# PRINCIPLES OF ELECTRIC CIRCUITS

Fifth Edition

# PRINCIPLES OF ELECTRIC CIRCUITS

THOMAS L. FLOYD

Prentice Hall
Upper Saddle River, New Jersey ■ Columbus, Ohio

**Library of Congress Cataloging-Publication Data**

Floyd, Thomas L.
Principles of electric circuits / Thomas L. Floyd.—5th ed.
p. cm.
Includes index.
ISBN 0-13-232224-2
1. Electric circuits. I. Title.
TK454.F57 1997
621.3815—dc20 96-13443
CIP

Cover photo: © Superstock
Editor: Linda Ludewig
Developmental Editor: Carol Hinklin Robison
Production Editor: Rex Davidson
Text Designer: Anne Flanagan
Cover Designer: Brian Deep
Production Buyer: Patricia A. Tonneman
Marketing Manager: Debbie Yarnell
Illustrations: Jane Lopez and Steve Botts

This book was set in Times Roman by The Clarinda Company and was printed and bound by Von Hoffmann Press, Inc. The cover was printed by Von Hoffmann Press, Inc.

Simon & Schuster/A Viacom Company
Upper Saddle River, New Jersey 07458



Printed in the United States of America

10 9 8 7 6 5 4 3 2

ISBN 0-13-232224-2

Prentice-Hall International (UK) Limited, *London*
Prentice-Hall of Australia Pty. Limited, *Sydney*
Prentice-Hall Canada Inc., *Toronto*
Prentice-Hall Hispanoamericana, S. A., *Mexico*
Prentice-Hall of India Private Limited, *New Delhi*
Prentice-Hall of Japan, Inc., *Tokyo*
Simon & Schuster Asia Pte. Ltd., *Singapore*
Editora Prentice-Hall do Brasil, Ltda., *Rio de Janeiro*

*In memory of my father, V. R. Floyd*

# PREFACE

This Fifth Edition of *Principles of Electric Circuits* provides a comprehensive, practical coverage of basic electrical concepts and circuits presented in an easy-to-read writing style with numerous illustrations. As in the previous editions, troubleshooting and applications are emphasized. This book has been thoroughly reviewed, and every effort has been made to ensure that the coverage is accurate and up-to-date.

## Features

- New **PSpice sections** in all chapters
- New **full-color** format
- **Technology Theory Into Practice (TECH TIP)** in all chapters except Chapters 1 and 22
- An overview and list of **objectives** at the opening of **each chapter**
- An introduction and **objectives** at the beginning of **each section** in a chapter
- A **related exercise** at the end of each worked example
- **Review questions** at the end of each section of text
- **Chapter glossaries**
- **Comprehensive glossary** at the end of the book
- End-of-chapter multiple-choice **self-tests**
- A **formula list** at the end of each chapter
- Extensive end-of-chapter **problem sets**
- Extensive **ancillary package** that includes
  - Two lab manuals

    Buchla, *Experiments in Basic Circuits: Theory and Application,* Third Edition.

    Stanley, *Experiments in Electric Circuits,* Fifth Edition
  - Transparency Masters and Transparencies (4-color acetates)
  - Test Item File
  - DOS PH Custom Test (IBM)
  - Instructor's Resource Manual (includes PSpice Version 6.3 eval CD-ROM)
  - PSpice/Electronics Workbench Data Disk
  - Bergwall Videos (to qualified adopters)
  - Electronics Workbench full software (to qualified adopters)

## Illustration of Chapter Features

***Chapter Opener*** Each chapter begins with a two-page opener as shown in Figure A. The left page includes a listing of the sections within the chapter, and a chapter overview.

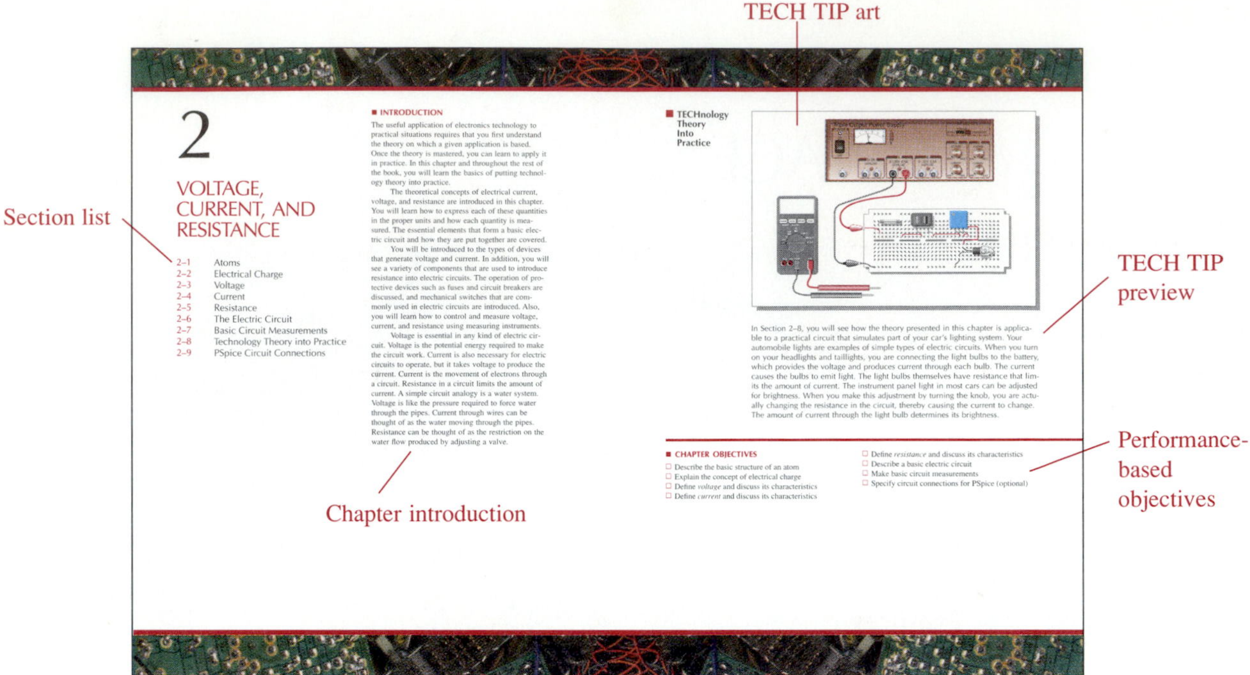


**FIGURE A**
*Chapter opener.*

The right page presents a preview of the Technology Theory Into Practice (TECH TIP) section and the chapter objectives.

***Section Opener*** Each section in a chapter begins with a brief introduction that provides a general overview of the material to be covered with a list of section objectives. This is illustrated in Figure B.

***Section Review*** Each section ends with a review consisting of questions or exercises that focus on the main concepts presented in the section. Answers to these section reviews are at the end of the chapter. This is also illustrated in Figure B.

***Worked Examples and Related Exercises*** Frequent examples help to demonstrate and clarify basic concepts or illustrate specific procedures. Each example concludes with a related exercise that reinforces or expands on the example. Some related exercises require a repetition of the example using different parameters or conditions. Others focus on a more limited part of the example or encourage further thought. Answers to the related exercises are at the end of the chapter. A typical worked example and related exercise are shown in Figure C.

***Troubleshooting Section*** Many chapters include troubleshooting sections that emphasize troubleshooting techniques and the use of test instruments as applied to situations that are related to topics covered in the chapter. These sections are optional and can be omitted without affecting the rest of the material.

***TECHnology Theory Into Practice (TECH TIP)*** The next-to-last section of each chapter (except Chapters 1 and 22) provides a practical application of material presented in the chapter. A series of activities involves comparing circuit board layouts with schematics, analyzing circuits, using test measurements to determine circuit operation, and in some cases, developing test procedures. Results and answers to the TECH TIPs are found in the Instructor's Resource Manual. A representative TECH TIP section is shown in Figure D.

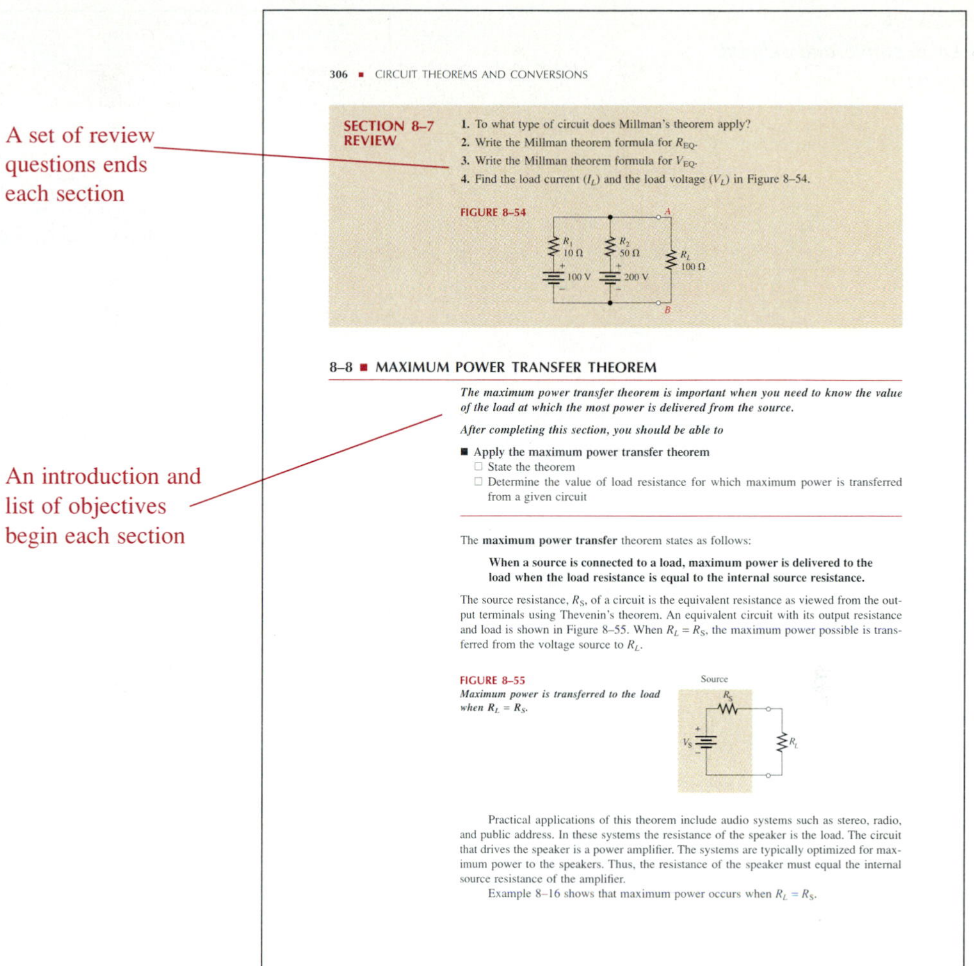


**FIGURE B**
*Section opener and section review.*

***PSpice*** The last section of each chapter provides an optional coverage of circuit analysis using PSpice. The PSpice material that is introduced in each chapter relates closely to the topics in the chapter. Each PSpice section incrementally builds on the material in previous chapters as you progress through the book.

***Chapter End Matter*** At the end of each chapter is a summary, glossary, formula list, multiple-choice self-test, and sectionalized problem set. Also, answers to section reviews and related exercises for examples are found at the end of each chapter. Terms that appear boldface in the text are defined in the glossaries.

## Suggestions for Use

This textbook is intended primarily for use in a two-term sequence in which dc topics (Chapters 1 through 10) are covered in the first term and ac topics (Chapters 11 through 22) are covered in the second term. Although not recommended, a one-term course is also possible with very selective coverage including the omission of many topics.

If time limitations or course emphasis restricts the topics that can be covered, as is generally the case, several possibilities for selective coverage exist. Suggestions for light treatment or omission do not necessarily imply that a given topic is less important than

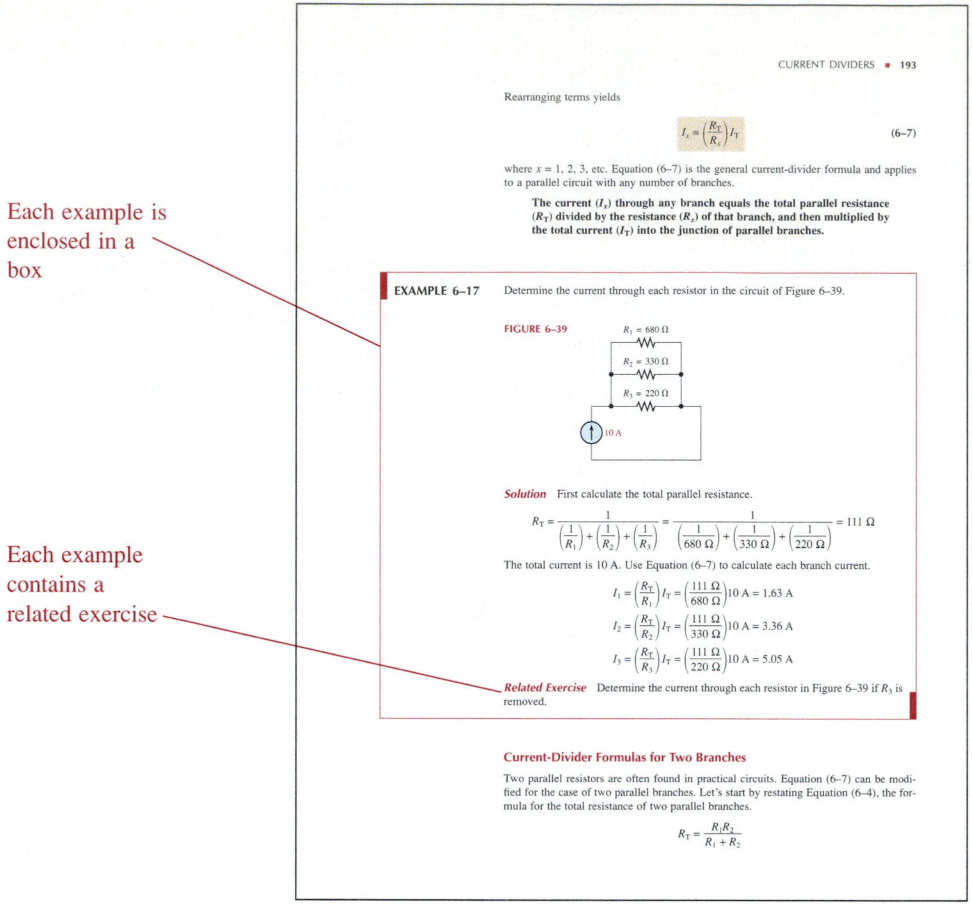


CURRENT DIVIDERS ■ 193

Rearranging terms yields

$$I_x = \left(\frac{R_T}{R_x}\right) I_T \qquad (6\text{–}7)$$

where $x = 1, 2, 3$, etc. Equation (6–7) is the general current-divider formula and applies to a parallel circuit with any number of branches.

**The current ($I_x$) through any branch equals the total parallel resistance ($R_T$) divided by the resistance ($R_x$) of that branch, and then multiplied by the total current ($I_T$) into the junction of parallel branches.**

**EXAMPLE 6–17** Determine the current through each resistor in the circuit of Figure 6–39.

FIGURE 6–39



***Solution*** First calculate the total parallel resistance.

$$R_T = \frac{1}{\left(\frac{1}{R_1}\right) + \left(\frac{1}{R_2}\right) + \left(\frac{1}{R_3}\right)} = \frac{1}{\left(\frac{1}{680\ \Omega}\right) + \left(\frac{1}{330\ \Omega}\right) + \left(\frac{1}{220\ \Omega}\right)} = 111\ \Omega$$

The total current is 10 A. Use Equation (6–7) to calculate each branch current.

$$I_1 = \left(\frac{R_T}{R_1}\right) I_T = \left(\frac{111\ \Omega}{680\ \Omega}\right) 10\ \text{A} = 1.63\ \text{A}$$

$$I_2 = \left(\frac{R_T}{R_2}\right) I_T = \left(\frac{111\ \Omega}{330\ \Omega}\right) 10\ \text{A} = 3.36\ \text{A}$$

$$I_3 = \left(\frac{R_T}{R_3}\right) I_T = \left(\frac{111\ \Omega}{220\ \Omega}\right) 10\ \text{A} = 5.05\ \text{A}$$

***Related Exercise*** Determine the current through each resistor in Figure 6–39 if $R_3$ is removed.

**Current-Divider Formulas for Two Branches**

Two parallel resistors are often found in practical circuits. Equation (6–7) can be modified for the case of two parallel branches. Let's start by restating Equation (6–4), the formula for the total resistance of two parallel branches.

$$R_T = \frac{R_1 R_2}{R_1 + R_2}$$

**FIGURE C**
***An example and related exercise.***

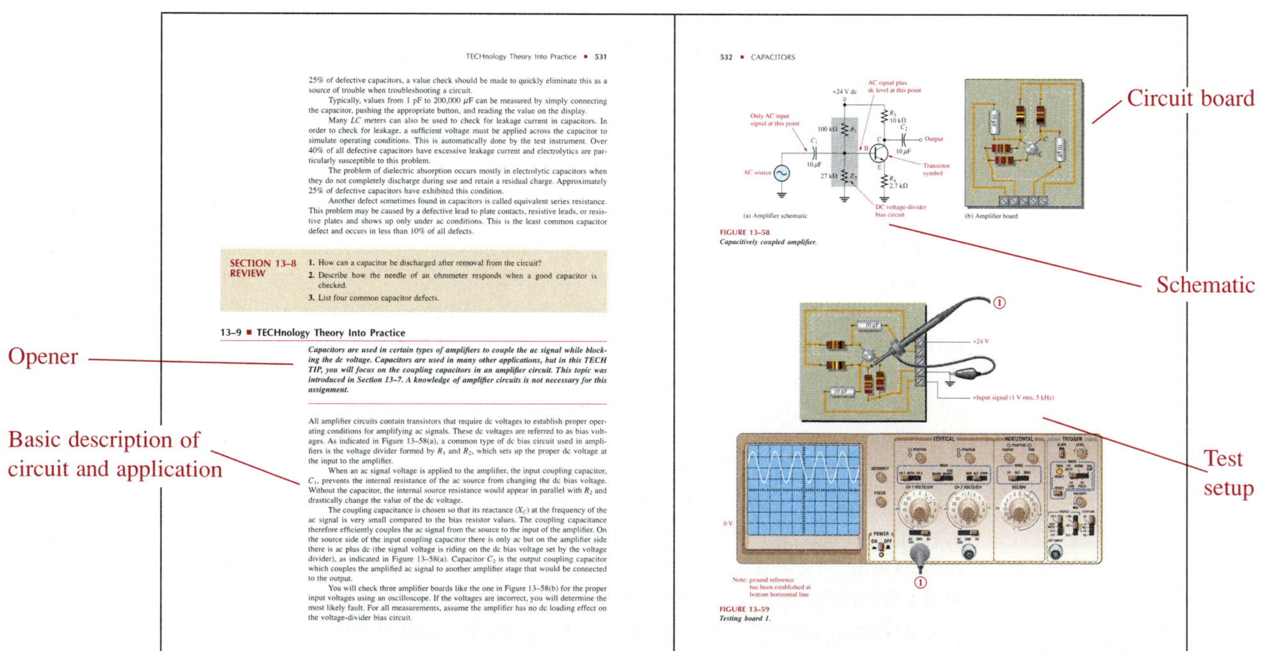


TECHnology Theory Into Practice ■ 531

25% of defective capacitors, a value check should be made to quickly eliminate this as a source of trouble when troubleshooting a circuit.

Typically, values from 1 pF to 200,000 μF can be measured by simply connecting the capacitor, pushing the appropriate button, and reading the value on the display.

Many *LC* meters can also be used to check for leakage current in capacitors. In order to check for leakage, a sufficient voltage must be applied across the capacitor to simulate operating conditions. This is automatically done by the test instrument. Over 40% of all defective capacitors have excessive leakage current and electrolytics are particularly susceptible to this problem.

The problem of dielectric absorption occurs mostly in electrolytic capacitors when they do not completely discharge during use and retain a residual charge. Approximately 25% of defective capacitors have exhibited this condition.

Another defect sometimes found in capacitors is called equivalent series resistance. This problem may be caused by a defective lead to plate contacts, resistive leads, or resistive plates and shows up only under ac conditions. This is the least common capacitor defect and occurs in less than 10% of all defects.

**SECTION 13–8 REVIEW**

1. How can a capacitor be discharged after removal from the circuit?
2. Describe how the needle of an ohmmeter responds when a good capacitor is checked.
3. List four common capacitor defects.

**13–9 ■ TECHnology Theory Into Practice**

***Capacitors are used in certain types of amplifiers to couple the ac signal while blocking the dc voltage. Capacitors are used in many other applications, but in this TECH TIP, you will focus on the coupling capacitors in an amplifier circuit. This topic was introduced in Section 13–7. A knowledge of amplifier circuits is not necessary for this assignment.***

All amplifier circuits contain transistors that require dc voltages to establish proper operating conditions for amplifying ac signals. These dc voltages are referred to as bias voltages. As indicated in Figure 13–58(a), a common type of dc bias circuit used in amplifiers is the voltage divider formed by $R_1$ and $R_2$, which sets up the proper dc voltage at the input to the amplifier.

When an ac signal voltage is applied to the amplifier, the input coupling capacitor, $C_1$, prevents the internal resistance of the ac source from changing the dc bias voltage. Without the capacitor, the internal source resistance would appear in parallel with $R_2$ and drastically change the value of the dc voltage.

The coupling capacitance is chosen so that its reactance ($X_C$) at the frequency of the ac signal is very small compared to the bias resistor values. The coupling capacitance therefore efficiently couples the ac signal from the source to the input of the amplifier. On the source side of the input coupling capacitor there is only ac but on the amplifier side there is ac plus dc (the signal voltage is riding on the dc bias voltage set by the voltage divider), as indicated in Figure 13–58(a). Capacitor $C_2$ is the output coupling capacitor which couples the amplified ac signal to another amplifier stage that would be connected to the output.

You will check three amplifier boards like the one in Figure 13–58(b) for the proper input voltages using an oscilloscope. If the voltages are incorrect, you will determine the most likely fault. For all measurements, assume the amplifier has no dc loading effect on the voltage-divider bias circuit.

532 ■ CAPACITORS



FIGURE 13–58
*Capacitively coupled amplifier.*



FIGURE 13–59
*Testing board 1.*

**FIGURE D**
***Representative pages from a typical TECHnology Theory Into Practice section.***

others but that, in the context of a specific program, the topic does not require the emphasis that the more fundamental topics do. Since course emphasis, level, and available time vary from one program to another, the omission or lighter treatment of selected topics must be made on an individual basis and, therefore, the following suggestions are intended only as a general guide.

1. Chapters that may be considered for omission or selective coverage:

   Chapter 8 Circuit Theorems and Conversions

   Chapter 9 Branch, Mesh, and Node Analysis

   Chapter 10 Magnetism and Electromagnetism

   Chapter 19 Basic Filters

   Chapter 20 Circuit Theorems in AC Analysis

   Chapter 21 Pulse Response of Reactive Circuits

   Chapter 22 Polyphase Systems in Power Applications
2. Troubleshooting sections, TECH TIP sections, and/or PSpice sections can be omitted. Other specific topics may be omitted or treated lightly on a section-by-section basis at the discretion of the instructor.
3. The order in which certain topics appear in the text can be altered at the instructor's discretion. For example, the topics of capacitors and inductors (Chapters 13 and 14) can be covered at the end of the dc course in the first term by delaying coverage of the ac topics in Sections 13–6, 13–7, 14–6, and 14–7 until the ac course in the second term. Another possibility is to cover Chapters 13 and 14 in the second term but cover Chapter 16 (*RC* Circuits) immediately after Chapter 13 and cover Chapter 17 (*RL* Circuits) immediately after Chapter 14.

***TECH TIPs*** These sections are very useful for motivation and as an introduction to applications of basic concepts and components. Possible uses are

1. An integral part of the chapter to illustrate how the concepts and components can be applied in a practical situation. The activities can be assigned for homework or as a class miniproject.
2. Extra credit assignments
3. In-class activities to promote discussion and interaction and to help answer the "need-to-know" questions that many students have.

## For the Student

All of the material in this preface is intended to provide both you and your instructor with information that will facilitate the most effective use of this textbook as both a teaching and a learning tool.

Acquiring knowledge and skills in any discipline is hard work, and perhaps even more so in the field of electronics. You must use this book as more than just a reference. You must really dig in by reading, thinking, and doing. Don't expect every concept or procedure to become clear immediately. Most of the topics are not overly difficult but some may require several readings, working many problems, and help from your instructor before you really understand them.

In order to master the material in each section, you should first read the material, then go through each example step-by-step, work the related exercises, and complete the review exercises at the end of the section. After completing a chapter, you should first work through the self-test and then, as a minimum, work the problems assigned by your instructor. Check your answers at the end of the chapter or book. Be sure to ask questions in class about anything that you do not understand.

The problems at the end of each chapter provide varying degrees of difficulty. In any technical field, it is important that you work lots of problems. Working through a problem gives you a level of insight and understanding that reading or classroom lectures alone do not provide. Never think that you can fully understand a concept or procedure by simply watching or listening to someone else. In the final analysis, you must do it yourself.

## A Brief History of Electronics

It is always good to have a sense of the history of your career field. So, before you begin your study of electric circuits, let's briefly look at the beginnings of electronics and some of the important developments that have led to the electronics technology that we have today. The names of many of the early pioneers in electricity and electromagnetics still live on in terms of familiar units and quantities. Names such as Ohm, Ampere, Volta, Farad, Henry, Coulomb, Oersted, and Hertz are some of the better known examples. More widely known names such as Franklin and Edison are also very significant in the history of electricity and electronics because of their tremendous contributions.

***The Beginning of Electronics*** The early experiments in electronics involved electric currents in glass vacuum tubes. One of the first to conduct such experiments was a German named Heinrich Geissler (1814–1879). Geissler removed most of the air from a glass tube and found that the tube glowed when there was an electric current through it. Around 1878, British scientist Sir William Crookes (1832–1919) experimented with tubes similar to those of Geissler. In his experiments, Crookes found that the current in the vacuum tubes seemed to consist of particles.

Thomas Edison (1847–1931), experimenting with the carbon-filament light bulb he had invented, made another important finding. He inserted a small metal plate in the bulb. When the plate was positively charged, there was a current from the filament to the plate. This device was the first thermionic diode. Edison patented it but never used it.

The electron was discovered in the 1890s. The French physicist Jean Baptiste Perrin (1870–1942) demonstrated that the current in a vacuum tube consisted of the movement of negatively charged particles in a given direction. Some of the properties of these particles were measured by Sir Joseph Thompson (1856–1940), a British physicist, in experiments he performed between 1895 and 1897. These negatively charged particles later became known as electrons. The charge on the electron was accurately measured by an American physicist, Robert A. Millikan (1868–1953), in 1909. As a result of these discoveries, electrons could be controlled, and the electronic age was ushered in.

***Putting the Electron to Work*** In 1904 a British scientist, John A. Fleming, constructed a vacuum tube that allowed electric current in only one direction. The tube was used to detect electromagnetic waves. Called the Fleming valve, it was the forerunner of the more recent vacuum diode tubes. Major progress in electronics, however, awaited the development of a device that could boost, or amplify, a weak electromagnetic wave or radio signal. This device was the audion, patented in 1907 by Lee deForest, an American. It was a triode vacuum tube capable of amplifying small electrical ac signals.

Two other Americans, Harold Arnold and Irving Langmuir, made great improvements in the triode vacuum tube between 1912 and 1914. About the same time, deForest and Edwin Armstrong, an electrical engineer, used the triode tube in an oscillator circuit. In 1914, the triode was incorporated in the telephone system and made the transcontinental telephone network possible. In 1916 Walter Schottky, a German, invented the tetrode tube. The tetrode, along with the pentode (invented in 1926 by Dutch engineer Tellegen), greatly improved the triode. The first television picture tube, called the kinescope, was developed in the 1920s by Vladimir Sworykin, an American researcher.

During World War II, several types of microwave tubes were developed that made possible modern microwave radar and other communications systems. In 1939, the magnetron was invented in Britain by Henry Boot and John Randall. In that same year, the klystron microwave tube was developed by two Americans, Russell Varian and his brother Sigurd Varian. The traveling-wave tube (TWT) was invented in 1943 by Rudolf Komphner, an Austrian-American.

***Solid-State Electronics*** The crystal detectors used in early radios were the forerunners of modern solid-state devices. However, the era of solid-state electronics began with the invention of the transistor in 1947 at Bell Labs. The inventors were Walter Brattain, John Bardeen, and William Shockley.

In the early 1960s, the integrated circuit (IC) was developed. It incorporated many transistors and other components on a single small chip of semiconductor material. Integrated circuit technology has been continuously developed and improved, allowing increasingly more complex circuits to be built on smaller chips.

Around 1965, the first integrated general-purpose operational amplifier was introduced. This low-cost, highly versatile device incorporated nine transistors and twelve resistors in a small package. It proved to have many advantages over comparable discrete component circuits in terms of reliability and performance. Since this introduction, the IC operational amplifier has become a basic building block for a wide variety of linear systems.

## Applications

Electricity and electronics are very broad and diverse fields and almost anything you can think of somehow uses electricity or electronics. The following are some of the major application areas.

***Computers*** As everyone knows, computers are used just about everywhere and their applications are almost unlimited. For example, computers are used in business for record keeping, accounting, payroll, inventory control, market analysis, and statistics. In industry, computers are used for controlling and monitoring manufacturing processes. Information, communications, navigation, medical, military, law enforcement, and domestic applications are some of the other broad areas where computers are found.

***Communications*** Electronic communications involves a wide variety of specialized fields that include telephone systems, satellites, radio, television, data communications, and radar, to name a few.

***Automation*** Electronic systems are used extensively in the control of manufacturing processes. For example, the robotic systems used in the assembly of automobiles are electronically controlled. A few other applications are the control of ingredient mixes in food processing, the operation of machine tools, distribution of power, and printing.

***Transportation*** Electronics is an integral part of most types of transportation equipment and systems. One example is the electronics found in most modern automobiles for entertainment, brake controls, ignition controls, engine monitoring, and communications. The airline industry would be completely shut down without electronics for navigation, communication, and scheduling purposes. Also, trucks, trains, and boats depend on electronics for much of their operation.

***Medicine*** Electronic systems are used in the medical field for many things. For example, the monitoring of patient functions such as heart rate, blood pressure, temperature, and respiration is accomplished electronically. Diagnostic tools such as the CAT scan, EKG, X-ray, and ultrasound are all electronic-based systems.

***Consumer Electronics*** Products used directly by the consumer constitute a large segment of the electronics market. There are systems for entertainment and information such as radios, televisions, VCRs, computers, sound systems, and electronic games. There are systems for communications such as cellular phones, pagers, recorders, intercoms, CB radio, short-wave radio, and computers. Also, many products that do work in the home such as washing machines, dryers, refrigerators, ranges, and security systems have electronic controls.

## Acknowledgements

I hope you will agree that this Fifth Edition of *Principles of Electric Circuits* is the best yet. As always, many people have contributed valuable ideas and constructive criticism for this new edition. It has been thoroughly reviewed for both content and accuracy.

Again, it has been a pleasure to work with the people at Prentice Hall. Their enthusiasm and dedication to quality continues to be a source of inspiration to me. My appreciation goes to Dave Garza, Linda Ludewig, Carol Robison, and Rex Davidson for their efforts on this project.

A significant amount of material used in the PSpice sections was contributed by Gary Snyder of Bently-Nevada Corporation. These PSpice sections, as well as the rest of the text, were thoroughly reviewed and checked for accuracy by Chuck Garbinsky, whose work produced many overall enhancements.

Lois Porter deserves special praise for, once again, doing such a thorough job of copy editing the manuscript. Also, Dave Buchla of Yuba College has provided many valuable suggestions for improvements to the book.

I am grateful to the following reviewers, whose comments and suggestions have been of tremendous help in the development of this Fifth Edition:

| | |
|---|---|
| Dwight Bechtel | Gwinnett Tech |
| Thomas Cress | Belleville Area College |
| Roger Harlow, Jr. | Mesa Community College |
| Richard E. Hoover | Owens Community College |
| Theodore Ingram | Texas Southern University |
| Daniel J. Kirsch | Kankakee Community College |
| Roger Peterson | Northwest Technical College |
| Bill Pulte | Metro Community College |
| Jereld R. Reeder | IVY Tech State College |
| Gordon F. Snyder, Jr. | Springfield Technical Community College |
| Richard K. Sturtevant | Springfield Technical Community College |
| Tim Woo | Durham Technical College |
| Stan Zeidell | Hallmark Institute |

The photographs used in this book were provided by Coilcraft, Fluke, Maxtec International Corp., Sencore Inc., Superior Electric Company, and Tektronix, Inc. The data sheets are courtesy of Texas Instruments.

Last, but not least, a special thanks to my wife Sheila.

*Thomas L. Floyd*

# CONTENTS

# 1

# COMPONENTS, QUANTITIES, AND UNITS



## ■ INTRODUCTION

This chapter presents a basic introduction to the field of electronics. An overview of electrical and electronic components and instruments gives you a preview of the types of things you will study throughout this book.

You must be familiar with the units used in electronics and know how to express electrical quantities in various ways using metric prefixes. Scientific notation is an indispensable tool whether you use a computer, a calculator, or do computations the old-fashioned way.

To help get you started using the computer, an introduction to PSpice is presented in this chapter. In each of the following chapters, additional aspects of PSpice are introduced so that it can be applied to the material covered. PSpice coverage is optional and its omission will not affect the rest of the coverage.

Fluke 867 graphical multimeter (courtesy of John Fluke Manufacturing Co.)

## CHAPTER OBJECTIVES

- Recognize some common components and measuring instruments
- List the electrical and magnetic quantities and their units
- Use scientific notation (powers of ten) to express quantities
- Use metric prefixes to express large and small numbers
- Convert from one metric unit to another
- Define SPICE and express electrical quantities in PSpice format (optional)

## 1–1 ■ ELECTRICAL COMPONENTS AND MEASURING INSTRUMENTS

***In this book, you will study many types of electrical components and measuring instruments. A thorough background in dc and ac fundamentals provides the foundation for understanding complex electronic devices and circuits. A preview of the basic types of electrical and electronic components and instruments that you will be studying in detail later in this and in other courses is provided in this section.***

***After completing this section, you should be able to***

- **Recognize some common components and measuring instruments**
  - ☐ State the purpose of a resistor
  - ☐ State the purpose of a capacitor
  - ☐ State the purpose of an inductor
  - ☐ State the purpose of a transformer
  - ☐ List some basic types of electronic test and measuring instruments

### Resistors

Resistors resist, or limit, electric current in a circuit. Several common types of resistors are shown in Figure 1–1.

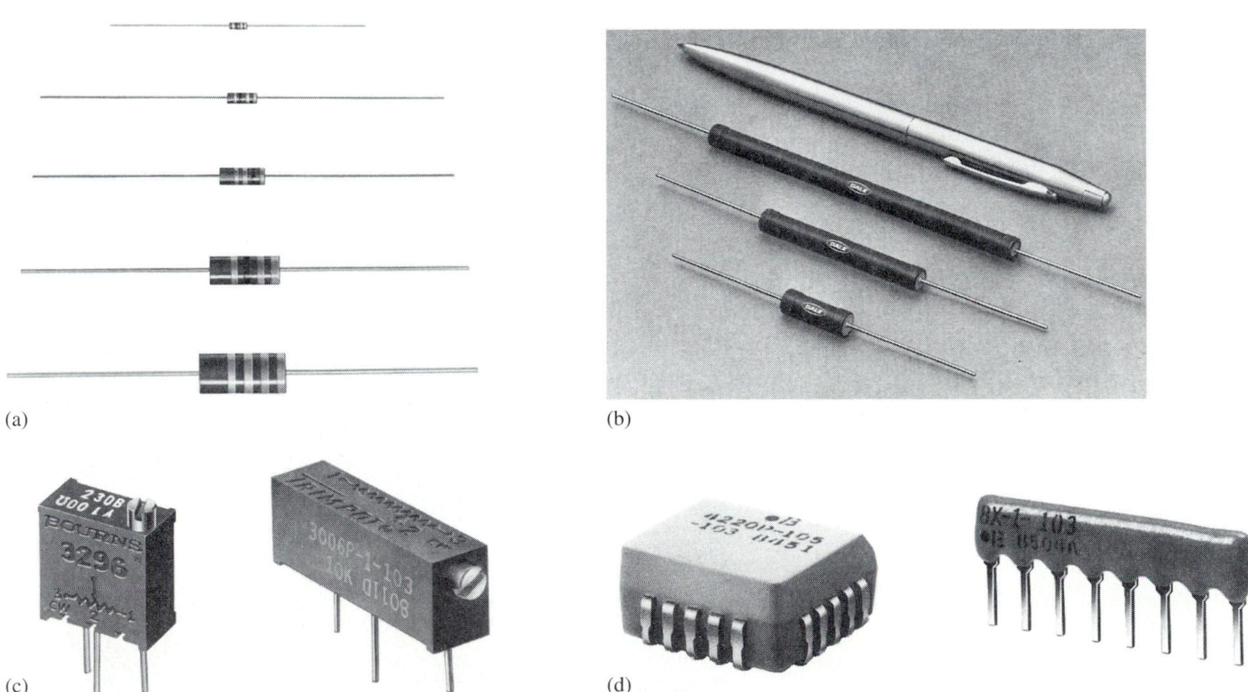


**FIGURE 1–1**
*Typical fixed and variable resistors. (a) Carbon-composition resistors with standard power ratings of 1/8 W, 1/4 W, 1/2 W, 1 W, and 2 W (courtesy of Allen-Bradley Co.). (b) Wirewound resistors (courtesy of Dale Electronics, Inc.). (c) Variable resistors or potentiometers (courtesy of Bourns, Inc.). (d) Resistor networks (courtesy of Bourns, Inc.).*

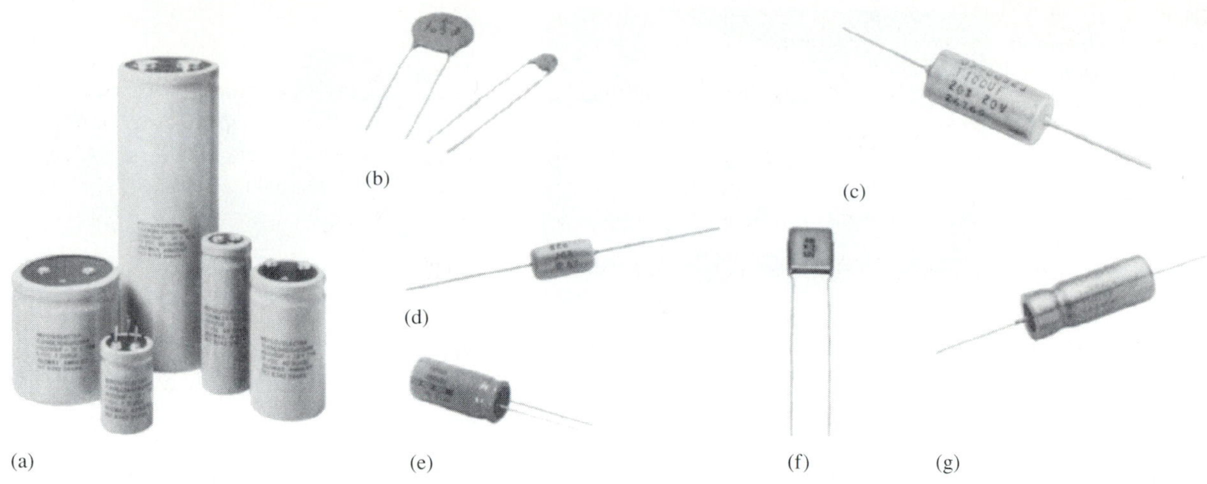


**FIGURE 1–2**
*Typical capacitors. Parts (a, c, d, and g) courtesy of Mepco/Centralab. Part (b) courtesy of Murata Erie, North America. Parts (e and f) courtesy of Sprague Electric Co.*

## Capacitors

Capacitors store electrical charge; they are found in most types of electronic circuits. Figure 1–2 shows several typical capacitors.

## Inductors

Inductors, also known as *coils,* are used to store energy in an electromagnetic field; they serve many useful functions in an electrical circuit. Figure 1–3 shows several typical inductors.

**FIGURE 1–3**
*Typical inductors. Surface mount, leaded both tuneable and fixed inductors (courtesy of Coilcraft).*

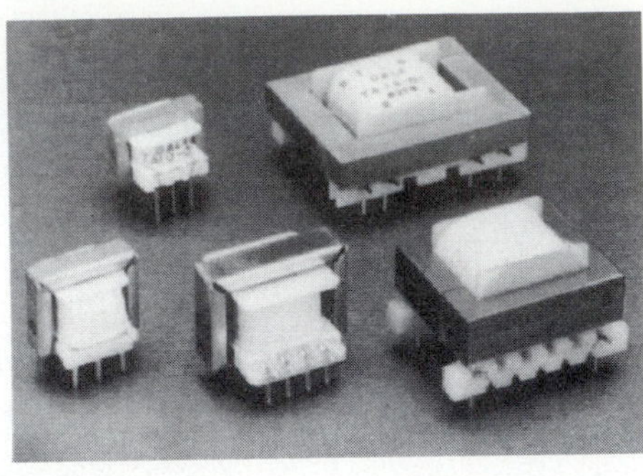

(a)

(b)

**FIGURE 1–4**
*Typical transformers (courtesy of Dale Electronics, Inc.).*

## Transformers

Transformers are used to magnetically couple ac voltages from one point in a circuit to another, or to increase or decrease the ac voltage. Several types of transformers are shown in Figure 1–4.

## Semiconductor Devices

Several varieties of diodes, transistors, and integrated circuits are shown in Figure 1–5.

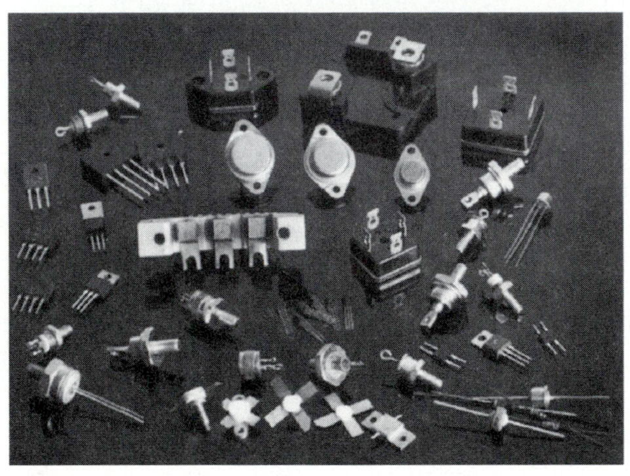

(a)

(b)

**FIGURE 1–5**
*A grouping of typical semiconductor devices (courtesy of Motorola, Inc. Used by permission.).*

## Electronic Instruments

Figure 1–6 shows a variety of instruments that are discussed throughout the text. Typical instruments include the power supply, for providing voltage and current; the voltmeter, for measuring voltage; the ammeter, for measuring current; the ohmmeter, for measuring resistance; the wattmeter, for measuring power; and the oscilloscope for observing and measuring ac voltages. The voltmeter, ammeter, and ohmmeter are available in a single instrument called a *multimeter.*

## SECTION 1–1 REVIEW

1. Name four types of common electrical components.
2. What instrument is used for measuring electrical current?
3. What instrument is used for measuring resistance?
4. What instrument is used for measuring voltage?
5. What is a multimeter?

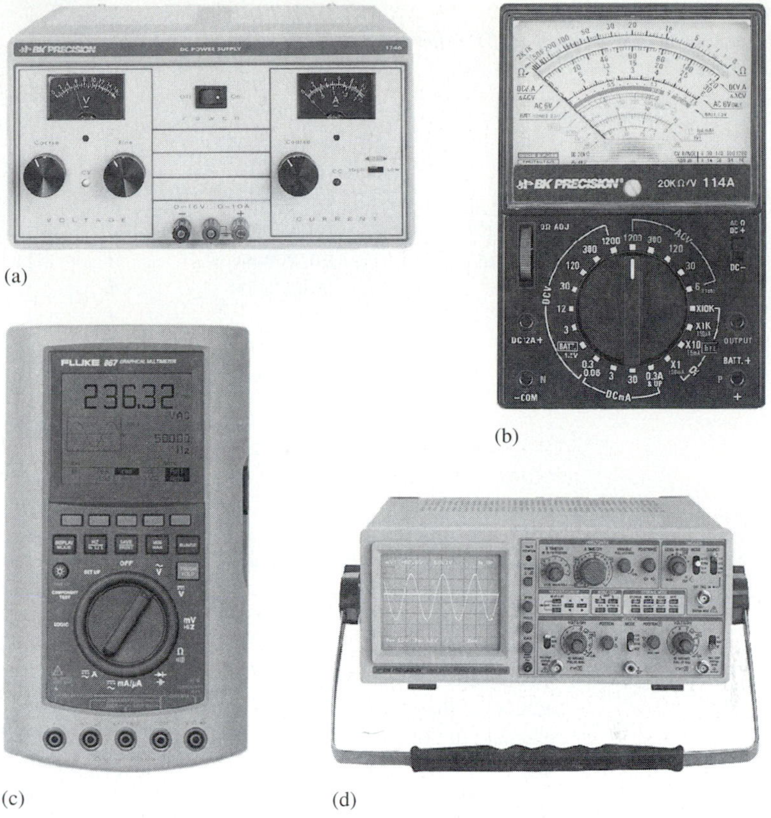


(a) (b) (c) (d)

**FIGURE 1–6**

***Typical instruments. (a) DC power supply. (b) Analog multimeter. (c) Digital multimeter (courtesy of John Fluke Manufacturing Co.). (d) Digital storage oscilloscope. (Parts a, b, and d courtesy of BK Precision, Maxtec International Corp.)***

## 1–2 ■ ELECTRICAL AND MAGNETIC UNITS

***In electronics work, you must deal with measurable quantities. For example, you must be able to express how many volts are measured at a certain point in a circuit, how much current there is through a wire, or how much power a certain amplifier produces. In this section, you are introduced to the units and symbols for most of the electrical and magnetic quantities that are used in this book.***

***After completing this section, you should be able to***

- **List the electrical and magnetic quantities and their units**
  - Specify the symbol for each quantity
  - Specify the symbol for each unit

Letter symbols are used in electronics to represent both quantities and their units. One symbol is used to represent the name of the quantity, and another symbol is used to represent the unit of measurement of that quantity. For example, *P* stands for *power,* and *W* stands for *watt,* which is the unit of power. Table 1–1 lists the most important electrical quantities, along with their SI units and symbols. The term *SI* is the French abbreviation for *International System* (*Système International* in French). Table 1–2 lists magnetic quantities, along with their SI units and symbols.

**TABLE 1–1**
***Electrical quantities and units with SI symbols***

| Quantity | Symbol | Unit | Symbol |
|---|---|---|---|
| capacitance | $C$ | farad | F |
| charge | $Q$ | coulomb | C |
| conductance | $G$ | siemens | S |
| current | $I$ | ampere | A |
| energy | $W$ | joule | J |
| frequency | $f$ | hertz | Hz |
| impedance | $Z$ | ohm | Ω |
| inductance | $L$ | henry | H |
| power | $P$ | watt | W |
| reactance | $X$ | ohm | Ω |
| resistance | $R$ | ohm | Ω |
| time | $t$ | second | s |
| voltage | $V$ | volt | V |

**TABLE 1–2**
***Magnetic quantities and units with SI symbols***

| Quantity | Symbol | Unit | Symbol |
|---|---|---|---|
| flux density | $B$ | tesla | T |
| magnetic flux | $\phi$ | weber | Wb |
| magnetizing force | $H$ | ampere-turns/meter | At/m |
| magnetomotive force | $F_m$ | ampere-turn | At |
| permeability | $\mu$ | webers/ampere-turns-meter | Wb/Atm |
| reluctance | $\mathcal{R}$ | ampere-turns/weber | At/Wb |

**SECTION 1–2 REVIEW**

1. What does *SI* stand for?
2. Without referring to Table 1–1, list as many electrical quantities as possible, including their symbols, units, and unit symbols.
3. Without referring to Table 1–2, list as many magnetic quantities as possible, including their symbols, units, and unit symbols.

## 1–3 ■ SCIENTIFIC NOTATION

***Working in the field of electronics, you will encounter both very small and very large quantities. For example, it is common to have electric current values of only a few thousandths or even a few millionths of an ampere. On the other hand, you will find resistance values of several thousand or several million ohms. This range of values is typical of many other electrical quantities also.***

***After completing this section, you should be able to***

- **Use scientific notation (powers of ten) to express quantities**
    - ☐ Express any number using a power of ten
    - ☐ Perform calculations with powers of ten

Scientific notation provides an easy method to express large and small numbers and to perform calculations involving such numbers. In scientific notation, a quantity is expressed as a product of a positive number (usually, but not necessarily, between 1 and 10) and a power of ten. For example, the quantity 150,000 would be expressed in scientific notation as $1.5 \times 10^5$ or $150 \times 10^3$.

## Powers of Ten

Table 1–3 lists some powers of ten, both positive and negative, and the corresponding decimal numbers. The power of ten is expressed as an exponent of the base 10 in each case. The exponent indicates the number of places that the decimal point is moved to the right or left to produce the decimal number. If the power of ten is positive, the decimal point is moved to the right to get the equivalent decimal number. For example,

$$10^4 = 1 \times 10^4 = 1.0000. = 10{,}000$$

If the power of ten is negative, the decimal point is moved to the left to get the equivalent decimal number. For example,

$$10^{-4} = 1 \times 10^{-4} = .0001. = 0.0001$$

**TABLE 1–3**
***Some positive and negative powers of ten***

| | |
|---|---|
| $10^6 = 1{,}000{,}000$ | $10^{-6} = 0.000001$ |
| $10^5 = 100{,}000$ | $10^{-5} = 0.00001$ |
| $10^4 = 10{,}000$ | $10^{-4} = 0.0001$ |
| $10^3 = 1{,}000$ | $10^{-3} = 0.001$ |
| $10^2 = 100$ | $10^{-2} = 0.01$ |
| $10^1 = 10$ | $10^{-1} = 0.1$ |
| $10^0 = 1$ | |

**EXAMPLE 1–1**

Express each number in scientific notation as a number between 1 and 10 times a positive power of ten:
**(a)** 200 **(b)** 5000 **(c)** 85,000 **(d)** 3,000,000

***Solution***

**(a)** $200 = 2 \times 10^2$ **(b)** $5000 = 5 \times 10^3$
**(c)** $85{,}000 = 8.5 \times 10^4$ **(d)** $3{,}000{,}000 = 3 \times 10^6$

***Related Exercise*** Express 4750 in scientific notation as a number between 1 and 10 times a positive power of ten.

**EXAMPLE 1–2**

Express each number in scientific notation as a number between 1 and 10 times a negative power of ten:
**(a)** 0.2 **(b)** 0.005 **(c)** 0.00063 **(d)** 0.000015

***Solution***

**(a)** $0.2 = 2 \times 10^{-1}$ **(b)** $0.005 = 5 \times 10^{-3}$
**(c)** $0.00063 = 6.3 \times 10^{-4}$ **(d)** $0.000015 = 1.5 \times 10^{-5}$

***Related Exercise*** Express 0.00738 in scientific notation as a number between 1 and 10 times a negative power of ten.

**EXAMPLE 1–3**

Express each of the following as a regular decimal number:
**(a)** $1 \times 10^5$ **(b)** $2 \times 10^3$ **(c)** $3.2 \times 10^{-2}$ **(d)** $250 \times 10^{-6}$

***Solution***
**(a)** $1 \times 10^5 = 100{,}000$ **(b)** $2 \times 10^3 = 2000$
**(c)** $3.2 \times 10^{-2} = 0.032$ **(d)** $250 \times 10^{-6} = 0.000250$

***Related Exercise*** Express $9.12 \times 10^3$ as a regular decimal number.

## Calculating with Powers of Ten

The advantage of scientific notation is in addition, subtraction, multiplication, and division of very small or very large numbers.

***Rules of Addition*** The rules for adding numbers using powers of ten are as follows:

1. Express the numbers to be added in the same power of ten.
2. Add the numbers without their powers of ten to get the sum.
3. Bring down the common power of ten, which is the power of ten of the sum.

**EXAMPLE 1–4**

Add $2 \times 10^6$ and $5 \times 10^7$.

***Solution***
1. Express both numbers in the same power of ten: $(2 \times 10^6) + (50 \times 10^6)$.
2. Add $2 + 50 = 52$.
3. Bring down the common power of ten ($10^6$), and the sum is $52 \times 10^6$.

***Related Exercise*** Add $3.1 \times 10^3$ and $0.55 \times 10^4$.

***Rules for Subtraction*** The rules for subtracting numbers using powers of ten are as follows:

1. Express the numbers to be subtracted in the same power of ten.
2. Subtract the numbers without their powers of ten to get the difference.
3. Bring down the common power of ten, which is the power of ten of the difference.

**EXAMPLE 1–5**

Subtract $25 \times 10^{-12}$ from $75 \times 10^{-11}$.

***Solution***
1. Express each number in the same power of ten: $(750 \times 10^{-12}) - (25 \times 10^{-12})$.
2. Subtract $750 - 25 = 725$.
3. Bring down the common power of ten ($10^{-12}$), and the difference is $725 \times 10^{-12}$.

***Related Exercise*** Subtract $98 \times 10^{-2}$ from $1530 \times 10^{-3}$.

***Rules for Multiplication*** The rules for multiplying numbers using powers of ten are as follows:

1. Multiply the numbers directly without their powers of ten.
2. Add the powers of ten algebraically (the powers do not have to be the same).

**EXAMPLE 1–6** Multiply $5 \times 10^{12}$ and $3 \times 10^{-6}$.

***Solution*** Multiply the numbers, and algebraically add the powers:

$$(5 \times 10^{12})(3 \times 10^{-6}) = 15 \times 10^{12 + (-6)} = 15 \times 10^{6}$$

***Related Exercise*** Multiply $3.2 \times 10^{6}$ and $1.5 \times 10^{-3}$.

***Rules for Division*** The rules for dividing numbers using powers of ten are as follows:

1. Divide the numbers directly without their powers of ten.
2. Subtract the power of ten in the denominator from the power of ten in the numerator (the powers do not have to be the same).

**EXAMPLE 1–7** Divide $50 \times 10^{8}$ by $25 \times 10^{3}$.

***Solution*** The division problem is written with a numerator and denominator as

$$\frac{50 \times 10^{8}}{25 \times 10^{3}}$$

Divide the numbers and subtract 3 from 8:

$$\frac{50 \times 10^{8}}{25 \times 10^{3}} = 2 \times 10^{8-3} = 2 \times 10^{5}$$

***Related Exercise*** Divide $100 \times 10^{12}$ by $4 \times 10^{6}$.

**SECTION 1–3 REVIEW**

1. Scientific notation uses powers of ten (T or F).
2. Express 100 as a power of ten.
3. Do the following operations:
   **(a)** $(1 \times 10^{5}) + (2 \times 10^{5})$ **(b)** $(3 \times 10^{6})(2 \times 10^{4})$
   **(c)** $(8 \times 10^{3}) \div (4 \times 10^{2})$ **(d)** $(25 \times 10^{-6}) - (130 \times 10^{-7})$

## 1–4 ■ METRIC PREFIXES

***In electrical and electronics applications, certain powers of ten are used more often than others. It is common practice to use metric prefixes to represent these quantities. You can think of the metric prefix as a shorthand way to express a large or small number.***

***After completing this section, you should be able to***

- **Use metric prefixes to express large and small numbers**
  - ☐ List the metric prefixes
  - ☐ Change a power of ten to a metric prefix
  - ☐ Use a calculator for numbers with metric prefixes
  - ☐ Use engineering notation

Table 1–4 lists the metric prefix for each of the commonly used powers of ten.

TABLE 1–4

*Commonly used metric prefixes and their symbols*

| Metric Prefix | Metric Symbol | Power of Ten | Value |
|---|---|---|---|
| tera | T | $10^{12}$ | one trillion |
| giga | G | $10^{9}$ | one billion |
| mega | M | $10^{6}$ | one million |
| kilo | k | $10^{3}$ | one thousand |
| milli | m | $10^{-3}$ | one-thousandth |
| micro | $\mu$ | $10^{-6}$ | one-millionth |
| nano | n | $10^{-9}$ | one-billionth |
| pico | p | $10^{-12}$ | one-trillionth |

## Use of Metric Prefixes

Now some examples will illustrate the use of metric prefixes. The number 2000 can be expressed in scientific notation as $2 \times 10^3$. Suppose you wish to represent 2000 watts (W) with a metric prefix. Since $2000 = 2 \times 10^3$, the metric prefix *kilo* (k) is used for $10^3$. So you can express 2000 W as 2 kW (2 kilowatts).

As another example, 0.015 ampere (A) can be expressed as $15 \times 10^{-3}$ A. The metric prefix *milli* (m) is used for $10^{-3}$. So 0.015 A becomes 15 mA (15 milliamperes).

### EXAMPLE 1–8

Express each quantity using a metric prefix:

**(a)** 50,000 V **(b)** 25,000,000 Ω **(c)** 0.000036 A

***Solution***

**(a)** $50{,}000 \text{ V} = 50 \times 10^3 \text{ V} = 50 \text{ kV}$

**(b)** $25{,}000{,}000\ \Omega = 25 \times 10^6\ \Omega = 25 \text{ M}\Omega$

**(c)** $0.000036 \text{ A} = 36 \times 10^{-6} \text{ A} = 36\ \mu\text{A}$

***Related Exercise*** Express using metric prefixes:

**(a)** 56,000,000 Ω **(b)** 0.000470 A

## Entering Numbers with Metric Prefixes on a Calculator

To enter a number expressed in scientific notation on a calculator, use the EXP key (the EE key on some calculators). Example 1–9 shows how to enter numbers with metric prefixes on a typical scientific calculator. Consult your user's manual for your particular calculator.

### EXAMPLE 1–9

**(a)** Enter 3.3 kΩ ($3.3 \times 10^3\ \Omega$) on your calculator.

**(b)** Enter 450 μA ($450 \times 10^{-6}$ A) on your calculator.

***Solution***

**(a)** Step 1: Enter [3] [.] [3]. The display shows 3.3.
Step 2: Press [EXP]. The display shows 3.3 00.
Step 3: Enter [3]. The display shows 3.3 03.

**(b)** Step 1: Enter [4] [5] [0]. The display shows 450.
Step 2: Press [EXP]. The display shows 450 00.
Step 3: Press [+/−]. Enter [6]. The display shows 450–06.

***Related Exercise*** Enter 259 mA on your calculator.

### Engineering Notation

*Engineering notation* is a term that refers to the application of scientific notation in which the powers of ten are limited to multiples of three, such as $10^3$, $10^{-3}$, $10^6$, $10^{-6}$, $10^9$, $10^{-9}$, $10^{12}$, and $10^{-12}$. The reason for this is that all units used in the fields of engineering and technology, such as ohms, amps, volts, and watts are expressed with prefixes of k (kilo, $10^3$), m (milli, $10^{-3}$), M (mega, $10^6$), $\mu$ (micro, $10^{-6}$), n (nano, $10^{-9}$), T (tera, $10^{12}$), and p (pico, $10^{-12}$).

**SECTION 1–4 REVIEW**

1. List the metric prefix for each of the following powers of ten: $10^6$, $10^3$, $10^{-3}$, $10^{-6}$, $10^{-9}$, and $10^{-12}$.
2. Use an appropriate metric prefix to express 0.000001 A.
3. Use an appropriate metric prefix to express 250,000 W.

## 1–5 ■ METRIC UNIT CONVERSIONS

***It is often necessary or convenient to convert a quantity from one metric-prefixed unit to another, such as from milliamperes (mA) to microamperes (μA). A metric prefix conversion is accomplished by moving the decimal point in the number an appropriate number of places to the left or to the right, depending on the particular conversion.***

***After completing this section, you should be able to***

- **Convert from one metric unit to another**
  - Convert between milli, micro, nano, and pico
  - Convert between kilo and mega

The following basic rules apply to metric unit conversions:

1. When converting from a larger unit to a smaller unit, move the decimal point to the right.
2. When converting from a smaller unit to a larger unit, move the decimal point to the left.
3. Determine the number of places that the decimal point is moved by finding the difference in the powers of ten of the units being converted.

For example, when converting from milliamperes (mA) to microamperes ($\mu$A), move the decimal point three places to the right because there is a three-place difference between the two units (mA is $10^{-3}$ A and $\mu$A is $10^{-6}$ A). The following examples illustrate a few conversions.

**EXAMPLE 1–10** Convert 0.15 milliampere (0.15 mA) to microamperes ($\mu$A).

***Solution*** Move the decimal point three places to the right.

$$0.15\text{ mA} = 0.15 \times 10^{-3}\text{ A} = 150 \times 10^{-6}\text{ A} = 150\ \mu\text{A}$$

***Related Exercise*** Convert 1 mA to microamperes.

**EXAMPLE 1–11** Convert 4500 microvolts (4500 $\mu$V) to millivolts (mV).

***Solution*** Move the decimal point three places to the left.

$$4500\ \mu\text{V} = 4500 \times 10^{-6}\ \text{V} = 4.5 \times 10^{-3}\ \text{V} = 4.5\ \text{mV}$$

***Related Exercise*** Convert 1000 $\mu$V to millivolts.

**EXAMPLE 1–12** Convert 5000 nanoamperes (5000 nA) to microamperes ($\mu$A).

***Solution*** Move the decimal point three places to the left.

$$5000\ \text{nA} = 5000 \times 10^{-9}\ \text{A} = 5 \times 10^{-6}\ \text{A} = 5\ \mu\text{A}$$

***Related Exercise*** Convert 893 nA to microamperes.

**EXAMPLE 1–13** Convert 47,000 picofarads (47,000 pF) to microfarads ($\mu$F).

***Solution*** Move the decimal point six places to the left.

$$47{,}000\ \text{pF} = 47{,}000 \times 10^{-12}\ \text{F} = 0.047 \times 10^{-6}\ \text{F} = 0.047\ \mu\text{F}$$

***Related Exercise*** Convert 0.0022 $\mu$F to picofarads.

**EXAMPLE 1–14** Convert 0.00022 microfarad (0.00022 $\mu$F) to picofarads (pF).

***Solution*** Move the decimal point six places to the right.

$$0.00022\ \mu\text{F} = 0.00022 \times 10^{-6}\ \text{F} = 220 \times 10^{-12}\ \text{F} = 220\ \text{pF}$$

***Related Exercise*** Convert 10,000 pF to microfarads.

**EXAMPLE 1–15** Convert 1800 kilohms (1800 k$\Omega$) to megohms (M$\Omega$).

***Solution*** Move the decimal point three places to the left.

$$1800\ \text{k}\Omega = 1800 \times 10^{3}\ \Omega = 1.8 \times 10^{6}\ \Omega = 1.8\ \text{M}\Omega$$

***Related Exercise*** Convert 2.2 k$\Omega$ to megohms.

When adding (or subtracting) quantities with different metric prefixes, first convert one of the quantities to the same prefix as the other as the next example shows.

**EXAMPLE 1–16** Add 15 mA and 8000 $\mu$A and express the result in milliamperes.

***Solution*** Convert 8000 $\mu$A to 8 mA and add.

$$\begin{aligned} 15\ \text{mA} + 8000\ \mu\text{A} &= 15 \times 10^{-3}\ \text{A} + 8000 \times 10^{-6}\ \text{A} \\ &= 15 \times 10^{-3}\ \text{A} + 8 \times 10^{-3}\ \text{A} = 15\ \text{mA} + 8\ \text{mA} = 23\ \text{mA} \end{aligned}$$

***Related Exercise*** Add 2873 mA to 10,000 $\mu$A.

**SECTION 1–5 REVIEW**

1. Convert 0.01 MV to kilovolts (kV).
2. Convert 250,000 pA to milliamperes (mA).
3. Add 0.05 MW and 75 kW.

## 1–6 ■ INTRODUCTION TO PSpice

***The speed and power of the personal computer make it suitable as a simulation and analysis tool. Using the proper simulation programs, savings in time and cost can usually be realized in the design, building, and testing of electronic circuits. With simulation software, you can analyze the operation of a prototype circuit and identify design flaws or performance problems before the circuit is actually constructed.***

***It is important to keep in mind that as powerful as modern simulation software is, you must still have a thorough understanding of circuit theory and the concepts and techniques of circuit analysis. The software is simply an additional tool that can be used to extend your analysis capabilities; it cannot interpret analysis results or design circuits for you.***

***After completing this section, you should be able to***

- **Define SPICE and express electrical quantities in PSpice format**
  - ☐ Describe a PSpice circuit file
  - ☐ Use the proper PSpice designations for several circuit components
  - ☐ Express component values in three different notations

### PSpice

The standard for analog electronic circuit simulation and analysis is SPICE (Simulation Program with Integrated Circuit Emphasis). PSpice from MicroSim Corporation is the PC-based version of SPICE. Although the capabilities of PSpice go far beyond what most technicians require, it performs many technician-oriented tasks very well, while requiring only a basic understanding of the program.

With PSpice, you can solve for voltages and currents in simple resistive and reactive circuits, or you can analyze the detailed performance of complex circuits. PSpice allows you to concentrate on learning circuit concepts rather than spending a lot of time on tedious calculations. It is, however, important that you know how to do circuit calculations.

PSpice, like most versions of SPICE, allows for text-based or graphical input and output. This means that you must provide PSpice with a text file that describes the circuit or schematic that shows a picture of the circuit. Text files must be ASCII files; that is, the file must not contain formatting codes. These files can be prepared using an ASCII text editor such as EDIT in DOS. You can also use a word processor that can save text as an ASCII or "DOS text" file. On computer systems with schematic capture capability, the schematic data are, at some point, converted to an ASCII file for input to PSpice; this usually goes unnoticed by the user. Consult the PSpice documentation for details about using PSpice on your computer.

### Circuit File

PSpice uses the ASCII text file as its source of information about a circuit. The circuit file describes the components in the circuit, how they are connected together, and what you wish to know about the circuit.

## Component Names

A *component* is a device in a circuit that affects some electrical characteristics of that circuit. Components include resistors, capacitors, inductors, diodes, transistors, voltage sources, and current sources. A circuit file describes the components of a circuit and how they are connected. The following rules are used in PSpice to identify some basic components (also known as *devices*) listed in a circuit file:

1. Resistor names begin with R.
2. Capacitor names begin with C.
3. Inductor names begin with L.
4. Independent voltage source names begin with V.
5. Independent current source names begin with I.

Notice that in each case these letters are the standard symbols for the electrical quantity related to the component. PSpice, of course, uses other component names, such as Q for transistor and D for diode, but the ones listed above are the primary components used in this course. Any others will be introduced as needed. Unless indicated otherwise, a voltage or current source shall imply an *independent* voltage or current source.

Component names may contain any combination of letters, numbers, and underscores provided that the first letter follows the rules stated previously and each component name in the circuit file is unique. PSpice considers uppercase and lowercase letters to be the same. For example, RTOT and rtot refer to the same device. Component names may be virtually any length, but longer names take more time to type and are easier to misspell. Consequently, names such as R1 are preferable to RESISTOR1.

A few examples of valid component names are as follows:

R1, r2, C1, c2, CAP1, VIN, Vout, VOLTAGE1, ir, I_L

Some examples of invalid component names and the reasons are as follows:

| | |
|---|---|
| R_#1 | (# not allowed in names) |
| C 2 | (spaces not allowed in names) |
| TEN_OHM_RESISTOR | (resistor names must start with R) |
| _VIN | (name cannot begin with underscore) |

## Component Values

In addition to names, the circuit file must contain the values of all components in the circuit. These values must be in the same line as the component name, so that PSpice knows which electrical unit to assign to each value. For example, if the component value 100 is in the description line for component VIN, PSpice knows that the component value is 100 V because VIN is the name of a voltage source. If, on the other hand, the component value 100 is in the description line with R13, PSpice interprets the component value to be 100 Ω because R13 is the name for a resistor. PSpice recognizes three different notations for component values: fixed, scientific, and metric.

***Fixed Notation*** In fixed notation, the component value is written as you normally would except any commas are omitted. For example, fifteen hundred is written as 1500 (not as 1,500). A fractional number is written with an optional zero before the decimal point as, for example, 0.153. Fixed notation is convenient as long as the values are not too large or too small.

***Scientific Notation*** In this notation, powers of ten and associated values are written similar to the way numbers are entered on a calculator with the letter E separating the number (mantissa) from the exponent (power of ten). For example, the number one thousand five hundred is written as 1.5E03 and the number one hundred fifty thousandths is written as 1.5E–01.

Specifying leading zeros in the exponent, as well as an explicit "+" sign for positive exponents, is optional, but it is considered good style. Thus, 1.5E03, 1.5E3, and 1.5E+03 are all equivalent and acceptable; however, the latter form is preferred.

***Metric Notation*** In this notation, component values are entered as they would be expressed using metric suffixes. The metric suffix abbreviations used in PSpice along with the corresponding metric prefixes and scale factors they represent are

F for femto ($10^{-15}$) K for kilo ($10^3$)
P for pico ($10^{-12}$) MEG for mega ($10^6$)
N for nano ($10^{-9}$) G for giga ($10^9$)
U for micro ($10^{-6}$) T for tera ($10^{12}$)
M for milli ($10^{-3}$)

Note that since PSpice is not case sensitive, MEG is used to avoid ambiguity with M. The value one thousand five hundred ($1.5 \times 10^3$) is written as 1.5K. The value one hundred fifty thousandths ($150 \times 10^{-3}$) is written as 1.5M. There is no space between the number and the suffix, and for values between 1 and 1000 no suffix is used.

Examples of component values and ways to express them in PSpice are shown in Table 1–5.

**TABLE 1–5**

| Value | Fixed | Scientific | Metric |
|---|---|---|---|
| 0.000022 | 0.000022 | 2.2E–05 | 22U |
| 0.47 | 0.47 | 4.7E–01 | 470M |
| 18 | 18 | 1.8E01 | 18 |
| 1,250 | 1250 | 1.25E02 | 1.25K |
| 680,000 | 680000 | 6.8E05 | 680K |
| 10,000,000 | 10000000 | 1.0E07 | 10MEG |

**SECTION 1–6 REVIEW**

1. Determine whether or not each of the following names is valid and, if so, identify the type of component.
   **(a)** Inductor_1 **(b)** L1 **(c)** cap12 **(d)** C 5
   **(e)** 3V **(f)** RESA **(g)** Value1 **(h)** r LOAD
   **(i)** R#2 **(j)** IS **(k)** Five Amps **(l)** R_E4
2. Which one of the following component names is invalid for PSpice?
   **(a)** CAP1 **(b)** V2 **(c)** L_300 **(d)** Amp5
   **(e)** R_5 **(f)** RTENOHMS **(g)** res_10 **(h)** Volt2
3. Identify the type of component for each PSpice name:
   **(a)** Current3 **(b)** LOAD **(c)** R2 **(d)** I8
   **(e)** VAR **(f)** INDUCTOR **(g)** V_4 **(h)** C_R
4. Express each of the following values in fixed notation, scientific notation, and metric notation for PSpice:
   **(a)** 0.0068 **(b)** 0.22 **(c)** 1000 **(d)** 180
   **(e)** 560,000 **(f)** 82,000,000

## ■ SUMMARY

- Resistors limit electric current.
- Capacitors store electrical charge.
- Inductors store energy electromagnetically.
- Inductors are also known as *coils*.
- Transformers magnetically couple ac voltages.
- Semiconductor devices include diodes, transistors, and integrated circuits.
- Power supplies provide current and voltage.
- Voltmeters measure voltage.
- Ammeters measure current.
- Ohmmeters measure resistance.
- A multimeter measures voltage, current, and resistance.
- Metric prefixes are a convenient method of expressing both large and small quantities.

## ■ SELF-TEST

**1.** Which of the following is not an electrical quantity?
**(a)** current **(b)** voltage **(c)** time **(d)** power

**2.** The unit of current is
**(a)** volt **(b)** watt **(c)** ampere **(d)** joule

**3.** The unit of voltage is
**(a)** ohm **(b)** watt **(c)** volt **(d)** farad

**4.** The unit of resistance is
**(a)** ampere **(b)** henry **(c)** hertz **(d)** ohm

**5.** 15,000 W is the same as
**(a)** 15 mW **(b)** 15 kW **(c)** 15 MW **(d)** 15 μW

**6.** The quantity $4.7 \times 10^3$ is the same as
**(a)** 470 **(b)** 4700 **(c)** 47,000 **(d)** 0.0047

**7.** The quantity $56 \times 10^{-3}$ is the same as
**(a)** 0.056 **(b)** 0.560 **(c)** 560 **(d)** 56,000

**8.** The number 3,300,000 can be expressed as
**(a)** $3300 \times 10^3$ **(b)** $3.3 \times 10^{-6}$ **(c)** $3.3 \times 10^6$ **(d)** either answer (a) or (c)

**9.** Ten milliamperes can be expressed as
**(a)** 10 MA **(b)** 10 μA **(c)** 10 kA **(d)** 10 mA

**10.** Five thousand volts can be expressed as
**(a)** 5000 V **(b)** 5 MV **(c)** 5 kV **(d)** either answer (a) or (c)

**11.** Twenty million ohms can be expressed as
**(a)** 20 mΩ **(b)** 20 MW **(c)** 20 MΩ **(d)** 20 μΩ

**12.** Hertz is the unit of
**(a)** power **(b)** inductance **(c)** frequency **(d)** time

**13.** A valid PSpice designation for a resistor is
**(a)** R#1 **(b)** ARESISTOR **(c)** RES_1 **(d)** R 1

## ■ PROBLEMS

### SECTION 1–3 Scientific Notation

**1.** Express each of the following numbers in scientific notation as a number between 1 and 10 times a positive power of ten:
**(a)** 3000 **(b)** 75,000 **(c)** 2,000,000

**2.** Express each number in scientific notation as a number between 1 and 10 times a negative power of ten:
**(a)** 1/500 **(b)** 1/2000 **(c)** 1/5,000,000

**3.** Express each of the following numbers in three ways, using $10^3$, $10^4$, and $10^5$:
**(a)** 8400 **(b)** 99,000 **(c)** $0.2 \times 10^6$

**4.** Express each of the following numbers in three ways, using $10^{-3}$, $10^{-4}$, and $10^{-5}$:
**(a)** 0.0002 **(b)** 0.6 **(c)** $7.8 \times 10^{-2}$

**5.** Express each of the following in regular decimal form:
**(a)** $2.5 \times 10^{-6}$ **(b)** $50 \times 10^{2}$ **(c)** $3.9 \times 10^{-1}$

**6.** Express each of the following in regular decimal form:
**(a)** $45 \times 10^{-6}$ **(b)** $8 \times 10^{-9}$ **(c)** $40 \times 10^{-12}$

**7.** Add the following numbers:
**(a)** $(92 \times 10^{6}) + (3.4 \times 10^{7})$ **(b)** $(5 \times 10^{-3}) + (85 \times 10^{-2})$
**(c)** $(560 \times 10^{-8}) + (460 \times 10^{-9})$

**8.** Perform the following subtractions:
**(a)** $(3.2 \times 10^{12}) - (1.1 \times 10^{12})$ **(b)** $(26 \times 10^{8}) - (1.3 \times 10^{9})$
**(c)** $(150 \times 10^{-12}) - (8 \times 10^{-11})$

**9.** Perform the following multiplications:
**(a)** $(5 \times 10^{3})(4 \times 10^{5})$ **(b)** $(12 \times 10^{12})(3 \times 10^{2})$
**(c)** $(2.2 \times 10^{-9})(7 \times 10^{-6})$

**10.** Divide the following:
**(a)** $(10 \times 10^{3}) \div (2.5 \times 10^{2})$ **(b)** $(250 \times 10^{-6}) \div (50 \times 10^{-8})$
**(c)** $(4.2 \times 10^{8}) \div (2 \times 10^{-5})$

**11.** Express each of the following numbers as a number having a power of ten of $10^{-6}$:
**(a)** $2.37 \times 10^{-3}$ **(b)** $0.001856 \times 10^{-2}$
**(c)** $5743.89 \times 10^{-12}$ **(d)** $100 \times 10^{3}$

**12.** Perform the following indicated operations:
**(a)** $(2.8 \times 10^{3})(3 \times 10^{2})/(2 \times 10^{2})$ **(b)** $(46)(10^{-3})(10^{5})/10^{6}$
**(c)** $(7.35)(0.5 \times 10^{12})/[(2)(10 \times 10^{10})]$ **(d)** $(30)^{2}(5)^{3}/10^{-2}$

**13.** Perform each operation:
**(a)** $(49 \times 10^{6})/(4 \times 10^{-8})$ **(b)** $(3 \times 10^{3})^{2}/(1.8 \times 10^{3})$
**(c)** $(1.5 \times 10^{-6})^{2}(4.7 \times 10^{6})$ **(d)** $1/[(5 \times 10^{-3})(0.01 \times 10^{-6})]$

## SECTION 1–4 METRIC PREFIXES

**14.** Express each of the following as a quantity having a metric prefix:
**(a)** $31 \times 10^{-3}$ A **(b)** $5.5 \times 10^{3}$ V **(c)** $200 \times 10^{-12}$ F

**15.** Express the following using metric prefixes:
**(a)** $3 \times 10^{-6}$ F **(b)** $3.3 \times 10^{6}$ Ω **(c)** $350 \times 10^{-9}$ A

**16.** Express the following quantities using powers of ten:
**(a)** 258 mA **(b)** 0.022 μF **(c)** 1200 kV

**17.** Express each of the following quantities as a metric unit:
**(a)** $2.4 \times 10^{-5}$ A **(b)** $970 \times 10^{4}$ Ω **(c)** $0.003 \times 10^{-9}$ W

**18.** Complete the following operations:
**(a)** (50 μA)(6.8 kΩ) = ________ V **(b)** 24 V/1.2 MΩ = ________ A
**(c)** 12 kV/20 mA = ______ Ω

**19.** Complete the following operations and express the results using metric prefixes:
**(a)** $(100\ \mu A)^{2}(8.2\ k\Omega)$ = ________ W **(b)** $(30\ kV)^{2}/2\ M\Omega$ = ________ W

## SECTION 1–5 METRIC UNIT CONVERSIONS

**20.** Perform the indicated conversions:
**(a)** 5 mA to microamperes **(b)** 3200 μW to milliwatts
(c) 5000 kV to megavolts **(d)** 10 MW to kilowatts

**21.** Determine the following:
**(a)** The number of microamperes in 1 milliampere
**(b)** The number of millivolts in 0.05 kilovolt
**(c)** The number of megohms in 0.02 kilohm
**(d)** The number of kilowatts in 155 milliwatts

## ■ ANSWERS TO SECTION REVIEWS

### Section 1–1

1. Common electrical components are resistors, capacitors, inductors, and transformers.
2. The ammeter measures current.
3. The ohmmeter measures resistance.
4. The voltmeter measures voltage.
5. A multimeter is an instrument that measures voltage, current, and resistance.

### Section 1–2

1. SI is the abbreviation for Système International.
2. Refer to Table 1–1 after you have compiled your list of electrical quantities.
3. Refer to Table 1–2 after you have compiled your list of magnetic quantities.

### Section 1–3

1. True
2. $100 = 1 \times 10^2$
3. **(a)** $(1 \times 10^5) + (2 \times 10^5) = 3 \times 10^5$ **(b)** $(3 \times 10^6)(2 \times 10^4) = 6 \times 10^{10}$
   **(c)** $(8 \times 10^3) \div (4 \times 10^2) = 2 \times 10^1$ **(d)** $(25 \times 10^{-6}) - (130 \times 10^{-7}) = 12 \times 10^{-6}$

### Section 1–4

1. $10^6$ = mega (M), $10^3$ = kilo (k), $10^{-3}$ = milli (m), $10^{-6}$ = micro ($\mu$), $10^{-9}$ = nano (n), and $10^{-12}$ = pico (p)
2. 0.000001 A = 1 $\mu$A (one microampere)
3. 250,000 W = 250 kW (250 kilowatts)

### Section 1–5

1. 0.01 MV = 10 kV **2.** 250,000 pA = 0.00025 mA
3. 0.05 MW + 75 kW = 50 kW + 75 kW = 125 kW

### Section 1–6

1. **(a)** Valid; current source **(b)** Valid; inductor
   **(c)** Valid; capacitor **(d)** Not valid; there is a space.
   **(e)** Not valid; begins with a number **(f)** Valid; resistor
   **(g)** Valid; voltage source **(h)** Not valid; there is a space.
   **(i)** Not valid; uses a space **(j)** Valid; current source
   **(k)** Not valid; there is a space. **(l)** Valid; resistor
2. **(d)** Amp5
3. **(a)** capacitor **(b)** inductor **(c)** resistor **(d)** current source
   **(e)** voltage source **(f)** current source **(g)** voltage source **(h)** capacitor
4. **(a)** 0.0068; 6.8E–03; 6.8M **(b)** 0.22; 2.2E–01, 220M
   **(c)** 1000; 1E+03; 1K **(d)** 180; 1.8E+02; 180
   **(e)** 560000; 5.6E+05; 560K **(f)** 82000000; 82E+06, 82MEG

## ■ ANSWERS TO RELATED EXERCISES FOR EXAMPLES

**1–1** $4.75 \times 10^3$
**1–2** $7.38 \times 10^{-3}$
**1–3** 9120
**1–4** $8.6 \times 10^3$
**1–5** $55 \times 10^{-2}$
**1–6** $4.8 \times 10^3$
**1–7** $25 \times 10^6$
**1–8** **(a)** 56 MΩ **(b)** 470 $\mu$A
**1–9** [2] [5] [9] [EXP] [+/–] [3]
**1–10** 1000 $\mu$A
**1–11** 1 mV
**1–12** 0.893 $\mu$A
**1–13** 2200 pF
**1–14** 0.010 $\mu$F
**1–15** 0.0022 MΩ
**1–16** 2883 mA

# 2

# VOLTAGE, CURRENT, AND RESISTANCE



## ■ INTRODUCTION

The useful application of electronics technology to practical situations requires that you first understand the theory on which a given application is based. Once the theory is mastered, you can learn to apply it in practice. In this chapter and throughout the rest of the book, you will learn the basics of putting technology theory into practice.

The theoretical concepts of electrical current, voltage, and resistance are introduced in this chapter. You will learn how to express each of these quantities in the proper units and how each quantity is measured. The essential elements that form a basic electric circuit and how they are put together are covered.

You will be introduced to the types of devices that generate voltage and current. In addition, you will see a variety of components that are used to introduce resistance into electric circuits. The operation of protective devices such as fuses and circuit breakers are discussed, and mechanical switches that are commonly used in electric circuits are introduced. Also, you will learn how to control and measure voltage, current, and resistance using measuring instruments.

Voltage is essential in any kind of electric circuit. Voltage is the potential energy required to make the circuit work. Current is also necessary for electric circuits to operate, but it takes voltage to produce the current. Current is the movement of electrons through a circuit. Resistance in a circuit limits the amount of current. A simple circuit analogy is a water system. Voltage is like the pressure required to force water through the pipes. Current through wires can be thought of as the water moving through the pipes. Resistance can be thought of as the restriction on the water flow produced by adjusting a valve.

## TECHnology Theory Into Practice

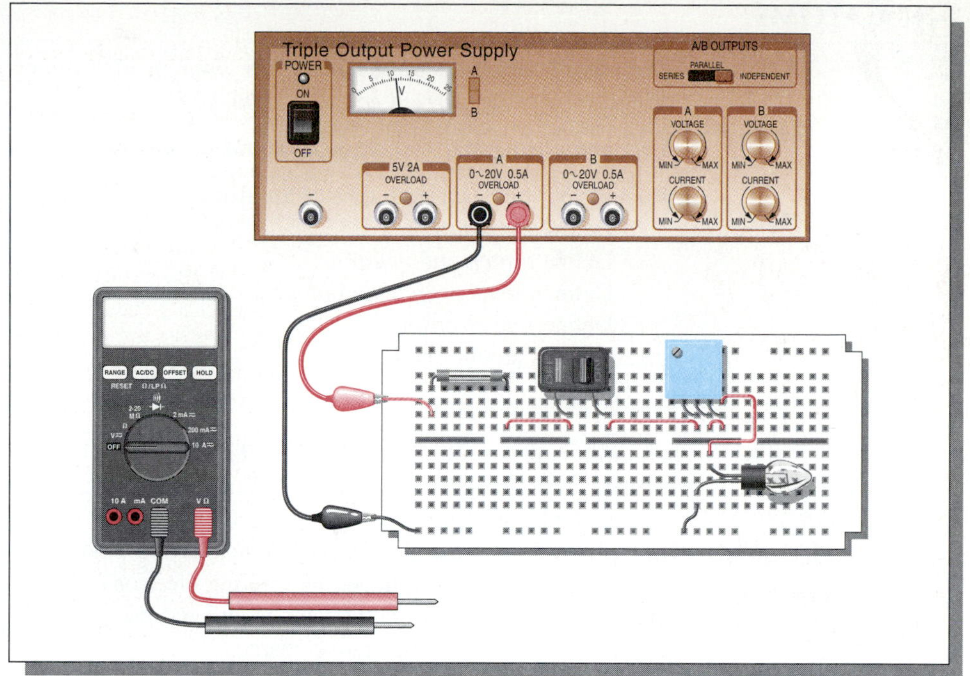


In Section 2–8, you will see how the theory presented in this chapter is applicable to a practical circuit that simulates part of your car's lighting system. Your automobile lights are examples of simple types of electric circuits. When you turn on your headlights and taillights, you are connecting the light bulbs to the battery, which provides the voltage and produces current through each bulb. The current causes the bulbs to emit light. The light bulbs themselves have resistance that limits the amount of current. The instrument panel light in most cars can be adjusted for brightness. When you make this adjustment by turning the knob, you are actually changing the resistance in the circuit, thereby causing the current to change. The amount of current through the light bulb determines its brightness.

## CHAPTER OBJECTIVES

- ☐ Describe the basic structure of an atom
- ☐ Explain the concept of electrical charge
- ☐ Define *voltage* and discuss its characteristics
- ☐ Define *current* and discuss its characteristics
- ☐ Define *resistance* and discuss its characteristics
- ☐ Describe a basic electric circuit
- ☐ Make basic circuit measurements
- ☐ Specify circuit connections for PSpice (optional)

## 2–1 ■ ATOMS

***The atomic structure of the materials used in electronics determines, among other things, how easily current can be established.***

***After completing this section, you should be able to***

- **Describe the basic structure of an atom**
  - ☐ Define *electron*
  - ☐ Define *proton* and *neutron*
  - ☐ Define *atomic weight* and *atomic number*
  - ☐ Define *free electron*
  - ☐ Discuss atomic shells

An **atom** is the smallest particle of an **element** that still retains the characteristics of that element. Different elements have different types of atoms. In fact, every element has a unique atomic structure.

According to the Bohr theory, atoms have a planetary type of structure, consisting of a central nucleus surrounded by orbiting electrons. The nucleus consists of positively charged particles called **protons** and uncharged particles called **neutrons.** The **electrons** are the basic particles of negative charge.

Each type of atom has a certain number of electrons and protons that distinguishes the atom from atoms of all other elements. For example, the simplest atom is that of hydrogen. It has one proton and one electron, as pictured in Figure 2–1(a). The helium atom, shown in Figure 2–1(b), has two protons and two neutrons in the nucleus, which is orbited by two electrons.

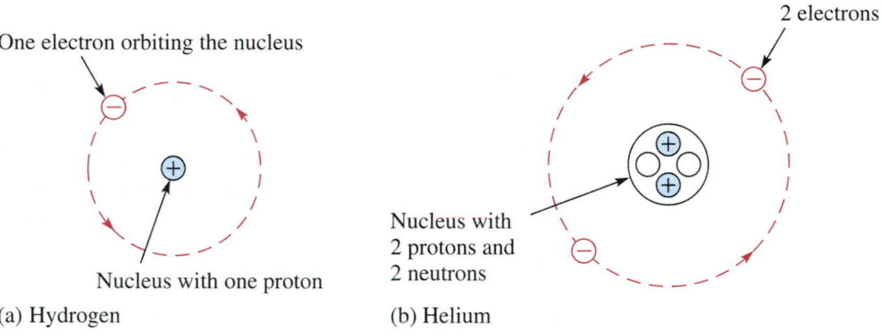


FIGURE 2–1
***Hydrogen and helium atoms.***

### Atomic Weight and Atomic Number

All elements are arranged in the periodic table of the elements in order according to their **atomic number,** which is the number of protons in the nucleus. The **atomic weight** is approximately the number of protons and neutrons in the nucleus. For example, hydrogen has an atomic number of one and an atomic weight of one. The atomic number of helium is two, and its atomic weight is four.

In their normal (or neutral) state, all atoms of a given element have the same number of electrons as protons. So the positive charges cancel the negative charges, and the atom has a net charge of zero.

## The Copper Atom

Since copper is the most commonly used metal in **electrical** applications, let's examine its atomic structure. The copper atom has 29 electrons in orbit around the nucleus. They do not all occupy the same orbit, however. They move in orbits at varying distances from the nucleus.

The orbits in which the electrons revolve are called **shells.** The number of electrons in each shell follows a predictable pattern according to the formula, $2N^2$, where $N$ is the number of the shell. The first shell of any atom can have up to 2 electrons, the second shell up to 8 electrons, the third shell up to 18 electrons, and the fourth shell up to 32 electrons.

A copper atom is shown in Figure 2–2. Notice that the fourth or outermost shell, called the **valence** shell, has only 1 electron, called the **valence electron.**

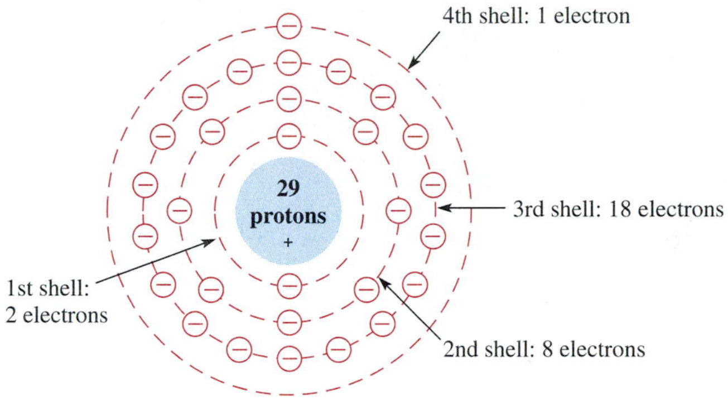

**FIGURE 2–2**
***The copper atom.***

4th shell: 1 electron
29 protons +
3rd shell: 18 electrons
1st shell: 2 electrons
2nd shell: 8 electrons

## Free Electrons

When the electron in the outer shell of the copper atom gains sufficient energy from the surrounding medium, it can break away from the parent atom and become what is called a **free electron.** The free electrons in the copper material are capable of moving from one atom to another in the material. In other words, they drift randomly from atom to atom within the copper. As you will see, the free electrons make electric current possible.

Three categories of materials are used in electronics: conductors, semiconductors, and insulators.

## Conductors

**Conductors** are materials that allow current to flow easily. They have a large number of free electrons and are characterized by one to three valence electrons in their structure. Most metals are good conductors. Silver is the best conductor, and copper is next. Copper is the most widely used conductive material because it is less expensive than silver. Copper wire is commonly used as a conductor in electric circuits.

## Semiconductors

**Semiconductors** are classed below the conductors in their ability to carry current because they have fewer free electrons than do conductors. Semiconductors have four valence electrons in their atomic structures. However, because of their unique characteristics, certain semiconductor materials are the basis for modern **electronic** devices such as the diode, transistor, and integrated circuit. Silicon and germanium are common semiconductor materials.

### Insulators

Insulating materials are poor conductors of electric current. In fact, **insulators** are used to prevent current where it is not wanted. Compared to conductive materials, insulators have very few free electrons and are characterized by more than four valence electrons in their atomic structures.

**SECTION 2–1 REVIEW**

1. What is the basic particle of negative charge?
2. Define *atom.*
3. What does a typical atom consist of?
4. Define *atomic weight* and *atomic number.*
5. Do all elements have the same types of atoms?
6. What is a free electron?
7. What is a shell in the atomic structure?

## 2–2 ■ ELECTRICAL CHARGE

***As you learned in the last section, the two types of charge are positive charge and negative charge. The electron is the smallest particle that exhibits negative electrical charge. When an excess of electrons exists in a material, there is a net negative electrical charge. When a deficiency of electrons exists, there is a net positive electrical charge.***

***After completing this section, you should be able to***

- **Explain the concept of electrical charge**
  - ☐ Name the unit of charge
  - ☐ Name the types of charge
  - ☐ Discuss attractive and repulsive forces
  - ☐ Determine the amount of charge on a given number of electrons

The **charge** of an electron and that of a proton are equal in magnitude. Electrical charge is symbolized by $Q$. Static electricity is the presence of a net positive or negative charge in a material. Everyone has experienced the effects of static electricity from time to time, for example, when attempting to touch a metal surface or another person or when the clothes in a dryer cling together.

Materials with charges of opposite polarity are attracted to each other, and materials with charges of the same polarity are repelled, as indicated in Figure 2–3. A force acts

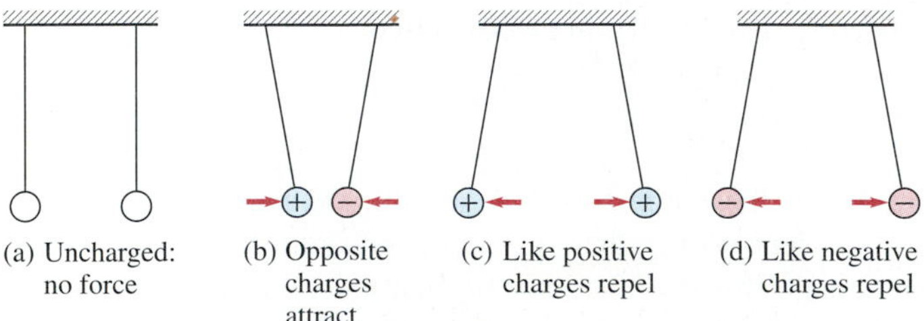


FIGURE 2–3
*Attraction and repulsion of electrical charges.*

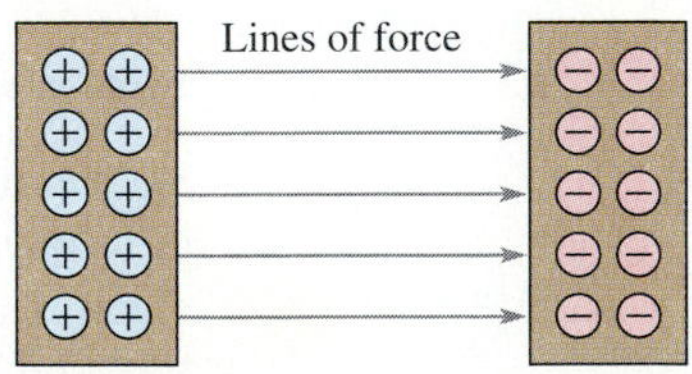


FIGURE 2–4
*Electric field between oppositely charged surfaces.*

between charges, as evidenced by the attraction or repulsion. This force, called an *electric field,* consists of invisible lines of force, as shown in Figure 2–4.

## Coulomb: The Unit of Charge

Electrical charge ($Q$) is measured in **coulombs,** abbreviated C.

**One coulomb is the total charge possessed by $6.25 \times 10^{18}$ electrons.**

A single electron has a charge of $1.6 \times 10^{-19}$ C. The unit of charge is named for Charles Coulomb (1736–1806), a French scientist. The total charge in a given number of electrons is stated in the following formula:

$$Q = \frac{\text{number of electrons}}{6.25 \times 10^{18}\ \text{electrons/C}} \qquad (2\text{–}1)$$

## Positive and Negative Charge

Consider a neutral atom—that is, one that has the same number of electrons and protons and thus has no net charge. If a valence electron is pulled away from the atom by the application of energy, the atom is left with a net positive charge (more protons than electrons) and becomes a positive ion. If an atom acquires an extra electron in its outer shell, it has a net negative charge and becomes a negative ion.

The amount of energy required to free a valence electron is related to the number of electrons in the outer shell. An atom can have up to eight valence electrons. The more complete the outer shell, the more stable the atom and thus the more energy is required to release an electron. Figure 2–5 illustrates the creation of a positive and a negative ion when sodium chloride dissolves and the sodium atom gives up its single valence electron to the chlorine atom.

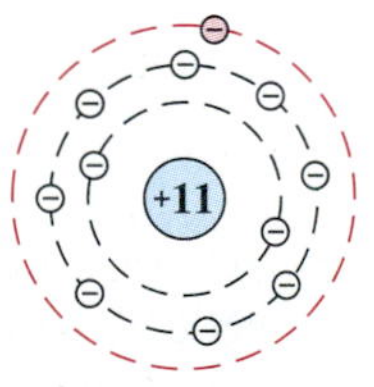


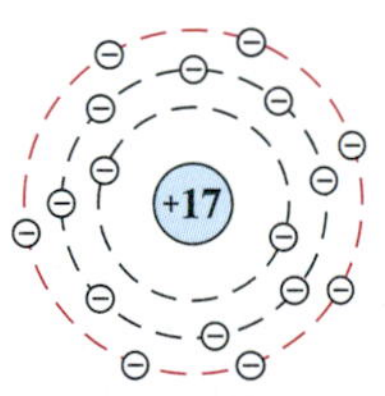


Sodium atom
(11 protons, 11 electrons)

Chlorine atom
(17 protons, 17 electrons)

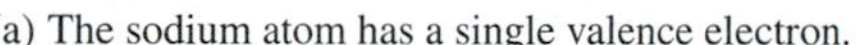

(a) The sodium atom has a single valence electron.

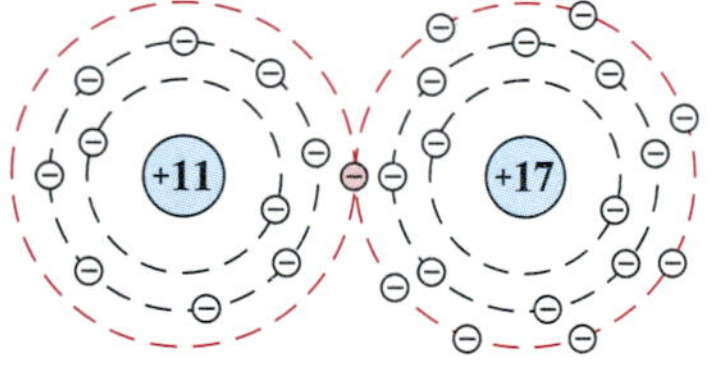


(b) The atoms combine by sharing the valence electron to form sodium chloride (table salt).

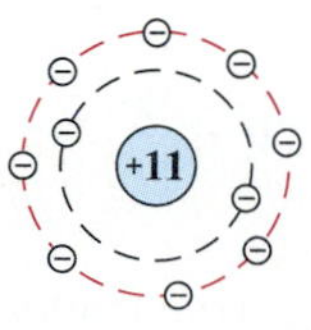


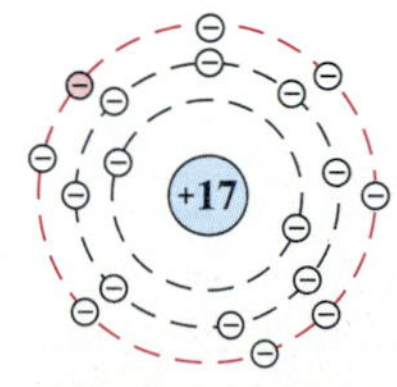


Positive sodium ion
(11 protons, 10 electrons)

Negative chlorine ion
(17 protons, 18 electrons)

(c) When dissolved, the sodium atom gives up the valence electron to become a positive ion, and the chlorine atom retains the extra valence electron to become a negative ion.

**FIGURE 2–5**
***Example of the formation of positive and negative ions.***

**EXAMPLE 2–1** How many coulombs do $93.75 \times 10^{16}$ electrons represent?

*Solution*

$$Q = \frac{\text{number of electrons}}{6.25 \times 10^{18}\ \text{electrons/C}} = \frac{93.75 \times 10^{16}\ \text{electrons}}{6.25 \times 10^{18}\ \text{electrons/C}} = 15 \times 10^{-2}\ \text{C} = 0.15\ \text{C}$$

*Related Exercise* How many electrons does it take to have 3 C of charge?

**SECTION 2–2 REVIEW**

1. What is the symbol for charge?
2. What is the unit of charge, and what is the unit symbol?
3. What causes positive and negative charge?
4. How much charge, in coulombs, is there in $10 \times 10^{12}$ electrons?

## 2–3 ■ VOLTAGE

***As you have seen, a force of attraction exists between a positive and a negative charge. A certain amount of energy must be exerted in the form of work to overcome the force and move the charges a given distance apart. All opposite charges possess a certain potential energy because of the separation between them. The difference in potential energy of the charges is the potential difference or voltage. Voltage is the driving force in electric circuits and is what establishes current.***

***After completing this section, you should be able to***

- **Define *voltage* and discuss its characteristics**
  - ☐ State the formula for voltage
  - ☐ Name and define the unit of voltage
  - ☐ Describe basic sources of voltage

Consider a water tank that is supported several feet above the ground. A given amount of energy must be exerted in the form of work to pump water up to fill the tank. Once the water is stored in the tank, it has a certain potential energy which, if released, can be used to perform work. For example, the water can be allowed to fall down a chute to turn a water wheel.

The difference in potential energy in electrical terms is called **voltage** ($V$) and is expressed as energy or work ($W$) per unit charge ($Q$).

$$V = \frac{W}{Q} \qquad \textbf{(2–2)}$$

where $W$ is expressed in **joules** (J) and $Q$ is in coulombs (C).

### Volt: The Unit of Voltage

The unit of voltage is the **volt,** symbolized by V.

> **One volt is the potential difference (voltage) between two points when one joule of energy is used to move one coulomb of charge from one point to the other.**

**EXAMPLE 2–2** If 50 J of energy are available for every 10 C of charge, what is the voltage?

***Solution***

$$V = \frac{W}{Q} = \frac{50\text{ J}}{10\text{ C}} = 5\text{ V}$$

***Related Exercise*** How much energy is used to move 50 C from one point to another when the voltage between the two points is 12 V?

## Sources of Voltage

***The Battery*** A voltage **source** is a source of potential energy that is also called *electromotive force* (emf). The **battery** is one type of voltage source that converts chemical energy into electrical energy. A voltage exists between the electrodes (terminals) of a battery, as shown by a voltaic cell in Figure 2–6. One electrode is positive and the other negative as a result of the separation of charges caused by the chemical action when two different conducting materials are dissolved in the electrolyte.

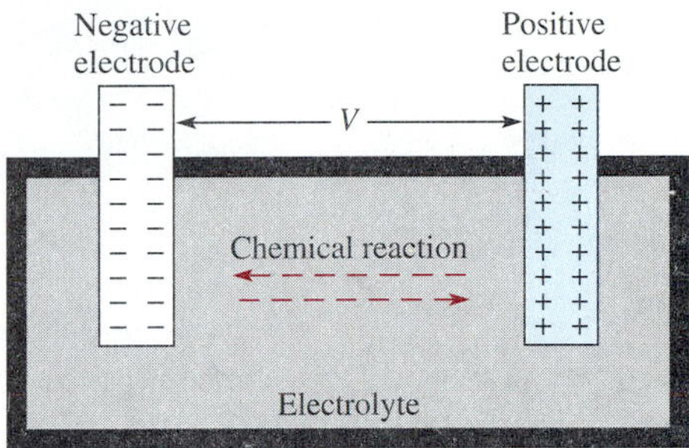


FIGURE 2–6
*A voltaic cell converts chemical energy into electrical energy.*

Batteries are generally classified as primary cells, which cannot be recharged, and secondary cells, which can be recharged by reversal of the chemical action. The amount of voltage provided by a battery varies. For example, a flashlight battery is 1.5 V and an automobile battery is 12 V. Some typical batteries are shown in Figure 2–7. For additional coverage of batteries, see Appendix B.

FIGURE 2–7
*Typical batteries (courtesy of Gould, Inc.).*

***The Electronic Power Supply*** Electronic **power supplies** convert the ac voltage from the wall outlet to a constant (dc) voltage that is available across two terminals, as indicated in Figure 2–8(a). Typical commercial power supplies are shown in Figure 2–8(b).

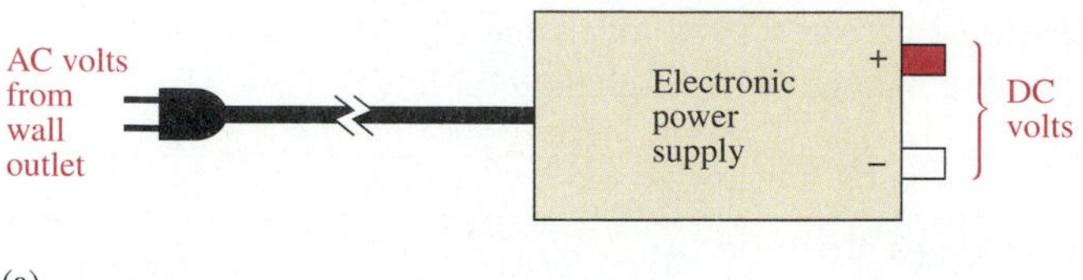


(a)

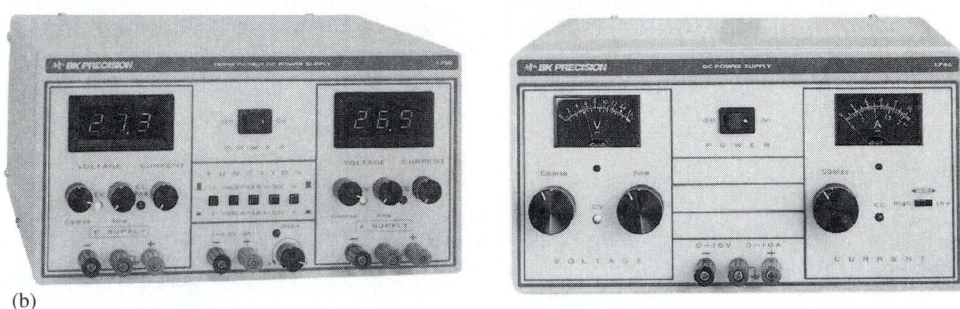

(b)

FIGURE 2–8
***Electronic power supplies. Part (b) courtesy of BK Precision, Maxtec International Corp.***

***The Solar Cell*** The operation of solar cells is based on the principle of *photovoltaic action,* which is the process whereby light energy is converted directly into electrical energy. A basic solar cell consists of two layers of different semiconductive materials joined together to form a junction. When one layer is exposed to light, many electrons acquire enough energy to break away from their parent atoms and cross the junction. This process forms negative ions on one side of the junction and positive ions on the other, and thus a potential difference (voltage) is developed. Figure 2–9 shows the construction of a basic solar cell.

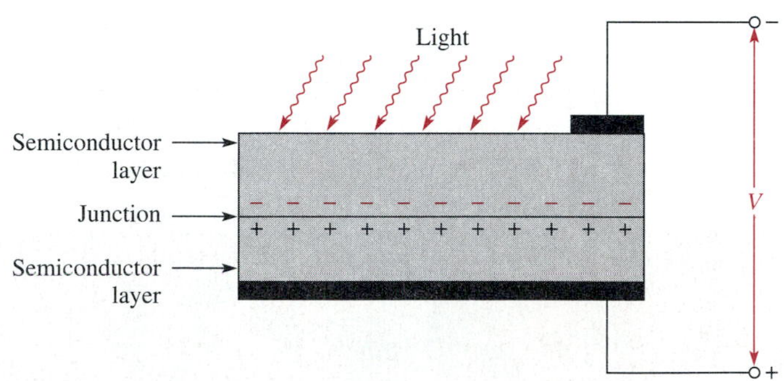


FIGURE 2–9
***Construction of a basic solar cell.***

***The Generator*** **Generators** convert mechanical energy into electrical energy using a principle called *electromagnetic induction* (see Chapter 10). A conductor is rotated through a magnetic field, and a voltage is produced across the conductor. A typical generator is pictured in Figure 2–10.

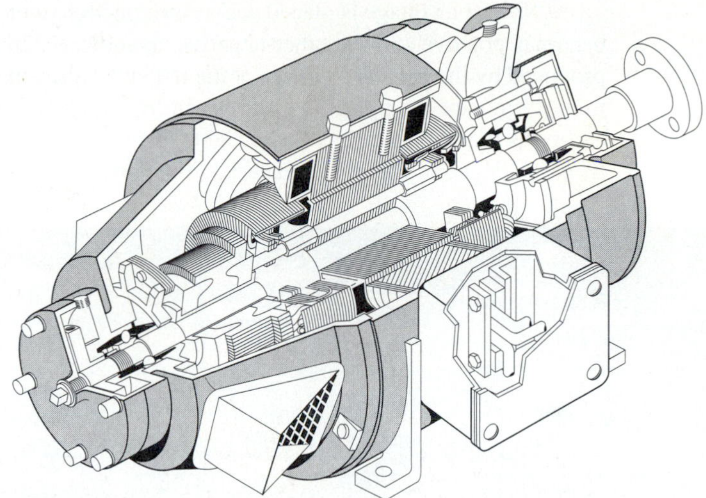

FIGURE 2–10
*Cutaway view of a dc generator.*

**SECTION 2–3 REVIEW**

1. Define *voltage.*
2. What is the unit of voltage?
3. How much is the voltage when there are 24 joules of energy for 10 coulombs of charge?
4. List four sources of voltage.

## 2–4 ■ CURRENT

***Voltage provides energy to electrons which allows them to move through a circuit. This movement of electrons is the current, which results in work being done in an electric circuit.***

***After completing this section, you should be able to***

- **Define *current* and discuss its characteristics**
  - ☐ Explain the movement of electrons
  - ☐ State the formula for current
  - ☐ Name and define the unit of current

As you have learned, free electrons are available in all conductive and semiconductive materials. These electrons drift randomly in all directions, from atom to atom, within the structure of the material, as indicated in Figure 2–11.

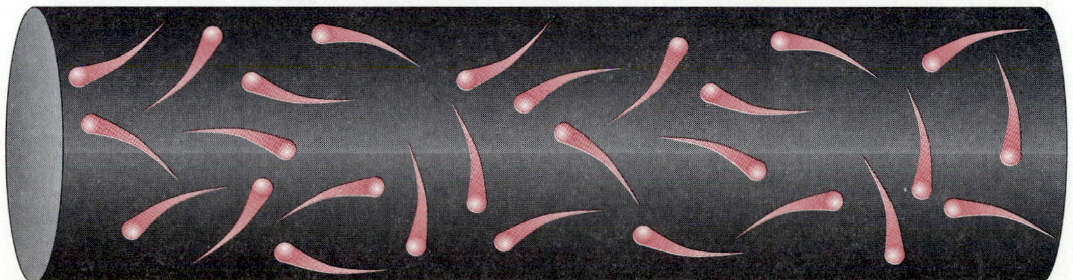

Randomly drifting free electron

FIGURE 2–11
*Random motion of free electrons in a material.*

Now, if a voltage is placed across the conductive or semiconductive material, one end becomes positive and the other negative, as indicated in Figure 2–12. The repulsive force produced by the negative voltage at the left end causes the free electrons (negative charges) to move toward the right. The attractive force produced by the positive voltage at the right end pulls the free electrons to the right. The result is a net movement of the free electrons from the negative end of the material to the positive end, as shown in Figure 2–12.

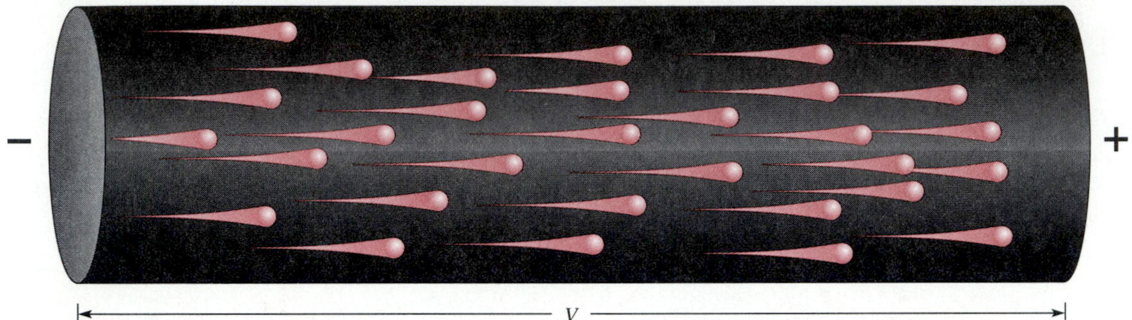


FIGURE 2–12
***Electrons flow from negative to positive when a voltage is applied across a conductive or semiconductive material.***

The movement of these free electrons from the negative end of the material to the positive end is the electrical **current,** symbolized by $I$.

**Electrical current is the rate of flow of charge.**

Current in a conductive material is measured by the number of electrons (amount of charge, $Q$) that flow past a point in a unit of time.

$$I = \frac{Q}{t} \qquad \textbf{(2–3)}$$

where $I$ is current in amperes, $Q$ is the charge of the electrons in coulombs, and $t$ is the time in seconds.

## Ampere: The Unit of Current

Current is measured in a unit called the **ampere** or *amp* for short, symbolized by A. It is named after André Ampère (1775–1836), a French physicist whose work contributed to the understanding of electrical current and its effects.

**One ampere (1 A) is the amount of current that exists when a number of electrons having a total charge of one coulomb (1 C) move through a given cross-sectional area in one second (1 s).**

See Figure 2–13. Remember, one coulomb is the charge carried by $6.25 \times 10^{18}$ electrons.

**EXAMPLE 2–3** Ten coulombs of charge flow past a given point in a wire in 2 s. What is the current in amperes?

***Solution***

$$I = \frac{Q}{t} = \frac{10\text{ C}}{2\text{ s}} = 5\text{ A}$$

***Related Exercise*** If there are 8 A of current through the filament of a lamp, how many coulombs of charge move through the filament in 1.5 s?

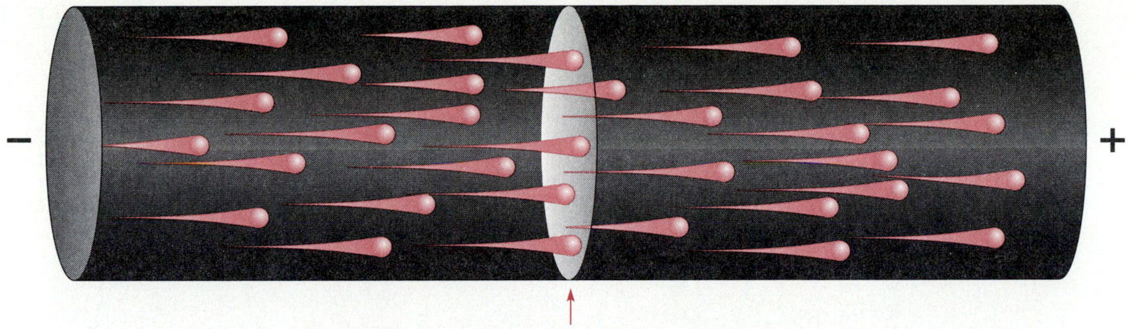

−

+

When a number of electrons having 1 coulomb of charge pass through this cross-sectional area in 1 second, there is 1 ampere of current.

FIGURE 2–13
*Illustration of one ampere of current in a material (1 C/s).*

**SECTION 2–4 REVIEW**

1. Define *current* and state its unit.
2. How many electrons make up one coulomb of charge?
3. What is the current in amperes when 20 C flow past a point in a wire in 4 s?

## 2–5 ■ RESISTANCE

***When there is current in a material, the free electrons move through the material and occasionally collide with atoms. These collisions cause the electrons to lose some of their energy, and thus their movement is restricted. The more collisions, the more the flow of electrons is restricted. This restriction varies and is determined by the type of material. The property of a material that restricts the flow of electrons is called resistance.***

***After completing this section, you should be able to***

- **Define *resistance* and discuss its characteristics**
  - ☐ Name and define the unit of resistance
  - ☐ Describe basic types of resistors
  - ☐ Identify resistance by color code or labeling

**Resistance is the opposition to current.**

The schematic symbol for **resistance** is shown in Figure 2–14.

FIGURE 2–14
***Resistance/resistor symbol.***

When there is current through any material that has resistance, heat is produced by the collisions of free electrons and atoms. Therefore, wire, which typically has a very small resistance, becomes warm when there is sufficient current through it.

### Ohm: The Unit of Resistance

Resistance, *R,* is expressed in **ohms,** named after Georg Simon Ohm (1787–1854) and symbolized by the Greek letter omega (Ω).

**One ohm (1 Ω) of resistance exists if there is one ampere (1 A) of current in a material when one volt (1 V) is applied across the material.**

***Conductance*** The reciprocal of resistance is **conductance,** symbolized by $G$. It is a measure of the ease with which current is established. The formula is

$$G = \frac{1}{R} \tag{2–4}$$

The unit of conductance is the siemens, abbreviated S. Occasionally, the earlier unit of mho is still used.

## Resistors

Components that are specifically designed to have a certain amount of resistance are called **resistors.** The principal applications of resistors are to limit current, to divide voltage, and, in certain cases, to generate heat. Although a variety of different types of resistors come in many shapes and sizes, they can all be placed in one of two main categories: fixed or variable.

***Fixed Resistors*** Fixed resistors are available with a large selection of resistance values that are set during manufacturing and cannot be changed easily. They are constructed using various methods and materials. Figure 2–15 shows several common types.

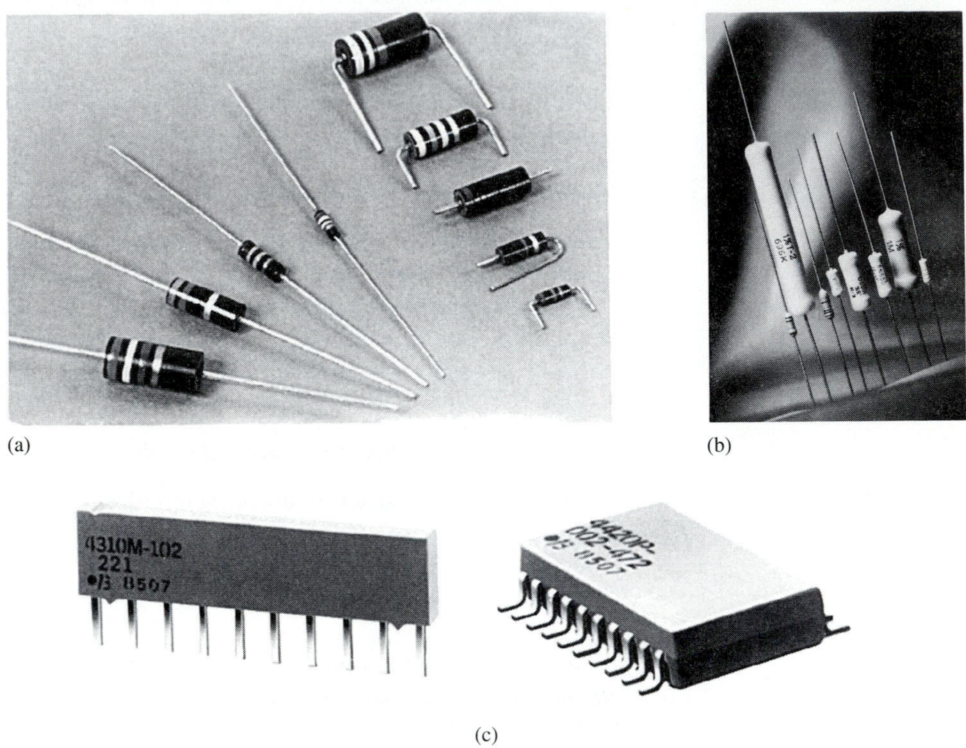


FIGURE 2–15
***Typical fixed resistors. Parts (a) and (b) courtesy of Stackpole Carbon Co. Part (c) courtesy of Bourns, Inc.***

One common fixed resistor is the carbon-composition type, which is made with a mixture of finely ground carbon, insulating filler, and a resin binder. The ratio of carbon to insulating filler sets the resistance value. The mixture is formed into rods, and conductive lead connections are made. The entire resistor is then encapsulated in an insulated coating for protection. Figure 2–16(a) shows the construction of a typical carbon-composition resistor. The chip resistor is another type of fixed resistor and is in the cate-

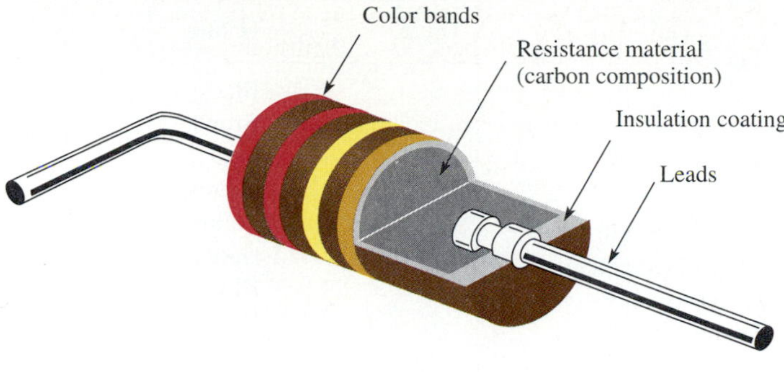


(a) Cutaway view of a carbon-composition resistor

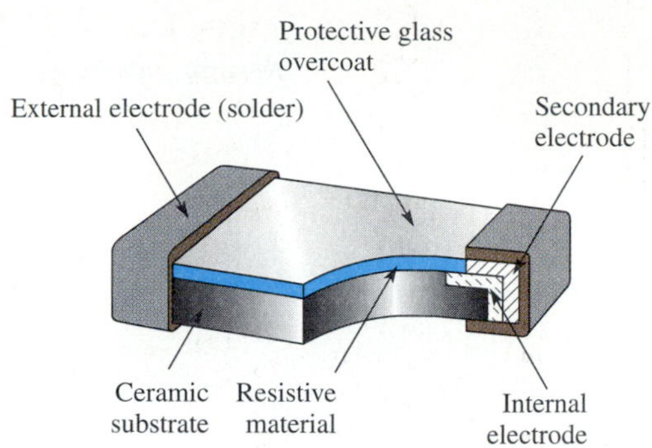


(b) Cutaway view of a chip resistor

**FIGURE 2–16**
***Two types of fixed resistors (not to scale).***

gory of SMT (surface mount technology) components. It has the advantage of a very small size for compact assemblies. Figure 2–16(b) shows the construction of a chip resistor.

Other types of fixed resistors include carbon film, metal film, and wirewound. In film resistors, a resistive material is deposited evenly onto a high-grade ceramic rod. The resistive film may be carbon (carbon film) or nickel chromium (metal film). In these types of resistors, the desired resistance value is obtained by removing part of the resistive material in a helical pattern along the rod using a spiraling technique as shown in Figure 2–17(a). Very close **tolerance** can be achieved with this method. Film resistors are also available in the form of resistor networks as shown in Figure 2–17(b).

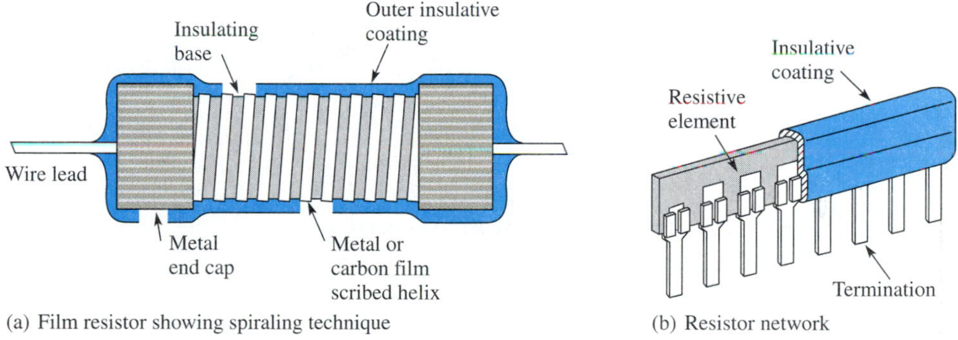


(a) Film resistor showing spiraling technique

(b) Resistor network

**FIGURE 2–17**
***Construction views of typical film resistors.***

Wirewound resistors are constructed with resistive wire wound around an insulating rod and then sealed. Normally, wirewound resistors are used because of their relatively high power ratings.

***Resistor Color Codes*** Fixed resistors with value tolerances of 5%, 10%, or 20% are color coded with four bands to indicate the resistance value and the tolerance. This color-code band system is shown in Figure 2–18, and the color code is listed in Table 2–1.

**FIGURE 2–18**
***Color-code bands on a resistor.***

1st digit
2nd
digit
Percent tolerance
Multiplier
(Number of zeros
following 2nd digit)

**TABLE 2–1**
***Resistor color code***

| | Digit | Color |
|---|---|---|
| Resistance value, first three bands: First band—1st digit Second band—2nd digit Third band—number of zeros following 2nd digit | 0 | Black |
| | 1 | Brown |
| | 2 | Red |
| | 3 | Orange |
| | 4 | Yellow |
| | 5 | Green |
| | 6 | Blue |
| | 7 | Violet |
| | 8 | Gray |
| | 9 | White |
| Tolerance, fourth band | 5% | Gold |
| | 10% | Silver |
| | 20% | No band |

The color code is read as follows:

1. Start with the band closest to one end of the resistor. The first band is the first digit of the resistance value. If it is not clear which is the banded end, start from the end that does not begin with a gold or silver band.
2. The second band is the second digit of the resistance value.
3. The third band is the number of zeros following the second digit, or multiplier.
4. The fourth band indicates the tolerance and is usually gold or silver.

For example, a 5% tolerance means that the *actual* resistance value is within ±5% of the color-coded value. Thus, a 100 Ω resistor with a tolerance of ±5% can have acceptable values as low as 95 Ω and as high as 105 Ω.

For resistance values less than 10 Ω, the third band is either gold or silver. Gold represents a multiplier of 0.1, and silver represents 0.01. For example, a color code of red, violet, gold, and silver represents 2.7 Ω with a tolerance of ±10%. A table of standard resistance values is in Appendix A.

**EXAMPLE 2–4** Find the resistance value in ohms and the percent tolerance for each of the color-coded resistors shown in Figure 2–19.

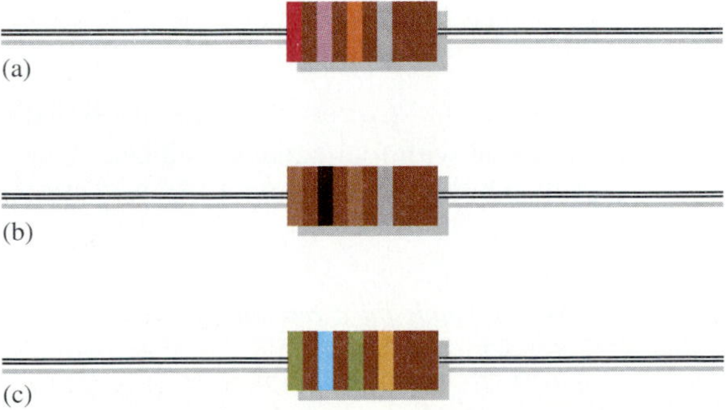


FIGURE 2–19

***Solution***

(a) First band is red = 2, second band is violet = 7, third band is orange = 3 zeros, fourth band is silver = 10% tolerance.

$$R = 27{,}000\ \Omega \pm 10\%$$

(b) First band is brown = 1, second band is black = 0, third band is brown = 1 zero, fourth band is silver = 10% tolerance.

$$R = 100\ \Omega \pm 10\%$$

(c) First band is green = 5, second band is blue = 6, third band is green = 5 zeros, fourth band is gold = 5% tolerance.

$$R = 5{,}600{,}000\ \Omega \pm 5\%$$

***Related Exercise*** A certain resistor has a yellow first band, a violet second band, a red third band, and no fourth band. Determine its value in ohms.

Certain precision resistors with tolerances of 1% or 2% are color coded with five bands. Beginning at the banded end, the first band is the first digit of the resistance value, the second band is the second digit, the third band is the third digit, the fourth band is the multiplier, and the fifth band indicates the tolerance. Table 2–1 applies, except that brown indicates 1% and red indicates 2%.

Numerical labels are often used on certain types of resistors where the resistance value and tolerance are stamped on the body of the resistor. For example, a common system uses R to designate the decimal point and letters to indicate tolerance as follows:

F = ±1%, G = ±2%, J = ±5%, K = ±10%, M = ±20%

For values above 100 Ω, three digits are used to indicate resistance value, followed by a fourth digit that specifies the number of zeros. For values less than 100 Ω, R indicates the decimal point.

Some examples are as follows: 6R8M is a 6.8 Ω ±20% resistor; 3301F is a 3300 Ω ±1% resistor; and 2202J is a 22,000 Ω ±5% resistor.

***Resistor Reliability Band*** The fifth band on some color-coded resistors indicates the resistor's reliability in percent of failures per 1000 hours of use. The fifth-band reliability color code is listed in Table 2–2. For example, a brown fifth band means that if a group of like resistors are operated under standard conditions for 1000 hours, 1% of the resistors in that group will fail.

**TABLE 2–2**
***Fifth-band reliability color code***

| Color | Failures (%) during 1000 Hours of Operation |
|---|---|
| Brown | 1.0% |
| Red | 0.1% |
| Orange | 0.01% |
| Yellow | 0.001% |

***Variable Resistors*** Variable resistors are designed so that their resistance values can be changed easily with a manual or an automatic adjustment.

Two basic uses for variable resistors are to divide voltage and to control current. The variable resistor used to divide voltage is called a **potentiometer.** The variable resis-

1 3 2
(a) Potentiometer

1 2
(b) Rheostat

1 3 2
(c) Potentiometer connected as a rheostat

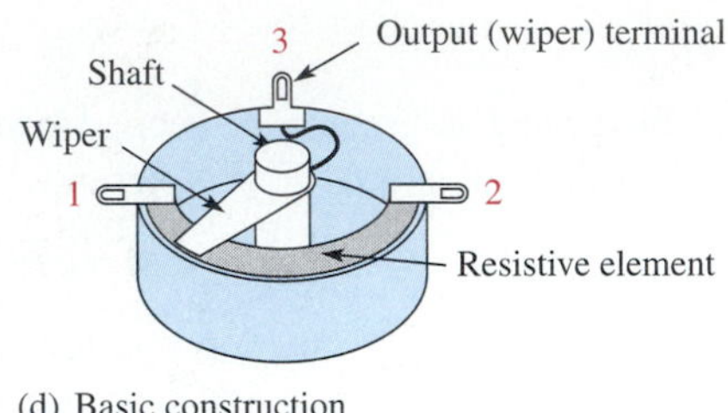


(d) Basic construction

FIGURE 2–20
***Potentiometer and rheostat symbols and basic construction of one type of potentiometer.***

tor used to control current is called a **rheostat.** Schematic symbols for these types are shown in Figure 2–20. The potentiometer is a three-terminal device, as indicated in part (a). Terminals 1 and 2 have a fixed resistance between them, which is the total resistance. Terminal 3 is connected to a moving contact **(wiper).** You can vary the resistance between 3 and 1 or between 3 and 2 by moving the contact up or down.

Figure 2–20(b) shows the rheostat as a two-terminal variable resistor. Part (c) shows how you can use a potentiometer as a rheostat by connecting terminal 3 to either terminal 1 or terminal 2. Parts (b) and (c) are equivalent symbols. Part (d) shows a simplified construction diagram of a potentiometer (which can also be configured as a rheostat). Some typical potentiometers are pictured in Figure 2–21.

Potentiometers and rheostats can be classified as linear or tapered, as shown in Figure 2–22, where a potentiometer with a total resistance of 100 $\Omega$ is used as an example. As shown in part (a), in a linear potentiometer, the resistance between either terminal and the moving contact varies linearly with the position of the moving contact. For example, one-half of a turn results in one-half the total resistance. Three-quarters of a turn results in three-quarters of the total resistance between the moving contact and one terminal, or one-quarter of the total resistance between the other terminal and the moving contact.

In the **tapered** potentiometer, the resistance varies nonlinearly with the position of the moving contact, so that one-half of a turn does not necessarily result in one-half the total resistance. This concept is illustrated in Figure 2–22(b), where the nonlinear values are arbitrary.

The potentiometer is used as a voltage-control device because when a fixed voltage is applied across the end terminals, a variable voltage is obtained at the wiper contact with respect to either end terminal. The rheostat is used as a current-control device because the current can be changed by changing the wiper position.

***Two Types of Automatically Variable Resistors*** A **thermistor** is a type of variable resistor that is temperature-sensitive. When its temperature coefficient is negative, the resistance changes inversely with temperature. When its temperature coefficient is positive, the resistance changes directly with temperature.

The resistance of a **photoconductive cell** changes with a change in light intensity. This cell also has a negative temperature coefficient. Symbols for both of these devices are shown in Figure 2–23 on page 38.

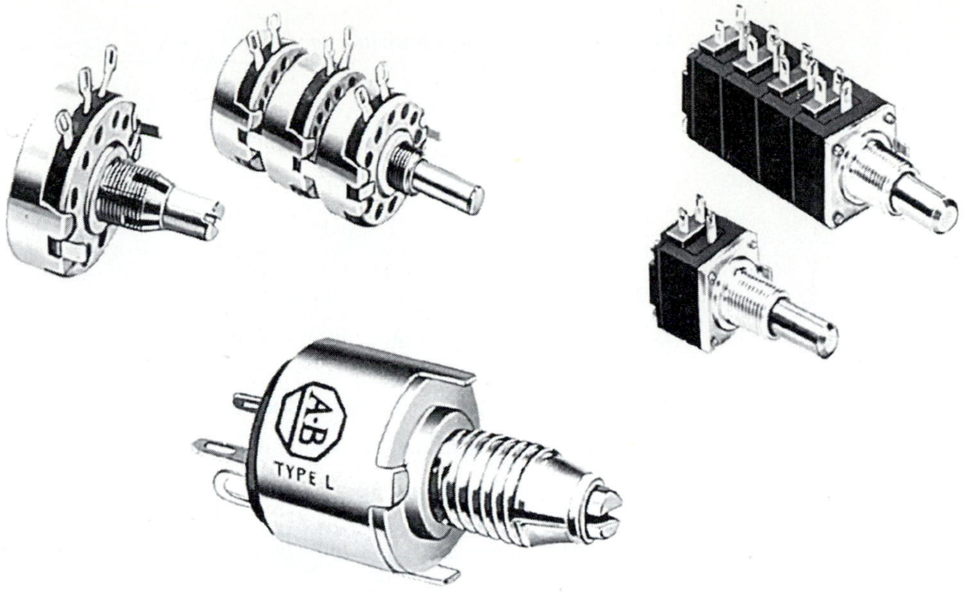

(a)

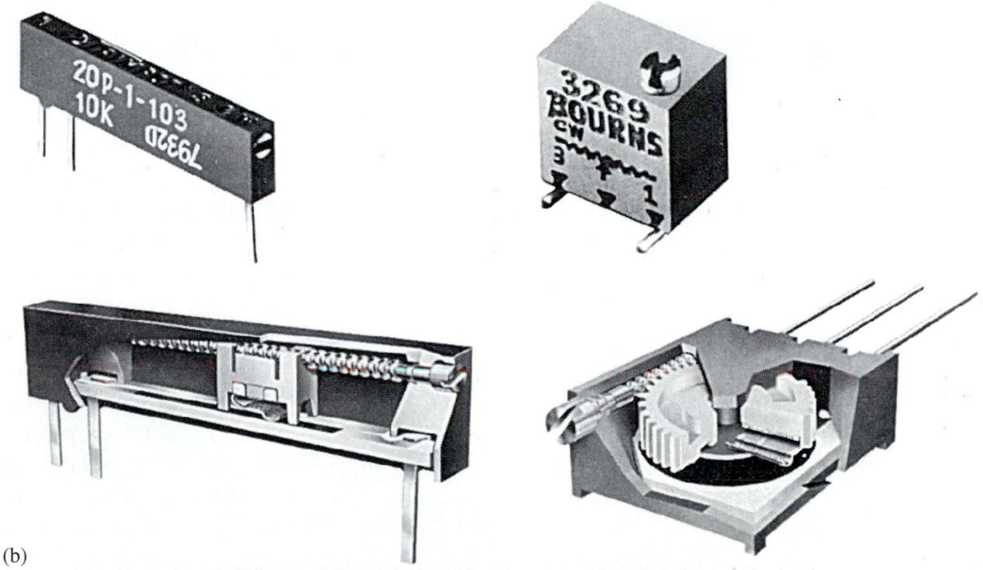

(b)

**FIGURE 2–21**
***(a) Typical potentiometers (courtesy of Allen-Bradley Co.). (b) Trimmer potentiometers with construction views (courtesy of Bourns Trimpot).***

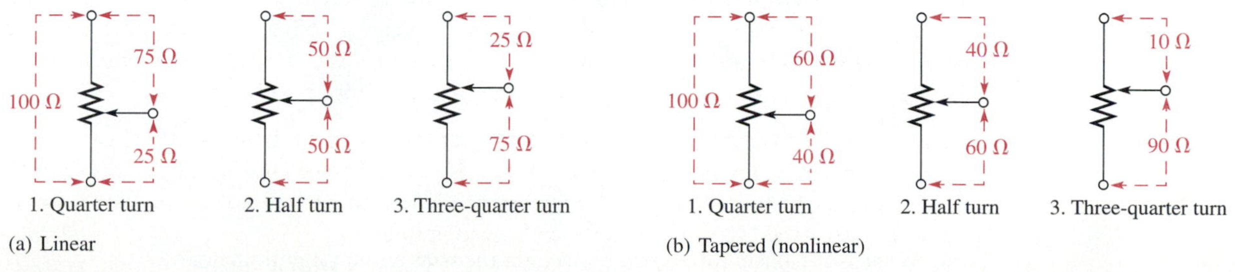


**FIGURE 2–22**
***Examples of (a) linear and (b) tapered potentiometers.***

FIGURE 2–23

***Resistive devices with sensitivities to temperature and light.***

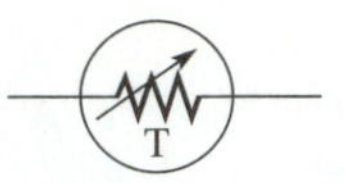


(a) Thermistor

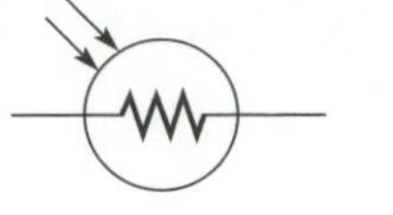

(b) Photoconductive cell

## SECTION 2–5 REVIEW

1. Define *resistance* and name its unit.
2. What are the two main categories of resistors? Briefly explain the difference between them.
3. In the resistor color code, what does each band represent?
4. Determine the resistance and tolerance for each of the resistors in Figure 2–24.

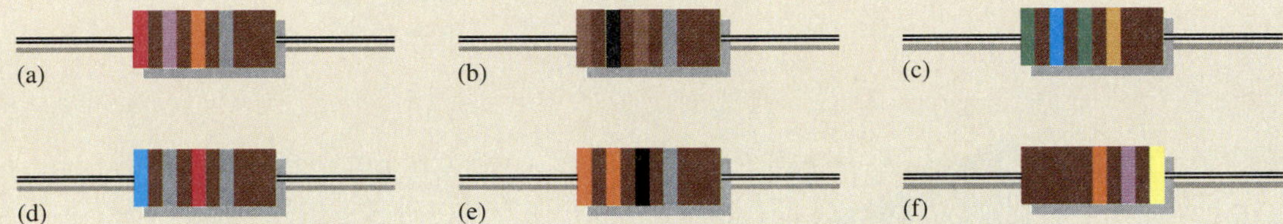


FIGURE 2–24

5. From the selection of resistors in Figure 2–25, select the following values: 330 Ω, 2.2 kΩ, 56 kΩ, 100 kΩ, and 39 kΩ.

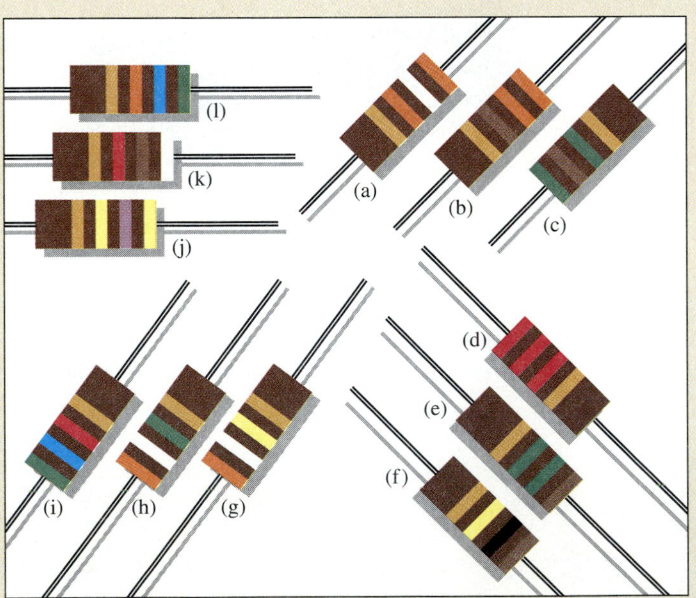


FIGURE 2–25

6. What is the basic difference between a rheostat and a potentiometer?
7. What is a thermistor?

## 2–6 ■ THE ELECTRIC CIRCUIT

***A basic electric circuit is an arrangement of physical components which use voltage, current, and resistance to perform some useful function.***

***After completing this section, you should be able to***

- **Describe a basic electric circuit**
  - ☐ Relate a schematic to a physical circuit
  - ☐ Define *open circuit* and *closed circuit*
  - ☐ Describe various types of protective devices
  - ☐ Describe various types of switches
  - ☐ Explain how wire sizes are related to gage numbers
  - ☐ Define *ground*

### Direction of Current

For a few years after the discovery of electricity, people assumed all current consisted of moving positive charges. However, in the 1890s, the electron was identified as the charge carrier for current in conductive materials.

Today, there are two accepted conventions for the direction of electrical current. *Electron flow direction,* preferred by many in the fields of electrical and electronics technology, assumes for analysis purposes that current is out of the negative terminal of a voltage source, through the circuit, and into the positive terminal of the source. *Conventional current direction* assumes for analysis purposes that current is out of the positive terminal of a voltage source, through the circuit, and into the negative terminal of the source. By following the direction of conventional current, there is a rise in voltage across a source (negative to positive) and a drop in voltage across a resistor (positive to negative).

It actually makes no difference which direction of current is assumed as long as it is used *consistently.* The results of electric circuit analysis are not affected by the direction of current that is assumed for analytical purposes. The direction used for analysis is largely a matter of preference, and there are many proponents for each approach.

Conventional current direction is used widely in electronics technology and is used almost exclusively at the engineering level. Conventional current direction is used throughout this text. An alternate version of this text that uses electron flow direction is also available.

### The Basic Circuit

Basically, an electric **circuit** consists of a voltage source, a load, and a path for current between the source and the **load.** Figure 2–26 shows in pictorial form an example of a simple electric circuit: a battery connected to a lamp with two conductors (wires). The battery is the voltage source, the lamp is the load on the battery because it draws current from the battery, and the two wires provide the current path from the positive terminal of the battery to the lamp and back to the negative terminal of the battery, as shown in part (b). Current goes through the filament of the lamp (which has a resistance), causing it to emit visible light. Current through the battery occurs by chemical action. In many practical cases, one terminal of the battery is connected to a common or ground point. For example, in most automobiles, the negative battery terminal is connected to the metal chassis of the car. The chassis is the ground for the automobile electrical system and acts as a conductor which completes the circuit.

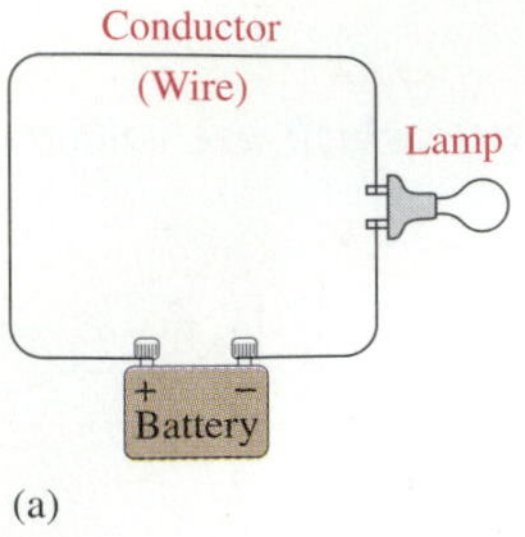


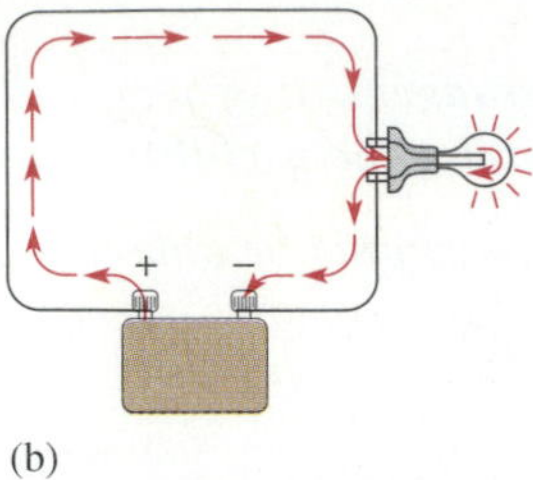


FIGURE 2–26
*A simple electric circuit.*

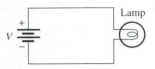


FIGURE 2–27
*Schematic for the circuit in Figure 2–26(a).*

## The Electric Circuit Schematic

An electric circuit can be represented by a **schematic** using standard symbols for each element, as shown in Figure 2–27 for the simple circuit in Figure 2–26(a). The purpose of a schematic is to show in an organized manner how the various components in a given circuit are interconnected so that the operation of the circuit can be determined.

## Closed and Open Circuits

The example circuit in Figure 2–26 illustrated a **closed circuit**—that is, a circuit in which the current has a complete path. When the current path is broken, the circuit is called an **open circuit.**

***Switches*** **Switches** are commonly used for controlling the opening or closing of circuits by either mechanical or electronic means. For example, a switch is used to turn a lamp on or off as illustrated in Figure 2–28. Each circuit pictorial is shown with its associated schematic. The type of switch indicated is a single-pole–single-throw (SPST) toggle switch.

Figure 2–29 shows a somewhat more complicated circuit using a single-pole–double-throw (SPDT) type of switch to control the current to two different lamps. When one lamp is on, the other is off, and vice versa, as illustrated by the two schematics in parts (b) and (c), which represent each of the switch positions.

FIGURE 2–28
*Basic closed and open circuits using an SPST switch for control.*

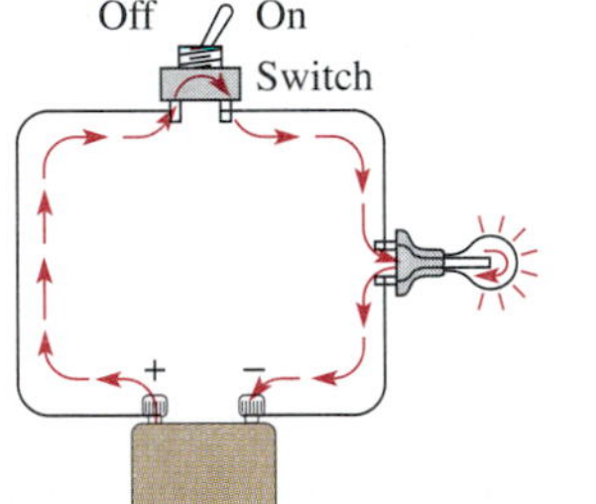


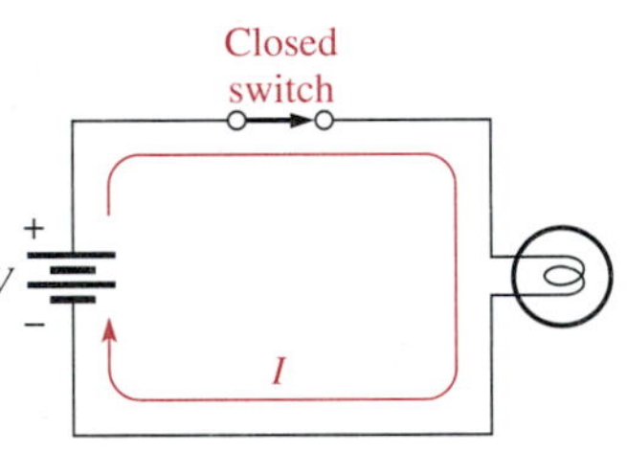


(a) There is current in a *closed* circuit (switch is ON or in the *closed* position).

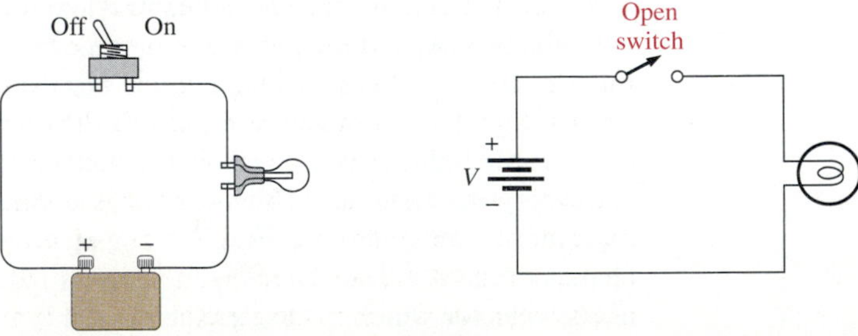


(b) There is no current in an *open* circuit (switch is OFF or in the *open* position).

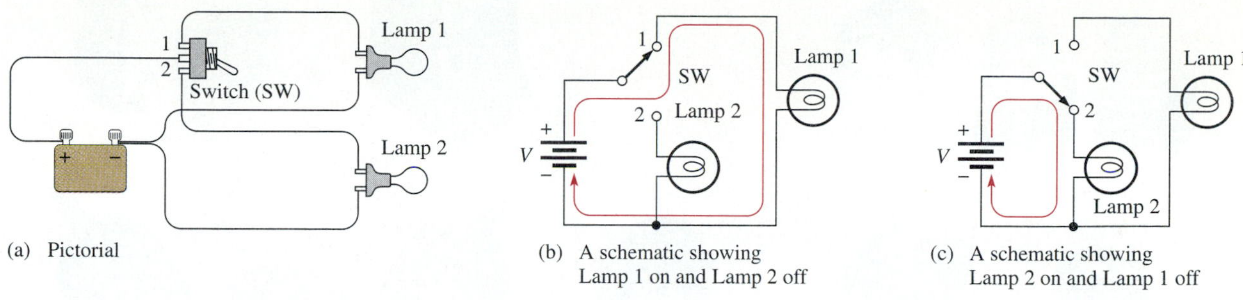


**FIGURE 2–29**
***An example of an SPDT switch controlling two lamps.***

The term *pole* refers to the movable arm in a switch, and the term *throw* indicates the number of contacts that are affected (either opened or closed) by a single switch action (a single movement of a pole).

In addition to the SPST and the SPDT switches (symbols are shown in Figure 2–30(a) and (b)), the following other types are important:

- *Double-pole–single throw (DPST).* The DPST switch permits simultaneous opening or closing of two sets of contacts. The symbol is shown in Figure 2–30(c). The dashed line indicates that the contact arms are mechanically linked so that both move with a single switch action.
- *Double-pole–double-throw (DPDT).* The DPDT switch provides connection from one set of contacts to either of two other sets. The schematic symbol is shown in Figure 2–30(d).
- *Push-button (PB).* In the normally open push-button switch (NOPB), shown in Figure 2–30(e), connection is made between two contacts when the button is depressed, and connection is broken when the button is released. In the normally closed push-button switch (NCPB), shown in Figure 2–30(f), connection between the two contacts is broken when the button is depressed.
- *Rotary.* In a rotary switch, a knob is turned to make connection between one contact and any one of several others. A symbol for a simple six-position rotary switch is shown in Figure 2–30(g).

Figure 2–31 shows several varieties of switches.

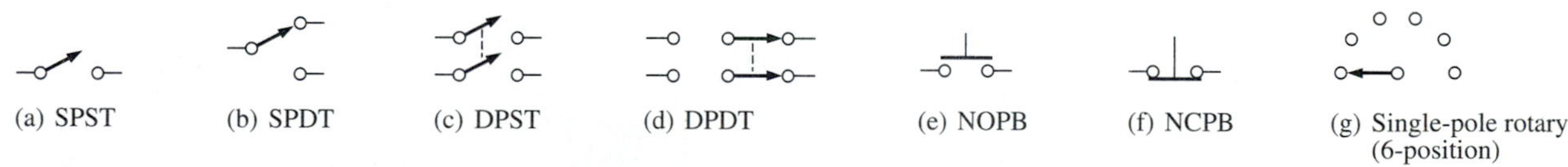


**FIGURE 2–30**
***Switch symbols.***

**Protective Devices** **Fuses** and **circuit breakers** are used to deliberately create an open circuit when the current exceeds a specified number of amperes due to a malfunction or other abnormal condition in a circuit. For example, a 20 A fuse or circuit breaker will open a circuit when the current exceeds 20 A.

The basic difference between a fuse and a circuit breaker is that when a fuse is "blown," it must be replaced; but when a circuit breaker opens, it can be reset and reused repeatedly. Both of these devices protect against damage to a circuit due to excess current or prevent a hazardous condition created by the overheating of wires and other components when the current is too great. Several typical fuses and circuit breakers, along with their schematic symbols, are shown in Figure 2–32.

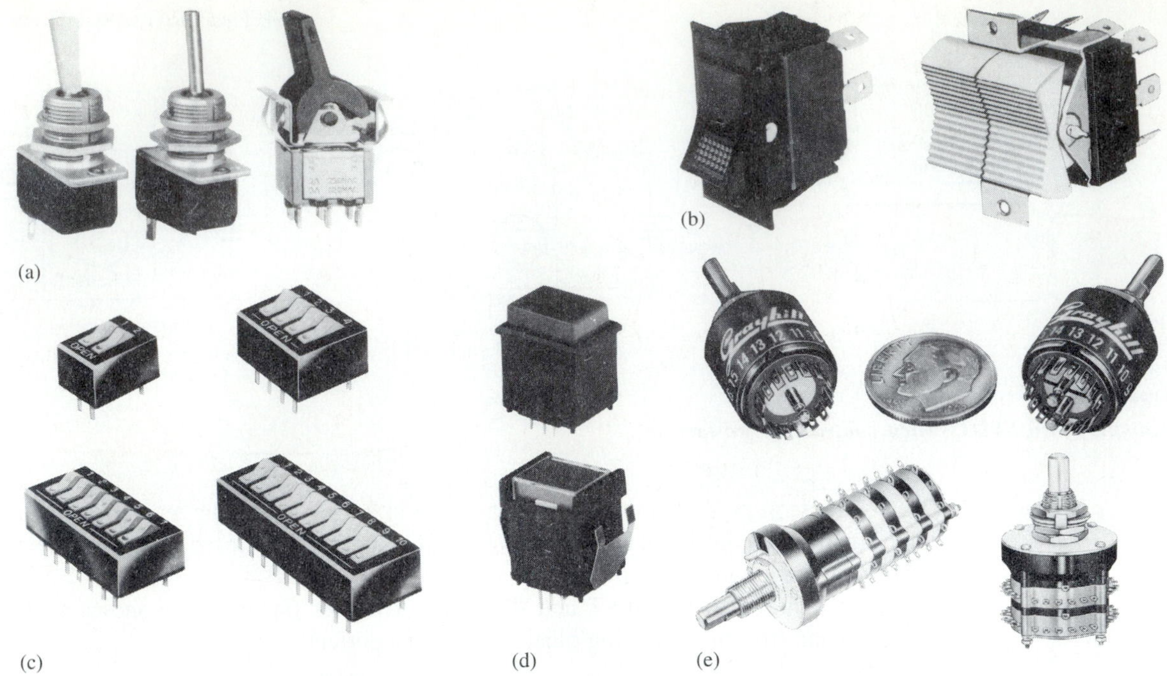


**FIGURE 2–31**

*Switches. (a) Typical toggle-lever switches (courtesy of Eaton Corp.). (b) Rocker switches (courtesy of Eaton Corp.). (c) Rocker DIP (dual-in-line package) switches (courtesy of Amp, Inc. and Grayhill, Inc.). (d) Push-button switches (courtesy of Eaton Corp.). (e) Rotary-position switches (courtesy of Grayhill, Inc.).*

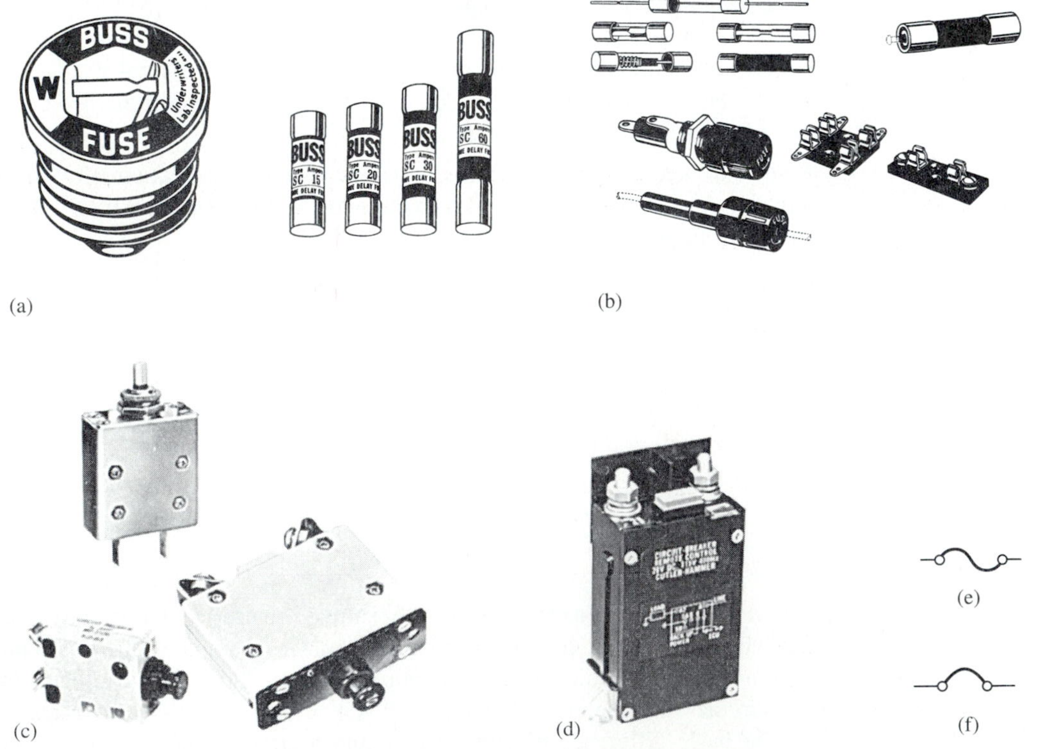


**FIGURE 2–32**

*Fuses and circuit breakers. (a) Power fuses (courtesy of Bussman Manufacturing Corp.). (b) Fuses and fuse holders (courtesy of Bussman Manufacturing Corp.). (c) Circuit breakers (courtesy of Bussman Manufacturing Corp.). (d) Remote control circuit breaker (courtesy of Eaton Corp.). (e) Fuse symbol. (f) Circuit breaker symbol.*

## Wires

Wires are the most common form of conductive material used in electrical applications. They vary in diameter and are arranged according to standard gage numbers, called **American Wire Gage (AWG)** sizes. The larger the gage number is, the smaller the wire diameter is. The size of a wire is also specified in terms of its cross-sectional area, as illustrated in Figure 2–33. The unit of cross-sectional area is the **circular mil,** abbreviated CM. One circular mil is the area of a wire with a diameter of 0.001 inch (1 mil). You can find the cross-sectional area by expressing the diameter in thousandths of an inch (mils) and squaring it, as follows:

$$A = d^2 \tag{2–5}$$

where $A$ is the cross-sectional area in circular mils and $d$ is the diameter in mils. Table 2–3 lists the AWG sizes with their corresponding cross-sectional area and resistance in ohms per 1000 ft at 20°C.

**FIGURE 2–33**
*Cross-sectional area of a wire.*

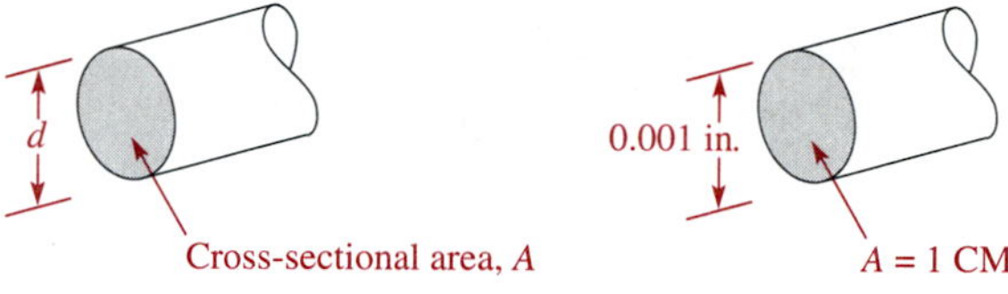


**TABLE 2–3**
*American Wire Gage (AWG) sizes for solid round copper*

| AWG # | Area (CM) | Ω/1000 ft at 20°C | AWG # | Area (CM) | Ω/1000 ft at 20°C |
|---|---|---|---|---|---|
| 0000 | 211,600 | 0.0490 | 19 | 1,288.1 | 8.051 |
| 000 | 167,810 | 0.0618 | 20 | 1,021.5 | 10.15 |
| 00 | 133,080 | 0.0780 | 21 | 810.10 | 12.80 |
| 0 | 105,530 | 0.0983 | 22 | 642.40 | 16.14 |
| 1 | 83,694 | 0.1240 | 23 | 509.45 | 20.36 |
| 2 | 66,373 | 0.1563 | 24 | 404.01 | 25.67 |
| 3 | 52,634 | 0.1970 | 25 | 320.40 | 32.37 |
| 4 | 41,742 | 0.2485 | 26 | 254.10 | 40.81 |
| 5 | 33,102 | 0.3133 | 27 | 201.50 | 51.47 |
| 6 | 26,250 | 0.3951 | 28 | 159.79 | 64.90 |
| 7 | 20,816 | 0.4982 | 29 | 126.72 | 81.83 |
| 8 | 16,509 | 0.6282 | 30 | 100.50 | 103.2 |
| 9 | 13,094 | 0.7921 | 31 | 79.70 | 130.1 |
| 10 | 10,381 | 0.9989 | 32 | 63.21 | 164.1 |
| 11 | 8,234.0 | 1.260 | 33 | 50.13 | 206.9 |
| 12 | 6,529.0 | 1.588 | 34 | 39.75 | 260.9 |
| 13 | 5,178.4 | 2.003 | 35 | 31.52 | 329.0 |
| 14 | 4,106.8 | 2.525 | 36 | 25.00 | 414.8 |
| 15 | 3,256.7 | 3.184 | 37 | 19.83 | 523.1 |
| 16 | 2,582.9 | 4.016 | 38 | 15.72 | 659.6 |
| 17 | 2,048.2 | 5.064 | 39 | 12.47 | 831.8 |
| 18 | 1,624.3 | 6.385 | 40 | 9.89 | 1049.0 |

**EXAMPLE 2–5** What is the cross-sectional area of a wire with a diameter of 0.005 inch?

***Solution***

$$d = 0.005 \text{ in.} = 5 \text{ mils}$$
$$A = d^2 = 5^2 = 25 \text{ CM}$$

***Related Exercise*** What is the cross-sectional area of a 0.0015 in. diameter wire?

## Wire Resistance

Although copper wire conducts electricity extremely well, it still has some resistance, as do all conductors. The resistance of a wire depends on four factors: (a) type of material, (b) length of wire, (c) cross-sectional area, and (d) temperature.

Each type of conductive material has a characteristic called its *resistivity,* $\rho$. For each material, $\rho$ is a constant value at a given temperature. The formula for the resistance of a wire of length $l$ and cross-sectional area $A$ is

$$R = \frac{\rho l}{A} \quad \textbf{(2–6)}$$

This formula shows that resistance increases with resistivity and length and decreases with cross-sectional area. For resistance to be calculated in ohms, the length must be in feet, the cross-sectional area in circular mils, and the resistivity in CM-Ω/ft.

**EXAMPLE 2–6** Find the resistance of a 100 ft length of copper wire with a cross-sectional area of 810.1 CM. The resistivity of copper is 10.4 CM-Ω/ft.

***Solution***

$$R = \frac{\rho l}{A} = \frac{(10.4 \text{ CM-}\Omega\text{/ft})(100 \text{ ft})}{810.1 \text{ CM}} = 1.284 \ \Omega$$

***Related Exercise*** What is the resistance of a 1000 ft length of copper wire that has a diameter of 0.0015 in.?

As mentioned, Table 2–3 lists the resistance of the various standard wire sizes in ohms per 1000 feet at 20°C. For example, a 1000 ft length of 14 gage copper wire has a resistance of 2.525 Ω. A 1000 ft length of 22 gage wire has a resistance of 16.14 Ω. For a given length, the smaller wire has more resistance. Thus, for a given voltage, larger wires can carry more current than smaller ones.

## Ground

The term *ground* comes from the method used in ac power distribution where one side of the power line is neutralized by connecting it to a metal rod driven into the ground. This method of grounding is called *earth ground.*

In electrical and electronic systems, the metal chassis that houses the assembly or a large conductive area on a printed circuit board is used as the electrical reference point and is called *chassis ground* or *circuit ground.* Circuit ground may or may not be connected to earth ground. For example, the negative terminal of the battery and one side of all the electrical circuits in most cars are connected to the metal chassis.

**Ground** is the reference point in electric circuits and has a potential of 0 V with respect to other points in the circuit. All of the ground points in a circuit are electrically the same and are therefore common points. Two ground symbols are shown in Figure

2–34. The symbol in part (a) is commonly used in schematic drawings to represent a reference ground, and the one in part (b) represents a chassis ground. The one in part (a) will be used throughout this textbook.

Figure 2–35 illustrates a simple circuit with ground connections. The current is from the positive terminal of the 12 V source, through the lamp, and back to the negative terminal of the source through the ground connection. Ground provides a return path for the current back to the source because all of the ground points are electrically the same point. The voltage at the top of the circuit is +12 V with respect to ground.

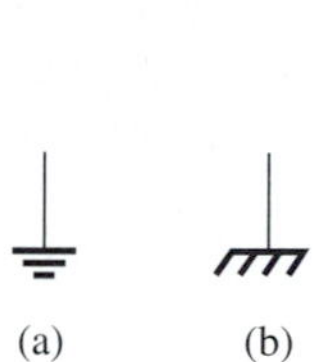


**FIGURE 2–34**
*Symbols for ground.*

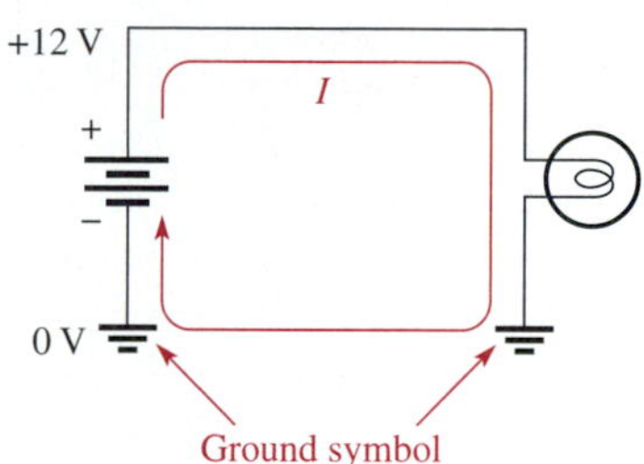


**FIGURE 2–35**
*A simple circuit with ground connections.*

**SECTION 2–6 REVIEW**

1. What are the basic elements of an electric circuit?
2. What is an open circuit?
3. What is a closed circuit?
4. What is the resistance across an open switch? Ideally, what is the resistance across a closed switch?
5. What is the difference between a fuse and a circuit breaker?
6. Which wire is larger in diameter, AWG 3 or AWG 22?
7. What is ground in an electric circuit?

## 2–7 ■ BASIC CIRCUIT MEASUREMENTS

***An electronics technician cannot function without knowing how to measure voltage, current, and resistance.***

***After completing this section, you should be able to***

- **Make basic circuit measurements**
  - ☐ Properly place a voltmeter in a circuit
  - ☐ Properly place an ammeter in a circuit
  - ☐ Properly connect an ohmmeter to measure resistance
  - ☐ Set up and read basic meters

Voltage, current, and resistance measurements are commonly required in electronics work. Special types of instruments are used to measure these basic electrical quantities.

The instrument used to measure voltage is a **voltmeter,** the instrument used to measure current is an **ammeter,** and the instrument used to measure resistance is an **ohmmeter.** Commonly, all three instruments are combined into a single instrument such as a **multimeter** or a VOM (volt-ohm-milliammeter), in which you can choose what specific quantity to measure by selecting the switch setting.

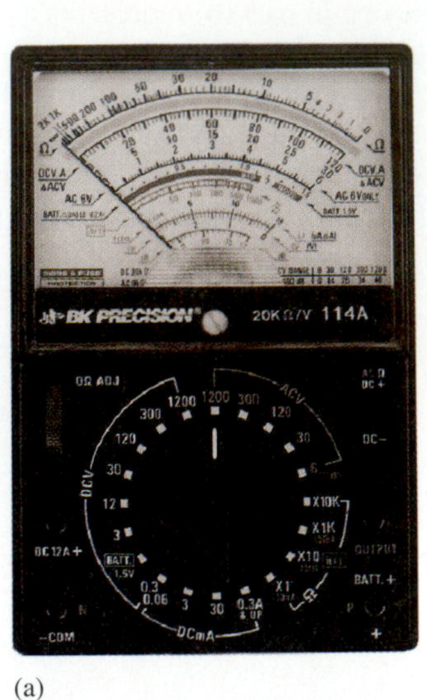


(a)

(b)

**FIGURE 2–36**
***Typical portable multimeters. (a) Analog (courtesy of BK Precision, Maxtec International Corp.). (b) Digital (courtesy of John Fluke Manufacturing Co.).***

Figure 2–36 shows typical multimeters. Part (a) shows an analog meter with a needle pointer, and part (b) shows a digital multimeter (DMM), which provides a digital readout of the measured quantity plus graphing capability.

## Meter Symbols

Throughout this book, certain symbols will be used to represent the different meters, as shown in Figure 2–37. You may see any of three types of symbols for voltmeters, ammeters, and ohmmeters, depending on which symbol most effectively conveys the information required. Generally, the pictorial analog symbol is used when relative measurements or changes in quantities are to be depicted by the position or movement of the needle. The pictorial digital symbol is used when specific values are to be indicated in a circuit. The

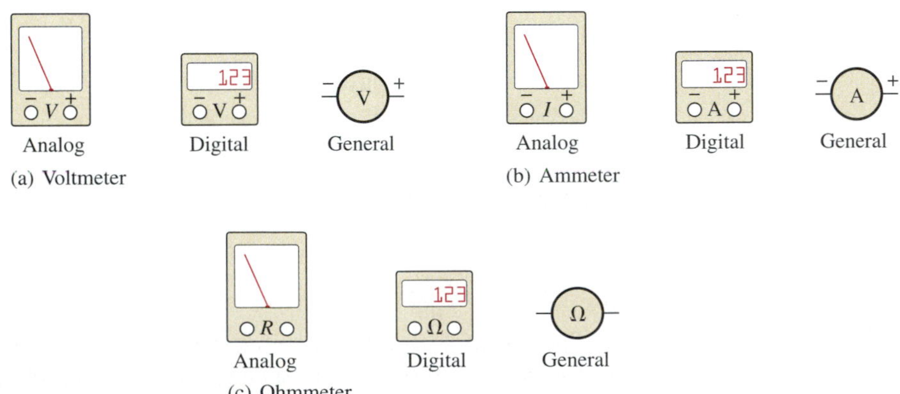


**FIGURE 2–37**
***Meter symbols used in this book.***

general schematic symbol is used to indicate placement of meters in a circuit when no values or value changes need to be shown.

## Measuring Current with an Ammeter

Figure 2–38 illustrates how to measure current with an ammeter. Part (a) shows the simple circuit in which the current through the resistor is to be measured. Connect the ammeter in the current path by first opening the circuit, as shown in part (b). Then insert the meter as shown in part (c). Such a connection is a series connection. The polarity of the meter must be such that the current is in at the positive terminal and out at the negative terminal.

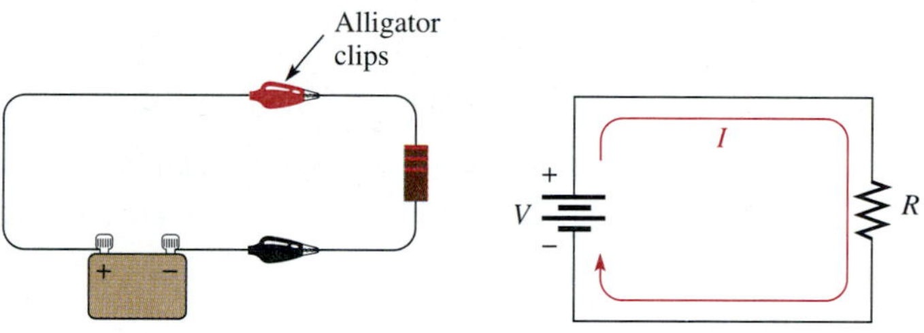


(a) Circuit in which the current is to be measured

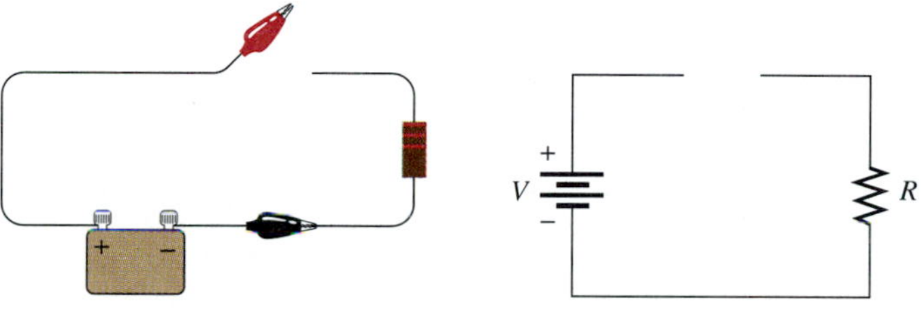


(b) Open the circuit either between the resistor and the positive terminal or between the resistor and the negative terminal of source.

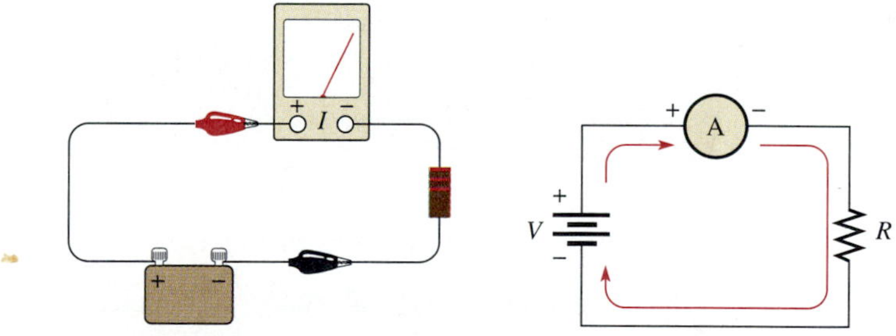


(c) Install the ammeter in the current path with polarity as shown (negative to negative, positive to positive).

**FIGURE 2–38**
***Example of an ammeter connection.***

## Measuring Voltage with a Voltmeter

To measure voltage, connect the voltmeter across the component for which the voltage is to be found. Such a connection is a parallel connection. The negative terminal of the meter must be connected to the negative side of the circuit, and the positive terminal of the meter to the positive side of the circuit. Figure 2–39 shows a voltmeter connected to measure the voltage across the resistor.

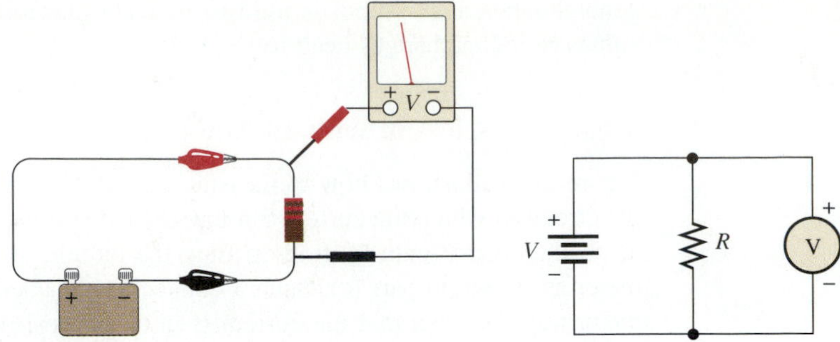


FIGURE 2–39
***Example of a voltmeter connection.***

## Measuring Resistance with an Ohmmeter

To measure resistance, connect the ohmmeter across the resistor. *The resistor must first be removed or disconnected from the circuit.* This procedure is shown in Figure 2–40.

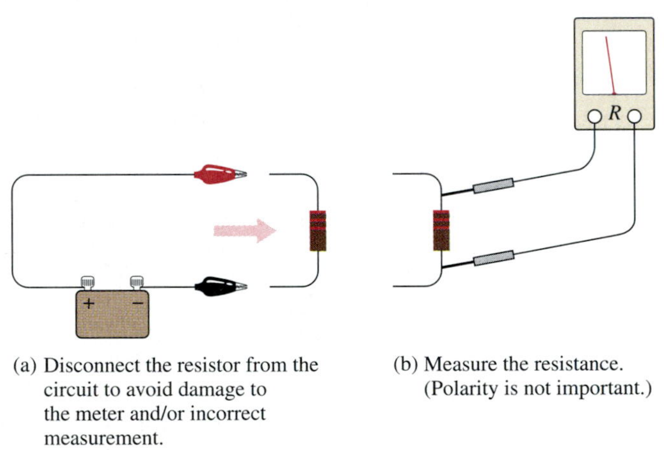


(a) Disconnect the resistor from the circuit to avoid damage to the meter and/or incorrect measurement.

(b) Measure the resistance. (Polarity is not important.)

FIGURE 2–40
***Example of using an ohmmeter.***

## Reading Analog Multimeters

A representation of a typical analog multimeter is shown in Figure 2–41. This particular instrument can be used to measure both direct current (dc) and alternating current (ac) quantities as well as resistance values. It has four selectable functions: dc volts (DC VOLTS), dc milliamperes (DC mA), ac volts (AC VOLTS), and OHMS. Most analog multimeters are similar to this one.

Within each function there are several ranges, as indicated by the brackets around the selector switch. For example, the DC VOLTS function has 0.3 V, 3 V, 12 V, 60 V, 300 V, and 600 V ranges. Thus, dc voltages from 0.3 V full-scale to 600 V full-scale can be measured. On the DC mA function, direct currents from 0.06 mA full-scale to 120 mA full-scale can be measured. On the ohm scale, the settings are ×1, ×10, ×100, ×1000, and ×100,000.

***The Ohm Scale*** Ohms are read on the top scale of the meter. This scale is nonlinear; that is, the values represented by each division (large or small) vary as you go across the scale. In Figure 2–41, notice how the scale becomes more compressed as you go from right to left.

**FIGURE 2–41**
*A typical analog multimeter.*

To read the actual value in ohms, multiply the number on the scale as indicated by the pointer by the factor selected by the switch. For example, when the switch is set at ×100 and the pointer is at 20, the reading is 20 × 100 = 2000 Ω.

As another example, assume that the switch is at ×10 and the pointer is at the seventh small division between the 1 and 2 marks, indicating 17 Ω (1.7 × 10). Now, if the meter remains connected to the same resistance and the switch setting is changed to ×1, the pointer will move to the second small division between the 15 and 20 marks. This, of course, is also a 17 Ω reading, illustrating that a given resistance value can often be read at more than one switch setting.

***The AC-DC Scales*** The second, third, and fourth scales from the top, labeled "AC" and "DC," are used in conjunction with the DC VOLTS, DC mA, and AC VOLTS functions. The upper ac-dc scale ends at the 300 mark and is used with the range settings that are multiples of three, such as 0.3, 3, and 300. For example, when the switch is at 3 on the DC VOLTS function, the 300 scale has a full-scale value of 3 V. At the range setting of 300, the full-scale value is 300 V, and so on.

The middle ac-dc scale ends at 60. This scale is used in conjunction with range settings that are multiples of 6, such as 0.06, 60, and 600. For example, when the switch is at 60 on the DC VOLTS function, the full-scale value is 60 V.

The lower ac-dc scale ends at 12 and is used in conjunction with switch settings that are multiples of 12, such as 1.2, 12, and 120.

**EXAMPLE 2–7**

In parts (a), (b), and (c) of Figure 2–42 on page 50, determine the quantity that is being measured and its value.

***Solution***

**(a)** The switch in Figure 2–42(a) is set on the DC VOLTS function and the 60 V range. The reading taken from the middle ac-dc scale is 18 V.

**(b)** The switch in Figure 2–42(b) is set on the DC mA function and the 12 mA range. The reading taken from the lower ac-dc scale is approximately 7.2 mA.

**(c)** The switch in Figure 2–42(c) is set on the OHMS function and the ×1000 range. The reading taken from the ohm scale (top scale) is approximately 7 kΩ.

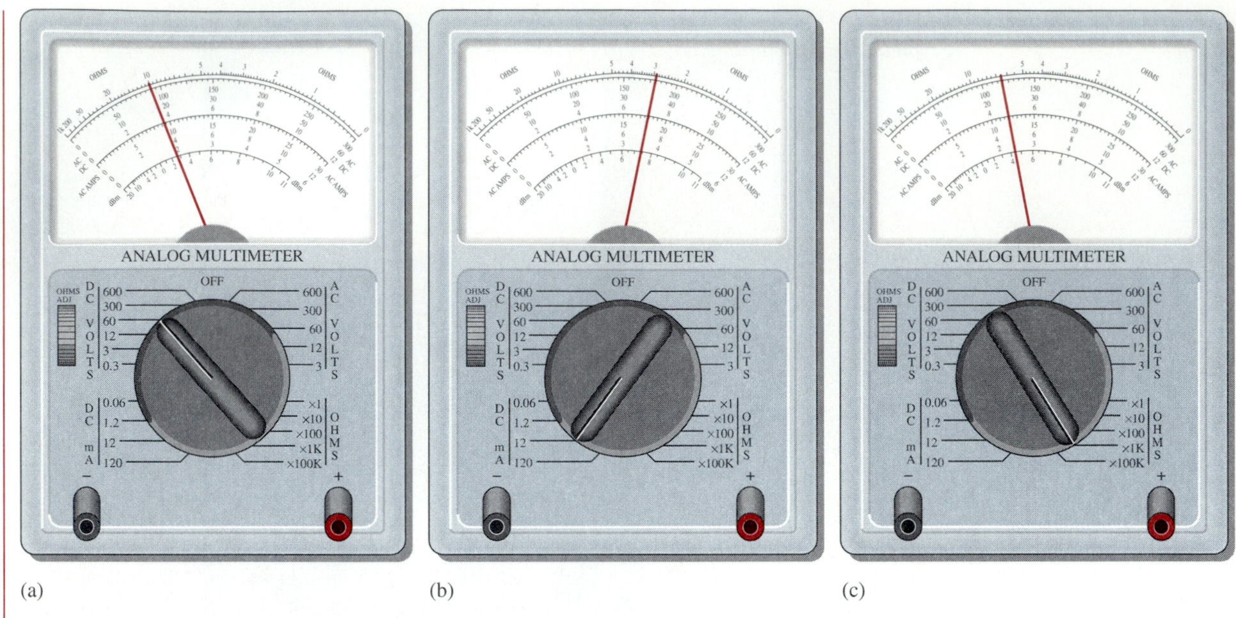


FIGURE 2–42

***Related Exercise*** In Figure 2–42 (c) the switch is moved to the ×100 setting. Assuming that the same resistance is being measured, what will the needle do?

## Digital Multimeters (DMMs)

DMMs are the most widely used type of electronic measuring instrument. Generally, DMMs provide more functions, better accuracy, greater ease of reading, and greater reliability than do many analog meters. Analog meters have at least one advantage over DMMs, however. They can track short-term variations and trends in a measured quantity that many DMMs are too slow to respond to. Figure 2–43 shows typical DMMs.

***DMM Functions*** The basic functions found on most DMMs include the following:

- Ohms
- DC voltage and current
- AC voltage and current

Some DMMs provide special functions such as transistor or diode tests, power measurement, and decibel measurement for audio amplifier tests. Some meters require manual selection of the ranges for the functions. Many meters now provide automatic range selection and are called *autoranging.*

***DMM Displays*** DMMs are available with either LCD (liquid-crystal display) or LED (light-emitting diode) readouts. The LCD is the most commonly used readout in battery-powered instruments because it requires only very small amounts of current. A typical battery-powered DMM with an LCD readout operates on a 9 V battery that will last from a few hundred hours to 2000 hours and more. The disadvantages of LCD readouts are that (a) they are difficult or impossible to see in low-light conditions and (b) they are relatively slow to respond to measurement changes. LEDs, on the other hand, can be seen in the dark and respond quickly to changes in measured values. LED displays require much more current than LCDs, and, therefore, battery life is shortened when they are used in portable equipment.

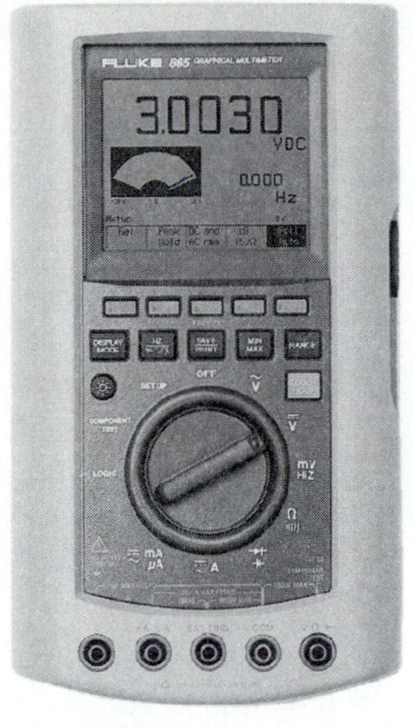


(a)

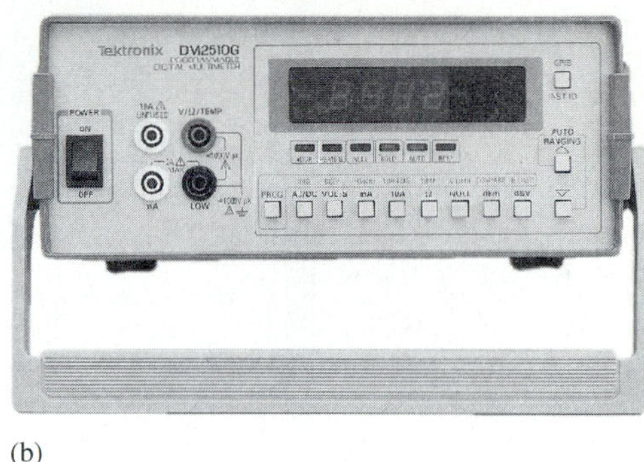


(b)

**FIGURE 2–43**
***Typical digital multimeters (DMMs). Part (a) courtesy of John Fluke Manufacturing Co. Part (b) courtesy of Tektronix, Inc.***

Both LCD and LED DMM displays are in a 7-segment format. Each digit in a display consists of seven separate segments as shown in Figure 2–44 (a). Each of the ten decimal digits is formed by activation of appropriate segments, as illustrated in Figure 2–44 (b). In addition to the seven segments, there is also a decimal point.

(a) (b)

**FIGURE 2–44**
***Seven-segment display.***

***Resolution*** The resolution of a meter is the smallest increment of a quantity that the meter can measure. The smaller the increment, the better the resolution. One factor that determines the resolution of a meter is the number of digits in the display.

Because many meters have 3½ digits in their display, we will use this case for illustration. A 3½-digit multimeter has three digit positions that can indicate from 0 through 9, and one digit position that can indicate only a value of 1. This latter digit, called the *half-digit,* is always the most significant digit in the display. For example, suppose that a DMM is reading 0.999 V, as shown in Figure 2–45(a). If the voltage increases by 0.001 V to 1 V, the display correctly shows 1.000 V, as shown in part (b). The "1" is the half-digit. Thus, with 3½ digits, a variation of 0.001 V, which is the resolution, can be observed.

Now, suppose that the voltage increases to 1.999 V. This value is indicated on the meter as shown in Figure 2–45(c). If the voltage increases by 0.001 V to 2 V, the half-digit cannot display the "2," so the display shows 2.00. The half-digit is blanked and only

(a) Resolution: 0.001 V

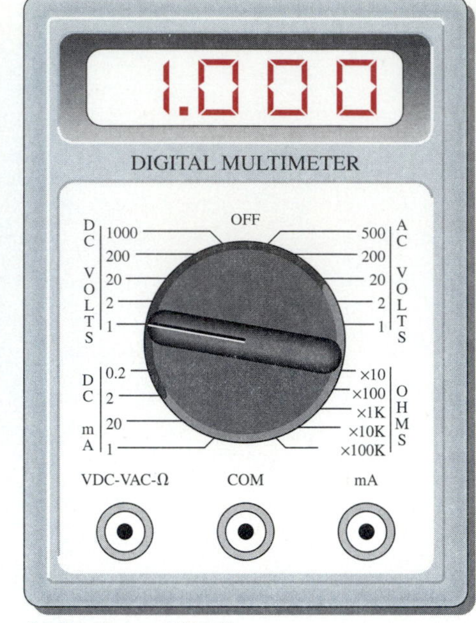

(b) Resolution: 0.001 V

(c) Resolution: 0.001 V

(d) Resolution: 0.01 V

**FIGURE 2–45**

***A $3\frac{1}{2}$-digit DMM illustrates how the resolution changes with the number of digits in use.***

three digits are active, as indicated in part (d). With only three digits active, the resolution is 0.01 V rather than 0.001 V as it is with $3\frac{1}{2}$ active digits. The resolution remains 0.01 V up to 19.99 V. The resolution goes to 0.1 V for readings of 20.0 V to 199.9 V. At 200 V, the resolution goes to 1 V, and so on.

The resolution capability of a DMM is also determined by the internal circuitry and the rate at which the measured quantity is sampled. DMMs with displays of $4\frac{1}{2}$ through $8\frac{1}{2}$ digits are also available.

***Accuracy*** The accuracy is the degree to which a measured value represents the true or accepted value of a quantity. The accuracy of a DMM is established strictly by its internal circuitry. For typical meters, accuracies range from 0.01% to 0.5%, with some precision laboratory-grade meters going to 0.002%.

**SECTION 2–7 REVIEW**

1. Name the meters for measurement of (a) current, (b) voltage, and (c) resistance.
2. Place two ammeters in the circuit of Figure 2–29 to measure the current through either lamp (be sure to observe the polarities). How can the same measurements be accomplished with only one ammeter?
3. Show how to place a voltmeter to measure the voltage across lamp 2 in Figure 2–29.
4. The multimeter in Figure 2–41 is set on the 3 V range to measure dc voltage. The pointer is at 150 on the upper ac-dc scale. What voltage is being measured?
5. How do you set up the meter in Figure 2–41 to measure 275 V dc, and on what scale do you read the voltage?
6. If you expect to measure a resistance in excess of 20 kΩ, where do you set the switch?
7. List two common types of DMM displays, and discuss the advantages and disadvantages of each.
8. Define *resolution* in a DMM.

## 2–8 ■ TECHnology Theory Into Practice

***In this TECH TIP section, a dc voltage is applied to a circuit in order to make current flow through a lamp and produce light. You will see how the current is controlled by resistance. The circuit that you will be working with simulates the instrument illumination circuit in your car, which allows you to increase or decrease the amount of light on the instruments.***

The instrument panel illumination circuit in an automobile operates from the 12 V battery that is the voltage source for the circuit. The circuit uses a potentiometer connected as a rheostat, controlled by a knob on the instrument panel, which is used to set the amount of current through the lamp to back-light the instruments. The brightness of the lamp is proportional to the amount of current through the lamp. The switch used to turn the lamp on and off is the same one used for the head lights. There is a fuse for circuit protection in case of a short.

Figure 2–46 shows the schematic for the illumination circuit. Figure 2–47 shows a breadboarded circuit which simulates the illumination circuit by using components that are functionally equivalent but not physically the same as those in a car. A laboratory dc power supply is used in the place of an actual automobile battery. The circuit board in Figure 2–47 is a type that is commonly used for constructing circuits on the test bench.

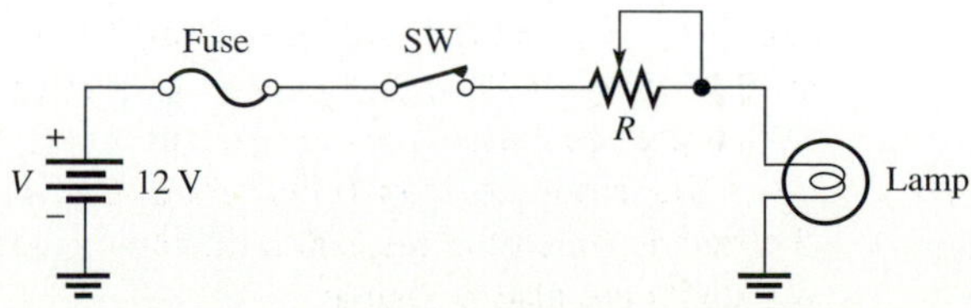


FIGURE 2–46
***Basic automobile panel illumination circuit schematic.***

### The Test Bench

Figure 2–47 shows the breadboarded circuit, a dc power supply, and a digital multimeter. The power supply is connected to provide 12 V to the circuit. The multimeter is used to measure current, voltage, and resistance in the circuit.

- ☐ Identify each component in the circuit and check the breadboarded circuit to make sure it is connected as the schematic in Figure 2–46 indicates.
- ☐ Explain the purpose of each component in the circuit.

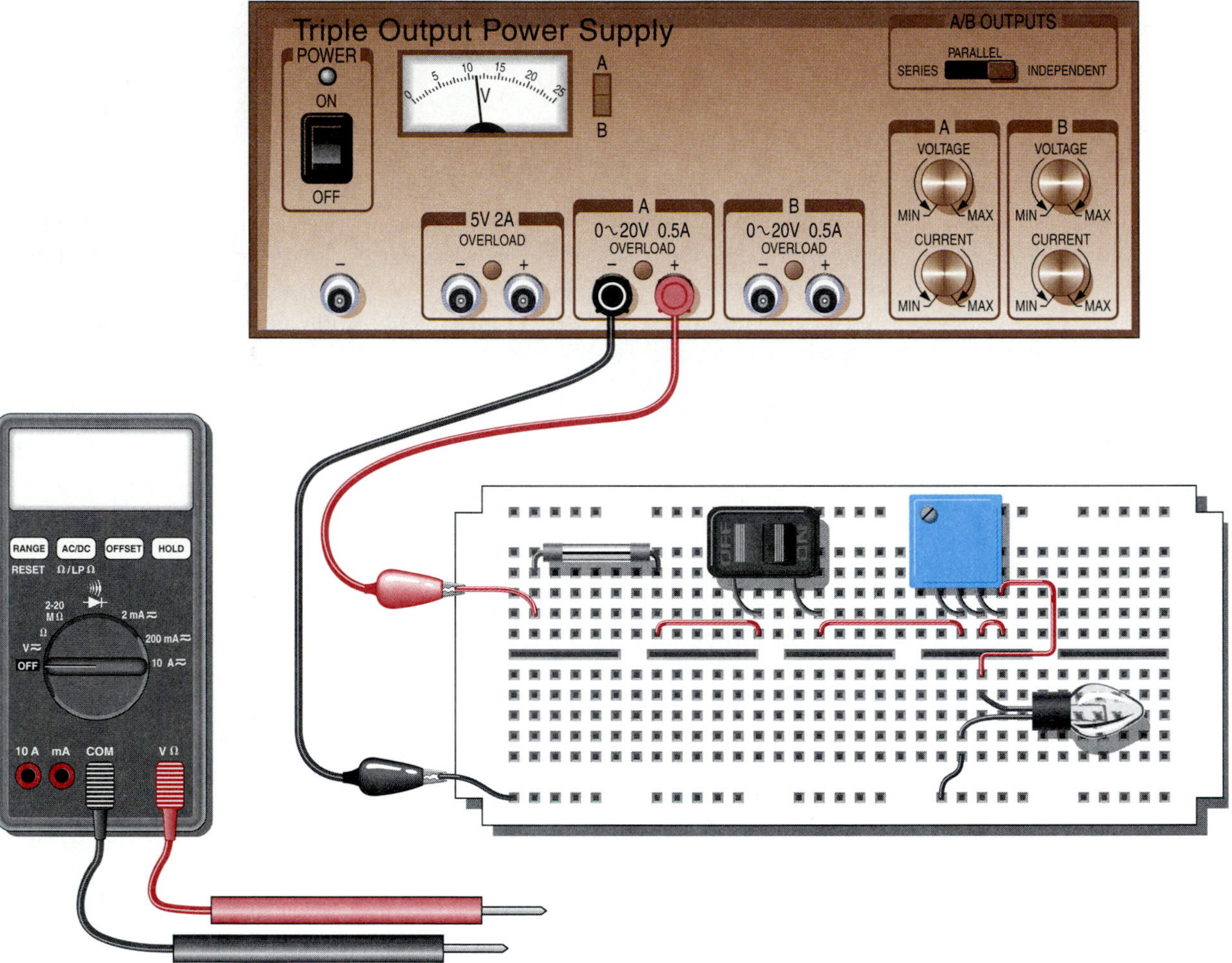


V≂ indicates dc/ac function

**FIGURE 2–47**
***Test bench setup for simulating the automobile panel illumination circuit.***

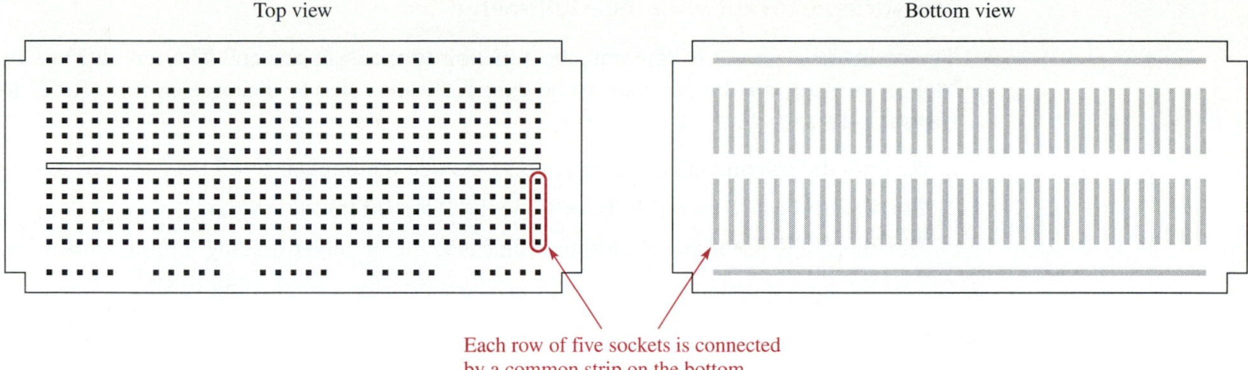
Top view
Bottom view
Each row of five sockets is connected
by a common strip on the bottom.

**FIGURE 2–48**
***A typical circuit board used for breadboarding.***

As shown in Figure 2–48, the typical circuit board consists of rows of small sockets into which component leads and wires are inserted. In this particular configuration, all five sockets in each row are connected together and are effectively one electrical point as shown in the bottom view. All sockets arranged on the outer edges of the board are typically connected together as shown.

### Measuring Current with the Multimeter

The meter must be set to the ammeter function to measure current. The circuit must be broken in order for the ammeter to be connected in series to measure current. Refer to Figure 2–49.

- ☐ Redraw the schematic in Figure 2–46 to include the ammeter.
- ☐ For which measurement (A, B, or C) is the lamp brightest? Explain.
- ☐ List the change(s) in the circuit that can cause the ammeter reading to go from A to B.
- ☐ List the circuit condition(s) that will produce the ammeter reading in C.

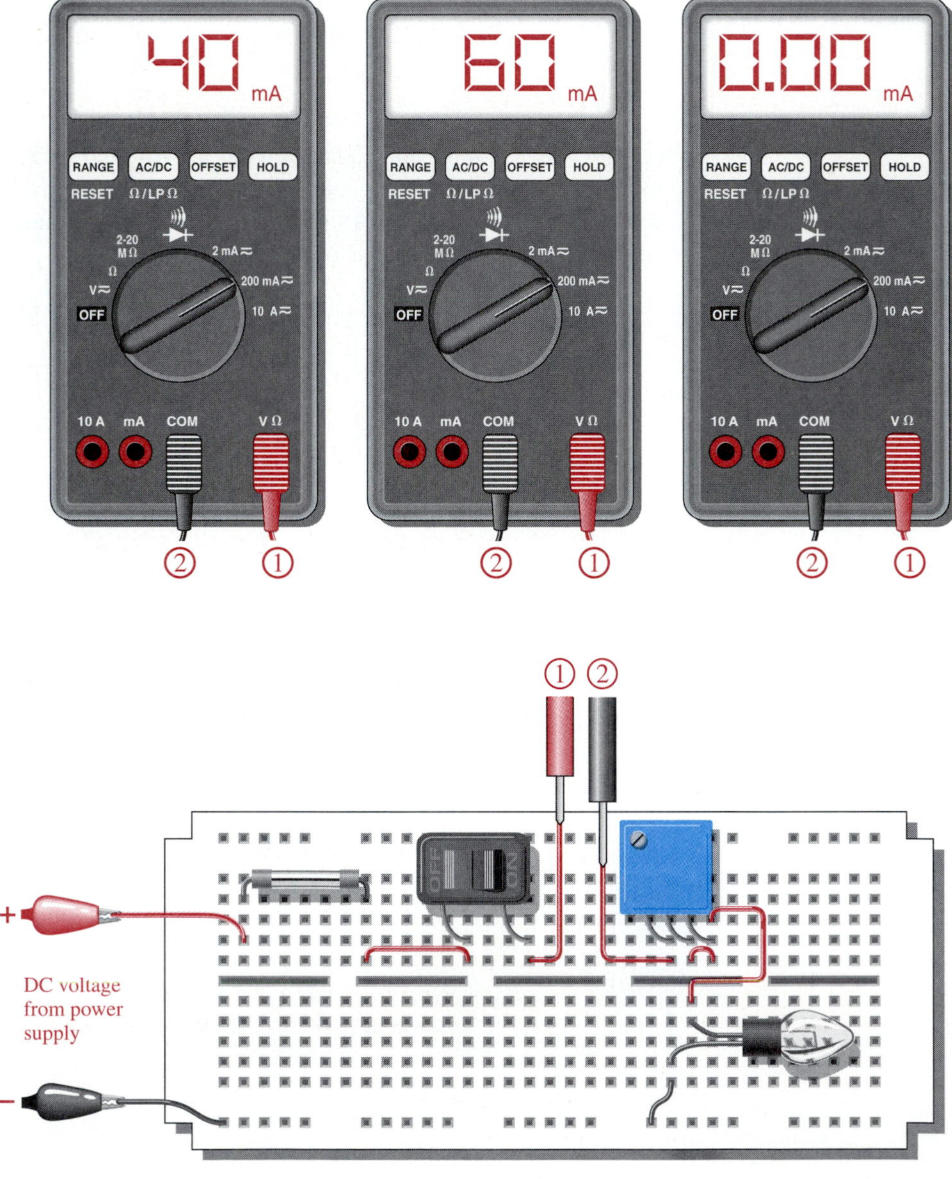


**FIGURE 2–49**

***Current measurements. The circled numbers indicate the meter-to-circuit connections.***

## Measuring Voltage with the Multimeter

The meter must be set to the voltmeter function to measure voltage. The voltmeter must be connected to the two points across which the voltage is to be measured. Refer to Figure 2–50.

- ☐ Redraw the schematic in Figure 2–46 to include the voltmeter.
- ☐ Across which component is the voltage measured?
- ☐ For which measurement (A or B) is the lamp brighter? Explain.
- ☐ List the change(s) in the circuit that can cause the voltmeter reading to go from A to B.

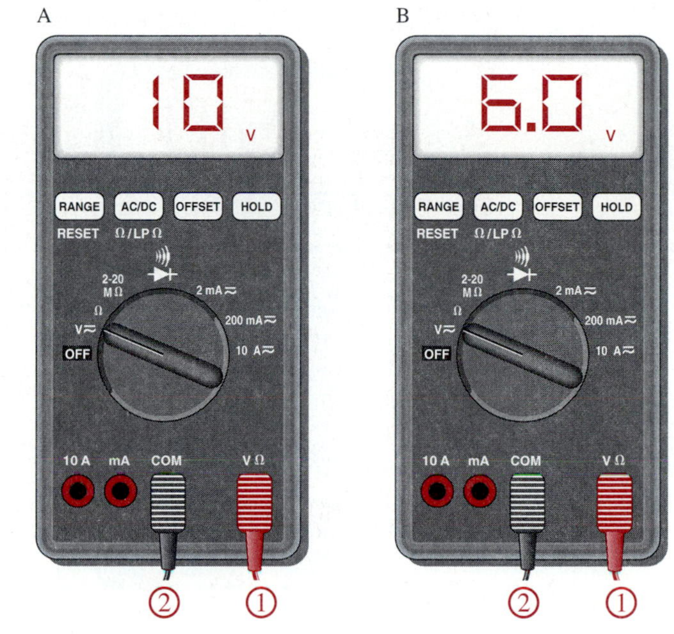


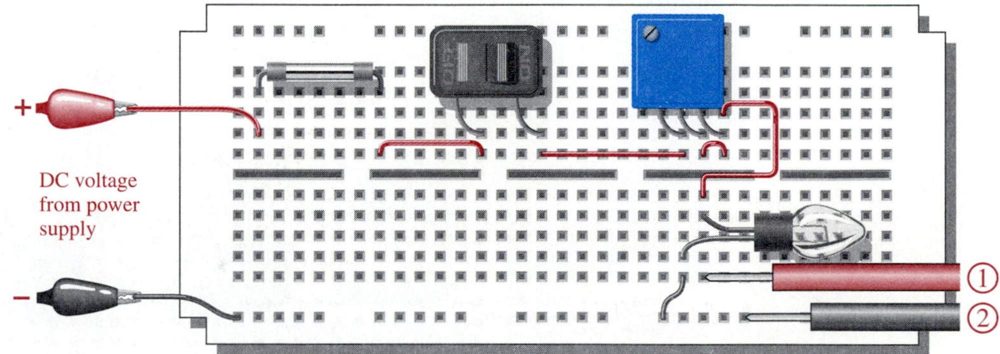


**FIGURE 2–50**
*Voltage measurements.*

### Measuring Resistance with the Multimeter

The meter must be set to the ohmmeter function to measure current. Before the ohmmeter is connected, the resistance to be measured must be disconnected from the circuit. Before disconnecting any component, first turn the power supply off. Refer to Figure 2–51.

- ☐ For which component is the resistance measured?
- ☐ For which measurement (A or B) will the lamp be brighter when the circuit is reconnected and the power turned on? Explain.

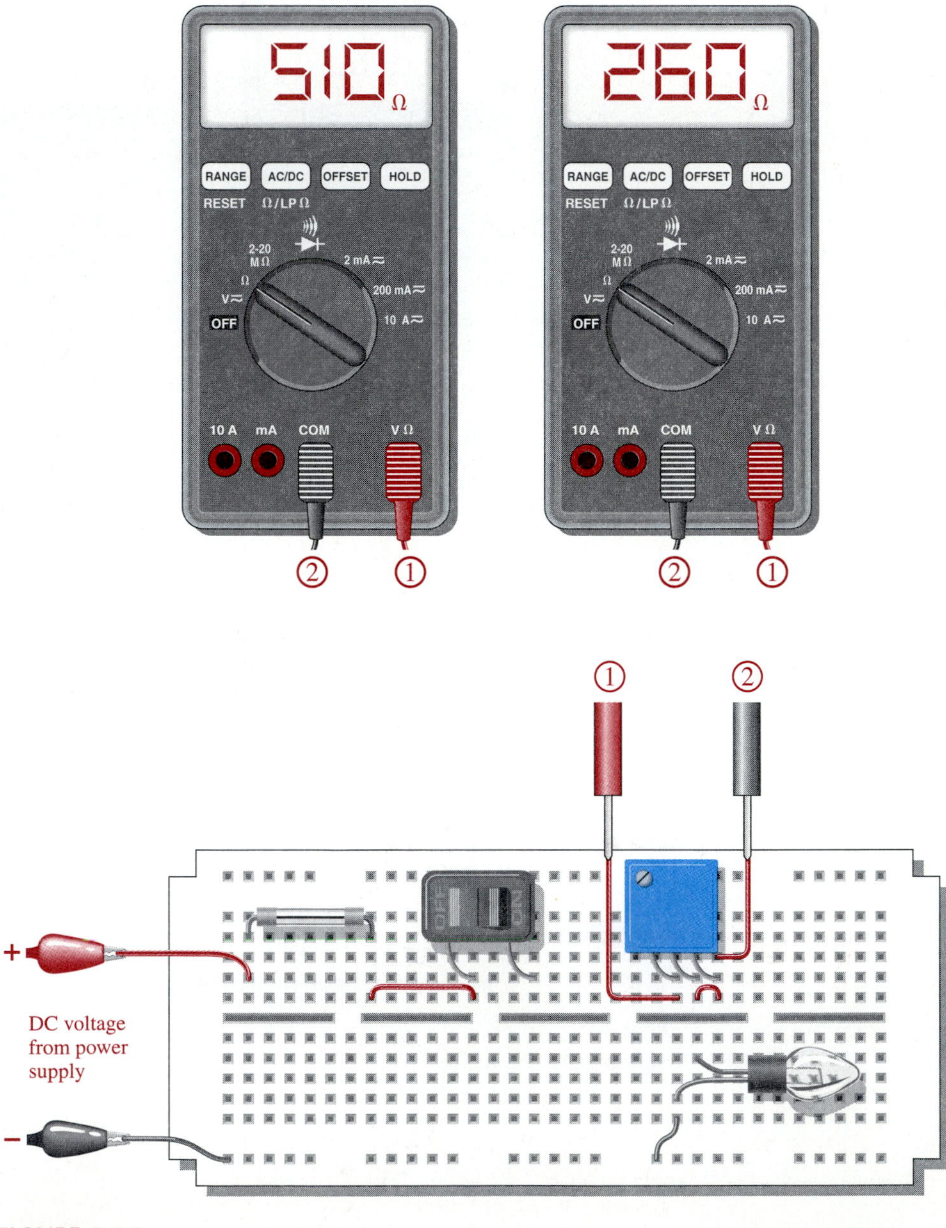


**FIGURE 2–51**
*Resistance measurements.*

**SECTION 2–8 REVIEW**

1. If the dc supply voltage in the panel illumination circuit is reduced, how is the amount of light produced by the lamp affected? Explain.
2. Should the potentiometer be adjusted to a higher or lower resistance for the circuit to produce more light?

## 2–9 ■ PSpice CIRCUIT CONNECTIONS

***In the last chapter, you learned how to properly identify circuit components (devices) and express their values in PSpice formats. In this section, you will learn to specify how components are connected together in a circuit.***

***After completing this section, you should be able to***

- **Specify circuit connections for PSpice**
  - Define *node*
  - Use node labels to indicate component connections
  - Write a description line for each component in a circuit

### Node Information

PSpice determines how components are connected together in a circuit by reading the node labels specified for each component description.

A **node** is a unique point in a circuit where two or more components are connected together. A node is to a circuit file what a wire is to an actual circuit. Each component in a circuit has at least two nodes associated with it because each component has at least two leads physically connected to other components in a circuit or to ground.

By keeping track of the node labels and the components connected to each one, PSpice can determine how the components connect to each other to form a circuit.

### Assigning Nodes

The simple circuit in Figure 2–52 consists of a 1.5 V battery, $V_S$, and a 1 kΩ resistor, $R_1$. There are two nodes in this circuit. The first node is at the point where the negative terminal of the battery connects to one end of the resistor by way of the ground. The second node is the point where the positive terminal of the battery connects to the other end of the resistor.

FIGURE 2–52

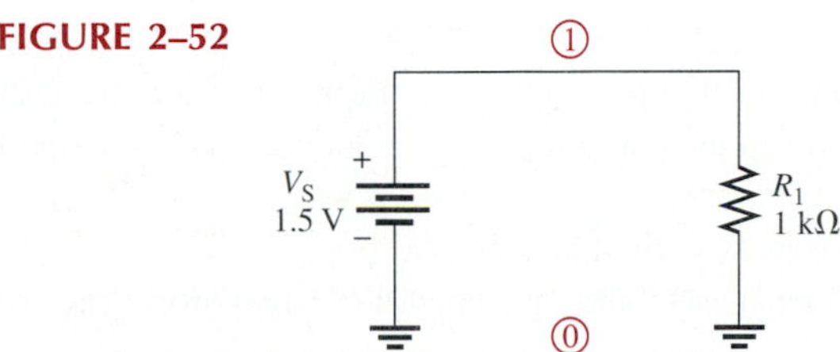


Although all nodes except ground can use any text string as a label, the usual convention is to use positive integers. In the figure, the nodes are labeled 0 and 1. In PSpice, the ground node must always have 0 as its label. All voltages are referenced relative to the ground node. In this circuit, both the battery and the resistor connect to nodes 0 and 1.

### Writing the Component Description

The general PSpice format for a component with $n$ nodes is

```
Component_name  node_1  node_2  . . .  node_n  Component_value
```

where each item is separated by one or more spaces. The component information for the circuit of Figure 2–52 is written in PSpice as follows:

```
VS  1  0  1.5
R1  1  0  1K
```

The first line tells PSpice that between nodes 1 and 0 there is a 1.5 V voltage source. The second line indicates that between nodes 1 and 0 there is a 1 kΩ resistor. Recall that PSpice knows that VS specifies a voltage source and it automatically assigns the unit of volts. It also knows that R1 specifies a resistor and it automatically assigns the unit of ohms.

The order in which the nodes are listed may or may not make a difference. If changing the leads of the component in the actual circuit makes no difference, then the order in which nodes are listed makes no difference. In the circuit in Figure 2–52, reversing the resistor leads will make no difference in the actual circuit, but reversing the battery will reverse the direction of current.

*PSpice assumes that for a voltage source the value given is for the first node referenced to the second node.* The voltage source description line

```
VS  1  0  1.5
```

is interpreted by PSpice to mean that node 1 is 1.5 V more positive than node 0. If the order of nodes is reversed as follows,

```
VS  0  1  1.5
```

then PSpice will interpret it to mean that node 0 is 1.5 V more positive than node 1; or, stated another way, node 1 is 1.5 V more negative than node 0.

### SECTION 2–9 REVIEW

**1.** Interpret each of the following component description lines:

**(a)** V1  0  1  10
**(b)** C1  1  0  4.7E-06
**(c)** RES1  0  1  2.2E3

**2.** Write the component description lines for the circuit in Figure 2–52 if the voltage source is changed to 9 V and the resistor is changed to 10 kΩ.

## ■ SUMMARY

- An atom is the smallest particle of an element that retains the characteristics of that element.
- When electrons in the outer orbit of an atom (valence electrons) break away, they become free electrons.
- Free electrons make current possible.
- Like charges repel each other, and opposite charges attract each other.
- Voltage must be applied to a circuit before current can flow.
- Resistance limits the current.
- Basically, an electric circuit consists of a source, a load, and a current path.
- An open circuit is one in which the current path is broken.
- A closed circuit is one which has a complete current path.
- An ammeter is connected in line with the current path.
- A voltmeter is connected across the current path.

- An ohmmeter is connected across a resistor (resistor must be removed from circuit).
- One coulomb is the charge of $6.25 \times 10^{18}$ electrons.
- One volt is the potential difference (voltage) between two points when one joule of energy is used to move one coulomb from one point to the other.
- One ampere is the amount of current that exists when one coulomb of charge moves through a given cross-sectional area of a material in one second.
- One ohm is the resistance when there is one ampere of current in a material with one volt applied across the material.
- Figure 2–53 shows the electrical symbols introduced in this chapter.

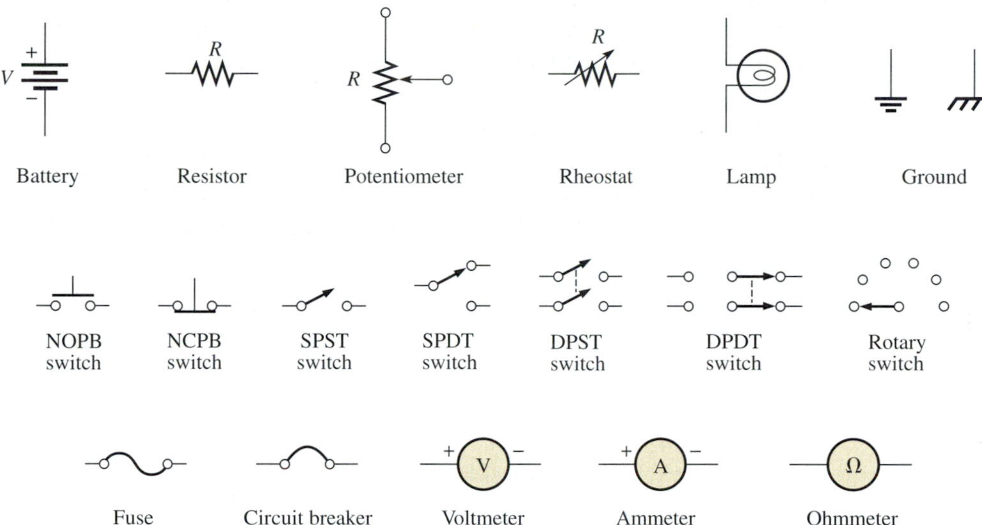


FIGURE 2–53

## ■ GLOSSARY

**American wire gage (AWG)** A standardization based on wire diameter.

**Ammeter** An electrical instrument used to measure current.

**Ampere (A)** The unit of electrical current.

**Atom** The smallest particle of an element possessing the unique characteristics of that element.

**Atomic number** The number of protons in a nucleus.

**Atomic weight** The number of protons and neutrons in the nucleus of an atom.

**Battery** An energy source that uses a chemical reaction to convert chemical energy into electrical energy.

**Charge** An electrical property of matter that exists because of an excess or a deficiency of electrons. Charge can be either positive or negative.

**Circuit** An interconnection of electrical components designed to produce a desired result. A basic circuit consists of a source, a load, and an interconnecting current path.

**Circuit breaker** A resettable protective device used for interrupting excessive current in an electric circuit.

**Circular mil (CM)** The unit of the cross-sectional area of a wire.

**Closed circuit** A circuit with a complete current path.

**Conductance** The ability of a circuit to allow current. The unit is the siemens (S).

**Conductor** A material in which electric current is easily established. An example is copper.

**Coulomb (C)** The unit of electrical charge.

**Current** The rate of flow of charge (electrons).

**Electrical** Related to the use of electrical voltage and current to achieve desired results.

**Electron** The basic particle of electrical charge in matter. The electron possesses negative charge.

**Electronic** Related to the movement and control of free electrons in semiconductors or vacuum devices.

**Element** One of the unique substances that make up the known universe. Each element is characterized by a unique atomic structure.

**Free electron** A valence electron that has broken away from its parent atom and is free to move from atom to atom within the atomic structure of a material.

**Fuse** A protective device that burns open when there is excessive current in a circuit.

**Generator** An energy source that produces electrical signals.

**Ground** The common or reference point in a circuit.

**Insulator** A material that does not allow current under normal conditions.

**Joule (J)** The unit of energy.

**Load** The device in a circuit upon which work is done.

**Multimeter** An instrument that measures voltage, current, and resistance.

**Neutron** An atomic particle having no electrical charge.

**Node** A unique point in a circuit where two or more components are connected.

**Ohm (Ω)** The unit of resistance.

**Ohmmeter** An instrument for measuring resistance.

**Open circuit** A circuit in which there is not a complete current path.

**Photoconductive cell** A type of variable resistor that is light-sensitive.

**Potentiometer** A three-terminal variable resistor.

**Power supply** An electronic instrument that produces voltage, current, and power from the ac power line or batteries in a form suitable for use in powering electronic equipment.

**Proton** A positively charged atomic particle.

**Resistance** Opposition to current. The unit is the ohm (Ω).

**Resistor** An electrical component designed specifically to provide resistance.

**Rheostat** A two-terminal variable resistor.

**Schematic** A symbolized diagram of an electrical or electronic circuit.

**Semiconductor** A material that has a conductance value between that of a conductor and an insulator. Silicon and germanium are examples.

**Shell** The orbit in which an electron revolves.

**Source** A device that produces electrical energy.

**Switch** An electrical device for opening and closing a current path.

**Tapered** Nonlinear, such as a tapered potentiometer.

**Thermistor** A type of variable resistor.

**Tolerance** The limits of variation in the value of a component.

**Valence** Related to the outer shell or orbit of an atom.

**Valence electron** An electron that is present in the outermost shell of an atom.

**Volt** The unit of voltage or electromotive force.

**Voltage** The amount of energy available to move a certain number of electrons from one point to another in an electric circuit.

**Voltmeter** An instrument used to measure voltage.

**Wiper** The sliding contact in a potentiometer.

## FORMULAS

**(2–1)** $$Q = \frac{\text{number of electrons}}{6.25 \times 10^{18}\ \text{electrons/C}}$$ Charge

**(2–2)** $$V = \frac{W}{Q}$$ Voltage equals energy divided by charge.

**(2–3)** $$I = \frac{Q}{t}$$ Current equals charge divided by time.

(2–4) $G = \frac{1}{R}$ Conductance is the reciprocal of resistance.

(2–5) $A = d^2$ Cross-sectional area equals the diameter squared.

(2–6) $R = \frac{\rho l}{A}$ Resistance is resistivity times length divided by cross-sectional area.

## ■ SELF-TEST

**1.** A neutral atom with an atomic number of three has how many electrons?
**(a)** 1 **(b)** 3 **(c)** none **(d)** depends on the type of atom

**2.** Electron orbits are called
**(a)** shells **(b)** nuclei **(c)** waves **(d)** valences

**3.** Materials in which no current can flow are called
**(a)** filters **(b)** conductors **(c)** insulators **(d)** semiconductors

**4.** When placed close together, a positively charged material and a negatively charged material will
**(a)** repel **(b)** become neutral **(c)** attract **(d)** exchange charges

**5.** The charge on a single electron is
**(a)** $6.25 \times 10^{-18}$ C **(b)** $1.6 \times 10^{-19}$ C **(c)** $1.6 \times 10^{-19}$ J **(d)** $3.14 \times 10^{-6}$ C

**6.** *Potential difference* is another term for
**(a)** energy
**(b)** voltage
**(c)** distance of an electron from the nucleus
**(d)** charge

**7.** The unit of energy is the
**(a)** watt **(b)** coulomb **(c)** joule **(d)** volt

**8.** Which one of the following is not a type of energy source?
**(a)** battery **(b)** solar cell **(c)** generator **(d)** potentiometer

**9.** Which one of the following is not a possible condition in an electric circuit?
**(a)** voltage and no current **(b)** current and no voltage
**(c)** voltage and current **(d)** no voltage and no current

**10.** Electrical current is defined as
**(a)** free electrons
**(b)** the rate of flow of free electrons
**(c)** the energy required to move electrons
**(d)** the charge on free electrons

**11.** There is no current in a circuit when
**(a)** a switch is closed **(b)** a switch is open **(c)** there is no voltage
**(d)** answers (a) and (c) **(e)** answers (b) and (c)

**12.** The primary purpose of a resistor is to
**(a)** increase current **(b)** limit current
**(c)** produce heat **(d)** resist current change

**13.** Potentiometers and rheostats are types of
**(a)** voltage sources **(b)** variable resistors **(c)** fixed resistors **(d)** circuit breakers

**14.** The current in a given circuit is not to exceed 22 A. Which value of fuse is best?
**(a)** 10 A **(b)** 25 A **(c)** 20 A **(d)** a fuse is not necessary

## ■ PROBLEMS

### SECTION 2–2 Electrical Charge

**1.** How many coulombs of charge do $50 \times 10^{31}$ electrons possess?

**2.** How many electrons does it take to make 80 $\mu$C (microcoulombs) of charge?

## SECTION 2–3 Voltage

**3.** Determine the voltage in each of the following cases:

**(a)** 10 J/C **(b)** 5 J/2 C **(c)** 100 J/25 C

**4.** Five hundred joules of energy are used to move 100 C of charge through a resistor. What is the voltage across the resistor?

**5.** What is the voltage of a battery that uses 800 J of energy to move 40 C of charge through a resistor?

**6.** How much energy does a 12 V battery use to move 2.5 C through a circuit?

**7.** If a resistor with a current of 2 A through it converts 1000 J of electrical energy into heat energy in 15 s, what is the voltage across the resistor?

## SECTION 2–4 Current

**8.** Determine the current in each of the following cases:

**(a)** 75 C in 1 s **(b)** 10 C in 0.5 s **(c)** 5 C in 2 s

**9.** Six-tenths coulomb passes a point in 3 s. What is the current in amperes?

**10.** How long does it take 10 C to flow past a point if the current is 5 A?

**11.** How many coulombs pass a point in 0.1 s when the current is 1.5 A?

**12.** $574 \times 10^{15}$ electrons flow through a wire in 250 ms. What is the current in amperes?

## SECTION 2–5 Resistance

**13.** Determine the resistance values for the following:

**(a)** Red, violet, orange, gold **(b)** Brown, gray, red, silver

**14.** Find the minimum and the maximum resistance within the tolerance limits for each resistor in Problem 13.

**15.** Determine the color bands for each of the following values: 330 Ω, 2.2 kΩ, 56 kΩ, 100 kΩ, and 39 kΩ.

**16.** The adjustable contact of a linear potentiometer is set at the mechanical center of its adjustment. If the total resistance is 1000 Ω, what is the resistance between each end terminal and the adjustable contact?

**17.** Find the conductance for each of the following resistance values:

**(a)** 5 Ω **(b)** 25 Ω **(c)** 100 Ω

**18.** Find the resistance corresponding to the following conductances:

**(a)** 0.1 S **(b)** 0.5 S **(c)** 0.02 S

**19.** A 120 V source is to be connected to a 1500 Ω resistive load by two lengths of wire as shown in Figure 2–54. The voltage source is to be located 50 ft from the load. Determine the gage number of the *smallest* wire that can be used if the *total* resistance of the two lengths of wire is not to exceed 6 Ω. Refer to Table 2–3.

FIGURE 2–54

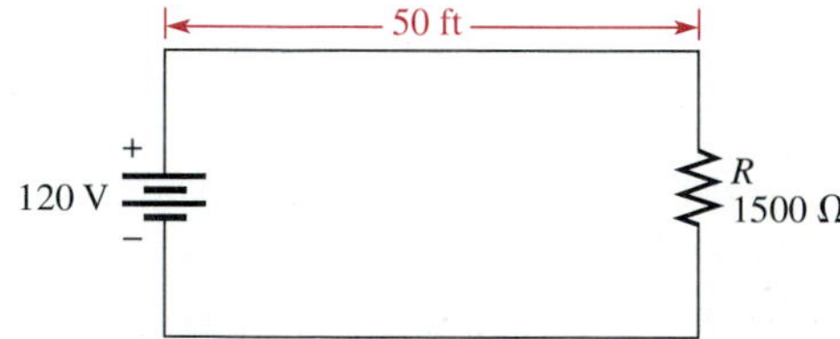


**20.** Determine the resistance and tolerance of each resistor labeled as follows:

**(a)** 4R7J **(b)** 5602M **(c)** 1501F

## SECTION 2–6 The Electric Circuit

**21.** Trace the current path in Figure 2–55(a) with the switch in position 2.

**22.** With the switch in either position, redraw the circuit in Figure 2–55(d) with a fuse connected to protect the circuit against excessive current.

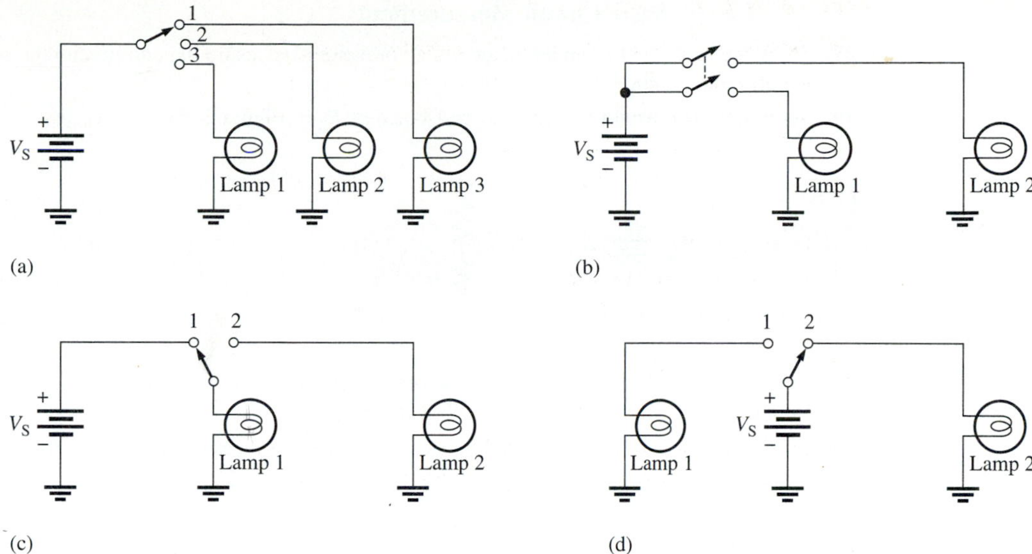


FIGURE 2–55

**23.** There is only one circuit in Figure 2–55 in which it is possible to have all lamps on at the same time. Determine which circuit it is.

**24.** Through which resistor in Figure 2–56 is there always current, regardless of the position of the switches?

FIGURE 2–56

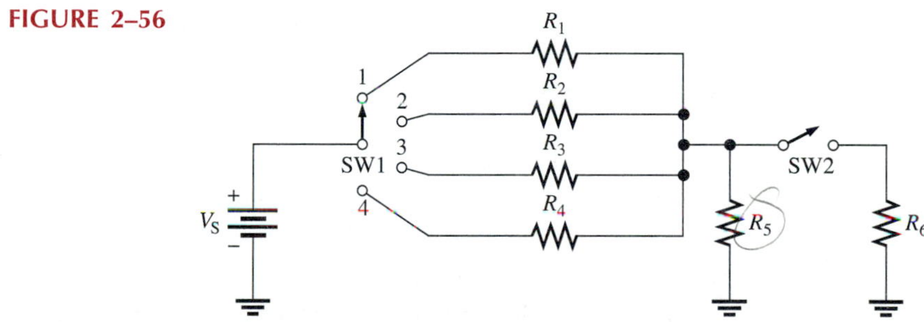


**25.** Devise a switch arrangement whereby two voltage sources ($V_{S1}$ and $V_{S2}$) can be connected simultaneously to either of two resistors ($R_1$ and $R_2$) as follows:

$V_{S1}$ connected to $R_1$ and $V_{S2}$ connected to $R_2$

or $V_{S1}$ connected to $R_2$ and $V_{S2}$ connected to $R_1$

**26.** The different sections of a stereo system are represented by the blocks in Figure 2–57. Show how a single switch can be used to connect the phonograph, the CD (compact disk) player, the tape deck, the AM tuner, or the FM tuner to the amplifier by a single knob control. Only one section can be connected to the amplifier at any time.

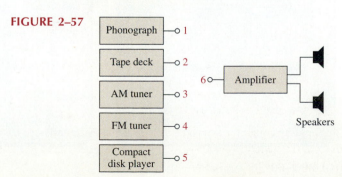

FIGURE 2–57

Phonograph 1
Tape deck 2
AM tuner 3
FM tuner 4
Compact disk player 5
6 Amplifier
Speakers

## SECTION 2–7 Basic Circuit Measurements

**27.** Show the placement of an ammeter and a voltmeter to measure the current and the source voltage in Figure 2–58.

**28.** Show how you would measure the resistance of $R_2$ in Figure 2–58.

FIGURE 2–58

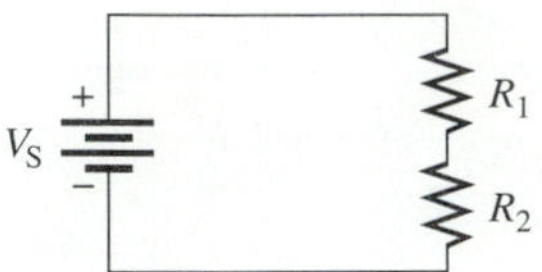


**29.** In Figure 2–59, how much voltage does each meter indicate when the switch is in position 1? In position 2?

**30.** In Figure 2–59, indicate how to connect an ammeter to measure the current from the voltage source regardless of the switch position.

FIGURE 2–59

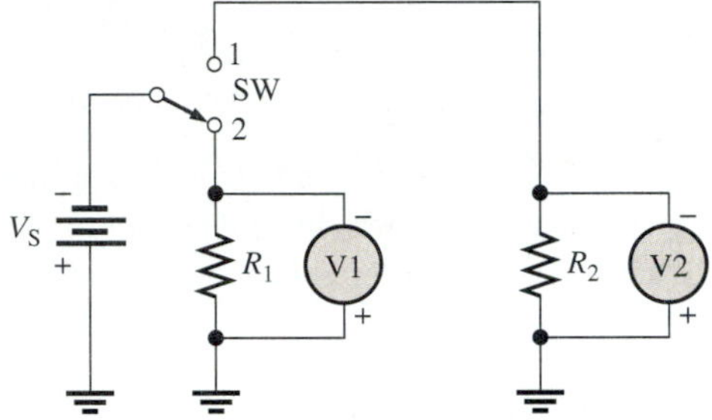


**31.** In Figure 2–56, show the proper placement of ammeters to measure the current through each resistor and the current out of the battery.

**32.** Show the proper placement of voltmeters to measure the voltage across each resistor in Figure 2–56.

**33.** What are the voltage readings in Figures 2–60(a) and 2–60(b)?

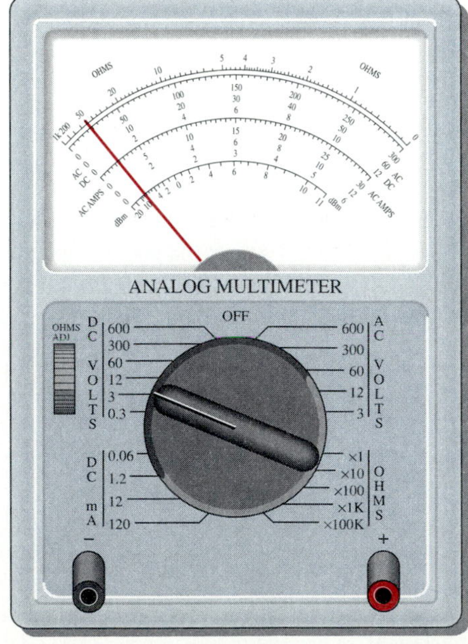


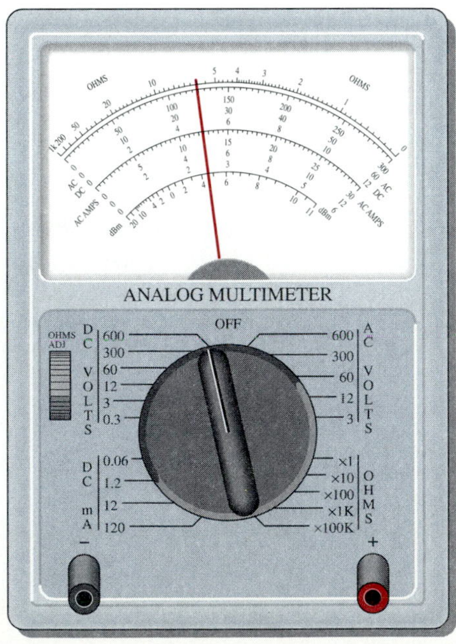


FIGURE 2–60

**34.** How much resistance is the ohmmeter in Figure 2–61 measuring?

FIGURE 2–61

**35.** Determine the resistance indicated by each of the following ohmmeter readings and range settings:

**(a)** pointer at 2, range setting at ×10

**(b)** pointer at 15, range setting at ×100,000

**(c)** pointer at 45, range setting at ×100

**36.** What is the maximum resolution of a 4½-digit DMM?

**37.** Indicate schematically how you would connect the multimeter in Figure 2–61 to the circuit in Figure 2–62 to measure each of the following quantities. In each case indicate the appropriate function and range.

**(a)** $I_1$ **(b)** $V_1$ **(c)** $R_1$

FIGURE 2–62

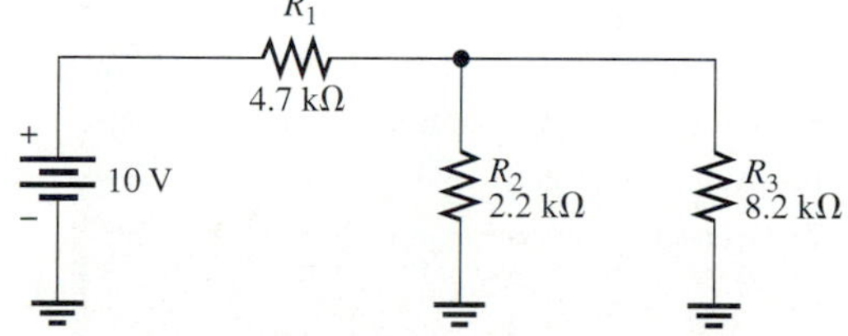


## ■ ANSWERS TO SECTION REVIEWS

### Section 2–1

**1.** The electron is the basic particle of negative charge.

**2.** An atom is the smallest particle of an element that retains the unique characteristics of the element.

**3.** An atom is a positively charged nucleus surrounded by orbiting electrons.

**4.** *Atomic weight:* approximately the number of protons and neutrons; *Atomic number:* number of protons

5. No, each element has a different type of atom.
6. A free electron is an outer-shell electron that has drifted away from the parent atom.
7. Shells are electron orbits.

## Section 2–2

1. $Q$ = charge
2. Unit of charge is the coulomb; C
3. Positive or negative charge is caused by the loss or acquisition respectively of an outer-shell (valence) electron.
4. $Q = \dfrac{10 \times 10^{12} \text{ electrons}}{6.25 \times 10^{18} \text{ electrons/C}} = 1.6 \times 10^{-6} \text{ C} = 1.6\ \mu\text{C}$

## Section 2–3

1. Voltage is energy per unit charge.
2. Volt is the unit of voltage.
3. $V = W/Q = 24 \text{ J}/10 \text{ C} = 2.4 \text{ V}$
4. Battery, power supply, solar cell, and generator are voltage sources.

## Section 2–4

1. Current is the rate of flow of electrons; its unit is the ampere (A).
2. electrons/coulomb = $6.25 \times 10^{18}$
3. $I = Q/t = 20 \text{ C}/4 \text{ s} = 5 \text{ A}$

## Section 2–5

1. Resistance is opposition to current; its unit is the ohm (Ω)
2. Two resistor categories are fixed and variable. The value of a fixed resistor cannot be changed, but that of a variable resistor can.
3. *First band:* first digit of resistance value. *Second band:* second digit of resistance value. *Third band:* number of zeros following the second digit. *Fourth band:* % tolerance.
4. **(a)** 27 kΩ ± 10% **(b)** 100 Ω ± 10%
   **(c)** 5.6 MΩ ± 5% **(d)** 6.8 kΩ ± 10%
   **(e)** 33 Ω ± 10% **(f)** 47 kΩ ± 20%
5. 330 Ω: (b), 2.2 kΩ: (d), 56 kΩ: (e), 100 kΩ: (f), 39 kΩ: (a)
6. A rheostat has two terminals; a potentiometer has three terminals.
7. A thermistor is a temperature-sensitive resistor.

## Section 2–6

1. An electric circuit consists of a source, load, and current path between source and load.
2. An open circuit is one that has no path for current.
3. A closed circuit is one that has a complete path for current.
4. Infinite resistance across an open switch; zero resistance across a closed switch
5. A fuse is not resettable, a circuit breaker is.
6. AWG 3 is larger.
7. Ground is the common or reference point.

## Section 2–7

1. **(a)** An ammeter measures current.
   **(b)** A voltmeter measures voltage.
   **(c)** An ohmmeter measures resistance.
2. See Figure 2–63.

FIGURE 2–63

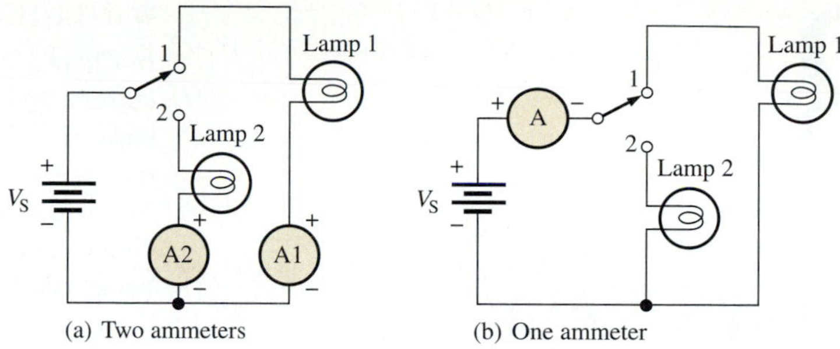


(a) Two ammeters (b) One ammeter

FIGURE 2–64

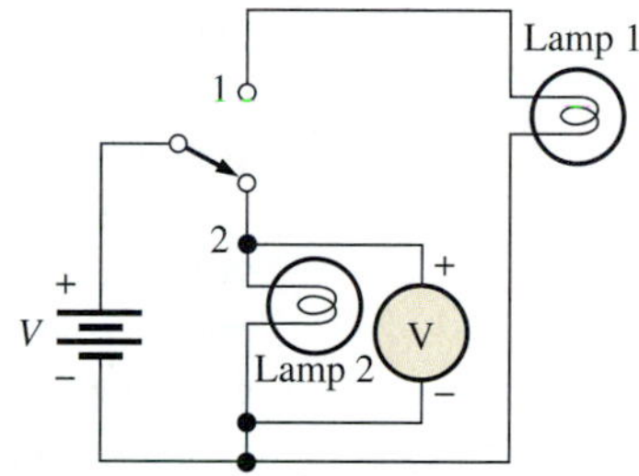


3. See Figure 2–64.
4. 1.5 V
5. Set the range switch to 600 and read on the middle ac-dc scale. The number read is multiplied by 10.
6. ×1000 range
7. Two types of DMM displays are liquid-crystal display (LCD) and light-emitting display (LED). The LCD requires little current, but it is difficult to see in low light and is slow to respond. The LED can be seen in the dark, and it responds quickly. However, it requires much more current than does the LCD.
8. Resolution is the smallest increment of a quantity that the meter can measure.

## Section 2–8

1. Less voltage causes less light because the current is reduced.
2. A lower resistance will result in more light.

## Section 2–9

1. **(a)** A 10 V source $V_1$ is connected between node 0 and node 1 with the positive at node 0.
   **(b)** A 4.7 $\mu$F capacitor $C_1$ is connected from node 1 to node 0.
   **(c)** A 2.2 kΩ resistor RES1 is connected from node 0 to node 1.
2. 
```
VS  1  0  9
R1  1  0  10K
```

## ■ ANSWERS TO RELATED EXERCISES FOR EXAMPLES

**2–1** $1.88 \times 10^{19}$ electrons

**2–2** 600 J

**2–3** 12 C

**2–4** 4700 Ω ± 20%

**2–5** 2.25 CM

**2–6** 4.62 kΩ

**2–7** The needle will move left to the "70" mark.

# 3

# OHM'S LAW



## ■ INTRODUCTION

In Chapter 2, you studied the concept of voltage, current, and resistance. You also were introduced to a basic electric circuit. In this chapter, you will learn how voltage, current, and resistance are interrelated. You will also learn how to analyze a simple electric circuit. In this chapter and throughout the rest of the book, you will learn the basics of putting technology theory into practice.

Ohm's law is perhaps the single most important tool for the analysis of electric circuits. There are many other laws, theorems, and rules—some of which you can live without; however, you *must* know and be able to apply Ohm's law.

In 1826 Georg Simon Ohm found that current, voltage, and resistance are related in a specific and predictable way. Ohm expressed this relationship with a formula that is known today as Ohm's law. In this chapter, you will learn Ohm's law and how to use it in solving circuit problems. Ohm's law is one of the basic foundation elements upon which the rest of your study and work in electronics will be built.

## TECHnology Theory Into Practice

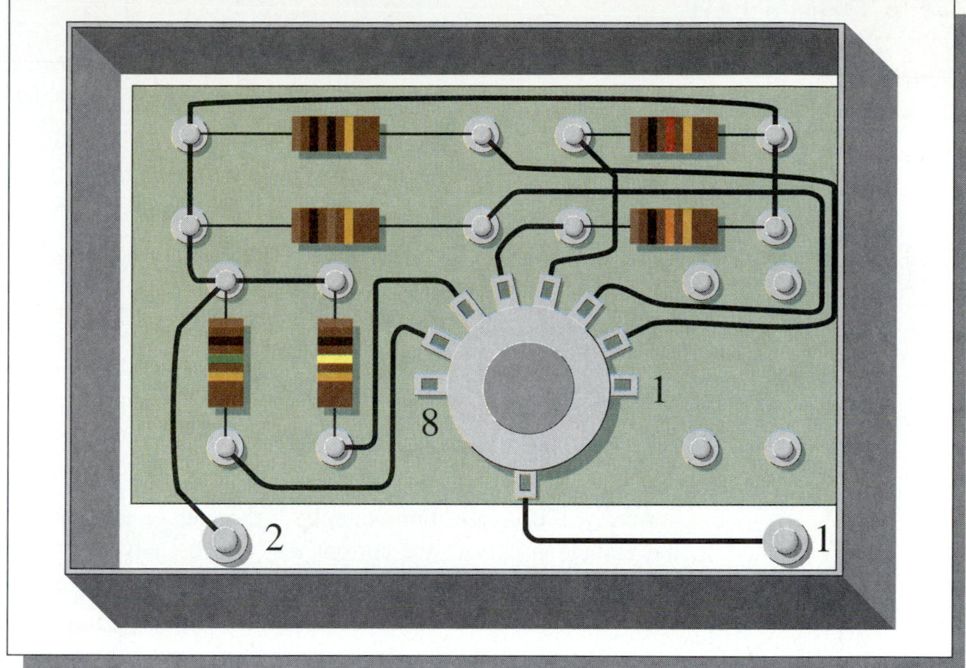


In Section 3–6, you will see how Ohm's law is applicable to a practical circuit. Visualize the following situation: On the job as a technician, you are assigned to test a resistance box that has been constructed for use in the lab. The resistance box provides several decades of resistance (10 Ω, 100 Ω, 1 kΩ, etc.) that are switch selectable. This is a rush job and the resistance box is urgently needed in a product test setup. Checking out the box is simple enough. All you have to do is measure the resistance between the two terminals for each setting of the selector switch and verify that the resistance is correct. After connecting the multimeter and selecting the ohmmeter function, you discover that it is not functioning as an ohmmeter. This is the only meter in the lab that is available immediately. The boss wants this done right away, so what do you do?

## CHAPTER OBJECTIVES

- ☐ Explain Ohm's law
- ☐ Calculate current in a circuit
- ☐ Calculate voltage in a circuit
- ☐ Calculate resistance in a circuit
- ☐ Explain the proportional relationship of current, voltage, and resistance
- ☐ Create PSpice circuit files and discuss the output file (optional)

## 3–1 ■ OHM'S LAW

***Ohm's law describes mathematically how voltage, current, and resistance in a circuit are related. Ohm's law is used in three equivalent forms depending on which quantity you need to determine. In this section, you will learn each of these forms.***

***After completing this section, you should be able to***

- **Explain Ohm's law**
  - ☐ Describe how *V, I,* and *R* are related
  - ☐ Express *I* as a function of *V* and *R*
  - ☐ Express *V* as a function of *I* and *R*
  - ☐ Express *R* as a function of *V* and *I*

Ohm determined experimentally that if the voltage across a resistor is increased, the current through the resistor will also increase; and, likewise, if the voltage is decreased, the current will decrease. For example, if the voltage is doubled, the current will double. If the voltage is halved, the current will also be halved. This relationship is illustrated in Figure 3–1, with relative meter indications of voltage and current.

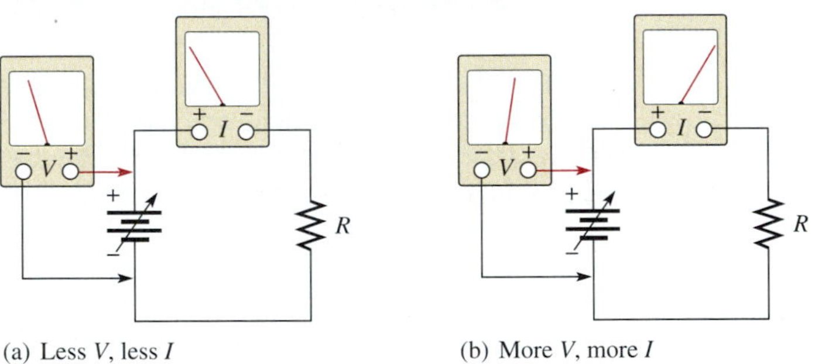


**FIGURE 3–1**
***Effect of changing the voltage with the resistance at a constant value.***

**Ohm's law** also states that if the voltage is kept constant, less resistance results in more current, and, also, more resistance results in less current. For example, if the resistance is halved, the current doubles. If the resistance is doubled, the current is halved. This concept is illustrated by the meter indications in Figure 3–2, where the resistance is increased and the voltage is held constant.

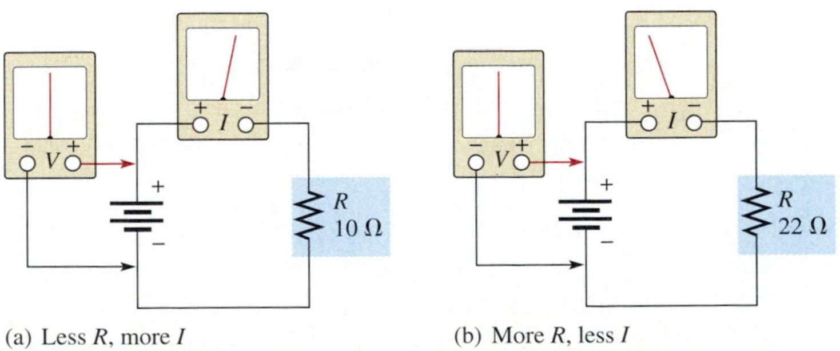


**FIGURE 3–2**
***Effect of changing the resistance with the voltage at a constant value.***

## Formula for Current

Ohm's law can be stated as follows:

$$I = \frac{V}{R} \tag{3–1}$$

This formula describes what was indicated by the circuits of Figures 3–1 and 3–2. For a constant value of $R$, if the value of $V$ is increased, the value of $I$ increases; if $V$ is decreased, $I$ decreases. Also notice in Equation (3–1) that if $V$ is constant and $R$ is increased, $I$ decreases. Similarly, if $V$ is constant and $R$ is decreased, $I$ increases.

Using Equation (3–1), you can calculate the current if the values of voltage and resistance are known.

## Formula for Voltage

Ohm's law can also be stated another way. By multiplying both sides of Equation (3–1) by $R$ and transposing terms, you obtain an equivalent form of Ohm's law, as follows:

$$V = IR \tag{3–2}$$

With this equation, you can calculate voltage if the current and resistance are known.

## Formula for Resistance

There is a third equivalent way to state Ohm's law. By dividing both sides of Equation (3–2) by $I$ and transposing terms, you obtain

$$R = \frac{V}{I} \tag{3–3}$$

This form of Ohm's law is used to determine resistance if voltage and current values are known.

Remember, the three formulas—Equations (3–1), (3–2) and (3–3)—are all equivalent. They are simply three different ways of expressing Ohm's law.

## SECTION 3–1 REVIEW

1. Ohm's law defines how three basic quantities are related. What are these quantities?
2. Write the Ohm's law formula for current.
3. Write the Ohm's law formula for voltage.
4. Write the Ohm's law formula for resistance.
5. If the voltage across a fixed-value resistor is tripled, does the current increase or decrease, and by how much?
6. If the voltage across a fixed resistor is cut in half, how much will the current change?
7. There is a fixed voltage across a resistor, and you measure a current of 1 A. If you replace the resistor with one that has twice the resistance value, how much current will you measure?
8. In a circuit the voltage is doubled and the resistance is cut in half. Would the current increase or decrease, and if so, by how much?

## 3–2 ■ CALCULATING CURRENT

***In this section, you will learn to determine current values when you know the values of voltage and resistance. You will also see how to use quantities expressed with metric prefixes in circuit calculations.***

***After completing this section, you should be able to***

- **Calculate current in a circuit**
  - ☐ Use Ohm's law to find current when voltage and resistance are known
  - ☐ Use voltage and resistance values expressed with metric prefixes

In the following examples, the formula $I = V/R$ is used. In order to get current in amperes, you must express the value of voltage in volts and the value of resistance in ohms.

**EXAMPLE 3–1**

How many amperes of current are in the circuit of Figure 3–3?

FIGURE 3–3

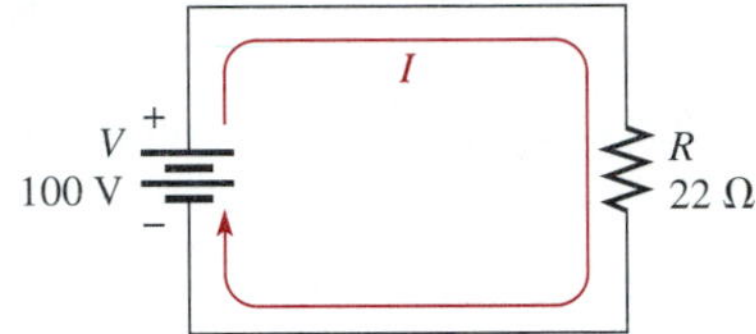


***Solution*** Use the formula $I = V/R$, and substitute 100 V for $V$ and 22 Ω for $R$.

$$I = \frac{V}{R} = \frac{100\text{ V}}{22\ \Omega} = 4.55\text{ A}$$

There are 4.55 A of current in this circuit.

***Related Exercise*** If $R$ is changed to 33 Ω in Figure 3–3, what is the current?

**EXAMPLE 3–2**

If the resistance in Figure 3–3 is changed to 47 Ω and the voltage to 50 V, what is the new value of current?

***Solution*** Substitute $V = 50$ V and $R = 47$ Ω into the formula $I = V/R$.

$$I = \frac{V}{R} = \frac{50\text{ V}}{47\ \Omega} = 1.06\text{ A}$$

***Related Exercise*** If $V = 5$ V and $R = 1000$ Ω, what is the current?

### Larger Units of Resistance (kΩ and MΩ)

In electronics, resistance values of thousands of **ohms** or even millions of ohms are common. As you learned in Chapter 1, large values of resistance are indicated by the metric system prefixes *kilo* (k) and *mega* (M). Thus, thousands of ohms are expressed in kilohms (kΩ), and millions of ohms in megohms (MΩ). The following examples illustrate how to use kilohms and megohms when you calculate current.

**EXAMPLE 3–3** Calculate the current in Figure 3–4.

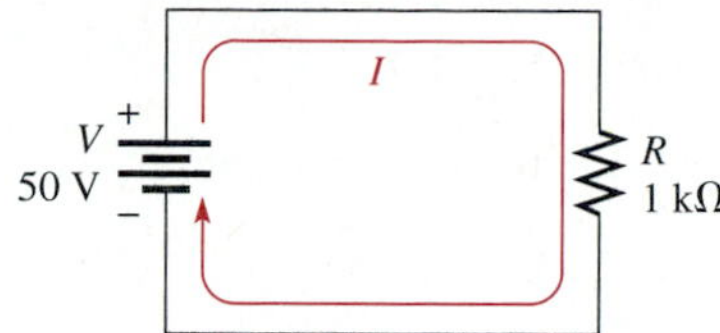

FIGURE 3–4

*Solution* Remember that 1 kΩ is the same as $1 \times 10^3$ Ω. Use the formula $I = V/R$ and substitute 50 V for $V$ and $1 \times 10^3$ Ω for $R$.

$$I = \frac{V}{R} = \frac{50\text{ V}}{1\text{ k}\Omega} = \frac{50\text{ V}}{1 \times 10^3\ \Omega} = 50 \times 10^{-3}\text{ A} = 50\text{ mA}$$

*Related Exercise* Calculate the current in Figure 3–4 if $R$ is changed to 10 kΩ.

In Example 3–3, $50 \times 10^{-3}$ A is expressed as 50 milliamperes (50 mA). This can be used to advantage when you divide volts by kilohms. The current will be in milliamperes, as Example 3–4 illustrates.

**EXAMPLE 3–4** How many milliamperes are in the circuit of Figure 3–5?

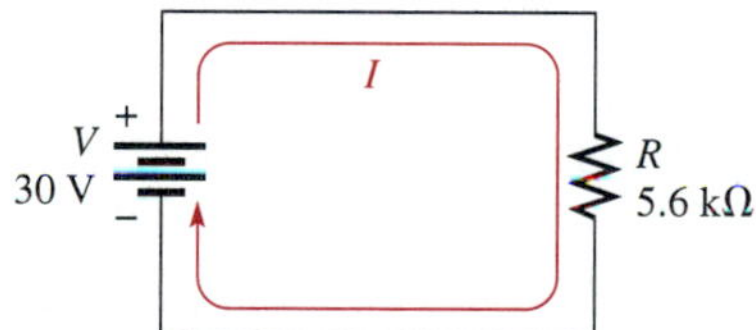

FIGURE 3–5

*Solution* When you divide volts by kilohms, you get current in milliamperes.

$$I = \frac{V}{R} = \frac{30\text{ V}}{5.6\text{ k}\Omega} = 5.36\text{ mA}$$

*Related Exercise* What is the current in milliamperes if $R$ is changed to 2.2 kΩ?

If volts are applied when resistance values are in megohms, the current is in microamperes ($\mu$A), as Example 3–5 shows.

**EXAMPLE 3–5** Determine the amount of current in the circuit of Figure 3–6.

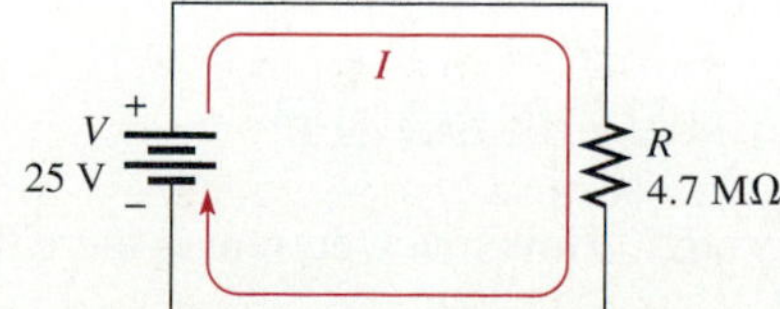

FIGURE 3–6

***Solution*** Recall that 4.7 MΩ equals $4.7 \times 10^6$ Ω. Substitute 25 V for $V$ and $4.7 \times 10^6$ Ω for $R$.

$$I = \frac{V}{R} = \frac{25\text{ V}}{4.7\text{ M}\Omega} = \frac{25\text{ V}}{4.7 \times 10^6\ \Omega} = 5.32 \times 10^{-6}\text{ A} = 5.32\ \mu\text{A}$$

***Related Exercise*** What is the current if $V$ is increased to 100 V in Figure 3–6?

**EXAMPLE 3–6**

Change the value of $R$ in Figure 3–6 to 1.8 MΩ. What is the new value of current?

***Solution*** When you divide volts by megohms, you get current in microamperes.

$$I = \frac{V}{R} = \frac{25\text{ V}}{1.8\text{ M}\Omega} = 13.9\ \mu\text{A}$$

***Related Exercise*** If $R$ is doubled in the circuit of Figure 3–6, what is the new value of current?

## Larger Units of Voltage (kV)

Small voltages, usually less than 50 V are common in semiconductor circuits. Occasionally, however, large voltages are encountered. For example, the high-voltage supply in a television receiver is around 20,000 V (20 kilovolts, or 20 kV), and transmission voltages generated by the power companies may be as high as 345,000 V (345 kV). The following two examples illustrate how to use voltage values in the kilovolt range when you calculate current.

**EXAMPLE 3–7**

How much current is produced by a voltage of 24 kV across a 12 kΩ resistance?

***Solution*** Since kilovolts are divided by kilohms, the prefixes cancel; therefore, the current is in amperes.

$$I = \frac{V}{R} = \frac{24\text{ kV}}{12\text{ k}\Omega} = \frac{24 \times 10^3\text{ V}}{12 \times 10^3\ \Omega} = 2\text{ A}$$

***Related Exercise*** What is the current in mA produced by 1 kV across a 27 kΩ resistance?

**EXAMPLE 3–8**

How much current is there through a 100 MΩ resistor when 50 kV are applied?

***Solution*** In this case, divide 50 kV by 100 MΩ to get the current. Substitute $50 \times 0^3$ V for 50 kV and $100 \times 10^6$ Ω for 100 MΩ.

$$I = \frac{V}{R} = \frac{50\text{ kV}}{100\text{ M}\Omega} = \frac{50 \times 10^3\text{ V}}{100 \times 10^6\ \Omega} = 0.5 \times 10^{-3}\text{ A} = 0.5\text{ mA}$$

Remember that the power of ten in the denominator is subtracted from the power of ten in the numerator. So 50 was divided by 100, giving 0.5, and 6 was subtracted from 3, giving $10^{-3}$.

[5] [0] [EXP] [3] [÷] [1] [0] [0] [EXP] [6] [=]

***Related Exercise*** How much current is there through a 6.8 MΩ resistor when 10 kV are applied?

**SECTION 3–2 REVIEW**

In Problems 1–4, calculate $I$ when

1. $V = 10$ V and $R = 5.6\ \Omega$.
2. $V = 100$ V and $R = 560\ \Omega$.
3. $V = 5$ V and $R = 2.2$ k$\Omega$.
4. $V = 15$ V and $R = 4.7$ M$\Omega$.
5. If a 4.7 M$\Omega$ resistor has 20 kV across it, how much current is there?
6. How much current will 10 kV across 2.2 k$\Omega$ produce?

## 3–3 ■ CALCULATING VOLTAGE

***In this section, you will learn to determine voltage values when you know the values of current and resistance. You will also see how to use quantities expressed with metric prefixes in circuit calculations. After completing this section, you should be able to***

- **Calculate voltage in a circuit**
  - ☐ Use Ohm's law to find voltage when current and resistance are known
  - ☐ Use current and resistance values expressed with metric prefixes

In the following examples, the formula $V = IR$ is used. To obtain voltage in volts, you must express the value of $I$ in amperes and the value of $R$ in ohms.

**EXAMPLE 3–9** In the circuit of Figure 3–7, how much voltage is needed to produce 5 A of current?

FIGURE 3–7

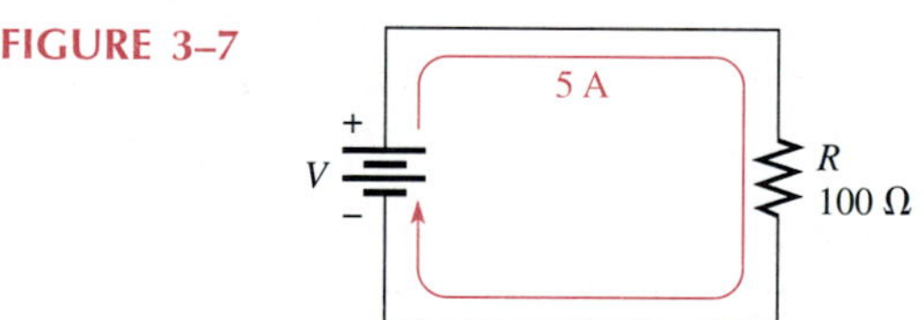


*Solution* Substitute 5 A for $I$ and 100 $\Omega$ for $R$ into the formula $V = IR$.

$$V = IR = (5\text{ A})(100\ \Omega) = 500\text{ V}$$

Thus, 500 V are required to produce 5 A of current through a 100 $\Omega$ resistor.

*Related Exercise* In Figure 3–7, how much voltage is required to produce 12 A of current?

### Smaller Units of Current (mA and $\mu$A)

The following two examples illustrate how to use current values in the milliampere (mA) and microampere ($\mu$A) ranges when you calculate voltage.

**EXAMPLE 3–10** How much voltage will be measured across the resistor in Figure 3–8?

FIGURE 3–8

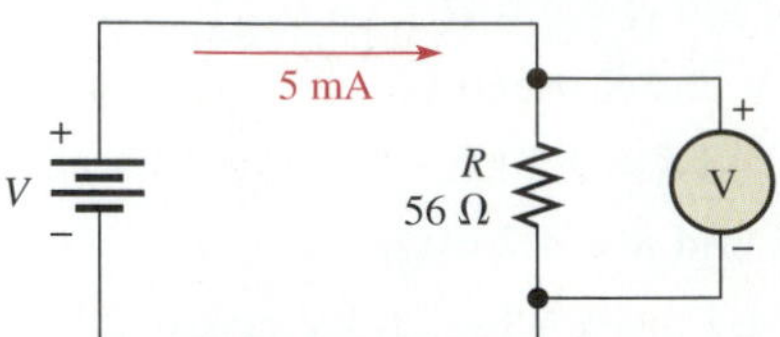


***Solution*** Five milliamperes equals $5 \times 10^{-3}$ A. Substitute the values for $I$ and $R$ into the formula $V = IR$.

$$V = IR = (5\text{ mA})(56\ \Omega) = (5 \times 10^{-3}\text{ A})(56\ \Omega) = 280 \times 10^{-3}\text{ V} = 280\text{ mV}$$

When milliamperes are multiplied by ohms, you get millivolts.

***Related Exercise*** How much voltage is measured across $R$ if $R = 33\ \Omega$ and $I = 1.5$ mA in Figure 3–8?

**EXAMPLE 3–11** Suppose that there is a current of 8 $\mu$A through a 10 Ω resistor. How much voltage is across the resistor?

***Solution*** Eight microamperes equals $8 \times 10^{-6}$ A. Substitute the values for $I$ and $R$ into the formula $V = IR$.

$$V = IR = (8\ \mu\text{A})(10\ \Omega) = (8 \times 10^{-6}\text{ A})(10\ \Omega) = 80 \times 10^{-6}\text{ V} = 80\ \mu\text{V}$$

When microamperes are multiplied by ohms, you get microvolts.

***Related Exercise*** If there are 3.2 $\mu$A through a 47 Ω resistor, what is the voltage across the resistor?

These examples have demonstrated that when you multiply milliamperes and ohms, you get millivolts. When you multiply microamperes and ohms, you get microvolts.

## Larger Units of Resistance (kΩ and MΩ)

The following two examples illustrate how to use resistance values in the kilohm (kΩ) and megohm (MΩ) ranges when you calculate voltage.

**EXAMPLE 3–12** The circuit in Figure 3–9 has a current of 10 mA. What is the voltage?

FIGURE 3–9

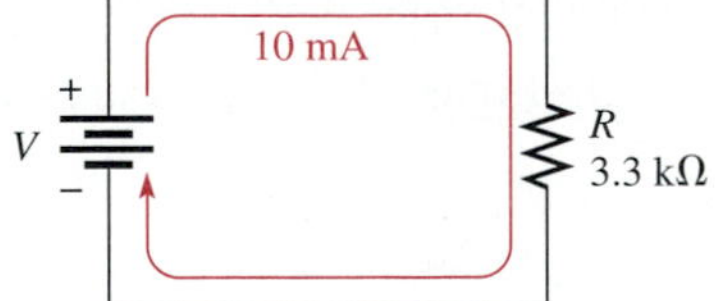


***Solution*** Ten milliamperes equals $10 \times 10^{-3}$ A and 3.3 kΩ equals $3.3 \times 10^{3}$ Ω. Substitute these values into the formula $V = IR$.

$$V = IR = (10\text{ mA})(3.3\text{ k}\Omega) = (10 \times 10^{-3}\text{ A})(3.3 \times 10^{3}\ \Omega) = 33\text{ V}$$

Notice that $10^{-3}$ and $10^{3}$ cancel. Therefore, milliamperes cancel kilohms when multiplied, and the result is volts.

***Related Exercise*** If the current in Figure 3–9 is 25 mA, what is the voltage?

**EXAMPLE 3–13** If there is a current of 50 $\mu$A through a 4.7 M$\Omega$ resistor, what is the voltage?

***Solution*** Fifty microamperes equals $50 \times 10^{-6}$ A and 4.7 M$\Omega$ is $4.7 \times 10^{6}$ $\Omega$. Substitute these values into the formula $V = IR$.

$$V = IR = (50\ \mu\text{A})(4.7\ \text{M}\Omega) = (50 \times 10^{-6}\ \text{A})(4.7 \times 10^{6}\ \Omega) = 235\ \text{V}$$

Notice that $10^{-6}$ and $10^{6}$ cancel. Therefore, microamperes cancel megohms when multiplied, and the result is volts.

***Related Exercise*** If there are 450 $\mu$A through a 3.9 M$\Omega$ resistor, what is the voltage?

**SECTION 3–3 REVIEW**

In Problems 1–7, calculate $V$ when

1. $I = 1$ A and $R = 10\ \Omega$.
2. $I = 8$ A and $R = 470\ \Omega$.
3. $I = 3$ mA and $R = 100\ \Omega$.
4. $I = 25\ \mu$A and $R = 56\ \Omega$.
5. $I = 2$ mA and $R = 1.8$ k$\Omega$.
6. $I = 5$ mA and $R = 100$ M$\Omega$.
7. $I = 10\ \mu$A and $R = 2.2$ M$\Omega$.
8. How much voltage is required to produce 100 mA through 4.7 k$\Omega$?
9. What voltage do you need to cause 3 mA of current in a 3.3 k$\Omega$ resistance?
10. A battery produces 2 A of current into a 6.8 $\Omega$ resistive load. What is the battery voltage?

## 3–4 ■ CALCULATING RESISTANCE

***In this section, you will learn to determine resistance values when you know the values of current and voltage. You will also see how to use quantities expressed with metric prefixes in circuit calculations.***

***After completing this section, you should be able to***

- **Calculate resistance in a circuit**
  - ☐ Use Ohm's law to find resistance when voltage and current are known
  - ☐ Use current and voltage values expressed with metric prefixes

In the following examples, the formula $R = V/I$ is used. To get resistance in ohms, you must express the value of $I$ in amperes and the value of $V$ in volts.

**EXAMPLE 3–14** In the circuit of Figure 3–10, how much resistance is needed to draw 3.08 A of current from the battery?

FIGURE 3–10

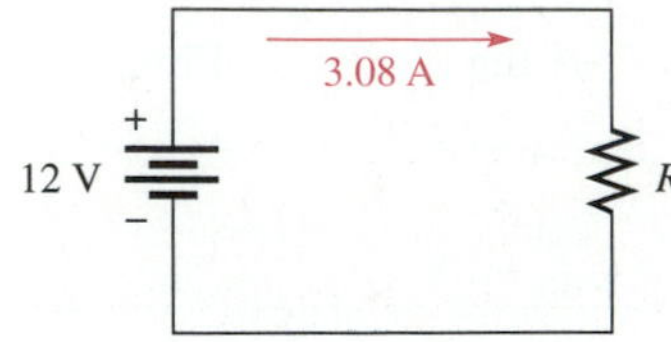


***Solution*** Substitute 12 V for $V$ and 3.08 A for $I$ into the formula $R = V/I$.

$$R = \frac{V}{I} = \frac{12\text{ V}}{3.08\text{ A}} = 3.90\ \Omega$$

***Related Exercise*** In Figure 3–10, to what value must $R$ be changed for a current of 5.45 A?

## Smaller Units of Current (mA and $\mu$A)

The following two examples illustrate how to use current values in the milliampere (mA) and microampere ($\mu$A) ranges when you calculate resistance.

**EXAMPLE 3–15** Suppose that the ammeter in Figure 3–11 indicates 4.55 mA of current and the voltmeter reads 150 V. What is the value of $R$?

FIGURE 3–11

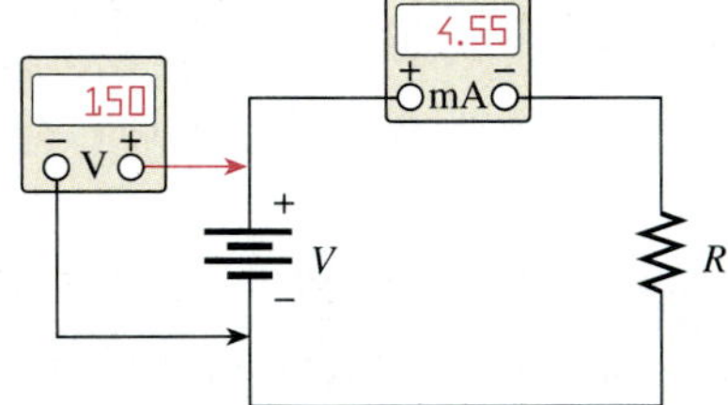


***Solution*** 4.55 mA equals $4.55 \times 10^{-3}$ A. Substitute the voltage and current values into the formula $R = V/I$.

$$R = \frac{V}{I} = \frac{150\text{ V}}{4.55\text{ mA}} = \frac{150\text{ V}}{4.55 \times 10^{-3}\text{ A}} = 33 \times 10^{3}\ \Omega = 33\text{ k}\Omega$$

When volts are divided by milliamperes, the resistance is in kilohms.

***Related Exercise*** If the ammeter indicates 1.10 mA and the voltmeter reads 75 V, what is the value of $R$?

**EXAMPLE 3–16** Suppose that the value of the resistor in Figure 3–11 is changed. If the battery voltage is still 150 V and the ammeter reads 68.2 $\mu$A, what is the new resistor value?

***Solution*** 68.2 $\mu$A equals $68.2 \times 10^{-6}$ A. Substitute $V$ and $I$ values into the equation for $R$.

$$R = \frac{V}{I} = \frac{150\text{ V}}{68.2\ \mu\text{A}} = \frac{150\text{ V}}{68.2 \times 10^{-6}\text{ A}} = 2.2 \times 10^{6}\ \Omega = 2.2\text{ M}\Omega$$

When volts are divided by microamperes, the resistance has units of megohms.

***Related Exercise*** If the resistor is changed in Figure 3–11 so that the ammeter reads 45.5 $\mu$A, what is the new resistor value? Assume $V = 150$ V.

**SECTION 3–4 REVIEW**

In Problems, 1–5, calculate $R$ when

1. $V = 10$ V and $I = 2.13$ A.
2. $V = 270$ V and $I = 10$ A.
3. $V = 20$ kV and $I = 5.13$ A.
4. $V = 15$ V and $I = 2.68$ mA.
5. $V = 5$ V and $I = 2.27$ $\mu$A.
6. You have a resistor across which you measure 25 V, and your ammeter indicates 53.2 mA of current. What is the resistor's value in kilohms? In ohms?

## 3–5 ■ THE RELATIONSHIP OF CURRENT, VOLTAGE, AND RESISTANCE

***Ohm's law describes how current is related to voltage and resistance. Current and voltage are linearly proportional; current and resistance are inversely related. Because voltage is the "driving force," it is not dependent on the resistance when used as a source. When it is a voltage drop, it is directly proportional to the resistance for a given current.***

***After completing this section, you should be able to***

- **Explain the proportional relationship of current, voltage, and resistance**
  - ☐ Show graphically that $I$ and $V$ are directly proportional
  - ☐ Show graphically that $I$ and $R$ are inversely proportional
  - ☐ Explain why $I$ and $V$ are linearly proportional

### The Linear Relationship of Current and Voltage

Current and voltage are **linearly** proportional; that is, if one is increased or decreased by a certain percentage, the other will increase or decrease by the same percentage, assuming that the resistance is constant in value. For example, if the voltage across a resistor is tripled, the current will triple.

**EXAMPLE 3–17** Show that if the voltage in the circuit of Figure 3–12 is increased to three times its present value, the current will triple in value.

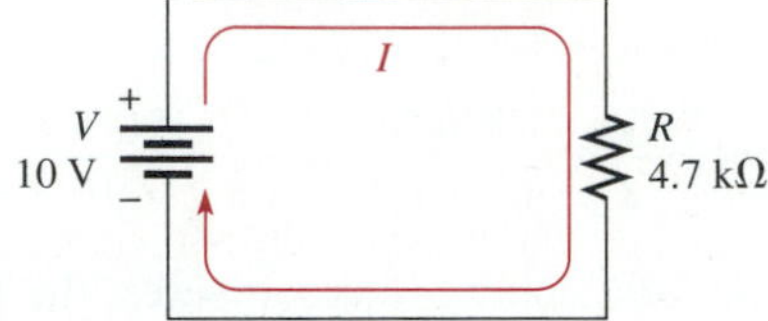


FIGURE 3–12

***Solution*** With 10 V, the current is

$$I = \frac{V}{R} = \frac{10\text{ V}}{4.7\text{ k}\Omega} = 2.13\text{ mA}$$

If the voltage is increased to 30 V, the current will be

$$I = \frac{V}{R} = \frac{30\text{ V}}{4.7\text{ k}\Omega} = 6.38\text{ mA}$$

The current went from 2.13 mA to 6.38 mA when the voltage was tripled to 30 V.

***Related Exercise*** If the voltage in Figure 3–12 is quadrupled, will the current also quadruple?

## A Graph of Current Versus Voltage

Let's take a constant value of resistance, for example, 10 Ω, and calculate the current for several values of voltage ranging from 10 V to 100 V. The current values obtained are shown in Figure 3–13(a). The graph of the $I$ values versus the $V$ values is shown in Figure 3–13(b). Note that it is a straight line graph. This graph tells us that a change in voltage results in a linearly proportional change in current. No matter what value $R$ is, assuming that $R$ is constant, the graph of $I$ versus $V$ will always be a straight line.

| $V$ | $I$ |
|---|---|
| 10 V | 1 A |
| 20 V | 2 A |
| 30 V | 3 A |
| 40 V | 4 A |
| 50 V | 5 A |
| 60 V | 6 A |
| 70 V | 7 A |
| 80 V | 8 A |
| 90 V | 9 A |
| 100 V | 10 A |

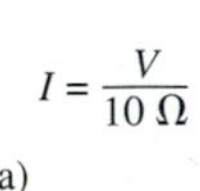

$$I = \frac{V}{10\ \Omega}$$

(a)

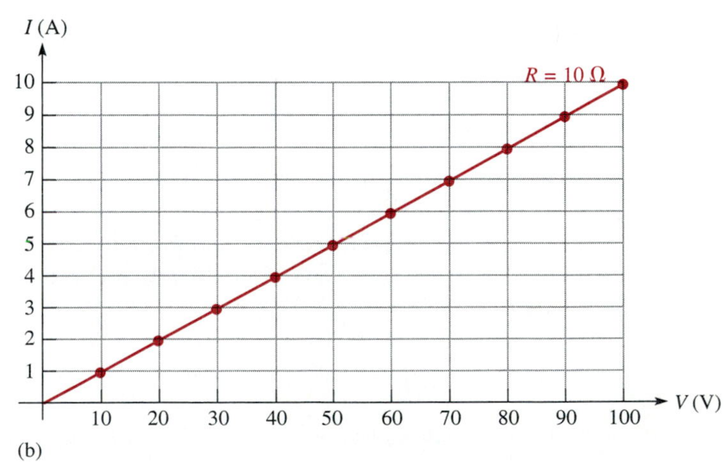


(b)

FIGURE 3–13
*Graph of current versus voltage for R = 10 Ω.*

Example 3–18 illustrates a use for the linear relationship between voltage and current in a resistive circuit.

**EXAMPLE 3–18**

Assume that you are measuring the current in a circuit that is operating with 25 V. The ammeter reads 50 mA. Later, you notice that the current has dropped to 40 mA. Assuming that the resistance did not change, you must conclude that the voltage source has changed. How much has the voltage changed, and what is its new value?

***Solution*** The current has dropped from 50 mA to 40 mA, which is a decrease of 20%. Since the voltage is linearly proportional to the current, the voltage has decreased by the same percentage that the current did. Taking 20% of 25 V, you get

$$\text{Change in voltage} = (0.2)(25\text{ V}) = 5\text{ V}$$

Subtract this change from the original voltage to get the new voltage.

$$\text{New voltage} = 25\text{ V} - 5\text{ V} = 20\text{ V}$$

Notice that you did not need the resistance value in order to find the new voltage.

***Related Exercise*** If the current drops to 0 A under the same conditions stated in the example, what is the voltage?

## Current and Resistance Are Inversely Related

As you have seen, current varies inversely with resistance as expressed by Ohm's law, $I = V/R$. When the resistance is reduced, the current goes up; when the resistance is increased, the current goes down. For example, if the source voltage is held constant and the resistance is halved, the current doubles in value; when the resistance is doubled, the current is reduced by half.

Let's take a constant value of voltage, for example, 10 V, and calculate the current for several values of resistance ranging from 10 Ω to 100 Ω. The values obtained are shown in Figure 3–14(a). The graph of the $I$ values versus the $R$ values is shown in Figure 3–14(b).

**FIGURE 3–14**
***Graph of current versus resistance for $V = 10$ V.***

| $R$ (Ω) | $I$ (A) |
|---|---|
| 10 | 1.000 |
| 20 | 0.500 |
| 30 | 0.333 |
| 40 | 0.250 |
| 50 | 0.200 |
| 60 | 0.167 |
| 70 | 0.143 |
| 80 | 0.125 |
| 90 | 0.111 |
| 100 | 0.100 |

(a)

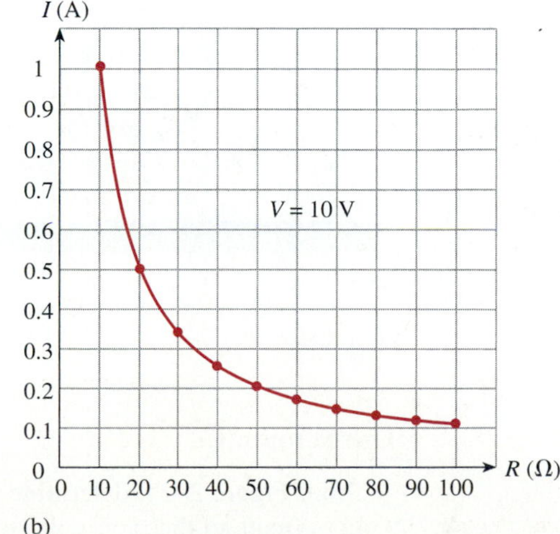


(b)

**SECTION 3–5 REVIEW**

1. What does *linearly proportional* mean?
2. In a circuit, $V = 2$ V and $I = 10$ mA. If $V$ is changed to 1 V, what will $I$ equal?
3. If $I = 3$ A at a certain voltage, what will it be if the voltage is doubled?
4. By how many volts must you increase a 12 V source in order to increase the current in a circuit by 50%?

## 3–6 ■ TECHnology Theory Into Practice

***In this TECH Tip, an existing resistance box that is to be used as part of a test setup in the lab is to be checked out and modified. Your task is to modify the circuit so that it will meet the requirements of the new application. You will have to apply your knowledge of Ohm's law in order to complete this assignment.***

***The specifications are as follows:***

1. ***Each resistor is switch selectable and only one resistor is selected at a time.***
2. ***The lowest resistor value is to be 10 Ω.***
3. ***Each successively higher resistance in the switch sequence must be a decade (10 times) increase over the previous one.***
4. ***The maximum resistor value must be 1 MΩ.***
5. ***The maximum voltage across any resistor in the box will be 4 V.***
6. ***Two additional resistors are required, one to limit the current to 10 mA ± 10% and the other to limit the current to 5 mA ± 10% with a 4 V drop.***

### The Existing Resistor Circuit

The existing resistance box is shown in both top and bottom views in Figure 3–15. The switch is a rotary type.

(a) Top view

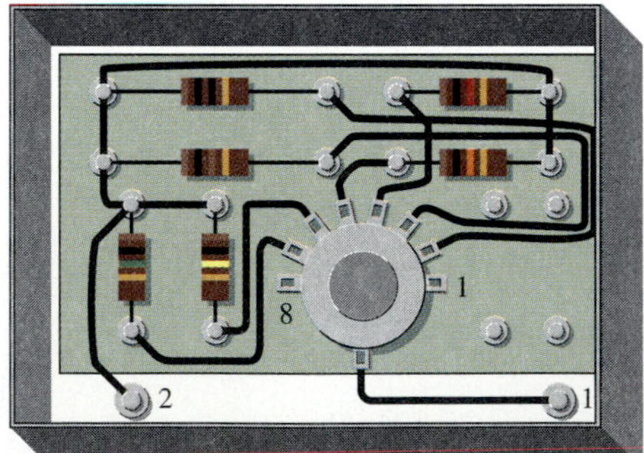


(b) Bottom view

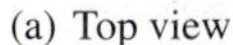

FIGURE 3–15

### The Schematic

- ☐ From Figure 3–15, determine the resistor values and draw the schematic for the existing circuit so that you will know what you have to work with. Determine the resistor numbering from the $R$ labels on the top view.

### The Schematic for the New Requirements

□ Draw the schematic for a circuit that will accomplish the following:

1. One resistor at a time is to be connected by the switch between terminals 1 and 2 of the box.
2. Provide switch selectable resistor values beginning with 10 Ω and increasing in decade increments to 1 MΩ.
3. Each of the resistors must be selectable by a sequence of adjacent switch positions in ascending order.
4. There must be two switch-selectable resistors, one is in switch position 1 (shown in Figure 3–15, bottom view) and must limit the current to 10 mA ± 10% with a 4 V drop and the other is in switch position 8 and must limit the current to 5 mA ± 10% with a 4 V drop.
5. All the resistors must be standard values with 10% tolerance.

□ Determine the modifications that must be made to the existing circuit to meet the specifications and develop a detailed list of the changes including resistance values, wiring, and new components. You should number each point in the schematic for easy reference.

### A Test Procedure

□ After the resistance box has been modified to meet the new specifications, it must be tested to see if it is working properly. Determine how you would test the resistance box and what instruments you would use. Then detail your test procedure in a step-by-step format.

### Troubleshooting the Circuit

□ When an ohmmeter is connected across terminals 1 and 2 of the resistance box, determine the most likely fault in each of the following cases:

1. The ohmmeter shows an infinitely high resistance when the switch is in position 3.
2. The ohmmeter shows an infinitely high resistance in all switch positions.
3. The ohmmeter shows an incorrect value of resistance when the switch is in position 6.

**SECTION 3–6 REVIEW**

1. Explain how you applied Ohm's law to this application assignment.
2. Determine the current through each resistor when 4 V is applied across it.

## 3–7 ■ PSpice FILES

***In the previous chapter, you learned how to describe components with PSpice description lines that show the type of component, its value, and how it is connected in a circuit. In this section, you will learn how these lines are put together to form a circuit file for PSpice to process. The basic output file that PSpice generates is also discussed.***

***After completing this section, you should be able to***

- **Create PSpice circuit files and discuss the output file**
  - □ Describe a circuit file
  - □ List the types of lines in a circuit file

☐ Write a circuit file for a simple circuit
☐ Describe an output file

## The Circuit (.CIR) File

The input file that PSpice processes is called a circuit file or .CIR file; it describes the circuit to be simulated and analyzed. PSpice will assume the file extension .CIR if one is not specified. For example, if you specify the file name MYFILE when running PSpice, it will look for a file called MYFILE.CIR.

Circuit files contain a title line, component description lines (discussed in Chapter 2), comments, and control statements that allow you to examine specific characteristics of a circuit.

***Title Line*** The title line is always the first line in the circuit file and, as the name suggests, describes what the circuit file is about. The title line may be left blank, but the circuit description cannot start until the second line. PSpice puts this title in the output file, so you can identify which circuit file produces a given output file.

***Description Lines*** Each component description line identifies a component, the nodes to which it is connected, and its value. For example,

```
R1  1  0  1K
```

identifies resistor $R_1$ connected between node 1 and node 0 with a value of 1 kΩ.

***Comments*** Comments are for use by you or other users and are ignored by PSpice. Comments allow you to clarify what the circuit file is simulating or what the circuit is supposed to do. A comment can take up a whole line or just the trailing portion of one. A full-line comment starts with an asterisk (*) in column one. An in-line comment starts with a semicolon (;) and extends to the end of the line. The following examples will illustrate:

```
*This is a comment line
R1   1   0   1K   ;This is an in-line comment
```

***Control Statements*** The .END control statement is the first one you will learn. Others are introduced in later chapters as necessary. Typically .END is the last statement in the circuit file and tells PSpice when it has all the information for a particular circuit and can begin analyzing. In certain cases, there can be more than one circuit in a given circuit file; thus, more than one .END statement must be used at appropriate points throughout the file.

A complete PSpice circuit file for the simple circuit in Figure 3–16 is as follows:

```
SIMPLECIRCUIT
*1 July 1997 by George Ohm
VS   1   0   10    ;Source voltage
R1   1   0   3.3K  ;The only resistor
.END
```

The comments, of course, could be omitted without affecting the simulation.

**FIGURE 3–16**

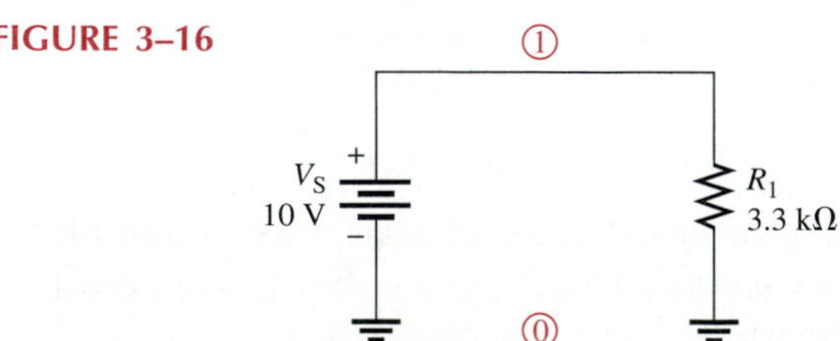

### The Output (.OUT) File

When PSpice analyzes a circuit file, it creates a file called the output file. When you run PSpice, it will prompt you for the name of the output file. You may give any file name you wish, but if you do not specify one, PSpice will name the output file the same as the input file except with an .OUT extension. For example, if the name of your circuit file is SIMPLE.CIR and you do not specify the name for the output file, PSpice will assign the name SIMPLE.OUT.

The output file consists of two sections. The first is the circuit description section, which contains the contents of the circuit file and any errors detected. The second section of the output file is the analysis section, which contains the results of the circuit analysis. PSpice will generate the analysis section only if no errors were found in the circuit file. The output file for the circuit in Figure 3–16 would contain the values of the node voltages and device currents.

**SECTION 3–7 REVIEW**

1. Name the types of lines in a PSpice circuit file.
2. Identify each of the following PSpice circuit file lines by its type.
   (a) RESCKT.CIR
   (b) R1 0 1 100
   (c) *This is a basic circuit
   (d) ;R1 is 100 ohms
3. Name the two sections of a PSpice output file.

## ■ SUMMARY

- Voltage and current are linearly proportional.
- Ohm's law gives the relationship of voltage, current, and resistance.
- Current is directly proportional to voltage.
- Current is inversely proportional to resistance.
- A kilohm (kΩ) is one thousand ohms.
- A megohm (MΩ) is one million ohms.
- A microampere ($\mu$A) is one-millionth of an ampere.
- A milliampere (mA) is one-thousandth of an ampere.
- Use $I = V/R$ when calculating the current.
- Use $V = IR$ when calculating the voltage.
- Use $R = V/I$ when calculating the resistance.
- Figure 3–17 illustrates the Ohm's law relationships.

FIGURE 3–17

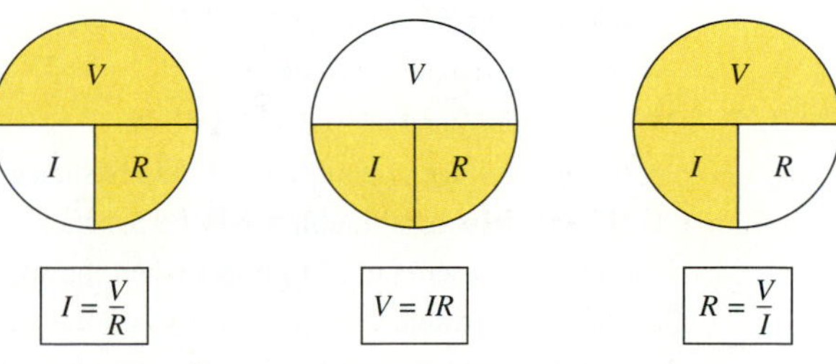


## ■ GLOSSARY

**Linear** Characterized by a straight-line relationship.

**Ohm (Ω)** The unit of resistance.

**Ohm's law** A law stating that current is directly proportional to voltage and inversely proportional to resistance.

## ■ FORMULAS

(3–1) $I = \frac{V}{R}$ Ohm's law for current

(3–2) $V = IR$ Ohm's law for voltage

(3–3) $R = \frac{V}{I}$ Ohm's law for resistance

## ■ SELF-TEST

**1.** Ohm's law states that
**(a)** current equals voltage times resistance
**(b)** voltage equals current times resistance
**(c)** resistance equals current divided by voltage
**(d)** voltage equals current squared times resistance

**2.** When the voltage across a resistor is doubled, the current will
**(a)** triple **(b)** halve **(c)** double **(d)** not change

**3.** When 10 V are applied across a 20 Ω resistor, the current is
**(a)** 10 A **(b)** 0.5 A **(c)** 200 A **(d)** 2 A

**4.** When there are 10 mA of current through a 1 kΩ resistor, the voltage across the resistor is
**(a)** 100 V **(b)** 0.1 V **(c)** 10 kV **(d)** 10 V

**5.** If 20 V are applied across a resistor and there are 6.06 mA of current, the resistance is
**(a)** 3.3 kΩ **(b)** 33 kΩ **(c)** 330 Ω **(d)** 3.03 kΩ

**6.** A current of 250 $\mu$A through a 4.7 kΩ resistor produces a voltage drop of
**(a)** 53.2 V **(b)** 1.18 mV **(c)** 18.8 V **(d)** 1.18 V

**7.** A resistance of 2.2 MΩ is connected across a 1 kV source. The resulting current is approximately
**(a)** 2.2 mA **(b)** 0.455 mA **(c)** 45.5 $\mu$A **(d)** 0.455 A

**8.** How much resistance is required to limit the current from a 10 V battery to 1 mA?
**(a)** 100 Ω **(b)** 1 kΩ **(c)** 10 Ω **(d)** 10 kΩ

**9.** An electric heater draws 2.5 A from a 110 V source. The resistance of the heating element is
**(a)** 275 Ω **(b)** 22.7 mΩ **(c)** 44 Ω **(d)** 440 Ω

**10.** The current through a flashlight bulb is 20 mA and the total battery voltage is 4.5 V. The resistance of the bulb is
**(a)** 90 Ω **(b)** 225 Ω **(c)** 4.44 Ω **(d)** 45 Ω

## ■ PROBLEMS

### SECTION 3–1 Ohm's law

**1.** In a circuit consisting of a voltage source and a resistor, describe what happens to the current when
**(a)** the voltage is tripled
**(b)** the voltage is reduced by 75%
**(c)** the resistance is doubled
**(d)** the resistance is reduced by 35%
**(e)** the voltage is doubled and the resistance is cut in half
**(f)** the voltage is doubled and the resistance is doubled

**2.** State the formula used to find $I$ when the values of $V$ and $R$ are known.

**3.** State the formula used to find $V$ when the values of $I$ and $R$ are known.

**4.** State the formula used to find $R$ when the values of $V$ and $I$ are known.

### SECTION 3–2 Calculating Current

**5.** Determine the current in each case:
**(a)** $V = 5$ V, $R = 1$ Ω **(b)** $V = 15$ V, $R = 10$ Ω
**(c)** $V = 50$ V, $R = 100$ Ω **(d)** $V = 30$ V, $R = 15$ kΩ
**(e)** $V = 250$ V, $R = 5.6$ MΩ

6. Determine the current in each case:
   (a) $V = 9$ V, $R = 2.7$ kΩ  (b) $V = 5.5$ V, $R = 10$ kΩ
   (c) $V = 40$ V, $R = 68$ kΩ  (d) $V = 1$ kV, $R = 2.2$ kΩ
   (e) $V = 66$ kV, $R = 10$ MΩ
7. A 10 Ω resistor is connected across a 12 V battery. What is the current through the resistor?
8. A certain resistor has the following color code: orange, orange, red, gold. Determine the maximum and minimum currents you should expect to measure when a 12 V source is connected across the resistor.
9. A resistor is connected across the terminals of a 25 V source. Determine the current in the resistor if the color code is yellow, violet, orange, silver.
10. The potentiometer connected as a rheostat in Figure 3–18 is used to control the current to a heating element. When the rheostat is adjusted to a value of 8 Ω or less, the heating element can burn out. What is the rated value of the fuse needed to protect the circuit if the voltage across the heating element at the point of maximum current is 100 V?

FIGURE 3–18

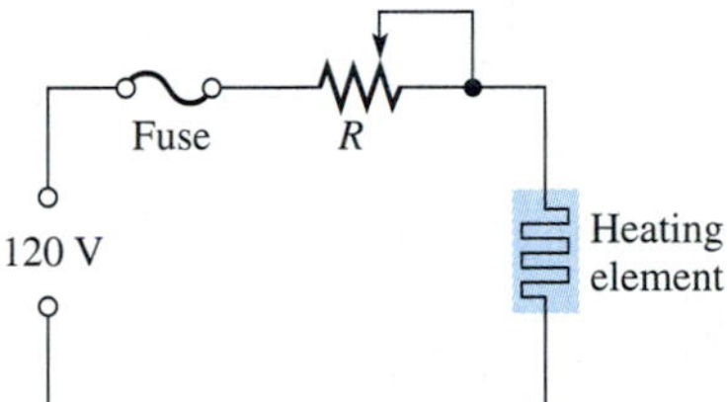


## SECTION 3–3 Calculating Voltage

11. Calculate the voltage for each value of $I$ and $R$:
   (a) $I = 2$ A, $R = 18$ Ω  (b) $I = 5$ A, $R = 56$ Ω
   (c) $I = 2.5$ A, $R = 680$ Ω  (d) $I = 0.6$ A, $R = 47$ Ω
   (e) $I = 0.1$ A, $R = 560$ Ω
12. Calculate the voltage for each value of $I$ and $R$:
   (a) $I = 1$ mA, $R = 10$ Ω  (b) $I = 50$ mA, $R = 33$ Ω
   (c) $I = 3$ A, $R = 5.6$ kΩ  (d) $I = 1.6$ mA, $R = 2.2$ kΩ
   (e) $I = 250$ μA, $R = 1$ kΩ  (f) $I = 500$ mA, $R = 1.5$ MΩ
   (g) $I = 850$ μA, $R = 10$ MΩ  (h) $I = 75$ μA, $R = 47$ Ω
13. Three amperes of current are measured through a 27 Ω resistor connected across a voltage source. How much voltage does the source produce?
14. Assign a voltage value to each source in the circuits of Figure 3–19 to obtain the indicated amounts of current.

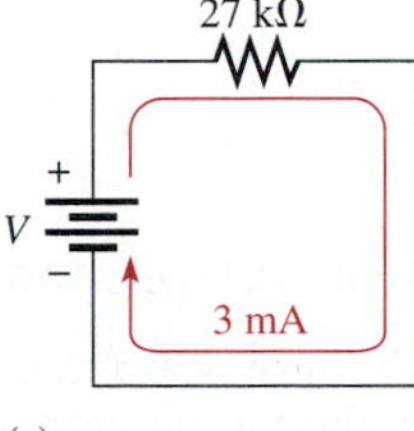

(a)

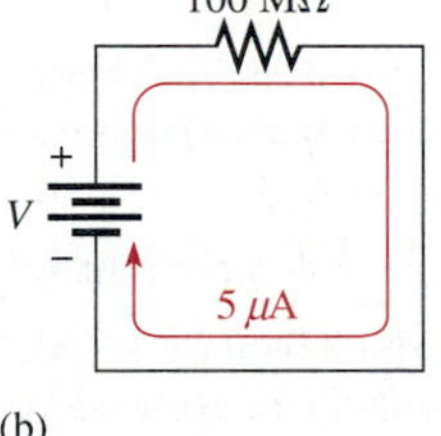

(b)

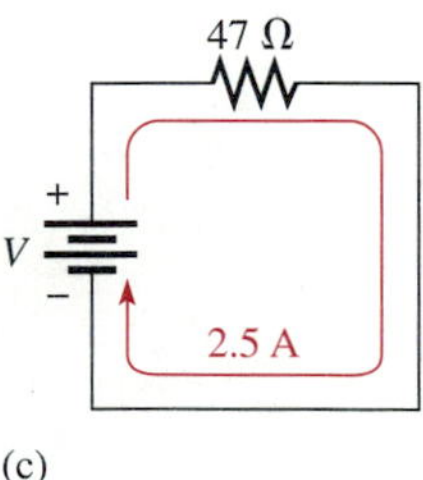

(c)

FIGURE 3–19

15. A 6 V source is connected to a 100 Ω resistor by two 12 ft lengths of 18 gage copper wire. Determine the following:
   (a) Current
   (b) Resistor voltage drop
   (c) Voltage drop across each length of wire

## SECTION 3–4 Calculating Resistance

**16.** Calculate the resistance of a rheostat for each value of $V$ and $I$:

**(a)** $V = 10$ V, $I = 2$ A  **(b)** $V = 90$ V, $I = 45$ A
**(c)** $V = 50$ V, $I = 5$ A  **(d)** $V = 5.5$ V, $I = 10$ A
**(e)** $V = 150$ V, $I = 0.5$ A

**17.** Calculate the resistance of a rheostat for each set of $V$ and $I$ values:

**(a)** $V = 10$ kV, $I = 5$ A  **(b)** $V = 7$ V, $I = 2$ mA
**(c)** $V = 500$ V, $I = 250$ mA  **(d)** $V = 50$ V, $I = 500\ \mu$A
**(e)** $V = 1$ kV, $I = 1$ mA

**18.** Six volts are applied across a resistor. A current of 2 mA is measured. What is the value of the resistor?

**19.** The filament of a light bulb in the circuit of Figure 3–20(a) has a certain amount of resistance, represented by an equivalent resistance in Figure 3–20(b). If the bulb operates with 120 V and 0.8 A of current, what is the resistance of its filament?

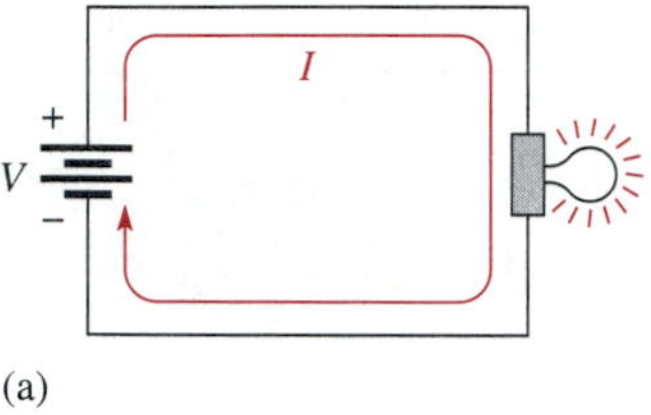

(a)

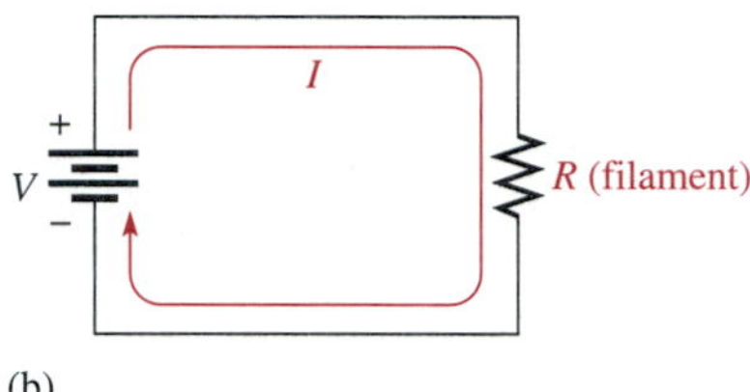

(b)

FIGURE 3–20

**20.** A certain electrical device has an unknown resistance. You have available a 12 V battery and an ammeter. How would you determine the value of the unknown resistance? Draw the necessary circuit connections.

**21.** By varying the rheostat (variable resistor) in the circuit of Figure 3–21, you can change the amount of current. The setting of the rheostat is such that the current is 750 mA. What is the resistance value of this setting? To adjust the current to 1 A, to what resistance value must you set the rheostat?

FIGURE 3–21

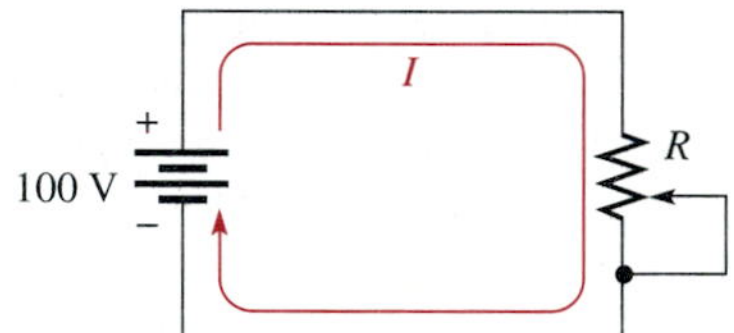


**22.** A 120 V lamp-dimming circuit is controlled by a rheostat and protected from excessive current by a 2 A fuse. To what minimum resistance value can the rheostat be set without blowing the fuse? Assume a lamp resistance of 15 Ω.

## SECTION 3–5 The Relationship of Current, Voltage, and Resistance

**23.** A variable voltage source is connected to the circuit of Figure 3–22. Start at 0 V and increase the voltage in 10 V steps up to 100 V. Determine the current at each voltage point, and plot a graph of $V$ versus $I$. Is the graph a straight line? What does the graph indicate?

FIGURE 3–22

24. In a certain circuit, $I = 5$ mA when $V = 1$ V. Determine the current for each of the following voltages in the same circuit:
   **(a)** $V = 1.5$ V **(b)** $V = 2$ V **(c)** $V = 3$ V
   **(d)** $V = 4$ V **(e)** $V = 10$ V

25. Figure 3–23 is a graph of voltage versus current for three resistance values. Determine $R_1$, $R_2$, and $R_3$.

FIGURE 3–23

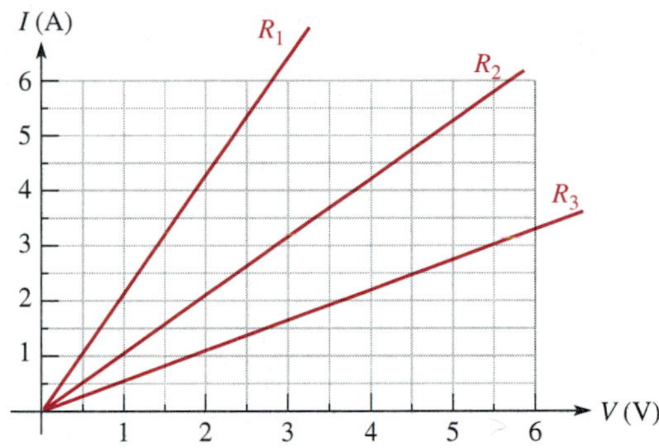


26. Which circuit in Figure 3–24 has the most current? The least current?

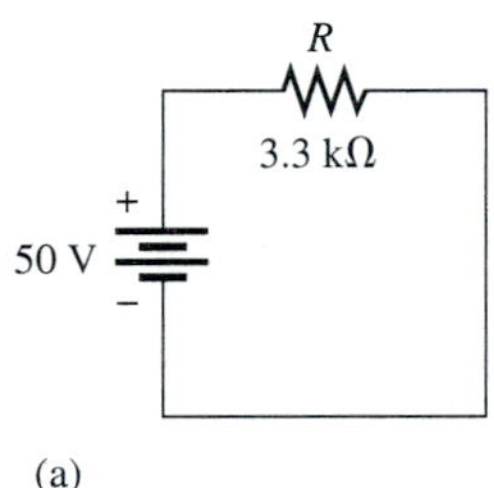


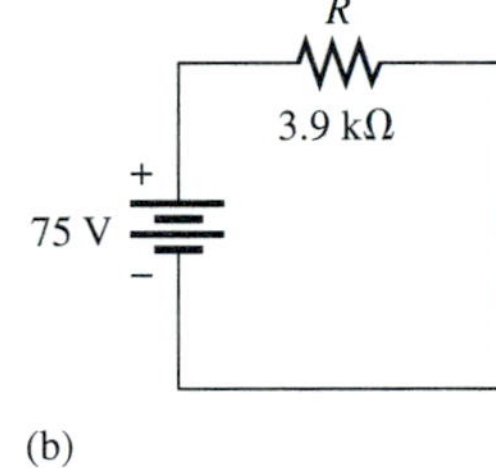


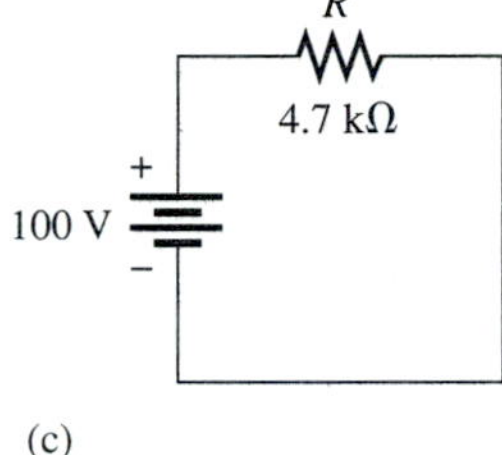


FIGURE 3–24

27. You are measuring the current in a circuit that is operated on a 10 V battery. The ammeter reads 50 mA. Later, you notice that the current has dropped to 30 mA. Eliminating the possibility of a resistance change, you must conclude that the voltage has changed. How much has the voltage of the battery changed, and what is its new value?

28. If you wish to increase the amount of current in a resistor from 100 mA to 150 mA by changing the 20 V source, by how many volts should you change the source? To what new value should you set it?

29. Plot a graph of current versus voltage for voltage values ranging from 10 V to 100 V in 10 V steps for each of the following resistance values:
   **(a)** 1 Ω **(b)** 5 Ω **(c)** 20 Ω **(d)** 100 Ω

## ■ ANSWERS TO SECTION REVIEWS

### Section 3–1

1. Current, voltage, and resistance
2. $I = V/R$
3. $V = IR$
4. $R = V/I$
5. When voltage is tripled, current increases by three times.
6. When voltage is halved, current reduces to one-half of original value.
7. 0.5 A
8. The current would increase by four times if the voltage doubles and the resistance is halved.

### Section 3–2

1. $I = 10\text{ V}/5.6\ \Omega = 1.79\text{ A}$
2. $I = 100\text{ V}/560\ \Omega = 179\text{ mA}$
3. $I = 5\text{ V}/2.2\text{ k}\Omega = 2.27\text{ mA}$
4. $I = 15\text{ V}/4.7\text{ M}\Omega = 3.19\ \mu\text{A}$
5. $I = 20\text{ kV}/4.7\text{ M}\Omega = 4.26\text{ mA}$
6. $I = 10\text{ kV}/2.2\text{ k}\Omega = 4.55\text{ A}$

### Section 3–3

1. $V = (1\text{ A})(10\ \Omega) = 10\text{ V}$
2. $V = (8\text{ A})(470\ \Omega) = 3.76\text{ kV}$
3. $V = (3\text{ mA})(100\ \Omega) = 300\text{ mV}$
4. $V = (25\ \mu\text{A})(56\ \Omega) = 1.4\text{ mV}$
5. $V = (2\text{ mA})(1.8\text{ k}\Omega) = 3.6\text{ V}$
6. $V = (5\text{ mA})(100\text{ M}\Omega) = 500\text{ kV}$
7. $V = (10\ \mu\text{A})(2.2\text{ M}\Omega) = 22\text{ V}$
8. $V = (100\text{ mA})(4.7\text{ k}\Omega) = 470\text{ V}$
9. $V = (3\text{ mA})(3.3\text{ k}\Omega) = 9.9\text{ V}$
10. $V = (2\text{ A})(6.8\ \Omega) = 13.6\text{ V}$

### Section 3–4

1. $R = 10\text{ V}/2.13\text{ A} = 4.7\ \Omega$
2. $R = 270\text{ V}/10\text{ A} = 27\ \Omega$
3. $R = 20\text{ kV}/5.13\text{ A} = 3.9\text{ k}\Omega$
4. $R = 15\text{ V}/2.68\text{ mA} = 5.6\text{ k}\Omega$
5. $R = 5\text{ V}/2.27\ \mu\text{A} = 2.2\text{ M}\Omega$
6. $R = 25\text{ V}/53.2\text{ mA} = 0.47\text{ k}\Omega = 470\ \Omega$

### Section 3–5

1. Linearly proportional means that the same percentage change occurs in two quantities.
2. $I = 5\text{ mA}$
3. $I = 6\text{ A}$
4. $\Delta V = 6\text{ V}$

### Section 3–6

1. For the new resistors, $R = V/I$.
2. $I = 4\text{ V}/10\ \Omega = 400\text{ mA}$; $I = 4\text{ V}/100\ \Omega = 40\text{ mA}$; $I = 4\text{ V}/1\text{ k}\Omega = 4\text{ mA}$; $I = 4\text{ V}/10\text{ k}\Omega = 400\ \mu\text{A}$; $I = 4\text{ V}/100\text{ k}\Omega = 40\ \mu\text{A}$; $I = 4\text{ V}/1\text{ M}\Omega = 4\ \mu\text{A}$.

### Section 3–7

1. Types of lines in a circuit file are title, component description, comment, and control.
2. **(a)** title **(b)** description **(c)** full-line comment **(d)** in-line comment
3. Two sections of a PSpice output file are circuit description and analysis.

## ▪ ANSWERS TO RELATED EXERCISES FOR EXAMPLES

**3–1** 3.03 A
**3–2** 0.005 A
**3–3** 0.005 A
**3–4** 13.6 mA
**3–5** 21.3 μA
**3–6** 2.66 μA
**3–7** 37.0 mA
**3–8** 1.47 mA
**3–9** 1200 V
**3–10** 49.5 mV
**3–11** 0.150 mV
**3–12** 82.5 V
**3–13** 1755 V
**3–14** 2.20 Ω
**3–15** 68.2 kΩ
**3–16** 3.30 MΩ
**3–17** Yes
**3–18** 0 V

# 4

# ENERGY AND POWER



## ■ INTRODUCTION

From Chapter 3, you know the relationship of current, voltage, and resistance as stated by Ohm's law. The existence of these three quantities in an electric circuit results in the fourth basic quantity known as power. A specific relationship exists between power and *I, V,* and *R,* as you will learn. In this chapter and throughout the rest of the book, you will learn the basics of putting technology theory into practice.

Energy is the fundamental ability to do work, and power is the rate at which energy is used. Current carries electrical energy through a circuit. As the free electrons pass through the resistance of the circuit, they give up their energy when they collide with atoms in the resistive material. The electrical energy given up by the electrons is converted into heat energy. The rate at which the electrical energy is used is the power in the circuit.

## TECHnology Theory Into Practice

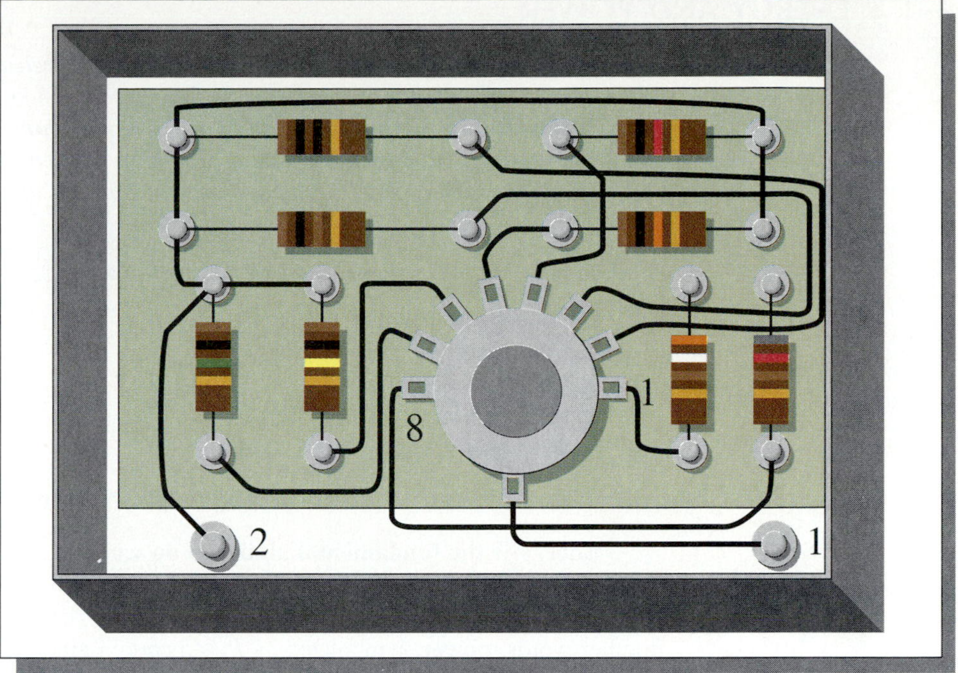


In Section 4–6, you will see how the theory learned in this chapter is applicable to the resistance box introduced in the last chapter. Suppose your supervisor tells you that the resistance box is to be used in testing a circuit in which there will be a maximum of 4 V across all the resistors. You are asked to evaluate the power rating of each resistor and, if it is not sufficient, to replace the resistor with one that has an adequate power rating.

## CHAPTER OBJECTIVES

- Define *energy* and *power*
- Calculate power in a circuit
- Properly select resistors based on power consideration
- Explain energy conversion and voltage drop
- Discuss power supplies and their characteristics
- Perform basic dc analysis using PSpice (optional)

## 4–1 ■ ENERGY AND POWER

***When there is current through a resistance, energy is released in the form of heat. A common example of this is a light bulb that becomes too hot to touch. The current through the filament that produces light also produces unwanted heat because the filament has resistance. Power is a measure of how fast energy is being used; electrical components must be able to dissipate a certain amount of energy in a given period of time.***

***After completing this section, you should be able to***

- **Define energy and power**
  - ☐ Express power in terms of energy
  - ☐ State the unit of power
  - ☐ State the common units of energy
  - ☐ Perform energy and power calculations

**Energy is the fundamental ability to do work.**

**Power is the rate at which energy is used.**

In other words, **power,** symbolized by *P,* is a certain amount of **energy** used in a certain length of time, expressed as follows:

$$\text{Power} = \frac{\text{energy}}{\text{time}}$$

$$P = \frac{W}{t} \qquad \textbf{(4–1)}$$

Energy is measured in **joules** (J), time is measured in seconds (s), and power is measured in **watts** (W). Note that an italic *W* is used to represent energy in the form of work and a nonitalic W is used for watts, the unit of power.

Energy in joules divided by time in seconds gives power in watts. For example, if 50 J of energy are used in 2 s, the power is 50 J/2 s = 25 W. By definition,

**One watt is the amount of power when one joule of energy is used in one second.**

Thus, the number of joules used in one second is always equal to the number of watts. For example, if 75 J are used in 1 s, the power is 75 W.

**EXAMPLE 4–1** An amount of energy equal to 100 J is used in 5 s. What is the power in watts?

***Solution***

$$P = \frac{\text{energy}}{\text{time}} = \frac{W}{t} = \frac{100\ \text{J}}{5\ \text{s}} = 20\ \text{W}$$

***Related Exercise*** If 100 W of power occurs for 30 s, how much energy, in joules, is used?

Amounts of power much less than one watt are common in certain areas of electronics. As with small current and voltage values, metric prefixes are used to designate small amounts of power. Thus, milliwatts (mW), microwatts ($\mu$W), and even picowatts (pW) are commonly found in some applications.

In the electrical utilities field, kilowatts (kW) and megawatts (MW) are common units. Radio and television stations also use large amounts of power to transmit signals.

**EXAMPLE 4–2**

Express the following values of electrical power using appropriate metric prefixes:
**(a)** 0.045 W **(b)** 0.000012 W **(c)** 3500 W **(d)** 10,000,000 W

***Solution***

**(a)** 0.045 W = 45 mW **(b)** 0.000012 W = 12 $\mu$W
**(c)** 3500 W = 3.5 kW **(d)** 10,000,000 W = 10 MW

***Related Exercise*** Express the following amounts of power in watts without metric prefixes:
**(a)** 1 mW **(b)** 1800 $\mu$W **(c)** 1000 mW **(d)** 1 $\mu$W

## The Kilowatt-hour (kWh) Unit of Energy

Since power is the rate at which energy is used, as expressed in Equation (4–1), power utilized over a period of time represents energy consumption. If you multiply power and time, you have energy, $W$.

$$\text{Energy} = \text{power} \times \text{time}$$

$$W = Pt \tag{4–2}$$

Previously, the joule was defined as a unit of energy. However, there is another way to express energy. Since power is expressed in watts and time in seconds, units of energy called the watt-second (Ws), watt-hour (Wh), and kilowatt-hour (kWh) can be used.

When you pay your electric bill, you are charged on the basis of the amount of energy you use, not the power. Because power companies deal in huge amounts of energy, the most practical unit is the **kilowatt-hour.** You use a kilowatt-hour of energy when you use one thousand watts of power for one hour. For example, a 100 W light bulb burning for 10 h uses 1 kWh of energy.

$$W = Pt = (100 \text{ W})(10 \text{ h}) = 1000 \text{ Wh} = 1 \text{ kWh}$$

**EXAMPLE 4–3**

Determine the number of kilowatt-hours for each of the following energy consumptions:
**(a)** 1400 W for 1 h **(b)** 2500 W for 2 h **(c)** 100,000 W for 5 h

***Solution***

**(a)** 1400 W = 1.4 kW
$W = Pt = (1.4 \text{ kW})(1 \text{ h}) = 1.4 \text{ kWh}$

**(b)** 2500 W = 2.5 kW
$W = (2.5 \text{ kW})(2 \text{ h}) = 5 \text{ kWh}$

**(c)** 100,000 W = 100 kW
$W = (100 \text{ kW})(5 \text{ h}) = 500 \text{ kWh}$

***Related Exercise*** How many kilowatt-hours are used by a 250 W bulb burning for 8 h?

**SECTION 4–1 REVIEW**

1. Define *power.*
2. Write the formula for power in terms of energy and time.
3. Define *watt.*
4. Express each of the following values of power in the most appropriate units:
   **(a)** 68,000 W **(b)** 0.005 W **(c)** 0.000025 W
5. If you use 100 W of power for 10 h, how much energy (in kilowatt-hours) have you used?
6. Convert 2000 Wh to kilowatt-hours.
7. Convert 360,000 Ws to kilowatt-hours.

## 4–2 ■ POWER IN AN ELECTRIC CIRCUIT

***The generation of heat in an electric circuit is often an unwanted by-product of current through the resistance in the circuit. In some cases, however, the generation of heat is the primary purpose of a circuit as, for example, in an electric resistive heater. In any case, you must always deal with power in electrical and electronic circuits.***

***After completing this section, you should be able to***

- **Calculate power in a circuit**
  - ☐ Determine power knowing $I$ and $R$
  - ☐ Determine power knowing $V$ and $I$
  - ☐ Determine power knowing $V$ and $R$

When there is current through resistance, the collisions of the electrons give off heat, resulting in a conversion of energy as indicated in Figure 4–1. There is always a certain amount of power in an electric circuit, and it is dependent on the amount of resistance and on the amount of current, expressed as follows:

$$P = I^2R \quad \textbf{(4–3)}$$

We can get an equivalent expression for power by substituting $V$ for $IR$ ($I^2$ is $I \times I$).

$$P = I^2R = (I \times I)R = I(IR) = (IR)I$$

$$P = VI \quad \textbf{(4–4)}$$

We obtain another equivalent expression by substituting $V/R$ for $I$ (Ohm's law) as follows:

$$P = VI = V\left(\frac{V}{R}\right)$$

$$P = \frac{V^2}{R} \quad \textbf{(4–5)}$$

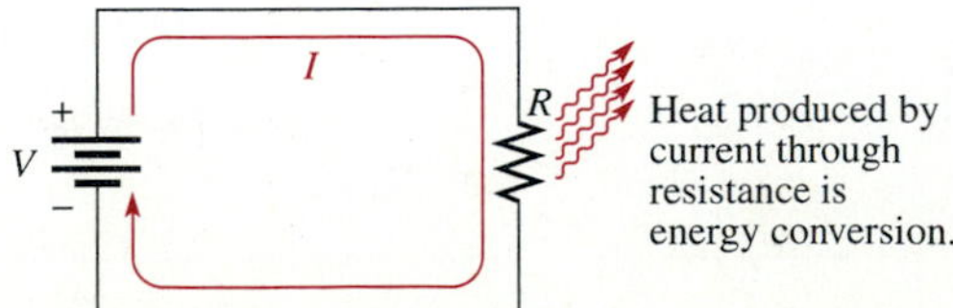
I
+
V
–
R
Heat produced by current through resistance is energy conversion.

**FIGURE 4–1**
***Power in an electric circuit is seen as heat given off by the resistance.***

The relationships expressed in the preceding formulas are known as **Watt's law.** In each case, $I$ must be in amps, $V$ in volts, and $R$ in ohms.

## Using the Appropriate Power Formula

To calculate the power in a resistance, you can use any one of the three power formulas, depending on what information you have. For example, assume that you know the values of current and voltage. In this case you calculate the power with the formula $P = VI$. If you know $I$ and $R$, use the formula $P = I^2R$. If you know $V$ and $R$, use the formula $P = V^2/R$.

**EXAMPLE 4–4** Calculate the power in each of the three circuits of Figure 4–2.

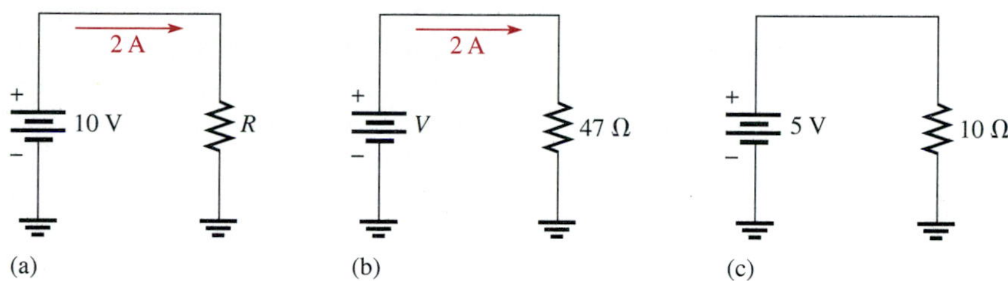


FIGURE 4–2

***Solution*** In circuit (a), $V$ and $I$ are known. Therefore, use Equation (4–4).

$$P = VI = (10\text{ V})(2\text{ A}) = 20\text{ W}$$

In circuit (b), $I$ and $R$ are known. Therefore, use Equation (4–3).

$$P = I^2R = (2\text{ A})^2(47\ \Omega) = 188\text{ W}$$

In circuit (c), $V$ and $R$ are known. Therefore, use Equation (4–5).

$$P = \frac{V^2}{R} = \frac{(5\text{ V})^2}{10\ \Omega} = 2.5\text{ W}$$

***Related Exercise*** Determine $P$ in each circuit of Figure 4–2 for the following changes:
Circuit (a): $I$ doubled and $V$ remains the same
Circuit (b): $R$ doubled and $I$ remains the same
Circuit (c): $V$ halved and $R$ remains the same

**EXAMPLE 4–5** A 100 W light bulb operates on 120 V. How much current does it require?

***Solution*** Use the formula $P = VI$ and solve for $I$ by first transposing the terms to get $I$ on the left of the equation.

$$VI = P$$

Divide both sides of the equation by $V$ to get $I$ by itself.

$$\frac{\cancel{V}I}{\cancel{V}} = \frac{P}{V}$$

The $V$'s cancel on the left, leaving

$$I = \frac{P}{V}$$

Substituting 100 W for $P$ and 120 V for $V$ yields

$$I = \frac{P}{V} = \frac{100\text{ W}}{120\text{ V}} = 0.833\text{ A} = 833\text{ mA}$$

***Related Exercise*** A light bulb draws 545 mA from a 110 V source. What is the power?

**SECTION 4–2 REVIEW**

1. If there are 10 V across a resistor and a current of 3 A through it, what is the power?
2. How much power does the source in Figure 4–3 generate? What is the power in the resistor? Are the two values the same? Why?

**FIGURE 4–3**

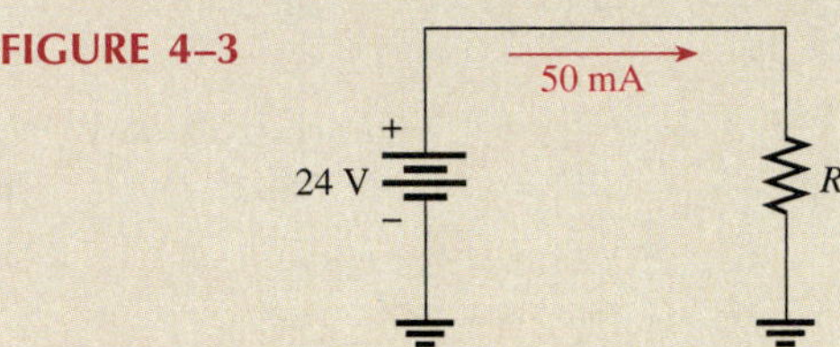


3. If there is a current of 5 A through a 56 Ω resistor, what is the power?
4. How much power is dissipated by 20 mA through a 4.7 kΩ resistor?
5. Five volts are applied to a 10 Ω resistor. What is the power?
6. How much power does a 2.2 kΩ resistor with 8 V across it dissipate?
7. What is the resistance of a 75 W bulb that takes 0.5 A?

## 4–3 ■ RESISTOR POWER RATINGS

***As you already know, a resistor gives off heat when there is current through it. The limit to the amount of heat that a resistor can give off is specified by its power rating.***

***After completing this section, you should be able to***

- **Properly select resistors based on power consideration**
  - ☐ Define *power rating*
  - ☐ Explain how physical characteristics of resistors determine their power rating
  - ☐ Check for resistor failure with an ohmmeter

The **power rating** is the maximum amount of power that a resistor can dissipate without being damaged by excessive heat buildup. The power rating is not related to the ohmic value (resistance) but rather is determined mainly by the physical composition, size, and shape of the resistor. All else being equal, the larger the surface area of a resistor, the more power it can dissipate. *The surface area of a cylindrically shaped resistor is approximately equal to the length (l) times the circumference (c),* as indicated in Figure 4–4.

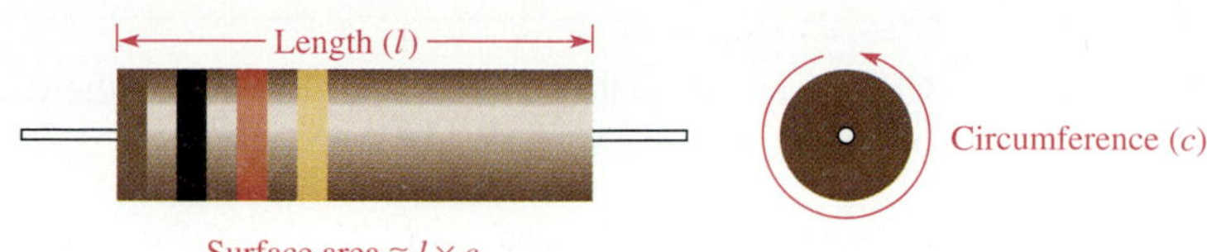


FIGURE 4–4
***The power rating of a resistor is directly related to its surface area.***

Carbon-composition resistors are available in standard power ratings from ⅛ W to 2 W, as shown in Figure 4–5. Available power ratings for other types of resistors vary. For example, carbon-film and metal-film resistors have ratings up to 10 W, and wire-wound resistors have ratings up to 225 W or greater. Figure 4–6 shows some of these resistors.

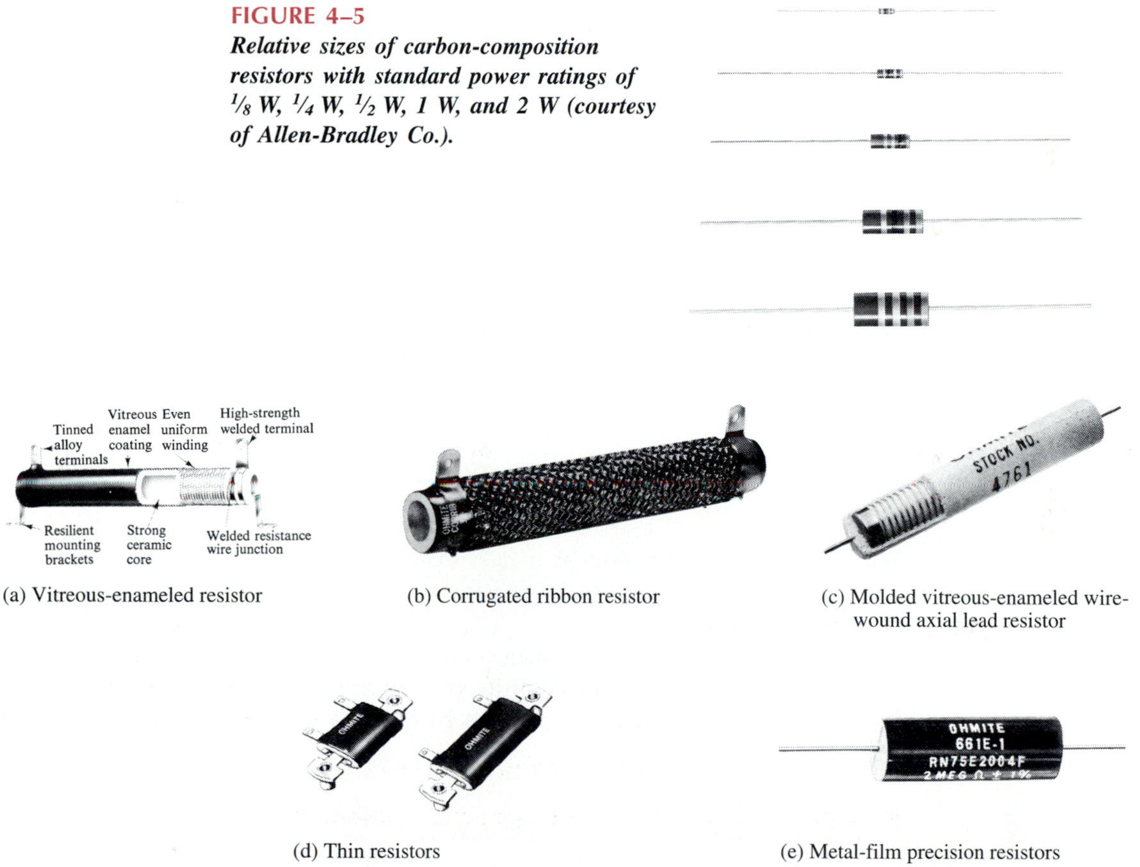


FIGURE 4–5
***Relative sizes of carbon-composition resistors with standard power ratings of ⅛ W, ¼ W, ½ W, 1 W, and 2 W (courtesy of Allen-Bradley Co.).***

(a) Vitreous-enameled resistor

(b) Corrugated ribbon resistor

(c) Molded vitreous-enameled wire-wound axial lead resistor

(d) Thin resistors

(e) Metal-film precision resistors

FIGURE 4–6
***Typical resistors with high power ratings (courtesy of Ohmite Manufacturing Co.).***

## Selecting the Proper Power Rating for an Application

When a resistor is used in a circuit, its power rating must be greater than the maximum power that it will have to handle. For example, if a carbon-composition resistor is to dissipate 0.75 W in a circuit application, its rating should be at least the next higher standard value which is 1 W. *Ideally, a rating that is approximately twice the actual power should be used when possible.*

**EXAMPLE 4–6** Choose an adequate power rating for each of the carbon-composition resistors in Figure 4–7 (⅛ W, ¼ W, ½ W, 1 W, or 2 W).

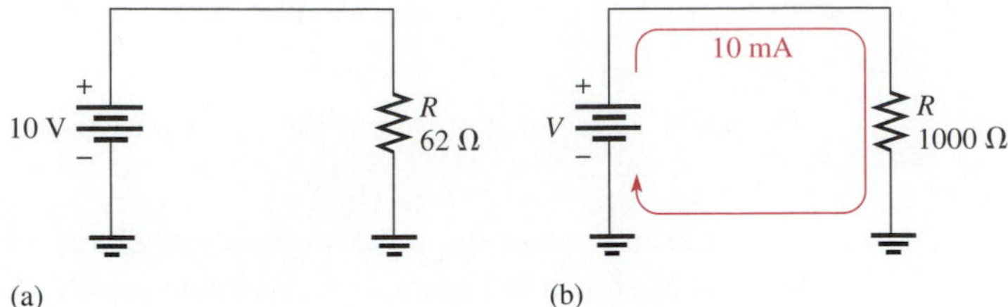


FIGURE 4–7

***Solution*** In Figure 4–7(a), the actual power is

$$P = \frac{V^2}{R} = \frac{(10\text{ V})^2}{62\ \Omega} = \frac{100\text{ V}^2}{62\ \Omega} = 1.61\text{ W}$$

Select a resistor with a power rating higher than the actual power. In this case, at least a 2 W resistor should be used.

In Figure 4–7(b), the actual power is

$$P = I^2R = (10\text{ mA})^2(1000\ \Omega) = (10 \times 10^{-3}\text{ A})^2(1000\ \Omega) = 0.1\text{ W}$$

At least a ⅛ W (0.125 W) resistor should be used in this case.

***Related Exercise*** A certain resistor is required to dissipate 0.25 W. What standard rating should be used?

## Resistor Failures

When the power in a resistor is greater than its rating, the resistor will become excessively hot. As a result, either the resistor will burn open or its resistance value will be greatly altered.

A resistor that has been damaged because of overheating can often be detected by the charred or altered appearance of its surface. If there is no visual evidence, a resistor that is suspected of being damaged can be checked with an ohmmeter for an open or incorrect resistance value.

## Checking a Resistor with an Ohmmeter

A typical analog multimeter (VOM) and a digital multimeter are shown in Figures 4–8(a) and 4–8(b), respectively. The large switch on the analog meter is called a range switch. Notice the resistance (OHMS) settings on both meters. For the analog meter in part (a), each setting indicates the amount by which the ohms scale (top scale) on the meter is to be multiplied. For example, if the pointer is at 50 on the ohms scale and the range switch is set at ×100, the resistance being measured is 50 × 100 Ω = 5000 Ω. *If the resistor is open, the pointer will stay at full left scale (∞ means infinite) regardless of the range switch setting.*

For the digital meter in Figure 4–8(b), you use the range switch to select the appropriate setting for the value of resistance being measured. You do not have to multiply to get the correct reading because you have a direct digital readout of the resistance value.

(a)

(b)

**FIGURE 4–8**

***Typical portable multimeters. Part (a) courtesy of BK Precision, Maxtec International Corp. Part (b) courtesy of John Fluke Manufacturing Co.***

**EXAMPLE 4–7** Determine whether the resistor in each circuit of Figure 4–9 has possibly been damaged by overheating.

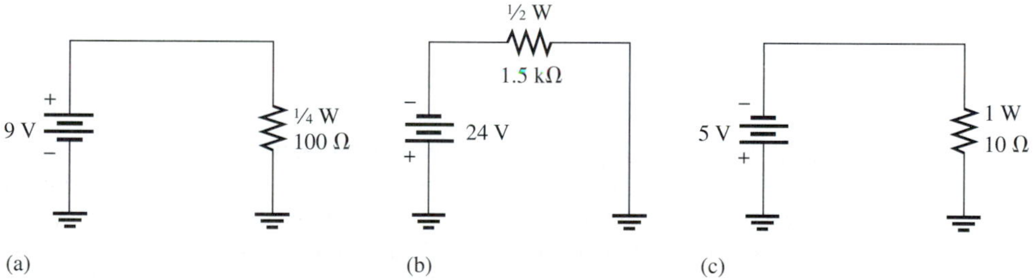


**FIGURE 4–9**

***Solution*** In the circuit in Figure 4–9(a),

$$P = \frac{V^2}{R} = \frac{(9\text{ V})^2}{100\ \Omega} = 0.810\text{ W}$$

The rating of the resistor is ¼ W (0.25 W), which is insufficient to handle the power. The resistor has been overheated and may be burned out, making it an open.

In the circuit of Figure 4–9(b),

$$P = \frac{V^2}{R} = \frac{(24\text{ V})^2}{1.5\text{ k}\Omega} = 0.384\text{ W}$$

The rating of the resistor is ½ W (0.5 W), which is sufficient to handle the power.

In the circuit of Figure 4–9(c),

$$P = \frac{V^2}{R} = \frac{(5\text{ V})^2}{10\ \Omega} = 2.5\text{ W}$$

The rating of the resistor is 1 W, which is insufficient to handle the power. The resistor has been overheated and may be burned out, making it an open.

***Related Exercise*** A 0.25 W, 1 kΩ resistor is connected across a 12 V battery. Will it burn out?

**SECTION 4–3 REVIEW**

1. Name two important values associated with a resistor.
2. How does the physical size of a resistor determine the amount of power that it can handle?
3. List the standard power ratings of carbon-composition resistors.
4. A resistor must handle 0.3 W. What minimum power rating of a carbon resistor should be used to dissipate the energy properly?

## 4–4 ■ ENERGY CONVERSION AND VOLTAGE DROP IN RESISTANCE

***As you have seen, when there is current through a resistance, electrical energy is converted to heat energy. This heat is caused by collisions of the free electrons within the atomic structure of the resistive material. When a collision occurs, heat is given off; and the electron loses some of its acquired energy.***

***After completing this section, you should be able to***

- **Explain energy conversion and voltage drop**
  - ☐ Discuss the cause of energy conversion in a circuit
  - ☐ Define *voltage drop*
  - ☐ Explain the relationship between energy conversion and voltage drop

In Figure 4–10, electrons are flowing out of the negative terminal of the battery. They have acquired energy from the battery and are at their highest energy level at the negative side of the circuit. As the electrons move through the resistor, they lose energy. The electrons emerging from the upper end of the resistor are at a lower energy level than those entering the lower end because some of the energy they had has been converted to heat. The drop in energy level of the electrons as they move through the resistor creates a potential difference, or **voltage drop,** across the resistor having the polarity shown in Figure 4–10. Notice that the upper end of the resistor in Figure 4–10 is less negative (more positive) than the lower end.

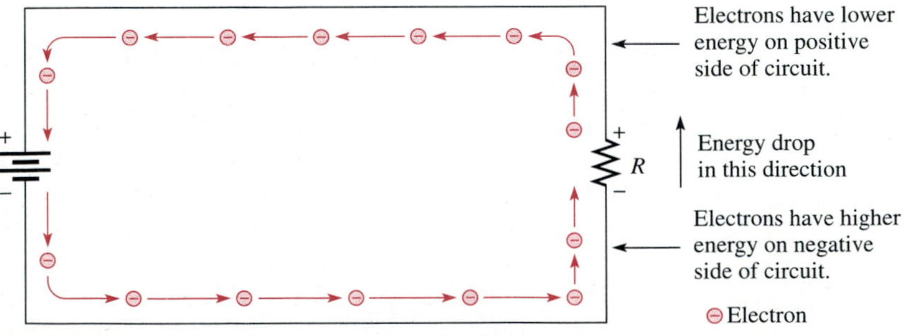


FIGURE 4–10
*Electron flow in a simple circuit.*

**SECTION 4–4 REVIEW**

1. What is the basic reason for energy conversion in a resistor?
2. What is a voltage drop?

## 4–5 ■ POWER SUPPLIES

***A power supply is a device that provides power to a load. A load is any electrical device or circuit that is connected to the output of the power supply and draws current from the supply.***

***After completing this section, you should be able to***

- **Discuss power supplies and their characteristics**
  - Define *ampere-hour rating*
  - Discuss power supply efficiency

Figure 4–11 shows a block diagram of a power supply with a loading device connected to it. The load can be anything from a light bulb to a computer. The power supply produces a voltage across its two output terminals and provides current through the load, as indicated in the figure. The product $IV_{OUT}$ is the amount of power produced by the supply and consumed by the load. For a given output voltage ($V_{OUT}$), more current drawn by the load means more power from the supply.

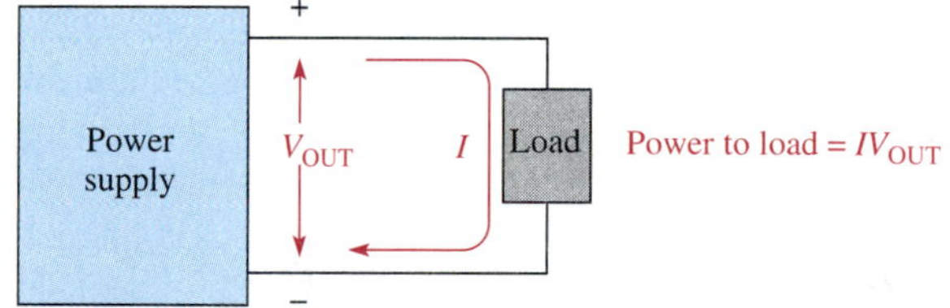


FIGURE 4–11
*Block diagram of power supply and load.*

Power supplies range from simple batteries to regulated electronic circuits where an accurate output voltage is automatically maintained. A battery is a dc power supply that converts chemical energy into electrical energy. Electronic power supplies normally convert 110 V ac (alternating current) from a wall outlet into a regulated dc (direct current) or ac voltage.

### Ampere-hour Ratings of Batteries

Batteries convert chemical energy into electrical energy. Because of their limited source of chemical energy, batteries have a certain capacity that limits the amount of time over which they can produce a given power level. This capacity is measured in ampere-hours (Ah). The **ampere-hour rating** determines the length of time that a battery can deliver a certain amount of current to a load at the rated voltage.

A rating of one ampere-hour means that a battery can deliver one ampere of current to a load for one hour at the rated voltage output. This same battery can deliver two amperes for one-half hour. The more current the battery is required to deliver, the shorter is the life of the battery. In practice, a battery usually is rated for a specified current level and output voltage. For example, a 12 V automobile battery may be rated for 70 Ah at 3.5 A. This means that it can produce 3.5 A for 20 h.

**EXAMPLE 4–8** For how many hours can a battery deliver 2 A if it is rated at 70 Ah?

***Solution*** The ampere-hour rating is the current times the hours.

$$70 \text{ Ah} = (2 \text{ A})(x \text{ h})$$

Solving for $x$ h yields

$$x = \frac{70 \text{ Ah}}{2 \text{ A}} = 35 \text{ h}$$

***Related Exercise*** A certain battery delivers 10 A for 6 h. What is its Ah rating?

## Power Supply Efficiency

An important characteristic of electronic power supplies is efficiency. **Efficiency** is the ratio of the output power to the input power.

$$\text{Efficiency} = \frac{\text{output power}}{\text{input power}}$$

$$\text{Efficiency} = \frac{P_{OUT}}{P_{IN}} \qquad \textbf{(4–6)}$$

Efficiency is often expressed as a percentage. For example, if the input power is 100 W and the output power is 50 W, the efficiency is (50 W/100 W) × 100% = 50%.

All power supplies require that power be put into them. For example, an electronic power supply might use the ac power from a wall outlet as its input. Its output may be regulated dc or ac. The output power is *always* less than the input power because some of the total power must be used internally to operate the power supply circuitry. This amount is normally called the *power loss.* The output power is the input power minus the amount of internal power loss.

$$P_{OUT} = P_{IN} - P_{LOSS} \qquad \textbf{(4–7)}$$

High efficiency means that little power is lost and there is a higher proportion of output power for a given input power.

**EXAMPLE 4–9** A certain power supply unit requires 25 W of input power. It can produce an output power of 20 W. What is its efficiency, and what is the power loss?

***Solution***

$$\text{Efficiency} = \left(\frac{P_{OUT}}{P_{IN}}\right)100\% = \left(\frac{20 \text{ W}}{25 \text{ W}}\right)100\% = 80\%$$

$$P_{LOSS} = P_{IN} - P_{OUT} = 25 \text{ W} - 20 \text{ W} = 5 \text{ W}$$

***Related Exercise*** A power supply has an efficiency of 92%. If $P_{IN}$ is 50 W, what is $P_{OUT}$?

**SECTION 4–5 REVIEW**

1. When a loading device draws an increased amount of current from a power supply, does this change represent a greater or a smaller load on the supply?
2. A power supply produces an output voltage of 10 V. If the supply provides 0.5 A to a load, what is the power to a load?
3. If a battery has an ampere-hour rating of 100 Ah, how long can it provide 5 A to a load?
4. If the battery in Question 3 is a 12 V device, what is its power to a load for the specified value of current?
5. An electronic power supply used in the lab operates with an input power of 1 W. It can provide an output power of 750 mW. What is its efficiency? Determine the power loss.

## 4–6 ■ TECHnology Theory Into Practice

***In this TECH TIP section, the resistance box that you worked with in Chapter 3 is back on your bench. The last time, you verified that all the resistor values were correct. This time there is a new requirement. The box must work with a maximum of 4 V, so you must make sure each resistor has a sufficient power rating; and if the power rating is insufficient, replace the resistor with one that is adequate.***

### Power Ratings

Assume the power rating of each resistor in the resistance box as modified in Chapter 3 is ⅛ W. The box is shown in Figure 4–12.

(a) Top view

(b) Bottom view

FIGURE 4–12

- ☐ Determine if the power rating of each resistor is adequate for a maximum of 4 V.
- ☐ If a rating is not adequate, determine the lowest rating required to handle the maximum power. Choose from standard ratings of ¼ W, ½ W, 1 W, 2 W, and 5 W.
- ☐ Add the power rating of each resistor to the schematic developed in Chapter 3.

**SECTION 4–6 REVIEW**

1. How many resistors were replaced because of inadequate power ratings?
2. If the resistance must operate with 10 V maximum, which resistors must be changed and to what minimum power ratings?

## 4–7 ■ PSpice DC ANALYSIS

***In the last chapter, you saw that PSpice could determine the voltage at every node in a circuit and the current from the voltage source. PSpice will also calculate voltage and current for every component in a circuit. You tell PSpice what you want to know about a circuit using control statements in the circuit file. In addition to the .END control statement that you already know, two new ones are introduced in this section: the .DC statement and the .PRINT statement.***

***After completing this section, you should be able to***

- **Perform basic dc analysis using PSpice**
  - ☐ Use the .DC statement in a circuit file
  - ☐ Use the .PRINT statement in a circuit file

### The .DC Control Statement

Generally, the .DC control statement tells PSpice to calculate dc values for a circuit. More precisely, it tells PSpice to vary a dc parameter over some range of values and collect information about the circuit at specific intervals. This information lets you determine the dc characteristics of a circuit, just as you could determine how a circuit behaves by connecting it to a dc voltage source and taking circuit measurements for specific dc voltage source values. Although, the .DC statement can be used for a variety of parameters, only dc sources are considered for now.

The general format of the .DC control statement is

```
.DC  <parameter>  <start value>  <end value>  <increment>
```

The parameter is the name of the dc source, the start and end values define the range to be swept, and the increment is the value of each step within the range. PSpice will begin with the parameter at the start value and continue the dc analysis for each increment until the end value is exceeded.

As an example, suppose you have a circuit with a dc current source labeled $I_S$ and want to know the circuit values when $I_S$ is 0 mA, 1 mA, 2 mA, 3 mA, and 4 mA. Since you want PSpice to sweep $I_S$ from 0 mA to 4 mA in 1 mA increments, the .DC control statement for this is

```
.DC  IS  0  4M  1M
```

Similarly, the statement

```
.DC  VIN  2  –3  –0.5
```

will sweep the voltage source $V_{IN}$ from 2 V down to –3 V in increments of 0.5 V. The negative increment in the statement is required because PSpice must sweep from a positive value to a negative value; that is, it is sweeping down rather than up.

### The .PRINT (Print Output) Control Statement

The .PRINT statement tells PSpice to print specific values in the output file and, thus, requires some sort of analysis. Unless PSpice performs some type of analysis on the circuit (such as a dc analysis), there are no values for it to print in the output file.

The general format of the .PRINT control statement is

```
.PRINT  <analysis type>  <variable 1>  <variable n>
```

The analysis type can be, for example, dc analysis and the variables can be the current, voltage, or power associated with some component in the circuit. The variables themselves are of the forms V(component) or V(node1, node2) for voltage and I(component) for current. The components are identified by their names in the circuit file. If, for example, you want PSpice to determine the currents through resistors $R_1$, $R_2$, and $R_3$ in a given circuit, you would put the control statement

```
.PRINT  DC  I(R1)  I(R2)  I(R3)
```

in the circuit file. Similarly, the statement

```
.PRINT  DC  V(R1)  I(R1)
```

would put the voltage and current values for $R_1$ in the output file.

You should realize at this point that PSpice must have a .DC control statement in the circuit file before it can use a .PRINT statement. Although the .DC statement specifies a range of values, it can be used for a fixed value by simply making the start and end values the same and using any increment value. For example,

```
.DC  V1  10  10  1
```

tells PSpice to begin calculating circuit values at V1 = 10 V and to calculate additional circuit values every 1 volt above this until V1 exceeds 10 V. The only value of V1 that satisfies these conditions is 10 V because the next point (11 V) is beyond the end value of 10 V.

**EXAMPLE 4–10** Write a PSpice circuit file for the circuit in Figure 4–13 and tell PSpice to calculate the values of the voltages and currents for $R_1$ and $R_2$. The dc source voltage has a fixed value of 8 V.

**FIGURE 4–13**

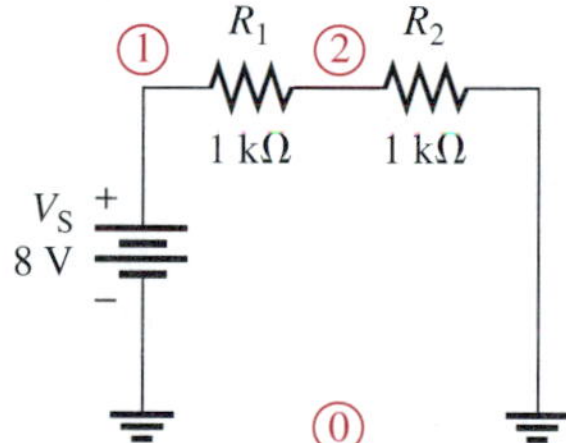


***Solution*** The circuit file is as follows with nodes assigned as indicated in the figure:

```
Basic Circuit
*2 February 1997 by D.O. Watt
VS  1  0  8
R1  1  2  1K
R2  2  0  1K
.DC  VS  8  8  1
.PRINT  DC  V(R1)  V(R2)
.PRINT  DC  I(R1)  I(R2)
.END
```

Save the file and run PSpice. Next view the .OUT file or print the output file on your printer. Among other things, the output file will show the following results:

| VS | V(R1) | V(R2) |
|---|---|---|
| 8.00E+00 | 4.000E+00 | 4.000E+00 |

| VS | I(R1) | I(R2) |
|---|---|---|
| 8.00E+00 | 4.000E-03 | 4.000E-03 |

***Related Exercise*** Rework the circuit file for the circuit in Figure 4–13 for a variation in the source voltage from 4 V to 8 V in 1 V increments.

**SECTION 4–7 REVIEW**

1. Explain each of the following .DC control statements:
   (a) .DC V1 1 8 1 (b) .DC I1 1M 5M 2.5M
2. Explain each of the following .PRINT statements:
   (a) .PRINT DC V(R2) V(R3)
   (b) .PRINT DC I(R1) I(R2) I(R3)

## ■ SUMMARY

- One watt equals one joule per second.
- Watt is the unit of power, joule is a unit of energy, and second is a unit of time.
- The power rating of a resistor determines the maximum power that it can handle safely.
- Resistors with a larger physical size can dissipate more power in the form of heat than smaller ones.
- A resistor should have a power rating higher than the maximum power that it is expected to handle in the circuit.
- Power rating is not related to resistance value.
- A resistor normally opens when it burns out.
- Energy is equal to power multiplied by time.
- The kilowatt-hour is a unit of energy.
- One kilowatt-hour is one thousand watts used for one hour.
- A power supply is an energy source used to operate electrical and electronic devices.
- A battery is one type of power supply that converts chemical energy into electrical energy.
- An electronic power supply converts commercial energy (ac from the power company) to regulated dc or ac at various voltage levels.
- The output power of a supply is the output voltage times the load current.
- A load is a device that draws current from the power supply.
- The capacity of a battery is measured in ampere-hours (Ah).
- One ampere-hour equals one ampere used for one hour, or any other combination of amperes and hours that has a product of one.
- A power supply with a high efficiency wastes less power than one with a lower efficiency.

## ■ GLOSSARY

**Ampere-hour rating** A number given in ampere-hours (Ah) determined by multiplying the current (A) times the length of time (h) a battery can deliver that current to a load.

**Efficiency** The ratio of the output power to the input power of a circuit, expressed as a percent.

**Energy** The fundamental ability to do work.

**Joule (J)** The unit of energy.

**Kilowatt-hour (kWh)** A common unit of energy used mainly by utility companies.

**Power** The rate of energy usage.

**Power rating** The maximum amount of power that a resistor can dissipate without being damaged by excessive heat buildup.

**Voltage drop** The drop in energy level through a resistor.

**Watt (W)** The unit of power. One watt is the power when 1 J of energy is used in 1 s.

**Watt's law** A law that states the relationships of power to current, voltage, and resistance.

## ■ FORMULAS

| | | |
|---|---|---|
| **(4–1)** | $P = \frac{W}{t}$ | Power equals energy divided by time. |
| **(4–2)** | $W = Pt$ | Energy equals power multiplied by time. |
| **(4–3)** | $P = I^2R$ | Power equals current squared times resistance. |

| | | |
|---|---|---|
| **(4–4)** | $P = VI$ | Power equals voltage times current. |
| **(4–5)** | $P = \frac{V^2}{R}$ | Power equals voltage squared divided by resistance. |
| **(4–6)** | $\text{Efficiency} = \frac{P_{OUT}}{P_{IN}}$ | Power supply efficiency |
| **(4–7)** | $P_{OUT} = P_{IN} - P_{LOSS}$ | Output power is input power less power loss. |

## ■ SELF-TEST

**1.** Power can be defined as
**(a)** energy **(b)** heat
**(c)** the rate at which energy is used **(d)** the time required to use energy

**2.** Two hundred joules of energy are consumed in 10 s. The power is
**(a)** 2000 W **(b)** 10 W **(c)** 20 W **(d)** 2 W

**3.** If it takes 300 ms to use 10,000 J of energy, the power is
**(a)** 33.3 kW **(b)** 33.3 W **(c)** 33.3 mW

**4.** In 50 kW, there are
**(a)** 500 W **(b)** 5000 W **(c)** 0.5 MW **(d)** 50,000 W

**5.** In 0.045 W, there are
**(a)** 45 kW **(b)** 45 mW **(c)** 4,500 μW **(d)** 0.00045 MW

**6.** For 10 V and 50 mA, the power is
**(a)** 500 mW **(b)** 0.5 W **(c)** 500,000 μW **(d)** answers (a), (b), and (c)

**7.** When the current through a 10 kΩ resistor is 10 mA, the power is
**(a)** 1 W **(b)** 10 W **(c)** 100 mW **(d)** 1000 μW

**8.** A 2.2 kΩ resistor dissipates 0.5 W. The current is
**(a)** 15.1 mA **(b)** 0.227 mA **(c)** 1.1 mA **(d)** 4.4 mA

**9.** A 330 Ω resistor dissipates 2 W. The voltage is
**(a)** 2.57 V **(b)** 660 V **(c)** 6.6 V **(d)** 25.7 V

**10.** If you used 500 W of power for 24 h, you have used
**(a)** 0.5 kWh **(b)** 2400 kWh **(c)** 12,000 kWh **(d)** 12 kWh

**11.** How many watt-hours represent 75 W used for 10 h?
**(a)** 75 Wh **(b)** 750 Wh **(c)** 0.75 Wh **(d)** 7500 Wh

**12.** A 100 Ω resistor must carry a maximum current of 35 mA. Its rating should be at least
**(a)** 35 W **(b)** 35 mW **(c)** 123 mW **(d)** 3500 mW

**13.** The power rating of a carbon-composition resistor that is to handle up to 1.1 W should be
**(a)** 0.25 W **(b)** 1 W **(c)** 2 W **(d)** 5 W

**14.** A 22 Ω half-watt resistor and a 220 Ω half-watt resistor are connected across a 10 V source. Which one(s) will overheat?
**(a)** 22 Ω **(b)** 220 Ω **(c)** both **(d)** neither

**15.** When the needle of an analog ohmmeter indicates infinity, the resistor being measured is
**(a)** overheated **(b)** shorted **(c)** open **(d)** reversed

**16.** A 12 V battery is connected to a 600 Ω load. Under these conditions, it is rated at 50 Ah. How long can it supply current to the load?
**(a)** 2500 h **(b)** 50 h **(c)** 25 h **(d)** 4.16 h

**17.** A given power supply is capable of providing 8 A for 2.5 h. Its ampere-hour rating is
**(a)** 2.5 Ah **(b)** 20 Ah **(c)** 8 Ah

**18.** A power supply produces a 0.5 W output with an input of 0.6 W. Its percentage of efficiency is
**(a)** 50% **(b)** 60% **(c)** 83.3% **(d)** 45%

## ■ PROBLEMS

### SECTION 4–1 Energy and Power

1. What is the power when energy is consumed at the rate of 350 J/s?
2. How many watts are used when 7500 J of energy are consumed in 5 h?
3. How many watts does 1000 J in 50 ms equal?
4. Convert the following to kilowatts:
   **(a)** 1000 W **(b)** 3750 W **(c)** 160 W **(d)** 50,000 W
5. Convert the following to megawatts:
   **(a)** 1,000,000 W **(b)** $3 \times 10^6$ W **(c)** $15 \times 10^7$ W **(d)** 8700 kW
6. Convert the following to milliwatts:
   **(a)** 1 W **(b)** 0.4 W **(c)** 0.002 W **(d)** 0.0125 W
7. Convert the following to microwatts:
   **(a)** 2 W **(b)** 0.0005 W **(c)** 0.25 mW **(d)** 0.00667 mW
8. Convert the following to watts:
   **(a)** 1.5 kW **(b)** 0.5 MW **(c)** 350 mW **(d)** 9000 $\mu$W
9. A particular electronic device uses 100 mW of power. If it runs for 24 h, how many joules of energy does it consume?
10. A certain appliance uses 300 W. If it is allowed to run continuously for 30 days, how many kilowatt-hours of energy does it consume?
11. At the end of a 31 day period, your utility bill shows that you have used 1500 kWh. What is your average daily power?
12. Convert $5 \times 10^6$ watt-minutes to kWh.
13. Convert 6700 watt-seconds to kWh.
14. For how many seconds must there be 5 A of current through a 47 Ω resistor in order to consume 25 J?

### SECTION 4–2 Power in an Electric Circuit

15. If a 75 V source is supplying 2 A to a load, what is the resistance value of the load?
16. If a resistor has 5.5 V across it and 3 mA through it, what is the power?
17. An electric heater works on 120 V and draws 3 A of current. How much power does it use?
18. How much power is produced by 500 mA of current through a 4.7 kΩ resistor?
19. Calculate the power handled by a 10 kΩ resistor carrying 100 $\mu$A.
20. If there are 60 V across a 680 Ω resistor, what is the power?
21. A 56 Ω resistor is connected across the terminals of a 1.5 V battery. What is the power dissipation in the resistor?
22. If a resistor is to carry 2 A of current and handle 100 W of power, how many ohms must it be? Assume that the voltage can be adjusted to any required value.
23. A 12 V source is connected across a 10 Ω resistor.
    **(a)** How much energy is used in two minutes?
    **(b)** If the resistor is disconnected after one minute, is the power greater than, less than, or equal to the power during the two minute interval?

### SECTION 4–3 Resistor Power Ratings

24. A 6.8 kΩ resistor has burned out in a circuit. You must replace it with another resistor with the same resistance value. If the resistor carries 10 mA, what should its power rating be? Assume that you have available carbon-composition resistors in all the standard power ratings.
25. A certain type of power resistor comes in the following ratings: 3 W, 5 W, 8 W, 12 W, 20 W. Your particular application requires a resistor that can handle approximately 8 W. Which rating would you use? Why?

## SECTION 4–4 Energy Conversion and Voltage Drop in Resistance

**26.** For each circuit in Figure 4–14, assign the proper polarity for the voltage drop across the resistor.

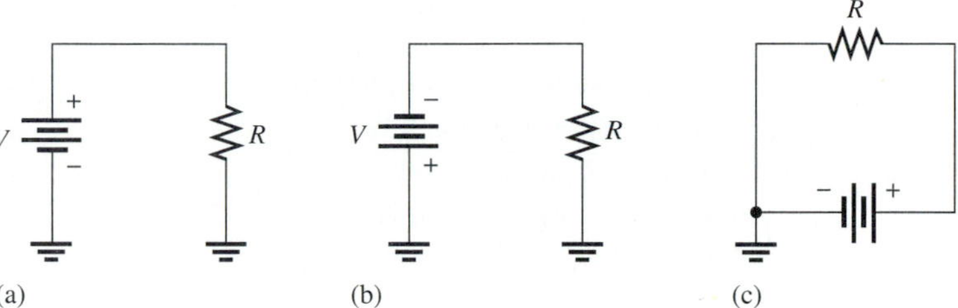


FIGURE 4–14

## SECTION 4–5 Power Supplies

**27.** A 50 Ω load uses 1 W of power. What is the output voltage of the power supply?

**28.** A battery can provide 1.5 A of current for 24 h. What is its ampere-hour rating?

**29.** How much continuous current can be drawn from an 80 Ah battery for 10 h?

**30.** If a battery is rated at 650 mAh, how much current will it provide for 48 h?

**31.** If the input power is 500 mW and the output power is 400 mW, how much power is lost? What is the efficiency of this power supply?

**32.** To operate at 85% efficiency, how much output power must a source produce if the input power is 5 W?

**33.** A certain power supply provides a continuous 2 W to a load. It is operating at 60% efficiency. In a 24 h period, how many kilowatt-hours does the power supply use?

# ■ ANSWERS TO SECTION REVIEWS

### Section 4–1

**1.** Power is the rate at which energy is used.

**2.** $P = W/t$

**3.** Watt is the unit of power. One watt is the power when 1 J of energy is used in 1 s.

**4.** **(a)** 68,000 W = 68 kW **(b)** 0.005 W = 5 mW **(c)** 0.000025 W = 25 $\mu$W

**5.** $W$ = 100 W/10 h = 1 kWh

**6.** 2000 Wh = 2 kWh

**7.** 360,000 Ws = 0.1 kWh

### Section 4–2

**1.** $P$ = (10 V)(3 A) = 30 W

**2.** $P$ = (24 V)(50 mA) = 1.2 W; 1.2 W; the values are the same because all energy generated by the source is dissipated by the resistance.

**3.** $P = (5\text{ A})^2(56\ \Omega) = 1400\text{ W}$

**4.** $P = (20\text{ mA})^2(4.7\text{ k}\Omega) = 1.88\text{ W}$

**5.** $P = (5\text{ V})^2/10\ \Omega = 2.5\text{ W}$

**6.** $P = (8\text{ V})^2/2.2\text{ k}\Omega = 29.1\text{ mW}$

**7.** $R = 75\text{ W}/(0.5\text{ A})^2 = 300\ \Omega$

### Section 4–3

**1.** Resistors have resistance and a power rating.

**2.** A larger surface area of a resistor dissipates more power.

**3.** 0.125 W, 0.25 W, 0.5 W, 1 W, 2 W

**4.** A 0.5 W rating should be used for 0.3 W.

### Section 4–4

1. Energy conversion in a resistor is caused by collisions of the free electrons within the atomic structure.
2. A voltage drop is the potential difference between two points due to energy conversion.

### Section 4–5

1. More current means a greater load.
2. $P = (10\text{ V})(0.5\text{ A}) = 5\text{ W}$
3. $t = 100\text{ Ah}/5\text{ A} = 20\text{ h}$
4. $P = (12\text{ V})(5\text{ A}) = 60\text{ W}$
5. Eff = (0.75 W/1 W)100% = 75%; $P_{LOSS} = 1000\text{ mW} - 750\text{ mW} = 250\text{ mW}$

### Section 4–6

**1.** Two  **2.** 10 Ω, 10 W; 100 Ω, 1 W; 400 Ω, ¼ W

### Section 4–7

1. **(a)** Vary voltage source $V_1$ from 1 V to 8 V in steps of 1 V.
   **(b)** Vary current $I_1$ from 1 mA to 3.5 mA.
2. **(a)** Print the values of the voltages across $R_2$ and $R_3$ in the output file.
   **(b)** Print the values of the currents through $R_1$, $R_2$, and $R_3$ in the output file.

## ■ ANSWERS TO RELATED EXERCISES FOR EXAMPLES

**4–1** 3000 J
**4–2** **(a)** 0.001 W **(b)** 0.0018 W **(c)** 1 W **(d)** 0.000001 W
**4–3** 2 kWh
**4–4** **(a)** 40 W **(b)** 376 W **(c)** 625 mW
**4–5** 60 W
**4–6** 0.5 W
**4–7** No
**4–8** 60 Ah
**4–9** 46 W
**4–10** Change the .DC statement as follows: `.DC VS 4 8 1`

# 5

# SERIES CIRCUITS



## ■ INTRODUCTION

In Chapter 3 you learned about Ohm's law, and in Chapter 4 you learned about power in resistors. In this chapter, those concepts are applied to circuits in which resistors are connected in a series arrangement.

Resistive circuits can be of two basic forms: series and parallel. In this chapter, series circuits are studied. Parallel circuits are covered in Chapter 6, and combinations of series and parallel resistors are examined in Chapter 7. In this chapter, you will see how Ohm's law is used in series circuits; and you will learn another important circuit law, Kirchhoff's voltage law. Also, several applications of series circuits, including voltage dividers, are presented. In this chapter and throughout the rest of the book, you will learn the basics of putting technology theory into practice.

When resistors are connected in series and a voltage is applied across the series connection, there is only one path for current; and, therefore, each resistor in series has the same amount of current through it. All of the resistances in series add together to produce a total resistance. The voltage drops across each of the resistors add up to the voltage applied across the entire series connection.

■ **TECHnology Theory Into Practice**

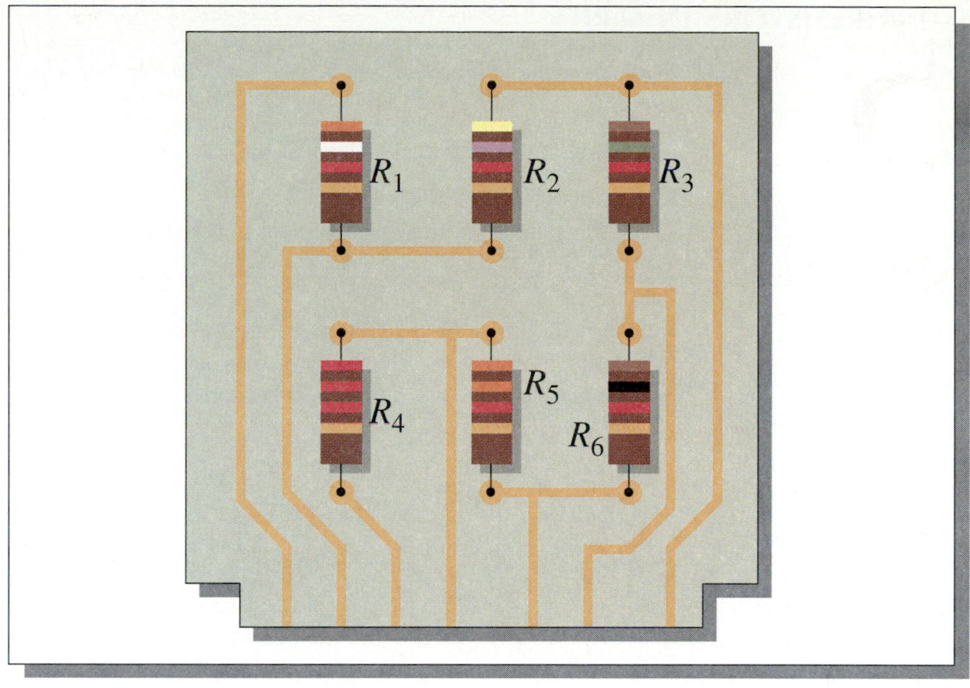


Your supervisor asks you to evaluate a voltage-divider circuit board that will be connected to a 12 V battery to provide a selection of fixed reference voltages for use with an electronic instrument.

## ■ CHAPTER OBJECTIVES

- Identify a series circuit
- Determine the current in a series circuit
- Determine total series resistance
- Apply Ohm's law in series circuits
- Determine the total effect of voltage sources in series
- Apply Kirchhoff's voltage law
- Use a series circuit as a voltage divider
- Determine power in a series circuit
- Determine and identify ground in a circuit
- Troubleshoot series circuits
- Use PSpice to analyze series circuits and to graph the results (optional)

## 5–1 ■ RESISTORS IN SERIES

***When connected in series, resistors form a "string" in which there is only one path for current.***

***After completing this section, you should be able to***

- **Identify a series circuit**
  - Translate a physical arrangement of resistors into a schematic

Figure 5–1(a) shows two resistors connected in **series** between point $A$ and point $B$. Part (b) of the figure shows three resistors in series, and part (c) shows four in series. Of course, there can be any number of resistors in a series circuit.

FIGURE 5–1
*Resistors in series.*

The only way for electrons to get from point $A$ to point $B$ in any of the connections of Figure 5–1 is to go through each of the resistors. The following is an important way to identify a series circuit.

**A series circuit provides only one path for current between two points in a circuit so that the current is the same through each series resistor.**

### Identifying Series Circuits

In an actual circuit diagram, a series circuit may not always be as easy to identify as those in Figure 5–1. For example, Figure 5–2 shows series resistors drawn in other ways. Remember, if there is only one current path between two points, the resistors between those two points are in series, no matter how they appear in a diagram.

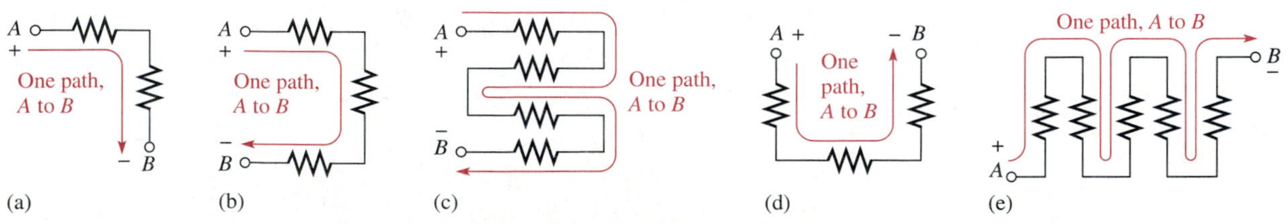


FIGURE 5–2
*Some examples of series circuits. Notice that the current is the same at all points.*

**EXAMPLE 5–1** Suppose that there are five resistors positioned on a circuit board as shown in Figure 5–3. Wire them together in series so that, starting from the positive (+) terminal, $R_1$ is first, $R_2$ is second, $R_3$ is third, and so on. Draw a schematic showing this connection.

FIGURE 5–3

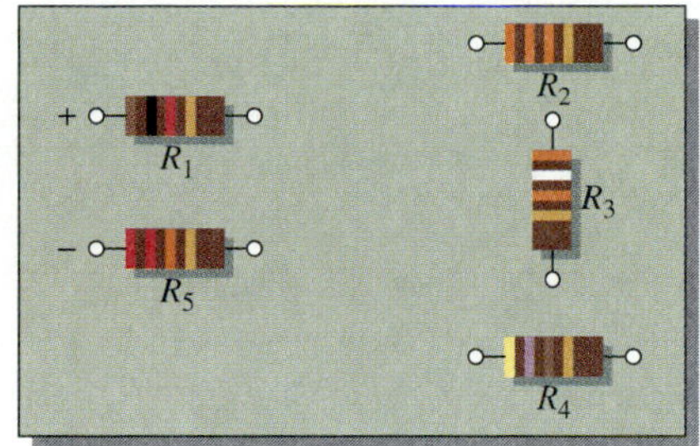


***Solution*** The wires are connected as shown in Figure 5–4(a), which is the assembly diagram. The schematic is shown in Figure 5–4(b). Note that the schematic does not necessarily show the actual physical arrangement of the resistors as does the assembly diagram. The schematic shows how components are connected electrically; the assembly diagram shows how components are arranged and interconnected physically.

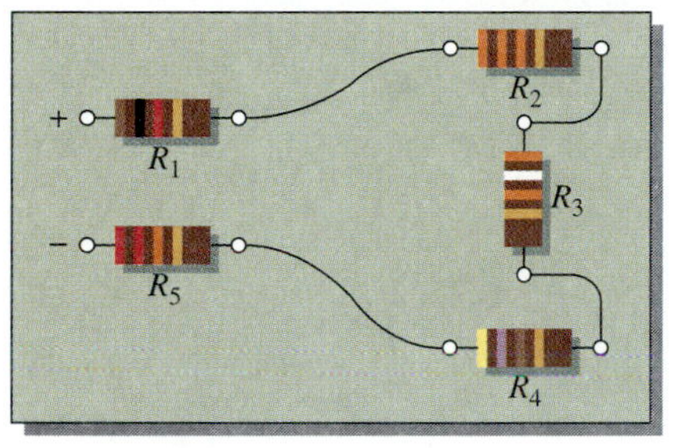


(a) Assembly diagram

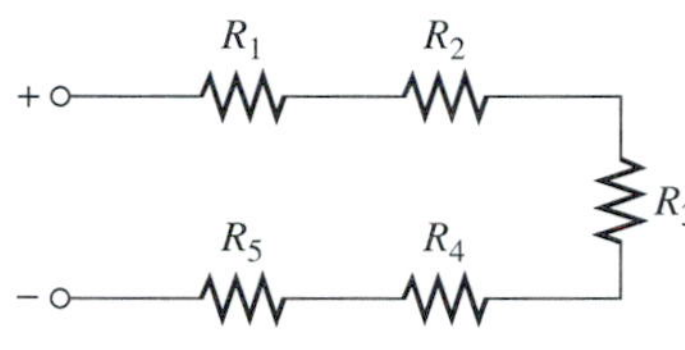


(b) Schematic

FIGURE 5–4

***Related Exercise*** (a) Show how you would rewire the circuit board in Figure 5–4(a) so that all the odd-numbered resistors come first followed by the even-numbered ones. (b) Determine the value of each resistor.

**EXAMPLE 5–2** Describe how the resistors on the printed circuit (PC) board in Figure 5–5 are related electrically. Determine the resistance value of each resistor.

FIGURE 5–5

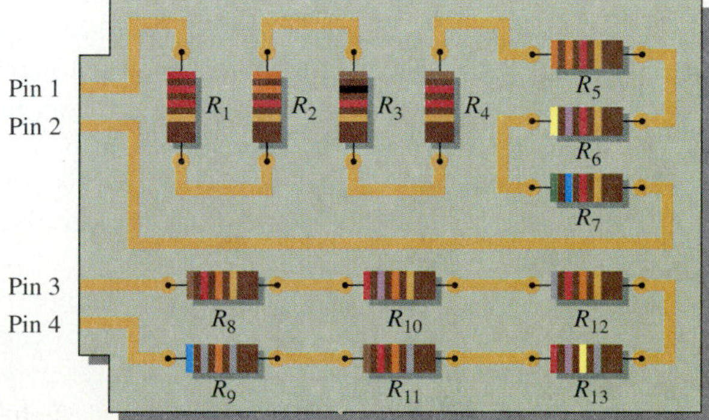

***Solution*** Resistors $R_1$ through $R_7$ are in series with each other. This series combination is connected between pins 1 and 2 on the PC board.

Resistors $R_8$ through $R_{13}$ are in series with each other. This series combination is connected between pins 3 and 4 on the PC board.

The values of the resistors are $R_1$ = 2.2 kΩ, $R_2$ = 3.3 kΩ, $R_3$ = 1 kΩ, $R_4$ = 1.2 kΩ, $R_5$ = 3.3 kΩ, $R_6$ = 4.7 kΩ, $R_7$ = 5.6 kΩ, $R_8$ = 12 kΩ, $R_9$ = 68 kΩ, $R_{10}$ = 27 kΩ, $R_{11}$ = 12 kΩ, $R_{12}$ = 82 kΩ, and $R_{13}$ = 270 kΩ.

***Related Exercise*** How is the circuit changed when pin 2 and pin 3 in Figure 5–5 are connected?

## SECTION 5–1 REVIEW

1. How are the resistors connected in a series circuit?
2. How can you identify a series circuit?
3. Complete the schematics for the circuits in each part of Figure 5–6 by connecting each group of resistors in series in numerical order from terminal *A* to *B*.
4. Connect each group of series resistors in Figure 5–6 in series with each other.

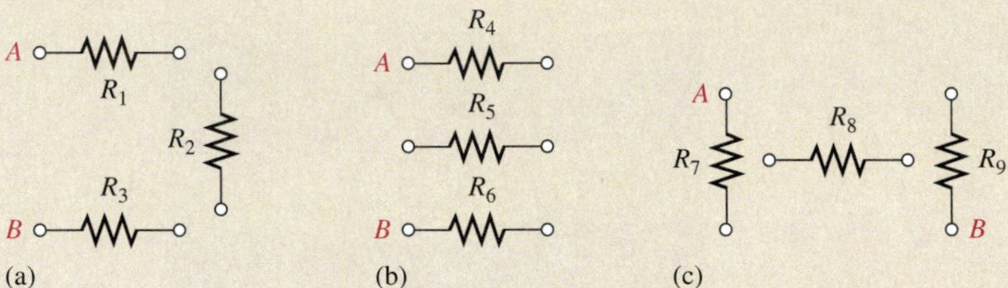


FIGURE 5–6

## 5–2 ■ CURRENT IN A SERIES CIRCUIT

***There is the same amount of current through all points in a series circuit. The current through each resistor in a series circuit is the same as the current through all the other resistors that are in series with it.***

***After completing this section, you should be able to***

- **Determine the current in a series circuit**
  - Show that the current is the same at all points in a series circuit

Figure 5–7 shows three resistors connected in series to a voltage source. *At any point in this circuit, the current into that point must equal the current out of that point,* as illustrated by the current directional arrows at points *A, B, C,* and *D* in part (b). Notice also that the current out of each of the resistors must equal the current in because there is no place where part of the current can branch off and go somewhere else. Therefore, the current in each section of the circuit is the same as the current in all other sections. It has only one path going from the positive (+) side of the source to the negative (−) side.

Let's assume that the battery in Figure 5–7 supplies two amperes of current to the series resistance. There are two amperes of current out of the battery's positive terminal. When ammeters are connected at several points in the circuit as shown in Figure 5–8, each meter reads two amperes.

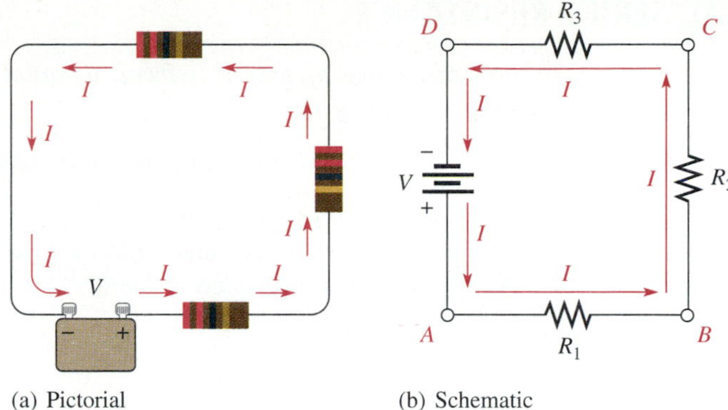


FIGURE 5–7
***Current into any point in a series circuit is the same as the current out of that point.***

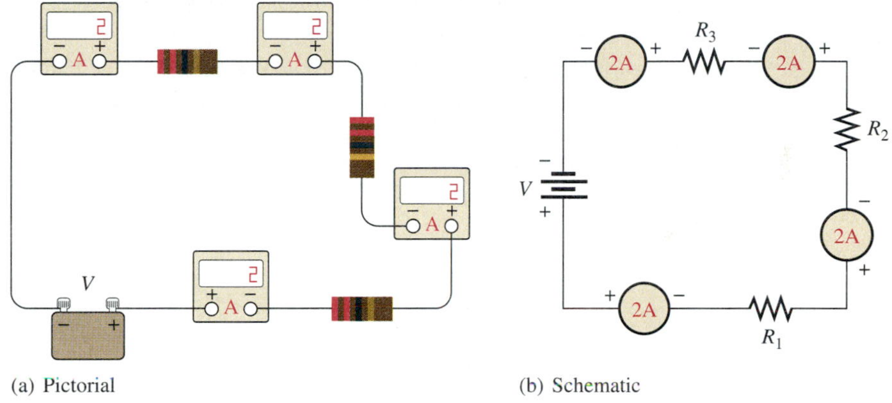


FIGURE 5–8
***Current is the same at all points in a series circuit.***

## SECTION 5–2 REVIEW

1. In a circuit with a 10 Ω and a 4.7 Ω resistor in series, there is 1 A of current through the 10 Ω resistor. How much current is through the 4.7 Ω resistor?
2. A milliammeter is connected between points *A* and *B* in Figure 5–9. It measures 50 mA. If you move the meter and connect it between points *C* and *D*, how much current will it indicate? Between *E* and *F?*
3. In Figure 5–10, how much current does ammeter 1 indicate? How much current does ammeter 2 indicate?
4. What statement can you make about the amount of current in a series circuit?

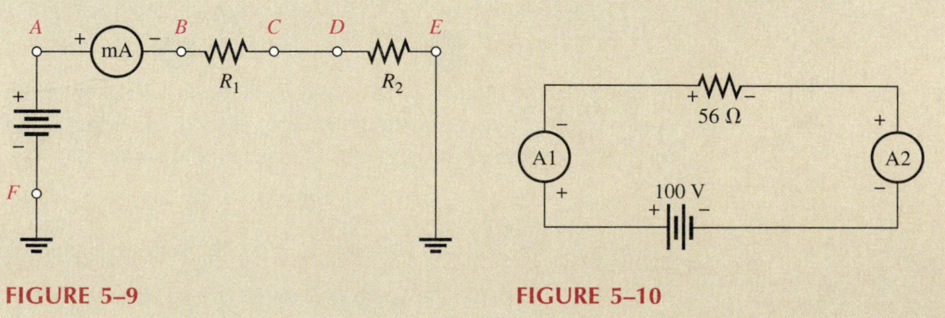


FIGURE 5–9

FIGURE 5–10

## 5–3 ■ TOTAL SERIES RESISTANCE

***The total resistance of a series circuit is equal to the sum of the resistances of each individual series resistor.***

***After completing this section, you should be able to***

- **Determine total series resistance**
  - ☐ Explain why resistance values add in series
  - ☐ Apply the series resistance formula

### Series Resistor Values Add

When resistors are connected in series, the resistor values add because each resistor offers opposition to the current in direct proportion to its resistance. A greater number of resistors connected in series creates more opposition to current. More opposition to current implies a higher value of resistance. Thus, every time a resistor is added in series, the total resistance increases.

Figure 5–11 illustrates how series resistances add to increase the total resistance. Figure 5–11(a) has a single 10 Ω resistor. Figure 5–11(b) shows another 10 Ω resistor connected in series with the first one, making a total resistance of 20 Ω. If a third 10 Ω resistor is connected in series with the first two, as shown in Figure 5–11(c), the total resistance becomes 30 Ω.

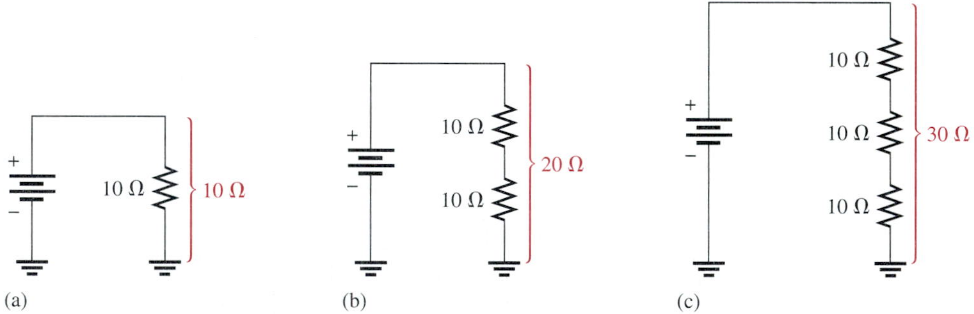


**FIGURE 5–11**
***Total resistance increases with each additional series resistor.***

### Series Resistance Formula

For any number of individual resistors connected in series, the total resistance is the sum of each of the individual values.

$$R_T = R_1 + R_2 + R_3 + \cdots + R_n \qquad \textbf{(5–1)}$$

where $R_T$ is the total resistance and $R_n$ is the last resistor in the series string ($n$ can be any positive integer equal to the number of resistors in series). For example, if there are four resistors in series ($n = 4$), the total resistance formula is

$$R_T = R_1 + R_2 + R_3 + R_4$$

If there are six resistors in series ($n = 6$), the total resistance formula is

$$R_T = R_1 + R_2 + R_3 + R_4 + R_5 + R_6$$

To illustrate the calculation of total series resistance, let's determine $R_T$ in the circuit of Figure 5–12, where $V_S$ is the source voltage. The circuit has five resistors in series. To get the total resistance, simply add the values as follows:

$$R_T = 56\ \Omega + 100\ \Omega + 27\ \Omega + 10\ \Omega + 5.6\ \Omega = 198.6\ \Omega$$

Note in Figure 5–12 that the order in which the resistances are added does not matter. You can physically change the positions of the resistors in the circuit without affecting the total resistance.

**FIGURE 5–12**
***Example of five resistors in series.***

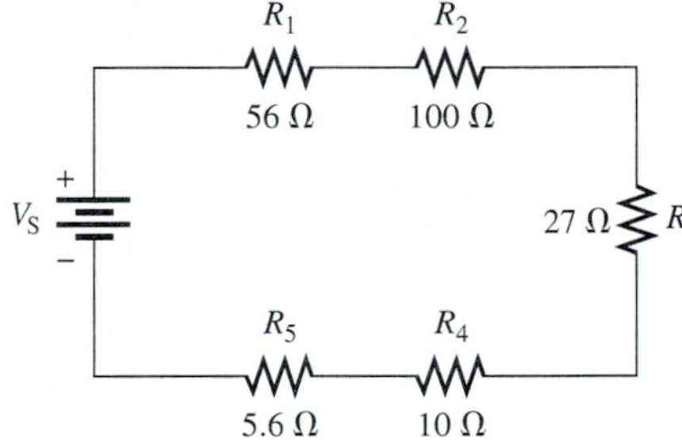


**EXAMPLE 5–3** Connect the resistors in Figure 5–13 in series, and determine the total resistance, $R_T$.

**FIGURE 5–13**

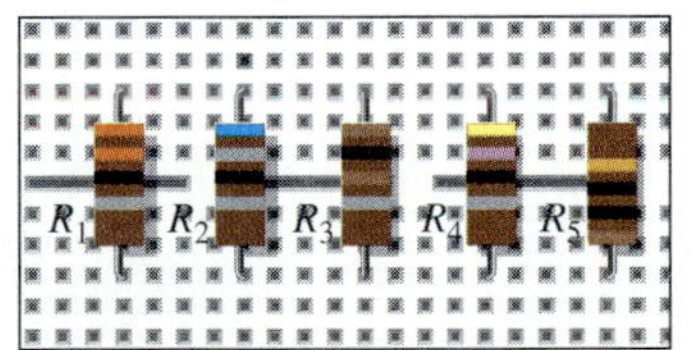


***Solution*** The resistors are connected as shown in Figure 5–14. Find the total resistance by adding all the values as follows:

$$R_T = R_1 + R_2 + R_3 + R_4 + R_5 = 33\ \Omega + 68\ \Omega + 100\ \Omega + 47\ \Omega + 10\ \Omega = 258\ \Omega$$

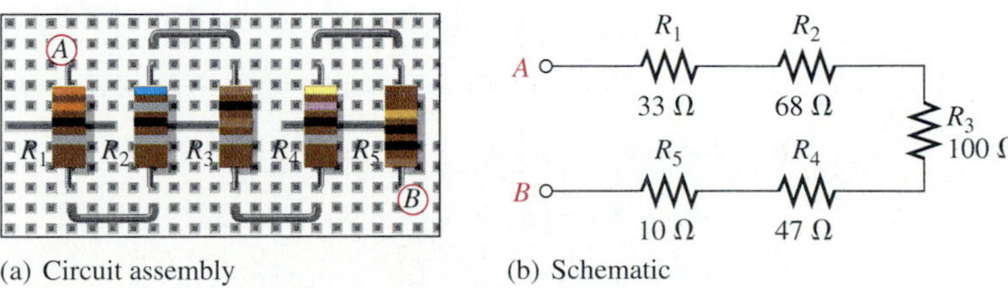


(a) Circuit assembly (b) Schematic

**FIGURE 5–14**

***Related Exercise*** Determine the total resistance in Figure 5–14(a) if the positions of $R_2$ and $R_4$ are interchanged.

**EXAMPLE 5–4** What is the total resistance ($R_T$) in the circuit of Figure 5–15?

FIGURE 5–15

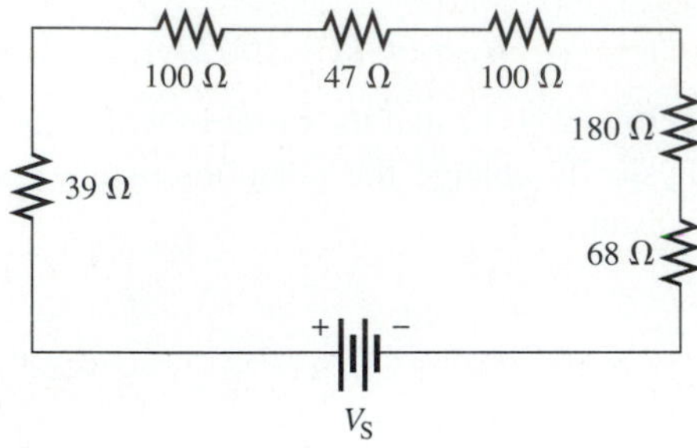


***Solution*** Sum all the values:

$$R_T = 39\ \Omega + 100\ \Omega + 47\ \Omega + 100\ \Omega + 180\ \Omega + 68\ \Omega = 534\ \Omega$$

***Related Exercise*** What is the total resistance for the following series resistors: 1 kΩ, 2.2 kΩ, 3.3 kΩ, and 5.6 kΩ?

**EXAMPLE 5–5** Determine the value of $R_4$ in the circuit of Figure 5–16.

FIGURE 5–16

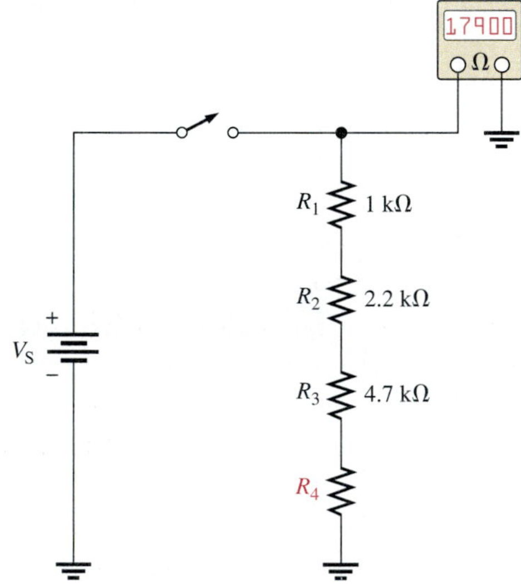


***Solution*** From the ohmmeter reading, $R_T = 17.9$ kΩ.

$$R_T = R_1 + R_2 + R_3 + R_4$$

Solving for $R_4$ yields

$$R_4 = R_T - (R_1 + R_2 + R_3) = 17.9\ \text{k}\Omega - (1\ \text{k}\Omega + 2.2\ \text{k}\Omega + 4.7\ \text{k}\Omega) = 10\ \text{k}\Omega$$

The calculator sequence for determining $R_4$ is

1 7 • 9 EXP 3 − ( 1 EXP 3 + 2 • 2 EXP 3 + 4 • 7 EXP 3 ) =

***Related Exercise*** Determine the value of $R_4$ in Figure 5–16 if the ohmmeter reading is 14,700 Ω.

### Equal-value Series Resistors

When a circuit has more than one resistor of the same value in series, there is a shortcut method to obtain the total resistance: Simply multiply the resistance value by the number of equal-value resistors that are in series. This method is essentially the same as adding the values. For example, five 100 Ω resistors in series have an $R_T$ of 5(100 Ω) = 500 Ω. In general, the formula is expressed as

$$R_T = nR \qquad (5\text{–}2)$$

where $n$ is the number of equal-value resistors and $R$ is the resistance value.

**EXAMPLE 5–6**

Find the $R_T$ of eight 22 Ω resistors in series.

***Solution*** Find $R_T$ by adding the values:

$$R_T = 22\ \Omega + 22\ \Omega + 22\ \Omega + 22\ \Omega + 22\ \Omega + 22\ \Omega + 22\ \Omega + 22\ \Omega = 176\ \Omega$$

However, it is much easier to multiply.

$$R_T = 8(22\ \Omega) = 176\ \Omega$$

***Related Exercise*** Find $R_T$ for three 1 kΩ resistors and two 720 Ω resistors in series.

**SECTION 5–3 REVIEW**

1. The following resistors (one each) are in series: 1 Ω, 2.2 Ω, 3.3 Ω, and 4.7 Ω. What is the total resistance?
2. The following resistors are in series: one 100 Ω, two 56 Ω, four 12 Ω, and one 330 Ω. What is the total resistance?
3. Suppose that you have one resistor each of the following values: 1 kΩ, 2.7 kΩ, 5.6 kΩ, and 560 Ω. To get a total resistance of 13.76 kΩ, you need one more resistor. What should its value be?
4. What is the $R_T$ for twelve 56 Ω resistors in series?
5. What is the $R_T$ for twenty 5.6 kΩ resistors and thirty 8.2 kΩ resistors in series?

## 5–4 ■ OHM'S LAW IN SERIES CIRCUITS

***The use of Ohm's law and the basic concepts of series circuits are applied in several examples.***

***After completing this section, you should be able to***

- **Apply Ohm's law in series circuits**
  - ☐ Find the current in a series circuit
  - ☐ Find the voltage across each resistor in series

**EXAMPLE 5–7** Find the current in the circuit of Figure 5–17.

FIGURE 5–17

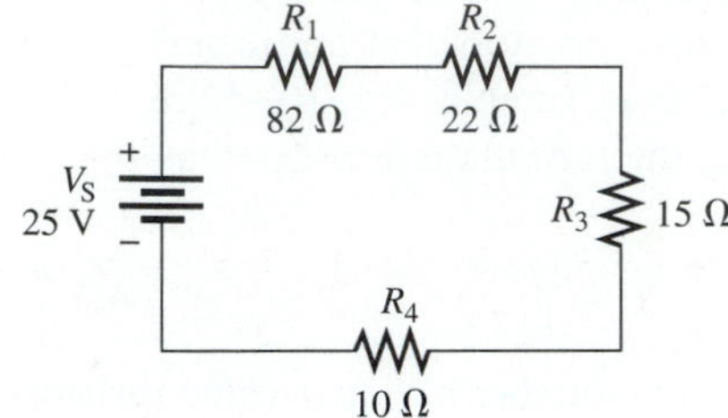


*Solution* The current is determined by the voltage and the total resistance. First, calculate the total resistance:

$$R_T = R_1 + R_2 + R_3 + R_4 = 82\ \Omega + 22\ \Omega + 15\ \Omega + 10\ \Omega = 129\ \Omega$$

Next, use Ohm's law to calculate the current:

$$I = \frac{V_S}{R_T} = \frac{25\ \text{V}}{129\ \Omega} = 0.194\ \text{A} = 194\ \text{mA}$$

Remember, the same current exists at all points in the circuit. Thus, each resistor has 194 mA through it.

*Related Exercise* What is the current in Figure 5–17 if $R_4$ is changed to 100 Ω?

**EXAMPLE 5–8** The current in the circuit of Figure 5–18 is 1 mA. For this amount of current, what must the source voltage $V_S$ be?

FIGURE 5–18

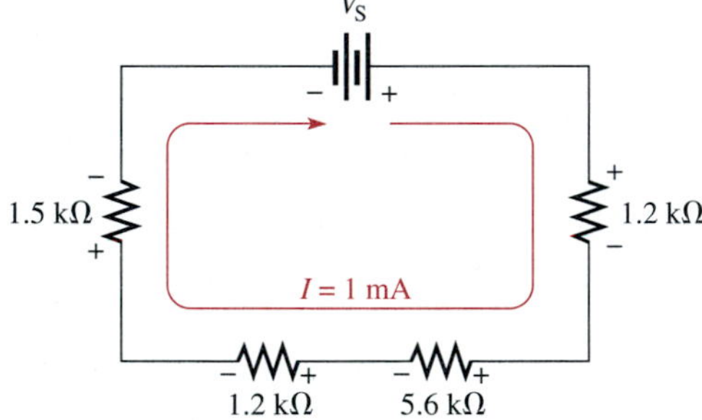


*Solution* In order to calculate $V_S$, first determine $R_T$:

$$R_T = 1.2\ \text{k}\Omega + 5.6\ \text{k}\Omega + 1.2\ \text{k}\Omega + 1.5\ \text{k}\Omega = 9.5\ \text{k}\Omega$$

Now use Ohm's law to get $V_S$:

$$V_S = IR_T = (1\ \text{mA})(9.5\ \text{k}\Omega) = 9.5\ \text{V}$$

*Related Exercise* Calculate $V_S$ if the 5.6 kΩ resistor is changed to 3.9 kΩ with the current the same.

**EXAMPLE 5–9** Calculate the voltage across each resistor in Figure 5–19, and find the value of $V_S$. To what maximum value can $V_S$ be raised before the fuse blows?

FIGURE 5–19

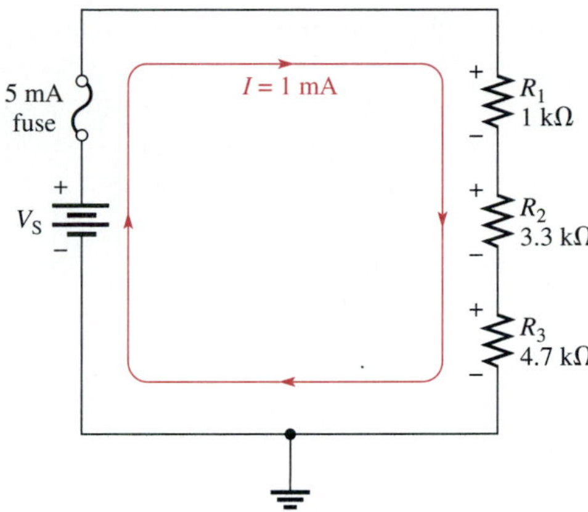


***Solution*** By Ohm's law, the voltage across each resistor is equal to its resistance multiplied by the current through it. Use the Ohm's law formula $V = IR$ to determine the voltage across each of the resistors. Keep in mind that there is the same current through each series resistor. The fuse is effectively like a wire and has negligible resistance. The voltage across $R_1$ (designated $V_1$) is

$$V_1 = IR_1 = (1\text{ mA})(1\text{ k}\Omega) = 1\text{ V}$$

The voltage across $R_2$ is

$$V_2 = IR_2 = (1\text{ mA})(3.3\text{ k}\Omega) = 3.3\text{ V}$$

The voltage across $R_3$ is

$$V_3 = IR_3 = (1\text{ mA})(4.7\text{ k}\Omega) = 4.7\text{ V}$$

To find the value of $V_S$, first determine $R_T$:

$$R_T = 1\text{ k}\Omega + 3.3\text{ k}\Omega + 4.7\text{ k}\Omega = 9\text{ k}\Omega$$

The source voltage $V_S$ is equal to the current times the total resistance.

$$V_S = IR_T = (1\text{ mA})(9\text{ k}\Omega) = 9\text{ V}$$

Notice that if you add the voltage drops of the resistors, they total 9 V, which is the same as the source voltage.

The fuse has a rating of 5 mA; thus $V_S$ can be increased to a value where $I =$ 5 mA. Calculate the maximum value of $V_S$ as follows:

$$V_{S(max)} = IR_T = (5\text{ mA})(9\text{ k}\Omega) = 45\text{ V}$$

***Related Exercise*** Repeat the calculations for $V_1$, $V_2$, $V_3$, $V_S$, and $V_{S(max)}$ if $R_3 =$ 2.2 kΩ and $I$ is maintained at 1 mA.

**EXAMPLE 5–10** As you have learned, some resistors are not color coded with bands but have the values stamped on the resistor body. When the circuit board shown in Figure 5–20 was assembled, the resistors were erroneously mounted with the labels turned down, and there is no documentation showing the resistor values. Without removing the resistors from the board, use Ohm's law to determine the resistance of each one. Assume that a voltmeter, ammeter, and power supply are immediately available, but no ohmmeter is available.

FIGURE 5–20

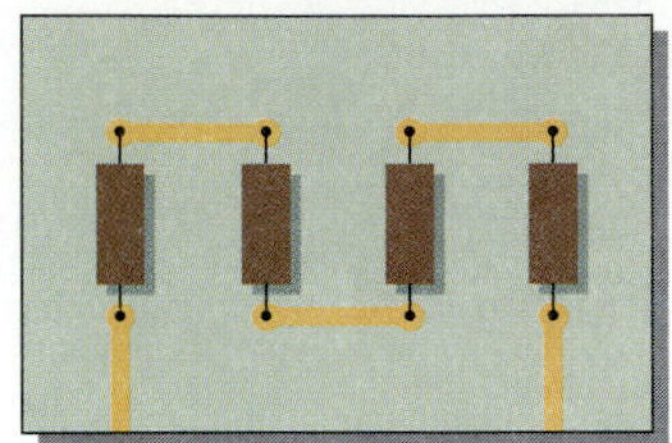

***Solution*** The resistors are all in series, so the current is the same through each one. Measure the current by connecting a 12 V source (arbitrary value) and an ammeter as shown in Figure 5–21. Measure the voltage across each resistor by placing the voltmeter across the first resistor. Then repeat this measurement for the other three resistors. For illustration, the voltage values indicated in color are assumed to be the measured values.

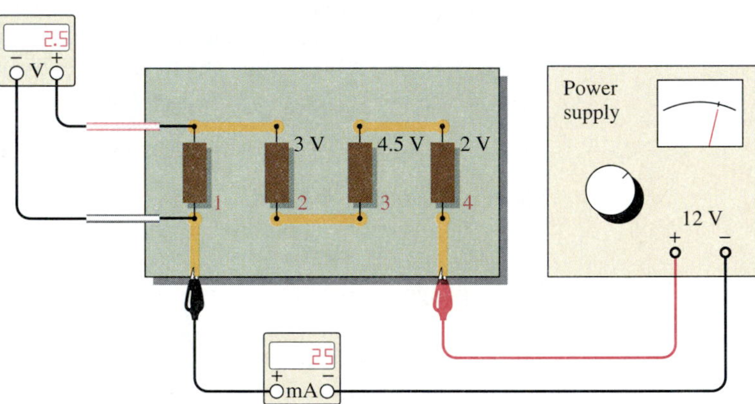


FIGURE 5–21
***The voltmeter readings across each resistor are indicated.***

Determine the resistance of each resistor by substituting the measured values of current and voltage into the Ohm's law formula as follows:

$$R_1 = \frac{V_1}{I} = \frac{2.5\text{ V}}{25\text{ mA}} = 100\ \Omega$$

$$R_2 = \frac{V_2}{I} = \frac{3\text{ V}}{25\text{ mA}} = 120\ \Omega$$

$$R_3 = \frac{V_3}{I} = \frac{4.5\text{ V}}{25\text{ mA}} = 180\ \Omega$$

$$R_4 = \frac{V_4}{I} = \frac{2\text{ V}}{25\text{ mA}} = 80\ \Omega$$

The calculator sequence for $R_1$ is

***Related Exercise*** What is an easier way to determine the resistance values?

**SECTION 5–4 REVIEW**

1. A 10 V battery is connected across three 100 Ω resistors in series. What is the current through each resistor?
2. How much voltage is required to produce 5 A through the circuit of Figure 5–22?

FIGURE 5–22

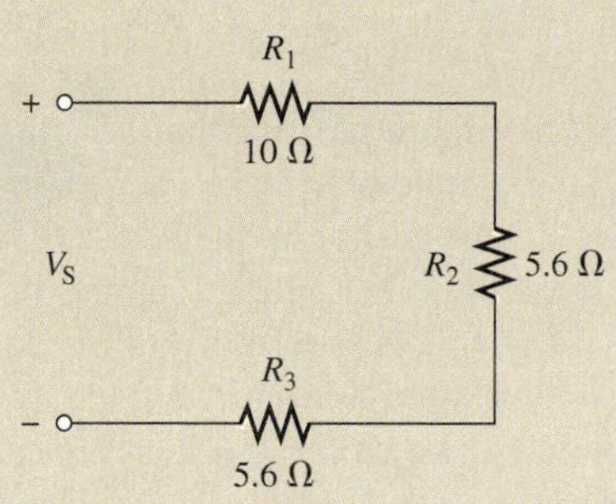

3. How much voltage is dropped across each resistor in Figure 5–22 when the current is 5 A?
4. There are four equal-value resistors connected in series with a 5 V source. A current of 4.63 mA is measured. What is the value of each resistor?

## 5–5 ■ VOLTAGE SOURCES IN SERIES

***A voltage source is an energy source that provides a constant voltage to a load. Batteries and power supplies are practical examples.***

***After completing this section, you should be able to***

- **Determine the total effect of voltage sources in series**
  - Determine the total voltage of series sources with the same polarities
  - Determine the total voltage of series sources with opposite polarities

When two or more voltage sources are in series, the total voltage is equal to the algebraic sum of the individual source voltages. The algebraic sum means that the polarities of the sources must be included when the sources are combined in series. Sources with opposite polarities have voltages with opposite signs.

$$V_{S(tot)} = V_{S1} + V_{S2} + \cdots + V_{Sn}$$

When the voltage sources are all in the same direction in terms of their polarities, as in Figure 5–23(a), all of the voltages have the same sign when added; there is a total of 4.5 V from terminal $A$ to terminal $B$ with $A$ more positive than $B$.

$$V_{AB} = 1.5\text{ V} + 1.5\text{ V} + 1.5\text{ V} = +4.5\text{ V}$$

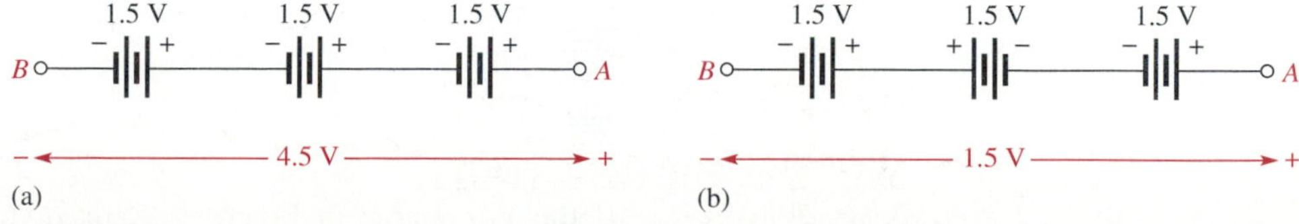


FIGURE 5–23
*Voltage sources in series add algebraically.*

The voltage has a double subscript, $AB$, to indicate that it is the voltage at point $A$ with respect to point $B$.

In Figure 5–23(b), the middle voltage source is opposite to the other two; so its voltage has an opposite sign when added to the others. For this case the total voltage from $A$ to $B$ is

$$V_{AB} = +1.5\text{ V} - 1.5\text{ V} + 1.5\text{ V} = +1.5\text{ V}$$

Terminal $A$ is 1.5 V more positive than terminal $B$.

A familiar example of voltage sources in series is the flashlight. When you put two 1.5 V batteries in your flashlight, they are connected in series, giving a total of 3 V. When connecting batteries or other voltage sources in series to increase the total voltage, always connect from the positive (+) terminal of one to the negative (−) terminal of another. Such a connection is illustrated in Figure 5–24.

The following two examples illustrate calculation of total source voltage.

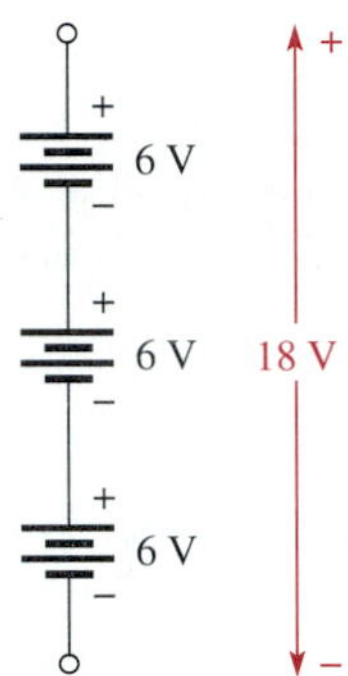


**FIGURE 5–24**
***Connection of three 6 V batteries to obtain 18 V.***

## EXAMPLE 5–11

What is the total source voltage ($V_{S(tot)}$) in Figure 5–25?

**FIGURE 5–25**

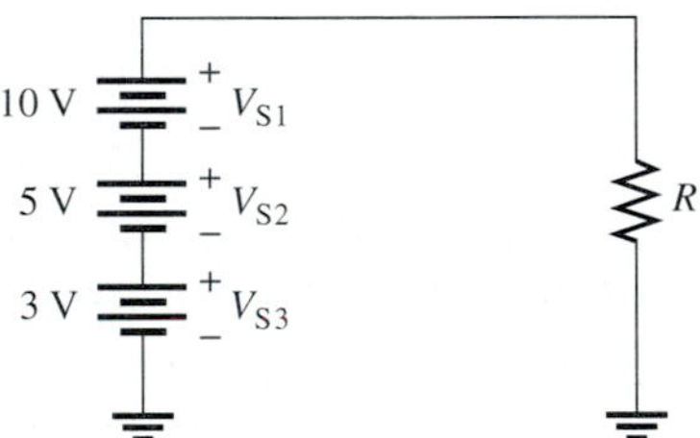


***Solution*** The polarity of each source is the same (the sources are connected in the same direction in the circuit). So add the three voltages to get the total.

$$V_{S(tot)} = V_{S1} + V_{S2} + V_{S3} = 10\text{ V} + 5\text{ V} + 3\text{ V} = 18\text{ V}$$

The three individual sources can be replaced by a single equivalent source of 18 V with its polarity as shown in Figure 5–26.

**FIGURE 5–26**

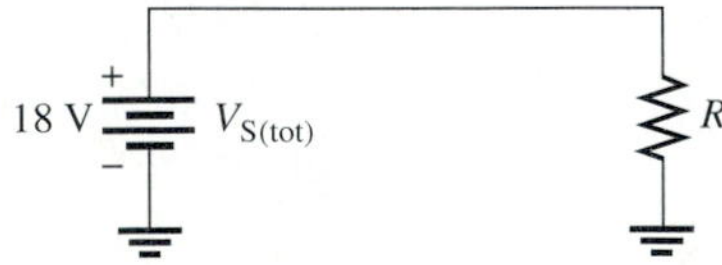


***Related Exercise*** If the $V_{S3}$ source in Figure 5–25 is reversed, what is the total source voltage?

**EXAMPLE 5–12** Determine $V_{S(tot)}$ in Figure 5–27.

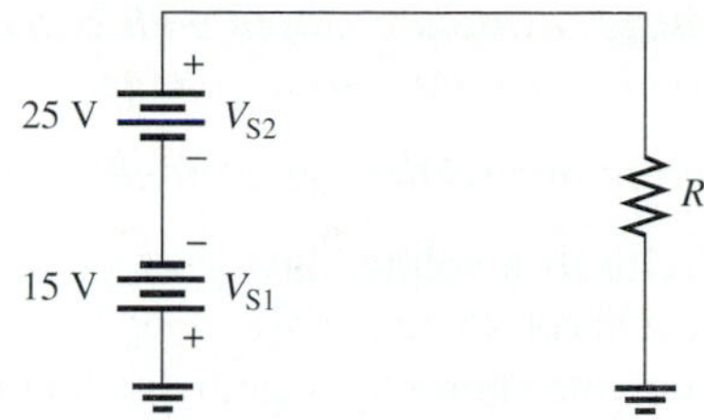


FIGURE 5–27

***Solution*** These sources are connected in opposing directions. If you go clockwise around the circuit, you go from plus to minus through $V_{S1}$, and minus to plus through $V_{S2}$. The total voltage is the difference of the two source voltages (algebraic sum of oppositely signed values). The total voltage has the same polarity as the larger-value source. Here we will choose $V_{S2}$ to be positive:

$$V_{S(tot)} = V_{S2} - V_{S1} = 25 \text{ V} - 15 \text{ V} = 10 \text{ V}$$

The two sources in Figure 5–27 can be replaced by a single 10 V equivalent source with polarity as shown in Figure 5–28.

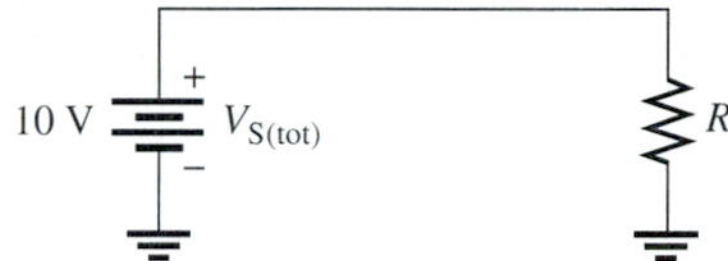


FIGURE 5–28

***Related Exercise*** If an 8 V source in the direction of $V_{S1}$ is added in series in Figure 5–27, what is $V_{S(tot)}$?

## SECTION 5–5 REVIEW

1. Four 1.5 V flashlight batteries are connected in series plus to minus. What is the total voltage of all four cells?
2. How many 12 V batteries must be connected in series to produce 60 V? Sketch a schematic that shows the battery connections.
3. The resistive circuit in Figure 5–29 is used to bias a transistor amplifier. Show how to connect two 15 V power supplies in order to get 30 V across the two resistors.
4. Determine the total source voltage in each circuit of Figure 5–30.
5. Sketch the equivalent single-source circuit for each circuit of Figure 5–30.

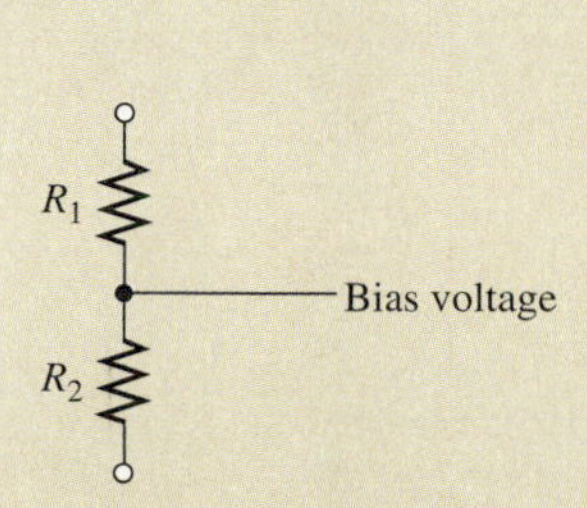


FIGURE 5–29

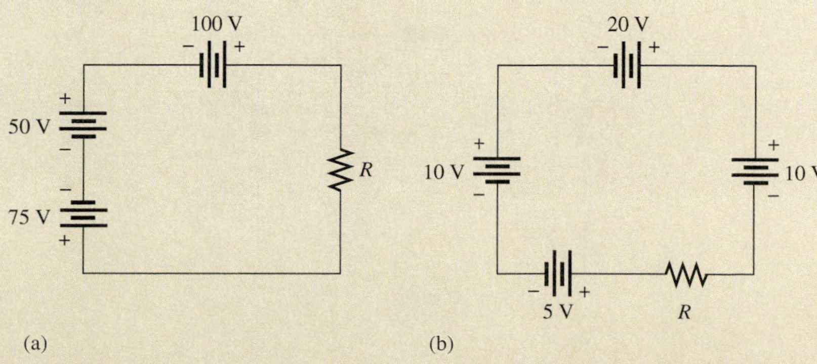


FIGURE 5–30

## 5–6 ■ KIRCHHOFF'S VOLTAGE LAW

***Kirchhoff's voltage law is a fundamental circuit law that states that the algebraic sum of all the voltages around a closed path is zero or, in other words, the sum of the voltage drops equals the total source voltage.***

***After completing this section, you should be able to***

- **Apply Kirchhoff's voltage law**
  - ☐ State Kirchhoff's voltage law
  - ☐ Determine the source voltage by adding the voltage drops
  - ☐ Determine an unknown voltage drop

In an electric circuit, the voltages across the resistors (voltage drops) *always* have polarities opposite to the source voltage polarity. For example, in Figure 5–31, follow a clockwise loop around the circuit and note that the source polarity is minus-to-plus and each voltage drop is plus-to-minus.

Also notice in Figure 5–31 that the current is out of the positive side of the source and through the resistors as indicated by the arrows. The current is into the positive side of each resistor and out the negative side. The drop in energy level across a resistor creates a potential difference, or voltage drop, with a plus-to-minus polarity in the direction of the current.

Notice that the voltage from point *A* to point *B* in the circuit of Figure 5–31 equals the source voltage, $V_S$. Also, the voltage from *A* to *B* is the sum of the series resistor voltage drops. Therefore, the source voltage is equal to the sum of the three voltage drops.

This discussion is an example of **Kirchhoff's voltage law,** which is generally stated as:

> **The sum of all the voltage drops around a single closed loop in a circuit is equal to the total source voltage in that loop.**

The general concept of Kirchhoff's voltage law is illustrated in Figure 5–32 and expressed by Equation (5–3):

$$V_S = V_1 + V_2 + V_3 + \cdots + V_n \qquad \textbf{(5–3)}$$

where the subscript *n* represents the number of voltage drops.

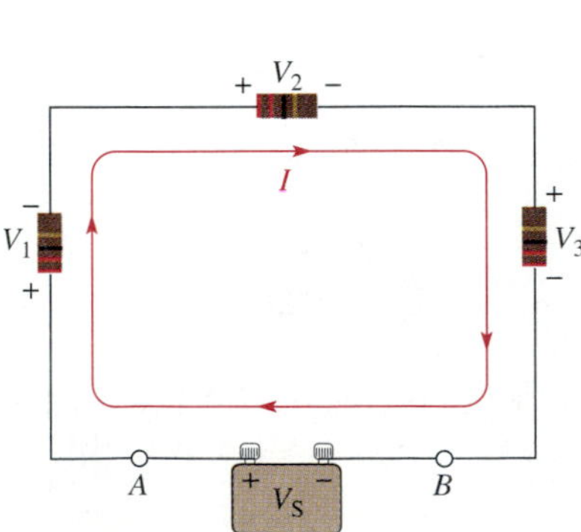


**FIGURE 5–31**
***Kirchhoff's voltage law: The sum of the voltage drops equals the source voltage.***

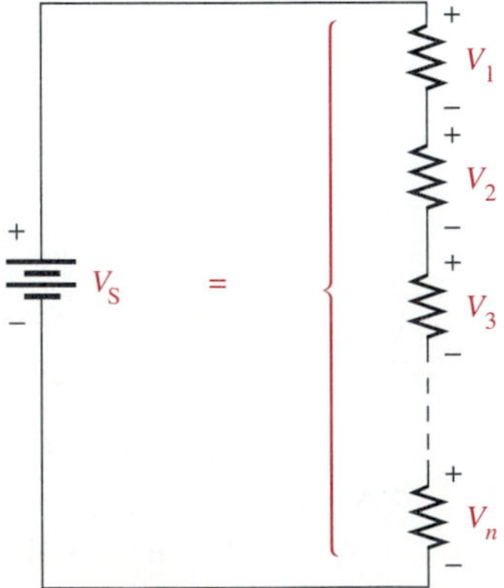


**FIGURE 5–32**
***Sum of n voltage drops equals the source voltage.***

## Another Way to State Kirchhoff's Voltage Law

If all the voltage drops around a closed loop are added and then this total is subtracted from the source voltage, the result is zero. This result occurs because the sum of the voltage drops always equals the source voltage.

**The algebraic sum of all voltages (both source and drops) around a closed path is zero.**

Therefore, another way of expressing Kirchhoff's voltage law is as follows:

$$V_S - V_1 - V_2 - V_3 = 0 \quad \textbf{(5–4)}$$

You can verify Kirchhoff's voltage law by connecting a circuit and measuring each resistor voltage and the source voltage as illustrated in Figure 5–33. When the resistor voltages are added together, their sum will equal the source voltage. Any number of resistors can be added.

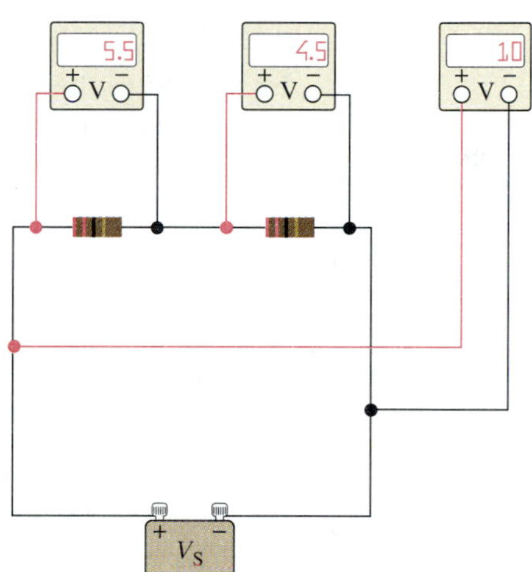


FIGURE 5–33
***Experimental verification of Kirchhoff's voltage law.***

The following three examples use Kirchhoff's voltage law to solve circuit problems.

**EXAMPLE 5–13** Determine the source voltage $V_S$ in Figure 5–34 where the two voltage drops are given.

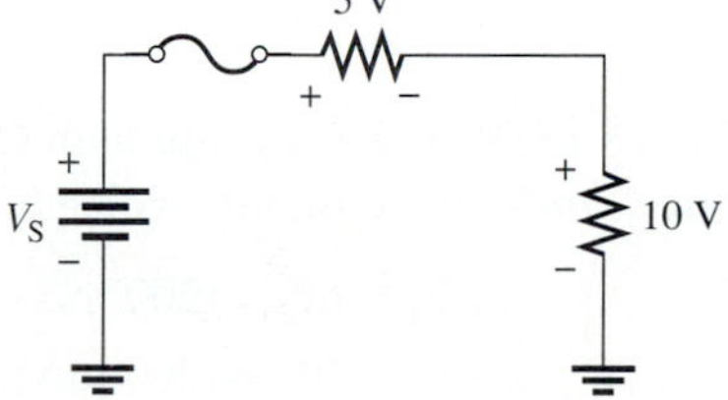


FIGURE 5–34

***Solution*** By Kirchhoff's voltage law, (Eq. 5–3), the source voltage (applied voltage) must equal the sum of the voltage drops. Adding the voltage drops gives the value of the source voltage.

$$V_S = 5\text{ V} + 10\text{ V} = 15\text{ V}$$

***Related Exercise*** If $V_S$ is increased to 30 V, determine the two voltage drops.

**EXAMPLE 5–14** Determine the unknown voltage drop, $V_3$, in Figure 5–35.

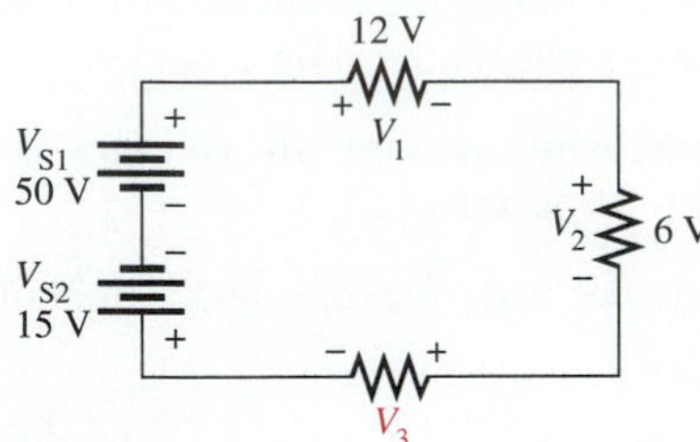

FIGURE 5–35

***Solution*** By Kirchhoff's voltage law (Eq. 5–4), the algebraic sum of all the voltages around the circuit is zero. The value of each voltage drop except $V_3$ is known. Substitute these values into the equation.

$$V_{S1} - V_{S2} - V_1 - V_2 - V_3 = 0$$
$$50\text{ V} - 15\text{ V} - 12\text{ V} - 6\text{ V} - V_3 = 0\text{ V}$$

Next combine the known values. Transpose 17 V to the right side of the equation, and cancel the minus signs.

$$17\text{ V} - V_3 = 0\text{ V}$$
$$-V_3 = -17\text{ V}$$
$$V_3 = 17\text{ V}$$

The voltage drop across $R_3$ is 17 V, and its polarity is as shown in Figure 5–35.

***Related Exercise*** Determine $V_3$ if the polarity of $V_{S2}$ is reversed in Figure 5–35.

**EXAMPLE 5–15** Find the value of $R_4$ in Figure 5–36.

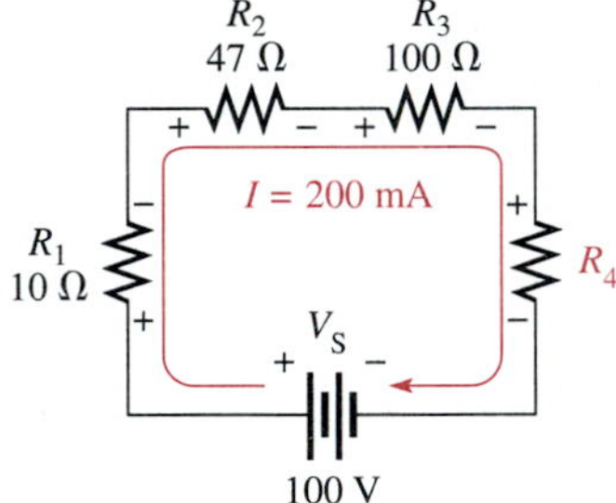

FIGURE 5–36

***Solution*** In this problem we will use both Ohm's law and Kirchhoff's voltage law.

First, use Ohm's law to find the voltage drop across each of the known resistors.

$$V_1 = IR_1 = (200\text{ mA})(10\ \Omega) = 2\text{ V}$$
$$V_2 = IR_2 = (200\text{ mA})(47\ \Omega) = 9.4\text{ V}$$
$$V_3 = IR_3 = (200\text{ mA})(100\ \Omega) = 20\text{ V}$$

Next, use Kirchhoff's voltage law to find $V_4$, the voltage drop across the unknown resistor.

$$V_S - V_1 - V_2 - V_3 - V_4 = 0\text{ V}$$
$$100\text{ V} - 2\text{ V} - 9.4\text{ V} - 20\text{ V} - V_4 = 0\text{ V}$$
$$68.6\text{ V} - V_4 = 0\text{ V}$$
$$V_4 = 68.6\text{ V}$$

Now that you know $V_4$, use Ohm's law to calculate $R_4$ as follows:

$$R_4 = \frac{V_4}{I} = \frac{68.6\text{ V}}{200\text{ mA}} = 343\ \Omega$$

This is most likely a 330 Ω resistor because 343 Ω is within a standard tolerance range (+5%) of 330 Ω.

**Related Exercise** Determine $R_4$ in Figure 5–36 for $V_S = 150$ V and $I = 200$ mA.

**SECTION 5–6 REVIEW**

1. State Kirchhoff's voltage law in two ways.
2. A 50 V source is connected to a series resistive circuit. What is the total of the voltage drops in this circuit?
3. Two equal-value resistors are connected in series across a 10 V battery. What is the voltage drop across each resistor?
4. In a series circuit with a 25 V source, there are three resistors. One voltage drop is 5 V, and the other is 10 V. What is the value of the third voltage drop?
5. The individual voltage drops in a series string are as follows: 1 V, 3 V, 5 V, 8 V, and 7 V. What is the total voltage applied across the series string?

## 5–7 ■ VOLTAGE DIVIDERS

***A series circuit acts as a voltage divider. You will learn what this term means and why voltage dividers are an important application of series circuits.***

***After completing this section, you should be able to***

- **Use a series circuit as a voltage divider**
  - Apply the voltage-divider formula
  - Use the potentiometer as an adjustable voltage divider
  - Describe some voltage-divider applications.

To illustrate how a series string of resistors acts as a voltage divider, we will examine Figure 5–37 where there are two resistors in series. There are two voltage drops: one across $R_1$ and one across $R_2$. These voltage drops are $V_1$ and $V_2$, respectively, as indicated in the diagram.

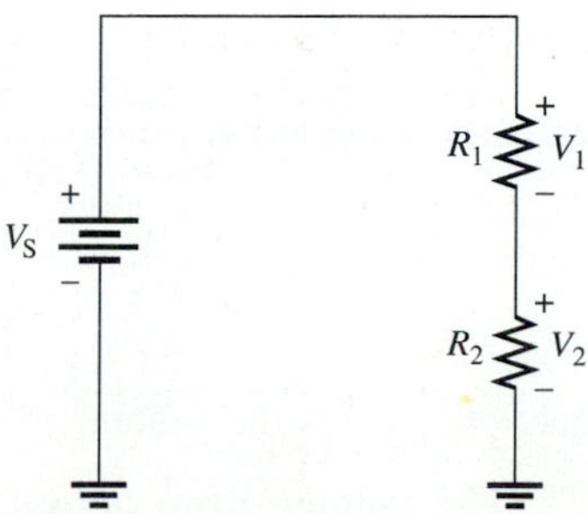


FIGURE 5–37
*Two-resistor voltage divider.*

Since each resistor has the same current, the voltage drops are proportional to the resistance values. For example, if the value of $R_2$ is twice that of $R_1$, then the value of $V_2$

is twice that of $V_1$. In other words, *the total voltage drop divides among the series resistors in amounts directly proportional to the resistance values.*

For example, in Figure 5–37, if $V_S$ is 10 V, $R_1$ is 50 Ω, and $R_2$ is 100 Ω, then $V_1$ is one-third the total voltage, or 3.33 V, because $R_1$ is one-third the total resistance (150 Ω). Likewise, $V_2$ is two-thirds $V_S$, or 6.67 V.

## Voltage-Divider Formula

With a few steps, a formula for determining how the voltages divide among series resistors can be developed. Let's assume that we have several resistors in series as shown in Figure 5–38. This figure shows five resistors as an example, but there can be any number.

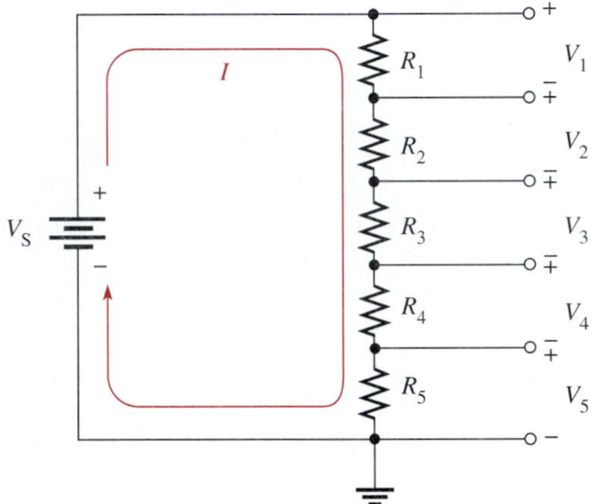


**FIGURE 5–38**
***Five-resistor voltage divider.***

Let's call the voltage drop across any one of the resistors $V_x$, where $x$ represents the number of a particular resistor (1, 2, 3, and so on). By Ohm's law, the voltage drop across any of the resistors in Figure 5–38 can be written as follows:

$$V_x = IR_x$$

where $x$ = 1, 2, 3, 4, or 5.

The current is equal to the source voltage divided by the total resistance ($I = V_S/R_T$). For the example circuit of Figure 5–38, the total resistance is $R_1 + R_2 + R_3 + R_4 + R_5$. Substituting $V_S/R_T$ for $I$ in the expression for $V_x$ results in

$$V_x = \left(\frac{V_S}{R_T}\right)R_x$$

Rearranging the terms yields

$$V_x = \left(\frac{R_x}{R_T}\right)V_S \qquad \textbf{(5–5)}$$

Equation (5–5) is the general voltage-divider formula. It can be stated as follows:

**The voltage drop across any resistor or combination of resistors in a series circuit is equal to the ratio of that resistance value to the total resistance, multiplied by the source voltage.**

The following three examples illustrate use of the voltage-divider formula.

**EXAMPLE 5–16** Determine $V_1$ (the voltage across $R_1$) and $V_2$ (the voltage across $R_2$) in the voltage divider in Figure 5–39.

FIGURE 5–39

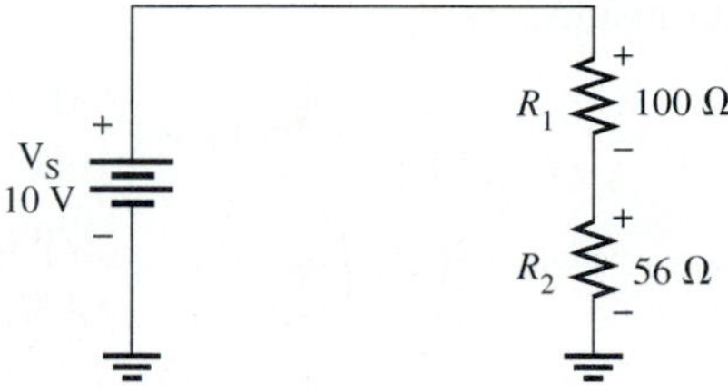


*Solution* To determine $V_1$, use the voltage-divider formula, $V_x = (R_x/R_T)V_S$, where $x = 1$. The total resistance is

$$R_T = R_1 + R_2 = 100\ \Omega + 56\ \Omega = 156\ \Omega$$

$R_1$ is 100 Ω and $V_S$ is 10 V. Substitute these values into the voltage-divider formula.

$$V_1 = \left(\frac{R_1}{R_T}\right)V_S = \left(\frac{100\ \Omega}{156\ \Omega}\right)10\ \text{V} = 6.41\ \text{V}$$

There are two ways to find the value of $V_2$: Kirchhoff's voltage law or the voltage-divider formula. If you use Kirchhoff's voltage law ($V_S = V_1 + V_2$), substitute the values for $V_S$ and $V_1$ as follows:

$$V_2 = V_S - V_1 = 10\ \text{V} - 6.41\ \text{V} = 3.59\ \text{V}$$

A second way to find $V_2$ is to use the voltage-divider formula where $x = 2$.

$$V_2 = \left(\frac{R_2}{R_T}\right)V_S = \left(\frac{56\ \Omega}{156\ \Omega}\right)10\ \text{V} = 3.59\ \text{V}$$

You get the same result either way.

*Related Exercise* Find the voltages across $R_1$ and $R_2$ in Figure 5–39 if $R_2$ is changed to 180 Ω.

**EXAMPLE 5–17** Calculate the voltage drop across each resistor in the voltage divider of Figure 5–40.

FIGURE 5–40

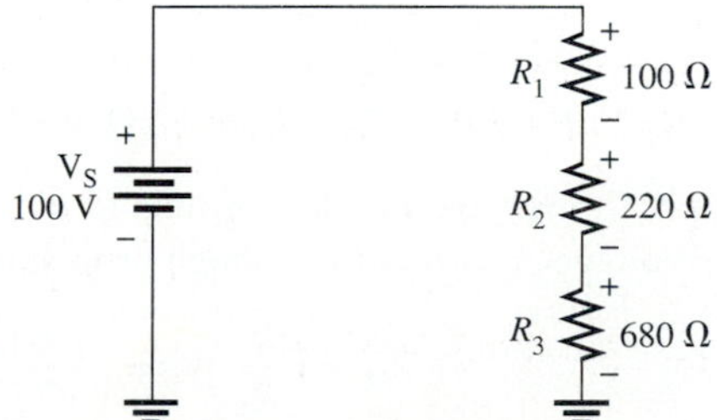


*Solution* Look at the circuit for a moment and consider the following: The total resistance is 1000 Ω. Ten percent of the total voltage is across $R_1$ because it is 10% of the total resistance (100 Ω is 10% of 1000 Ω). Likewise, 22% of the total voltage is dropped across $R_2$ because it is 22% of the total resistance (220 Ω is 22% of 1000 Ω). Finally, $R_3$ drops 68% of the total voltage because 680 Ω is 68% of 1000 Ω.

Because of the convenient values in this problem, it is easy to figure the voltages mentally. ($V_1 = 0.10 \times 100\ \text{V} = 10\ \text{V}$, $V_2 = 0.22 \times 100\ \text{V} = 22\ \text{V}$, and $V_3 = 0.68 \times 100\ \text{V} = 68\ \text{V}$). Such is not always the case, but sometimes a little thinking will produce a result more efficiently and eliminate some calculating. This is also a good way to roughly estimate what your results should be so that you will recognize an unreasonable answer as a result of a calculation error.

Although you have already reasoned through this problem, the calculations will verify your results.

$$V_1 = \left(\frac{R_1}{R_T}\right)V_S = \left(\frac{100\ \Omega}{1000\ \Omega}\right)100\ \text{V} = 10\ \text{V}$$

$$V_2 = \left(\frac{R_2}{R_T}\right)V_S = \left(\frac{220\ \Omega}{1000\ \Omega}\right)100\ \text{V} = 22\ \text{V}$$

$$V_3 = \left(\frac{R_3}{R_T}\right)V_S = \left(\frac{680\ \Omega}{1000\ \Omega}\right)100\ \text{V} = 68\ \text{V}$$

Notice that the sum of the voltage drops is equal to the source voltage, in accordance with Kirchhoff's voltage law. This check is a good way to verify your results.

***Related Exercise*** If $R_1$ and $R_2$ in Figure 5–40 are changed to 680 Ω, what are the voltage drops?

**EXAMPLE 5–18** Determine the voltages between the following points in the voltage divider of Figure 5–41:

**(a)** *A* to *B* **(b)** *A* to *C* **(c)** *B* to *C* **(d)** *B* to *D* **(e)** *C* to *D*

**FIGURE 5–41**

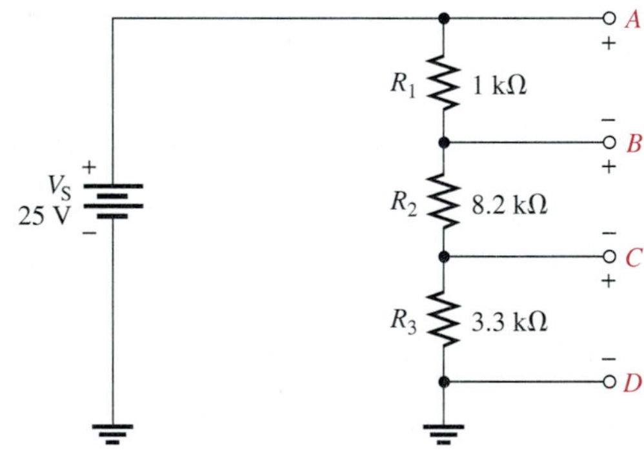


***Solution*** First determine $R_T$.

$$R_T = R_1 + R_2 + R_3 = 1\ \text{k}\Omega + 8.2\ \text{k}\Omega + 3.3\ \text{k}\Omega = 12.5\ \text{k}\Omega$$

Now apply the voltage-divider formula to obtain each required voltage.

**(a)** The voltage *A* to *B* is the voltage drop across $R_1$.

$$V_{AB} = \left(\frac{R_1}{R_T}\right)V_S = \left(\frac{1\ \text{k}\Omega}{12.5\ \text{k}\Omega}\right)25\ \text{V} = 2\ \text{V}$$

**(b)** The voltage from *A* to *C* is the combined voltage drop across both $R_1$ and $R_2$. In this case, $R_x$ in the general formula given in Equation (5–5) is $R_1 + R_2$.

$$V_{AC} = \left(\frac{R_1 + R_2}{R_T}\right)V_S = \left(\frac{9.2\ \text{k}\Omega}{12.5\ \text{k}\Omega}\right)25\ \text{V} = 18.4\ \text{V}$$

**(c)** The voltage from *B* to *C* is the voltage drop across $R_2$.

$$V_{BC} = \left(\frac{R_2}{R_T}\right)V_S = \left(\frac{8.2\ \text{k}\Omega}{12.5\ \text{k}\Omega}\right)25\ \text{V} = 16.4\ \text{V}$$

**(d)** The voltage from $B$ to $D$ is the combined voltage drop across both $R_2$ and $R_3$. In this case, $R_x$ in the general formula is $R_2 + R_3$.

$$V_{BD} = \left(\frac{R_2 + R_3}{R_T}\right)V_S = \left(\frac{11.5\text{ k}\Omega}{12.5\text{ k}\Omega}\right)25\text{ V} = 23\text{ V}$$

**(e)** Finally, the voltage from $C$ to $D$ is the voltage drop across $R_3$.

$$V_{CD} = \left(\frac{R_3}{R_T}\right)V_S = \left(\frac{3.3\text{ k}\Omega}{12.5\text{ k}\Omega}\right)25\text{ V} = 6.6\text{ V}$$

If you connect this voltage divider in the lab, you can verify each of the calculated voltages by connecting a voltmeter between the appropriate points in each case.

***Related Exercise*** Determine each of the previously calculated voltages if $V_S$ is doubled.

## The Potentiometer as an Adjustable Voltage Divider

Recall from Chapter 2 that a potentiometer is a variable resistor with three terminals. A potentiometer connected to a voltage source is shown in Figure 5–42(a) with the schematic shown in part (b). Notice that the two end terminals are labeled 1 and 2. The adjustable terminal or wiper is labeled 3. The potentiometer acts as a voltage divider, which can be illustrated by separating the total resistance into two parts, as shown in Figure 5–42(c). The resistance between terminal 1 and terminal 3 ($R_{13}$) is one part, and the resistance between terminal 3 and terminal 2 ($R_{32}$) is the other part. So this potentiometer actually is a two-resistor voltage divider that can be manually adjusted.

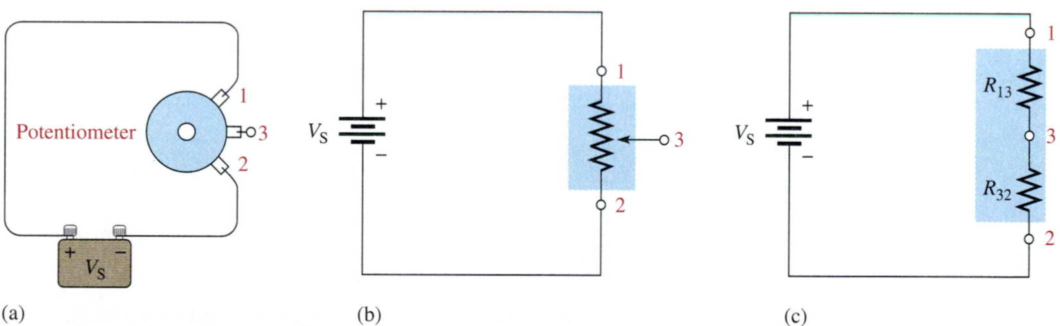


FIGURE 5–42
***The potentiometer as a voltage divider.***

Figure 5–43 shows what happens when the wiper terminal (3) is moved. In part (a) of Figure 5–43, the wiper is exactly centered, making the two resistances equal. If we measure the voltage across terminals 3 to 2 as indicated by the voltmeter symbol, we have one-half of the total source voltage. When the wiper is moved up, as in Figure 5–43(b), the resistance between terminals 3 and 2 increases, and the voltage across it increases proportionally. When the wiper is moved down, as in Figure 5–43(c), the resistance between terminals 3 and 2 decreases, and the voltage decreases proportionally.

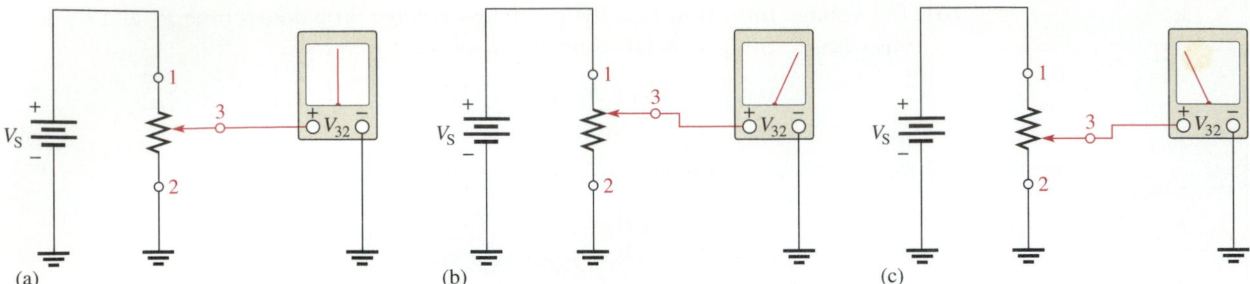


**FIGURE 5–43**
***Adjusting the voltage divider.***

## Applications of Voltage Dividers

The volume control of radio or TV receivers is a common application of a potentiometer used as a voltage divider. Since the loudness of the sound is dependent on the amount of voltage associated with the audio signal, you can increase or decrease the volume by adjusting the potentiometer, that is, by turning the knob of the volume control on the set. The block diagram in Figure 5–44 shows how a potentiometer can be used for volume control in a typical receiver.

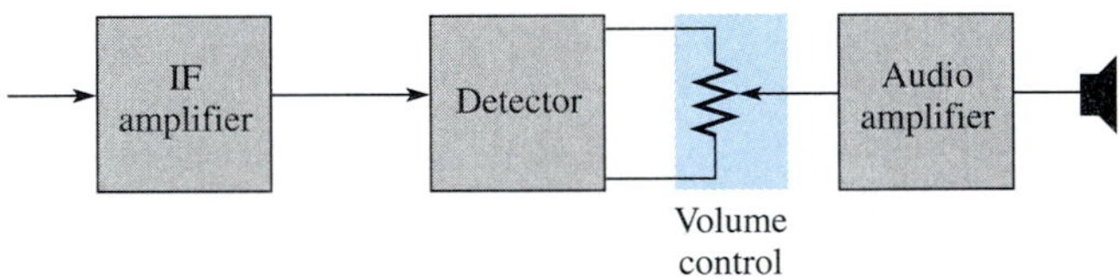


**FIGURE 5–44**
***A voltage divider used for volume control.***

Another application of a voltage divider is illustrated in Figure 5–45, which depicts a potentiometer voltage divider as a fuel-level sensor in an automobile gas tank. As shown in part (a), the float moves up as the tank is filled and moves down as the tank empties. The float is mechanically linked to the wiper arm of a potentiometer, as shown in part (b). The output voltage varies proportionally with the position of the wiper arm. As the fuel in the tank decreases, the sensor output voltage also decreases. The output voltage goes to the indicator circuitry, which controls the fuel gauge or digital readout to show the fuel level. The schematic of this system is shown in part (c).

Still another application for voltage dividers is in setting the dc operating voltage (bias) in transistor amplifiers. Figure 5–46 shows a voltage divider used for this purpose. You will study transistor amplifiers and biasing later, so it is important that you understand the basics of voltage dividers at this point.

These examples are only three out of many possible applications of voltage dividers.

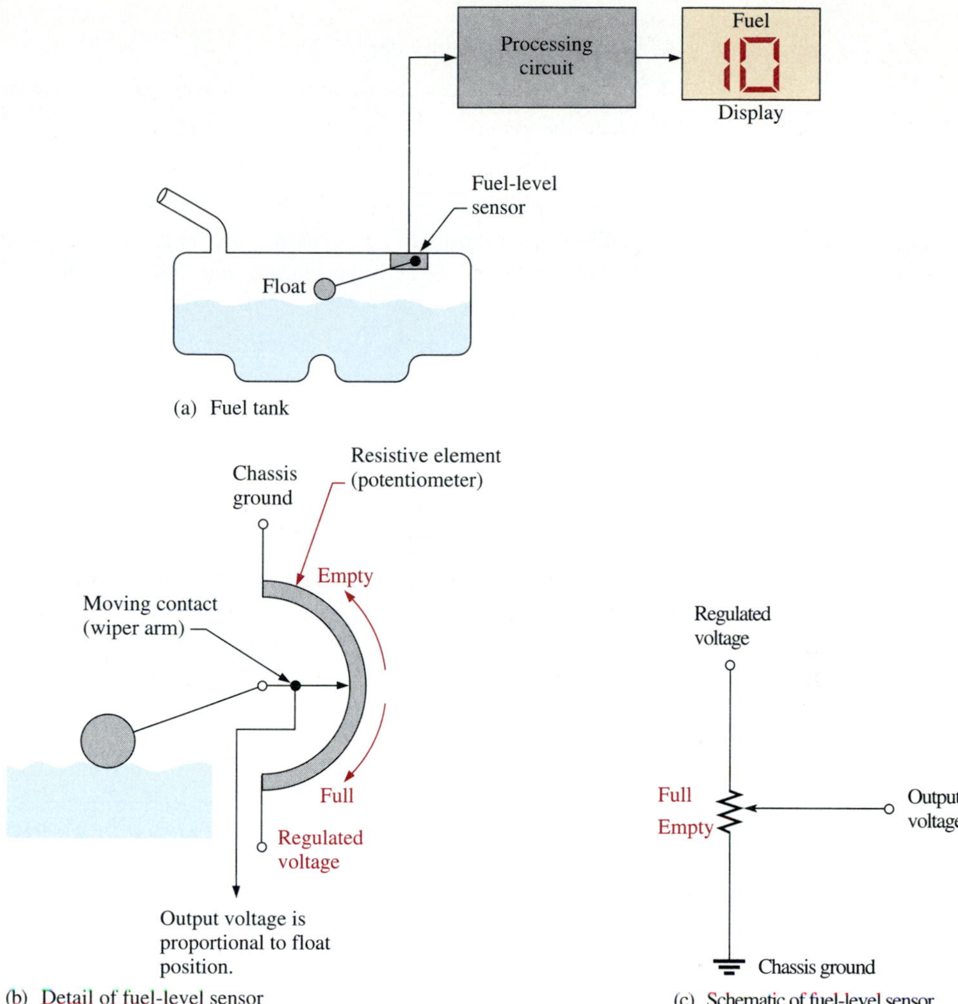


**FIGURE 5–45**

***A potentiometer voltage divider used as an automotive fuel-level sensor.***

**FIGURE 5–46**

***The voltage divider as a bias circuit for a transistor amplifier.***

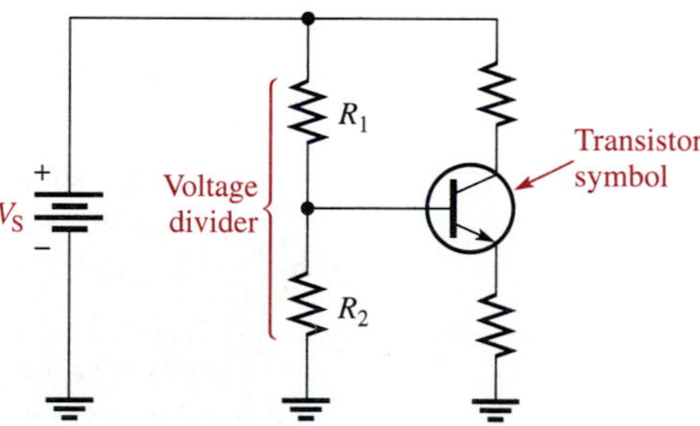

**SECTION 5–7 REVIEW**

1. What is a voltage divider?
2. How many resistors can there be in a series voltage-divider circuit?
3. Write the general formula for voltage dividers.
4. If two series resistors of equal value are connected across a 10 V source, how much voltage is there across each resistor?
5. A 47 Ω resistor and an 82 Ω resistor are connected as a voltage divider. The source voltage is 100 V. Sketch the circuit, and determine the voltage across each of the resistors.
6. The circuit of Figure 5–47 is an adjustable voltage divider. If the potentiometer is linear, where would you set the wiper in order to get 5 V from *A* to *B* and 5 V from *B* to *C*?

FIGURE 5–47

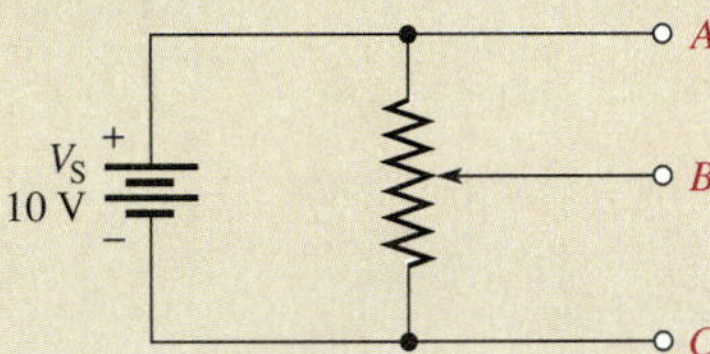


## 5–8 ■ POWER IN A SERIES CIRCUIT

***The power dissipated by each individual resistor in a series circuit contributes to the total power in the circuit. The individual powers are additive.***

***After completing this section, you should be able to***

- **Determine power in a series circuit**

The total amount of power in a series resistive circuit is equal to the sum of the powers in each resistor in series.

$$P_T = P_1 + P_2 + P_3 + \cdots + P_n \qquad \textbf{(5–6)}$$

where $P_T$ is the total power and $P_n$ is the power in the last resistor in series. In other words, the powers are additive.

The power formulas that you learned in Chapter 4 are, of course, directly applicable to series circuits. Since there is the same current through each resistor in series, the following formulas are used to calculate the total power:

$$P_T = V_S I$$
$$P_T = I^2 R_T$$
$$P_T = \frac{V_S^2}{R_T}$$

where $V_S$ is the total source voltage across the series connection and $R_T$ is the total resistance. Example 5–19 illustrates how to calculate total power in a series circuit.

**EXAMPLE 5–19** Determine the total amount of power in the series circuit in Figure 5–48.

**FIGURE 5–48**

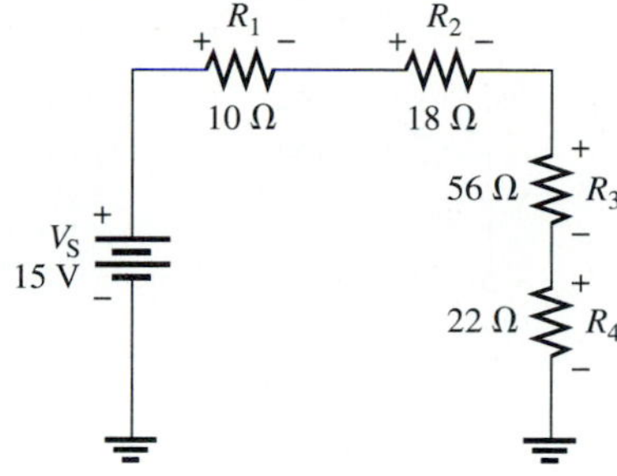


***Solution*** The source voltage is 15 V. The total resistance is

$$R_T = R_1 + R_2 + R_3 + R_4 = 10\ \Omega + 18\ \Omega + 56\ \Omega + 22\ \Omega = 106\ \Omega$$

The easiest formula to use is $P_T = V_S^2/R_T$ since you know both $V_S$ and $R_T$.

$$P_T = \frac{V_S^2}{R_T} = \frac{(15\text{ V})^2}{106\ \Omega} = \frac{225\text{ V}^2}{106\ \Omega} = 2.12\text{ W}$$

If you determine the power in each resistor separately and all of these powers are added, you obtain the same result. Another calculation will illustrate. First, find the current.

$$I = \frac{V_S}{R_T} = \frac{15\text{ V}}{106\ \Omega} = 142\text{ mA}$$

Next, calculate the power for each resistor using $P = I^2R$.

$$P_1 = I^2R_1 = (142\text{ mA})^2(10\ \Omega) = 0.202\text{ W}$$
$$P_2 = I^2R_2 = (142\text{ mA})^2(18\ \Omega) = 0.363\text{ W}$$
$$P_3 = I^2R_3 = (142\text{ mA})^2(56\ \Omega) = 1.13\text{ W}$$
$$P_4 = I^2R_4 = (142\text{ mA})^2(22\ \Omega) = 0.444\text{ W}$$

Now, add these powers to get the total power.

$$P_T = P_1 + P_2 + P_3 + P_4 = 0.202\text{ W} + 0.363\text{ W} + 1.13\text{ W} + 0.444\text{ W} = 2.14\text{ W}$$

This result compares closely to the total power as determined previously by the formula $P_T = V_S^2/R_T$. The small difference is due to rounding.

***Related Exercise*** What is the power in the circuit of Figure 5–48 if $V_S$ is increased to 30 V?

As you have learned, the amount of power in a resistor is important because the power rating of the resistors must be high enough to handle the expected power in the circuit. The following example illustrates practical considerations relating to power in a series circuit.

**EXAMPLE 5–20** Determine if the indicated power rating (½ W) of each resistor in Figure 5–49 is sufficient to handle the actual power. If a rating is not adequate, specify the required minimum rating.

FIGURE 5–49

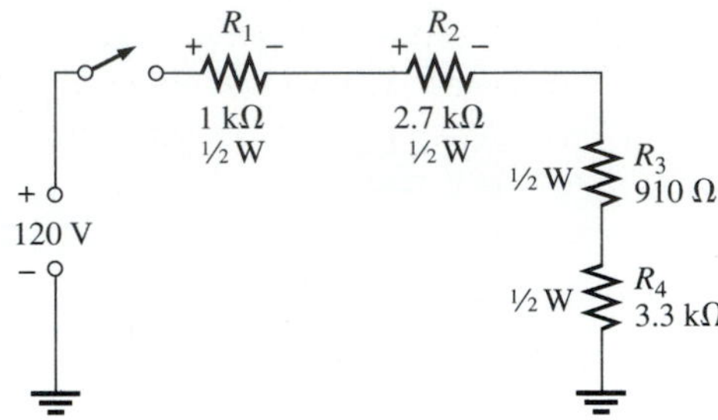


***Solution*** First determine the total resistance. Then use Ohm's law to calculate the current.

$$R_T = R_1 + R_2 + R_3 + R_4 = 1\text{ k}\Omega + 2.7\text{ k}\Omega + 910\ \Omega + 3.3\text{ k}\Omega = 7.91\text{ k}\Omega$$

$$I = \frac{V_S}{R_T} = \frac{120\text{ V}}{7.91\text{ k}\Omega} = 15.2\text{ mA}$$

The power in each resistor is

$$P_1 = I^2R_1 = (15.2\text{ mA})^2(1\text{ k}\Omega) = 231\text{ mW}$$
$$P_2 = I^2R_2 = (15.2\text{ mA})^2(2.7\text{ k}\Omega) = 624\text{ mW}$$
$$P_3 = I^2R_3 = (15.2\text{ mA})^2(910\ \Omega) = 210\text{ mW}$$
$$P_4 = I^2R_4 = (15.2\text{ mA})^2(3.3\text{ k}\Omega) = 762\text{ mW}$$

$R_2$ and $R_4$ do not have a rating sufficient to handle the actual power, which exceeds ½ W in each of these two resistors, and they will burn out if the switch is closed. These resistors must be replaced by 1 W resistors.

***Related Exercise*** Determine the minimum power rating required for each resistor in Figure 5–49 if the source voltage is increased to 240 V.

**SECTION 5–8 REVIEW**

1. If you know the power in each resistor in a series circuit, how can you find the total power?
2. The resistors in a series circuit dissipate the following powers: 2 W, 5 W, 1 W, and 8 W. What is the total power in the circuit?
3. A circuit has a 100 Ω, a 330 Ω, and a 680 Ω resistor in series. There is a current of 1 A through the circuit. What is the total power?

## 5–9 ■ CIRCUIT GROUND

***Voltage is relative. That is, the voltage at one point in a circuit is always measured relative to another point. For example, if we say that there are +100 V at a certain point in a circuit, we mean that the point is 100 V more positive than some designated reference point in the circuit. This reference point is usually the ground point.***

***After completing this section, you should be able to***

■ **Determine and identify ground in a circuit**
- ☐ Measure voltage with respect to ground
- ☐ Define the term *circuit ground*

The concept of **ground** was introduced in Chapter 2. In most electronic equipment, a large conductive area on a printed circuit board or the metal housing is used as the common or reference point, called **circuit ground** or *chassis ground*, as illustrated in Figure 5–50.

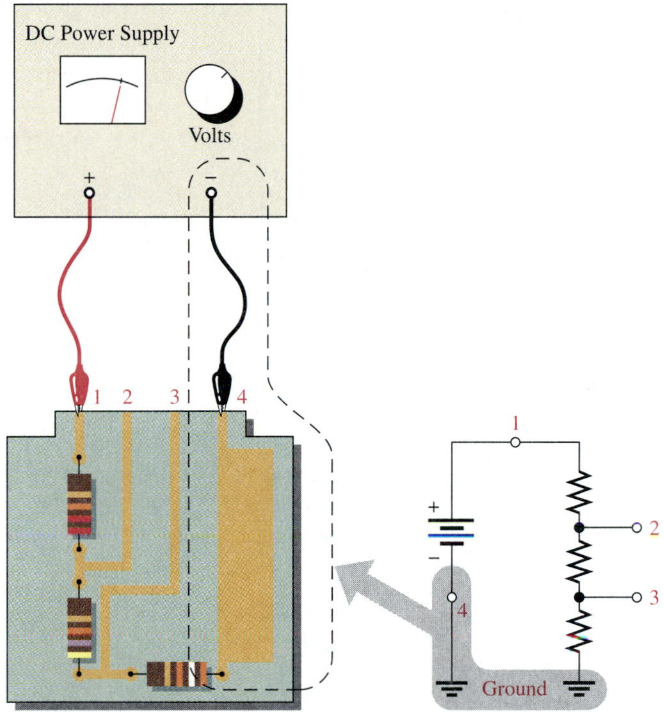


FIGURE 5–50
*Simple illustration of circuit ground.*

Ground has a potential of zero volts (0 V) with respect to all other points in the circuit that are referenced to it, as illustrated in Figure 5–51. In part (a), the negative side of the source is grounded, and all voltages indicated are positive with respect to ground. In part (b), the positive side of the source is ground. The voltages at all other

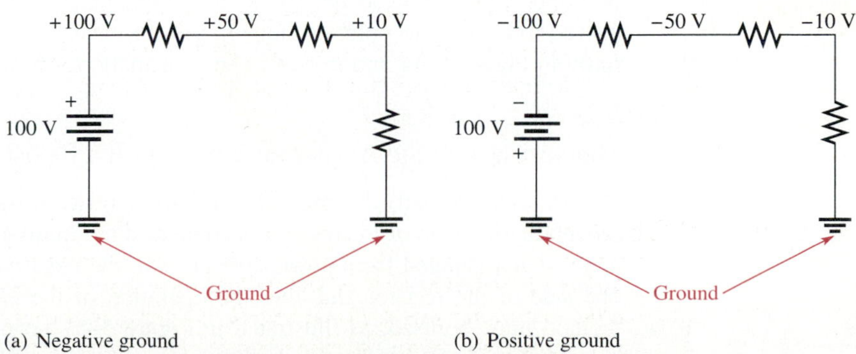


FIGURE 5–51
*Example of negative and positive grounds.*

points are therefore negative with respect to ground. Recall that all points shown grounded in a circuit are connected together through ground and are effectively the same point electrically.

## Measuring Voltages with Respect to Ground

When voltages are measured with respect to ground in a circuit, one meter lead is connected to the circuit ground, and the other to the point at which the voltage is to be measured. In a negative ground circuit, the negative meter terminal is connected to the circuit ground. The positive terminal of the voltmeter is then connected to the positive voltage point. Measurement of positive voltage is illustrated in Figure 5–52, where the meter reads the voltage at point *A* with respect to ground.

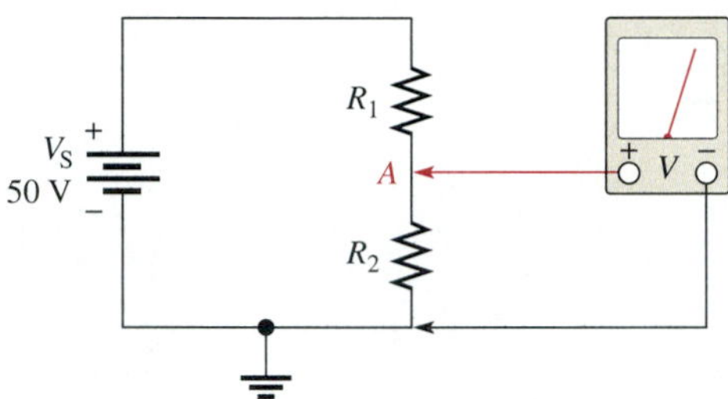


**FIGURE 5–52**
***Measuring a voltage with respect to negative ground.***

For a circuit with a positive ground, the positive voltmeter lead is connected to ground, and the negative lead is connected to the negative voltage point, as indicated in Figure 5–53. Here the meter reads the voltage at point *A* with respect to ground.

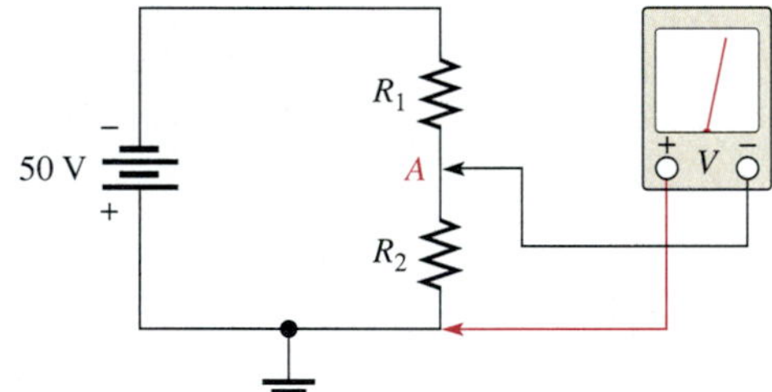


**FIGURE 5–53**
***Measuring a voltage with respect to positive ground.***

When voltages must be measured at several points in a circuit, the ground lead can be clipped to ground at one point in the circuit and left there. The other lead is then moved from point to point as the voltages are measured. This method is illustrated pictorially in Figure 5–54 and in equivalent schematic form in Figure 5–55.

## Measuring Voltage Across an Ungrounded Resistor

Voltage can normally be measured across a resistor, as shown in Figure 5–56, even though neither side of the resistor is connected to circuit ground. In some cases, when the meter is not isolated from power line ground, the negative lead of the meter will ground one side of the resistor and alter the operation of the circuit. In this situation, another method must be used, as illustrated in Figure 5–57 on page 148. The voltages on each side of the resistor are measured with respect to ground. The difference of these two measurements is the voltage drop across the resistor.

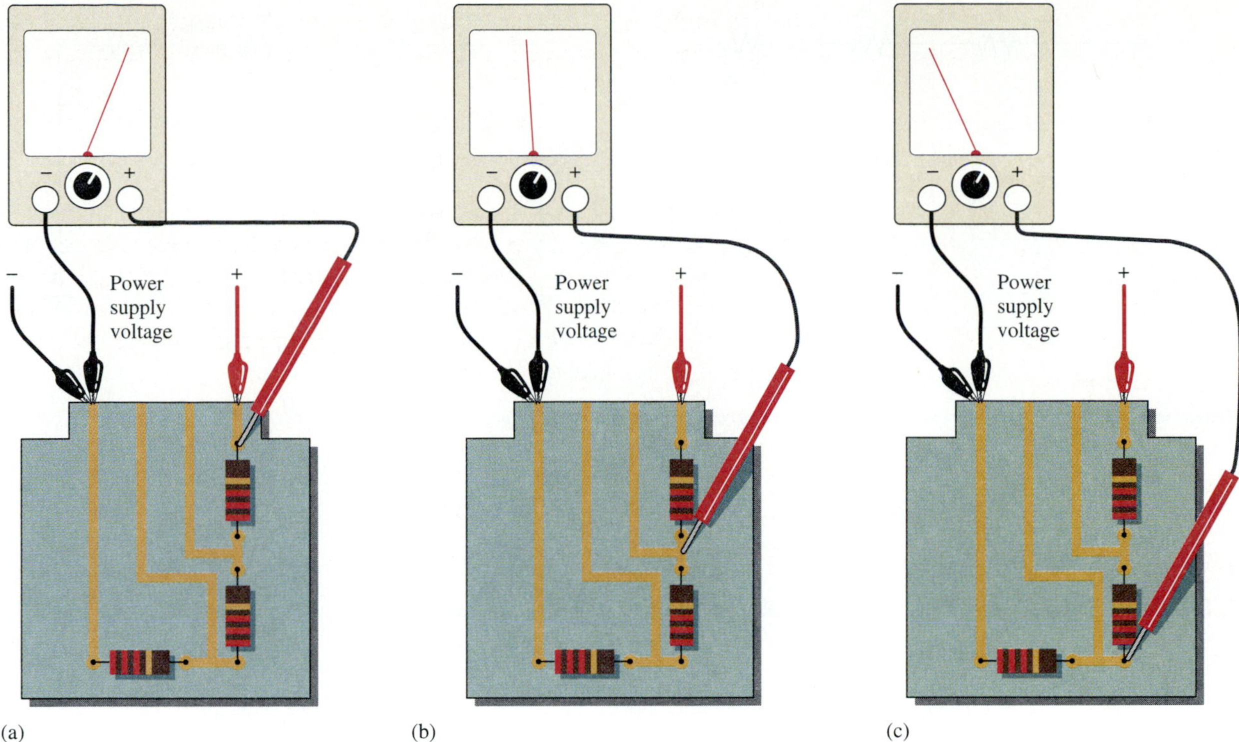


**FIGURE 5–54**
*Measuring voltages at several points in a circuit with respect to ground.*

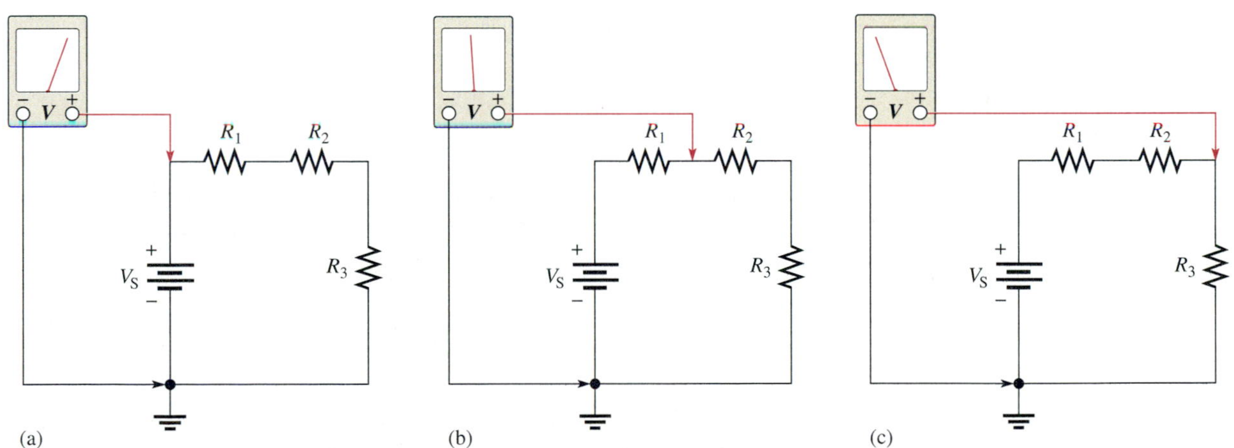


**FIGURE 5–55**
*Equivalent schematics for Figure 5–54.*

**FIGURE 5–56**
*Measuring voltage across a resistor.*

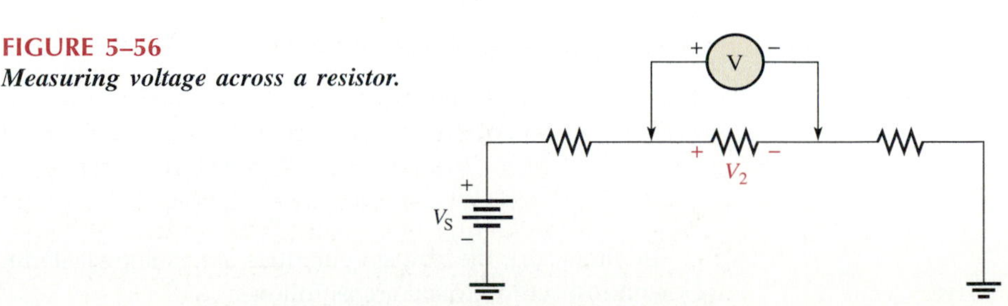

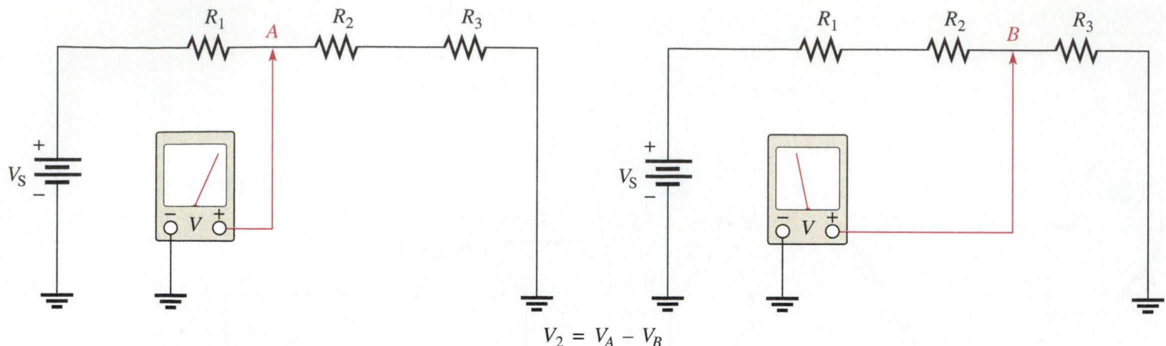


FIGURE 5–57
***Measuring voltage across a resistor with two separate measurements to ground.***

**EXAMPLE 5–21** Determine the voltages of each of the indicated points in each circuit of Figure 5–58. Assume that 25 V are dropped across each resistor.

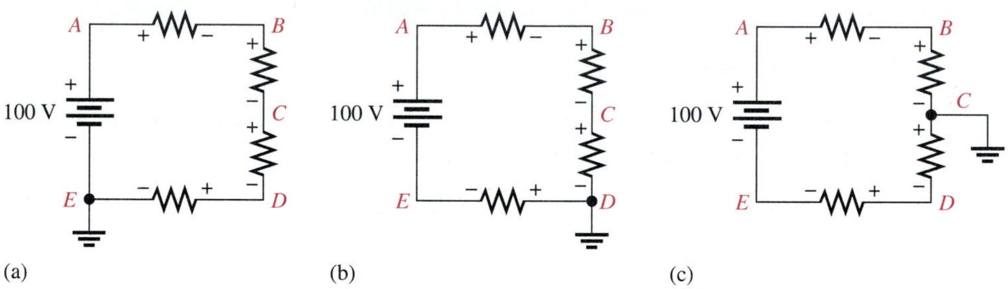


FIGURE 5–58

***Solution*** In circuit (a), the voltage polarities are as shown. Point $E$ is ground. Single-letter subscripts denote voltage at a point with respect to ground. The voltages with respect to ground are as follows:

$$V_E = 0\text{ V}$$
$$V_D = +25\text{ V}$$
$$V_C = +50\text{ V}$$
$$V_B = +75\text{ V}$$
$$V_A = +100\text{ V}$$

In circuit (b), the voltage polarities are as shown. Point $D$ is ground. The voltages with respect to ground are as follows:

$$V_E = -25\text{ V}$$
$$V_D = 0\text{ V}$$
$$V_C = +25\text{ V}$$
$$V_B = +50\text{ V}$$
$$V_A = +75\text{ V}$$

In circuit (c), the voltage polarities are as shown. Point $C$ is ground. The voltages with respect to ground are as follows:

$$V_E = -50\text{ V}$$
$$V_D = -25\text{ V}$$
$$V_C = 0\text{ V}$$
$$V_B = +25\text{ V}$$
$$V_A = +50\text{ V}$$

***Related Exercise*** If the ground is at point $A$ in the circuit in Figure 5–58, what are the voltages at each of the points with respect to ground?

**SECTION 5–9 REVIEW**

1. What is the common reference point in a circuit called?
2. Voltages in a circuit are generally referenced to ground (T or F).
3. The housing or chassis is often used as circuit ground (T or F).
4. What is the symbol for ground?

## 5–10 ■ TROUBLESHOOTING

***Open resistors or contacts and one point shorted to another are common problems in all circuits including series circuits.***

***After completing this section, you should be able to***

- **Troubleshoot series circuits**
  - ☐ Check for an open circuit
  - ☐ Check for a short circuit
  - ☐ Identify primary causes of shorts

### Open Circuit

The most common failure in a series circuit is an open. For example, when a resistor or a lamp burns out, it causes a break in the current path and creates an **open circuit** as illustrated in Figure 5–59.

**An open in a series circuit prevents current.**

***Checking for an Open Element*** Sometimes a visual check will reveal a charred resistor or an open lamp filament. However, it is possible for a resistor to open without showing visible signs of damage. In this situation, a voltage check of the series circuit is required. The general procedure is as follows: *Measure the voltage across each resistor in series. The voltage across all of the good resistors will be zero. The voltage across the open resistor will equal the source voltage.*

This condition occurs because an open resistor will prevent current through the series circuit. With no current, there can be no voltage drop across any of the good resistors. Since $IR = (0\text{ A})R = 0\text{ V}$, in accordance with Ohm's law, the voltage on each side of a good resistor is the same. The total voltage must then appear across the open resistor in accordance with Kirchhoff's voltage law, as illustrated in Figure 5–60. To fix the circuit, replace the open resistor.

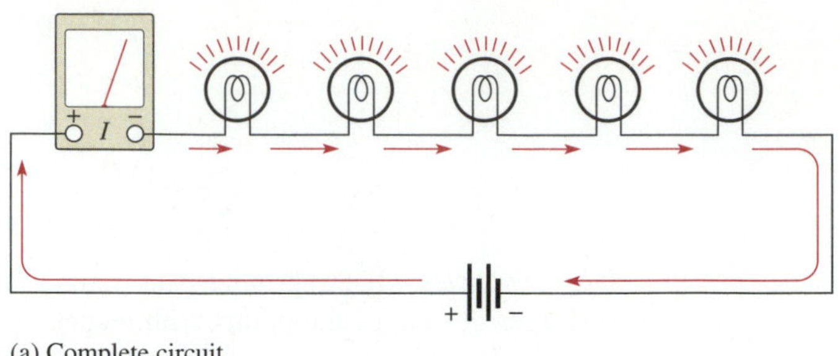


(a) Complete circuit

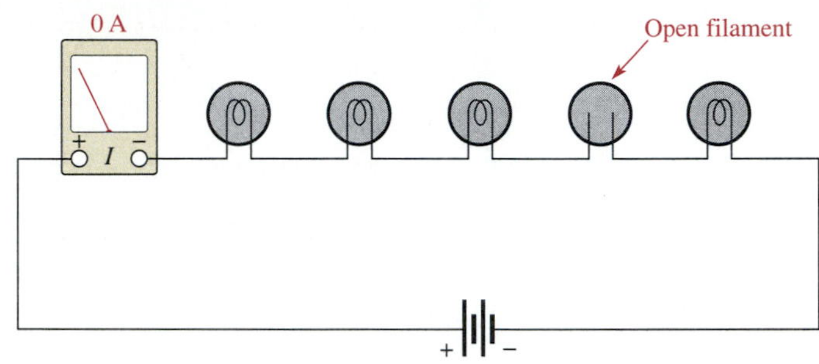


(b) Open circuit

FIGURE 5–59

***An open circuit prevents current.***

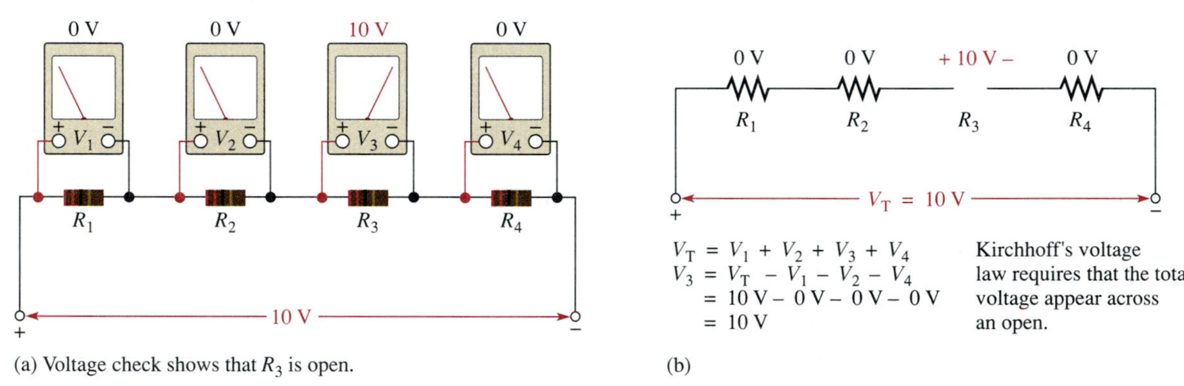


(a) Voltage check shows that $R_3$ is open.

(b)

FIGURE 5–60

***Troubleshooting a series circuit for an open element.***

## Short Circuit

Sometimes an unwanted **short circuit** occurs when two conductors touch or a foreign object such as solder or a wire clipping accidentally connects two sections of a circuit together. This situation is particularly common in circuits with a high component density. Several potential causes of short circuits are illustrated on the PC board in Figure 5–61.

When there is a short, a portion of the series resistance is bypassed (all of the current goes through the short), thus reducing the total resistance as illustrated in Figure 5–62. Notice that the current increases as a result of the short.

**A short in a series circuit causes more current.**

**FIGURE 5–61**
*Examples of shorts on a PC board.*

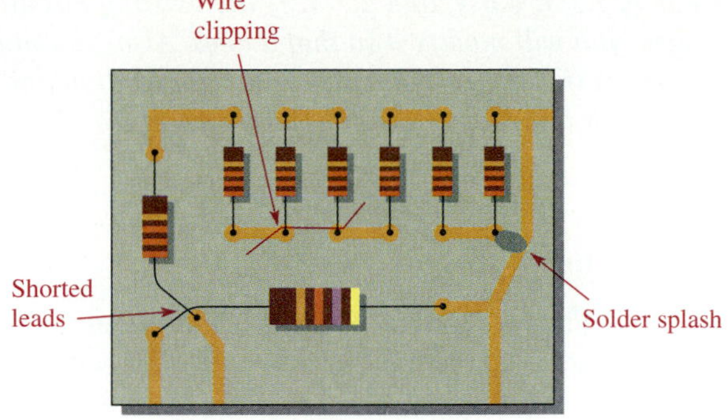


**FIGURE 5–62**
*Example of the effect of a short in a series circuit.*

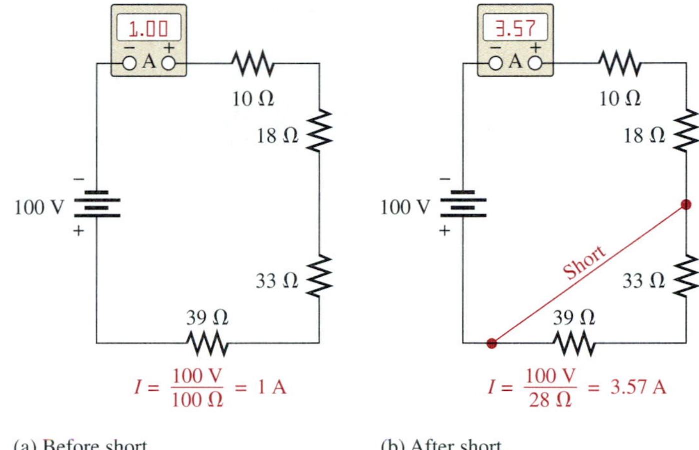


**SECTION 5–10 REVIEW**

1. Define *short circuit.*
2. Define *open circuit.*
3. What happens when a series circuit opens?
4. Name two general ways in which an open circuit can occur in practice. What may cause a short circuit to occur?
5. When a resistor fails, it will normally fail open (T or F).
6. The total voltage across a string of series resistors is 24 V. If one of the resistors is open, how much voltage is there across it? How much is there across each of the good resistors?

## 5–11 ■ TECHnology Theory Into Practice

***For this TECH TIP assignment, your supervisor has given you a voltage-divider board to evaluate and modify if necessary. It will be used to obtain five different voltage levels from a 12 V battery that has a 6.5 Ah rating. The voltage divider is to be used to provide positive reference voltages to an electronic circuit in an analog-to-digital converter. Your task will be to check the circuit to see if it provides the following voltages within a tolerance of ±5% with respect to the negative side of the battery: 10.4 V,***

***8.0 V, 7.3 V, 6.0 V, and 2.7 V. If the existing circuit does not provide the specified voltages, you will modify it so that it does. Also, you must make sure that the power ratings of the resistors are adequate for the application and determine how long the battery will last with the voltage divider connected to it.***

## The Schematic of the Circuit

☐ Use Figure 5–63 to determine the resistor values and draw the schematic of the voltage-divider circuit so you will know what you are working with. All the resistors on the board are ¼ W.

FIGURE 5–63

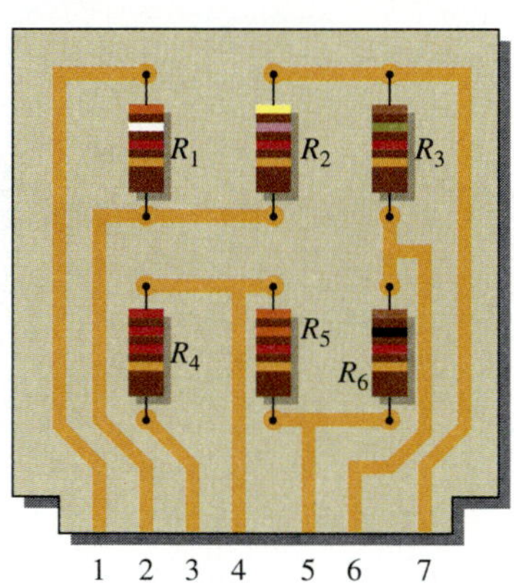


## The Voltages

☐ Determine the voltage at each pin on the circuit board with respect to the negative side of the battery when the positive side of the 12 V battery is connected to pin 3 and the negative side is connected to pin 1. Compare the existing voltages to the following specifications:

Pin 1: negative terminal of 12 V battery

Pin 2: 2.7 V ± 5%

Pin 3: positive terminal of 12 V battery

Pin 4: 10.4 V ± 5%

Pin 5: 8.0 V ± 5%

Pin 6: 7.3 V ± 5%

Pin 7: 6.0 V ± 5%

☐ If the output voltages of the existing circuit are not the same as those stated in the specifications, make the necessary changes in the circuit to meet the specifications. Draw a schematic of the modified circuit showing resistor values and adequate power ratings.

## The Battery

☐ Find the total current drawn from the 12 V battery when the voltage-divider circuit is connected and determine how many days the 6.5 Ah battery will last.

### A Test Procedure

☐ Determine how you would test the voltage-divider board and what instruments you would use. Then detail your test procedure in a step-by-step format.

### Troubleshooting

☐ Determine the most likely fault for each of the following cases (voltages are with respect to the negative battery terminal (pin 1 on the circuit board):

1. No voltage at any of the pins on the circuit board
2. 12 V at pins 3 and 4. All other pins have 0 V.
3. 12 V at all pins except 0 V at pin 1
4. 12 V at pin 6 and 0 V at pin 7
5. 3.3 V at pin 2

**SECTION 5–11 REVIEW**

1. What is the total power dissipated by the voltage-divider circuit with a 12 V battery?
2. What are the output voltages from the voltage divider if a 6 V battery is used?
3. When the voltage-divider board is connected to the electronic circuit to which it is providing positive reference voltages, which pin on the board should be connected to the ground of the electronic circuit?

## 5–12 ■ PSpice DC ANALYSIS

***In this section, you will learn two additional PSpice control statements: the .OP statement and the .PLOT statement. You will also learn how to break long statement lines into several shorter ones using continuation lines.***

***After completing this section, you should be able to***

- **Use PSpice to analyze series circuits and to graph the results**
  - ☐ Use the .OP statement in a circuit file
  - ☐ Use the .PLOT statement in a circuit file
  - ☐ Use the line continuation symbol in a statement
  - ☐ Write a circuit file to analyze a voltage-divider circuit

### The .OP (DC Operating or Bias Point) Control Statement

PSpice will always calculate all dc operating (bias point) voltages and voltage source currents for a specified circuit, but it will only list these in the output file when there are no other analysis control statements such as .DC in the circuit file. For example, if your circuit file asks for the voltage and current for only one resistor in a circuit with several resistors, PSpice will list only that single voltage and current in the output file although it calculates all of the dc operating voltages.

Sometimes you may want to see all dc node voltages in addition to the values specified in .DC statements. The .OP statement is used to tell PSpice to include the dc operating voltages in the output file. For now you can consider the dc operating voltages to be the dc node voltages, although this isn't entirely correct as you will learn when ac circuits are discussed.

For the voltage-divider circuit in Figure 5–64, the following circuit file will result in the voltages across each of the resistors being printed in the output file.

```
Voltage Divider
*Figure 5–64
VS  1  0  12
R1  1  2  1K
R2  2  3  560
R3  3  4  2.2K
R4  4  0  1.5K
.DC  VS  12  12  1
.PRINT  DC  V(R1)  V(R2)  V(R3)  V(R4)
.END
```

If the circuit file is modified to include

```
.OP
```

PSpice will also print the values of V(1), V(2), V(3), and V(4) in the output file. These are the voltages at the specified nodes with respect to ground. PSpice will also print the value of the current through the voltage source and power dissipation of the circuit.

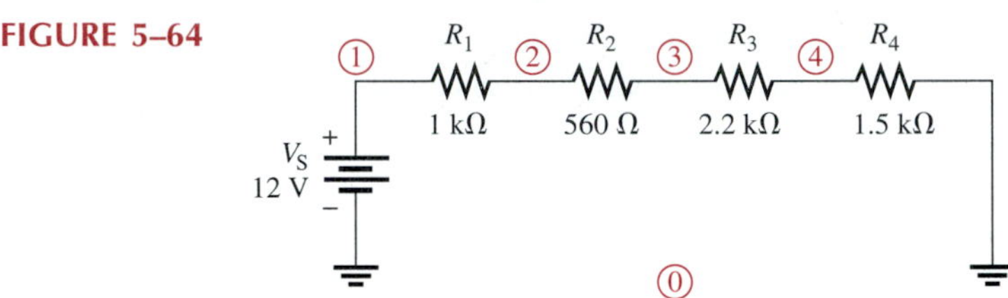


**FIGURE 5–64**

## The .PLOT Control Statement

Sometimes you may need to know how a circuit responds to sources that are not constant or the values of the circuit voltages and currents for several different source values. While you can use .PRINT to find the values, long tabular listings are not very convenient. This is especially true if you want to determine how the voltages and currents vary as a function of the source value. For example, suppose you want to find how the current changes as you vary the voltage across a resistor. If you make a table of current and voltage values, you may be able to see that the relationship is linear, but a graph will make it immediately obvious.

In PSpice, the .PRINT statement provides tables of data, while the .PLOT statement shows this data as a text-based graph or plot. The format of the basic .PLOT statement is similar to the general .PRINT statement:

```
.PLOT  <analysis type>  <variable 1>. . .<variable n>
```

The analysis type can be DC (among other options), and the variables are the voltages or currents that you want to plot. One difference between the .PRINT and .PLOT statements is that with .PLOT you may specify no more than eight variables on one .PLOT statement. However, you may use any number of .PLOT statements. For example, the following statements in a circuit file will tell PSpice to vary the source voltage $V_S$ from 0 V to 10 V in 1 V increments, calculate the currents through resistors $R_1$ and $R_2$, and then plot the results in the output file.

```
.DC  VS  0  10  1
.PLOT  DC  I(R1)  I(R2)
```

In some cases, you may want a plot for only part of the plot variables' range so the start and end values for the data can be specified using the general form of the .PLOT statement as follows:

```
.PLOT  <analysis type>  <variable list>  (<start>,<end>)
```

The *variable list* is all the variables to be plotted, *start* is the starting value, and *end* is the ending value. For example, if you need only the currents through $R_1$ and $R_2$ from 1 mA to 5 mA, the .PLOT statement is as follows:

```
.PLOT  DC  I(R1)  I(R2)  (1M,5M)
```

## The + (Line Continuation) Symbol

So far, the examples of the .PRINT and .PLOT statement lines have been fairly short. It is possible to have a sufficient number of variables in a single description line so that the line covers several text lines. To let PSpice know that information on one line belongs with information on the previous line, a plus (+) symbol is placed in the first column of the continuation line. For example, to print the current and voltage for resistors $R_1$ through $R_8$, the statement can be written as follows:

```
.PRINT  DC  V(R1)  V(R2)  V(R3)  V(R4)  V(R5)  V(R6)
+           V(R7)  V(R8)  I(R1)  I(R2)  I(R3)  I(R4)
+           I(R5)  I(R6)  I(R7)  I(R8)
```

Note that the + symbol only links information on a line with the information on the line immediately before it. You cannot place any other information (other than a comment or blank lines) between linked lines.

### SECTION 5–12 REVIEW

1. Explain how the .OP statement is used.
2. Explain each of the following sets of statement lines:
   **(a)** .DC VS 5 15 2
   .PLOT DC V(R1) V(R2) V(R3)
   **(b)** .DC VS 0 20 0.5
   .PLOT DC I(R1) I(R2) (1M,10M)
3. Explain how the + symbol is used.

## ■ SUMMARY

- The current is the same at all points in a series circuit.
- The total series resistance is the sum of all resistors in the series circuit.
- The total resistance between any two points in a series circuit is equal to the sum of all resistors connected in series between those two points.
- If all of the resistors in a series circuit are of equal value, the total resistance is the number of resistors multiplied by the resistance value.
- Voltage sources in series add algebraically.
- First statement of Kirchhoff's voltage law: The sum of the voltage drops equals the total source voltage.
- Second statement of Kirchhoff's voltage law: The sum of all the voltages around a closed path is zero.
- The voltage drops in a circuit are always opposite in polarity to the total source voltage.
- Current is out of the positive side of a source in the conventional direction and into the negative side.
- Current is into the positive side of each resistor and out of the more negative (less positive) side.
- A voltage drop is considered to be from a more positive voltage to a more negative voltage.

- A voltage divider is a series arrangement of resistors.
- A voltage divider is so named because the voltage drop across any resistor in the series circuit is divided down from the total voltage by an amount proportional to that resistance value in relation to the total resistance.
- A potentiometer can be used as an adjustable voltage divider.
- The total power in a resistive circuit is the sum of all the individual powers of the resistors making up the series circuit.
- All voltages in a circuit are referenced to ground unless otherwise specified.
- Ground is zero volts with respect to all points referenced to it in the circuit.
- *Negative ground* is the term used when the negative side of the source is grounded.
- *Positive ground* is the term used when the positive side of the source is grounded.
- The voltage across an open series element equals the source voltage.

## GLOSSARY

**Circuit ground** A method of grounding whereby the metal chassis that houses the assembly or a large conductive area on a printed circuit board is used as the common or reference point; also called chassis ground.

**Ground** In electric circuits, the common or reference point.

**Open circuit** A circuit in which the current path is broken.

**Series** In an electric circuit, a relationship of components in which the components are connected such that they provide a single current path between two points.

**Short circuit** A circuit in which there is a zero or abnormally low resistance path between two points; usually an inadvertent condition.

## FORMULAS

| | | |
|---|---|---|
| **(5–1)** | $R_T = R_1 + R_2 + R_3 + \cdots + R_n$ | Total resistance of $n$ resistors in series |
| **(5–2)** | $R_T = nR$ | Total resistance of $n$ equal-value resistors in series |
| **(5–3)** | $V_S = V_1 + V_2 + V_3 + \cdots + V_n$ | Kirchhoff's voltage law |
| **(5–4)** | $V_S - V_1 - V_2 - V_3 = 0$ | Kirchhoff's voltage law |
| **(5–5)** | $V_x = \left(\frac{R_x}{R_T}\right)V_S$ | Voltage-divider formula |
| **(5–6)** | $P_T = P_1 + P_2 + P_3 + \cdots + P_n$ | Total power |

## SELF-TEST

1. Five resistors are connected in series and there is a current of 2 A into the first resistor. The amount of current out of the second resistor is

   **(a)** 2 A **(b)** 1 A **(c)** 4 A **(d)** 0.4 A

2. To measure the current out of the third resistor in a circuit consisting of four series resistors, an ammeter can be placed

   **(a)** between the third and fourth resistors **(b)** between the second and third resistors
   **(c)** at the positive terminal of the source **(d)** at any point in the circuit

3. When a third resistor is connected in series with two series resistors, the total resistance

   **(a)** remains the same **(b)** increases
   **(c)** decreases **(d)** increases by one-third

4. When one of four series resistors is removed from a circuit and the circuit reconnected, the current

   **(a)** decreases by the amount of current through the removed resistor
   **(b)** decreases by one-fourth
   **(c)** quadruples
   **(d)** increases

5. A series circuit consists of three resistors with values of 100 Ω, 220 Ω, and 330 Ω. The total resistance is
   **(a)** less than 100 Ω **(b)** the average of the values **(c)** 550 Ω **(d)** 650 Ω
6. A 9 V battery is connected across a series combination of 68 Ω, 33 Ω, 100 Ω, and 47 Ω resistors. The amount of current is
   **(a)** 36.3 mA **(b)** 27.6 A **(c)** 22.3 mA **(d)** 363 mA
7. While putting four 1.5 V batteries in a flashlight, you accidentally put one of them in backward. The light will be
   **(a)** brighter than normal **(b)** dimmer than normal **(c)** off **(d)** the same
8. If you measure all the voltage drops and the source voltage in a series circuit and add them together, taking into consideration the polarities, you will get a result equal to
   **(a)** the source voltage
   **(b)** the total of the voltage drops
   **(c)** zero
   **(d)** the total of the source voltage and the voltage drops
9. There are six resistors in a given series circuit and each resistor has 5 V dropped across it. The source voltage is
   **(a)** 5 V **(b)** 30 V
   **(c)** dependent on the resistor values **(d)** dependent on the current
10. A series circuit consists of a 4.7 kΩ, a 5.6 kΩ, and a 10 kΩ resistor. The resistor that has the most voltage across it is
    **(a)** the 4.7 kΩ **(b)** the 5.6 kΩ
    **(c)** the 10 kΩ **(d)** impossible to determine from the given information
11. Which of the following series combinations dissipates the most power when connected across a 100 V source?
    **(a)** One 100 Ω resistor **(b)** Two 100 Ω resistors
    **(c)** Three 100 Ω resistors **(d)** Four 100 Ω resistors
12. The total power in a certain circuit is 10 W. Each of the five equal-value series resistors making up the circuit dissipates
    **(a)** 10 W **(b)** 50 W **(c)** 5 W **(d)** 2 W
13. When you connect an ammeter in a series-resistive circuit and turn on the source voltage, the meter reads zero. You should check for
    **(a)** a broken wire **(b)** a shorted resistor
    **(c)** an open resistor **(d)** answers (a) and (c)
14. While checking out a series-resistive circuit, you find that the current is higher than it should be. You should look for
    **(a)** an open circuit **(b)** a short
    **(c)** a low resistor value **(d)** answers (b) and (c)

## ■ PROBLEMS

### SECTION 5–1 Resistors in Series

1. Connect each set of resistors in Figure 5–65 in series between points *A* and *B*.

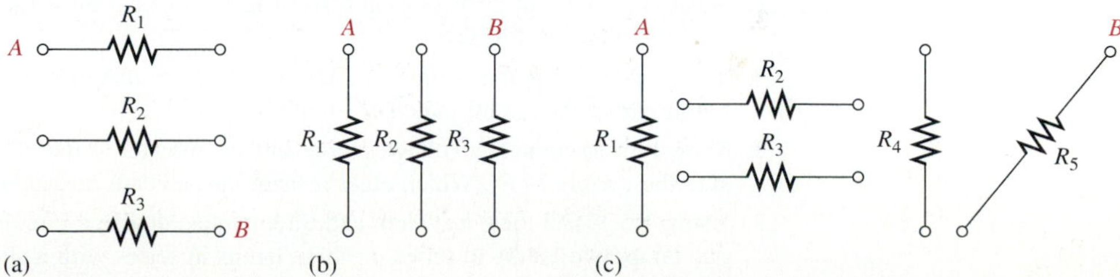


FIGURE 5–65

2. Determine which resistors in Figure 5–66 are in series. Show how to interconnect the pins to put all the resistors in series.

FIGURE 5–66

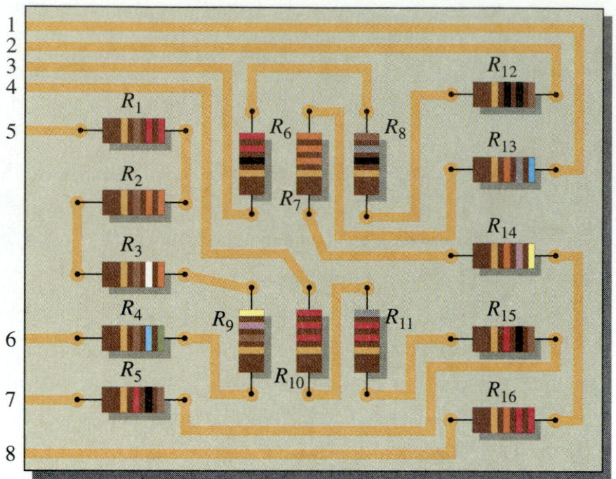


3. On the double-sided PC board in Figure 5–67, identify each group of series resistors. Note that many of the interconnections feed through the board from the top side to the bottom side.

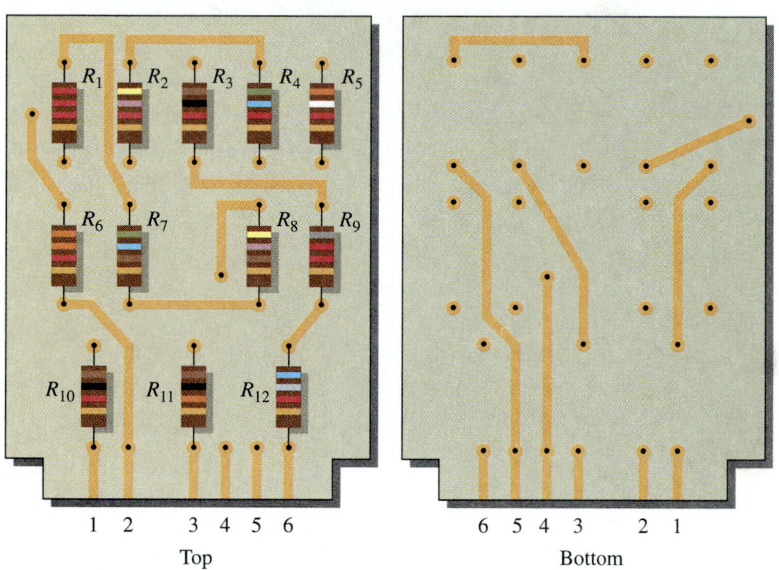


FIGURE 5–67

## SECTION 5–2 Current in a Series Circuit

4. What is the current through each resistor in a series circuit if the total voltage is 12 V and the total resistance is 120 Ω?
5. The current from the source in Figure 5–68 is 5 mA. How much current does each milliammeter in the circuit indicate?
6. Show how to connect a voltage source and an ammeter to the PC board in Figure 5–66 to measure the current in $R_1$. Which other resistor currents are measured by this setup?
7. Using 1.5 V batteries, a switch, and three lamps, devise a circuit to apply 4.5 V across either one lamp, two lamps in series, or three lamps in series with a single-control switch. Draw the schematic.

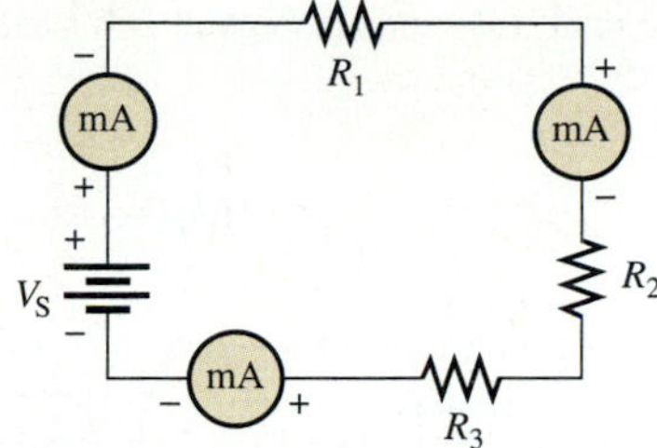


FIGURE 5–68

## SECTION 5–3 Total Series Resistance

**8.** The following resistors (one each) are connected in a series circuit: 1 Ω, 2.2 Ω, 5.6 Ω, 12 Ω, and 22 Ω. Determine the total resistance.

**9.** Find the total resistance of each of the following groups of series resistors:

**(a)** 560 Ω and 1000 Ω **(b)** 47 Ω and 56 Ω

**(c)** 1.5 kΩ, 2.2 kΩ, and 10 kΩ **(d)** 1 MΩ, 470 kΩ, 1 kΩ, 2.2 MΩ

**10.** Calculate $R_T$ for each circuit of Figure 5–69.

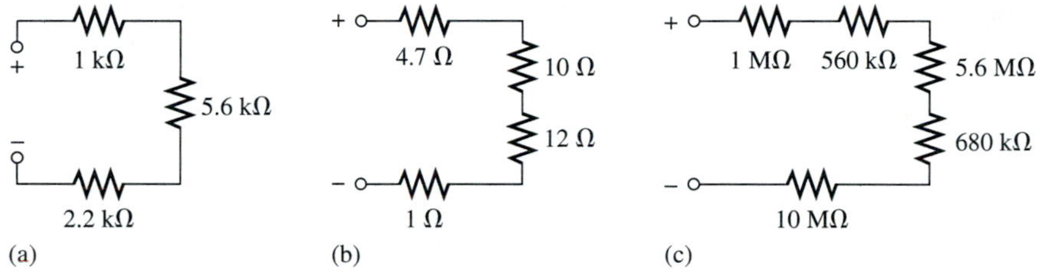


FIGURE 5–69

**11.** What is the total resistance of twelve 5.6 kΩ resistors in series?

**12.** Six 56 Ω resistors, eight 100 Ω resistors, and two 22 Ω resistors are all connected in series. What is the total resistance?

**13.** If the total resistance in Figure 5–70 is 17.4 kΩ, what is the value of $R_5$?

FIGURE 5–70

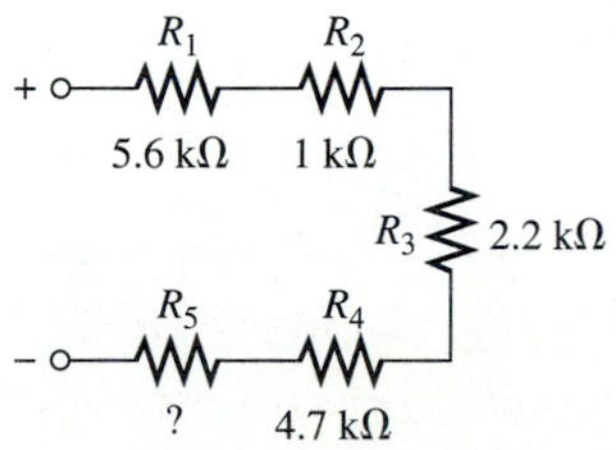

$R_1$ 5.6 kΩ, $R_2$ 1 kΩ, $R_3$ 2.2 kΩ, $R_4$ 4.7 kΩ, $R_5$ ?

**14.** You have the following resistor values available to you in the lab in unlimited quantities: 10 Ω, 100 Ω, 470 Ω, 560 Ω, 680 Ω, 1 kΩ, 2.2 kΩ, and 5.6 kΩ. All of the other standard values are out of stock. A project that you are working on requires an 18 kΩ resistance. What combinations of the available values would you use in series to achieve this total resistance?

**15.** Find the total resistance in Figure 5–69 if all three circuits are connected in series.

**16.** What is the total resistance from $A$ to $B$ for each switch position in Figure 5–71?

FIGURE 5–71

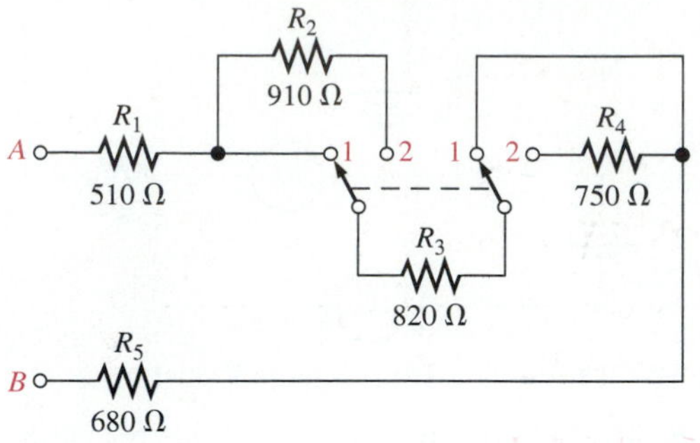

## SECTION 5–4 Ohm's Law in Series Circuits

**17.** What is the current in each circuit of Figure 5–72?

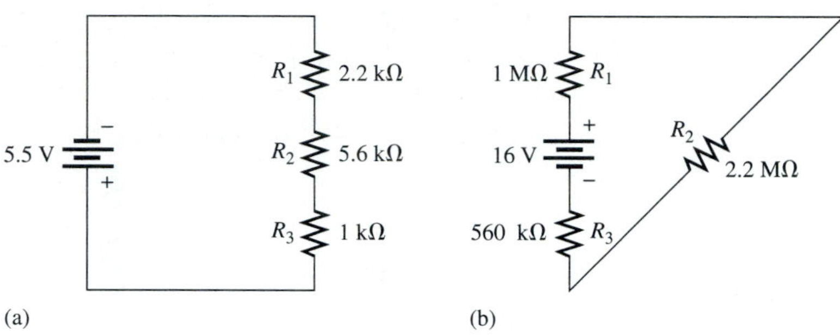


FIGURE 5–72

**18.** Determine the voltage drop across each resistor in Figure 5–72.

**19.** Three 470 Ω resistors are connected in series with a 500 V source. How much current is in the circuit?

**20.** Four equal-value resistors are in series with a 5 V battery, and 2.23 mA are measured. What is the value of each resistor?

**21.** What is the value of each resistor in Figure 5–73?

**22.** Determine $V_{R1}$, $R_2$, and $R_3$ in Figure 5–74.

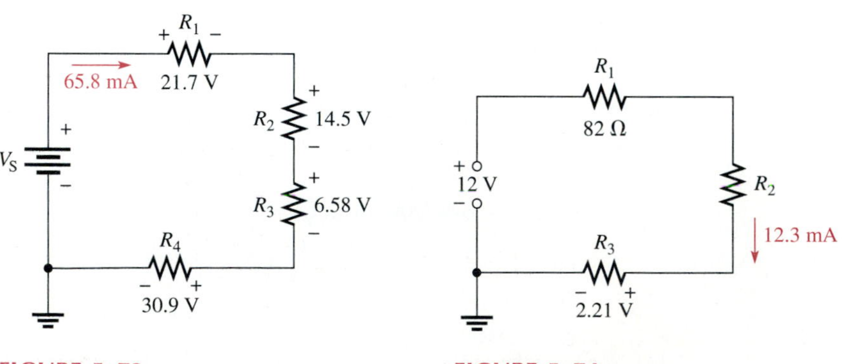


FIGURE 5–73

FIGURE 5–74

**23.** Determine the current measured by the meter in Figure 5–75 for each switch position.

**24.** Determine the current measured by the meter in Figure 5–76 for each position of the ganged switch.

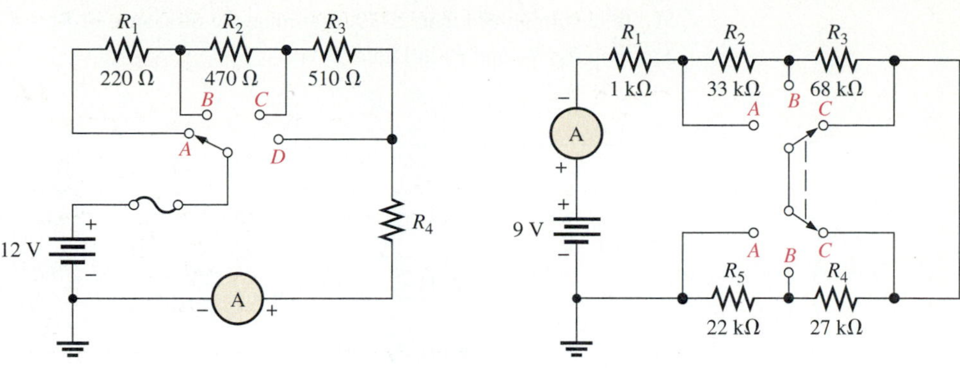


FIGURE 5–75

FIGURE 5–76

## SECTION 5–5 Voltage Sources in Series

**25.** *Series aiding* is a term sometimes used to describe voltage sources of the same polarity in series. If a 5 V and a 9 V source are connected in this manner, what is the total voltage?

**26.** The term *series opposing* means that sources are in series with opposite polarities. If a 12 V and a 3 V battery are series opposing, what is the total voltage?

**27.** Determine the total source voltage in each circuit of Figure 5–77.

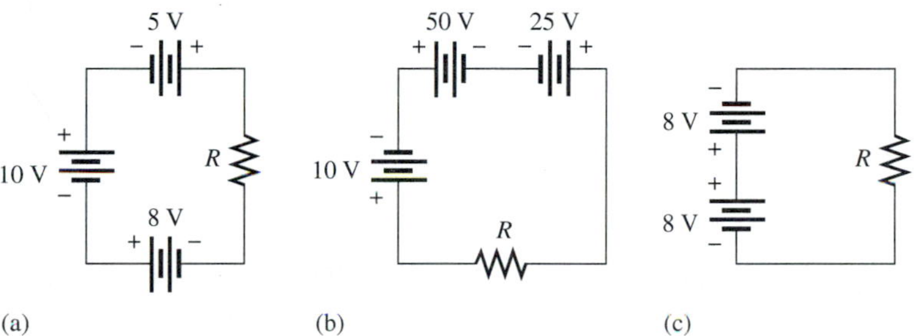


FIGURE 5–77

## SECTION 5–6 Kirchhoff's Voltage Law

**28.** The following voltage drops are measured across three resistors in series: 5.5 V, 8.2 V, and 12.3 V. What is the value of the source voltage to which these resistors are connected?

**29.** Five resistors are in series with a 20 V source. The voltage drops across four of the resistors are 1.5 V, 5.5 V, 3 V, and 6 V. How much voltage is dropped across the fifth resistor?

**30.** Determine the unspecified voltage drop(s) in each circuit of Figure 5–78. Show how to connect a voltmeter to measure each unknown voltage drop.

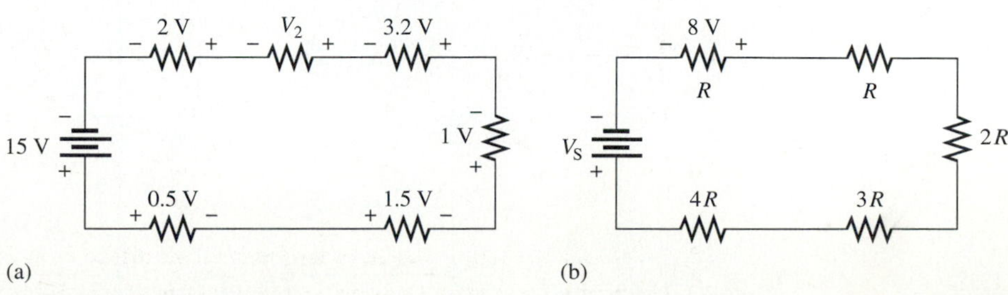


FIGURE 5–78

31. In the circuit of Figure 5–79, determine the resistance of $R_4$.
32. Find $R_1$, $R_2$, and $R_3$ in Figure 5–80.

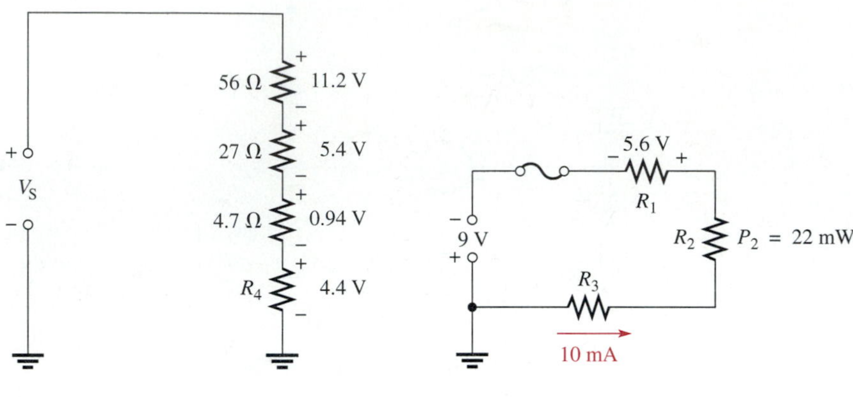


FIGURE 5–79

FIGURE 5–80

33. Determine the voltage across $R_5$ for each position of the switch in Figure 5–81. The current in each position is as follows: *A*, 3.35 mA; *B*, 3.73 mA; *C*, 4.50 mA; *D*, 6.00 mA.

FIGURE 5–81

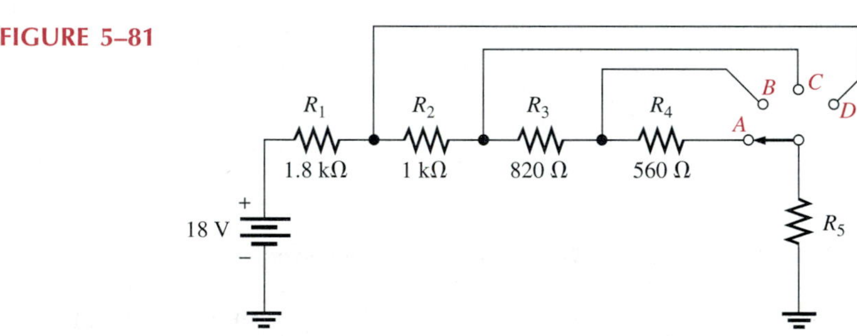


34. Determine the voltage across each resistor in Figure 5–81 for each switch position.

## SECTION 5–7 Voltage Dividers

35. The total resistance of a circuit is 560 Ω. What percentage of the total voltage appears across a 27 Ω resistor that makes up part of the total series resistance?
36. Determine the voltage between points *A* and *B* in each voltage divider of Figure 5–82.

FIGURE 5–82

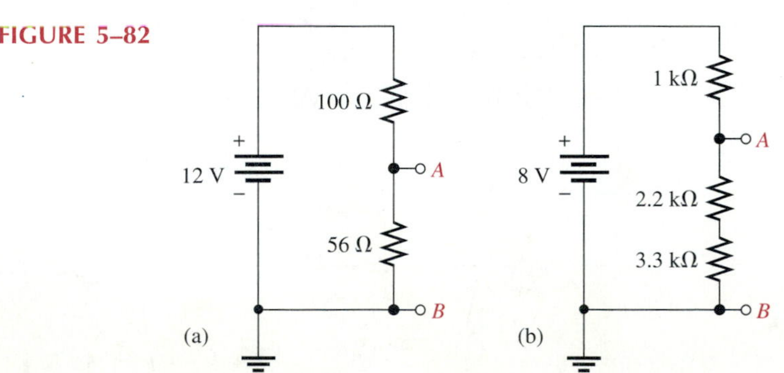

**37.** What is the voltage across each resistor in Figure 5–83? $R$ is the lowest-value resistor, and all others are multiples of that value as indicated.

**38.** Determine the voltage at each point in Figure 5–84 with respect to the negative side of the battery.

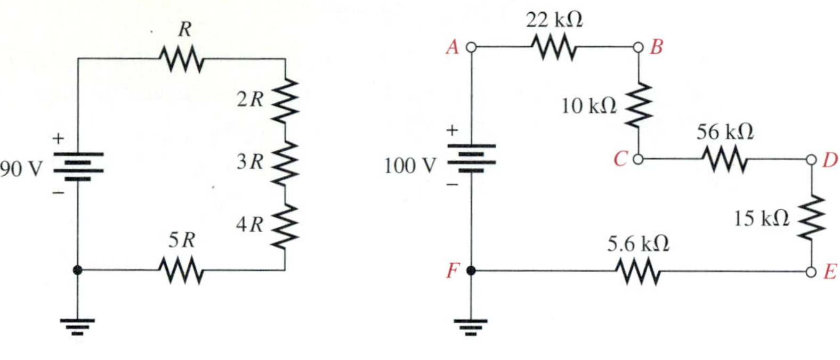


FIGURE 5–83

FIGURE 5–84

**39.** If there are 10 V across $R_1$ in Figure 5–85, what is the voltage across each of the other resistors?

FIGURE 5–85

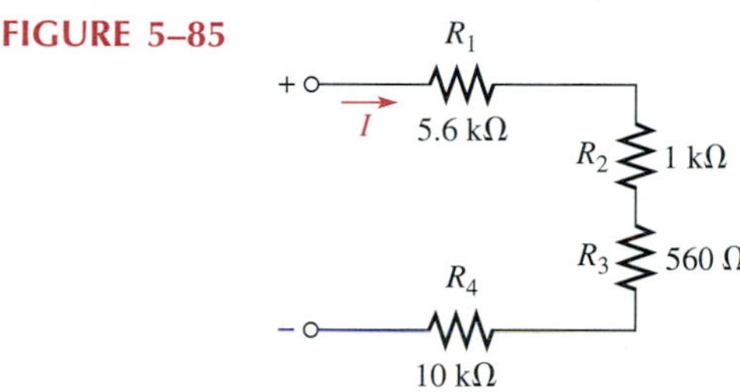


**40.** With the table of standard resistor values given in Appendix A, design a voltage divider to provide the following approximate voltages with respect to ground using a 30 V source: 8.18 V, 14.7 V, and 24.6 V. The current drain on the source must be limited to no more than 1 mA. The number of resistors, their values, and their wattage ratings must be specified. A schematic showing the circuit arrangement and resistor placement must be provided.

**41.** Design a variable voltage divider to provide an output voltage adjustable from a minimum of 10 V to a maximum of 100 V within ±1% using a 120 V source. The maximum voltage must occur at the maximum resistance setting of the potentiometer, and the minimum voltage must occur at the minimum resistance (zero) setting. The maximum current is to be 10 mA.

## SECTION 5–8 Power in a Series Circuit

**42.** Five series resistors each handle 50 mW. What is the total power?

**43.** What is the total power in the circuit in Figure 5–85? Use the results of Problem 39.

**44.** The following ¼ W resistors are in series: 1.2 kΩ, 2.2 kΩ, 3.9 kΩ, and 5.6 kΩ. What is the maximum voltage that can be applied across the series resistors without exceeding a power rating? Which resistor will burn out first if excessive voltage is applied?

**45.** Find $R_T$ in Figure 5–86.

FIGURE 5–86

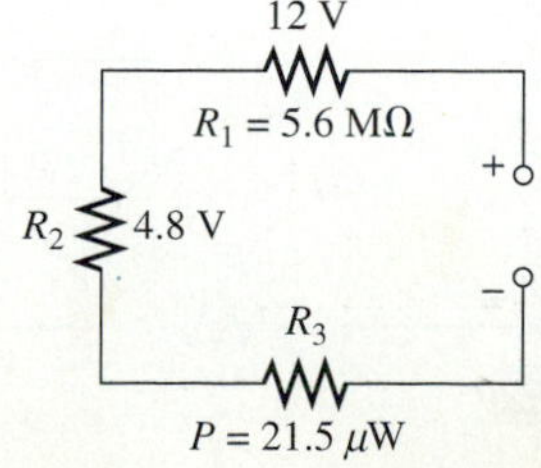

46. A certain series circuit consists of a ⅛ W resistor, a ¼ W resistor and a ½ W resistor. The total resistance is 2400 Ω. If each of the resistors is operating in the circuit at its maximum power dissipation, determine the following:
(a) $I$ (b) $V_T$ (c) The value of each resistor

## SECTION 5–9 Circuit Ground

47. Determine the voltage at each point with respect to ground in Figure 5–87.
48. In Figure 5–88, how would you determine the voltage across $R_2$ by measuring, without connecting a meter directly across the resistor?
49. Determine the voltage at each point with respect to ground in Figure 5–88.

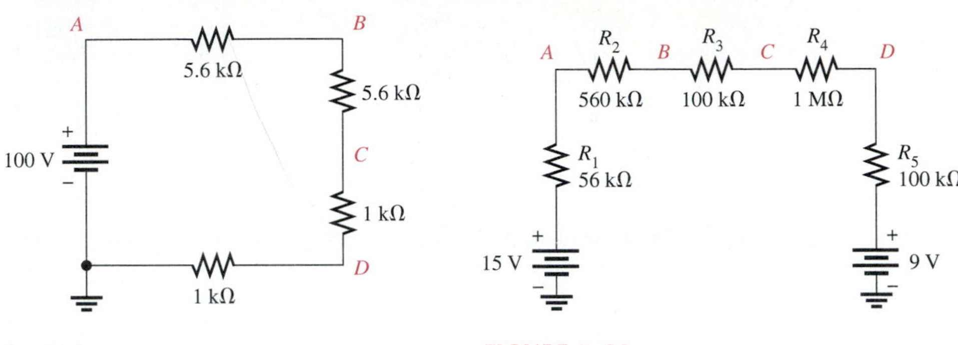


FIGURE 5–87

FIGURE 5–88

## SECTION 5–10 Troubleshooting

50. A string of five series resistors is connected across a 12 V battery. Zero volts is measured across all of the resistors except $R_2$. What is wrong with the circuit? What voltage will be measured across $R_2$?
51. By observing the meters in Figure 5–89, determine the types of failures in the circuits and which components have failed.
52. What current would you measure in Figure 5–89(b) if only $R_2$ were shorted?

FIGURE 5–89

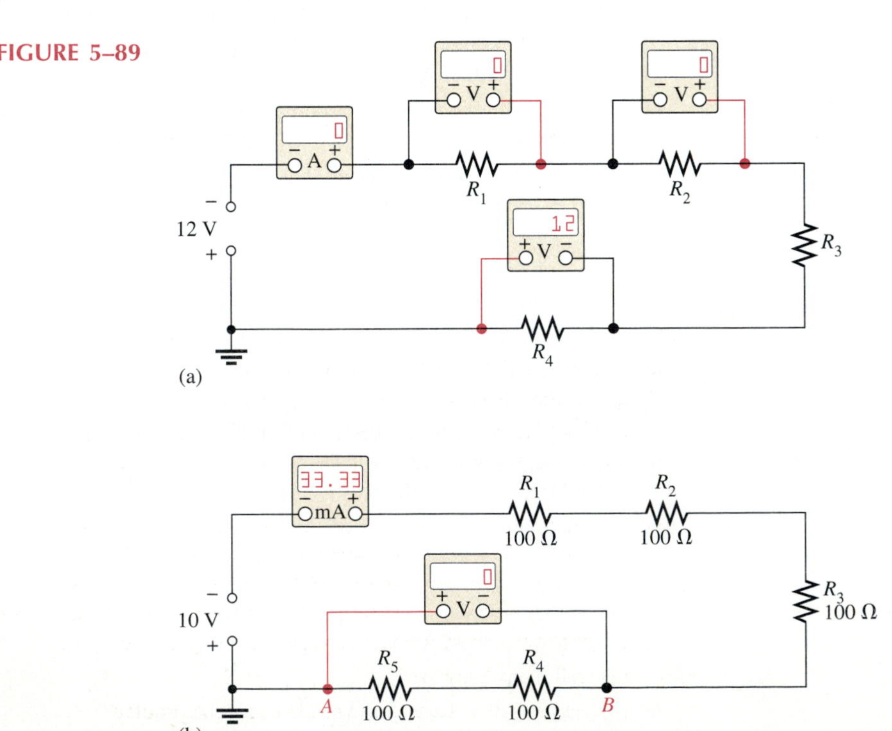

**53.** Table 5–1 shows the results of resistance measurements on the PC circuit board in Figure 5–90. Are these results correct? If not, identify the possible problems.

TABLE 5–1

| Between Pins | Resistance |
|---|---|
| 1 and 2 | ∞ |
| 1 and 3 | ∞ |
| 1 and 4 | 4.23 kΩ |
| 1 and 5 | ∞ |
| 1 and 6 | ∞ |
| 2 and 3 | 23.6 kΩ |
| 2 and 4 | ∞ |
| 2 and 5 | ∞ |
| 2 and 6 | ∞ |
| 3 and 4 | ∞ |
| 3 and 5 | ∞ |
| 3 and 6 | ∞ |
| 4 and 5 | ∞ |
| 4 and 6 | ∞ |
| 5 and 6 | 19.9 kΩ |

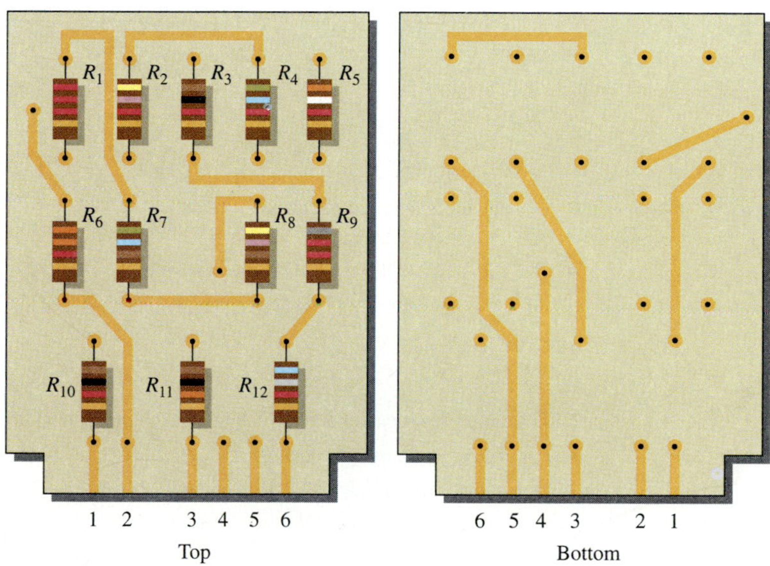


FIGURE 5–90

**54.** You measure 15 kΩ between pins 5 and 6 on the PC board in Figure 5–90. Does this indicate a problem? If so, identify it.

**55.** In checking out the PC board in Figure 5–90, you measure 17.83 kΩ between pins 1 and 2. Also, you measure 13.6 kΩ between pins 2 and 4. Does this indicate a problem on the PC board? If so, identify the fault.

**56.** The three groups of series resistors on the PC board in Figure 5–90 are connected in series with each other to form a single series circuit by connecting pin 2 to pin 4 and pin 3 to pin 5. A voltage source is connected across pins 1 and 6 and an ammeter is placed in series. As you increase the source voltage, you observe the corresponding increase in current. Suddenly, the current drops to zero and you smell smoke. All resistors are ½ W.

**(a)** What has happened?

**(b)** Specifically, what must you do to fix the problem?

**(c)** At what voltage did the failure occur?

## ANSWERS TO SECTION REVIEWS

### Section 5–1

1. Series resistors are connected end-to-end in a "string" with each lead of a given resistor connected to a different resistor.
2. There is a single current path in a series circuit.
3. See Figure 5–91.

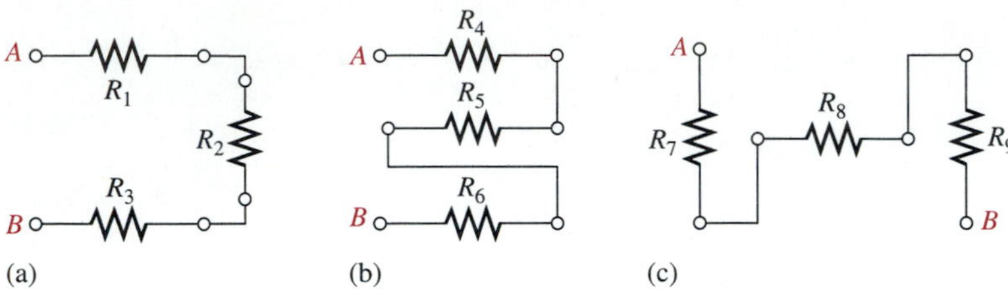


FIGURE 5–91

4. See Figure 5–92.

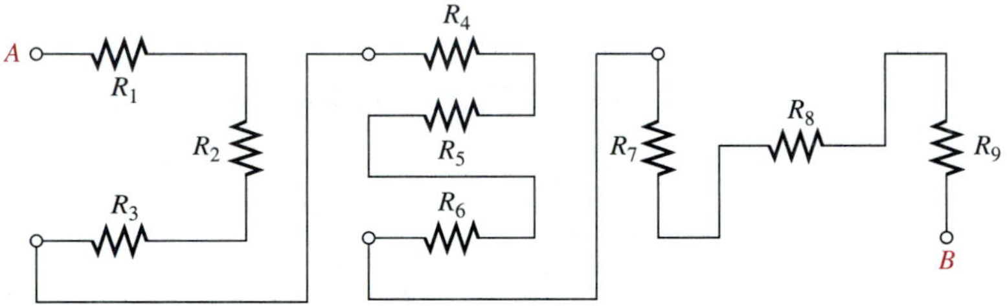


FIGURE 5–92

### Section 5–2

1. $I = 1$ A
2. The ammeter measures 50 mA between $C$ and $D$ and 50 mA between $E$ and $F$.
3. $I = 100\text{ V}/56\ \Omega = 1.79$ A; 1.79 A
4. In a series circuit, current is the same at all points.

### Section 5–3

1. $R_T = 1\ \Omega + 2.2\ \Omega + 3.3\ \Omega + 4.7\ \Omega = 11.2\ \Omega$
2. $R_T = 100\ \Omega + 2(56\ \Omega) + 4(12\ \Omega) + 330\ \Omega = 590\ \Omega$
3. $R = 13.76\text{ k}\Omega - (1\text{ k}\Omega + 2.7\text{ k}\Omega + 5.6\text{ k}\Omega + 560\ \Omega) = 3.9\text{ k}\Omega$
4. $R_T = 12(56\ \Omega) = 672\ \Omega$
5. $R_T = 20(5.6\text{ k}\Omega) + 30(8.2\text{ k}\Omega) = 358\text{ k}\Omega$

### Section 5–4

1. $I = 10\text{ V}/300\ \Omega = 33.3$ mA
2. $V_S = (5\text{ A})(21.2\ \Omega) = 106$ V
3. $V_1 = (5\text{ A})(10\ \Omega) = 50$ V; $V_2 = (5\text{ A})(5.6\ \Omega) = 28$ V; $V_3 = (5\text{ A})(5.6\ \Omega) = 28$ V
4. $R = (\frac{1}{4})(5\text{ V}/4.63\text{ mA}) = 270\ \Omega$

### Section 5–5

1. $V_T = 4(1.5\text{ V}) = 6.0$ V
2. 60 V/12 V = 5; see Figure 5–93.
3. See Figure 5–94.

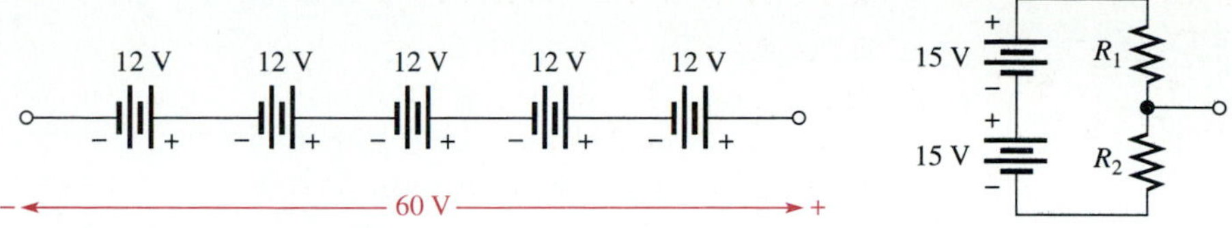


FIGURE 5–93

FIGURE 5–94

4. **(a)** $V_{S(tot)} = 100\text{ V} + 50\text{ V} - 75\text{ V} = 75\text{ V}$
   **(b)** $V_{S(tot)} = 20\text{ V} + 10\text{ V} - 10\text{ V} - 5\text{ V} = 15\text{ V}$
5. See Figure 5–95.

FIGURE 5–95

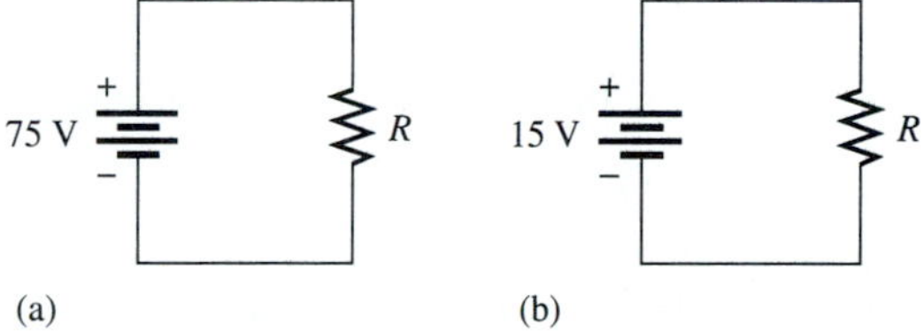


## Section 5–6

1. **(a)** Kirchhoff's law states the sum of the voltages around a closed path is zero;
   **(b)** Kirchhoff's law states the sum of the voltage drops equals the total source voltage.
2. $V_T = V_S = 50\text{ V}$
3. $V_{R1} = V_{R2} = 5\text{ V}$
4. $V_R = 25\text{ V} - 10\text{ V} - 5\text{ V} = 10\text{ V}$
5. $V_S = 1\text{ V} + 3\text{ V} + 5\text{ V} + 8\text{ V} + 7\text{ V} = 24\text{ V}$

## Section 5–7

1. A voltage divider is a circuit with two or more series resistors in which the voltage taken across any resistor or combination or resistors is proportional to the value of that resistance.
2. Two or more resistors form a voltage divider.
3. $V_x = (R_x/R_T)V_S$
4. $V_R = 10\text{ V}/2 = 5\text{ V}$
5. $V_{47} = (47\ \Omega/129\ \Omega)100\text{ V} = 36.4\text{ V}$; $V_{82} = (82\ \Omega/129\ \Omega)100\text{ V} = 63.6\text{ V}$; see Figure 5–96.
6. Set the wiper at the midpoint.

FIGURE 5–96

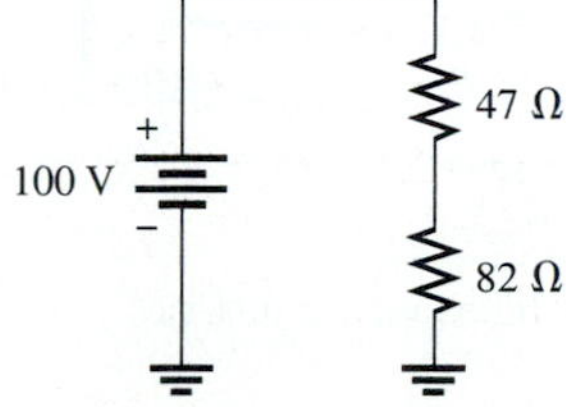


## Section 5–8

1. Add the power in each resistor to get total power.
2. $P_T = 2\text{ W} + 5\text{ W} + 1\text{ W} + 8\text{ W} = 16\text{ W}$
3. $P_T = (1\text{ A})^2(1110\ \Omega) = 1110\text{ W}$

## Section 5–9

1. The common reference point in a circuit is ground.
2. True

3. True
4. See Figure 5–97.

FIGURE 5–97

### Section 5–10

1. A short circuit is a zero resistance path that bypasses a portion of a circuit.
2. An open circuit is a break in the current path.
3. When a circuit opens, current ceases.
4. An open can be created by a switch or by a component failure. A short can be created by a switch or, unintentionally, by a wire clipping or solder splash.
5. True, a resistor normally fails open.
6. 24 V across the open $R$; 0 V across the other $R$s

### Section 5–11

1. $P_T = (12\text{ V})^2/16.6\text{ k}\Omega = 8.67\text{ mW}$
2. Pin 2: 4.59 V; Pin 6: 2.35 V; Pin 5: 1.99 V; Pin 4: 795 mV; Pin 7: 2.89 V
3. Pin 3 connects to ground.

### Section 5–12

1. The .OP statement puts detailed information about the dc operating point in the output file, including node voltages, source currents, and total power dissipation.
2. **(a)** Voltages across $R_1$, $R_2$, and $R_3$ are graphed as the source voltage is varied from 5 V to 15 V in 2 V steps.
   **(b)** Currents through $R_1$ and $R_2$ between 1 mA and 10 mA are graphed as source voltage is varied from 0 V to 20 V in 0.5 V steps.
3. The plus symbol + is used to continue a statement from a preceding line.

## ■ ANSWERS TO RELATED EXERCISES FOR EXAMPLES

**5–1** **(a)** See Figure 5–98.
**(b)** $R_1 = 1\text{ k}\Omega$, $R_2 = 33\text{ k}\Omega$, $R_3 = 39\text{ k}\Omega$, $R_4 = 470\ \Omega$, $R_5 = 22\text{ k}\Omega$

FIGURE 5–98

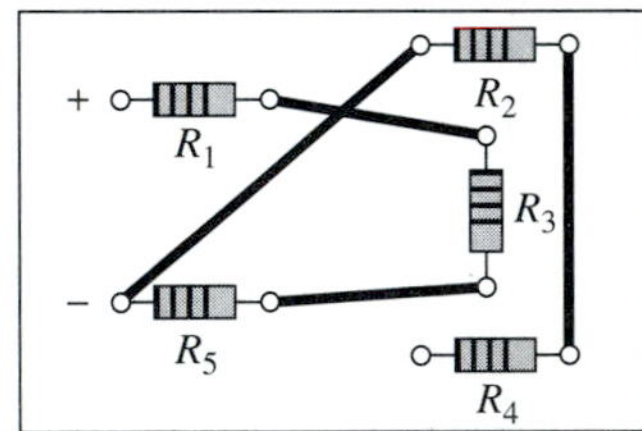


**5–2** All resistors on the board are in series.
**5–3** 258 Ω
**5–4** 12.1 kΩ
**5–5** 6.8 kΩ
**5–6** 4440 Ω
**5–7** 114 mA
**5–8** 7.8 V
**5–9** $V_1 = 1$ V, $V_2 = 3.3$ V, $V_3 = 2.2$ V; $V_S = 6.5$ V; $V_{S(max)} = 32.5$ V
**5–10** Use an ohmmeter.
**5–11** 12 V

**5–12** 2 V

**5–13** 10 V and 20 V

**5–14** 47 V

**5–15** 593 Ω

**5–16** $V_1 = 3.57$ V; $V_2 = 6.43$ V

**5–17** $V_1 = V_2 = V_3 = 33.3$ V

**5–18** $V_{AB} = 4$ V; $V_{AC} = 36.8$ V; $V_{BC} = 32.8$ V; $V_{BD} = 46$ V; $V_{CD} = 13.2$ V

**5–19** 8.49 W

**5–20** $P_1 = 0.92$ W (1 W); $P_2 = 2.49$ W (5 W); $P_3 = 0.838$ W (1 W); $P_4 = 3.04$ W (5 W)

**5–21** $V_A = 0$ V; $V_B = -25$ V; $V_C = -50$ V; $V_D = -75$ V; $V_E = -100$ V

# 6

# PARALLEL CIRCUITS



## ■ INTRODUCTION

In Chapter 5, you learned about series circuits and how to apply Ohm's law and Kirchhoff's voltage law. You also saw how a series circuit can be used as a voltage divider to obtain several specified voltages from a single source voltage. The effects of opens and shorts in series circuits were also examined.

In this chapter, you will see how Ohm's law is used in parallel circuits; and you will learn Kirchhoff's current law. Also, several applications of parallel circuits, including automotive lighting, residential wiring, and analog ammeters are presented. You will learn how to determine total parallel resistance and how to troubleshoot for open resistors. In this chapter and throughout the rest of the book, you will learn the basics of putting technology theory into practice.

When resistors are connected in parallel and a voltage is applied across the parallel circuit, each resistor provides a separate path for current. The total resistance of a parallel circuit is reduced as more resistors are connected in parallel. The voltage across each of the parallel resistors is equal to the voltage applied across the entire parallel circuit.

## TECHnology Theory Into Practice

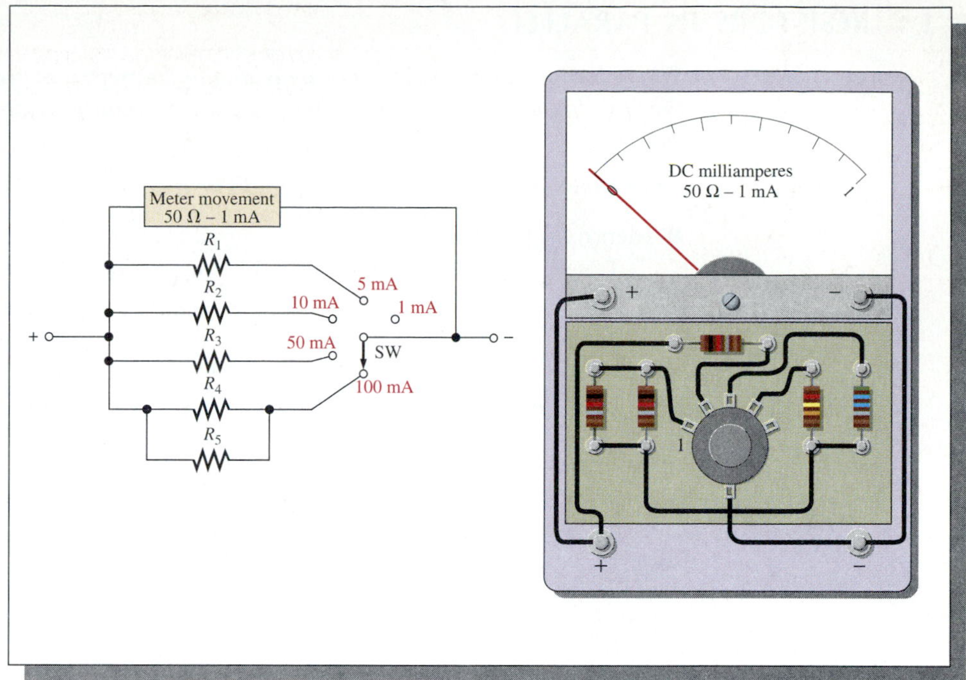


You are employed by an electronic instruments company, and your first job is to troubleshoot any defective instrument that fails the routine line check. In this particular assignment, you must determine the problem(s) with a defective five-range milliammeter so that it can be repaired. The knowledge of parallel circuits and of basic ammeters that you will acquire in this chapter plus your understanding of Ohm's law, current dividers, and the resistor color code will be put to good use.

## CHAPTER OBJECTIVES

- ☐ Identify a parallel circuit
- ☐ Determine the voltage across each parallel branch
- ☐ Apply Kirchhoff's current law
- ☐ Determine total parallel resistance
- ☐ Apply Ohm's law in a parallel circuit
- ☐ Determine the total effect of current sources in parallel
- ☐ Use a parallel circuit as a current divider
- ☐ Determine power in a parallel circuit
- ☐ Describe some basic applications of parallel circuits
- ☐ Troubleshoot parallel circuits
- ☐ Use PSpice to analyze parallel circuits and graph current values (optional)

## 6–1 ■ RESISTORS IN PARALLEL

***When two or more resistors are individually connected between the same two points, they are in parallel with each other. A parallel circuit provides more than one path for current.***

***After completing this section, you should be able to***

- **Identify a parallel circuit**
    - ☐ Translate a physical arrangement of resistors into a schematic

Each parallel current path is called a **branch.** Two resistors connected in parallel are shown in Figure 6–1(a). As shown in part (b), the current out of the source divides when it gets to point *A*. Part of it goes through $R_1$ and part through $R_2$. If additional resistors are connected in parallel with the first two, more current paths are provided between point *A* and point *B,* as shown in Figure 6–1(c). All points along the top are electrically the same as point *A,* and all the points along the bottom are electrically the same as point *B.*

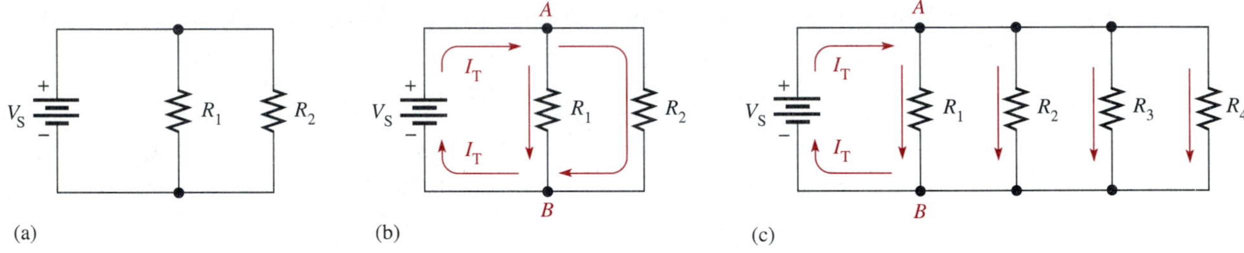


FIGURE 6–1
*Resistors in parallel.*

### Identifying Parallel Circuits

In Figure 6–1, it is obvious that the resistors are connected in **parallel.** Often, in actual circuit diagrams, the parallel relationship is not as clear. It is important that you learn to recognize parallel circuits regardless of how they may be drawn.

A rule for identifying parallel circuits is as follows:

**If there is more than one current path (branch) between two points, and if the voltage between those two points also appears across each of the branches, then there is a parallel circuit between those two points.**

Figure 6–2 shows parallel resistors drawn in different ways between two points labeled *A* and *B.* Notice that in each case, the current "travels" two paths going from *A* to *B,* and the voltage across each branch is the same. Although these examples in Figure 6–2 show only two parallel paths, there can be any number of resistors in parallel.

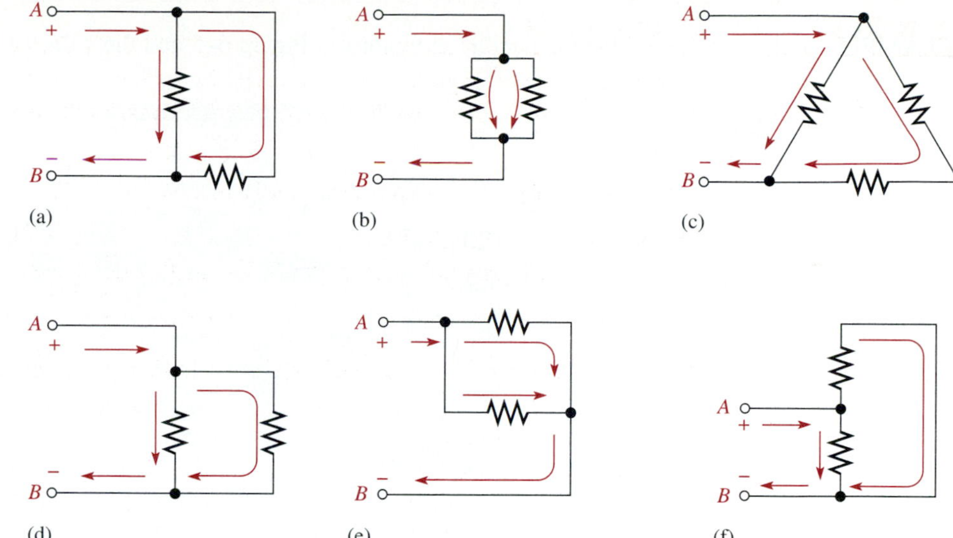


**FIGURE 6–2**
***Examples of circuits with two parallel paths.***

## EXAMPLE 6–1

Suppose that there are five resistors positioned on a circuit board as shown in Figure 6–3. Show the wiring required to connect all of the resistors in parallel. Draw a schematic and label each of the resistors with its value.

**FIGURE 6–3**

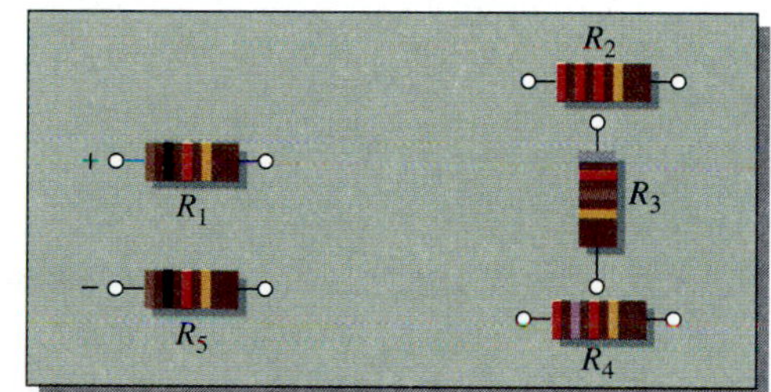


***Solution*** Wires are connected as shown in the assembly diagram of Figure 6–4(a). The schematic is shown in Figure 6–4(b). Again, note that the schematic does not necessarily have to show the actual physical arrangement of the resistors. The schematic shows how components are connected electrically.

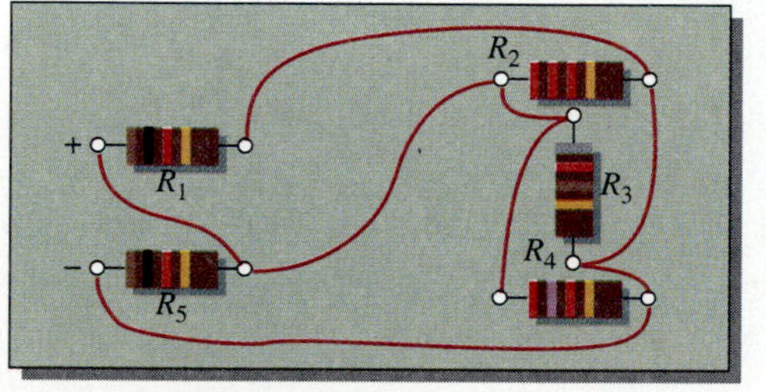


(a) Assembly wiring diagram

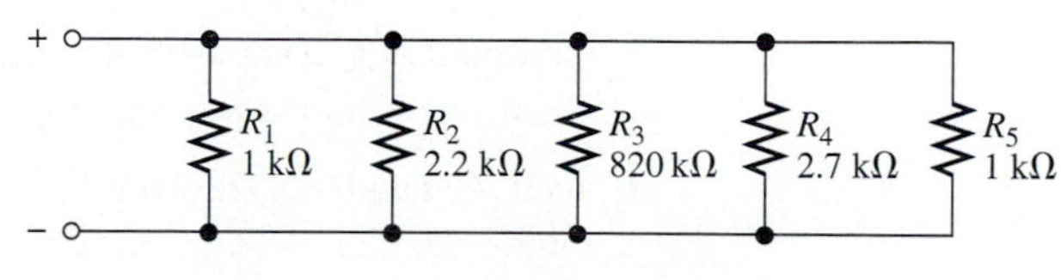


(b) Schematic

**FIGURE 6–4**

***Related Exercise*** How would the circuit have to be rewired if $R_2$ is removed?

**EXAMPLE 6–2** Determine the parallel groupings in Figure 6–5 and the value of each resistor.

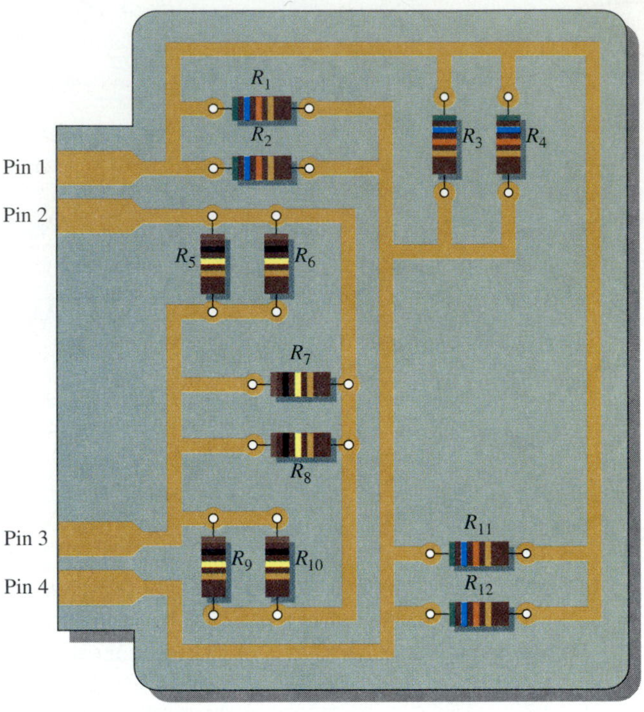


FIGURE 6–5

***Solution*** Resistors $R_1$ through $R_4$ and $R_{11}$ and $R_{12}$ are all in parallel. This parallel combination is connected to pins 1 and 4. Each resistor in this group is 56 kΩ.

Resistors $R_5$ through $R_{10}$ are all in parallel. This combination is connected to pins 2 and 3. Each resistor in this group is 100 kΩ.

***Related Exercise*** How would you connect all of the resistors in Figure 6–5 in parallel?

**SECTION 6–1 REVIEW**

1. How are the resistors connected in a parallel circuit?
2. How do you identify a parallel circuit?
3. Complete the schematics for the circuits in each part of Figure 6–6 by connecting the resistors in parallel between points *A* and *B*.
4. Now connect each group of parallel resistors in Figure 6–6 in parallel with each other.

FIGURE 6–6

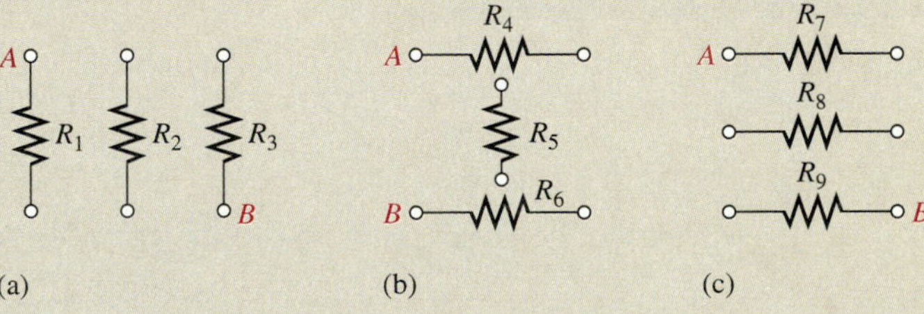

## 6–2 ■ VOLTAGE DROP IN PARALLEL CIRCUITS

***As mentioned in the previous section, each current path in a parallel circuit is called a branch. The voltage across any given branch of a parallel circuit is equal to the voltage across each of the other branches in parallel.***

***After completing this section, you should be able to***

- **Determine the voltage across each parallel branch**
  - Explain why the voltage is the same across all parallel resistors

To illustrate voltage drop in a parallel circuit, let's examine Figure 6–7(a). Points *A, B, C,* and *D* along the left side of the parallel circuit are electrically the same point because the voltage is the same along this line. You can think of all of these points as being connected by a single wire to the negative terminal of the battery. The points *E, F, G,* and *H* along the right side of the circuit are all at a voltage equal to that of the positive terminal of the source. Thus, voltage across each parallel resistor is the same, and each is equal to the source voltage.

Figure 6–7(b) is the same circuit as in part (a), drawn in a slightly different way. Here the left side of each resistor is connected to a single point, which is the negative battery terminal. The right side of each resistor is connected to a single point, which is the positive battery terminal. The resistors are still all in parallel across the source.

In Figure 6–8, a 12 V battery is connected across three parallel resistors. When the voltage is measured across the battery and then across each of the resistors, the readings are the same. As you can see, the same voltage appears across each branch in a parallel circuit.

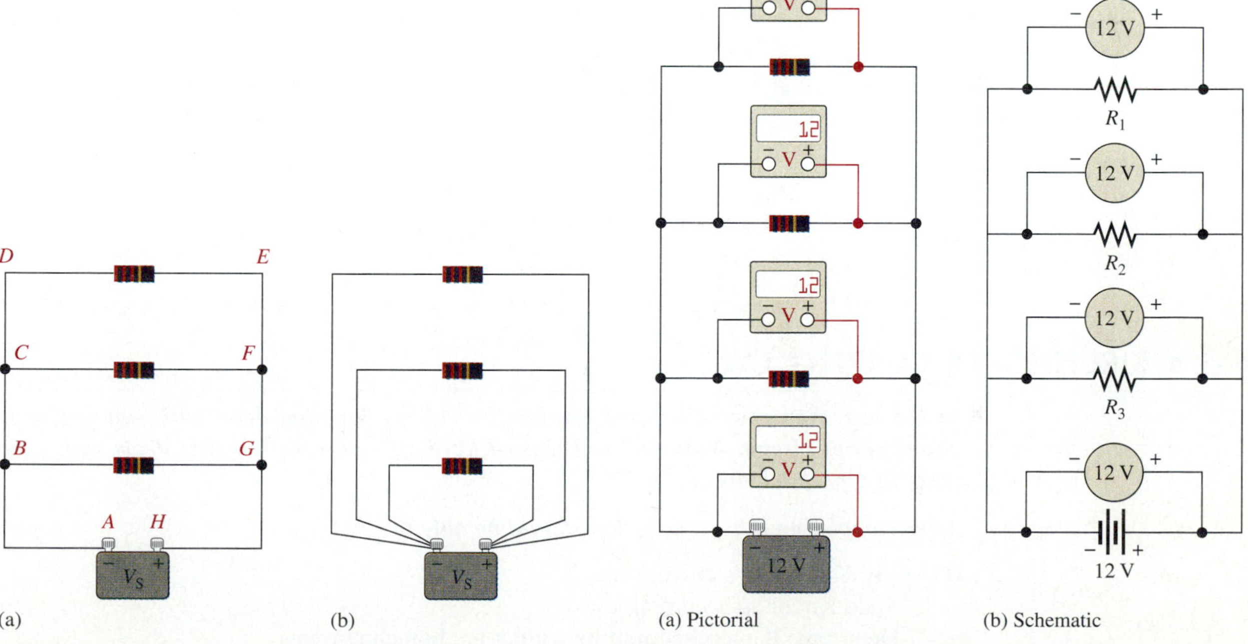


**FIGURE 6–7**
***Voltage across parallel branches is the same.***

**FIGURE 6–8**
***The same voltage appears across each resistor in parallel.***

**EXAMPLE 6–3** Determine the voltage across each resistor in Figure 6–9.

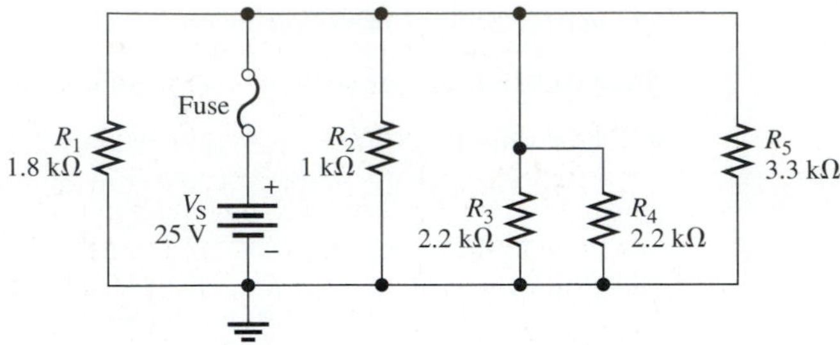


FIGURE 6–9

***Solution*** The five resistors are in parallel, so the voltage drop across each one is equal to the applied source voltage. There is no voltage drop across the fuse.

$$V_1 = V_2 = V_3 = V_4 = V_5 = V_S = 25\text{ V}$$

***Related Exercise*** If $R_4$ is removed from the circuit, what is the voltage across $R_3$?

**SECTION 6–2 REVIEW**

1. A 10 Ω and a 22 Ω resistor are connected in parallel with a 5 V source. What is the voltage across each of the resistors?
2. A voltmeter is connected across $R_1$ in Figure 6–10. It measures 118 V. If you move the meter and connect it across $R_2$, how much voltage will it indicate? What is the source voltage?
3. In Figure 6–11, how much voltage does voltmeter 1 indicate? Voltmeter 2?
4. How are voltages across each branch of a parallel circuit related?

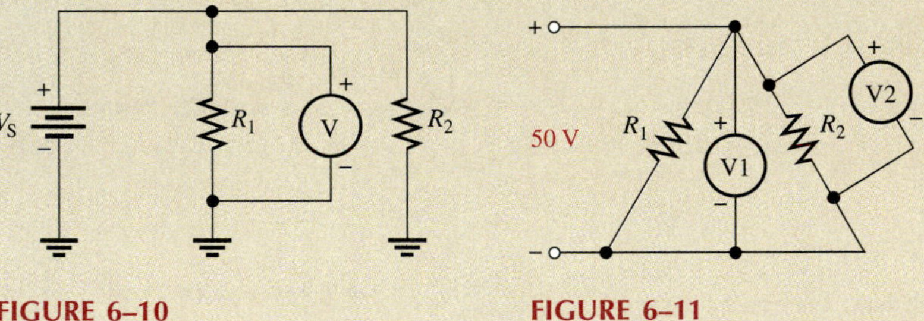


FIGURE 6–10

FIGURE 6–11

## 6–3 ■ KIRCHHOFF'S CURRENT LAW

***In the last chapter, you learned Kirchhoff's voltage law that dealt with voltages in a closed series circuit. Now, you will learn Kirchhoff's current law that deals with currents in a parallel circuit.***

***After completing this section, you should be able to***

- **Apply Kirchhoff's current law**
  - ☐ State Kirchhoff's current law
  - ☐ Determine the total current by adding the branch currents
  - ☐ Determine an unknown branch current

**Kirchhoff's current law,** often abbreviated KCL, is stated as follows:

> **The sum of the currents into a junction (total current in) is equal to the sum of the currents out of that junction (total current out).**

A **junction** is any point in a circuit where two or more components are connected. So, in a parallel circuit, a junction is where the parallel branches come together. For example, in the circuit of Figure 6–12, point *A* is one junction and point *B* is another. Let's start at the positive terminal of the source and follow the current. The total current $I_T$ from the source is *into* the junction at point *A*. At this point, the current splits up among the three branches as indicated. Each of the three branch currents ($I_1$, $I_2$, and $I_3$) is *out of* junction *A*. Kirchhoff's current law says that the total current into junction *A* is equal to the total current out of junction *A;* that is,

$$I_T = I_1 + I_2 + I_3$$

Now, following the currents in Figure 6–12 through the three branches, you see that they come back together at point *B*. Currents $I_1$, $I_2$, and $I_3$ are into junction *B*, and $I_T$ is out of junction *B*. Kirchhoff's current law formula at this junction is therefore the same as at junction *A*.

$$I_T = I_1 + I_2 + I_3$$

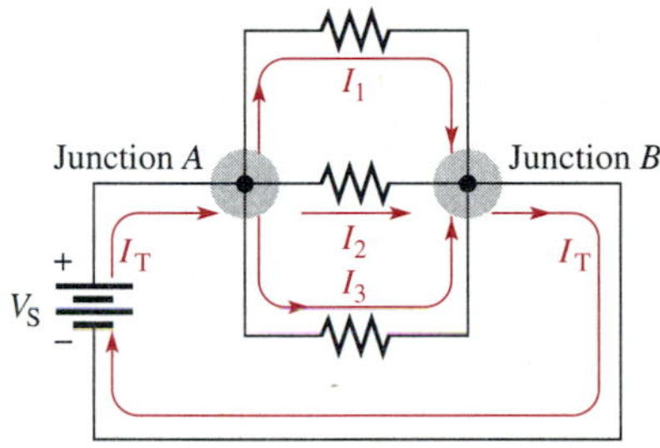


FIGURE 6–12
***Total current into junction A equals the sum of currents out of junction A. The sum of currents into junction B equals the total current out of junction B.***

## General Formula for Kirchhoff's Current Law

The previous discussion used a specific example to illustrate Kirchhoff's current law. Figure 6–13 shows a generalized circuit junction where a number of branches are connected

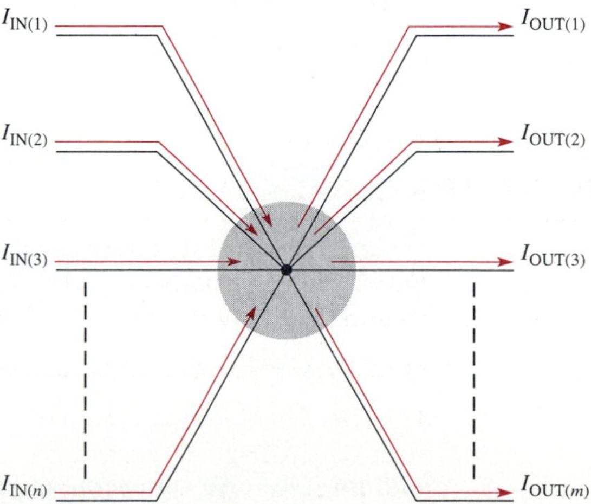


FIGURE 6–13
***Generalized circuit junction illustrating Kirchhoff's current law.***

to a point in the circuit. Currents $I_{IN(1)}$ through $I_{IN(n)}$ are into the junction ($n$ can be any number). Currents $I_{OUT(1)}$ through $I_{OUT(m)}$ are out of the junction ($m$ can be any number, but not necessarily equal to $n$). By Kirchhoff's current law, the sum of the currents into a junction must equal the sum of the currents out of the junction. With reference to Figure 6–13, the general formula for Kirchhoff's current law is

$$I_{IN(1)} + I_{IN(2)} + \cdots + I_{IN(n)} = I_{OUT(1)} + I_{OUT(2)} + \cdots + I_{OUT(m)} \quad \textbf{(6–1)}$$

If all the terms on the right side of Equation (6–1) are brought over to the left side, their signs change to negative, and a zero is left on the right side as follows:

$$I_{IN(1)} + I_{IN(2)} + \cdots + I_{IN(n)} - I_{OUT(1)} - I_{OUT(2)} - \cdots - I_{OUT(m)} = 0$$

Kirchhoff's current law can also be stated in this way:

**The algebraic sum of all the currents entering and leaving a junction is equal to zero.**

You can verify Kirchhoff's current law by connecting a circuit and measuring each branch current and the total current from the source, as illustrated in Figure 6–14. When the branch currents are added together, their sum will equal the total current. This rule applies for any number of branches.

The following three examples illustrate use of Kirchhoff's current law.

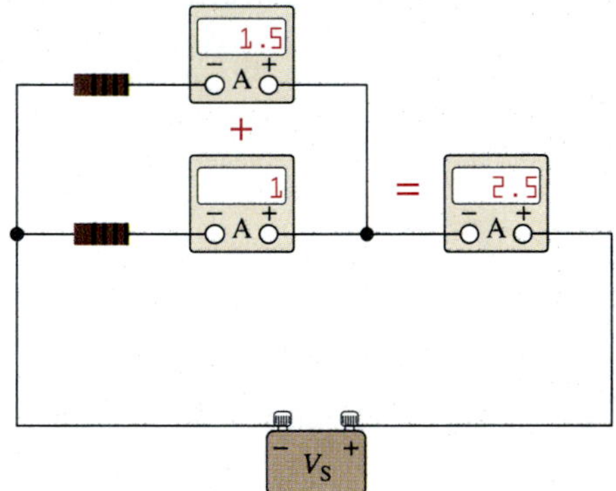


**FIGURE 6–14**
***Experimental verification of Kirchhoff's current law.***

## EXAMPLE 6–4

The branch currents in the circuit of Figure 6–15 are known. Determine the total current entering junction $A$ and the total current leaving junction $B$.

**FIGURE 6–15**

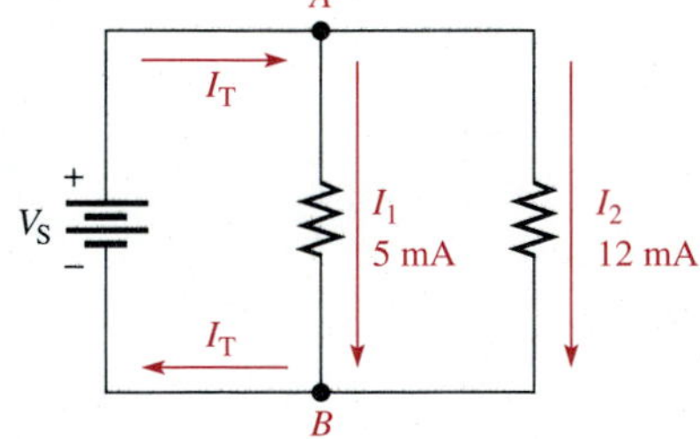


***Solution*** The total current out of junction $A$ is the sum of the two branch currents. So the total current into $A$ is

$$I_T = I_1 + I_2 = 5\text{ mA} + 12\text{ mA} = 17\text{ mA}$$

The total current entering junction $B$ is the sum of the two branch currents. So the total current out of $B$ is

$$I_T = I_1 + I_2 = 5 \text{ mA} + 12 \text{ mA} = 17 \text{ mA}$$

The calculator sequence is

5 EXP +/– 3 + 1 2 EXP +/– 3 =

***Related Exercise*** If a third branch is added to the circuit in Figure 6–15, and its current is 3 mA, what is the total current into junction $A$ and out of junction $B$?

## EXAMPLE 6–5

Determine the current $I_2$ through $R_2$ in Figure 6–16.

FIGURE 6–16

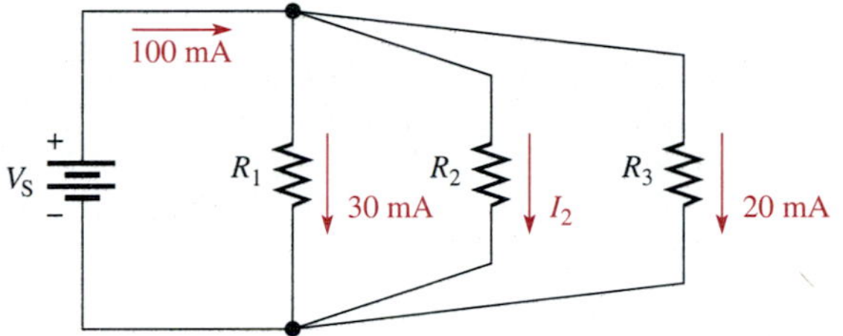


***Solution*** The total current into the junction of the three branches is $I_T = I_1 + I_2 + I_3$. From Figure 6–16, you know the total current and the branch currents through $R_1$ and $R_3$. Solve for $I_2$ as follows:

$$I_2 = I_T - I_1 - I_3 = 100 \text{ mA} - 30 \text{ mA} - 20 \text{ mA} = 50 \text{ mA}$$

***Related Exercise*** Determine $I_T$ and $I_2$ if a fourth branch is added to the circuit in Figure 6–16 and it has 12 mA through it.

## EXAMPLE 6–6

Use Kirchhoff's current law to find the current measured by ammeters A1 and A2 in Figure 6–17.

FIGURE 6–17

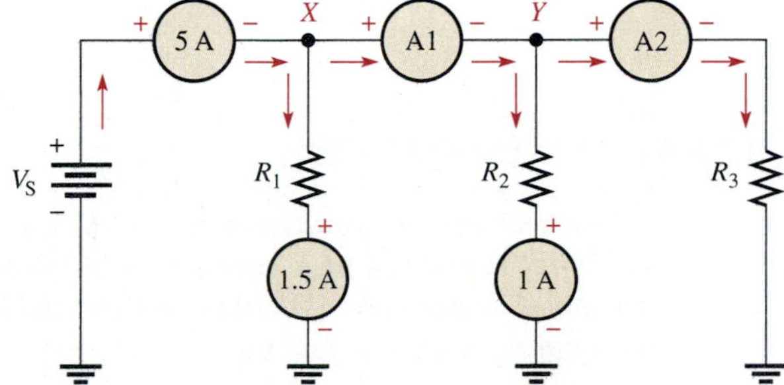


***Solution*** The total current into junction $X$ is 5 A. Two currents are out of junction $X$: 1.5 A through resistor $R_1$ and the current through A1. Kirchhoff's current law applied at junction $X$ gives

$$5 \text{ A} = 1.5 \text{ A} + I_{A1}$$

Solving for $I_{A1}$ yields

$$I_{A1} = 5\text{ A} - 1.5\text{ A} = 3.5\text{ A}$$

The total current into junction $Y$ is $I_{A1} = 3.5$ A. Two currents are out of junction $Y$: 1 A through resistor $R_2$ and the current through A2 and $R_3$. Kirchhoff's current law applied at junction $Y$ gives

$$3.5\text{ A} = 1\text{ A} + I_{A2}$$

Solving for $I_{A2}$ yields

$$I_{A2} = 3.5\text{ A} - 1\text{ A} = 2.5\text{ A}$$

***Related Exercise*** How much current will an ammeter measure when it is placed in the circuit right below $R_3$? Below the negative battery terminal?

**SECTION 6–3 REVIEW**

1. State Kirchhoff's current law in two ways.
2. There is a total current of 2.5 A into the junction of three parallel branches. What is the sum of all three branch currents?
3. In Figure 6–18, 100 mA and 300 mA are into the junction. What is the amount of current out of the junction?
4. Determine $I_1$ in the circuit of Figure 6–19.
5. Two branch currents enter a junction, and two branch currents leave the same junction. One of the currents entering the junction is 1 A, and one of the currents leaving the junction is 3 A. The total current entering and leaving the junction is 8 A. Determine the value of the unknown current entering the junction and the value of the unknown current leaving the junction.

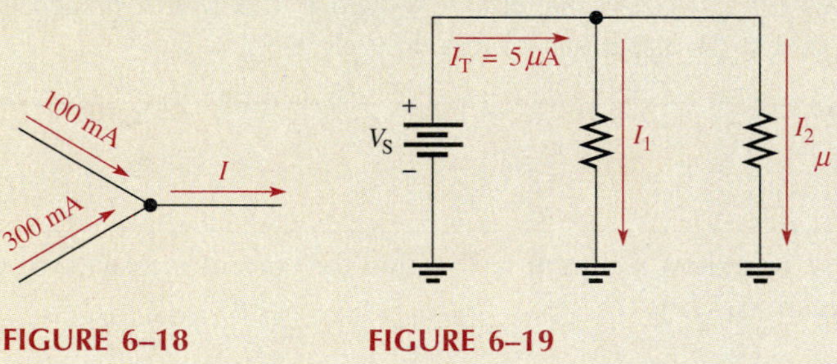


FIGURE 6–18 FIGURE 6–19

## 6–4 ■ TOTAL PARALLEL RESISTANCE

***When resistors are connected in parallel, the total resistance of the circuit decreases. The total resistance of a parallel circuit is always less than the value of the smallest resistor. For example, if a 10 Ω resistor and a 100 Ω resistor are connected in parallel, the total resistance is less than 10 Ω.***

*After completing this section, you should be able to*

- **Determine total parallel resistance**
  - ☐ Explain why resistance decreases as resistors are connected in parallel
  - ☐ Apply the parallel-resistance formula

## The Number of Current Paths Affects Total Resistance

As you know, when resistors are connected in parallel, the current has more than one path. The number of current paths is equal to the number of parallel branches.

For example, in Figure 6–20(a), there is only one current path since it is a series circuit. There is a certain amount of current, $I_1$, through $R_1$. If resistor $R_2$ is connected in parallel with $R_1$, as shown in Figure 6–20(b), there is an additional amount of current, $I_2$, through $R_2$. The total current from the source has increased with the addition of the parallel resistor. Assuming that the source voltage is constant, an increase in the total current from the source means that the total resistance has decreased, in accordance with Ohm's law. Additional resistors connected in parallel will further reduce the resistance and increase the total current.

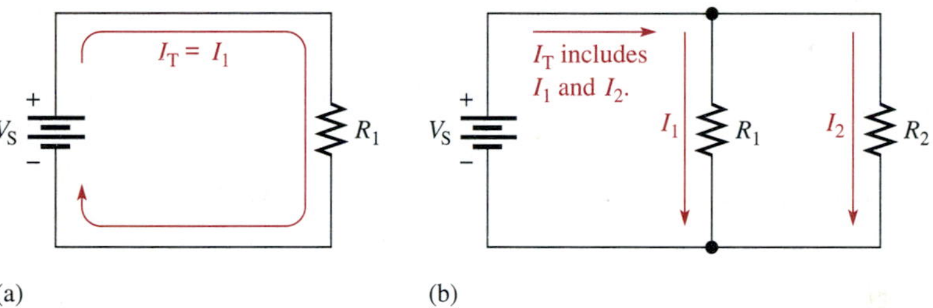


**FIGURE 6–20**
***Addition of resistors in parallel reduces total resistance and increases total current.***

## Formula for Total Parallel Resistance

The circuit in Figure 6–21 shows a general case of $n$ resistors in parallel ($n$ can be any number). From Kirchhoff's current law, the current equation is

$$I_T = I_1 + I_2 + I_3 + \cdots + I_n$$

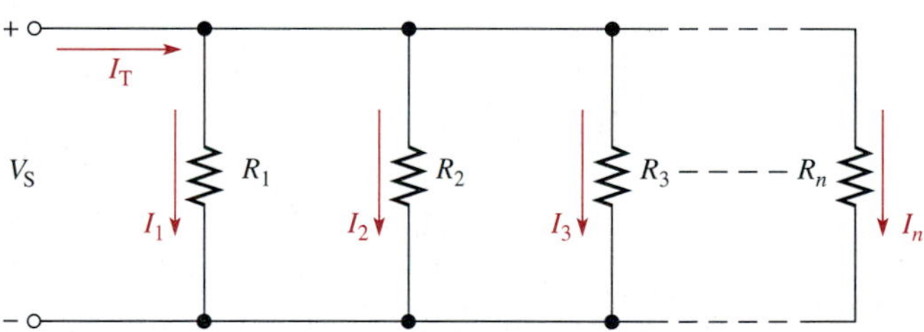


**FIGURE 6–21**
***Circuit with n resistors in parallel.***

Since $V_S$ is the voltage across each of the parallel resistors, by Ohm's law, $I_1 = V_S/R_1$, $I_2 = V_S/R_2$, and so on. By substitution into the current equation,

$$\frac{V_S}{R_T} = \frac{V_S}{R_1} + \frac{V_S}{R_2} + \frac{V_S}{R_3} + \cdots + \frac{V_S}{R_n}$$

The term $V_S$ can be factored out of the right side of the equation and canceled with $V_S$ on the left side, leaving only the resistance terms.

$$\frac{1}{R_T} = \frac{1}{R_1} + \frac{1}{R_2} + \frac{1}{R_3} + \cdots + \frac{1}{R_n} \qquad \textbf{(6–2)}$$

Solve for $R_T$ in Equation (6–2) by taking the reciprocal of (that is, by inverting) both sides of the equation.

$$R_T = \frac{1}{\left(\frac{1}{R_1}\right) + \left(\frac{1}{R_2}\right) + \left(\frac{1}{R_3}\right) + \cdots + \left(\frac{1}{R_n}\right)} \qquad \textbf{(6–3)}$$

Equation (6–3) shows that to find the total parallel resistance, add all the $1/R$ terms and then take the reciprocal of the sum. Example 6–7 shows how to use the formula in a specific case, and Example 6–8 shows how to use a calculator to determine parallel resistances. Recall that the reciprocal of resistance ($1/R$) is called *conductance* and is symbolized by $G$. The unit of conductance is the siemens (S).

**EXAMPLE 6–7** Calculate the total parallel resistance between points $A$ and $B$ of the circuit in Figure 6–22.

**FIGURE 6–22**

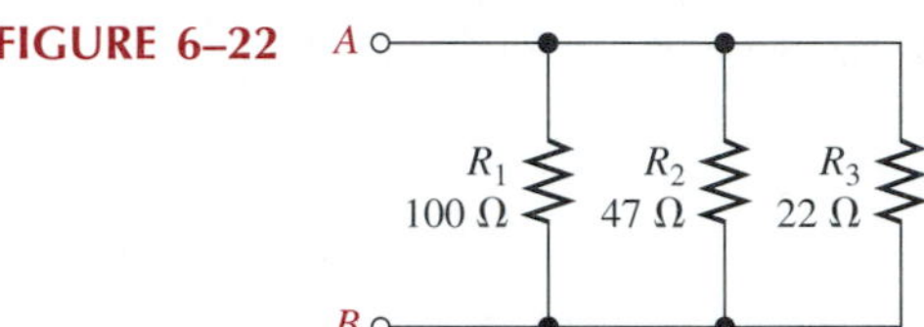


***Solution*** Use Equation (6–3) to calculate the total parallel resistance when you know the individual resistances. First, find the reciprocal of each of the three resistors:

$$\frac{1}{R_1} = \frac{1}{100\ \Omega} = 10\text{ mS}$$

$$\frac{1}{R_2} = \frac{1}{47\ \Omega} = 21.3\text{ mS}$$

$$\frac{1}{R_3} = \frac{1}{22\ \Omega} = 45.5\text{ mS}$$

Next, calculate $R_T$ by adding $1/R_1$, $1/R_2$, and $1/R_3$ and taking the reciprocal of the sum:

$$R_T = \frac{1}{10\text{ mS} + 21.3\text{ mS} + 45.5\text{ mS}} = \frac{1}{76.8\text{ mS}} = 13.0\ \Omega$$

For a quick accuracy check, notice that the value of $R_T$ (13.0 Ω) is smaller than the smallest value in parallel, which is $R_3$ (22 Ω), as it should be.

***Related Exercise*** If a 33 Ω resistor is connected in parallel in Figure 6–22, what is $R_T$?

## Calculator Solution

The parallel-resistance formula is easily solved on a calculator. The general procedure is to enter the value of $R_1$ and then take its reciprocal by pressing the [2nd] [1/x] keys. ([1/x] is not a secondary function on some calculators.) Next press the [+] key; then enter the value of $R_2$ and take its reciprocal. Repeat this procedure until all of the resistor values have been entered and the reciprocal of each has been added. The final step is to press the [2nd] [1/x] keys to convert $1/R_T$ to $R_T$. The total parallel resistance is now on the display.

This calculator procedure is illustrated in Example 6–8.

**EXAMPLE 6–8** Show the steps required for a calculator solution of Example 6–7.

***Solution***

Step 1: Enter 100. Display shows 100.

Step 2: Press 2nd 1/x. Display shows 0.01.

Step 3: Press + key. Display shows 0.01.

Step 4: Enter 47. Display shows 47.

Step 5: Press 2nd 1/x. Display shows 0.021276596.

Step 6: Press + key. Display shows 0.031276596.

Step 7: Enter 22. Display shows 22.

Step 8: Press 2nd 1/x. Display shows 0.045454545.

Step 9: Press = key. Display shows 0.076731141.

Step 10: Press 2nd 1/x. Display shows 13.03251828.

The number displayed in Step 10 is the total resistance in ohms. Round it to 13.0 Ω.

***Related Exercise*** Show the additional calculator steps for $R_T$ when a 33 Ω resistor is placed in parallel in the circuit of Example 6–7.

## The Case of Two Resistors in Parallel

Equation (6–3) is a general formula for finding the total resistance for any number of resistors in parallel. It is often useful to consider only two resistors in parallel because this setup occurs commonly in practice. Also, any number of resistors in parallel can be broken down into pairs as an alternate way to find the $R_T$. Based on Equation (6–3), the formula for the total resistance of two resistors in parallel is

$$R_T = \frac{1}{\left(\frac{1}{R_1}\right) + \left(\frac{1}{R_2}\right)}$$

Combining the terms in the denominator yields

$$R_T = \frac{1}{\left(\frac{R_1 + R_2}{R_1 R_2}\right)}$$

which can be rewritten as follows:

$$R_T = \frac{R_1 R_2}{R_1 + R_2} \qquad \textbf{(6–4)}$$

Equation (6–4) states

**The total resistance for two resistors in parallel is equal to the product of the two resistors divided by the sum of the two resistors.**

This equation is sometimes referred to as the "product over the sum" formula. Example 6–9 illustrates how to use it.

**EXAMPLE 6–9** Calculate the total resistance connected to the voltage source of the circuit in Figure 6–23.

FIGURE 6–23

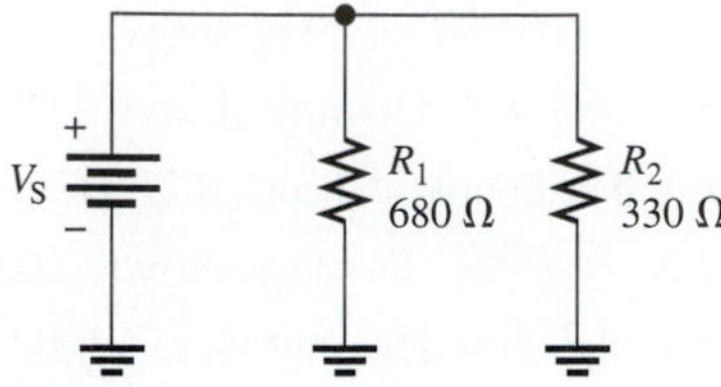


***Solution*** Use Equation (6–4) as follows:

$$R_T = \frac{R_1R_2}{R_1 + R_2} = \frac{(680\ \Omega)(330\ \Omega)}{680\ \Omega + 330\ \Omega} = \frac{224{,}400\ \Omega^2}{1010\ \Omega} = 222\ \Omega$$

The calculator sequence is

6 8 0 × 3 3 0 ÷ ( 6 8 0 + 3 3 0 ) =

***Related Exercise*** Determine $R_T$ if a 220 Ω resistor is added in parallel in Figure 6–23.

## The Case of Equal-Value Resistors in Parallel

Another special case of parallel circuits is the parallel connection of several resistors each having the same resistance value. There is a shortcut method of calculating $R_T$ when this case occurs.

If several resistors in parallel have the same resistance, they can be assigned the same symbol $R$. For example, $R_1 = R_2 = R_3 = \cdots = R_n = R$. Starting with Equation (6–3), we can develop a special formula for finding $R_T$.

$$R_T = \frac{1}{\left(\frac{1}{R}\right) + \left(\frac{1}{R}\right) + \left(\frac{1}{R}\right) + \cdots + \left(\frac{1}{R}\right)}$$

Notice that in the denominator, the same term, $1/R$, is added $n$ times ($n$ is the number of equal-value resistors in parallel). Therefore, the formula can be written as

$$R_T = \frac{1}{n/R}$$

or

$$R_T = \frac{R}{n} \qquad \textbf{(6–5)}$$

Equation (6–5) says that when any number of resistors ($n$), all having the same resistance ($R$), are connected in parallel, $R_T$ is equal to the resistance divided by the number of resistors in parallel. Example 6–10 shows how to use this formula.

**EXAMPLE 6–10** Four 8 Ω speakers are connected in parallel to the output of an amplifier. What is the total resistance across the output of the amplifier?

***Solution*** There are four 8 Ω resistors in parallel. Use Equation (6–5) as follows:

$$R_{\mathrm{T}} = \frac{R}{n} = \frac{8\ \Omega}{4} = 2\ \Omega$$

***Related Exercise*** If two of the speakers are removed, what is the resistance across the output?

## Determining an Unknown Parallel Resistor

Sometimes you need to determine the values of resistors that are to be combined to produce a desired total resistance. For example, two parallel resistors are used to obtain a known total resistance. If one resistor value is known or arbitrarily chosen, then the second resistor value can be calculated using Equation (6–4), the formula for two parallel resistors. The formula for determining the value of an unknown resistor $R_x$ is developed as follows:

$$R_{\mathrm{T}} = \frac{R_x R_A}{R_x + R_A}$$

$$R_{\mathrm{T}}(R_x + R_A) = R_x R_A$$

$$R_{\mathrm{T}}R_x + R_{\mathrm{T}}R_A = R_x R_A$$

$$R_A R_x - R_{\mathrm{T}}R_x = R_{\mathrm{T}}R_A$$

$$R_x(R_A - R_{\mathrm{T}}) = R_{\mathrm{T}}R_A$$

$$R_x = \frac{R_A R_{\mathrm{T}}}{R_A - R_{\mathrm{T}}} \qquad \textbf{(6–6)}$$

where $R_x$ is the unknown resistor and $R_A$ is the known or selected value. Example 6–11 illustrates use of this formula.

**EXAMPLE 6–11** Suppose that you wish to obtain a resistance as close to 150 Ω as possible by combining two resistors in parallel. There is a 330 Ω resistor available. What other value do you need?

***Solution*** $R_{\mathrm{T}} = 150\ \Omega$ and $R_A = 330\ \Omega$. Therefore,

$$R_x = \frac{R_A R_{\mathrm{T}}}{R_A - R_{\mathrm{T}}} = \frac{(330\ \Omega)(150\ \Omega)}{330\ \Omega - 150\ \Omega} = 275\ \Omega$$

The closest standard value is 270 Ω.

***Related Exercise*** If you need to obtain a total resistance of 130 Ω, what value can you add in parallel to the parallel combination of 330 Ω and 270 Ω? First find the value of 330 Ω and 270 Ω in parallel and treat that value as a single resistor.

### Notation for Parallel Resistors

Sometimes, for convenience, parallel resistors are designated by two parallel vertical marks. For example, $R_1$ in parallel with $R_2$ can be rewritten as $R_1 \parallel R_2$. Also, when several resistors are in parallel with each other, this notation can be used. For example,

$$R_1 \parallel R_2 \parallel R_3 \parallel R_4 \parallel R_5$$

indicates that $R_1$ through $R_5$ are all in parallel.

This notation is also used with resistance values. For example,

$$10\ \text{k}\Omega \parallel 5\ \text{k}\Omega$$

means that a 10 kΩ resistor is in parallel with a 5 kΩ resistor.

**SECTION 6–4 REVIEW**

1. Does the total resistance increase or decrease as more resistors are connected in parallel?
2. The total parallel resistance is always less than _______.
3. Write the general formula for $R_T$ with any number of resistors in parallel.
4. Write the special formula for two resistors in parallel.
5. Write the special formula for any number of equal-value resistors in parallel.
6. Calculate $R_T$ for Figure 6–24.
7. Determine $R_T$ for Figure 6–25.
8. Find $R_T$ for Figure 6–26.

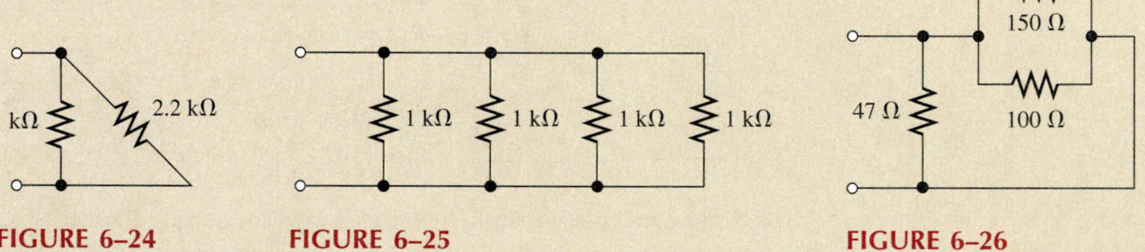


FIGURE 6–24 FIGURE 6–25 FIGURE 6–26

## 6–5 ■ OHM'S LAW IN PARALLEL CIRCUITS

***In this section, examples illustrate how Ohm's law can be applied to parallel circuit analysis.***

***After completing this section, you should be able to***

- **Apply Ohm's law in a parallel circuit**
  - ☐ Find the total current in a parallel circuit
  - ☐ Find each branch current in a parallel circuit
  - ☐ Find the voltage across a parallel circuit
  - ☐ Find the resistance in a parallel circuit

The following examples illustrate how to apply Ohm's law to determine the total current, branch currents, voltage, and resistance in parallel circuits.

**EXAMPLE 6–12** Find the total current produced by the battery in Figure 6–27.

FIGURE 6–27

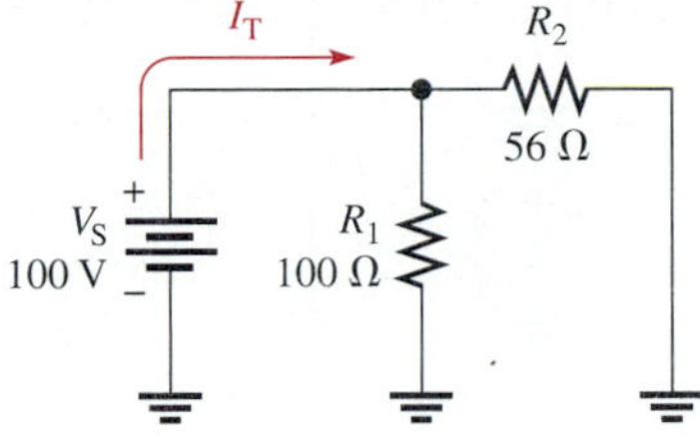


***Solution*** The battery "sees" a total parallel resistance which determines the amount of current that it generates. First, calculate $R_T$.

$$R_T = \frac{R_1R_2}{R_1 + R_2} = \frac{(100\ \Omega)(56\ \Omega)}{100\ \Omega + 56\ \Omega} = \frac{5600\ \Omega^2}{156\ \Omega} = 35.9\ \Omega$$

The battery voltage is 100 V. Use Ohm's law to find $I_T$.

$$I_T = \frac{100\ \text{V}}{35.9\ \Omega} = 2.79\ \text{A}$$

***Related Exercise*** What is $I_T$ in Figure 6–27 if $R_2$ is changed on 120 Ω? What is the current through $R_1$?

**EXAMPLE 6–13** Determine the current through each resistor in the parallel circuit of Figure 6–28.

FIGURE 6–28

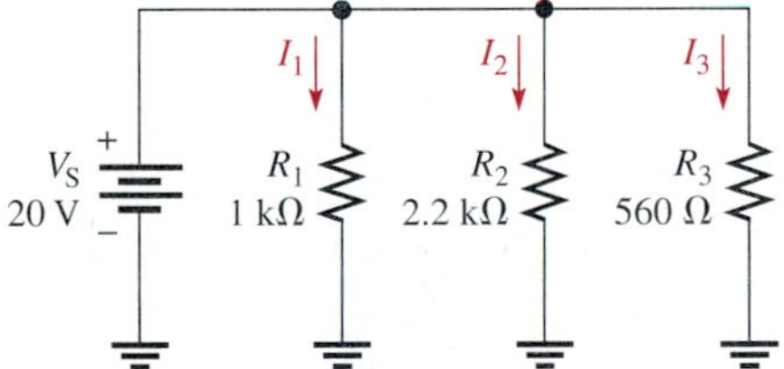


***Solution*** The voltage across each resistor (branch) is equal to the source voltage. That is, the voltage across $R_1$ is 20 V, the voltage across $R_2$ is 20 V, and the voltage across $R_3$ is 20 V. The current through each resistor is determined as follows:

$$I_1 = \frac{V_S}{R_1} = \frac{20\ \text{V}}{1\ \text{k}\Omega} = 20\ \text{mA}$$

$$I_2 = \frac{V_S}{R_2} = \frac{20\ \text{V}}{2.2\ \text{k}\Omega} = 9.09\ \text{mA}$$

$$I_3 = \frac{V_S}{R_3} = \frac{20\ \text{V}}{560\ \Omega} = 35.7\ \text{mA}$$

***Related Exercise*** If an additional resistor of 910 Ω is connected in parallel to the circuit in Figure 6–28, determine all of the branch currents.

**EXAMPLE 6–14** Find the voltage across the parallel circuit in Figure 6–29.

**FIGURE 6–29**

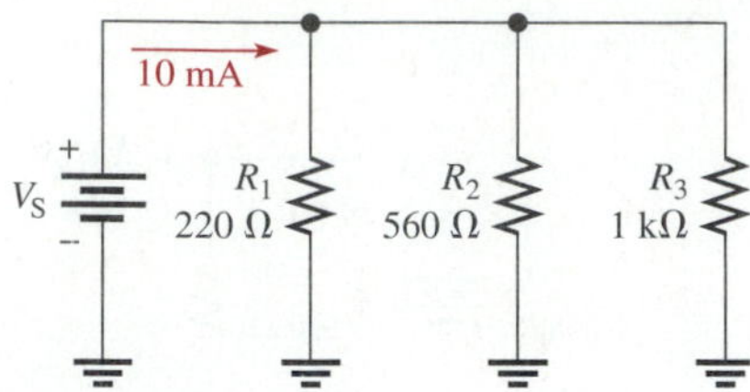


***Solution*** The total current into the parallel circuit is 10 mA. If you know the total resistance, than you can apply Ohm's law to get the voltage. The total resistance is

$$R_T = \frac{1}{\left(\frac{1}{R_1}\right) + \left(\frac{1}{R_2}\right) + \left(\frac{1}{R_3}\right)} = \frac{1}{\left(\frac{1}{220\ \Omega}\right) + \left(\frac{1}{560\ \Omega}\right) + \left(\frac{1}{1\ \text{k}\Omega}\right)}$$

$$= \frac{1}{4.55\ \text{mS} + 1.79\ \text{mS} + 1\ \text{mS}} = \frac{1}{7.34\ \text{mS}} = 136\ \Omega$$

Therefore, the source voltage is

$$V_S = I_T R_T = (10\ \text{mA})(136\ \Omega) = 1.36\ \text{V}$$

***Related Exercise*** Find the voltage if $R_3$ is decreased to 680 Ω in Figure 6–29 and $I_T$ is 10 mA.

**EXAMPLE 6–15** The circuit board in Figure 6–30 has three resistors in parallel used for setting the gain of an instrumentation amplifier. The values of two of the resistors are known from the color codes, but the top resistor is not clearly marked (maybe the bands are worn off from handling). Determine the value of the unknown resistor using only an ammeter and a dc power supply.

**FIGURE 6–30**

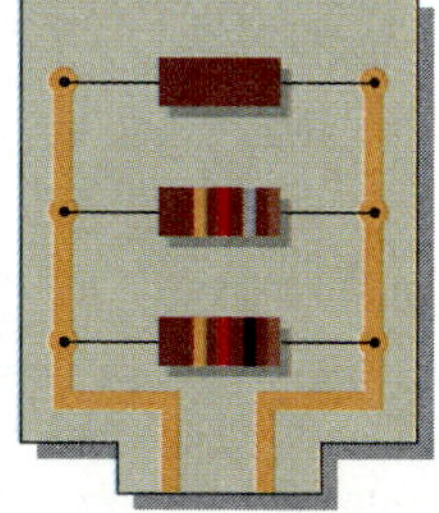

***Solution*** If you can determine the total resistance of the three resistors in parallel, then you can use the parallel-resistance formula to calculate the unknown resistance. You can use Ohm's law to find the total resistance if voltage and total current are known.

In Figure 6–31, a 10 V source (arbitrary value) is connected across the resistors, and the total current is measured. Using these measured values, find the total resistance.

$$R_T = \frac{V}{I_T} = \frac{10\ \text{V}}{20.1\ \text{mA}} = 498\ \Omega$$

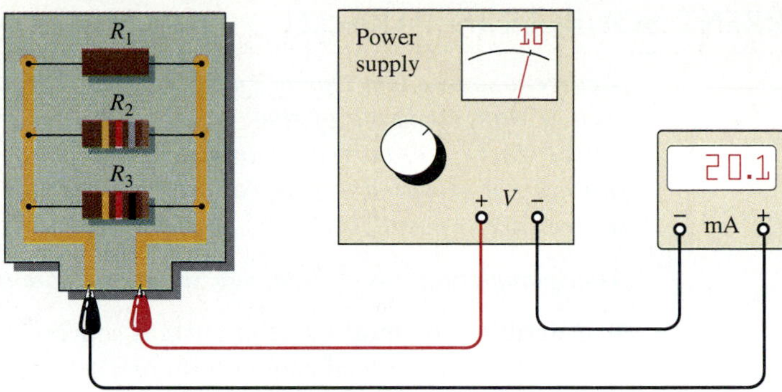


FIGURE 6–31

Use Equation (6–2) to find the unknown resistance as follows:

$$\frac{1}{R_T} = \frac{1}{R_1} + \frac{1}{R_2} + \frac{1}{R_3}$$

$$\frac{1}{R_1} = \frac{1}{R_T} - \frac{1}{R_2} - \frac{1}{R_3} = \frac{1}{498\ \Omega} - \frac{1}{1.8\ \text{k}\Omega} - \frac{1}{1\ \text{k}\Omega} = 453\ \mu\text{S}$$

$$R_1 = \frac{1}{453\ \mu\text{S}} = 2.21\ \text{k}\Omega$$

The calculator sequence for $R_1$ is

4 9 8 2nd 1/x − 1 • 8 EXP 3 2nd 1/x − 1 EXP 3 2nd 1/x = 2nd 1/x

***Related Exercise*** Is there an easier way to determine the value of $R_1$? If so, what is it?

## SECTION 6–5 REVIEW

1. A 10 V battery is connected across three 68 Ω resistors that are in parallel. What is the total current from the battery?
2. How much voltage is required to produce 2 A of current through the circuit of Figure 6–32?
3. How much current is there through each resistor of Figure 6–32?
4. There are four equal-value resistors in parallel with a 12 V source, and 5.85 mA of current from the source. What is the value of each resistor?
5. A 1 kΩ and a 2.2 kΩ resistor are connected in parallel. There is a total of 100 mA through the parallel combination. How much voltage is dropped across the resistors?

FIGURE 6–32

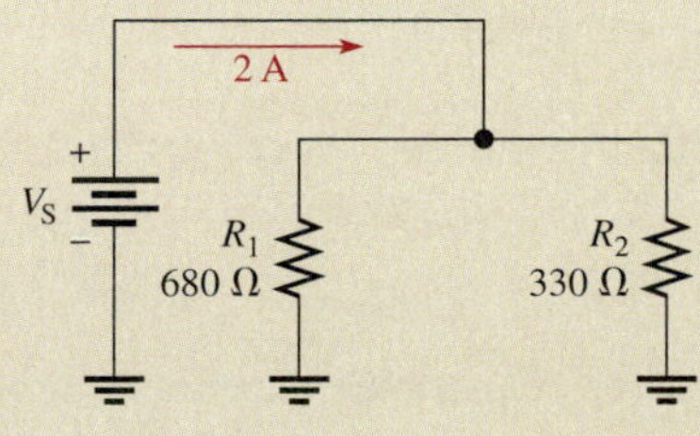

## 6–6 ■ CURRENT SOURCES IN PARALLEL

***A current source is a type of energy source that provides a constant current to a load even if the resistance of that load changes. A transistor can be used as a current source; therefore, current sources are important in electronic circuits. At this point, you are not prepared to study transistors, but you do need to understand how current sources act in parallel.***

***After completing this section, you should be able to***

- **Determine the total effect of current sources in parallel**
  - ☐ Determine the total current from parallel sources having the same direction
  - ☐ Determine the total current from parallel sources having opposite directions

In general, the total current produced by current sources in parallel is equal to the algebraic sum of the individual current sources. The algebraic sum means that you must consider the direction of current when you combine the sources in parallel. For example, in Figure 6–33(a), the three current sources in parallel provide current in the same direction (into point $A$). So the total current into point $A$ is

$$I_T = 1\text{ A} + 2\text{ A} + 2\text{ A} = 5\text{ A}$$

In Figure 6–33(b), the 1 A source provides current in a direction opposite to the other two. The total current into point $A$ in this case is

$$I_T = 2\text{ A} + 2\text{ A} - 1\text{ A} = 3\text{ A}$$

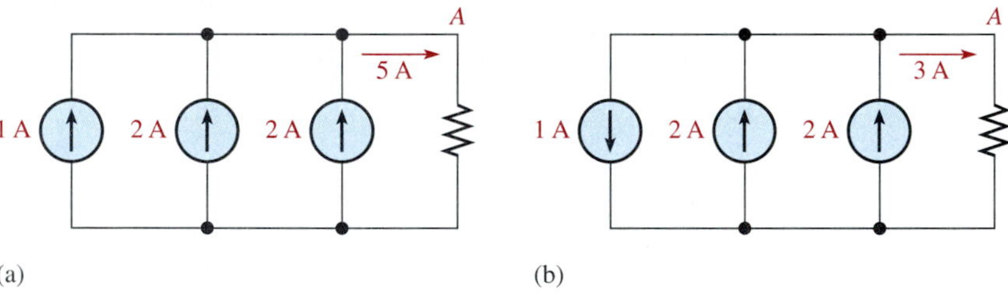


FIGURE 6–33

**EXAMPLE 6–16** Determine the current through $R_L$ in Figure 6–34.

FIGURE 6–34

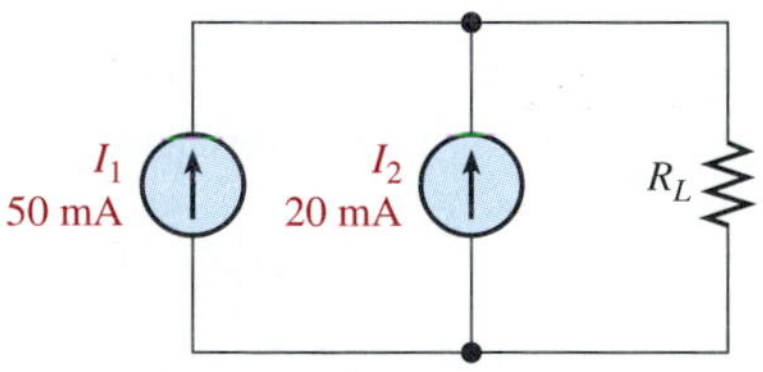

***Solution*** The two current sources are in the same direction; so the current through $R_L$ is

$$I_L = I_1 + I_2 = 50\text{ mA} + 20\text{ mA} = 70\text{ mA}$$

***Related Exercise*** Determine the current through $R_L$ if the direction of $I_2$ is reversed.

**SECTION 6–6 REVIEW**

1. Four 0.5 A current sources are connected in parallel in the same direction. What current will be produced through a load resistor?
2. How many 1 A current sources must be connected in parallel to produce a total current output of 3 A? Sketch a schematic showing the sources connected.
3. In a transistor amplifier circuit, the transistor can be represented by a 10 mA current source, as shown in Figure 6–35. The transistors act in parallel, as in the case of a differential amplifier. How much current is there through the resistor $R_E$?

FIGURE 6–35

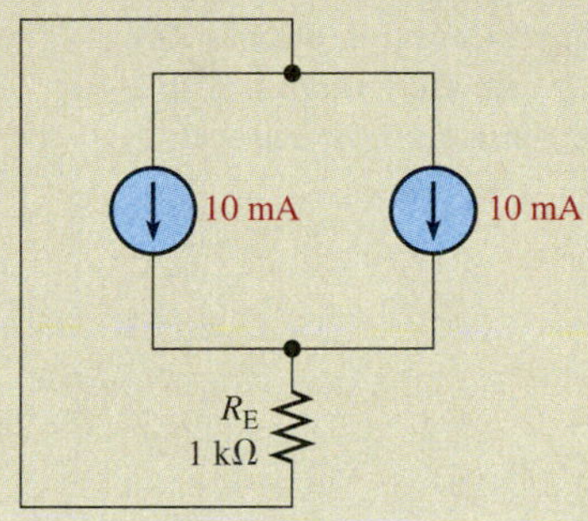


## 6–7 ■ CURRENT DIVIDERS

***A parallel circuit acts as a current divider because the current entering the junction of parallel branches "divides" up into several individual branch currents.***

***After completing this section, you should be able to***

- **Use a parallel circuit as a current divider**
  - Apply the current-divider formula
  - Determine an unknown branch current

In a parallel circuit, the total current into the junction of parallel branches divides among the branches. Thus, a parallel circuit acts as a current divider. This current-divider principle is illustrated in Figure 6–36 for a two-branch parallel circuit in which part of the total current $I_T$ goes through $R_1$ and part through $R_2$.

FIGURE 6–36
*Total current divides between two branches.*

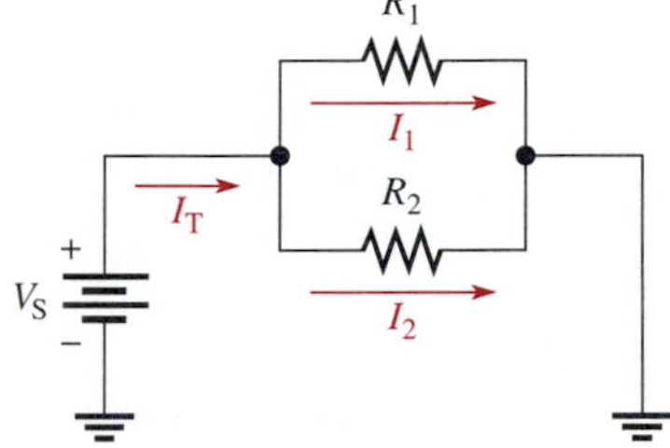


Since the same voltage is across each of the resistors in parallel, the branch currents are inversely proportional to the values of the resistors. For example, if the value of $R_2$ is twice that of $R_1$, then the value of $I_2$ is one-half that of $I_1$. In other words,

**The total current divides among parallel resistors in a manner inversely proportional to the resistance values.**

The branches with higher resistance have less current, and the branches with lower resistance have more current, in accordance with Ohm's law. If all the branches have the same resistance, the branch currents are all equal.

Figure 6–37 shows specific values to demonstrate how the currents divide according to the branch resistances. Notice that in this case the resistance of the upper branch is one-tenth the resistance of the lower branch, but the upper branch current is ten times the lower branch current.

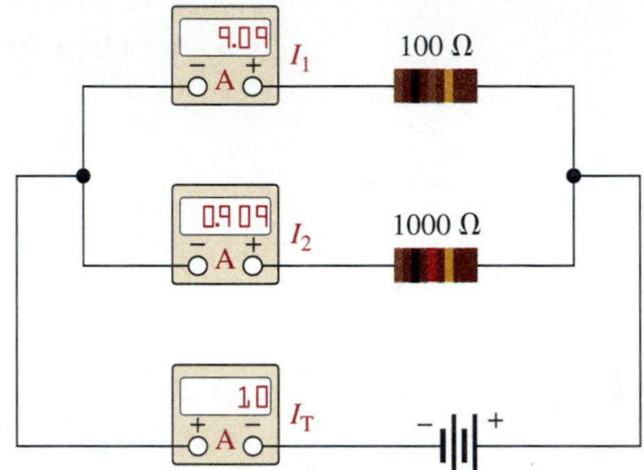


**FIGURE 6–37**
***The branch with the lowest resistance has the most current, and the branch with the highest resistance has the least current.***

## General Current-Divider Formula for Any Number of Parallel Branches

With a few steps, a formula for determining how currents divide among parallel resistors can be developed. Let's assume that we have $n$ resistors in parallel, as shown in Figure 6–38, where $n$ can be any number.

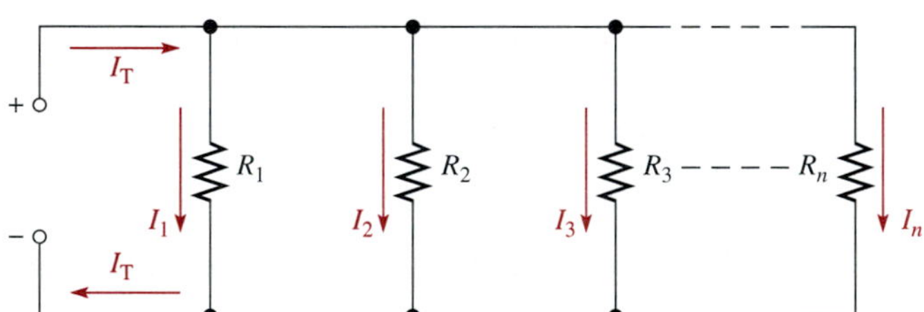


**FIGURE 6–38**
***Generalized parallel circuit with n branches.***

Let's call the current through any one of the parallel resistors $I_x$, where $x$ represents the number of a particular resistor (1, 2, 3, and so on). By Ohm's law, the current through any one of the resistors in Figure 6–38 can be written as follows:

$$I_x = \frac{V_S}{R_x}$$

The source voltage, $V_S$, appears across each of the parallel resistors, and $R_x$ represents any one of the parallel resistors. The total source voltage $V_S$ is equal to the total current times the total parallel resistance.

$$V_S = I_T R_T$$

Substituting $I_T R_T$ for $V_S$ in the expression for $I_x$ results in

$$I_x = \frac{I_T R_T}{R_x}$$

Rearranging terms yields

$$I_x = \left(\frac{R_T}{R_x}\right)I_T \quad \textbf{(6–7)}$$

where $x = 1, 2, 3$, etc. Equation (6–7) is the general current-divider formula and applies to a parallel circuit with any number of branches.

**The current ($I_x$) through any branch equals the total parallel resistance ($R_T$) divided by the resistance ($R_x$) of that branch, and then multiplied by the total current ($I_T$) into the junction of parallel branches.**

**EXAMPLE 6–17** Determine the current through each resistor in the circuit of Figure 6–39.

FIGURE 6–39

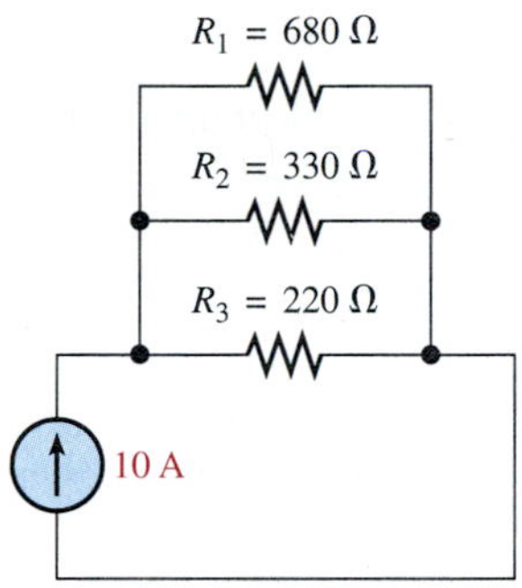

$R_1 = 680\ \Omega$, $R_2 = 330\ \Omega$, $R_3 = 220\ \Omega$, 10 A

***Solution*** First calculate the total parallel resistance.

$$R_T = \frac{1}{\left(\frac{1}{R_1}\right) + \left(\frac{1}{R_2}\right) + \left(\frac{1}{R_3}\right)} = \frac{1}{\left(\frac{1}{680\ \Omega}\right) + \left(\frac{1}{330\ \Omega}\right) + \left(\frac{1}{220\ \Omega}\right)} = 111\ \Omega$$

The total current is 10 A. Use Equation (6–7) to calculate each branch current.

$$I_1 = \left(\frac{R_T}{R_1}\right)I_T = \left(\frac{111\ \Omega}{680\ \Omega}\right)10\text{ A} = 1.63\text{ A}$$

$$I_2 = \left(\frac{R_T}{R_2}\right)I_T = \left(\frac{111\ \Omega}{330\ \Omega}\right)10\text{ A} = 3.36\text{ A}$$

$$I_3 = \left(\frac{R_T}{R_3}\right)I_T = \left(\frac{111\ \Omega}{220\ \Omega}\right)10\text{ A} = 5.05\text{ A}$$

***Related Exercise*** Determine the current through each resistor in Figure 6–39 if $R_3$ is removed.

## Current-Divider Formulas for Two Branches

Two parallel resistors are often found in practical circuits. Equation (6–7) can be modified for the case of two parallel branches. Let's start by restating Equation (6–4), the formula for the total resistance of two parallel branches.

$$R_T = \frac{R_1R_2}{R_1 + R_2}$$

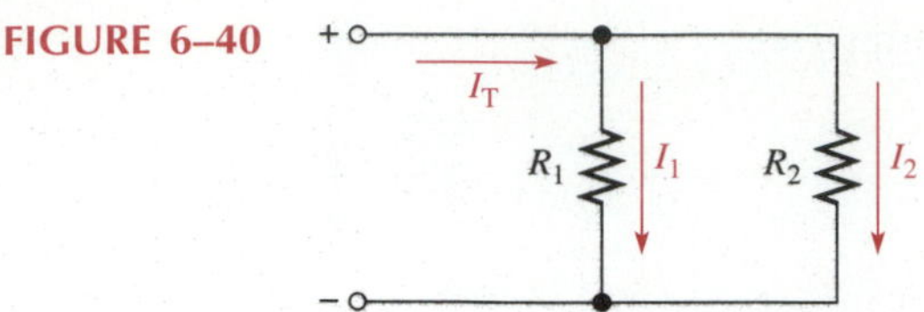


FIGURE 6–40

When there are two parallel resistors, as in Figure 6–40, you may want to find the current through either or both of the branches. To do this, you need two special formulas.

Based on Equation (6–7), the formulas for $I_1$ and $I_2$ can be written as follows:

$$I_1 = \left(\frac{R_T}{R_1}\right)I_T \quad \text{and} \quad I_2 = \left(\frac{R_T}{R_2}\right)I_T$$

Substituting $R_1R_2/(R_1 + R_2)$ for $R_T$ and canceling result in

$$I_1 = \frac{\left(\frac{\cancel{R_1}R_2}{R_1 + R_2}\right)}{\cancel{R_1}}I_T \quad \text{and} \quad I_2 = \frac{\left(\frac{R_1\cancel{R_2}}{R_1 + R_2}\right)}{\cancel{R_2}}I_T$$

Therefore, the current-divider formulas for two branches are the following:

$$I_1 = \left(\frac{R_2}{R_1 + R_2}\right)I_T \tag{6–8}$$

$$I_2 = \left(\frac{R_1}{R_1 + R_2}\right)I_T \tag{6–9}$$

When there are only two resistors in parallel, these equations are a little easier to use than Equation (6–7) because it is not necessary to know $R_T$.

Note that in Equations (6–8) and (6–9), the current in one of the branches is equal to the opposite branch resistance divided by the sum of the two resistors, all multiplied by the total current. In all applications of the current-divider equations, you must know the total current into the parallel branches. Example 6–18 illustrates application of these special current-divider formulas.

**EXAMPLE 6–18** Find $I_1$ and $I_2$ in Figure 6–41.

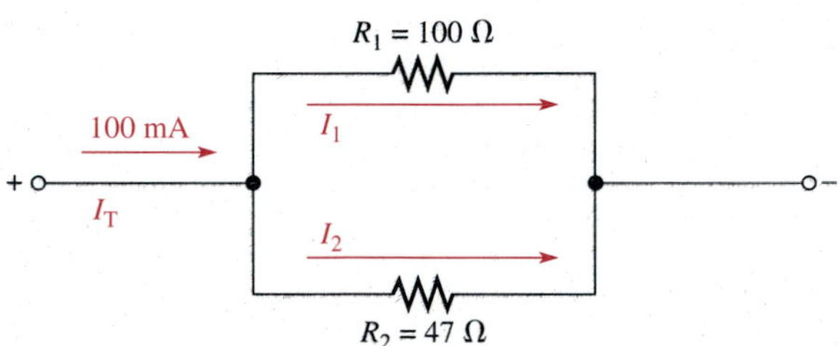


FIGURE 6–41

***Solution*** Use Equation (6–8) to determine $I_1$.

$$I_1 = \left(\frac{R_2}{R_1 + R_2}\right)I_T = \left(\frac{47\ \Omega}{147\ \Omega}\right)100\text{ mA} = 32.0\text{ mA}$$

Use Equation (6–9) to determine $I_2$.

$$I_2 = \left(\frac{R_2}{R_1 + R_2}\right)I_T = \left(\frac{100\ \Omega}{147\ \Omega}\right)100\ \text{mA} = 68.0\ \text{mA}$$

***Related Exercise*** If $R_1 = 56\ \Omega$, and $R_2 = 82\ \Omega$ in Figure 6–41 and $I_T$ stays the same, what will each branch current be?

**SECTION 6–7 REVIEW**

1. Write the general current-divider formula.
2. Write the two special formulas for calculating each branch current for a two-branch circuit.
3. A parallel circuit has the following resistors in parallel: 220 Ω, 100 Ω, 82 Ω, 47 Ω, and 22 Ω. Which resistor has the most current through it? The least current?
4. Determine the current through $R_3$ in Figure 6–42.
5. Find $I_1$ and $I_2$ in the circuit of Figure 6–43.

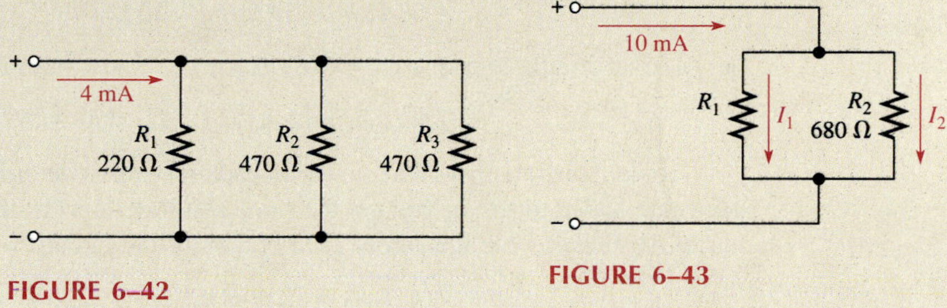


FIGURE 6–42

FIGURE 6–43

## 6–8 ■ POWER IN PARALLEL CIRCUITS

***Total power in a parallel circuit is found by adding up the powers of all the individual resistors, the same as you did for series circuits.***

***After completing this section, you should be able to***

■ **Determine power in a parallel circuit**

Equation (6–10) states the formula for finding total power in a concise way for any number of resistors in parallel.

$$P_T = P_1 + P_2 + P_3 + \cdots + P_n \qquad \textbf{(6–10)}$$

where $P_T$ is the total power and $P_n$ is the power in the last resistor in parallel. As you can see, the powers are additive, just as in the series circuit.

The power formulas in Chapter 4 are directly applicable to parallel circuits. The following formulas are used to calculate the total power $P_T$:

$$P_T = VI_T$$

$$P_T = I_T^2 R_T$$

$$P_T = \frac{V^2}{R_T}$$

where $V$ is the voltage across the parallel circuit, $I_T$ is the total current into the parallel circuit, and $R_T$ is the total resistance of the parallel circuit. Example 6–19 shows how total power can be calculated in a parallel circuit.

**EXAMPLE 6–19** Determine the total amount of power in the parallel circuit in Figure 6–44.

FIGURE 6–44

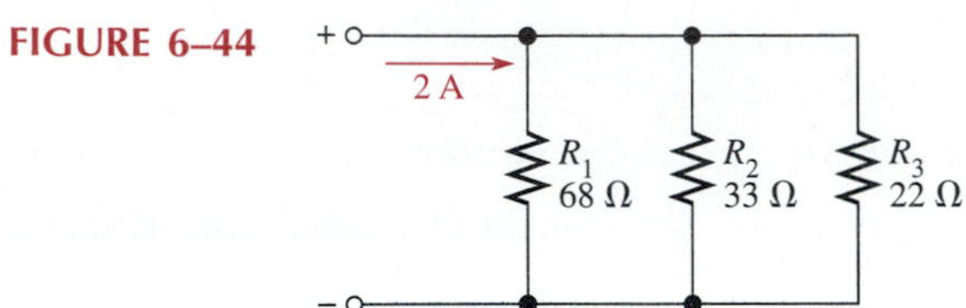


***Solution*** The total current is 2 A. The total resistance is

$$R_T = \frac{1}{\left(\frac{1}{68\ \Omega}\right) + \left(\frac{1}{33\ \Omega}\right) + \left(\frac{1}{22\ \Omega}\right)} = 11.1\ \Omega$$

The easiest formula to use is $P_T = I_T^2R_T$ since you know both $I_T$ and $R_T$. Thus,

$$P_T = I_T^2R_T = (2\ \text{A})^2(11.1\ \Omega) = 44.4\ \text{W}$$

To demonstrate that if the power in each resistor is determined and if all of these values are added together, you get the same result, let's work through another calculation. First, find the voltage across each branch of the circuit.

$$V_S = I_TR_T = (2\ \text{A})(11.1\ \Omega) = 22.2\ \text{V}$$

Remember that the voltage across all branches is the same.

Next, use $P = V^2/R$ to calculate the power for each resistor.

$$P_1 = \frac{(22.2\ \text{V})^2}{68\ \Omega} = 7.25\ \text{W}$$

$$P_2 = \frac{(22.2\ \text{V})^2}{33\ \Omega} = 14.9\ \text{W}$$

$$P_3 = \frac{(22.2\ \text{V})^2}{22\ \Omega} = 22.4\ \text{W}$$

Now, add these powers to get the total power.

$$P_T = 7.25\ \text{W} + 14.9\ \text{W} + 22.4\ \text{W} = 44.6\ \text{W}$$

This calculation shows that the sum of the individual powers is equal (approximately) to the total power as determined by one of the power formulas. Rounding to three significant figures accounts for the difference.

***Related Exercise*** Find the total power in Figure 6–44 if the total current is doubled.

**EXAMPLE 6–20** The amplifier in one channel of a stereo system as shown in Figure 6–45 drives four speakers. If the maximum voltage to the speakers is 15 V, how much power must the amplifier be able to deliver to the speakers?

**FIGURE 6–45**

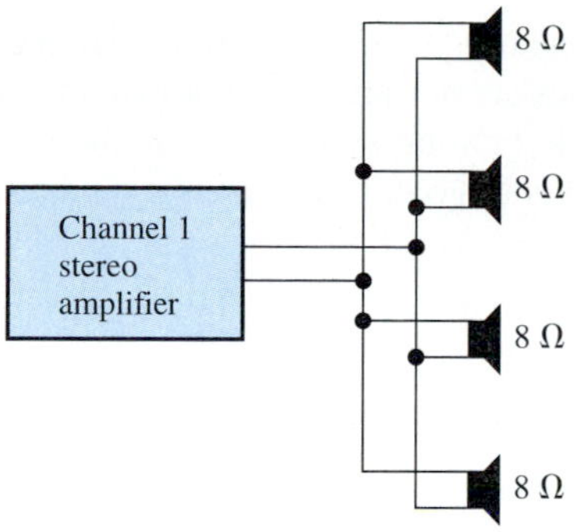

*Solution* The speakers are connected in parallel to the amplifier output, so the voltage across each is the same. The maximum power to each speaker is

$$P_{max} = \frac{V_{max}^2}{R} = \frac{(15\text{ V})^2}{8\ \Omega} = 28.1\text{ W}$$

The total power that the amplifier must be capable of delivering to the speaker system is four times the power in an individual speaker because the total power is the sum of the individual powers.

$$P_{T(max)} = P_{(max)} + P_{(max)} + P_{(max)} + P_{(max)} = 4P_{(max)} = 4(28.1\text{ W}) = 112.4\text{ W}$$

*Related Exercise* If the amplifier can produce a maximum of 18 V, what is the maximum total power to the speakers?

**SECTION 6–8 REVIEW**

1. If you know the power in each resistor in a parallel circuit, how can you find the total power?
2. The resistors in a parallel circuit dissipate the following powers: 2.38 W, 5.12 W, 1.09 W, and 8.76 W. What is the total power in the circuit?
3. A circuit has a 1 kΩ, a 2.7 kΩ, and a 3.9 kΩ resistor in parallel. There is a total current of 1 A into the parallel circuit. What is the total power?

## 6–9 ■ EXAMPLES OF PARALLEL CIRCUIT APPLICATIONS

***Parallel circuits are found in some form in virtually every electronic system. In many of these applications, the parallel relationship of components may not be obvious until you have covered some advanced topics that you will study later. For now, we will look at some common and familiar applications of parallel circuits.***

***After completing this section, you should be able to***

- **Describe some basic applications of parallel circuits**
  - ☐ Discuss the lighting system in automobiles
  - ☐ Discuss residential wiring

□ Describe how current through a heating element is controlled
□ Explain basically how a multiple-range ammeter works

## Automotive

One advantage of a parallel circuit over a series circuit is that when one branch opens, the other branches are not affected. For example, Figure 6–46 shows a simplified diagram of an automobile lighting system. When one headlight on a car goes out, it does not cause the other lights to go out, because they are all in parallel.

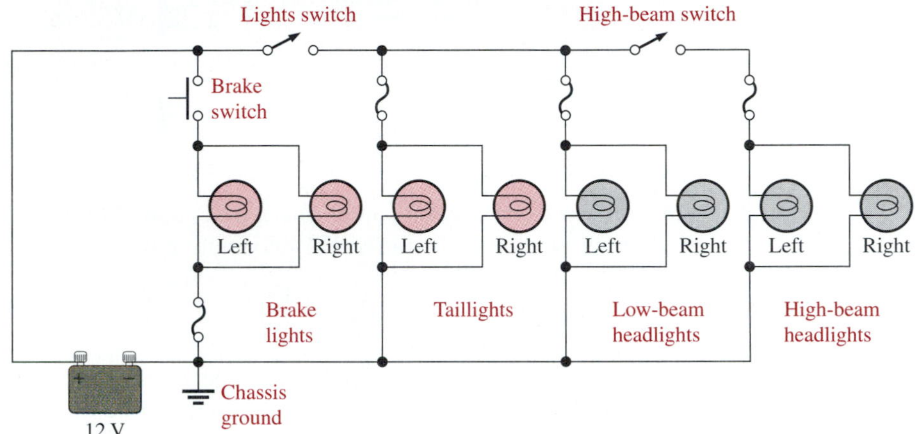


**FIGURE 6–46**
***Simplified diagram of the exterior light system of an automobile.***

Notice that the brake lights are switched on independently of the head and taillights. They come on only when the driver closes the brake light switch by depressing the brake pedal. When the lights switch is closed, both low-beam headlights and both taillights are on. The high-beam headlights are on only when both the lights switch and the high-beam switch are closed. If any one of the lights burns out (opens), there is still current in each of the other lights.

## Residential

Another common use of parallel circuits is in residential electrical systems. All the lights and appliances in a home are wired in parallel. Figure 6–47(a) shows a typical room wiring arrangement with two switch-controlled lights and three wall outlets in parallel.

Figure 6–47(b) shows a simplified parallel arrangement of four heating elements on an electric range. The four-position switches in each branch allow the user to control the amount of current through the heating elements by selecting the appropriate limiting resistor. The lowest resistor value (H setting) allows the highest amount of current for maximum heat. The highest resistor value (L setting) allows the least amount of current for minimum heat; M designates the medium settings.

## Ammeters

Another example in which parallel circuits are used is the familiar analog (needle-type) ammeter or milliammeter. Parallel circuits are an important part of the operation of the ammeter because they allow the user to select various ranges in order to measure many different current values.

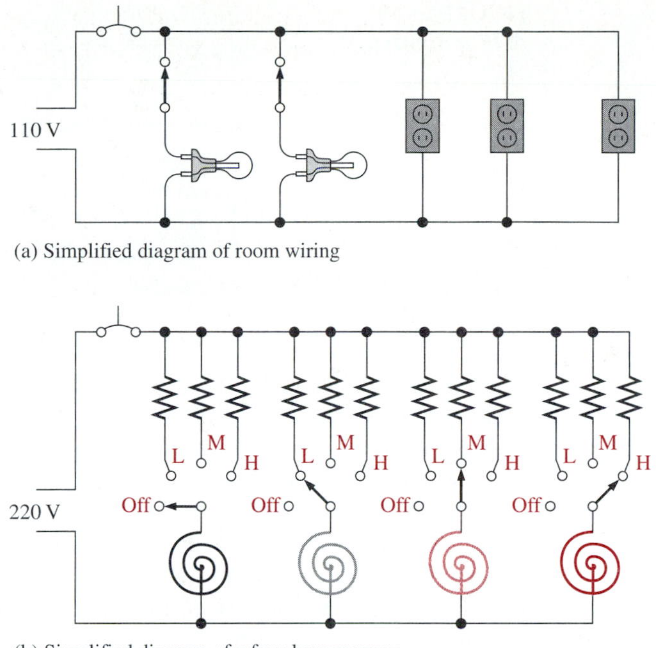


(a) Simplified diagram of room wiring

(b) Simplified diagram of a four-burner range

**FIGURE 6–47**
***Examples of parallel circuits in residential wiring and appliances.***

The mechanism in an ammeter that causes the pointer to move in proportion to the current is called the *meter movement,* which is based on a magnetic principle that you will learn later. Right now, it is sufficient to know that a meter movement has a certain resistance and a maximum current. This maximum current, called the *full-scale deflection current,* causes the pointer to go all the way to the end of the scale. For example, a certain meter movement has a 50 Ω resistance and a full-scale deflection current of 1 mA. A meter with this particular movement can measure currents of 1 mA or less as indicated in Figure 6–48(a) and (b). Currents greater than 1 mA will cause the pointer to "peg" (or stop) at full scale as indicated in part (c).

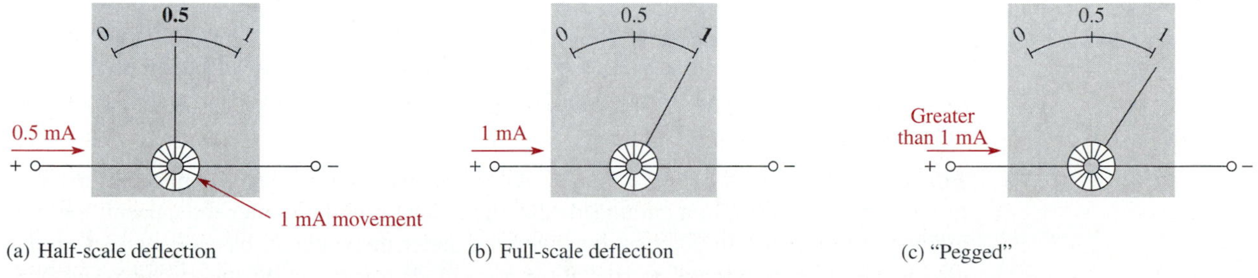


(a) Half-scale deflection

(b) Full-scale deflection

(c) "Pegged"

**FIGURE 6–48**
***A 1 mA meter.***

Figure 6–49 shows a simple ammeter with a resistor in parallel with the meter movement; this resistor is called a *shunt resistor.* Its purpose is to bypass a portion of current around the meter movement to extend the range of currents that can be measured. The figure specifically shows 9 mA through the shunt resistor and 1 mA through the meter movement. Thus, up to 10 mA can be measured. To find the actual current value, simply multiply the reading on the scale by 10.

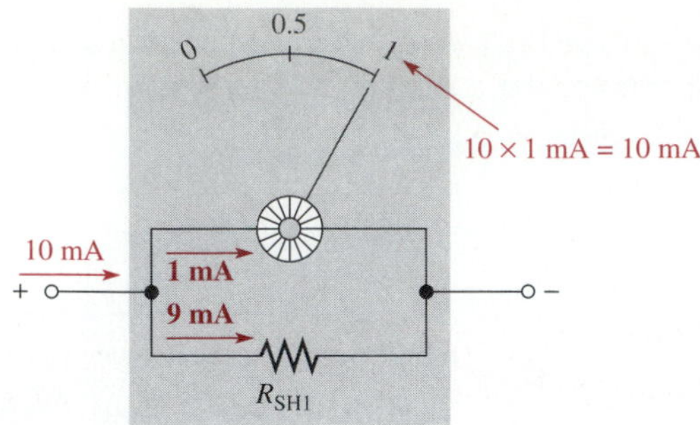


**FIGURE 6–49**
*A 10 mA meter.*

A practical ammeter has a range switch that permits the selection of several full-scale current settings. In each switch position, a certain amount of current is bypassed through a parallel resistor as determined by the resistance value. In our example, the current though the movement is never greater than 1 mA.

Figure 6–50 illustrates a meter with three ranges: 1 mA, 10 mA, and 100 mA. When the range switch is in the 1 mA position, all of the current into the meter goes through the meter movement. In the 10 mA setting, up to 9 mA goes through $R_{SH1}$ and up to 1 mA through the movement. In the 100 mA setting, up to 99 mA goes through $R_{SH2}$, and the movement can still have only 1 mA for full-scale.

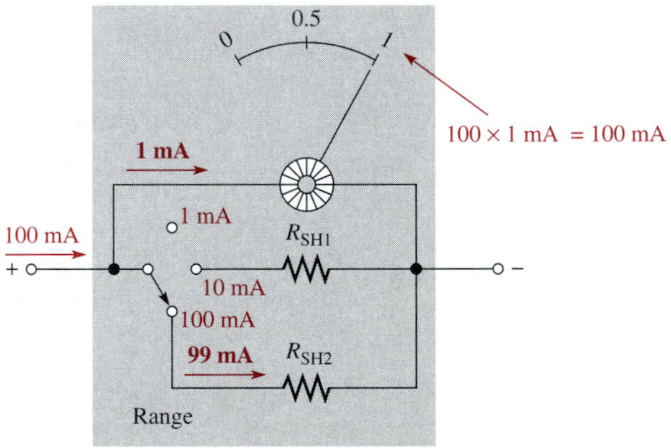


**FIGURE 6–50**
*A milliammeter with three ranges.*

For example, in Figure 6–50, if 50 mA of current are being measured, the needle points at the 0.5 mark on the scale; you must multiply 0.5 by 100 to find the current value. In this situation, 0.5 mA is through the movement (half-scale deflection) and 49.5 mA are through $R_{SH2}$.

***Effect of the Ammeter on a Circuit*** As you know, an ammeter is connected in series to measure the current in a circuit. Ideally, the meter should not alter the current that it is intended to measure. In practice, however, the meter unavoidably has some effect on the circuit, because its internal resistance is connected in series with the circuit resistance. However, in most cases, the meter's internal resistance is so small compared to the circuit resistance that it can be neglected.

For example, if the meter has a 50 Ω movement ($R_M$) and a 100 μA full-scale current ($I_M$), the voltage dropped across the movement is

$$V_M = I_M R_M = (100\ \mu\text{A})(50\ \Omega) = 5\text{ mV}$$

The shunt resistance ($R_{SH}$) for the 10 mA range, for example, is

$$R_{SH} = \frac{V_M}{I_{SH}} = \frac{5\text{ mV}}{9.9\text{ mA}} = 0.505\ \Omega$$

As you can see, the total resistance of the ammeter on the 10 mA range is the resistance of the movement in parallel with the shunt resistance.

$$R_M \parallel R_{SH} = 50\ \Omega \parallel 0.505\ \Omega = 0.5\ \Omega$$

**EXAMPLE 6–21** How much does an ammeter with a 100 $\mu$A, 50 Ω movement affect the current in the circuit of Figure 6–51?

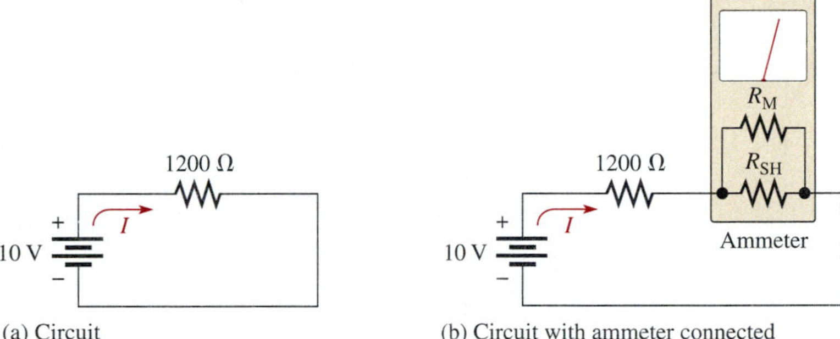


**FIGURE 6–51**

***Solution*** The true current in the circuit is

$$I_{true} = \frac{10\ \text{V}}{1200\ \Omega} = 8.3333\ \text{mA}$$

The meter is set on the 10 mA range in order to measure this particular amount of current. It was found that the meter's resistance on the 10 mA range is 0.5 Ω. When the meter is connected in the circuit, its resistance is in series with the 1200 Ω resistor. Thus, there is a total of 1200.5 Ω.

The current actually measured by the ammeter is

$$I_{meas} = \frac{10\ \text{V}}{1200.5\ \Omega} = 8.3299\ \text{mA}$$

This current differs from the true circuit current by only 0.04%. Therefore, the meter does not significantly alter the current value, a situation which, of course, is necessary because the measuring instrument should not change the quantity that is to be measured accurately.

***Related Exercise*** How much will the measured current differ from the true current if the circuit resistance is 12 kΩ rather than 1200 Ω?

## SECTION 6–9 REVIEW

1. For the ammeter in Figure 6–51, what is the maximum resistance that the meter will have when connected in a circuit? What is the maximum current that can be measured at the setting?
2. Do the shunt resistors have resistance values considerably less than or more than that of the meter movement? Why?

## 6–10 ■ TROUBLESHOOTING

***In this section, you will see how an open in a parallel branch affects the parallel circuit.***

***After completing this section, you should be able to***

- **Troubleshoot parallel circuits**
  - ☐ Check for an open circuit

### Open Branches

Recall that an open circuit is one in which the current path is interrupted and there is no current. In this section we examine what happens when a branch of a parallel circuit opens.

If a switch is connected in a branch of a parallel circuit, as shown in Figure 6–52, an open or a closed path can be made by the switch. When the switch is closed, as in Figure 6–52(a), $R_1$ and $R_2$ are in parallel. The total resistance is 50 Ω (two 100 Ω resistors in parallel). Current is through both resistors. If the switch is opened, as in Figure 6–52(b), $R_1$ is effectively removed from the circuit, and the total resistance is 100 Ω. Current is now only through $R_2$.

**When an open circuit occurs in a parallel branch, the total resistance increases, the total current decreases, and the same current continues through each of the remaining parallel paths.**

The decrease in total current equals the amount of current that was previously in the open branch. The other branch currents remain the same.

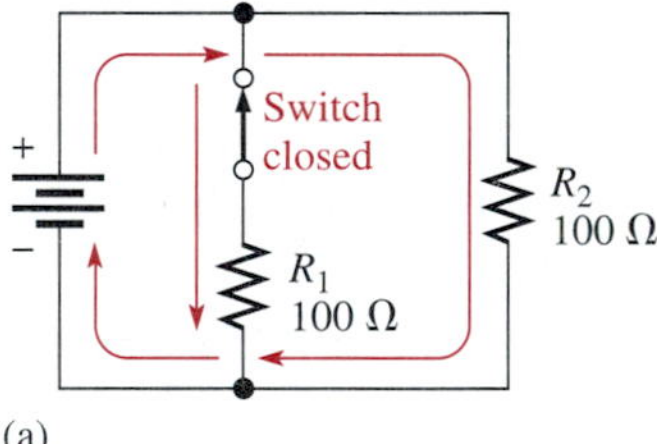


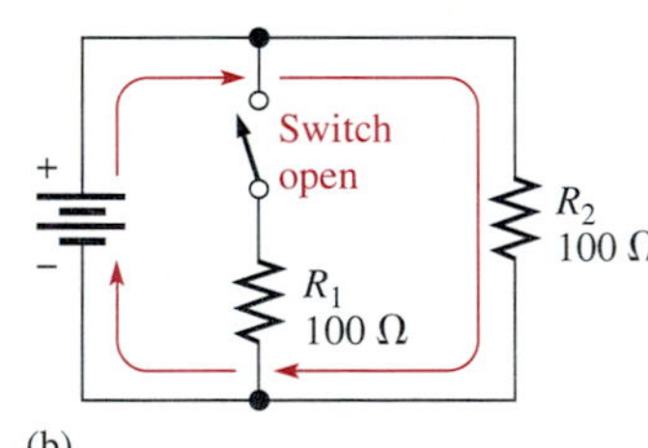


FIGURE 6–52
***When switch opens, total current decreases and current through $R_2$ remains unchanged.***

Consider the lamp circuit in Figure 6–53. There are four bulbs in parallel with a 120 V source. In part (a), there is current through each bulb. Now suppose that one of the bulbs burns out, creating an open path as shown in Figure 6–53(b). This light will go out because there is no current through the open path. Notice, however, that current continues through all the other parallel bulbs, and they continue to glow. The open branch does not change the voltage across the parallel branches; it remains at 120 V.

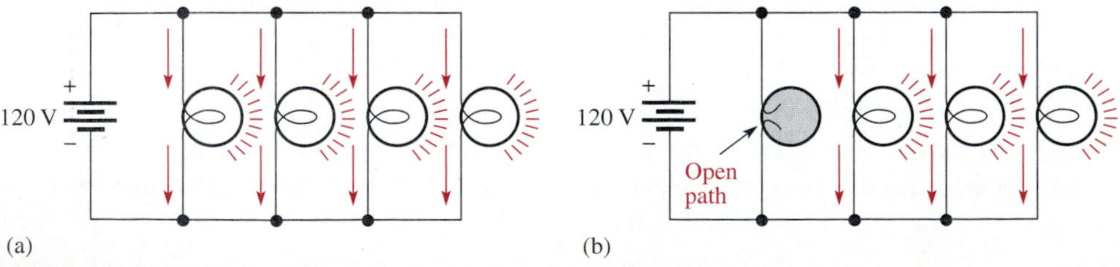


FIGURE 6–53
***When one lamp opens, total current decreases and other branch currents remain unchanged.***

You can see that a parallel circuit has an advantage over a series circuit in lighting systems because if one or more of the parallel bulbs burn out, the others will stay on. In a series circuit, when one bulb goes out, all of the others go out also because the current path is completely interrupted.

When a resistor in a parallel circuit opens, the open resistor cannot be located by measurement of the voltage across the branches because the same voltage exists across all the branches. Thus, there is no way to tell which resistor is open by simply measuring voltage. The good resistors will always have the same voltage as the open one, as illustrated in Figure 6–54 (note that the middle resistor is open).

If a visual inspection does not reveal the open resistor, it must be located by current measurements. In practice, measuring current is more difficult than measuring voltage because you must insert the ammeter in series to measure the current. Thus, a wire or a printed circuit board connection must be cut or disconnected, or one end of a component must be lifted off the circuit board, in order to connect the ammeter in series. This procedure, of course, is not required when voltage measurements are made, because the meter leads are simply connected across a component.

**FIGURE 6–54**
***Parallel branches (open or not) have the same voltage.***

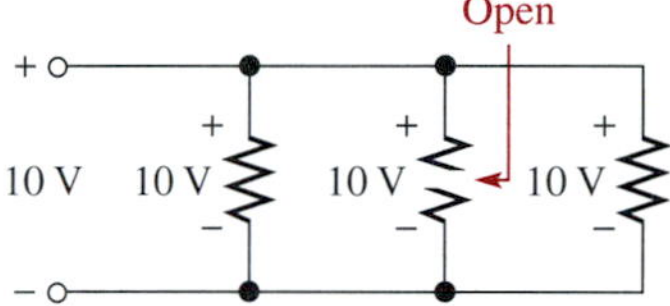


## Finding an Open Branch by Current Measurement

In a parallel circuit, the total current should be measured. *When a parallel resistor opens, the total current $I_T$, is always less than its normal value.* Once $I_T$ and the voltage across the branches are known, a few calculations will determine the open resistor when all the resistors are of different resistance values.

Consider the two-branch circuit in Figure 6–55(a). If one of the resistors opens, the total current will equal the current in the good resistor. Ohm's law quickly tells you what the current in each resistor should be.

$$I_1 = \frac{50\text{ V}}{560\ \Omega} = 89.3\text{ mA}$$

$$I_2 = \frac{50\text{ V}}{100\ \Omega} = 500\text{ mA}$$

If $R_2$ is open, the total current is 89.3 mA, as indicated in Figure 6–55(b). If $R_1$ is open, the total current is 500 mA, as indicated in Figure 6–55(c).

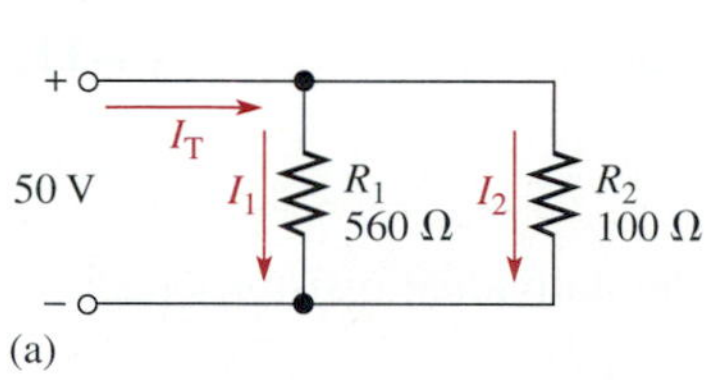


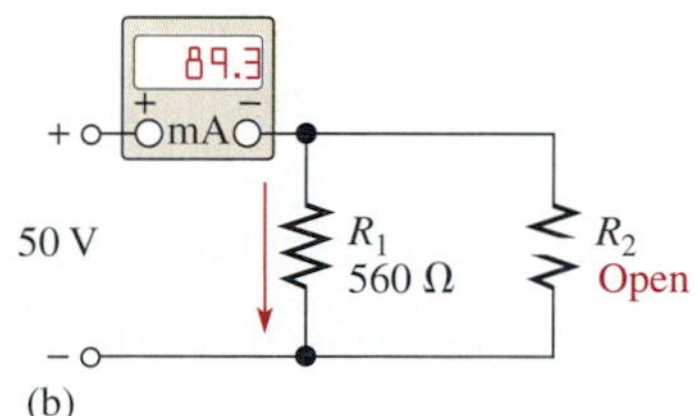


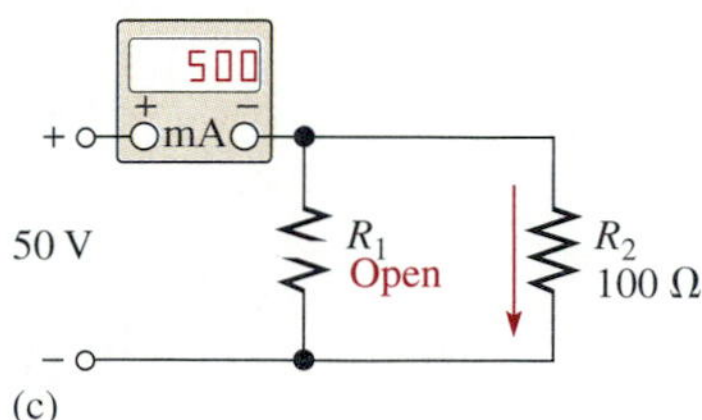


**FIGURE 6–55**
***Finding an open path by current measurement.***

This procedure can be extended to any number of branches having unequal resistances. If the parallel resistances are all equal, the current in each branch must be checked until a branch is found with no current. This is the open resistor.

**EXAMPLE 6–22** In Figure 6–56, there is a total current of 31.09 mA, and the voltage across the parallel branches is 20 V. Is there an open resistor, and, if so, which one is it?

FIGURE 6–56

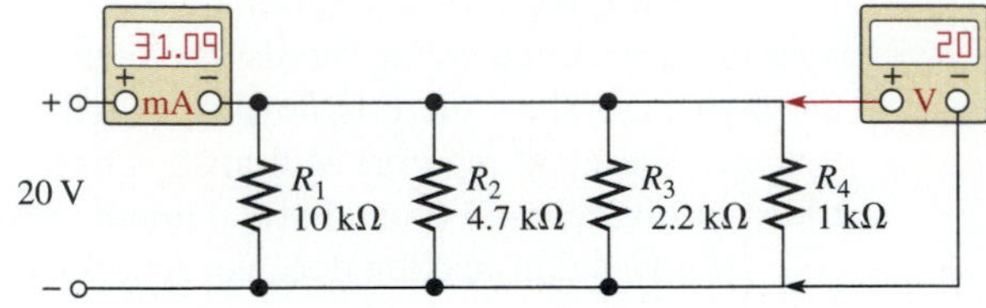


***Solution*** Calculate the current in each branch.

$$I_1 = \frac{V}{R_1} = \frac{20\text{ V}}{10\text{ k}\Omega} = 2\text{ mA}$$

$$I_2 = \frac{V}{R_2} = \frac{20\text{ V}}{4.7\text{ k}\Omega} = 4.26\text{ mA}$$

$$I_3 = \frac{V}{R_3} = \frac{20\text{ V}}{2.2\text{ k}\Omega} = 9.09\text{ mA}$$

$$I_4 = \frac{V}{R_4} = \frac{20\text{ V}}{1\text{ k}\Omega} = 20\text{ mA}$$

The total current should be

$$I_T = I_1 + I_2 + I_3 + I_4 = 2\text{ mA} + 4.26\text{ mA} + 9.09\text{ mA} + 20\text{ mA} = 35.35\text{ mA}$$

The actual measured current is 31.09 mA, as stated, which is 4.26 mA less than normal, indicating that the branch carrying 4.26 mA is open. Thus, $R_2$ must be open.

***Related Exercise*** What is the total current measured in Figure 6–56 if $R_4$ and not $R_2$ is open?

## Finding an Open Branch by Resistance Measurement

If the parallel circuit to be checked can be disconnected from its voltage source and from any other circuit to which it may be connected, a measurement of the total resistance can be used to locate an open branch.

Conductance, $G$, is the reciprocal of resistance ($1/R$) and its unit is the siemens (S). Based on Equation (6–2), the total conductance of a parallel circuit is the sum of the conductances of all the resistors.

$$G_T = G_1 + G_2 + G_3 + \cdots + G_n \qquad \textbf{(6–11)}$$

To locate an open branch, do the following steps:

**1.** Calculate what the total conductance should be using the individual resistor values:

$$G_{T(calc)} = \frac{1}{R_1} + \frac{1}{R_2} + \frac{1}{R_3} + \cdots + \frac{1}{R_n}$$

**2.** Measure the total resistance with an ohmmeter and calculate the total measured conductance:

$$G_{T(meas)} = \frac{1}{R_{T(meas)}}$$

3. Subtract the measured total conductance (Step 2) from the calculated total conductance (Step 1). The result is the conductance of the open branch and the resistance is obtained by taking its reciprocal ($R = 1/G$).

$$R_{\text{open}} = \frac{1}{G_{\text{T(calc)}} - G_{\text{T(meas)}}} \tag{6–12}$$

**EXAMPLE 6–23** Check the printed circuit board in Figure 6–57 for open branches.

FIGURE 6–57

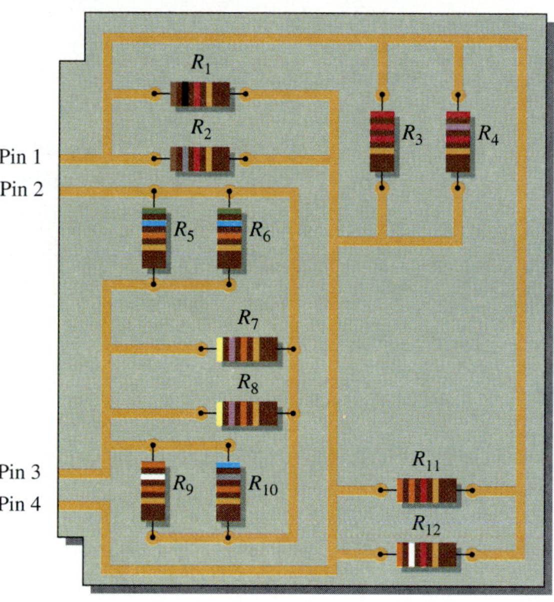


***Solution*** There are two separate parallel circuits on the board. The circuit between pin 1 and pin 4 is checked as follows (we will assume one of the resistors is open):

1. Calculate what the total conductance should be using the individual resistor values:

$$G_{\text{T(calc)}} = \frac{1}{1\ \text{k}\Omega} + \frac{1}{1.8\ \text{k}\Omega} + \frac{1}{2.2\ \text{k}\Omega} + \frac{1}{2.7\ \text{k}\Omega} + \frac{1}{3.3\ \text{k}\Omega} + \frac{1}{3.9\ \text{k}\Omega} = 2.94\ \text{mS}$$

2. Measure the total resistance with an ohmmeter and calculate the total measured conductance. Assume that your ohmmeter measures 402 Ω.

$$G_{\text{T(meas)}} = \frac{1}{402\ \Omega} = 2.49\ \text{mS}$$

3. Subtract the measured total conductance (Step 2) from the calculated total conductance (Step 1). The result is the conductance of the open branch and the resistance is obtained by taking its reciprocal.

$$G_{\text{open}} = G_{\text{T(calc)}} - G_{\text{T(meas)}} = 2.94\ \text{mS} - 2.49\ \text{mS} = 0.45\ \text{mS}$$

$$R_{\text{open}} = \frac{1}{G_{\text{open}}} = \frac{1}{0.45\ \text{mS}} = 2.2\ \text{k}\Omega$$

Resistor $R_3$ is open and must be replaced.

***Related Exercise*** Your ohmmeter indicates 9.6 kΩ between pin 2 and pin 3 on the PC board in Figure 6–57. Determine if this is correct and, if not, which resistor is open.

### Shorted Branches

When a branch in a parallel circuit shorts, the current increases to an excessive value, usually causing the resistor to burn open or a fuse or circuit breaker to blow. This results in a difficult troubleshooting problem because it is hard to isolate the shorted branch although sometimes a visual check may indicate a burned resistor that may be opened.

**SECTION 6–10 REVIEW**

1. If a parallel branch opens, what changes can be detected in the circuit's voltage and the currents, assuming that the parallel circuit is across a constant voltage source?
2. What happens to the total resistance if one branch opens?
3. If several light bulbs are connected in parallel and one of the bulbs opens (burns out), will the others continue to glow?
4. There is 1 A of current in each branch of a parallel circuit. If one branch opens, what is the current in each of the remaining branches?
5. A three-branch circuit normally has the following branch currents: 1 A, 2.5 A, and 1.2 A. If the total current measures 3.5 A, which branch is open?

## 6–11 ▪ TECHnology Theory Into Practice

***A five-range milliammeter has just come off the assembly line, but it has failed the line check. For any current in excess of 1 mA, the needle "pegs" (goes off scale to the right) on all the range settings except 1 mA and 50 mA. In these two ranges, the meter seems to work correctly. As you know, a multiple-range meter circuit is based on shunt resistance placed in parallel with the meter movement to achieve the specified full-scale deflection of the needle. Your task is to troubleshoot the meter and determine what repairs are necessary.***

### The Meter Circuit

The meter face is shown in Figure 6–58(a). The circuitry is exposed by removing the front of the meter as shown in part (b). The precision resistors are 1% tolerance with a five-band color code. The first three bands indicate resistance value. In the fourth band, gold indicates a multiplier of 0.1 and silver a multiplier of 0.01. The brown fifth band is for 1% tolerance.

### The Schematic

☐ The meter schematic is shown in Figure 6–59. Carefully check the circuit connections to make sure they agree with the schematic. If there is a wiring problem, specify how to repair it. Resistors $R_1$, $R_2$, etc. can be identified by the way they are wired in the circuit. The 100 mA switch position is labeled "1". Resistance values are not shown in the schematic.

### Troubleshooting

There was a wiring problem in the meter circuit which, let's assume, you have located and corrected. Now, assume that when you connect the meter to the test instrument to check for proper operation you find that the meter works properly on the 1 mA, 5 mA,

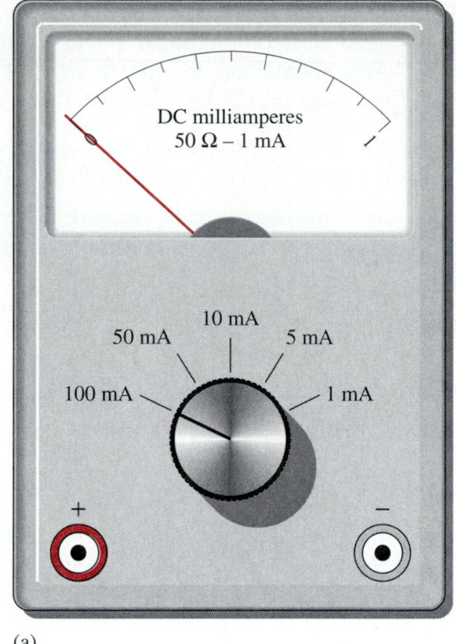


(a)

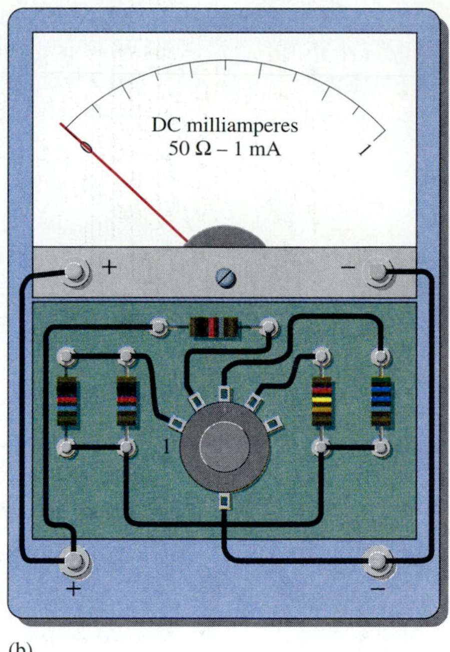


(b)

FIGURE 6–58

FIGURE 6–59

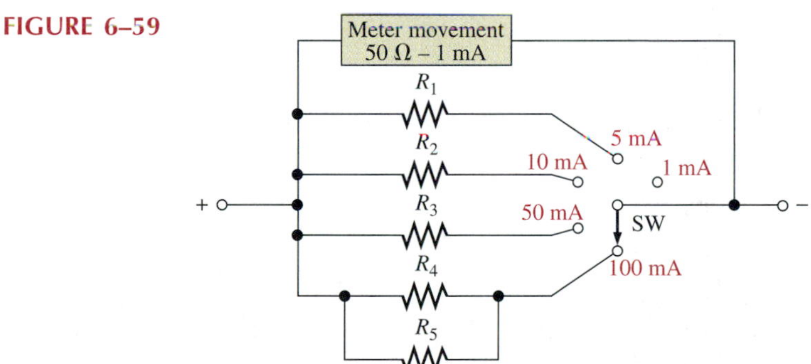


10 mA, and 50 mA ranges but the needle still "pegs" on the 100 mA range setting for a current of 50 mA or greater.

Since the needle pegs on the 100 mA range, all or most of the current must be going through the meter movement. Recall that a shunt resistor carries most of the current being measured so that the current through the meter movement does not exceed its 1 mA maximum. For this problem to occur, one of the shunt resistors for this range is either not connected, failed open, or has a resistance value that is much too high.

□ Check the circuit in Figure 6–58 again for another wiring error and check the color-coded values of $R_4$ and $R_5$. For some reason, the schematic does not show the resistor values so you will have to calculate what they should be using your knowledge of Ohm's law and current dividers. Refer to the table of standard resistor values in Appendix A.

**SECTION 6–11 REVIEW**

1. What is the total resistance between the positive and negative terminals of the milliammeter in Figure 6–58 when the range switch is in the 1 mA position? The 5 mA position? The 10 mA position?
2. When the switch is in the 5 mA position and the needle points to the middle mark on the scale, how much current is being measured?
3. When the meter is on the 5 mA range and is measuring the amount of current determined in Question 2, how much of the total current is through the meter movement and how much through the shunt resistor?

## 6–12 ■ PSpice DC ANALYSIS

***In this section, PSpice is used to solve a basic parallel circuit. So far, the circuits used to illustrate PSpice statements have been quite simple. They were meant to show how to create a circuit file using the basic statements so that you can use PSpice for more complex circuits. As you will see in future chapters, the power of PSpice is in its ability to quickly analyze very complex circuits that would take much effort or be impractical to do manually.***

***After completing this section, you should be able to***

- **Use PSpice to analyze parallel circuits and graph current values**
  - ☐ Write a circuit file to analyze a parallel circuit
  - ☐ Use all of the command statements covered so far

### A Parallel Circuit

Suppose you want to analyze a parallel circuit with a 24 V dc voltage source and eight resistors in parallel and solve for all the branch currents and the total resistance. Such a circuit with its two node assignments is shown in Figure 6–60.

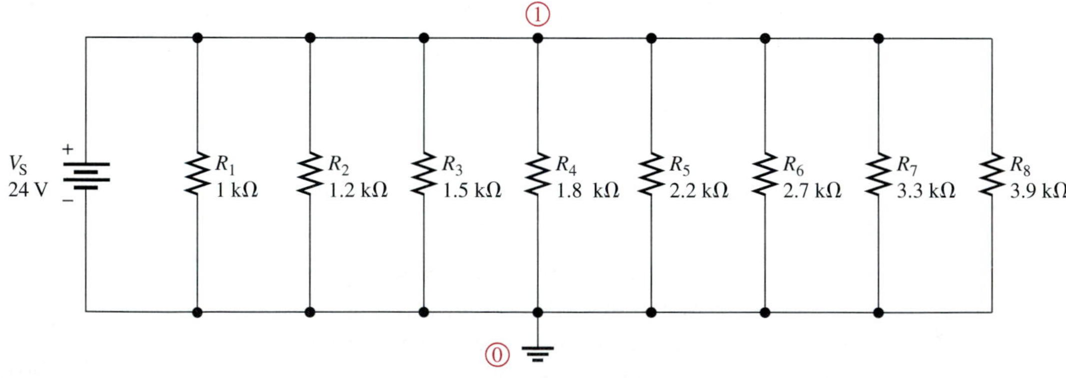


FIGURE 6–60

PSpice will not solve for resistance directly, since (as a rule) the circuit resistances are fixed by design and the voltages and currents are usually the unknown values. This is not a problem, however, because the .OP statement can be used to show the total current in a circuit in the output file. Then you calculate the total resistance by dividing the total voltage by the total current. The circuit file for Figure 6–60 is as follows:

```
Parallel Circuit
*Figure 6–60
VS  1  0  24
R1  1  0  1K
R2  1  0  1.2K
R3  1  0  1.5K
R4  1  0  1.8K
R5  1  0  2.2K
R6  1  0  2.7K
R7  1  0  3.3K
R8  1  0  3.9K
.OP
.DC  VS  24  24  1
.PRINT  DC  I(R1)  I(R2)  I(R3)  I(R4)  I(R5)  I(R6)  I(R7)  I(R8)
.END
```

If you want to plot the values of the currents in the circuit file, add the following .PLOT statement to the circuit file:

```
.PLOT  DC  I(R1)  I(R2)  I(R3)  I(R4)  I(R5)  I(R6)  I(R7)  I(R8)
```

If you want to plot the value of only the current in $R_8$ as the source voltage is varied from 0 V to 24 V in 1 V increments, change the .DC and .PLOT statements as follows:

```
.DC  VS  0  24  1
.PLOT  DC  I(R8)
```

If you want to plot the value of the current in $R_8$ as the source voltage is varied from 0 V to 24 V in 1 V increments but only for current values greater than 1 mA and less than 3.5 mA, change the .DC and .PLOT statements as follows:

```
.DC  VS  0  24  1
.PLOT  DC  I(R8)  (1M, 3.5M)
```

## SECTION 6–12 REVIEW

1. Write a circuit file for a parallel circuit that has ten resistors connected to a 100 V source. There are two 1 MΩ resistors, two 1.5 MΩ resistors, three 100 kΩ resistors, two 560 kΩ resistors, and one 820 kΩ resistor. You need the values of all the branch currents.
2. Modify the circuit file in question 1 to plot only the currents in the 1 MΩ resistors between 25 $\mu$A and 75 $\mu$A as the source is varied from 30 V to 80 V in 10 V increments.

## ■ SUMMARY

- Resistors in parallel are connected across the same points.
- A parallel combination has more than one path for current.
- The number of current paths equals the number of resistors in parallel.
- The total parallel resistance is less than the lowest-value resistor.
- The voltages across all branches of a parallel circuit are the same.
- Current sources in parallel add algebraically.
- One way to state Kirchhoff's current law: The sum of the currents into a junction (total current in) equals the sum of the currents out of the junction (total current out).
- Another way to state Kirchhoff's current law: The algebraic sum of all the currents at a junction is zero.
- A parallel circuit is a current divider, so called because the total current entering the junction of parallel branches divides up into each of the branches.

- If all of the branches of a parallel circuit have equal resistance, the currents through all of the branches are equal.
- The total power in a parallel-resistive circuit is the sum of all of the individual powers of the resistors making up the parallel circuit.
- The total power for a parallel circuit can be calculated with the power formulas using values of total current, total resistance, or total voltage.
- If one of the branches of a parallel circuit opens, the total resistance increases, and therefore the total current decreases.
- If a branch of a parallel circuit opens, there is still current through the remaining branches.

## GLOSSARY

**Branch** One current path in a parallel circuit.

**Junction** A point at which two or more components are connected.

**Kirchhoff's current law** A law stating that the total current into a junction equals the total current out of the junction.

**Parallel** The relationship in electric circuits in which two or more current paths are connected between the same two points.

## FORMULAS

| | | |
|---|---|---|
| **(6–1)** | $I_{IN(1)} + I_{IN(2)} + \cdots + I_{IN(n)} = I_{OUT(1)} + I_{OUT(2)} + \cdots + I_{OUT(m)}$ | Kirchhoff's current law |
| **(6–2)** | $\frac{1}{R_T} = \frac{1}{R_1} + \frac{1}{R_2} + \frac{1}{R_3} + \cdots + \frac{1}{R_n}$ | Reciprocal for total parallel resistance |
| **(6–3)** | $R_T = \frac{1}{\left(\frac{1}{R_1}\right) + \left(\frac{1}{R_2}\right) + \left(\frac{1}{R_3}\right) + \cdots + \left(\frac{1}{R_n}\right)}$ | Total parallel resistance |
| **(6–4)** | $R_T = \frac{R_1R_2}{R_1 + R_2}$ | Special case for two resistors in parallel |
| **(6–5)** | $R_T = \frac{R}{n}$ | Special case for *n* equal-value resistors in parallel |
| **(6–6)** | $R_x = \frac{R_A R_T}{R_A - R_T}$ | Unknown parallel resistor |
| **(6–7)** | $I_x = \left(\frac{R_T}{R_x}\right)I_T$ | General current-divider formula |
| **(6–8)** | $I_1 = \left(\frac{R_2}{R_1 + R_2}\right)I_T$ | Two-branch current-divider formula |
| **(6–9)** | $I_2 = \left(\frac{R_1}{R_1 + R_2}\right)I_T$ | Two-branch current-divider formula |
| **(6–10)** | $P_T = P_1 + P_2 + P_3 + \cdots + P_n$ | Total power |
| **(6–11)** | $G_T = G_1 + G_2 + G_3 + \cdots + G_n$ | Total conductance |
| **(6–12)** | $R_{open} = \frac{1}{G_{T(calc)} - G_{T(meas)}}$ | Open branch resistance |

## ■ SELF-TEST

1. In a parallel circuit, each resistor has
   **(a)** the same current **(b)** the same voltage
   **(c)** the same power **(d)** all of the above
2. When a 1.2 kΩ resistor and a 100 Ω resistor are connected in parallel, the total resistance is
   **(a)** greater than 1.2 kΩ
   **(b)** greater than 100 Ω but less than 1.2 kΩ
   **(c)** less than 100 Ω but greater than 90 Ω
   **(d)** less than 90 Ω
3. A 330 Ω resistor, a 270 Ω resistor, and a 68 Ω resistor are all in parallel. The total resistance is approximately
   **(a)** 668 Ω **(b)** 47 Ω **(c)** 68 Ω **(d)** 22 Ω
4. Eight resistors are in parallel. The two lowest-value resistors are both 1 kΩ. The total resistance
   **(a)** is less than 8 kΩ **(b)** is greater than 1 kΩ
   **(c)** is less than 1 kΩ **(d)** is less than 500 Ω
5. When an additional resistor is connected across an existing parallel circuit, the total resistance
   **(a)** decreases **(b)** increases
   **(c)** remains the same **(d)** increases by the value of the added resistor
6. If one of the resistors in a parallel circuit is removed, the total resistance
   **(a)** decreases by the value of the removed resistor **(b)** remains the same
   **(c)** increases **(d)** doubles
7. The currents into a junction flow along two paths. One current is 5 A and the other is 3 A. The total current out of the junction is
   **(a)** 2 A **(b)** unknown **(c)** 8 A **(d)** the larger of the two
8. The following resistors are in parallel across a voltage source: 390 Ω, 560 Ω, and 820 Ω. The resistor with the least current is
   **(a)** 390 Ω **(b)** 560 Ω
   **(c)** 820 Ω **(d)** impossible to determine without knowing the voltage
9. A sudden decrease in the total current into a parallel circuit may indicate
   **(a)** a short **(b)** an open resistor
   **(c)** a drop in source voltage **(d)** either (b) or (c)
10. In a four-branch parallel circuit, there are 10 mA of current in each branch. If one of the branches opens, the current in each of the other three branches is
    **(a)** 13.3 mA **(b)** 10 mA **(c)** 0 A **(d)** 30 mA
11. In a certain three-branch parallel circuit, $R_1$ has 10 mA through it, $R_2$ has 15 mA through it, and $R_3$ has 20 mA through it. After measuring a total current of 35 mA, you can say that
    **(a)** $R_1$ is open **(b)** $R_2$ is open
    **(c)** $R_3$ is open **(d)** the circuit is operating properly
12. If there are a total of 100 mA into a parallel circuit consisting of three branches and two of the branch currents are 40 mA and 20 mA, the third branch current is
    **(a)** 60 mA **(b)** 20 mA **(c)** 160 mA **(d)** 40 mA
13. A complete short develops across one of five parallel resistors on a PC board. The most likely result is
    **(a)** the shorted resistor will burn out
    **(b)** one or more of the other resistors will burn out
    **(c)** the fuse in the power supply will blow
    **(d)** the resistance values will be altered
14. The power dissipation in each of four parallel branches is 1 W. The total power dissipation is
    **(a)** 1 W **(b)** 4 W **(c)** 0.25 W **(d)** 16 W

## ■ PROBLEMS

### SECTION 6–1 Resistors in Parallel

1. Show how to connect the resistors in Figure 6–61(a) in parallel across the battery.

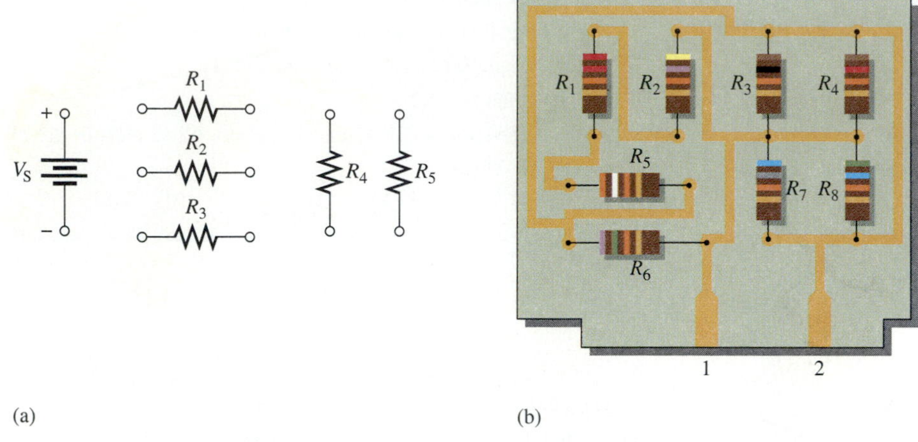


FIGURE 6–61

2. Determine whether or not all the resistors in Figure 6–61(b) are connected in parallel on the printed circuit (PC) board.
3. Identify which groups of resistors are in parallel on the double-sided PC board in Figure 6–62.

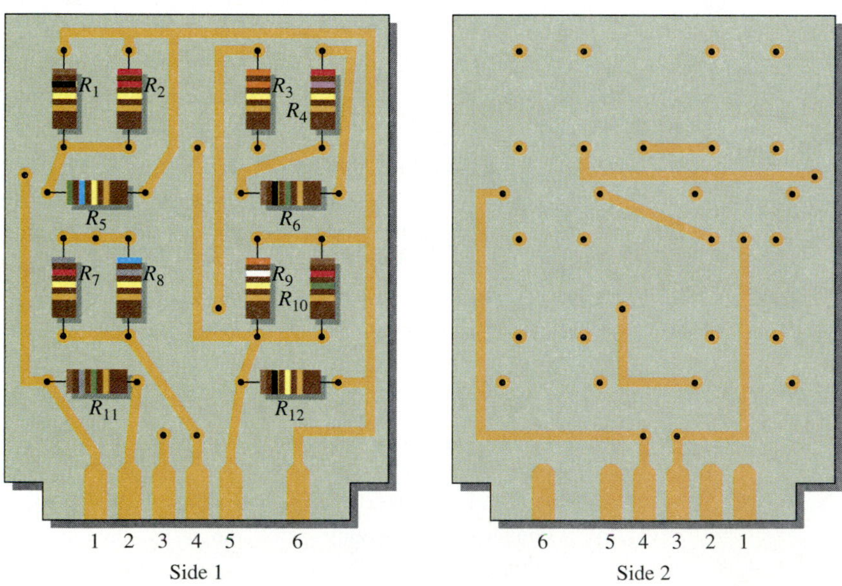


FIGURE 6–62

### SECTION 6–2 Voltage Drop in Parallel Circuits

4. What is the voltage across and the current through each parallel resistor if the total voltage is 12 V and the total resistance is 550 Ω? There are four resistors, all of equal value.
5. The source voltage in Figure 6–63 is 100 V. How much voltage does each of the meters read?
6. What is the voltage across each resistor in Figure 6–64 for each switch position?

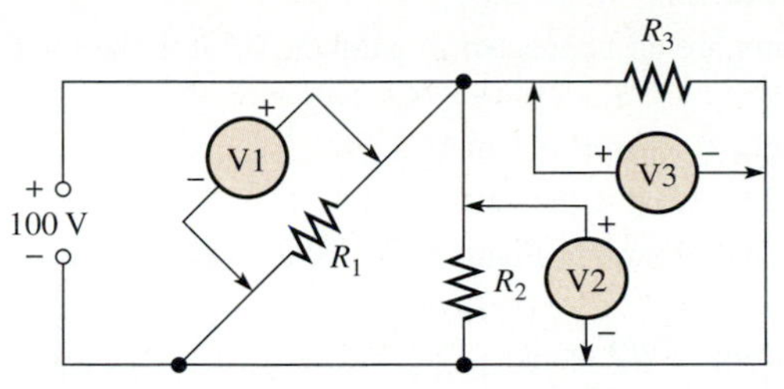


FIGURE 6–63

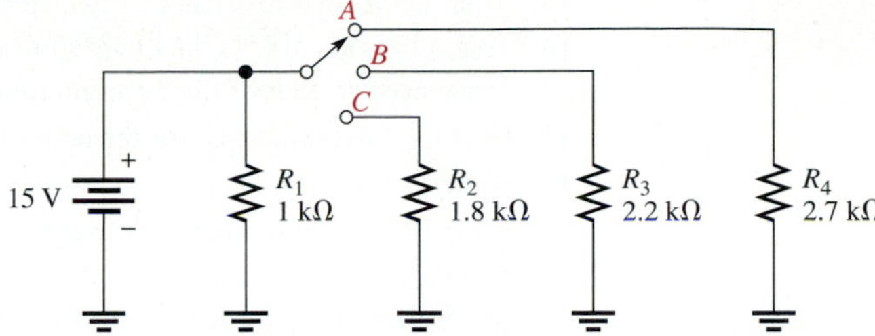


FIGURE 6–64

## SECTION 6–3 Kirchhoff's Current Law

**7.** The following currents are measured in the same direction in a three-branch parallel circuit: 250 mA, 300 mA, and 800 mA. What is the value of the current into the junction of these three branches?

**8.** There is a total of 500 mA of current into five parallel resistors. The currents through four of the resistors are 50 mA, 150 mA, 25 mA, and 100 mA. What is the current through the fifth resistor?

**9.** In the circuit of Figure 6–65, determine the resistance $R_2$, $R_3$, and $R_4$.

FIGURE 6–65

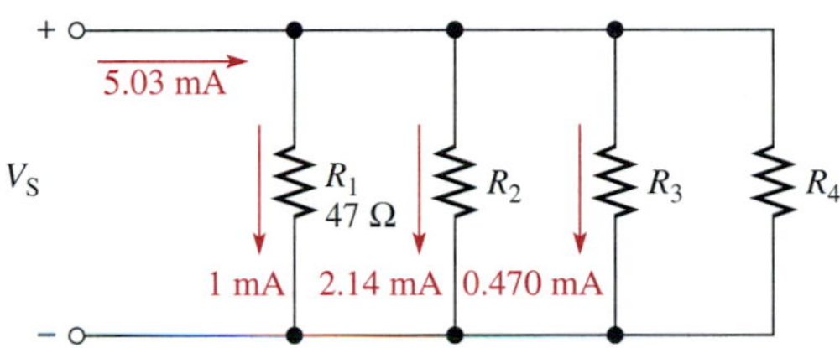


**10.** The electrical circuit in a room has a ceiling lamp that draws 1.25 A and four wall outlets. Two table lamps that each draw 0.833 A are plugged into two outlets, and a TV set that draws 1 A is connected to the third outlet. When all of these items are in use, how much current is in the main line serving the room? If the main line is protected by a 5 A circuit breaker, how much current can be drawn from the fourth outlet? Draw a schematic of this wiring.

**11.** The total resistance of a parallel circuit is 25 Ω. What is the current through a 220 Ω resistor that makes up part of the parallel circuit if the total current is 100 mA?

## SECTION 6–4 Total Parallel Resistance

**12.** The following resistors are connected in parallel: 1 MΩ, 2.2 MΩ, 5.6 MΩ, 12 MΩ, and 22 MΩ. Determine the total resistance.

**13.** Find the total resistance for each following group of parallel resistors:

**(a)** 560 Ω and 1000 Ω **(b)** 47 Ω and 56 Ω

**(c)** 1.5 kΩ, 2.2 kΩ, 10 kΩ **(d)** 1 MΩ, 470 kΩ, 1 kΩ, 2.7 MΩ

**14.** Calculate $R_T$ for each circuit in Figure 6–66.

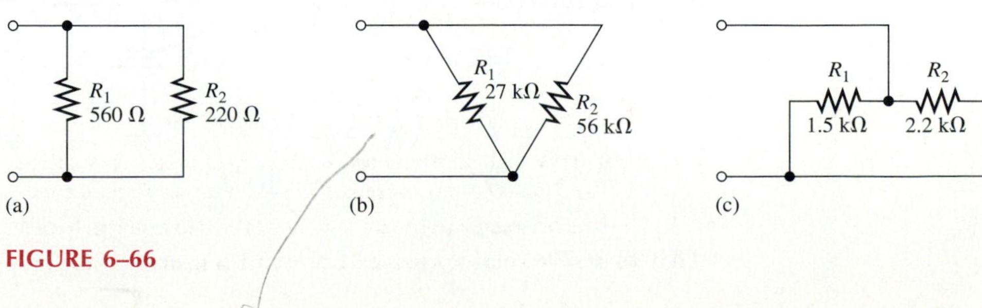


FIGURE 6–66

**15.** What is the total resistance of twelve 6.8 kΩ resistors in parallel?

**16.** Five 47 Ω, ten 100 Ω, and two 10 Ω resistors are all connected in parallel. What is the total resistance for each of the three groupings?

**17.** Find the total resistance for the entire parallel circuit in Problem 16.

**18.** If the total resistance in Figure 6–67 is 389.2 Ω, what is the value of $R_2$?

**19.** What is the total resistance between point *A* and ground in Figure 6–68 for the following conditions?

**(a)** SW1 and SW2 open **(b)** SW1 closed, SW2 open

**(c)** SW1 open, SW2 closed **(d)** SW1 and SW2 closed

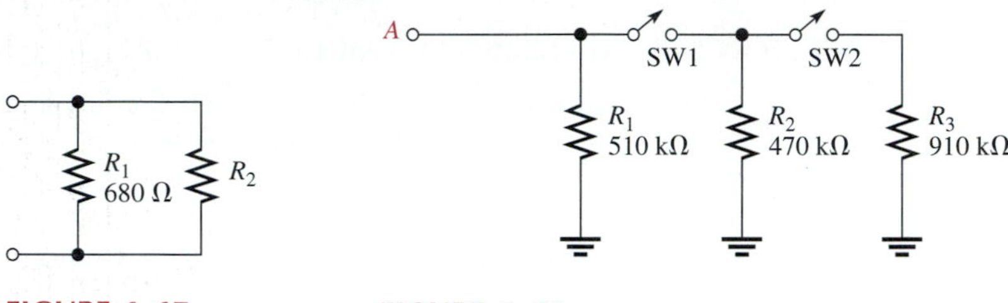


FIGURE 6–67

FIGURE 6–68

## SECTION 6–5 Ohm's Law in Parallel Circuits

**20.** What is the total current in each circuit of Figure 6–69?

FIGURE 6–69

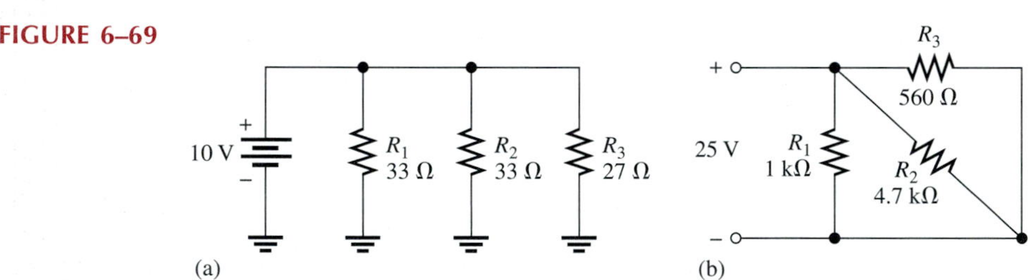


**21.** Three 33 Ω resistors are connected in parallel with a 110 V source. What is the current from the source?

**22.** Four equal-value resistors are connected in parallel. Five volts are applied across the parallel circuit, and 1.11 mA are measured from the source. What is the value of each resistor?

**23.** Christmas tree lights are usually connected in parallel. If a set of lights is connected to a 110 V source and the filament of each bulb has a resistance of 2.2 kΩ, what is the current through each bulb? Why is it better to have these bulbs in parallel rather than in series?

**24.** Find the values of the unspecified labeled quantities in each circuit of Figure 6–70.

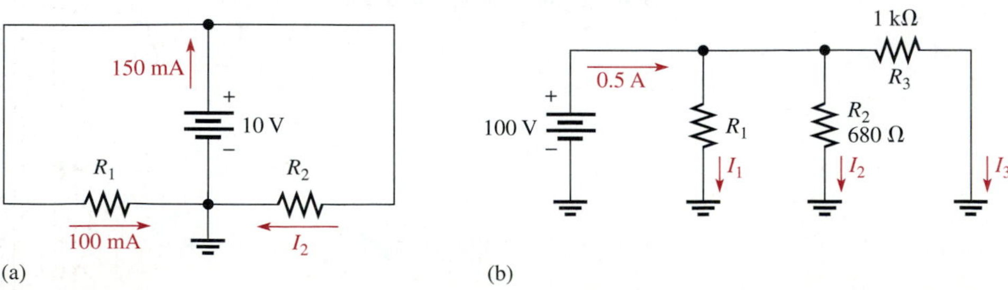


FIGURE 6–70

**25.** To what minimum value can the 100 Ω rheostat in Figure 6–71 be adjusted before the 0.5 A fuse blows?

**FIGURE 6–71**

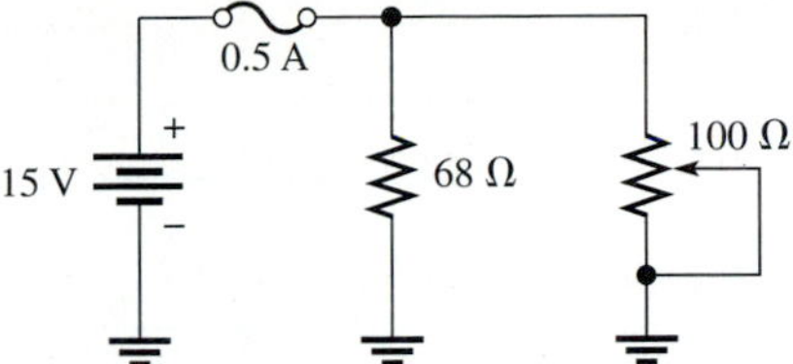


**26.** Determine the total current from the source and the current through each resistor for each switch position in Figure 6–72.

**FIGURE 6–72**

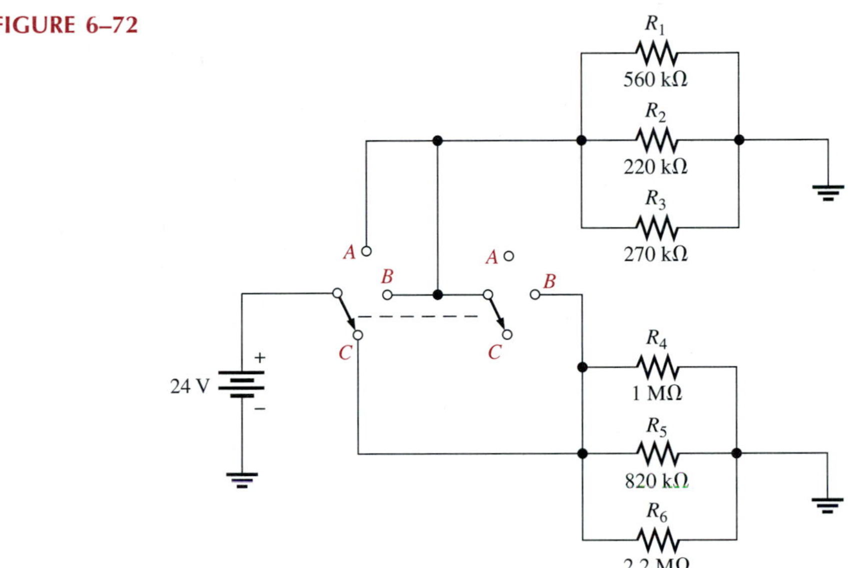


**27.** Find the values of the unspecified quantities in Figure 6–73.

**FIGURE 6–73**

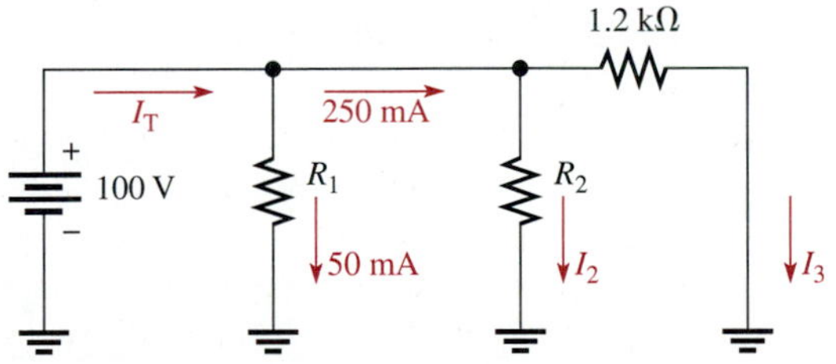

## SECTION 6–6 Current Sources in Parallel

**28.** Determine the current through $R_L$ in each circuit in Figure 6–74.

**FIGURE 6–74**

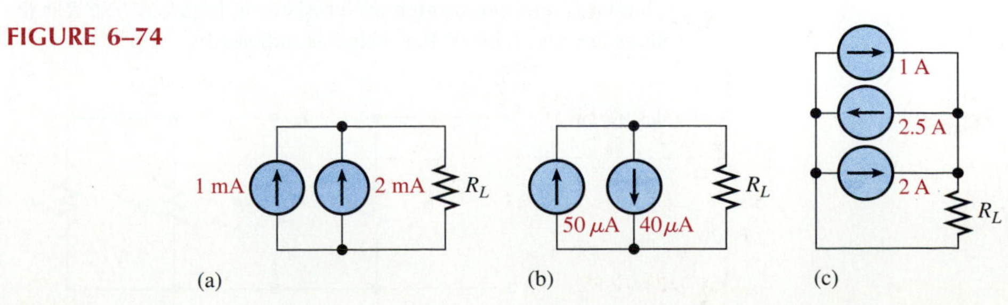

**29.** Find the current through the resistor for each position of the ganged switch in Figure 6–75.

FIGURE 6–75

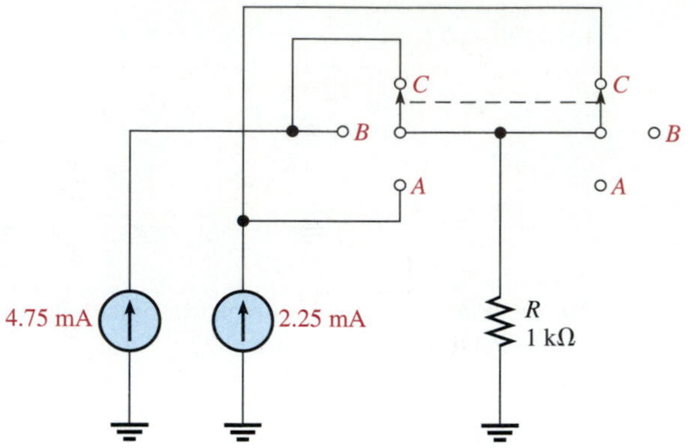

## SECTION 6–7 Current Dividers

**30.** How much branch current should each meter in Figure 6–76 indicate?

FIGURE 6–76

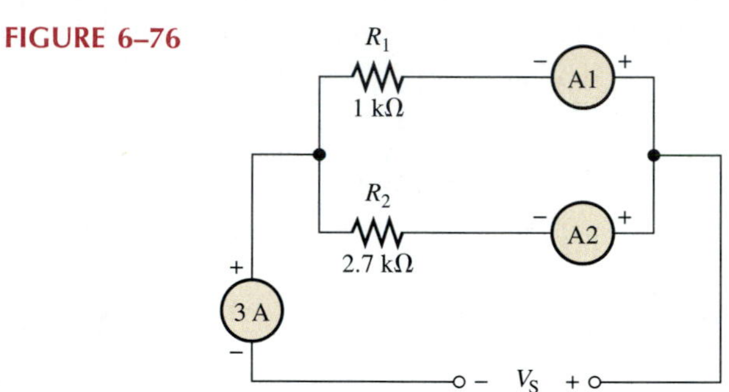


**31.** Determine the current in each branch of the current dividers of Figure 6–77.

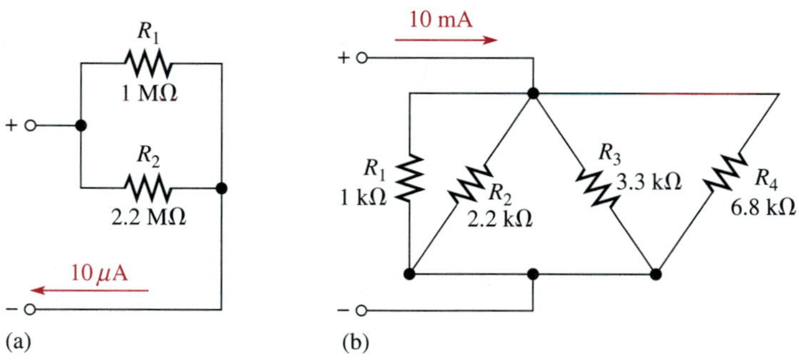


FIGURE 6–77

**32.** What is the current through each resistor in Figure 6–78? $R$ is the lowest-value resistor, and all others are multiples of that value as indicated.

FIGURE 6–78

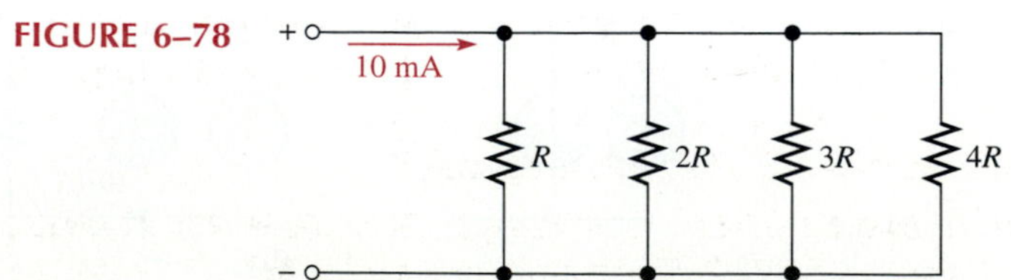

**33.** Determine all of the resistor values in Figure 6–79. $R_T = 773\ \Omega$.

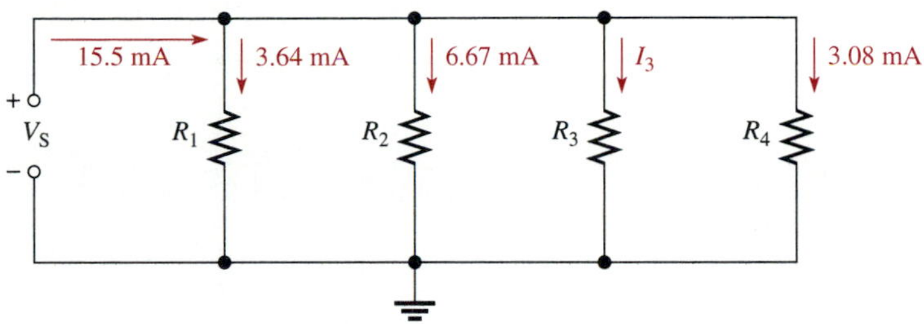


FIGURE 6–79

**34.** **(a)** Determine the required value of the shunt resistor $R_{SH1}$ in the ammeter of Figure 6–49 if the resistance of the meter movement is 50 Ω.

**(b)** Find the required value for $R_{SH2}$ in the meter circuit of Figure 6–50 ($R_M = 50\ \Omega$).

**35.** Add a fourth range position to the meter in Figure 6–50 so that currents up to 1 A (1000 mA) can be measured. Specify the value of the additional shunt resistor. Assume a 1 mA, 50 Ω meter movement.

## SECTION 6–8 Power in Parallel Circuits

**36.** Five parallel resistors each handle 40 mW. What is the total power?

**37.** Determine the total power in each circuit of Figure 6–77.

**38.** Six light bulbs are connected in parallel across 110 V. Each bulb is rated at 75 W. What is the current through each bulb, and what is the total current?

**39.** Find the values of the unspecified quantities in Figure 6–80.

FIGURE 6–80

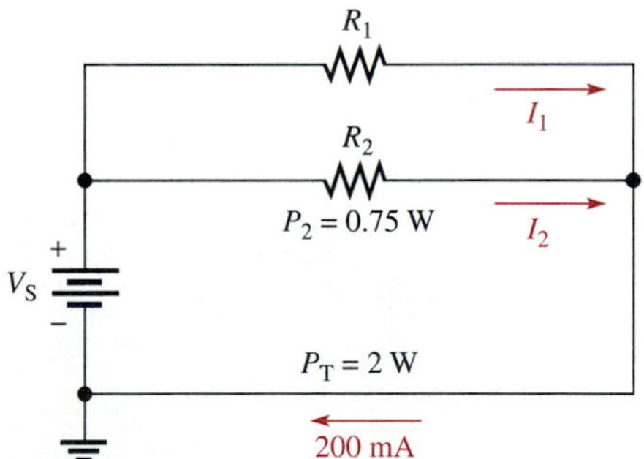


**40.** A certain parallel circuit consists of only ½ W resistors. The total resistance is 1 kΩ, and the total current is 50 mA. If each resistor is operating at one-half its maximum power level, determine the following:

**(a)** The number of resistors  **(b)** The value of each resistor

**(c)** The current in each branch  **(d)** The applied voltage

## SECTION 6–10 Troubleshooting

**41.** If one of the bulbs burns out in Problem 38, how much current will be through each of the remaining bulbs? What will the total current be?

**42.** In Figure 6–81, the current and voltage measurements are indicated. Has a resistor opened, and, if so, which one?

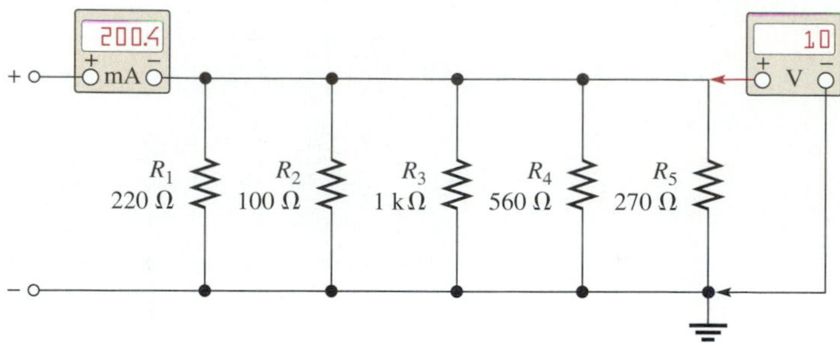


FIGURE 6–81

**43.** What is wrong with the circuit in Figure 6–82?

FIGURE 6–82

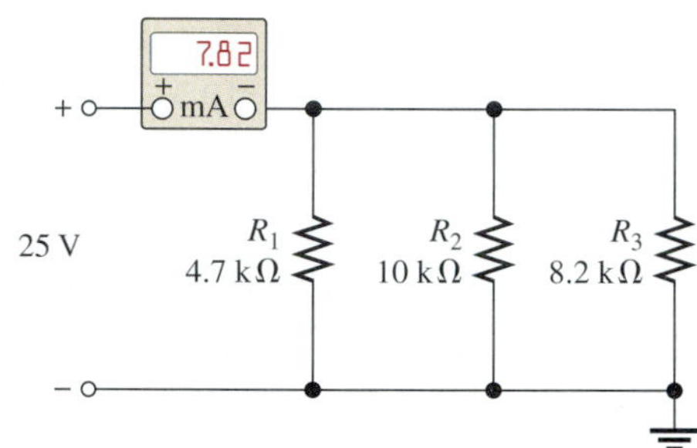


**44.** What is wrong with the circuit in Figure 6–82 if the meter reads 5.55 mA?

**45.** Develop a test procedure to check the circuit board in Figure 6–83 to make sure that there are no open components. You must do this test without removing a component from the board. List the procedure in a detailed step-by-step format.

FIGURE 6–83

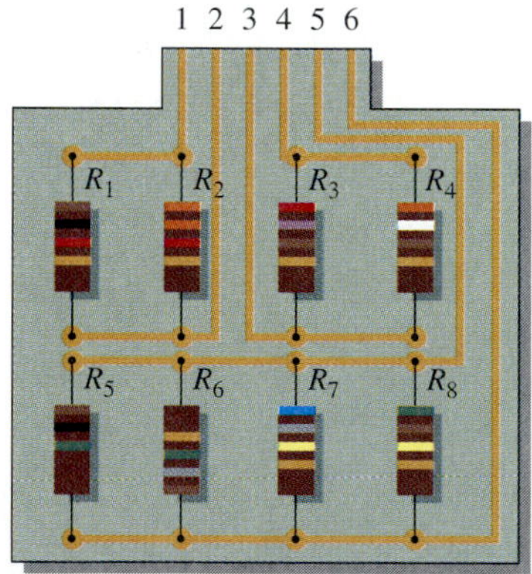


**46.** For the circuit board shown in Figure 6–84, determine the resistance between the following pins if there is a short between pins 2 and 4:

**(a)** 1 and 2 **(b)** 2 and 3 **(c)** 3 and 4 **(d)** 1 and 4

**47.** For the circuit board shown in Figure 6–84, determine the resistance between the following pins if there is a short between pins 3 and 4:

**(a)** 1 and 2 **(b)** 2 and 3 **(c)** 2 and 4 **(d)** 1 and 4

FIGURE 6–84

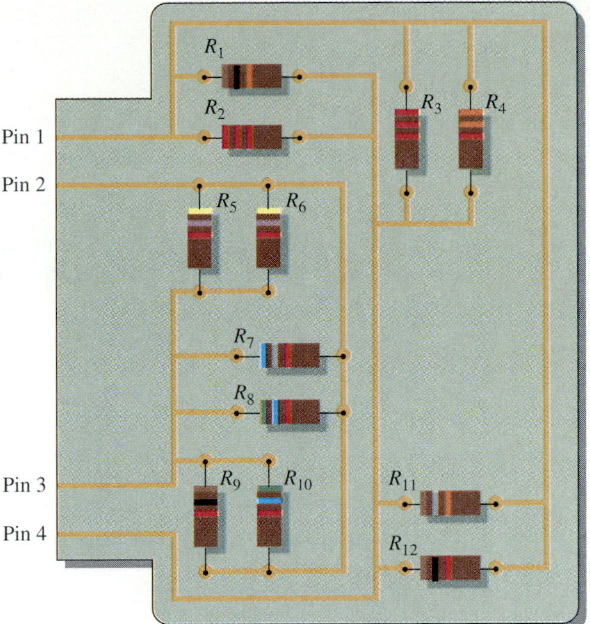


## ANSWERS TO SECTION REVIEWS

### Section 6–1

1. Parallel resistors are connected between the same two points.
2. A parallel circuit has more than one current path between two given points.
3. See Figure 6–85.

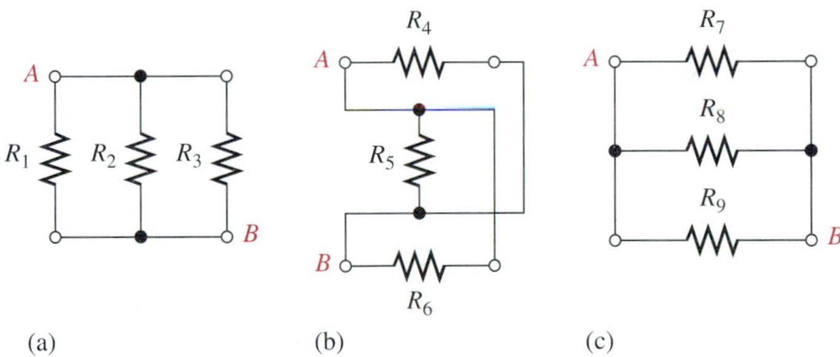


FIGURE 6–85

4. See Figure 6–86.

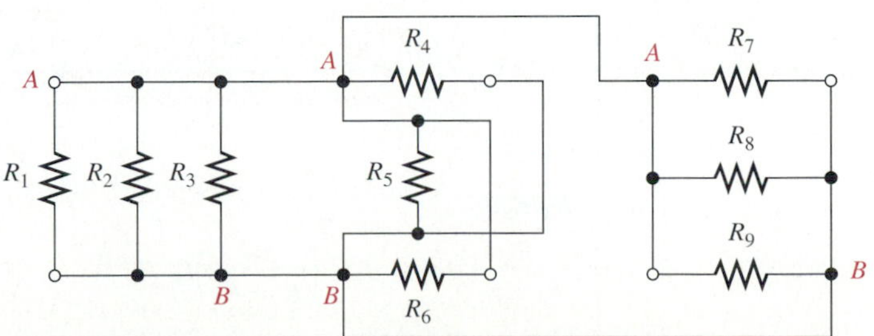


FIGURE 6–86

## Section 6–2

1. $V_{10\Omega} = V_{22\Omega} = 5$ V
2. $V_{R2} = 118$ V; $V_S = 118$ V
3. $V_1 = 50$ V and $V_2 = 50$ V
4. Voltage is the same across all parallel branches.

## Section 6–3

1. Kirchhoff's law: The algebraic sum of all the currents at a junction is zero; The sum of the currents entering a junction equals the sum of the currents leaving that junction.
2. $I_1 = I_2 = I_3 = I_T = 2.5$ A
3. $I_{OUT} = 100$ mA + 300 mA = 400 mA
4. $I_1 = I_T - I_2 = 3\ \mu$A
5. $I_{IN} = 8$ A − 1 A = 7 A; $I_{OUT} = 8$ A − 3 A = 5 A

## Section 6–4

1. $R_T$ decreases with more resistors in parallel.
2. The total parallel resistance is less than the smallest branch resistance.
3. $R_T = \dfrac{1}{(1/R_1) + (1/R_2) + \cdots + (1/R_n)}$
4. $R_T = R_1R_2/(R_1 + R_2)$
5. $R_T = R/n$
6. $R_T = (1\text{ k}\Omega)(2.2\text{ k}\Omega)/3.2\text{ k}\Omega = 688\ \Omega$
7. $R_T = 1\text{ k}\Omega/4 = 250\ \Omega$
8. $R_T = \dfrac{1}{1/47\ \Omega + 1/150\ \Omega + 1/100\ \Omega} = 26.4\ \Omega$

## Section 6–5

1. $I_T = 10\text{ V}/22.7\ \Omega = 441$ mA
2. $V_S = (2\text{ A})(222\ \Omega) = 444$ V
3. $I_1 = 444\text{ V}/680\ \Omega = 653$ mA; $I_2 = 444\text{ V}/330\ \Omega = 1.35$ A
4. $R_T = 12\text{ V}/5.85\text{ mA} = 2.05\text{ k}\Omega$; $R = (2.05\text{ k}\Omega)(4) = 8.2\text{ k}\Omega$
5. $V = (100\text{ mA})(688\ \Omega) = 68.8$ V

## Section 6–6

1. $I_T = 4(0.5\text{ A}) = 2$ A
2. Three sources; See Figure 6–87.
3. $I_{R_E} = 10$ mA + 10 mA = 20 mA

FIGURE 6–87

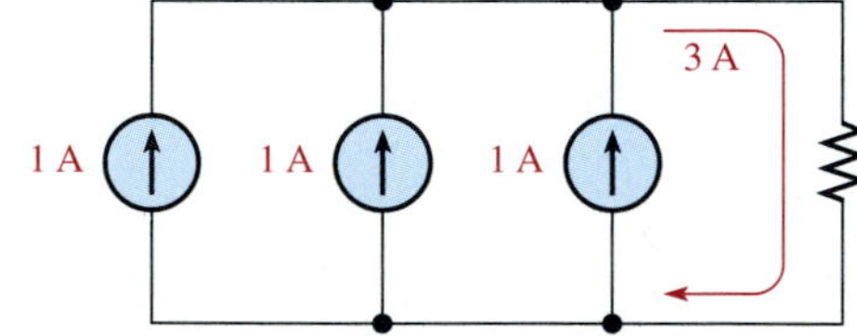


## Section 6–7

1. $I_x = (R_T/R_x)I_T$
2. $I_1 = \left(\dfrac{R_2}{R_1 + R_2}\right)I_T \qquad I_2 = \left(\dfrac{R_1}{R_1 + R_2}\right)I_T$
3. The 22 Ω has the most current; the 220 Ω has the least current.
4. $I_3 = (114\ \Omega/470\ \Omega)4\text{ mA} = 970\ \mu$A
5. $I_1 = (680\ \Omega/1010\ \Omega)10\text{ mA} = 6.73$ mA; $I_2 = (330\ \Omega/1010\ \Omega)10\text{ mA} = 3.27$ mA

### Section 6–8

1. Add the power of each resistor to get total power.
2. $P_T = 2.38\ W + 5.12\ W + 1.09\ W + 8.76\ W = 17.4\ W$
3. $P_T = (1\ A)^2(615\ \Omega) = 615\ W$

### Section 6–9

1. $R_{max} = 50\ \Omega$; $I_{max} = 1\ mA$
2. $R_{SH}$ is less than $R_M$ because the shunt resistors must allow currents much greater than the current through the meter movement.

### Section 6–10

1. When a branch opens, there is no change in voltage; the total current decreases.
2. If a branch opens, total parallel resistance increases.
3. The remaining bulbs continue to glow.
4. All remaining branch currents are 1 A.
5. The branch with 1.2 A is open.

### Section 6–11

1. $R_T = 50\ \Omega$ in 1 mA position; $R_T = 9.94\ \Omega$ in 5 mA position; $R_T = 5.08\ \Omega$ in 10 mA position.
2. $I = 2.5\ mA$
3. $I_M = 0.5\ mA$; $I_{R1} = 2\ mA$

### Section 6–12

1. 
```
VS   1  0  100
R1   1  0  1MEG
R2   1  0  1MEG
R3   1  0  1.5MEG
R4   1  0  1.5MEG
R5   1  0  100K
R6   1  0  100K
R7   1  0  100K
R8   1  0  560K
R9   1  0  560K
R10  1  0  820K
.OP
.DC  VS  100  100  1
.PRINT  DC  I(R1)  I(R2)  I(R3)  I(R4)  I(R5)  I(R6)  I(R7)  I(R8)  I(R9)  I(R10)
.END
```
2. Change the .DC statement to

```
.DC  VS  30  80  10
```

Replace the .PRINT statement with

```
.PLOT  DC  I(R1)  I(R2)  (25U, 75U)
```

## ■ ANSWERS TO RELATED EXERCISES FOR EXAMPLES

**6–1** See Figure 6–88.

**6–2** Connect pin 1 to pin 2 and pin 3 to pin 4.

**6–3** 25 V

**6–4** 20 mA into junction *A* and out of junction *B*

**6–5** $I_T = 112\ mA$, $I_2 = 50\ mA$

**6–6** 2.5 A; 5 A

**6–7** 9.34 Ω

**6–8** Replace Step 9 with "Press [+] key." Display shows 0.076731141.
Replace Step 10 with "Enter 33." Display shows 33.
Step 11: Press [2nd] [1/x]. Display shows 0.03030303.
Step 12: Press [−] key. Display shows 0.107034172.
Step 13: Press [2nd] [1/x]. Display shows 9.342810674

**6–9** 111 Ω

**6–10** 4 Ω

**6–11** 1044 Ω

**6–12** 1.83 A; 1 A

**6–13** $I_1 = 20$ mA; $I_2 = 9.09$ mA; $I_3 = 35.7$ mA, $I_4 = 22.0$ mA

**6–14** 1.28 V

**6–15** Measure $R_T$ with an ohmmeter and calculate $R_1$ using $R_1 = \dfrac{1}{(1/R_T) - (1/R_2) - (1/R_3)}$

**6–16** 30 mA

**6–17** $I_1 = 3.27$ A; $I_2 = 6.73$ A

**6–18** $I_1 = 59.4$ mA; $I_2 = 40.6$ mA

**6–19** 178 W

**6–20** 162 W

**6–21** 0.3 μA

**6–22** 15.4 mA

**6–23** Not correct, $R_{10}$ (68 kΩ) must be open.

FIGURE 6–88

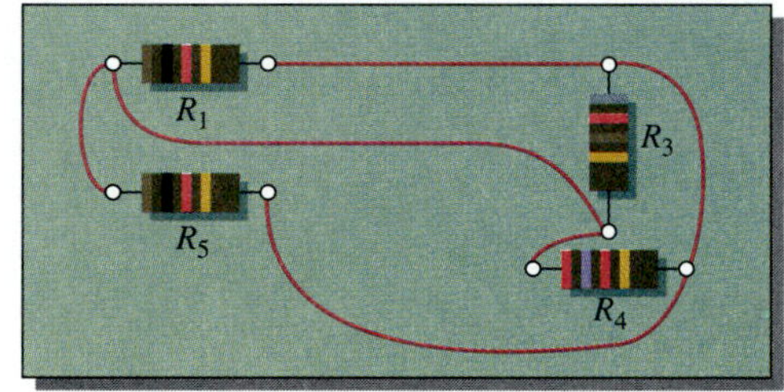

# 7

# SERIES-PARALLEL CIRCUITS



## ■ INTRODUCTION

In Chapters 5 and 6, series circuits and parallel circuits were studied individually. In this chapter, both series and parallel resistors are combined into series-parallel circuits. In many practical situations, you will have both series and parallel combinations within the same circuit, and the methods you learned for series circuits and for parallel circuits will apply.

Important types of series-parallel circuits are introduced in this chapter. These circuits include the voltage divider with a resistive load, the ladder network, and the Wheatstone bridge. In this chapter and throughout the rest of the book, you will learn the basics of putting technology theory into practice.

The analysis of series-parallel circuits requires the use of Ohm's law, Kirchhoff's voltage and current laws, and the methods for finding total resistance and power that you learned in the last two chapters. The topic of loaded voltage dividers is very important because this type of circuit is found in many practical situations. One example is the voltage-divider bias circuit for a transistor amplifier, which you will study in a later course. Ladder networks are important in several areas including a major type of digital-to-analog conversion, which you will study in the digital fundamentals course. The Wheatstone bridge is used in many systems for the measurement of unknown parameters.

## TECHnology Theory Into Practice

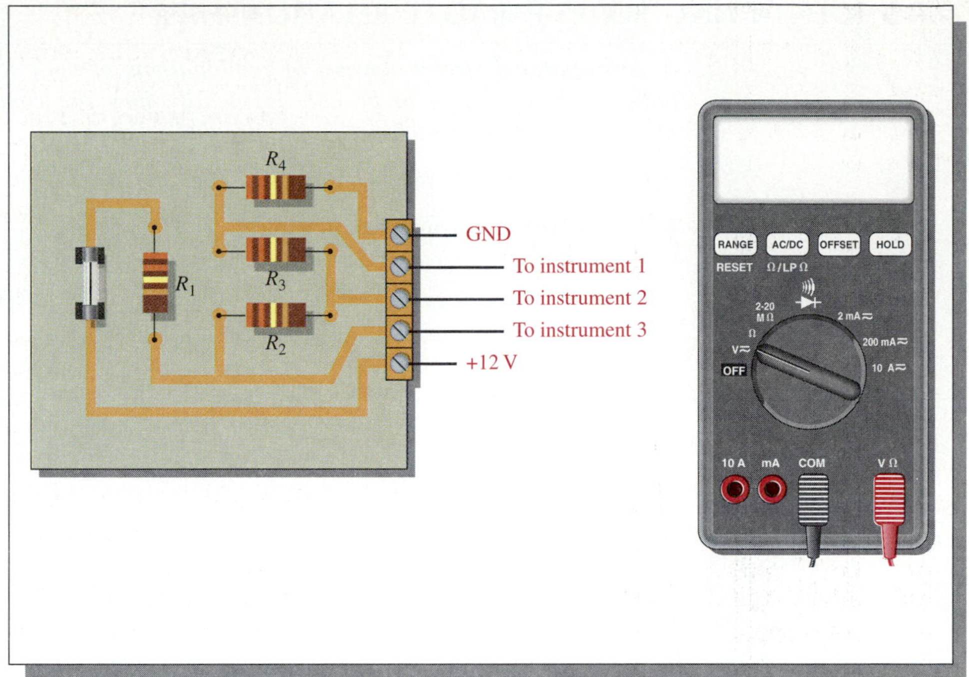


In the TECH TIP assignment in Section 7–8, you will evaluate a voltage-divider circuit board used in a portable power supply by applying your knowledge of loaded voltage dividers gained in this chapter as well as skills developed in previous chapters. The voltage divider in this application is designed to provide reference voltages to three different instruments that act as loads on the circuit. You will also be required to troubleshoot the circuit board for various common faults.

## CHAPTER OBJECTIVES

- ☐ Identify series-parallel relationships
- ☐ Analyze series-parallel circuits
- ☐ Analyze loaded voltage dividers
- ☐ Determine the loading effect of a voltmeter on a circuit
- ☐ Analyze ladder networks
- ☐ Analyze a Wheatstone bridge
- ☐ Troubleshoot series-parallel circuits
- ☐ Use PSpice to analyze series-parallel circuits (optional)

## 7–1 ■ IDENTIFYING SERIES-PARALLEL RELATIONSHIPS

***A series-parallel circuit consists of combinations of both series and parallel current paths. It is important to be able to identify how the components in a circuit are arranged in terms of their series and parallel relationships.***

***After completing this section, you should be able to***

- **Identify series-parallel relationships**
  - Recognize how each resistor in a given circuit is related to the other resistors
  - Determine series and parallel relationships on a PC board

Figure 7–1(a) shows a simple series-parallel combination of resistors. Notice that the resistance from point $A$ to point $B$ is $R_1$. The resistance from point $B$ to point $C$ is $R_2$ and $R_3$ in parallel ($R_2 \parallel R_3$). The resistance from point $A$ to point $C$ is $R_1$ in series with the parallel combination of $R_2$ and $R_3$, as indicated in Figure 7–1(b).

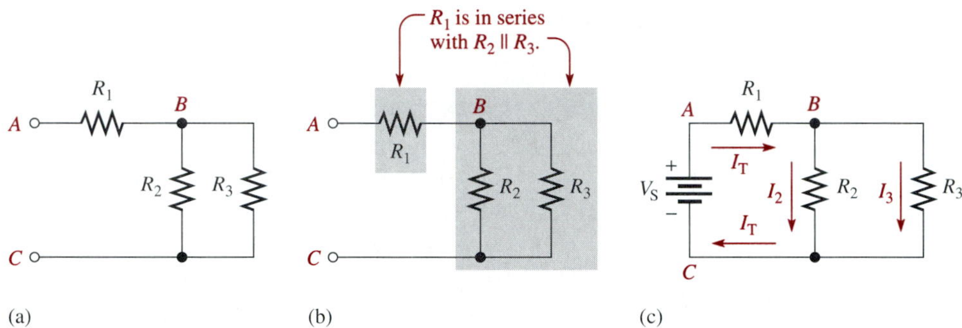


**FIGURE 7–1**
*A simple series-parallel circuit.*

When the circuit of Figure 7–1(a) is connected to a voltage source as shown in Figure 7–1(c), the total current is through $R_1$ and divides at point $B$ into the two parallel paths. These two branch currents then recombine, and the total current is into the negative source terminal as shown.

Now, to illustrate series-parallel relationships, we will increase the complexity of the circuit in Figure 7–1(a) step-by-step. In Figure 7–2(a), another resistor ($R_4$) is connected in series with $R_1$. The resistance between points $A$ and $B$ is now $R_1 + R_4$, and this combination is in series with the parallel combination of $R_2$ and $R_3$, as illustrated in Figure 7–2(b).

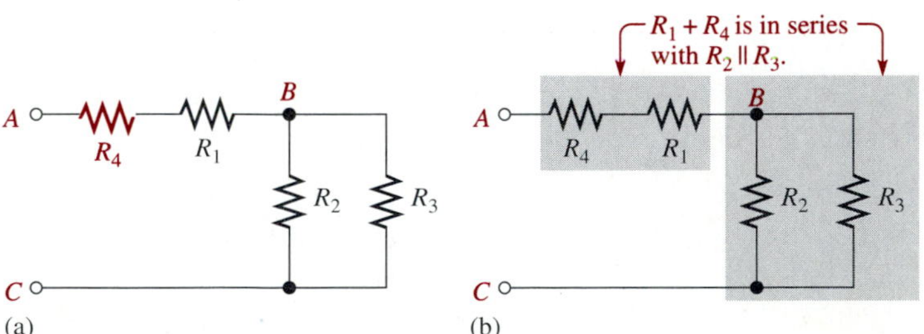


**FIGURE 7–2**
*$R_4$ is added to the circuit in series with $R_1$.*

In Figure 7–3(a), $R_5$ is connected in series with $R_2$. The series combination of $R_2$ and $R_5$ is in parallel with $R_3$. This entire series-parallel combination is in series with the $R_1 + R_4$ combination, as illustrated in Figure 7–3(b).

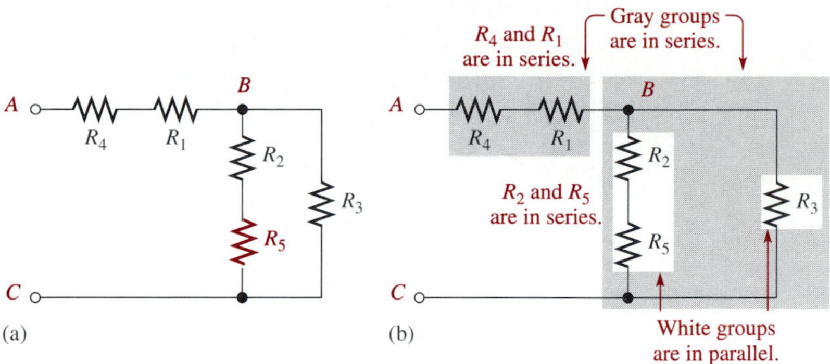


**FIGURE 7–3**

***$R_5$ is added to the circuit in series with $R_2$.***

In Figure 7–4(a), $R_6$ is connected in parallel with the series combination of $R_1$ and $R_4$. The series-parallel combination of $R_1$, $R_4$, and $R_6$ is in series with the series-parallel combination of $R_2$, $R_3$, and $R_5$, as indicated in Figure 7–4(b).

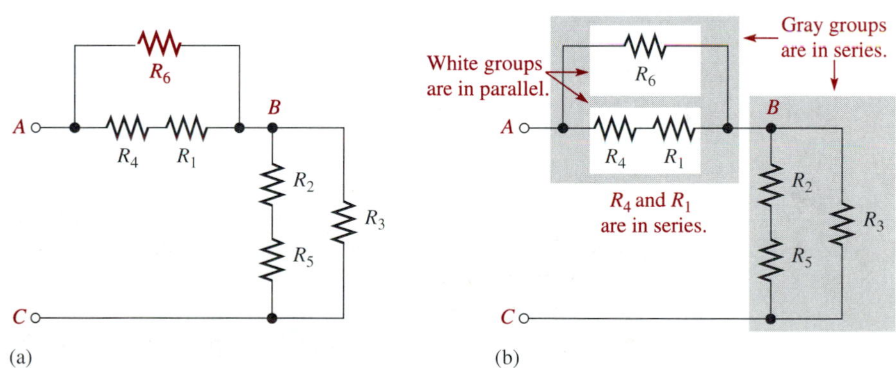


**FIGURE 7–4**

***$R_6$ is added to the circuit in parallel with the series combination of $R_1$ and $R_4$.***

**EXAMPLE 7–1** Identify the series-parallel relationships in Figure 7–5.

**FIGURE 7–5**

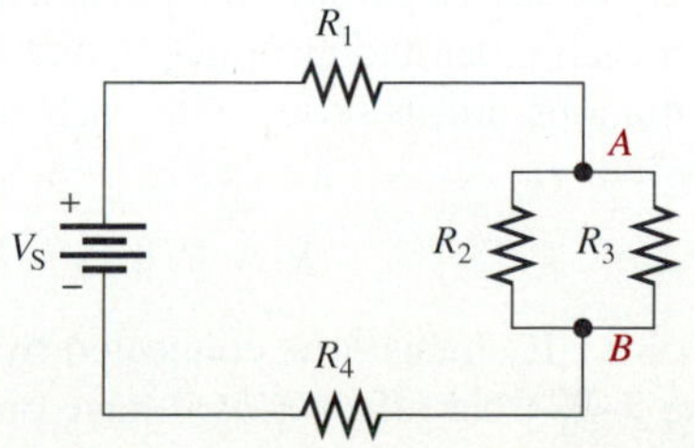

***Solution*** Starting at the positive terminal of the source, follow the current paths. All of the current produced by the source must go through $R_1$, which is in series with the rest of the circuit.

The total current takes two paths when it gets to point $A$. Part of it is through $R_2$, and part of it through $R_3$. Resistors $R_2$ and $R_3$ are in parallel with each other, and this parallel combination is in series with $R_1$.

At point $B$, the currents through $R_2$ and $R_3$ come together again. Thus, the total current is through $R_4$. Resistor $R_4$ is in series with $R_1$ and the parallel combination of $R_2$ and $R_3$. The currents are shown in Figure 7–6, where $I_T$ is the total current.

**FIGURE 7–6**

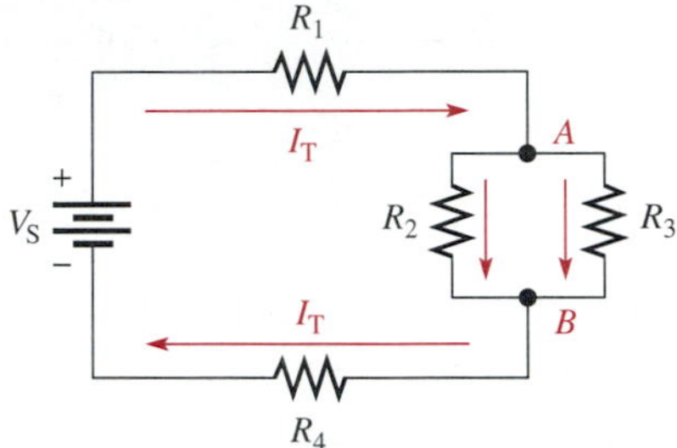


In summary, $R_1$ and $R_4$ are in series with the parallel combination of $R_2$ and $R_3$ as stated by the following expression:

$$R_1 + R_2 \parallel R_3 + R_4$$

***Related Exercise*** If another resistor, $R_5$, is connected from point $A$ to the negative side of the source in Figure 7–6, what is its relationship to the other resistors?

## EXAMPLE 7–2

Identify the series-parallel relationships in Figure 7–7.

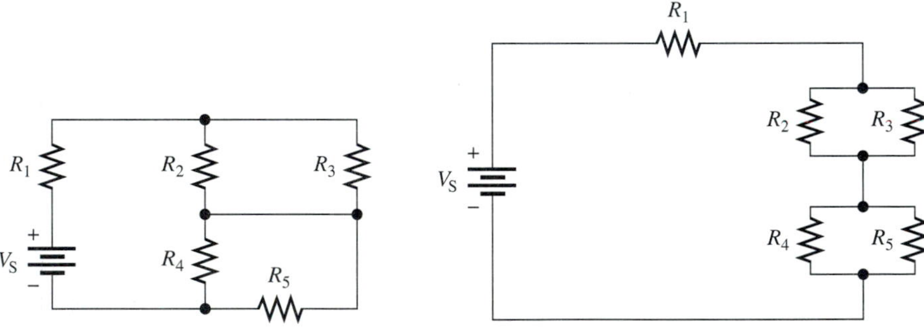


**FIGURE 7–7** **FIGURE 7–8**

***Solution*** Sometimes it is easier to see a particular circuit arrangement if it is drawn in a different way. In this case, the circuit schematic is redrawn in Figure 7–8, which better illustrates the series-parallel relationships. Now you can see that $R_2$ and $R_3$ are in parallel with each other and also that $R_4$ and $R_5$ are in parallel with each other. Both parallel combinations are in series with each other and with $R_1$ as stated by the following expression:

$$R_1 + R_2 \parallel R_3 + R_4 \parallel R_5$$

***Related Exercise*** If a resistor is connected from the bottom end of $R_3$ to the top end of $R_5$ in Figure 7–8, what effect does it have on the circuit?

**EXAMPLE 7–3** Describe the series-parallel combination between points $A$ and $D$ in Figure 7–9.

FIGURE 7–9

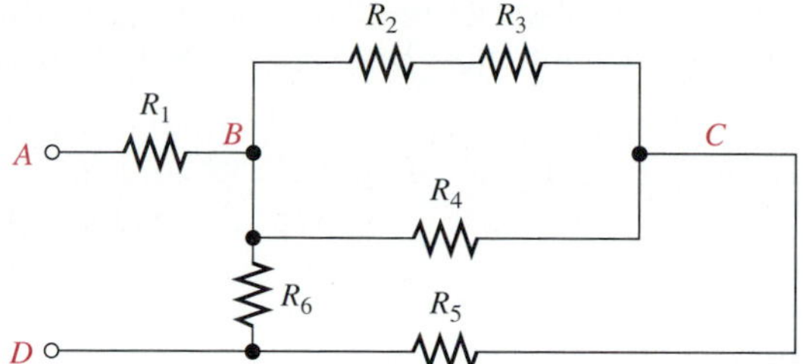


***Solution*** Between points $B$ and $C$, there are two parallel paths. The lower path consists of $R_4$, and the upper path consists of a series combination of $R_2$ and $R_3$. This parallel combination is in series with $R_5$. The $R_2$, $R_3$, $R_4$, $R_5$ combination is in parallel with $R_6$. Resistor $R_1$ is in series with this entire combination as stated by the following expression:

$$R_1 + R_6 \parallel (R_5 + (R_4 \parallel (R_2 + R_3)))$$

***Related Exercise*** If a resistor is connected from point $C$ to point $D$ in Figure 7–9, describe its parallel relationship.

**EXAMPLE 7–4** Describe the total resistance between each pair of points in Figure 7–10.

FIGURE 7–10

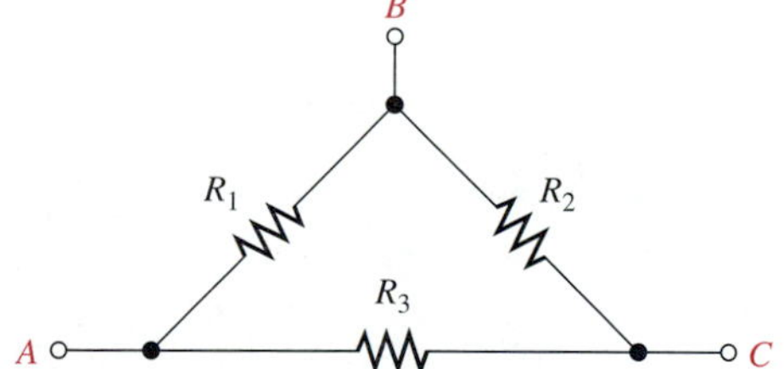


***Solution***

1. From point $A$ to $B$: $R_1$ is in parallel with the series combination of $R_2$ and $R_3$.

$$R_1 \parallel (R_2 + R_3)$$

2. From point $A$ to $C$: $R_3$ is in parallel with the series combination of $R_1$ and $R_2$.

$$R_3 \parallel (R_1 + R_2)$$

3. From point $B$ to $C$: $R_2$ is in parallel with the series combination of $R_1$ and $R_3$.

$$R_2 \parallel (R_1 + R_3)$$

***Related Exercise*** Describe the total resistance between each point in Figure 7–10 and ground if a resistor, $R_4$, is connected from point $C$ to ground.

## Determining Relationships on a Printed Circuit (PC) Board

Usually, the physical arrangement of components on a PC board bears no resemblance to the actual circuit relationships. By tracing out the circuit on the PC board and rearranging the components on paper into a recognizable form, you can determine the series-parallel relationships. Example 7–5 will illustrate.

### EXAMPLE 7–5

Determine the relationships of the resistors on the PC board in Figure 7–11.

FIGURE 7–11

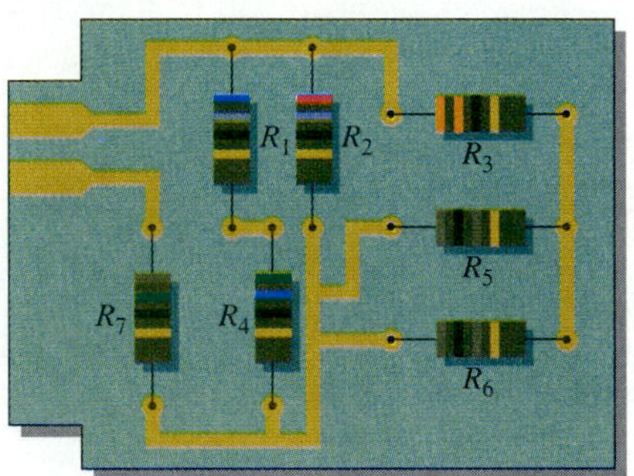


***Solution*** In Figure 7–12(a), the schematic is drawn in the same arrangement as that of the resistors on the board. In part (b), the resistors are reoriented so that the series-parallel relationships are obvious. Resistors $R_1$ and $R_4$ are in series: $R_1 + R_4$ is in parallel with $R_2$; $R_5$ and $R_6$ are in parallel and this combination is in series with $R_3$. The $R_3$, $R_5$, and $R_6$ series-parallel combination is in parallel with both $R_2$ and the $R_1 + R_4$ combination. This entire series-parallel combination is in series with $R_7$, as Figure 7–12(c) illustrates.

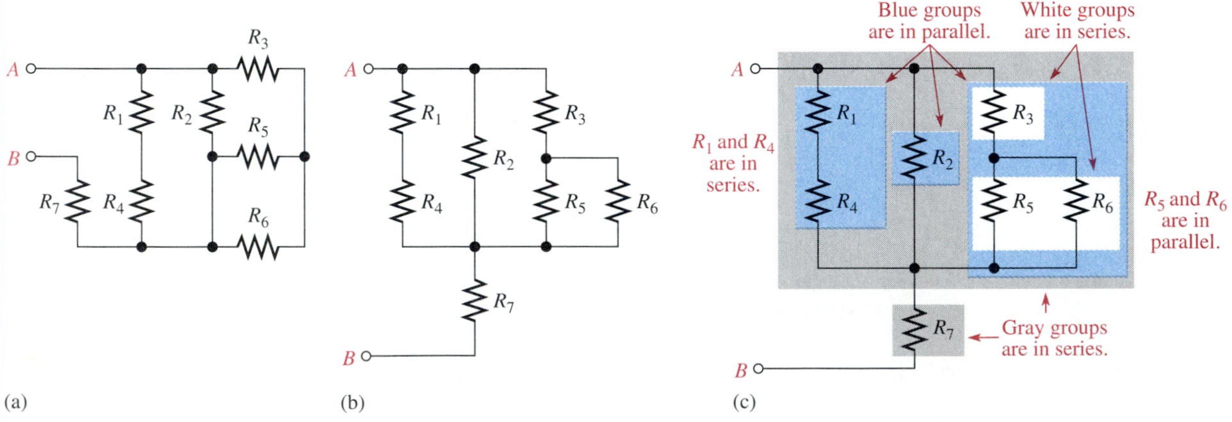


FIGURE 7–12

***Related Exercise*** If $R_5$ is removed from the circuit, what is the relationship of $R_3$ and $R_6$?

**SECTION 7–1 REVIEW**

1. Define *series-parallel resistive circuit.*
2. A certain series-parallel circuit is described as follows: $R_1$ and $R_2$ are in parallel. This parallel combination is in series with another parallel combination of $R_3$ and $R_4$. Sketch the circuit.
3. In the circuit of Figure 7–13, describe the series-parallel relationships of the resistors.
4. Which resistors are in parallel in Figure 7–14?

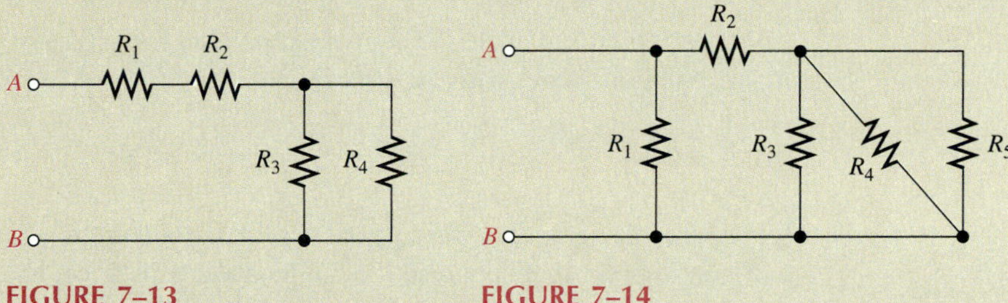


FIGURE 7–13

FIGURE 7–14

5. Describe the parallel arrangements in Figure 7–15.
6. Are the parallel combinations in Figure 7–15 in series?

FIGURE 7–15

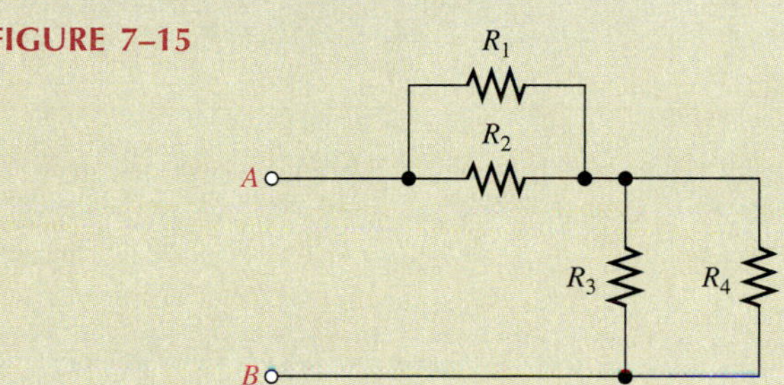


## 7–2 ■ ANALYSIS OF SERIES-PARALLEL CIRCUITS

***Several quantities are important when you have a circuit that is a series-parallel configuration of resistors.***

***After completing this section, you should be able to***

- **Analyze series-parallel circuits**
  - ☐ Determine total resistance
  - ☐ Determine all the currents
  - ☐ Determine all the voltage drops

### Total Resistance

In Chapter 5, you learned how to determine total series resistance. In Chapter 6, you learned how to determine total parallel resistance. To find the total resistance ($R_T$) of a series-parallel combination, simply define the series and parallel relationships; then perform the calculations that you have previously learned. The following two examples illustrate this general approach.

**EXAMPLE 7–6** Determine $R_T$ of the circuit in Figure 7–16 between points $A$ and $B$.

FIGURE 7–16

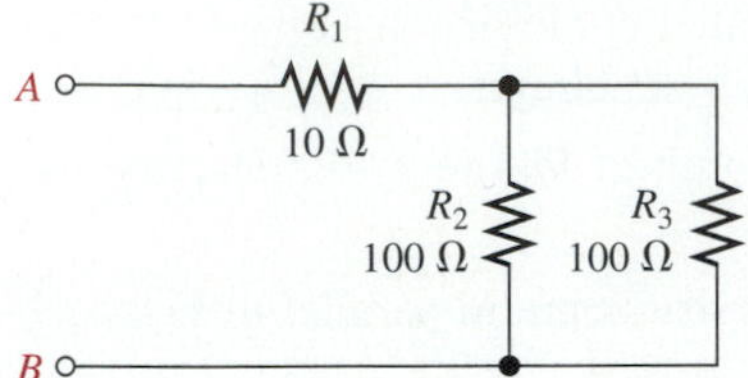


***Solution*** First, calculate the equivalent parallel resistance of $R_2$ and $R_3$. Since $R_2$ and $R_3$ are equal in value, you can use Equation (6–5).

$$R_{2\text{-}3} = \frac{R}{n} = \frac{100\ \Omega}{2} = 50\ \Omega$$

Notice that the term $R_{2\text{-}3}$ was used here to designate the total resistance of a portion of a circuit in order to distinguish it from the total resistance, $R_T$, of the complete circuit.

Now, since $R_1$ is in series with $R_{2\text{-}3}$, add their values as follows:

$$R_T = R_1 + R_{2\text{-}3} = 10\ \Omega + 50\ \Omega = 60\ \Omega$$

***Related Exercise*** Determine $R_T$ in Figure 7–16 if $R_3$ is changed to 82 Ω.

**EXAMPLE 7–7** Find the total resistance between the positive and negative terminals of the battery in Figure 7–17.

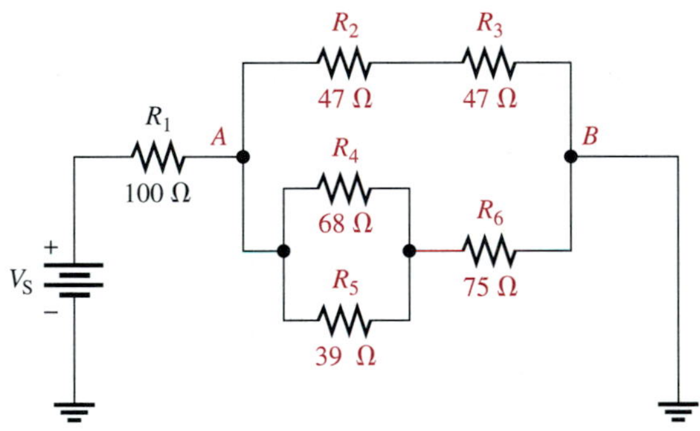


FIGURE 7–17

***Solution*** In the upper branch, $R_2$ is in series with $R_3$. This series combination is designated $R_{2\text{-}3}$ and is equal to $R_2 + R_3$.

$$R_{2\text{-}3} = R_2 + R_3 = 47\ \Omega + 47\ \Omega = 94\ \Omega$$

In the lower branch, $R_4$ and $R_5$ are in parallel with each other. This parallel combination is designated $R_{4\text{-}5}$ and is calculated as follows:

$$R_{4\text{-}5} = \frac{R_4 R_5}{R_4 + R_5} = \frac{(68\ \Omega)(39\ \Omega)}{68\ \Omega + 39\ \Omega} = 24.8\ \Omega$$

Also in the lower branch, the parallel combination of $R_4$ and $R_5$ is in series with $R_6$. This series-parallel combination is designated $R_{4\text{-}5\text{-}6}$ and is calculated as follows:

$$R_{4\text{-}5\text{-}6} = R_6 + R_{4\text{-}5} = 75\ \Omega + 24.8\ \Omega = 99.8\ \Omega$$

Figure 7–18 shows the original circuit in a simplified equivalent form.

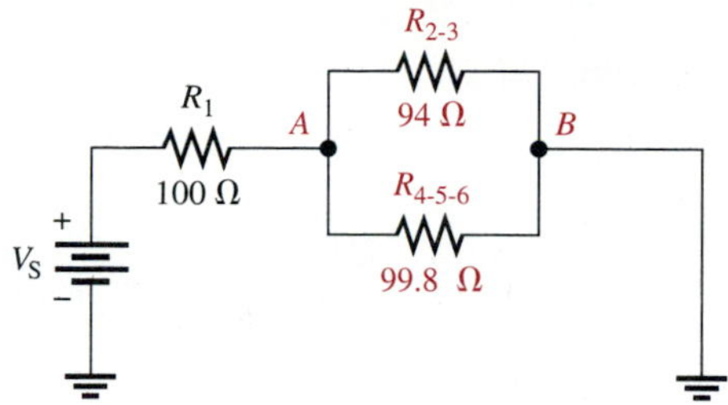


**FIGURE 7–18**

Now you can find the equivalent resistance points $A$ and $B$. It is $R_{2\text{-}3}$ in parallel with $R_{4\text{-}5\text{-}6}$. The equivalent resistance is calculated as follows:

$$R_{AB} = \frac{1}{(1/94\ \Omega) + (1/99.8\ \Omega)} = 48.4\ \Omega$$

Finally, the total resistance is $R_1$ in series with $R_{AB}$.

$$R_T = R_1 + R_{AB} = 100\ \Omega + 48.4\ \Omega = 148.4\ \Omega$$

***Related Exercise*** Determine $R_T$ if a 68 Ω resistor is added in parallel from point $A$ to point $B$ in Figure 7–17.

## Total Current

Once the total resistance and the source voltage are known, you can find total current in a circuit by applying Ohm's law. Total current is the total source voltage divided by the total resistance.

$$I_T = \frac{V_S}{R_T}$$

For example, let's find the total current in the circuit of Example 7–7 (Figure 7–17). Assume that the source voltage is 30 V. The calculation is

$$I_T = \frac{V_S}{R_T} = \frac{30\ \text{V}}{148.4\ \Omega} = 202\ \text{mA}$$

## Branch Currents

Using the current-divider formula, Kirchhoff's current law, Ohm's law, or combinations of these, you can find the current in any branch of a series-parallel circuit. In some cases, it may take repeated application of the formula to find a given current. The following two examples will help you understand the procedure. (Notice that the subscripts for the current variables ($I$) match the $R$ subscripts; for example, current through $R_1$ is referred to as $I_1$.)

**EXAMPLE 7–8** Find the current through $R_2$ and the current through $R_3$ in Figure 7–19.

FIGURE 7–19

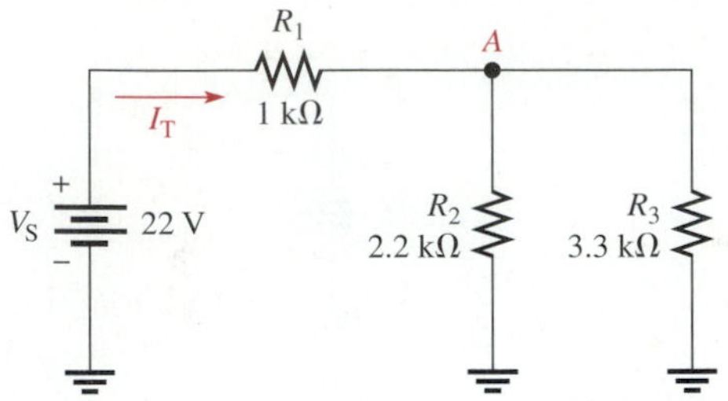

***Solution*** First you need to know how much current is into the junction (point $A$) of the parallel branches. This is the total circuit current. To find $I_T$, you need to know $R_T$.

$$R_T = R_1 + \frac{R_2R_3}{R_2 + R_3} = 1\text{ k}\Omega + \frac{(2.2\text{ k}\Omega)(3.3\text{ k}\Omega)}{2.2\text{ k}\Omega + 3.3\text{ k}\Omega} = 1\text{ k}\Omega + 1.32\text{ k}\Omega = 2.32\text{ k}\Omega$$

$$I_T = \frac{V_S}{R_T} = \frac{22\text{ V}}{2.32\text{ k}\Omega} = 9.48\text{ mA}$$

Use the current-divider rule for two branches as given in Chapter 6 to find the current through $R_2$ as follows:

$$I_2 = \left(\frac{R_3}{R_2 + R_3}\right)I_T = \left(\frac{3.3\text{ k}\Omega}{5.5\text{ k}\Omega}\right)9.48\text{ mA} = 5.69\text{ mA}$$

Now you can use Kirchhoff's current law to find the current through $R_3$ as follows:

$$I_T = I_2 + I_3$$

$$I_3 = I_T - I_2 = 9.48\text{ mA} - 5.69\text{ mA} = 3.79\text{ mA}$$

***Related Exercise*** A 4.7 kΩ resistor is connected in parallel with $R_3$ in Figure 7–19. Determine the current through the new resistor.

**EXAMPLE 7–9** Determine the current through $R_4$ in Figure 7–20 if $V_S = 50$ V.

FIGURE 7–20

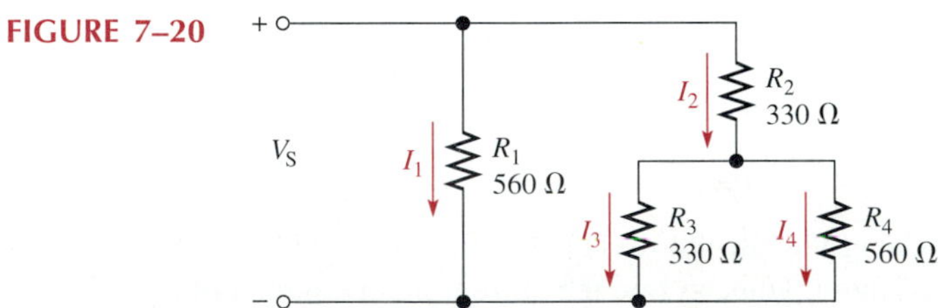


***Solution*** First find the current ($I_2$) into the junction of $R_3$ and $R_4$. Once you know this current, use the current-divider formula to find $I_4$.

Notice that there are two main branches in the circuit. The left-most branch consists of only $R_1$. The right-most branch has $R_2$ in series with the parallel combination of $R_3$ and $R_4$. The voltage across both of these main branches is the same and equal to 50 V. Find the current ($I_2$) into the junction of $R_3$ and $R_4$ by calculating the equivalent resistance ($R_{2\text{-}3\text{-}4}$) of the right-most main branch and then applying Ohm's law; $I_2$ is the total current through this main branch. Thus,

$$R_{2\text{-}3\text{-}4} = R_2 + \frac{R_3R_4}{R_3 + R_4} = 330\ \Omega + \frac{(330\ \Omega)(560\ \Omega)}{890\ \Omega} = 538\ \Omega$$

$$I_2 = \frac{V_S}{R_{2\text{-}3\text{-}4}} = \frac{50\ \text{V}}{538\ \Omega} = 93\ \text{mA}$$

Use the current-divider formula to calculate $I_4$ as follows:

$$I_4 = \left(\frac{R_3}{R_3 + R_4}\right)I_2 = \left(\frac{330\ \Omega}{890\ \Omega}\right)93\ \text{mA} = 34.5\ \text{mA}$$

***Related Exercise*** Determine the current through $R_1$ and $R_3$ in Figure 7–20 if $V_S$ = 20 V.

## Voltage Drops

It is often necessary to find the voltages across certain parts of a series-parallel circuit. You can find these voltages by using the voltage-divider formula given in Chapter 5, Kirchhoff's voltage law, Ohm's law, or combinations of each. The following three examples illustrate use of the formulas.

**EXAMPLE 7–10** Determine the voltage drop from point $A$ to ground in Figure 7–21. Then find the voltage ($V_1$) across $R_1$.

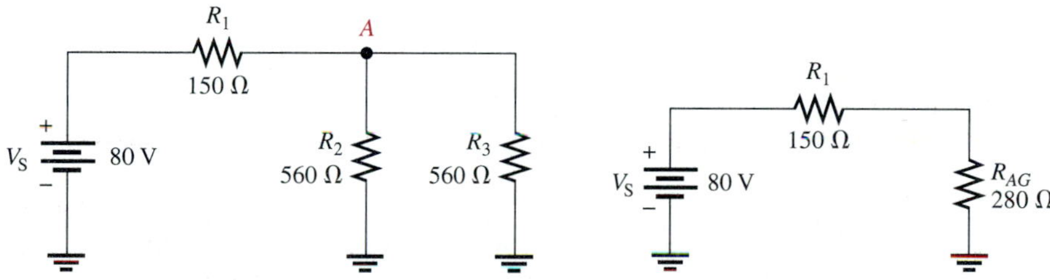


FIGURE 7–21

FIGURE 7–22

***Solution*** Note that $R_2$ and $R_3$ are in parallel in this circuit. Since they are equal in value, their equivalent resistance from point $A$ to ground is

$$R_{AG} = \frac{560\ \Omega}{2} = 280\ \Omega$$

In the equivalent circuit shown in Figure 7–22, $R_1$ is in series with $R_{AG}$. The total circuit resistance as seen from the source is

$$R_T = R_1 + R_{AG} = 150\ \Omega + 280\ \Omega = 430\ \Omega$$

Use the voltage-divider formula to find the voltage across the parallel combination of Figure 7–21 (between point $A$ and ground).

$$V_{AG} = \left(\frac{R_{AG}}{R_T}\right)V_S = \left(\frac{280\ \Omega}{430\ \Omega}\right)80\ \text{V} = 52.1\ \text{V}$$

Now use Kirchhoff's voltage law to find $V_1$.

$$V_S = V_1 + V_{AG}$$
$$V_1 = V_S - V_{AG} = 80\ \text{V} - 52.1\ \text{V} = 27.9\ \text{V}$$

***Related Exercise*** Determine $V_{AG}$ and $V_{R1}$ if $R_1$ is changed to 220 Ω in Figure 7–21.

**EXAMPLE 7–11** Determine the voltage across each resistor in the circuit of Figure 7–23.

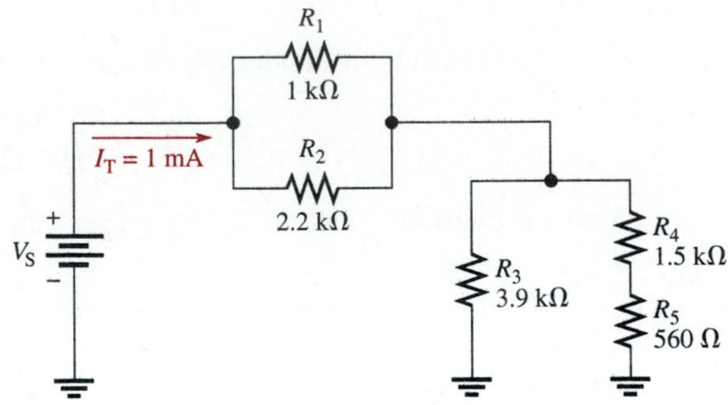


**FIGURE 7–23**

***Solution*** The source voltage is not given, but you know the total current from the figure. Since $R_1$ and $R_2$ are in parallel, they each have the same voltage. The current through $R_1$ is

$$I_1 = \left(\frac{R_2}{R_1 + R_2}\right)I_T = \left(\frac{2.2\text{ k}\Omega}{3.2\text{ k}\Omega}\right)1\text{ mA} = 688\ \mu\text{A}$$

The voltages across $R_1$ and $R_2$ are

$$V_1 = I_1R_1 = (688\ \mu\text{A})(1\text{ k}\Omega) = 688\text{ mV}$$
$$V_2 = V_1 = 688\text{ mV}$$

The current through $R_3$ is found by applying the current-divider formula.

$$I_3 = \left[\frac{R_4 + R_5}{R_3 + (R_4 + R_5)}\right]I_T = \left(\frac{2.06\text{ k}\Omega}{5.96\text{ k}\Omega}\right)1\text{ mA} = 346\ \mu\text{A}$$

The voltage across $R_3$ is

$$V_3 = I_3R_3 = (346\ \mu\text{A})(3.9\text{ k}\Omega) = 1.35\text{ V}$$

The currents through $R_4$ and $R_5$ are the same because these resistors are in series.

$$I_4 = I_5 = I_T - I_3 = 1\text{ mA} - 346\ \mu\text{A} = 654\ \mu\text{A}$$

The voltages across $R_4$ and $R_5$ are calculated as follows:

$$V_4 = I_4R_4 = (654\ \mu\text{A})(1.5\text{ k}\Omega) = 981\text{ mV}$$
$$V_5 = I_5R_5 = (654\ \mu\text{A})(560\ \Omega) = 366\text{ mV}$$

***Related Exercise*** What is the source voltage, $V_S$, in the circuit of Figure 7–23?

**EXAMPLE 7–12** Determine the voltage drop across each resistor in Figure 7–24.

FIGURE 7–24

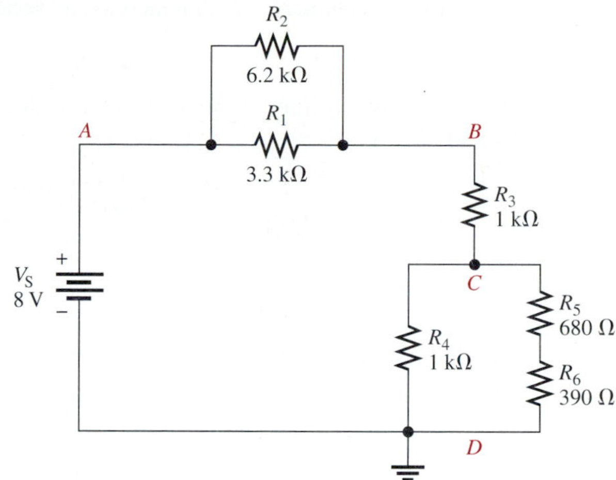


***Solution*** Because the total voltage is given in the figure, you can solve this problem using the voltage-divider formula.

First you need to reduce each parallel combination to an equivalent resistance. Since $R_1$ and $R_2$ are in parallel between points $A$ and $B$, combine their values as follows:

$$R_{AB} = \frac{R_1R_2}{R_1 + R_2} = \frac{(3.3\text{ k}\Omega)(6.2\text{ k}\Omega)}{9.5\text{ k}\Omega} = 2.15\text{ k}\Omega$$

Since $R_4$ is in parallel with the series combination of $R_5$ and $R_6$ between points $C$ and $D$, combine these values to obtain

$$R_{CD} = \frac{R_4(R_5 + R_6)}{R_4 + R_5 + R_6} = \frac{(1\text{ k}\Omega)(1.07\text{ k}\Omega)}{2.07\text{ k}\Omega} = 517\ \Omega$$

The equivalent circuit is drawn in Figure 7–25. The total circuit resistance is

$$R_T = R_{AB} + R_3 + R_{CD} = 2.15\text{ k}\Omega + 1\text{ k}\Omega + 517\ \Omega = 3.67\text{ k}\Omega$$

FIGURE 7–25

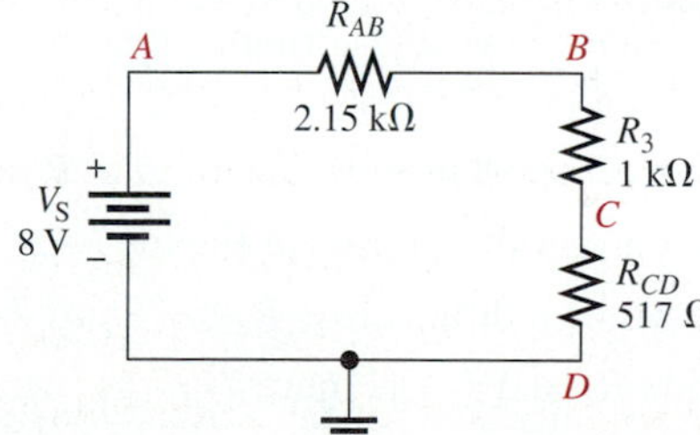


Now use the voltage-divider formula to determine the voltages in the equivalent circuit as follows:

$$V_{AB} = \left(\frac{R_{AB}}{R_T}\right)V_S = \left(\frac{2.15\text{ k}\Omega}{3.67\text{ k}\Omega}\right)8\text{ V} = 4.69\text{ V}$$

$$V_{CD} = \left(\frac{R_{CD}}{R_T}\right)V_S = \left(\frac{517\ \Omega}{3.67\text{ k}\Omega}\right)8\text{ V} = 1.13\text{ V}$$

$$V_3 = \left(\frac{R_3}{R_T}\right)V_S = \left(\frac{1\text{ k}\Omega}{3.67\text{ k}\Omega}\right)8\text{ V} = 2.18\text{ V}$$

Refer to Figure 7–24. $V_{AB}$ equals the voltage across both $R_1$ and $R_2$, so

$$V_1 = V_2 = V_{AB} = 4.69 \text{ V}$$

$V_{CD}$ is the voltage across $R_4$ and across the series combination of $R_5$ and $R_6$. Therefore,

$$V_4 = V_{CD} = 1.13 \text{ V}$$

Now apply the voltage-divider formula to the series combination of $R_5$ and $R_6$ to get $V_5$ and $V_6$.

$$V_5 = \left(\frac{R_5}{R_5 + R_6}\right)V_{CD} = \left(\frac{680\ \Omega}{1070\ \Omega}\right)1.13 \text{ V} = 718 \text{ mV}$$

$$V_6 = \left(\frac{R_6}{R_5 + R_6}\right)V_{CD} = \left(\frac{390\ \Omega}{1070\ \Omega}\right)1.13 \text{ V} = 412 \text{ mV}$$

***Related Exercise*** $R_2$ is removed from the circuit in Figure 7–24. Calculate $V_{AB}$, $V_{BC}$, and $V_{CD}$.

As you have seen in this section, the analysis of series-parallel circuits can be approached in many ways, depending on what information you need and what circuit values you know. The examples in this section do not represent an exhaustive coverage, but they give you an idea of how to approach series-parallel circuit analysis.

If you know Ohm's law, Kirchhoff's laws, the voltage-divider formula, and the current-divider formula, and if you know how to apply these laws, you can solve most resistive circuit analysis problems. The ability to recognize series and parallel combinations is, of course, essential.

## SECTION 7–2 REVIEW

1. List the circuit laws and formulas that may be necessary in the analysis of series-parallel circuits.
2. Find the total resistance between $A$ and $B$ in the circuit of Figure 7–26.
3. Find the current through $R_3$ in Figure 7–26.
4. Find the voltage drop across $R_2$ in Figure 7–26.
5. Determine $R_T$ and $I_T$ in Figure 7–27 as "seen" by the source.

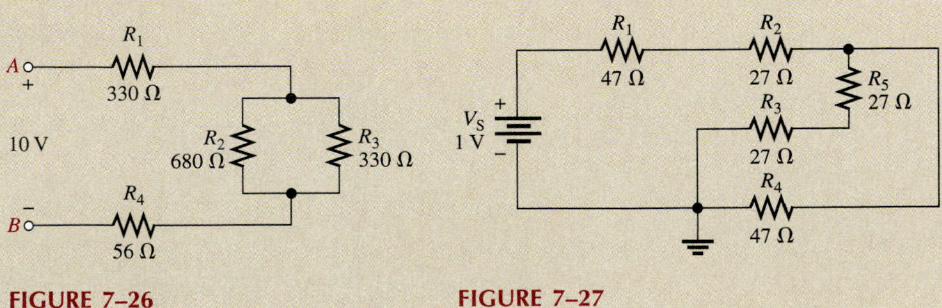


FIGURE 7–26

FIGURE 7–27

## 7–3 ■ VOLTAGE DIVIDERS WITH RESISTIVE LOADS

***Voltage dividers were introduced in Chapter 5. In this section, you will learn how resistive loads affect the operation of voltage-divider circuits.***

***After completing this section, you should be able to***

- **Analyze loaded voltage dividers**
  - □ Determine the effect of a resistive load on a voltage-divider circuit
  - □ Explain how a voltmeter can alter the value of the voltage being measured

The voltage divider in Figure 7–28(a) produces an output voltage ($V_{OUT}$) of 5 V because the two resistors are of equal value. This voltage is the *unloaded output voltage*. When a load resistor, $R_L$, is connected from the output to ground as shown in Figure 7–28(b), the output voltage is reduced by an amount that depends on the value of $R_L$. The load resistor is in parallel with $R_2$, reducing the resistance from point $A$ to ground and, as a result, also reducing the voltage across the parallel combination. This is one effect of loading a voltage divider. Another effect of a **load** is that more current is drawn from the source because the total resistance of the circuit is reduced.

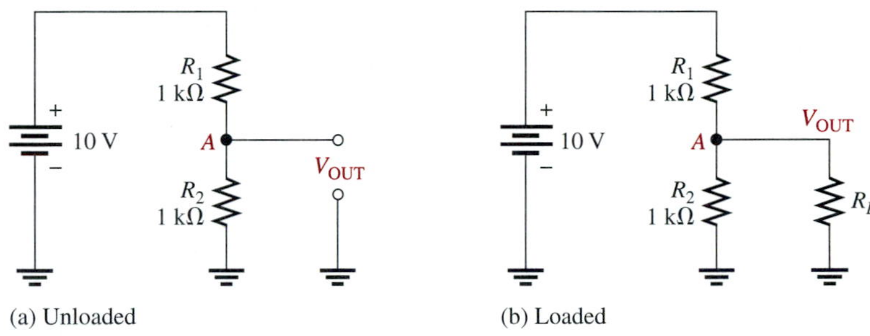


**FIGURE 7–28**
***A voltage divider with both unloaded and loaded outputs.***

The larger $R_L$ is compared to $R_2$, the less the output voltage is reduced from its unloaded value, as illustrated in Figure 7–29. When two resistors are connected in parallel and one of the resistors is much greater than the other, the total resistance is close to the value of the smaller resistance.

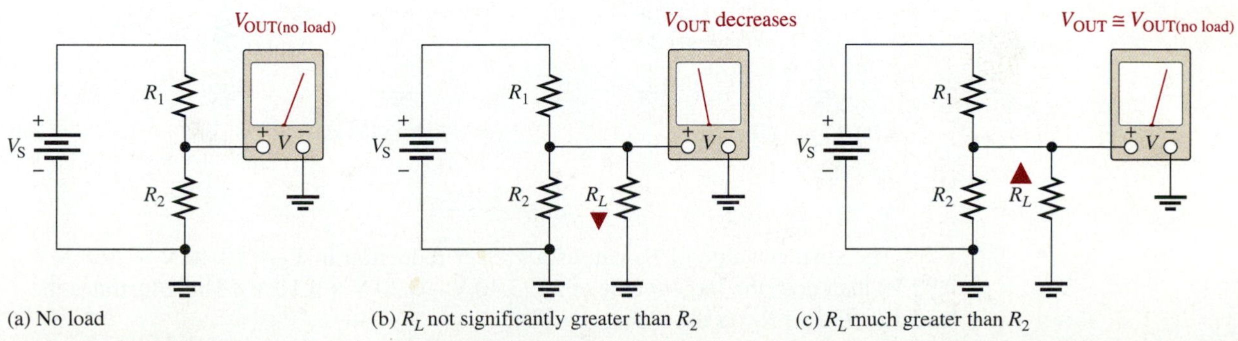


**FIGURE 7–29**
***The effect of a load resistor.***

**EXAMPLE 7–13**

**(a)** Determine the unloaded output voltage of the voltage divider in Figure 7–30.

**(b)** Find the loaded output voltages of the voltage divider in Figure 7–30 for the following two values of load resistance: $R_L = 10\ \text{k}\Omega$ and $R_L = 100\ \text{k}\Omega$.

FIGURE 7–30

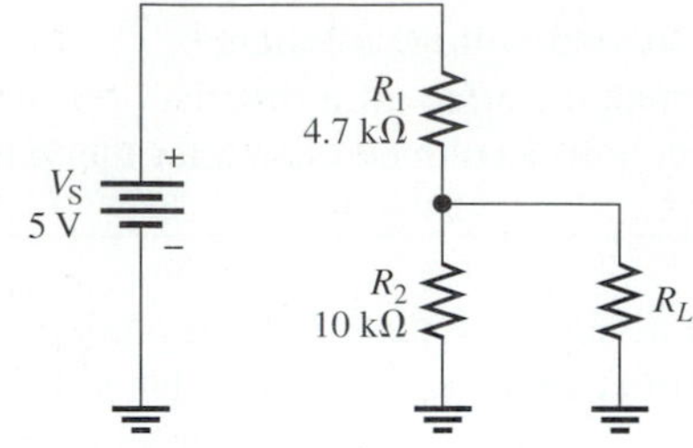


### *Solution*

**(a)** The unloaded output voltage is

$$V_{\text{OUT(unloaded)}} = \left(\frac{R_2}{R_1 + R_2}\right)V_S = \left(\frac{10\ \text{k}\Omega}{14.7\ \text{k}\Omega}\right)5\ \text{V} = 3.40\ \text{V}$$

**(b)** With the 10 kΩ load resistor connected, $R_L$ is in parallel with $R_2$, which gives

$$R_2 \parallel R_L = \frac{R_2 R_L}{R_2 + R_L} = \frac{100\ \text{k}\Omega}{20} = 5\ \text{k}\Omega$$

The equivalent circuit is shown in Figure 7–31(a). The loaded output voltage is

$$V_{\text{OUT(loaded)}} = \left(\frac{R_2 \parallel R_L}{R_1 + R_2 \parallel R_L}\right)V_S = \left(\frac{5\ \text{k}\Omega}{9.7\ \text{k}\Omega}\right)5\ \text{V} = 2.58\ \text{V}$$

With the 100 kΩ load, the resistance from output to ground is

$$R_2 \parallel R_L = \frac{R_2 R_L}{R_2 + R_L} = \frac{(10\ \text{k}\Omega)(100\ \text{k}\Omega)}{110\ \text{k}\Omega} = 9.1\ \text{k}\Omega$$

The equivalent circuit is shown in Figure 7–31(b). The loaded output voltage is

$$V_{\text{OUT(loaded)}} = \left(\frac{R_2 \parallel R_L}{R_1 + R_2 \parallel R_L}\right)V_S = \left(\frac{9.1\ \text{k}\Omega}{13.8\ \text{k}\Omega}\right)5\ \text{V} = 3.30\ \text{V}$$

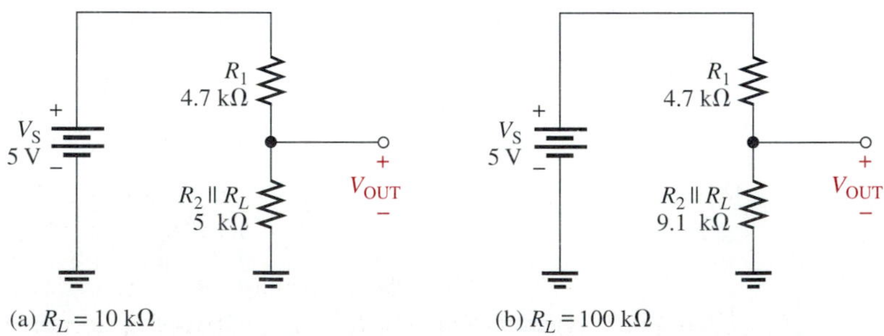


FIGURE 7–31

The smaller value of $R_L$ causes a greater reduction in $V_{OUT}$ (3.40 V – 2.58 V = 0.82 V) than does the larger value of $R_L$ (3.40 V – 3.30 V = 0.10 V). This illustrates the loading effect of $R_L$ on the voltage divider.

***Related Exercise*** Determine $V_{OUT}$ in Figure 7–30 for a 1 MΩ load resistance.

## Bleeder Current

Voltage dividers are sometimes useful in obtaining various voltages from a fixed power supply. For example, suppose that you wish to derive 12 V and 6 V from a 24 V supply. To do so requires a voltage divider with two taps, as shown in Figure 7–32. In this example, $R_1$ must equal $R_2 + R_3$, and $R_2$ must equal $R_3$. The actual values of the resistors are set by the amount of current that is to be drawn from the source under unloaded conditions. This current, called the **bleeder current,** represents a continuous drain on the source. With these ideas in mind, in Example 7–14 a voltage divider is designed to meet certain specified requirements.

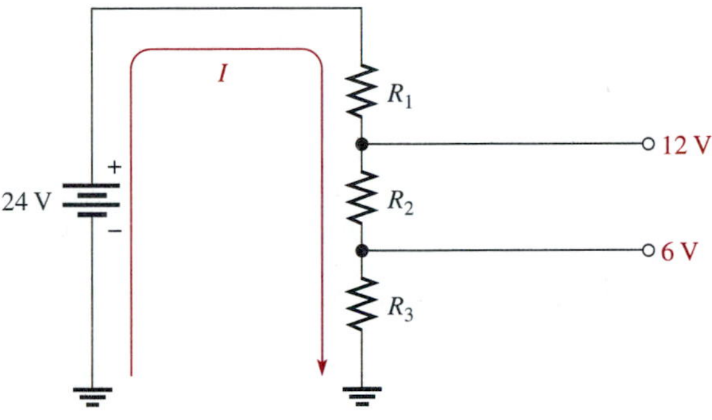


**FIGURE 7–32**
*Voltage divider with two output taps.*

**EXAMPLE 7–14**

A power supply requires 12 V and 6 V from a 24 V battery. The unloaded current drain on this battery is not to exceed 1 mA. Determine the values of the resistors. Also determine the output voltage at the 12 V tap when both outputs are loaded with 100 kΩ each.

***Solution*** A circuit as shown in Figure 7–32 is required. In order to have an unloaded current of 1 mA, the total resistance must be

$$R_T = \frac{V_S}{I} = \frac{24\text{ V}}{1\text{ mA}} = 24\text{ k}\Omega$$

To get exactly 12 V, $R_1 = R_2 + R_3 = 12\text{ k}\Omega$. To get exactly 6 V, $R_2 = R_3 = 6\text{ k}\Omega$.

The 12 kΩ is a standard value, but the closest standard value to 6 kΩ is 6.04 kΩ in a 1% resistor and 6.2 kΩ in a 5% resistor. We will use 6.2 kΩ for $R_2$ and $R_3$, although this will result in small differences from the desired unloaded output voltages as follows:

$$V_{12V} = \left(\frac{R_2 + R_3}{R_1 + R_2 + R_3}\right)V_S = \left(\frac{12.4\text{ k}\Omega}{24.4\text{ k}\Omega}\right)24\text{ V} = 12.2\text{ V}$$

$$V_{6V} = \left(\frac{R_3}{R_1 + R_2 + R_3}\right)V_S = \left(\frac{6.2\text{ k}\Omega}{24.4\text{ k}\Omega}\right)24\text{ V} = 6.1\text{ V}$$

The 100 kΩ loads are connected to the outputs as shown in Figure 7–33. The loaded output voltage at the 12 V tap is determined as follows. First, the equivalent resistance from the 12 V tap to ground is the 100 kΩ load resistor $R_{L1}$ in parallel with the combination of $R_2$ in series with the parallel combination of $R_3$ and $R_{L2}$. This equivalent resistance is determined as follows: For $R_3$ in parallel with $R_{L2}$,

$$R_{EQ1} = \frac{R_3R_{L2}}{R_3 + R_{L2}} = \frac{(6.2\text{ k}\Omega)(100\text{ k}\Omega)}{106.2\text{ k}\Omega} = 5.84\text{ k}\Omega$$

For $R_2$ in series with $R_{EQ1}$,

$$R_{EQ2} = R_2 + R_{EQ1} = 6.2\text{ k}\Omega + 5.84\text{ k}\Omega = 12.0\text{ k}\Omega$$

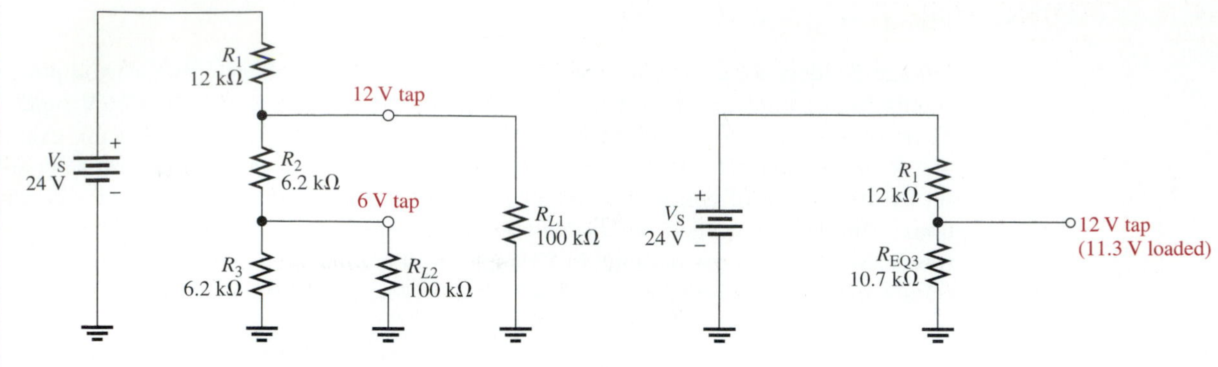


FIGURE 7–33

FIGURE 7–34

For $R_{L1}$ in parallel with $R_{EQ2}$,

$$R_{EQ3} = \frac{R_{L1}R_{EQ2}}{R_{L1} + R_{EQ2}} = \frac{(100\text{ k}\Omega)(12.0\text{ k}\Omega)}{112\text{ k}\Omega} = 10.7\text{ k}\Omega$$

$R_{EQ3}$ is the equivalent resistance from the 12 V tap to ground. The equivalent circuit from the 12 V tap to ground is shown in Figure 7–34. Using this equivalent circuit, calculate the voltage at the 12 V tap of the loaded voltage divider as follows:

$$V_{12V} = \left(\frac{R_{EQ3}}{R_1 + R_{EQ3}}\right)V_S = \left(\frac{10.7\text{ k}\Omega}{22.7\text{ k}\Omega}\right)24\text{ V} = 11.3\text{ V}$$

As you can see, the output voltage at the 12 V tap decreases slightly from its unloaded value when the 100 kΩ loads are connected. Smaller values of load resistance would result in a greater decrease in the output voltage.

***Related Exercise*** Determine the output voltage at the 12 V tap when both outputs are loaded with 10 kΩ each.

## SECTION 7–3 REVIEW

1. A load resistor is connected to an output tap on a voltage divider. What effect does the load resistor have on the output voltage at this tap?
2. A larger-value load resistor will cause the output voltage to change less than a smaller-value one will (T or F).
3. For the voltage divider in Figure 7–35, determine the unloaded output voltage. Also determine the output voltage with a 10 kΩ load resistor connected across the output.

FIGURE 7–35

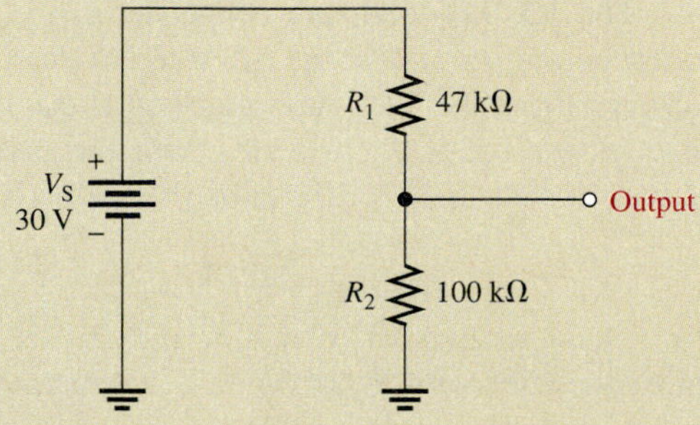

## 7–4 ■ LOADING EFFECT OF A VOLTMETER

***As you have learned, voltmeters must be connected in parallel with a resistor in order to measure the voltage across the resistor. Because of its internal resistance, a voltmeter puts a load on the circuit and will affect, to a certain extent, the voltage that is being measured. Until now, we have ignored the loading effect because the internal resistance of a voltmeter is very high, and normally it has negligible effect on the circuit that is being measured. However, if the internal resistance of the voltmeter is not sufficiently greater than the circuit resistance across which it is connected, the loading effect will cause the measured voltage to be less than its actual value. You should always be aware of this effect.***

***After completing this section, you should be able to***

- **Determine the loading effect of a voltmeter on a circuit**
  - ☐ Explain why a voltmeter can load a circuit
  - ☐ Discuss the internal resistance of a voltmeter

### Why a Voltmeter Can Load a Circuit

When a voltmeter is connected to a circuit as shown, for example, in Figure 7–36(a), its internal resistance appears in parallel with $R_3$, as shown in part (b). The resistance from point $A$ to point $B$ is altered by the loading effect of the voltmeter's internal resistance $R_M$ and is equal to $R_3 \parallel R_M$, as indicated in part (c).

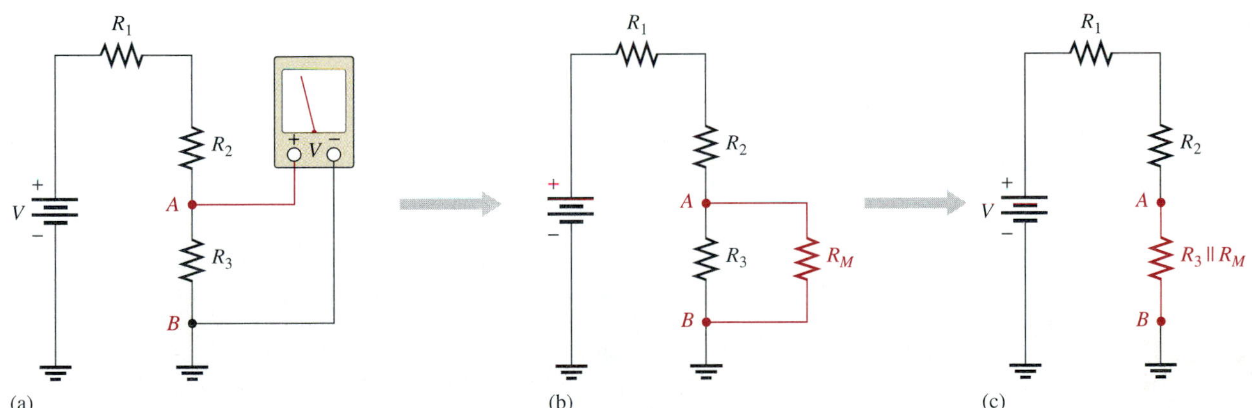


**FIGURE 7–36**
*The loading effect of a voltmeter.*

If $R_M$ is much greater than $R_3$, the resistance from $A$ to $B$ changes very little, and the meter indicates the actual voltage. If $R_M$ is not sufficiently greater than $R_3$, the resistance from $A$ to $B$ is reduced significantly, and the voltage across $R_3$ is altered by the loading effect of the meter. A good rule of thumb is that *if the meter resistance is at least ten times greater than the resistance across which it is connected, the loading effect can be neglected.*

### Internal Resistance of a Voltmeter

Three basic categories of voltmeters are the electromagnetic analog voltmeter, whose internal resistance is determined by its **sensitivity factor;** the electronic analog voltmeter, whose internal resistance is set by an input amplifier and is typically at least 10 MΩ; and the digital voltmeter (the most commonly used type), whose internal resistance is also typically at least 10 MΩ. The electronic analog voltmeter and the digital voltmeter present fewer loading problems than the electromagnetic type because their internal resistances are much higher.

As mentioned, the internal resistance of an electromagnetic voltmeter is specified by the sensitivity factor. A common sensitivity value is 20,000 Ω/V. The internal resistance depends on the dc voltage range switch setting of the meter and is determined by multiplying the sensitivity by the range setting. For example, if the dc voltage range switch is set at 10 V, the internal resistance of the meter is

$$R_M = 20{,}000\ \Omega/\text{V} \times 10\ \text{V} = 200{,}000\ \Omega$$

The lower the range setting, the less the internal resistance and the greater the loading effect of the voltmeter on a circuit. Always use the highest possible range setting that will allow an accurate reading from the meter scale. However, you would not use a 60 V setting to measure 1 V because you could not read 1 V accurately on the 60 V meter scale. The following example shows how the range setting affects the loading.

**EXAMPLE 7–15** How much does the electromagnetic voltmeter affect the voltage being measured for each range switch setting indicated in Figure 7–37? Assume the meter has a sensitivity factor of 20,000 Ω/V.

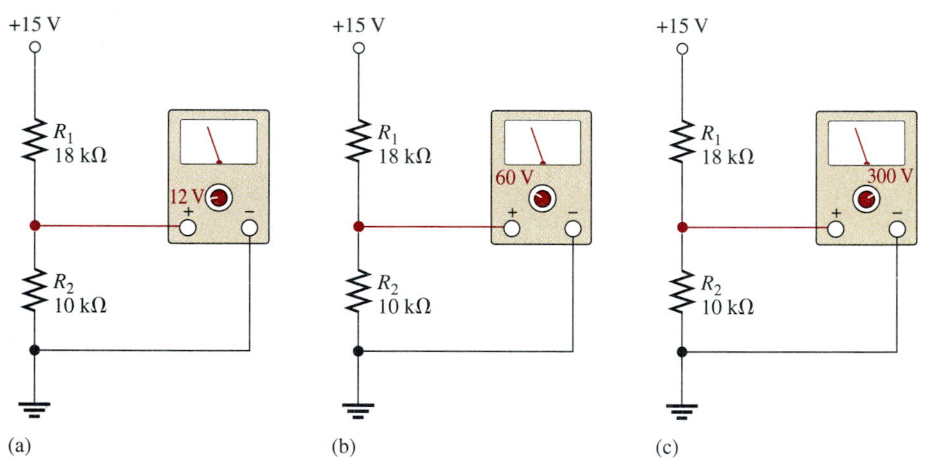


FIGURE 7–37

***Solution*** The unloaded voltage across $R_2$ in the voltage-divider circuit is

$$V_{R2} = \left(\frac{R_2}{R_1 + R_2}\right)V = \left(\frac{10\ \text{k}\Omega}{28\ \text{k}\Omega}\right)15\ \text{V} = 5.36\ \text{V}$$

Refer to Figure 7–37(a). The internal resistance of the meter is

$$R_M = 20{,}000\ \Omega/\text{V} \times 12\ \text{V} = 240\ \text{k}\Omega$$

The meter's resistance in parallel with $R_2$ is

$$R_2 \parallel R_M = \left(\frac{R_2 \parallel R_M}{R_2 + R_M}\right) = \frac{(10\ \text{k}\Omega)(240\ \text{k}\Omega)}{250\ \text{k}\Omega} = 9.6\ \text{k}\Omega$$

The voltage actually measured by the meter is

$$V_{R2} = \left(\frac{R_2 \parallel R_M}{R_1 + R_2 \parallel R_M}\right)V = \left(\frac{9.6\ \text{k}\Omega}{27.6\ \text{k}\Omega}\right)15\ \text{V} = 5.22\ \text{V}$$

Refer to Figure 7–37(b). The internal resistance of the meter is

$$R_M = 20{,}000\ \Omega/\text{V} \times 60\ \text{V} = 1.2\ \text{M}\Omega$$

and

$$R_2 \parallel R_M = \frac{R_2 R_M}{R_2 + R_M} = \frac{(10\ \text{k}\Omega)(1.2\ \text{M}\Omega)}{1.21\ \text{M}\Omega} = 9.92\ \text{k}\Omega$$

The voltage actually measured is

$$V_{R2} = \left(\frac{R_2 \parallel R_M}{R_1 + R_2 \parallel R_M}\right)V = \left(\frac{9.92\ \text{k}\Omega}{27.9\ \text{k}\Omega}\right)15\ \text{V} = 5.33\ \text{V}$$

Refer to Figure 7–37(c). The internal resistance of the meter is

$$R_M = 20{,}000\ \Omega/\text{V} \times 300\ \text{V} = 6\ \text{M}\Omega$$

and

$$R_2 \parallel R_M = \frac{R_2 R_M}{R_2 + R_M} = \frac{(10\ \text{k}\Omega)(6\ \text{M}\Omega)}{6.01\ \text{M}\Omega} = 9.98\ \text{k}\Omega$$

The voltage actually measured is

$$V_{R2} = \left(\frac{R_2 \parallel R_M}{R_1 + R_2 \parallel R_M}\right)V = \left(\frac{9.98\ \text{k}\Omega}{27.98\ \text{k}\Omega}\right)15\ \text{V} = 5.35\ \text{V}$$

As you can see, the loading decreases as you increase the range setting on the voltmeter because the measured voltage is closer to the unloaded voltage. However, the 12 V setting is the best choice in this case because, although the loading effect is greater, you can read the expected voltage much more accurately than you can on the 60 V range. Of course, using the 300 V range is not a good choice in this case.

***Related Exercise*** Calculate the voltage across $R_2$ in Figure 7–37 if the range switch is set on 100 V.

**SECTION 7–4 REVIEW**

1. What is meant when we say a voltmeter loads a circuit?
2. If a voltmeter with a 10 MΩ internal resistance is measuring the voltage across a 1 kΩ resistor, should you normally be concerned about the loading effect?
3. A certain electromagnetic-type voltmeter is set on the 12 V range to measure 2.5 V. If you switch it down to the 3 V range, will there be an increase or a decrease in the measured voltage? What is the advantage of using the 3 V range instead of the 12 V range in this situation?
4. What is the internal resistance of a 30,000 Ω/V voltmeter on its 50 V range setting?

## 7–5 ■ LADDER NETWORKS

***A resistive ladder network is a special type of series-parallel circuit. One form of ladder network is commonly used to scale down voltages to certain weighted values for digital-to-analog conversion, a process that you will study in another course.***

***After completing this section, you should be able to***

- **Analyze ladder networks**
  - ☐ Determine the voltages in a three-step ladder network
  - ☐ Analyze an $R/2R$ ladder

In this section, we examine basic resistive ladder networks of limited complexity, beginning with the one shown in Figure 7–38. One approach to the analysis of a ladder network is to simplify it one step at a time, starting at the side farthest from the source. In this way, the current in any branch or the voltage at any point can be determined, as illustrated in Example 7–16.

**FIGURE 7–38**
***Basic three-step ladder network.***

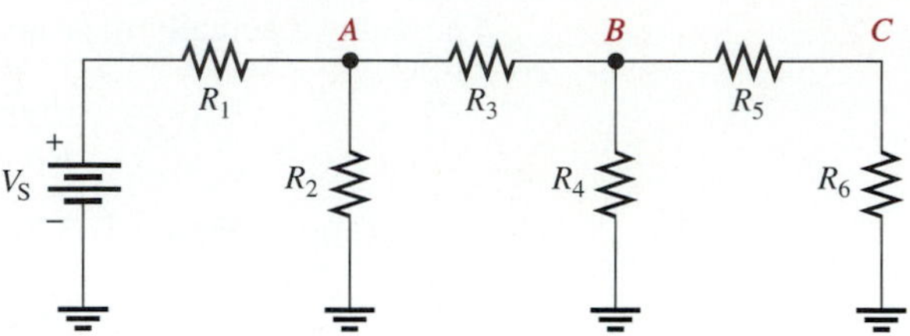


## EXAMPLE 7–16

Determine the current through each resistor and the voltage at each point in the ladder network of Figure 7–39.

**FIGURE 7–39**

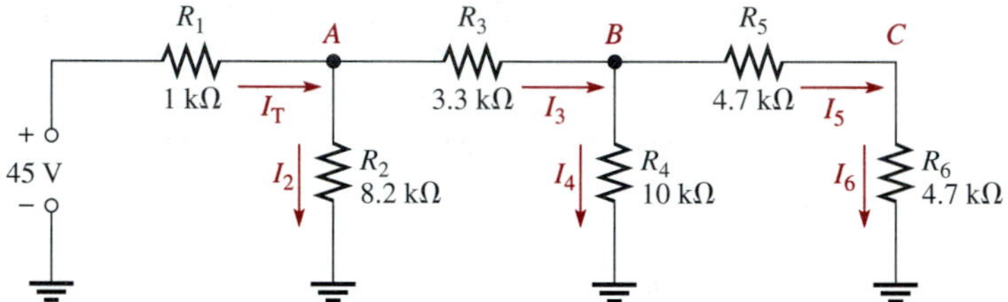


***Solution*** To find the current through each resistor, you must know the total current from the source ($I_T$). To obtain $I_T$, you must find the total resistance "seen" by the source.

Determine $R_T$ in a step-by-step process, starting at the right of the circuit diagram. First notice that $R_5$ and $R_6$ are in series across $R_4$. So the resistance from point $B$ to ground is

$$R_{BG} = \frac{R_4(R_5 + R_6)}{R_4 + (R_5 + R_6)} = \frac{(10\text{ k}\Omega)(9.4\text{ k}\Omega)}{19.4\text{ k}\Omega} = 4.85\text{ k}\Omega$$

Using $R_{BG}$ (the resistance from point $B$ to ground), you can draw the equivalent circuit as shown in Figure 7–40.

**FIGURE 7–40**

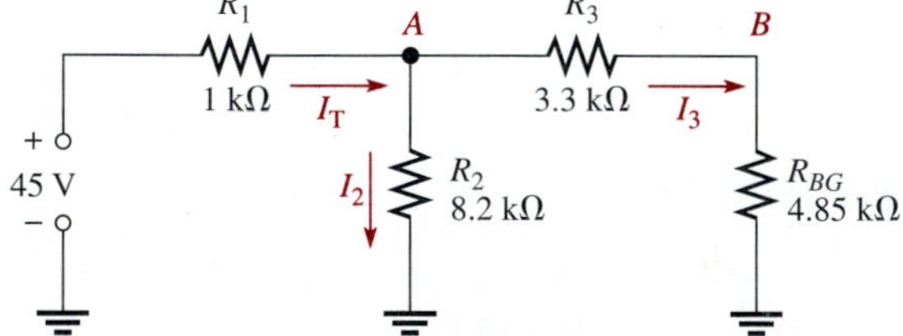


Next, the resistance from point $A$ to ground ($R_{AG}$) is $R_2$ in parallel with the series combination of $R_3$ and $R_{BG}$. Resistance $R_{AG}$ is calculated as follows:

$$R_{AG} = \frac{R_2(R_3 + R_{BG})}{R_2 + (R_3 + R_{BG})} = \frac{(8.2\text{ k}\Omega)(8.15\text{ k}\Omega)}{16.35\text{ k}\Omega} = 4.09\text{ k}\Omega$$

Using $R_{AG}$, you can further simplify the equivalent circuit of Figure 7–40 as shown in Figure 7–41.

**FIGURE 7–41**

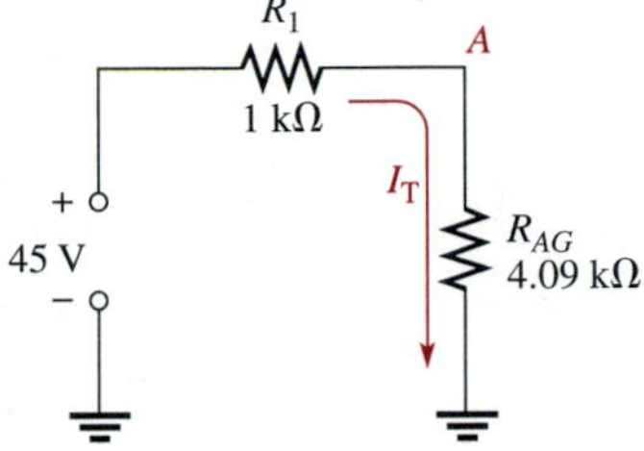


Finally, the total resistance "seen" by the source is $R_1$ in series with $R_{AG}$.

$$R_T = R_1 + R_{AG} = 1\text{ k}\Omega + 4.09\text{ k}\Omega = 5.09\text{ k}\Omega$$

The total circuit current is

$$I_T = \frac{V_S}{R_T} = \frac{45\text{ V}}{5.09\text{ k}\Omega} = 8.84\text{ mA}$$

As indicated in Figure 7–40, $I_T$ is into point $A$ and divides between $R_2$ and the branch containing $R_3 + R_{BG}$. Since the branch resistances are approximately equal in this particular example, half the total current is through $R_2$ and half into point $B$. So the currents through $R_2$ and $R_3$ are

$$I_2 = 4.42\text{ mA}$$
$$I_3 = 4.42\text{ mA}$$

If the branch resistances are not equal, use the current-divider formula. As indicated in Figure 7–39, $I_3$ is into point $B$ and is divided between $R_4$ and the branch containing $R_5 + R_6$. Therefore, the currents through $R_4$, $R_5$, and $R_6$ can be calculated.

$$I_4 = \frac{R_5 + R_6}{R_4 + (R_5 + R_6)}I_3 = \left(\frac{9.4\text{ k}\Omega}{19.4\text{ k}\Omega}\right)4.42\text{ mA} = 2.14\text{ mA}$$
$$I_5 = I_6 = I_3 - I_4 = 4.42\text{ mA} - 2.14\text{ mA} = 2.28\text{ mA}$$

To determine $V_A$, $V_B$, and $V_C$, apply Ohm's law as follows:

$$V_A = I_2R_2 = (4.42\text{ mA})(8.2\text{ k}\Omega) = 36.2\text{ V}$$
$$V_B = I_4R_4 = (2.14\text{ mA})(10\text{ k}\Omega) = 21.4\text{ V}$$
$$V_C = I_6R_6 = (2.28\text{ mA})(4.7\text{ k}\Omega) = 10.7\text{ V}$$

***Related Exercise*** Recalculate the currents through each resistor and the voltages at each point in Figure 7–39 if $R_1$ is increased to 2.2 kΩ.

## The *R*/2*R* Ladder Network

A basic *R*/2*R* ladder network circuit is shown in Figure 7–42. As you can see, the name comes from the relationship of the resistor values (one set of resistors has twice the value of the others). This type of ladder network circuit is used in applications where digital codes are converted to speech, music, or other types of analog signals as found, for example, in the area of digital recording and reproduction. This application area is called *digital-to-analog* (*D/A*) *conversion.*

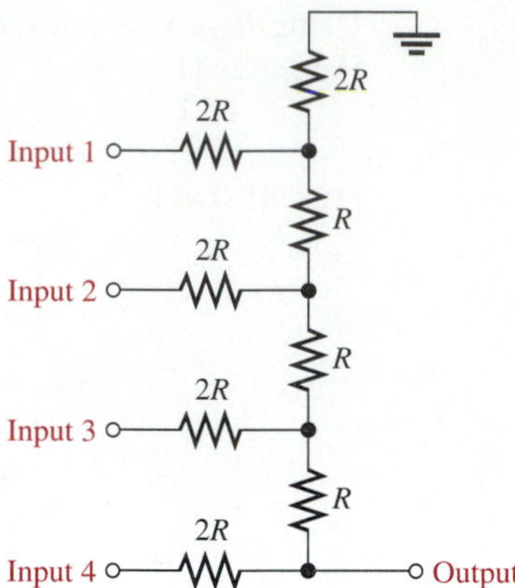


**FIGURE 7–42**
*A basic four-step R/2R ladder network.*

Let's examine the general operation of a basic $R/2R$ ladder using the four-step circuit in Figure 7–43. In a later course in digital fundamentals, you will learn specifically how this type of circuit is used in D/A conversion.

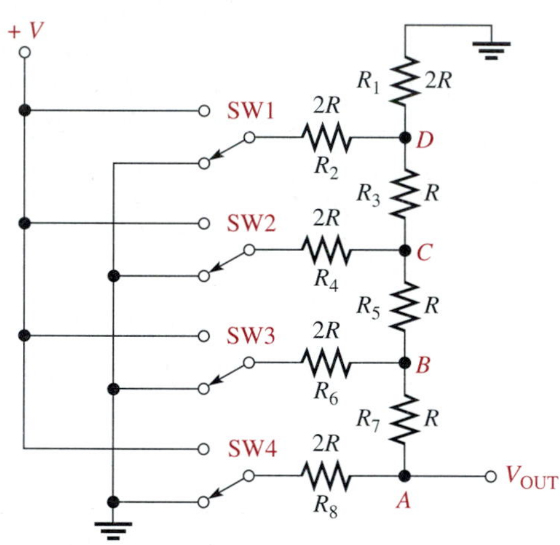


**FIGURE 7–43**
*R/2R ladder with switch inputs to simulate a two-level (digital) code.*

The switches are used in this illustration to simulate the digital (two-level) inputs. One switch position is connected to ground (0 V) and the other position is connected to a positive voltage ($V$). The analysis is as follows: Start by assuming that switch SW4 in Figure 7–43 is at the $V$ position and the others are at ground so that the inputs are as shown in Figure 7–44(a).

The total resistance from point $A$ to ground is found by first combining $R_1$ and $R_2$ in parallel from point $D$ to ground to simplify the circuit as shown in Figure 7–44(b).

$$R_1 \parallel R_2 = \frac{2R}{2} = R$$

$R_1 \parallel R_2$ is in series with $R_3$ from point $C$ to ground as illustrated in part (c).

$$R_1 \parallel R_2 + R_3 = R + R = 2R$$

Next, the above combination is in parallel with $R_4$ from point $C$ to ground as shown in part (d).

$$(R_1 \parallel R_2 + R_3) \parallel R_4 = 2R \parallel 2R = \frac{2R}{2} = R$$

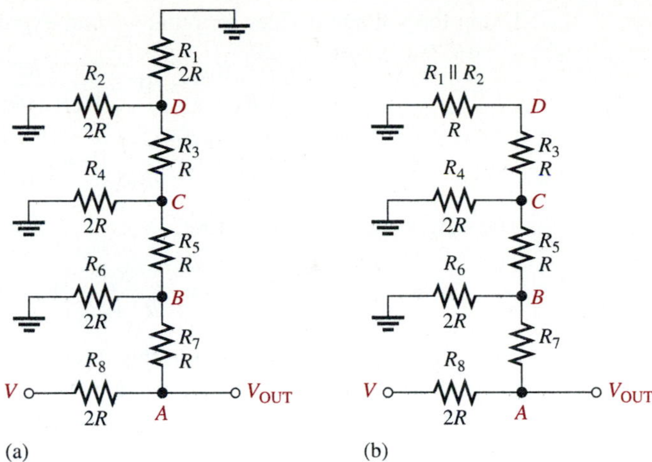


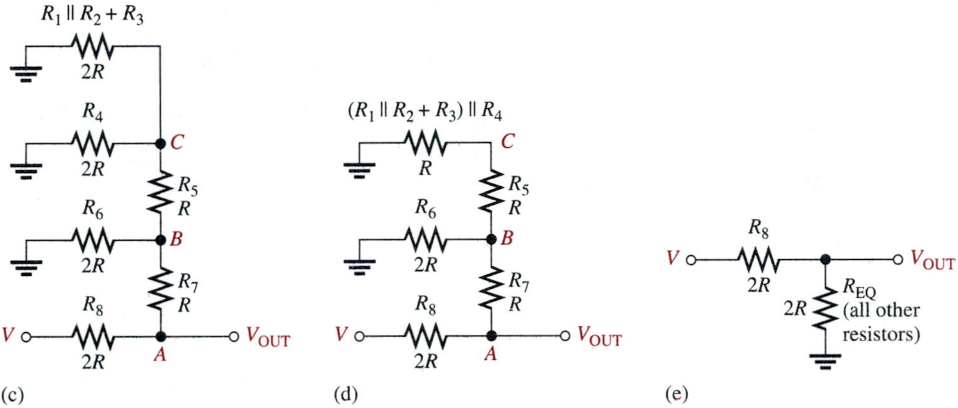


**FIGURE 7–44**
***Simplification of R/2R ladder for analysis.***

Continuing this simplification process results in the circuit in part (e) in which the output voltage can be expressed as

$$V_{OUT} = \left(\frac{2R}{4R}\right)V = \frac{V}{2}$$

A similar analysis, except with switch SW3 in Figure 7–43 connected to $V$ and the other switches connected to ground, results in the simplified circuit shown in Figure 7–45. The analysis for this case is as follows: The resistance from point $B$ to ground is

$$R_{BG} = (R_7 + R_8) \parallel 2R = 3R \parallel 2R = \frac{6R}{5}$$

**FIGURE 7–45**
***Simplified ladder with V input at SW3 in Figure 7–43.***

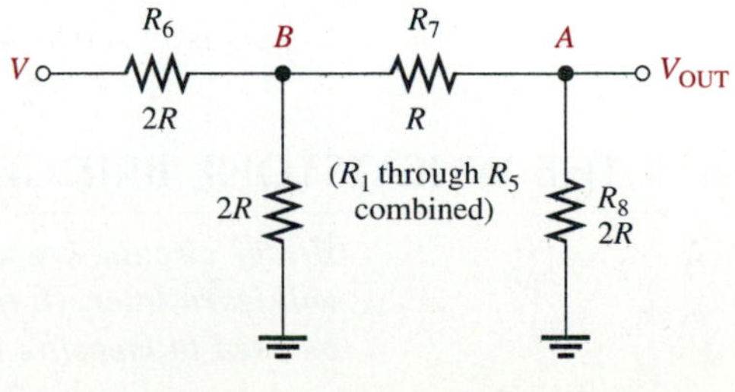

Using the voltage-divider formula, we can express the voltage at point $B$ as

$$V_B = \left(\frac{R_{BG}}{R_6 + R_{BG}}\right)V = \left(\frac{6R/5}{2R + 6R/5}\right)V$$
$$= \left(\frac{6R/5}{10R/5 + 6R/5}\right)V = \left(\frac{6R/5}{16R/5}\right)V = \left(\frac{6R}{16R}\right)V = \frac{3V}{8}$$

The output voltage is, therefore,

$$V_{\text{OUT}} = \left(\frac{R_8}{R_7 + R_8}\right)V_B = \left(\frac{2R}{3R}\right)\left(\frac{3V}{8}\right) = \frac{V}{4}$$

Notice that the output voltage in this case ($V/4$) is one half the output voltage ($V/2$) for the case where $V$ is connected at switch SW4.

A similar analysis for each of the remaining switch inputs in Figure 7–43 results in output voltages as follows: For SW2 connected to $V$ and the other switches connected to ground,

$$V_{\text{OUT}} = \frac{V}{8}$$

For SW1 connected to $V$ and the other switches connected to ground,

$$V_{\text{OUT}} = \frac{V}{16}$$

When more than one input at a time are connected to $V$, the total output is the sum of the individual outputs. These particular relationships among the output voltages for the various levels of inputs are very important in the application of $R/2R$ ladder networks to digital-to-analog conversion.

**SECTION 7–5 REVIEW**

1. Sketch a basic four-step ladder network.
2. Determine the total circuit resistance presented to the source by the ladder network of Figure 7–46.
3. What is the total current in Figure 7–46?
4. What is the current through $R_2$ in Figure 7–46?
5. What is the voltage at point $A$ with respect to ground in Figure 7–46?

FIGURE 7–46

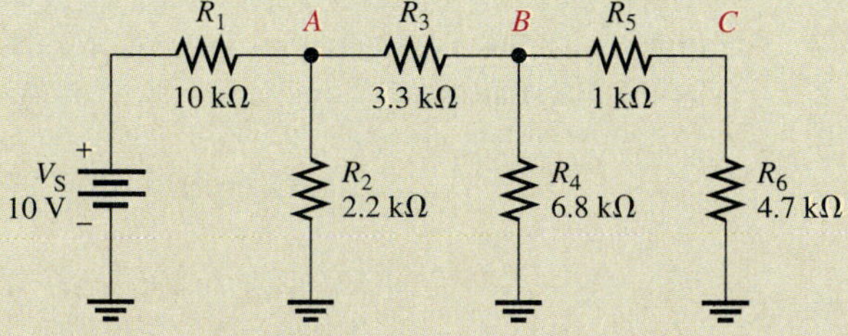


## 7–6 ■ THE WHEATSTONE BRIDGE

***Bridge circuits are widely used in measurement devices and other applications that you will learn later. In this section, you will study the balanced resistive bridge, which can be used to measure unknown resistance values.***

***After completing this section, you should be able to***

- **Analyze a Wheatstone bridge**
  - ☐ Determine when a bridge is balanced
  - ☐ Determine an unknown resistance using a balanced bridge
  - ☐ Describe how a Wheatstone bridge can be used for temperature measurement

The circuit shown in Figure 7–47(a) is known as a *Wheatstone bridge.* Figure 7–47(b) is the same circuit electrically, but it is drawn in a different way.

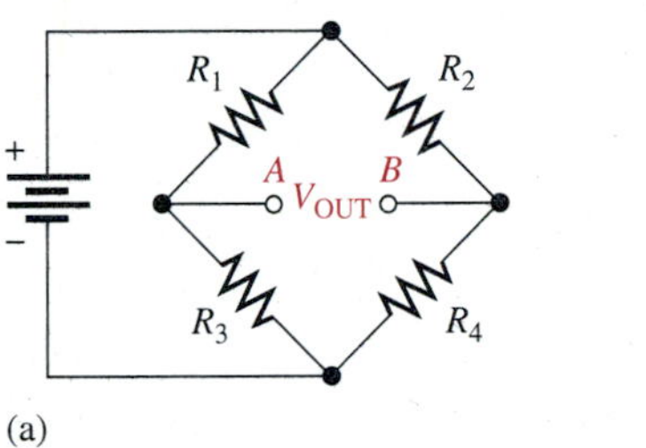

(a)

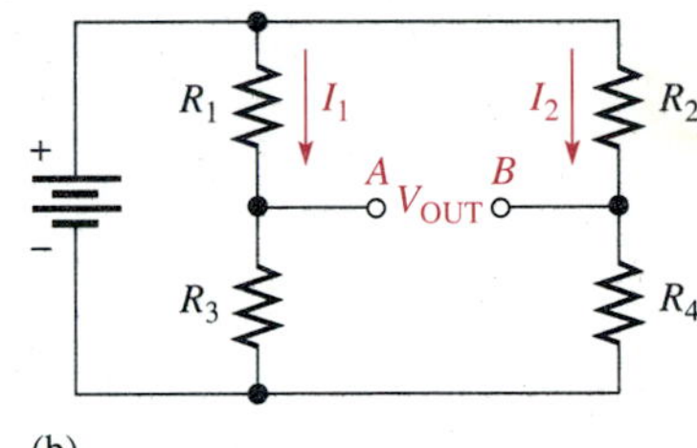

(b)

**FIGURE 7–47**
***Wheatstone bridge.***

A bridge is said to be balanced when the voltage ($V_{OUT}$) across the output terminals $A$ and $B$ is zero; that is, $V_A = V_B$. In Figure 7–47(b), if $V_A$ equals $V_B$, then $V_{R1} = V_{R2}$, since the top sides of both $R_1$ and $R_2$ are connected to the same point. Also $V_{R3} = V_{R4}$, since the bottom sides of both $R_3$ and $R_4$ connect to the same point. The voltage ratios can be written as

$$\frac{V_1}{V_3} = \frac{V_2}{V_4}$$

Substituting by Ohm's law yields

$$\frac{I_1R_1}{I_1R_3} = \frac{I_2R_2}{I_2R_4}$$

The currents cancel to give

$$\frac{R_1}{R_3} = \frac{R_2}{R_4}$$

Solving for $R_1$ yields the following formula:

$$R_1 = R_3\left(\frac{R_2}{R_4}\right)$$

This formula can be used to determine an unknown resistance. First, make $R_3$ a variable resistor and call it $R_V$. Also, set the ratio $R_2/R_4$ to a known value. If $R_V$ is adjusted until the bridge is balanced, the product of $R_V$ and the ratio $R_2/R_4$ is equal to $R_1$, which is the unknown resistor ($R_{UNK}$).

$$R_{UNK} = R_V\left(\frac{R_2}{R_4}\right) \quad \textbf{(7–1)}$$

The bridge is balanced when the voltage across the output terminals equals zero ($V_A = V_B$). A *galvanometer* (a meter that measures small currents in either direction and is zero at center scale) is connected between the output terminals. Then $R_V$ is adjusted until the galvanometer shows zero current ($V_A = V_B$), indicating a balanced condition. The setting of $R_V$ multiplied by the ratio $R_2/R_4$ gives the value of $R_{UNK}$. Figure 7–48 shows this arrangement. For example, if $R_2/R_4 = 1/10$ and $R_V = 680\ \Omega$, then $R_{UNK} = (680\ \Omega)(1/10) = 68\ \Omega$.

FIGURE 7–48
***Balanced Wheatstone bridge.***

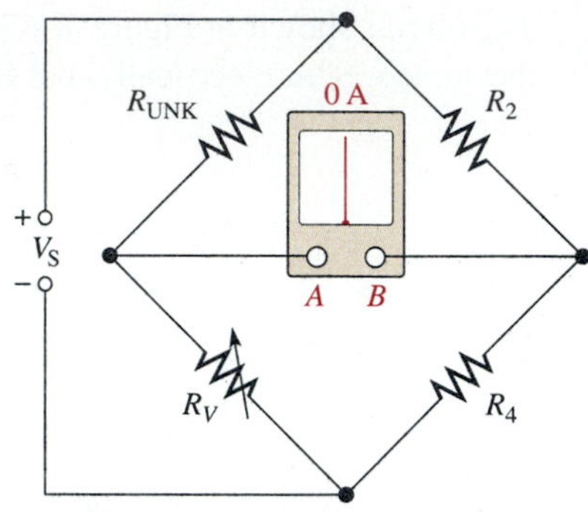


**EXAMPLE 7–17** What is $R_{UNK}$ under the balanced bridge conditions shown in Figure 7–49?

FIGURE 7–49

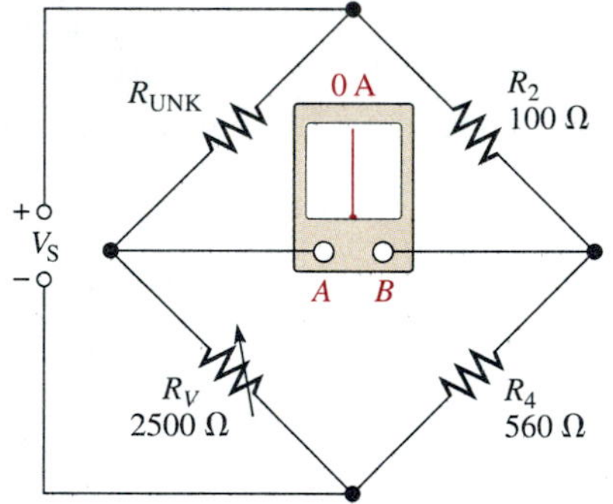


***Solution***

$$R_{UNK} = R_V\left(\frac{R_2}{R_4}\right) = 2500\ \Omega\left(\frac{100\ \Omega}{560\ \Omega}\right) = 446\ \Omega$$

***Related Exercise*** If $R_V$ must be adjusted to 1.8 kΩ in order to balance the bridge, what is $R_{UNK}$?

## A Bridge Application

Bridge circuits are used for many measurements other than for determining an unknown resistance. One application of the Wheatstone bridge is in accurate temperature measurement. A temperature-sensitive element such as a thermistor is connected in a Wheatstone bridge as shown in Figure 7–50. An amplifier is connected across the output from $A$ to $B$ in order to increase the output voltage from the bridge to a usable value. The bridge is calibrated so that it is balanced at a specified reference temperature. As the temperature changes, the resistance of the sensing element changes proportionately, and the bridge becomes unbalanced. As a result, $V_{AB}$ changes and is amplified (increased) and converted to a form for direct temperature readout on a gauge or a digital-type display.

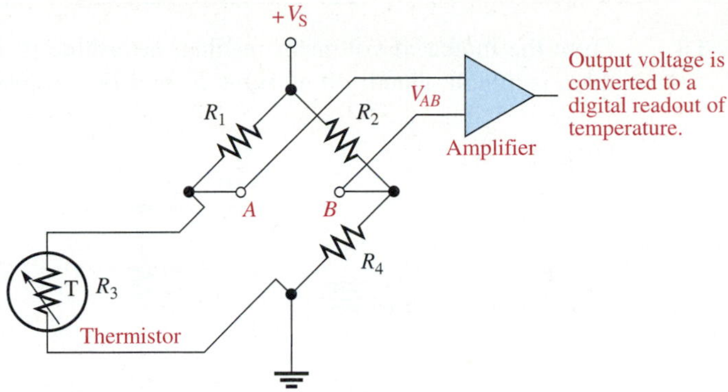


**FIGURE 7–50**
***A simplified circuit for temperature measurement.***

**SECTION 7–6 REVIEW**

1. Sketch a basic Wheatstone bridge circuit.
2. Under what condition is the bridge balanced?
3. What formula is used to determine the value of the unknown resistance when the bridge is balanced?
4. What is the unknown resistance for the values shown in Figure 7–51?

**FIGURE 7–51**

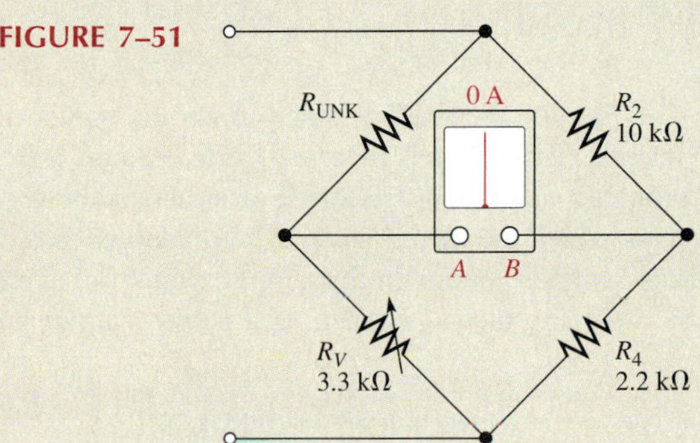


## 7–7 ■ TROUBLESHOOTING

***Troubleshooting is the process of identifying and locating a failure or problem in a circuit. Some troubleshooting techniques have already been discussed in relation to both series circuits and parallel circuits. These methods are now extended to the series-parallel circuit.***

***After completing this section, you should be able to***

- **Troubleshoot series-parallel circuits**
  - ☐ Determine the effects of an open circuit
  - ☐ Determine the effects of a short circuit
  - ☐ Locate opens and shorts

Opens and shorts are typical problems that occur in electric circuits. As mentioned in Chapter 5, if a resistor burns out, it will normally produce an open circuit. Bad solder connections, broken wires, and poor contacts can also be causes of open paths. Pieces of foreign material, such as solder splashes, broken insulation on wires, and so on, can often lead to shorts in a circuit. A short is a zero resistance path between two points. The following three examples illustrate troubleshooting in series-parallel resistive circuits.

**EXAMPLE 7–18** From the indicated voltmeter reading, determine if there is a fault in Figure 7–52. If there is a fault, identify it as either a short or an open.

FIGURE 7–52

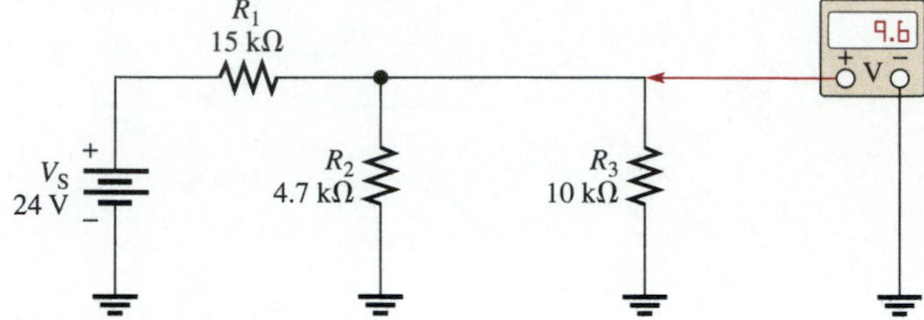


***Solution*** First determine what the voltmeter should be indicating. Since $R_2$ and $R_3$ are in parallel, their equivalent resistance is

$$R_{2\text{-}3} = \frac{R_2R_3}{R_2 + R_3} = \frac{(4.7\text{ k}\Omega)(10\text{ k}\Omega)}{14.7\text{ k}\Omega} = 3.20\text{ k}\Omega$$

The voltage across the equivalent parallel combination is determined by the voltage-divider formula as follows:

$$V_{2\text{-}3} = \left(\frac{R_{2\text{-}3}}{R_1 + R_{2\text{-}3}}\right)V_S = \left(\frac{3.20\text{ k}\Omega}{18.2\text{ k}\Omega}\right)24\text{ V} = 4.22\text{ V}$$

Thus, 4.22 V is the voltage reading that you should get on the meter. But the meter reads 9.6 V instead. This value is incorrect, and, because it is higher than it should be, $R_2$ or $R_3$ is probably open. Why? Because if either of these two resistors is open, the resistance across which the meter is connected is larger than expected. A higher resistance will drop a higher voltage in this circuit, which is, in effect, a voltage divider.

Start by assuming that $R_2$ is open. If it is, the voltage across $R_3$ is

$$V_3 = \left(\frac{R_3}{R_1 + R_3}\right)V_S = \left(\frac{10\text{ k}\Omega}{25\text{ k}\Omega}\right)24\text{ V} = 9.6\text{ V}$$

Since the measured voltage is also 9.6 V, this calculation shows that $R_2$ is open. Replace $R_2$ with a new resistor.

***Related Exercise*** What would be the voltmeter reading if $R_3$ were open? If $R_1$ were open?

**EXAMPLE 7–19** Suppose that you measure 24 V with the voltmeter in Figure 7–53. Determine if there is a fault, and, if there is, isolate it.

FIGURE 7–53

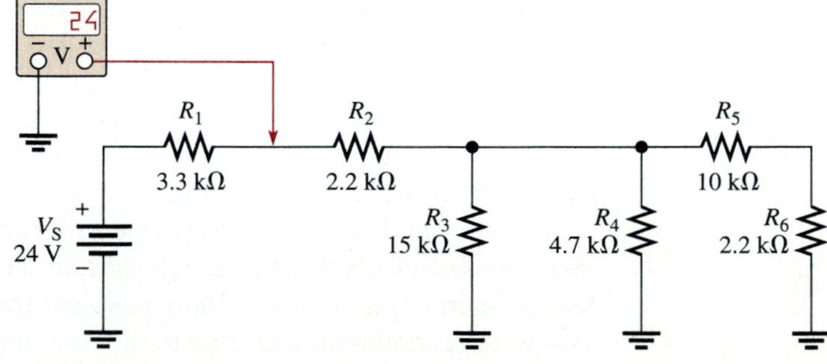

***Solution*** There is no voltage drop across $R_1$ because both sides of the resistor are at +24 V. Either there is no current through $R_1$ from the source, which tells you that $R_2$ is open in the circuit, or $R_1$ is shorted.

If $R_1$ were open, the meter in Figure 7–53 would not read 24 V. The most logical failure is in $R_2$. If $R_2$ is open, then there will be no current from the source and thus no voltage is dropped across $R_1$. To verify this, measure across $R_2$ with the voltmeter as shown in Figure 7–54. If $R_2$ is open, the meter will indicate 24 V. The right side of $R_2$ will be at zero volts because there is no current through any of the other resistors to cause a voltage drop across them.

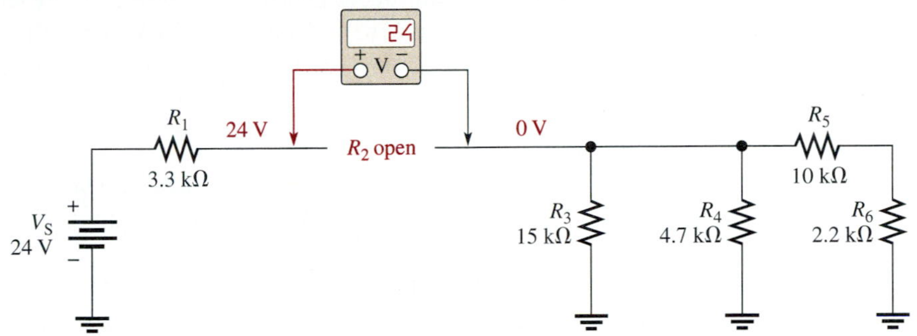


FIGURE 7–54

***Related Exercise*** What would be the voltage across an open $R_5$ in Figure 7–53, assuming no other faults?

**EXAMPLE 7–20** The two voltmeters in Figure 7–55 indicate the voltage shown. Determine if there are any opens or shorts in the circuit and, if so, where they are located.

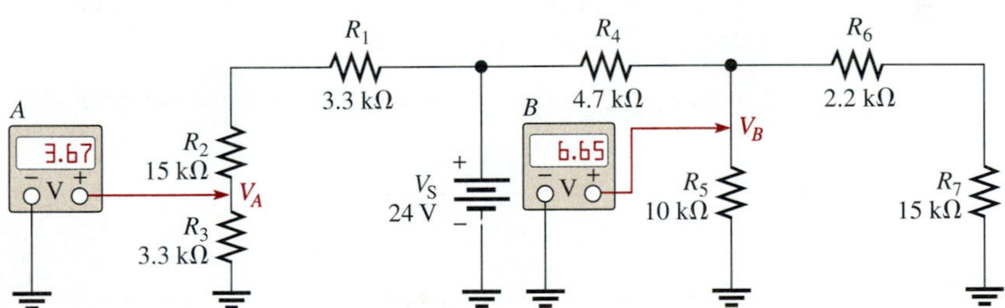


FIGURE 7–55

***Solution*** First determine if the voltmeter readings are correct. $R_1$, $R_2$, and $R_3$ act as a voltage divider on the left side of the source. The voltage ($V_A$) across $R_3$ is calculated as follows:

$$V_A = \left(\frac{R_3}{R_1 + R_2 + R_3}\right)V_S = \left(\frac{3.3\text{ k}\Omega}{21.6\text{ k}\Omega}\right)24\text{ V} = 3.67\text{ V}$$

The voltmeter A reading is correct.

Now see if the voltmeter B reading is correct. The part of the circuit to the right of the source also acts as a voltage divider. The series-parallel combination of $R_5$, $R_6$, and $R_7$ is in series with $R_4$. The equivalent resistance of the $R_5$, $R_6$, and $R_7$ combination is calculated on the next page.

$$R_{5\text{-}6\text{-}7} = \frac{(R_6 + R_7)R_5}{R_5 + R_6 + R_7} = \frac{(17.2\ \text{k}\Omega)(10\ \text{k}\Omega)}{27.2\ \text{k}\Omega} = 6.32\ \text{k}\Omega$$

where $R_5$ is in parallel with $R_6$ and $R_7$ in series. $R_{5\text{-}6\text{-}7}$ and $R_4$ form a voltage divider. Voltmeter B is measuring the voltage across $R_{5\text{-}6\text{-}7}$. Is it correct? Check as follows:

$$V_B = \left(\frac{R_{5\text{-}6\text{-}7}}{R_4 + R_{5\text{-}6\text{-}7}}\right)V_S = \left(\frac{6.32\ \text{k}\Omega}{11\ \text{k}\Omega}\right)24\ \text{V} = 13.8\ \text{V}$$

Thus, the actual measured voltage (6.65 V) at this point is incorrect. Some further thought will help to isolate the problem.

Resistor $R_4$ is not open, because if it were, the meter would read 0 V. If there were a short across it, the meter would read 24 V. Since the actual voltage is much less than it should be, $R_{5\text{-}6\text{-}7}$ must be less than the calculated value of 6.32 kΩ. The most likely problem is a short across $R_7$. If there is a short from the top of $R_7$ to ground, $R_6$ is effectively in parallel with $R_5$. In this case, $R_{5\text{-}6}$ is

$$R_{5\text{-}6} = \frac{R_5 R_6}{R_5 + R_6} = \frac{(10\ \text{k}\Omega)(2.2\ \text{k}\Omega)}{12.2\ \text{k}\Omega} = 1.80\ \text{k}\Omega$$

Then $V_B$ is

$$V_B = \left(\frac{R_{5\text{-}6}}{R_4 + R_{5\text{-}6}}\right)V_S = \left(\frac{1.80\ \text{k}\Omega}{6.5\ \text{k}\Omega}\right)24\ \text{V} = 6.65\ \text{V}$$

This value for $V_B$ agrees with the voltmeter B reading. So there is a short across $R_7$. If this were an actual circuit, you would try to find the physical cause of the short.

***Related Exercise*** If $R_2$ in Figure 7–55 were shorted, what would voltmeter A read? What would voltmeter B read?

## SECTION 7–7 REVIEW

1. Name two types of common circuit faults.
2. In Figure 7–56, one of the resistors in the circuit is open. Based on the meter reading, determine which is the bad resistor.

FIGURE 7–56

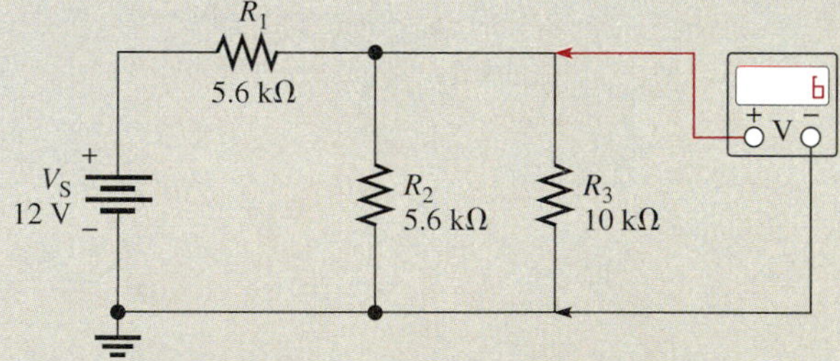


3. For the following faults in Figure 7–57, what voltage would be measured at point $A$ with respect to ground?
   (a) No faults (b) $R_1$ open (c) Short across $R_5$ (d) $R_3$ and $R_4$ open
   (e) $R_2$ open

FIGURE 7–57

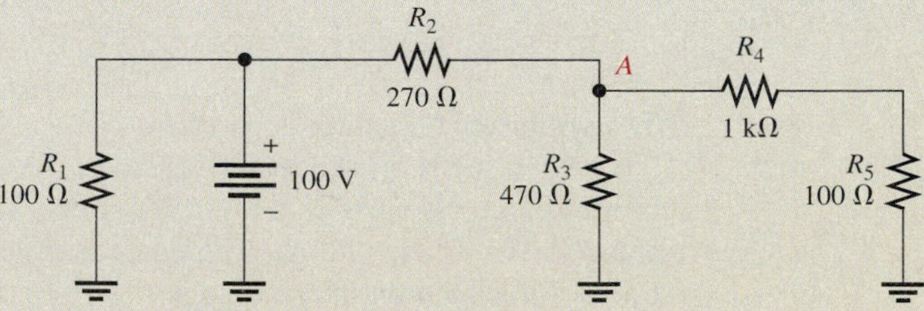

## 7–8 ■ TECHnology Theory Into Practice

***A voltage divider with three output voltages has been designed and constructed on a PC board. The voltage divider is to be used as part of a portable power supply unit for supplying up to three different reference voltages to measuring instruments in the field. The power supply unit contains a battery pack combined with a voltage regulator that produces a constant +12 V to the voltage-divider circuit board. In this assignment, you will apply your knowledge of loaded voltage dividers, Kirchhoff's laws and Ohm's law to determine the operating parameters of the voltage divider in terms of voltages and currents for all possible load configurations. You will also troubleshoot the circuit for various malfunctions.***

### The Schematic of the Voltage Divider

☐ Draw the schematic and label the resistor values for the circuit board in Figure 7–58.

**FIGURE 7–58**
*Voltage-divider circuit board.*

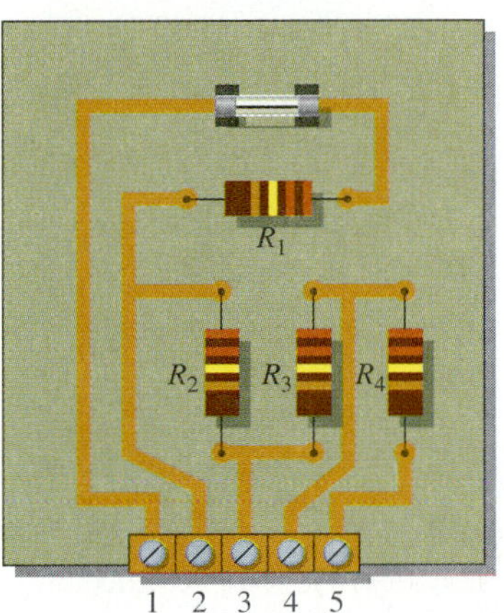


### The 12 V Power Supply

☐ Specify how to connect a 12 V power supply to the circuit board so that all resistors are in series and pin 2 has the highest output voltage.

### The Unloaded Output Voltages

☐ Calculate each of the output voltages with no loads connected. Add these voltage values to a copy of the table in Figure 7–59.

### The Loaded Output Voltages

The instruments to be connected to the voltage divider each have a 10 MΩ input resistance. This means that when an instrument is connected to a voltage-divider output there is effectively a 10 MΩ resistor from that output to ground (negative side of source).

☐ Determine the output voltage across each load for the following load configurations and add these voltage values to a copy of the table in Figure 7–59.

1. A 10 MΩ load connected to pin 2.
2. A 10 MΩ load connected to pin 3.
3. A 10 MΩ load connected to pin 4.
4. 10 MΩ loads connected to pin 2 and pin 3.

<table>
<tr><th>10 MΩ Load</th><th>$V_{OUT\,(2)}$</th><th>$V_{OUT\,(3)}$</th><th>$V_{OUT\,(4)}$</th><th>% Deviation</th><th>$I_{LOAD\,(2)}$</th><th>$I_{LOAD\,(3)}$</th><th>$I_{LOAD\,(4)}$</th></tr>
<tr><td>None</td><td></td><td></td><td></td><td></td><td></td><td></td><td></td></tr>
<tr><td>Pin 2</td><td></td><td></td><td></td><td></td><td></td><td></td><td></td></tr>
<tr><td>Pin 3</td><td></td><td></td><td></td><td></td><td></td><td></td><td></td></tr>
<tr><td>Pin 4</td><td></td><td></td><td></td><td></td><td></td><td></td><td></td></tr>
<tr><td rowspan="2">Pins 2 and 3</td><td rowspan="2"></td><td rowspan="2"></td><td rowspan="2"></td><td>2</td><td rowspan="2"></td><td rowspan="2"></td><td rowspan="2"></td></tr>
<tr><td>3</td></tr>
<tr><td rowspan="2">Pins 2 and 4</td><td rowspan="2"></td><td rowspan="2"></td><td rowspan="2"></td><td>2</td><td rowspan="2"></td><td rowspan="2"></td><td rowspan="2"></td></tr>
<tr><td>4</td></tr>
<tr><td rowspan="2">Pins 3 and 4</td><td rowspan="2"></td><td rowspan="2"></td><td rowspan="2"></td><td>3</td><td rowspan="2"></td><td rowspan="2"></td><td rowspan="2"></td></tr>
<tr><td>4</td></tr>
<tr><td rowspan="3">Pins 2, 3, and 4</td><td rowspan="3"></td><td rowspan="3"></td><td rowspan="3"></td><td>2</td><td rowspan="3"></td><td rowspan="3"></td><td rowspan="3"></td></tr>
<tr><td>3</td></tr>
<tr><td>4</td></tr>
</table>

**FIGURE 7–59**
***Table of operating parameters for the power supply voltage divider.***

5. 10 MΩ loads connected to pin 2 and pin 4.
6. 10 MΩ loads connected to pin 3 and pin 4.
7. 10 MΩ loads connected to pin 2, pin 3, and pin 4.

## Percent Deviation of the Output Voltages

☐ Calculate how much each loaded output voltage deviates from its unloaded value for each of the load configurations listed above and express each as a percentage using the following formula:

$$\text{Percent deviation} = \left(\frac{V_{OUT(unloaded)} - V_{OUT(loaded)}}{V_{OUT(unloaded)}}\right)100\%$$

Add the values to a copy of the table of Figure 7–59.

## The Load Currents

☐ Calculate the current to each 10 MΩ load for each of the load configurations listed above. Add these values to a copy of the table in Figure 7–59.

## Troubleshooting

The voltage-divider circuit board is connected to a 12 V power supply and to the three instruments to which it provides reference voltages, as shown in Figure 7–60. Voltages at each of the numbered test points are measured with a voltmeter in each of eight different cases.

☐ For each case in Figure 7–60, determine the fault indicated by the voltage measurements.

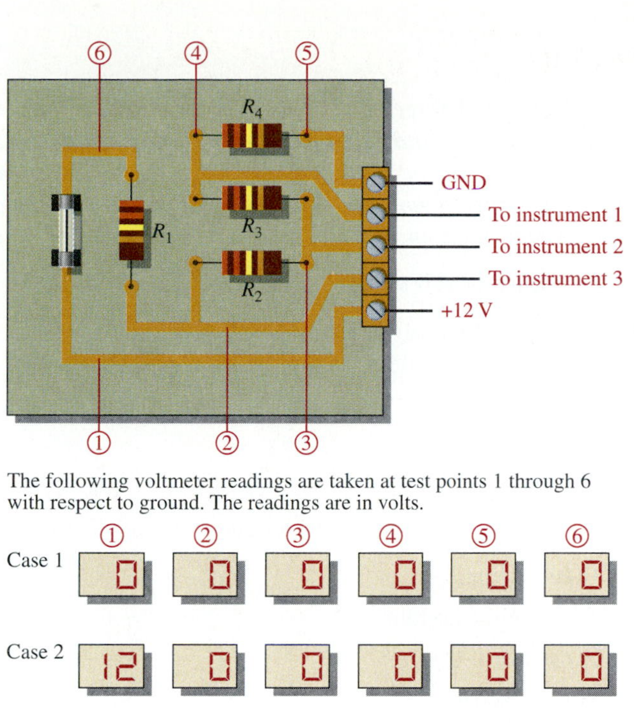

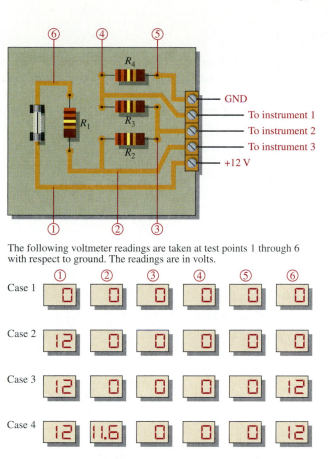

The following voltmeter readings are taken at test points 1 through 6 with respect to ground. The readings are in volts.

| | ① | ② | ③ | ④ | ⑤ | ⑥ |
|---|---|---|---|---|---|---|
| Case 1 | 0 | 0 | 0 | 0 | 0 | 0 |
| Case 2 | 12 | 0 | 0 | 0 | 0 | 0 |
| Case 3 | 12 | 0 | 0 | 0 | 0 | 12 |
| Case 4 | 12 | 11.6 | 0 | 0 | 0 | 12 |
| Case 5 | 12 | 11.3 | 10.9 | 0 | 0 | 12 |
| Case 6 | 12 | 11 | 10.3 | 10 | 0 | 12 |
| Case 7 | 12 | 5.9 | 0 | 0 | 0 | 12 |
| Case 8 | 12 | 7.8 | 3.8 | 0 | 0 | 12 |

**FIGURE 7–60**

## SECTION 7–8 REVIEW

1. If the portable unit covered in this section is to supply reference voltages to all three instruments, how many days can a 100 mAh battery be used as the power supply?
2. Can ⅛ W resistors be used on the voltage-divider board?
3. If ⅛ W resistors are used, will an output shorted to ground cause any of the resistors to overheat due to excessive power?

## 7–9 ■ PSpice DC ANALYSIS

***In this section, PSpice is used to analyze a series-parallel circuit. By now you have enough knowledge and experience to do a more complex problem. Although the circuit in this section can be analyzed manually, it is of sufficient complexity that you will begin to see the advantages of PSpice. The main challenge is to set up the circuit file so that PSpice can analyze the circuit and provide the needed results.***

***After completing this section, you should be able to***

- **Use PSpice to analyze series-parallel circuits**
  - ☐ Write a circuit file to analyze a series-parallel circuit

### A Series-Parallel Circuit

Suppose you want to analyze the circuit in Figure 7–61 for the following values:

**(a)** Total resistance

**(b)** Total source current

**(c)** Current through $R_6$

**(d)** Voltage across $R_7$

**FIGURE 7–61**

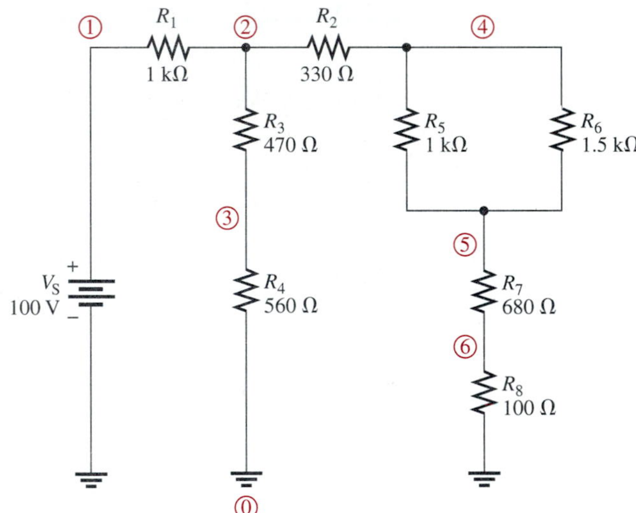


First, the problem asks for the total resistance. PSpice does not solve for resistance directly since, generally, the circuit resistances are fixed by design; the voltages and currents are the unknown quantities. This is not a major problem, however, because the .OP statement can be used to find the current from the voltage source, and then the total resistance is simply the source voltage (100 V) divided by the source current. Finally, the voltage and current for the specific resistors may be provided by .DC and .PRINT statements.

Using the given node assignments, the circuit file for the required analysis of the circuit in Figure 7–61 is as follows:

```
Series-Parallel Circuit
*Figure 7–61
VS  1  0  100
R1  1  2  1K
R2  2  4  330
R3  2  3  470
R4  3  0  560
```

```
R5  4  5  1K
R6  4  5  1.5K
R7  5  6  680
R8  6  0  100
.OP
.DC  VS  100  100  1
.PRINT  DC  I(R6)  V(R7)
.END
```

After you have written the circuit file, save it under some filename (such as PRACT1.CIR) and run PSpice on it. The resulting output file will provide the following values:

$$\text{Total current from } V_S = -60.87 \text{ mA}$$
$$I_{R6} = 9.153 \text{ mA}$$
$$V_{R7} = 15.56 \text{ V}$$

The negative source current reflects the PSpice convention of treating positive current as being into the + node of a source.

The total resistance is then determined using your calculator as follows, ignoring the negative sign on the source current:

$$R_T = \frac{100 \text{ V}}{60.87 \text{ mA}} = 1.64 \text{ k}\Omega$$

## SECTION 7–9 REVIEW

1. Rewrite the circuit file for Figure 7–61 if VS is changed to 50 V and you want to find the currents through $R_2$, $R_3$, $R_4$, and $R_5$ and the voltages across $R_1$ and $R_7$.
2. Write a circuit file to find the voltage across and the current through each resistor in the circuit of Figure 7–62.

FIGURE 7–62

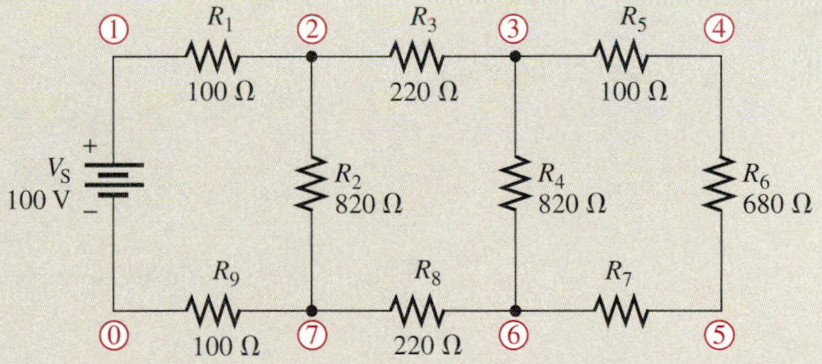


## ■ SUMMARY

- A series-parallel circuit is a combination of both series paths and parallel paths.
- To determine total resistance in a series-parallel circuit, identify the series and parallel relationships, and then apply the formulas for series resistance and parallel resistance from Chapters 5 and 6.
- To find the total current, divide the total voltage by the total resistance.
- To determine branch currents, apply the current-divider formula, Kirchhoff's current law, or Ohm's law. Consider each circuit problem individually to determine the most appropriate method.
- To determine voltage drops across any portion of a series-parallel circuit, use the voltage-divider formula, Kirchhoff's voltage law, or Ohm's law. Consider each circuit problem individually to determine the most appropriate method.
- When a load resistor is connected across a voltage-divider output, the output voltage decreases.

- The load resistor should be large compared to the resistance across which it is connected, in order that the loading effect may be minimized. A *10-times* value is sometimes used as a rule of thumb, but the value depends on the accuracy required for the output voltage.
- To find total resistance of a ladder network, start at the point farthest from the source and reduce the resistance in steps.
- A Wheatstone bridge can be used to measure an unknown resistance.
- A bridge is balanced when the output voltage is zero. The balanced condition produces zero current through a load connected across the output terminals of the bridge.
- Open circuits and short circuits are typical circuit faults.
- Resistors normally open when they burn out.

## ■ GLOSSARY

**Bleeder current** The current left after the total load current is subtracted from the total current into the circuit.

**Load** An element (resistor or other component) connected across the output terminals of a circuit that draws current from the circuit.

**Sensitivity factor** The ohms-per-volt rating of a voltmeter.

## ■ FORMULA

(7–1) $R_{UNK} = R_V\left(\frac{R_2}{R_4}\right)$ Unknown resistance in a Wheatstone bridge

## ■ SELF-TEST

1. Which of the following statements are true concerning Figure 7–63?
   (a) $R_1$ and $R_2$ are in series with $R_3$, $R_4$, and $R_5$
   (b) $R_1$ and $R_2$ are in series
   (c) $R_3$, $R_4$, and $R_5$ are in parallel
   (d) The series combination of $R_1$ and $R_2$ is in parallel with the series combination of $R_3$, $R_4$, and $R_5$
   (e) answers (b) and (d)

FIGURE 7–63

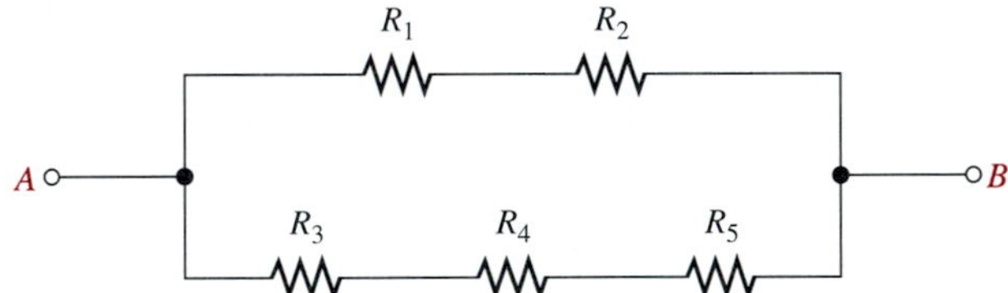


2. The total resistance of Figure 7–63 can be found with which of the following formulas?
   (a) $R_1 + R_2 + R_3 \parallel R_4 \parallel R_5$ (b) $R_1 \parallel R_2 + R_3 \parallel R_4 \parallel R_5$
   (c) $(R_1 + R_2) \parallel (R_3 + R_4 + R_5)$ (d) none of these answers
3. If all of the resistors in Figure 7–63 have the same value, when voltage is applied across terminals *A* and *B*, the current is
   (a) greatest in $R_5$ (b) greatest in $R_3$, $R_4$, and $R_5$
   (c) greatest in $R_1$ and $R_2$ (d) the same in all the resistors
4. Two 1 kΩ resistors are in series and this series combination is in parallel with a 2.2 kΩ resistor. The voltage across one of the 1 kΩ resistors is 6 V. The voltage across the 2.2 kΩ resistor is
   (a) 6 V (b) 3 V (c) 12 V (d) 13.2 V
5. The parallel combination of a 330 Ω resistor and a 470 Ω resistor is in series with the parallel combination of four 1 kΩ resistors. A 100 V source is connected across the circuit. The resistor with the most current has a value of
   (a) 1 kΩ (b) 330 Ω (c) 470 Ω

6. In the circuit described in Question 5, the resistor(s) with the most voltage has (have) a value of
   (a) 1 kΩ  (b) 470 Ω  (c) 330 Ω
7. In the circuit of Question 5, the percentage of the total current through any single 1 kΩ resistor is
   (a) 100%  (b) 25%  (c) 50%  (d) 31.3%
8. The output of a certain voltage divider is 9 V with no load. When a load is connected, the output voltage
   (a) increases  (b) decreases
   (c) remains the same  (d) becomes zero
9. A certain voltage divider consists of two 10 kΩ resistors in series. Which of the following load resistors will have the most effect on the output voltage?
   (a) 1 MΩ  (b) 20 kΩ  (c) 100 kΩ  (d) 10 kΩ
10. When a load resistance is connected to the output of a voltage-divider circuit, the current drawn from the source
    (a) decreases  (b) increases  (c) remains the same  (d) is cut off
11. In a ladder network, simplification should begin at
    (a) the source  (b) the resistor farthest from the source
    (c) the center  (d) the resistor closest to the source
12. In a certain four-step $R/2R$ ladder network, the smallest resistor value is 10 kΩ. The largest value is
    (a) indeterminable  (b) 20 kΩ  (c) 50 kΩ  (d) 100 kΩ
13. The output voltage of a balanced Wheatstone bridge is
    (a) equal to the source voltage
    (b) equal to zero
    (c) dependent on all of the resistor values in the bridge
    (d) dependent on the value of the unknown resistor
14. A certain Wheatstone bridge has the following resistor values: $R_V = 8$ kΩ, $R_2 = 680$ Ω, and $R_4 = 2.2$ kΩ. The unknown resistance is
    (a) 2473 Ω  (b) 25.9 kΩ  (c) 187 Ω  (d) 2890 Ω
15. You are measuring the voltage at a given point in a circuit that has very high resistance values and the measured voltage is a little lower than it should be. This is possibly because of
    (a) one or more of the resistance values being off
    (b) the loading effect of the voltmeter
    (c) the source voltage is too low
    (d) all the above

## ■ PROBLEMS

### SECTION 7–1 Identifying Series-Parallel Relationships

1. Visualize and sketch the following series-parallel combinations:
   (a) $R_1$ in series with the parallel combination of $R_2$ and $R_3$
   (b) $R_1$ in parallel with the series combination of $R_2$ and $R_3$
   (c) $R_1$ in parallel with a branch containing $R_2$ in series with a parallel combination of four other resistors
2. Visualize and sketch the following series-parallel circuits:
   (a) A parallel combination of three branches, each containing two series resistors
   (b) A series combination of three parallel circuits, each containing two resistors

3. In each circuit of Figure 7–64, identify the series and parallel relationships of the resistors viewed from the source.

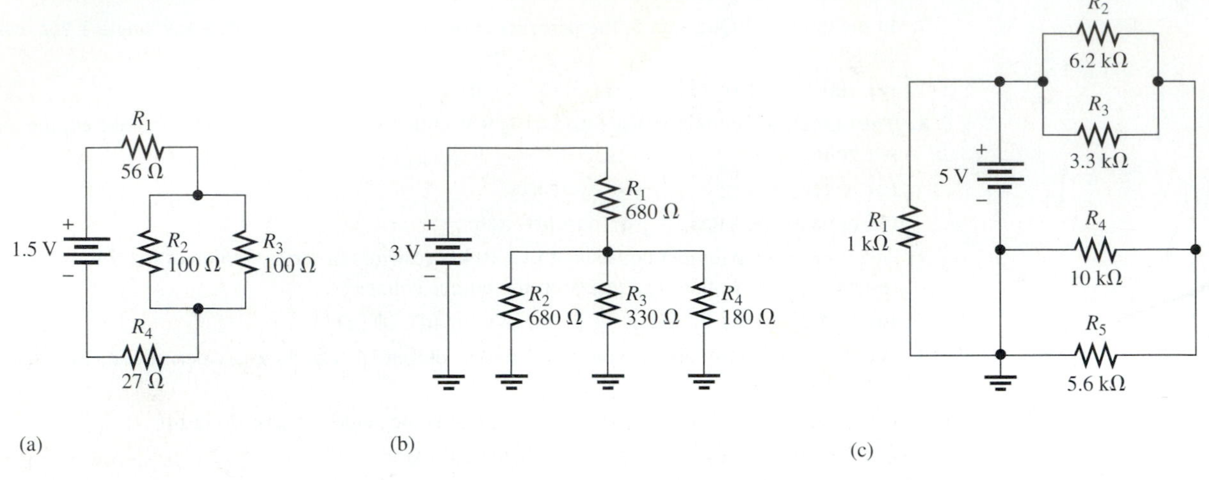


FIGURE 7–64

4. For each circuit in Figure 7–65, identify the series and parallel relationships of the resistors viewed from the source.

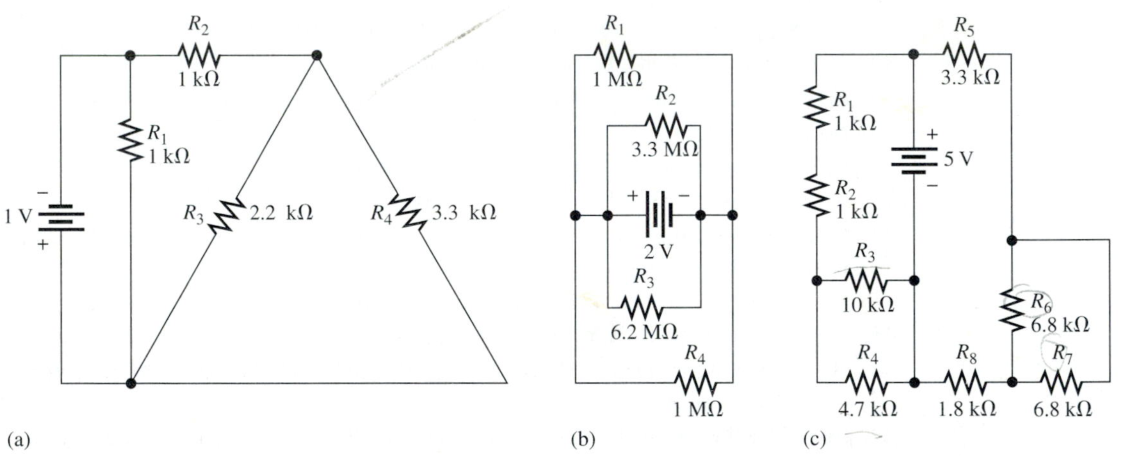


FIGURE 7–65

5. Draw the schematic of the PC board layout in Figure 7–66 showing resistor values and identify the series-parallel relationships.

FIGURE 7–66

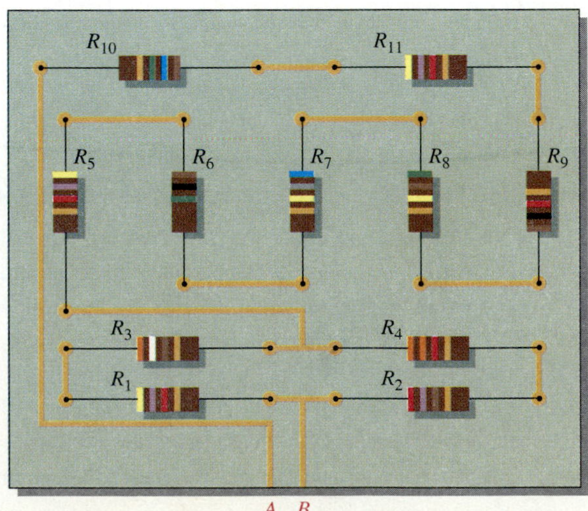

6. Develop a schematic for the double-sided PC board in Figure 7–67 and label the resistor values.

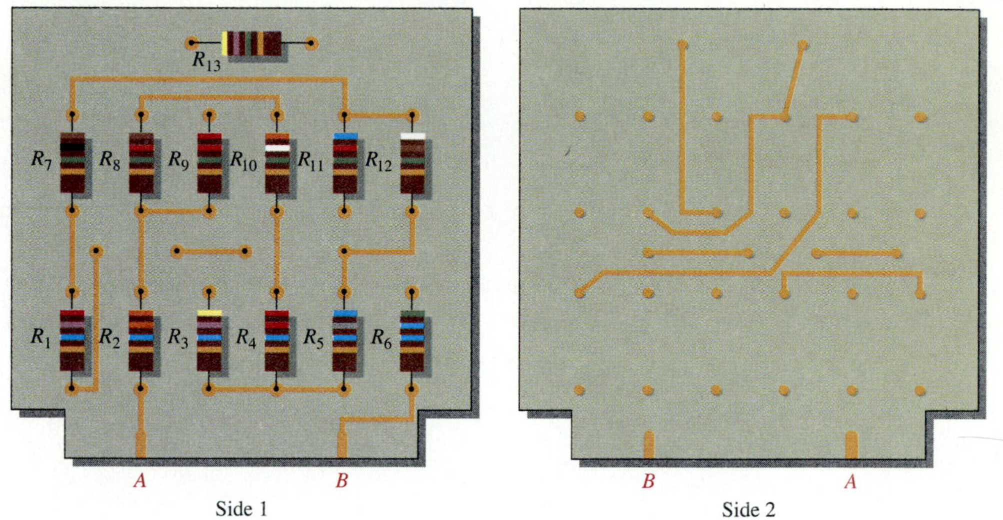


FIGURE 7–67

7. Lay out a PC board for the circuit in Figure 7–65(c). The battery is to be connected external to the board.

## SECTION 7–2 Analysis of Series-Parallel Circuits

8. A certain circuit is composed of two parallel resistors. The total resistance is 667 Ω. One of the resistors is 1 kΩ. What is the other resistor?
9. For each circuit in Figure 7–64, determine the total resistance presented to the source.
10. Repeat Problem 9 for each circuit in Figure 7–65.
11. Determine the current through each resistor in each circuit in Figure 7–64; then calculate each voltage drop.
12. Determine the current through each resistor in each circuit in Figure 7–65; then calculate each voltage drop.
13. Find $R_T$ for all combinations of the switches in Figure 7–68.

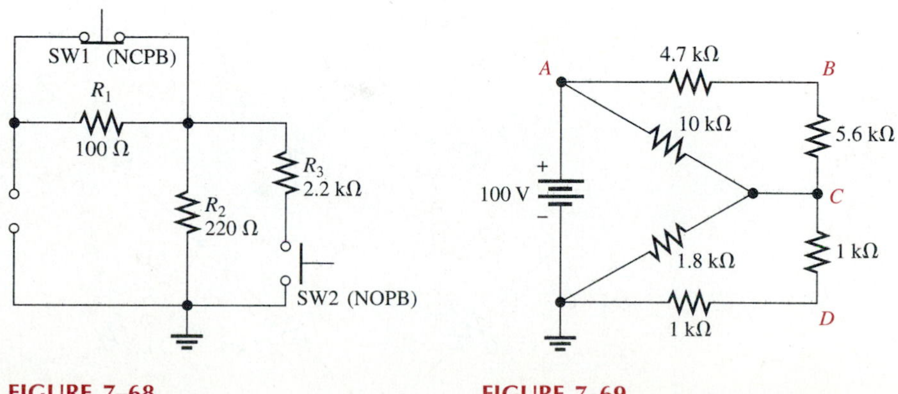


FIGURE 7–68

FIGURE 7–69

14. Determine the resistance between $A$ and $B$ in Figure 7–69 with the source removed.
15. Determine the voltage at each point with respect to ground in Figure 7–69.

16. Determine the voltage at each point with respect to ground in Figure 7–70.

17. In Figure 7–70, how would you determine the voltage across $R_2$ by measuring without connecting a meter directly across the resistor?

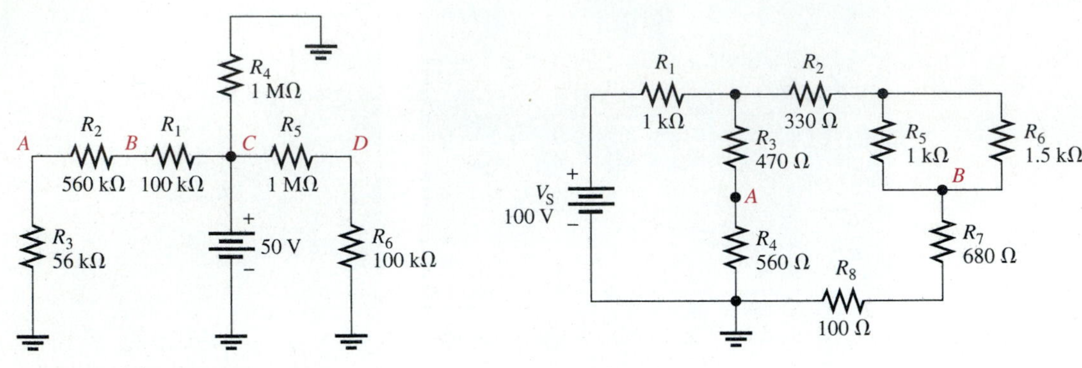


FIGURE 7–70

FIGURE 7–71

18. Determine the voltage, $V_{AB}$, in Figure 7–71.

19. Find the value of $R_2$ in Figure 7–72.

FIGURE 7–72

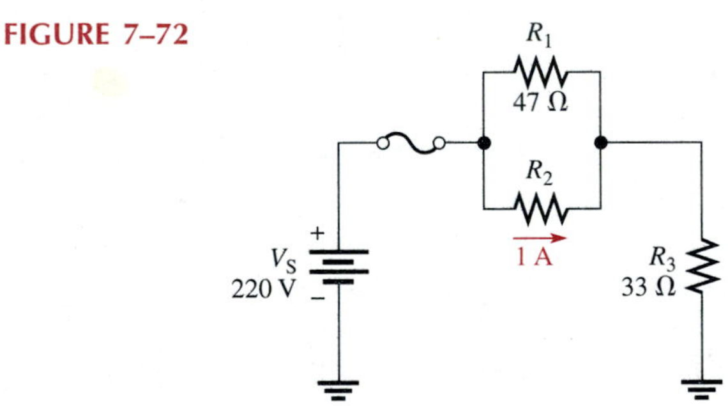


20. Find the resistance between point $A$ and each of the other points ($R_{AB}$, $R_{AC}$, $R_{AD}$, $R_{AE}$, $R_{AF}$, and $R_{AG}$) in Figure 7–73.

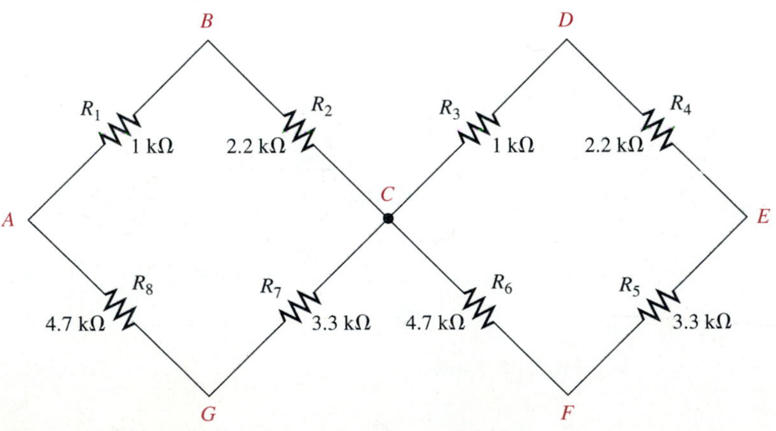


FIGURE 7–73

**21.** Find the resistance between each of the following sets of points in Figure 7–74: *AB, BC,* and *CD.*

**FIGURE 7–74**

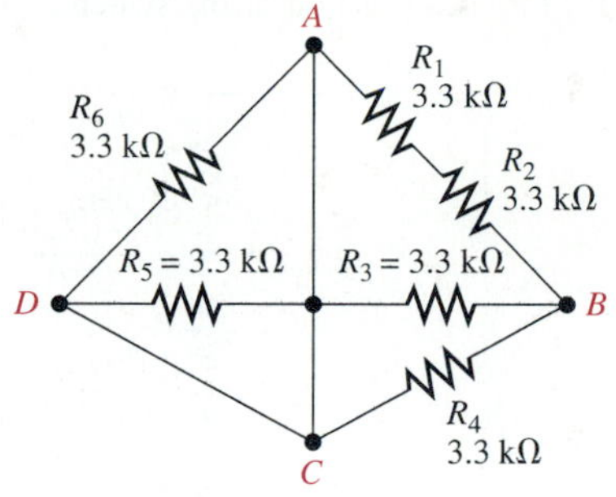


**22.** Determine the value of each resistor in Figure 7–75.

**FIGURE 7–75**

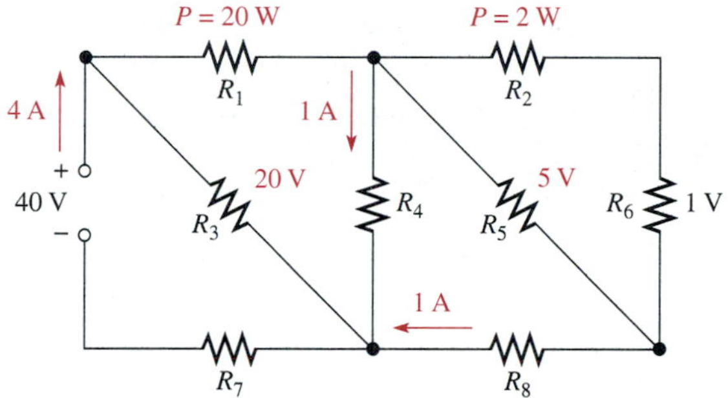


## SECTION 7–3 Voltage Dividers with Resistive Loads

**23.** A voltage divider consists of two 56 kΩ resistors and a 15 V source. Calculate the unloaded output voltage. What will the output voltage be if a load resistor of 1 MΩ is connected to the output?

**24.** A 12 V battery output is divided down to obtain two output voltages. Three 3.3 kΩ resistors are used to provide the two taps. Determine the output voltages. If a 10 kΩ load is connected to the higher of the two outputs, what will its loaded value be?

**25.** Which will cause a smaller decrease in output voltage for a given voltage divider, a 10 kΩ load or a 47 kΩ load?

**26.** In Figure 7–76, determine the continuous current drain on the battery with no load across the output terminals. With a 10 kΩ load, what is the battery current?

**FIGURE 7–76**

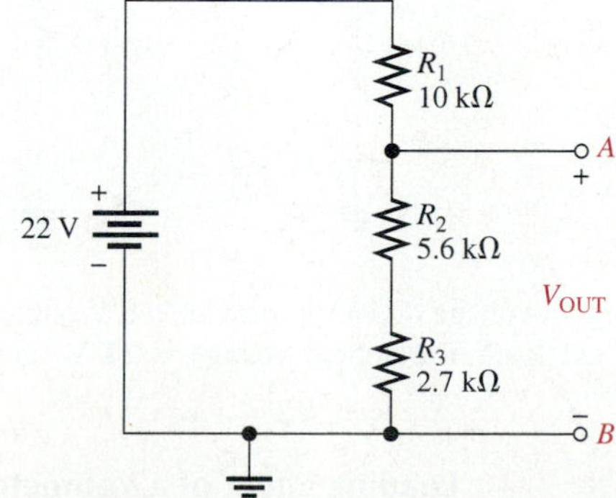


**27.** Determine the resistance values for a voltage divider that must meet the following specifications: The current drain under unloaded condition is not to exceed 5 mA. The source voltage is to be 10 V, and the required outputs are to be 5 V and 2.5 V. Sketch the circuit. Determine the effect on the output voltages if a 1 kΩ load is connected to each tap one at a time.

**28.** The voltage divider in Figure 7–77 has a switched load. Determine the voltage at each tap ($V_1$, $V_2$, and $V_3$) for each position of the switch.

FIGURE 7–77

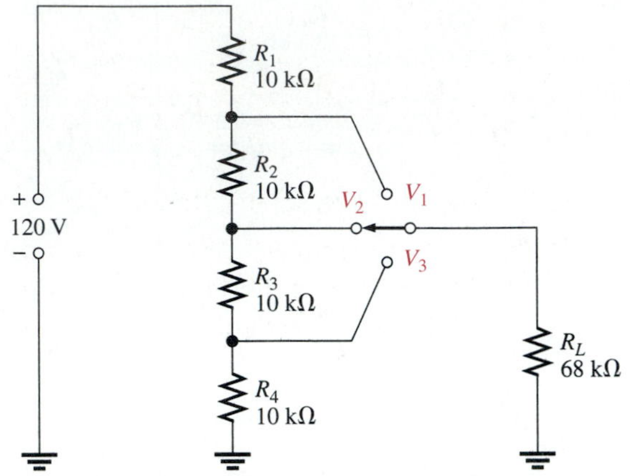

**29.** Figure 7–78 shows a dc biasing arrangement for a field-effect transistor amplifier. Biasing is a common method for setting up certain dc voltage levels required for proper amplifier operation. Although you are not expected to be familiar with transistor amplifiers at this point, the dc voltages and currents in the circuit can be determined using methods that you already know.

**(a)** Find $V_G$ and $V_S$ **(b)** Determine $I_1$, $I_2$, $I_D$, and $I_S$ **(c)** Find $V_{DS}$ and $V_{DG}$

FIGURE 7–78

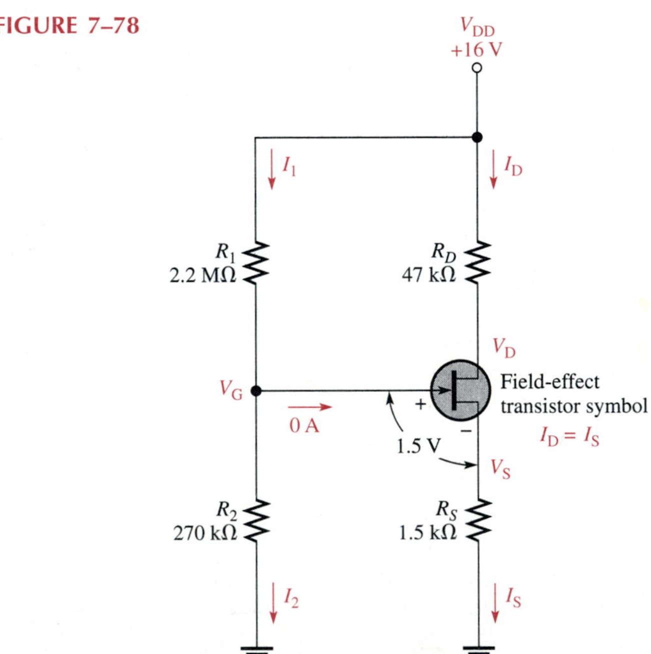


**30.** Design a voltage divider to provide a 6 V output with no load and a minimum of 5.5 V across a 1 kΩ load. The source voltage is 24 V, and the unloaded current drain is not to exceed 100 mA.

## SECTION 7–4 Loading Effect of a Voltmeter

**31.** On which one of the following voltage range settings will a voltmeter present the minimum load on a circuit?

**(a)** 1 V **(b)** 10 V **(c)** 100 V **(d)** 1000 V

**32.** Determine the internal resistance of a 20,000 Ω/V voltmeter on each of the following range settings.

**(a)** 0.5 V **(b)** 1 V **(c)** 5 V **(d)** 50 V **(e)** 100 V **(f)** 1000 V

**33.** The voltmeter described in Problem 32 is used to measure the voltage across $R_4$ in Figure 7–64(a).

**(a)** What range should be used?

**(b)** How much less is the voltage measured by the meter than the actual voltage?

**34.** Repeat Problem 33 if the voltmeter is used to measure the voltage across $R_4$ in the circuit of Figure 7–64(b).

## SECTION 7–5 Ladder Networks

**35.** For the circuit shown in Figure 7–79, calculate the following:

**(a)** Total resistance across the source **(b)** Total current from the source

**(c)** Current through the 910 Ω resistor **(d)** Voltage from point $A$ to point $B$

FIGURE 7–79

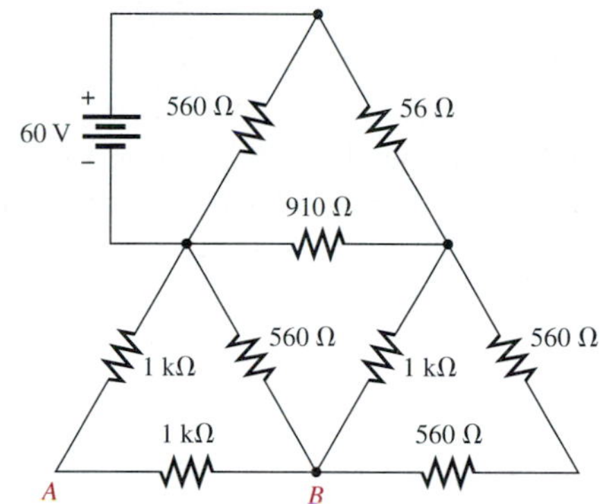


**36.** Determine the total resistance and the voltage at points $A$, $B$, and $C$ in the ladder network of Figure 7–80.

FIGURE 7–80

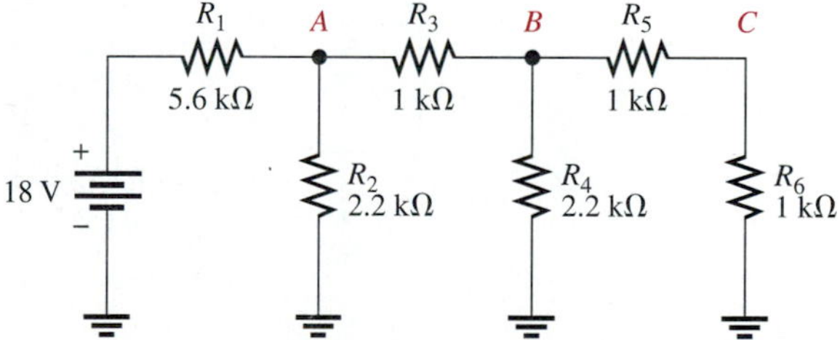


**37.** Determine the total resistance between terminals $A$ and $B$ of the ladder network in Figure 7–81. Also calculate the current in each branch with 10 V between $A$ and $B$.

FIGURE 7–81

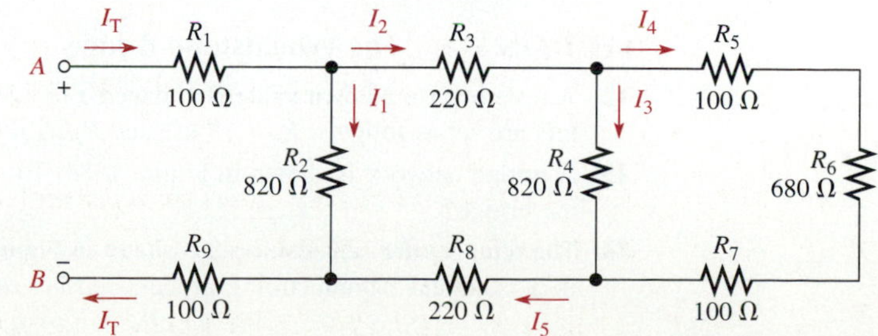

**38.** What is the voltage across each resistor in Figure 7–81 with 10 V between *A* and *B*?

**39.** Find $I_T$ and $V_{OUT}$ in Figure 7–82.

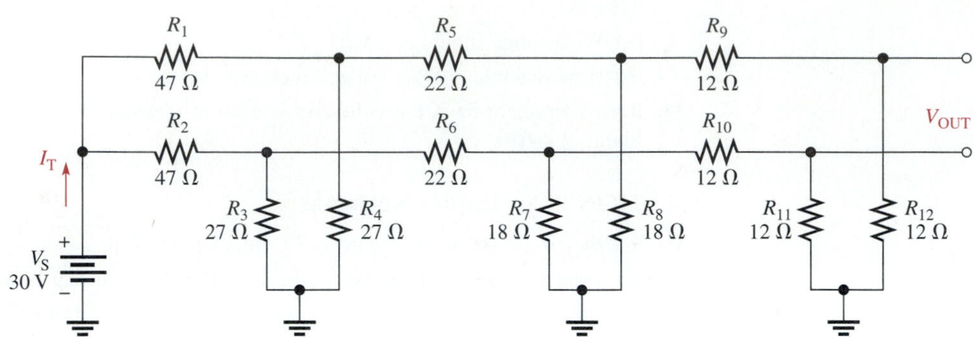


FIGURE 7–82

**40.** Determine $V_{OUT}$ for the *R*/2*R* ladder network in Figure 7–83 for the following conditions:

**(a)** Switch SW2 connected to +12 V and the others connected to ground

**(b)** Switch SW1 connected to +12 V and the others connected to ground

FIGURE 7–83

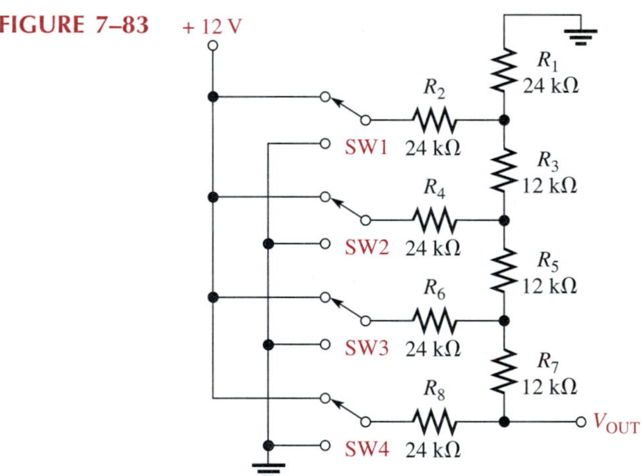


**41.** Repeat Problem 40 for the following conditions:

**(a)** SW3 and SW4 to +12 V, SW1 and SW2 to ground

**(b)** SW3 and SW1 to +12 V, SW2 and SW4 to ground

**(c)** All switches to +12 V

## SECTION 7–6 The Wheatstone Bridge

**42.** A resistor of unknown value is connected to a Wheatstone bridge circuit. The bridge parameters are set as follows: $R_V = 18$ kΩ and $R_2/R_4 = 0.02$. What is $R_{UNK}$?

**43.** A bridge network is shown in Figure 7–84. To what value must $R_V$ be set in order to balance the bridge?

**44.** The temperature-sensitive bridge circuit in Figure 7–50 is used to detect when the temperature in a chemical manufacturing process reaches 100°C. The resistance of the thermistor drops

FIGURE 7–84

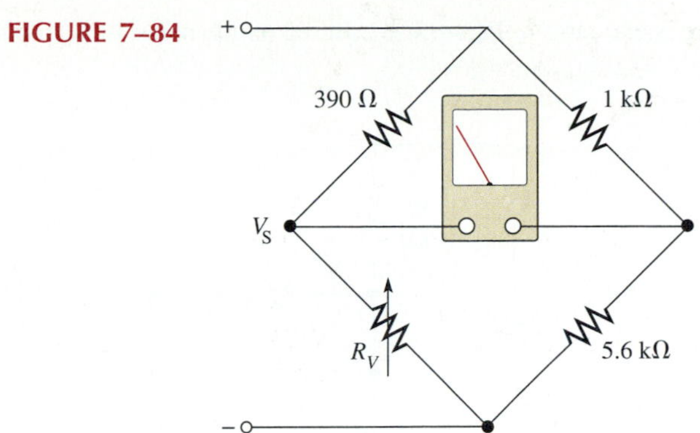


from 5 kΩ at a nominal 20°C to 100 Ω at 100°C. If $R_1 = 1$ kΩ and $R_2 = 2.2$ kΩ, to what value must $R_4$ be set to produce a balanced bridge when the temperature reaches 100°C?

## SECTION 7–7 Troubleshooting

**45.** Is the voltmeter reading in Figure 7–85 correct?

FIGURE 7–85

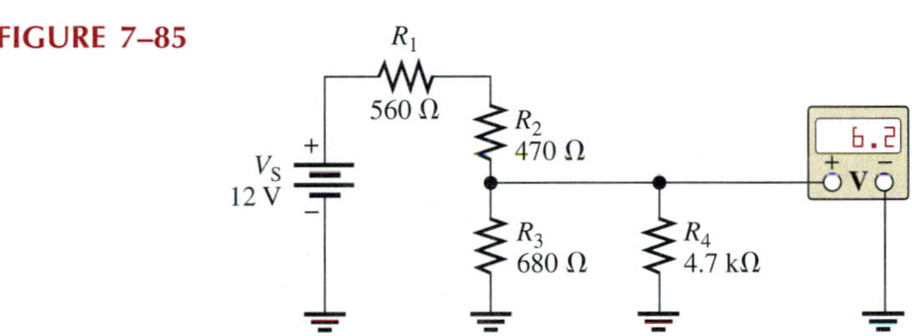


**46.** Are the meter readings in Figure 7–86 correct?

FIGURE 7–86

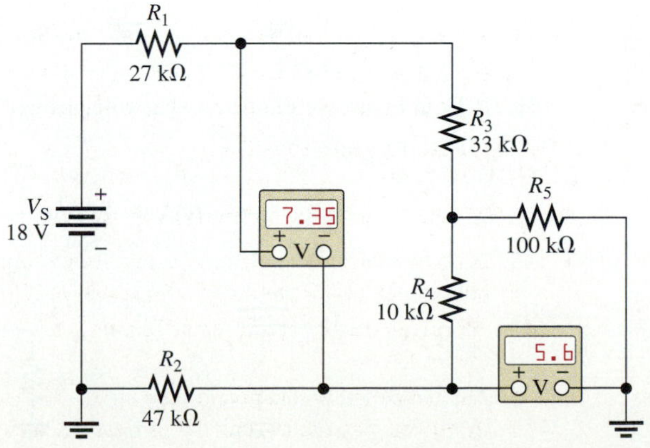

**47.** There is one fault in Figure 7–87. Based on the meter indications, determine what the fault is.

FIGURE 7–87

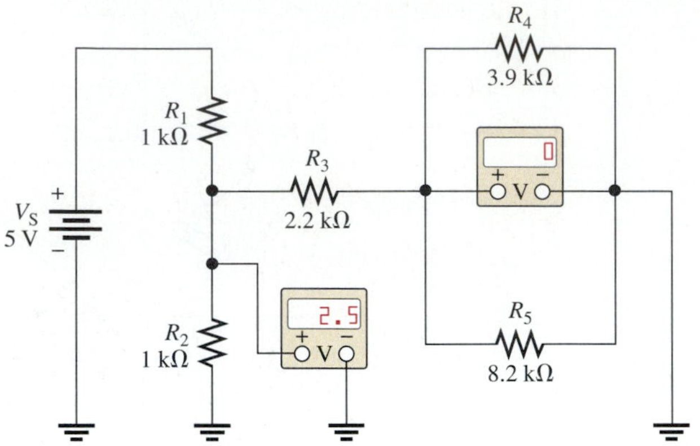

**48.** Look at the meters in Figure 7–88 and determine if there is a fault in the circuit. If there is a fault, identify it.

FIGURE 7–88

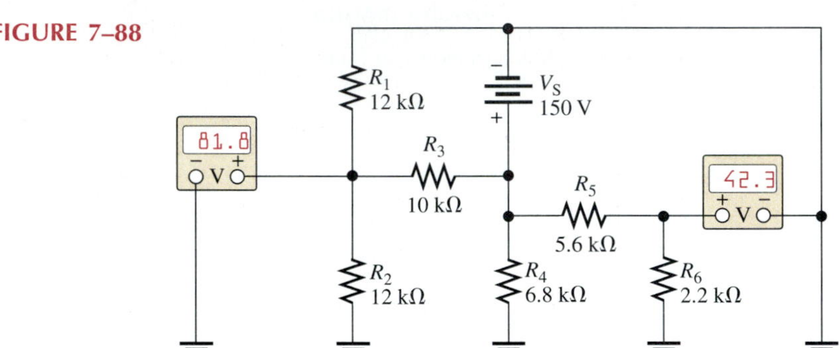


**49.** Check the meter readings in Figure 7–89 and locate any fault that may exist.

FIGURE 7–89

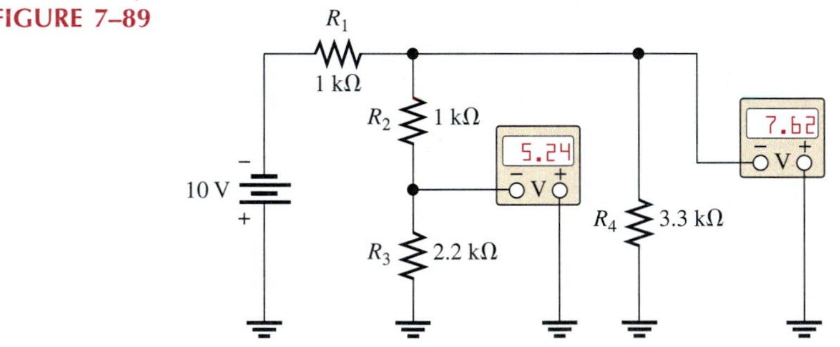


**50.** If $R_2$ in Figure 7–90 opens, what voltages will be read at points $A$, $B$, and $C$?

FIGURE 7–90

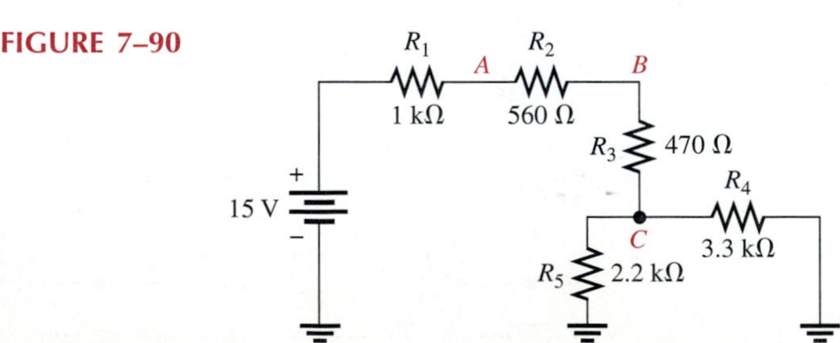

## ANSWERS TO SECTION REVIEWS

### Section 7–1

1. A series-parallel resistive circuit is a circuit consisting of both series and parallel connections.
2. See Figure 7–91.
3. Resistors $R_1$ and $R_2$ are in series with the parallel combination of $R_3$ and $R_4$.
4. All the resistors are in parallel.
5. Resistors $R_1$ and $R_2$ are in parallel; $R_3$ and $R_4$ are in parallel.
6. Yes, the parallel combinations are in series.

FIGURE 7–91

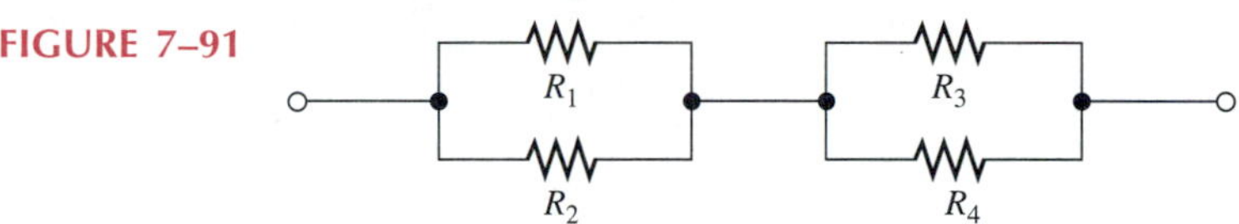


### Section 7–2

1. Voltage-divider and current-divider formulas, Kirchhoff's laws, and Ohm's law can be used in series-parallel analysis.
2. $R_T = R_1 + R_2 \parallel R_3 + R_4 = 608\ \Omega$
3. $I_3 = [R_2/(R_2 + R_3)]I_T = 11.1$ mA
4. $V_2 = I_2R_2 = 3.65$ V
5. $R_T = 47\ \Omega + 27\ \Omega + (27\ \Omega + 27\ \Omega) \parallel 47\ \Omega = 99.1\ \Omega$; $I_T = 1\text{ V}/99.1\ \Omega = 10.1$ mA

### Section 7–3

1. The load resistor decreases the output voltage.
2. True
3. $V_{OUT(unloaded)} = (100\text{ k}\Omega/147\text{ k}\Omega)30\text{ V} = 20.4$ V; $V_{OUT(loaded)} = (9.1\text{ k}\Omega/56.1\text{ k}\Omega)30\text{ V} = 4.87$ V

### Section 7–4

1. A voltmeter loads a circuit because the internal resistance of the meter appears in parallel with the circuit resistance across which it is connected, reducing the resistance between those two points of the circuit and drawing current from the circuit.
2. No, because the meter resistance is much larger than 1 kΩ.
3. There is a decrease in measured voltage because $R_M$ is smaller. You can read the scale more accurately on the 3 V range.
4. $R_M = (30{,}000\ \Omega/\text{V})5\text{ V} = 1.5\text{ M}\Omega$

### Section 7–5

1. See Figure 7–92.
2. $R_T = 11.6\text{ k}\Omega$
3. $I_T = 10\text{ V}/11.6\text{ k}\Omega = 859\ \mu$A
4. $I_2 = 640\ \mu$A
5. $V_A = 141$ V

FIGURE 7–92

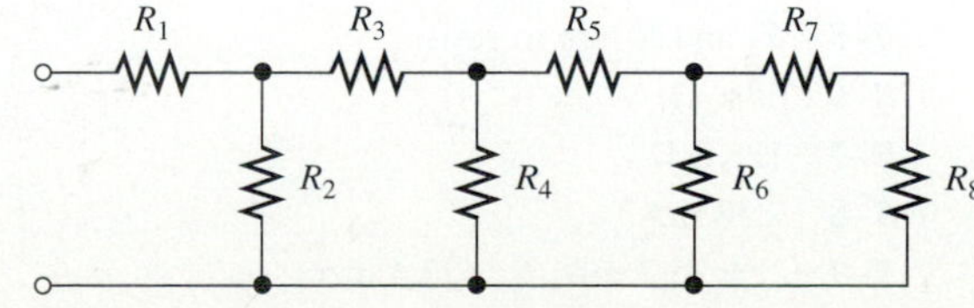

### Section 7–6

1. See Figure 7–93.
2. The bridge is balanced when $V_A = V_B$.
3. $R_{UNK} = R_V(R_2/R_4)$
4. $R_{UNK} = 15\ k\Omega$

FIGURE 7–93

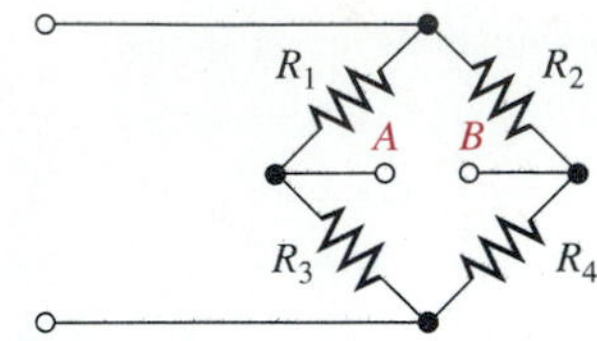


### Section 7–7

1. Common circuit faults are opens and shorts.
2. The 10 kΩ resistor ($R_3$) is open.
3. **(a)** $V_A = 55$ V **(b)** $V_A = 54.9$ V **(c)** $V_A = 54.2$ V **(d)** $V_A = 100$ V **(e)** $V_A = 0$ V

### Section 7–8

1. The battery will last 417 days.
2. Yes, ⅛ W resistors can be used.
3. No, none of the resistors will overheat.

### Section 7–9

1.
```
VS  1  0  50
.PRINT  DC  I(R2)  I(R3)  I(R4)  I(R5)  V(R1)  V(R7)
```
2.
```
VS  1  0  10
R1  1  2  100
R2  2  7  820
R3  2  3  220
R4  3  6  820
R5  3  4  100
R6  4  5  680
R7  5  6  100
R8  6  7  220
R9  7  0  100
.DC  VS  10  10  1
.PRINT  I(R1)  I(R2)  I(R3)  I(R3)  I(R4)  I(R5)  I(R6)  I(R7)  I(R8)  I(R9)
.PRINT  V(R1)  V(R2)  V(R3)  V(R4)  V(R5)  V(R6)  V(R7)  V(R8)  V(R9)
.END
```

## ■ ANSWERS TO RELATED EXERCISES FOR EXAMPLES

**7–1** The new resistor is in parallel with $R_4 + R_2 \parallel R_3$.

**7–2** The resistor has no effect because it is shorted.

**7–3** The new resistor is in parallel with $R_5$.

**7–4** $A$ to gnd: $R_T = R_4 + R_3 \parallel (R_1 + R_2)$
$B$ to gnd: $R_T = R_4 + R_2 \parallel (R_1 + R_3)$
$C$ to gnd: $R_T = R_4$

**7–5** $R_3$ and $R_6$ are in series.

**7–6** 55.1 Ω

**7–7** 128.3 Ω

**7–8** 2.38 mA

**7–9** $I_1 = 35.7$ mA; $I_3 = 23.4$ mA

**7–10** $V_A = 44.8$ V; $V_1 = 35.2$ V

**7–11** 2.04 V

**7–12** $V_{AB} = 5.48$ V; $V_{BC} = 1.66$ V; $V_{CD} = 0.86$ V

**7–13** 3.39 V

**7–14** 7.07 V

**7–15** 5.34 V

**7–16** $I_1 = 7.16$ mA; $I_2 = 3.57$ mA; $I_3 = 3.57$ mA; $I_4 = 1.74$ mA; $I_5 = 1.85$ mA; $I_6 = 1.85$ mA; $V_A = 29.3$ V; $V_B = 17.4$ V; $V_C = 8.70$ V

**7–17** 321 Ω

**7–18** 5.73 V; 0 V

**7–19** 9.46 V

**7–20** $V_A = 12$ V; $V_B = 13.8$ V

# 8

# CIRCUIT THEOREMS AND CONVERSIONS



## ■ INTRODUCTION

In previous chapters, you saw how to analyze various types of circuits using Ohm's law and Kirchhoff's laws. Some types of circuits are difficult to analyze using only those basic laws and require additional methods in order to simplify the analysis.

The theorems and conversions in this chapter make analysis easier for certain types of circuits. These methods do not replace Ohm's law and Kirchhoff's laws, but are normally used in conjunction with them in certain situations. In this chapter and throughout the rest of the book, you will learn the basics of putting technology theory into practice.

Because all electric circuits are driven by either voltage sources or current sources, it is important to understand how to work with these elements. The superposition theorem will help you to deal with circuits that have multiple sources. Thevenin's, Norton's, and Millman's theorems provide methods for reducing a circuit to a simple equivalent form for ease of analysis. The maximum power transfer theorem is used in applications where it is important for a given circuit to provide maximum power to a load. An example of this is an audio amplifier that provides maximum power to a speaker. Delta-to-wye and wye-to-delta conversions are sometimes useful when analyzing bridge circuits that are commonly found in systems that measure physical parameters such as temperature, pressure, and strain.

## TECHnology Theory Into Practice

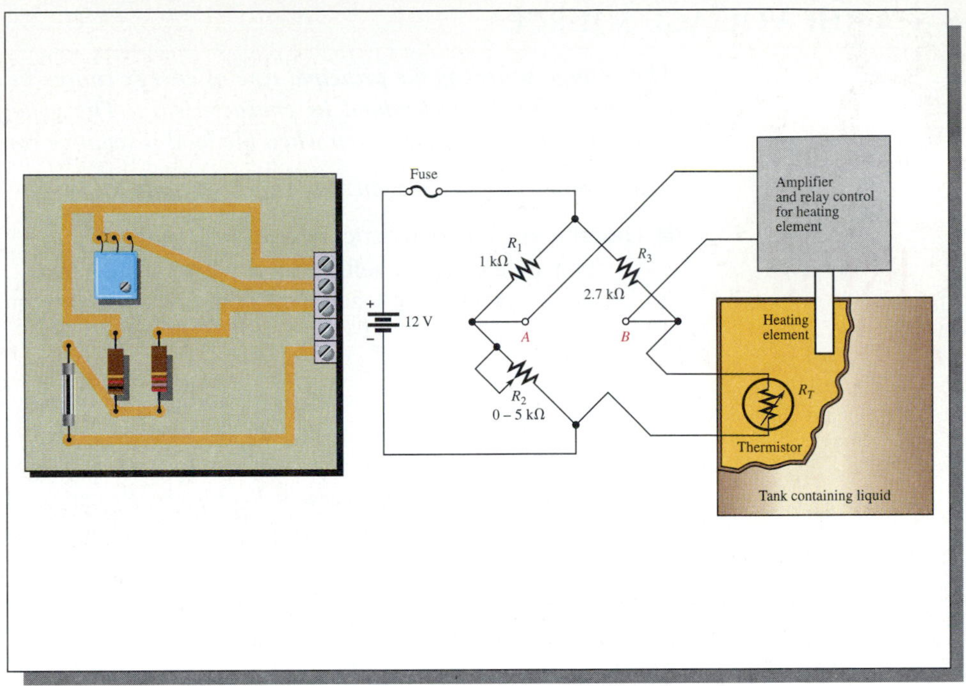


In the TECH TIP assignment in Section 8–10, you will be working with a temperature measurement and control circuit that uses a Wheatstone bridge, which you studied in Chapter 7. You will utilize Thevenin's theorem as well as other techniques in the evaluation of this circuit.

## CHAPTER OBJECTIVES

- ☐ Describe the characteristics of a voltage source
- ☐ Describe the characteristics of a current source
- ☐ Perform source conversions
- ☐ Apply the superposition theorem to circuit analysis
- ☐ Apply Thevenin's theorem to simplify a circuit for analysis
- ☐ Apply Norton's theorem to simplify a circuit
- ☐ Apply Millman's theorem to parallel sources
- ☐ Apply the maximum power transfer theorem
- ☐ Perform Δ-to-Y and Y-to-Δ conversions
- ☐ Use PSpice to determine equivalent circuits (optional)

## 8–1 ■ THE VOLTAGE SOURCE

***The voltage source is the principal type of energy source in electronic applications, so it is important to understand its characteristics. The voltage source ideally provides constant voltage to a load even when the load resistance varies.***

***After completing this section, you should be able to***

- **Describe the characteristics of a voltage source**
  - ☐ Compare a practical voltage source to an ideal source
  - ☐ Discuss the effect of loading on a practical voltage source

Figure 8–1(a) is the familiar symbol for an ideal dc **voltage source.** The voltage across its terminals $A$ and $B$ remains fixed regardless of the value of load resistance that may be connected across its output. Figure 8–1(b) shows a load resistor, $R_L$, connected. All of the source voltage, $V_S$, is dropped across $R_L$. Ideally, $R_L$ can be changed to any value except zero, and the voltage will remain fixed. The ideal voltage source has an internal resistance of zero.

**FIGURE 8–1**
***Ideal dc voltage source.***

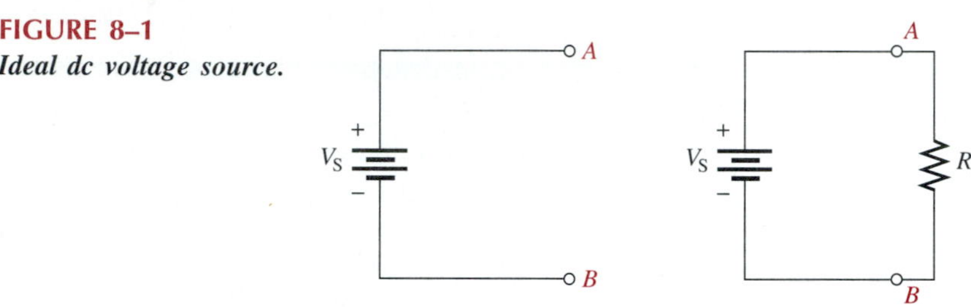


In reality, no voltage source is ideal. That is, all voltage sources have some inherent internal resistance as a result of their physical and/or chemical makeup, which can be represented by a resistor in series with an ideal source, as shown in Figure 8–2(a). $R_S$ is the internal source resistance and $V_S$ is the source voltage. With no load, the output voltage (voltage from $A$ to $B$) is $V_S$. This voltage is sometimes called the *open circuit voltage.*

**FIGURE 8–2**
***Practical voltage source.***

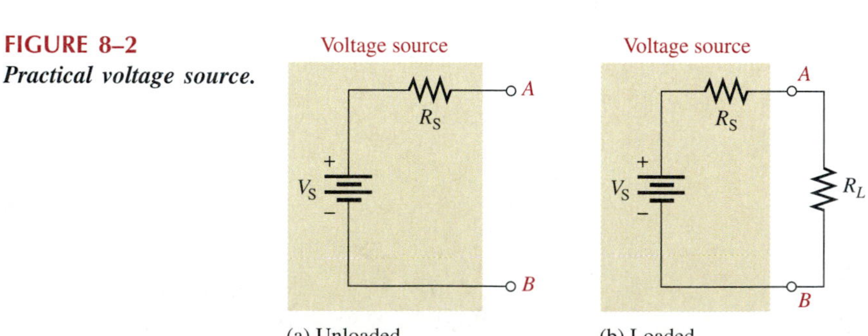


### Loading of the Voltage Source

When a load resistor is connected across the output terminals, as shown in Figure 8–2(b), all of the source voltage does not appear across $R_L$. Some of the voltage is dropped across $R_S$ because $R_S$ and $R_L$ are in series.

If $R_S$ is very small compared to $R_L$, the source approaches ideal because almost all of the source voltage, $V_S$, appears across the larger resistance, $R_L$. Very little voltage is dropped across the internal resistance, $R_S$. If $R_L$ changes, most of the source voltage remains across the output as long as $R_L$ is much larger than $R_S$. As a result, very little change occurs in the output voltage. The larger $R_L$ is compared to $R_S$, the less change there is in the output voltage. As a rule, before it can be neglected, $R_L$ should be at least ten times $R_S$ ($R_L \geq 10R_S$).

Example 8–1 illustrates the effect of changes in $R_L$ on the output voltage when $R_L$ is much greater than $R_S$. Example 8–2 shows the effect of smaller load resistances.

## EXAMPLE 8–1

Calculate the voltage output of the source in Figure 8–3 for the following values of $R_L$: 100 Ω, 560 Ω, and 1 kΩ.

FIGURE 8–3

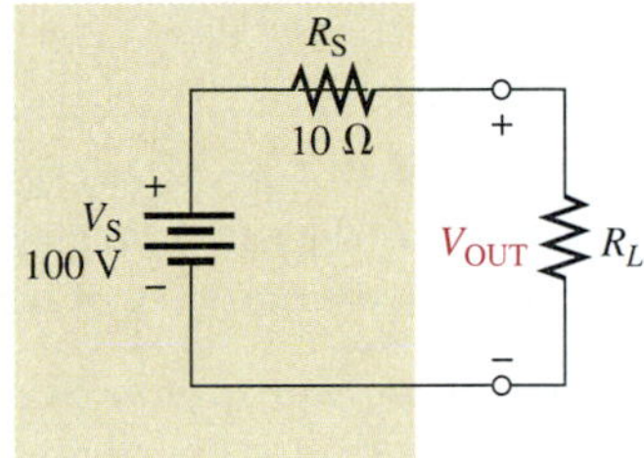


***Solution*** For $R_L = 100\ \Omega$, the voltage output is

$$V_{OUT} = \left(\frac{R_L}{R_S + R_L}\right)V_S = \left(\frac{100\ \Omega}{110\ \Omega}\right)100\ \text{V} = 90.9\ \text{V}$$

For $R_L = 560\ \Omega$,

$$V_{OUT} = \left(\frac{560\ \Omega}{570\ \Omega}\right)100\ \text{V} = 98.2\ \text{V}$$

For $R_L = 1\ \text{k}\Omega$,

$$V_{OUT} = \left(\frac{1000\ \Omega}{1010\ \Omega}\right)100\ \text{V} = 99.0\ \text{V}$$

Notice that the output voltage is within 10% of the source voltage, $V_S$, for all three values of $R_L$, because $R_L$ is at least ten times $R_S$.

***Related Exercise*** Determine $V_{OUT}$ in Figure 8–3 if $R_S = 50\ \Omega$ and $R_L = 10\ \text{k}\Omega$.

## EXAMPLE 8–2

Determine $V_{OUT}$ for $R_L = 10\ \Omega$ and for $R_L = 1\ \Omega$ in Figure 8–3.

***Solution*** For $R_L = 10\ \Omega$, the voltage output is

$$V_{OUT} = \left(\frac{R_L}{R_S + R_L}\right)V_S = \left(\frac{10\ \Omega}{20\ \Omega}\right)100\ \text{V} = 50\ \text{V}$$

For $R_L = 1\ \Omega$,

$$V_{OUT} = \left(\frac{1\ \Omega}{11\ \Omega}\right)100\ \text{V} = 9.09\ \text{V}$$

***Related Exercise*** What is $V_{OUT}$ with no load resistor in Figure 8–3?

Notice in Example 8–2 that the output voltage decreases significantly as $R_L$ is made smaller compared to $R_S$. This example illustrates the requirement that $R_L$ must be much larger than $R_S$ in order to maintain the output voltage near its open circuit value.

**SECTION 8–1 REVIEW**

1. What is the symbol for the ideal voltage source?
2. Sketch a practical voltage source.
3. What is the internal resistance of the ideal voltage source?
4. What effect does the load have on the output voltage of the practical voltage source?

## 8–2 ■ THE CURRENT SOURCE

***The current source is another type of energy source that ideally provides a constant current to a load even when the resistance of the load varies. The concept of the current source is important in certain types of transistor circuits.***

***After completing this section, you should be able to***

- **Describe the characteristics of a current source**
  - ☐ Compare a practical current source to an ideal source
  - ☐ Discuss the effect of loading on a practical current source

Figure 8–4(a) shows a symbol for the ideal **current source.** The arrow indicates the direction of current, and $I_S$ is the value of the source current. An ideal current source produces a fixed or constant value of current through a load, regardless of the value of the load. This concept is illustrated in Figure 8–4(b), where a load resistor is connected to the current source between terminals *A* and *B*. The ideal current source has an infinitely large internal resistance.



FIGURE 8–4
*Ideal current source.*

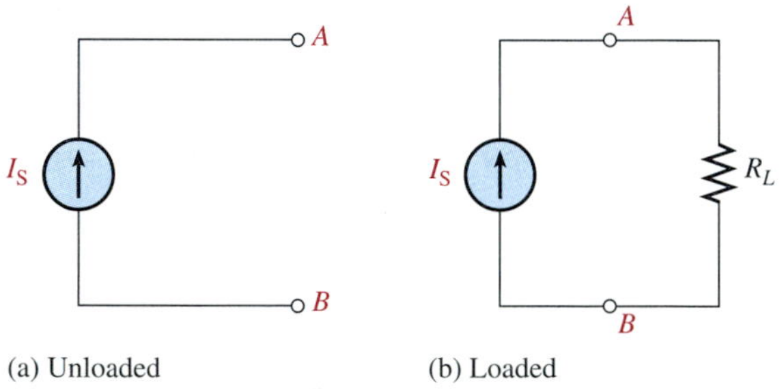

Transistors act basically as current sources, and for this reason, knowledge of the current source concept is important. You will find that the equivalent model of a transistor does contain a current source.

Although the ideal current source can be used in most analysis work, no actual device is ideal. A practical current source representation is shown in Figure 8–5. Here the internal resistance appears in parallel with the ideal current source.

If the internal source resistance, $R_S$ is much larger than a load resistor, the practical source approaches ideal. The reason is illustrated in the practical current source shown in Figure 8–5. Part of the current $I_S$ is through $R_S$, and part through $R_L$. Resistors $R_S$ and $R_L$ act as a current divider. If $R_S$ is much larger than $R_L$, most of the current is through $R_L$ and

**FIGURE 8–5**
***Practical current source with load.***

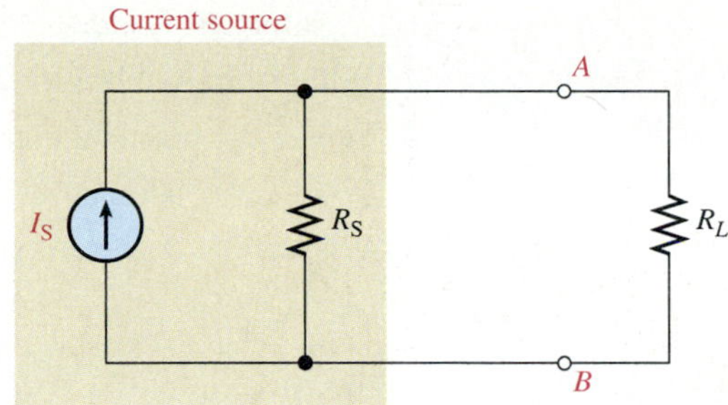


very little through $R_S$. As long as $R_L$ remains much smaller than $R_S$, the current through it will stay almost constant, no matter how much $R_L$ changes.

If there is a constant-current source, you can normally assume that $R_S$ is so much larger than the load that $R_S$ can be neglected. This simplifies the source to ideal, making the analysis easier.

Example 8–3 illustrates the effect of changes in $R_L$ on the load current when $R_L$ is much smaller than $R_S$. Generally, $R_L$ should be at least ten times smaller than $R_S$ ($10R_L \leq R_S$).

## EXAMPLE 8–3

Calculate the load current ($I_L$) in Figure 8–6 for the following values of $R_L$: 100 Ω, 560 Ω, and 1 kΩ.

**FIGURE 8–6**

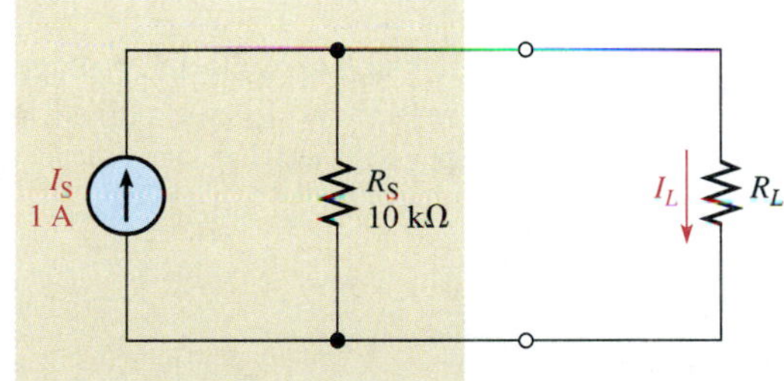


***Solution*** For $R_L = 100\ \Omega$, the load current is

$$I_L = \left(\frac{R_S}{R_S + R_L}\right)I_S = \left(\frac{10\text{ k}\Omega}{10.1\text{ k}\Omega}\right)1\text{ A} = 990\text{ mA}$$

For $R_L = 560\ \Omega$,

$$I_L = \left(\frac{10\text{ k}\Omega}{10.56\text{ k}\Omega}\right)1\text{ A} = 947\text{ mA}$$

For $R_L = 1\text{ k}\Omega$,

$$I_L = \left(\frac{10\text{ k}\Omega}{11\text{ k}\Omega}\right)1\text{ A} = 909\text{ mA}$$

Notice that the load current, $I_L$, is within 10% of the source current for each value of $R_L$ because $R_L$ is at least ten times smaller than $R_S$.

***Related Exercise*** At what value of $R_L$ in Figure 8–6 will the load current equal 750 mA?

**SECTION 8–2 REVIEW**

1. What is the symbol for an ideal current source?
2. Sketch the practical current source.
3. What is the internal resistance of the ideal current source?
4. What effect does the load have on the load current of the practical current source?

## 8–3 ▪ SOURCE CONVERSIONS

***In circuit analysis, it is sometimes useful to convert a voltage source to an equivalent current source, or vice versa.***

***After completing this section, you should be able to***

- **Perform source conversions**
  - ☐ Convert a voltage source to a current source
  - ☐ Convert a current source to a voltage source
  - ☐ Define *terminal equivalency*

### Converting a Voltage Source into a Current Source

The source voltage, $V_S$, divided by the internal source resistance, $R_S$, gives the value of the equivalent source current.

$$I_S = \frac{V_S}{R_S}$$

The value of $R_S$ is the same for both the voltage and current sources. As illustrated in Figure 8–7, the directional arrow for the current points from minus to plus. The equivalent current source is in parallel with $R_S$.

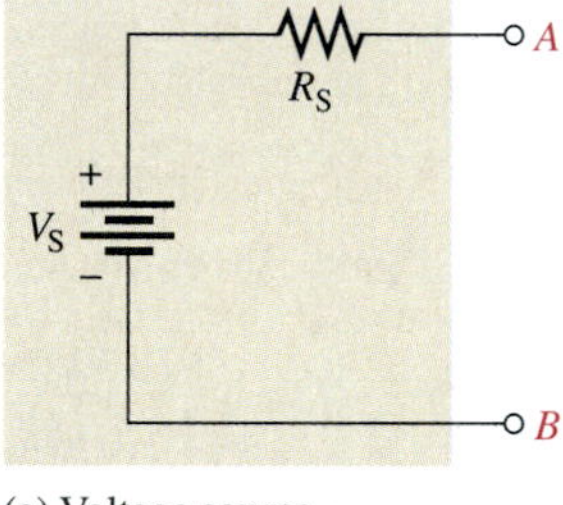


(a) Voltage source

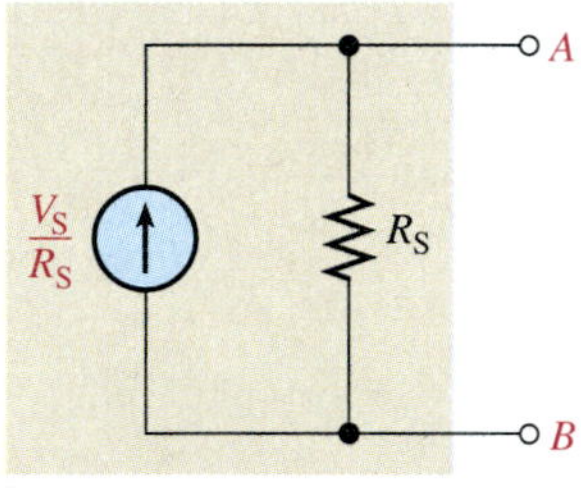


(b) Current source

FIGURE 8–7
*Conversion of voltage source to equivalent current source.*

Equivalency of two sources means that for any given load resistance connected to the two sources, the same load voltage and load current are produced by both sources. This concept is called **terminal equivalency.**

You can show that the voltage source and the current source in Figure 8–7 are equivalent by connecting a load resistor to each, as shown in Figure 8–8, and then calculating the load current. For the voltage source, the load current is

$$I_L = \frac{V_S}{R_S + R_L}$$

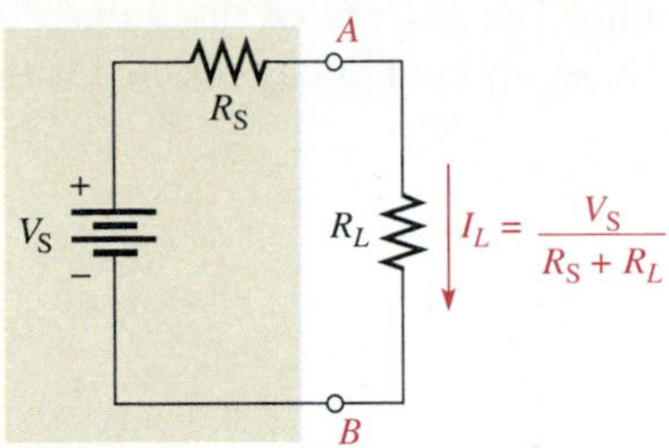


(a) Loaded voltage source

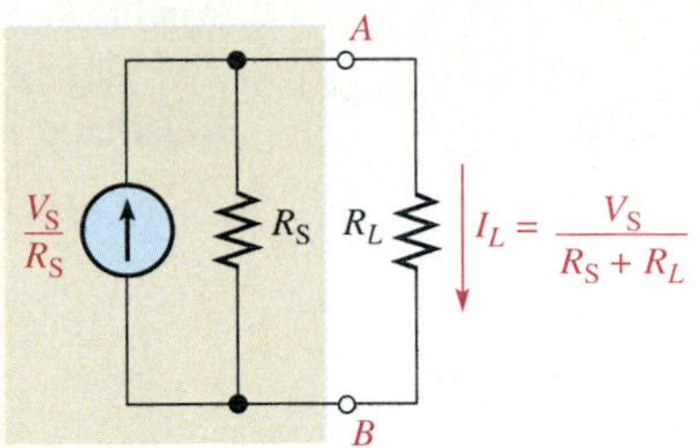


(b) Loaded current source

FIGURE 8–8
*Equivalent sources with loads.*

For the current source,

$$I_L = \left(\frac{R_S}{R_S + R_L}\right)\frac{V_S}{R_S} = \frac{V_S}{R_S + R_L}$$

As you see, both expressions for $I_L$ are the same. These equations prove that the sources are equivalent as far as the load or terminals $AB$ are concerned.

**EXAMPLE 8–4** Convert the voltage source in Figure 8–9 to an equivalent current source and show the equivalent circuit.

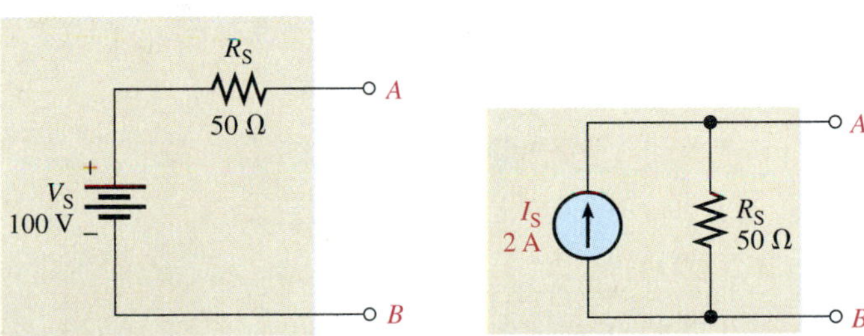


FIGURE 8–9 FIGURE 8–10

***Solution*** The value of $R_S$ is the same as with a voltage source. Therefore, the equivalent current source is

$$I_S = \frac{V_S}{R_S} = \frac{100\text{ V}}{50\ \Omega} = 2\text{ A}$$

Figure 8–10 shows the equivalent circuit.

***Related Exercise*** Determine $I_S$ and $R_S$ of a current source equivalent to a voltage source with $V_S = 12$ V and $R_S = 10\ \Omega$.

## Converting a Current Source into a Voltage Source

The source current, $I_S$, multiplied by the internal source resistance, $R_S$, gives the value of the equivalent source voltage.

$$V_S = I_S R_S$$

Again, $R_S$ remains the same. The polarity of the voltage source is minus to plus in the direction of the current. The equivalent voltage source is the voltage in series with $R_S$, as illustrated in Figure 8–11.

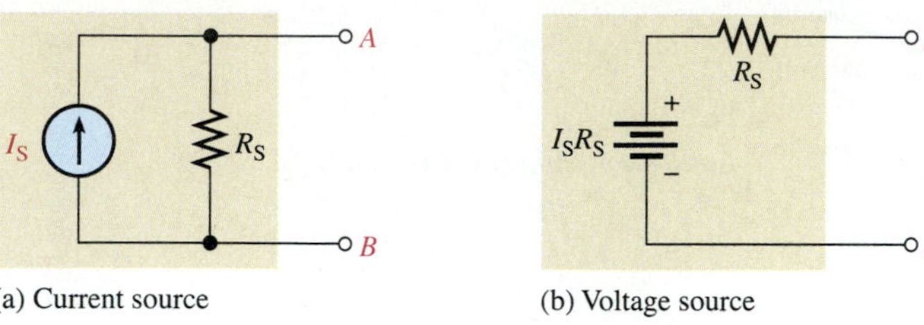


(a) Current source (b) Voltage source

FIGURE 8–11
***Conversion of current source to equivalent voltage source.***

**EXAMPLE 8–5** Convert the current source in Figure 8–12 to an equivalent voltage source and show the equivalent circuit.

FIGURE 8–12

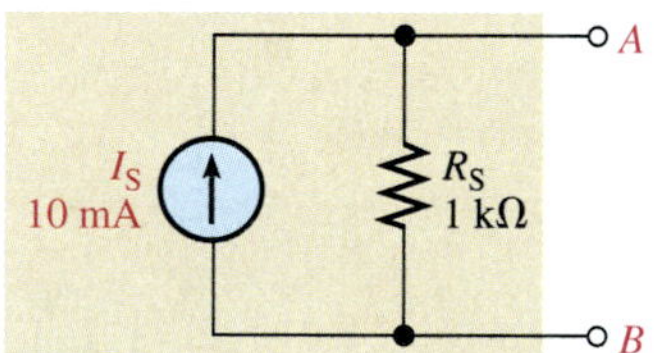


***Solution*** The value of $R_S$ is the same as with a current source. Therefore, the equivalent voltage source is

$$V_S = I_SR_S = (10 \text{ mA})(1 \text{ k}\Omega) = 10 \text{ V}$$

Figure 8–13 shows the equivalent circuit.

FIGURE 8–13

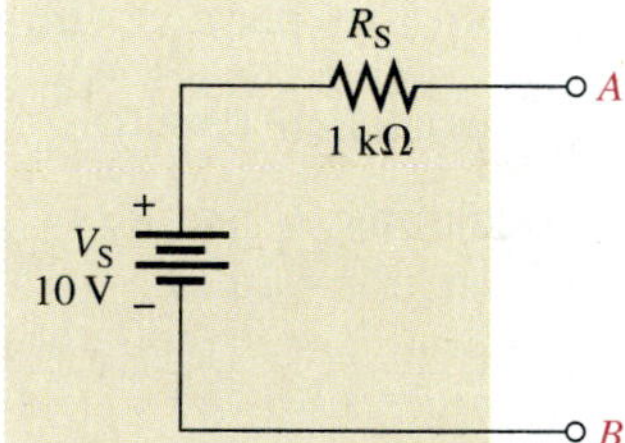


***Related Exercise*** Determine $V_S$ and $R_S$ of a voltage source equivalent to a current source with $I_S = 500$ mA and $R_S = 600\ \Omega$.

**SECTION 8–3 REVIEW**

1. Write the formula for converting a voltage source to a current source.
2. Write the formula for converting a current source to a voltage source.
3. Convert the voltage source in Figure 8–14 to an equivalent current source.
4. Convert the current source in Figure 8–15 to an equivalent voltage source.

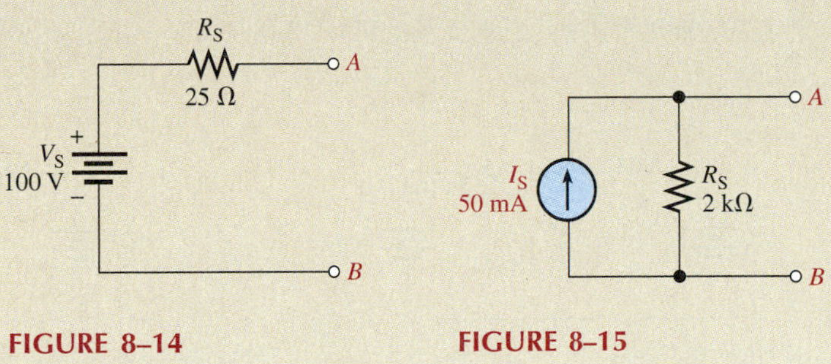


FIGURE 8–14

FIGURE 8–15

## 8–4 ■ THE SUPERPOSITION THEOREM

***Some circuits require more than one voltage or current source. For example, certain types of amplifiers require both a positive and a negative voltage source for proper operation.***

***After completing this section, you should be able to***

- **Apply the superposition theorem to circuit analysis**
  - ☐ State the superposition theorem
  - ☐ List the steps in applying the theorem

The superposition method is a way to determine currents and voltages in a circuit that has multiple sources by taking one source at a time. The other sources are replaced by their internal resistances. Recall that the ideal voltage source has a zero internal resistance. In this section, all voltage sources will be treated as ideal in order to simplify the coverage.

A general statement of the **superposition theorem** is as follows:

**The current in any given branch of a multiple-source circuit can be found by determining the currents in that particular branch produced by each source acting alone, with all other sources replaced by their internal resistances. The total current in the branch is the algebraic sum of the individual source currents in that branch.**

The steps in applying the superposition method are as follows:

**Step 1:** Take one voltage (or current) source at a time and replace each of the other voltage (or current) sources with either a short for a voltage source or an open for a current source (a short represents zero internal resistance and an open represents infinite internal resistance).

**Step 2:** Determine the particular current or voltage that you want just as if there were only one source in the circuit.

**Step 3:** Take the next source in the circuit and repeat Steps 1 and 2 for each source.

**Step 4:** To find the actual current in a given branch, add or subtract the currents due to each individual source. If the currents are in the same direction, add them. If the currents are in opposite directions, subtract them with the direction of the resulting current the same as the larger of the original quantities. Once the current is found, voltage can be determined.

An example of the approach to superposition is demonstrated in Figure 8–16 for a series-parallel circuit with two voltage sources. Study the steps in this figure. Examples 8–6 through 8–9 will clarify this procedure.

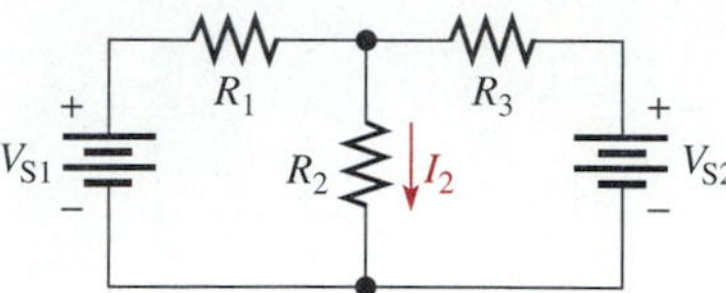


(a) Problem: Find $I_2$.

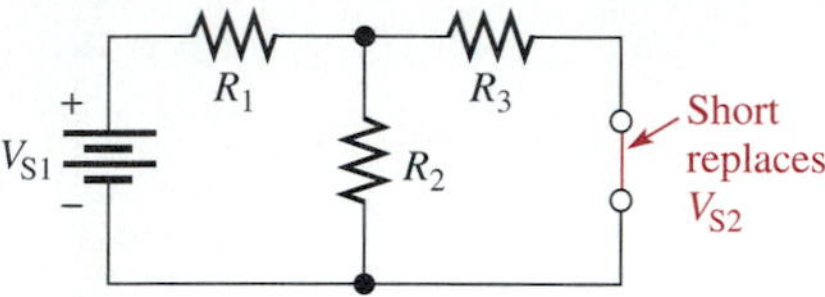


(b) Replace $V_{S2}$ with zero resistance (short).

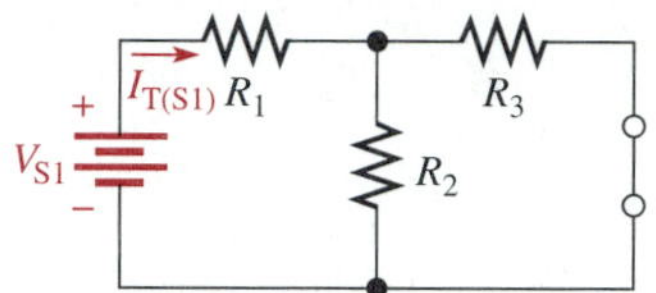


(c) Find $R_T$ and $I_T$ looking from $V_{S1}$:
$R_{T(S1)} = R_1 + R_2 \parallel R_3$
$I_{T(S1)} = V_{S1}/R_{T(S1)}$

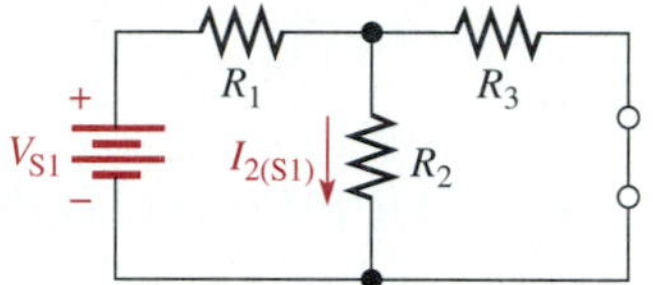


(d) Find $I_2$ due to $V_{S1}$ (current divider):

$$I_{2(S1)} = \left(\frac{R_3}{R_2 + R_3}\right) I_{T(S1)}$$

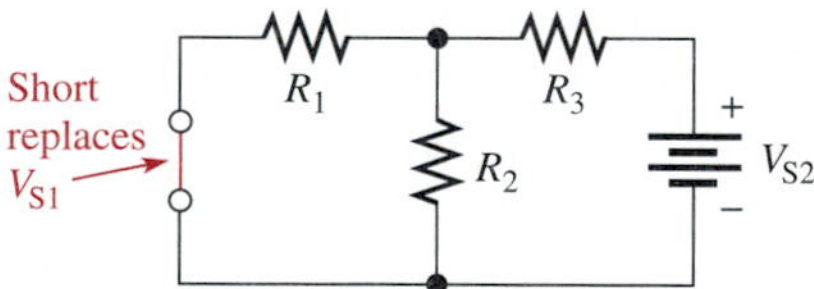


(e) Replace $V_{S1}$ with zero resistance (short).

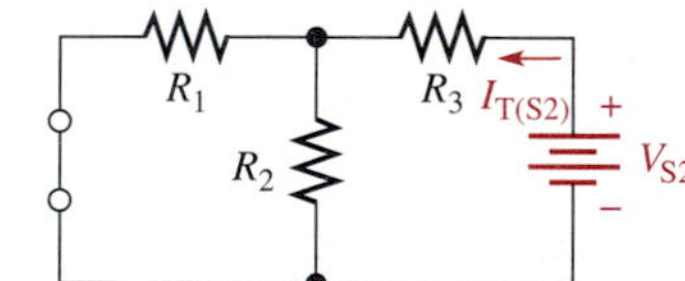


(f) Find $R_T$ and $I_T$ looking from $V_{S2}$:
$R_{T(S2)} = R_3 + R_1 \parallel R_2$
$I_{T(S2)} = V_{S2}/R_{T(S2)}$

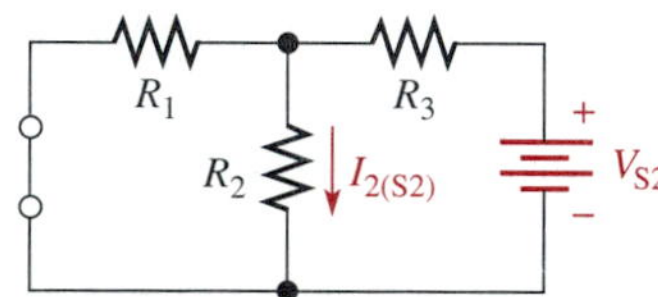


(g) Find $I_2$ due to $V_{S2}$:

$$I_{2(S2)} = \left(\frac{R_1}{R_1 + R_2}\right) I_{T(S2)}$$

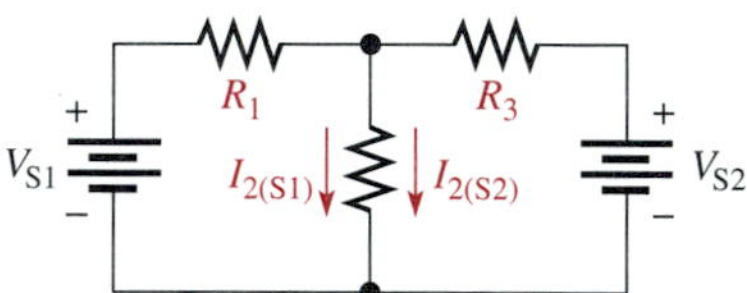


(h) Add $I_{2(S1)}$ and $I_{2(S2)}$ to get the actual $I_2$ (they are in same direction):
$I_2 = I_{2(S1)} + I_{2(S2)}$

**FIGURE 8–16**
***Demonstration of the superposition method.***

**EXAMPLE 8–6** Find the current in $R_2$ of Figure 8–17 by using the superposition theorem.

FIGURE 8–17

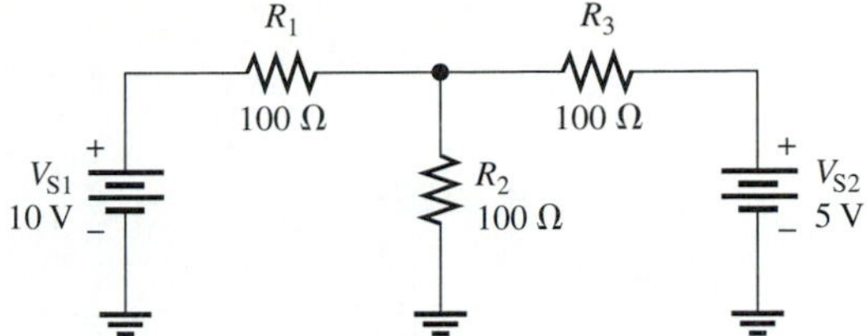


***Solution***

**Step 1:** Replace $V_{S2}$ with a short and find the current in $R_2$ due to voltage source $V_{S1}$, as shown in Figure 8–18. To find $I_2$, use the current-divider formula (Equation 6–7). Looking from $V_{S1}$,

$$R_{T(S1)} = R_1 + \frac{R_3}{2} = 100\ \Omega + 50\ \Omega = 150\ \Omega$$

$$I_{T(S1)} = \frac{V_{S1}}{R_{T(S1)}} = \frac{10\ \text{V}}{150\ \Omega} = 66.7\ \text{mA}$$

The current in $R_2$ due to $V_{S1}$ is

$$I_{2(S1)} = \left(\frac{R_3}{R_2 + R_3}\right) I_{T(S1)} = \left(\frac{100\ \Omega}{200\ \Omega}\right) 66.7\ \text{mA} = 33.3\ \text{mA}$$

Note that this current is downward through $R_2$.

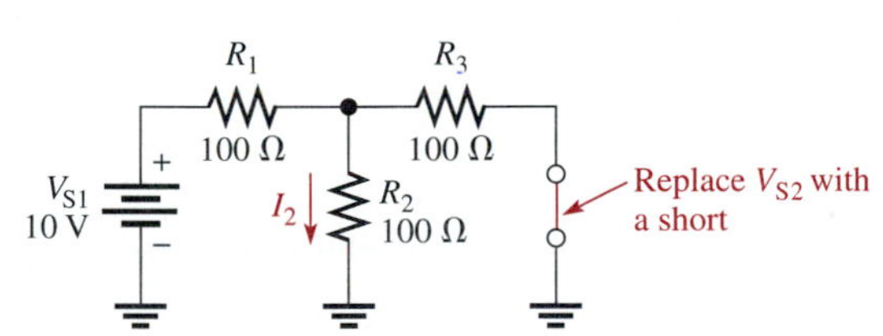


FIGURE 8–18

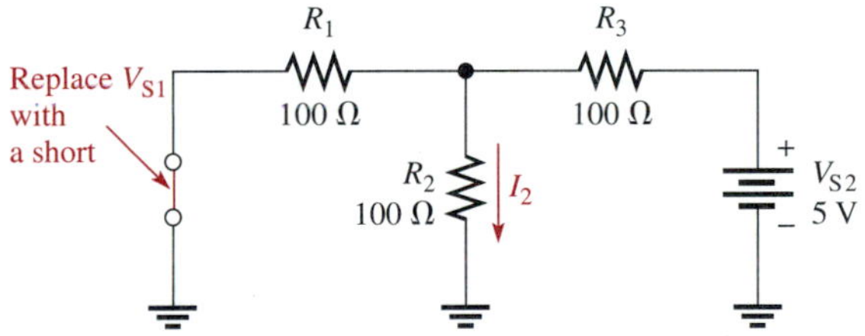

FIGURE 8–19

**Step 2:** Find the current in $R_2$ due to voltage source $V_{S2}$ by replacing $V_{S1}$ with a short, as shown in Figure 8–19. Looking from $V_{S2}$,

$$R_{T(S2)} = R_3 + \frac{R_1}{2} = 100\ \Omega + 50\ \Omega = 150\ \Omega$$

$$I_{T(S2)} = \frac{V_{S2}}{R_{T(S2)}} = \frac{5\ \text{V}}{150\ \Omega} = 33.3\ \text{mA}$$

The current in $R_2$ due to $V_{S2}$ is

$$I_{2(S2)} = \left(\frac{R_1}{R_1 + R_2}\right) I_{T(S2)} = \left(\frac{100\ \Omega}{200\ \Omega}\right) 33.3\ \text{mA} = 16.7\ \text{mA}$$

Note that this current is downward through $R_2$.

**Step 3:** Both component currents are downward through $R_2$, so they have the same algebraic sign. Therefore, add the values to get the total current through $R_2$.

$$I_{2(tot)} = I_{2(S1)} + I_{2(S2)} = 33.3\ \text{mA} + 16.7\ \text{mA} = 50\ \text{mA}$$

***Related Exercise*** Determine the total current through $R_2$ if the polarity of $V_{S2}$ in Figure 8–17 is reversed.

**EXAMPLE 8–7** Find the current through $R_2$ in the circuit of Figure 8–20.

FIGURE 8–20

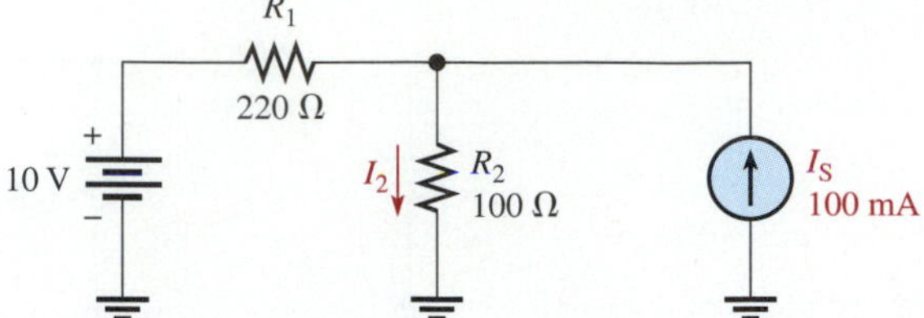


***Solution***

**Step 1:** Find the current in $R_2$ due to $V_S$ by replacing $I_S$ with an open, as shown in Figure 8–21.

FIGURE 8–21

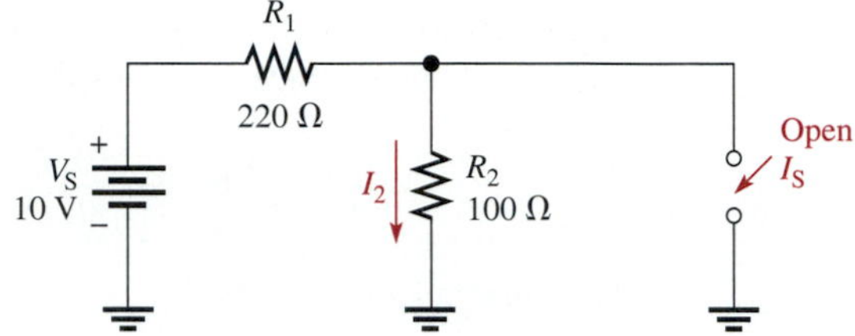


Notice that all of the current produced by $V_S$ is through $R_2$. Looking from $V_S$,

$$R_T = R_1 + R_2 = 320\ \Omega$$

The current through $R_2$ due to $V_S$ is

$$I_{2(V_S)} = \frac{V_S}{R_T} = \frac{10\text{ V}}{320\ \Omega} = 31.2\text{ mA}$$

Note that this current is downward through $R_2$.

**Step 2:** Find the current through $R_2$ due to $I_S$ by replacing $V_S$ with a short, as shown in Figure 8–22.

FIGURE 8–22

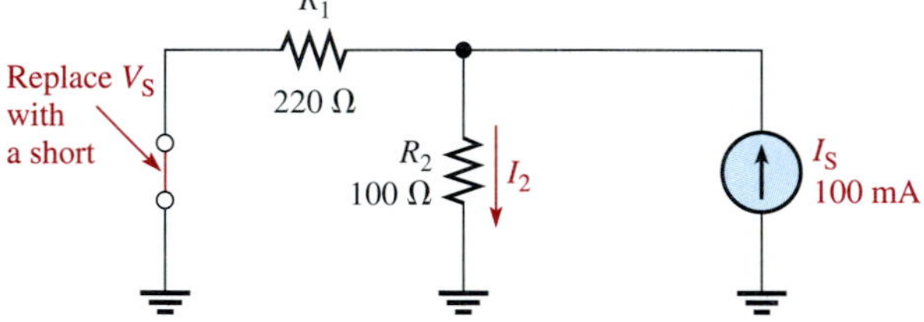


Using the current-divider formula, determine the current through $R_2$ due to $I_S$ as follows:

$$I_{2(I_S)} = \left(\frac{R_1}{R_1 + R_2}\right)I_S = \left(\frac{220\ \Omega}{320\ \Omega}\right)100\text{ mA} = 68.8\text{ mA}$$

Note that this current also is downward through $R_2$.

**Step 3:** Both currents are in the same direction through $R_2$, so add them to get the total.

$$I_{2(\text{tot})} = I_{2(V_S)} + I_{2(I_S)} = 31.2\text{ mA} + 68.8\text{ mA} = 100\text{ mA}$$

***Related Exercise*** If the polarity of $V_S$ in Figure 8–20 is reversed, how is the value of $I_S$ affected?

**EXAMPLE 8–8** Find the current through the 100 Ω resistor in Figure 8–23.

FIGURE 8–23

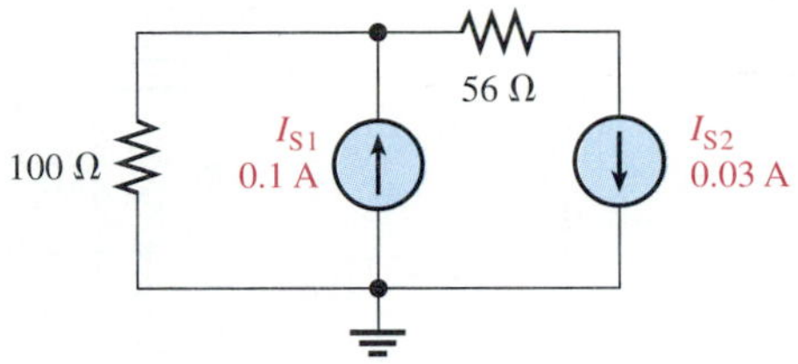

*Solution*

**Step 1:** Find the current through the 100 Ω resistor due to current source $I_{S1}$ by replacing source $I_{S2}$ with an open, as shown in Figure 8–24. As you can see, the entire 0.1 A from the current source $I_{S1}$ is downward through the 100 Ω resistor.

FIGURE 8–24

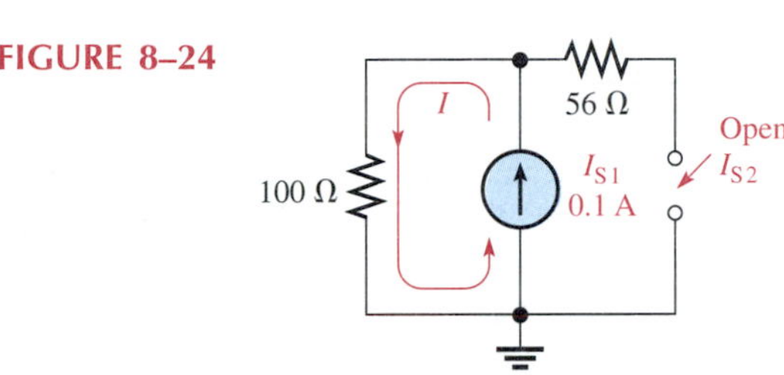


**Step 2:** Find the current through the 100 Ω resistor due to source $I_{S2}$ by replacing source $I_{S1}$ with an open, as indicated in Figure 8–25. Notice that all of the 0.03 A from source $I_{S2}$ is upward through the 100 Ω resistor.

FIGURE 8–25

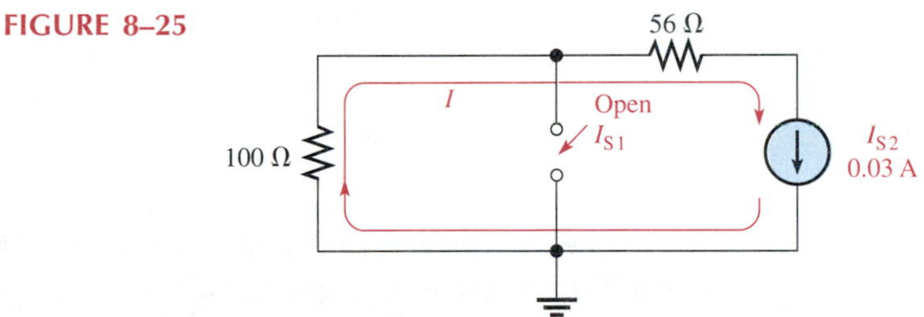


**Step 3:** To get the total current through the 100 Ω resistor, subtract the smaller current from the larger because they are in opposite directions. The resulting total current is in the direction of the larger current from source $I_{S1}$.

$$I_{100\Omega(\text{tot})} = I_{100\Omega(I_{S1})} - I_{100\Omega(I_{S2})}$$
$$= 0.1\text{ A} - 0.03\text{ A} = 0.07\text{ A}$$

The resulting current is downward through the resistor.

***Related Exercise*** If the 100 Ω resistor in Figure 8–23 is changed to 68 Ω, what will be the current through it?

**EXAMPLE 8–9** Find the total current through $R_3$ in Figure 8–26.

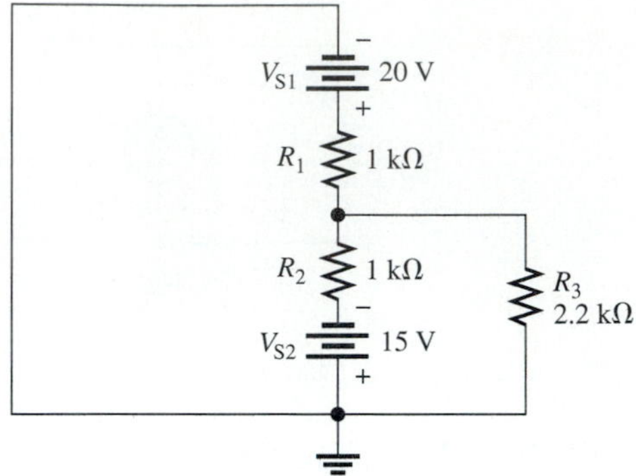

FIGURE 8–26

***Solution***

**Step 1:** Find the current through $R_3$ due to source $V_{S1}$ by replacing source $V_{S2}$ with a short, as shown in Figure 8–27.

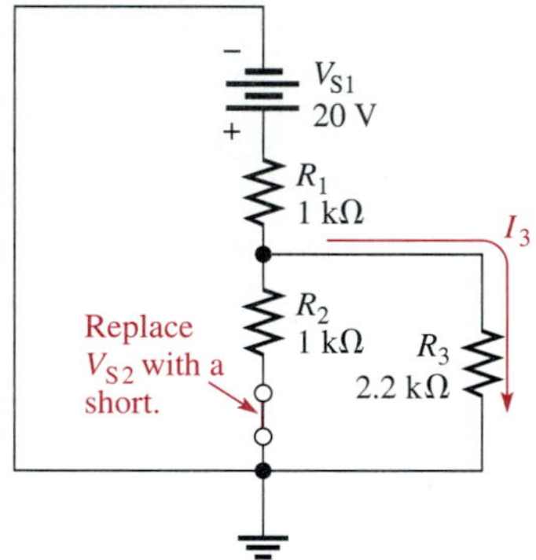

FIGURE 8–27

Looking from $V_{S1}$,

$$R_{T(S1)} = R_1 + \frac{R_2R_3}{R_2 + R_3} = 1\text{ k}\Omega + \frac{(1\text{ k}\Omega)(2.2\text{ k}\Omega)}{3.2\text{ k}\Omega} = 1.69\text{ k}\Omega$$

$$I_{T(S1)} = \frac{V_{S1}}{R_{T(S1)}} = \frac{20\text{ V}}{1.69\text{ k}\Omega} = 11.8\text{ mA}$$

Now apply the current-divider formula to get the current through $R_3$ due to source $V_{S1}$ as follows:

$$I_{3(S1)} = \left(\frac{R_2}{R_2 + R_3}\right)I_{T(S1)} = \left(\frac{1\text{ k}\Omega}{3.2\text{ k}\Omega}\right)11.8\text{ mA} = 3.69\text{ mA}$$

Notice that this current is downward through $R_3$.

**Step 2:** Find $I_3$ due to source $V_{S2}$ by replacing source $V_{S1}$ with a short, as shown in Figure 8–28. Looking from $V_{S2}$,

$$R_{T(S2)} = R_2 + \frac{R_1R_3}{R_1 + R_3} = 1\text{ k}\Omega + \frac{(1\text{ k}\Omega)(2.2\text{ k}\Omega)}{3.2\text{ k}\Omega} = 1.69\text{ k}\Omega$$

$$I_{T(S2)} = \frac{V_{S2}}{R_{T(S2)}} = \frac{15\text{ V}}{1.69\text{ k}\Omega} = 8.88\text{ mA}$$

FIGURE 8–28

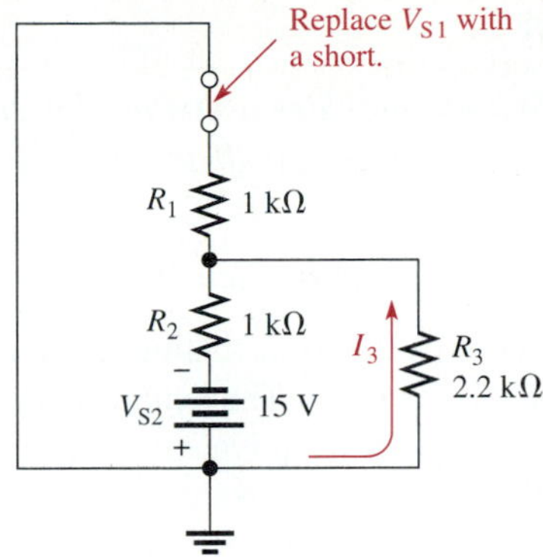


Now apply the current-divider formula to find the current through $R_3$ due to source $V_{S2}$ as follows:

$$I_{3(S2)} = \left(\frac{R_1}{R_1 + R_3}\right)I_{T(S2)} = \left(\frac{1\text{ k}\Omega}{3.2\text{ k}\Omega}\right)8.88\text{ mA} = 2.78\text{ mA}$$

Notice that this current is upward through $R_3$.

**Step 3:** Calculate the total current through $R_3$ as follows:

$$I_{3(tot)} = I_{3(S1)} - I_{3(S2)} = 3.69\text{ mA} - 2.78\text{ mA} = 0.91\text{ mA} = 910\ \mu\text{A}$$

This current is downward through $R_3$.

***Related Exercise*** Find the total current through $R_3$ in Figure 8–26 if $V_{S1}$ is changed to 12 V and its polarity reversed.

## SECTION 8–4 REVIEW

1. State the superposition theorem.
2. Why is the superposition theorem useful for analysis of multiple-source linear circuits?
3. Why is a voltage source shorted and a current source opened when the superposition theorem is applied?
4. Using the superposition theorem, find the current through $R_1$ in Figure 8–29.
5. If, as a result of applying the superposition theorem, two currents are in opposing directions through a branch of a circuit, in which direction is the net current?

FIGURE 8–29

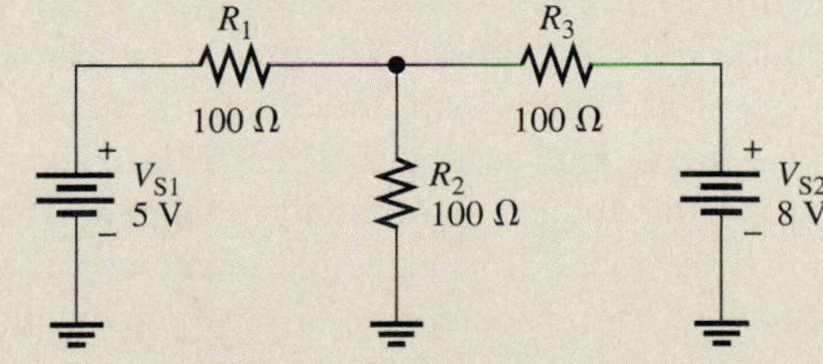

## 8–5 ■ THEVENIN'S THEOREM

***Thevenin's theorem provides a method for simplifying a circuit to a standard equivalent form. In many cases, this theorem can be used to simplify the analysis of complex circuits.***

***After completing this section, you should be able to***

- **Apply Thevenin's theorem to simplify a circuit for analysis**
  - Describe the form of a Thevenin equivalent circuit
  - Obtain the Thevenin equivalent voltage source
  - Obtain the Thevenin equivalent resistance
  - Explain terminal equivalency in the context of Thevenin's theorem
  - Thevenize a portion of a circuit
  - Thevenize a bridge circuit

The Thevenin equivalent form of any resistive circuit consists of an equivalent voltage source ($V_{TH}$) and an equivalent resistance ($R_{TH}$), arranged as shown in Figure 8–30. The values of the equivalent voltage and resistance depend on the values in the original circuit. Any resistive circuit can be simplified regardless of its complexity.

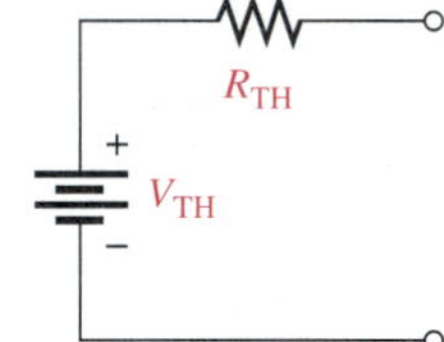


FIGURE 8–30
***The general form of a Thevenin equivalent circuit is simply a nonideal voltage source. Any resistive circuit can be reduced to this form.***

### Thevenin's Equivalent Voltage ($V_{TH}$) and Equivalent Resistance ($R_{TH}$)

The equivalent voltage, $V_{TH}$, is one part of the complete Thevenin equivalent circuit. The other part is $R_{TH}$.

> **The Thevenin equivalent voltage ($V_{TH}$) is the open circuit voltage between two points in a circuit.**

Any component connected between these two points effectively "sees" $V_{TH}$ in series with $R_{TH}$. As defined by **Thevenin's theorem,**

> **The Thevenin equivalent resistance ($R_{TH}$) is the total resistance appearing between two terminals in a given circuit with all sources replaced by their internal resistances.**

### Terminal Equivalency

Although a Thevenin equivalent circuit is not the same as its original circuit, it acts the same in terms of the output voltage and current. Try the following demonstration as illustrated in Figure 8–31. Place a resistive circuit of any complexity in a box with only the output terminals exposed. Then place the Thevenin equivalent of that circuit in an identical box with, again, only the output terminals exposed. Connect identical load resistors across the output terminals of each box. Next connect a voltmeter and an ammeter to measure the voltage and current for each load as shown in the figure. The measured values will be identical (neglecting tolerance variations), and you will not be able to determine which box contains the original circuit and which contains the Thevenin equivalent. That is, in terms of your observations, both circuits are the same. This condition is sometimes known as *terminal equivalency,* because both circuits look the same from the "viewpoint" of the two output terminals.

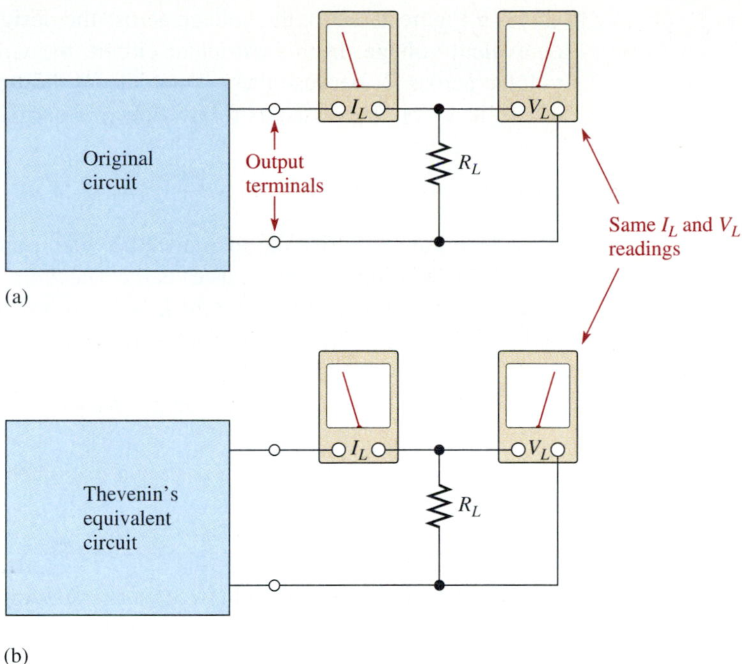


**FIGURE 8–31**
***Which box contains the original circuit and which contains the Thevenin equivalent circuit? You cannot tell by observing the meters.***

## The Thevenin Equivalent of a Circuit

To find the Thevenin equivalent of any circuit, determine the equivalent voltage, $V_{TH}$, and the equivalent resistance, $R_{TH}$. As an example, the Thevenin equivalent for the circuit between points $A$ and $B$ is developed in Figure 8–32.

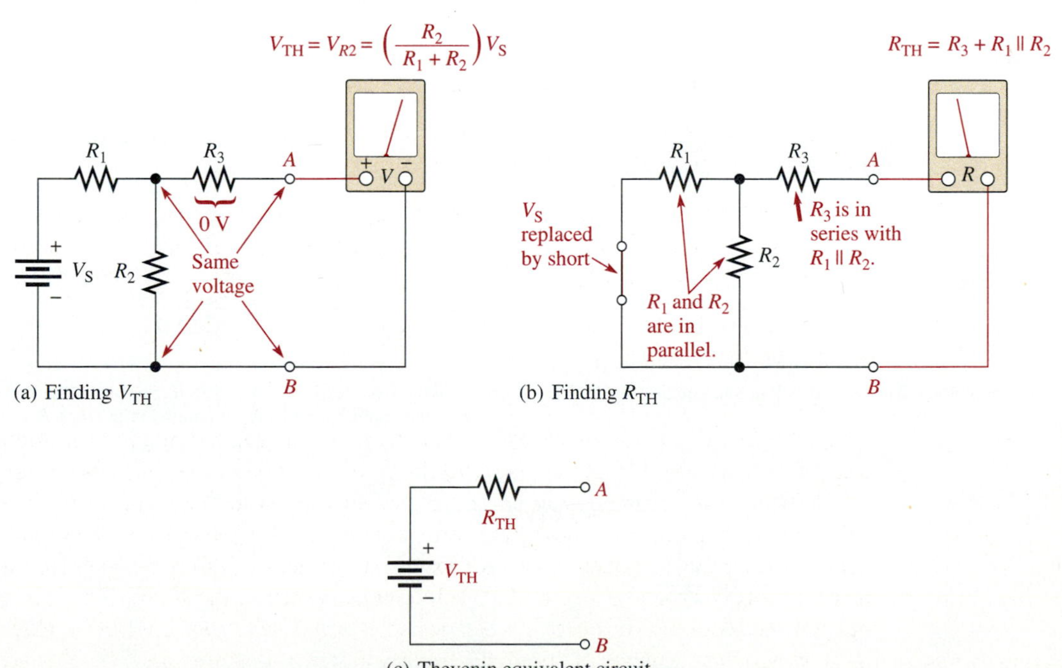


**FIGURE 8–32**
***Example of the simplification of a circuit by Thevenin's theorem.***

In Figure 8–32(a), the voltage across the designated points $A$ and $B$ is the Thevenin equivalent voltage. In this particular circuit, the voltage from $A$ to $B$ is the same as the voltage across $R_2$ because there is no current through $R_3$ and, therefore, no voltage drop across it. $V_{TH}$ is expressed as follows for this particular example:

$$V_{TH} = \left(\frac{R_2}{R_1 + R_2}\right)V_S$$

In Figure 8–32(b), the resistance between points $A$ and $B$ with the source replaced by a short (zero internal resistance) is the Thevenin equivalent resistance. In this particular circuit, the resistance from $A$ to $B$ is $R_3$ in series with the parallel combination of $R_1$ and $R_2$. Therefore, $R_{TH}$ is expressed as follows:

$$R_{TH} = R_3 + \frac{R_1R_2}{R_1 + R_2}$$

The Thevenin equivalent circuit is shown in Figure 8–32(c).

## EXAMPLE 8–10

Find the Thevenin equivalent between terminals $A$ and $B$ of the circuit in Figure 8–33.

FIGURE 8–33

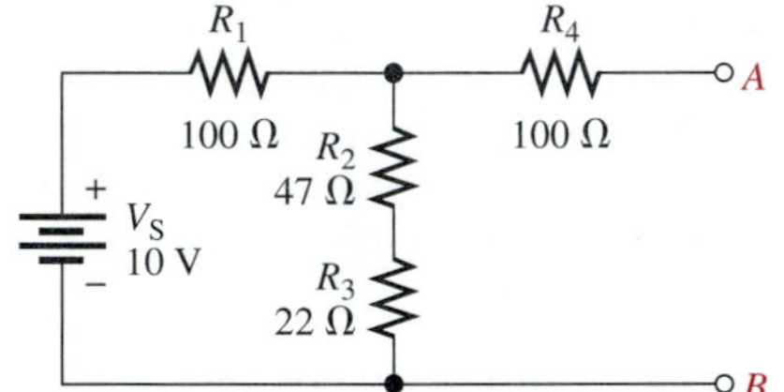


***Solution*** $V_{TH}$ equals the voltage across $R_2 + R_3$ as shown in Figure 8–34(a). Use the voltage-divider principle to find $V_{TH}$.

$$V_{TH} = \left(\frac{R_2 + R_3}{R_1 + R_2 + R_3}\right)V_S = \left(\frac{69\ \Omega}{169\ \Omega}\right)10\text{ V} = 4.08\text{ V}$$

(a) The voltage from $A$ to $B$ is $V_{TH}$ and equals $V_{2\text{-}3}$.

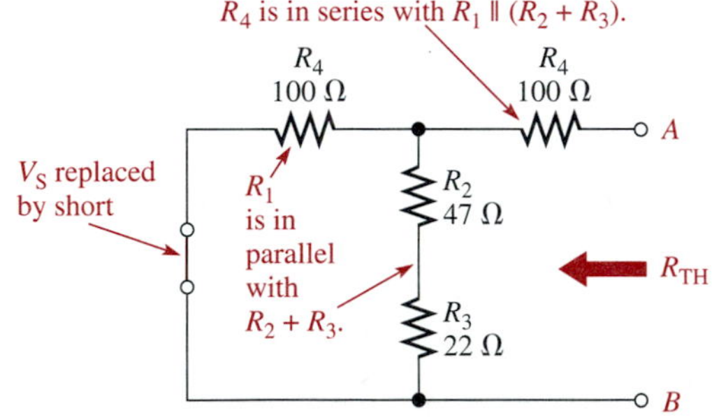


(b) Looking from terminals $A$ and $B$, $R_4$ appears in series with the combination of $R_1$ in parallel with $(R_2 + R_3)$.

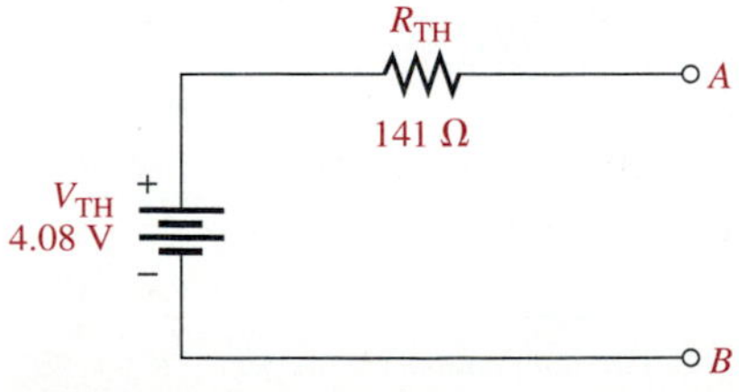


(c) Thevenin equivalent circuit

FIGURE 8–34

To find $R_{TH}$, first replace the source with a short to simulate a zero internal resistance. Then $R_1$ appears in parallel with $R_2 + R_3$, and $R_4$ is in series with the series-parallel combination of $R_1$, $R_2$, and $R_3$ as indicated in Figure 8–34(b).

$$R_{TH} = R_4 + \frac{R_1(R_2 + R_3)}{R_1 + R_2 + R_3} = 100\ \Omega + \frac{(100\ \Omega)(69\ \Omega)}{169\ \Omega} = 141\ \Omega$$

The resulting Thevenin equivalent circuit is shown in Figure 8–34(c).

***Related Exercise*** Determine $V_{TH}$ and $R_{TH}$ if a 56 Ω resistor is connected in parallel across $R_2$ and $R_3$.

## Thevenin Equivalency Depends on the Viewpoint

The Thevenin equivalent for any circuit depends on the location of the two points from between which the circuit is "viewed." In Figure 8–33, you viewed the circuit from between the two points labeled *A* and *B*. Any given circuit can have more than one Thevenin equivalent, depending on how the viewpoints are designated. For example, if you view the circuit in Figure 8–35 from between points *A* and *C*, you obtain a completely different result than if you viewed it from between points *A* and *B* or from between points *B* and *C*.

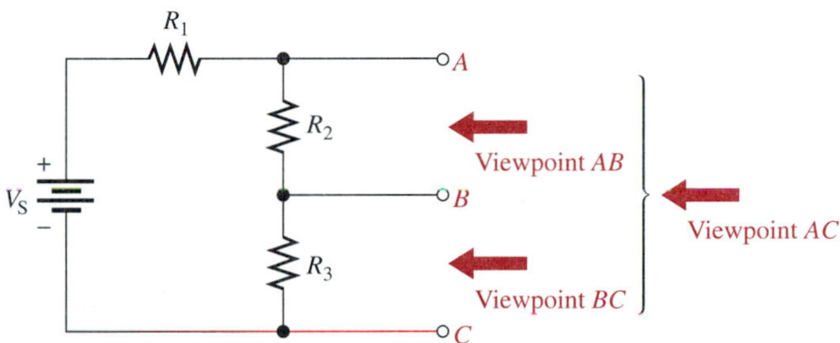


**FIGURE 8–35**
***Thevenin's equivalent depends on viewpoint.***

In Figure 8–36(a), when viewed from between points *A* and *C*, $V_{TH}$ is the voltage across $R_2 + R_3$ and can be expressed using the voltage-divider formula as

$$V_{TH} = \left(\frac{R_2 + R_3}{R_1 + R_2 + R_3}\right)V_S$$

Also, as shown in Figure 8–36(b), the resistance between points *A* and *C* is $R_2 + R_3$ in parallel with $R_1$ (the source is replaced by a short) and can be expressed as

$$R_{TH} = \frac{R_1(R_2 + R_3)}{R_1 + R_2 + R_3}$$

The resulting Thevenin equivalent circuit is shown in Figure 8–36(c).

When viewed from between points *B* and *C* as indicated in Figure 8–36(d), $V_{TH}$ is the voltage across $R_3$ and can be expressed as

$$V_{TH} = \left(\frac{R_3}{R_1 + R_2 + R_3}\right)V_S$$

As shown in Figure 8–36(e), the resistance between points *B* and *C* is $R_3$ in parallel with the series combination of $R_1$ and $R_2$.

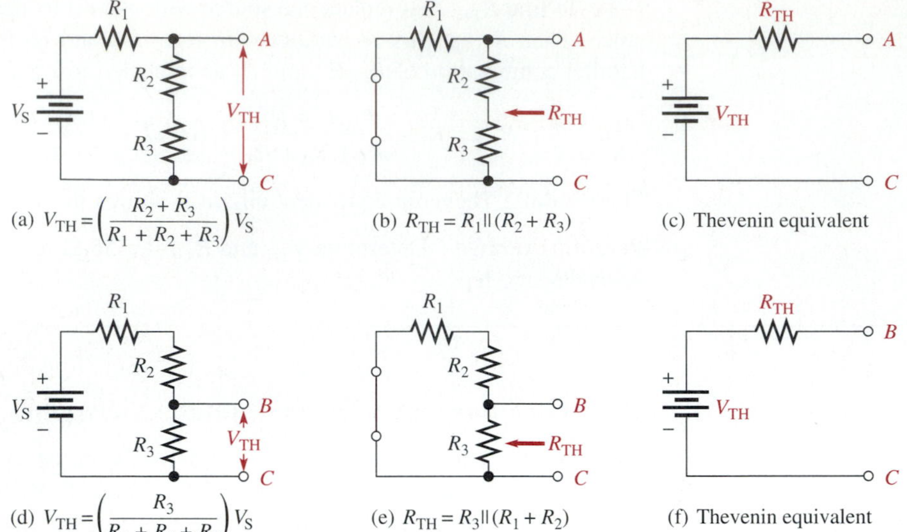


**FIGURE 8–36**

***Example of circuit thevenized from two viewpoints, resulting in two different equivalent circuits. (The $V_{TH}$ and $R_{TH}$ values are different.)***

$$R_{TH} = \frac{R_3(R_1 + R_2)}{R_1 + R_2 + R_3}$$

The resulting Thevenin equivalent is shown in Figure 8–36(f).

## Thevenizing a Portion of a Circuit

In many cases it helps to thevenize only a portion of a circuit. For example, when you need to know the equivalent circuit as viewed by one particular resistor in the circuit, you can remove the resistor and apply Thevenin's theorem to the remaining part of the circuit as viewed from the points between which that resistor was connected. Figure 8–37 illustrates the thevenizing of part of a circuit.

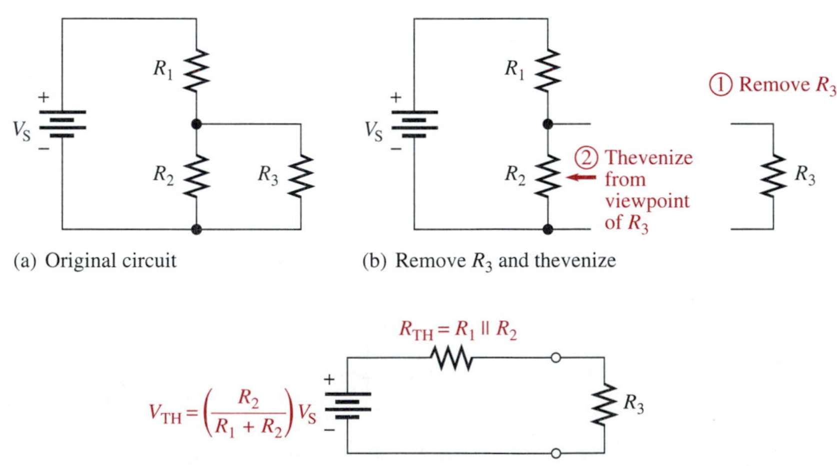


**FIGURE 8–37**

***Example of thevenizing a portion of a circuit. In this case, the circuit is thevenized from the viewpoint of $R_3$.***

Using this type of approach, you can easily find the voltage and current for a specified resistor for any number of resistor values using only Ohm's law. This method eliminates the necessity of reanalyzing the original circuit for each different resistance value.

## Thevenizing a Bridge Circuit

The usefulness of Thevenin's theorem is perhaps best illustrated when it is applied to a Wheatstone bridge circuit. For example, when a load resistor is connected to the output terminals of a Wheatstone bridge, as shown in Figure 8–38, the circuit is very difficult to analyze because it is not a straightforward series-parallel arrangement. If you doubt that this analysis is difficult, try to identify which resistors are in parallel and which are in series.

FIGURE 8–38
***Wheatstone bridge with load resistor is not a series-parallel circuit.***

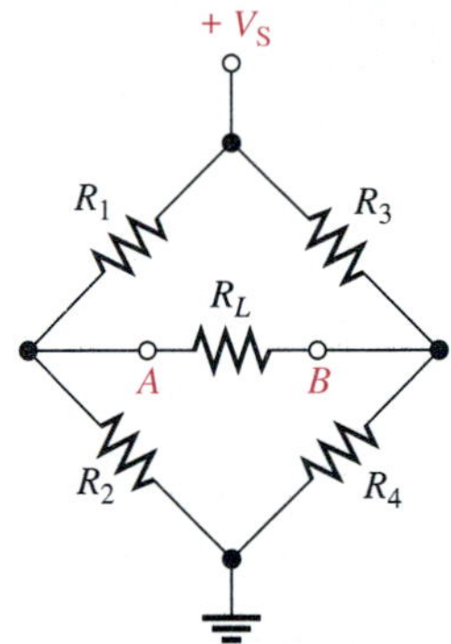


Using Thevenin's theorem, you can simplify the bridge circuit to an equivalent circuit viewed from the load resistor as shown step-by-step in Figure 8–39. Study carefully the steps in this figure. Once the equivalent circuit for the bridge is found, the voltage and current for any value of load resistor can easily be determined.

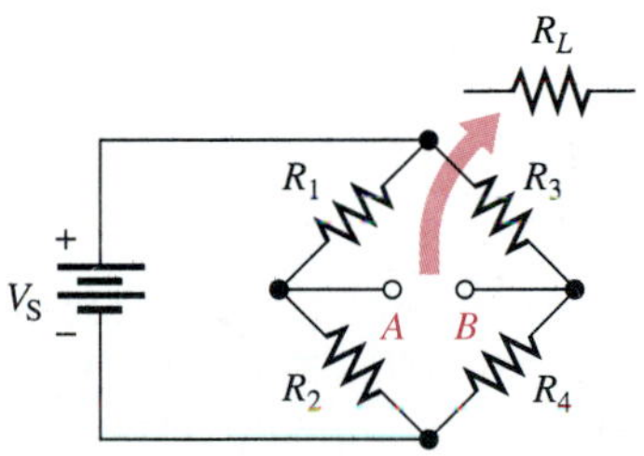


(a) Remove $R_L$.

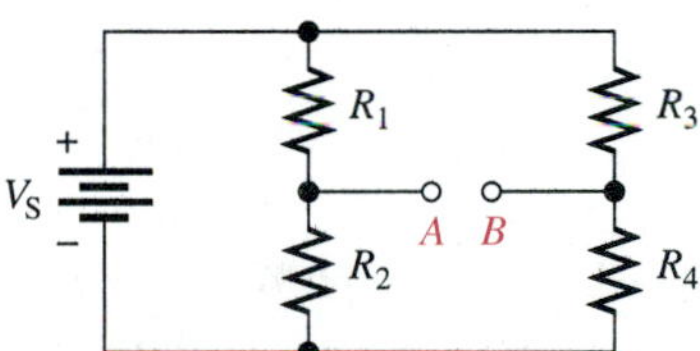


(b) Redraw to find $V_{TH}$.

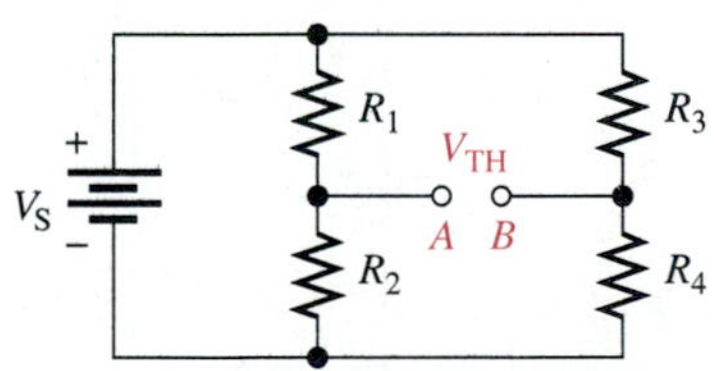


(c) $V_{TH} = V_A - V_B = \left(\frac{R_2}{R_1 + R_2}\right)V_S - \left(\frac{R_4}{R_3 + R_4}\right)V_S$

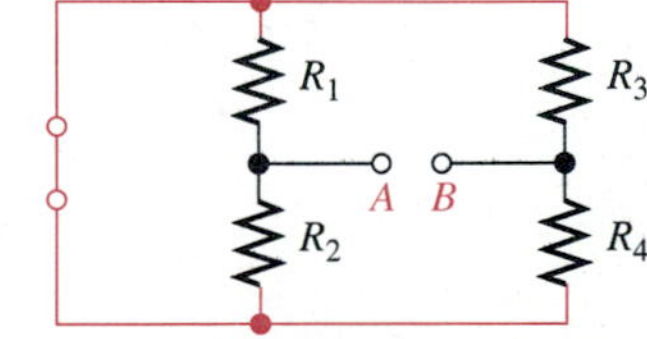


(d) Replace $V_S$ with a short. *Note*: The colored lines represent the same electrical point as the colored lines in Part (e).

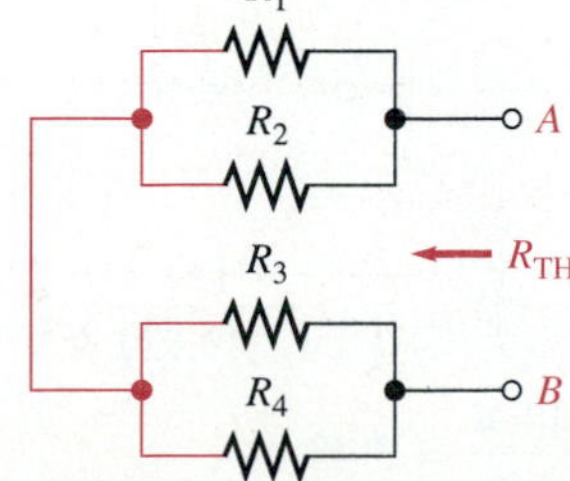


(e) Redraw to find $R_{TH}$:
$R_{TH} = R_1 \parallel R_2 + R_3 \parallel R_4$

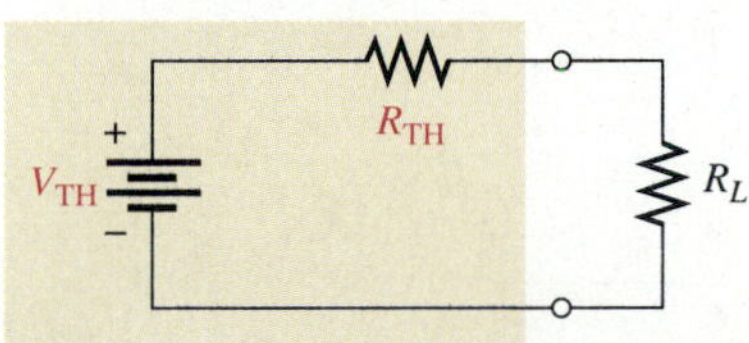


(f) Thevenin's equivalent with $R_L$ reconnected

FIGURE 8–39
***Simplifying a Wheatstone bridge with Thevenin's theorem.***

**EXAMPLE 8–11** Determine the voltage and current for the load resistor, $R_L$, in the bridge circuit of Figure 8–40.

FIGURE 8–40

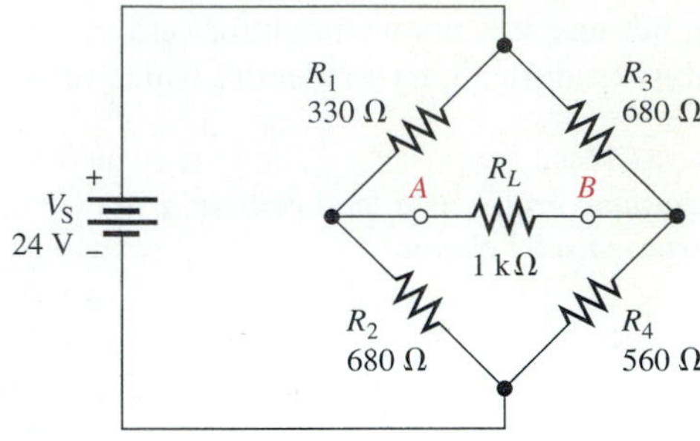


***Solution***

**Step 1:** Remove $R_L$.

**Step 2:** To thevenize the bridge as viewed from between points *A* and *B*, as was shown in Figure 8–39, first determine $V_{TH}$.

$$V_{TH} = V_A - V_B = \left(\frac{R_2}{R_1 + R_2}\right)V_S - \left(\frac{R_4}{R_3 + R_4}\right)V_S$$

$$= \left(\frac{680\ \Omega}{1010\ \Omega}\right)24\text{ V} - \left(\frac{560\ \Omega}{1240\ \Omega}\right)24\text{ V} = 16.16\text{ V} - 10.84\text{ V} = 5.32\text{ V}$$

**Step 3:** Determine $R_{TH}$.

$$R_{TH} = \frac{R_1R_2}{R_1 + R_2} + \frac{R_3R_4}{R_3 + R_4}$$

$$= \frac{(330\ \Omega)(680\ \Omega)}{1010\ \Omega} + \frac{(680\ \Omega)(560\ \Omega)}{1240\ \Omega} = 222\ \Omega + 307\ \Omega = 529\ \Omega$$

**Step 4:** Place $V_{TH}$ and $R_{TH}$ in series to form the Thevenin equivalent circuit.

**Step 5:** Connect the load resistor from points *A* and *B* of the equivalent circuit, and determine the load voltage and current as illustrated in Figure 8–41.

FIGURE 8–41

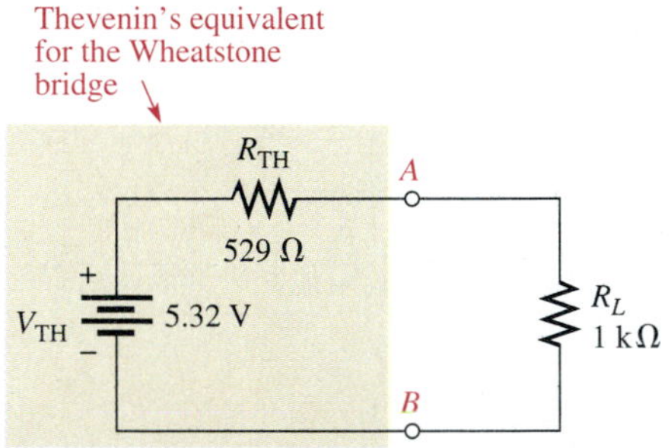

Thevenin's equivalent for the Wheatstone bridge

$R_{TH}$ 529 Ω; A; + $V_{TH}$ – 5.32 V; B; $R_L$ 1 kΩ

$$V_L = \left(\frac{R_L}{R_L + R_{TH}}\right)V_{TH} = \left(\frac{1\text{ k}\Omega}{1.529\text{ k}\Omega}\right)5.32\text{ V} = 3.48\text{ V}$$

$$I_L = \frac{V_L}{R_L} = \frac{3.48\text{ V}}{1\text{ k}\Omega} = 3.48\text{ mA}$$

***Related Exercise*** Calculate $I_L$ for $R_1 = 2.2$ kΩ, $R_2 = 3.3$ kΩ, $R_3 = 3.9$ kΩ, and $R_4 = 2.7$ kΩ.

## Summary of Thevenin's Theorem

Remember, the Thevenin equivalent circuit is *always* an equivalent voltage source in series with an equivalent resistance regardless of the original circuit that it replaces. The significance of Thevenin's theorem is that the equivalent circuit can replace the original circuit as far as any external load is concerned. Any load resistor connected between the terminals of a Thevenin equivalent circuit will have the same current through it and the same voltage across it as if it were connected to the terminals of the original circuit.

A summary of steps for applying Thevenin's theorem is as follows:

**Step 1:** Open the two terminals (remove any load) between which you want to find the Thevenin equivalent circuit.

**Step 2:** Determine the voltage ($V_{TH}$) across the two open terminals.

**Step 3:** Determine the resistance ($R_{TH}$) between the two terminals with all voltage sources shorted and all current sources opened.

**Step 4:** Connect $V_{TH}$ and $R_{TH}$ in series to produce the complete Thevenin equivalent for the original circuit.

**Step 5:** Place the load resistor removed in Step 1 across the terminals of the Thevenin equivalent circuit. The load current can now be calculated using only Ohm's law, and it has the same value as the load current in the original circuit.

## Determining $V_{TH}$ and $R_{TH}$ by Measurement

Thevenin's theorem is largely an analytical tool that is applied theoretically in order to simplify circuit analysis. However, in many cases, Thevenin's equivalent can be found for an actual circuit by the following general measurement methods. These steps are illustrated in Figure 8–42.

**Step 1:** Remove any load from the output terminals of the circuit.

**Step 2:** Measure the open terminal voltage. The voltmeter used must have an internal resistance much greater (at least 10 times greater) than the $R_{TH}$ of the circuit. ($V_{TH}$ is the open terminal voltage.)

**Step 3:** Connect a variable resistor (rheostat) across the output terminals. Its maximum value must be greater than $R_{TH}$.

**Step 4:** Adjust the rheostat and measure the terminal voltage. When the terminal voltage equals $0.5V_{TH}$, the resistance of the rheostat is equal to $R_{TH}$.

**Step 5:** Disconnect the rheostat from the terminals and measure its resistance with an ohmmeter. This measured resistance is equal to $R_{TH}$.

This procedure for determining $R_{TH}$ differs from the theoretical procedure because it is impractical to short voltage sources or open current sources in an actual circuit. Also, when measuring $R_{TH}$, be certain that the circuit is capable of providing the required current to the variable resistor load and that the variable resistor can handle the required power. These considerations may make the procedure impractical in some cases.

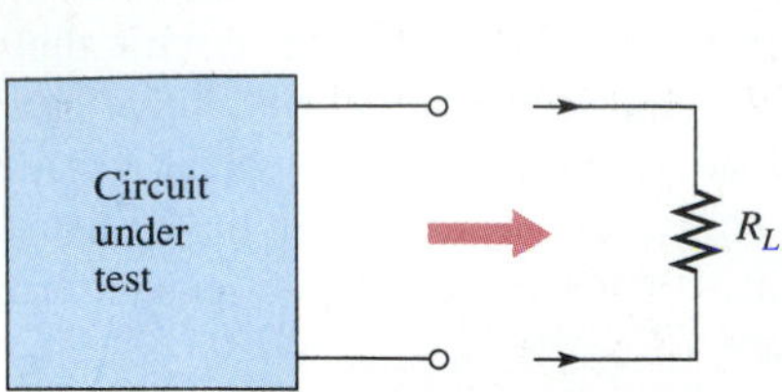


Step 1: Open the terminals (remove load).

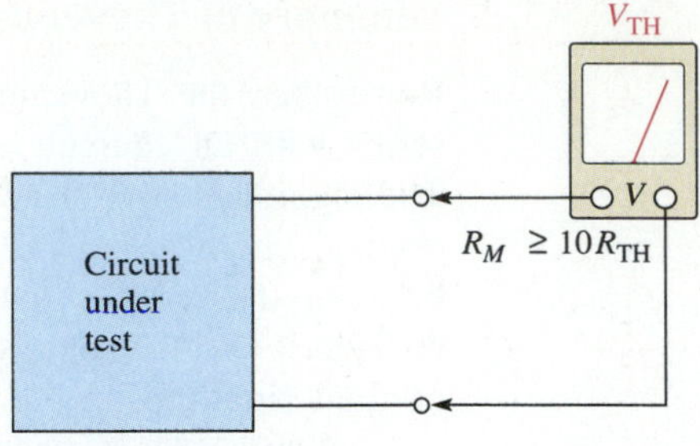


Step 2: Measure $V_{TH}$.

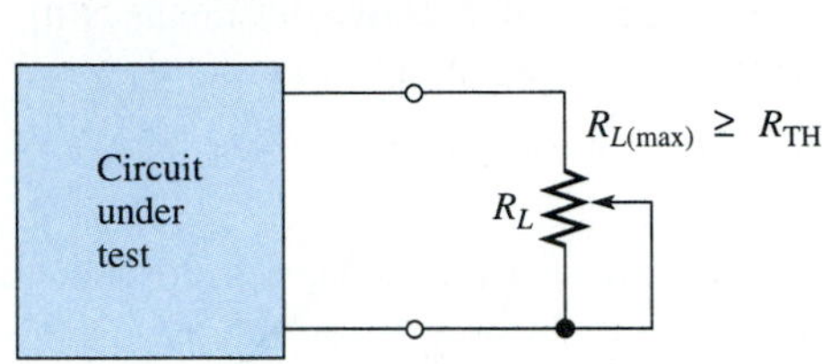


Step 3: Connect variable load resistance across the terminals.

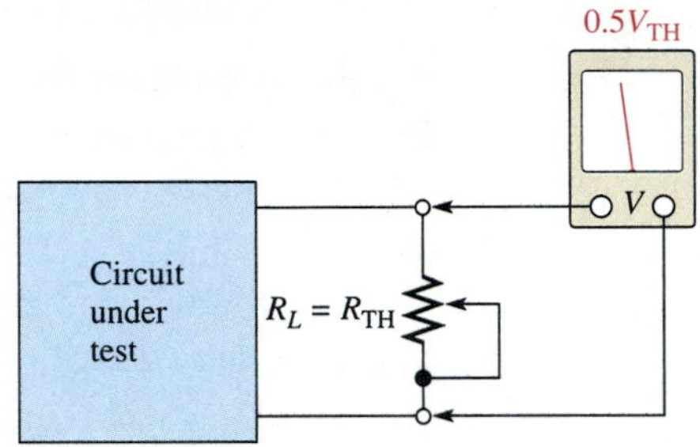


Step 4: Adjust $R_L$ until $V_L = 0.5V_{TH}$.
When $V_L = 0.5V_{TH}$, $R_L = R_{TH}$.

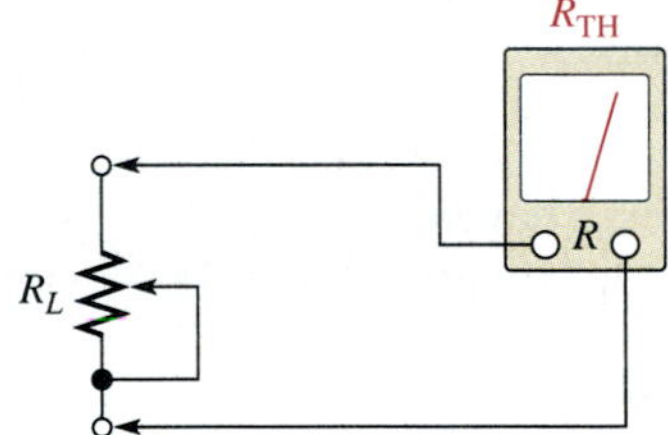


Step 5: Remove $R_L$ and measure its resistance to get $R_{TH}$.

**FIGURE 8–42**
***Determination of Thevenin's equivalent by measurement.***

## SECTION 8–5 REVIEW

1. What are the two components of a Thevenin equivalent circuit?
2. Draw the general form of a Thevenin equivalent circuit.
3. How is $V_{TH}$ defined?
4. How is $R_{TH}$ defined?
5. For the original circuit in Figure 8–43, draw the Thevenin equivalent circuit as viewed by $R_L$.

**FIGURE 8–43**

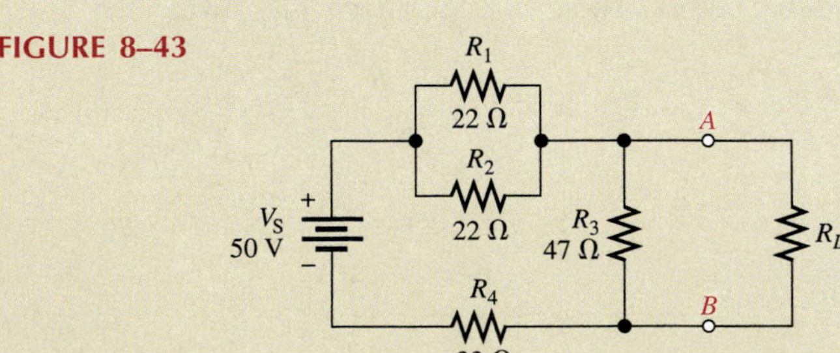

## 8–6 ■ NORTON'S THEOREM

***Like Thevenin's theorem, Norton's theorem provides a method of reducing a more complex circuit to a simpler equivalent form. The basic difference is that Norton's theorem results in an equivalent current source in parallel with an equivalent resistance.***

***After completing this section, you should be able to***

- **Apply Norton's theorem to simplify a circuit**
  - ☐ Describe the form of a Norton equivalent circuit
  - ☐ Obtain the Norton equivalent current source
  - ☐ Obtain the Norton equivalent resistance

The form of Norton's equivalent circuit is shown in Figure 8–44. Regardless of how complex the original circuit is, it can always be reduced to this equivalent form. The equivalent current source is designated $I_N$, and the equivalent resistance is designated $R_N$.

FIGURE 8–44
*Form of Norton's equivalent circuit.*

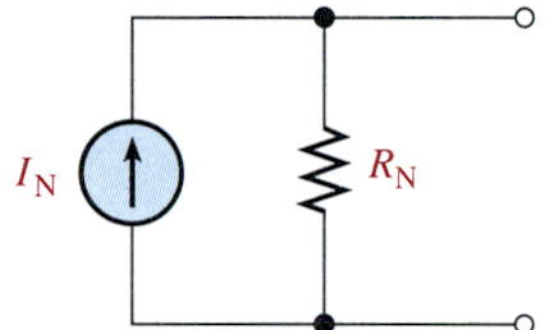


To apply **Norton's theorem,** you must know how to find the two quantities $I_N$ and $R_N$. Once you know them for a given circuit, simply connect them in parallel to get the complete Norton circuit.

### Norton's Equivalent Current ($I_N$)

As stated, $I_N$ is one part of the complete Norton equivalent circuit; $R_N$ is the other part.

> **Norton's equivalent current ($I_N$) is the short-circuit current between two points in a circuit.**

Any component connected between these two points effectively "sees" a current source of value $I_N$ in parallel with $R_N$.

To illustrate, suppose that a resistive circuit of some kind has a resistor ($R_L$) connected between two points in the circuit, as shown in Figure 8–45(a). You want to find the Norton circuit that is equivalent to the one shown as "seen" by $R_L$. To find $I_N$, calculate the current between points $A$ and $B$ with these two points shorted, as shown in Figure 8–45(b). Example 8–12 demonstrates how to find $I_N$.

FIGURE 8–45
*Determining the Norton equivalent current, $I_N$.*

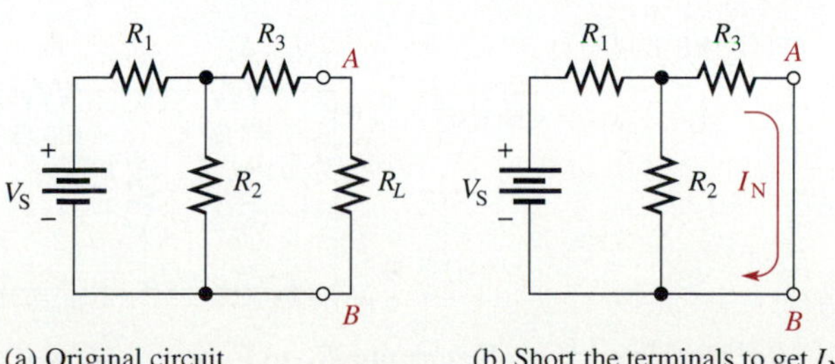


(a) Original circuit (b) Short the terminals to get $I_N$.

**EXAMPLE 8–12** Determine $I_N$ for the circuit within the beige area in Figure 8–46(a).

FIGURE 8–46

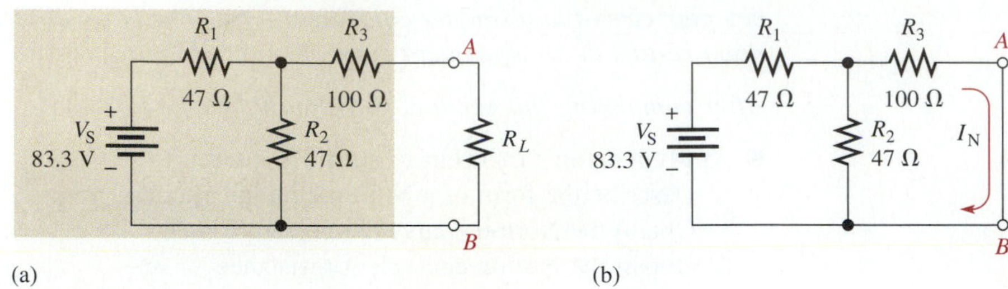


***Solution*** Short terminals $A$ and $B$ as shown in Figure 8–46(b). $I_N$ is the current through the short and is calculated as follows. First, the total resistance seen by the voltage source is

$$R_T = R_1 + \frac{R_2R_3}{R_2 + R_3} = 47\ \Omega + \frac{(47\ \Omega)(100\ \Omega)}{147\ \Omega} = 79\ \Omega$$

The total current from the source is

$$I_T = \frac{V_S}{R_T} = \frac{83.3\ \text{V}}{79\ \Omega} = 1.05\ \text{A}$$

Now apply the current-divider formula to find $I_N$ (the current through the short).

$$I_N = \left(\frac{R_2}{R_2 + R_3}\right)I_T = \left(\frac{47\ \Omega}{147\ \Omega}\right)1.05\ \text{A} = 336\ \text{mA}$$

This is the value for the equivalent Norton current source.

***Related Exercise*** Determine $I_N$ in Figure 8–46(a) if all the resistor values are doubled.

## Norton's Equivalent Resistance ($R_N$)

Norton's equivalent resistance ($R_N$) is defined in the same way as $R_{TH}$:

> **$R_N$ is the total resistance appearing between two terminals in a given circuit with all sources replaced by their internal resistances.**

Example 8–13 demonstrates how to find $R_N$.

**EXAMPLE 8–13** Find $R_N$ for the circuit within the beige area of Figure 8–46(a) (see Example 8–12).

***Solution*** First reduce $V_S$ to zero by shorting it, as shown in Figure 8–47. Looking in at terminals $A$ and $B$, you can see that the parallel combination of $R_1$ and $R_2$ is in series with $R_3$. Thus,

$$R_N = R_3 + \frac{R_1}{2} = 100\ \Omega + \frac{47\ \Omega}{2} = 124\ \Omega$$

FIGURE 8–47

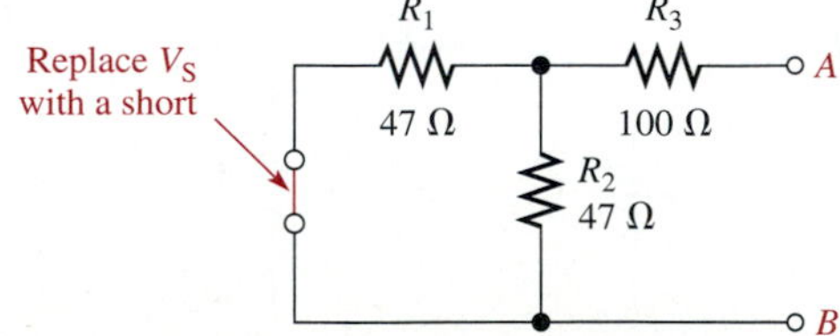


***Related Exercise*** Determine $R_N$ in Figure 8–46(a) if all the resistor values are doubled.

Examples 8–12 and 8–13 showed how to find the two equivalent components of a Norton equivalent circuit, $I_N$ and $R_N$. Keep in mind that these values can be found for any linear circuit. Once these are known, they must be connected in parallel to form the Norton equivalent circuit, as illustrated in Example 8–14.

**EXAMPLE 8–14** Draw the complete Norton equivalent circuit for the original circuit in Figure 8–46(a) (Example 8–12).

***Solution*** In Examples 8–12 and 8–13 you found that $I_N = 336$ mA and $R_N = 124\ \Omega$. The Norton equivalent circuit is shown in Figure 8–48.

FIGURE 8–48

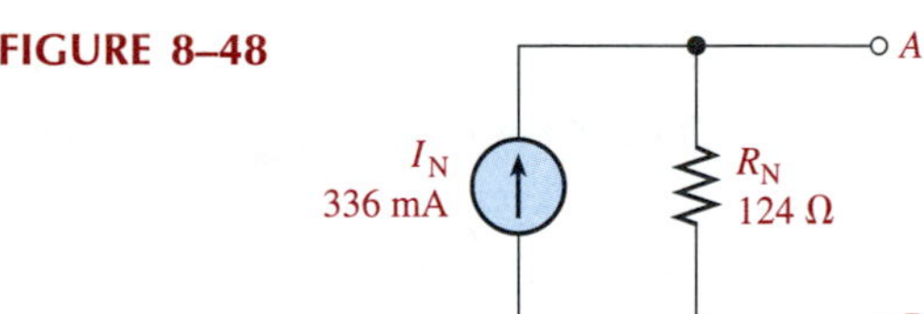


***Related Exercise*** Find $I_N$ and $R_N$ for the circuit in Figure 8–46(a) if all the resistor values are doubled.

## Summary of Norton's Theorem

Any load resistor connected between the terminals of a Norton equivalent circuit will have the same current through it and the same voltage across it as if it were connected to the terminals of the original circuit. A summary of steps for theoretically applying Norton's theorem is as follows:

**Step 1:** Short the two terminals between which you want to find the Norton equivalent circuit.

**Step 2:** Determine the current ($I_N$) through the shorted terminals.

**Step 3:** Determine the resistance ($R_N$) between the two terminals (opened) with all voltage sources shorted and all current sources opened ($R_N = R_{TH}$).

**Step 4:** Connect $I_N$ and $R_N$ in parallel to produce the complete Norton equivalent for the original circuit.

Norton's equivalent circuit can also be derived from Thevenin's equivalent circuit by use of the source conversion method discussed in Section 8–3.

**SECTION 8–6 REVIEW**

1. What are the two components of a Norton equivalent circuit?
2. Draw the general form of a Norton equivalent circuit.
3. How is $I_N$ defined?
4. How is $R_N$ defined?
5. Find the Norton circuit as seen by $R_L$ in Figure 8–49.

FIGURE 8–49

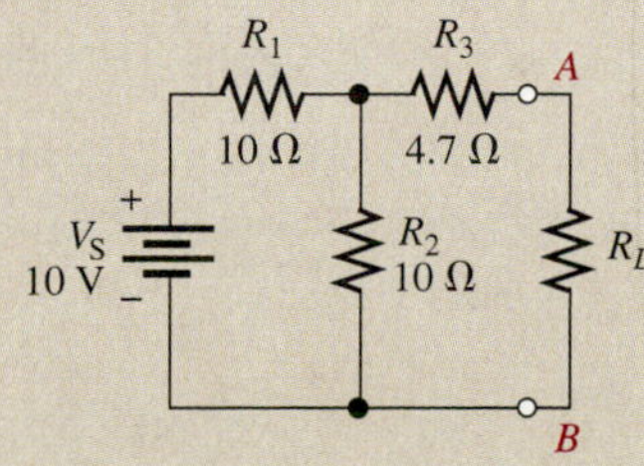

## 8–7 ■ MILLMAN'S THEOREM

***Millman's theorem provides a way to reduce any number of parallel voltage sources to a single equivalent voltage source. It simplifies finding the voltage across or current through a load. Millman's theorem gives the same results as Thevenin's theorem for the special case of parallel voltage sources.***

***After completing this section, you should be able to***

- **Apply Millman's theorem to parallel sources**
  - ☐ Determine the Millman equivalent voltage
  - ☐ Determine the Millman equivalent resistance

A conversion by **Millman's theorem** is illustrated in Figure 8–50.

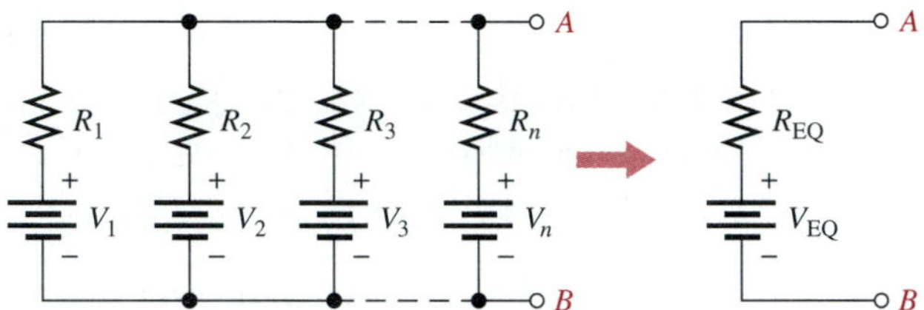


FIGURE 8–50
*Reduction of parallel voltage sources to a single equivalent voltage source.*

### Millman's Equivalent Voltage ($V_{EQ}$) and Equivalent Resistance ($R_{EQ}$)

Millman's theorem provides formulas for calculating the equivalent resistance, $R_{EQ}$, and the equivalent voltage, $V_{EQ}$, for circuits with the general form shown in Figure 8–51(a). To see how each of these formulas is derived, first convert each of the parallel voltage sources into current sources, as shown in Figure 8–51.

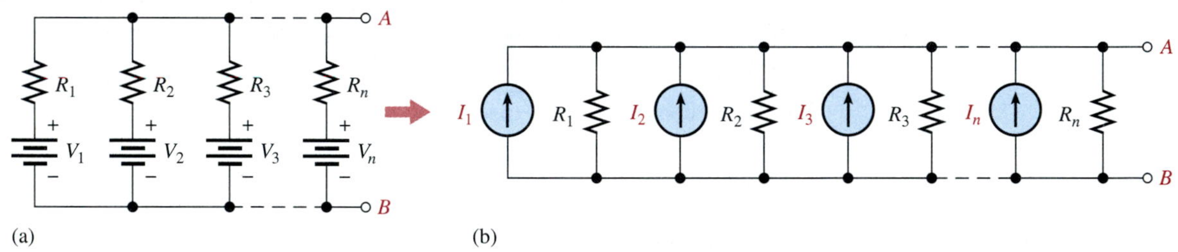


FIGURE 8–51
*Parallel voltage sources converted to current sources.*

In Figure 8–51(b), the total conductance between terminals $A$ and $B$ is

$$G_T = G_1 + G_2 + G_3 + \cdots + G_n$$

where $G_T = 1/R_T$, $G_1 = 1/R_1$, and so on. Remember, the current sources are effectively open. The Millman equivalent resistance $R_{EQ}$ is the total resistance $R_T$.

$$R_{EQ} = \frac{1}{G_T} = \frac{1}{(1/R_1) + (1/R_2) + (1/R_3) + \cdots + (1/R_n)} \quad \textbf{(8–1)}$$

The Millman equivalent voltage $V_{EQ}$ is $I_T R_{EQ}$. Since $I = V/R$, $I_T$ can be expressed as follows:

$$I_T = I_1 + I_2 + I_3 + \cdots + I_n = \frac{V_1}{R_1} + \frac{V_2}{R_2} + \frac{V_3}{R_3} + \cdots + \frac{V_n}{R_n}$$

The formulas for $R_{EQ}$ and $I_T$ are used to obtain the equivalent voltage formula.

$$V_{EQ} = \frac{(V_1/R_1) + (V_2/R_2) + (V_3/R_3) + \cdots + (V_n/R_n)}{(1/R_1) + (1/R_2) + (1/R_3) + \cdots + (1/R_n)} \quad \textbf{(8–2)}$$

Equations (8–1) and (8–2) are the two Millman formulas. The equivalent voltage source has a polarity such that the total current through a load will be in the same direction as in the original circuit.

**EXAMPLE 8–15** Use Millman's formulas to find the current through $R_L$ and the voltage across $R_L$ in Figure 8–52.

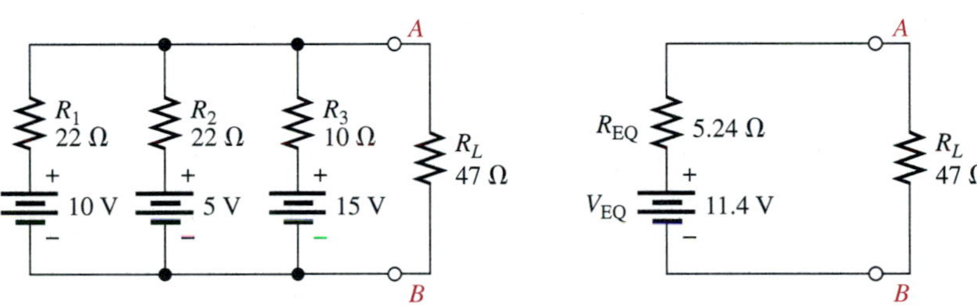


FIGURE 8–52

FIGURE 8–53

***Solution*** Apply Millman's theorem to reduce the circuit to an equivalent resistance and voltage source.

$$R_{EQ} = \frac{1}{G_T} = \frac{1}{(1/R_1) + (1/R_2) + (1/R_3)}$$

$$= \frac{1}{(1/22\ \Omega) + (1/22\ \Omega) + (1/10\ \Omega)} = \frac{1}{0.191\ \text{S}} = 5.24\ \Omega$$

$$V_{EQ} = I_T R_{EQ} = \frac{(V_1/R_1) + (V_2/R_2) + (V_3/R_3)}{(1/R_1) + (1/R_2) + (1/R_3)}$$

$$= \frac{(10\ \text{V}/22\ \Omega) + (5\ \text{V}/22\ \Omega) + (15\ \text{V}/10\ \Omega)}{(1/22\ \Omega) + (1/22\ \Omega) + (1/10\ \Omega)} = \frac{2.18\ \text{A}}{0.191\ \text{S}} = 11.4\ \text{V}$$

The single equivalent voltage source and equivalent resistance are shown in Figure 8–53.

Now calculate the current ($I_L$) through $R_L$ and the voltage ($V_L$) across $R_L$.

$$I_L = \frac{V_{EQ}}{R_{EQ} + R_L} = \frac{11.4\ \text{V}}{52.24\ \Omega} = 218\ \text{mA}$$

$$V_L = I_L R_L = (218\ \text{mA})(47\ \Omega) = 10.3\ \text{V}$$

***Related Exercise*** Find the current through $R_L$ if another branch with a 12 V source and an 18 Ω resistor is added to the circuit in Figure 8–52.

**SECTION 8–7 REVIEW**

1. To what type of circuit does Millman's theorem apply?
2. Write the Millman theorem formula for $R_{EQ}$.
3. Write the Millman theorem formula for $V_{EQ}$.
4. Find the load current ($I_L$) and the load voltage ($V_L$) in Figure 8–54.

FIGURE 8–54

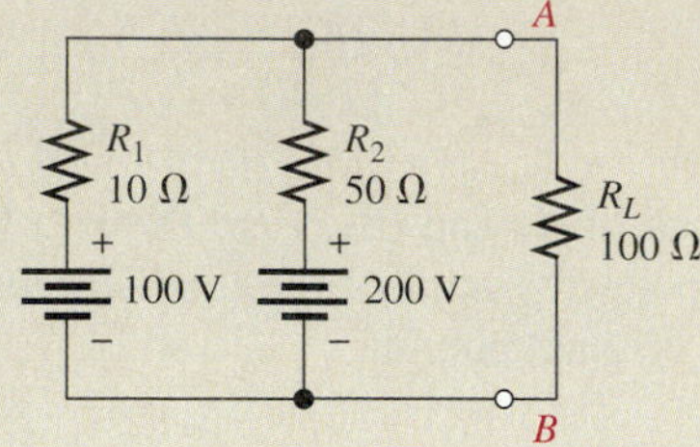


## 8–8 ■ MAXIMUM POWER TRANSFER THEOREM

***The maximum power transfer theorem is important when you need to know the value of the load at which the most power is delivered from the source.***

***After completing this section, you should be able to***

- **Apply the maximum power transfer theorem**
  - State the theorem
  - Determine the value of load resistance for which maximum power is transferred from a given circuit

The **maximum power transfer** theorem states as follows:

> **When a source is connected to a load, maximum power is delivered to the load when the load resistance is equal to the internal source resistance.**

The source resistance, $R_S$, of a circuit is the equivalent resistance as viewed from the output terminals using Thevenin's theorem. An equivalent circuit with its output resistance and load is shown in Figure 8–55. When $R_L = R_S$, the maximum power possible is transferred from the voltage source to $R_L$.

FIGURE 8–55
***Maximum power is transferred to the load when $R_L = R_S$.***

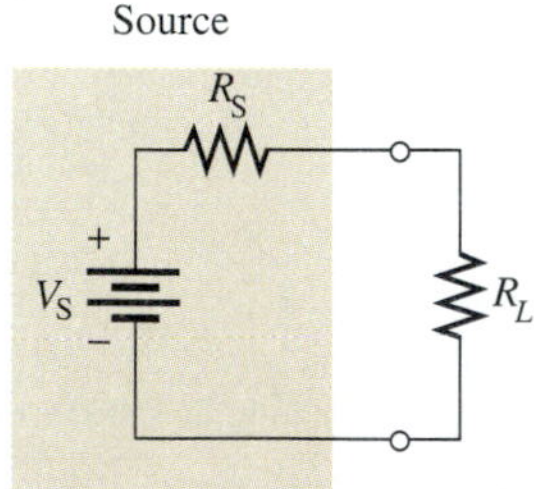


Practical applications of this theorem include audio systems such as stereo, radio, and public address. In these systems the resistance of the speaker is the load. The circuit that drives the speaker is a power amplifier. The systems are typically optimized for maximum power to the speakers. Thus, the resistance of the speaker must equal the internal source resistance of the amplifier.

Example 8–16 shows that maximum power occurs when $R_L = R_S$.

**EXAMPLE 8–16** The source in Figure 8–56 has an internal source resistance of 75 Ω. Determine the load power for each of the following values of load resistance:

**(a)** 25 Ω **(b)** 50 Ω **(c)** 75 Ω **(d)** 100 Ω **(e)** 125 Ω

Draw a graph showing the load power versus the load resistance.

**FIGURE 8–56**

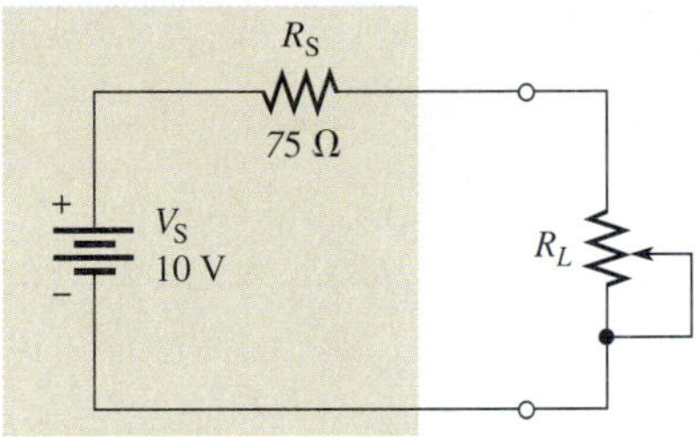


***Solution*** Use Ohm's law ($I = V/R$) and the power formula ($P = I^2R$) to find the load power, $P_L$, for each value of load resistance.

**(a)** For $R_L = 25\ \Omega$.

$$I = \frac{V_S}{R_S + R_L} = \frac{10\text{ V}}{75\ \Omega + 25\ \Omega} = 100\text{ mA}$$

$$P_L = I^2R_L = (100\text{ mA})^2(25\ \Omega) = 250\text{ mW}$$

**(b)** For $R_L = 50\ \Omega$,

$$I = \frac{V_S}{R_S + R_L} = \frac{10\text{ V}}{125\ \Omega} = 80\text{ mA}$$

$$P_L = I^2R_L = (80\text{ mA})^2(50\ \Omega) = 320\text{ mW}$$

**(c)** For $R_L = 75\ \Omega$,

$$I = \frac{V_S}{R_S + R_L} = \frac{10\text{ V}}{150\ \Omega} = 66.7\text{ mA}$$

$$P_L = I^2R_L = (66.7\text{ mA})^2(75\ \Omega) = 334\text{ mW}$$

**(d)** For $R_L = 100\ \Omega$,

$$I = \frac{V_S}{R_S + R_L} = \frac{10\text{ V}}{175\ \Omega} = 57.1\text{ mA}$$

$$P_L = I^2R_L = (57.1\text{ mA})^2(100\ \Omega) = 326\text{ mW}$$

**(e)** For $R_L = 125\ \Omega$.

$$I = \frac{V_S}{R_S + R_L} = \frac{10\text{ V}}{200\ \Omega} = 50\text{ mA}$$

$$P_L = I^2R_L = (50\text{ mA})^2(125\ \Omega) = 313\text{ mW}$$

Notice that the load power is greatest when $R_L = 75\ \Omega$, which is the same as the internal source resistance. When the load resistance is less than or greater than this value, the power drops off, as the curve in Figure 8–57 graphically illustrates.

The calculator sequence for $R_L = 25\ \Omega$ is

[1] [0] [÷] [(] [7] [5] [+] [2] [5] [)] [=] [$x^2$] [×] [2] [5] [=]

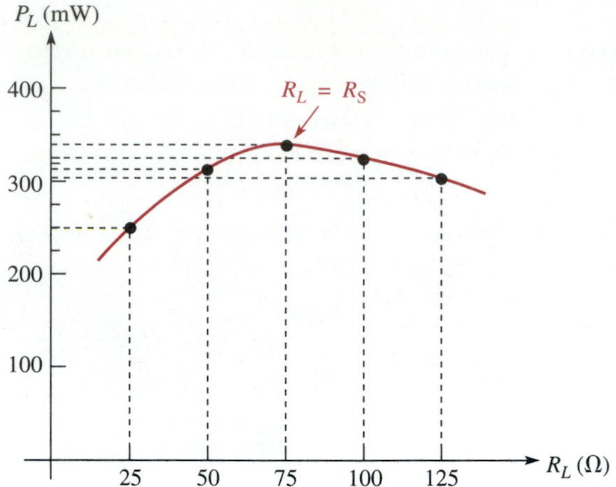


FIGURE 8–57
*Curve showing that the load power is maximum when* $R_L = R_S$.

***Related Exercise*** If the source resistance in Figure 8–56 is 600 Ω, what is the maximum power than can be delivered to a load?

**SECTION 8–8 REVIEW**

1. State the maximum power transfer theorem.
2. When is maximum power delivered from a source to a load?
3. A given circuit has an internal source resistance of 50 Ω. What will be the value of the load to which the maximum power is delivered?

## 8–9 ■ DELTA-TO-WYE (Δ-TO-Y) AND WYE-TO-DELTA (Y-TO-Δ) CONVERSIONS

***Conversions between delta-type and wye-type circuit arrangements are useful in certain specialized applications. One example is in the analysis of a loaded Wheatstone bridge circuit. In this section, the conversion formulas and rules for remembering them are given.***

***After completing this section, you should be able to***

- **Perform Δ-to-Y and Y-to-Δ conversions**
  - ☐ Apply Δ-to-Y conversion to a bridge circuit

A resistive delta (Δ) circuit has the form shown in Figure 8–58(a). A wye (Y) circuit is shown in Figure 8–58(b). Notice that letter subscripts are used to designate resistors in

FIGURE 8–58
*Delta and wye circuits.*

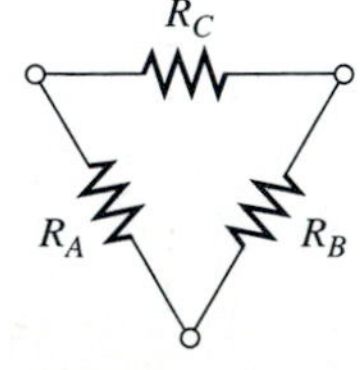


(a) Delta

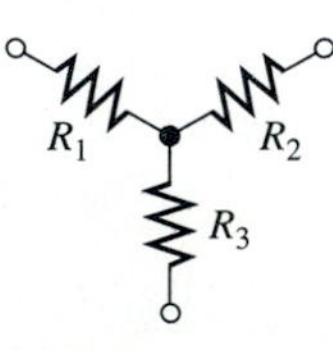


(b) Wye

the delta circuit and that numerical subscripts are used to designate resistors in the wye circuit.

## Δ-to-Y Conversion

It is convenient to think of the wye positioned within the delta, as shown in Figure 8–59. To convert from delta to wye, you need $R_1$, $R_2$, and $R_3$ in terms of $R_A$, $R_B$, and $R_C$. The conversion rule is as follows:

**Each resistor in the wye is equal to the product of the resistors in two adjacent delta branches, divided by the sum of all three delta resistors.**

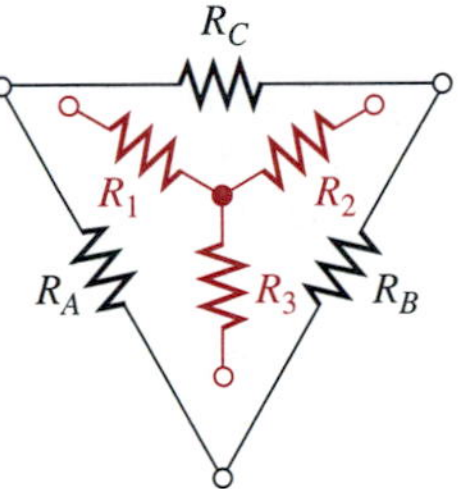


**FIGURE 8–59**
*"Y within Δ" aid for conversion formulas.*

In Figure 8–59, $R_A$ and $R_C$ are adjacent to $R_1$; therefore,

$$R_1 = \frac{R_A R_C}{R_A + R_B + R_C} \quad \textbf{(8–3)}$$

Also, $R_B$ and $R_C$ are adjacent to $R_2$, so

$$R_2 = \frac{R_B R_C}{R_A + R_B + R_C} \quad \textbf{(8–4)}$$

and $R_A$ and $R_B$ are adjacent to $R_3$, so

$$R_3 = \frac{R_A R_B}{R_A + R_B + R_C} \quad \textbf{(8–5)}$$

## Y-to-Δ Conversion

To convert from wye to delta, you need $R_A$, $R_B$, and $R_C$ in terms of $R_1$, $R_2$, and $R_3$. The conversion rule is as follows:

**Each resistor in the delta is equal to the sum of all possible products of wye resistors taken two at a time, divided by the opposite wye resistor.**

In Figure 8–59, $R_2$ is opposite to $R_A$; therefore,

$$R_A = \frac{R_1R_2 + R_1R_3 + R_2R_3}{R_2} \quad \textbf{(8–6)}$$

Also, $R_1$ is opposite to $R_B$, so

$$R_B = \frac{R_1R_2 + R_1R_3 + R_2R_3}{R_1} \quad \textbf{(8–7)}$$

and $R_3$ is opposite to $R_C$, so

$$R_C = \frac{R_1R_2 + R_1R_3 + R_2R_3}{R_3} \qquad \textbf{(8-8)}$$

The following two examples illustrate conversion between these two forms of circuits.

**EXAMPLE 8–17** Convert the delta circuit in Figure 8–60 to a wye circuit.

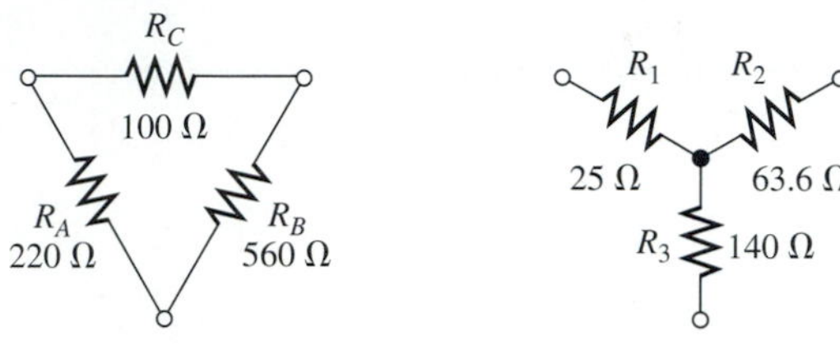


FIGURE 8–60

FIGURE 8–61

***Solution*** Use Equations (8–3), (8–4), and (8–5).

$$R_1 = \frac{R_AR_C}{R_A + R_B + R_C} = \frac{(220\ \Omega)(100\ \Omega)}{220\ \Omega + 560\ \Omega + 100\ \Omega} = 25\ \Omega$$

$$R_2 = \frac{R_BR_C}{R_A + R_B + R_C} = \frac{(560\ \Omega)(100\ \Omega)}{880\ \Omega} = 63.6\ \Omega$$

$$R_3 = \frac{R_AR_B}{R_A + R_B + R_C} = \frac{(220\ \Omega)(560\ \Omega)}{880\ \Omega} = 140\ \Omega$$

The resulting wye circuit is shown in Figure 8–61.

***Related Exercise*** Convert the delta circuit to a wye circuit for $R_A = 2.2\ \text{k}\Omega$, $R_B = 1\ \text{k}\Omega$, and $R_C = 1.8\ \text{k}\Omega$.

**EXAMPLE 8–18** Convert the wye circuit in Figure 8–62 to a delta circuit.

FIGURE 8–62

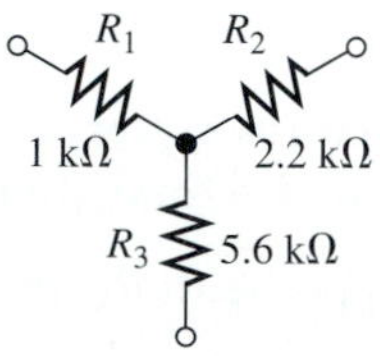

***Solution*** Use Equations (8–6), (8–7), and (8–8).

$$R_A = \frac{R_1R_2 + R_1R_3 + R_2R_3}{R_2}$$

$$= \frac{(1\ \text{k}\Omega)(2.2\ \text{k}\Omega) + (1\ \text{k}\Omega)(5.6\ \text{k}\Omega) + (2.2\ \text{k}\Omega)(5.6\ \text{k}\Omega)}{2.2\ \text{k}\Omega} = 9.15\ \text{k}\Omega$$

$$R_B = \frac{R_1R_2 + R_1R_3 + R_2R_3}{R_1}$$

$$= \frac{(1\text{ k}\Omega)(2.2\text{ k}\Omega) + (1\text{ k}\Omega)(5.6\text{ k}\Omega) + (2.2\text{ k}\Omega)(5.6\text{ k}\Omega)}{1\text{ k}\Omega} = 20.1\text{ k}\Omega$$

$$R_C = \frac{R_1R_2 + R_1R_3 + R_2R_3}{R_3}$$

$$= \frac{(1\text{ k}\Omega)(2.2\text{ k}\Omega) + (1\text{ k}\Omega)(5.6\text{ k}\Omega) + (2.2\text{ k}\Omega)(5.6\text{ k}\Omega)}{5.6\text{ k}\Omega} = 3.59\text{ k}\Omega$$

The resulting delta circuit is shown in Figure 8–63.

FIGURE 8–63

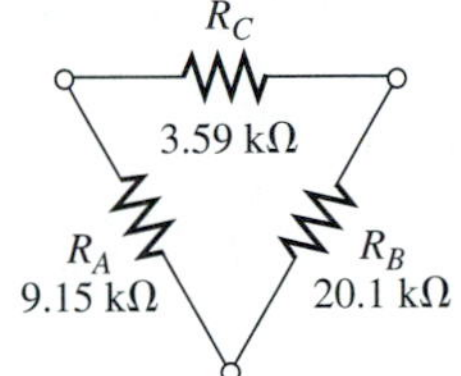


***Related Exercise*** Convert the wye circuit to a delta circuit for $R_1$ = 100 Ω, $R_2$ = 330 Ω, and $R_3$ = 470 Ω.

## Application of Δ-to-Y Conversion to the Simplification of a Bridge Circuit

In Section 8–5 you learned how Thevenin's theorem can be used to simplify a bridge circuit. Now you will see how Δ-to-Y conversion can be used for converting a bridge circuit to a series-parallel form for easier analysis.

Figure 8–64 illustrates how the delta (Δ) formed by $R_A$, $R_B$, and $R_C$ can be converted to a wye (Y), thus creating an equivalent series-parallel circuit. Equations (8–3), (8–4), and (8–5) are used in this conversion.

In a bridge circuit, the load is connected across points *C* and *D*. In Figure 8–64(a), $R_C$ represents the load resistor. When voltage is applied across points *A* and *B*, the

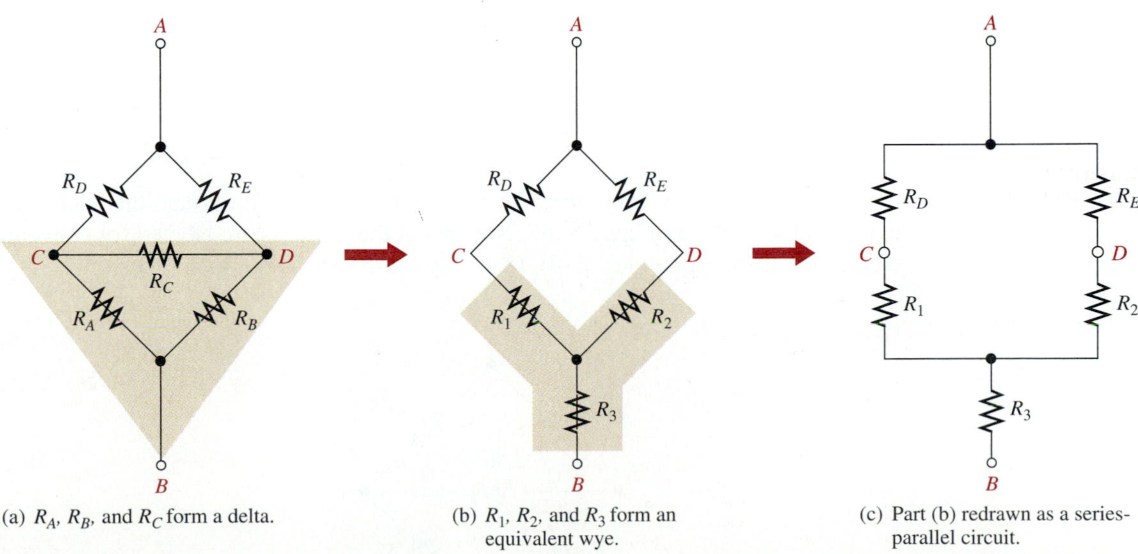


(a) $R_A$, $R_B$, and $R_C$ form a delta.

(b) $R_1$, $R_2$, and $R_3$ form an equivalent wye.

(c) Part (b) redrawn as a series-parallel circuit.

**FIGURE 8–64**
***Conversion of a bridge circuit to a series-parallel configuration.***

voltage from $C$ to $D$ ($V_{CD}$) can be determined using the equivalent series-parallel circuit in Figure 8–64(c) as follows. The total resistance from point $A$ to point $B$ is

$$R_T = \frac{(R_1 + R_D)(R_2 + R_E)}{(R_1 + R_D) + (R_2 + R_E)} + R_3$$

Then,

$$I_T = \frac{V_{AB}}{R_T}$$

The resistance of the parallel portion of the circuit in Figure 8–64(c) is

$$R_{T(p)} = \frac{(R_1 + R_D)(R_2 + R_E)}{(R_1 + R_D) + (R_2 + R_E)}$$

The current through the left branch is

$$I_{AC} = \left(\frac{R_{T(p)}}{R_1 + R_D}\right)I_T$$

The current through the right branch is

$$I_{AD} = \left(\frac{R_{T(p)}}{R_2 + R_E}\right)I_T$$

The voltage at point $C$ with respect to point $A$ is

$$V_{CA} = V_A - I_{AC}R_D$$

The voltage at point $D$ with respect to point $A$ is

$$V_{DA} = V_A - I_{AD}R_E$$

The voltage from point $C$ to point $D$ is

$$V_{CD} = V_{CA} - V_{DA}$$
$$= (V_A - I_{AC}R_D) - (V_A - I_{AD}R_E) = I_{AD}R_E - I_{AC}R_D$$

$V_{CD}$ is the voltage across the load ($R_C$) in the bridge circuit of Figure 8–64(a).

The load current through $R_C$ can be found by Ohm's law.

$$I_{R_C} = \frac{V_{CD}}{R_C}$$

**EXAMPLE 8–19** Determine the voltage across the load resistor and the current through the load resistor in the bridge circuit in Figure 8–65. Notice that the resistors are labeled for convenient conversion using Equations (8–3), (8–4), and (8–5). $R_C$ is the load resistor.

FIGURE 8–65

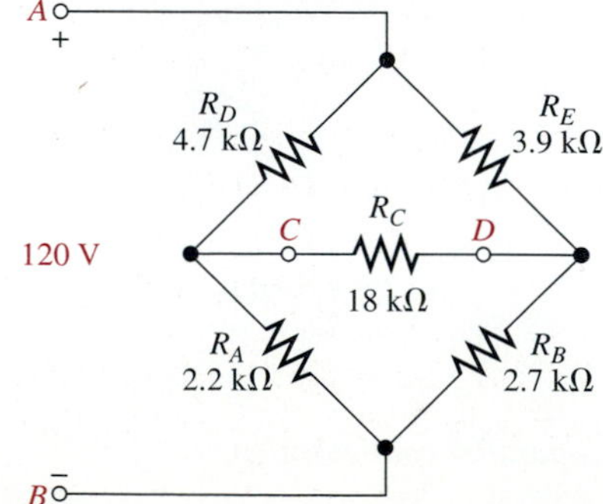

***Solution*** First, convert the delta formed by $R_A$, $R_B$, and $R_C$ to a wye.

$$R_1 = \frac{R_A R_C}{R_A + R_B + R_C} = \frac{(2.2\text{ k}\Omega)(18\text{ k}\Omega)}{2.2\text{ k}\Omega + 2.7\text{ k}\Omega + 18\text{ k}\Omega} = 1.73\text{ k}\Omega$$

$$R_2 = \frac{R_B R_C}{R_A + R_B + R_C} = \frac{(2.7\text{ k}\Omega)(18\text{ k}\Omega)}{22.9\text{ k}\Omega} = 2.12\text{ k}\Omega$$

$$R_3 = \frac{R_A R_B}{R_A + R_B + R_C} = \frac{(2.2\text{ k}\Omega)(2.7\text{ k}\Omega)}{22.9\text{ k}\Omega} = 259\ \Omega$$

The resulting equivalent series-parallel circuit is shown in Figure 8–66.

**FIGURE 8–66**

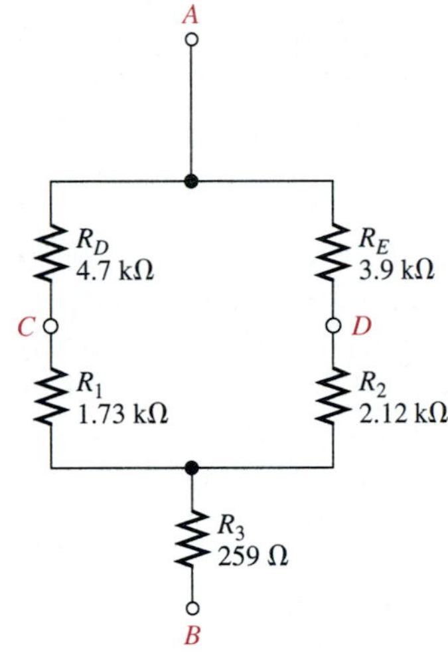


Now, determine $R_T$ and the branch currents in Figure 8–66.

$$R_T = \frac{(R_1 + R_D)(R_2 + R_E)}{(R_1 + R_D) + (R_2 + R_E)} + R_3$$

$$= \frac{(6.43\text{ k}\Omega)(6.02\text{ k}\Omega)}{6.43\text{ k}\Omega + 6.02\text{ k}\Omega} + 259\ \Omega = 3.11\text{ k}\Omega + 259\ \Omega = 3.37\text{ k}\Omega$$

$$I_T = \frac{V_{AB}}{R_T} = \frac{120\text{ V}}{3.37\text{ k}\Omega} = 35.6\text{ mA}$$

The total resistance of the parallel part of the circuit, $R_{T(p)}$, is 3.11 kΩ.

$$I_{AC} = \left(\frac{R_{T(p)}}{R_1 + R_D}\right) I_T = \left(\frac{3.11\text{ k}\Omega}{1.73\text{ k}\Omega + 4.7\text{ k}\Omega}\right) 35.6\text{ mA} = 17.2\text{ mA}$$

$$I_{AD} = \left(\frac{R_{T(p)}}{R_2 + R_E}\right) I_T = \left(\frac{3.11\text{ k}\Omega}{2.12\text{ k}\Omega + 3.9\text{ k}\Omega}\right) 35.6\text{ mA} = 18.4\text{ mA}$$

The voltage from point $C$ to point $D$ is

$$V_{CD} = I_{AD} R_E - I_{AC} R_D = (18.4\text{ mA})(3.9\text{ k}\Omega) - (17.2\text{ mA})(4.7\text{ k}\Omega)$$
$$= 71.8\text{ V} - 80.8\text{ V} = -9\text{ V}$$

$V_{CD}$ is the voltage across the load ($R_C$) in the bridge circuit shown in Figure 8–65.

The load current through $R_C$ is

$$I_{R_C} = \frac{V_{CD}}{R_C} = \frac{-9\text{ V}}{18\text{ k}\Omega} = -500\ \mu\text{A}$$

***Related Exercise*** Determine the load current, $I_{R_C}$, in Figure 8–65 for the following resistor values: $R_A = 27$ kΩ, $R_B = 33$ kΩ, $R_D = 39$ kΩ, $R_E = 47$ kΩ, and $R_C = 100$ kΩ.

**SECTION 8–9 REVIEW**

1. Sketch a delta circuit.
2. Sketch a wye circuit.
3. Write the formulas for delta-to-wye conversion.
4. Write the formulas for wye-to-delta conversion.

## 8–10 ■ TECHnology Theory Into Practice

***The Wheatstone bridge circuit was covered in Chapter 7. In this section, you will work with a bridge that is to be used in a temperature-measuring circuit in which a device called a thermistor is the temperature sensor. You will be doing a preliminary analysis of the circuit in which Thevenin's theorem can be used to advantage.***

The Wheatstone bridge circuit will be used in a temperature-sensing application where the temperature of a liquid that is used in a certain industrial process is monitored using a thermistor in one leg of the bridge. The **thermistor** used in this application is a temperature-sensing resistor with a negative temperature coefficient in which the resistance decreases as temperature increases.

When a certain preset temperature is reached, the bridge becomes balanced and its output voltage is zero. This zero-voltage condition is detected by a high-gain amplifier circuit that operates a relay to turn off the heating element. As the temperature decreases below the preset value, the bridge again becomes unbalanced causing the amplifier to close the relay and turn the heating element back on. This process maintains the temperature of the liquid in a tank within defined limits.

The amplifier effectively has an internal resistance of 10 kΩ between its input terminals. You do not need to know any additional details of the amplifier circuit in this assignment because you are to concentrate only on the bridge circuit. The study of amplifiers will come in a later course.

The Wheatstone bridge control and temperature-measuring circuit is shown in Figure 8–67. The amplifier and relay circuitry is to be connected across the bridge between

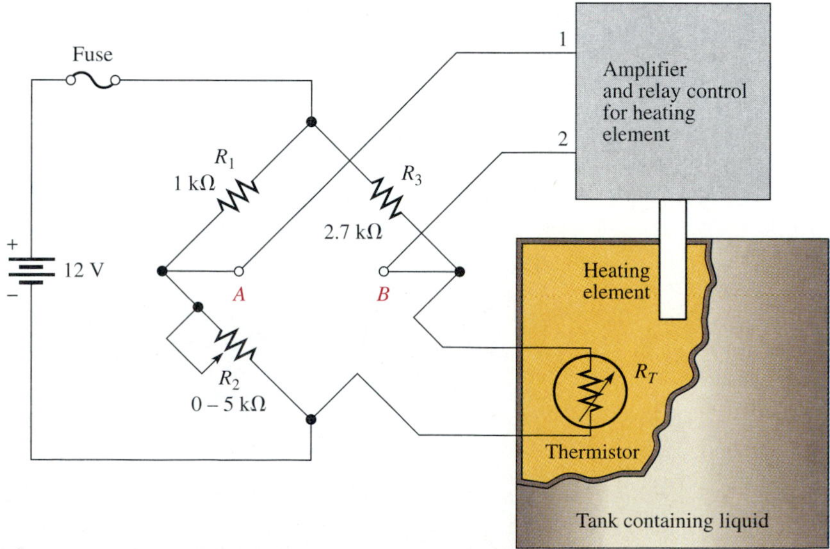


**FIGURE 8–67**
*Wheatstone bridge control and temperature-measuring circuit.*

points $A$ and $B$ as indicated, so there will be a 10 kΩ load between these points. The thermistor is connected in one leg of the bridge but is remotely located in the tank and away from the rest of the circuit. The variable resistor, $R_2$, is used to set the desired temperature at which the liquid in the tank will be maintained. The temperature characteristics of the thermistor, shown in the graph of Figure 8–68, indicates how the resistance of the thermistor changes with temperature.

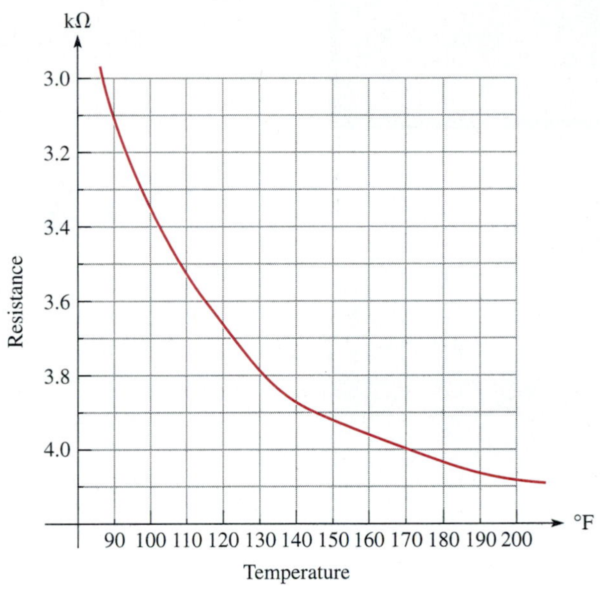


**FIGURE 8–68**
***Graph of thermistor resistance versus temperature.***

The Wheatstone bridge is built on a printed circuit board as shown in Figure 8–69(a). A probe-type thermistor is inserted through the wall of the tank and into the liquid as indicated in part (b) and the circuit is powered by a 12 V battery shown in part (c). The amplifier and relay circuitry is housed in a separate module that is not shown.

## The Printed Circuit Board

□ Check the printed circuit board to make sure that it agrees with the schematic in Figure 8–67. Relate each input, output, and component on the board to the schematic.

## Wiring List

□ Develop a point-to-point wiring list to properly interconnect the elements in Figure 8–69. Although it is not shown, include the amplifier inputs (call them Amp 1 and Amp 2).

## Balancing the Bridge

□ From the graph in Figure 8–68, calculate the resistance value to which $R_2$ must be set in order to balance the bridge at 170°F.

## Analysis

□ Assume that $R_2$ is adjusted for balance at 170°F. Using Thevenin's theorem, determine the voltage across the thermistor and the current through the thermistor for each of the following temperatures: 90°, 100°, 110°, 120°, 130°, 140°, 150°, 160°, 170°, 180°, 190°, and 200°. These data could be used to evaluate the power dissipation requirements for the thermistor based on a derating formula.

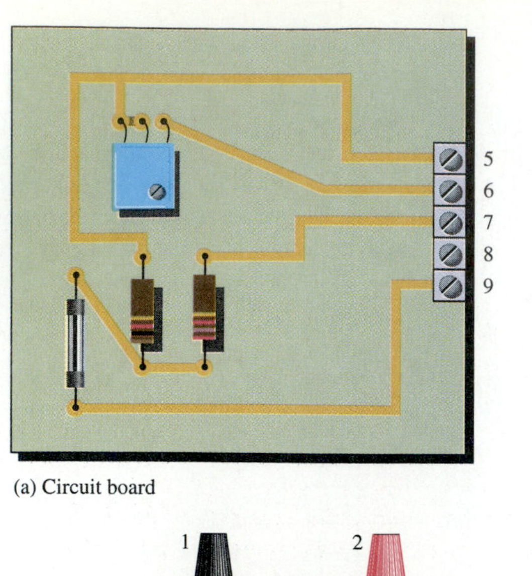


(a) Circuit board

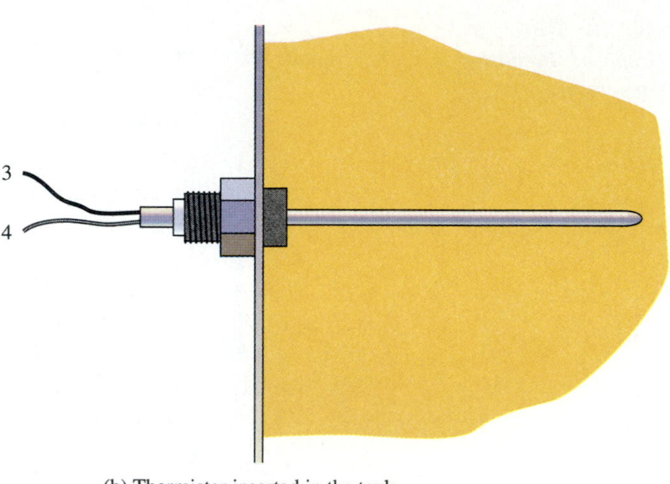

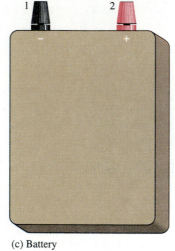

(b) Thermister inserted in the tank

**FIGURE 8–69**

(c) Battery

**FIGURE 8–70**

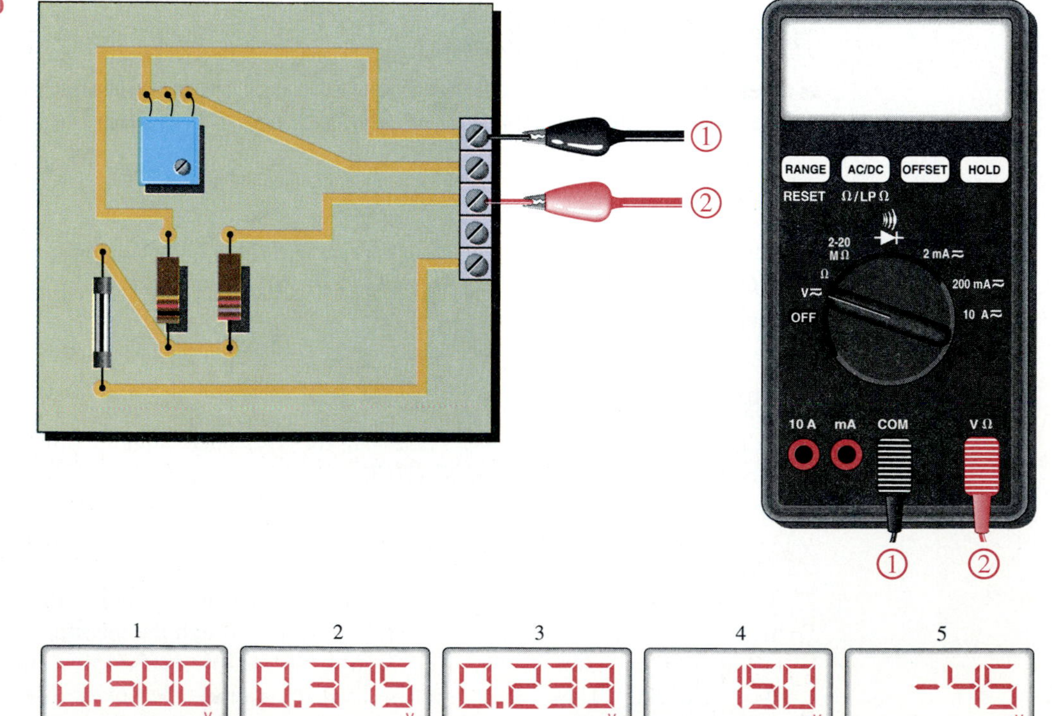

☐ In Figure 8–70, determine the approximate temperature in the tank for each of the voltages indicated by the voltmeter. $R_2$ is set at 1.5 kΩ. The board is completely connected according to the wiring list that you developed although, for simplicity, the wiring is not shown. Assume the meter presents no load to the bridge circuit.

**SECTION 8–10 REVIEW**

1. Discuss the purpose of the thermistor.
2. In the system diagram of Figure 8–67, when does the heating element turn on?

## 8–11 ■ PSpice DC ANALYSIS

***Now that you have learned how PSpice simplifies circuit analysis, you may think that Thevenin's theorem and other circuit conversion theorems are no longer useful. Although PSpice solves for currents and voltages without using an equivalent circuit, simplification theorems are still useful. Because an equivalent circuit is simpler than the original circuit, it helps you to better understand basic circuit operation and provides more insight into how different loads affect circuit performance. Also, equivalent circuits make PSpice circuit files shorter and simpler to write. In this section, PSpice is used to find the Thevenin and Norton equivalents for a resistive circuit.***

***After completing this section, you should be able to***

- **Use PSpice to determine equivalent circuits**
  - ☐ Write a circuit file to determine the Thevenin equivalent voltage
  - ☐ Write a circuit file to determine the Norton equivalent current
  - ☐ Use a multiple circuit file for more than one equivalent circuit

### Thevenin and Norton Sources

As you have learned, the Thevenin and Norton equivalents for a resistive circuit consist of a single equivalent source and a single equivalent resistor. The Thevenin equivalent voltage source is determined by removing the load from the circuit to be thevenized and finding the resulting open circuit voltage. Similarly, the Norton equivalent current source is determined by replacing the load with a short and finding the resulting short circuit current. The circuit in Figure 8–71 is used as an example.

FIGURE 8–71

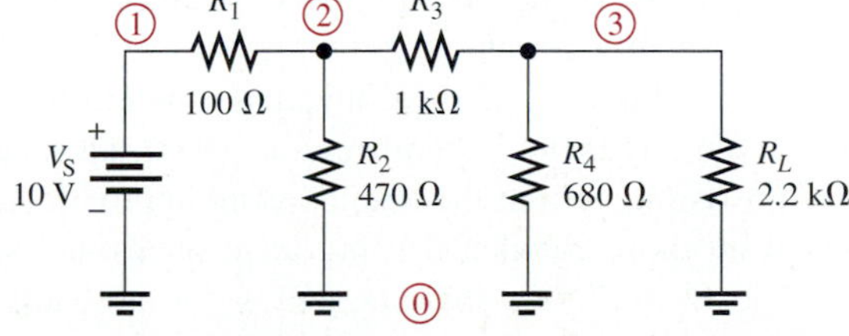


The circuit file for the original circuit in Figure 8–71 is as follows:

```
Circuit Equivalent Demonstration
*Original Circuit for Figure 8-71
VS  1  0  10
R1  1  2  100
R2  2  0  470
R3  2  3  1K
R4  3  0  680
RL  3  0  2.2K
.END
```

To determine the Thevenin equivalent looking from $R_L$, the first thought might be to simply remove $R_L$ from the circuit and delete the associated line in the circuit file. This approach will work only if another resistor in the circuit has the open circuit voltage across it. In the case of the circuit in Figure 8–71, the open circuit voltage appears across $R_4$. One difficulty with this approach is that an original circuit will not always have a single resistor across which the open circuit voltage appears, and PSpice can only give voltages and currents associated with single components.

The best approach to determining the Thevenin equivalent voltage is to replace $R_L$ with another resistor with an extremely large value (100 GΩ, for example) that simulates an effective open. For the Norton equivalent current, $R_L$ is replaced with another resistor with an extremely small value (1 pΩ, for example) that simulates an effective short. For the circuit in Figure 8–71, the circuit file to find the Thevenin voltage would be as follows:

```
Circuit Equivalent Demonstration
*Thevenin Equivalent Voltage for Figure 8–71
VS  1  0  10
R1  1  2  100
R2  2  0  470
R3  2  3  1K
R4  3  0  680
RL  3  0  100G
.DC  VS  10  10  1
.PRINT  DC  V(RL)
.END
```

The circuit file to find the Norton current would be as follows:

```
Circuit Equivalent Demonstration
*Norton Equivalent Current for Figure 8–71
VS  1  0  10
R1  1  2  100
R2  2  0  470
R3  2  3  1K
R4  3  0  680
RL  3  0  1P
.DC  VS  10  10  1
.PRINT  DC  I(RL)
.END
```

Notice that the circuit files are identical except for the comment line, the value of RL, and the .PRINT line. V(RL) is the Thevenin equivalent voltage and I(RL) is the Thevenin equivalent current.

This is a valid technique for determining the Thevenin or Norton equivalent of a given circuit for several reasons. First, all measurements and calculations are limited in precision, so that the results using approximate opens and shorts are not much different from those calculated using ideal opens and shorts. In fact, for the four digits typically shown in PSpice answers, you would probably see no difference.

The second reason that makes the approach valid is that circuit analysis is not meant to be just a mental exercise using fictitious circuits made of ideal components. It is meant to deal with actual circuits built from components that have tolerances and variations. The accuracy of PSpice is more than adequate for analyzing real-world circuits. So, calculations using an approximate open or an approximate short are well within the tolerances of actual circuits with real opens and shorts.

The third reason that makes the approach valid is that all practical meters will affect a circuit when a measurement is being made. The best voltmeters have several gigohms of resistance and the best ammeters have several microohms of resistance. Therefore, 100 GΩ is much closer to an actual open and 1 pΩ is much closer to an actual short than the resistance of measuring instruments.

## Thevenin and Norton Resistances

The techniques previously discussed will allow you to find the Thevenin equivalent voltage and the Norton equivalent current for any resistive circuit because PSpice is designed to calculate voltages and currents in a circuit. Since PSpice will not calculate resistance directly, you must use the values of $V_{TH}$ and $I_N$ to calculate $R_{TH}$ and $R_N$. As you know, $R_{TH} = R_N$ for a given circuit. You can find the equivalent resistance using the Ohm's law formula and the values for V(RL) and I(RL) determined by PSpice.

$$R_{TH} = R_N = \frac{\text{V(RL)}}{\text{I(RL)}}$$

## Multiple Circuit Files

So far, only single circuit files have been used in PSpice. Now, we need PSpice to evaluate two separate circuit files: the open load circuit file for the Thevenin equivalent voltage and the shorted load circuit file for the Norton equivalent current because both values are needed to calculate the equivalent resistance. You could run each circuit file separately, but it is easier and faster to combine both circuit files in one text file and run PSpice only once.

You can place several circuit files in a single text file provided that each circuit file has its own .END statement. Although it is not required, it is best to give each circuit file a title line because PSpice puts everything into the same output file and the title line at the top of each page makes it clear which output belongs to which file. There must be no blank lines between the title line of one circuit file and the .END statement of the previous circuit file.

A multiple circuit file for the Thevenin equivalent and the Norton equivalent of the circuit in Figure 8–71 is as follows:

```
Thevenin Equivalent Circuit
*Thevenin Equivalent Voltage for Figure 8–71
VS  1  0  10
R1  1  2  100
R2  2  0  470
R3  2  3  1K
R4  3  0  680
RL  3  0  100G
.DC  VS  10  10  1
.PRINT  DC  V(RL)
.END
Norton Equivalent Circuit
*Norton Equivalent Current for Figure 8–71
VS  1  0  10
R1  1  2  100
R2  2  0  470
R3  2  3  1K
R4  3  0  680
RL  3  0  1P
.DC  VS  10  10  1
.PRINT  DC  I(RL)
.END
```

Save these files as a single text file under a filename such as EQUIV.CIR, run PSpice, and use your text editor to look at the output file or print it out. PSpice will give the value of 3.181 V for V(RL) and the value of 7.618 mA for I(RL). Dividing the equivalent voltage by the equivalent current gives the equivalent resistance.

$$R_{TH} = R_N = \frac{\text{V(RL)}}{\text{I(RL)}} = \frac{3.181 \text{ V}}{7.618 \text{ mA}} = 417.6 \ \Omega$$

**SECTION 8–11 REVIEW**

1. Why is it necessary to substitute an extremely high resistance to simulate an open in order to determine $V_{TH}$ using PSpice?
2. Why is it necessary to substitute an extremely low resistance to simulate a short in order to determine $I_N$ using PSpice?
3. What are the two criteria for a multiple circuit file when a title line is used for each circuit file?

## ■ SUMMARY

- An ideal voltage source has zero internal resistance. It provides a constant voltage across its terminals regardless of the load resistance.
- A practical voltage source has a nonzero internal resistance. Its terminal voltage is essentially constant when $R_L \geq 10R_S$ (rule of thumb).
- An ideal current source has infinite internal resistance. It provides a constant current regardless of the load resistance.
- A practical current source has a finite internal resistance. Its current is essentially constant when $10R_L \leq R_S$.
- The superposition theorem is useful for multiple-source circuits.
- Thevenin's theorem provides for the reduction of any linear resistive circuit to an equivalent form consisting of an equivalent voltage source in series with an equivalent resistance.
- The term *equivalency,* as used in Thevenin's and Norton's theorems, means that when a given load resistance is connected to the equivalent circuit, it will have the same voltage across it and the same current through it as when it was connected to the original circuit.
- Norton's theorem provides for the reduction of any linear resistive circuit to an equivalent form consisting of an equivalent current source in parallel with an equivalent resistance.
- Millman's theorem provides for the reduction of parallel voltage sources to a single equivalent voltage source consisting of an equivalent voltage and an equivalent series resistance.
- Maximum power is transferred to a load from a source when the load resistance equals the internal source resistance.

## ■ GLOSSARY

**Current source** A device that ideally provides a constant value of current regardless of load.

**Maximum power transfer theorem** A theorem that states the maximum power is transferred from a source to a load when the load resistance equals the internal source resistance.

**Millman's theorem** A method for reducing parallel voltage sources to a single equivalent voltage source.

**Norton's theorem** A method for simplifying a given circuit to an equivalent circuit with a current source in parallel with a resistance.

**Superposition theorem** A method for the analysis of circuits with more than one source.

**Terminal equivalency** The concept that when any given load resistance is connected to two sources, the same load voltage and load current are produced by both sources.

**Thermistor** A temperature-sensitive resistor.

**Thevenin's theorem** A method for simplifying a given circuit to an equivalent circuit with a voltage source in series with a resistance.

**Voltage source** A device that ideally provides a constant value of voltage regardless of load.

## ■ FORMULAS

### Millman's Theorem

**(8–1)** $$R_{EQ} = \frac{1}{G_T} = \frac{1}{(1/R_1) + (1/R_2) + (1/R_3) + \cdots + (1/R_n)}$$

**(8–2)** $$V_{EQ} = \frac{(V_1/R_1) + (V_2/R_2) + (V_3/R_3) + \cdots + (V_n/R_n)}{(1/R_1) + (1/R_2) + (1/R_3) + \cdots + (1/R_n)}$$

### Δ-to-Y Conversions

(8–3) $$R_1 = \frac{R_A R_C}{R_A + R_B + R_C}$$

(8–4) $$R_2 = \frac{R_B R_C}{R_A + R_B + R_C}$$

(8–5) $$R_3 = \frac{R_A R_B}{R_A + R_B + R_C}$$

### Y-to-Δ Conversions

(8–6) $$R_A = \frac{R_1R_2 + R_1R_3 + R_2R_3}{R_2}$$

(8–7) $$R_B = \frac{R_1R_2 + R_1R_3 + R_2R_3}{R_1}$$

(8–8) $$R_C = \frac{R_1R_2 + R_1R_3 + R_2R_3}{R_3}$$

## ■ SELF-TEST

**1.** A 100 Ω load is connected across an ideal voltage source with $V_S = 10$ V. The voltage across the load is

**(a)** 0 V **(b)** 10 V **(c)** 100 V

**2.** A 100 Ω load is connected across a voltage source with $V_S = 10$ V and $R_S = 10$ Ω. The voltage across the load is

**(a)** 10 V **(b)** 0 V **(c)** 9.09 V **(d)** 0.909 V

**3.** A certain voltage source has the values $V_S = 25$ V and $R_S = 5$ Ω. The values for an equivalent current source are

**(a)** 5 A, 5 Ω **(b)** 25 A, 5 Ω **(c)** 5 A, 125 Ω

**4.** A certain current source has the values $I_S = 3$ $\mu$A and $R_S = 1$ MΩ. The values for an equivalent voltage source are

**(a)** 3 $\mu$V, 1 MΩ **(b)** 3 V, 1 MΩ **(c)** 1 V, 3 MΩ

**5.** In a two-source circuit, one source acting alone produces 10 mA through a given branch. The other source acting alone produces 8 mA in the opposite direction through the same branch. The actual current through the branch is

**(a)** 10 mA **(b)** 18 mA **(c)** 8 mA **(d)** 2 mA

**6.** Thevenin's theorem converts a circuit to an equivalent form consisting of

**(a)** a current source and a series resistance

**(b)** a voltage source and a parallel resistance

**(c)** a voltage source and a series resistance

**(d)** a current source and parallel resistance

**7.** The Thevenin equivalent voltage for a given circuit is found by

**(a)** shorting the output terminals

**(b)** opening the output terminals

**(c)** shorting the voltage source

**(d)** removing the voltage source and replacing it with a short

**8.** A certain circuit produces 15 V across its open output terminals, and when a 10 kΩ load is connected across its output terminals, it produces 12 V. The Thevenin equivalent for this circuit is

**(a)** 15 V in series with 10 kΩ **(b)** 12 V in series with 10 kΩ

**(c)** 12 V in series with 2.5 kΩ **(d)** 15 V in series with 2.5 kΩ

**9.** A circuit can be reduced to an equivalent current source in parallel with an equivalent resistance using

**(a)** Millman's theorem **(b)** Thevenin's theorem

**(c)** Norton's theorem **(d)** superposition theorem

**10.** Maximum power is transferred from a source to a load when

**(a)** the load resistance is very large

**(b)** the load resistance is very small

**(c)** the load resistance is twice the source resistance

**(d)** the load resistance equals the source resistance

**11.** For the circuit described in Question 8, maximum power is transferred to a

**(a)** 10 kΩ load **(b)** 2.5 kΩ load **(c)** an infinitely large resistance load

## ■ PROBLEMS

### SECTION 8–3 Source Conversions

**1.** A voltage source has the values $V_S = 300$ V and $R_S = 50\ \Omega$. Convert it to an equivalent current source.

**2.** Convert the practical voltage sources in Figure 8–72 to equivalent current sources.

FIGURE 8–72

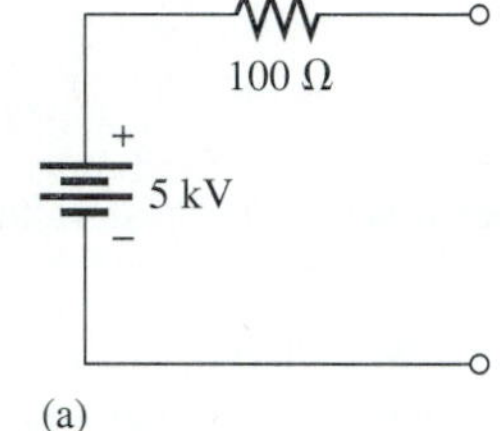


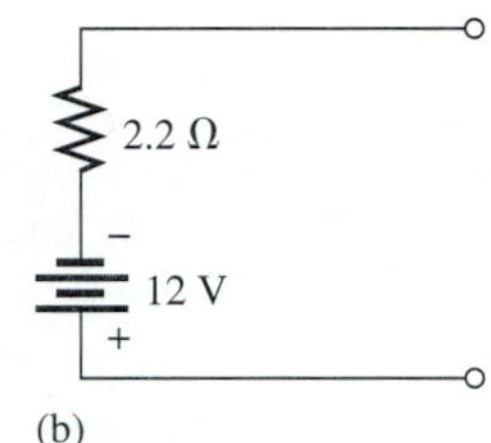


**3.** A current source has an $I_S$ of 600 mA and an $R_S$ of 1.2 kΩ. Convert it to an equivalent voltage source.

**4.** Convert the practical current sources in Figure 8–73 to equivalent voltage sources.

FIGURE 8–73

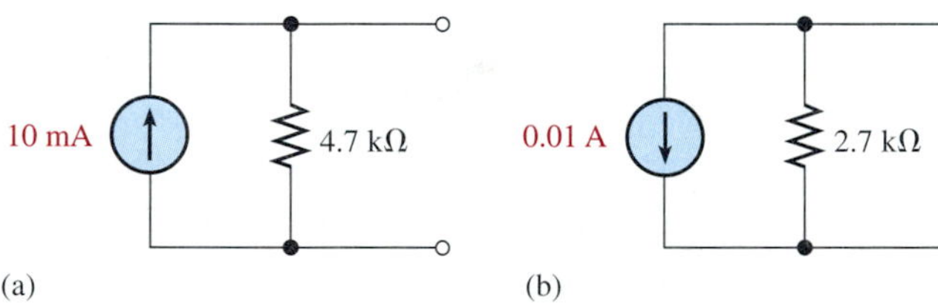


### SECTION 8–4 The Superposition Theorem

**5.** Using the superposition method, calculate the current through $R_5$ in Figure 8–74.

FIGURE 8–74

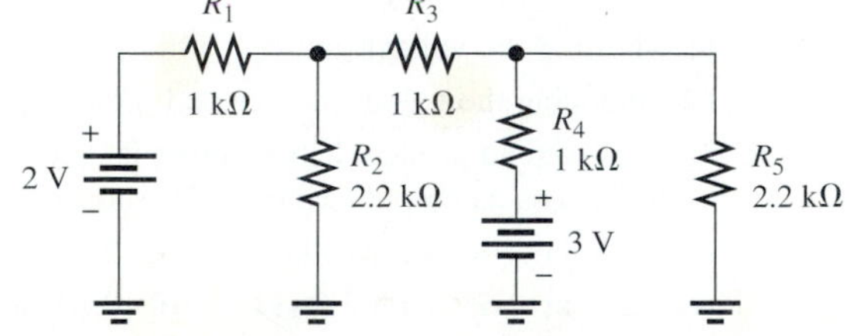


**6.** Use the superposition theorem to find the current in and the voltage across the $R_2$ branch of Figure 8–74.

**7.** Using the superposition theorem, solve for the current through $R_3$ in Figure 8–75.

**FIGURE 8–75**

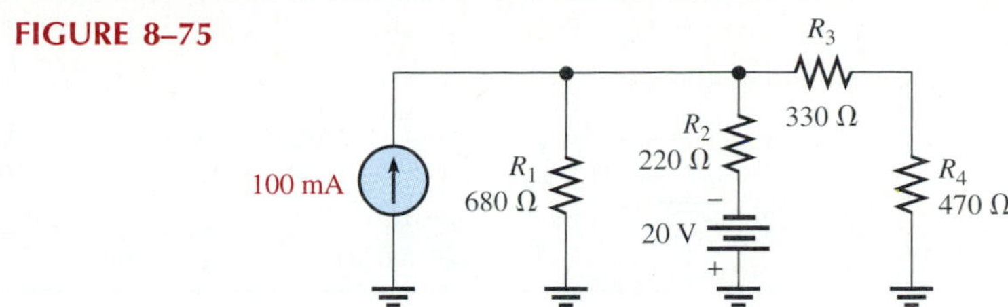


**8.** Using the superposition theorem, find the load current in each circuit of Figure 8–76.

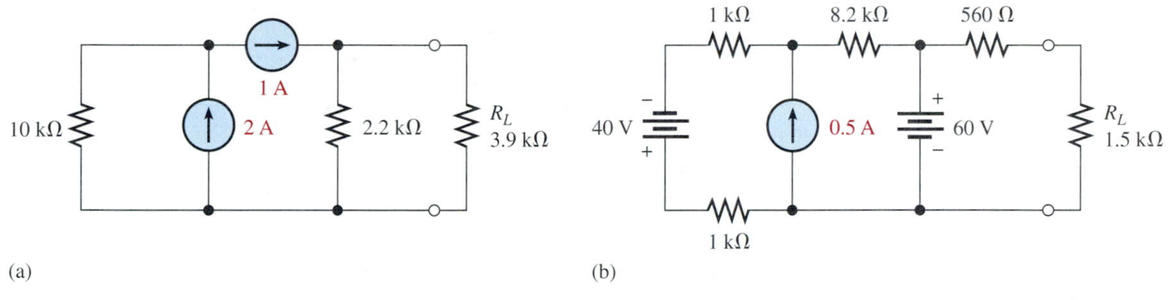


**FIGURE 8–76**

**9.** Determine the voltage from point $A$ to point $B$ in Figure 8–77.

**FIGURE 8–77**

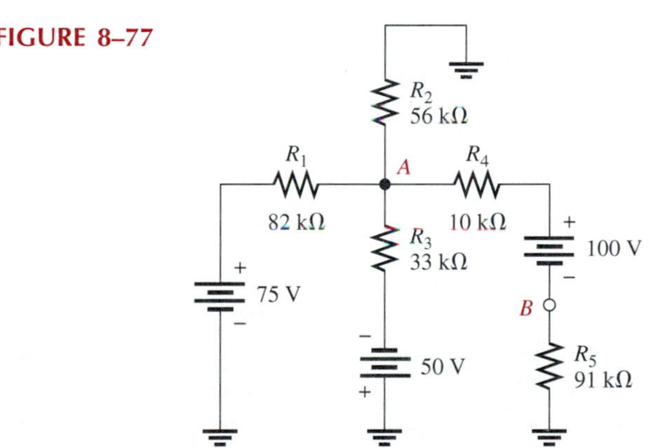


**10.** The switches in Figure 8–78 are closed in sequence, SW1 first. Find the current through $R_4$ after each switch closure.

**FIGURE 8–78**

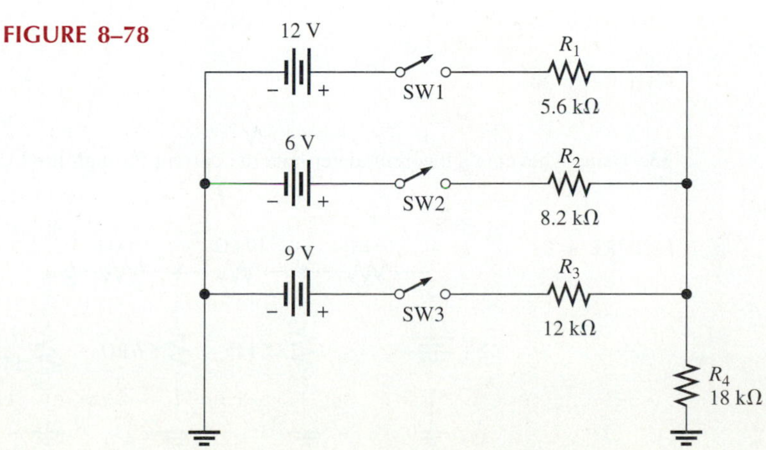

11. Figure 8–79 shows two ladder networks. Determine the current provided by each of the batteries when terminals $A$ are connected ($A$ to $A$) and terminals $B$ are connected ($B$ to $B$).

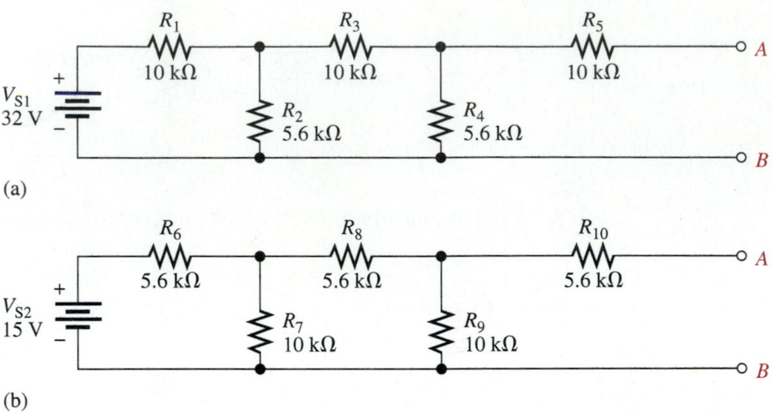


FIGURE 8–79

## SECTION 8–5 Thevenin's Theorem

12. For each circuit in Figure 8–80, determine the Thevenin equivalent as seen by $R_L$.

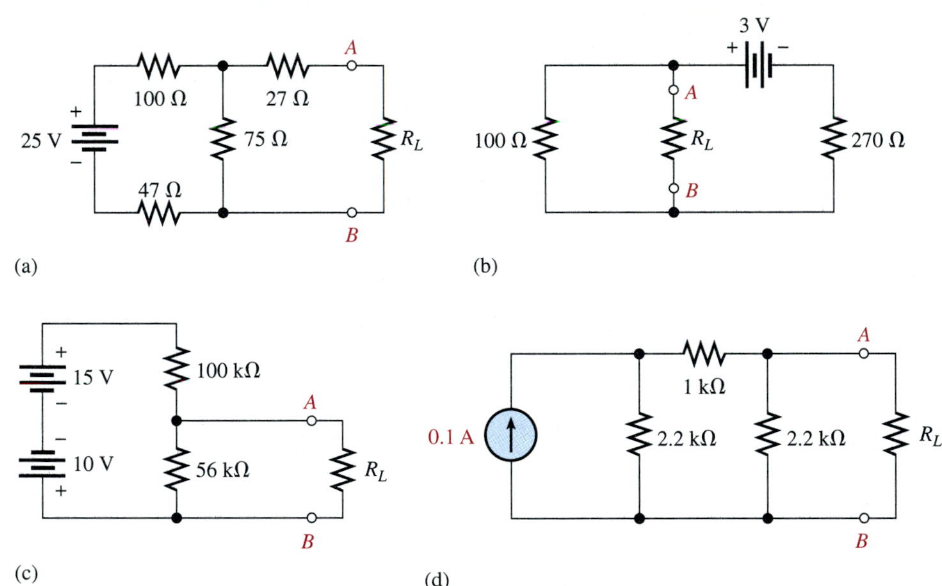


FIGURE 8–80

13. Using Thevenin's theorem, determine the current through the load $R_L$ in Figure 8–81.

FIGURE 8–81

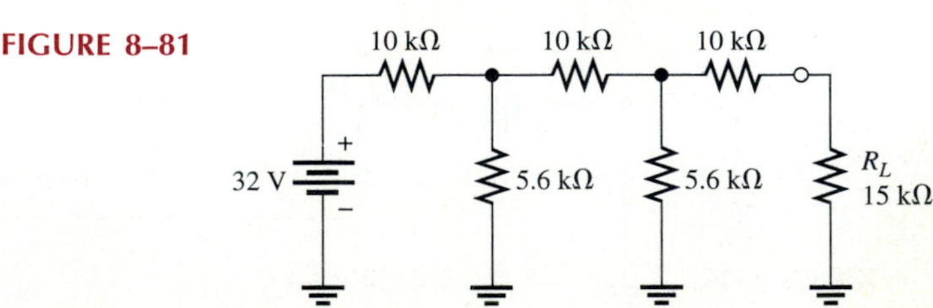

**14.** Using Thevenin's theorem, find the voltage across $R_4$ in Figure 8–82.

**15.** Find the Thevenin equivalent for the circuit external to the amplifier in Figure 8–83.

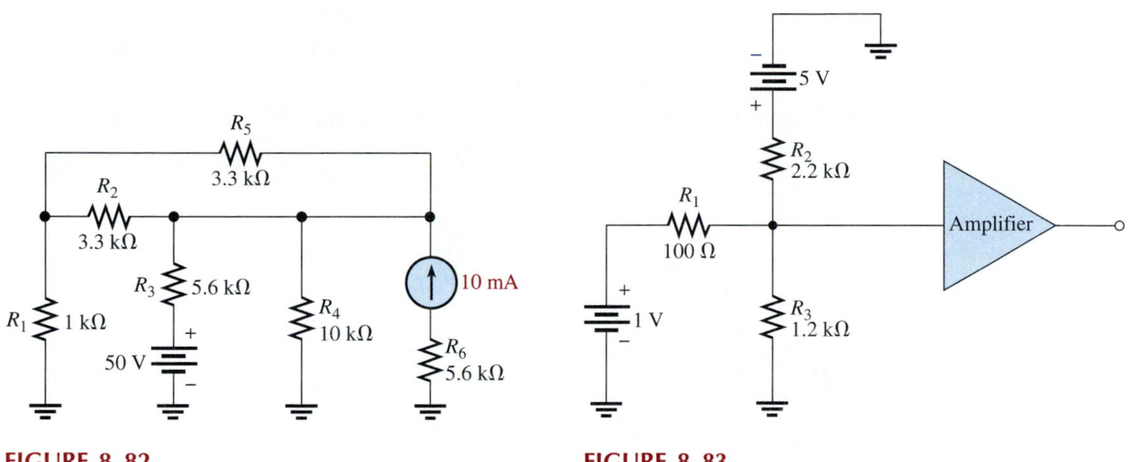


FIGURE 8–82

FIGURE 8–83

**16.** Determine the current into point $A$ when $R_8$ is 1 kΩ, 5 kΩ, and 10 kΩ in Figure 8–84.

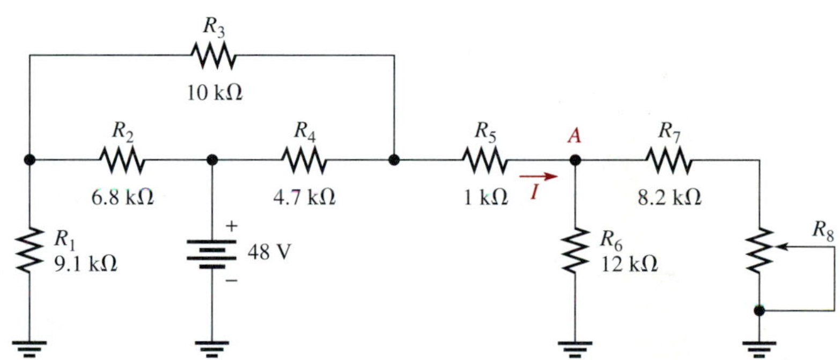


FIGURE 8–84

**17.** Find the current through the load resistor in the bridge circuit of Figure 8–85.

**18.** Determine the Thevenin equivalent looking from terminals $AB$ for the circuit in Figure 8–86.

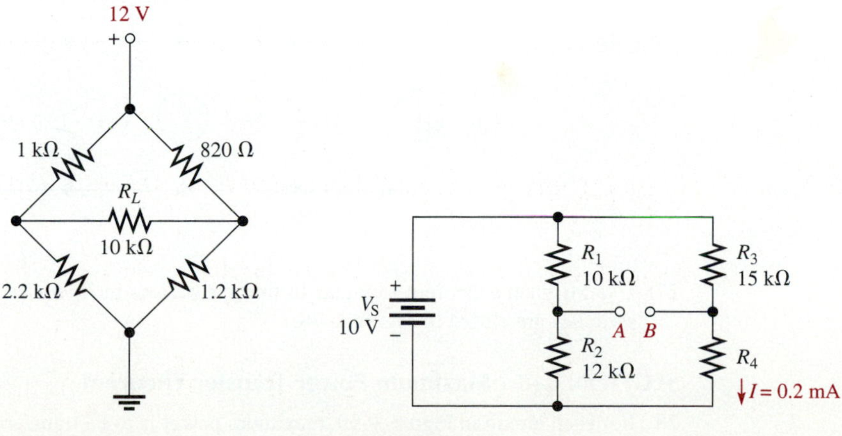


FIGURE 8–85

FIGURE 8–86

## SECTION 8–6 Norton's Theorem

**19.** For each circuit in Figure 8–80, determine the Norton equivalent as seen by $R_L$.

**20.** Using Norton's theorem, find the current through the load resistor $R_L$ in Figure 8–81.

**21.** Using Norton's theorem, find the voltage across $R_5$ in Figure 8–82.

**22.** Using Norton's theorem, find the current through $R_1$ in Figure 8–84 when $R_8$ = 8 kΩ.

**23.** Determine the Norton equivalent circuit for the bridge in Figure 8–85 with $R_L$ removed.

**24.** Reduce the circuit between terminals $A$ and $B$ in Figure 8–87 to its Norton equivalent.

FIGURE 8–87

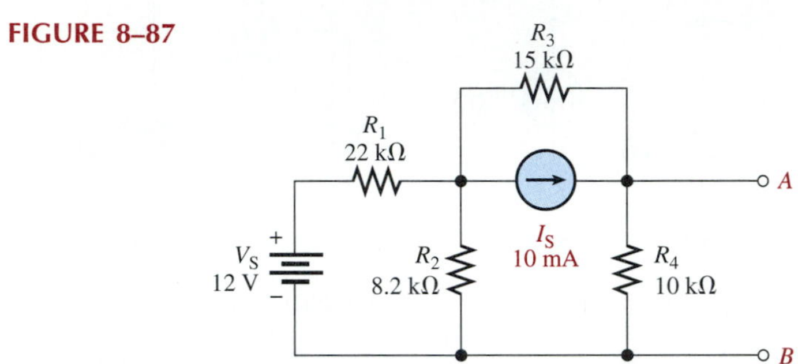


## SECTION 8–7 Millman's Theorem

**25.** Apply Millman's theorem to the circuit of Figure 8–88.

FIGURE 8–88

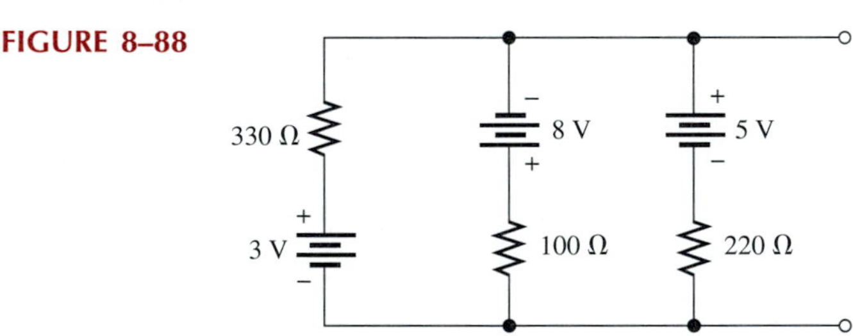


**26.** Use a Millman's theorem formula to reduce the circuit in Figure 8–89 to a single voltage source.

FIGURE 8–89

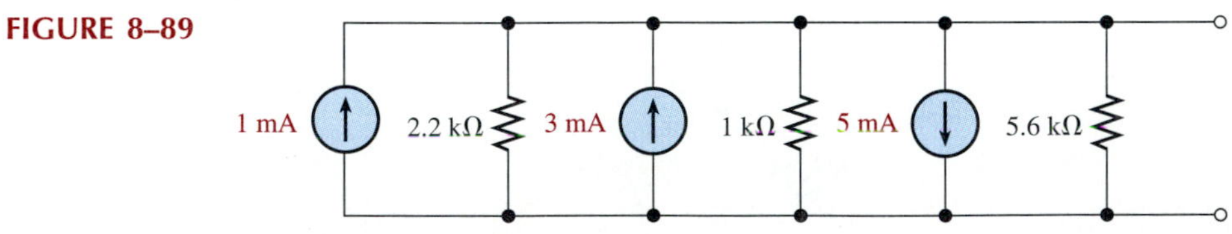


**27.** Use Millman's theorem formulas to find the current in $R_L$ for each case in which at least two switches are closed in Figure 8–78.

## SECTION 8–8 Maximum Power Transfer Theorem

**28.** For each circuit in Figure 8–90, maximum power is to be transferred to the load $R_L$. Determine the appropriate value for $R_L$ in each case.

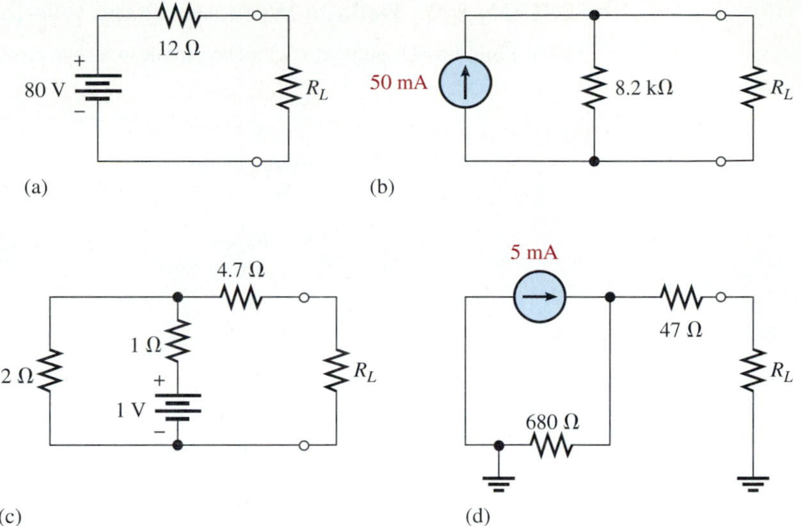


**FIGURE 8–90**

**29.** Determine the value of $R_L$ for maximum power in Figure 8–91.

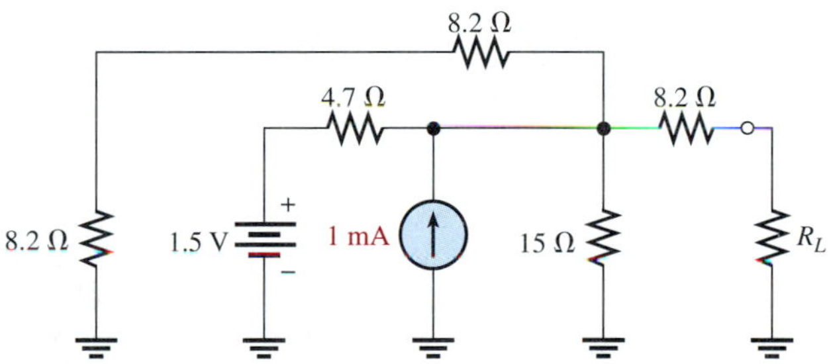


**FIGURE 8–91**

**30.** How much power is delivered to the load when $R_L$ is 10% higher than its value for maximum power in Figure 8–91.

**31.** What are the values of $R_4$ and $R_{TH}$ when the maximum power is transferred from the thevenized source to the ladder network in Figure 8–92?

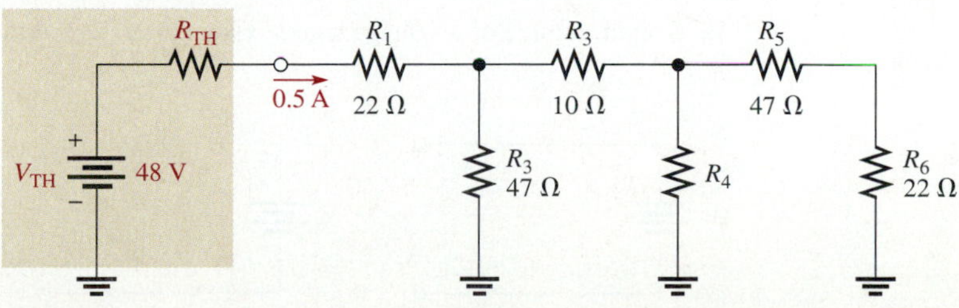


**FIGURE 8–92**

## SECTION 8–9 Delta-to-Wye (Δ-to-Y) and Wye-to-Delta (Y-to-Δ) Conversions

**32.** In Figure 8–93, convert each delta circuit to a wye circuit.

FIGURE 8–93

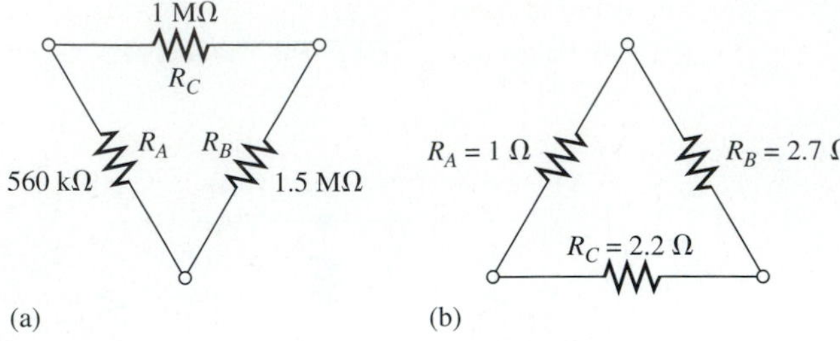


**33.** In Figure 8–94, convert each wye circuit to a delta circuit.

FIGURE 8–94

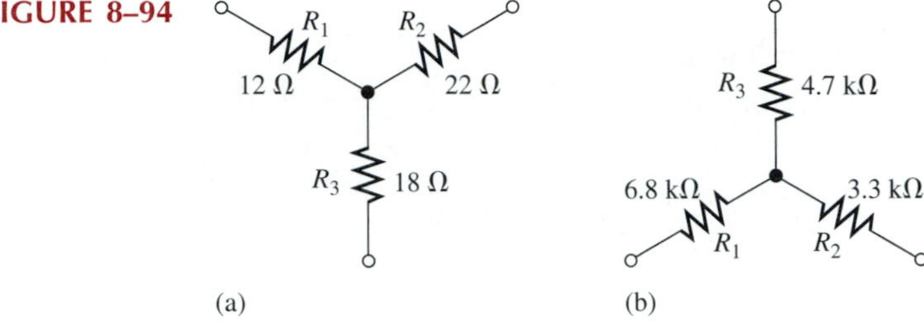


**34.** Find all currents in the circuit of Figure 8–95.

FIGURE 8–95

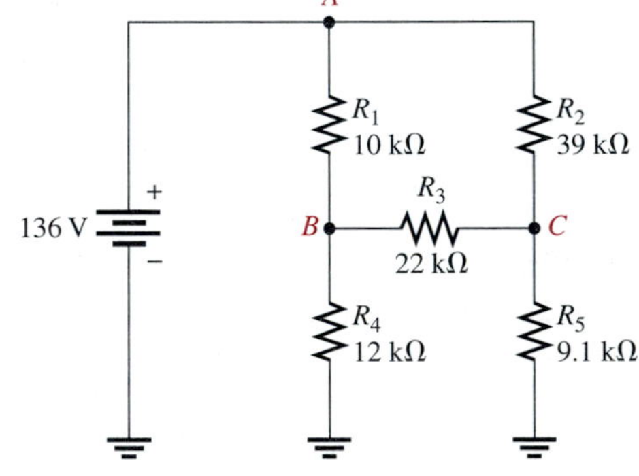


## ▪ ANSWERS TO SECTION REVIEWS

### Section 8–1

1. For ideal voltage source, see Figure 8–96.
2. For practical voltage source, see Figure 8–97.
3. The internal resistance of an ideal voltage source is zero ohms.
4. Output voltage of a voltage source varies directly with load resistance.

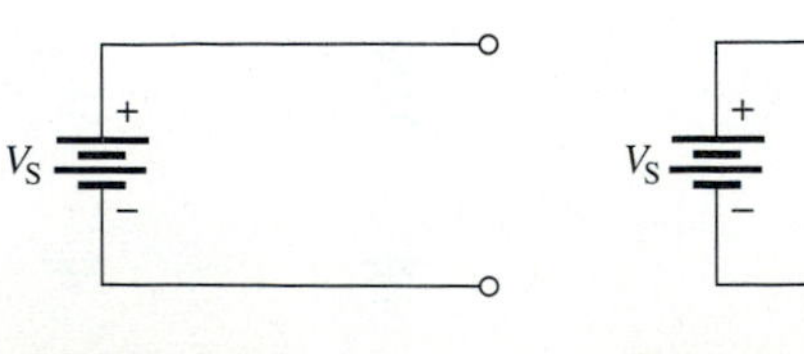


FIGURE 8–96 FIGURE 8–97

## Section 8–2

1. For ideal current source, see Figure 8–98.
2. For practical current source, see Figure 8–99.
3. An ideal current source has infinite internal resistance.
4. Load current from a current source varies inversely with load resistance.

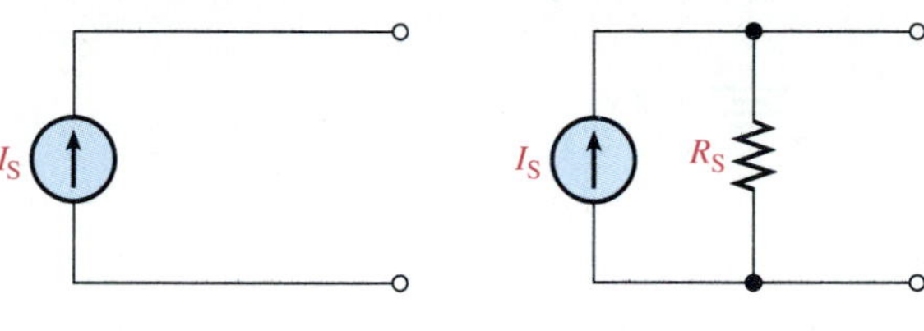


FIGURE 8–98 FIGURE 8–99

## Section 8–3

1. $I_S = V_S/R_S$
2. $V_S = I_S R_S$
3. See Figure 8–100.
4. See Figure 8–101.

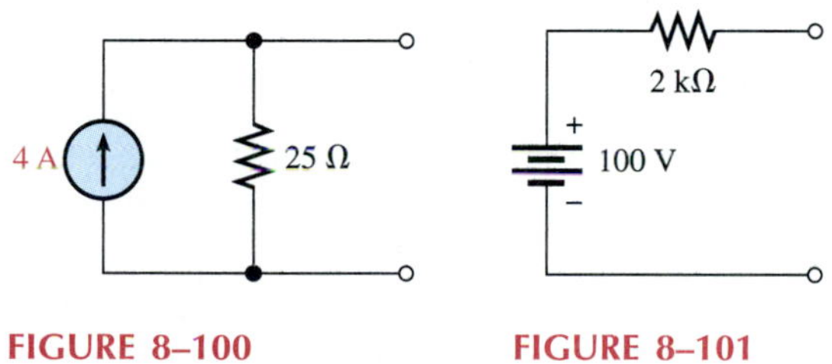


FIGURE 8–100 FIGURE 8–101

## Section 8–4

1. The superposition theorem states that the total current in any branch of a multiple-source linear circuit is equal to the algebraic sum of the currents due to the individual sources acting alone, with the other sources replaced by their internal resistances.
2. The superposition theorem allows each source to be treated independently.
3. A short simulates the internal resistance of an ideal voltage source; an open simulates the internal resistance of an ideal current source.
4. $I_{R1} = 6.67$ mA
5. The net current is in the direction of the larger current.

## Section 8–5

1. A Thevenin equivalent circuit consists of $V_{TH}$ and $R_{TH}$.
2. See Figure 8–102 for the general form of a Thevenin equivalent circuit.

FIGURE 8–102

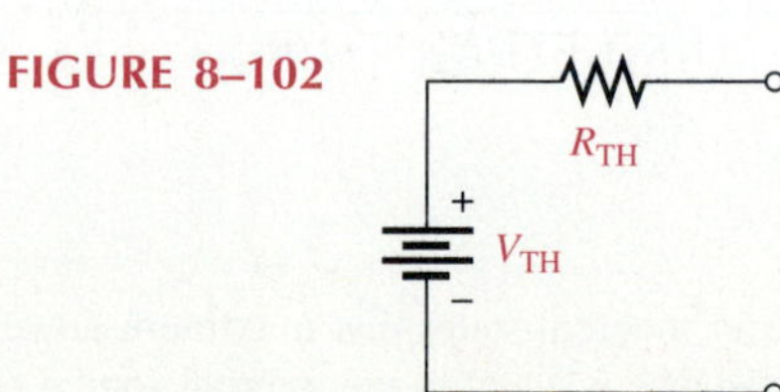

3. $V_{TH}$ is the open circuit voltage between two terminals in a circuit.
4. $R_{TH}$ is the resistance as viewed from two terminals in a circuit, with all sources replaced by their internal resistances.
5. See Figure 8–103.

FIGURE 8–103

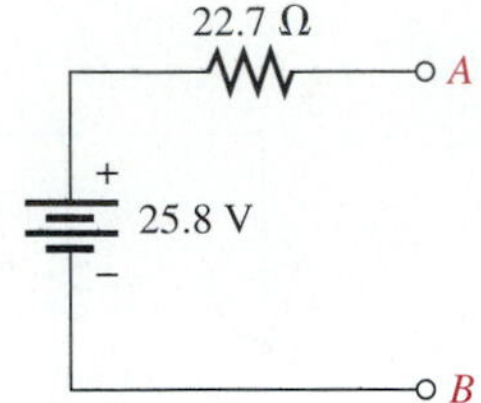


## Section 8–6

1. A Norton equivalent circuit consists of $I_N$ and $R_N$.
2. See Figure 8–104 for the general form of a Norton equivalent circuit.

FIGURE 8–104

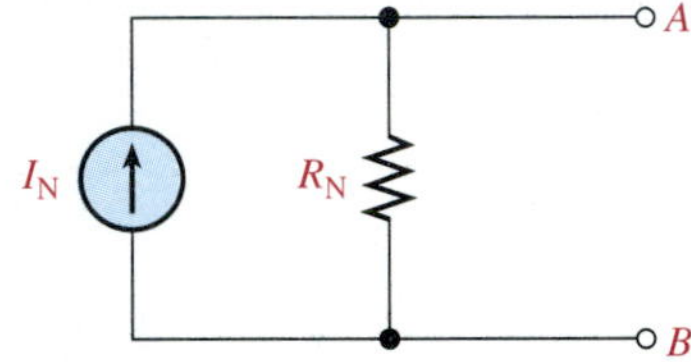


3. $I_N$ is the short circuit current between two terminals in a circuit.
4. $R_N$ is the resistance as viewed from the two open terminals in a circuit.
5. See Figure 8–105.

FIGURE 8–105

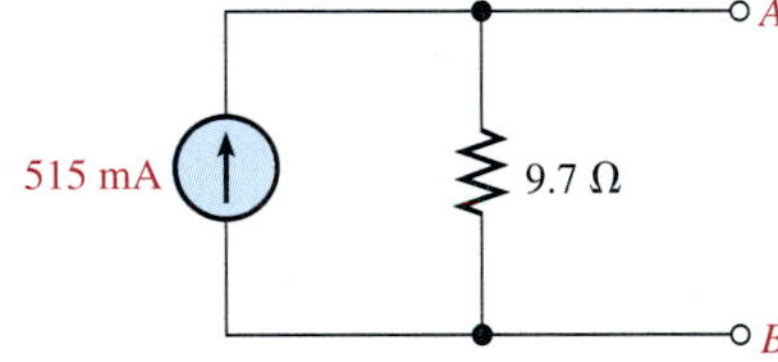


## Section 8–7

1. Millman's theorem applies to parallel voltage sources.
2. The Millman equation for $R_{EQ}$ is as follows:

$$R_{EQ} = \frac{1}{(1/R_1) + (1/R_2) + (1/R_3) + \cdots + (1/R_n)}$$

3. The Millman equation for $V_{EQ}$ is as follows:

$$V_{EQ} = \frac{(V_1/R_1) + (V_2/R_2) + (V_3/R_3) + \cdots + (V_n/R_n)}{(1/R_1) + (1/R_2) + (1/R_3) + \cdots + (1/R_n)}$$

4. $I_L = 1.08$ A; $V_L = 108$ V

## Section 8–8

1. The maximum power transfer theorem states that maximum power is transferred from a source to a load when the load resistance is equal to the internal source resistance.

2. Maximum power is delivered to a load when $R_L = R_S$.
3. $R_L = R_S = 50\ \Omega$

## Section 8–9

1. For a delta circuit see Figure 8–106.
2. For a wye circuit, see Figure 8–107.

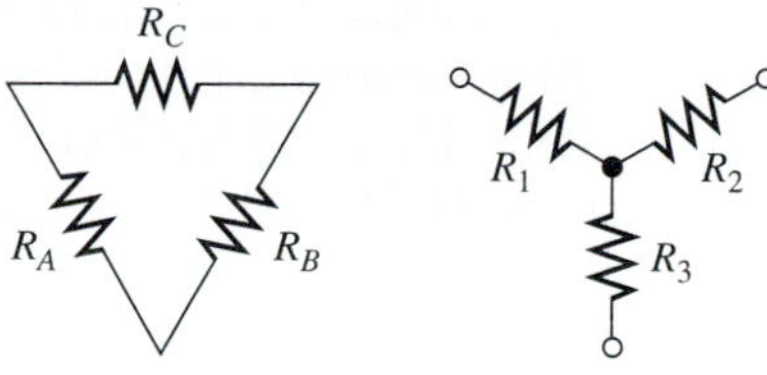


FIGURE 8–106 FIGURE 8–107

3. The delta-to-wye conversion equations are

$$R_1 = \frac{R_A R_C}{R_A + R_B + R_C} \qquad R_2 = \frac{R_B R_C}{R_A + R_B + R_C} \qquad R_3 = \frac{R_A R_B}{R_A + R_B + R_C}$$

4. The wye-to-delta conversion equations are

$$R_A = \frac{R_1R_2 + R_1R_3 + R_2R_3}{R_2} \qquad R_B = \frac{R_1R_2 + R_1R_3 + R_2R_3}{R_1} \qquad R_C = \frac{R_1R_2 + R_1R_3 + R_2R_3}{R_3}$$

## Section 8–10

1. A thermister senses a change in temperature and produces a resulting change in resistance.
2. The heating element turns on when the bridge becomes unbalanced.

## Section 8–11

1. PSpice can calculate voltage only when associated with a component.
2. PSpice can calculate current only when associated with a component.
3. Each circuit file must have its own .END statement; there must be no blank lines between the title line of one circuit and the .END statement of the previous circuit file.

## ■ ANSWERS TO RELATED EXERCISES FOR EXAMPLES

**8–1** 99.5 V
**8–2** 100 V
**8–3** 3.33 kΩ
**8–4** 1.2 A; 10 Ω
**8–5** 300 V; 600 Ω
**8–6** 16.6 mA
**8–7** $I_S$ is not affected.
**8–8** 70 mA
**8–9** 5 mA
**8–10** 2.36 V; 124 Ω
**8–11** 1.17 mA
**8–12** 169 mA
**8–13** 247 Ω
**8–14** $I_N = 169$ mA; $R_N = 247\ \Omega$
**8–15** 227 mA
**8–16** 41.7 mW
**8–17** $R_1 = 792\ \Omega$, $R_2 = 360\ \Omega$, $R_3 = 440\ \Omega$
**8–18** $R_A = 712\ \Omega$, $R_B = 2.35\ \text{k}\Omega$, $R_C = 500\ \Omega$
**8–19** 3.02 μA

# 9

# BRANCH, MESH, AND NODE ANALYSIS



## ■ INTRODUCTION

In the last chapter, you learned about the superposition theorem, Thevenin's theorem, Norton's theorem, Millman's theorem, maximum power transfer theorem, and several types of conversion methods. These theorems and conversion methods are useful in solving circuit problems.

In this chapter, three more circuit analysis methods are introduced. These methods are based on Ohm's law and Kirchhoff's laws and are particularly useful in the analysis of multiple loop circuits having two or more voltage or current sources. The methods presented here can be used alone or in conjunction with the techniques covered in the previous chapters. With experience, you will learn which method is best for a particular problem or you may develop a preference for one of them. In this chapter and throughout the rest of the book, you will learn the basics of putting technology theory into practice.

In the branch current method, Kirchhoff's laws are applied to solve for current in various branches of a multiple loop circuit. A loop is a complete current path within a circuit. The method of determinants is useful in solving simultaneous equations that occur in multiple loop analysis. In the mesh current method, you will solve for loop currents rather than branch currents. In the node voltage method, the voltages at the independent nodes in a circuit are found. A node is the junction of two or more components.

## TECHnology Theory Into Practice

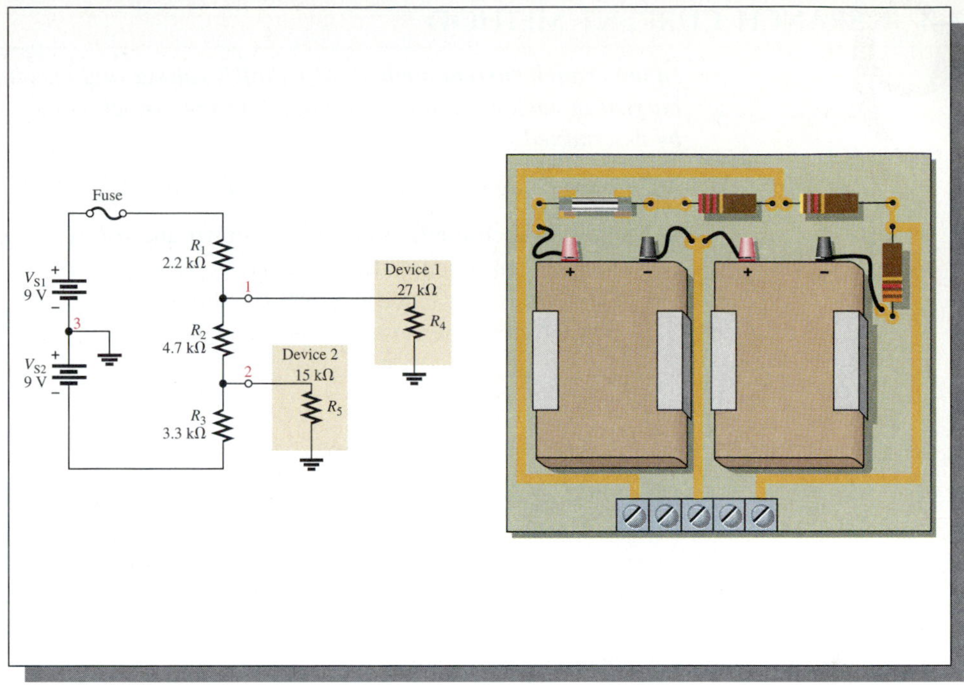


For this chapter, the TECH TIP assignment is to analyze a dual-polarity loaded voltage divider to determine if certain voltage measurements are correct. One of the methods covered in this chapter will be used for the analysis.

## CHAPTER OBJECTIVES

- ☐ Use the branch current method to find unknown quantities in a circuit
- ☐ Use determinants to solve simultaneous equations
- ☐ Use mesh analysis to find unknown quantities in a circuit
- ☐ Use node analysis to find unknown quantities in a circuit
- ☐ Use PSpice to analyze multiple loop circuits (optional)

## 9–1 ■ BRANCH CURRENT METHOD

***In the branch current method, Kirchhoff's voltage and current laws are used to find the current in each branch of a circuit. Once the branch currents are known, voltages can be determined.***

***After completing this section, you should be able to***

- **Use the branch method to find unknown quantities in a circuit**
  - ☐ Identify loops and nodes in a circuit
  - ☐ Develop a set of branch current equations
  - ☐ Solve the equations for an unknown current

### Loops, Nodes, and Branches

Figure 9–1 shows a circuit with two voltage sources and two loops. This circuit will be used as the basic model throughout the chapter to illustrate each of the three circuit analysis methods. In this circuit, there are two nonredundant closed loops, as indicated by the arrows. A **loop** is a complete current path within a circuit. Also, there are four nodes as indicated by the letters *A, B, C,* and *D.* A **node** is a point where two or more components are connected. A **branch** is a path that connects two nodes.

The following are the general steps used in applying the branch current method. These steps are demonstrated with the aid of Figure 9–2.

**Step 1:** Assign a current in each circuit branch in an arbitrary direction.

**Step 2:** Show the polarities of the resistor voltages according to the assigned branch current directions.

**Step 3:** Apply Kirchhoff's voltage law around each closed loop (sum of voltages is equal to zero).

**Step 4:** Apply Kirchhoff's current law at the minimum number of nodes so that all branch currents are included (sum of currents at a node equals zero).

**Step 5:** Solve the equations resulting from Steps 3 and 4 for the branch current values.

First, the **branch currents** $I_{R1}$, $I_{R2}$, and $I_{R3}$ are assigned in the direction shown in Figure 9–2. Do not worry about the actual current directions at this point.

Second, the polarities of the voltage drops across $R_1$, $R_2$, and $R_3$ are indicated in the figure according to the current directions.

Third, Kirchhoff's voltage law applied to the two loops gives the following equations:

Equation 1: $R_1I_{R1} + R_2I_{R2} - V_{S1} = 0$ for loop 1

Equation 2: $R_2I_{R2} + R_3I_{R3} - V_{S2} = 0$ for loop 2

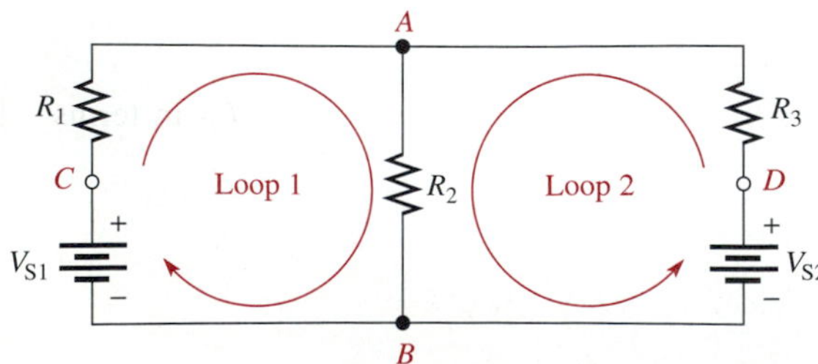


**FIGURE 9–1**
***Basic multiple-loop circuit showing loops and nodes.***

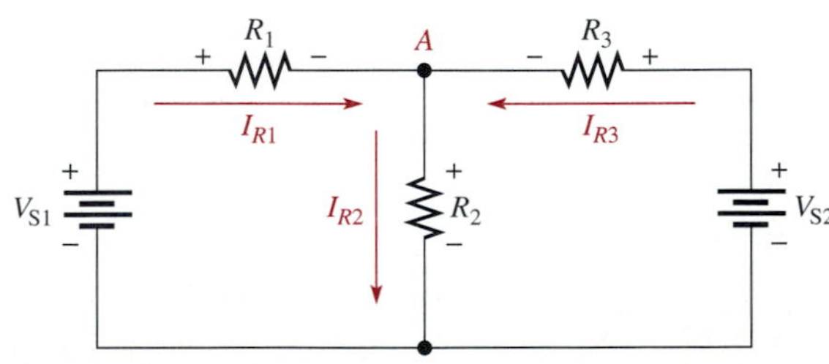


**FIGURE 9–2**
***Circuit for demonstrating branch current analysis***

Fourth, Kirchhoff's current law is applied to node *A,* including all branch currents as follows:

$$\text{Equation 3:} \quad I_{R1} - I_{R2} + I_{R3} = 0$$

The negative sign indicates that $I_{R2}$ is out of the junction.

Fifth and last, the three equations must be solved for the three unknown currents, $I_{R1}$, $I_{R2}$, and $I_{R3}$. The three equations in the above steps are called *simultaneous equations* and can be solved in two ways: by substitution or by determinants. Example 9–1 shows how to solve equations by substitution. In Section 9–2, you will study determinants and how to use them to find branch currents; in the following sections, you will use determinants in two other methods of circuit analysis.

**EXAMPLE 9–1** Use the branch current method to find each branch current in Figure 9–3.

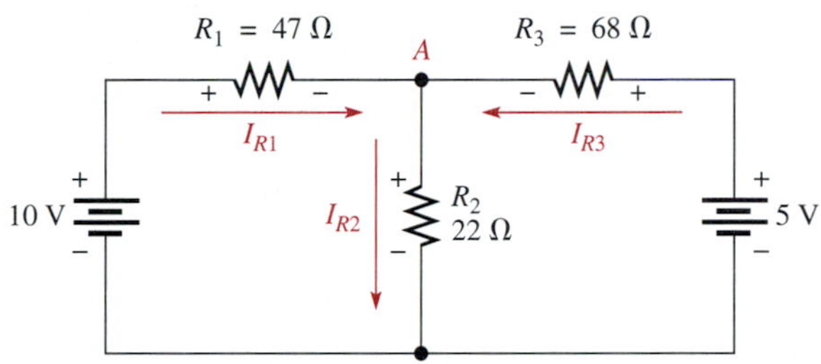


FIGURE 9–3

***Solution***

**Step 1:** Assign branch currents as shown in Figure 9–3. Keep in mind that you can assume any current direction at this point and that the final solution will have a negative sign if the actual current is opposite to the assigned current.

**Step 2:** Mark the polarities of the resistor voltage drops as shown in the figure.

**Step 3:** Applying Kirchhoff's voltage law around the left loop gives

$$47I_{R1} + 22I_{R2} - 10 = 0$$

Around the right loop gives

$$22I_{R2} + 68I_{R3} - 5 = 0$$

where all resistance values are in ohms and voltage values are in volts. For simplicity, the units are not shown.

**Step 4:** At node *A,* the current equation is

$$I_{R1} - I_{R2} + I_{R3} = 0$$

**Step 5:** The equations are solved by substitution as follows. First, find $I_{R1}$ in terms of $I_{R2}$ and $I_{R3}$.

$$I_{R1} = I_{R2} - I_{R3}$$

Now, substitute $I_{R2} - I_{R3}$ for $I_{R1}$ in the left loop equation.

$$47(I_{R2} - I_{R3}) + 22I_{R2} = 10$$
$$47I_{R2} - 47I_{R3} + 22I_{R2} = 10$$
$$69I_{R2} - 47I_{R3} = 10$$

Next, take the right loop equation and solve for $I_{R2}$ in terms of $I_{R3}$.

$$22I_{R2} = 5 - 68I_{R3}$$

$$I_{R2} = \frac{5 - 68I_{R3}}{22}$$

Substituting this expression for $I_{R2}$ into $69I_{R2} - 47I_{R3} = 10$ yields

$$69\left(\frac{5 - 68I_{R3}}{22}\right) - 47I_{R3} = 10$$

$$\frac{345 - 4692I_{R3}}{22} - 47I_{R3} = 10$$

$$15.68 - 213.27I_{R3} - 47I_{R3} = 10$$

$$-260.27I_{R3} = -5.68$$

$$I_{R3} = \frac{5.68}{260.27} = 0.0218 \text{ A} = 21.8 \text{ mA}$$

Now, substitute this value of $I_{R3}$ into the right loop equation.

$$22I_{R2} + 68(0.0218) = 5$$

Solve for $I_{R2}$.

$$I_{R2} = \frac{5 - 68(0.0218)}{22} = \frac{3.52}{22} = 0.16 \text{ A} = 160 \text{ mA}$$

Substituting $I_{R2}$ and $I_{R3}$ values into the current equation at node $A$ yields

$$I_{R1} - 0.16 + 0.0218 = 0$$

$$I_{R1} = 0.16 - 0.0218 = 0.138 \text{ A} = 138 \text{ mA}$$

***Related Exercise*** Determine the branch currents in Figure 9–3 with the polarity of the 5 V source reversed.

**SECTION 9–1 REVIEW**

1. What basic circuit laws are used in the branch current method?
2. When assigning branch currents, you should be careful of the directions (T or F).
3. What is a loop?
4. What is a node?

## 9–2 ■ DETERMINANTS

***When several unknown quantities are to be found, such as the three branch currents in Example 9–1, you must have a number of equations equal to the number of unknowns. In this section, you will learn how to solve for two and three unknowns using the systematic method of determinants. This method is an alternate to the substitution method, which you used in the previous section.***

***After completing this section, you should be able to***

- **Use determinants to solve simultaneous equations**
  - ☐ Set up second-order determinants to solve two simultaneous equations
  - ☐ Set up third-order determinants to solve three simultaneous equations
  - ☐ Evaluate determinants using either the expansion method or the cofactor method

## Solving Two Simultaneous Equations for Two Unknowns

To illustrate the method of second-order determinants, let's assume two loop equations as follows:

$$10I_1 + 5I_2 = 15$$
$$2I_1 + 4I_2 = 8$$

We want to find the value of $I_1$ and $I_2$. To do so, we form a **determinant** with the coefficients of the unknown currents. A **coefficient** is the number associated with an unknown. For example, 10 is the coefficient for $I_1$ in the first equation.

The first column in the determinant consists of the coefficients of $I_1$, and the second column consists of the coefficients of $I_2$. The resulting determinant appears as follows:

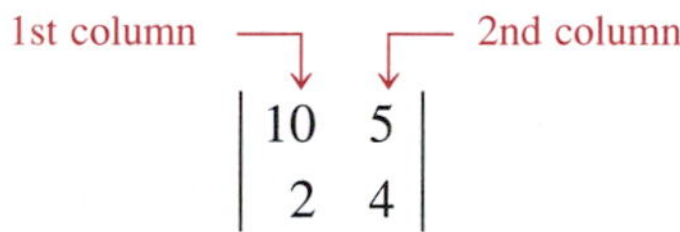


This is called the *characteristic determinant* for the set of equations.

Next, we form another determinant and use it in conjunction with the characteristic determinant to solve for $I_1$. We form this determinant for our example by replacing the coefficients of $I_1$ in the characteristic determinant with the constants on the right side of the equations. Doing this, we get the following determinant:

$$\begin{vmatrix} 15 & 5 \\ 8 & 4 \end{vmatrix}$$

Replace coefficients of $I_1$ with constants from right sides of equations.

We can now solve for $I_1$ by evaluating both determinants and then dividing by the characteristic determinant. To evaluate the determinants, we cross-multiply and subtract the resulting products. An evaluation of the characteristic determinant in this example is illustrated in the following two steps:

**Step 1:** Multiply the first number in the left column by the second number in the right column.

$$\begin{vmatrix} 10 & 5 \\ 2 & 4 \end{vmatrix} = 10 \times 4 = 40$$

**Step 2:** Multiply the second number in the left column by the first number in the right column and subtract from the product in Step 1. This result is the value of the determinant (30 in this case).

$$\begin{vmatrix} 10 & 5 \\ 2 & 4 \end{vmatrix} = 40 - (2 \times 5) = 40 - 10 = 30$$

Repeat the same procedure for the other determinant that was set up for $I_1$.

$$\begin{vmatrix} 15 & 5 \\ 8 & 4 \end{vmatrix} = 15 \times 4 = 60$$

$$\begin{vmatrix} 15 & 5 \\ 8 & 4 \end{vmatrix} = 60 - (8 \times 5) = 60 - 40 = 20$$

The value of this determinant is 20. Now we can solve for $I_1$ by dividing the $I_1$ determinant by the characteristic determinant as follows:

$$I_1 = \frac{\begin{vmatrix} 15 & 5 \\ 8 & 4 \end{vmatrix}}{\begin{vmatrix} 10 & 5 \\ 2 & 4 \end{vmatrix}} = \frac{20}{30} = 0.667 \text{ A}$$

To find $I_2$, we form another determinant by substituting the constants on the right side of the equations for the coefficients of $I_2$.

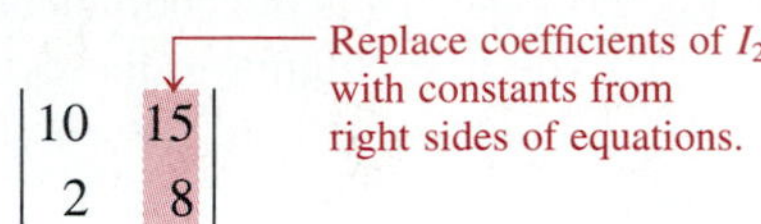
$\begin{vmatrix} 10 & 15 \\ 2 & 8 \end{vmatrix}$
Replace coefficients of $I_2$ with constants from right sides of equations.

We solve for $I_2$ by dividing this determinant by the characteristic determinant already evaluated.

$$I_2 = \frac{\begin{vmatrix} 10 & 15 \\ 2 & 8 \end{vmatrix}}{30} = \frac{(10 \times 8) - (2 \times 15)}{30} = \frac{80 - 30}{30} = \frac{50}{30} = 1.67 \text{ A}$$

**EXAMPLE 9–2** Solve the following set of equations for the unknown currents:

$$2I_1 - 5I_2 = 10$$
$$6I_1 + 10I_2 = 20$$

***Solution*** The characteristic determinant is evaluated as follows:

$$\begin{vmatrix} 2 & -5 \\ 6 & 10 \end{vmatrix} = (2)(10) - (-5)(6) = 20 - (-30) = 20 + 30 = 50$$

Solving for $I_1$ yields

$$I_1 = \frac{\begin{vmatrix} 10 & -5 \\ 20 & 10 \end{vmatrix}}{50} = \frac{(10)(10) - (-5)(20)}{50} = \frac{100 - (-100)}{50} = \frac{200}{50} = 4 \text{ A}$$

Solving for $I_2$ yields

$$I_2 = \frac{\begin{vmatrix} 2 & 10 \\ 6 & 20 \end{vmatrix}}{50} = \frac{(2)(20) - (6)(10)}{50} = \frac{40 - 60}{50} = -0.4 \text{ A}$$

In a circuit problem, a result with a negative sign indicates that the direction of actual current is opposite to the assigned direction.

Note that multiplication can be expressed either by the multiplication sign such as 2 × 10 or by parentheses such as (2)(10).

***Related Exercise*** Solve the following set of equations for $I_1$:

$$5I_1 + 3I_2 = 4$$
$$I_1 + 2I_2 = -6$$

## Solving Three Simultaneous Equations for Three Unknowns

Third-order determinants can be evaluated by either the expansion method or the cofactor method. First, we will illustrate the expansion method (which is good only for third order) using the following three equations:

$$1I_1 + 3I_2 - 2I_3 = 7$$
$$0I_1 + 4I_2 + 1I_3 = 8$$
$$-5I_1 + 1I_2 + 6I_3 = 9$$

The three-column characteristic determinant for this set of equations is formed in a similar way to that used earlier for the second-order determinant. The first column consists of the coefficients of $I_1$, the second column consists of the coefficients of $I_2$, and the third column consists of the coefficients of $I_3$, as shown below.

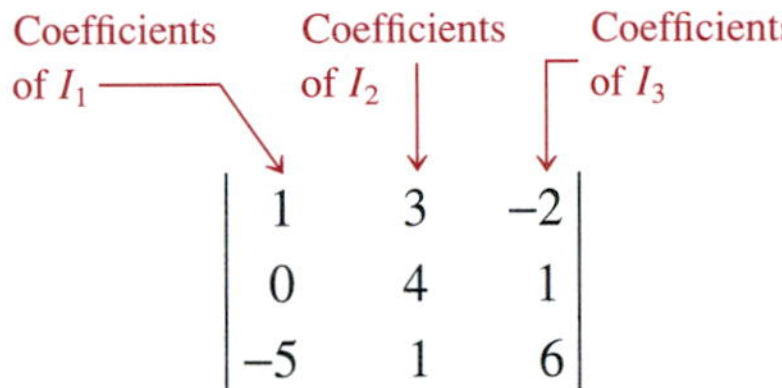


***The Expansion Method*** This third-order determinant is evaluated by the expansion method as shown in the following steps.

**Step 1:** Rewrite the first two columns immediately to the right of the determinant.

$$\begin{vmatrix} 1 & 3 & -2 \\ 0 & 4 & 1 \\ -5 & 1 & 6 \end{vmatrix} \begin{matrix} 1 & 3 \\ 0 & 4 \\ -5 & 1 \end{matrix}$$

**Step 2:** Identify the three downward diagonal groups of three coefficients each.

$$\begin{vmatrix} 1 & 3 & -2 \\ 0 & 4 & 1 \\ -5 & 1 & 6 \end{vmatrix} \begin{matrix} 1 & 3 \\ 0 & 4 \\ -5 & 1 \end{matrix}$$

**Step 3:** Multiply the numbers in each diagonal and add the products.

$$\begin{vmatrix} 1 & 3 & -2 \\ 0 & 4 & 1 \\ -5 & 1 & 6 \end{vmatrix} \begin{matrix} 1 & 3 \\ 0 & 4 \\ -5 & 1 \end{matrix}$$

$$(1)(4)(6) + (3)(1)(-5) + (-2)(0)(1) = 24 + (-15) + 0 = 9$$

**Step 4:** Repeat Steps 2 and 3 for the three upward diagonal groups of three coefficients.

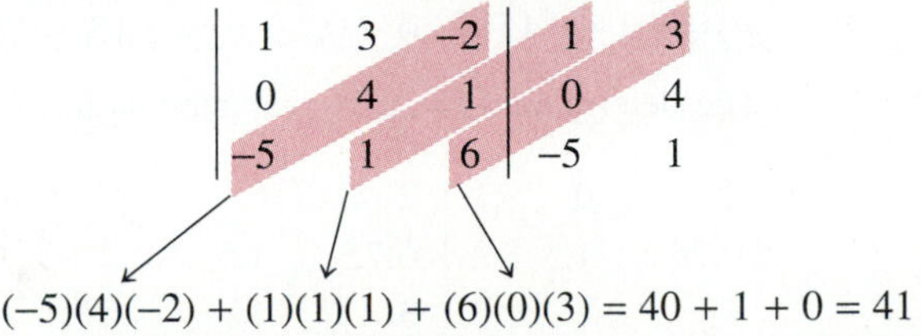


$$(-5)(4)(-2) + (1)(1)(1) + (6)(0)(3) = 40 + 1 + 0 = 41$$

**Step 5:** Subtract the result in Step 4 from the result in Step 3 to get the value of the characteristic determinant.

$$9 - 41 = -32$$

To solve for $I_1$ in the given set of three equations, a determinant is formed by substituting the constants on the right of the equations for the coefficients of $I_1$ in the characteristic determinant.

$$\begin{vmatrix} 7 & 3 & -2 \\ 8 & 4 & 1 \\ 9 & 1 & 6 \end{vmatrix}$$

This determinant is evaluated using the method described in the previous steps.

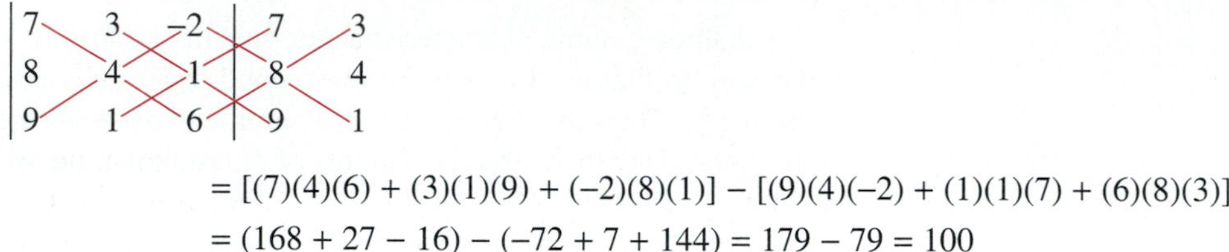

$$= [(7)(4)(6) + (3)(1)(9) + (-2)(8)(1)] - [(9)(4)(-2) + (1)(1)(7) + (6)(8)(3)]$$
$$= (168 + 27 - 16) - (-72 + 7 + 144) = 179 - 79 = 100$$

$I_1$ is found by dividing this determinant value by the value of the characteristic determinant.

$$I_1 = \frac{\begin{vmatrix} 7 & 3 & -2 \\ 8 & 4 & 1 \\ 9 & 1 & 6 \end{vmatrix}}{\begin{vmatrix} 1 & 3 & -2 \\ 0 & 4 & 1 \\ -5 & 1 & 6 \end{vmatrix}} = \frac{100}{-32} = -3.125 \text{ A}$$

$I_2$ and $I_3$ are found in a similar way.

**EXAMPLE 9–3** Determine the value of $I_2$ from the following set of equations:

$$2I_1 + 0.5I_2 + 1I_3 = 0$$
$$0.75I_1 + 0I_2 + 2I_3 = 1.5$$
$$3I_1 + 0.2I_2 + 0I_3 = -1$$

***Solution*** The characteristic determinant is evaluated as follows:

$$\begin{vmatrix} 2 & 0.5 & 1 \\ 0.75 & 0 & 2 \\ 3 & 0.2 & 0 \end{vmatrix} \begin{matrix} 2 & 0.5 \\ 0.75 & 0 \\ 3 & 0.2 \end{matrix}$$

$$= [(2)(0)(0) + (0.5)(2)(3) + (1)(0.75)(0.2)] - [(3)(0)(1) + (0.2)(2)(2) + (0)(0.75)(0.5)]$$
$$= (0 + 3 + 0.15) - (0 + 0.8 + 0) = 3.15 - 0.8 = 2.35$$

The determinant for $I_2$ is evaluated as follows:

$$\begin{vmatrix} 2 & 0 & 1 \\ 0.75 & 1.5 & 2 \\ 3 & -1 & 0 \end{vmatrix} \begin{matrix} 2 & 0 \\ 0.75 & 1.5 \\ 3 & -1 \end{matrix}$$

$$= [(2)(1.5)(0) + (0)(2)(3) + (1)(0.75)(-1)] - [(3)(1.5)(1) + (-1)(2)(2) + (0)(0.75)(0)]$$
$$= [0 + 0 + (-0.75)] - [4.5 + (-4) + 0] = -0.75 - 0.5 = -1.25$$

Finally,

$$I_2 = \frac{-1.25}{2.35} = -0.532\text{ A} = -532\text{ mA}$$

***Related Exercise*** Determine the value of $I_1$ in the set of equations used in Example 9–3.

***The Cofactor Method*** Unlike the expansion method, the cofactor method can be used to evaluate determinants with higher orders than three and is therefore more versatile. We will use a third-order determinant to illustrate the method, keeping in mind that fourth-, fifth-, and higher-order determinant can be evaluated in a similar way. The following specific determinant is used to demonstrate the cofactor method on a step-by-step basis.

$$\begin{vmatrix} 1 & 3 & -2 \\ 4 & 0 & -1 \\ 5 & 1.5 & 6 \end{vmatrix}$$

**Step 1:** Select any one column or row in the determinant. Each number in the selected column or row is used as a multiplying factor. For illustration, we will use the first column.

$$\begin{vmatrix} 1 & 3 & -2 \\ 4 & 0 & -1 \\ 5 & 1.5 & 6 \end{vmatrix}$$

**Step 2:** Determine the cofactor for each number in the selected column (or row). The cofactor for a given number is the determinant formed by all numbers that are *not* in the same column or row as the given number. This is illustrated as follows:

$$\begin{vmatrix} 1 & 3 & -2 \\ 4 & 0 & -1 \\ 5 & 1.5 & 6 \end{vmatrix} \qquad \begin{vmatrix} 1 & 3 & -2 \\ 4 & 0 & -1 \\ 5 & 1.5 & 6 \end{vmatrix} \qquad \begin{vmatrix} 1 & 3 & -2 \\ 4 & 0 & -1 \\ 5 & 1.5 & 6 \end{vmatrix}$$

↑ Cofactor for 1 ↑ Cofactor for 4 ↑ Cofactor for 5

**Step 3:** Assign the proper sign to each multiplying factor according to the following format (note the alternating pattern).

$$\begin{vmatrix} + & - & + \\ - & + & - \\ + & - & + \end{vmatrix}$$

**Step 4:** Sum all of the products of each multiplying factor and its associated cofactor using the appropriate sign.

$$1 \times \begin{vmatrix} 0 & -1 \\ 1.5 & 6 \end{vmatrix} - 4 \times \begin{vmatrix} 3 & -2 \\ 1.5 & 6 \end{vmatrix} + 5 \times \begin{vmatrix} 3 & -2 \\ 0 & -1 \end{vmatrix}$$

$$= 1[(0)(6) - (1.5)(-1)] - 4[(3)(6) + (1.5)(-2)] + 5[(3)(-1) + (0)(-2)]$$

$$= 1(1.5) - 4(15) + 5(-3) = 1.5 - 60 - 15 = -73.5$$

**EXAMPLE 9–4** Repeat Example 9–3 using the cofactor method to find $I_2$. The equations are repeated below.

$$2I_1 + 0.5I_2 + 1I_3 = 0$$
$$0.75I_1 + 0I_2 + 2I_3 = 1.5$$
$$3I_1 + 0.2I_2 + 0I_3 = -1$$

***Solution*** Evaluate the characteristic determinant as follows:

$$\begin{vmatrix} 2 & 0.5 & 1 \\ 0.75 & 0 & 2 \\ 3 & 0.2 & 0 \end{vmatrix} = 2\begin{vmatrix} 0 & 2 \\ 0.2 & 0 \end{vmatrix} - 0.75\begin{vmatrix} 0.5 & 1 \\ 0.2 & 0 \end{vmatrix} + 3\begin{vmatrix} 0.5 & 1 \\ 0 & 2 \end{vmatrix}$$

$$= 2[(0)(0) - (0.2)(2)] - 0.75[(0.5)(0) - (0.2)(1)] + 3[(0.5)(2) - (0)(1)]$$
$$= 2(-0.4) - 0.75(-0.2) + 3(1) = -0.8 + 0.15 + 3 = 2.35$$

The determinant for $I_2$ is

$$\begin{vmatrix} 2 & 0 & 1 \\ 0.75 & 1.5 & 2 \\ 3 & -1 & 0 \end{vmatrix} = 2\begin{vmatrix} 1.5 & 2 \\ -1 & 0 \end{vmatrix} - 0.75\begin{vmatrix} 0 & 1 \\ -1 & 0 \end{vmatrix} + 3\begin{vmatrix} 0 & 1 \\ 1.5 & 2 \end{vmatrix}$$

$$= 2[(1.5)(0) - (-1)(2)] - 0.75[(0)(0) - (-1)(1)] + 3[(0)(2) - (1.5)(1)]$$
$$= 2(2) - 0.75(1) + 3(-1.5) = 4 - 0.75 - 4.5 = -1.25$$

Then,

$$I_2 = \frac{-1.25}{2.35} = -0.532\text{ A} = -532\text{ mA}$$

***Related Exercise*** Find $I_1$ using the cofactor method. Use the set of equations in Example 9–4.

**SECTION 9–2 REVIEW**

1. Evaluate the following determinants:

**(a)** $\begin{vmatrix} 0 & -1 \\ 4 & 8 \end{vmatrix}$ **(b)** $\begin{vmatrix} 0.25 & 0.33 \\ -0.5 & 1 \end{vmatrix}$ **(c)** $\begin{vmatrix} 1 & 3 & 7 \\ 2 & -1 & 7 \\ -4 & 0 & -2 \end{vmatrix}$

2. Set up the characteristic determinant for the following set of simultaneous equations:

$$2I_1 + 3I_2 = 0$$
$$5I_1 + 4I_2 = 1$$

3. Find $I_2$ in Question 2.

## 9–3 ■ MESH CURRENT METHOD

***In the mesh current method, you will work with loop currents instead of branch currents. A branch current is the actual current through a branch. An ammeter placed in a given branch will measure the branch current. Loop currents are different because they are mathematical quantities that are used to make circuit analysis somewhat easier than with the branch current method. The term mesh comes from the fact that a multiple-loop circuit, when drawn, can be imagined to resemble a wire mesh.***

***After completing this section, you should be able to***

- **Use mesh analysis to find unknown quantities in a circuit**
  - ☐ Assign loop currents
  - ☐ Apply Kirchhoff's voltage law around each loop
  - ☐ Develop the loop (mesh) equations
  - ☐ Solve the loop equations

---

A systematic method of mesh analysis is given in the following steps and is illustrated in Figure 9–4, which is the same circuit configuration as in Figure 9–1 used in the branch current analysis. It demonstrates the basic principles well.

**FIGURE 9–4**
***Circuit for mesh analysis.***

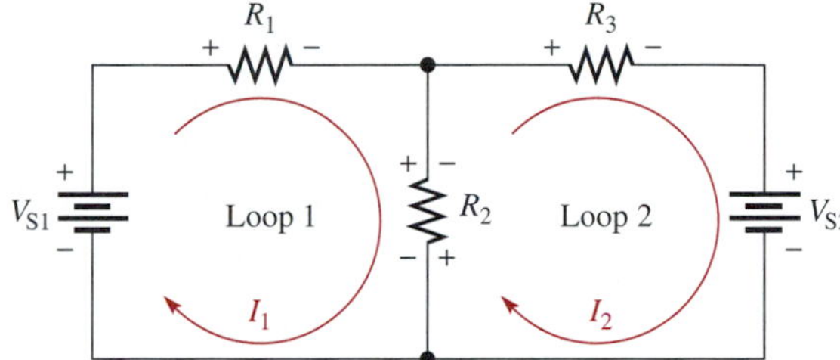


**Step 1:** Assign a current in the clockwise (CW) direction around each closed loop. This may not be the actual current direction, but it does not matter. The number of current assignments must be sufficient to include current through all components in the circuit. No redundant current assignments should be made. The direction does not have to be clockwise, but we will use CW for consistency.

**Step 2:** Indicate the voltage drop polarities in each loop based on the assigned current directions.

**Step 3:** Apply Kirchhoff's voltage law around each closed loop. When more than one loop current passes through a component, include its voltage drop. This results in one equation for each loop.

**Step 4:** Using substitution or determinants, solve the resulting equations for the loop currents.

First, the loop currents $I_1$ and $I_2$ are assigned in the CW direction as shown in Figure 9–4. A loop current could be assigned around the outer perimeter of the circuit, but this information would be redundant since $I_1$ and $I_2$ already pass through all of the components.

Second, the polarities of the voltage drops across $R_1$, $R_2$, and $R_3$ are shown based on the loop current directions. Notice that $I_1$ and $I_2$ are in opposite directions through $R_2$ because $R_2$ is common to both loops. Therefore, two voltage polarities are indicated. In reality, $R_2$ currents cannot be separated into two parts, but remember that the loop currents are basically mathematical quantities used for analysis purposes. The polarities of the voltage sources are fixed and are not affected by the current assignments.

Third, Kirchhoff's voltage law applied to the two loops results in the following two equations:

$$R_1I_1 + R_2(I_1 - I_2) = V_{S1} \qquad \text{for loop 1}$$
$$R_3I_2 + R_2(I_2 - I_1) = -V_{S2} \qquad \text{for loop 2}$$

Fourth, the like terms in the equations are combined and rearranged for convenient solution so that they have the same position in each equation, that is, the $I_1$ term is first and the $I_2$ term is second. The equations are rearranged into the following form. Once the loop currents are evaluated, all of the branch currents can be determined.

$$(R_1 + R_2)I_1 - R_2I_2 = V_{S1} \qquad \text{for loop 1}$$
$$-R_2I_1 + (R_2 + R_3)I_2 = -V_{S2} \qquad \text{for loop 2}$$

Notice that in the mesh current method only two equations are required for the same circuit that required three equations in the branch current method. The last two equations (developed in the fourth step) follow a certain format to make mesh analysis easier. Referring to these last two equations, notice that for loop 1, the total resistance in the loop, $R_1 + R_2$, is multiplied by $I_1$ (its loop current). Also in the loop 1 equation, the resistance common to both loops, $R_2$, is multiplied by the other loop current, $I_2$, and subtracted from the first term. The same general form is seen in the loop 2 equation except that the terms have been rearranged. From these observations, the format for setting up the equation for a loop circuit can be stated as follows:

1. Sum the resistances around the loop, and multiply by the loop current.
2. Subtract the common resistance(s) times the adjacent loop current(s).
3. Set the terms in Steps 1 and 2 equal to the total source voltage in the loop. The sign of the source voltage is positive if the assigned loop current is out of its positive terminal. The sign is negative if the loop current is into its positive terminal.
4. Rearrange the terms so that like terms appear in the same position in each equation.

Example 9–5 illustrates the application of this format to the mesh current analysis of a circuit.

**EXAMPLE 9–5** Using the mesh current method, find the branch currents in Figure 9–5.

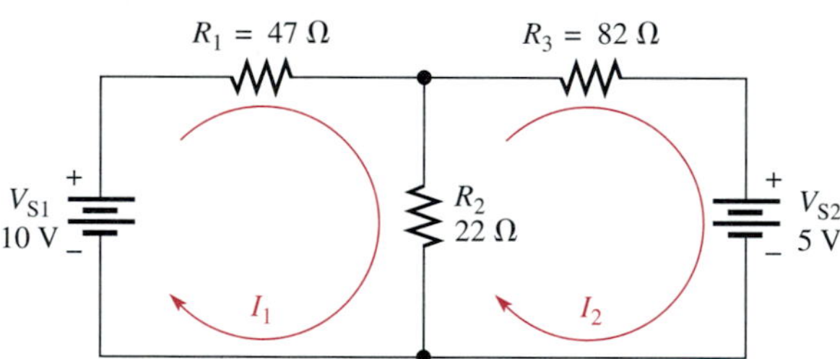


FIGURE 9–5

***Solution*** Assign the loop currents ($I_1$ and $I_2$) as shown in Figure 9–5; resistance values are in ohms and voltage values are in volts. Use the format described to set up the two loop equations.

$$(47 + 22)I_1 - 22I_2 = 10$$
$$69I_1 - 22I_2 = 10 \qquad \text{for loop 1}$$
$$-22I_1 + (22 + 82)I_2 = -5$$
$$-22I_1 + 104I_2 = -5 \qquad \text{for loop 2}$$

Use determinants to find $I_1$.

$$I_1 = \frac{\begin{vmatrix} 10 & -22 \\ -5 & 104 \end{vmatrix}}{\begin{vmatrix} 69 & -22 \\ -22 & 104 \end{vmatrix}} = \frac{(10)(104) - (-5)(-22)}{(69)(104) - (-22)(-22)} = \frac{1040 - 110}{7176 - 484} = 139 \text{ mA}$$

Solving for $I_2$ yields

$$I_2 = \frac{\begin{vmatrix} 69 & 10 \\ -22 & -5 \end{vmatrix}}{6692} = \frac{(69)(-5) - (-22)(10)}{6692} = \frac{-345 - (-220)}{6692} = -18.7 \text{ mA}$$

The negative sign on $I_2$ means that its direction must be reversed.

Now find the actual branch currents. Since $I_1$ is the only current through $R_1$, it is also the branch current $I_{R1}$.

$$I_{R1} = I_1 = 139 \text{ mA}$$

Since $I_2$ is the only current through $R_3$, it is also the branch current $I_{R3}$.

$$I_{R3} = I_2 = -18.7 \text{ mA} \qquad \text{(opposite direction of that originally assigned to } I_2\text{)}$$

Both loop currents $I_1$ and $I_2$ are through $R_2$ in the same direction. Remember, the negative $I_2$ value told you to reverse its assigned direction.

$$I_{R2} = I_1 - I_2 = 139 \text{ mA} - (-18.7 \text{ mA}) = 158 \text{ mA}$$

Keep in mind that once you know the branch currents, you can find the voltages by using Ohm's law.

***Related Exercise*** Find $I_1$ by the mesh current method if the 10 V source in Figure 9–5 is reversed.

## Circuits with More Than Two Loops

The mesh method also can be systematically applied to circuits with any number of loops. Of course, the more loops there are, the more difficult is the solution. However, the basic procedure still applies. For example, for a three-loop circuit, three simultaneous equations are required. Example 9–6 illustrates the analysis of a three-loop circuit.

**EXAMPLE 9–6** Find $I_3$ in Figure 9–6.

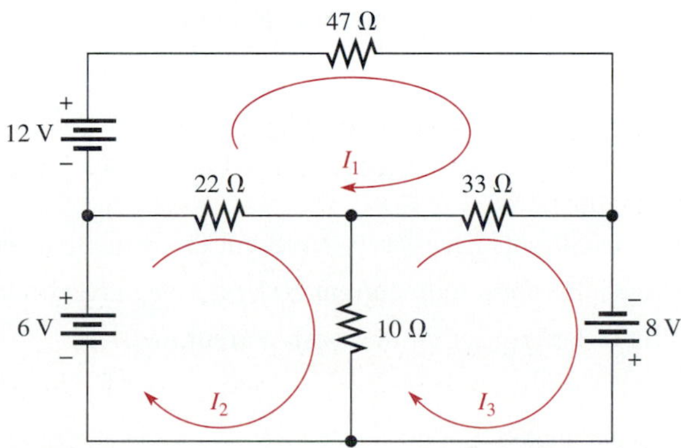


FIGURE 9–6

***Solution*** Assign three CW loop currents ($I_1$, $I_2$, and $I_3$) as shown in Figure 9–6. Then use the format procedure to set up each loop equation. The polarity of a voltage source is positive when the assigned mesh current is out of the positive terminal. A concise restatement of this procedure is as follows:

> **(Sum of resistors in loop) times (loop current) minus (each common resistor) times (associated adjacent loop current) equals (source voltage in the loop).**

The loop equations are as follows:

$$102I_1 - 22I_2 - 33I_3 = 12 \qquad \text{for loop 1}$$
$$-22I_1 + 32I_2 - 10I_3 = 6 \qquad \text{for loop 2}$$
$$-33I_1 - 10I_2 + 43I_3 = 8 \qquad \text{for loop 3}$$

These three equations can be solved for the loop currents by substitution or, more easily, with third-order determinants. $I_3$ is found using determinants as follows.

The characteristic determinant is evaluated as follows:

$$\begin{vmatrix} 102 & -22 & -33 \\ -22 & 32 & -10 \\ -33 & -10 & 43 \end{vmatrix} = 102\begin{vmatrix} 32 & -10 \\ -10 & 43 \end{vmatrix} - (-22)\begin{vmatrix} -22 & -33 \\ -10 & 43 \end{vmatrix} + (-33)\begin{vmatrix} -22 & -33 \\ 32 & -10 \end{vmatrix}$$

$$= 102[(32)(43) - (-10)(-10)] - (-22)[(-22)(43) - (-10)(-33)] + (-33)[(-22)(-10) - (32)(-33)]$$
$$= 102(1276) + 22(-1276) - 33(1276) = 130{,}152 - 28{,}072 - 42{,}108 = 59{,}972$$

The $I_3$ determinant is evaluated as follows:

$$\begin{vmatrix} 102 & -22 & 12 \\ -22 & 32 & 6 \\ -33 & -10 & 8 \end{vmatrix} = 102\begin{vmatrix} 32 & 6 \\ -10 & 8 \end{vmatrix} - (-22)\begin{vmatrix} -22 & 12 \\ -10 & 8 \end{vmatrix} + (-33)\begin{vmatrix} -22 & 12 \\ 32 & 6 \end{vmatrix}$$

$$= 102[(33)(8) - (-10)(6)] + 22[(-22)(8) - (-10)(12)] - 33[(-22)(6) - (32)(12)]$$
$$= 102(316) + 22(-56) - 33(-516) = 32{,}232 - 1232 + 17{,}028 = 48{,}028$$

$I_3$ is determined by dividing the value of the $I_3$ determinant by the value of the characteristic determinant.

$$I_3 = \frac{48{,}028}{59{,}972} = 801 \text{ mA}$$

The other loop currents are found similarly. The actual branch currents and voltages can be determined ones you know the loop currents.

***Related Exercise*** Find loop current $I_1$ in Figure 9–6.

**SECTION 9–3 REVIEW**

1. Do the loop currents necessarily represent the actual currents in the branches?
2. When you solve for a loop current and get a negative value, what does it mean?
3. What circuit law is used in the mesh current method?

## 9–4 ■ NODE VOLTAGE METHOD

***Another alternate method of analysis of multiple-loop circuits is called the node voltage method. It is based on finding the voltages at each node in the circuit using Kirchhoff's current law. A node is the junction of two or more components.***

***After completing this section, you should be able to***

- **Use node analysis to find unknown quantities in a circuit**
  - ☐ Select the nodes at which the voltage is unknown and assign currents
  - ☐ Apply Kirchhoff's current law at each node
  - ☐ Develop the node equations
  - ☐ Solve the node equations

---

The general steps for the node voltage method of circuit analysis are as follows:

**Step 1:** Determine the number of nodes.

**Step 2:** Select one node as a reference. All voltages will be relative to the reference node. Assign voltage designations to each node where the voltage is unknown.

**Step 3:** Assign currents at each node where the voltage is unknown, except at the reference node. The directions are arbitrary.

**Step 4:** Apply Kirchhoff's current law to each node where currents are assigned.

**Step 5:** Express the current equations in terms of voltages, and solve the equations for the unknown node voltages using Ohm's law.

We will use Figure 9–7 to illustrate the general approach to node voltage analysis. First, establish the nodes. In this case, there are four nodes, as indicated in the figure. Second, let's use node $B$ as reference. Think of it as circuit ground. Node voltages $C$ and $D$ are already known to be the source voltages. The voltage at node $A$ is the only unknown; it is designated as $V_A$. Third, arbitrarily assign the branch currents at node $A$ as indicated in the figure. Fourth, the Kirchhoff current equation at node $A$ is

$$I_{R1} - I_{R2} + I_{R3} = 0$$

Fifth, express the currents in terms of circuit voltages using Ohm's law as follows:

$$I_{R1} = \frac{V_1}{R_1} = \frac{V_{S1} - V_A}{R_1}$$

$$I_{R2} = \frac{V_2}{R_2} = \frac{V_A}{R_2}$$

$$I_{R3} = \frac{V_3}{R_3} = \frac{V_{S2} - V_A}{R_3}$$

Substituting these into the current equation yields

$$\frac{V_{S1} - V_A}{R_1} - \frac{V_A}{R_2} + \frac{V_{S2} - V_A}{R_3} = 0$$

The only unknown is $V_A$; so solve the single equation by combining and rearranging terms. Once the voltage is known, all branch currents can be calculated. Example 9–7 illustrates this method further.

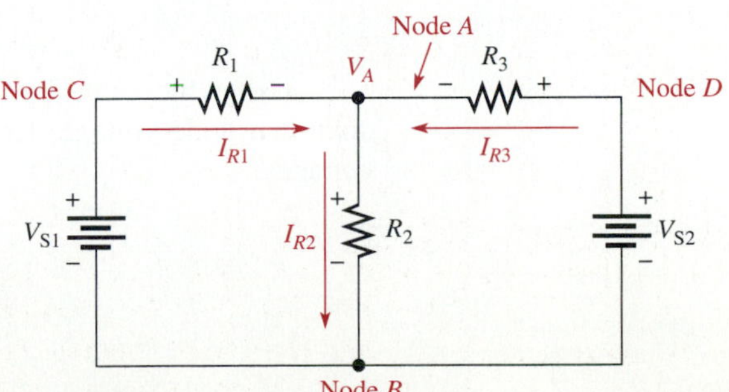


**FIGURE 9–7**
***Circuit for node voltage analysis.***

**EXAMPLE 9–7** Find the node voltages in Figure 9–8.

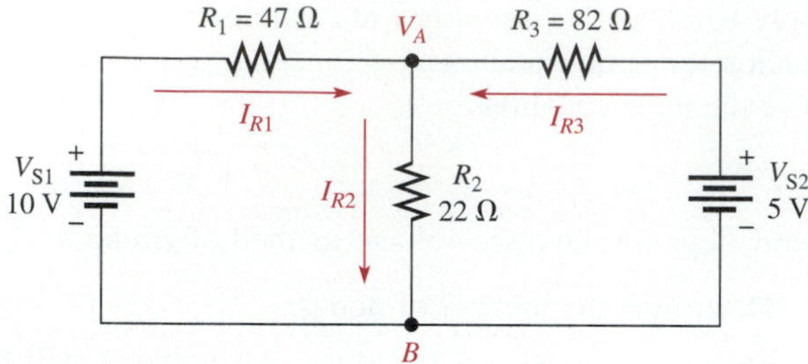


**FIGURE 9–8**

***Solution*** The reference node is chosen at $B$. The unknown node voltage is $V_A$, as indicated in Figure 9–8. This is the only unknown voltage. Branch currents are assigned at node $A$ as shown. The current equation is

$$I_{R1} - I_{R2} + I_{R3} = 0$$

Substitution for currents using Ohm's law gives the equation in terms of voltages.

$$\frac{10 - V_A}{47} - \frac{V_A}{22} + \frac{5 - V_A}{82} = 0$$

Rearranging the terms yields

$$\frac{10}{47} - \frac{V_A}{47} - \frac{V_A}{22} + \frac{5}{82} - \frac{V_A}{82} = 0$$

$$-\frac{V_A}{47} - \frac{V_A}{22} - \frac{V_A}{82} = -\frac{10}{47} - \frac{5}{82}$$

To solve for $V_A$, combine the terms on each side of the equation and find the common denominator.

$$\frac{1804V_A + 3854V_A + 1034V_A}{84{,}788} = \frac{820 + 235}{3854}$$

$$\frac{6692V_A}{84{,}788} = \frac{1055}{3854}$$

$$V_A = \frac{(1055)(84{,}788)}{(6692)(3854)} = 3.47\text{ V}$$

***Related Exercise*** Find $V_A$ in Figure 9–8 if the 5 V source is reversed.

Using the same basic procedure, you can analyze circuits with more than one unknown node voltage. Example 9–8 illustrates this calculation for two unknown node voltages.

**EXAMPLE 9–8** Using the node analysis method, calculate $V_A$ and $V_B$ in the circuit of Figure 9–9.

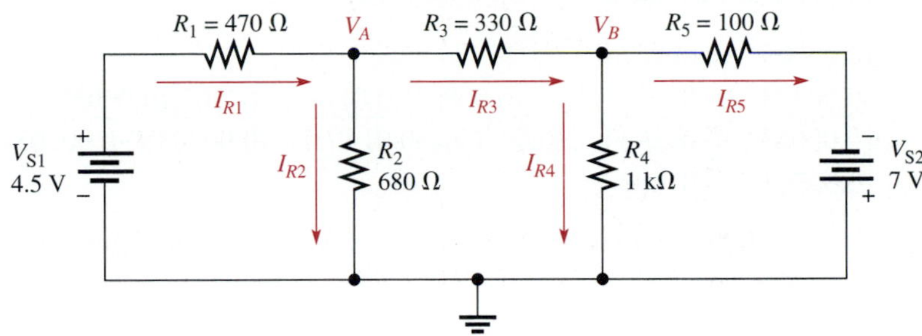


FIGURE 9–9

***Solution*** First, assign the branch currents as shown in Figure 9–9. Next, apply Kirchhoff's current law at each node. At node $A$,

$$I_{R1} - I_{R2} - I_{R3} = 0$$

Using Ohm's law substitution for the currents yields

$$\left(\frac{4.5 - V_A}{470}\right) - \left(\frac{V_A}{680}\right) - \left(\frac{V_A - V_B}{330}\right) = 0$$

$$\frac{4.5}{470} - \frac{V_A}{470} - \frac{V_A}{680} - \frac{V_A}{330} + \frac{V_B}{330} = 0$$

$$\left(\frac{1}{470} + \frac{1}{680} + \frac{1}{330}\right)V_A - \left(\frac{1}{330}\right)V_B = \frac{4.5}{470}$$

Now, using the 1/x key of your calculator, evaluate the coefficients and constant. The resulting equation for node $A$ is

$$0.00663V_A - 0.00303V_B = 0.00957$$

At node $B$,

$$I_{R3} - I_{R4} - I_{R5} = 0$$

Again use Ohm's law substitution:

$$\left(\frac{V_A - V_B}{330}\right) - \left(\frac{V_B}{1000}\right) - \left(\frac{V_B - (-7)}{100}\right) = 0$$

$$\frac{V_A}{330} - \frac{V_B}{330} - \frac{V_B}{1000} - \frac{V_B}{100} - \frac{7}{100} = 0$$

$$\left(\frac{1}{330}\right)V_A - \left(\frac{1}{330} + \frac{1}{1000} + \frac{1}{100}\right)V_B = \frac{7}{100}$$

Evaluate the coefficients and constant. The equation for node $B$ is

$$0.00303V_A - 0.01403V_B = 0.07$$

Now, these two node equations must be solved for $V_A$ and $V_B$. Use determinants to get the solutions.

$$V_A = \frac{\begin{vmatrix} 0.00957 & -0.00303 \\ 0.07 & -0.01403 \end{vmatrix}}{\begin{vmatrix} 0.00663 & -0.00303 \\ 0.00303 & -0.01403 \end{vmatrix}} = \frac{(0.00957)(-0.01403) - (0.07)(-0.00303)}{(0.00663)(-0.01403) - (0.00303)(-0.00303)} = -928 \text{ mV}$$

$$V_B = \frac{\begin{vmatrix} 0.00663 & 0.00957 \\ 0.00303 & 0.07 \end{vmatrix}}{\begin{vmatrix} 0.00663 & -0.00303 \\ 0.00303 & -0.01403 \end{vmatrix}} = \frac{(0.00663)(0.07) - (0.00303)(0.00957)}{(0.00663)(-0.01403) - (0.00303)(-0.00303)} = -5.19 \text{ V}$$

***Related Exercise*** What is the value of $V_A$ in Figure 9–9 if the 4.5 V source is reversed?

**SECTION 9–4 REVIEW**

1. What circuit law is the basis for the node voltage method?
2. What is the reference node?

## 9–5 ■ TECHnology Theory Into Practice

***Analysis of a dual-polarity loaded voltage divider provides an opportunity to apply one of the analysis methods covered in this chapter. The dual-polarity voltage divider circuit in this section operates from two voltage sources. One source is +9 V and the other is −9 V. These two voltage sources supply both positive and negative voltages that are divided down to produce reference voltages for two different devices.***

The voltage divider that you will check out in this TECH TIP section will be used to provide reference voltages to two devices using 9 V batteries. One of the devices requires a positive reference voltage and presents a 27 kΩ load to the voltage divider. The other device requires a negative reference voltage and presents a 15 kΩ load to the voltage divider. The schematic of the loaded dual-polarity voltage divider is shown in Figure 9–10.

FIGURE 9–10

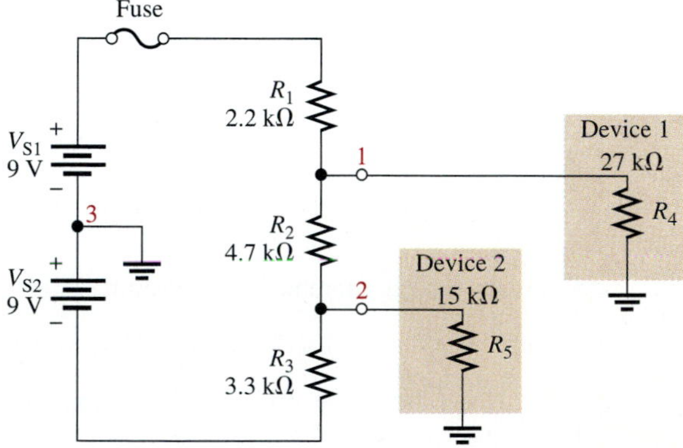

FIGURE 9–11

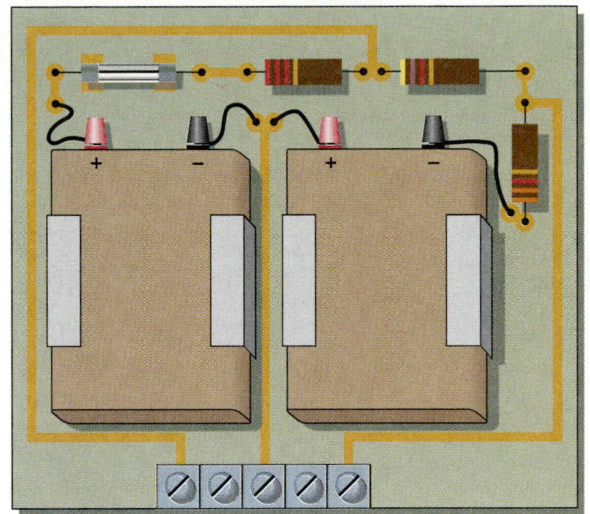


The self-contained dual-polarity voltage divider is constructed on the PC board shown in Figure 9–11. The two batteries are clip-mounted directly on the board and wired to the printed circuit pads as indicated. The two load devices can be connected to the terminal strip.

## The Printed Circuit Board and Schematic

□ Check the printed circuit board in Figure 9–11 to make sure that it agrees with the schematic in Figure 9–10. Relate each input, output, and component on the board to the schematic.

## Analysis and Troubleshooting

□ Referring to Figure 9–12, determine if the voltmeter readings are correct. The output voltages are measured without the loads connected.

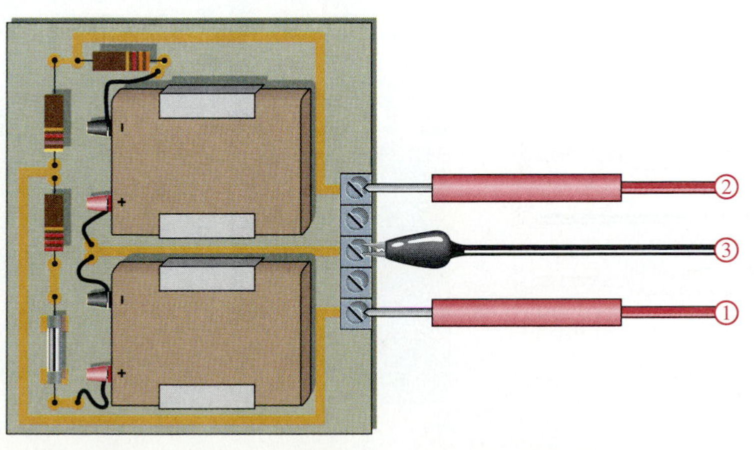


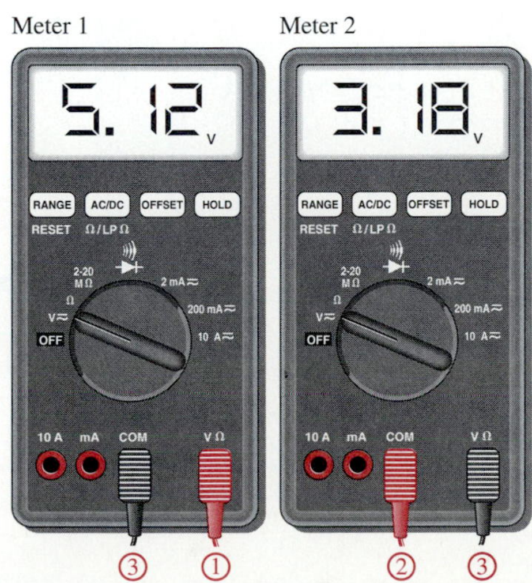


FIGURE 9–12
*Circled numbers show corresponding connections.*

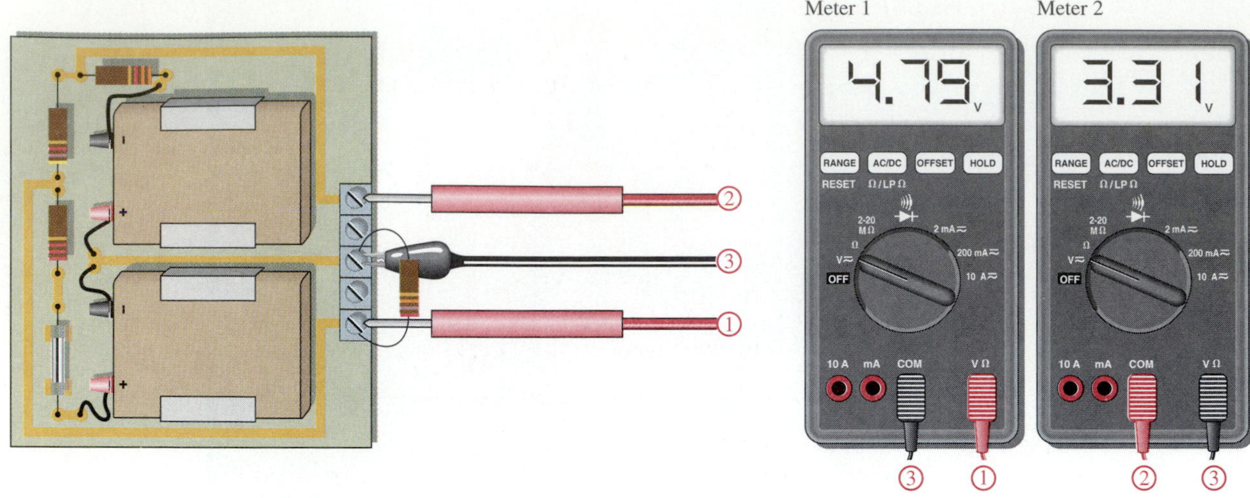


FIGURE 9–13
*Circled numbers show corresponding connections.*

- ☐ Referring to Figure 9–13, apply the node voltage method to determine if the voltmeter readings are correct. You may need to redraw the schematic in a more familiar form. The output voltages are measured with a 27 kΩ load connected from output 1 to ground to simulate one of the devices connected to the voltage divider.
- ☐ Referring to Figure 9–14, apply the node voltage method to determine if the voltmeter readings are correct. You may need to redraw the schematic in a more familiar form. The output voltages are measured with a 27 kΩ load connected from output 1 to ground and a 15 kΩ load from output 2 to ground to simulate both devices connected to the voltage divider.
- ☐ If meter 1 reads +7.50 V and meter 2 reads +5.71 V in Figure 9–14, what is the problem?
- ☐ If meter 1 reads +8.27 V and meter 2 reads −7.38 V in Figure 9–14, what is the problem?

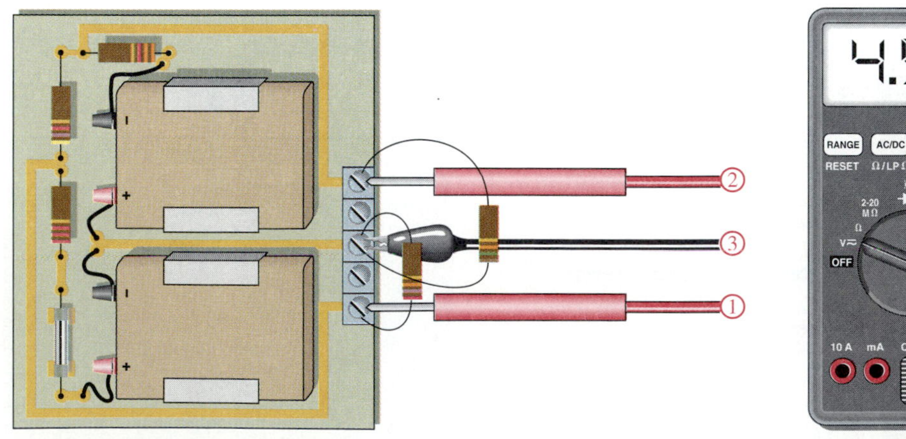


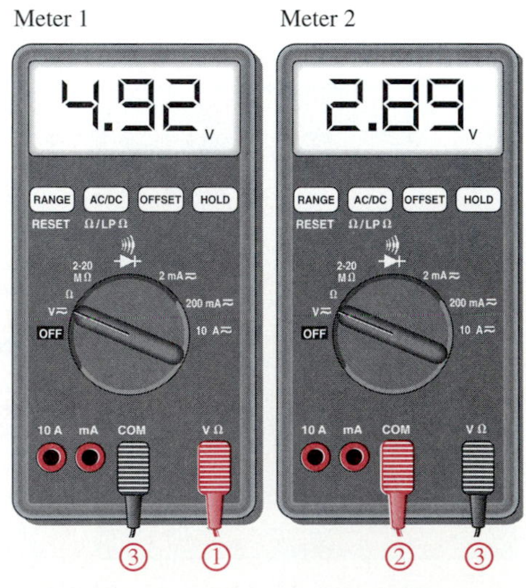


FIGURE 9–14
*Circled numbers show corresponding connections.*

**SECTION 9–5 REVIEW**

1. Determine the minimum value of fuse for the voltage-divider circuit.
2. If the batteries are rated at 10 Ah, how long will the voltage-divider circuit operate under unloaded conditions?

## 9–6 ■ PSpice DC ANALYSIS

***By now you should feel confident about using PSpice to analyze any resistive circuit, regardless of the number of components or how they are connected together. In this section, you will see how PSpice is used to analyze multiple-loop circuits.***

***After completing this section, you should be able to***

- **Use PSpice to analyze multiple loop circuits**
  - ☐ Write a circuit file to determine the current in a given resistor
  - ☐ Write a circuit file to determine the voltages at specified points

### Multiple-Loop Circuit Analysis 1

For the first example, we will write a circuit file to determine the current through $R_1$ in Figure 9–15 with the nodes labeled as shown.

FIGURE 9–15

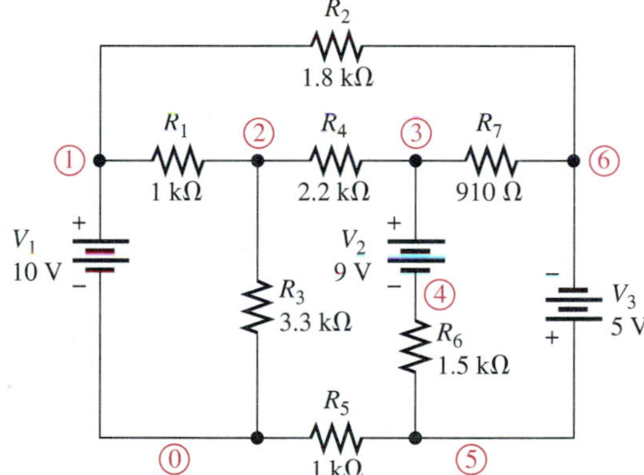


The circuit file for the circuit in Figure 9–15 is as follows:

```
Multiple-Loop Circuit Analysis 1
*Figure 9–15
V1  1  0  10
V2  3  4  9
V3  5  6  5
R1  1  2  1K
R2  1  6  1.8K
R3  2  0  3.3K
R4  2  3  2.2K
R5  0  5  1K
R6  4  5  1.5K
R7  3  6  910
.DC  V1  10  10  1
.PRINT  DC  I(R1)
.END
```

Remember that use of the .PRINT statement must include some sort of analysis control statement (in this case the .DC statement). In the preceding file, V1 was used in the .DC statement, but we could have used either V2 or V3 as follows:

```
.DC  V2  9  9  1
```

or

```
.DC  V3  5  5  1
```

Also, we could have used all three .DC control statements in the circuit file. However, this would cause PSpice to analyze the circuit three times and place the results for each in the output file. In this case the analyses would be redundant because the value of I(R1) would be the same for each analysis. It is best to use just one .DC control statement in a situation like this.

Notice in the voltage source lines, the more positive node is always listed first. This is because PSpice assumes that the first node listed is the more positive node.

After saving the circuit file, running PSpice, and examining the output file, you should find that I(R1) = 2.699 mA.

## Multiple Loop Circuit Analysis 2

For the second example, the circuit in Figure 9–16 is analyzed for the voltages at points A, B, and C. Up to now in PSpice analysis, only numbers have been used to designate nodes. However, PSpice also recognizes letters as well as numbers for node designations, except for the ground or reference node which must always be zero (0).

FIGURE 9–16

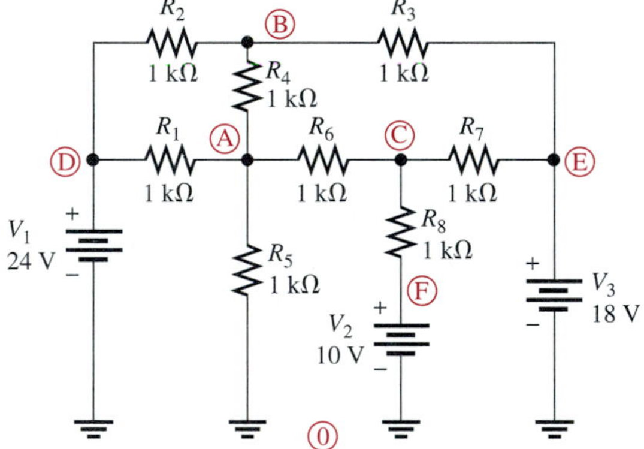


The file for the circuit in Figure 9–16 is as follows:

```
Multiple-Loop Circuit Analysis 2
*Figure 9–16
V1  D  0  24
V2  F  0  10
V3  E  0  18
R1  A  D  1K
R2  B  D  1K
R3  B  E  1K
R4  A  B  1K
R5  A  0  1K
R6  A  C  1K
R7  C  E  1K
R8  C  F  1K
.END
```

Recall that the .OP statement is not needed if the circuit file does not specify any type of circuit analysis other than voltages at the nodes.

After you run PSpice, the output file will show the following results as well as values for all the other node voltages.

V(A) = 14.200 V
V(B) = 18.733 V
V(C) = 14.06 V

## SECTION 9–6 REVIEW

1. Modify the circuit file for Figure 9–15 to also determine the value of V(R1) and V(R2).
2. Modify the circuit file for Figure 9–16 to also determine I(R1) and I(R2).

## ■ SUMMARY

- The branch current method is based on Kirchhoff's voltage law and Kirchhoff's current law.
- Simultaneous equations can be solved by substitution or by determinants.
- The number of equations must be equal to the number of unknowns.
- Second-order determinants are evaluated by adding the signed cross-products.
- Third-order determinants are evaluated by the expansion method or by the cofactor method.
- The mesh current method is based on Kirchhoff's voltage law.
- A loop current is not necessarily the actual current in a branch.
- The node voltage method is based on Kirchhoff's current law.

## ■ GLOSSARY

**Branch** One current path that connects two nodes.

**Branch current** The actual current in a branch.

**Coefficient** The constant number that appears in front of a variable.

**Determinant** An array of coefficients and constants in a given set of simultaneous equations.

**Loop** A closed current path in a circuit.

**Node** The junction of two or more components.

## ■ SELF-TEST

**1.** In Figure 9–1, there is/are
**(a)** 1 loop **(b)** 1 unknown node **(c)** 2 loops
**(d)** 2 unknown nodes **(e)** both answers (b) and (c)

**2.** In assigning the direction of branch currents,
**(a)** the directions are critical **(b)** they must all be in the same direction
**(c)** they must all point into a node **(d)** the directions are not critical

**3.** The branch current method uses
**(a)** Ohm's law and Kirchhoff's voltage law
**(b)** Kirchhoff's voltage and current laws
**(c)** the superposition theorem and Kirchhoff's current law
**(d)** Thevenin's theorem and Kirchhoff's voltage law

**4.** A determinant for two simultaneous equations will have
**(a)** 2 rows and 1 column **(b)** 1 row and 2 columns
**(c)** 2 rows and 2 columns

**5.** The first row of a certain determinant has the numbers 2 and 4. The second row has the numbers 6 and 1. The value of this determinant is
**(a)** 22 **(b)** 2 **(c)** –22 **(d)** 8

6. The expansion method for evaluating determinants is
   **(a)** only good for second-order determinants **(b)** only good for third-order determinants
   **(c)** good for any determinant **(d)** better than the cofactor method
7. The mesh current method is based on
   **(a)** Kirchhoff's current law **(b)** Ohm's law
   **(c)** the superposition theorem **(d)** Kirchhoff's voltage law
8. The node voltage method is based on
   **(a)** Kirchhoff's current law **(b)** Ohm's law
   **(c)** the superposition theorem **(d)** Kirchhoff's voltage law
9. In the node voltage method,
   **(a)** currents are assigned at each node
   **(b)** currents are assigned at the reference node
   **(c)** the current directions are arbitrary
   **(d)** currents are assigned only at the nodes where the voltage is unknown
   **(e)** both answers (c) and (d)
10. Generally, the node voltage method results in
   **(a)** more equations than the mesh current method
   **(b)** fewer equations than the mesh current method
   **(c)** the same number of equations as the mesh current method

## ■ PROBLEMS

### SECTION 9–1 Branch Current Method

1. Identify all possible loops in Figure 9–17.

FIGURE 9–17

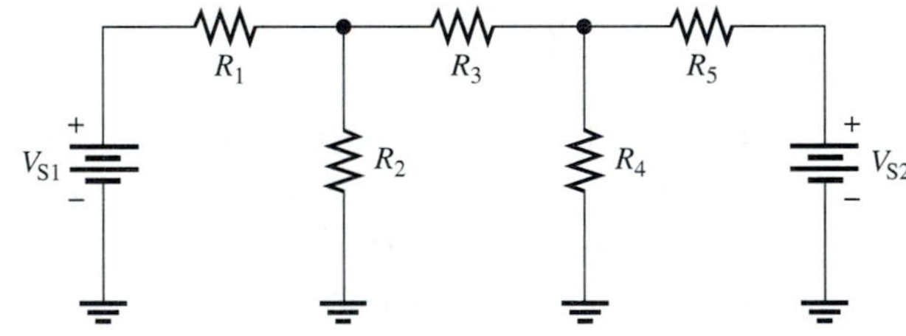


2. Identify all nodes in Figure 9–17. Which ones have a known voltage?
3. Write the Kirchhoff current equation for the current assignment shown at node $A$ in Figure 9–18.

FIGURE 9–18

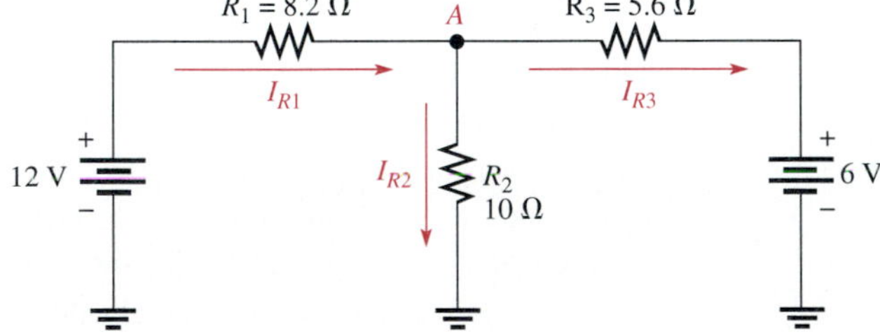


4. Solve for each of the branch currents in Figure 9–18.
5. Find the voltage drop across each resistor in Figure 9–18 and indicate its actual polarity.
6. Using the substitution method, solve the following set of equations for $I_{R1}$ and $I_{R2}$.

$$100I_{R1} + 50I_{R2} = 30$$
$$75I_{R1} + 90I_{R2} = 15$$

**7.** Using the substitution method, solve the following set of three equations for all currents:

$$5I_{R1} - 2I_{R2} + 8I_{R3} = 1$$
$$2I_{R1} + 4I_{R2} - 12I_{R3} = 5$$
$$10I_{R1} + 6I_{R2} + 9I_{R3} = 0$$

**8.** Find the current through each resistor in Figure 9–19.

FIGURE 9–19

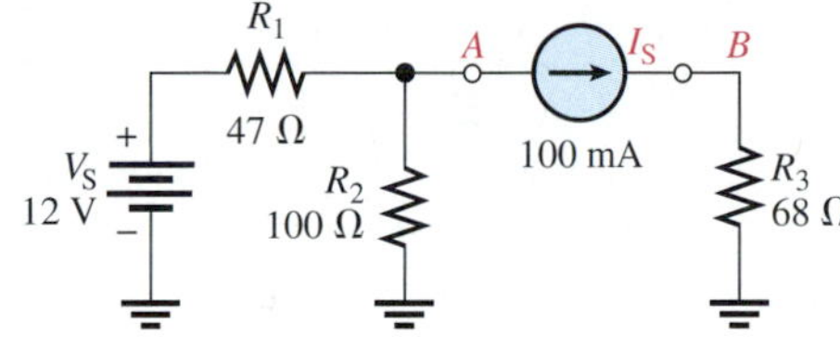


**9.** In Figure 9–19, determine the voltage across the current source (points *A* and *B*).

## SECTION 9–2 Determinants

**10.** Evaluate each determinant:

**(a)** $\begin{vmatrix} 4 & 6 \\ 2 & 3 \end{vmatrix}$ **(b)** $\begin{vmatrix} 9 & -1 \\ 0 & 5 \end{vmatrix}$ **(c)** $\begin{vmatrix} 12 & 15 \\ -2 & -1 \end{vmatrix}$ **(d)** $\begin{vmatrix} 100 & 50 \\ 30 & -20 \end{vmatrix}$

**11.** Using determinants, solve the following set of equations for both currents:

$$-I_1 + 2I_2 = 4$$
$$7I_1 + 3I_2 = 6$$

**12.** Evaluate each of the determinants using the expansion method:

**(a)** $\begin{vmatrix} 1 & 0 & -2 \\ 5 & 4 & 1 \\ 2 & 10 & 0 \end{vmatrix}$ **(b)** $\begin{vmatrix} 0.5 & 1 & -0.8 \\ 0.1 & 1.2 & 1.5 \\ -0.1 & -0.3 & 5 \end{vmatrix}$

**13.** Evaluate each of the determinants using the cofactor method:

**(a)** $\begin{vmatrix} 25 & 0 & -20 \\ 10 & 12 & 5 \\ -8 & 30 & -16 \end{vmatrix}$ **(b)** $\begin{vmatrix} 1.08 & 1.75 & 0.55 \\ 0 & 2.12 & -0.98 \\ 1 & 3.49 & -1.05 \end{vmatrix}$

**14.** Find $I_1$ and $I_3$ in Example 9–3.

**15.** Solve for $I_1$, $I_2$, $I_3$ in the following set of equations:

$$2I_1 - 6I_2 + 10I_3 = 9$$
$$3I_1 + 7I_2 - 8I_3 = 3$$
$$10I_1 + 5I_2 - 12I_3 = 0$$

**16.** Find $V_1$, $V_2$, $V_3$, and $V_4$ from the following set of equations:

$$16V_1 + 10V_2 - 8V_3 - 3V_4 = 15$$
$$2V_1 + 0V_2 + 5V_3 + 2V_4 = 0$$
$$-7V_1 - 12V_2 + 0V_3 + 0V_4 = 9$$
$$-1V_1 + 20V_2 - 18V_3 + 0V_4 = 10$$

## SECTION 9–3 Mesh Current Method

17. Using the mesh current method, find the loop currents in Figure 9–20.
18. Find the branch currents in Figure 9–20.
19. Determine the voltages and their proper polarities for each resistor in Figure 9–20.

FIGURE 9–20

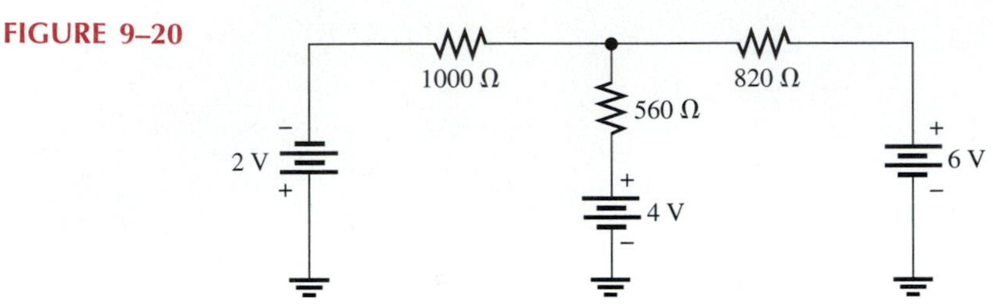


20. Write the loop equations for the circuit in Figure 9–21.
21. Solve for the loop currents in Figure 9–21.
22. Find the current through each resistor in Figure 9–21.

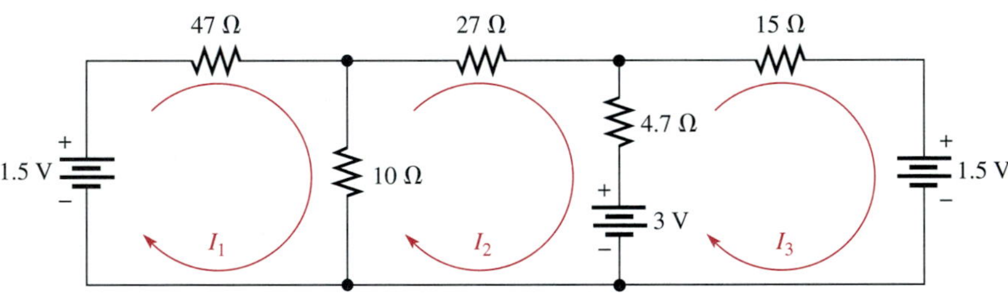


FIGURE 9–21

23. Determine the voltage across the open bridge terminals, *AB*, in Figure 9–22.
24. When a 10 Ω resistor is connected from point *A* to point *B* in Figure 9–22, what is the current through it?

FIGURE 9–22

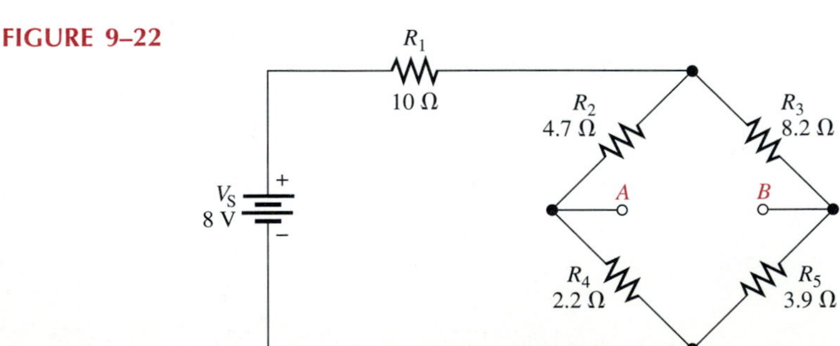

**25.** Find the current through $R_1$ in Figure 9–23.

FIGURE 9–23

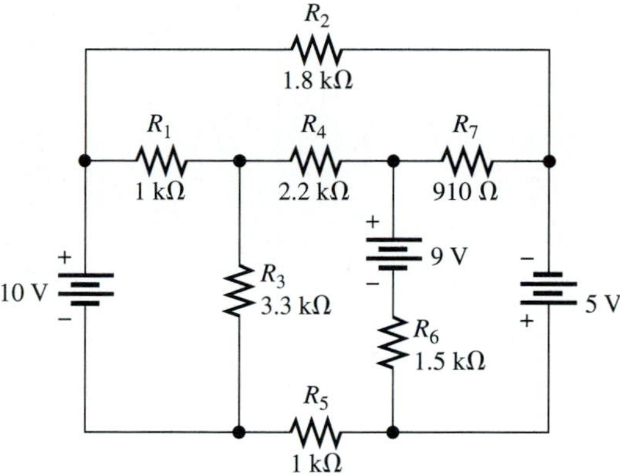


## SECTION 9–4 Node Voltage Method

**26.** In Figure 9–24, use the node voltage method to find the voltage at point $A$ with respect to ground.

**27.** What are the branch current values in Figure 9–24? Show the actual direction of current in each branch.

**28.** Write the node voltage equations for Figure 9–21.

FIGURE 9–24

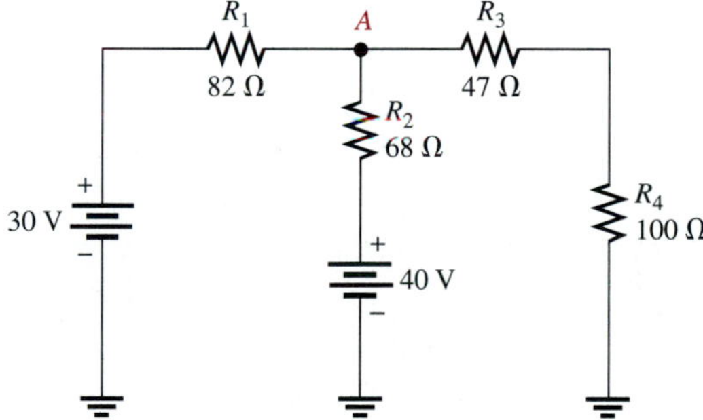


**29.** Use node analysis to determine the voltage at points $A$ and $B$ with respect to ground in Figure 9–25.

FIGURE 9–25

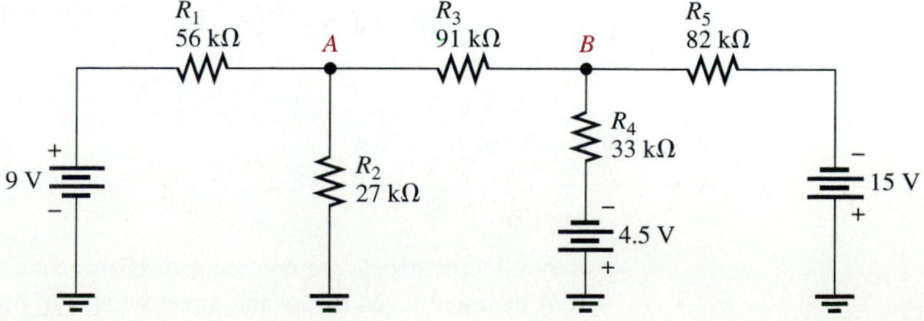

**30.** Find the voltage at points $A$, $B$, and $C$ in Figure 9–26.

FIGURE 9–26

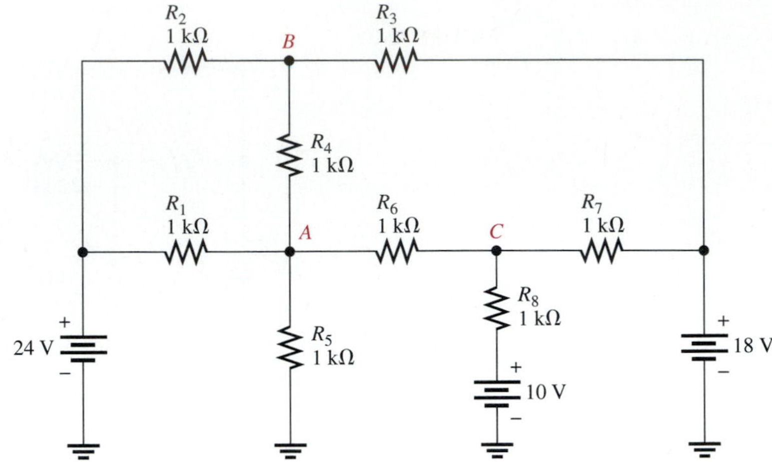


**31.** Use node analysis, mesh analysis, or any other procedure to find all currents and the voltages at each unknown node in Figure 9–27.

FIGURE 9–27

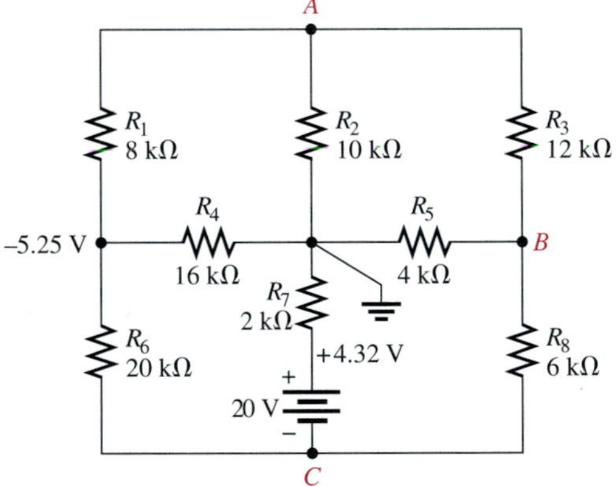


## ▪ ANSWERS TO SECTION REVIEWS

### Section 9–1

**1.** Kirchhoff's voltage law and Kirchhoff's current law are used in the branch current method.

**2.** False, but write the equations so that they are consistent with your assigned directions.

**3.** A loop is a closed path within a circuit.

**4.** A node is a junction of two or more components.

### Section 9–2

**1. (a)** 4 **(b)** 0.415 **(c)** –98

**2.** $\begin{vmatrix} 2 & 3 \\ 5 & 4 \end{vmatrix}$

**3.** –0.286 A

### Section 9–3

**1.** No, loop currents are not necessarily the same as branch currents.

**2.** A negative value means the direction should be reversed.

**3.** Kirchhoff's voltage law is used in mesh analysis.

### Section 9–4

**1.** Kirchhoff's current law is the basis for node analysis.
**2.** A reference node is the junction to which all circuit voltages are referenced.

### Section 9–5

**1.** $I_T$ = 1.90 mA; use smallest available fuse with a rating greater than 1.90 mA.
**2.** 10 Ah/1.76 mA = 5682 h

### Section 9–6

**1.** Change .PRINT statement to:

```
.PRINT   DC   I(R1)   V(R1)   V(R2)
```

**2.** Add:

```
.DC   V1   24   24   1
.PRINT   I(R1)   I(R2)
```

---

## ■ ANSWERS TO RELATED EXERCISES FOR EXAMPLES

**9–1** $I_1$ = 176 mA; $I_2$ = 77.8 mA; $I_3$ = −98.7 mA
**9–2** 3.71 A
**9–3** −298 mA
**9–4** −298 mA
**9–5** −172 mA
**9–6** 553 mA
**9–7** 1.92 V
**9–8** −4.13 V

# 10

# MAGNETISM AND ELECTROMAGNETISM



## ■ INTRODUCTION

This chapter is somewhat of a departure from the coverage of dc circuits in the previous nine chapters because two totally different concepts are introduced—magnetism and electromagnetism. The operation of many types of devices such as the relay, the solenoid, and the speaker is based partially on magnetic or electromagnetic principles.

The concept of electromagnetic induction is important in an electrical component called an inductor or coil that is covered in Chapter 14. At that time, a review of portions of this chapter may be helpful.

In this chapter and throughout the rest of the book, you will learn the basics of putting technology theory into practice.

Two types of magnets are the permanent magnet and the electromagnet. The permanent magnet maintains a constant magnetic field between its two poles with no external excitation. The electromagnet produces a magnetic field only when there is current through it. The electromagnet is basically a coil of wire wound around a magnetic core material.

## TECHnology Theory Into Practice

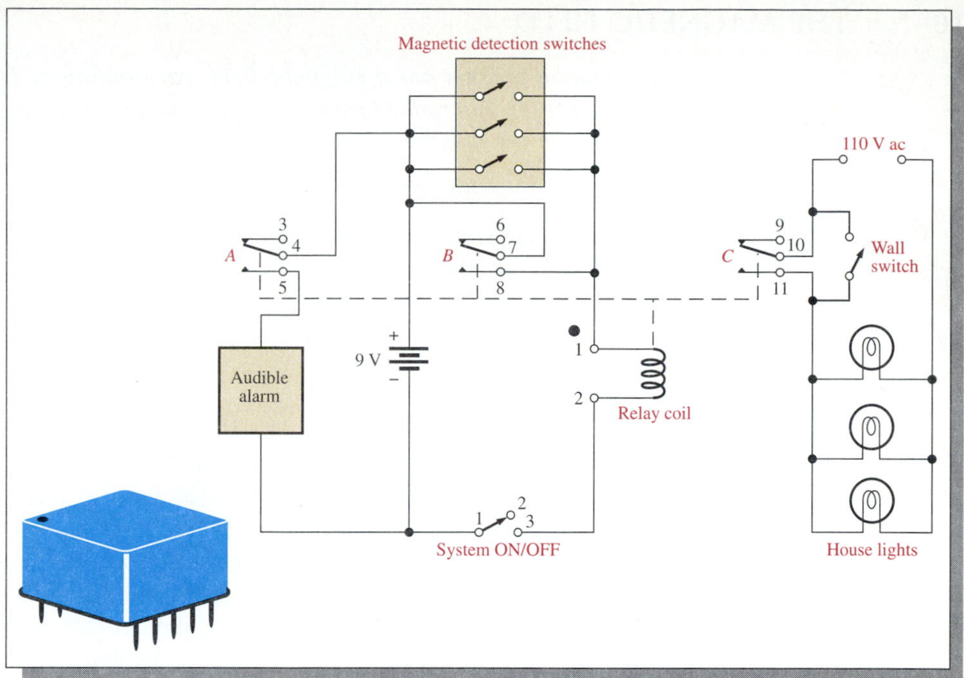


In the TECH TIP assignment in Section 10–7, you will learn how electromagnetic relays can be used in burglar alarm systems, and you will develop a procedure to check out the basic alarm system indicated by the above schematic.

## CHAPTER OBJECTIVES

- ☐ Explain the principles of the magnetic field
- ☐ Explain the principles of electromagnetism
- ☐ Describe the principle of operation for several types of electromagnetic devices
- ☐ Explain magnetic hysteresis
- ☐ Discuss the principle of electromagnetic induction
- ☐ Describe some applications of electromagnetic induction
- ☐ Use Probe in circuit analysis (optional)

## 10–1 ■ THE MAGNETIC FIELD

***A permanent magnet has a magnetic field surrounding it. The magnetic field consists of lines of force that radiate from the north pole to the south pole and back to the north pole through the magnetic material.***

***After completing this section, you should be able to***

- **Explain the principles of the magnetic field**
  - ☐ Define *magnetic flux*
  - ☐ Define *magnetic flux density*
  - ☐ Discuss how materials are magnetized
  - ☐ Explain how a magnetic switch works

A permanent magnet, such as the bar magnet shown in Figure 10–1, has a magnetic field surrounding it. The magnetic field consists of lines of force, or flux lines, that radiate from the north pole (N) to the south pole (S) and back to the north pole through the magnetic material. For clarity, only a few lines of force are shown in the figure. Imagine, however, that many lines surround the magnet in three dimensions. The lines shrink to the smallest possible size and blend together, although they do not touch. This effectively forms a continuous magnetic field surrounding the magnet.

FIGURE 10–1
***Magnetic lines of force around a bar magnet.***

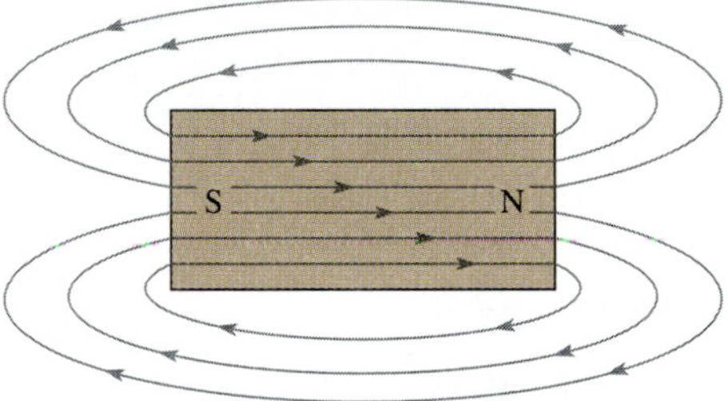


Gray lines represent a few magnetic lines of force in the magnetic field.

### Attraction and Repulsion of Magnetic Poles

When unlike poles of two permanent magnets are placed close together, an attractive force is produced by the magnetic fields, as indicated in Figure 10–2(a). When two like poles are brought close together, they repel each other, as shown in part (b).

### Altering a Magnetic Field

When a nonmagnetic material such as paper, glass, wood, or plastic is placed in a magnetic field, the lines of force are unaltered, as shown in Figure 10–3(a). However, when a magnetic material such as iron is placed in the magnetic field, the lines of force tend to change course and pass through the iron rather than through the surrounding air. They do so because the iron provides a magnetic path that is more easily established than that of air. Figure 10–3(b) illustrates this principle.

### Magnetic Flux ($\phi$)

The group of force lines going from the north pole to the south pole of a magnet is called the **magnetic flux,** symbolized by $\phi$ (the Greek letter phi). The number of lines of force in a magnetic field determines the value of the flux. The more lines of force, the greater the flux and the stronger the magnetic field.

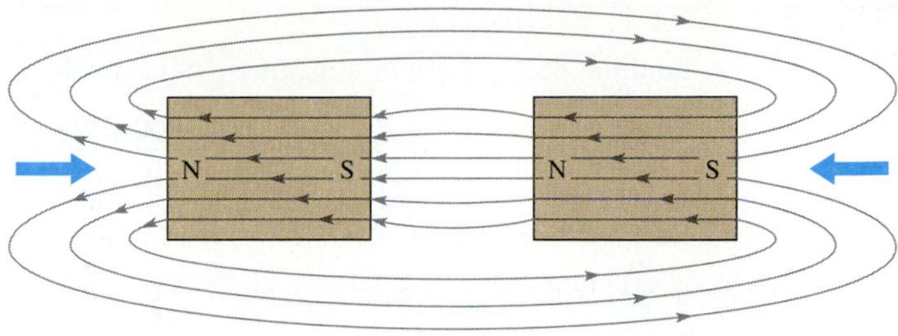


(a) Unlike poles attract.

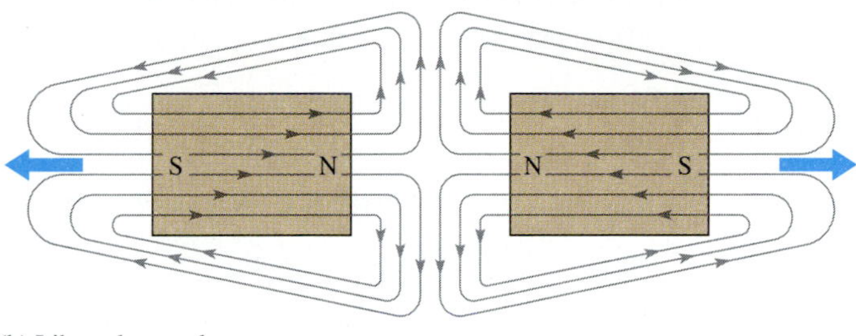


(b) Like poles repel.

FIGURE 10–2
***Magnetic attraction and repulsion.***

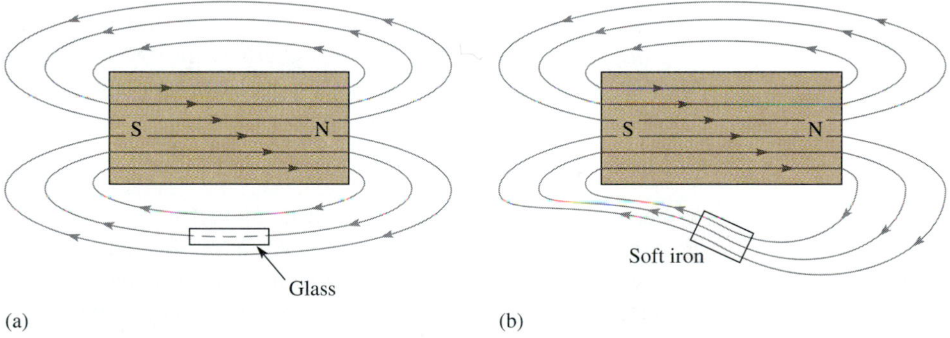


(a) (b)

FIGURE 10–3
***Effect of (a) nonmagnetic and (b) magnetic materials on a magnetic field.***

The unit of magnetic flux is the **weber** (Wb). One weber equals $10^8$ lines. The weber is a very large unit; thus, in most practical situations, the microweber ($\mu$Wb) is used. One microweber equals 100 lines of magnetic flux.

## Magnetic Flux Density *(B)*

The **magnetic flux density** is the amount of flux per unit area perpendicular to the magnetic field. Its symbol is *B*, and its unit is the **tesla** (T). One tesla equals one weber per square meter (Wb/m$^2$). The following formula expresses the flux density:

$$B = \frac{\phi}{A} \qquad \textbf{(10-1)}$$

where $\phi$ is the flux and $A$ is the cross-sectional area in square meters (m$^2$) of the magnetic field.

**EXAMPLE 10–1** Find the flux density in a magnetic field in which the flux in 0.1 $m^2$ is 800 $\mu$Wb.

***Solution***

$$B = \frac{\phi}{A} = \frac{800\ \mu\text{Wb}}{0.1\ \text{m}^2} = 8000\ \mu\text{T}$$

***Related Exercise*** Calculate $\phi$ if $B = 4700\ \mu$T and $A = 0.05\ \text{m}^2$.

**EXAMPLE 10–2** If the flux density in a certain magnetic material is 2.3 T and the area of the material is 0.38 in.$^2$, what is the flux through the material?

***Solution*** First, 0.38 in.$^2$ must be converted to square meters. 39.37 in. = 1 m; therefore,

$$A = 0.38\ \text{in}^2[1\ \text{m}^2/(39.37\ \text{in})^2] = 245.16 \times 10^{-6}\ \text{m}^2$$

The flux through the material is

$$\phi = BA = (2.3\ \text{T})(245.16 \times 10^{-6}\ \text{m}^2) = 564\ \mu\text{Wb}$$

***Related Exercise*** Calculate $B$ if $A = 0.05$ in.$^2$ and $\phi = 1000\ \mu$Wb.

***The Gauss*** Although the tesla (T) is the SI unit for flux density, another unit called the *gauss,* from the CGS (centimeter-gram-second) system, is sometimes used ($10^4$ gauss = 1 T). In fact, the instrument used to measure flux density is the gaussmeter.

## How Materials Become Magnetized

Ferromagnetic materials such as iron, nickel, and cobalt become magnetized when placed in the magnetic field of a magnet. We have all seen a permanent magnet pick up things like paper clips, nails, and iron filings. In these cases, the object becomes magnetized (that is, it actually becomes a magnet itself) under the influence of the permanent magnetic field and becomes attracted to the magnet. When removed from the magnetic field, the object tends to lose its magnetism.

Ferromagnetic materials have minute magnetic domains created within their atomic structure. These domains can be viewed as very small bar magnets with north and south poles. When the material is not exposed to an external magnetic field, the magnetic domains are randomly oriented, as shown in Figure 10–4(a). When the material is placed in a magnetic field, the domains align themselves as shown in part (b). Thus, the object itself effectively becomes a magnet.

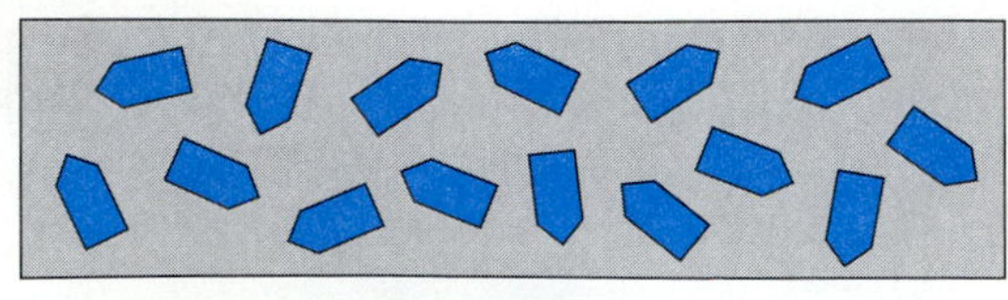

(a) The magnetic domains (N S) are randomly oriented in the unmagnetized material.

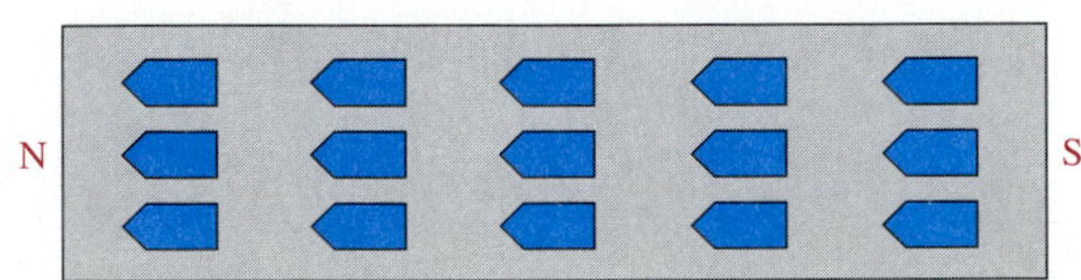


(b) The magnetic domains become aligned when the material is magnetized.

**FIGURE 10–4**
***Magnetic domains in (a) an unmagnetized and (b) a magnetized material.***

## An Application

Permanent magnets have almost endless applications, one of which is presented here as an illustration. Figure 10–5 shows a typical magnetically operated, normally closed (NC) magnetic switch. When the magnet is near the switch mechanism, the metallic arm is held in its NC position. When the magnet is moved away, the spring pulls the arm up, breaking the contact as shown in Figure 10–6.

**FIGURE 10–5**
***Magnetic switch set (courtesy of Tandy Corp.).***

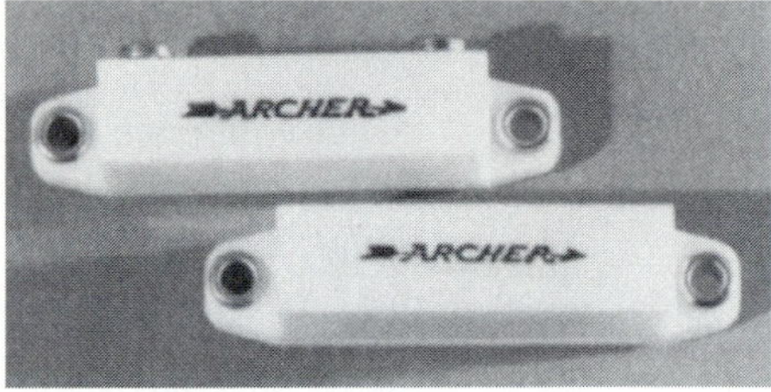


**FIGURE 10–6**
***Operation of a magnetic switch.***

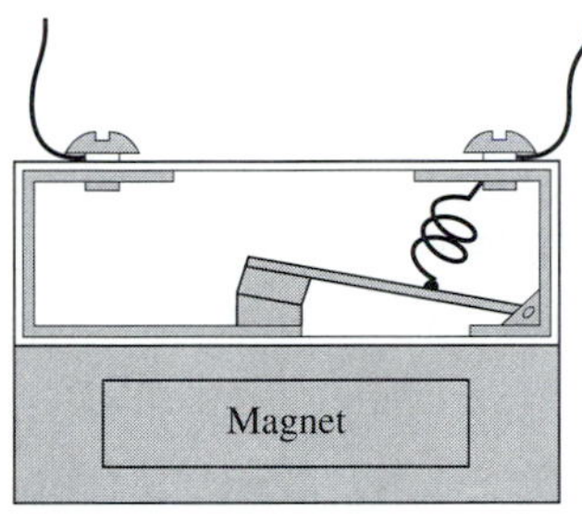


(a) Contact is closed.

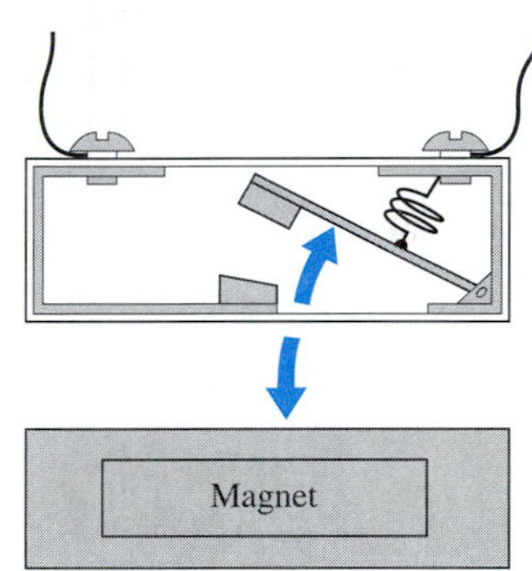


(b) Contact opens.

Switches of this type are commonly used in perimeter alarm systems to detect entry into a building through windows or doors. As Figure 10–7 shows, several openings can be protected by magnetic switches wired to a common transmitter. When any one of the switches opens, the transmitter is activated and sends a signal to a central receiver and alarm unit.

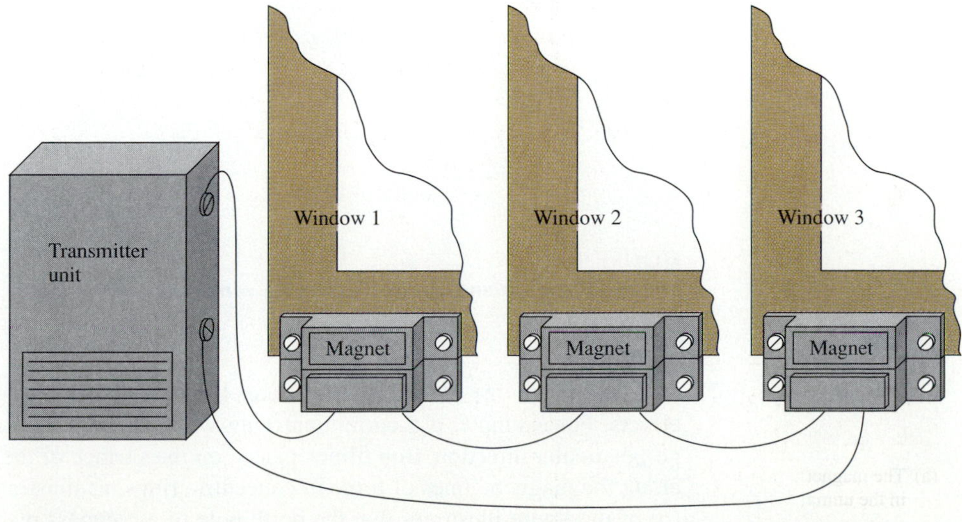


**FIGURE 10–7**
***Connection of a typical perimeter alarm system.***

**SECTION 10–1 REVIEW**

1. When the north poles of two magnets are placed close together, do they repel or attract each other?
2. What is magnetic flux?
3. What is the flux density when $\phi = 4.5\ \mu\text{Wb}$ and $A = 5 \times 10^{-3}\ \text{m}^2$?

## 10–2 ■ ELECTROMAGNETISM

***Electromagnetism is the production of a magnetic field by current in a conductor. Many types of useful devices such as tape recorders, electric motors, speakers, solenoids, and relays are based on electromagnetism.***

***After completing this section, you should be able to***

- **Explain the principles of electromagnetism**
  - ☐ Determine the direction of the magnetic lines of force
  - ☐ Define *permeability*
  - ☐ Define *reluctance*
  - ☐ Define *magnetomotive force*
  - ☐ Describe a basic electromagnet

Current produces a magnetic field, called an **electromagnetic field,** around a conductor, as illustrated in Figure 10–8. The invisible lines of force of the magnetic field form a concentric circular pattern around the conductor and are continuous along its length.

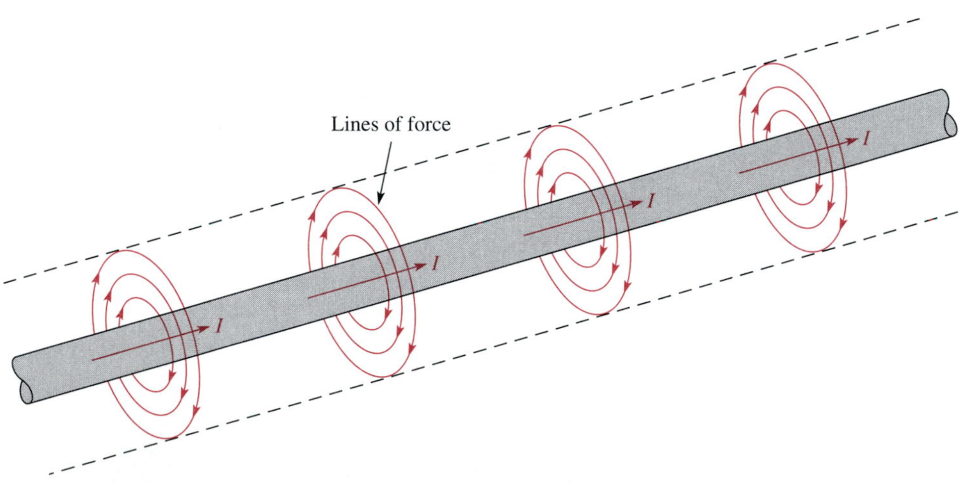


FIGURE 10–8
***Magnetic field around a current-carrying conductor.***

Although the magnetic field cannot be seen, it is capable of producing visible effects. For example, if a current-carrying wire is inserted through a sheet of paper in a perpendicular direction, iron filings placed on the surface of the paper arrange themselves along the magnetic lines of force in concentric rings, as illustrated in Figure 10–9(a). Part (b) of the figure illustrates that the north pole of a compass placed in the electromagnetic field will point in the direction of the lines of force. The field is stronger closer to the conductor and becomes weaker with increasing distance from the conductor.

FIGURE 10–9
*Visible effects of an electromagnetic field.*

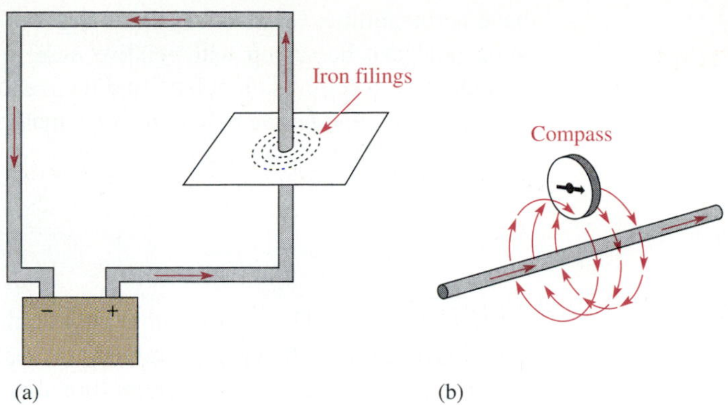


## Direction of the Lines of Force

The direction of the lines of force surrounding the conductor is indicated in Figure 10–10. When the direction of current is left to right, as in part (a), the lines are in a clockwise direction. When current is right to left, as in part (b), the lines are in a counterclockwise direction.

FIGURE 10–10
*Magnetic lines of force around a current-carrying conductor.*

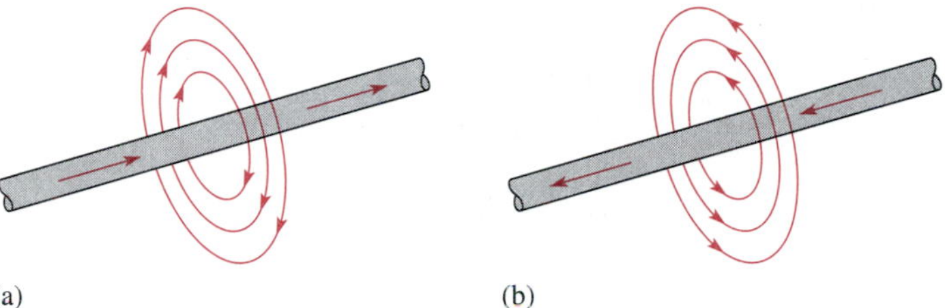


***Right-Hand Rule*** An aid to remembering the direction of the lines of force is illustrated in Figure 10–11. Imagine that you are grasping the conductor with your right hand, with your thumb pointing in the direction of current. Your fingers point in the direction of the magnetic lines of force.

FIGURE 10–11
*Illustration of right-hand rule.*

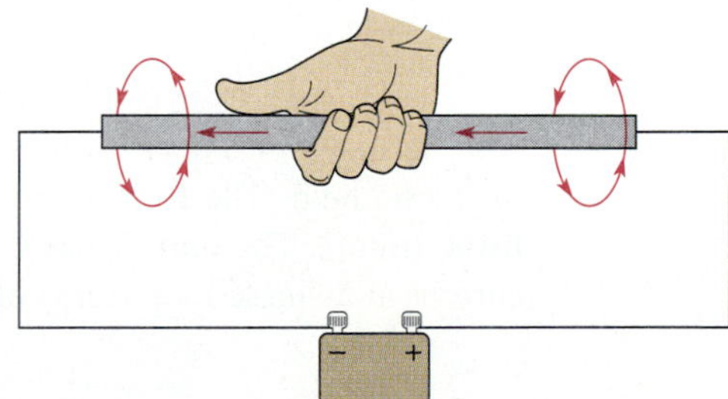

## Electromagnetic Properties

Several important properties relating to electromagnetic fields are now discussed.

***Permeability ($\mu$)*** The ease with which a magnetic field can be established in a given material is measured by the **permeability** of that material. The higher the permeability, the more easily a magnetic field can be established.

The symbol of permeability is $\mu$ (the Greek letter mu), and its value varies depending on the type of material. The permeability of a vacuum ($\mu_0$) is $4\pi \times 10^{-7}$ Wb/At·m (weber/ampere-turn·meter) and is used as a reference. Ferromagnetic materials typically have permeabilities hundreds of times larger than that of a vacuum, indicating that a magnetic field can be set up with relative ease in these materials. Ferromagnetic materials include iron, steel, nickel, cobalt, and their alloys.

The *relative permeability* ($\mu_r$) of a material is the ratio of its absolute permeability to the permeability of a vacuum.

$$\mu_r = \frac{\mu}{\mu_0} \quad \textbf{(10–2)}$$

***Reluctance ($\mathcal{R}$)*** The opposition to the establishment of a magnetic field in a material is called **reluctance.** The value of reluctance is directly proportional to the length ($l$) of the magnetic path, and inversely proportional to the permeability ($\mu$) and to the cross-sectional area ($A$) of the material as expressed by the following equation:

$$\mathcal{R} = \frac{l}{\mu A} \quad \textbf{(10–3)}$$

Reluctance in magnetic circuits is analogous to resistance in electric circuits. The unit of reluctance can be derived using $l$ in meters, $A$ (area) in square meters, and $\mu$ in Wb/At·m as follows:

$$\mathcal{R} = \frac{l}{\mu A} = \frac{\cancel{m}}{(\text{Wb/At}\cdot\cancel{m})(\cancel{m^2})} = \frac{\text{At}}{\text{Wb}}$$

At/Wb is ampere-turns/weber.

**EXAMPLE 10–3**

What is the reluctance of a material that has a length of 0.05 m, a cross-sectional area of 0.012 m$^2$, and a permeability of 3500 $\mu$Wb/At·m?

***Solution***

$$\mathcal{R} = \frac{l}{\mu A} = \frac{0.05\text{ m}}{(3500 \times 10^{-6}\text{ Wb/At}\cdot\text{m})(0.012\text{ m}^2)} = 1190\text{ At/Wb}$$

***Related Exercise*** What happens to the reluctance of the material in this example if $l$ is doubled and $A$ is halved?

***Magnetomotive Force (mmf)*** As you have learned, current in a conductor produces a magnetic field. The force that produces the magnetic field is called the **magnetomotive force** (mmf). The unit of mmf, the **ampere-turn** (At), is established on the basis of the current in a single loop (turn) of wire. The formula for mmf is as follows:

$$F_m = NI \quad \textbf{(10–4)}$$

where $F_m$ is the magnetomotive force, $N$ is the number of turns of wire, and $I$ is the current in amperes.

Figure 10–12 illustrates that a number of turns of wire carrying a current around a magnetic material creates a force that sets up flux lines through the magnetic path. The amount of flux depends on the magnitude of the mmf and on the reluctance of the material, as expressed by the following equation:

$$\phi = \frac{F_m}{\mathcal{R}} \quad \textbf{(10–5)}$$

Equation (10–5) is known as the *Ohm's law for magnetic circuits* because the flux ($\phi$) is analogous to current, the mmf ($F_m$) is analogous to voltage, and the reluctance ($\mathcal{R}$) is analogous to resistance.

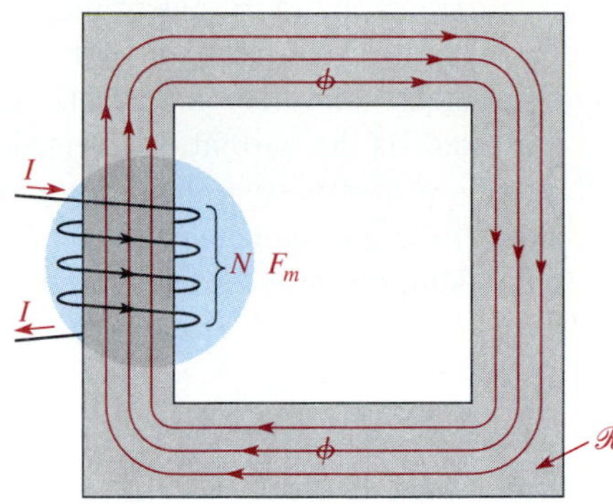


**FIGURE 10–12**
***A basic magnetic circuit.***

**EXAMPLE 10–4**

How much flux is established in the magnetic path of Figure 10–13 if the reluctance of the material is $28 \times 10^3$ At/Wb?

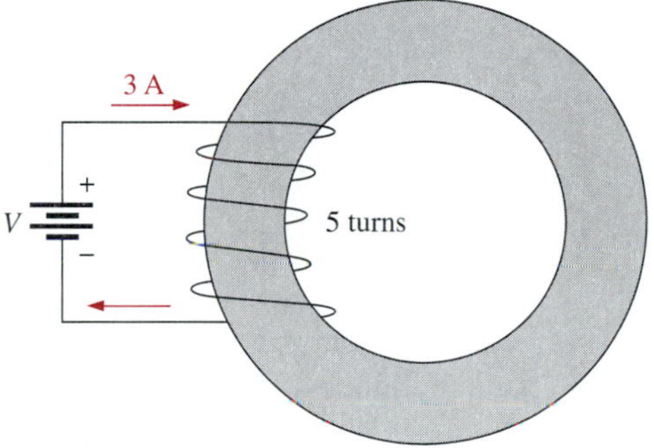


**FIGURE 10–13**

***Solution***

$$\phi = \frac{F_m}{\mathcal{R}} = \frac{NI}{\mathcal{R}} = \frac{(5\text{ t})(3\text{ A})}{28 \times 10^3\text{ At/Wb}} = 536\ \mu\text{Wb}$$

***Related Exercise*** How much flux is established in the magnetic path of Figure 10–13 if the reluctance is $7.5 \times 10^3$ At/Wb, the number of turns is 30, and the current is 1.8 A?

**EXAMPLE 10–5**

There are two amperes of current through a wire with 5 turns.

**(a)** What is the mmf?

**(b)** What is the reluctance of the circuit if the flux is 250 $\mu$Wb?

***Solution***

**(a)** $N = 5$ and $I = 2$ A

$F_m = NI = (5\text{ t})(2\text{ A}) = 10$ At

**(b)** $\mathcal{R} = \dfrac{F_m}{\phi} = \dfrac{10\text{ At}}{250\ \mu\text{Wb}} = 40 \times 10^3$ At/Wb

***Related Exercise*** Rework the example for $I = 850$ mA and $N = 50$. The flux is 500 $\mu$Wb.

## The Electromagnet

An electromagnet is based on the properties that you have just learned. A basic electromagnet is simply a coil of wire wound around a core material that can be easily magnetized.

The shape of the electromagnet can be designed for various applications. For example, Figure 10–14 shows a U-shaped magnetic core. When the coil of wire is connected to a battery and there is current, as shown in part (a), a magnetic field is established as indicated. If the current is reversed, as shown in part (b), the direction of the magnetic field is also reversed. The closer the north and south poles are brought together, the smaller the air gap between them becomes, and the easier it becomes to establish a magnetic field, because the reluctance is lessened.

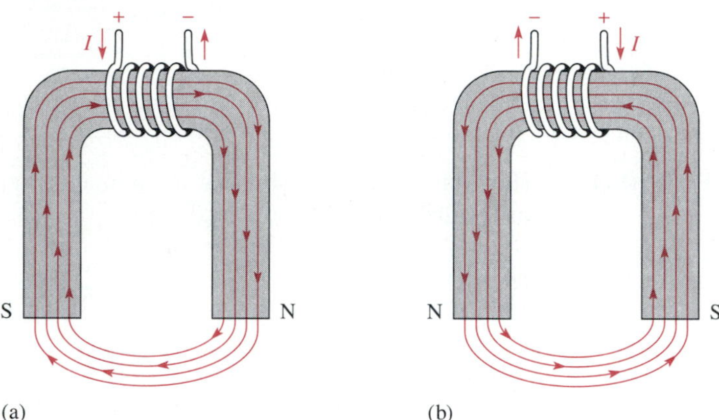


**FIGURE 10–14**
***Reversing the current in the coil causes the electromagnetic field to reverse.***

A good example of one application of an electromagnet is the process of recording on magnetic tape. In this situation, the recording head is an electromagnet with a narrow air gap, as shown in Figure 10–15. Current sets up a magnetic field across the air gap, and as the recording head passes over the magnetic tape, the tape is permanently magnetized. In digital recording, for example, the tape is magnetized in one direction for a binary 1 and in the other direction for a binary 0, as illustrated in the figure. This magnetization in different directions is accomplished by reversing the coil current in the recording head.

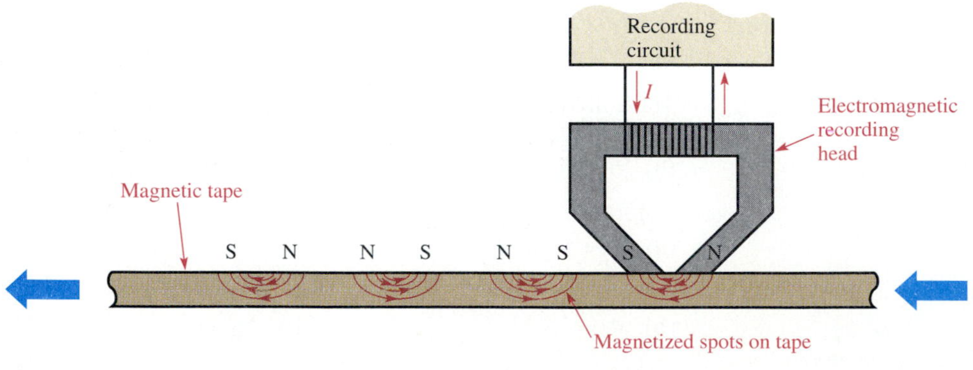


**FIGURE 10–15**
***An electromagnetic recording head recording information on magnetic tape by magnetizing the tape as it passes by.***

**SECTION 10–2 REVIEW**

1. Explain the difference between magnetism and electromagnetism.
2. What happens to the magnetic field in an electromagnet when the current through the coil is reversed?

## 10–3 ■ ELECTROMAGNETIC DEVICES

***In the last section, you learned that the recording head is a type of electromagnetic device. Now, several other common devices are introduced.***

***After completing this section, you should be able to***

- **Describe the principle of operation for several types of electromagnetic devices**
  - ☐ Discuss how a solenoid works
  - ☐ Discuss how a relay works
  - ☐ Discuss how a speaker works
  - ☐ Discuss the basic analog meter movement

The recording head illustrated in the last section is one example of an electromagnetic device. Several other examples are now presented.

### The Solenoid

The **solenoid** is a type of electromagnetic device that has a movable iron core called a *plunger.* The movement of this iron core depends on both an electromagnetic field and a mechanical spring force. The basic structure of a solenoid is shown in Figure 10–16. It consists of a cylindrical coil of wire wound around a nonmagnetic hollow form. A stationary iron core is fixed in position at the end of the shaft and a sliding iron core (plunger) is attached to the stationary core with a spring.

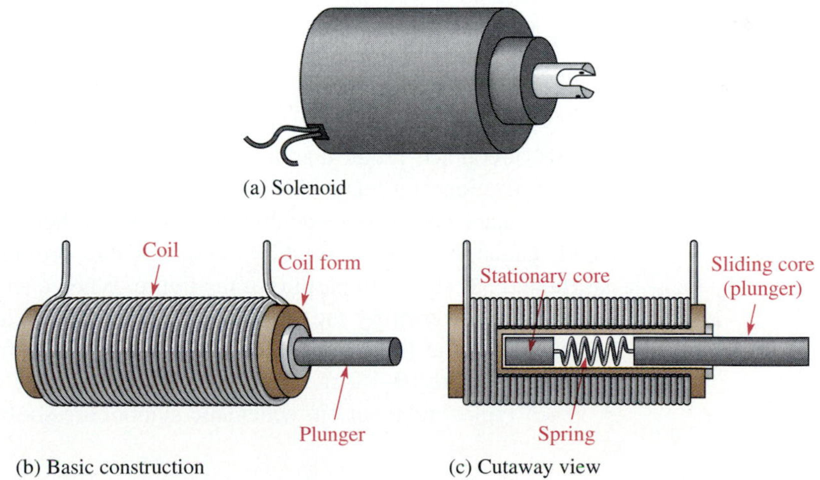


**FIGURE 10–16**
***Basic solenoid structure.***

In the at-rest, or unenergized, state, the plunger is extended. The solenoid is energized by current through the coil, which sets up an electromagnetic field that magnetizes both iron cores. The south pole of the stationary core attracts the north pole of the movable core causing it to slide inward, thus retracting the plunger and compressing the spring. As long as there is coil current, the plunger remains retracted by the attractive force of the magnetic fields. When the current is cut off, the magnetic fields collapse and the force of the compressed spring pushes the plunger back out. This basic solenoid operation is illustrated in Figure 10–17 for the unenergized and the energized conditions. The solenoid is used for applications such as opening and closing valves and automobile door locks.

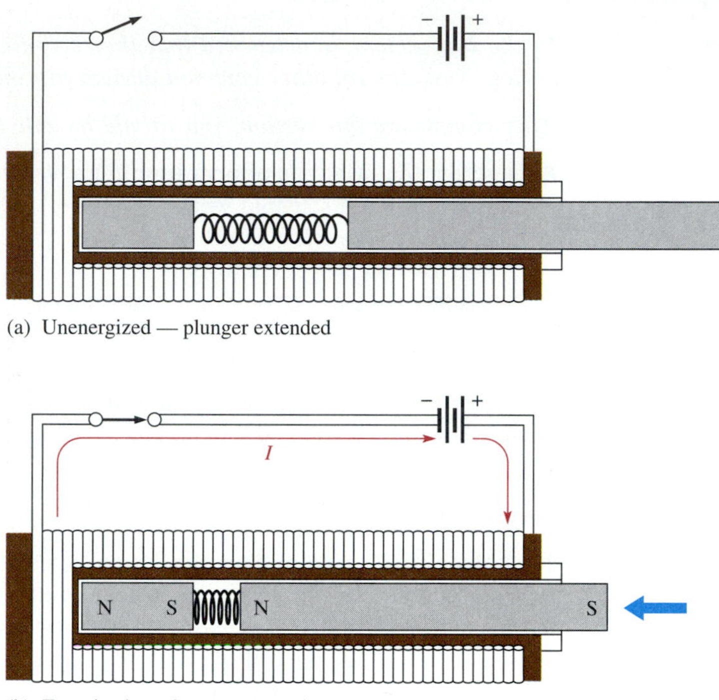


**FIGURE 10–17**
***Basic solenoid operation.***

## The Relay

**Relays** differ from solenoids in that the electromagnetic action is used to open or close electrical contacts rather than to provide mechanical movement. Figure 10–18 shows the basic operation of a relay with one normally open (NO) contact and one normally closed (NC) contact (single pole–double throw). When there is no coil current, the armature is held against the upper contact by the spring, thus providing continuity from terminal 1 to terminal 2, as shown in part (a) of the figure. When energized with coil current, the armature is pulled down by the attractive force of the electromagnetic field and makes connection with the lower contact to provide continuity from terminal 1 to terminal 3, as shown in Figure 10–18(b).

A typical relay and its schematic symbol are shown in Figure 10–19.

## The Speaker

Permanent-magnet speakers are commonly used in stereos, radios, and TVs, and their operation is based on the principle of electromagnetism. A typical speaker is constructed with a permanent magnet and an electromagnet, as shown in Figure 10–20(a). The cone of the speaker consists of a paper-like diaphragm to which is attached a hollow cylinder with a coil around it, forming an electromagnet. One of the poles of the permanent magnet is positioned within the cylindrical coil. When there is current through the coil in one

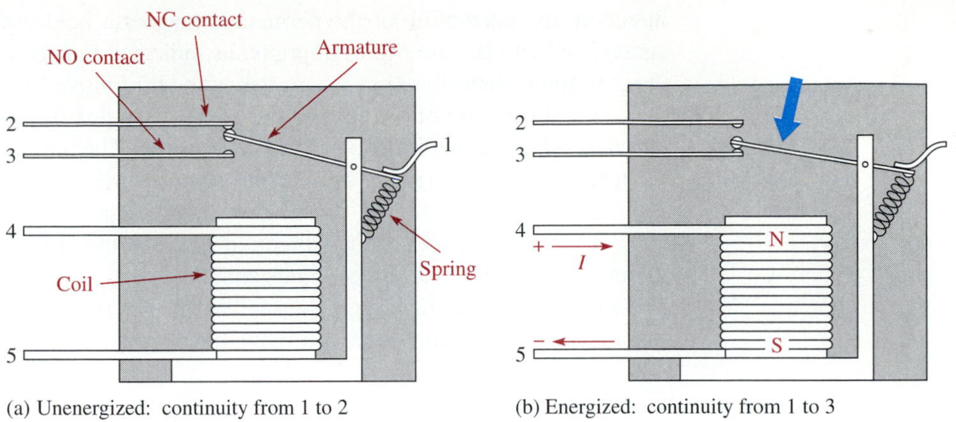


(a) Unenergized: continuity from 1 to 2

(b) Energized: continuity from 1 to 3

**FIGURE 10–18**
***Basic structure of a single-pole–double-throw relay.***

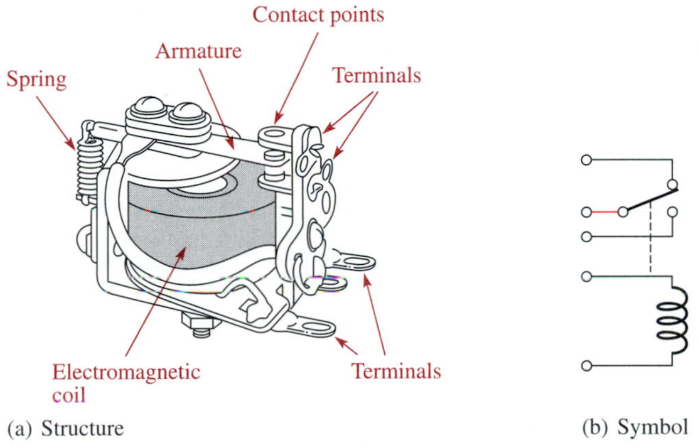


(a) Structure

(b) Symbol

**FIGURE 10–19**
***A typical relay.***

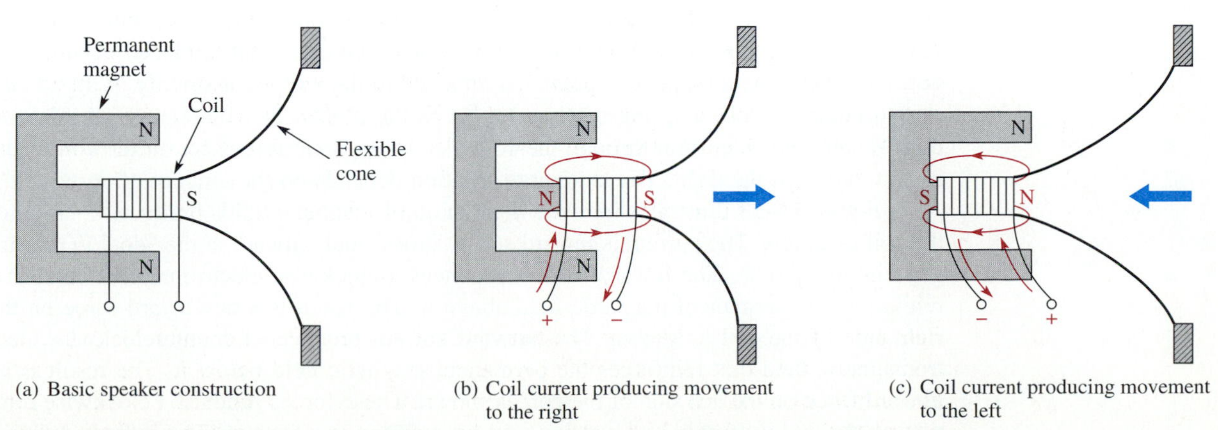


(a) Basic speaker construction

(b) Coil current producing movement to the right

(c) Coil current producing movement to the left

**FIGURE 10–20**
***Basic speaker operation.***

direction, the interaction of the permanent magnetic field with the electromagnetic field causes the cylinder to move to the right, as indicated in Figure 10–20(b). Current through the coil in the other direction causes the cylinder to move to the left, as shown in part (c).

The movement of the coil cylinder causes the flexible diaphragm also to move in or out, depending on the direction of the coil current. The amount of coil current determines the intensity of the magnetic field, which controls the amount that the diaphragm moves.

As shown in Figure 10–21, when an audio signal (voice or music) is applied to the coil, the current varies in both direction and amount. In response, the diaphragm will vibrate in and out by varying amounts and at varying rates corresponding to the audio signal. Vibration in the diaphragm causes the air that is in contact with it to vibrate in the same manner. These air vibrations move through the air as sound waves.

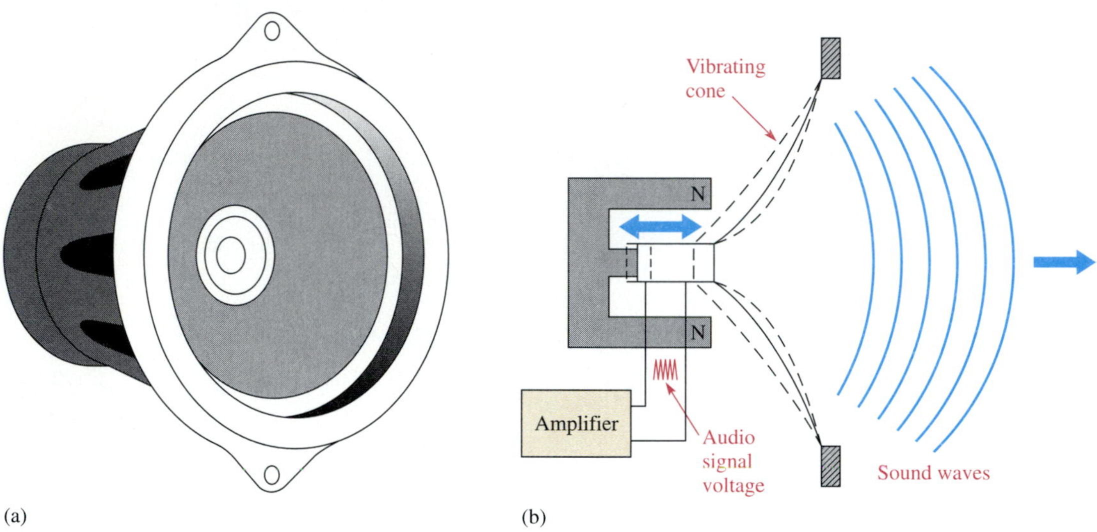


**FIGURE 10–21**
***The speaker converts audio signal voltages into sound waves.***

## Analog Meter Movement

The d'Arsonval meter movement is the most common type used in analog multimeters. In this type of meter movement, the pointer is deflected in proportion to the amount of current through a coil. Figure 10–22 shows a basic d'Arsonval meter movement. It consists of a coil of wire wound on a bearing-mounted assembly that is placed between the poles of a permanent magnet. A pointer is attached to the moving assembly. With no current through the coil, a spring mechanism keeps the pointer at its left-most (zero) position. When there is current through the coil, electromagnetic forces act on the coil, causing a rotation to the right. The amount of rotation depends on the amount of current.

Figure 10–23 illustrates how the interaction of magnetic fields produces rotation of the coil assembly. The current is inward at the "cross" and outward at the "dot" in the single winding shown. The inward current produces a clockwise electromagnetic field that reinforces the permanent magnetic field above it. The result is a downward force on the right side of the coil as shown. The outward current produces a counterclockwise electromagnetic field that reinforces the permanent magnetic field below it. The result is an upward force on the left side of the coil as shown. These forces produce a clockwise rotation of the coil assembly and are opposed by a spring mechanism. The indicated forces and the spring force are balanced at the value of the current. When current is removed, the spring force returns the pointer to its zero position.

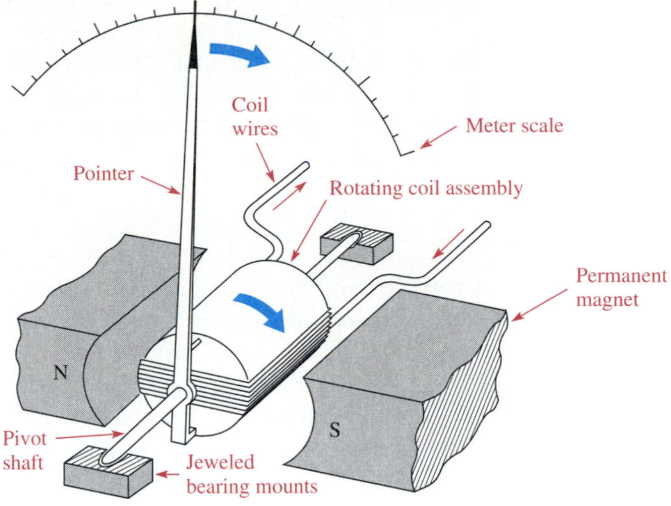


FIGURE 10–22
*The basic d'Arsonval meter movement.*

FIGURE 10–23
*When the electromagnetic field interacts with the permanent magnetic field, forces are exerted on the rotating coil assembly, causing it to move clockwise and thus deflecting the pointer.*

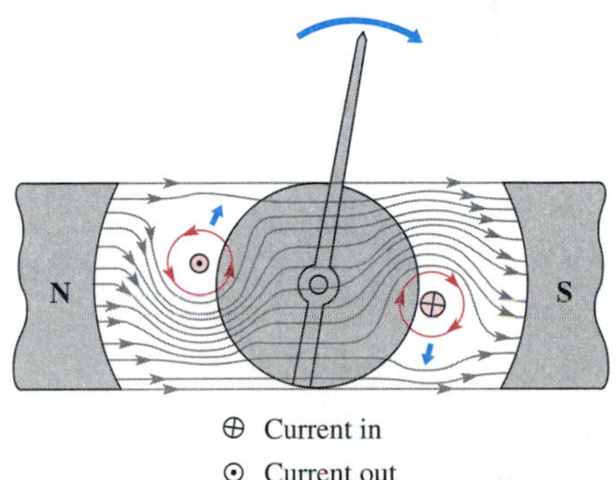


**SECTION 10–3 REVIEW**

1. Explain the difference between a solenoid and a relay.
2. What is the movable part of a solenoid called?
3. What is the movable part of a relay called?
4. Upon what basic principle is the d'Arsonval meter movement based?

## 10–4 ■ MAGNETIC HYSTERESIS

***When a magnetizing force is applied to a material, the flux density in the material changes in a certain way, which we will now examine.***

***After completing this section, you should be able to***

- **Explain magnetic hysteresis**
    - ☐ State the formula for magnetizing force
    - ☐ Discuss a hysteresis curve
    - ☐ Define *retentivity*

## Magnetizing Force *(H)*

The **magnetizing force** in a material is defined to be the magnetomotive force ($F_m$) per unit length ($l$) of the material, as expressed by the following equation. The unit of magnetizing force ($H$) is ampere-turns per meter (At/m).

$$H = \frac{F_m}{l} \tag{10–6}$$

where $F_m = NI$. Note that the magnetizing force depends on the number of turns ($N$) of the coil of wire, the current ($I$) through the coil, and the length ($l$) of the material. It does not depend on the type of material.

Since $\phi = F_m/\mathcal{R}$, as $F_m$ increases, the flux increases. Also, the magnetizing force ($H$) increases. Recall that the flux density ($B$) is the flux per unit cross-sectional area ($B = \phi/A$), so $B$ is also proportional to $H$. The curve showing how these two quantities ($B$ and $H$) are related is called the $B$-$H$ curve or the hysteresis curve. The parameters that influence both $B$ and $H$ are illustrated in Figure 10–24.

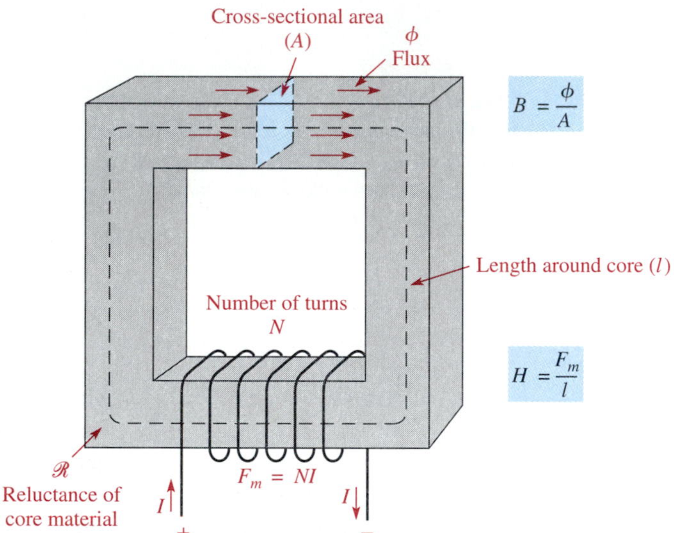


**FIGURE 10–24**
***Parameters that determine the magnetizing force (H) and the flux density (B).***

## The Hysteresis Curve and Retentivity

**Hysteresis** is a characteristic of a magnetic material whereby a change in magnetization lags the application of a magnetizing force. The magnetizing force ($H$) can be readily increased or decreased by varying the current through the coil of wire, and it can be reversed by reversing the voltage polarity across the coil.

Figure 10–25 illustrates the development of the hysteresis curve. We start by assuming a magnetic core is unmagnetized so that $B = 0$. As the magnetizing force ($H$) is increased from zero, the flux density ($B$) increases proportionally as indicated by the curve in Figure 10–25(a). When $H$ reaches a certain value, $B$ begins to level off. As $H$ continues to increase, $B$ reaches a saturation value ($B_{sat}$) when $H$ reaches a value ($H_{sat}$) as illustrated in Figure 10–25(b). Once saturation is reached, a further increase in $H$ will not increase $B$.

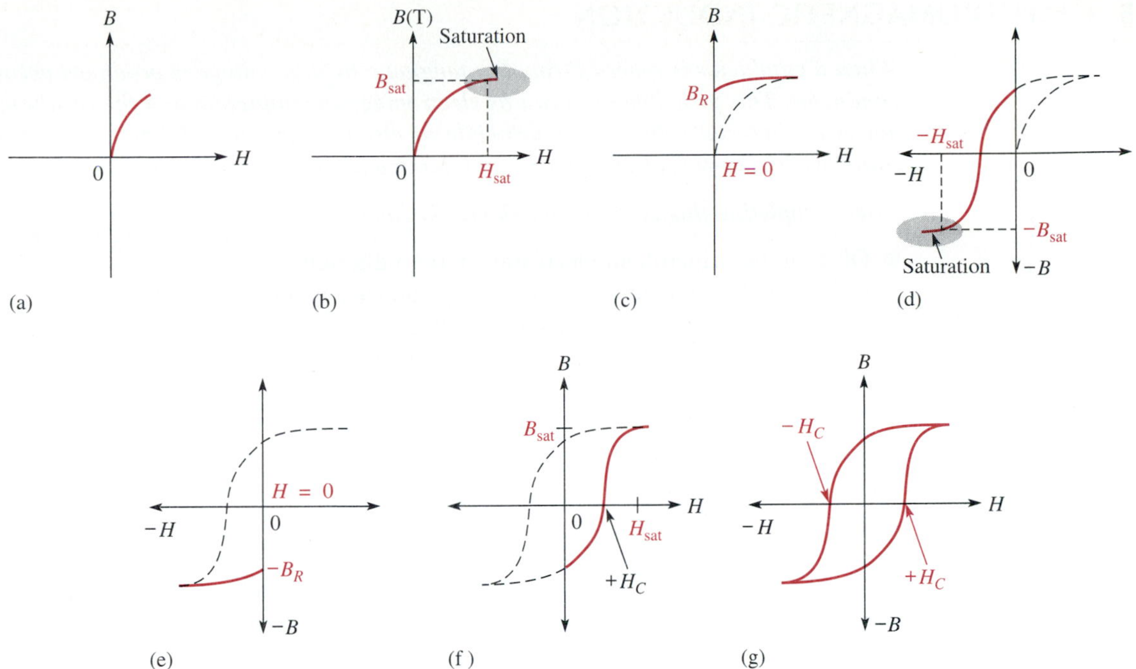


**FIGURE 10–25**
***Development of a magnetic hysteresis curve.***

Now, if $H$ is decreased to zero, $B$ will fall back along a different path to a residual value ($B_R$) as shown in Figure 10–25(c). This indicates that the material continues to be magnetized even with the magnetizing force removed ($H = 0$). The ability of a material to maintain a magnetized state without the presence of a magnetizing force is called **retentivity.** The retentivity of a material is indicated by the ratio of $B_R$ to $B_{sat}$.

Reversal of the magnetizing force is represented by negative values of $H$ on the curve and is achieved by reversing the current in the coil of wire. An increase in $H$ in the negative direction causes saturation to occur at a value ($-H_{sat}$) where the flux density is at its maximum negative value as indicated in Figure 10–25(d).

When the magnetizing force is removed ($H = 0$), the flux density goes to its negative residual value ($-B_R$) as shown in Figure 10–25(e). From the $-B_R$ value, the flux density follows the curve indicated in part (f) back to its maximum positive value when the magnetizing force equals $H_{sat}$ in the positive direction.

The complete $B$-$H$ curve is shown in Figure 10–25(g) and is called the *hysteresis curve*. The magnetizing force required to make the flux density zero is called the *coercive force,* $H_C$.

Materials with a low retentivity do not retain a magnetic field very well, while those with high retentivities exhibit values of $B_R$ very close to the saturation value of $B$. Depending on the application, retentivity in a magnetic material can be an advantage or a disadvantage. In permanent magnets and memory cores, for example, high retentivity is required. In ac motors, retentivity is undesirable because the residual magnetic field must be overcome each time the current reverses, thus wasting energy.

## SECTION 10–4 REVIEW

1. For a given wirewound core, how does an increase in current through the coil affect the flux density?
2. Define *retentivity*.

## 10–5 ■ ELECTROMAGNETIC INDUCTION

***When a conductor is moved through a magnetic field, a voltage is produced across the conductor. This principle is known as electromagnetic induction and the resulting voltage is an induced voltage. The principle of electromagnetic induction is what makes transformers, electrical generators, and many other devices possible.***

***After completing this section, you should be able to***

- **Discuss the principle of electromagnetic induction**
  - Explain how voltage is induced in a conductor in a magnetic field
  - Determine polarity of an induced voltage
  - Discuss forces on a conductor in a magnetic field
  - State Faraday's law
  - State Lenz's law

### Relative Motion

When a wire is moved across a magnetic field, there is a relative motion between the wire and the magnetic field. Likewise, when a magnetic field is moved past a stationary wire, there is also relative motion. In either case, this relative motion results in an induced voltage ($v_{ind}$) in the wire, as Figure 10–26 indicates. The lowercase $v$ stands for instantaneous voltage.

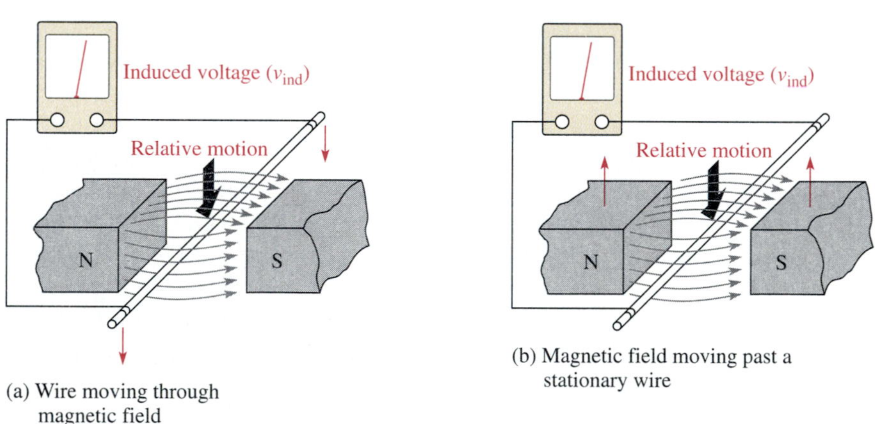


(a) Wire moving through magnetic field

(b) Magnetic field moving past a stationary wire

FIGURE 10–26
***Relative motion between a wire and a magnetic field.***

The amount of the induced voltage depends on the rate at which the wire and the magnetic field move with respect to each other. The faster the relative speed, the greater the induced voltage.

### Polarity of the Induced Voltage

If the conductor in Figure 10–26 is moved first one way and then another in the magnetic field, a reversal of the polarity of the induced voltage will be observed. As the wire is moved downward, a voltage is induced with the polarity indicated in Figure 10–27(a). As the wire is moved upward, the polarity is as indicated in part (b) of the figure.

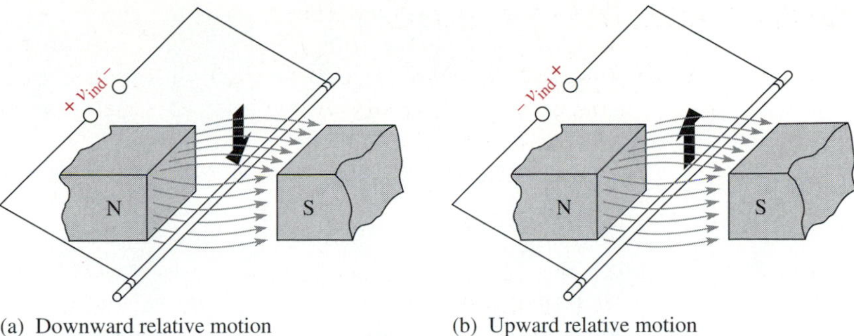


(a) Downward relative motion (b) Upward relative motion

**FIGURE 10–27**
***Polarity of induced voltage depends on direction of motion.***

## Induced Current

When a load resistor is connected to the wire in Figure 10–27, the voltage induced by the relative motion in the magnetic field will cause a current in the load, as shown in Figure 10–28. This current is called the induced current ($i_{ind}$). The lowercase $i$ stands for instantaneous current.

**FIGURE 10–28**
***Induced current in a load as the wire moves through the magnetic field.***

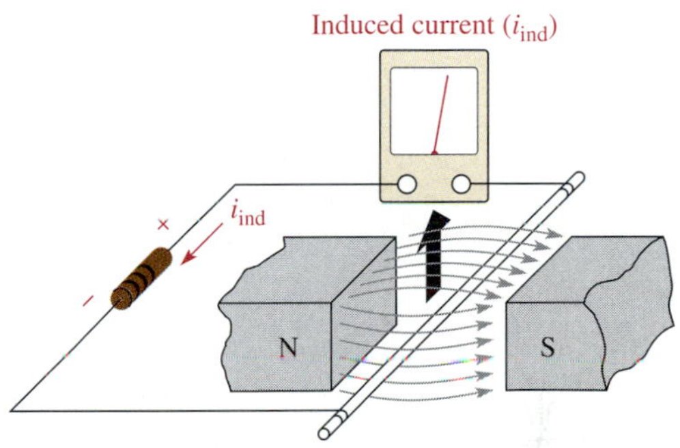


The action of producing a voltage and a resulting current in a load by moving a conductor across a magnetic field is the basis for electrical generators. The concept of a conductor existing in a moving magnetic field is the basis for inductance in an electric circuit.

## Forces on a Current-Carrying Conductor in a Magnetic Field (Motor Action)

Figure 10–29(a) shows current outward through a wire in a magnetic field. The electromagnetic field set up by the current interacts with the permanent magnetic field; as a result, the permanent lines of force above the wire tend to be deflected down under the wire, because they are opposite in direction to the electromagnetic lines of force. Therefore, the flux density above is reduced, and the magnetic field is weakened. The flux density below the conductor is increased, and the magnetic field is strengthened. An upward force on the conductor results, and the conductor tends to move toward the weaker magnetic field.

Figure 10–29(b) shows the current inward, resulting in a force on the conductor in the downward direction. This force is the basis for electrical motors.

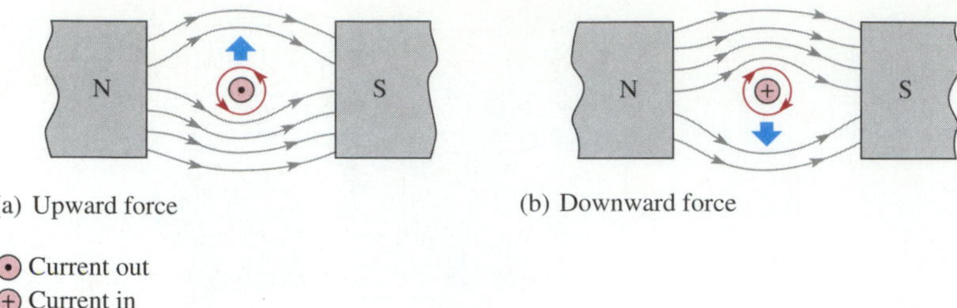


FIGURE 10–29
*Forces on a current-carrying conductor in a magnetic field (motor action).*

## Faraday's Law

Michael Faraday discovered the principle of **electromagnetic induction** in 1831. He found that moving a magnet through a coil of wire induced a voltage across the coil, and that when a complete path was provided, the induced voltage caused an induced current, as you have learned. Faraday's two observations are stated as follows:

1. The amount of voltage induced in a coil is directly proportional to the rate of change of the magnetic field with respect to the coil ($d\phi/dt$).
2. The amount of voltage induced in a coil is directly proportional to the number of turns of wire in the coil ($N$).

Faraday's first observation is demonstrated in Figure 10–30, where a bar magnet is moved through a coil, thus creating a changing magnetic field. In part (a) of the figure, the magnet is moved at a certain rate, and a certain induced voltage is produced as indicated. In part (b), the magnet is moved at a faster rate through the coil, creating a greater induced voltage.

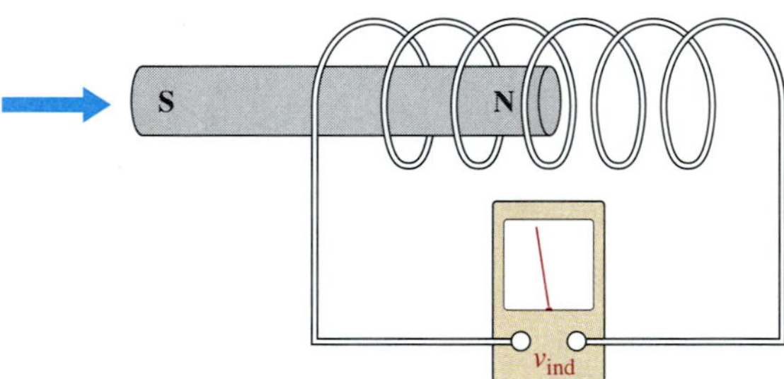


(a) As magnet moves slowly to the right, magnetic field is changing with respect to coil, and a voltage is induced.

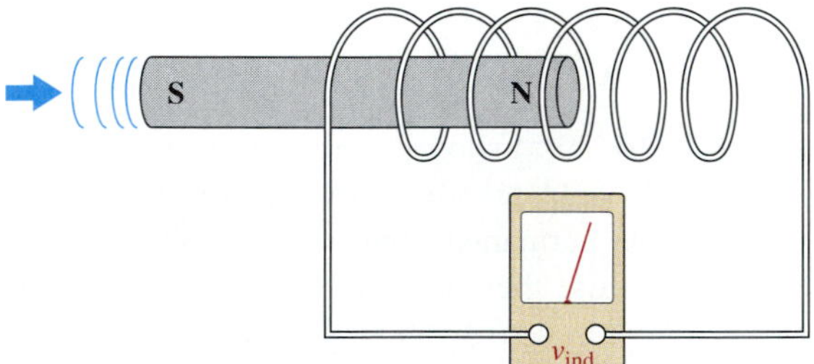


(b) As magnet moves more rapidly to the right, magnetic field is changing more rapidly with respect to coil, and a greater voltage is induced.

FIGURE 10–30
*A demonstration of Faraday's first observation: The amount of induced voltage is directly proportional to the rate of change of the magnetic field with respect to the coil.*

Faraday's second observation is demonstrated in Figure 10–31. In part (a), the magnet is moved through the coil and a voltage is induced as shown. In part (b), the magnet is moved at the same speed through a coil that has a greater number of turns. The greater number of turns creates a greater induced voltage.

**Faraday's law** is stated as follows:

**The voltage induced across a coil of wire equals the number of turns in the coil times the rate of change of the magnetic flux.**

Faraday's law is expressed in equation form as

$$v_{ind} = N\left(\frac{d\phi}{dt}\right) \qquad \textbf{(10–7)}$$

FIGURE 10–31

***A demonstration of Faraday's second observation: The amount of induced voltage is directly proportional to the number of turns in the coil.***

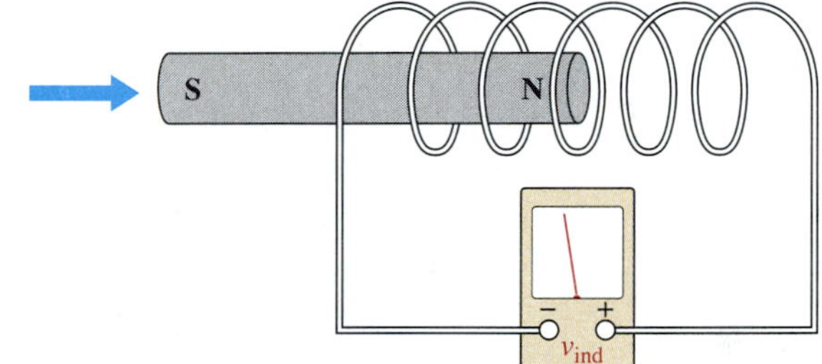


(a) Magnet moves through a coil and induces a voltage.

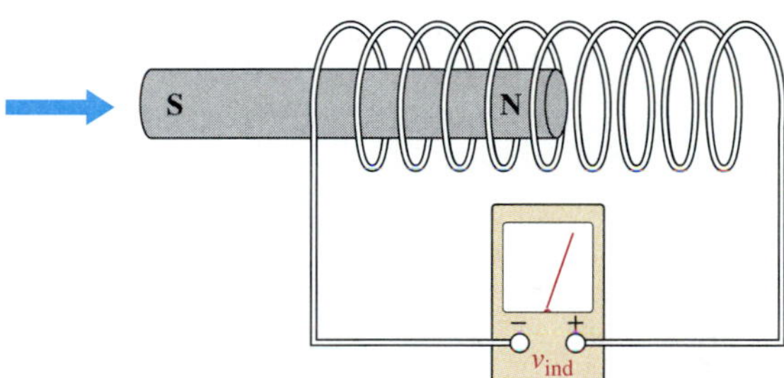


(b) Magnet moves at same rate through a coil with more turns (loops) and induces a greater voltage.

**EXAMPLE 10–6** Apply Faraday's law to find the induced voltage across a coil with 100 turns that is located in a magnetic field that is changing at a rate of 5 Wb/s.

***Solution***

$$v_{ind} = N\left(\frac{d\phi}{dt}\right) = (100\text{ t})(5\text{ Wb/s}) = 500\text{ V}$$

***Related Exercise*** Find the induced voltage across a 250 turn coil in a magnetic field that is changing at 50 $\mu$Wb/s.

## Lenz's Law

You have learned that a changing magnetic field induces a voltage in a coil that is directly proportional to the rate of change of the magnetic field and the number of turns in the coil. Lenz's law defines the polarity or direction of the induced voltage.

**When the current through a coil changes, an induced voltage is created as a result of the changing electromagnetic field, and the direction of the induced voltage is such that it always opposes the change in current.**

**SECTION 10–5 REVIEW**

1. What is the induced voltage across a stationary conductor in a stationary magnetic field?
2. When the speed at which a conductor is moved through a magnetic field is increased, does the induced voltage increase, decrease, or remain the same?
3. When there is current through a conductor in a magnetic field, what happens?

## 10–6 ■ APPLICATIONS OF ELECTROMAGNETIC INDUCTION

***In this section, two interesting applications of electromagnetic induction are discussed—an automotive crankshaft position sensor and a dc generator. Although there are many varied applications, these two are representative.***

***After completing this section, you should be able to***

- **Describe some applications of electromagnetic induction**
  - ☐ Explain how a crankshaft position sensor works
  - ☐ Explain how a dc generator works

### Automotive Crankshaft Position Sensor

An interesting automotive application is a type of engine sensor that detects the crankshaft position directly using electromagnetic induction. The electronic engine controller in many automobiles uses the position of the crankshaft to set ignition timing and, sometimes, to adjust the fuel control system. Figure 10–32 shows the basic concept. A steel disk is attached to the engine's crankshaft by an extension rod; the protruding tabs on the disk represent specific crankshaft positions.

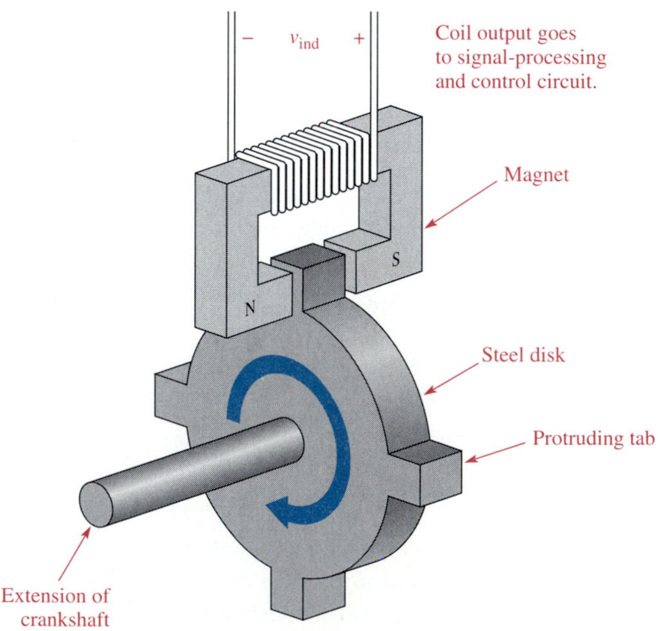


FIGURE 10–32
***A crankshaft position sensor that produces a voltage when a tab passes through the air gap of the magnet.***

As illustrated in Figure 10–32, as the disk rotates with the crankshaft, the tabs periodically pass through the air gap of the permanent magnet. Since steel has a much lower reluctance than does air (a magnetic field can be established in steel much more easily than in air), the magnetic flux suddenly increases as a tab comes into the air gap, causing a voltage to be induced across the coil. This process is illustrated in Figure 10–33. The electronic engine control circuit uses the induced voltage as an indicator of the crankshaft position.

**FIGURE 10–33**
***As the tab passes through the air gap of the magnet, the coil senses a change in the magnetic field, and a voltage is induced.***

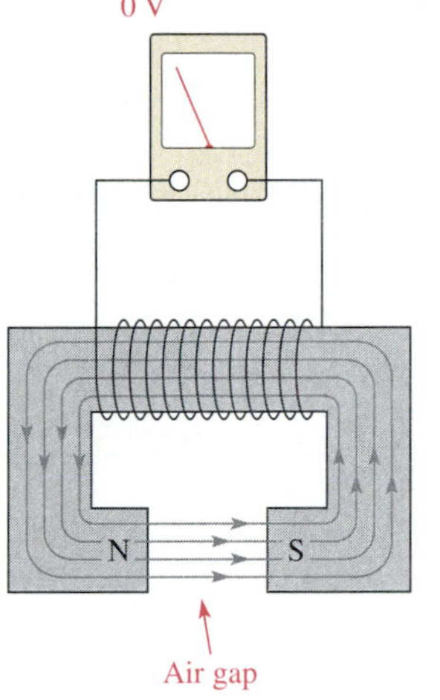


(a) There is no changing magnetic field, so there is no induced voltage.

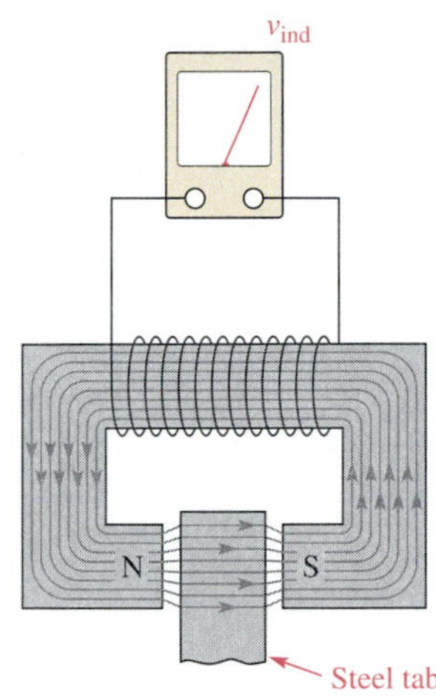


(b) Insertion of the steel tab reduces the reluctance of the air gap, causing the magnetic flux to increase and thus inducing a voltage.

## A DC Generator

Figure 10–34 shows a simplified dc generator consisting of a single loop of wire in a permanent magnetic field. Notice that each end of the loop is connected to a split-ring arrangement. This conductive metal ring is called a *commutator.* As the loop is rotated in the magnetic field, the split commutator ring also rotates. Each half of the split ring rubs against the fixed contacts, called *brushes,* and connects the loop to an external circuit.

**FIGURE 10–34**
***A basic dc generator.***

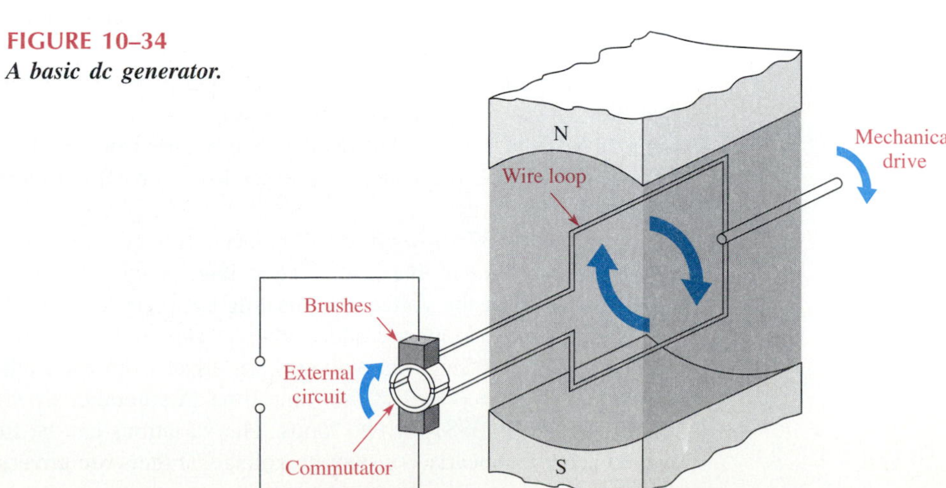

As the loop rotates through the magnetic field, it cuts through the flux lines at varying angles, as illustrated in Figure 10–35. At position *A* in its rotation, the loop of wire is effectively moving parallel with the magnetic field. Therefore, at this instant, the rate at which it is cutting through the magnetic flux lines is zero. As the loop moves from position *A* to position *B*, it cuts through the flux lines at an increasing rate. At position *B*, it is moving effectively perpendicular to the magnetic field and thus is cutting through a maximum number of lines. As the loop rotates from position *B* to position *C*, the rate at which it cuts the flux lines decreases to minimum (zero) at *C*. From position *C* to position *D*, the rate at which the loop cuts the flux lines increases to a maximum at *D* and then back to a minimum again at *A*.

**FIGURE 10–35**
***End view of loop cutting through the magnetic field.***

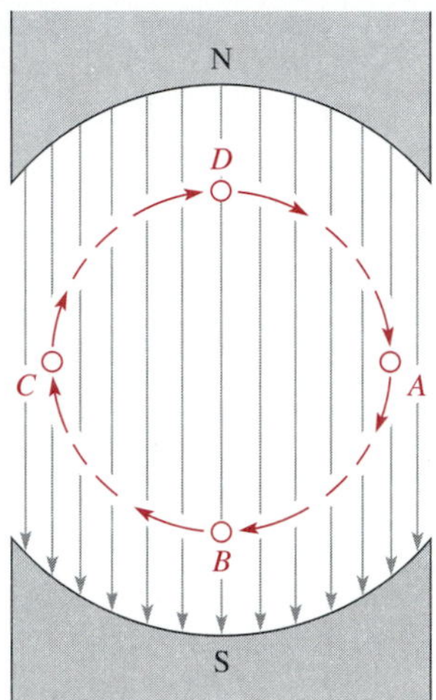


As you have learned, when a wire moves through a magnetic field, a voltage is induced, and by Faraday's law, the amount of induced voltage is proportional to the number of loops (turns) in the wire and the rate at which it is moving with respect to the magnetic field. Now you know that the angle at which the wire moves with respect to the magnetic flux lines determines the amount of induced voltage, because the rate at which the wire cuts through the flux lines depends on the angle of motion.

Figure 10–36 illustrates how a voltage is induced in the external circuit as the single loop rotates in the magnetic field. Assume that the loop is in its instantaneous horizontal position, so the induced voltage is zero. As the loop continues in its rotation, the induced voltage builds up to a maximum at position *B*, as shown in part (a) of the figure. Then, as the loop continues from *B* to *C*, the voltage decreases to zero at position *C*, as shown in part (b).

During the second half of the revolution, shown in Figure 10–36(c) and (d), the brushes switch to opposite commutator sections, so the polarity of the voltage remains the same across the output. Thus, as the loop rotates from position *C* to position *D* and then back to position *A*, the voltage increases from zero at *C* to a maximum at *D* and back to zero at *A*.

Figure 10–37 shows how the induced voltage varies as the loop goes through several rotations (three in this case). This voltage is a dc voltage because its polarities do not change. However, the voltage is pulsating between zero and its maximum value.

When more loops are added, the voltages induced across each loop are combined across the output. Since the voltages are offset from each other, they do not reach their maximum or zero values at the same time. A smoother dc voltage results, as shown in Figure 10–38 (p. 388) for two loops. The variations can be further smoothed out by filters to achieve a nearly constant dc voltage. (Filters are covered in Chapter 19.)

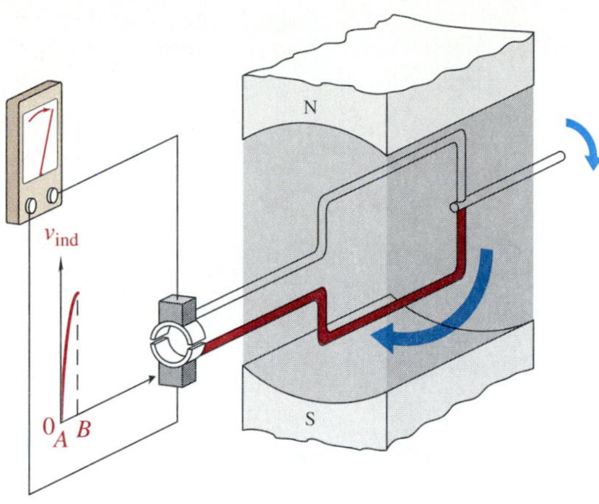


(a) Position $B$: Loop is moving perpendicular to flux lines, and voltage is maximum.

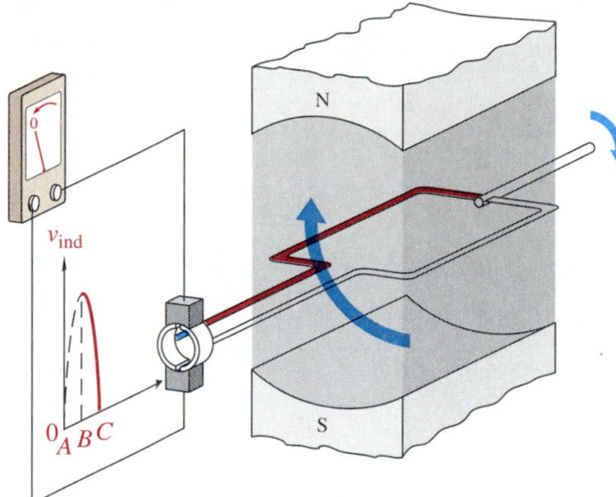


(b) Position $C$: Loop is moving parallel with flux lines, and voltage is zero.

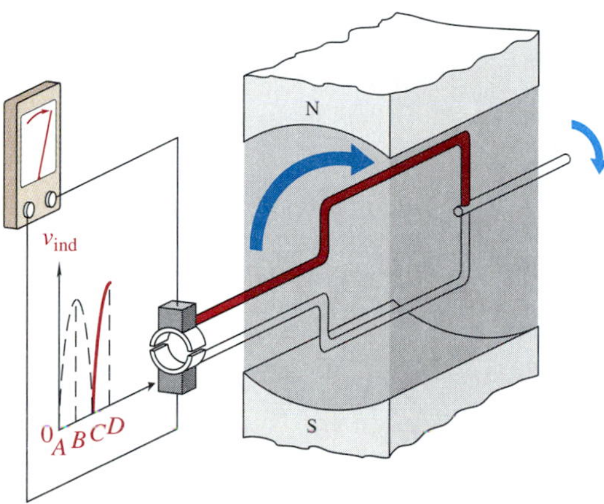


(c) Position $D$: Loop is moving perpendicular to flux lines, and voltage is maximum.

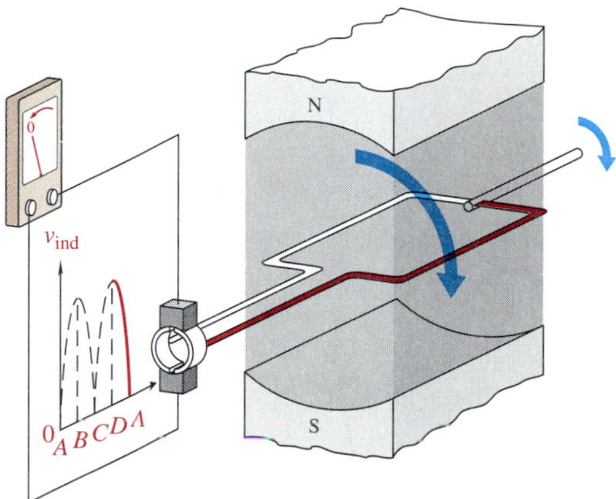


(d) Position $A$: Loop is moving parallel with flux lines, and voltage is zero.

**FIGURE 10–36**
***Operation of a basic dc generator.***

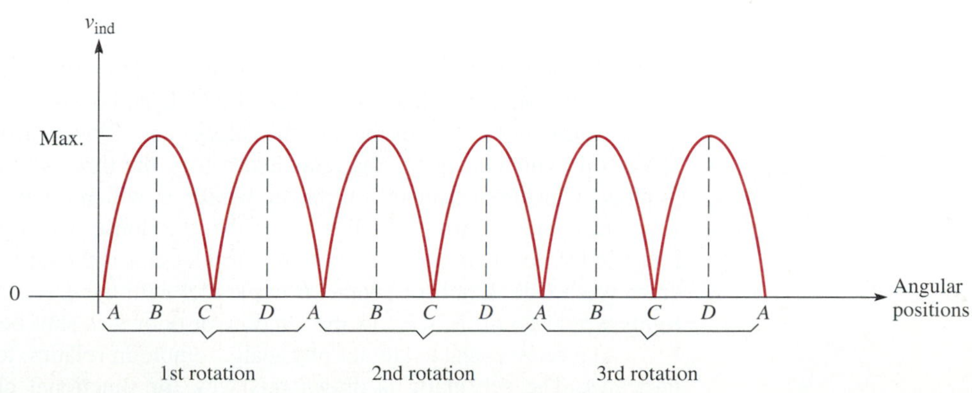


**FIGURE 10–37**
***Induced voltage over three rotations of the loop.***

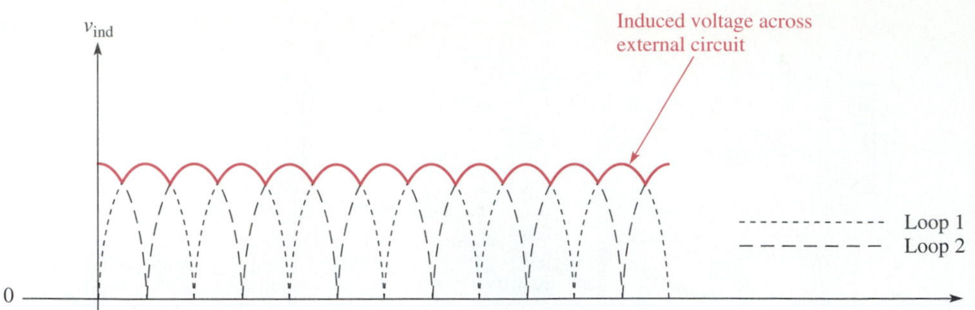


FIGURE 10–38
***The induced voltage for a two-loop generator. There is much less variation in the induced voltage.***

**SECTION 10–6 REVIEW**

1. If the steel disk in the crankshaft position sensor has stopped, with a tab in the magnet's air gap, what is the induced voltage?
2. What happens to the induced voltage if the loop in the basic dc generator suddenly begins rotating at a faster speed?

## 10–7 ■ TECHnology Theory Into Practice

***The relay is a common type of electromagnetic device that is used in many types of control applications. With a relay, a lower voltage, such as from a battery, can be used to switch a much higher voltage, such as the 110 V from an ac outlet. In this section, you will see how a relay can be used in a basic burglar alarm system.***

The schematic in Figure 10–39 shows a basic simplified intrusion alarm system that uses a relay to turn on an audible alarm (siren) and lights. The system operates from a 9 V battery so that even if power to the house is off, the audible alarm will still work.

The detection switches are normally open (NO) magnetic switches that are parallel connected and located in the windows and doors. The relay is a three-pole, double-throw device that operates with a coil voltage of 9 V dc and draws approximately 50 mA. When an intrusion occurs, one of the switches closes and allows current from the battery to the relay coil, which energizes the relay and causes the three sets of normally open contacts to close. Closure of contact *A* turns on the alarm, which draws 2 A from the battery. Closure of contact *C* turns on a light circuit in the house. Closure of contact *B* latches the relay and keeps it energized even if the intruder closes the door or window through which entry was made. If not for contact *B* in parallel with the detection switches, the alarm and lights would go off as soon as the window or door was shut behind the intruder.

The relay contacts are not physically remote in relation to the coil as the schematic indicates. The schematic is drawn this way for functional clarity. The entire relay is housed in the package shown in Figure 10–40. Also shown are the pin diagram and internal schematic for the relay.

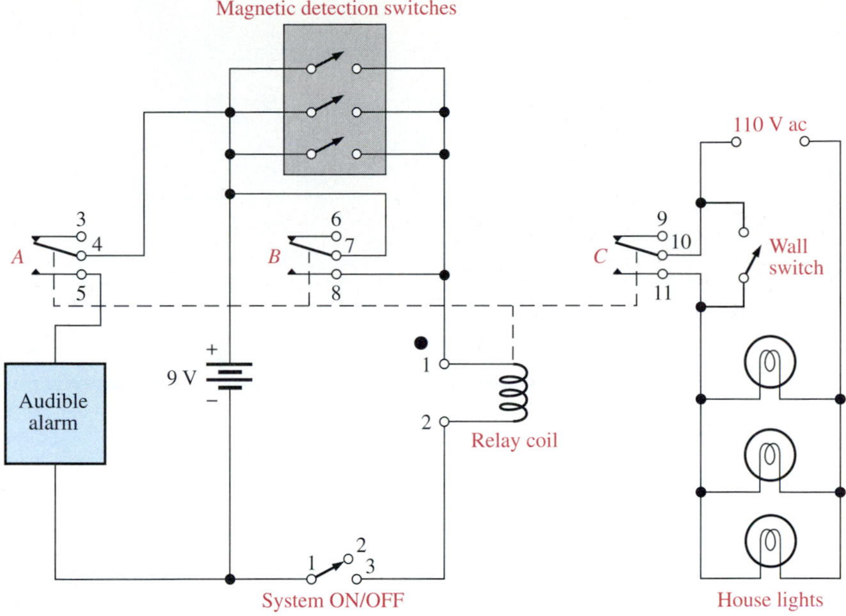


**FIGURE 10–39**
***Simplified burglar alarm system.***

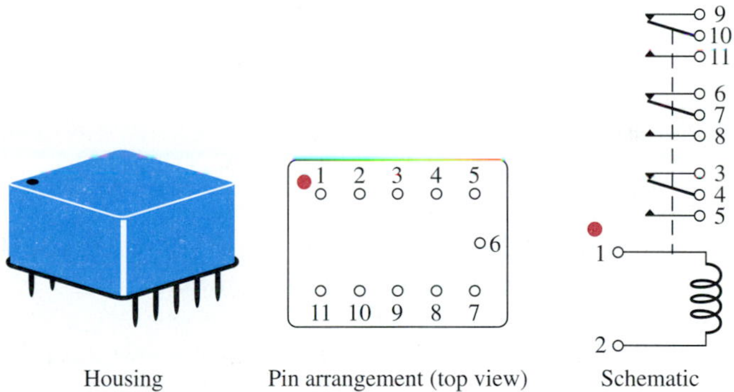


**FIGURE 10–40**
***Triple-pole–double-throw relay.***

## System Interconnections

☐ Develop a connection block diagram and point-to-point wire list for interconnecting the components in Figure 10–41 to create the alarm system shown in the schematic of Figure 10–39. The connection points on the components are indicated by letters.

## A Test Procedure

☐ Develop a detailed step-by-step procedure to check out the completely wired burglar alarm system.

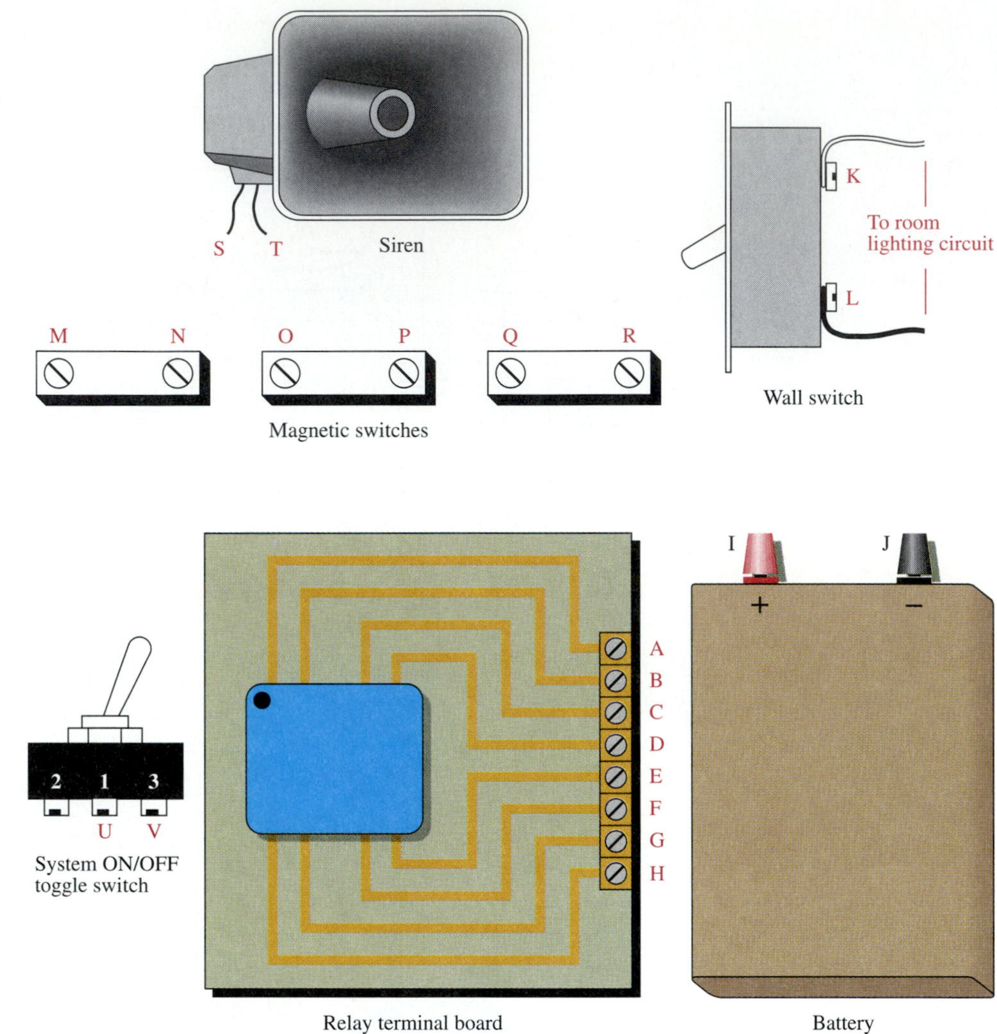


**FIGURE 10–41**
***Array of burglar alarm components.***

**SECTION 10–7 REVIEW**

1. What is the purpose of the detection switches?
2. What is the purpose of relay section B?

## 10–8 ■ INTRODUCTION TO PROBE

***In previous chapters you learned about the .PLOT statement that allows you to graph changing currents and voltages in a circuit. There is a more powerful analysis tool available called Probe that allows you to analyze varying circuit parameters in more detail. Unlike the .PLOT statement, which provides only crude "printer" plot capability directly from any standard SPICE simulator, Probe is a separate program that graphically displays data generated by PSpice. In this section, and in following chapters, various aspects of Probe are introduced.***

***After completing this section, you should be able to***

- **Use Probe in circuit analysis**
  - Write a circuit file using the .PROBE statement
  - Use Probe to graph any current or voltage in a circuit

## The .PROBE Control Statement

Just as PSpice requires a circuit file for input, Probe requires a data file for input. PSpice and Probe are written to allow for a very simple interface between the programs. You request data to be written to a Probe data file by including a .PROBE statement in the circuit file. Earlier versions of the two programs assumed the data file was always named PROBE.DAT, while later versions assume the data file will follow the name of the circuit file. This means that, for example, if PSpice was run successfully on a circuit file named TEST.CIR, and a .PROBE statement was found in the circuit file, a Probe data file named TEST.DAT would be generated. For those using older versions of PSpice, the PROBE.DAT file can be renamed to match the circuit file name, since even the older versions of Probe allow you to specify the Probe data file name on the command line when the program is called, just as the circuit file name is specified on the command line when PSpice is run. Since Probe is a graphical output program, there is also the additional need to specify the display device and hard copy output device to be used. Older versions of the software include a utility program called SETUPDEV.EXE for specifying the particular display and hard copy devices to use. Later versions provide a more integrated means of specifying such information, such as through environment variables, control file entries, and such. As was stated in Chapter 1, you should consult your local PSpice documentation for complete details about using PSpice and Probe on your particular computer system.

To compare the use of .PROBE with .PLOT, we will use the circuit in Figure 10–42 for analysis. The circuit file using the .PLOT statement to generate a plot showing how the currents through $R_1$, $R_2$, and $R_3$ change as $V_S$ changes from 0 V to 10 V is as follows:

```
Parallel Resistive Circuit
*.PLOT demonstration, Figure 10–42
VS  1  0  5
R1  1  0  330
R2  1  0  680
R3  1  0  1K
.DC  VS  0  10  1
.PLOT  DC  I(R1)  I(R2)  I(R3)
.END
```

**FIGURE 10–42**

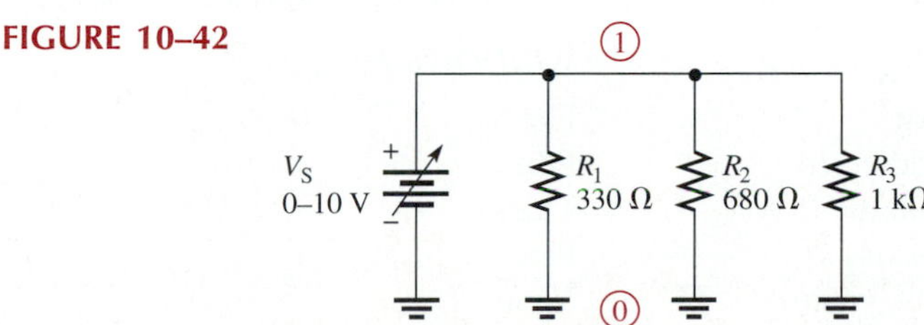


A circuit file to do the same analysis using Probe is almost identical. Instead of using the .PLOT statement, a .PROBE control statement is used and the comment line is changed, as follows:

```
Parallel Resistive Circuit
*.PROBE demonstration, Figure 10–42
VS  1  0  5
R1  1  0  330
R2  1  0  680
R3  1  0  1K
.DC  VS  0  10  1
.PROBE
.END
```

After running PSpice, you will find that PSpice has written a file with a .DAT extension. This is the file that Probe uses to display the circuit information that you may wish to see.

## Using Probe

Notice that in the circuit file containing the .PROBE statement, the currents you want to obtain are not specified as they are in the .PLOT statement. This is because Probe is an interactive program in which you tell it what you want to see as the program runs. As its name suggests, Probe lets you "probe" the circuit to examine currents and voltages, just as you might with an oscilloscope or other test instrument. Any current or voltage can be viewed, and the view may be changed at any time. This ability is very important in circuit analysis because what you see at one point in a circuit may determine what you wish to see next. Using .PLOT, you must change the circuit file and run PSpice again if you wish to see something new. With Probe, you simply change what you tell the program to display, and it retrieves the data per your request from the data file. For very long simulation runs, or for a very large number of nodes, the .PROBE statement can be specified with a specific list of circuit voltages and currents to record data on, so that excessively large data files can be avoided. For all our work in this book, however, and for most of the work you are likely to do in the near future, the default action of writing all voltages and currents to the data file with a plain .PROBE statement is the preferred usage.

Once a Probe data file has been created using PSpice, you run Probe by entering the program name followed by the data file name at the DOS prompt, or by selecting the program for execution via menu choices and mouse clicks if you are using Windows. When entering the program and data file names at the DOS prompt, you may leave off the .EXE and .DAT extensions, so that the following are equivalent:

```
C:\>PROBE.EXE MYFILE.DAT
C:\>PROBE MYFILE
```

Once Probe starts up and examines the data file, you will see a grid and a menu of commands or options. For the circuit file for Figure 10–42, the $x$-axis of the grid will range from 0 V to 10 V and show the label VS. Above or below the grid there will be a number of menu options. The options available can vary depending on the version of Probe, but the typical "main menu" selections are as follows:

| *DOS Format* | *WINDOWS Format* |
|---|---|
| Exit | File |
| Add_trace | Edit |
| X-axis | Trace |
| Y-axis | Plot |
| Plot_control | View |
| Display_control | Tools |
| Macros | Window |
| Hard_copy | Help |

An option that is active is always highlighted, and it will be selected when you press ENTER or click a mouse button. Maneuvering through menus and selecting options will vary from installation to installation; however, program operation is straightforward and largely intuitive.

## Viewing Information in Probe

When Probe starts, the graph is empty because it does not know the voltage or current trace you might want to see. Thus, you must tell the program what you wish to look at, by selecting the Add_trace or Trace menu option, and following through drop-down submenus or input prompts, depending on the program version. Most versions display a list of available voltages and currents, either automatically, or by pressing a function key on the computer keyboard. Traces then can be selected for display either by typing them directly or through use of a pointing device, such as a mouse.

For the example circuit in Figure 10–42, the currents through $R_1$, $R_2$, and $R_3$ are needed. So, you specify the variables I(R1), I(R2), I(R3) for display. Probe will display three lines on the graph, corresponding to the circuit variables (currents) you requested. Probe will label the $y$-axis with an appropriate range of values to display all the data within and will also label the $y$-axis with units (mA in this case) when all of the circuit variables are of the same electrical unit. Additional menu options may appear to allow you to add more traces to, remove traces from, or modify the appearance or ranges of the displayed data.

Probe provides many display and hard copy options for analyzing PSpice simulation data in a graphical form. As such, it is beyond the scope of this book to discuss the detailed operation of such a substantial program. Fortunately, the documentation supplied with the program is excellent, and while operation differs somewhat from platform to platform, users typically find the program is powerful, yet easy to learn.

One difference worth pointing out here, in using Probe versus the .PLOT statement, is in analyzing the voltage across components. Suppose you wish to see the voltages across $R_1$ in our example. In PSpice, you would simply specify V(R1) on the .PLOT statement. However, Probe does not accept trace specifications of voltages across component references. Voltages must be specified in terms of the circuit file nodes. The general format for specifying the voltage at a node X with respect to a node Y, in either PSpice or Probe is as follows:

```
V(X,Y)
```

In the example circuit of Figure 10–42, $R_1$ is connected between node 1 and node 0, so the voltage across $R_1$ may be specified as

```
V(1,0)
```

As a matter of notation, PSpice and Probe allow the ground node, node 0, to be omitted from voltage expressions, where node 0 is the reference node. Thus, for example, the voltage across $R_1$ may be abbreviated as

```
V(1)
```

## Printing Results from Probe

To print out the graphs displayed, select the Hard_copy or Print options from the available menus. Probe will provide options for plotting the data onto various page sizes or across more than one page, and perhaps with a choice of resolutions. Some experimentation may be required for this option to find a graph size and quality that is suitable.

## Exiting Probe

To exit Probe, select the Exit option when displayed. Non-Windows versions allow you to repeatedly select Exit while backing out of menus, while the Main menu exit option will end the program. The Windows versions of the program have an Exit option under the File main menu option.

## SECTION 10–8 REVIEW

1. Compare Probe with .PLOT.
2. Write a PSpice circuit file for the circuit in Figure 10–43 using Probe to display how voltages across $R_1$ and $R_2$ vary as $V_S$ is varied from 0 V to 10 V in 1 V increments.

FIGURE 10–43

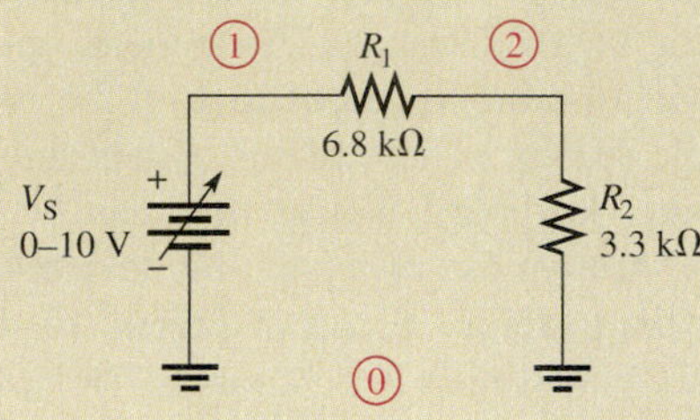


## ■ SUMMARY

- Unlike magnetic poles attract each other, and like poles repel each other.
- Materials that can be magnetized are called *ferromagnetic.*
- When there is current through a conductor, it produces an electromagnetic field around the conductor.
- You can use the right-hand rule to establish the direction of the electromagnetic lines of force around a conductor.
- An electromagnet is basically a coil of wire around a magnetic core.
- When a conductor moves within a magnetic field, or when a magnetic field moves relative to a conductor, a voltage is induced across the conductor.
- The faster the relative motion between a conductor and a magnetic field, the greater is the induced voltage.
- Table 10–1 summarizes the magnetic quantities.

TABLE 10–1

| Symbol | Quantity | Unit |
|---|---|---|
| $B$ | Magnetic flux density | Tesla (T) |
| $\phi$ | Magnetic flux | Weber (Wb) |
| $\mu$ | Permeability | Weber/ampere turn · meter (Wb/At · m) |
| $\mathcal{R}$ | Reluctance | At/Wb |
| $F_m$ | Magnetomotive force (mmf) | Ampere-turn (At) |
| $H$ | Magnetizing force | At/m |

## ■ GLOSSARY

**Ampere-turn** The unit of magnetomotive force (mmf).

**Electromagnetic field** A formation of a group of magnetic lines of force surrounding a conductor created by electrical current in the conductor.

**Electromagnetic induction** The phenomenon or process by which a voltage is produced in a conductor when there is relative motion between the conductor and a magnetic or electromagnetic field.

**Faraday's law** A law stating that the voltage induced across a coil of wire equals the number of turns in the coil times the rate of change of the magnetic flux.

**Hysteresis** A characteristic of a magnetic material whereby a change in magnetization lags the application of a magnetizing force.

**Magnetic flux** The lines of force between the north and south poles of a permanent magnet or an electromagnet.

**Magnetic flux density** The number of lines of force per unit area in a magnetic field.

**Magnetizing force** The amount of mmf per unit length of magnetic material.

**Magnetomotive force** The force that produces the magnetic field.

**Permeability** The measure of ease with which a magnetic field can be established in a material.

**Relay** An electromagnetically controlled mechanical device in which electrical contacts are opened or closed by a magnetizing current.

**Reluctance** The opposition to the establishment of a magnetic field in a material.

**Retentivity** The ability of a material, once magnetized, to maintain a magnetized state without the presence of a magnetizing force.

**Solenoid** An electromagnetically controlled device in which the mechanical movement of a shaft or plunger is activated by a magnetizing current.

**Tesla** The unit of flux density.

**Weber** The unit of magnetic flux.

## FORMULAS

| | | |
|---|---|---|
| **(10–1)** | $B = \frac{\phi}{A}$ | Magnetic flux density |
| **(10–2)** | $\mu_r = \frac{\mu}{\mu_0}$ | Relative permeability |
| **(10–3)** | $\mathcal{R} = \frac{l}{\mu A}$ | Reluctance |
| **(10–4)** | $F_m = NI$ | Magnetomotive force |
| **(10–5)** | $\phi = \frac{F_m}{\mathcal{R}}$ | Magnetic flux |
| **(10–6)** | $H = \frac{F_m}{l}$ | Magnetic flux intensity |
| **(10–7)** | $v_{\text{ind}} = N\left(\frac{d\phi}{dt}\right)$ | Faraday's law |

## SELF-TEST

**1.** When the south poles of two bar magnets are brought close together, there will be
**(a)** a force of attraction **(b)** a force of repulsion
**(c)** an upward force **(d)** no force

**2.** A magnetic field is made up of
**(a)** positive and negative charges **(b)** magnetic domains
**(c)** flux lines **(d)** magnetic poles

**3.** The direction of a magnetic field is from
**(a)** north pole to south pole **(b)** south pole to north pole
**(c)** inside to outside the magnet **(d)** front to back

**4.** Reluctance in a magnetic circuit is analogous to
**(a)** voltage in an electric circuit **(b)** current in an electric circuit
**(c)** power in an electric circuit **(d)** resistance in an electric circuit

**5.** The unit of magnetic flux is the
**(a)** tesla **(b)** weber **(c)** electron-volt **(d)** ampere-turn

**6.** The unit of magnetomotive force is the
**(a)** tesla **(b)** weber **(c)** ampere-turn **(d)** electron-volt

**7.** The unit of flux density is the
**(a)** tesla **(b)** weber **(c)** ampere-turn **(d)** electron-volt

**8.** The electromagnetic activation of a movable shaft is the basis for
**(a)** relays **(b)** circuit breakers **(c)** magnetic switches **(d)** solenoids

9. When there is current through a wire placed in a magnetic field,
   (a) the wire will overheat (b) the wire will become magnetized
   (c) a force is exerted on the wire (d) the magnetic field will be cancelled
10. A coil of wire is placed in a changing magnetic field. If the number of turns in the coil is increased, the voltage induced across the coil will
   (a) remain unchanged (b) decrease
   (c) increase (d) be excessive
11. If a conductor is moved back and forth at a constant rate in a constant magnetic field, the voltage induced in the conductor will
   (a) remain constant (b) reverse polarity
   (c) be reduced (d) be increased
12. In the crankshaft position sensor in Figure 10–32, the induced voltage across the coil is caused by
   (a) current in the coil
   (b) rotation of the disk
   (c) a tab passing through the magnetic field
   (d) acceleration of the disk's rotational speed

## ■ PROBLEMS

### SECTION 10–1 The Magnetic Field

1. The cross-sectional area of a magnetic field is increased, but the flux remains the same. Does the flux density increase or decrease?
2. In a certain magnetic field the cross-sectional area is 0.5 $m^2$ and the flux is 1500 $\mu$Wb. What is the flux density?
3. What is the flux in a magnetic material when the flux density is $2500 \times 10^{-6}$ T and the cross-sectional area is 150 $cm^2$?

### SECTION 10–2 Electromagnetism

4. What happens to the compass needle in Figure 10–9 when the current through the conductor is reversed?
5. What is the relative permeability of a ferromagnetic material whose absolute permeability is $750 \times 10^{-6}$ Wb/At · m?
6. Determine the reluctance of a material with a length of 0.28 m and a cross-sectional area of 0.08 $m^2$ if the absolute permeability is $150 \times 10^{-7}$ Wb/At·m.
7. What is the magnetomotive force in a 50 turn coil of wire when there are 3 A of current through it?

### SECTION 10–3 Electromagnetic Devices

8. Typically, when a solenoid is activated, is the plunger extended or retracted?
9. (a) What force moves the plunger when a solenoid is activated?
   (b) What force causes the plunger to return to its at-rest position?
10. Explain the sequence of events in the circuit of Figure 10–44 starting when switch 1 (SW1) is closed.
11. What causes the pointer in a d'Arsonval movement to deflect when there is current through the coil?

FIGURE 10–44

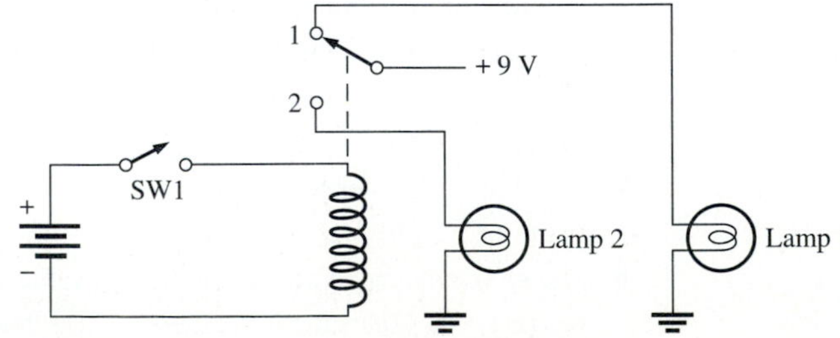

## SECTION 10–4 Magnetic Hysteresis

**12.** What is the magnetizing force in Problem 7 if the length of the core is 0.2 m?

**13.** How can the flux density in Figure 10–45 be changed without altering the physical characteristics of the core?

**14.** In Figure 10–45, determine the following:

**(a)** $H$ **(b)** $\phi$ **(c)** $B$

FIGURE 10–45

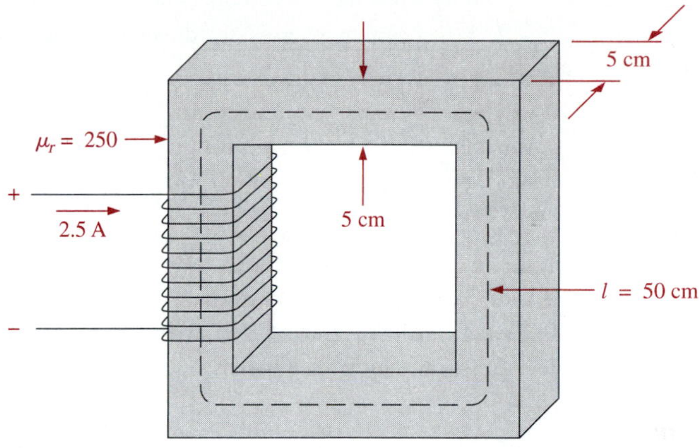


**15.** Determine from the hysteresis curves in Figure 10–46 which material has the most retentivity.

FIGURE 10–46

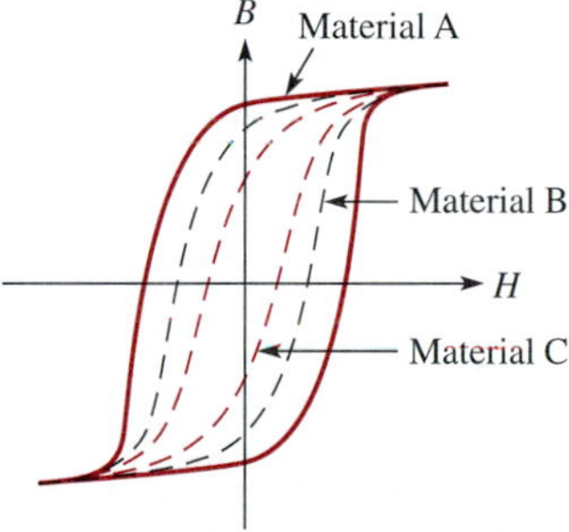


## SECTION 10–5 Electromagnetic Induction

**16.** According to Faraday's law, what happens to the induced voltage across a given coil if the rate of change of magnetic flux doubles?

**17.** The voltage induced across a certain coil is 100 mV. A 100 Ω resistor is connected to the coil terminals. What is the induced current?

**18.** A magnetic field is changing at a rate of $3500 \times 10^{-3}$ Wb/s. How much voltage is induced across a 50 turn coil that is placed in the magnetic field?

**19.** How does Lenz's law complement Faraday's law?

## SECTION 10–6 Applications of Electromagnetic Induction

**20.** In Figure 10–32 on page 384, why is there no induced voltage when the steel disk is not rotating?

**21.** Explain the purpose of the commutator and brushes in Figure 10–34 on page 385.

**22.** A basic one-loop dc generator is rotated at 60 rev/s. How many times each second does the dc output voltage peak (reach a maximum)?

**23.** Assume that another loop, 90 degrees from the first loop, is added to the dc generator in Problem 22. Make a graph of voltage versus time to show how the output voltage appears. Let the maximum voltage be 10 V.

## ANSWERS TO SECTION REVIEWS

### Section 10–1

1. North poles repel.
2. Magnetic flux is the group of lines of force that make up a magnetic field.
3. $B = \phi/A = 900\ \mu T$

### Section 10–2

1. Electromagnetism is produced by current through a conductor. An electromagnetic field exists only when there is current. A magnetic field exists independently of current.
2. When current reverses, the direction of the magnetic field also reverses.

### Section 10–3

1. A solenoid produces a movement only. A relay provides an electrical contact closure.
2. The movable part of a solenoid is the plunger.
3. The movable part of a relay is the armature.
4. The d'Arsonval movement is based on the interaction of magnetic fields.

### Section 10–4

1. An increase in current increases the flux density.
2. Retentivity is the ability of a material to remain magnetized after removal of the magnetizing force.

### Section 10–5

1. Zero voltage is induced.
2. Induced voltage increases.
3. A force is exerted on the conductor when there is current.

### Section 10–6

1. Zero voltage is induced in the air gap.
2. A faster rotation increases induced voltage.

### Section 10–7

1. The detection switches indicate an intrusion through a window or door when closed.
2. Section B latches the relay and keeps it energized when intrusion is detected.

### Section 10–8

1. Probe is interactive, .PLOT is not.
2. Series Resistive Circuit

```
VS  1  0  5
R1  1  2  6.8K
R2  2  0  3.3K
.DC  VS  O  10  1
.PROBE
.END
```

## ANSWERS TO RELATED EXERCISES FOR EXAMPLES

**10–1** 235 $\mu$Wb

**10–2** 31.0 T

**10–3** Reluctance increases to 4762 At/Wb.

**10–4** 7.2 mWb

**10–5** **(a)** $F_m = 42.5$ At
**(b)** $\mathcal{R} = 85 \times 10^3$ At/Wb

**10–6** 12.5 mV

# 11

# INTRODUCTION TO ALTERNATING CURRENT AND VOLTAGE



## ■ INTRODUCTION

In the preceding chapters, you have studied resistive circuits with dc currents and voltages.

This chapter provides an introduction to ac circuit analysis in which time-varying electrical signals, particularly the sine wave, are studied. An electrical signal is a voltage or current that changes in some consistent manner with time. In other words, the voltage or current fluctuates according to a certain pattern called a **waveform.**

Special emphasis is given to the sine wave because of its fundamental importance in ac circuit analysis. Other types of waveforms are also introduced, including pulse, triangular, and sawtooth. The use of the oscilloscope for displaying and measuring waveforms is introduced. In this chapter and throughout the rest of the book, you will learn the basics of putting technology theory into practice.

An alternating voltage is one that changes polarity at a certain rate and an alternating current is one that changes direction at a certain rate. The sinusoidal waveform is the most common and fundamental type because all other types of waveforms can be broken down into composite sine waves. The sine wave is a periodic type of waveform that repeats at fixed intervals. The time for each repetition is the *period* and the repetition rate is the *frequency.*

■ **TECHnology Theory Into Practice**

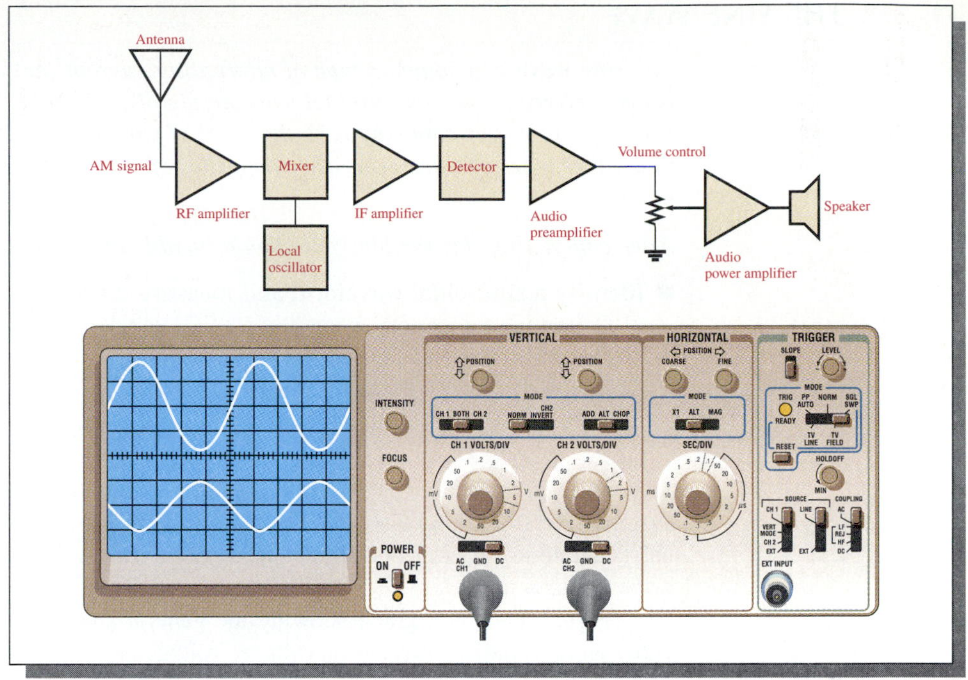


In the TECH TIP assignment in Section 11–10, you will measure voltage signals in an AM receiver using an oscilloscope.

## ■ CHAPTER OBJECTIVES

- ☐ Identify a sinusoidal waveform and measure its characteristics
- ☐ Describe how sine waves are generated
- ☐ Determine the various voltage and current values of a sine wave
- ☐ Describe angular relationships of sine waves
- ☐ Mathematically analyze a sinusoidal waveform
- ☐ Apply the basic circuit laws to ac resistive circuits
- ☐ Determine total voltages which have both ac and dc components
- ☐ Identify the characteristics of basic nonsinusoidal waveforms
- ☐ Use the oscilloscope to measure waveforms
- ☐ Use PSpice and Probe in ac circuit analysis (optional)

## 11–1 ■ THE SINE WAVE

***The sine wave is a common type of alternating current (ac) and alternating voltage. It is also referred to as a sinusoidal wave or, simply, sinusoid. The electrical service provided by the power company is in the form of sinusoidal voltage and current. In addition, other types of waveforms are composites of many individual sine waves called harmonics.***

***After completing this section, you should be able to***

- **Identify a sinusoidal waveform and measure its characteristics**
  - ☐ Determine the period
  - ☐ Determine the frequency
  - ☐ Relate the period and the frequency

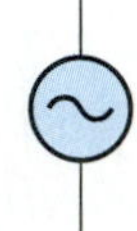

FIGURE 11–1
*Symbol for a sine wave voltage source.*

Sine waves are produced by two types of sources: rotating electrical machines (ac generators) or electronic oscillator circuits, which are in instruments known as electronic signal generators. Figure 11–1 shows the symbol used to represent either source of sine wave voltage.

Figure 11–2 is a graph showing the general shape of a sine wave, which can be either an **alternating current** or voltage. Voltage (or current) is displayed on the vertical axis and time ($t$) is displayed on the horizontal axis. Notice how the voltage (or current) varies with time. Starting at zero, the voltage (or current) increases to a positive maximum (peak), returns to zero, and then increases to a negative maximum (peak) before returning again to zero, thus completing one full cycle.

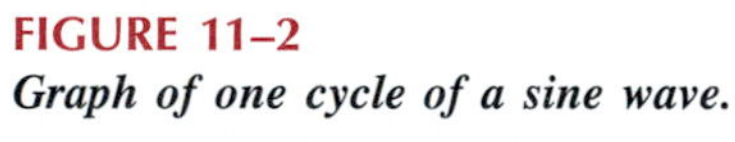

FIGURE 11–2
*Graph of one cycle of a sine wave.*

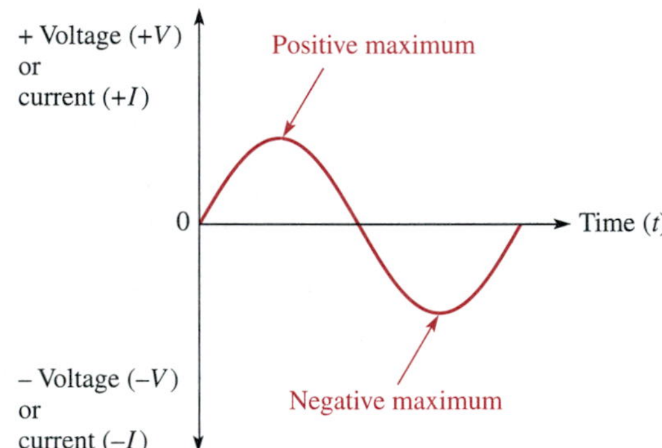


### Polarity of a Sine Wave

As you have learned, a sine wave changes polarity at its zero value; that is, it alternates between positive and negative values. When a sine wave voltage source ($V_s$) is applied to a resistive circuit, as in Figure 11–3, an alternating sine wave current results. When the voltage changes polarity, the current correspondingly changes direction as indicated.

During the positive alternation of the applied voltage $V_s$, the current is in the direction shown in Figure 11–3(a). During a negative alternation of the applied voltage, the current is in the opposite direction, as shown in Figure 11–3(b). The combined positive and negative alternations make up one **cycle** of a sine wave.

**FIGURE 11–3**
***Alternating current and voltage.***

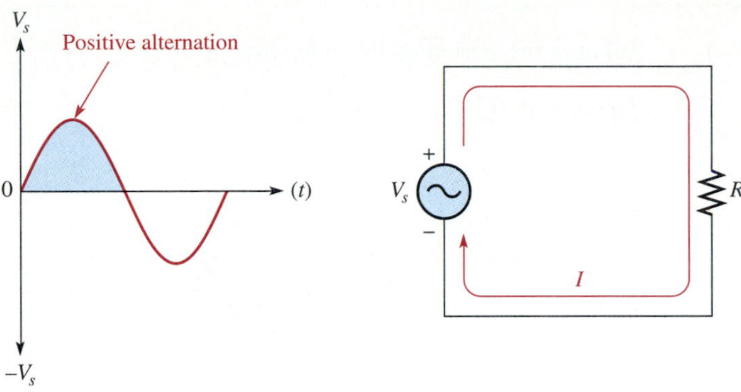


(a) Positive voltage: current direction as shown

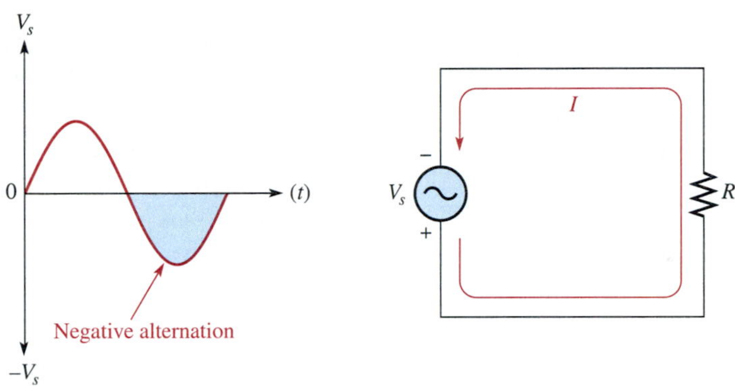


(b) Negative voltage: current reverses direction

## Period of a Sine Wave

A sine wave varies with time ($t$) in a definable manner.

> **The time required for a sine wave to complete one full cycle is called the period ($T$).**

Figure 11–4(a) illustrates the **period** of a sine wave. Typically, a sine wave continues to repeat itself in identical cycles, as shown in Figure 11–4(b). Since all cycles of a repetitive sine wave are the same, the period is always a fixed value for a given sine wave. The period of a sine wave does not necessarily have to be measured between the zero crossings at the beginning and end of a cycle. It can be measured from any peak in a given cycle to the corresponding peak in the next cycle.

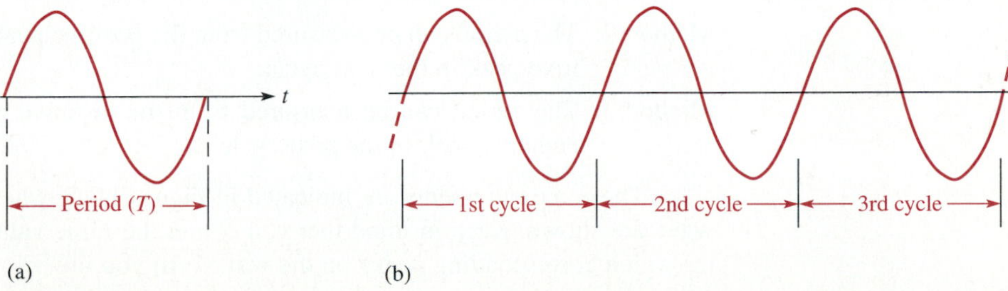


**FIGURE 11–4**
***The period of a sine wave is the same for each cycle.***

**EXAMPLE 11–1** What is the period of the sine wave in Figure 11–5?

FIGURE 11–5

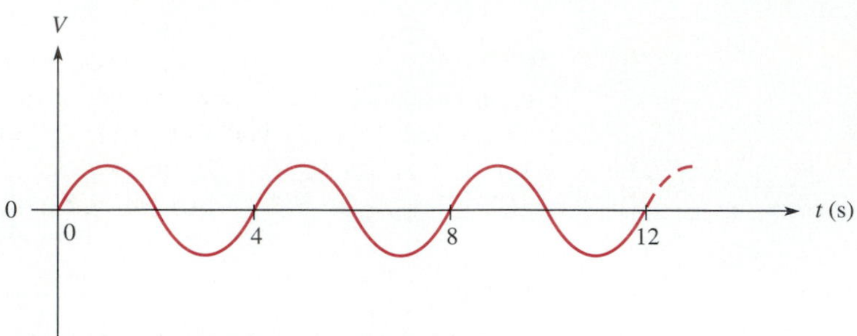


***Solution*** As shown in Figure 11–5, it takes four seconds (4 s) to complete each cycle. Therefore, the period is 4 s.

$$T = 4\text{ s}$$

***Related Exercise*** What is the period if the sine wave goes through five cycles in 12 s?

**EXAMPLE 11–2** Show three possible ways to measure the period of the sine wave in Figure 11–6. How many cycles are shown?

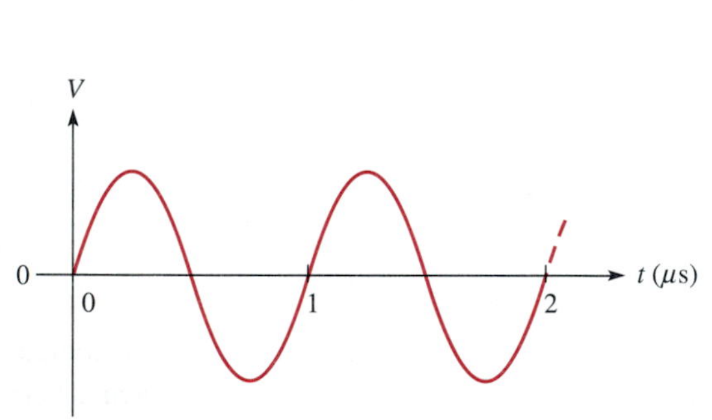


FIGURE 11–6

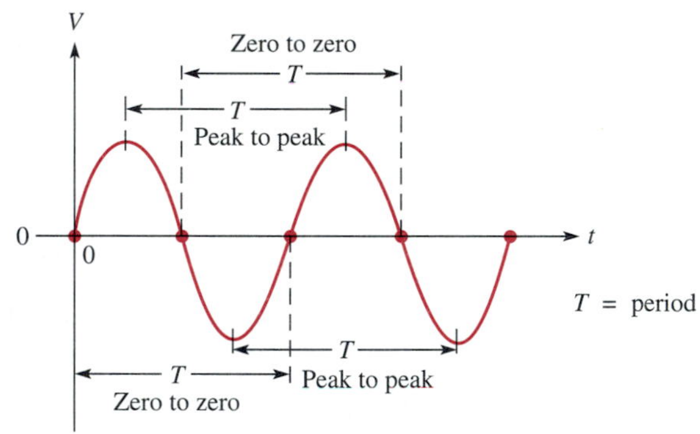


FIGURE 11–7
***Measurement of the period of a sine wave.***

***Solution***

Method 1: The period can be measured from one zero crossing to the corresponding zero crossing in the next cycle.

Method 2: The period can be measured from the positive peak in one cycle to the positive peak in the next cycle.

Method 3: The period can be measured from the negative peak in one cycle to the negative peak in the next cycle.

These measurements are indicated in Figure 11–7, where two cycles of the sine wave are shown. Keep in mind that you obtain the same value for the period no matter which corresponding peaks on the waveform you use.

***Related Exercise*** If a positive peak occurs at 1 ms and the next positive peak occurs at 2.5 ms, what is the period?

## Frequency of a Sine Wave

> **Frequency is the number of cycles that a sine wave completes in one second.**

The more cycles completed in one second, the higher the frequency. **Frequency** ($f$) is measured in units of **hertz,** Hz. One hertz is equivalent to one cycle per second; 60 Hz is 60 cycles per second; and so on. Figure 11–8 shows two sine waves. The sine wave in part (a) completes two full cycles in one second. The one in part (b) completes four cycles in one second. Therefore, the sine wave in part (b) has twice the frequency of the one in part (a).

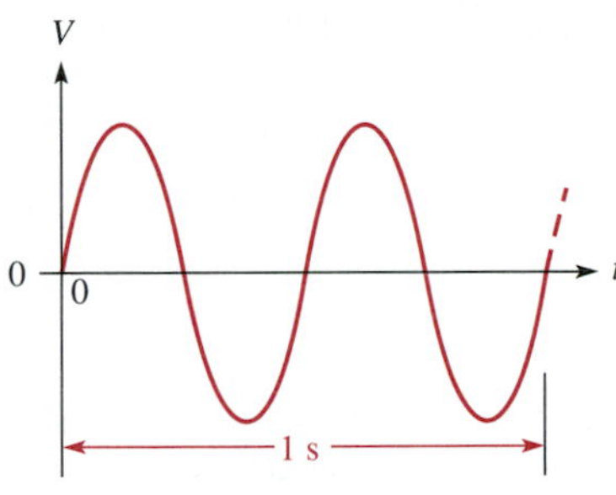


(a) Lower frequency: fewer cycles per second

(b) Higher frequency: more cycles per second

**FIGURE 11–8**
***Illustration of frequency.***

## Relationship of Frequency and Period

The formulas for the relationship between frequency ($f$) and period ($T$) are as follows:

$$f = \frac{1}{T} \quad \textbf{(11–1)}$$

$$T = \frac{1}{f} \quad \textbf{(11–2)}$$

There is a reciprocal relationship between $f$ and $T$. Knowing one, you can calculate the other with the 1/x key on your calculator. This inverse relationship makes sense because a sine wave with a longer period goes through fewer cycles in one second than one with a shorter period.

**EXAMPLE 11–3** Which sine wave in Figure 11–9 has a higher frequency? Determine the period and the frequency of both waveforms.

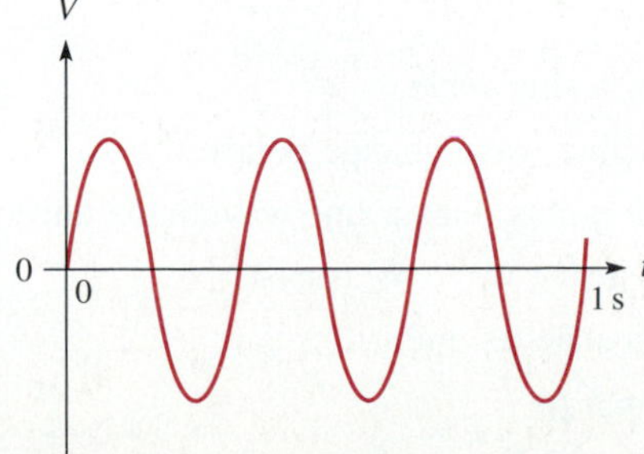


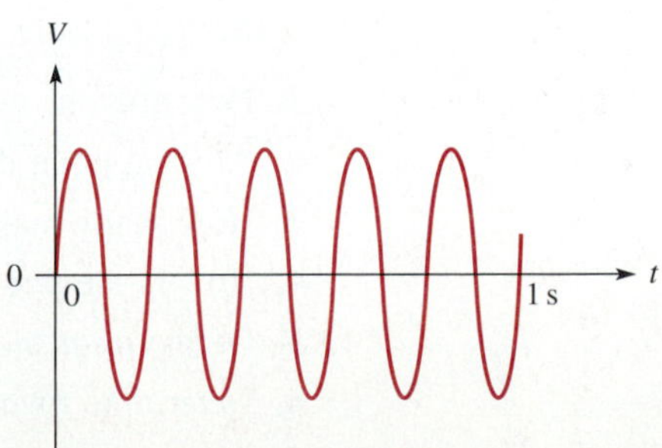


**FIGURE 11–9**

***Solution*** The sine wave in Figure 11–9(b) has the higher frequency because it completes more cycles in 1 s than does the sine wave in part (a).

In Figure 11–9(a), three cycles take 1 s. Therefore, one cycle takes 0.333 s (one-third second), and this is the period.

$$T = 0.333 \text{ s} = 333 \text{ ms}$$

The frequency is

$$f = \frac{1}{T} = \frac{1}{333 \text{ ms}} = 3 \text{ Hz}$$

In Figure 11–9(b), five cycles take 1 s. Therefore, one cycle takes 0.2 s (one-fifth second), and this is the period.

$$T = 0.2 \text{ s} = 200 \text{ ms}$$

The frequency is

$$f = \frac{1}{T} = \frac{1}{200 \text{ ms}} = 5 \text{ Hz}$$

***Related Exercise*** If the time between negative peaks of a given sine wave is 50 $\mu$s, what is the frequency?

**EXAMPLE 11–4**

The period of a certain sine wave is 10 ms. What is the frequency?

***Solution*** Use Equation (11–1).

$$f = \frac{1}{T} = \frac{1}{10 \text{ ms}} = \frac{1}{10 \times 10^{-3} \text{ s}} = 100 \text{ Hz}$$

***Related Exercise*** A certain sine wave goes through four cycles in 20 ms. What is the frequency?

**EXAMPLE 11–5**

The frequency of a sine wave is 60 Hz. What is the period?

***Solution*** Use Equation (11–2).

$$T = \frac{1}{f} = \frac{1}{60 \text{ Hz}} = 16.7 \text{ ms}$$

***Related Exercise*** If $T = 15\ \mu$s, what is $f$?

**SECTION 11–1 REVIEW**

1. Describe one cycle of a sine wave.
2. At what point does a sine wave change polarity?
3. How many maximum points does a sine wave have during one cycle?
4. How is the period of a sine wave measured?
5. Define *frequency,* and state its unit.
6. Determine $f$ when $T = 5\ \mu$s.
7. Determine $T$ when $f = 120$ Hz.

## 11–2 ■ SINE WAVE VOLTAGE SOURCES

***Two basic methods of generating sine wave voltages are electromagnetic and electronic. Sine waves are produced electromagnetically by ac generators and electronically by oscillator circuits.***

***After completing this section, you should be able to***

- **Describe how sine waves are generated**
  - ☐ Discuss the basic operation of an ac generator
  - ☐ Discuss factors that affect frequency in ac generators
  - ☐ Discuss factors that affect voltage in ac generators

### An AC Generator

Figure 11–10 shows a greatly simplified ac **generator** consisting of a single loop of wire in a permanent magnetic field. Notice that each end of the loop is connected to a separate solid conductive ring called a *slip ring*. As the loop rotates in the magnetic field, the slip rings also rotate and rub against the brushes that connect the loop to an external load. Compare this generator to the basic dc generator in Figure 10–34, and note the difference in the ring and brush arrangements.

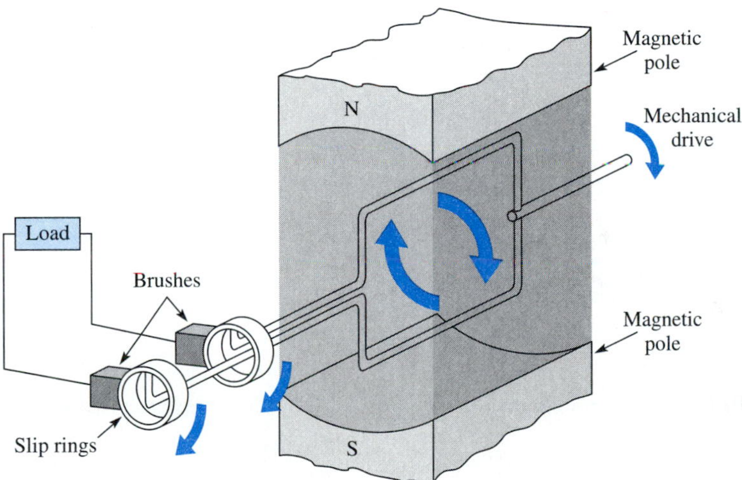


**FIGURE 11–10**
***A basic ac generator.***

As you learned in Chapter 10, when a conductor moves through a magnetic field, a voltage is induced. Figure 11–11 illustrates how a sine wave voltage is produced by the basic ac generator as the loop rotates. An oscilloscope is used to display the voltage waveform.

To begin, Figure 11–11(a) shows the loop rotating through the first quarter of a revolution. It goes from an instantaneous horizontal position, where the induced voltage is zero, to an instantaneous vertical position, where the induced voltage is maximum. At the horizontal position, the loop is instantaneously moving parallel with the flux lines; thus, no lines are being cut and the voltage is zero. As the loop rotates through the first quarter-cycle, it cuts through the flux lines at an increasing rate until it is instantaneously moving perpendicular to the flux lines at the vertical position and cutting through them at a maximum rate. Thus, the induced voltage increases from zero to a peak during the quarter-cycle. As shown on the display in part (a), this part of the rotation produces the first quarter of the sine wave cycle as the voltage builds up from zero to its positive maximum.

Figure 11–11(b) shows the loop completing the first half of a revolution. During this part of the rotation, the voltage decreases from its positive maximum back to zero as the rate at which the loop cuts through the flux lines decreases.

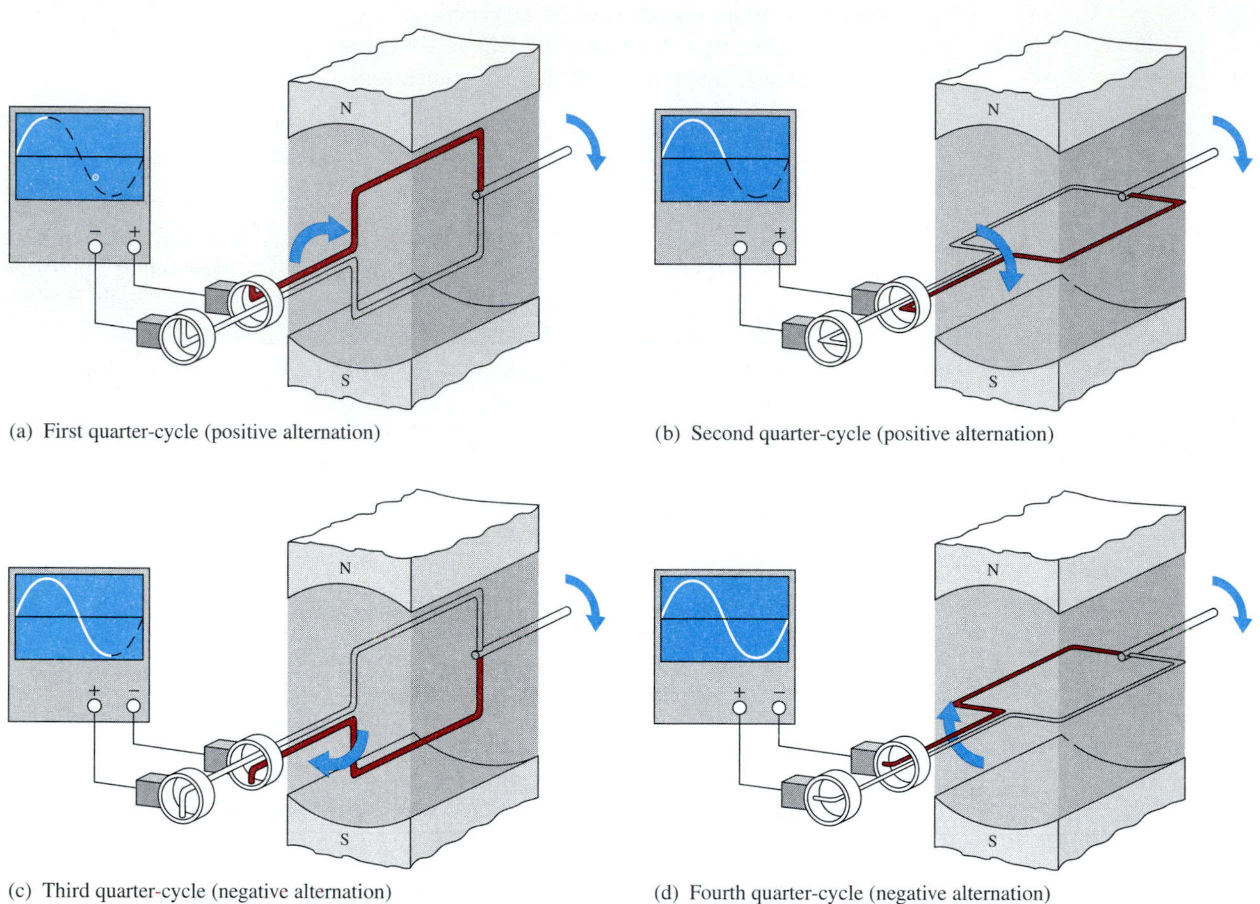


(a) First quarter-cycle (positive alternation)

(b) Second quarter-cycle (positive alternation)

(c) Third quarter-cycle (negative alternation)

(d) Fourth quarter-cycle (negative alternation)

**FIGURE 11–11**

***One revolution of the loop generates one cycle of the sine wave voltage.***

During the second half of the revolution, illustrated in Figures 11–11(c) and 11–11(d), the loop is cutting through the magnetic field in the opposite direction, so the voltage produced has a polarity opposite to that produced during the first half of the revolution. After one complete revolution of the loop, one full cycle of the sine wave voltage has been produced. As the loop continues to rotate, repetitive cycles of the sine wave are generated.

***Frequency*** You have seen that one revolution of the conductor through the magnetic field in the basic ac generator (also called an *alternator*) produces one cycle of induced sinusoidal voltage. It is obvious that the rate at which the conductor is rotated determines the time for completion of one cycle. For example, if the conductor completes 60 revolutions in one second (rps), the period of the resulting sine wave is 1/60 s, corresponding to a frequency of 60 Hz. Thus, the faster the conductor rotates, the higher the resulting frequency of the induced voltage, as illustrated in Figure 11–12.

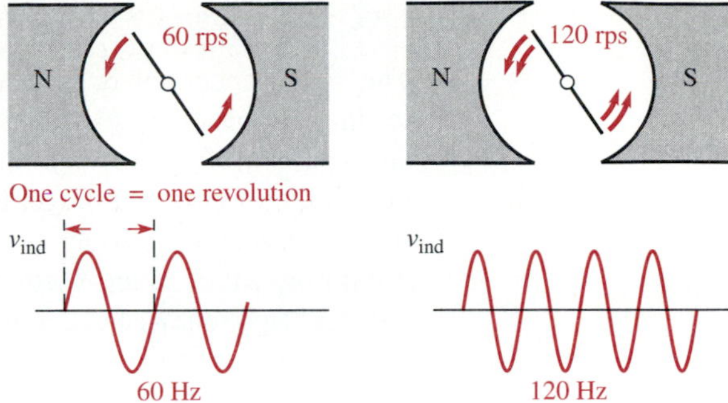


**FIGURE 11–12**
***Frequency is directly proportional to the rate of rotation of the loop.***

Another way of achieving a higher frequency is to increase the number of magnetic poles. In the previous discussion, two magnetic poles were used to illustrate the ac generator principle. During one revolution, the conductor passes under a north pole and a south pole, thus producing one cycle of a sine wave. When four magnetic poles are used instead of two, as shown in Figure 11–13, one cycle is generated during one-half a revolution. This doubles the frequency for the same rate of rotation.

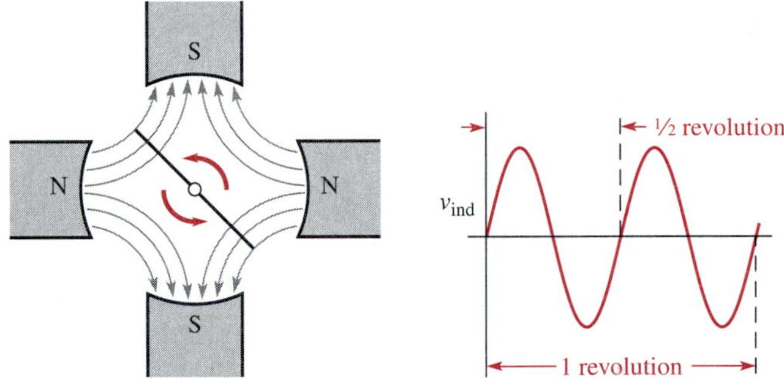


**FIGURE 11–13**
***Four poles achieve a higher frequency than two for the same rps.***

An expression for frequency in terms of the number of pole pairs and the number of revolutions per second (rps) is as follows:

$$f = (\text{number of pole pairs})(\text{rps}) \qquad \textbf{(11–3)}$$

**EXAMPLE 11–6** A four-pole generator has a rotation speed of 100 rps. Determine the frequency of the output voltage.

***Solution*** $f = (\text{number of pole pairs})(\text{rps}) = 2(100) = 200\text{ Hz}$

***Related Exercise*** If the frequency of the output of a four-pole generator is 60 Hz, what is the rps?

***Voltage Amplitude*** Recall from Chapter 10 that the voltage induced in a conductor depends on the number of turns ($N$) and the rate of change with respect to the magnetic field. Therefore, when the speed of rotation of the conductor is increased, not only the frequency of the induced voltage increases—so also does the amplitude. Since the frequency value normally is fixed, the most practical method of increasing the amount of induced voltage is to increase the number of loops.

### Electronic Signal Generators

The signal generator is an instrument that electronically produces sine waves for use in testing or controlling electronic circuits and systems. There are a variety of signal generators, ranging from special-purpose instruments that produce only one type of waveform in a limited frequency range, to programmable instruments that produce a wide range of frequencies and a variety of waveforms. All signal generators consist basically of an **oscillator,** which is an electronic circuit that produces sine wave voltages whose amplitude and frequency can be adjusted. Typical signal generators are shown in Figure 11–14.

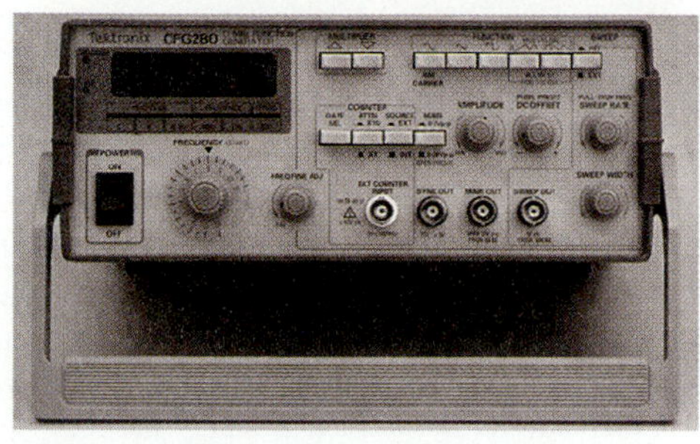


(a)

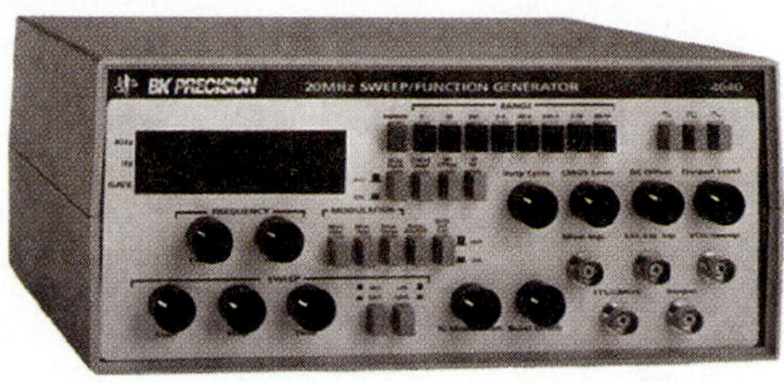


(b)

FIGURE 11–14
***Typical signal generators. Part (a) courtesy of Tektronix, Inc. Part (b) courtesy of BK Precision, Maxtec International Corp.***

**SECTION 11–2 REVIEW**

1. What two basic methods are used to generate sine wave voltages?
2. How are the speed of rotation and the frequency in an ac generator related?

## 11–3 ■ VOLTAGE AND CURRENT VALUES OF SINE WAVES

***There are several ways to express the value of a sine wave in terms of its voltage or its current magnitude. These are instantaneous, peak, peak-to-peak, rms, and average values.***

***After completing this section, you should be able to***

- **Determine the various voltage and current values of a sine wave**
  - ☐ Find the instantaneous value at any point
  - ☐ Find the peak value
  - ☐ Find the peak-to-peak value
  - ☐ Define *rms*
  - ☐ Explain why the average value is always zero over a complete cycle
  - ☐ Find the half-cycle average value

### Instantaneous Value

Figure 11–15 illustrates that at any point in time on a sine wave, the voltage (or current) has an **instantaneous value.** This instantaneous value is different at different points along the curve. Instantaneous values are positive during the positive alternation and negative during the negative alternation. Instantaneous values of voltage and current are symbol-

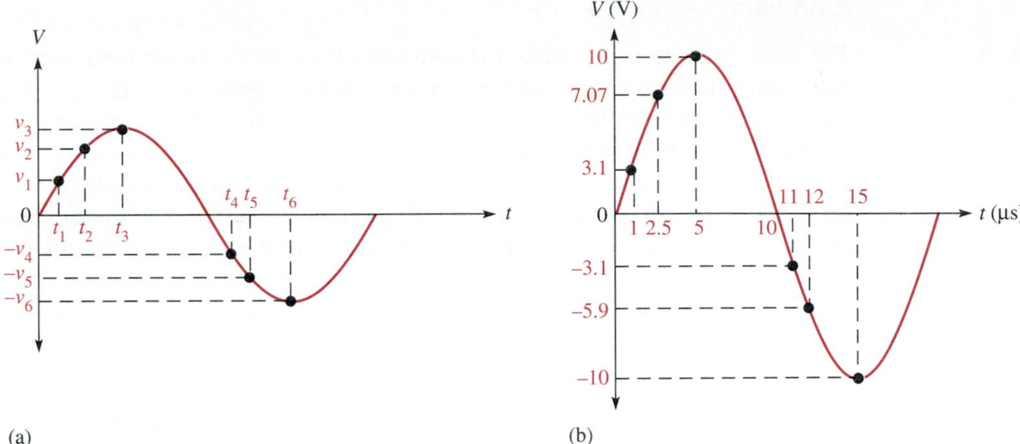


FIGURE 11–15
*Instantaneous values.*

ized by lowercase *v* and *i*, respectively, as shown in part (a). The curve is shown for voltage only, but it applies equally for current when the *v*'s are replaced with *i*'s. An example of instantaneous values is shown in part (b) where the instantaneous voltage is 3.1 V at 1 μs, 7.07 V at 2.5 μs, 10 V at 5 μs, 0 V at 10 μs, –3.1 V at 11 μs, and so on.

## Peak Value

The **peak value** of a sine wave is the value of voltage (or current) at the positive or the negative maximum (peaks) with respect to zero. Since the peaks are equal in **magnitude,** a sine wave is characterized by a single peak value. This is illustrated in Figure 11–16. For a given sine wave, the peak value is constant and is represented by $V_p$ or $I_p$.

## Peak-to-Peak Value

The **peak-to-peak value** of a sine wave, as shown in Figure 11–17, is the voltage or current from the positive peak to the negative peak. It is always twice the peak value as expressed in the following equations. Peak-to-peak voltage or current values are represented by $V_{pp}$ or $I_{pp}$.

$$V_{pp} = 2V_p \qquad \textbf{(11–4)}$$

$$I_{pp} = 2I_p \qquad \textbf{(11–5)}$$

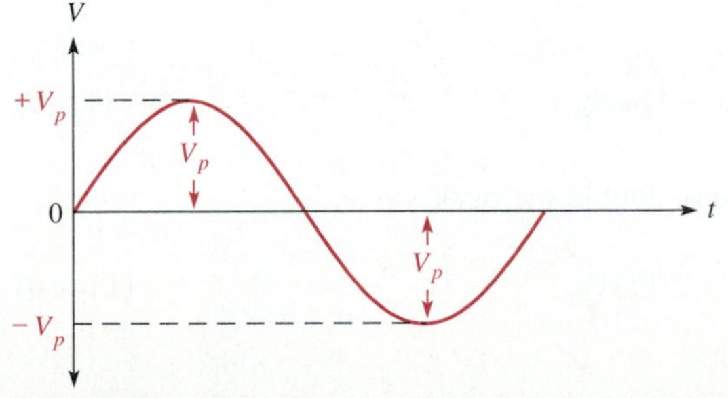


FIGURE 11–16
*Peak values.*

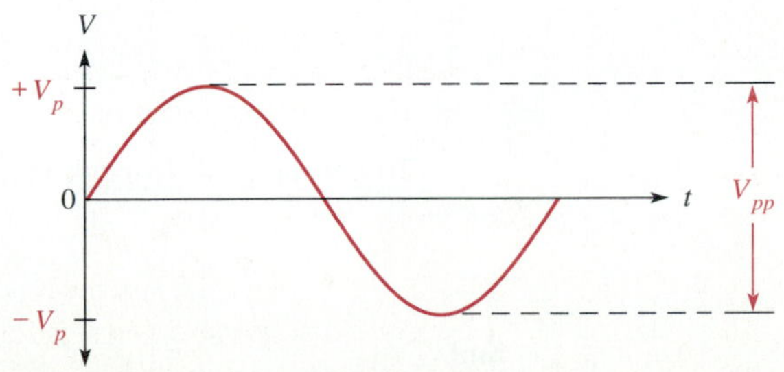


FIGURE 11–17
*Peak-to-peak value.*

## RMS Value

The term *rms* stands for **root mean square.** It refers to the mathematical process by which this value is derived (see Appendix C for the derivation). The rms value is also referred to as the **effective value.** Most ac voltmeters display rms voltage. The 110 volts at your wall outlet is an rms value.

The rms value of a sine wave is actually a measure of the heating effect of the sine wave. For example, when a resistor is connected across an ac (sine wave) voltage source, as shown in Figure 11–18(a), a certain amount of heat is generated by the power in the resistor. Figure 11–18(b) shows the same resistor connected across a dc voltage source. The value of the dc voltage can be adjusted so that the resistor gives off the same amount of heat as it does when connected to the ac source.

**The rms value of a sine wave is equal to the dc voltage that produces the same amount of heat in a resistance as does the sinusoidal voltage.**

FIGURE 11–18
***The meaning of the rms value of a sine wave.***

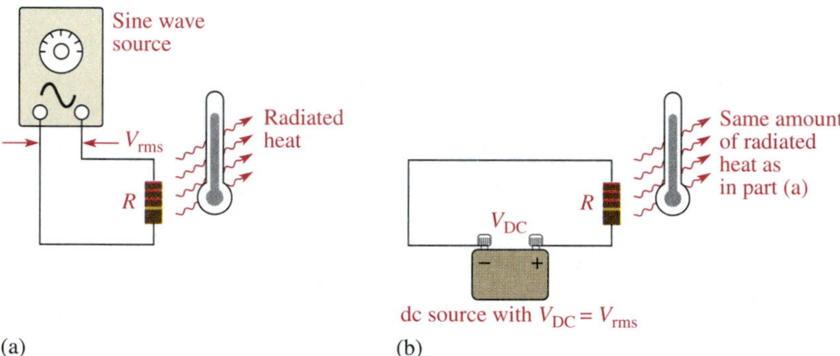


The peak value of a sine wave can be converted to the corresponding rms value using the following relationships, derived in Appendix C, for either voltage or current:

$$V_{rms} = 0.707V_p \qquad \textbf{(11–6)}$$

$$I_{rms} = 0.707I_p \qquad \textbf{(11–7)}$$

Using these formulas, you can also determine the peak value knowing the rms value as follows:

$$V_p = \frac{V_{rms}}{0.707}$$

$$V_p = 1.414V_{rms} \qquad \textbf{(11–8)}$$

Similarly,

$$I_p = 1.414I_{rms} \qquad \textbf{(11–9)}$$

To get the peak-to-peak value, simply double the peak value.

$$V_{pp} = 2.828V_{rms} \qquad \textbf{(11–10)}$$

and

$$I_{pp} = 2.828I_{rms} \qquad \textbf{(11–11)}$$

## Average Value

The average value of a sine wave taken over one complete cycle is always zero, because the positive values (above the zero crossing) offset the negative values (below the zero crossing).

To be useful for comparison purposes, the average value of a sine wave is defined over a half-cycle rather than over a full cycle. The **average value** is the total area under the half-cycle curve divided by the distance in radians of the curve along the horizontal axis. The result is derived in Appendix C and is expressed in terms of the peak value as follows for both voltage and current sine waves:

$$V_{avg} = \left(\frac{2}{\pi}\right)V_p$$

$$V_{avg} = 0.637V_p \qquad \textbf{(11–12)}$$

$$I_{avg} = \left(\frac{2}{\pi}\right)I_p$$

$$I_{avg} = 0.637I_p \qquad \textbf{(11–13)}$$

**EXAMPLE 11–7** Determine $V_p$, $V_{pp}$, $V_{rms}$, and the half-cycle $V_{avg}$ for the sine wave in Figure 11–19.

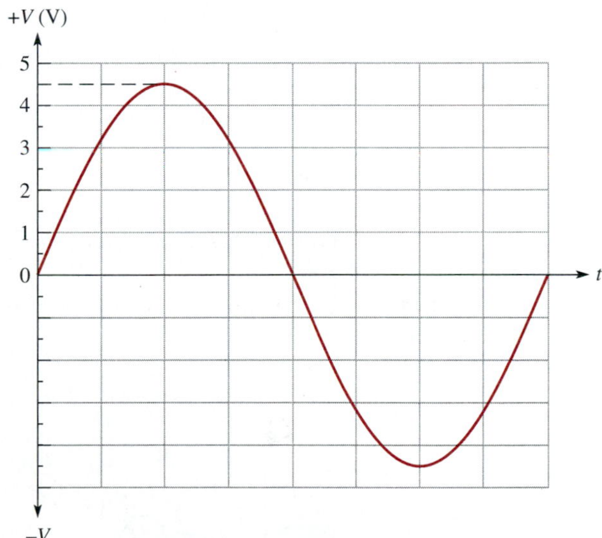


**FIGURE 11–19**

***Solution*** $V_p = 4.5$ V is taken directly from the graph. From this, calculate the other values.

$$V_{pp} = 2V_p = 2(4.5\text{ V}) = 9\text{ V}$$
$$V_{rms} = 0.707V_p = 0.707(4.5\text{ V}) = 3.18\text{ V}$$
$$V_{avg} = 0.637V_p = 0.637(4.5\text{ V}) = 2.87\text{ V}$$

***Related Exercise*** If $V_p = 25$ V, determine $V_{pp}$, $V_{rms}$, and $V_{avg}$ for a sine wave.

**SECTION 11–3 REVIEW**

1. Determine $V_{pp}$ in each case when
   (a) $V_p = 1$ V (b) $V_{rms} = 1.414$ V (c) $V_{avg} = 3$ V
2. Determine $V_{rms}$ in each case when
   (a) $V_p = 2.5$ V (b) $V_{pp} = 10$ V (c) $V_{avg} = 1.5$ V
3. Determine the half-cycle $V_{avg}$ in each case when
   (a) $V_p = 10$ V (b) $V_{rms} = 2.3$ V (c) $V_{pp} = 60$ V

## 11–4 ■ ANGULAR MEASUREMENT OF A SINE WAVE

***As you have seen, sine waves can be measured along the horizontal axis on a time basis; however, since the time for completion of one full cycle or any portion of a cycle is frequency-dependent, it is often useful to specify points on the sine wave in terms of an angular measurement expressed in degrees or radians. Angular measurement is independent of frequency.***

***After completing this section, you should be able to***

- **Describe angular relationships of sine waves**
  - ☐ Show how to measure a sine wave in terms of angles
  - ☐ Define *radian*
  - ☐ Convert radians to degrees
  - ☐ Determine the phase angle of a sine wave

A sine wave voltage can be produced by an ac generator. As the rotor of the ac generator goes through a full 360° of rotation, the resulting voltage output is one full cycle of a sine wave. Thus, the angular measurement of a sine wave can be related to the angular rotation of a generator, as shown in Figure 11–20.

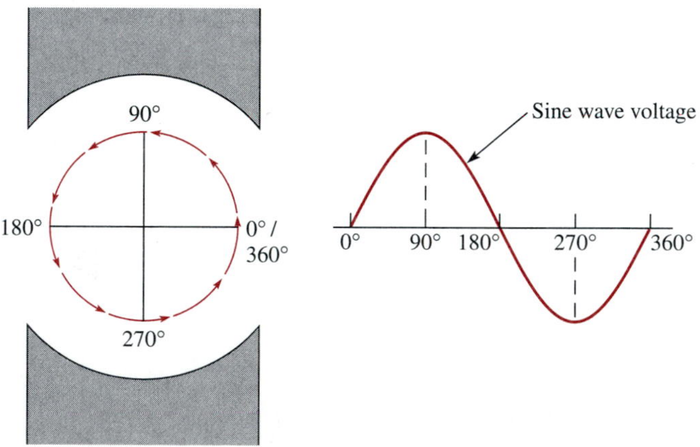


FIGURE 11–20
*Relationship of a sine wave to the rotational motion in an ac generator.*

### Angular Measurement

A **degree** is an angular measurement corresponding to 1/360 of a circle or a complete revolution. A **radian** (rad) is defined as the angular measurement along the circumference of a circle that is equal to the radius of the circle. One radian is equivalent to 57.3°, as illustrated in Figure 11–21. In a 360° revolution, there are $2\pi$ radians.

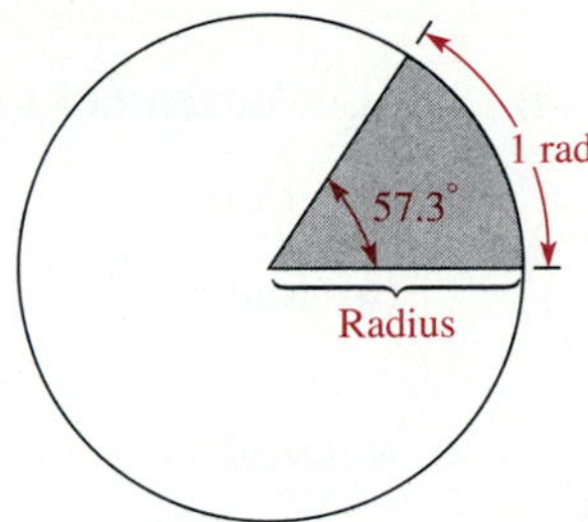


**FIGURE 11–21**
***Angular measurement showing relationship of radian (rad) to degrees (°).***

**The Greek letter π (pi) represents the ratio of the circumference of any circle to its diameter and has a constant value of approximately 3.1416.**

Most calculators have a [π] key so that the actual numerical value does not have to be entered.

Table 11–1 lists several values of degrees and the corresponding radian values. These angular measurements are illustrated in Figure 11–22.

**TABLE 11–1**

| Degrees (°) | Radians (rad) |
|---|---|
| 0 | 0 |
| 45 | π/4 |
| 90 | π/2 |
| 135 | 3π/4 |
| 180 | π |
| 225 | 5π/4 |
| 270 | 3π/2 |
| 315 | 7π/4 |
| 360 | 2π |

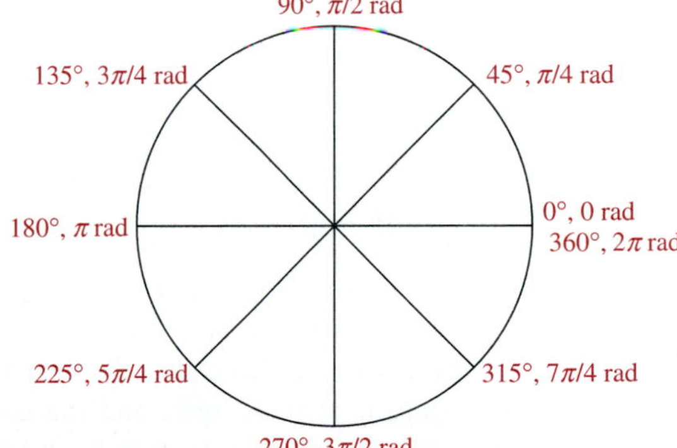


**FIGURE 11–22**
***Angular measurements.***

## Radian/Degree Conversion

Radians can be converted to degrees using Equation (11–14).

$$\text{rad} = \left(\frac{\pi\ \text{rad}}{180^\circ}\right) \times \text{degrees} \tag{11–14}$$

Similarly, degrees can be converted to radians with Equation (11–15).

$$\text{degrees} = \left(\frac{180^\circ}{\pi\ \text{rad}}\right) \times \text{rad} \tag{11–15}$$

**EXAMPLE 11–8** (a) Convert 60° to radians. (b) Convert π/6 rad to degrees.

***Solution***

(a) $\text{Rad} = \left(\dfrac{\pi \text{ rad}}{180°}\right)60° = \dfrac{\pi}{3} \text{ rad}$ (b) $\text{Degrees} = \left(\dfrac{180°}{\pi \text{ rad}}\right)\left(\dfrac{\pi}{6} \text{ rad}\right) = 30°$

***Related Exercise*** (a) Convert 15° to radians. (b) Convert 2π rad to degrees.

## Sine Wave Angles

The angular measurement of a sine wave is based on 360° or 2π rad for a complete cycle. A half-cycle is 180° or π rad; a quarter-cycle is 90° or π/2 rad; and so on. Figure 11–23(a) shows angles in degrees for a full cycle of a sine wave; part (b) shows the same points in radians.

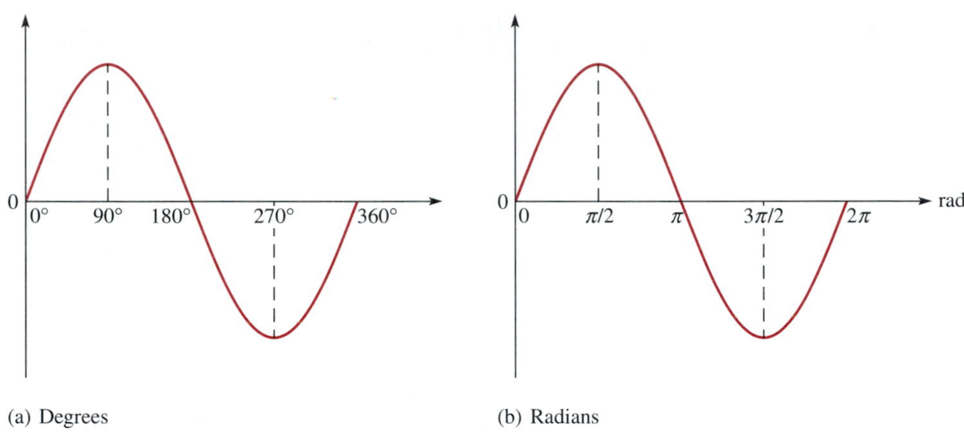


(a) Degrees (b) Radians

FIGURE 11–23
*Sine wave angles.*

## Phase of a Sine Wave

The **phase** of a sine wave is an angular measurement that specifies the position of that sine wave relative to a reference. Figure 11–24 shows one cycle of a sine wave to be used as the reference. Note that the first positive-going crossing of the horizontal axis (zero crossing) is at 0° (0 rad), and the positive peak is at 90° (π/2 rad). The negative-going zero crossing is at 180° (π rad), and the negative peak is at 270° (3π/2 rad). The cycle is completed at 360° (2π rad). When the sine wave is shifted left or right with respect to this reference, there is a phase shift.

Figure 11–25 illustrates phase shifts of a sine wave. In part (a), sine wave $B$ is shifted to the right by 90° (π/2 rad). Thus, there is a phase angle of 90° between sine wave $A$ and sine wave $B$. In terms of time, the positive peak of sine wave $B$ occurs later than the positive peak of sine wave $A$, because time increases to the right along the horizontal axis. In this case, sine wave $B$ is said to **lag** sine wave $A$ by 90° or π/2 radians. Stated another way, sine wave $A$ leads sine wave $B$ by 90°.

In Figure 11–25(b), sine wave $B$ is shown shifted left by 90°. Thus, again there is a phase angle of 90° between sine wave $A$ and sine wave $B$. In this case, the positive peak of sine wave $B$ occurs earlier in time than that of sine wave $A$; therefore, sine wave $B$ is said to **lead** by 90°. A sine wave shifted 90° to the left with respect to the reference is called a *cosine wave.*

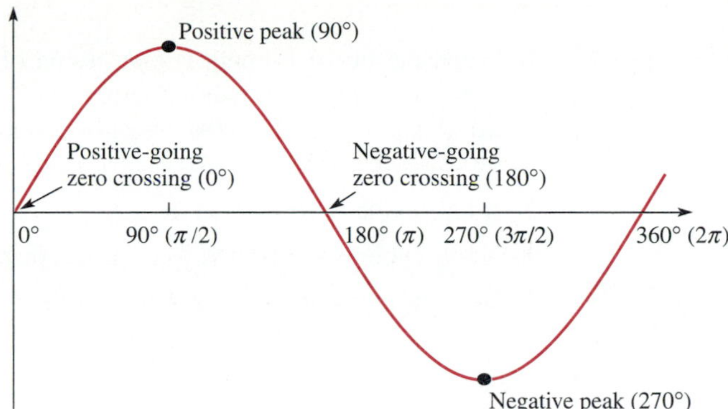


**FIGURE 11–24**
***Phase reference.***

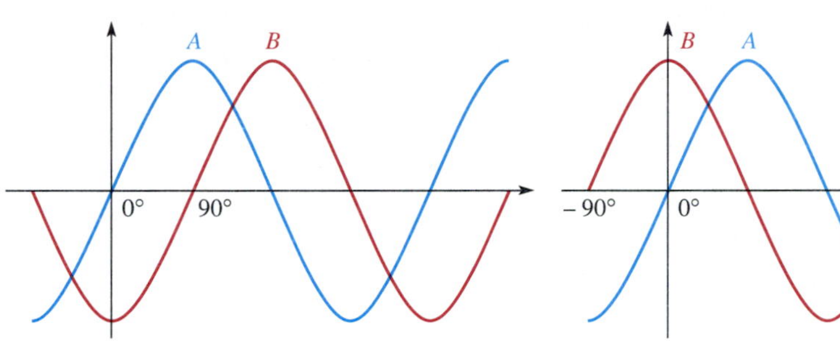

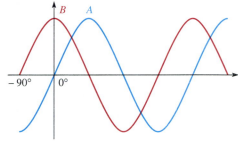

(a) *A* leads *B* by 90°, or *B* lags *A* by 90°.

(b) *B* leads *A* by 90°, or *A* lags *B* by 90°.

**FIGURE 11–25**
***Illustration of a phase shift.***

**EXAMPLE 11–9**

What are the phase angles between the two sine waves in parts (a) and (b) of Figure 11–26?

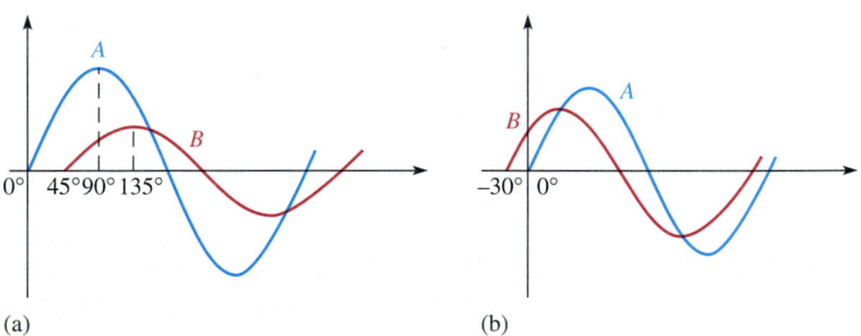


**FIGURE 11–26**

***Solution*** In Figure 11–26(a) the zero crossing of sine wave *A* is at 0°, and the corresponding zero crossing of sine wave *B* is at 45°. There is a 45° phase angle between the two waveforms with sine wave *A* leading.

In Figure 11–26(b) the zero crossing of sine wave *B* is at –30°, and the corresponding zero crossing of sine wave *A* is at 0°. There is a 30° phase angle between the two waveforms with sine wave *B* leading.

***Related Exercise*** If the positive-going zero crossing of one sine wave is at 15° and that of the second sine wave is at 23°, what is the phase angle between them?

**SECTION 11–4 REVIEW**

1. When the positive-going zero crossing of a sine-wave occurs at 0°, at what angle does each of the following points occur?
   **(a)** Positive peak **(b)** Negative-going zero crossing
   **(c)** Negative peak **(d)** End of first complete cycle
2. A half-cycle is completed in _______ degrees or _______ radians.
3. A full cycle is completed in _______ degrees or _______ radians.
4. Determine the phase angle between the two sine waves in Figure 11–27.

FIGURE 11–27

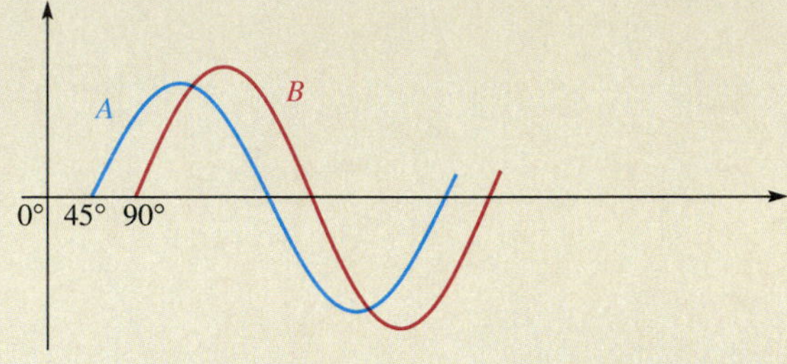


## 11–5 ■ THE SINE WAVE FORMULA

***A sine wave can be graphically represented by voltage or current values on the vertical axis and by angular measurement (degrees or radians) along the horizontal axis. This graph can be expressed mathematically, as you will see.***

***After completing this section, you should be able to***

- **Mathematically analyze a sinusoidal waveform**
  - ☐ State the sine wave formula
  - ☐ Find instantaneous values using the formula

A generalized graph of one cycle of a sine wave is shown in Figure 11–28. The **amplitude,** $A$, is the maximum value of the voltage or current on the vertical axis, and angular values run along the horizontal axis. The variable $y$ is an instantaneous value representing either voltage or current at a given angle, $\theta$.

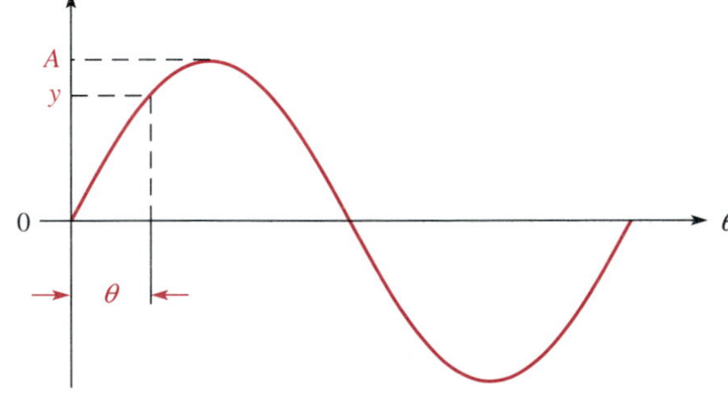


FIGURE 11–28
***One cycle of a generalized sine wave showing amplitude and phase.***

A sine wave curve follows a specific mathematical formula. The general expression for the sine wave curve in Figure 11–28 is

$$y = A \sin \theta \tag{11–16}$$

This expression states that any point on the sine wave, represented by an instantaneous value ($y$), is equal to the maximum value $A$ times the sine (sin) of the angle $\theta$ at that point. For example, a certain voltage sine wave has a peak value of 10 V. The instantaneous voltage at a point 60° along the horizontal axis can be calculated as follows, where $y = v$ and $A = V_p$:

$$v = V_p \sin \theta = 10 \sin 60° = 10(0.866) = 8.66 \text{ V}$$

Figure 11–29 shows this particular instantaneous value of the curve. You can find the sine of any angle on your calculator by first entering the value of the angle and then pressing the [sin] key. Verify that your calculator is in the degree mode, and, if not, use the [DRG] key.

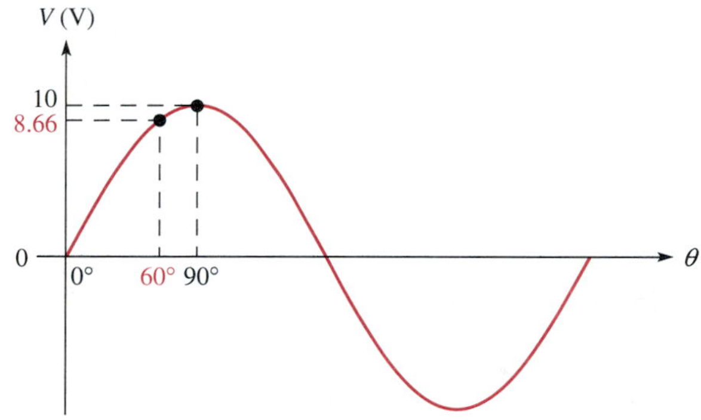


**FIGURE 11–29**
***Illustration of the instantaneous value of a sine wave voltage at $\theta = 60°$.***

## Expressions for Shifted Sine Waves

When a sine wave is shifted to the right of the reference (lagging) by a certain angle, $\phi$, as illustrated in Figure 11–30(a) where the reference is the vertical axis, the general expression is

$$y = A \sin(\theta - \phi) \tag{11–17}$$

When a sine wave is shifted to the left of the reference (leading) by a certain angle, $\phi$, as shown in Figure 11–30(b), the general expression is

$$y = A \sin(\theta + \phi) \tag{11–18}$$

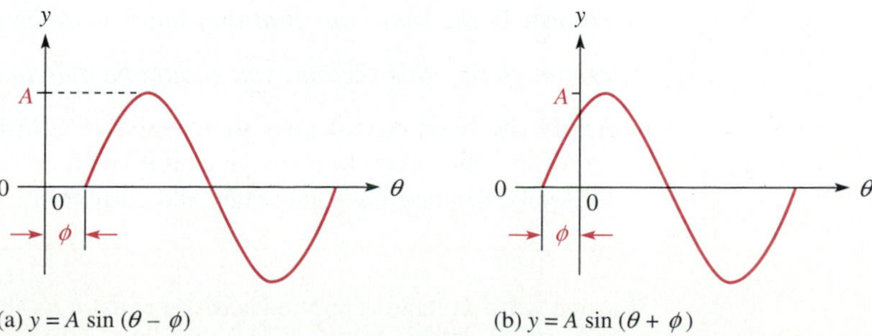


**FIGURE 11–30**
***Shifted sine waves.***

**EXAMPLE 11–10** Determine the instantaneous value at the 90° reference point on the horizontal axis for each sine wave voltage in Figure 11–31.

FIGURE 11–31

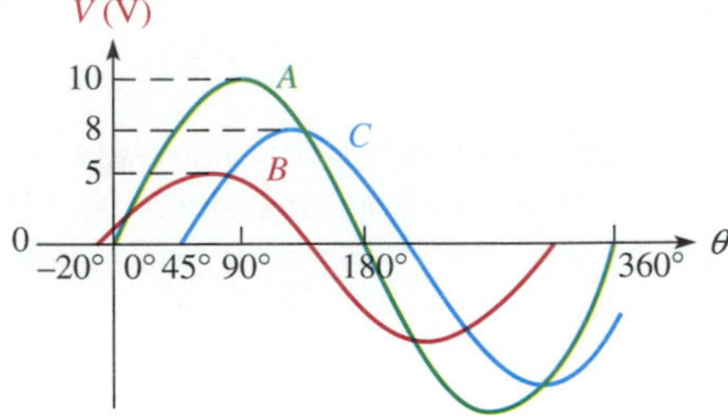


***Solution*** Sine wave $A$ is the reference. Sine wave $B$ is shifted left 20° with respect to $A$, so it leads. Sine wave $C$ is shifted right 45° with respect to $A$, so it lags.

$$v_A = V_p \sin\theta$$
$$= 10\sin(90°) = 10(1) = 10\text{ V}$$
$$v_B = V_p \sin(\theta + \phi_B)$$
$$= 5\sin(90° + 20°) = 5\sin(110°) = 5(0.9397) = 4.70\text{ V}$$
$$v_C = V_p \sin(\theta - \phi_C)$$
$$= 8\sin(90° - 45°) = 8\sin(45°) = 8(0.7071) = 5.66\text{ V}$$

***Related Exercise*** A sine wave has a peak value of 20 V. What is its instantaneous value at 65° from its zero crossing?

**SECTION 11–5 REVIEW**

1. Calculate the instantaneous value at 120° for the sine wave in Figure 11–29.
2. Determine the instantaneous value at the 45° point of a sine wave shifted 10° to the left of the zero reference ($V_p = 10$ V).
3. Find the instantaneous value of the 90° point of a sine wave shifted 25° to the right from the zero reference ($V_p = 5$ V).

## 11–6 ■ OHM'S LAW AND KIRCHHOFF'S LAWS IN AC CIRCUITS

***When time-varying ac voltages such as the sine wave are applied to a circuit, the circuit laws that you studied earlier still apply. Ohm's law and Kirchhoff's laws apply to ac circuits in the same way that they apply to dc circuits.***

***After completing this section, you should be able to***

- **Apply the basic circuit laws to ac resistive circuits**
  - ☐ Apply Ohm's law to resistive circuits with ac sources
  - ☐ Apply Kirchhoff's voltage law and current law to resistive circuits with ac sources

If a sine wave voltage is applied across a resistor as shown in Figure 11–32, there is a sine wave current. The current is zero when the voltage is zero and is maximum when the voltage is maximum. When the voltage changes polarity, the current reverses direction. As a result, the voltage and current are said to be in phase with each other.

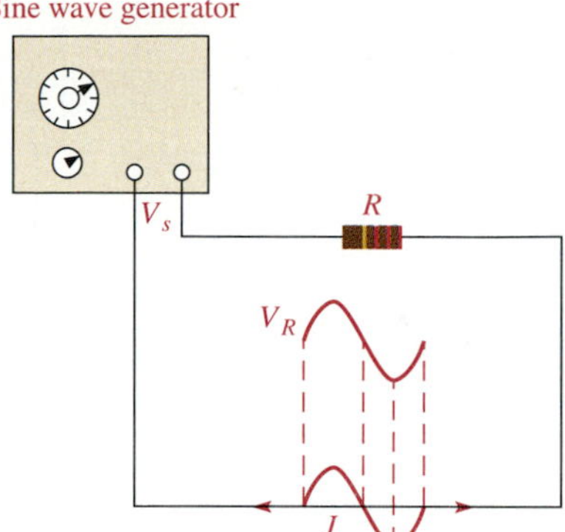


**FIGURE 11–32**
***A sine wave voltage produces a sine wave current.***

When using Ohm's law in ac circuits, remember that both the voltage and the current must be expressed consistently, that is, both as peak values, both as rms values, both as average values, and so on.

**EXAMPLE 11–11** Determine the rms voltage across each resistor and the rms current in Figure 11–33. The source voltage is given as an rms value.

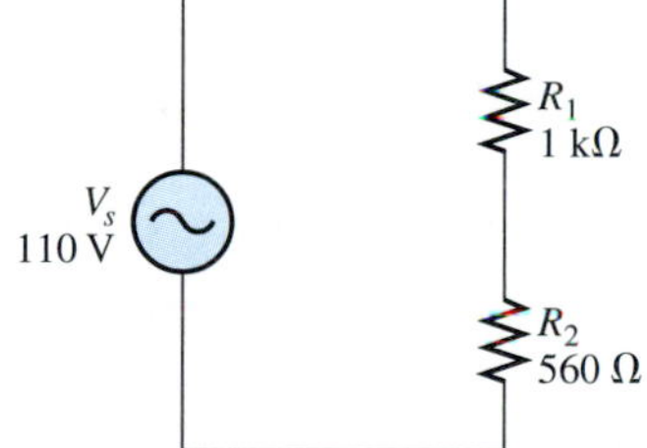


**FIGURE 11–33**

***Solution*** The total resistance of the circuit is

$$R_T = R_1 + R_2 = 1\text{ k}\Omega + 560\ \Omega = 1.56\text{ k}\Omega$$

Use Ohm's law to find the rms current.

$$I_{rms} = \frac{V_{s(rms)}}{R_T} = \frac{110\text{ V}}{1.56\text{ k}\Omega} = 70.5\text{ mA}$$

The rms voltage drop across each resistor is

$$V_{1(rms)} = I_{rms}R_1 = (70.5\text{ mA})(1\text{ k}\Omega) = 70.5\text{ V}$$

$$V_{2(rms)} = I_{rms}R_2 = (70.5\text{ mA})(560\ \Omega) = 39.5\text{ V}$$

***Related Exercise*** Repeat this example for a source voltage of 10 V peak.

Kirchhoff's voltage and current laws apply to ac circuits as well as to dc circuits. Figure 11–34 illustrates Kirchhoff's voltage law in a resistive circuit that has a sine wave voltage source. As you can see, the source voltage is the sum of all the voltage drops across the resistors, just as in a dc circuit.

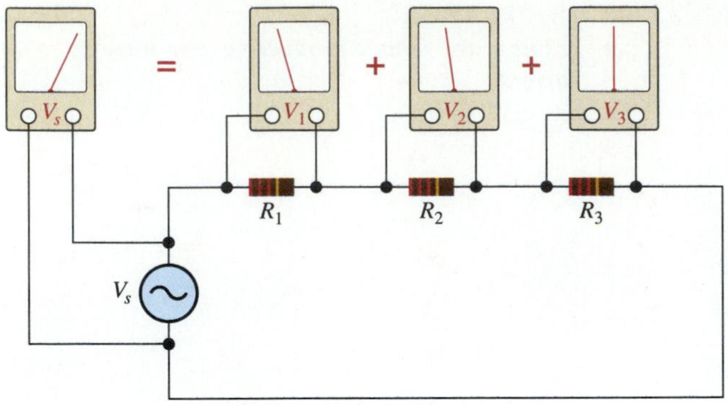

**FIGURE 11–34**
***Illustration of Kirchhoff's voltage law in an ac circuit.***

**EXAMPLE 11–12** All values in Figure 11–35 are given in rms.

**(a)** Find the unknown peak voltage drop in Figure 11–35(a).

**(b)** Find the total rms current in Figure 11–35(b).

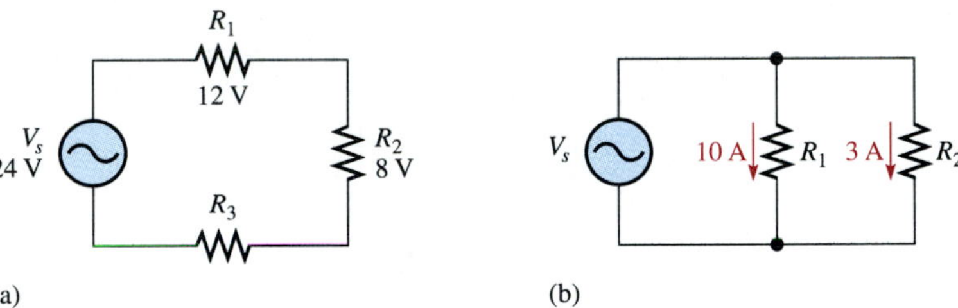


**FIGURE 11–35**

***Solution***

**(a)** Use Kirchhoff's voltage law to find $V_3$.

$$V_s = V_1 + V_2 + V_3$$

$$V_{3(\text{rms})} = V_{s(\text{rms})} - V_{1(\text{rms})} - V_{2(\text{rms})} = 24\text{ V} - 12\text{ V} - 8\text{ V} = 4\text{ V}$$

Convert rms to peak:

$$V_{3(p)} = 1.414V_{3(\text{rms})} = 1.414(4\text{ V}) = 5.66\text{ V}$$

**(b)** Use Kirchhoff's current law to find $I_{tot}$.

$$I_{tot(\text{rms})} = I_{1(\text{rms})} + I_{2(\text{rms})} = 10\text{ A} + 3\text{ A} = 13\text{ A}$$

***Related Exercise*** A series circuit has the following voltage drops: $V_{1(\text{rms})} = 3.50$ V, $V_{2(p)} = 4.25$ V, $V_{3(\text{avg})} = 1.70$ V. Determine the peak-to-peak source voltage.

**SECTION 11–6 REVIEW**

1. A sine wave voltage with a half-cycle average value of 12.5 V is applied to a circuit with a resistance of 330 Ω. What is the peak current in the circuit?
2. The peak voltage drops in a series resistive circuit are 6.2 V, 11.3 V, and 7.8 V. What is the rms value of the source voltage?

## 11–7 ■ SUPERIMPOSED DC AND AC VOLTAGES

***In many practical circuits, you will find both dc and ac voltages combined. An example of this is in amplifier circuits where ac signal voltages are superimposed on dc operating voltages.***

***After completing this section, you should be able to***

- **Determine total voltages which have both ac and dc components**

Figure 11–36 shows a dc source and an ac source in series. These two voltages will add algebraically to produce an ac voltage "riding" on a dc level, as measured across the resistor.

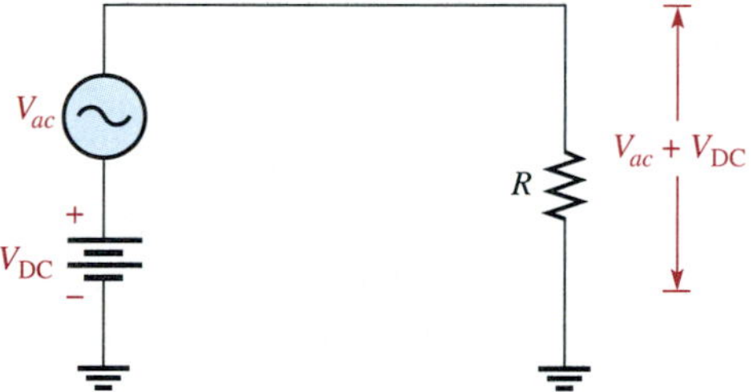


**FIGURE 11–36**
***Superimposed dc and ac voltages.***

If $V_{DC}$ is greater than the peak value of the sine wave, the combined voltage is a sine wave that never reverses polarity and is therefore nonalternating. That is, we have a sine wave riding on a dc level, as shown in Figure 11–37(a). If $V_{DC}$ is less than the peak value of the sine wave, the sine wave will be negative during a portion of its lower half-cycle, as illustrated in Figure 11–37(b), and is therefore alternating. In either case the sine wave will reach a maximum voltage equal to $V_{DC} + V_p$, and it will reach a minimum voltage equal to $V_{DC} - V_p$.

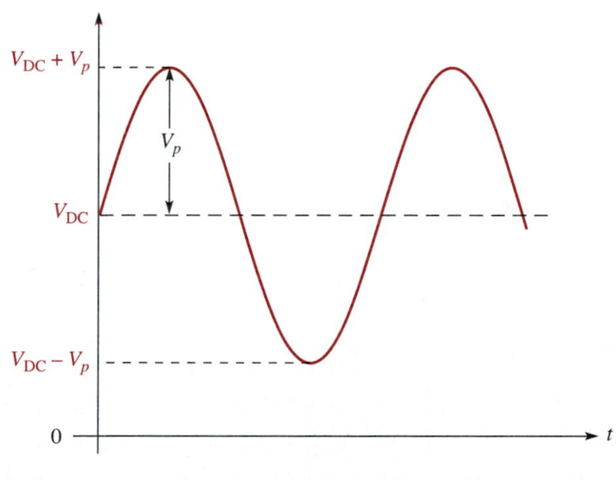


(a) $V_{DC} > V_p$. The sine wave never goes negative.

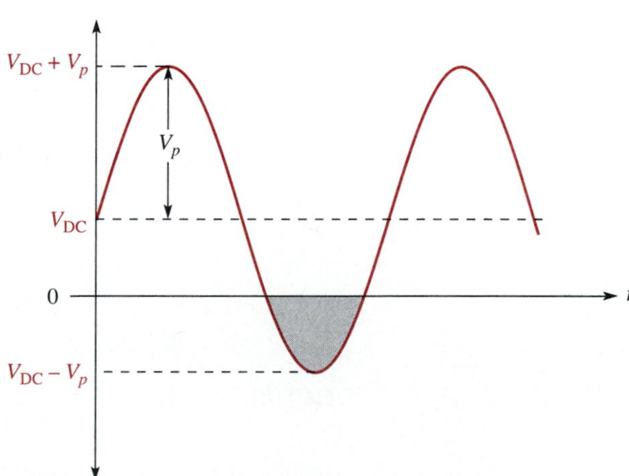


(b) $V_{DC} < V_p$. The sine wave reverses polarity during a portion of its cycle.

**FIGURE 11–37**
***Sine waves with dc levels.***

**EXAMPLE 11–13** Determine the total voltage across the resistor in each circuit of Figure 11–38.

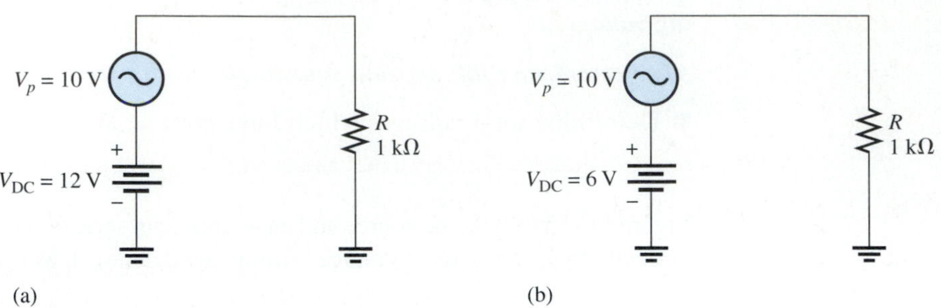


FIGURE 11–38

***Solution*** In Figure 11–38(a), the maximum voltage across $R$ is

$$V_{max} = V_{DC} + V_p = 12\text{ V} + 10\text{ V} = 22\text{ V}$$

The minimum voltage across $R$ is

$$V_{min} = V_{DC} - V_p = 12\text{ V} - 10\text{ V} = 2\text{ V}$$

Therefore, $V_{R(tot)}$ is a nonalternating sine wave that varies from +22 V to + 2 V, as shown in Figure 11–39(a).

In Figure 11–38(b), the maximum voltage across $R$ is

$$V_{max} = V_{DC} + V_p = 6\text{ V} + 10\text{ V} = 16\text{ V}$$

The minimum voltage across $R$ is

$$V_{min} = V_{DC} - V_p = -4\text{ V}$$

Therefore, $V_{R(tot)}$ is an alternating sine wave that varies from +16 V to –4 V, as shown in Figure 11–39(b).

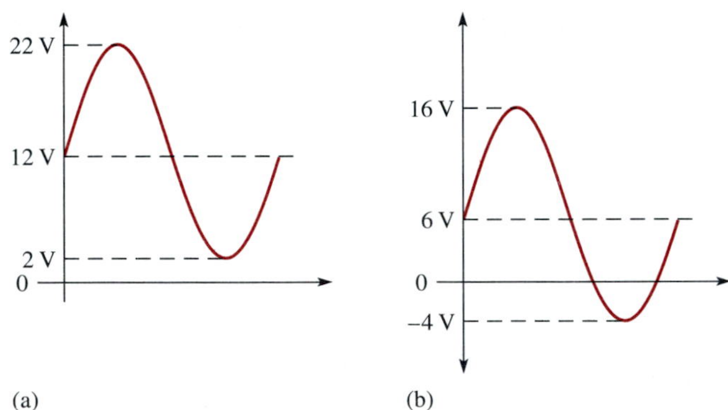


FIGURE 11–39

***Related Exercise*** Explain why the waveform in Figure 11–39(a) is nonalternating but the waveform in part (b) is considered to be alternating.

**SECTION 11–7 REVIEW**

1. What is the maximum positive value of the resulting total voltage when a sine wave with $V_p = 5$ V is added to a dc voltage of +2.5 V?
2. Will the resulting voltage in Question 1 alternate polarity?
3. If the dc voltage in Question 1 is −2.5 V, what is the maximum positive value of the resulting total voltage?

## 11–8 ■ NONSINUSOIDAL WAVEFORMS

***Sine waves are important in electronics, but they are by no means the only type of ac or time-varying waveform. Two other major types of waveforms, the pulse waveform and the triangular waveform, are discussed next.***

***After completing this section, you should be able to***

- **Identify the characteristics of basic nonsinusoidal waveforms**
  - Discuss the properties of a pulse waveform
  - Define *duty cycle*
  - Discuss the properties of triangular and sawtooth waveforms
  - Discuss the harmonic content of a waveform

### Pulse Waveforms

Basically, a **pulse** can be described as a very rapid transition **(leading edge)** from one voltage or current level **(baseline)** to an amplitude level, and then, after an interval of time, a very rapid transition **(trailing edge)** back to the original baseline level. The transitions in level are also called *steps.* An ideal pulse consists of two opposite-going steps of equal amplitude. When the leading or trailing edge is positive-going, it is called a **rising edge.** When the leading or trailing edge is negative-going, it is called a **falling edge.**

Figure 11–40(a) shows an ideal positive-going pulse consisting of two equal but opposite instantaneous steps separated by an interval of time called the **pulse width.** Part (b) of Figure 11–40 shows an ideal negative-going pulse. The height of the pulse measured from the baseline is its voltage (or current) amplitude.

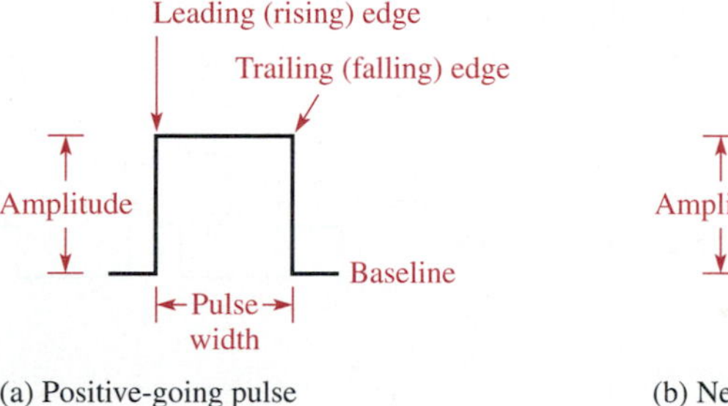


FIGURE 11–40
*Ideal pulses.*

In many applications, analysis is simplified by treating all pulses as ideal (composed of instantaneous steps and perfectly rectangular in shape). Actual pulses, however, are never ideal. All pulses possess certain characteristics that cause them to be different from the ideal.

In practice, pulses cannot change from one level to another instantaneously. Time is always required for a transition (step), as illustrated in Figure 11–41(a). As you can see, there is an interval of time during the rising edge in which the pulse is going from its lower value to its higher value. This interval is called the **rise time, $t_r$.**

**Rise time is the time required for the pulse to go from 10% of its full amplitude to 90% of its full amplitude.**

The interval of time during the falling edge in which the pulse is going from its higher value to its lower value is called the **fall time, $t_f$.**

**Fall time is the time required for the pulse to go from 90% of its full amplitude to 10% of its full amplitude.**

Pulse width, $t_W$, also requires a precise definition for the nonideal pulse because the rising and falling edges are not vertical.

**Pulse width is the time between the point on the rising edge, where the value is 50% of full amplitude, to the point on the falling edge, where the value is 50% of full amplitude.**

Pulse width is shown in Figure 11–41(b).

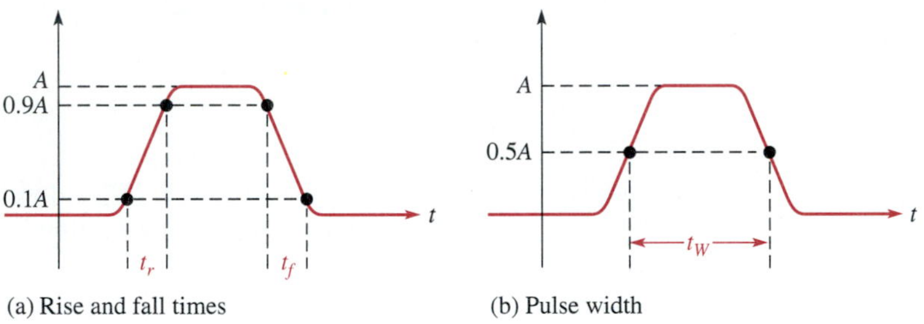


FIGURE 11–41
*Nonideal pulse.*

## Repetitive Pulses

Any waveform that repeats itself at fixed intervals is **periodic.** Some examples of periodic pulse waveforms are shown in Figure 11–42. Notice that, in each case, the pulses repeat at regular intervals. The rate at which the pulses repeat is the **pulse repetition frequency,** which is the fundamental frequency of the waveform. The frequency can be

FIGURE 11–42
*Repetitive pulse waveforms.*

$t_W$ $T$ (a)

$t_W$ $T$ (b)

$t_W$ $T$ (c)

expressed in hertz or in pulses per second. The time from one pulse to the corresponding point on the next pulse is the period, $T$. The relationship between frequency and period is the same as with the sine wave, $f = 1/T$.

A very important characteristic of repetitive pulse waveforms is the **duty cycle.**

**The duty cycle is the ratio of the pulse width ($t_W$) to the period ($T$) and is usually expressed as a percentage.**

$$\text{percent duty cycle} = \left(\frac{t_W}{T}\right)100\% \qquad (11\text{–}19)$$

**EXAMPLE 11–14** Determine the period, frequency, and duty cycle for the pulse waveform in Figure 11–43.

**FIGURE 11–43**

***Solution***

$$T = 10\ \mu\text{s}$$

$$f = \frac{1}{T} = \frac{1}{10\ \mu\text{s}} = 100\text{ kHz}$$

$$\text{percent duty cycle} = \left(\frac{1\ \mu\text{s}}{10\ \mu\text{s}}\right)100\% = 10\%$$

***Related Exercise*** A certain pulse waveform has a frequency of 200 kHz and a pulse width of 0.25 μs. Determine the duty cycle.

## Square Waves

A square wave is a pulse waveform with a duty cycle of 50%. Thus, the pulse width is equal to one-half of the period. A square wave is shown in Figure 11–44.

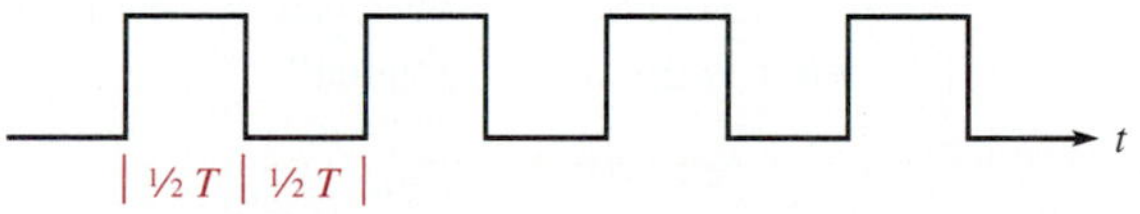

**FIGURE 11–44**
*Square wave.*



## The Average Value of a Pulse Waveform

The average value ($V_{avg}$) of a pulse waveform is equal to its baseline value plus its duty cycle times its amplitude. The lower level of a positive-going waveform or the upper level of a negative-going waveform is taken as the baseline. The formula is as follows:

$$V_{avg} = \text{baseline} + (\text{duty cycle})(\text{amplitude}) \qquad (11\text{–}20)$$

The following example illustrates the calculation of the average value.

**EXAMPLE 11–15** Determine the average value of each of the waveforms in Figure 11–45.

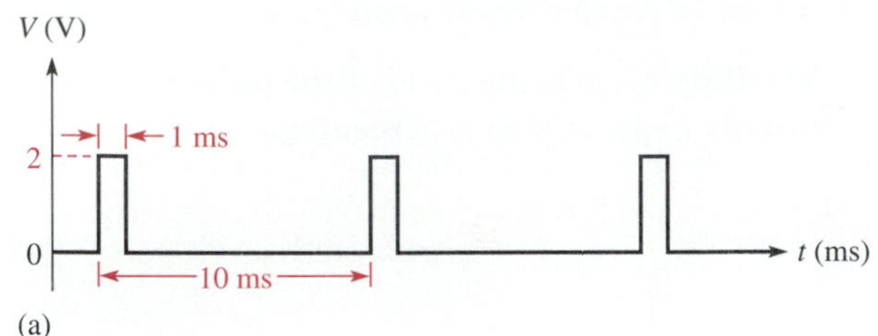


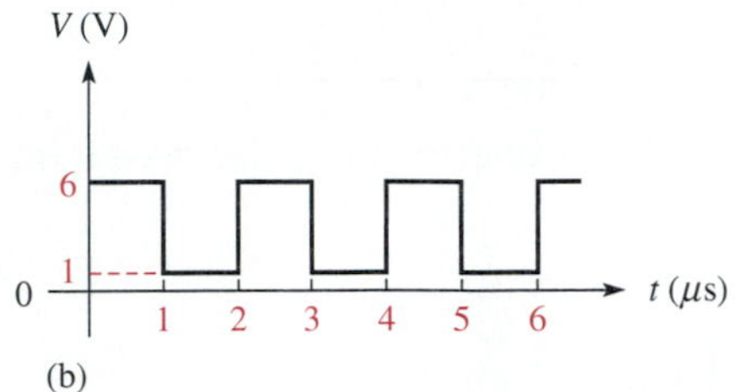


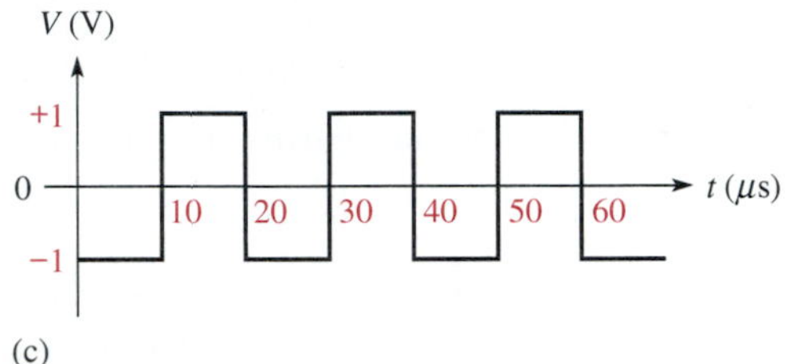


FIGURE 11–45

***Solution*** In Figure 11–45(a), the baseline is at 0 V, the amplitude is 2 V, and the duty cycle is 10%. The average value is

$$V_{avg} = \text{baseline} + (\text{duty cycle})(\text{amplitude})$$
$$= 0\text{ V} + (0.1)(2\text{ V}) = 0.2\text{ V}$$

The waveform in Figure 11–45(b) has a baseline of +1 V, an amplitude of 5 V, and a duty cycle of 50%. The average value is

$$V_{avg} = \text{baseline} + (\text{duty cycle})(\text{amplitude})$$
$$= 1\text{ V} + (0.5)(5\text{ V}) = 1\text{ V} + 2.5\text{ V} = 3.5\text{ V}$$

Figure 11–45(c) shows a square wave with a baseline of −1 V and an amplitude of 2 V. The average value is

$$V_{avg} = \text{baseline} + (\text{duty cycle})(\text{amplitude})$$
$$= -1\text{ V} + (0.5)(2\text{ V}) = -1\text{ V} + 1\text{ V} = 0\text{ V}$$

This is an alternating square wave, and, as with an alternating sine wave, it has an average of zero.

***Related Exercise*** If the baseline of the waveform in Figure 11–45(a) is shifted to 1 V, what is the average value?

## Triangular and Sawtooth Waveforms

Triangular and sawtooth waveforms are formed by voltage or current ramps. A **ramp** is a linear increase or decrease in the voltage or current. Figure 11–46 shows both positive- and negative-going ramps. In part (a), the ramp has a positive slope; in part (b), the ramp has a negative slope. The slope of a voltage ramp is $\pm V/t$ and is expressed in units of V/s. The slope of a current ramp is $\pm I/t$ and is expressed in units of A/s.

**FIGURE 11–46**
***Ramps.***

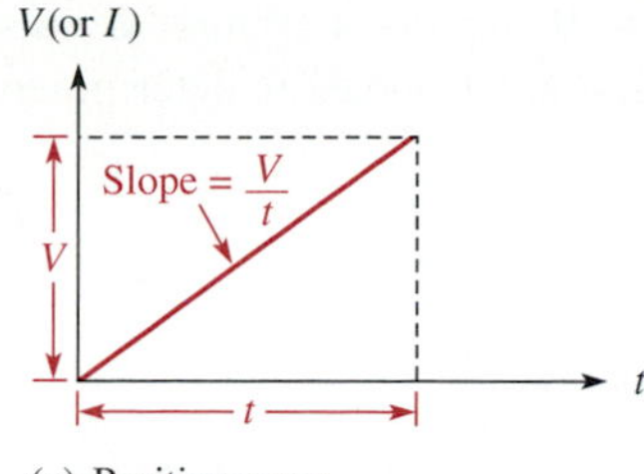


(a) Positive ramp

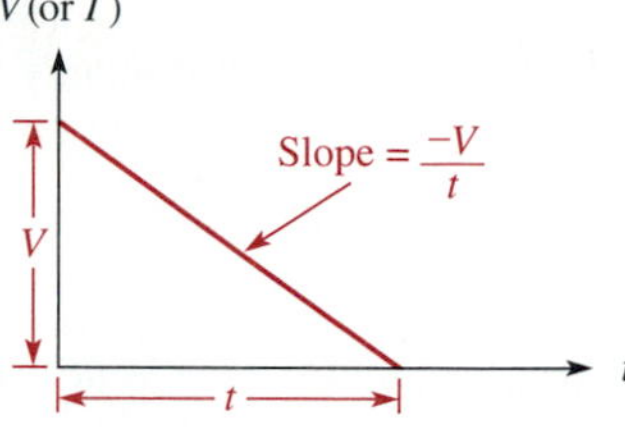


(b) Negative ramp

**EXAMPLE 11–16** What are the slopes of the voltage ramps in Figure 11–47?

**FIGURE 11–47**

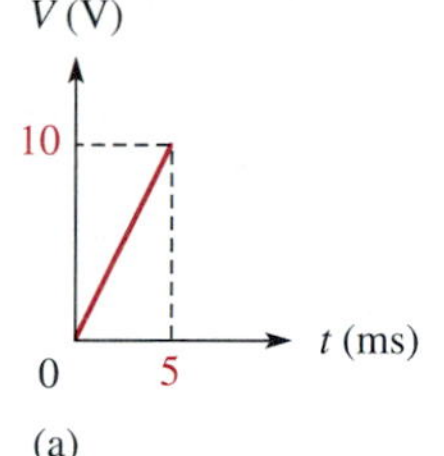


(a)

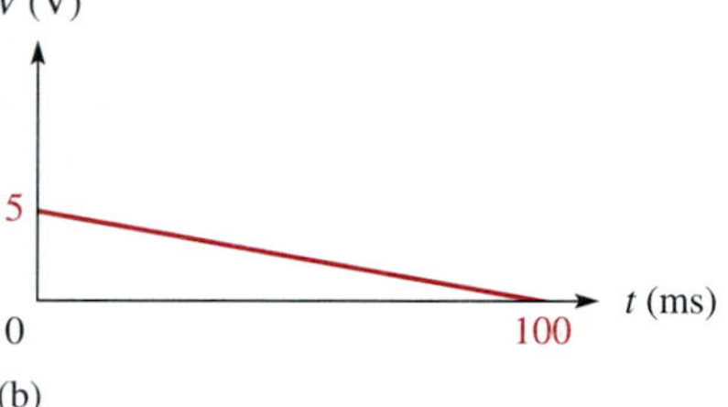


(b)

***Solution*** In Figure 11–47(a), the voltage increases from 0 V to +10 V in 5 ms. Thus, $V = 10$ V and $t = 5$ ms. The slope is

$$\frac{V}{t} = \frac{10\text{ V}}{5\text{ ms}} = 2\text{ V/ms} = 2\text{ kV/s}$$

In Figure 11–47(b), the voltage decreases from +5 V to 0 V in 100 ms. Thus, $V = -5$ V and $t = 100$ ms. The slope is

$$\frac{V}{t} = \frac{-5\text{ V}}{100\text{ ms}} = -0.05\text{ V/ms} = -50\text{ V/s}$$

***Related Exercise*** A certain voltage ramp has a slope of +12 V/$\mu$s. If the ramp starts at zero, what is the voltage at 0.01 ms?

***Triangular Waveforms*** Figure 11–48 shows that a **triangular waveform** is composed of positive-going and negative-going ramps having equal slopes. The period of this waveform is measured from one peak to the next corresponding peak, as illustrated. This particular triangular waveform is alternating and has an average value of zero.

**FIGURE 11–48**
***Alternating triangular waveform with a zero average value.***

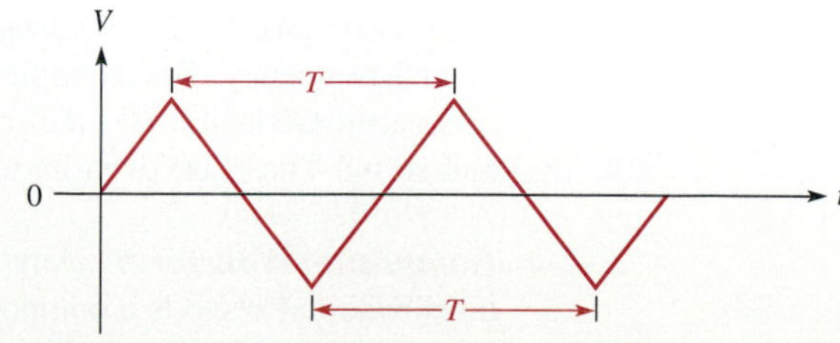

Figure 11–49 depicts a triangular waveform with a nonzero average value. The frequency for triangular waves is determined in the same way as for sine waves, that is, $f = 1/T$.

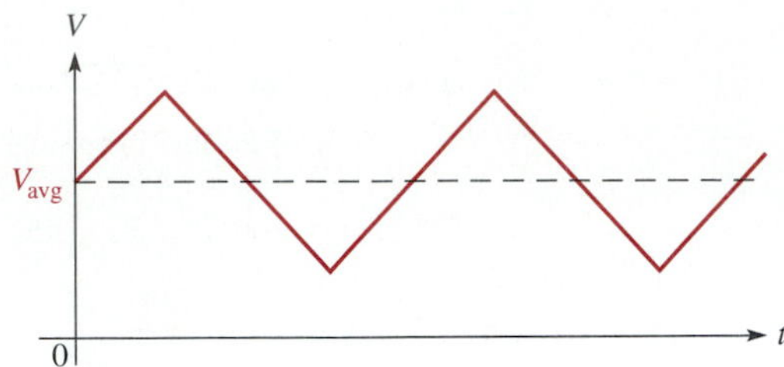


**FIGURE 11–49**
***Nonalternating triangular waveform with a nonzero average value.***

***Sawtooth Waveforms*** The **sawtooth waveform** is actually a special case of the triangular wave consisting of two ramps, one of much longer duration than the other. Sawtooth waveforms are used in many electronic systems. For example, the electron beam that sweeps across the screen of your TV receiver, creating the picture, is controlled by sawtooth voltages and currents. One sawtooth wave produces the horizontal beam movement, and the other produces the vertical beam movement. A sawtooth voltage is sometimes called a *sweep voltage*.

Figure 11–50 is an example of a sawtooth wave. Notice that it consists of a positive-going ramp of relatively long duration, followed by a negative-going ramp of relatively short duration.

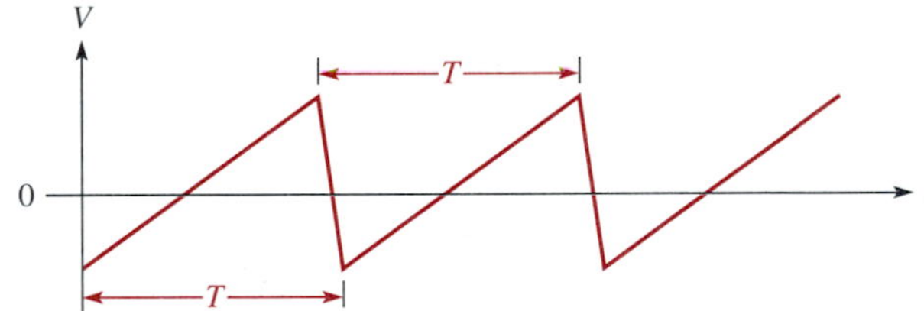


**FIGURE 11–50**
***Alternating sawtooth waveform.***

## Harmonics

A repetitive nonsinusoidal waveform is composed of a fundamental frequency and harmonic frequencies. The **fundamental frequency** is the repetition rate of the waveform, and the **harmonics** are higher frequency sine waves that are multiples of the fundamental.

***Odd Harmonics*** *Odd harmonics* are frequencies that are odd multiples of the fundamental frequency of a waveform. For example, a 1 kHz square wave consists of a fundamental of 1 kHz and odd harmonics of 3 kHz, 5 kHz, 7 kHz, and so on. The 3 kHz frequency in this case is called the third harmonic, the 5 kHz frequency is the fifth harmonic, and so on.

***Even Harmonics*** *Even harmonics* are frequencies that are even multiples of the fundamental frequency. For example, if a certain wave has a fundamental of 200 Hz, the second harmonic is 400 Hz, the fourth harmonic is 800 Hz, the sixth harmonic is 1200 Hz, and so on. These are even harmonics.

***Composite Waveform*** Any variation from a pure sine wave produces harmonics. A nonsinusoidal wave is a composite of the fundamental and the harmonics. Some types of waveforms have only odd harmonics, some have only even harmonics, and some contain both. The shape of the wave is determined by its harmonic content. Generally, only the fundamental and the first few harmonics are of significant importance in determining the wave shape.

A square wave is an example of a waveform that consists of a fundamental and only odd harmonics. When the instantaneous values of the fundamental and each odd harmonic are added algebraically at each point, the resulting curve will have the shape of a square wave, as illustrated in Figure 11–51. In part (a) of the figure, the fundamental and the third harmonic produce a wave shape that begins to resemble a square wave. In part (b), the fundamental, third, and fifth harmonics produce a closer resemblance. When the seventh harmonic is included, as in part (c), the resulting wave shape becomes even more like a square wave. As more harmonics are included, a square wave is approached.

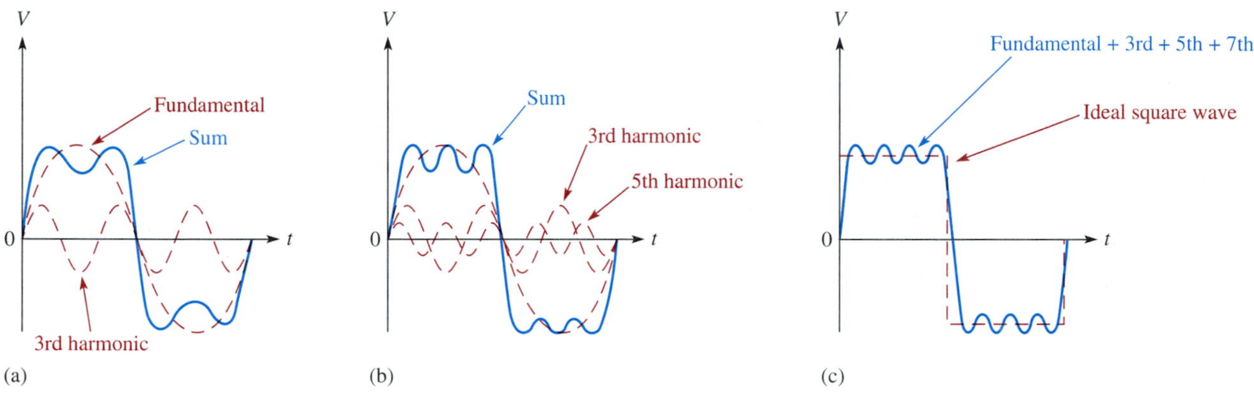


**FIGURE 11–51**
***Odd harmonics produce a square wave.***

**SECTION 11–8 REVIEW**

1. Define the following parameters:
   (a) rise time (b) fall time (c) pulse width
2. In a certain repetitive pulse waveform, the pulses occur once every millisecond. What is the frequency of this waveform?
3. Determine the duty cycle, amplitude, and average value of the waveform in Figure 11–52(a).
4. What is the period of the triangular wave in Figure 11–52(b)?
5. What is the frequency of the sawtooth wave in Figure 11–52(c)?

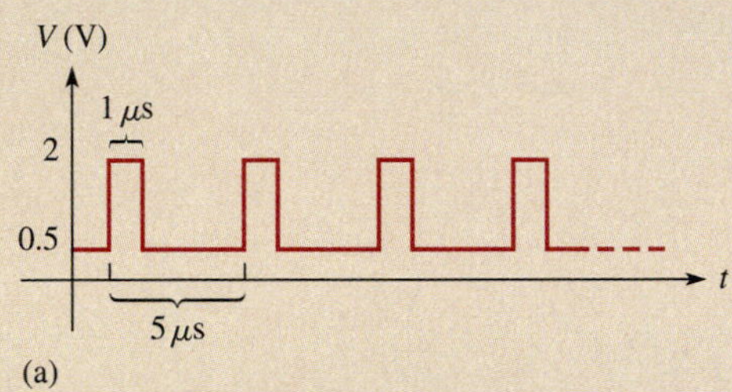


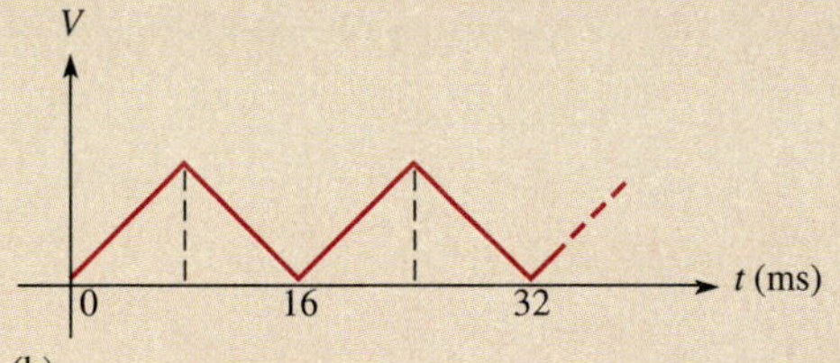


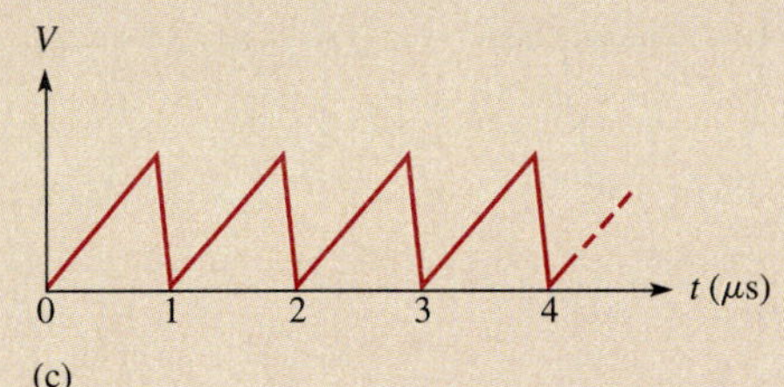


FIGURE 11–52

6. Define *fundamental frequency.*
7. What is the second harmonic of a fundamental frequency of 1 kHz?
8. What is the fundamental frequency of a square wave having a period of 10 μs?

## 11–9 ■ OSCILLOSCOPE MEASUREMENTS

***The oscilloscope, or scope for short, is one of the most widely used and versatile test instruments. It displays on a screen the actual shape of a voltage that is changing with time so that waveform measurements can be made.***

***After completing this section, you should be able to***

- **Use the oscilloscope to measure waveforms**
  - ☐ Describe a basic CRT
  - ☐ Identify basic oscilloscope controls
  - ☐ Explain how to measure amplitude
  - ☐ Explain how to measure period and frequency

Figure 11–53 shows two typical oscilloscopes. The one in part (a) is an analog oscilloscope. The one in part (b) is a digital instrument. You can think of the **oscilloscope** as essentially a "graphing machine" that graphs out a voltage and shows how it varies with time. The shape of the sine wave that you have seen throughout this chapter is simply a graph of voltage versus time. Of course, oscilloscopes can display any type of waveform, but we will use the sine wave to illustrate some basic concepts.

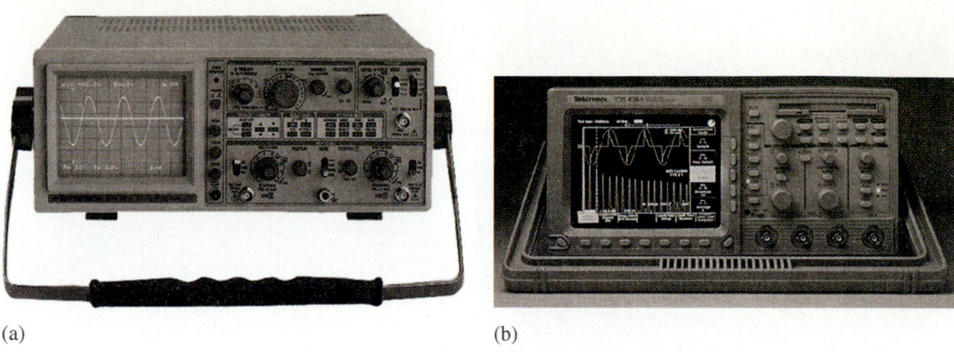

(a) (b)

FIGURE 11–53
***Oscilloscopes. Part (a) courtesy of BK Precision, Maxtec International Corp. Part (b) courtesy of Tektronix, Inc.***

## The Oscilloscope Screen

Generally, the screen of an oscilloscope is divided into ten horizontal divisions and eight vertical divisions, as shown in Figure 11–54. Each major division is divided into five small or minor divisions on both the horizontal and the vertical axes. The vertical axis on the scope screen is the voltage scale and the horizontal axis is the time scale. Values of voltage are read vertically and values of time are read horizontally.

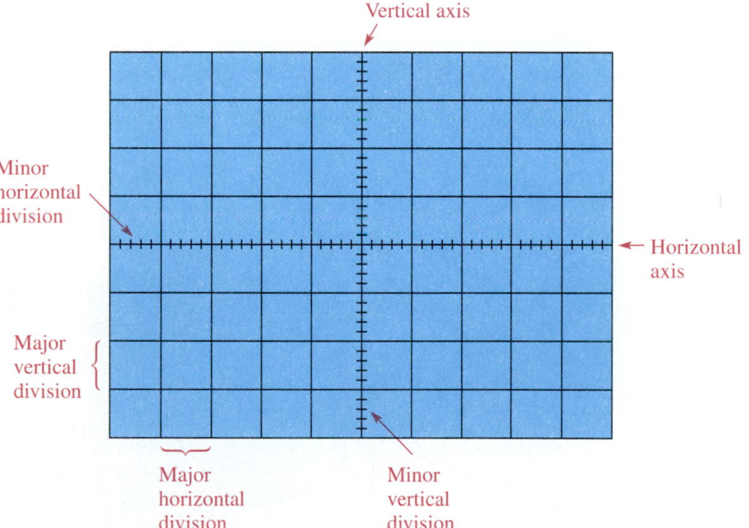


FIGURE 11–54
***The oscilloscope screen.***

## Cathode-Ray Tube (CRT)

The oscilloscope is built around the **cathode-ray tube (CRT),** which is the device that displays the waveforms. The screen of the scope is the front of the CRT.

The CRT is a vacuum tube device containing an electron gun that emits a narrow, focused beam of electrons. A phosphorescent coating on the face of the tube forms the screen. The beam is electronically focused and accelerated so that it strikes the screen, causing light to be emitted at the point of impact.

Figure 11–55 shows the basic construction of a CRT. The electron gun assembly contains a heater, a cathode, a control grid, and accelerating and focusing grids. The heater carries current that indirectly heats the cathode. The heated cathode emits electrons. The amount of voltage on the control grid determines the flow of electrons and thus the intensity of the beam. The electrons are accelerated by the accelerating grid and are

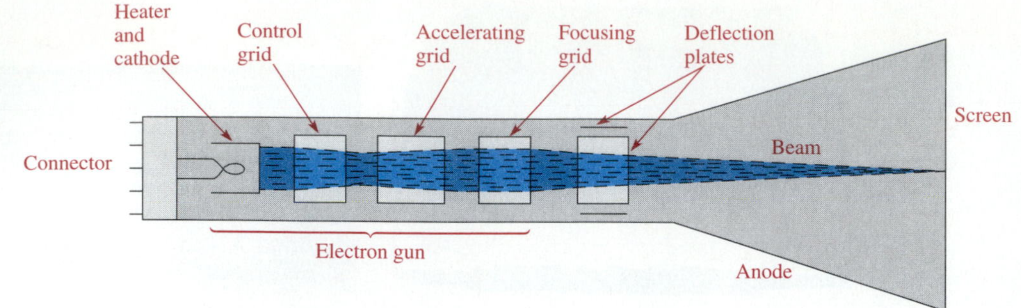


**FIGURE 11–55**
***Basic construction of a CRT.***

focused by the focusing grid into a narrow beam that converges at the screen. The beam is further accelerated to a high speed after it leaves the electron gun by a high voltage on the anode surfaces of the CRT.

The deflection plates in the CRT produce a "bending," or deflection, of the electron beam. This deflection allows the position of the point of impact on the screen to be varied. There are two sets of deflection plates: one set for vertical deflection, and the other set for horizontal deflection.

Figure 11–56 illustrates how a waveform (a sine wave in this case) is "plotted out" on the screen. A sawtooth voltage called the *horizontal sweep signal* is internally generated and applied to the horizontal deflection plates. Each cycle of the sawtooth causes the electron beam to sweep across the screen from left to right and then quickly return for another sweep. The external signal that is to be displayed on the screen is applied to the vertical deflection plates. This causes the electron beam to move up or down vertically as it is swept across the screen, tracing out a pattern that follows the variation of the external signal voltage over an interval of time.

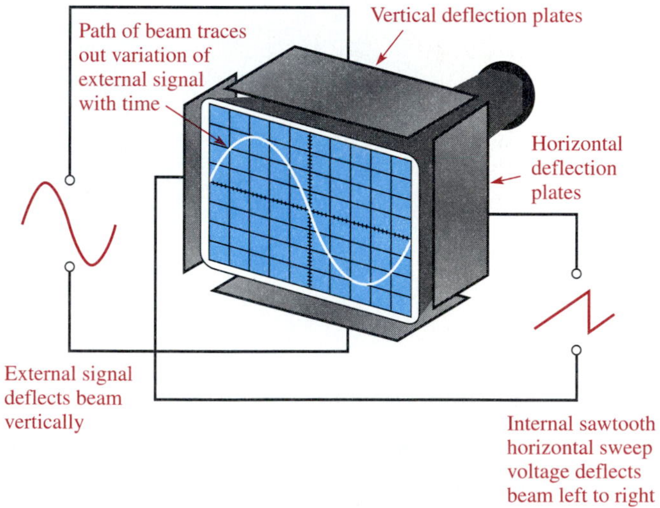


**FIGURE 11–56**
***How a waveform is traced out on the scope screen.***

## Operation of the Oscilloscope

Figure 11–57 shows a typical oscilloscope like the one used in the TECH TIP sections. This is a dual-channel instrument, as most scopes are. *Dual-channel* means that you can display two waveforms from two different inputs on the screen at the same time.

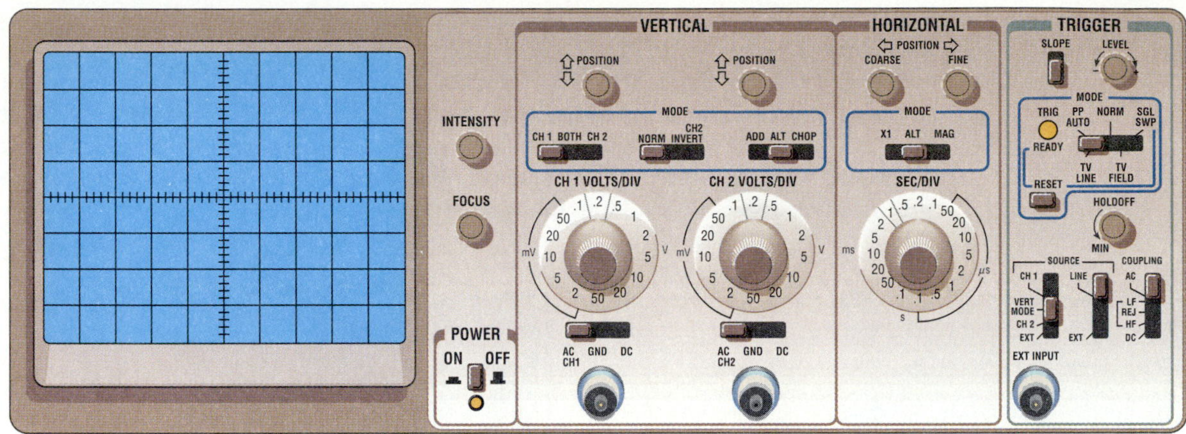


**FIGURE 11–57**
*A typical dual-channel oscilloscope.*

***The Volts/Division Control*** Notice that in Figure 11–57 there are two VOLTS/DIV switches, one for channel 1 (CH1) and one for channel 2 (CH2). The volts/division selector switch sets the number of volts to be represented by each major division on the vertical scale. For example, Figure 11–58(a) shows one cycle of a sine wave on the scope screen with the VOLTS/DIV switch set at 1 V. This means that each of the major vertical divisions is 1 V. The peak of the sine wave is three divisions high, and, since each division is 1 V, assuming a ×1 scope probe is used, the peak value of the sine wave is 3 V (3 divisions × 1 V/division) = 3 V.

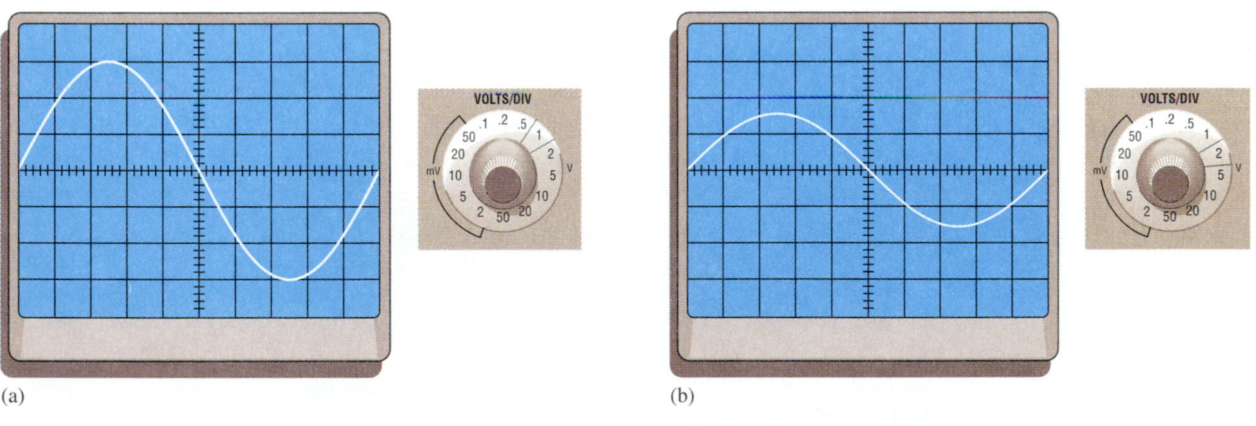


(a) (b)

**FIGURE 11–58**
*Using the VOLTS/DIV control.*

For the same sine wave, if the VOLTS/DIV setting is changed to 2 V, the sine wave will appear as shown in Figure 11–58(b). Notice that now the peak is 1½ major divisions high. Even though each division is now 2 V, the peak value is still 3 V because 1.5 divisions × 2 V/division = 3 V.

So, in summary, to measure the peak value of a sine wave on an oscilloscope, count the number of major vertical divisions from the zero crossing to the peak of the waveform and multiply by the VOLTS/DIV setting.

***The Seconds/Division Control*** The SEC/DIV selector switch sets the number of seconds, milliseconds, or microseconds to be represented by each major division on the horizontal scale. It actually controls how fast the electron beam sweeps horizontally across the screen.

For example, Figure 11–59(a) shows one cycle of a sine wave displayed on the screen. In this case, the entire cycle covers ten horizontal divisions. The SEC/DIV switch is set at 10 μs. This means that each major horizontal division is 10 μs. Since a full cycle of the sine wave covers 10 divisions, the period of the sine wave is 100 μs (10 divisions × 10 μs/division) = 100 μs. From this, $f = 1/100\ \mu s = 10$ kHz.

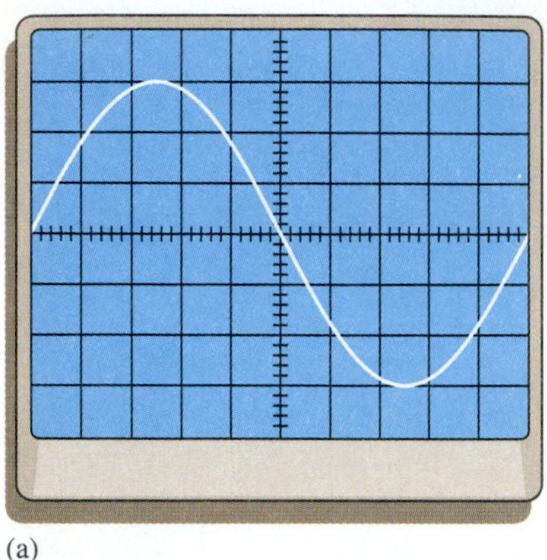

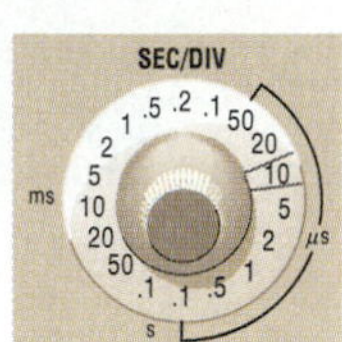


(a)

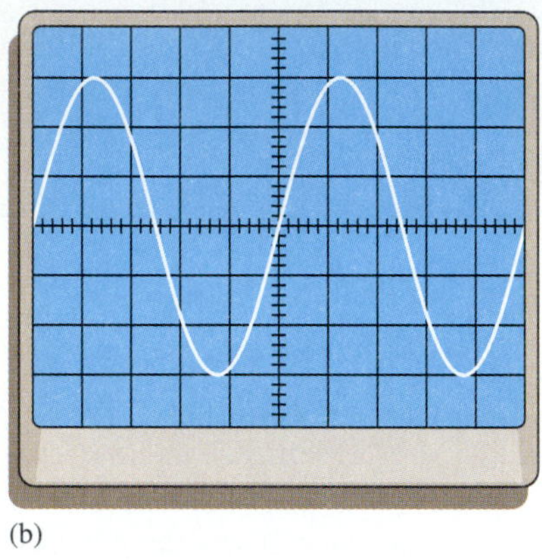

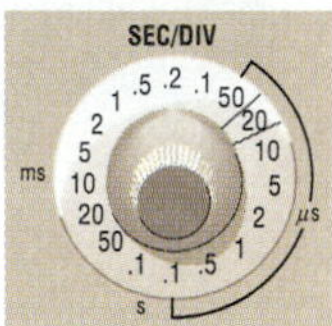


(b)

**FIGURE 11–59**
*Using the SEC/DIV control.*

For the same sine wave, if the SEC/DIV setting is changed to 20 μs, the sine wave will appear as shown in Figure 11–59(b). Notice that now a full cycle covers only five horizontal divisions and there are two cycles on the screen. Even though each division is now 20 μs, the period is still 100 μs because 5 divisions × 20 μs/division = 100 μs.

So, in summary, to measure the period of a waveform on an oscilloscope, count the number of major divisions covered by one cycle and then multiply by the SEC/DIV setting. Use the formula $f = 1/T$ to calculate the frequency.

**EXAMPLE 11–17** Determine the peak value and period of each sine wave in Figure 11–60 from the scope displays and the indicated settings for VOLTS/DIV and SEC/DIV.

***Solution*** Looking at the vertical scale in Figure 11–60(a),

$$V_p = 3 \text{ divisions} \times 0.5 \text{ V/division} = 1.5 \text{ V}$$

From the horizontal scale (one cycle covers ten divisions),

$$T = 10 \text{ divisions} \times 2 \text{ ms/division} = 20 \text{ ms}$$

Looking at the vertical scale in Figure 11–60(b),

$$V_p = 2.5 \text{ divisions} \times 50 \text{ mV/division} = 125 \text{ mV}$$

From the horizontal scale (one cycle covers six divisions),

$$T = 6 \text{ divisions} \times 0.1 \text{ ms/division} = 0.6 \text{ ms} = 600\ \mu\text{s}$$

Looking at the vertical scale in Figure 11–60(c),

$$V_p = 3.4 \text{ divisions} \times 2 \text{ V/division} = 6.8 \text{ V}$$

From the horizontal scale (one-half cycle covers ten divisions),

$$T = 20 \text{ divisions} \times 10\ \mu\text{s/division} = 200\ \mu\text{s}$$

Looking at the vertical scale in Figure 11–60(d),

$$V_p = 2 \text{ divisions} \times 5 \text{ V/division} = 10 \text{ V}$$

From the horizontal scale (one cycle covers two divisions),

$$T = 2 \text{ divisions} \times 2 \ \mu\text{s/division} = 4 \ \mu\text{s}$$

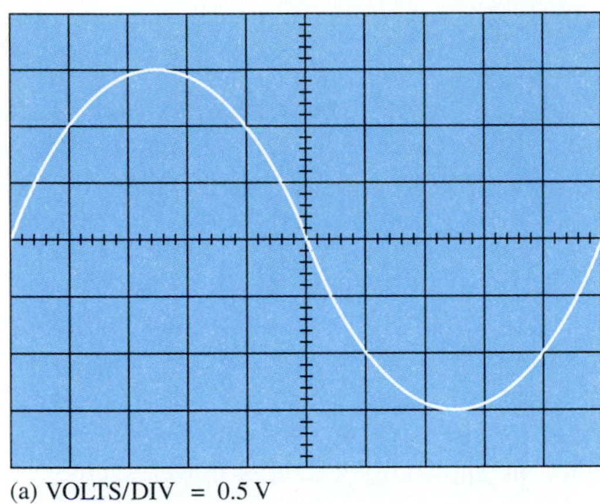

(a) VOLTS/DIV = 0.5 V
SEC/DIV = 2 ms

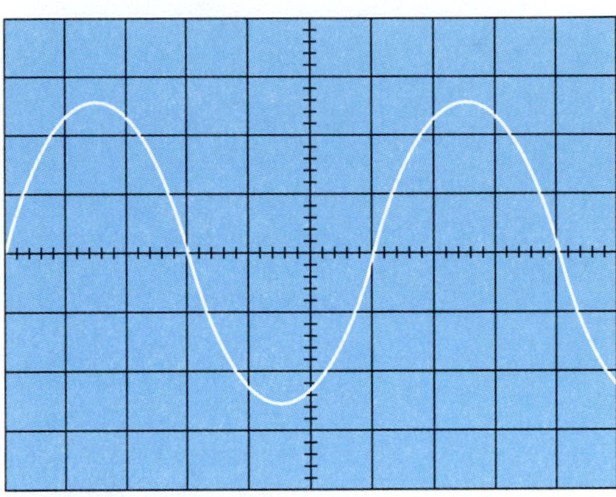

(b) VOLTS/DIV = 50 mV
SEC/DIV = 0.1 ms

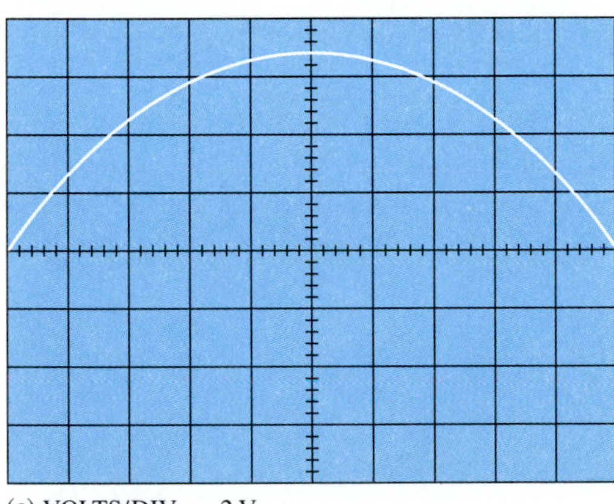

(c) VOLTS/DIV = 2 V
SEC/DIV = 10 $\mu$s
Sine wave is centered vertically on the screen.

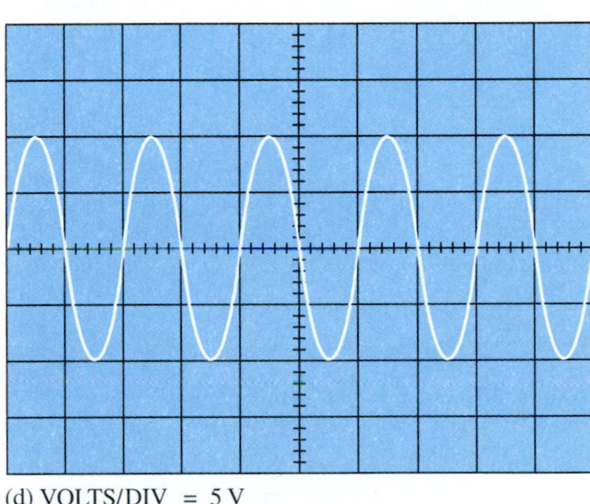

(d) VOLTS/DIV = 5 V
SEC/DIV = 2 $\mu$s

FIGURE 11–60

***Related Exercise*** Determine the rms value and the frequency for each waveform displayed in Figure 11–60.

## Other Oscilloscope Controls

The following descriptions refer to the oscilloscope in Figure 11–57 but apply also to most general-purpose scopes.

***Power Switch*** The power switch turns the power to the scope on and off. A light indicates when the power is on.

***Intensity*** The intensity control knob varies the brightness of the trace on the screen. Caution should be used so that the intensity is not left too high for an extended period of time, especially when the beam forms a motionless dot on the screen. Damage to the screen can result from excessive intensity.

***Focus*** This control focuses the beam so that it converges to a tiny point at the screen. An out-of-focus condition results in a fuzzy trace.

***Horizontal Position*** These control knobs (coarse and fine) adjust the neutral horizontal position of the beam. They are used to reposition horizontally a waveform display for more convenient viewing or measurement.

***Vertical Position*** The two vertical position controls move each trace up or down for easier measurement or observation.

***AC-GND-DC Switch*** This switch, located below the VOLTS/DIV control, allows the input signal to be ac coupled, dc coupled, or grounded. The ac coupling eliminates any dc component on the input signal. The dc coupling permits dc values to be displayed. The ground position allows a 0 V reference to be established on the screen.

***Signal Inputs*** The signals to be displayed are connected into the channel 1 (CH1) and/or channel 2 (CH2) input connectors. These connections are normally done with a special probe that minimizes the loading effect of the scope's input resistance on the circuit being measured. Oscilloscope voltage probes are generally either ×1 (nonattenuating) or ×10 (attenuates by 10). When a ×10 probe is used, the VOLTS/DIV setting must be multiplied by 10. To keep it simple, all applications in this book assume ×1 voltage probes.

***Mode Switches*** These switches provide for displaying either or both channel inputs, inverting channel 2 signal, adding two waveforms, and selecting between alternate and chopped mode of sweep.

***Trigger Control*** The **trigger** controls allow the beam to be triggered from various selected sources. The triggering of the beam causes it to begin its sweep across the screen. It can be triggered from an internally generated signal derived from an input signal, or from the line voltage, or from an externally applied trigger signal. The modes of triggering are auto, normal, single-sweep, and TV. In the auto mode, sweep occurs in the absence of an adequate trigger signal. In the normal mode, a trigger signal must be present for the sweep to occur. The TV mode provides triggering on the TV field or TV line signals. The slope switch allows the triggering to occur on either the positive-going slope or the negative-going slope of the trigger waveform. The level control selects the voltage level on the trigger signal at which the triggering occurs.

Basically, the trigger controls provide for synchronization of the horizontal sweep waveform and the input signal waveform. As a result, the display of the input signal is stable on the screen, rather than appearing to drift across the screen.

**SECTION 11–9 REVIEW**

1. What does *CRT* stand for?
2. On an oscilloscope, voltage is measured (horizontally, vertically) on the screen and time is measured (horizontally, vertically).
3. What can an oscilloscope do that a multimeter cannot?

## 11–10 ■ TECHnology Theory Into Practice

***As you learned in this chapter, nonsinusoidal waveforms contain a combination of various harmonic frequencies. Each of these harmonics is a sinusoidal waveform with a certain frequency. Certain sine wave frequencies are audible; that is, they can be heard by the human ear. A single audible frequency, or pure sine wave, is called a tone and generally falls in the frequency range from about 300 Hz to about 15 kHz. When you hear a tone reproduced through a speaker, its loudness, or volume, depends on its volt-***

***age amplitude. In this section, you will use your knowledge of sine wave characteristics and the operation of an oscilloscope to measure the frequency and amplitude of signals at various points in a basic radio receiver.***

Actual voice or music signals that are picked up by a radio receiver contain many harmonic frequencies with different voltage values. A voice or music signal is continuously changing, so its harmonic content is also changing. However, if a single sinusoidal frequency is transmitted and picked up by the receiver, you will hear a constant tone from the speaker.

Although, at this point you do not have the background to study amplifiers and receiver systems in detail, you can observe the signals at various points in the receiver. A block diagram of a typical AM receiver is shown in Figure 11–61. AM stands for amplitude modulation, a topic which will be covered in another course. For now, all you need to know is what a basic AM signal looks like and this is shown in Figure 11–62. As you can see, the amplitude of the sinusoidal waveform is changing. The higher radio frequency (RF) signal is called the *carrier* and its amplitude is varied or modulated by a lower frequency signal which is the audio (a tone in this case). Normally, however, the audio signal is a complex voice or music waveform.

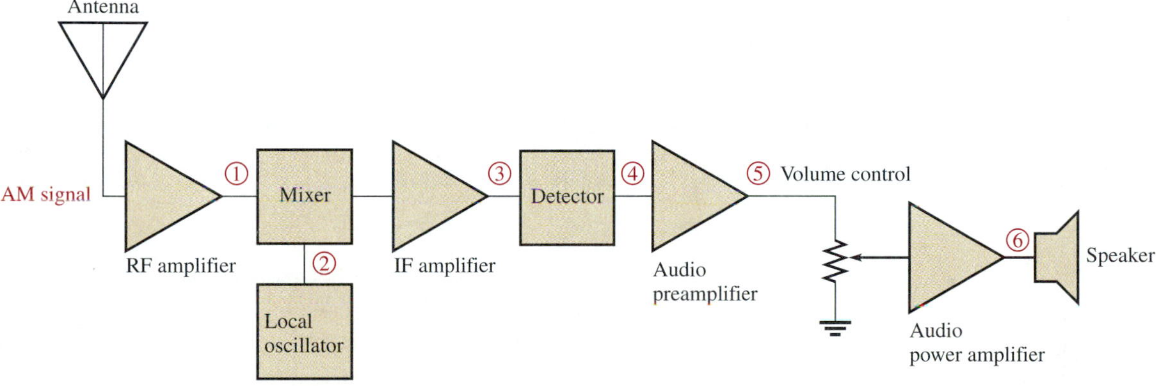


**FIGURE 11–61**
***Simplified block diagram of a basic radio receiver.***

**FIGURE 11–62**
***Example of an amplitude modulated (AM) signal.***

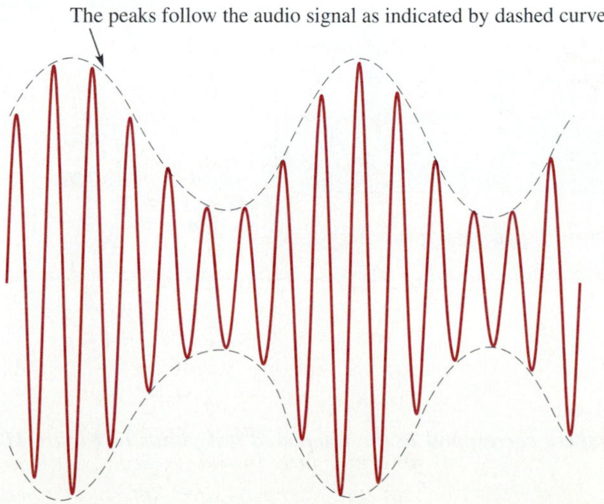

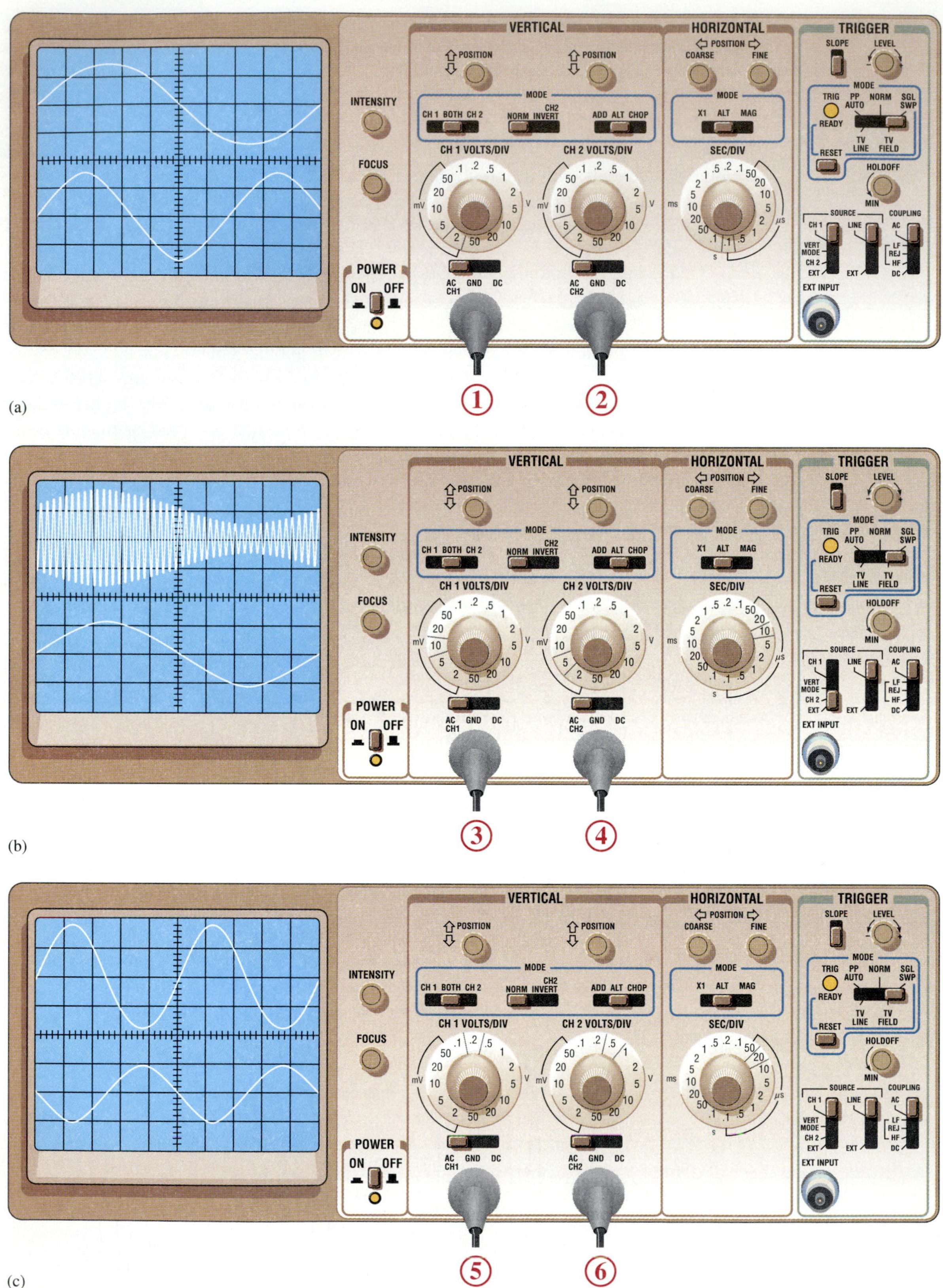


**FIGURE 11–63**

*Circled numbers correspond to the numbered test points in Figure 11–61.*

### Oscilloscope Measurements

Signals that are indicated by circled numbers at several test points on the receiver block diagram in Figure 11–61 are measured on the oscilloscope in Figure 11–63 on the channels with the same corresponding circled numbers. In all cases, the upper waveform on the screen is channel 1 and the lower waveform is channel 2.

The signal at point 1 is an AM signal but you can't see the amplitude variation because of the short time base. The waveform is spread out too much too see the modulating audio signal, which causes amplitude variations; so what you see is just one cycle of the carrier. At point 3, the higher carrier frequency is difficult to determine because the time base was selected to allow viewing of one full cycle of the modulating signal. In an AM receiver, this frequency is 455 kHz. In actual practice, the modulated carrier signal at point 3 cannot easily be viewed on the scope because it contains two frequencies which make it difficult to synchronize in order to obtain a stable pattern. A stable pattern is shown in this case to illustrate what the modulated waveform looks like.

- □ For each waveform in Figure 11–63, except point 3, determine the frequency and rms value. The signal at point 4 is the modulating tone extracted by the detector from the higher intermediate frequency (455 kHz).

### Amplifier Analysis

- □ All voltage amplifiers have a characteristic known as voltage gain. The voltage gain is the amount by which the amplitude of the output signal is greater than the amplitude of the input signal. Using this definition and the appropriate scope measurements, determine the gain of the audio preamplifier in this particular receiver.
- □ When an electrical signal is converted to sound by a speaker, the loudness of the sound depends on the amplitude of the signal applied to the speaker. Based on this, explain how the volume control potentiometer is used to adjust the loudness (volume) of the sound and determine the rms amplitude at the speaker.

### SECTION 11–10 REVIEW

1. What does RF stand for?
2. What does IF stand for?
3. Which frequency is higher, the carrier or the audio?
4. What is the variable in a given AM signal?

## 11–11 ■ INTRODUCTION TO AC ANALYSIS WITH PSpice AND PROBE

***Although PSpice and Probe have been used only for dc analysis up to this point, they are also capable of ac analysis. Probe is especially suited for ac circuit analysis because it permits you to see what happens in a circuit as the source value changes and, as you know, ac circuits have constantly varying sources.***

***After completing this section, you should be able to***

- **Use PSpice and Probe in ac circuit analysis**
  - □ Specify ac sources
  - □ Write circuit files for simple ac circuits
  - □ Use Probe to graph ac quantities
  - □ Use the .TRAN control statement

## Specification of AC Sources

First, let's review briefly. Earlier, you learned that the format for a current or voltage source in a PSpice circuit file is

```
<Source Name> <First Node> <Second Node> <Value>
```

where the source name begins with a "V" for a voltage source or an "I" for a current source. The value is assumed to be volts for a voltage source or amperes for a current source. You also learned that the voltage source value referenced the first node to the second, so that

```
VS 1 0 5
```

means that node 1 is 5 V more positive than node 0, while

```
VS 1 0 –5
```

means that node 1 is 5 V more negative than node 0. In contrast,

```
IS 1 0 1
```

means that 1 A of current is *into* node 1 and *out of* node 0, while

```
IS 1 0 –1
```

means that 1 A of current is *out of* node 1 and *into* node 0. This seeming contradiction can be explained by the fact that, by convention, PSpice regards positive current as being from the first node through the source to the second node. This may seem a bit confusing at first but, with usage, remembering the correct form becomes routine.

For ac circuits, additional information must be specified to describe an ac source. The general format for an independent voltage or current source in PSpice then, is

```
<Source Name> <First Node> <Second Node> [ [ DC ] <Value> ]
+ [ AC <Value> [ <Phase> ] ]
+ [ <Transient Specification> ]
```

where anything between matching square brackets [ ] is optional and may be omitted. Anything between matching angle brackets < > means to substitute the appropriate alphabetic or numeric value. Anything else is to be included literally, although it need not be in capital letters. Finally, the [ ] and < > brackets are *never* included; they are only shown here to describe the format.

Notice that the general format and the format we have been using for dc-only sources are consistent and differ only by the addition of AC and Transient Specification information. Recall that the "+" character is a line continuation symbol. If all the information you wish to specify will fit on a single line, the "+" is not necessary. However, you may want to use separate lines as shown, for clarity. PSpice resolves unspecified information for sources by using the following rules:

1. If AC or DC does not precede <Value>, DC is assumed.
2. If any of <Value>, <Phase> or <Transient Specification> is not present, it is assumed to be 0.

Thus, the line

```
VS 1 0 5
```

is interpreted by PSpice to be the same as

```
VS 1 0 DC 5 AC 0 0     (No Transient Specification)
```

PSpice and Probe allow you to analyze ac circuit behavior as time varies, or as frequency varies, or both. Compare this to dc circuits which, by definition, do not vary in time and therefore have zero or no frequency. The AC <Value> <Phase> part of a source description tells PSpice what to use for frequency-based analysis, while the <Transient Specification> part tells it what to use for time-based analysis. PSpice assumes the source for frequency-based analysis is always a sinusoid, with <Value> specifying the magnitude. The <Phase> may be specified to express a phase difference in degrees (if any), compared to other AC sources in the circuit. The <Transient Specification> part of a source description is more complex and may be used to specify not just sinusoidal waveforms, but pulses, square waves, exponential waveforms, and more. For now, however, we need only concentrate on the format for a sinusoidal waveform, and we need only use some of the available options for it. The complete formats for sinusoidal and other types of <Transient Specification> waveforms will be introduced as needed.

We will specify a *transient* sinusoidal ac source as

```
<Source Name> <First Node> <Second Node> SIN ( <DC Offset> <Peak Value> <Frequency> )
```

where <DC Offset> is the dc value on which the ac waveform is riding, <Peak Value> is the amplitude of the ac waveform in volts or amps, and <Frequency> is the frequency of the ac source in hertz. The enclosing parentheses must be included. For example, a 5 V peak, 1 kHz transient ac voltage source called VS, between nodes 1 and 0 with no dc component, may be written in the PSpice circuit file as

```
VS  1  0  SIN( 0  5  1K )
```

Other transient value specifications for SIN will be discussed as the need arises.

## The .TRAN (Transient Analysis) Control Statement

The .TRAN statement tells PSpice to calculate the circuit's behavior over time. The output you see from a transient analysis, using Probe, is much like what you might see on an oscilloscope, which displays voltages on a time base. In fact, PSpice and Probe can display much more than just voltages obtained from transient analysis, as was discussed before. This is a distinct advantage over a real oscilloscope, which generally only works with voltage inputs.

Like most other PSpice control statements, the .TRAN statement allows for a number of options, some of which we will ignore at this time in order to make the discussion simpler and clearer. The general format for the .TRAN statement we will use is as follows:

```
.TRAN <Print Step> <Final Time> [ <No Print Time> [ <Max Step> ] ]
```

where <Print Step> is the interval used for printing or plotting the results of the transient analysis; <Final Time> tells PSpice when to stop simulating; <No Print Time> tells PSpice how far into the simulation to go before saving output for the .PRINT, .PLOT, or .PROBE statements; and <Max Step> tells PSpice what to use for the maximum time step for calculating circuit behavior during the simulation. Notice that the <No Print Time> and <Max Step> values are both optional. Also notice, per the format above, that you can specify the <No Print Time> without specifying the <Max Step>, but you must specify <No Print Time> if you wish to use <Max Step>. The default value for <No Print Time> is 0 seconds, which is where PSpice always starts the simulation. The default value for <Max Step> is <Final Time>/50, which is usually sufficient to accurately reproduce the waveforms, since PSpice automatically reduces the internal time step (sometimes far below <Max Step>) during busy intervals of the simulation.

For those using a schematic capture or DOS shell version of PSpice and Probe, there is a pull-down menu for specifying Transient Analysis data, with dialog data entry boxes. The labels on the boxes correspond with the <Options> specified above, so simply enter the data in the appropriate boxes, as we have discussed.

As an example, the following line will cause PSpice to take samples from 0 s to 1 s at 10 ms intervals.

```
.TRAN  10m  1
```

To take samples for output at 10 μs intervals, starting 50 μs into the simulation, while letting PSpice use its default maximum calculation time step, over a total simulation run ending at 100 μs, the following would be used:

```
.TRAN  10U  100U  50U
```

Generally speaking, the <No Print Time> and <Max Step> parameters can be left off with satisfactory results, so you should probably not be too concerned with these at first. The <Print Step> value, or output sampling interval, is not critical provided there are enough samples taken to accurately reproduce the waveforms without using too much disk space or memory.

## Example of AC Circuit Analysis

Now that you know how to specify an ac source, let's look at the simple ac circuit in Figure 11–64 and its corresponding PSpice circuit file below.

FIGURE 11–64

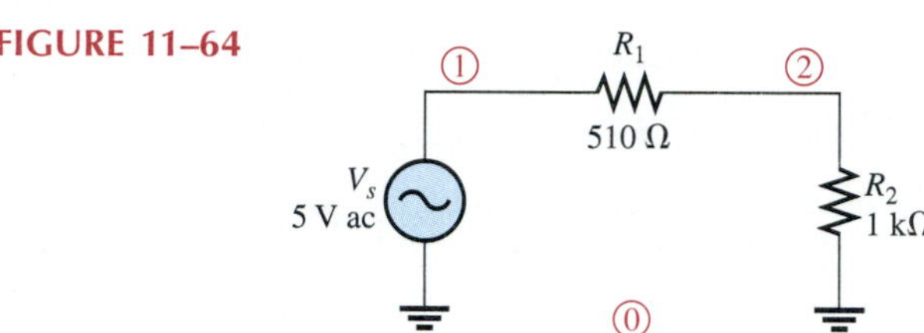


```
AC Voltage Divider
*.AC and .TRAN Demonstration Circuit
VS  1  0  SIN( 0  5  1K )
R1  1  2  510
R2  2  0  1K
.TRAN  20u  5m
.PROBE
.END
```

From the circuit file, we see that VS is a transient voltage source, consisting of a 5 V peak, 1 kHz sine wave with no dc offset, and that .TRAN will cause PSpice to sample data every 20 μs over a 5 ms interval.

When you run PSpice and Probe for this circuit, the opening Probe screen will show that the $x$-axis ranges from 0 s to 5 ms, just as the .TRAN control statement specified. As was stated before, you will see somewhat different menu commands and menu formats, depending on your installed platform. Equivalent functions and options are provided, however, regardless of platform, and the menus are easy to follow. So, select the Add_trace or Trace/Add option(s) from the displayed menu, and either type in or mouse-click on the following trace names when the prompt appears:

```
V(1)  V(1,2)  V(2)
```

Probe will graph three sine waves of different amplitudes, corresponding to $V_s$, $V_{R1}$, and $V_{R2}$. Each of the sine waves are in phase with each other because the circuit is purely resistive, and there will be five complete periods shown because the period of each sine wave is 1 ms, while the total simulation time is 5 ms.

## Changing the Scale of the *X*-Axis in Probe

You may find that five periods are too many to view conveniently or that you want a closer look at the data. To change the scale of the $x$-axis so that you see a smaller range, select the X_axis or Plot/X Axis Settings . . . menu option(s). You will then see either the original menu replaced with a new submenu of options, or perhaps see a pull-down or pop-up style menu appear over or along side the original one. Generally, you may do any or all of the following options, each of which affects the *x*-axis and how the data are displayed.

- □ Use logarithmic or linear scale (logarithmic scale must not include 0)
- □ Select a range for the plot axis
- □ Restrict the range of data to consider for plotting or processing
- □ Perform a Fourier transform to show the frequency content of the data
- □ Select the *x*-axis variable to plot the data against

For now, let's change the *x*-axis range so that only one period is displayed across the entire graph. To do this, select the option to set the range, and at the prompt or dialog data entry box, enter the values 1m and 2m, and press ENTER or mouse-click on OK to continue. Once the command is complete, Probe will redraw the graph so that only one period of the waveforms (traces), from 1 ms to 2 ms, is displayed. Probe gives you the capability of rescaling the plot to display all of the selected data with the Auto_Range option.

## A Note on Units in Probe

As you have seen, a lowercase "m" rather than an uppercase "M" is used in Probe to specify milliseconds. This is because Probe uses the standard one-letter abbreviations for metric prefixes, so that "m" stands for milli while "M" stands for mega. All other metric prefixes can be specified in either uppercase or lowercase, although for consistency it is best to use the standard abbreviations as follows: femto (f), pico (p), nano (n), micro (u), milli (m), kilo (k), mega (M), giga (G) and tera (T).

## SECTION 11–11 REVIEW

1. Specify a transient sinusoidal ac source voltage with a peak value of 10 V and a frequency of 15 kHz between nodes 1 and 0 in a circuit. Assume no phase shift or dc value.
2. Write a circuit file for the circuit in Figure 11–65 to view the voltages across each of the resistors at a rate of every 10 $\mu$s over a 5 ms interval.

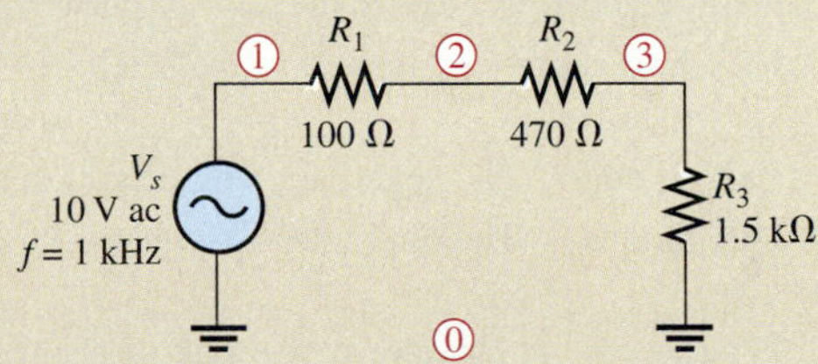


FIGURE 11–65

## ■ SUMMARY

- The sine wave is a time-varying, periodic waveform.
- The sine wave is a common form of ac (alternating current)
- Alternating current changes direction in response to changes in the polarity of the source voltage.
- One cycle of an alternating sine wave consists of a positive alternation and a negative alternation.
- Two common sources of sine waves are the electromagnetic ac generator and the electronic oscillator circuit.

- A full cycle of a sine wave is 360°, or $2\pi$ radians. A half-cycle is 180°, or $\pi$ radian. A quarter-cycle is 90°, or $\pi/2$ radians.
- A sine wave voltage can be generated by a conductor rotating in a magnetic field.
- Phase angle is the difference in degrees or radians between a given sine wave and a reference point.
- A pulse consists of a transition from a baseline level to an amplitude level, followed by a transition back to the baseline level.
- A triangle or sawtooth wave consists of positive- and negative-going ramps.
- Harmonic frequencies are odd or even multiples of the repetition rate of a nonsinusoidal waveform.
- Conversions of sine wave values are summarized in Table 11–2.

TABLE 11–2

| To Change From | To | Multiply By |
|---|---|---|
| Peak | rms | 0.707 |
| Peak | Peak-to-peak | 2 |
| Peak | Average | 0.637 |
| rms | Peak | 1.414 |
| Peak-to-peak | Peak | 0.5 |
| Average | Peak | 1.57 |

## ■ GLOSSARY

**Alternating current (ac)** Current that reverses direction in response to a change in source voltage polarity.

**Amplitude** The maximum value of a voltage or current.

**Average value** The average of a sine wave over one-half cycle. It is 0.637 times the peak value.

**Baseline** The normal level of a pulse waveform; the voltage level in the absence of a pulse.

**Cathode-ray tube (CRT)** A vacuum tube device containing an electron gun that emits a narrow focused beam of electrons onto a phosphor-coated screen.

**Cycle** One repetition of a periodic waveform.

**Degree** The unit of angular measure corresponding to 1/360 of a complete revolution.

**Duty cycle** A characteristic of a pulse waveform that indicates the percentage of time that a pulse is present during a cycle; the ratio of pulse width to period.

**Effective value** A measure of the heating effect of a sine wave; also known as the rms (root mean square) value.

**Falling edge** The negative-going transition of a pulse.

**Fall time ($t_f$)** The time interval required for a pulse to change from 90% to 10% of its full amplitude.

**Frequency** A measure of the rate of change of a periodic function; the number of cycles completed in 1 s. The unit of frequency is the hertz.

**Fundamental frequency** The repetition rate of a waveform.

**Generator** An energy source that produces electrical signals.

**Harmonics** The frequencies contained in a composite waveform, which are integer multiples of the pulse repetition frequency (fundamental).

**Hertz (Hz)** The unit of frequency. One hertz equals one cycle per second.

**Instantaneous value** The voltage or current value of a waveform at a given instant in time.

**Lag** A condition of the phase or time relationship of waveforms in which one waveform is behind the other in phase or time.

**Lead** A condition of the phase or time relationship of waveforms in which one waveform is ahead of the other in phase or time; also, a wire or cable connection to a device or instrument.

**Leading edge** The first step or transition of a pulse.

**Magnitude** The value of a quantity, such as the number of volts of voltage or the number of amperes of current.

**Oscillator** An electronic circuit that produces a time-varying signal without an external input signal using positive feedback.

**Oscilloscope** A measurement instrument that displays signal waveforms on a screen.

**Peak-to-peak value** The voltage or current value of a waveform measured from its minimum to its maximum points.

**Peak value** The voltage or current value of a waveform at its maximum positive or negative points.

**Period ($T$)** The time interval of one complete cycle of a periodic waveform.

**Periodic** Characterized by a repetition at fixed-time intervals.

**Phase** The relative displacement of a time-varying waveform in terms of its occurrence with respect to a reference.

**Pulse** A type of waveform that consists of two equal and opposite steps in voltage or current separated by a time interval.

**Pulse repetition frequency** The fundamental frequency of a repetitive pulse waveform; the rate at which the pulses repeat expressed in either hertz or pulses per second.

**Pulse width ($t_W$)** The time interval between the opposite steps of an ideal pulse. For a nonideal pulse, the time between the 50% points on the leading and trailing edges.

**Radian** A unit of angular measurement. There are $2\pi$ radians in one complete 360° revolution. One radian equals 57.3°.

**Ramp** A type of waveform characterized by a linear increase or decrease in voltage or current.

**Rise time ($t_r$)** The time interval required for a pulse to change from 10% to 90% of its amplitude.

**Rising edge** The positive-going transition of a pulse.

**Root mean square (rms)** The value of a sine wave that indicates its heating effect, also known as the effective value. It is equal to 0.707 times the peak value.

**Sawtooth waveform** A type of electrical waveform composed of ramps; a special case of a triangular waveform in which one ramp is much shorter than the other.

**Trailing edge** The second step or transition of a pulse.

**Triangular waveform** A type of electrical waveform that consists of two ramps.

**Trigger** The activating unit of some electronic devices or instruments.

**Waveform** The pattern of variations of a voltage or current showing how the quantity changes with time.

## ■ FORMULAS

| | | |
|---|---|---|
| **(11–1)** | $f = \dfrac{1}{T}$ | Frequency |
| **(11–2)** | $T = \dfrac{1}{f}$ | Period |
| **(11–3)** | $f =$ (number of pole pairs)(rps) | Output frequency of a generator |
| **(11–4)** | $V_{pp} = 2V_p$ | Peak-to-peak voltage (sine wave) |
| **(11–5)** | $I_{pp} = 2I_p$ | Peak-to-peak current (sine wave) |
| **(11–6)** | $V_{rms} = 0.707V_p$ | Root-mean-square voltage (sine wave) |
| **(11–7)** | $I_{rms} = 0.707I_p$ | Root-mean-square current (sine wave) |
| **(11–8)** | $V_p = 1.414V_{rms}$ | Peak voltage (sine wave) |
| **(11–9)** | $I_p = 1.414I_{rms}$ | Peak current (sine wave) |
| **(11–10)** | $V_{pp} = 2.828V_{rms}$ | Peak-to-peak voltage (sine wave) |
| **(11–11)** | $I_{pp} = 2.828I_{rms}$ | Peak to peak current (sine wave) |
| **(11–12)** | $V_{avg} = 0.637V_p$ | Half-cycle average voltage (sine wave) |

| | | |
|---|---|---|
| **(11–13)** | $I_{avg} = 0.637I_p$ | Half-cycle average current (sine wave) |
| **(11–14)** | $\text{rad} = \left(\frac{\pi \text{ rad}}{180°}\right) \times \text{degrees}$ | |
| **(11–15)** | $\text{degrees} = \left(\frac{180°}{\pi \text{ rad}}\right) \times \text{rad}$ | |
| **(11–16)** | $y = A \sin \theta$ | General formula for a sine wave |
| **(11–17)** | $y = A \sin(\theta - \phi)$ | Sine wave shifted to the right of the reference |
| **(11–18)** | $y = A \sin(\theta + \phi)$ | Sine wave shifted to the left of the reference |
| **(11–19)** | $\text{percent duty cycle} = \left(\frac{t_W}{T}\right)100\%$ | |
| **(11–20)** | $V_{avg} = \text{baseline} + (\text{duty cycle})(\text{amplitude})$ | Average value of a pulse waveform |

## ▪ SELF-TEST

**1.** The difference between alternating current (ac) and direct current (dc) is
**(a)** ac changes value and dc does not **(b)** ac changes direction and dc does not
**(c)** both answers (a) and (b) **(d)** neither answer (a) nor (b)

**2.** During each cycle, a sine wave reaches a peak value
**(a)** one time **(b)** two times
**(c)** four times **(d)** a number of times depending on the frequency

**3.** A sine wave with a frequency of 12 kHz is changing at a faster rate than a sine wave with a frequency of
**(a)** 20 kHz **(b)** 15,000 Hz **(c)** 10,000 Hz **(d)** 1.25 MHz

**4.** A sine wave with a period of 2 ms is changing at a faster rate than a sine wave with a period of
**(a)** 1 ms **(b)** 0.0025 s **(c)** 1.5 ms **(d)** 1000 $\mu$s

**5.** When a sine wave has a frequency of 60 Hz, in 10 s it goes through
**(a)** 6 cycles **(b)** 10 cycles **(c)** 1/16 cycle **(d)** 600 cycles

**6.** If the peak value of a sine wave is 10 V, the peak-to-peak value is
**(a)** 20 V **(b)** 5 V **(c)** 100 V **(d)** none of these

**7.** If the peak value of a sine wave is 20 V, the rms value is
**(a)** 14.14 V **(b)** 6.37 V **(c)** 7.07 V **(d)** 0.707 V

**8.** The average value of a 10 V peak sine wave over one complete cycle is
**(a)** 0 V **(b)** 6.37 V **(c)** 7.07 V **(d)** 5 V

**9.** The average half-cycle value of a sine wave with a 20 V peak is
**(a)** 0 V **(b)** 6.37 V **(c)** 12.74 V **(d)** 14.14 V

**10.** One sine wave has a positive-going zero crossing at 10° and another sine wave has a positive-going zero crossing at 45°. The phase angle between the two waveforms is
**(a)** 55° **(b)** 35° **(c)** 0° **(d)** none of these

**11.** The instantaneous value of a 15 A peak sine wave at a point 32° from its positive-going zero crossing is
**(a)** 7.95 A **(b)** 7.5 A **(c)** 2.13 A **(d)** 7.95 V

**12.** If the rms current through a 10 kΩ resistor is 5 mA, the rms voltage drop across the resistor is
**(a)** 70.7 V **(b)** 7.07 V **(c)** 5 V **(d)** 50 V

**13.** Two series resistors are connected to an ac source. If there are 6.5 V rms across one resistor and 3.2 V rms across the other, the peak source voltage is
**(a)** 9.7 V **(b)** 9.19 V **(c)** 13.72 V **(d)** 4.53 V

14. A 10 kHz pulse waveform consists of pulses that are 10 μs wide. Its duty cycle is
    **(a)** 100% **(b)** 10% **(c)** 1% **(d)** not determinable
15. The duty cycle of a square wave
    **(a)** varies with the frequency **(b)** varies with the pulse width
    **(c)** both answers (a) and (b) **(d)** is 50%

## ■ PROBLEMS

### SECTION 11–1 The Sine Wave

1. Calculate the frequency for each of the following values of period:
   **(a)** 1 s **(b)** 0.2 s **(c)** 50 ms
   **(d)** 1 ms **(e)** 500 μs **(f)** 10 μs
2. Calculate the period of each of the following values of frequency:
   **(a)** 1 Hz **(b)** 60 Hz **(c)** 500 Hz
   **(d)** 1 kHz **(e)** 200 kHz **(f)** 5 MHz
3. A sine wave goes through 5 cycles in 10 μs. What is its period?
4. A sine wave has a frequency of 50 kHz. How many cycles does it complete in 10 ms?

### SECTION 11–2 Sine Wave Voltage Sources

5. The conductive loop on the rotor of a simple two-pole, single-phase generator rotates at a rate of 250 rps. What is the frequency of the induced output voltage?
6. A certain four-pole generator has a speed of rotation of 3600 rpm. What is the frequency of the voltage produced by this generator?
7. At what speed of rotation must a four-pole generator be operated to produce a 400 Hz sine wave voltage?

### SECTION 11–3 Voltage and Current Values of Sine Waves

8. A sine wave has a peak value of 12 V. Determine the following values:
   **(a)** rms **(b)** peak-to-peak **(c)** average
9. A sinusoidal current has an rms value of 5 mA. Determine the following values:
   **(a)** peak **(b)** average **(c)** peak-to-peak
10. For the sine wave in Figure 11–66, determine the peak, peak-to-peak, rms, and average values.

FIGURE 11–66

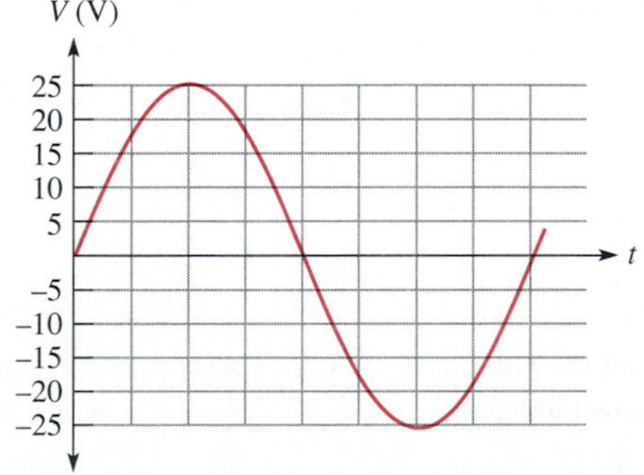


### SECTION 11–4 Angular Measurement of a Sine Wave

11. Convert the following angular values from degrees to radians:
    **(a)** 30° **(b)** 45° **(c)** 78°
    **(d)** 135° **(e)** 200° **(f)** 300°
12. Convert the following angular values from radians to degrees:
    **(a)** $\pi/8$ rad **(b)** $\pi/3$ rad **(c)** $\pi/2$ rad
    **(d)** $3\pi/5$ rad **(e)** $6\pi/5$ rad **(f)** $1.8\pi$ rad

13. Sine wave $A$ has a positive-going zero crossing at 30°. Sine wave $B$ has a positive-going zero crossing at 45°. Determine the phase angle between the two signals. Which signal leads?
14. One sine wave has a positive peak at 75°, and another has a positive peak at 100°. How much is each sine wave shifted in phase from the 0° reference? What is the phase angle between them?
15. Make a sketch of two sine waves as follows: Sine wave $A$ is the reference, and sine wave $B$ lags $A$ by 90°. Both have equal amplitudes.

## SECTION 11–5 The Sine Wave Formula

16. A certain sine wave has a positive-going zero crossing at 0° and an rms value of 20 V. Calculate its instantaneous value at each of the following angles:
    **(a)** 15° **(b)** 33° **(c)** 50° **(d)** 110°
    **(e)** 70° **(f)** 145° **(g)** 250° **(h)** 325°
17. For a particular 0° reference sinusoidal current, the peak value is 100 mA. Determine the instantaneous value at each of the following points:
    **(a)** 35° **(b)** 95° **(c)** 190° **(d)** 215° **(e)** 275° **(f)** 360°
18. For a 0° reference sine wave with an rms value of 6.37 V, determine its instantaneous value at each of the following points:
    **(a)** $\pi/8$ rad **(b)** $\pi/4$ rad **(c)** $\pi/2$ rad **(d)** $3\pi/4$ rad
    **(e)** $\pi$ rad **(f)** $3\pi/2$ rad **(g)** $2\pi$ rad
19. Sine wave $A$ lags sine wave $B$ by 30°. Both have peak values of 15 V. Sine wave $A$ is the reference with a positive-going crossing at 0°. Determine the instantaneous value of sine wave $B$ at 30°, 45°, 90°, 180°, 200°, and 300°.
20. Repeat Problem 19 for the case when sine wave $A$ leads sine wave $B$ by 30°.
21. A certain sine wave has a frequency of 2.2 kHz and an rms value of 25 V. Assuming a given cycle begins (zero crossing) at $t = 0$ s, what is the change in voltage from 0.12 ms to 0.2 ms?

## SECTION 11–6 Ohm's Law and Kirchhoff's Laws in AC Circuits

22. A sinusoidal voltage is applied to the resistive circuit in Figure 11–67. Determine the following:
    **(a)** $I_{rms}$ **(b)** $I_{avg}$ **(c)** $I_p$ **(d)** $I_{pp}$ **(e)** $i$ at the positive peak

FIGURE 11–67

23. Find the half-cycle average values of the voltages across $R_1$ and $R_2$ in Figure 11–68. All values shown are rms.

FIGURE 11–68

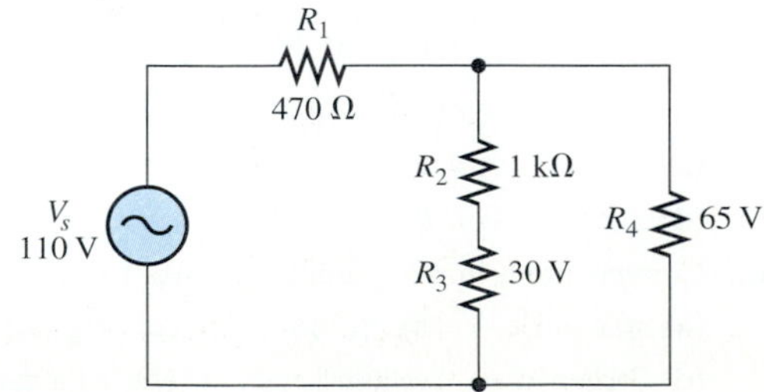

**24.** Determine the rms voltage across $R_3$ in Figure 11–69.

FIGURE 11–69

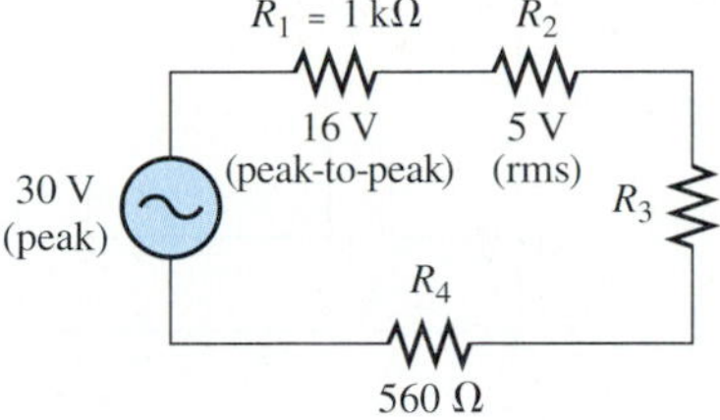


## SECTION 11–7 Superimposed DC and AC Voltages

**25.** A sine wave with an rms value of 10.6 V is riding on a dc level of 24 V. What are the maximum and minimum values of the resulting waveform?

**26.** How much dc voltage must be added to a 3 V rms sine wave in order to make the resulting voltage nonalternating (no negative values)?

**27.** A 6 V peak sine wave is riding on a dc voltage of 8 V. If the dc voltage is lowered to 5 V, how far negative will the sine wave go?

**28.** Figure 11–70 shows a sinusoidal voltage source in series with a dc source. Effectively, the two voltages are superimposed. Determine the power dissipation in the load resistor.

FIGURE 11–70

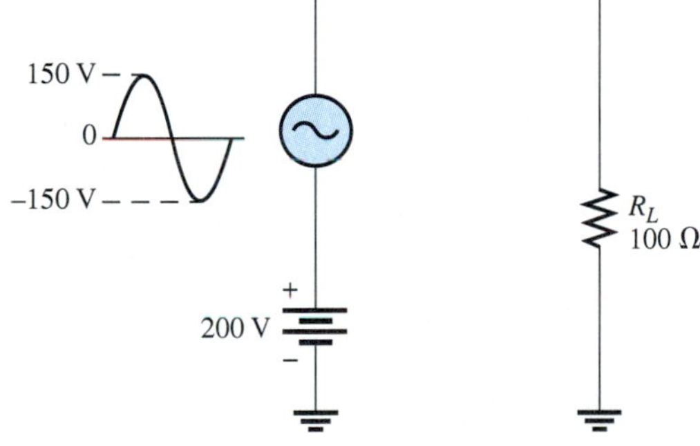


## SECTION 11–8 Nonsinusoidal Waveforms

**29.** From the graph in Figure 11–71, determine the approximate values of $t_r$, $t_f$, $t_W$, and amplitude.

FIGURE 11–71

**30.** The repetition frequency of a pulse waveform is 2 kHz, and the pulse width is 1 $\mu$s. What is the percent duty cycle?

**31.** Calculate the average value of the pulse waveform in Figure 11–72.

FIGURE 11–72

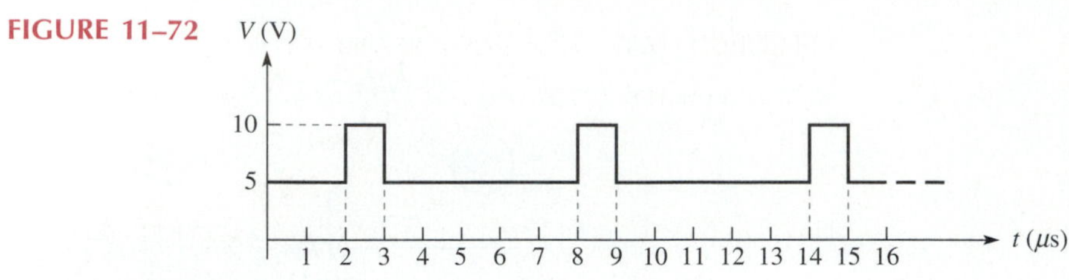


**32.** Determine the duty cycle for each waveform in Figure 11–73.

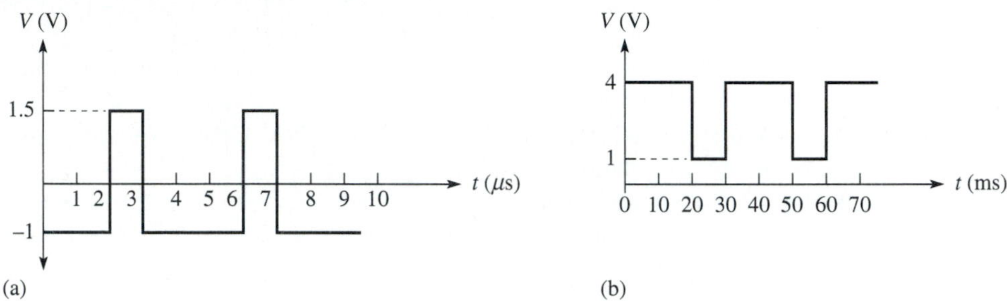


FIGURE 11–73

**33.** Find the average value of each pulse waveform in Figure 11–73.

**34.** What is the frequency of each waveform in Figure 11–73?

**35.** What is the frequency of each sawtooth waveform in Figure 11–74?

FIGURE 11–74

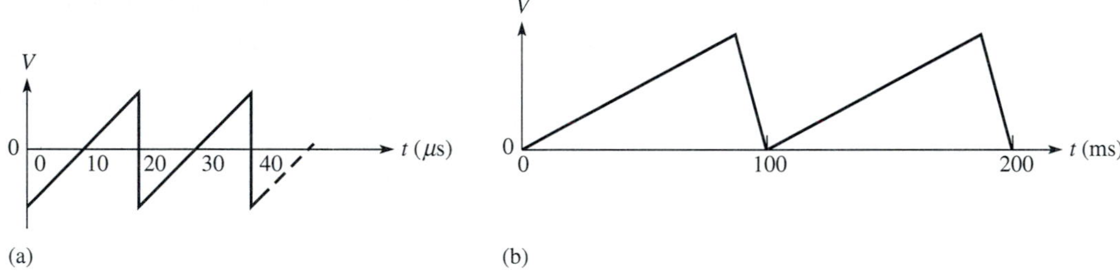


**36.** A nonsinusoidal waveform called a *stairstep* is shown in Figure 11–75. Determine its average value.

FIGURE 11–75

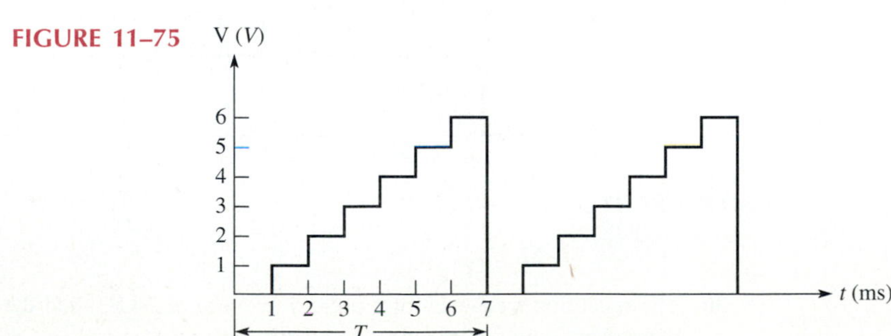

**37.** A square wave has a period of 40 μs. List the first six odd harmonics.

**38.** What is the fundamental frequency of the square wave mentioned in Problem 37?

## SECTION 11–9 Oscilloscope Measurements

**39.** Determine the peak value and the period of the sine wave displayed on the scope screen in Figure 11–76.

FIGURE 11–76

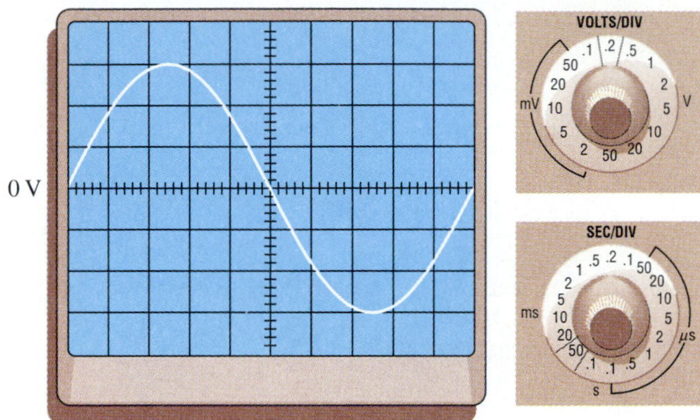


**40.** Determine the rms value and the frequency of the sine wave displayed on the scope screen in Figure 11–77.

FIGURE 11–77

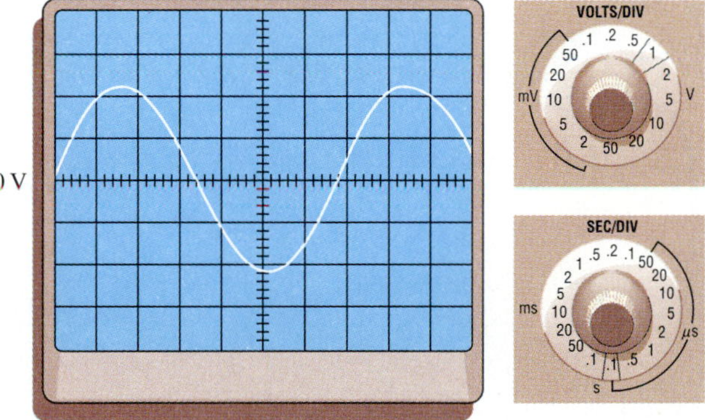


**41.** Find the amplitude, pulse width, and duty cycle for the pulse waveform displayed on the scope screen in Figure 11–78.

FIGURE 11–78

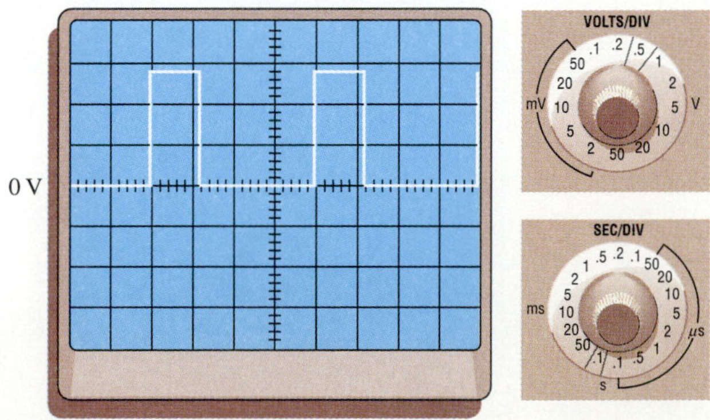

42. Based on the instrument settings and an examination of the scope display and the circuit board in Figure 11–79, determine the frequency and peak value of the input signal and output signal. The waveform shown is channel 1. Sketch the channel 2 waveform as it would appear on the scope with the indicated settings.

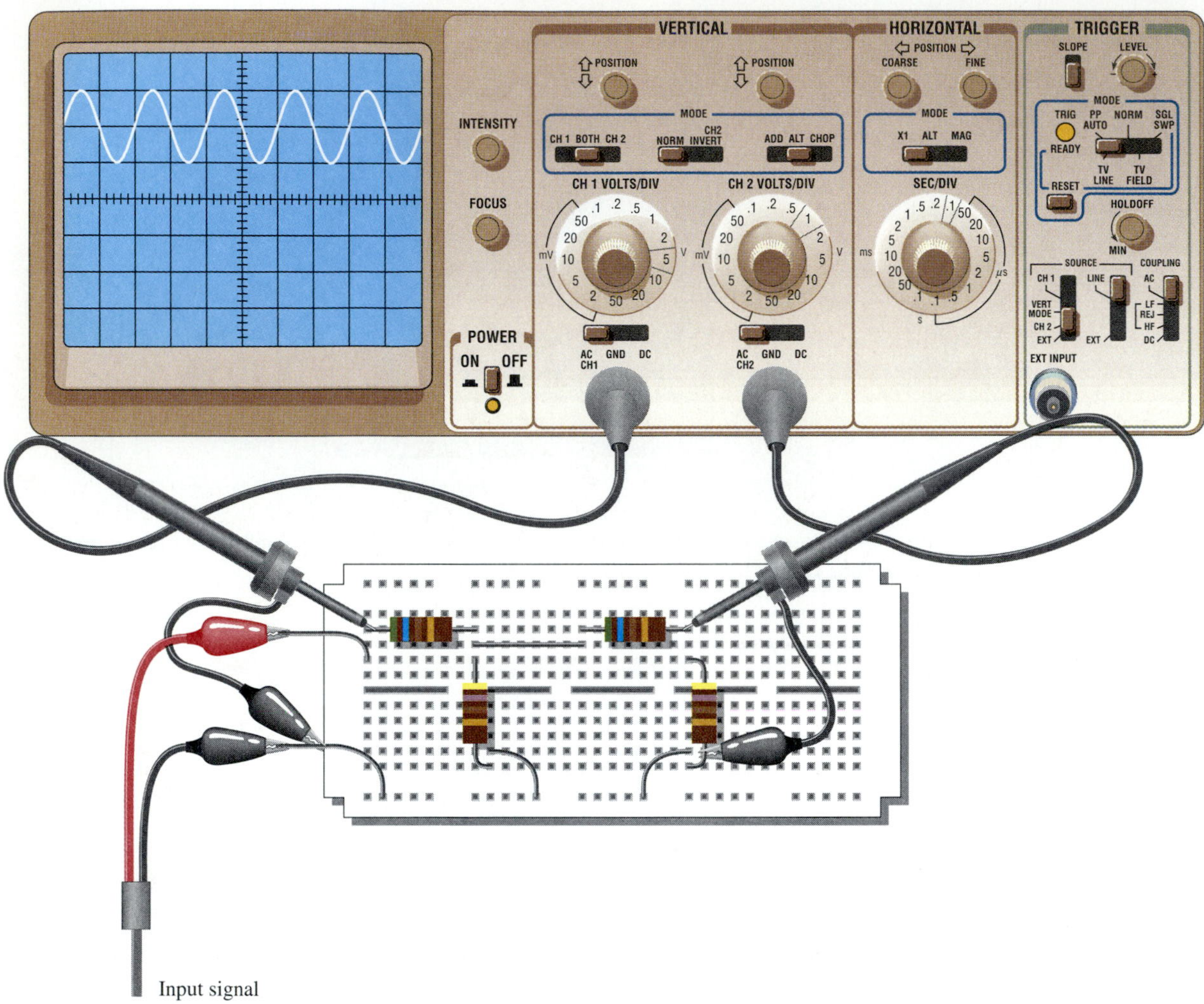


FIGURE 11–79

43. Examine the circuit board and the oscilloscope display in Figure 11–80 and determine the peak value and the frequency of the unknown input signal.

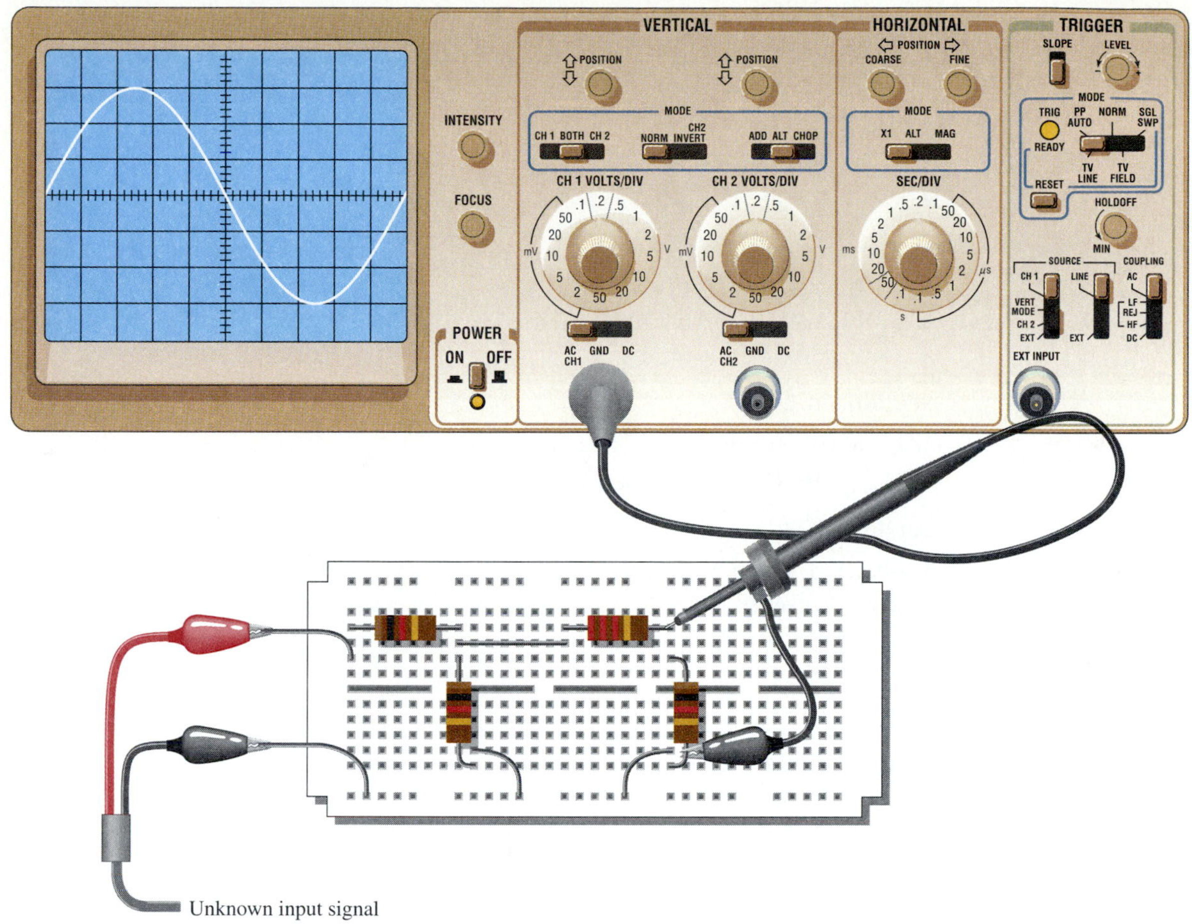


FIGURE 11–80

## ▪ ANSWERS TO SECTION REVIEWS

### Section 11–1

1. One cycle of a sine wave is from the zero crossing through a positive peak, then through zero to a negative peak and back to the zero crossing.
2. A sine wave changes polarity at the zero crossings.
3. A sine wave has two maximum points (peaks) per cycle.
4. The period is from one zero crossing to the next corresponding zero crossing, or from one peak to the next corresponding peak.
5. Frequency is the number of cycles completed in one second; the unit of frequency is the hertz.
6. $f = 1/T = 200$ kHz
7. $T = 1/f = 8.33$ ms

### Section 11–2

1. Sine waves are generated by electromagnetic and electronic methods.
2. Speed and frequency are directly proportional.

### Section 11–3

1. **(a)** $V_{pp} = 2(1\text{ V}) = 2\text{ V}$ **(b)** $V_{pp} = 2(1.414)(1.414\text{ V}) = 4\text{ V}$ **(c)** $V_{pp} = 2(1.57)(3\text{ V}) = 9.42\text{ V}$
2. **(a)** $V_{\text{rms}} = (0.707)(2.5\text{ V}) = 1.77\text{ V}$ **(b)** $V_{\text{rms}} = (0.5)(0.707)(10\text{ V}) = 3.54\text{ V}$
   **(c)** $V_{\text{rms}} = (0.707)(1.57)(1.5\text{ V}) = 1.66\text{ V}$
3. **(a)** $V_{\text{avg}} = (0.637)(10\text{ V}) = 6.37\text{ V}$ **(b)** $V_{\text{avg}} = (0.637)(1.414)(2.3\text{ V}) = 2.07\text{ V}$
   **(c)** $V_{\text{avg}} = (0.637)(0.5)(60\text{ V}) = 19.1\text{ V}$

## Section 11–4

1. **(a)** Positive peak at 90° **(b)** Negative-going zero crossing at 180°
   **(c)** Negative peak at 270° **(d)** End of cycle at 360°
2. Half-cycle: 180°; $\pi$
3. Full cycle: 360°; $2\pi$
4. $90° - 45° = 45°$

## Section 11–5

1. $v = 10 \sin(120°) = 8.66$ V
2. $v = 10 \sin(45° + 10°) = 8.19$ V
3. $v = 5 \sin(90° - 25°) = 4.53$ V

## Section 11–6

1. $I_p = V_p/R = (1.57)(12.5 \text{ V})/330\ \Omega = 59.5$ mA
2. $V_{s(rms)} = (0.707)(25.3 \text{ V}) = 17.9$ V

## Section 11–7

1. $+V_{max} = 5 \text{ V} + 2.5 \text{ V} = 7.5$ V
2. Yes, it will alternate.
3. $+V_{max} = 5 \text{ V} - 2.5 \text{ V} = 2.5$ V

## Section 11–8

1. **(a)** Rise time is the time interval from 10% to 90% of the rising pulse edge; **(b)** Fall time is the time interval from 90% to 10% of the falling pulse edge; **(c)** Pulse width is the time interval from 50% of the leading pulse edge to 50% of the trailing pulse edge.
2. $f = 1/1$ ms = 1 kHz
3. d.c. = (1/5)100% = 20%; Ampl. 1.5 V; $V_{avg} = 0.5 \text{ V} + 0.2(1.5 \text{ V}) = 0.8$ V
4. $T = 16$ ms
5. $f = 1/T = 1/1\ \mu$s = 1 MHz
6. Fundamental frequency is the repetition rate of the waveform.
7. 2nd harm.: 2 kHz
8. $f = 1/10\ \mu$s = 100 kHz

## Section 11–9

1. CRT is cathode-ray tube.
2. Voltage is measured vertically; time is measured horizontally.
3. An oscilloscope can display time-varying quantities.

## Section 11–10

1. RF is radio frequency.
2. IF is intermediate frequency.
3. Carrier frequency is higher than audio.
4. The amplitude varies in an AM signal.

## Section 11–11

1. `VS 1 0 SIN( 0 10 15K )`
2. Circuit for Fig. 11–65

```
VS  1  0  SIN( 0  10  1K )
R1  1  2  100
R2  2  3  470
R3  3  0  1.5K
.TRAN  10u  5 m
.PROBE
.END
```

## ■ ANSWERS TO RELATED EXERCISES FOR EXAMPLES

**11–1** 2.4 s

**11–2** 1.5 ms

**11–3** 20 kHz

**11–4** 200 Hz

**11–5** 66.7 kHz

**11–6** 30 rps

**11–7** $V_{pp} = 50$ V; $V_{rms} = 17.7$ V; $V_{avg} = 15.9$ V

**11–8** (a) $\pi/12$ rad (b) 360°

**11–9** 8°

**11–10** 18.1 V

**11–11** $I_{rms} = 4.53$ mA; $V_{1(rms)} = 4.53$ V; $V_{2(rms)} = 2.54$ V

**11–12** 23.7 V

**11–13** The waveform in part (a) never goes negative. The waveform in part (b) goes negative for a portion of its cycle.

**11–14** 5%

**11–15** 1.2 V

**11–16** 120 V

**11–17** Part (a) 1.06 V, 50 Hz;
part (b) 88.4 mV, 1.67 kHz
part (c) 4.81 V, 5 kHz;
part (d) 7.07 V, 250 kHz

# 12

# PHASORS AND COMPLEX NUMBERS



## ■ INTRODUCTION

In this chapter, two important tools for the analysis of ac circuits are introduced. These are phasors and complex numbers. You will see how phasors are a convenient, graphic way to represent sine wave voltages and currents in terms of their magnitude and phase angle. In later chapters, you will see how phasors can also represent other ac circuit quantities.

The complex number system is a means for expressing phasor quantities and for performing mathematical operations with those quantities. In this chapter and throughout the rest of the book, you will learn the basics of putting technology theory into practice.

Phasor diagrams are an abstract method of representing quantities that have both magnitude and direction. In the case of sinusoidal voltages and currents, the magnitude is the amplitude of the sine wave and the direction is its phase angle. Phasors provide a way to diagram sine waves and their phase relationships with other sine waves. The complex number system provides a way to mathematically express a phasor quantity and allows phasor quantities to be added, subtracted, multiplied, or divided.

## TECHnology Theory Into Practice

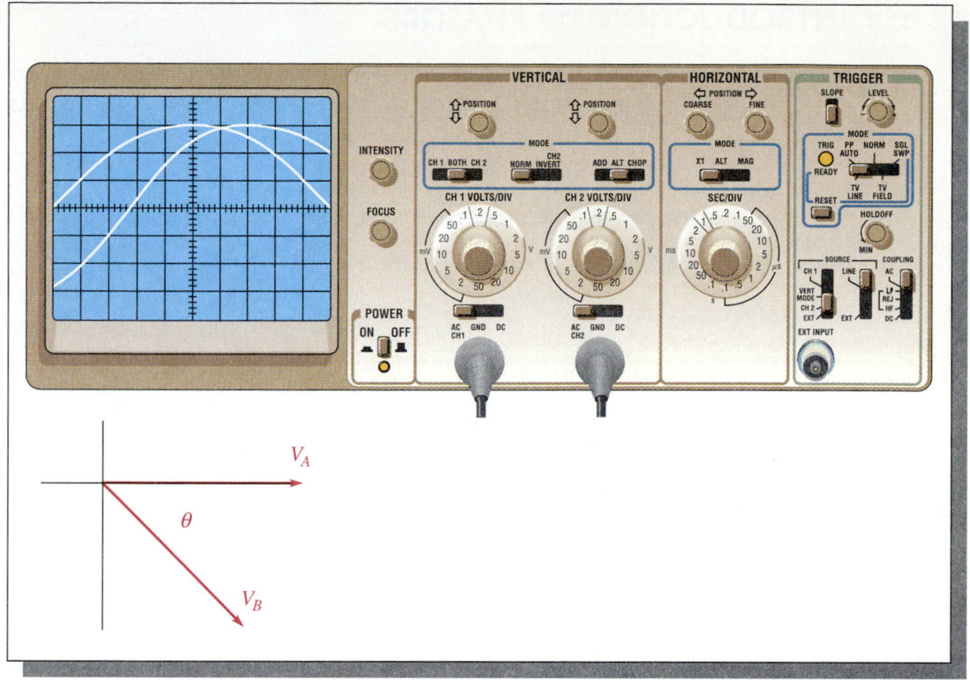


The phase relationship of two sine waves can be graphically represented by phasors and it can also be measured on an oscilloscope, as you will see in the TECH TIP in Section 12–5.

## CHAPTER OBJECTIVES

- ☐ Use a phasor to represent a sine wave
- ☐ Use complex numbers to express phasor quantities
- ☐ Represent phasors in two complex forms
- ☐ Do mathematical operations with complex numbers
- ☐ Use PSpice and Probe in phasor analysis (optional)

## 12–1 ▪ INTRODUCTION TO PHASORS

***Phasors provide a graphic means for representing quantities that have both magnitude and direction (angular position). Phasors are especially useful for representing sine waves in terms of their magnitude and phase angle and also for analysis of reactive circuits studied in later chapters.***

***After completing this section, you should be able to***

- **Use a phasor to represent a sine wave**
  - ☐ Define *phasor*
  - ☐ Explain how phasors are related to the sine wave formula
  - ☐ Draw a phasor diagram
  - ☐ Discuss angular velocity

A **phasor** is a graphic representation of the magnitude and angular position of a time-varying quantity. Examples of phasors are shown in Figure 12–1. The length of the phasor "arrow" represents the magnitude of a quantity. The angle, $\theta$ (relative to 0°), represents the angular position, as shown in part (a). The specific phasor example in part (b) has a magnitude of 2 and a phase angle of 45°. The phasor in part (c) has a magnitude of 3 and a phase angle of 180°. The phasor in part (d) has a magnitude of 1 and a phase angle of −45° (or +315°).

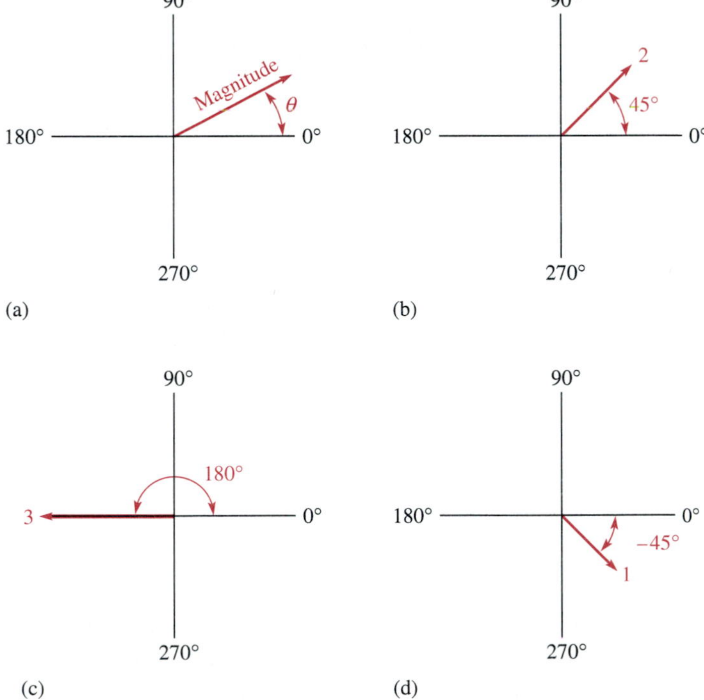


FIGURE 12–1
*Examples of phasors.*

### Phasor Representation of a Sine Wave

A full cycle of a sine wave can be represented by rotation of a phasor through 360 degrees.

> **The instantaneous value of the sine wave at any point is equal to the vertical distance from the tip of the phasor to the horizontal axis.**

Figure 12–2 shows how the phasor traces out the sine wave as it goes from 0° to 360°. You can relate this concept to the rotation in an ac generator (refer to Chapter 11).

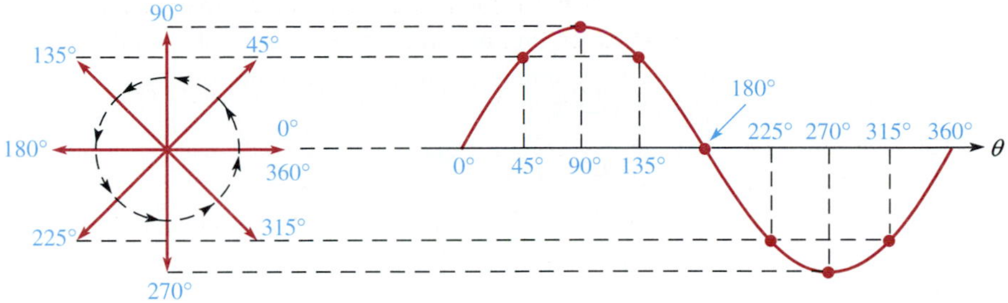


**FIGURE 12–2**
***Sine wave represented by rotational phasor motion.***

Notice in Figure 12–2 that the length of the phasor is equal to the peak value of the sine wave (observe the 90° and the 270° points). The angle of the phasor measured from 0° is the corresponding angular point on the sine wave.

## Phasors and the Sine Wave Formula

Let's examine a phasor representation at one specific angle. Figure 12–3 shows a voltage phasor at an angular position of 45° and the corresponding point on the sine wave. The instantaneous value of the sine wave at this point is related to both the position and the length of the phasor. As previously mentioned, the vertical distance from the phasor tip down to the horizontal axis represents the instantaneous value of the sine wave at that point.

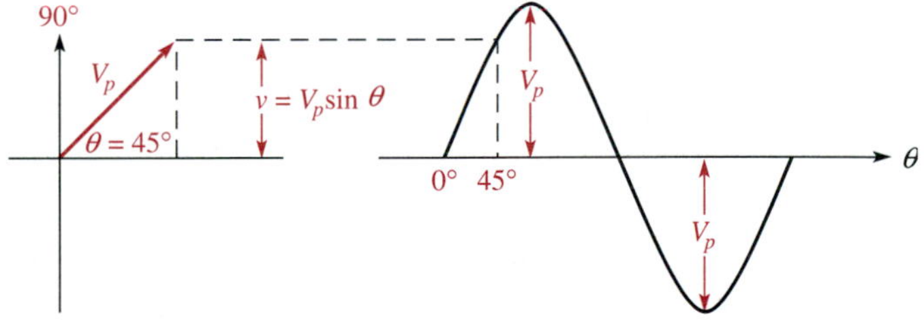


**FIGURE 12–3**
***Right triangle derivation of sine wave formula.***

Notice that when a vertical line is drawn from the phasor tip down to the horizontal axis, a right triangle is formed as shown in Figure 12–3. The length of the phasor is the hypotenuse of the triangle, and the vertical projection is the opposite side. From trigonometry,

> **The opposite side of a right triangle is equal to the hypotenuse times the sine of the angle $\theta$.**

In this case, the length of the phasor is the peak value of the sine wave voltage, $V_p$. Thus, the opposite side of the triangle, which is the instantaneous value, can be expressed as $v = V_p\sin\theta$. Recall that this formula is the one stated in Chapter 11 for calculating instantaneous sine wave values. Of course, this also applies to a sinusoidal current.

## Positive and Negative Phasor Angles

The position of a phasor at any instant can be expressed as a positive angle, as you have seen, or as an equivalent negative angle. Positive angles are measured counterclockwise from 0°. Negative angles are measured clockwise from 0°. For a given positive angle $\theta$, the corresponding negative angle is $\theta - 360°$, as illustrated in Figure 12–4(a). In part (b), a specific example is shown. The angle of the phasor in this case can be expressed as +225° or −135°.

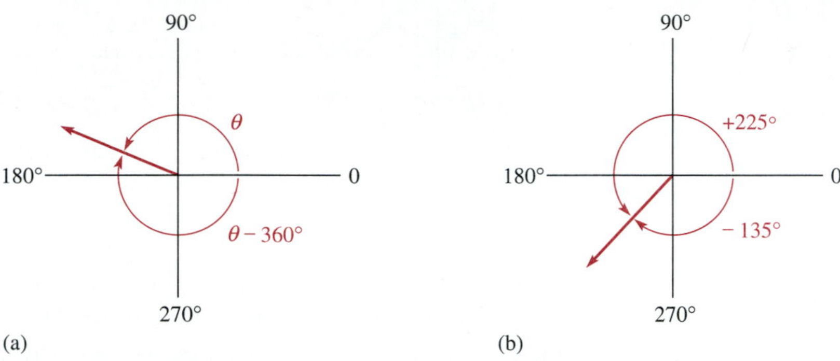


**FIGURE 12–4**
***Positive and negative phasor angles.***

**EXAMPLE 12–1** For the phasor in each part of Figure 12–5, determine the instantaneous sine wave value. Also express each positive angle shown as an equivalent negative angle. The length of each phasor represents the peak value of the sine wave.

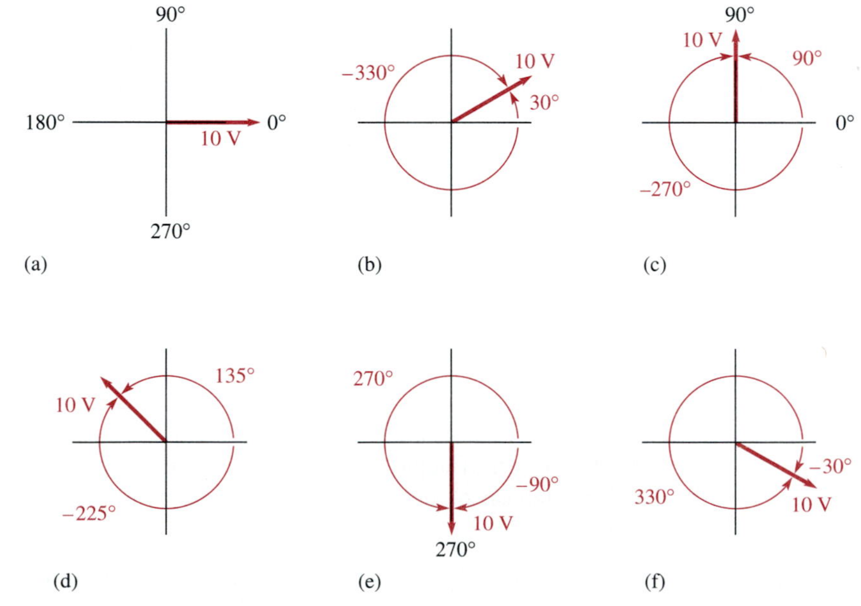


**FIGURE 12–5**

***Solution***

**(a)** $\theta = 0°$

$v = 10 \sin 0° = 10(0) = 0 \text{ V}$

**(b)** $\theta = 30° = \theta - 360° = 30° - 360° = -330°$
$v = 10 \sin 30° = 10(0.5) = 5\text{ V}$

**(c)** $\theta = 90° = -270°$
$v = 10 \sin 90° = 10(1) = 10\text{ V}$

**(d)** $\theta = 135° = -225°$
$v = 10 \sin 135° = 10(0.707) = 7.07\text{ V}$

**(e)** $\theta = 270° = -90°$
$v = 10 \sin 270° = 10(-1) = -10\text{ V}$

**(f)** $\theta = 330° = -30°$
$v = 10 \sin 330° = 10(-0.5) = -5\text{ V}$

The equivalent negative angles are shown in Figure 12–5.

***Related Exercise*** If a phasor is at 45° and its length represents 15 V rms, what is the instantaneous sine wave value?

## Phasor Diagrams

A phasor diagram can be used to show the relative relationship of two or more sine waves of the same frequency. A phasor in a fixed position represents a complete sine wave, because once the phase angle between two or more sine waves of the same frequency is established, it remains constant throughout the cycles. For example, the two sine waves in Figure 12–6(a) can be represented by a phasor diagram, as shown in part (b). As you can see, sine wave $B$ leads sine wave $A$ by 30°.

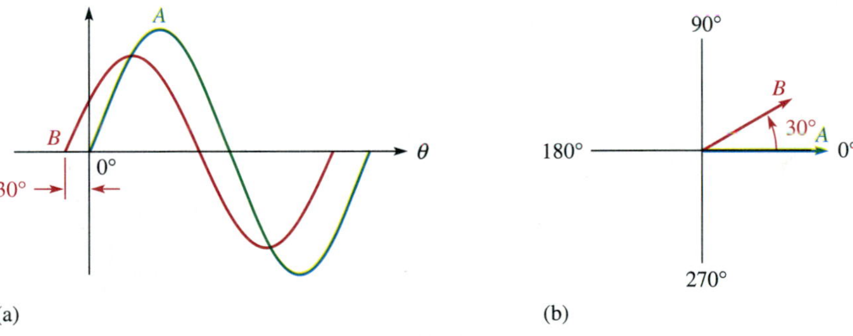


**FIGURE 12–6**
***Example of a phasor diagram.***

**EXAMPLE 12–2** Use a phasor diagram to represent the sine waves in Figure 12–7.

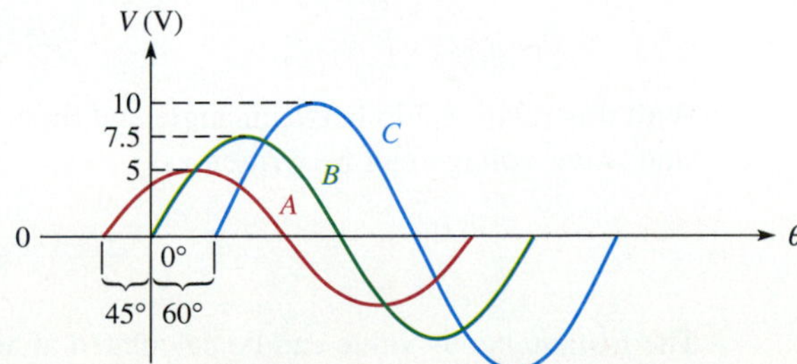


**FIGURE 12–7**

***Solution*** The phasor diagram representing the sine waves is shown in Figure 12–8. In this case, the length of each phasor represents the peak value of the sine wave.

FIGURE 12–8

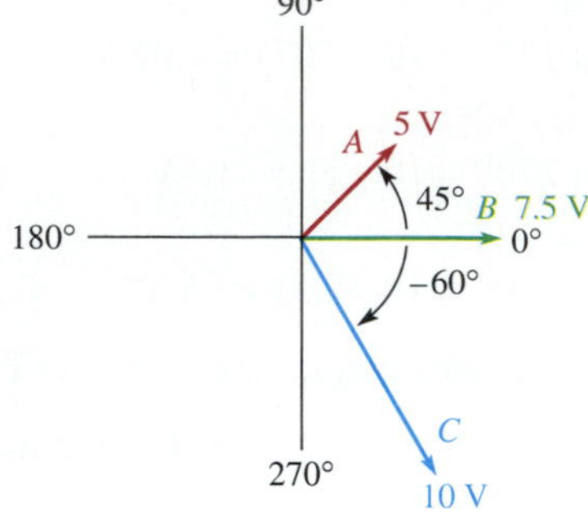


***Related Exercise*** Describe a phasor to represent a 5 V rms sine wave that lags sine wave $C$ in Figure 12–7 by 25°.

## Angular Velocity of a Phasor

As you have seen, one cycle of a sine wave is traced out when a phasor is rotated through 360 degrees. The faster it is rotated, the faster the sine wave cycle is traced out. Thus, the period and frequency are related to the velocity of rotation of the phasor. The velocity of rotation is called the **angular velocity** and is designated ω (the small Greek letter omega).

When a phasor rotates through 360 degrees or $2\pi$ radians, one complete cycle is traced out. Therefore, the time required for the phasor to go through $2\pi$ radians is the period of the sine wave. Because the phasor rotates through $2\pi$ radians in a time equal to the period $T$, the angular velocity can be expressed as

$$\omega = \frac{2\pi}{T}$$

Since $f = 1/T$,

$$\omega = 2\pi f \qquad \textbf{(12–1)}$$

When a phasor is rotated at an angular velocity $\omega$, then $\omega t$ is the angle through which the phasor has passed at any instant. Therefore, the following relationship can be stated:

$$\theta = \omega t \qquad \textbf{(12–2)}$$

With this relationship between angle and time, the equation for the instantaneous value of sine wave voltage can be written as

$$v = V_p \sin \omega t \qquad \textbf{(12–3)}$$

The instantaneous value can be calculated at any point in time along the sine wave curve if the frequency and peak value are known. The unit of $\omega t$ is the radian.

**EXAMPLE 12–3** What is the value of a sine wave voltage at 3 μs from the positive-going zero crossing when $V_p = 10$ V and $f = 50$ kHz?

***Solution***
$$v = V_p \sin \omega t = V_p \sin 2\pi ft$$
$$= 10 \sin[2\pi(50 \times 10^3 \text{ rad/s})(3 \times 10^{-6} \text{ s})] = 8.09 \text{ V}$$

***Related Exercise*** What is the value of a sine wave voltage at 12 μs from the positive-going zero crossing when $V_p = 50$ V and $f = 10$ kHz?

**SECTION 12–1 REVIEW**

1. What is a phasor?
2. What is the angular velocity of a phasor representing a sine wave with a frequency of 1500 Hz?
3. A certain phasor has an angular velocity of 628 rad/s. To what frequency does this correspond?
4. Sketch a phasor diagram to represent the two sine waves in Figure 12–9. Use peak values.

FIGURE 12–9

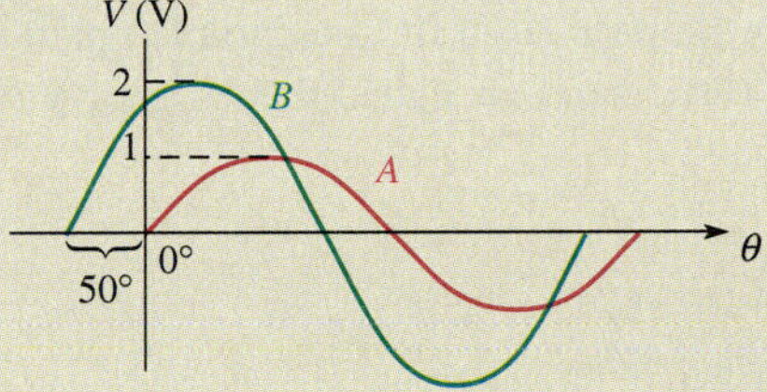


## 12–2 ■ THE COMPLEX NUMBER SYSTEM

***Complex numbers allow mathematical operations with phasor quantities and are useful in analysis of ac circuits. With the complex number system, you can add, subtract, multiply, and divide quantities that have both magnitude and angle, such as sine waves and other ac circuit quantities that will be studied later.***

***After completing this section, you should be able to***

- **Use complex numbers to express phasor quantities**
  - ☐ Describe the complex plane
  - ☐ Represent a point on the complex plane
  - ☐ Discuss real and imaginary numbers

### Positive and Negative Numbers

Positive numbers can be represented by points to the right of the origin on the horizontal axis of a graph, and negative numbers can be represented by points to the left of the origin, as illustrated in Figure 12–10(a). Also, positive numbers can be represented by points on the vertical axis above the origin, and negative numbers can be represented by points below the origin, as shown in Figure 12–10(b).

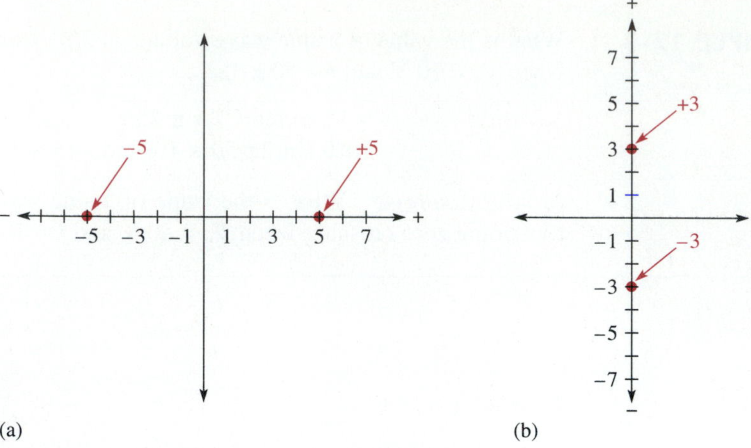


FIGURE 12–10
*Graphic representation of positive and negative numbers.*

## The Complex Plane

To distinguish between values on the horizontal axis and values on the vertical axis, a **complex plane** is used. In the complex plane, the horizontal axis is called the *real axis,* and the vertical axis is called the *imaginary axis,* as shown in Figure 12–11.

FIGURE 12–11
*The complex plane.*

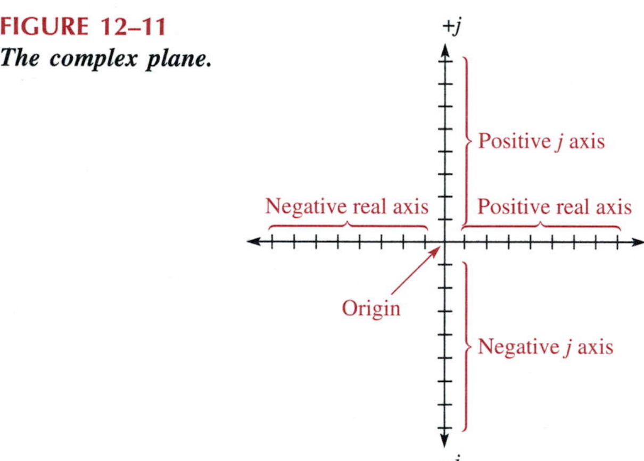


In electrical circuit work, a $\pm j$ prefix is used to designate numbers that lie on the imaginary axis in order to distinguish them from numbers lying on the real axis. This prefix is known as the *j operator.* In mathematics, an *i* is used instead of a *j*, but in electric circuits, the *i* can be confused with instantaneous current, so *j* is used.

## Angular Position on the Complex Plane

Angular positions can be represented on the complex plane, as shown in Figure 12–12. The positive real axis represents zero degrees. Proceeding counterclockwise, the $+j$ axis represents 90°, the negative real axis represents 180°, the $-j$ axis is the 270° point, and, after a full rotation of 360°, we are back to the positive real axis. Notice that the plane is sectioned into four quadrants.

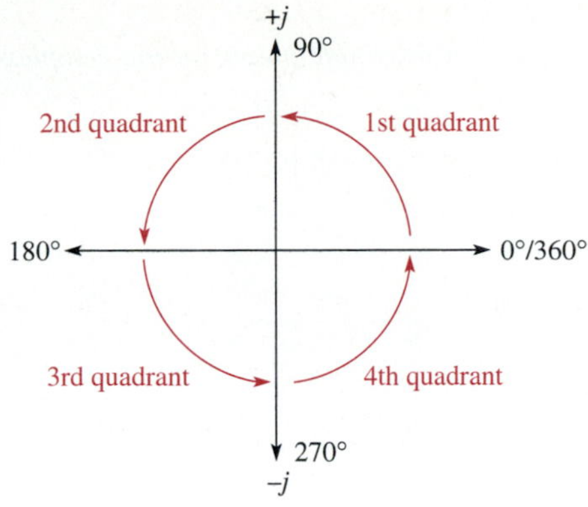


**FIGURE 12–12**
*Angles on the complex plane.*

## Representing a Point on the Complex Plane

A point located on the complex plane can be classified as real, imaginary ($\pm j$), or a combination of the two. For example, a point located 4 units from the origin on the positive real axis is the positive real number, +4, as shown in Figure 12–13(a). A point 2 units from the origin on the negative real axis is the negative real number, −2, as shown in part (b). A point on the $+j$ axis 6 units from the origin, as in part (c), is the positive **imaginary number,** $+j6$. Finally, a point 5 units along the $-j$ axis is the negative imaginary number, $-j5$, as in part (d).

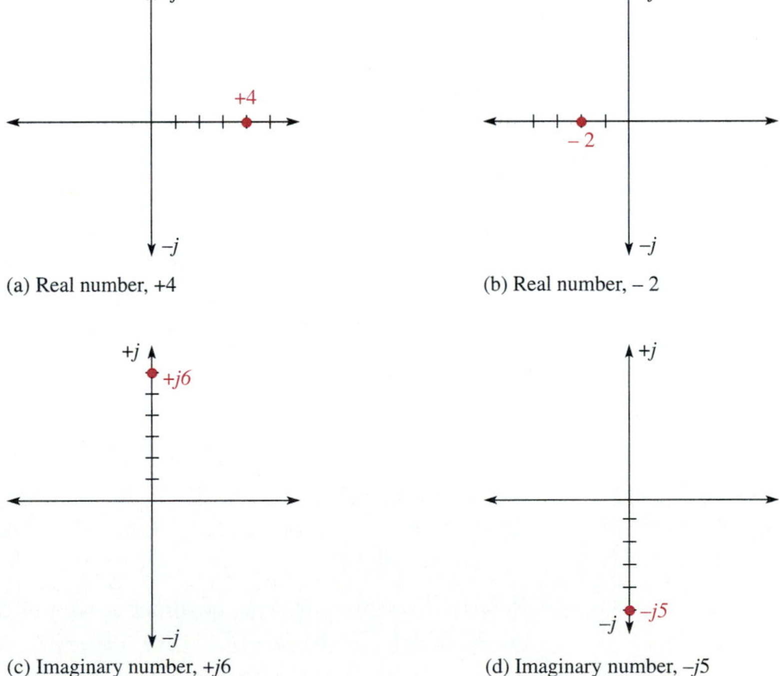


**FIGURE 12–13**
*Real and imaginary (j) numbers on the complex plane.*

When a point lies not on any axis but somewhere in one of the four quadrants, it is a complex number and can be defined by its coordinates. For example, in Figure 12–14, the point located in the first quadrant has a real value of +4 and a $j$ value of $+j4$ and is expressed as +4, $+j4$. The point located in the second quadrant has coordinates −3 and $+j2$. The point located in the third quadrant has coordinates −3 and $-j5$. The point located in the fourth quadrant has coordinates of +6 and $-j4$.

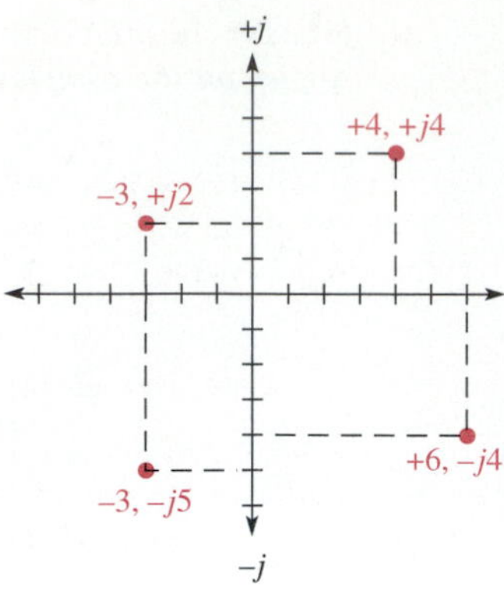


FIGURE 12–14
*Coordinate points on the complex plane.*

**EXAMPLE 12–4**

**(a)** Locate the following points on the complex plane: 7, *j*5; 5, −*j*2; −3.5, *j*1; and −5.5, −*j*6.5.

**(b)** Determine the coordinates for each point in Figure 12–15.

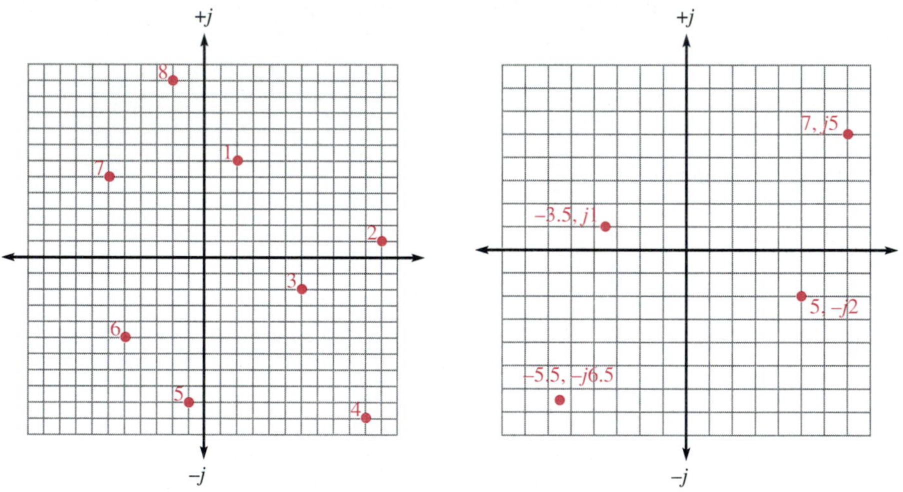


FIGURE 12–15

FIGURE 12–16

***Solution***

**(a)** See Figure 12–16.

**(b)**

| | | | |
|---|---|---|---|
| 1: 2, *j*6 | 2:11, *j*1 | 3: 6, −*j*2 | 4: 10, −*j*10 |
| 5: −1, −*j*9 | 6: −5, −*j*5 | 7: −6, *j*5 | 8: −2, *j*11 |

***Related Exercise*** In what quadrant is each of the following points located?

**(a)** +2.5, +*j*1  **(b)** 7, −*j*5  **(c)** −10, −*j*5  **(d)** −11, +*j*6.8

## Value of *j*

If we multiply the positive real value of +2 by *j*, the result is +*j*2. This multiplication has effectively moved the +2 through a 90° angle to the +*j* axis. Similarly, multiplying +2 by −*j* rotates it −90° to the −*j* axis.

Mathematically, the $j$ operator has a value of $\sqrt{-1}$. If $+j2$ is multiplied by $j$, we get

$$j^2 2 = (\sqrt{-1})(\sqrt{-1})(2) = (-1)(2) = -2$$

This calculation effectively places the value on the negative real axis. Therefore, multiplying a positive real number by $j^2$ converts it to a negative real number, which, in effect, is a rotation of 180° on the complex plane. These operations are illustrated in Figure 12–17.

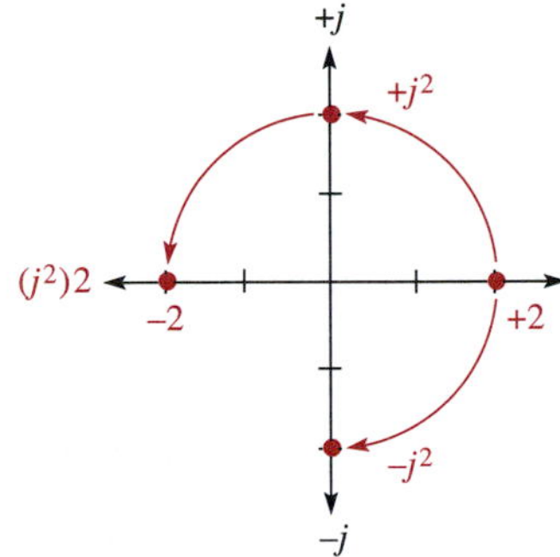


FIGURE 12–17
*Effect of the j operator on location of a number on the complex plane.*

SECTION 12–2 REVIEW

1. Locate the following points on the complex plane:
   (a) +3 (b) −4 (c) $+j1$
2. What is the angular difference between the following numbers:
   (a) +4 and $+j4$ (b) $+j6$ and −6 (c) $+j2$ and $-j2$

## 12–3 ■ RECTANGULAR AND POLAR FORMS

***The rectangular form and the polar form are two forms of complex numbers that are used to represent phasor quantities. Each has certain advantages when used in circuit analysis, depending on the particular application.***

***After completing this section, you should be able to***

- **Represent phasors in two complex forms**
  - ☐ Show how to represent a phasor in rectangular form
  - ☐ Show how to represent a phasor in polar form
  - ☐ Convert between rectangular and polar forms

As you know, a phasor quantity contains both magnitude and phase. In this text, italic letters such as $V$ and $I$ are used to represent magnitude only, and boldface nonitalic letters such as **V** and **I** are used to represent complete phasor quantities. Other circuit quantities that can be expressed in phasor form will be studied in later chapters.

### Rectangular Form

A phasor quantity is represented in **rectangular form** by the algebraic sum of the real value of the coordinate and the $j$ value of the coordinate. An "arrow" drawn from the origin to the coordinate point in the complex plane is used to represent graphically the phasor quantity. Examples of phasor quantities are $1 + j2$, $5 - j3$, $-4 + j4$, and $-2 - j6$, which are shown on the complex plane in Figure 12–18. As you can see, the rectangular coordinates describe the phasor in terms of its values projected onto the real axis and the $j$ axis.

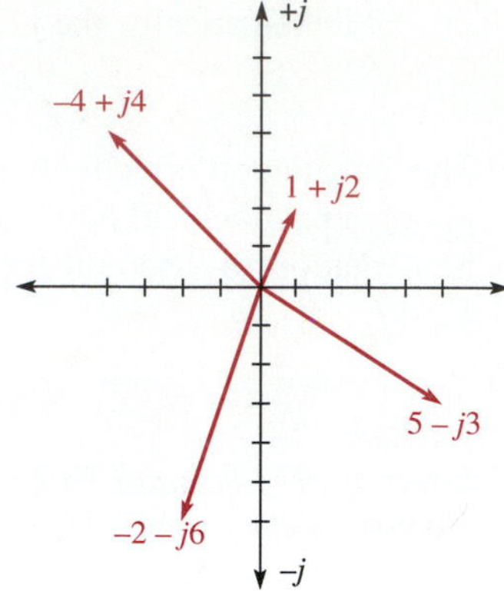


FIGURE 12–18
***Examples of phasors specified by rectangular coordinates.***

## Polar Form

Phasor quantities can also be expressed in **polar form,** which consists of the phasor magnitude and the angular position relative to the positive real axis. Examples are $2\angle 45°$, $5\angle 120°$, and $8\angle -30°$. The first number is the magnitude, and the symbol $\angle$ precedes the value of the angle. Figure 12–19 shows these phasors on the complex plane. The length of the phasor, of course, represents the magnitude of the quantity. Keep in mind that for every phasor expressed in polar form, there is also an equivalent expression in rectangular form.

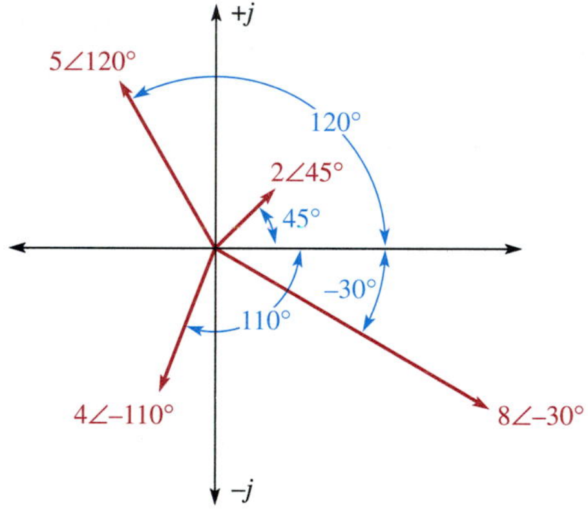


FIGURE 12–19
***Examples of phasors specified by polar values.***

## Conversion from Rectangular to Polar Form

Most scientific calculators have provisions for conversion between rectangular and polar forms. However, the basic conversion method is presented here so that you will understand the mathematical procedure.

A phasor can exist in any of the four quadrants of the complex plane, as indicated in Figure 12–20. The phase angle $\theta$ in each case is measured relative to the positive real axis (0°) as shown.

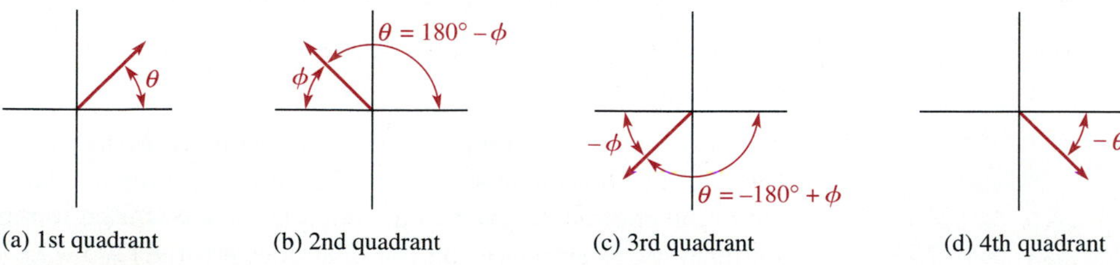


FIGURE 12–20
***All possible phasor quadrant locations. $\theta$ is the angle of the phasor relative to the positive real axis in each case, and $\phi$ is the angle in the 2nd and 3rd quadrants relative to the negative real axis.***

The first step to convert from rectangular form to polar form is to determine the magnitude of the phasor. A phasor can be visualized as forming a right triangle in the complex plane, as indicated in Figure 12–21, for each quadrant location. The horizontal side of the triangle is the real value, $A$, and the vertical side is the $j$ value, $B$. The hypotenuse of the triangle is the length of the phasor, $C$, representing the magnitude, and can be expressed, using the Pythagorean theorem, as

$$C = \sqrt{A^2 + B^2} \qquad \textbf{(12–4)}$$

Next, the angle $\theta$ indicated in parts (a) and (d) of Figure 12–21 is expressed as an inverse tangent function.

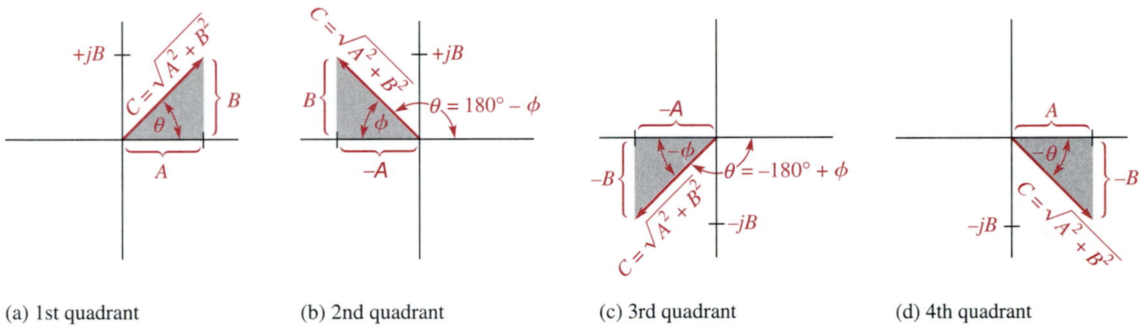


**FIGURE 12–21**
***Right angle relationships in the complex plane.***

$$\theta = \tan^{-1}\left(\frac{\pm B}{A}\right) \qquad \textbf{(12–5)}$$

The angle $\theta$ indicated in parts (b) and (c) of Figure 12–21 is

$$\theta = \pm 180^\circ \mp \phi$$

$$\theta = \pm 180^\circ \mp \tan^{-1}\left(\frac{B}{A}\right) \qquad \textbf{(12–6)}$$

In each case the appropriate signs must be used in the calculation. Note that $\tan^{-1}$ is INV TAN on some calculators.

The general formula for converting from rectangular to polar is as follows:

$$\pm A \pm jB = C\angle\pm\theta \qquad \textbf{(12–7)}$$

The following example illustrates the conversion procedure.

**EXAMPLE 12–5**

Convert the following complex numbers from rectangular form to polar form:
**(a)** $8 + j6$ **(b)** $10 - j5$ **(c)** $-12 - j18$ **(d)** $-7 + j10$

***Solution***

**(a)** The magnitude of the phasor represented by $8 + j6$ is

$$C = \sqrt{A^2 + B^2} = \sqrt{8^2 + 6^2} = \sqrt{100} = 10$$

Since the phasor is in the first quadrant, use Equation (12–5). The angle is

$$\theta = \tan^{-1}\left(\frac{\pm B}{A}\right) = \tan^{-1}\left(\frac{6}{8}\right) = 36.9°$$

$\theta$ is the angle relative to the positive real axis. The complete polar expression for this phasor is

$$\mathbf{C} = 10\angle 36.9°$$

Boldface, nonitalic letters represent phasor quantities.

**(b)** The magnitude of the phasor represented by $10 - j5$ is

$$C = \sqrt{10^2 + (-5)^2} = \sqrt{125} = 11.2$$

Since the phasor is in the fourth quadrant, use Equation (12–5). The angle is

$$\theta = \tan^{-1}\left(\frac{-5}{10}\right) = -26.6°$$

$\theta$ is the angle relative to the positive real axis. The complete polar expression for this phasor is

$$\mathbf{C} = 11.2\angle -26.6°$$

**(c)** The magnitude of the phasor represented by $-12 - j18$ is

$$C = \sqrt{(-12)^2 + (-18)^2} = \sqrt{468} = 21.6$$

Since the phasor is in the third quadrant, use Equation (12–6). The angle is

$$\theta = -180° + \tan^{-1}\left(\frac{18}{12}\right) = -180° + 56.3° = -123.7°$$

The complete polar expression for this phasor is

$$\mathbf{C} = 21.6\angle -123.7°$$

**(d)** The magnitude of the phasor represented by $-7 + j10$ is

$$C = \sqrt{(-7)^2 + 10^2} = \sqrt{149} = 12.2$$

Since the angle is in the second quadrant, use Equation (12–6). The angle is

$$\theta = 180° - \tan^{-1}\left(\frac{10}{7}\right) = 180° - 55° = 125°$$

The complete polar expression for this phasor is

$$\mathbf{C} = 12.2\angle 125°$$

Calculator sequences may vary depending on the type of calculator you use. The typical calculator sequences are

**(a)** [8] [x⇆y] [6] [2nd] [INV] [P–R] for angle, then [x⇆y] for magnitude
**(b)** [1] [0] [x⇆y] [5] [+/–] [2nd] [INV] [P–R] for angle, then [x⇆y] for magnitude
**(c)** [1] [2] [+/–] [x⇆y] [1] [8] [+/–] [2nd] [INV] [P–R] for angle, then [x⇆y] for magnitude
**(d)** [7] [+/–] [x⇆y] [1] [0] [2nd] [INV] [P–R] for angle, then [x⇆y] for magnitude

***Related Exercise*** Convert $18 + j23$ to polar form.

## Conversion from Polar to Rectangular Form

The polar form gives the magnitude and angle of a phasor quantity, as indicated in Figure 12–22.

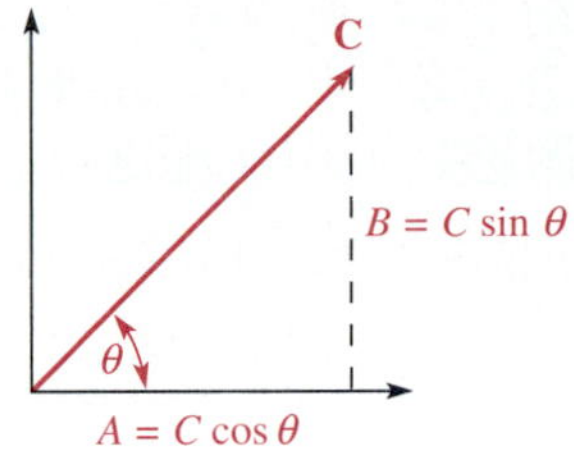


FIGURE 12–22
*Polar components of a phasor.*

To get the rectangular form, sides $A$ and $B$ of the triangle must be found, using the rules from trigonometry stated below:

$$A = C \cos(\pm\theta) \quad \textbf{(12–8)}$$

$$B = C \sin(\pm\theta) \quad \textbf{(12–9)}$$

The general polar-to-rectangular conversion formula is as follows:

$$C\angle\pm\theta = C \cos(\pm\theta) + jC \sin(\pm\theta) = \pm A \pm jB \quad \textbf{(12–10)}$$

The following example demonstrates this conversion.

**EXAMPLE 12–6** Convert the following polar quantities to rectangular form:
**(a)** $10\angle 30°$ **(b)** $200\angle -45°$ **(c)** $4\angle 135°$

*Solution*

**(a)** The real part of the phasor represented by $10\angle 30°$ is

$$A = C \cos(\pm\theta) = 10 \cos 30° = 10(0.866) = 8.66$$

The $j$ part of this phasor is

$$jB = jC \sin(\pm\theta) = j10 \sin 30° = j10(0.5) = j5$$

The complete rectangular expression is

$$A + jB = 8.66 + j5$$

**(b)** The real part of the phasor represented by $200\angle -45°$ is

$$A = 200 \cos(-45°) = 200(0.707) = 141$$

The $j$ part is

$$jB = j200 \sin(-45°) = j200(-0.707) = -j141$$

The complete rectangular expression is

$$A + jB = 141 - j141$$

**(c)** The real part of the phasor represented by $4\angle 135°$ is

$$A = 4 \cos 135° = 4(-0.707) = -2.83$$

The $j$ part is

$$jB = j4 \sin 135° = 4(0.707) = 2.83$$

The complete rectangular expression is

$$A + jB = -2.83 + j2.83$$

The typical calculator sequences are

**(a)** [1] [0] [x⇌y] [3] [0] [2nd] [P–R] for $j$ part, then [x⇌y] for real part
**(b)** [2] [0] [0] [x⇌y] [4] [5] [2nd] [P–R] for $j$ part, then [x⇌y] for real part
**(c)** [4] [x⇌y] [1] [3] [5] [2nd] [P–R] for $j$ part, then [x⇌y] for real part

***Related Exercise*** Convert $78\angle{-26^\circ}$ to rectangular form.

**SECTION 12–3 REVIEW**

1. Name the two parts of a complex number in rectangular form.
2. Name the two parts of a complex number in polar form.
3. Convert $2 + j2$ to polar form. In which quadrant does this phasor lie?
4. Convert $5\angle{-45^\circ}$ to rectangular form. In which quadrant does this phasor lie?

## 12–4 ▪ MATHEMATICAL OPERATIONS

***Complex numbers can be added, subtracted, multiplied, and divided.***

***After completing this section, you should be able to***

- **Do mathematical operations with complex numbers**
  - ☐ Add complex numbers
  - ☐ Subtract complex numbers
  - ☐ Multiply complex numbers
  - ☐ Divide complex numbers
  - ☐ Apply complex numbers to sine waves

### Addition

Complex numbers must be in rectangular form in order to add them. The rule is

**Add the real parts of each complex number to get the real part of the sum. Then add the $j$ parts of each complex number to get the $j$ part of the sum.**

**EXAMPLE 12–7** Add the following sets of complex numbers:
**(a)** $8 + j5$ and $2 + j1$  **(b)** $20 - j10$ and $12 + j6$

***Solution***
**(a)** $(8 + j5) + (2 + j1) = (8 + 2) + j(5 + 1) = 10 + j6$
**(b)** $(20 - j10) + (12 + j6) = (20 + 12) + j(-10 + 6) = 32 + j(-4) = 32 - j4$

***Related Exercise*** Add $5 - j11$ and $-6 + j3$.

### Subtraction

As in addition, the numbers must be in rectangular form to be subtracted. The rule is

**Subtract the real parts of the numbers to get the real part of the difference, and subtract the *j* parts of the numbers to get the *j* part of the difference.**

**EXAMPLE 12–8** Perform the following subtractions:
**(a)** Subtract $1 + j2$ from $3 + j4$. **(b)** Subtract $10 - j8$ from $15 + j15$.

*Solution*

**(a)** $(3 + j4) - (1 + j2) = (3 - 1) + j(4 - 2) = 2 + j2$

**(b)** $(15 + j15) - (10 - j8) = (15 - 10) + j[15 - (-8)] = 5 + j23$

*Related Exercise* Subtract $3.5 - j4.5$ from $-10 - j9$.

### Multiplication

Multiplication of two complex numbers in rectangular form is accomplished by multiplying, in turn, each term in one number by both terms in the other number and then combining the resulting real terms and the resulting *j* terms (recall that $j \times j = -1$). As an example,

$$(5 + j3)(2 - j4) = 10 - j20 + j6 + 12 = 22 - j14$$

Multiplication of two complex numbers is easier when both numbers are in polar form. The rule is

**Multiply the magnitudes, and add the angles algebraically.**

**EXAMPLE 12–9** Perform the following multiplications:
**(a)** $10\angle 45^\circ$ times $5\angle 20^\circ$ **(b)** $2\angle 60^\circ$ times $4\angle -30^\circ$

*Solution*

**(a)** $(10\angle 45^\circ)(5\angle 20^\circ) = (10)(5)\angle(45^\circ + 20^\circ) = 50\angle 65^\circ$

**(b)** $(2\angle 60^\circ)(4\angle -30^\circ) = (2)(4)\angle[60^\circ + (-30^\circ)] = 8\angle 30^\circ$

*Related Exercise* Multiply $50\angle 10^\circ$ times $30\angle -60^\circ$.

### Division

Division of two complex numbers in rectangular form is accomplished by multiplying both the numerator and the denominator by the complex conjugate of the denominator and then combining terms and simplifying. The complex conjugate of a number is found by changing the sign of the *j* term. As an example,

$$\frac{10 + j5}{2 + j4} = \frac{(10 + j5)(2 - j4)}{(2 + j4)(2 - j4)} = \frac{20 - j30 + 20}{4 + 16} = \frac{40 - j30}{20} = 2 - j1.5$$

Like multiplication, division is easier when the numbers are in polar form. The rule is

**Divide the magnitude of the numerator by the magnitude of the denominator to get the magnitude of the quotient, and subtract the denominator angle from the numerator angle to get the angle of the quotient.**

**EXAMPLE 12–10** Perform the following divisions:

**(a)** Divide $100\angle 50°$ by $25\angle 20°$. **(b)** Divide $15\angle 10°$ by $3\angle -30°$.

*Solution*

**(a)** $\dfrac{100\angle 50°}{25\angle 20°} = \left(\dfrac{100}{25}\right)\angle(50° - 20°) = 4\angle 30°$

**(b)** $\dfrac{15\angle 10°}{3\angle -30°} = \left(\dfrac{15}{3}\right)\angle[10° - (-30°)] = 5\angle 40°$

*Related Exercise* Divide $24\angle -30°$ by $6\angle 12°$.

## Application of Complex Numbers to Sine Waves

Since sine waves can be represented by phasors, they can be described in terms of complex numbers either in rectangular or polar form. For example, four series sine wave voltage sources all with the same frequency are shown in Figure 12–23. The sine waves are graphed in Figure 12–24(a), and the phasor representation is shown in part (b). The total voltage across the load in Figure 12–23 can be determined by first converting each phasor to rectangular form and then adding as follows:

$$\begin{aligned}
\mathbf{V}_{tot} &= \mathbf{V}_1 + \mathbf{V}_2 + \mathbf{V}_3 + \mathbf{V}_4 \\
&= 10\angle 120°\text{ V} + 4\angle 30°\text{ V} + 8\angle -30°\text{ V} + 6\angle -130°\text{ V} \\
&= (-5\text{ V} + j8.66\text{ V}) + (3.46\text{ V} + j2\text{ V}) + (6.93\text{ V} - j4\text{ V}) + (-3.86\text{ V} - j4.60\text{ V}) \\
&= 1.53\text{ V} + j2.06\text{ V} = 2.57\angle 53.4°\text{ V}
\end{aligned}$$

The total voltage has a peak value of 2.57 V and a phase angle of 53.4°.

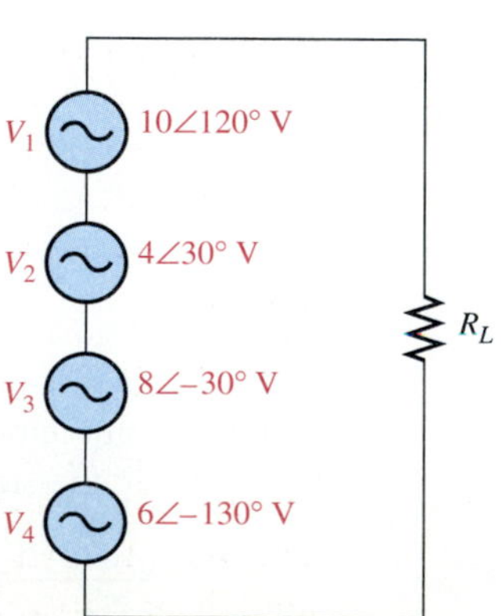


FIGURE 12–23
***Superimposed sine wave sources applied to a load.***

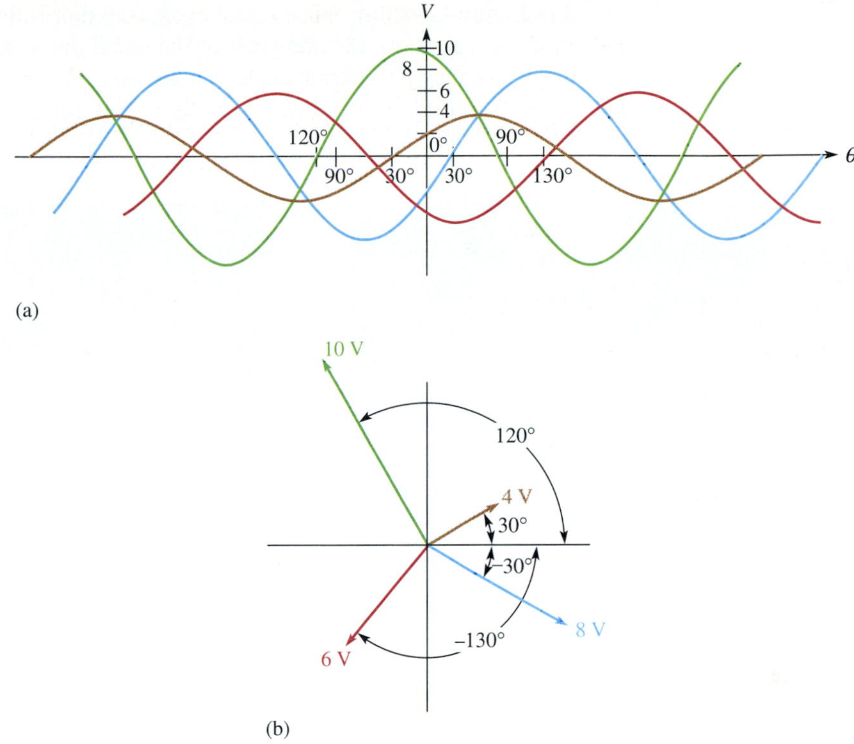


**FIGURE 12–24**

***A phasor diagram representing four out-of-phase sine waves in the circuit of Figure 12–23.***

As you have seen, sine waves can be represented in complex form and can be added, subtracted, multiplied, and divided using the rules that we have discussed. Also, as you will learn later, other electrical quantities, such as capacitive and inductive reactances, impedance, and power, can be described in complex form to ease many circuit analysis problems.

**SECTION 12–4 REVIEW**

1. Add $1 + j2$ and $3 - j1$.
2. Subtract $12 + j18$ from $15 + j25$.
3. Multiply $8\angle 45°$ times $2\angle 65°$.
4. Divide $30\angle 75°$ by $6\angle 60°$.

## 12–5 ■ TECHnology Theory Into Practice

***Phasors and complex numbers are mathematical concepts that are used for the purpose of ac circuit analysis. Although this Tech TIP does not deal directly with phasors or complex numbers, the angular relationships that are measured can be represented by phasors.***

One way to measure the approximate phase angle between two sine waves with the same frequency is use an oscilloscope and convert the horizontal axis into angular divisions. The seconds/division control can be switched off of the calibrated position and varied until there is exactly one-half cycle of the sine wave displayed across the screen as illustrated in Figure 12–25(a). Since a half cycle contains 180°, each of the ten main horizontal divisions represents 18° and each of the small divisions represents 3.6°, as indicated. For better accuracy, a quarter cycle can be used although it is difficult to establish the exact positive peak when the waveform is spread out.

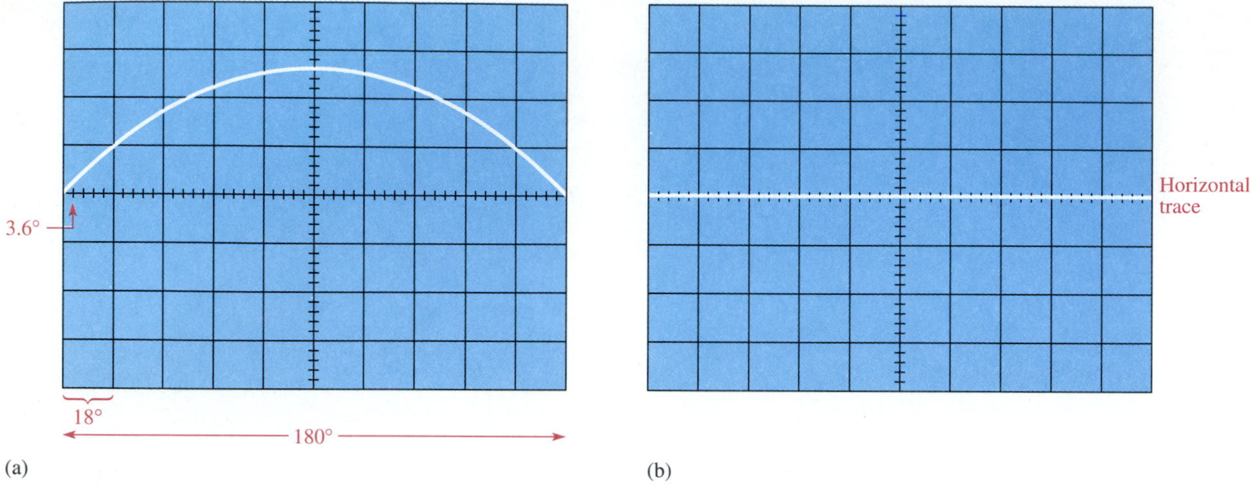


FIGURE 12–25

After the phase scale has been established using one of the sine waves, the traces for both oscilloscope channels must be superimposed on each other and aligned along the horizontal axis to prevent any vertical offset of the waveforms. This is done by touching channel 1 probe to ground (0 V) or by switching the AC/DC channel input switch to the ground (GND) position and adjusting the vertical position to bring it to the center line on the screen, as indicated in Figure 12–25(b). This is repeated for the channel 2 trace.

Next the probes are connected to the two signals and each is ac coupled (input switch set to AC) into the scope. The scope should be triggered on the reference waveform. You may wish to set the amplitudes to an approximately equal height by taking the volts/division controls off of the calibrate position and adjusting each one. This is illustrated in Figure 12–26. The phase angle between the two waveforms can be most easily measured from the zero crossings, as shown. In this example, the angle is 36°.

FIGURE 12–26

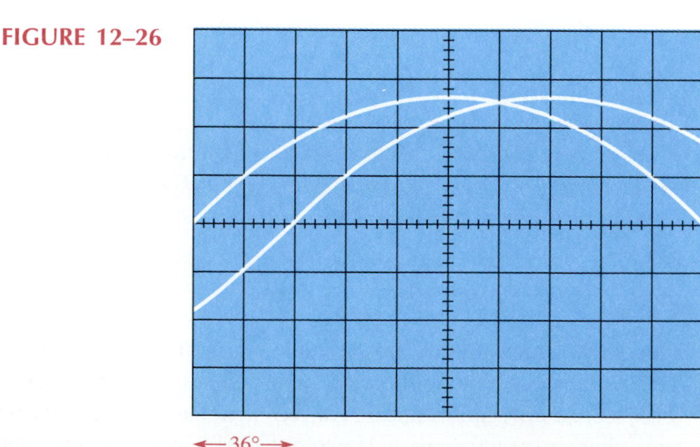

### Phase Angle Measurement

□ Determine the phase angle for each of the scope displays in Figure 12–27. A quarter cycle is shown in part (c).

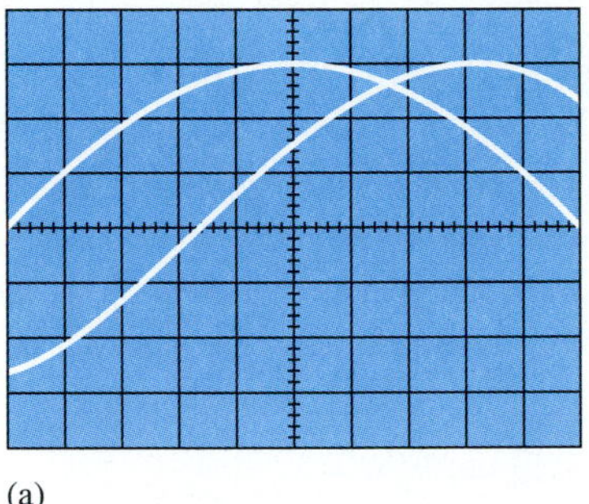
(a)

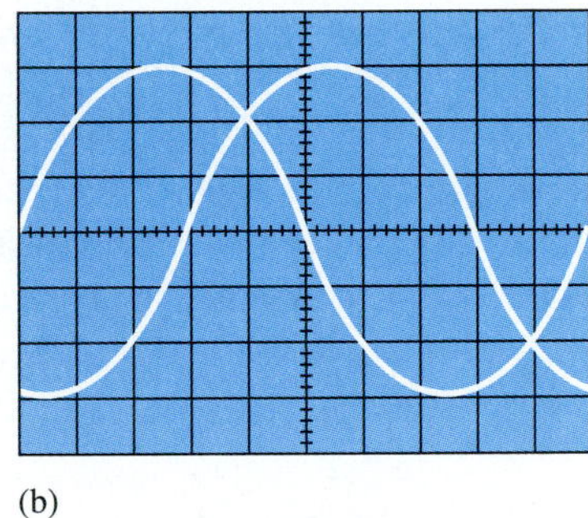
(b)

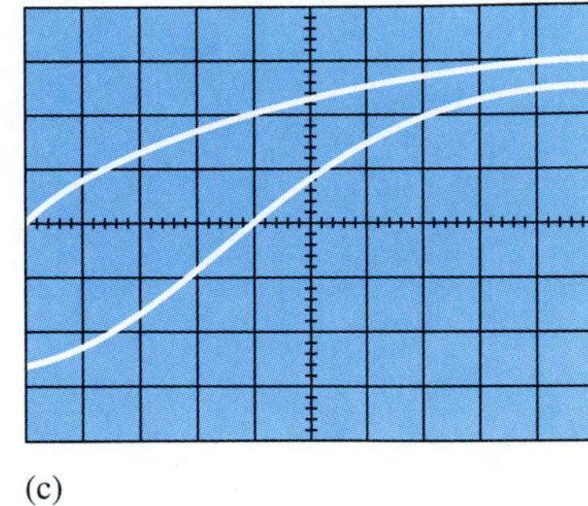
(c)

FIGURE 12–27

**SECTION 12–5 REVIEW**

1. When measuring the phase angle between two waveforms on an oscilloscope, is the seconds/division setting critical?
2. How do you prevent vertical offset of the waveforms when measuring the phase angle?

## 12–6 ■ PHASOR ANALYSIS WITH PSpice AND PROBE

***As you have learned, sinusoidal waveforms can be represented by phasors which contain magnitude and phase information. Phasors are useful for analyzing circuits where all currents and voltages have the same frequency, and PSpice and Probe can be used to obtain the phasor values in a circuit with relative ease.***

***After completing this section, you should be able to***

- **Use PSpice and Probe in phasor analysis**
  - □ Specify ac sources with phase shifts
  - □ Write circuit files that include waveforms with phase shifts
  - □ Use the cursor in Probe to measure ac waveform values

### Transient AC Voltages

In Chapter 11, you learned how to specify some transient values of an ac sinusoidal source so that Probe could display the ac waveform on a time scale. The three ac transient values that you learned are dc offset, peak amplitude, and frequency. In addition to these ac transient values, there are three more that can be specified: delay, damping factor, and phase.

The delay allows you to switch different waveforms on at different times after $t = 0$. The damping factor describes how quickly the waveform dies out, or decays, after it starts up. The larger the damping factor, the more quickly the waveform decays. The phase, expressed in degrees, indicates when the waveform has a positive-going zero crossing relative to a reference waveform that has a positive-going zero crossing at $t = 0$. For now, only the phase is important, so the other two values can be set to zero.

As an example, a 5 V, 1 kHz source labeled $V_1$ between node 1 and node 0 with no dc offset and a positive-going zero crossing at 90° is described by the following PSpice line:

```
V1  1  0  SIN(0  5  1K  0  0  90)
```

Similarly, a 3 V, 5 MHz source called Vin between node 3 and node 4 with no dc offset and a positive-going zero crossing at −30° is written as

```
VIN  3  4  SIN(0  3  5MEG  0  0  30)
```

Note that a positive angle as a transient value means the sine wave crosses zero before the reference waveform while a negative angle means that the sine wave crosses zero after the reference waveform. Thus, a positive phase angle as a transient value means the sine wave leads the reference waveform, while a negative phase angle means the sine wave lags the reference waveform.

## Phasor Analysis in an AC Circuit

In Figure 12–28, the 1 kHz voltage sources are expressed in polar form. From the superposition theorem, the voltage across $R_1$ is the phasor sum of the two source voltages.

**FIGURE 12–28**

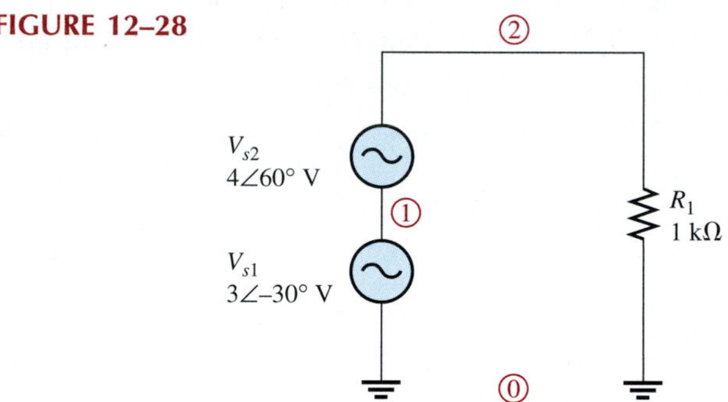


The PSpice circuit file is as follows:

```
Circuit with Two AC Sources
* Demonstration of Phasor Analysis
VS1  1  0  SIN(0  3  1K  0  0  –30)
VS2  2  1  SIN(0  4  1K  0  0  60)
R1  2  0  1K
.TRAN  20u  1m
.PROBE
.END
```

After running PSpice and starting Probe, add the following traces to the display:

```
V(1,0)  V(2,1)  V(2,0)
```

Probe will draw three sine waves showing the phase relationships of the two source voltages and the voltage across $R_1$, which is the phasor sum of $V_{S1}$ and $V_{S2}$.

## Using the Cursors

The magnitude of the phasor sum is 5 V and is easy to read. The phase shift, however, is more difficult to determine. The Cursor option in the menu allows you to take more precise readings than can be obtained by visual inspection. In addition, the Cursor option can be used for differential measurements to find the difference between two points on a waveform. (Windows Probe users should select Tools/Cursor/Display from a sequence of menus.)

When the Cursor option is selected, Probe responds by placing a small box around the marker for the first waveform it plotted and showing a box with the cursor position information in the following format:

```
C1 =    <X1>       <Y1>
C2 =    <X2>       <Y2>
dif =   <X1-X2>  <Y1-Y2>
```

<X1> and <Y1> are the coordinates of cursor C1 and <X2> and <Y2> are the coordinates of cursor C2. Some versions of Probe will label the cursors A1 and A2, but they work the same.

***Moving the Cursors*** Pressing the RIGHT arrow or the LEFT arrow key will cause the cursor C1 to track the waveform indicated by the boxed marker. At the same time, <X1> and <Y1> will change to indicate where on the waveform the cursor is (the $x$ and $y$ coordinates of the cursor). The cursor can be moved rapidly through the waveform by holding an arrow key down, or it can be moved in small increments with short taps on the arrow key.

To move cursor C2 on the waveform, hold down the SHIFT key while pressing the LEFT or RIGHT arrow key. Both cursors will follow the same waveform because both start on the first waveform when Cursor is first selected.

To move cursor C1 to another waveform, hold down the CONTROL key and press either the RIGHT or LEFT arrow key. With each press of the arrow key, the cursor will move to the next waveform in the display.

To move cursor C2 to another waveform, hold down the CONTROL and SHIFT keys and press either the RIGHT or LEFT arrow key. With each press of the arrow key, the cursor will move to the next waveform in the display.

***Changing the X-variable*** So far, Probe has used time as the $x$-axis variable because the .TRAN statement implies the time domain. If you select the X_axis or Plot/X Axis Settings option from the main menu, you will find an option on the submenu called X_variable or Axis Variable which can be used to change the $x$-axis variable. To change the $x$-axis to degrees in the preceding example, select this option, and at the prompt type

```
(Time * 360)/1 ms
```

and press ENTER or mouse-click on OK to get back to the plot. Probe will redraw the $x$-axis and number it from 0 to 400 corresponding to the phase angle in degrees. The phase angle of $V_{S1}$, $V_{S2}$, and the total voltage can be determined using the cursors.

## SECTION 12–6 REVIEW

1. Specify a sinusoidal ac source voltage with a peak value of 6 V, a frequency of 5 kHz, no dc value, and a phase shift of 45° between nodes 1 and 0 in a circuit. Assume the delay and damping factor are 0.
2. Write a circuit file for the circuit in Figure 12–29 to view the 2 kHz voltages across each of the resistors.

FIGURE 12–29

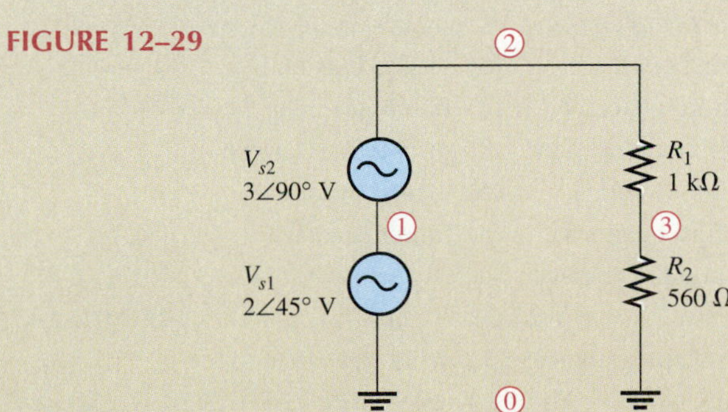

## ▪ SUMMARY

- A phasor represents a time-varying quantity in terms of both magnitude and direction.
- The angular position of a phasor represents the angle of the sine wave, and the length of a phasor represents the amplitude.
- A complex number represents a phasor quantity.
- Complex numbers can be added, subtracted, multiplied, and divided.
- The rectangular form of a complex number consists of a real part and a $j$ part.
- The polar form of a complex number consists of a magnitude and an angle.

## ▪ GLOSSARY

**Angular velocity** The rotational velocity of a phasor which is related to the frequency of the sine wave that the phasor represents.

**Complex plane** An area consisting of four quadrants on which a quantity containing both magnitude and direction can be represented.

**Imaginary number** A number that exists on the vertical axis of the complex plane.

**Phasor** A representation of a sine wave in terms of its magnitude (amplitude) and direction (phase angle).

**Polar form** One form of a complex number made up of a magnitude and an angle.

**Rectangular form** One form of a complex number made up of a real part and an imaginary part.

## ▪ FORMULAS

**(12–1)** $\omega = 2\pi f$

**(12–2)** $\theta = \omega t$

**(12–3)** $v = V_p \sin \omega t$

**(12–4)** $C = \sqrt{A^2 + B^2}$

**(12–5)** $\theta = \tan^{-1}\left(\frac{\pm B}{A}\right)$

**(12–6)** $\theta = \pm 180° \mp \tan^{-1}\left(\frac{B}{A}\right)$

**(12–7)** $\pm A \pm jB = C\angle \pm\theta$

**(12–8)** $A = C\cos(\pm\theta)$

**(12–9)** $B = C\sin(\pm\theta)$

**(12–10)** $C\angle\pm\theta = C\cos(\pm\theta) + jC\sin(\pm\theta) = \pm A \pm jB$

## ▪ SELF-TEST

1. A phasor represents
   **(a)** the magnitude of a quantity **(b)** the magnitude and direction of a quantity
   **(c)** the phase angle **(d)** the length of a quantity
2. A positive angle of 20° is equivalent to a negative angle of
   **(a)** −160° **(b)** −340° **(c)** −70° **(d)** −20°
3. In the complex plane, the number $3 + j4$ is located in the
   **(a)** first quadrant **(b)** second quadrant
   **(c)** third quadrant **(d)** fourth quadrant
4. In the complex plane, $12 - j6$ is located in the
   **(a)** first quadrant **(b)** second quadrant
   **(c)** third quadrant **(d)** fourth quadrant
5. The complex number $5 + j5$ is equivalent to
   **(a)** $5\angle 45°$ **(b)** $25\angle 0°$ **(c)** $7.07\angle 45°$ **(d)** $7.07\angle 135°$
6. The complex number $35\angle 60°$ is equivalent to
   **(a)** $35 + j35$ **(b)** $35 + j60$ **(c)** $17.5 + j30.3$ **(d)** $30.3 + j17.5$
7. $(4 + j7) + (-2 + j9)$ is equal to
   **(a)** $2 + j16$ **(b)** $11 + j11$ **(c)** $-2 + j16$ **(d)** $2 - j2$

8. $(16 - j8) - (12 + j5)$ is equal to
   **(a)** $28 - j13$ **(b)** $4 - j13$ **(c)** $4 - j3$ **(d)** $-4 + j13$
9. $(5\angle 45°)(2\angle 20°)$ is equal to
   **(a)** $7\angle 65°$ **(b)** $10\angle 25°$ **(c)** $10\angle 65°$ **(d)** $7\angle 25°$
10. $(50\angle 10°)/(25\angle 30°)$ is equal to
   **(a)** $25\angle 40°$ **(b)** $2\angle 40°$ **(c)** $25\angle -20°$ **(d)** $2\angle -20°$

## ■ PROBLEMS

### SECTION 12–1 Introduction to Phasors

1. Draw a phasor diagram to represent the sine waves in Figure 12–30.
2. Sketch the sine waves represented by the phasor diagram in Figure 12–31. The phasor lengths represent peak values.

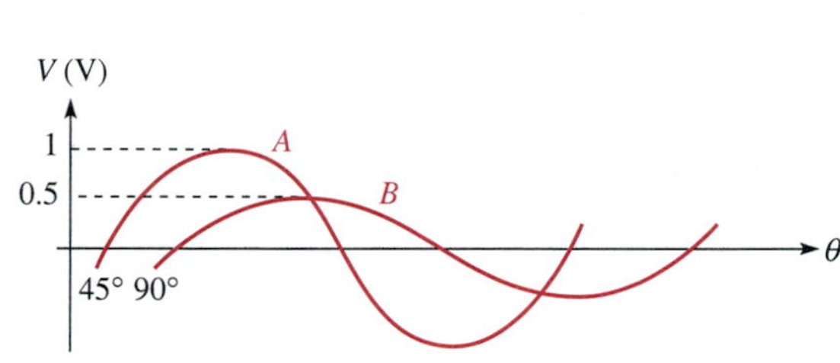


FIGURE 12–30

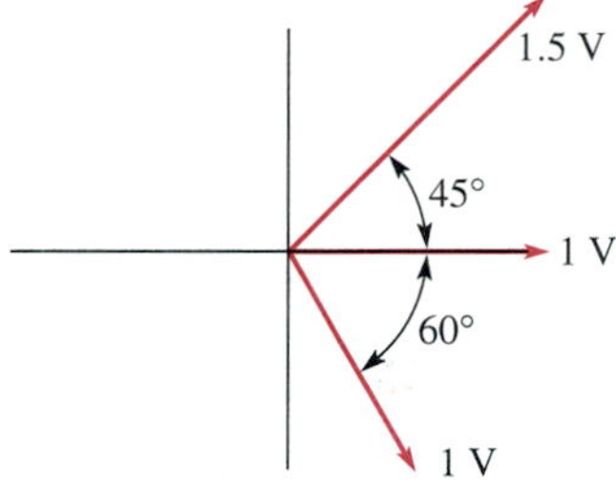


FIGURE 12–31

3. Determine the frequency for each angular velocity:
   **(a)** 60 rad/s **(b)** 360 rad/s **(c)** 2 rad/s **(d)** 1256 rad/s
4. Determine the value of sine wave $A$ in Figure 12–30 at each of the following times, measured from the positive-going zero crossing. Assume the frequency is 5 kHz.
   **(a)** 30 $\mu$s **(b)** 75 $\mu$s **(c)** 125 $\mu$s
5. In Figure 12–30, how many microseconds after the zero crossing does sine wave $A$ reach 0.8 V? Assume the frequency is 5 kHz.

### SECTION 12–2 The Complex Number System

6. Locate the following numbers on the complex plane:
   **(a)** $+6$ **(b)** $-2$ **(c)** $+j3$ **(d)** $-j8$
7. Locate the points represented by each of the following coordinates on the complex plane:
   **(a)** $3, j5$ **(b)** $-7, j1$ **(c)** $-10, -j10$
8. Determine the coordinates of each point having the same magnitude but located 180° away from each point in Problem 7.
9. Determine the coordinates of each point having the same magnitude but located 90° away from those in Problem 7.

### SECTION 12–3 Rectangular and Polar Forms

10. Points on the complex plane are described below. Express each point as a complex number in rectangular form:
    **(a)** 3 units to the right of the origin on the real axis, and up 5 units on the $j$ axis.
    **(b)** 2 units to the left of the origin on the real axis, and 1.5 units up on the $j$ axis.
    **(c)** 10 units to the left of the origin on the real axis, and down 14 units on the $-j$ axis.
11. What is the value of the hypotenuse of a right triangle whose sides are 10 and 15?
12. Convert each of the following rectangular numbers to polar form:
    **(a)** $40 - j40$ **(b)** $50 - j200$ **(c)** $35 - j20$ **(d)** $98 + j45$
13. Convert each of the following polar numbers to rectangular form:
    **(a)** $1000\angle -50°$ **(b)** $15\angle 160°$ **(c)** $25\angle -135°$ **(d)** $3\angle 180°$

14. Express each of the following polar numbers using a negative angle to replace the positive angle:
(a) $10\angle 120°$ (b) $32\angle 85°$ (c) $5\angle 310°$
15. Identify the quadrant in which each of the points in Problem 12 is located.
16. Identify the quadrant in which each point in Problem 14 is located.
17. Write the polar expressions using positive angles for each phasor in Figure 12–32.

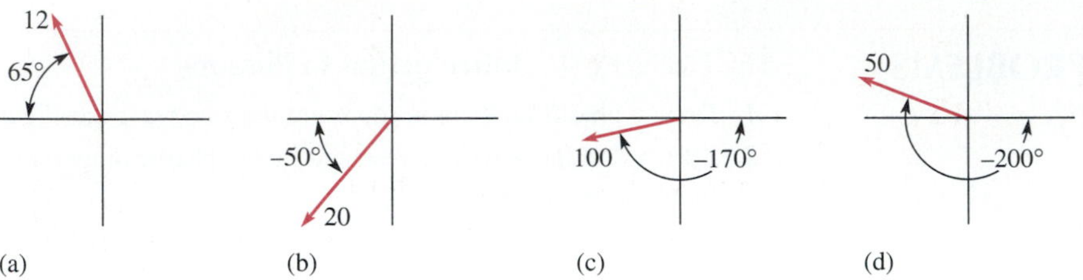


FIGURE 12–32

## SECTION 12–4 Mathematical Operations

18. Add the following sets of complex numbers:
(a) $9 + j3$ and $5 + j8$ (b) $3.5 - j4$ and $2.2 + j6$
(c) $-18 + j23$ and $30 - j15$ (d) $12\angle 45°$ and $20\angle 32°$
(e) $3.8\angle 75°$ and $1 + j1.8$ (f) $50 - j39$ and $60\angle -30°$
19. Perform the following subtractions:
(a) $(2.5 + j1.2) - (1.4 + j0.5)$ (b) $(-45 - j23) - (36 + j12)$
(c) $(8 - j4) - 3\angle 25°$ (d) $48\angle 135° - 33\angle -60°$
20. Multiply the following numbers:
(a) $4.5\angle 48°$ and $3.2\angle 90°$ (b) $120\angle -220°$ and $95\angle 200°$
(c) $-3\angle 150°$ and $4 - j3$ (d) $67 + j84$ and $102\angle 40°$
(e) $15 - j10$ and $-25 - j30$ (f) $0.8 + j0.5$ and $1.2 - j1.5$
21. Perform the following divisions:
(a) $\dfrac{8\angle 50°}{2.5\angle 39°}$ (b) $\dfrac{63\angle -91°}{9\angle 10°}$ (c) $\dfrac{28\angle 30°}{14 - j12}$ (d) $\dfrac{40 - j30}{16 + j8}$
22. Perform the following operations:
(a) $\dfrac{2.5\angle 65° - 1.8\angle -23°}{1.2\angle 37°}$ (b) $\dfrac{(100\angle 15°)(85 - j150)}{25 + j45}$
(c) $\dfrac{(250\angle 90° + 175\angle 75°)(50 - j100)}{(125 + j90)(35\angle 50°)}$ (d) $\dfrac{(1.5)^2(3.8)}{1.1} + j\left(\dfrac{8}{4} - j\dfrac{4}{2}\right)$
23. Three sine wave voltage sources are connected in series as shown in Figure 12–33. Determine the total voltage and current expressed as polar quantities. Resistance always has a zero phase angle as you will learn later, so $R_2 = 2.2\angle 0°$ kΩ.
24. What is the magnitude and phase of the voltages across each resistor in Figure 12–33?

FIGURE 12–33

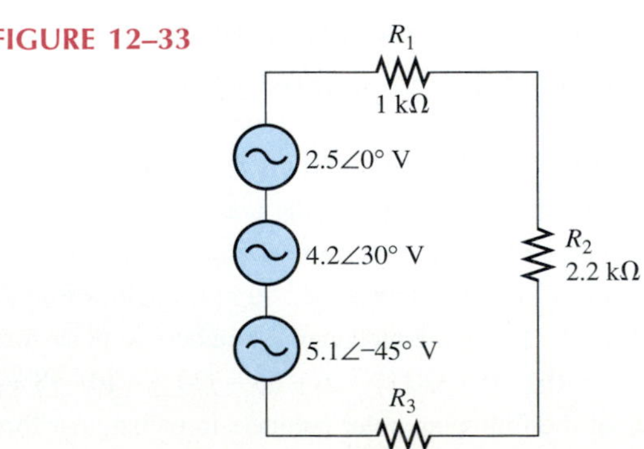

## ■ ANSWERS TO SECTION REVIEWS

### Section 12–1

1. A graphic representation of the magnitude and angular position of a time-varying quantity
2. 9425 rad/s
3. 100 Hz
4. See Figure 12–34.

FIGURE 12–34

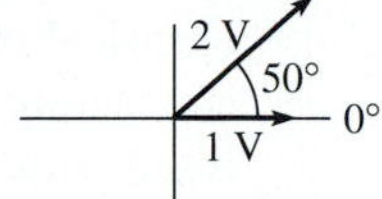


### Section 12–2

1. **(a)** 3 units right of the origin on real axis **(b)** 4 units left of the origin on real axis **(c)** 1 unit above origin on *j* axis
2. **(a)** 90° **(b)** 90° **(c)** 180°

### Section 12–3

1. Real part and *j* (imaginary) part
2. Magnitude and angle
3. $2.828\angle 45°$; first
4. $3.54 - j3.54$, fourth

### Section 12–4

1. $4 + j1$
2. $3 + j7$
3. $16\angle 110°$
4. $5\angle 15°$

### Section 12–5

1. No, the seconds/division setting is not critical.
2. Superimpose the traces by adjusting the vertical deflection (position).

### Section 12–6

1. `VS 1 0 SIN(0 6 5K 0 0 45)`
2. 
```
VS1  1  0  SIN(0  2  2K  0  0  45)
VS2  2  1  SIN(0  3  2K  0  0  90)
R1  2  3  1K
R2  3  0  560
.TRAN  5u  500u
.PROBE
.END
```

## ■ ANSWERS TO RELATED EXERCISES FOR EXAMPLES

**12–1** 15 V
**12–2** 7.07 V at −85°
**12–3** 34.2 V
**12–4** **(a)** 1st **(b)** 4th **(c)** 3rd **(d)** 2nd
**12–5** $29.2\angle 52°$
**12–6** $70.1 - j34.2$
**12–7** $-1 - j8$
**12–8** $-13.5 - j4.5$
**12–9** $1500\angle -50°$
**12–10** $4\angle -42°$

# 13

# CAPACITORS



## ■ INTRODUCTION

In previous chapters, the resistor has been the only passive electrical component that you have studied. The capacitor is the second type of basic passive electrical component.

In this chapter, you will learn about the capacitor and its characteristics. The basic construction and electrical properties are examined and the effects of connecting capacitors in series and in parallel are analyzed. How a capacitor works in both dc and ac circuits is an important part of this coverage and forms the basis for the study of reactive circuits in terms of both frequency response and time response. You will learn how to check for a faulty capacitor. In this chapter and throughout the rest of the book, you will learn the basics of putting technology theory into practice.

The capacitor is an electrical device that can store electrical charge, thereby creating an electric field that, in turn, stores energy. The measure of the energy-storing ability of a capacitor is its capacitance. When an electrical signal is applied to a capacitor, it reacts in a certain way and produces an opposition to current, which depends on the frequency of the applied signal. This opposition to current is called *capacitive reactance.*

## TECHnology Theory Into Practice

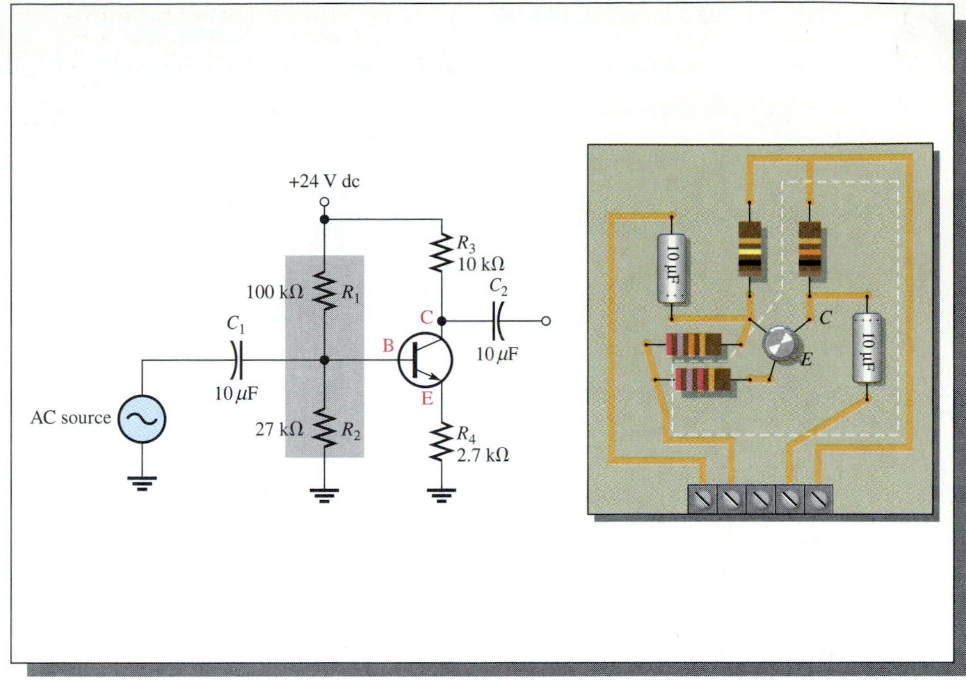


In the TECH TIP assignment in Section 13–9, you will see how a capacitor is used to couple signal voltages in an amplifier. You will also troubleshoot the amplifier using oscilloscope waveforms.

## CHAPTER OBJECTIVES

- ☐ Describe the basic structure and characteristics of a capacitor
- ☐ Discuss various types of capacitors
- ☐ Analyze series capacitors
- ☐ Analyze parallel capacitors
- ☐ Analyze capacitive dc switching circuits
- ☐ Analyze capacitive ac circuits
- ☐ Discuss some capacitor applications
- ☐ Test a capacitor
- ☐ Use PSpice and Probe in transient analysis (optional)

## 13–1 ■ THE BASIC CAPACITOR

***In this section, the basic structure and characteristics of capacitors are examined.***

***After completing this section, you should be able to***

- **Describe the basic structure and characteristics of a capacitor**
  - ☐ Explain how a capacitor stores charge
  - ☐ Define *capacitance* and state its unit
  - ☐ State Coulomb's law
  - ☐ Explain how a capacitor stores energy
  - ☐ Discuss voltage rating and temperature coefficient
  - ☐ Explain capacitor leakage
  - ☐ Specify how the physical characteristics affect the capacitance

### Basic Construction

In its simplest form, a **capacitor** is an electrical device constructed of two parallel conductive plates separated by an insulating material called the **dielectric.** Connecting leads are attached to the parallel plates. A basic capacitor is shown in Figure 13–1(a), and a schematic symbol is shown in part (b).

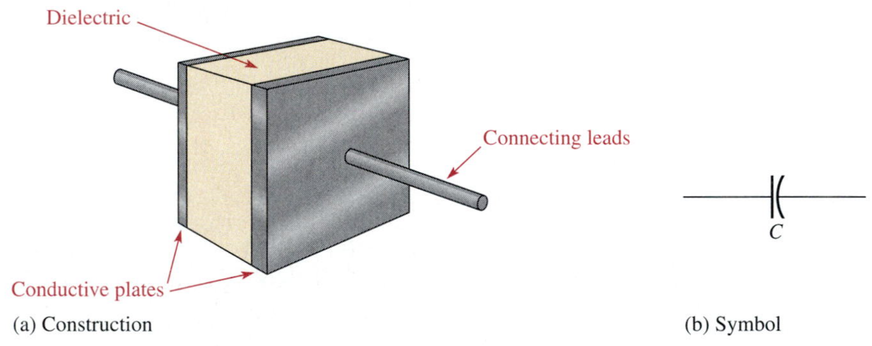


(a) Construction (b) Symbol

**FIGURE 13–1**
***The basic capacitor.***

### How a Capacitor Stores Charge

In the neutral state, both plates of a capacitor have an equal number of free electrons, as indicated in Figure 13–2(a). When the capacitor is connected to a voltage source through a resistor, as shown in part (b), electrons (negative charge) are removed from plate *A,* and an equal number are deposited on plate *B.* As plate *A* loses electrons and plate *B* gains electrons, plate *A* becomes positive with respect to plate *B.* During this charging process, electrons flow only through the connecting leads and the source. No electrons flow through the dielectric of the capacitor because it is an insulator. The movement of electrons ceases when the voltage across the capacitor equals the source voltage, as indicated in Figure 13–2(c). If the capacitor is disconnected from the source, it retains the stored charge for a long period of time (the length of time depends on the type of capacitor) and still has the voltage across it, as shown in Figure 13–2(d). Actually, the charged capacitor can be considered as a temporary battery.

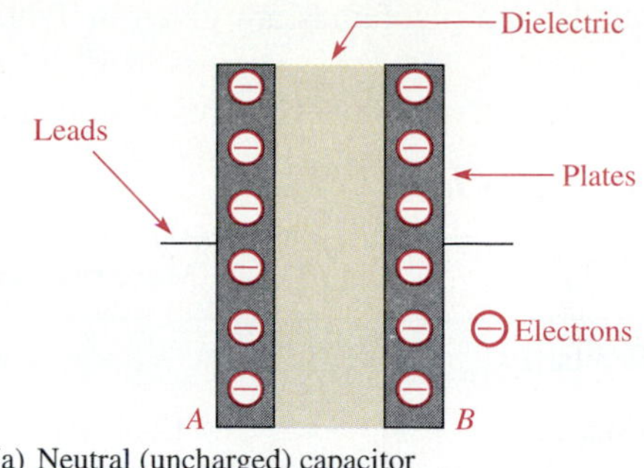

(a) Neutral (uncharged) capacitor (same charge on both plates)

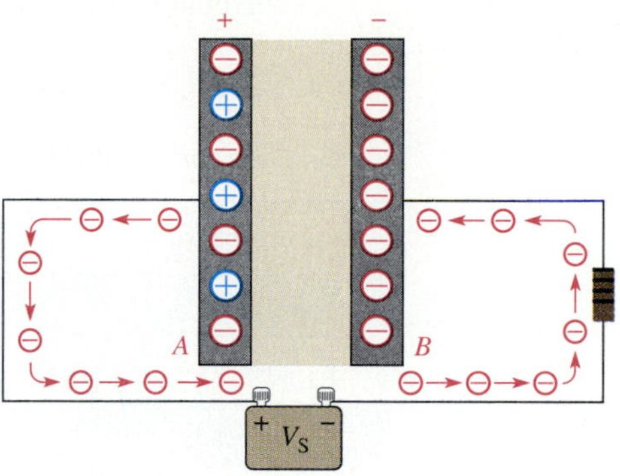

(b) Electrons flow from plate *A* to plate *B* as capacitor charges.

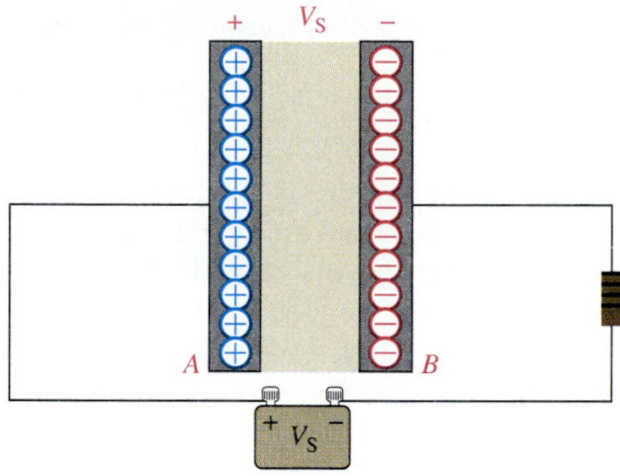

(c) Capacitor charged to $V_S$. No more electrons flow.

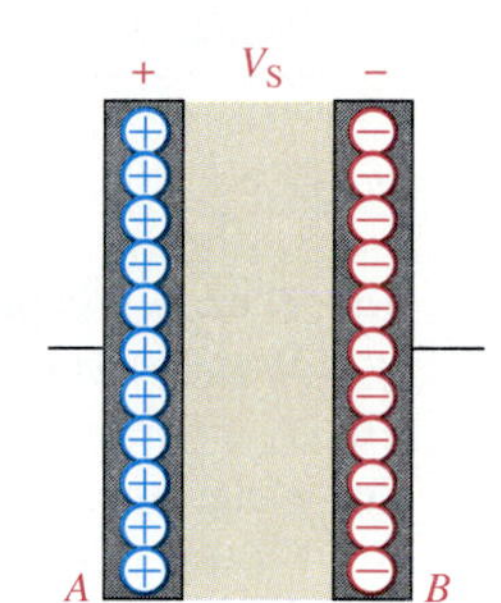

(d) Capacitor retains charge when disconnected from source.

FIGURE 13–2
***Illustration of a capacitor storing charge.***

## Capacitance

The amount of charge that a capacitor can store per unit of voltage across its plates is its **capacitance,** designated *C*. That is, capacitance is a measure of a capacitor's ability to store charge. The more charge per unit of voltage that a capacitor can store, the greater its capacitance, as expressed by the following formula:

$$C = \frac{Q}{V} \qquad \textbf{(13–1)}$$

where *C* is capacitance, *Q* is charge, and *V* is voltage.

By rearranging Equation (13–1), we obtain two other formulas as follows:

$$Q = CV \qquad \textbf{(13–2)}$$

$$V = \frac{Q}{C} \qquad \textbf{(13–3)}$$

***The Unit of Capacitance*** The **farad (F)** is the basic unit of capacitance, and the **coulomb (C)** is the unit of electrical charge. By definition, one farad is the amount of capacitance when one coulomb (C) of charge is stored with one volt across the plates.

Most capacitors that are used in electronics work have capacitance values in microfarads ($\mu$F) and picofarads (pF). A microfarad is one-millionth of a farad (1 $\mu$F = 1 × $10^{-6}$ F), and a picofarad is one-trillionth of a farad (1 pF = 1 × $10^{-12}$ F).

Conversions for farads, microfarads, and picofarads are given in Table 13–1.

**TABLE 13–1**
*Conversions for farads, microfarads, and picofarads*

| To Convert from | To | Move the Decimal Point |
|---|---|---|
| Farads | Microfarads | 6 places to right ($\times 10^6$) |
| Farads | Picofarads | 12 places to right ($\times 10^{12}$) |
| Microfarads | Farads | 6 places to left ($\times 10^{-6}$) |
| Microfarads | Picofarads | 6 places to right ($\times 10^6$) |
| Picofarads | Farads | 12 places to left ($\times 10^{-12}$) |
| Picofarads | Microfarads | 6 places to left ($\times 10^{-6}$) |

**EXAMPLE 13–1**

**(a)** A certain capacitor stores 50 microcoulombs (50 $\mu$C) with 10 V across its plates. What is its capacitance in units of microfarads?
**(b)** A 2 $\mu$F capacitor has 100 V across its plates. How much charge does it store?
**(c)** Determine the voltage across a 1000 pF capacitor that is storing 20 microcoulombs (20 $\mu$C) of charge.

***Solution***

**(a)** $C = \frac{Q}{V} = \frac{50\ \mu\text{C}}{10\ \text{V}} = 5\ \mu\text{F}$

**(b)** $Q = CV = (2\ \mu\text{F})(100\ \text{V}) = 200\ \mu\text{C}$

**(c)** $V = \frac{Q}{C} = \frac{20\ \mu\text{C}}{1000\ \text{pF}} = 20\ \text{kV}$

***Related Exercise*** Determine $V$ if $C = 1000$ pF and $Q = 100\ \mu$C.

**EXAMPLE 13–2**

Convert the following values to microfarads:
**(a)** 0.00001 F **(b)** 0.005 F **(c)** 1000 pF **(d)** 200 pF

***Solution***

**(a)** $0.00001\ \text{F} \times 10^6 = 10\ \mu\text{F}$ **(b)** $0.005\ \text{F} \times 10^6 = 5000\ \mu\text{F}$
**(c)** $1000\ \text{pF} \times 10^{-6} = 0.001\ \mu\text{F}$ **(d)** $200\ \text{pF} \times 10^{-6} = 0.0002\ \mu\text{F}$

***Related Exercise*** Convert 50,000 pF to microfarads.

**EXAMPLE 13–3**

Convert the following values to picofarads:
**(a)** $0.1 \times 10^{-8}$ F **(b)** 0.000025 F **(c)** 0.01 $\mu$F **(d)** 0.005 $\mu$F

***Solution***

**(a)** $0.1 \times 10^{-8}\ \text{F} \times 10^{12} = 1000\ \text{pF}$ **(b)** $0.000025\ \text{F} \times 10^{12} = 25 \times 10^6\ \text{pF}$
**(c)** $0.01\ \mu\text{F} \times 10^6 = 10{,}000\ \text{pF}$ **(d)** $0.005\ \mu\text{F} \times 10^6 = 5000\ \text{pF}$

***Related Exercise*** Convert 100 $\mu$F to picofarads.

## How a Capacitor Stores Energy

A capacitor stores energy in the form of an electric field that is established by the opposite charges on the two plates. The electric field is represented by lines of force between the positive and negative charges and concentrated within the dielectric, as shown in Figure 13–3.

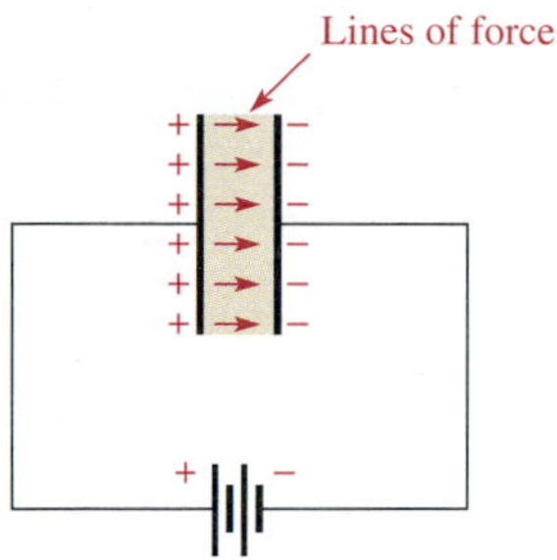


**FIGURE 13–3**
***The electric field stores energy in a capacitor.***

**Coulomb's law** states

**A force exists between two charged bodies that is directly proportional to the product of the two charges and inversely proportional to the square of the distance between the bodies.**

This relationship is expressed as

$$F = \frac{kQ_1Q_2}{d^2} \qquad \textbf{(13–4)}$$

where $F$ is the force in newtons, $Q_1$ and $Q_2$ are the charges in coulombs, $d$ is the distance between the charges in meters, and $k$ is a proportionality constant equal to $9 \times 10^9$.

Figure 13–4(a) illustrates the line of force between a positive and a negative charge. Figure 13–4(b) shows that many opposite charges on the plates of a capacitor create many lines of force, which form an electric field that stores energy within the dielectric.

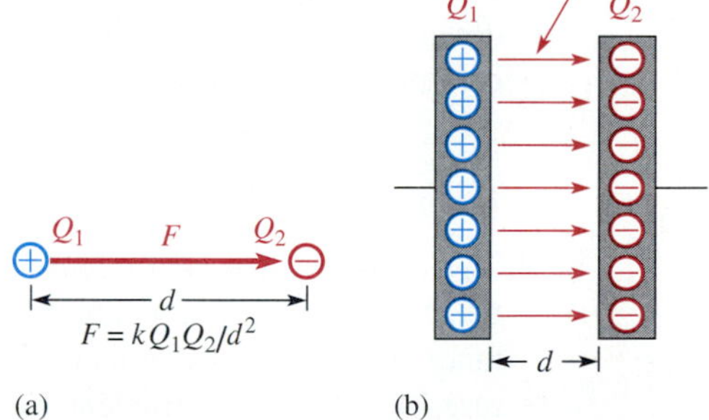


**FIGURE 13–4**
***Lines of force are created by opposite charges.***

The greater the forces between the charges on the plates of a capacitor, the more energy is stored. The amount of energy stored therefore is directly proportional to the capacitance because, from Coulomb's law, the more charge stored, the greater the force.

Also, from Equation (13–2), the amount of charge stored is directly related to the voltage as well as the capacitance. Therefore, the amount of energy stored is also dependent on the square of the voltage across the plates of the capacitor. The formula for the energy stored by a capacitor is as follows:

$$W = \frac{1}{2}CV^2 \qquad \textbf{(13–5)}$$

When capacitance ($C$) is in farads and voltage ($V$) is in volts, the energy ($W$) is in joules.

## Voltage Rating

Every capacitor has a limit on the amount of voltage that it can withstand across its plates. The voltage rating specifies the maximum dc voltage that can be applied without risk of damage to the device. If this maximum voltage, commonly called the *breakdown voltage* or *working voltage,* is exceeded, permanent damage to the capacitor can result.

Both the capacitance and the voltage rating must be taken into consideration before a capacitor is used in a circuit application. The choice of capacitance value is based on particular circuit requirements. The voltage rating should always be well above the maximum voltage expected in a particular application.

***Dielectric Strength*** The breakdown voltage of a capacitor is determined by the **dielectric strength** of the dielectric material used. The dielectric strength is expressed in V/mil (1 mil = 0.001 in.). Table 13–2 lists typical values for several materials. Exact values vary depending on the specific composition of the material.

TABLE 13–2
***Some common dielectric materials and their dielectric strengths***

| Material | Dielectric Strength (V/mil) |
|---|---|
| Air | 80 |
| Oil | 375 |
| Ceramic | 1000 |
| Paper (paraffined) | 1200 |
| Teflon® | 1500 |
| Mica | 1500 |
| Glass | 2000 |

The dielectric strength can best be explained by an example. Assume that a certain capacitor has a plate separation of 1 mil and that the dielectric material is ceramic. This particular capacitor can withstand a maximum voltage of 1000 V because its dielectric strength is 1000 V/mil. If the maximum voltage is exceeded, the dielectric may break down and conduct current, causing permanent damage to the capacitor. Similarly, if the ceramic capacitor has a plate separation of 2 mils, its breakdown voltage is 2000 V.

## Temperature Coefficient

The **temperature coefficient** indicates the amount and direction of a change in capacitance value with temperature. A positive temperature coefficient means that the capacitance increases with an increase in temperature or decreases with a decrease in temperature. A negative coefficient means that the capacitance decreases with an increase in temperature or increases with a decrease in temperature.

Temperature coefficients are typically specified in parts per million per Celsius degree (ppm/C°). For example, a negative temperature coefficient of 150 ppm/C° for a 1 $\mu$F capacitor means that for every degree rise in temperature, the capacitance decreases by 150 pF (there are one million picofarads in one microfarad).

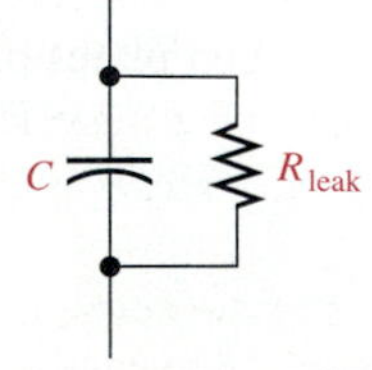

FIGURE 13–5
***Equivalent circuit for a nonideal capacitor.***

## Leakage

No insulating material is perfect. The dielectric of any capacitor will conduct some very small amount of current. Thus, the charge on a capacitor will eventually leak off. Some types of capacitors have higher leakages than others. An equivalent circuit for a nonideal capacitor is shown in Figure 13–5. The parallel resistor $R_{leak}$ represents the extremely high resistance of the dielectric material through which there is leakage current.

## Physical Characteristics of a Capacitor

The following parameters are important in establishing the capacitance and the voltage rating of a capacitor.

**Plate Area** *Capacitance is directly proportional to the physical size of the plates as determined by the plate area, A.* A larger plate area produces a larger capacitance, and a smaller plate area produces a smaller capacitance. Figure 13–6(a) shows that the plate area of a parallel plate capacitor is the area of one of the plates. If the plates are moved in relation to each other, as shown in Figure 13–6(b), the overlapping area determines the effective plate area. This variation in effective plate area is the basis for a certain type of variable capacitor.

FIGURE 13–6
***Capacitance is directly proportional to plate area (A).***

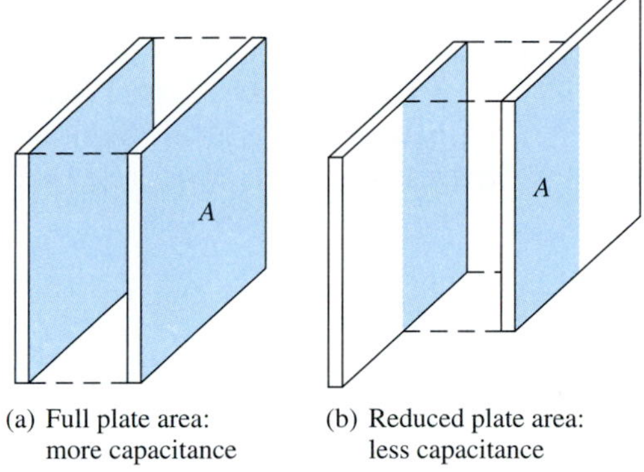


**Plate Separation** *Capacitance is inversely proportional to the distance between the plates.* The plate separation is designated *d,* as shown in Figure 13–7. A greater separation of the plates produces a smaller capacitance, as illustrated in the figure. The breakdown voltage is directly proportional to the plate separation. The further the plates are separated, the greater the breakdown voltage.

FIGURE 13–7
***Capacitance is inversely proportional to the distance between the plates.***

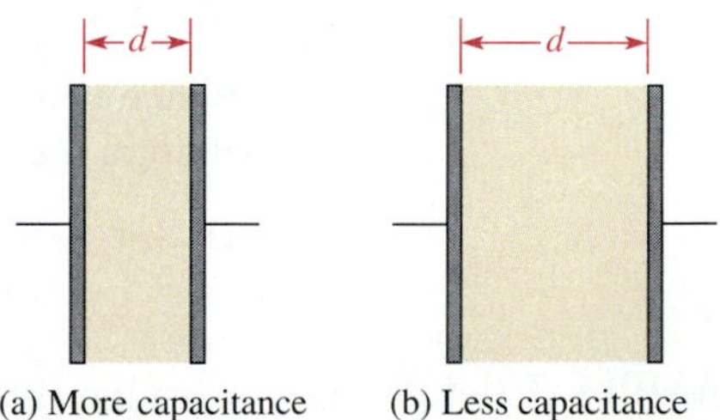


**Dielectric Constant** As you know, the insulating material between the plates of a capacitor is called the *dielectric.* Every dielectric material has the ability to concentrate the lines of force of the electric field existing between the oppositely charged plates of a capacitor and thus increase the capacity for energy storage. The measure of a material's ability to establish an electric field is called the **dielectric constant** or *relative permittivity,* symbolized by $\varepsilon_r$. ($\varepsilon$ is the Greek letter epsilon.)

*Capacitance is directly proportional to the dielectric constant.* The dielectric constant of a vacuum is defined as 1 and that of air is very close to 1. These values are used as a reference, and all other materials have values of $\varepsilon_r$ specified with respect to that of a vacuum of air. For example, a material with $\varepsilon_r = 8$ can result in a capacitance eight times greater than that of air with all other factors being equal.

Table 13–3 lists several common dielectric materials and typical dielectric constants for each. Values can vary because they depend on the specific composition of the material.

**TABLE 13–3**
***Some common dielectric materials and their dielectric constants***

| Material | Typical $\varepsilon_r$ Values |
|---|---|
| Air (vacuum) | 1.0 |
| Teflon® | 2.0 |
| Paper (paraffined) | 2.5 |
| Oil | 4.0 |
| Mica | 5.0 |
| Glass | 7.5 |
| Ceramic | 1200 |

The dielectric constant (relative permittivity) is dimensionless because it is a relative measure and is a ratio of the absolute permittivity of a material, $\varepsilon$, to the absolute permittivity of a vacuum, $\varepsilon_0$, as expressed by the following formula:

$$\varepsilon_r = \frac{\varepsilon}{\varepsilon_0} \qquad \textbf{(13–6)}$$

The value of $\varepsilon_0$ is $8.85 \times 10^{-12}$ F/m (farads per meter).

## Formula for Capacitance in Terms of Physical Parameters

You have seen how capacitance is directly related to plate area, $A$, and the dielectric constant, $\varepsilon_r$, and inversely related to plate separation, $d$. An exact formula for calculating the capacitance in terms of these three quantities is as follows:

$$C = \frac{A\varepsilon_r(8.85 \times 10^{-12}\text{ F/m})}{d} \qquad \textbf{(13–7)}$$

where $A$ is in square meters ($m^2$), $d$ is in meters (m), and $C$ is in farads (F). Recall that the absolute permittivity of a vacuum, $\varepsilon_0$, is $8.85 \times 10^{-12}$ F/m and that the absolute permittivity of a dielectric ($\varepsilon$), as derived from Equation (13–6), is $\varepsilon_r(8.85 \times 10^{-12}\text{ F/m})$.

**EXAMPLE 13–4**

Determine the capacitance of a parallel plate capacitor having a plate area of 0.01 $m^2$ and a plate separation of 0.02 m. The dielectric is mica, which has a dielectric constant of 5.0.

***Solution*** Use Equation (13–7).

$$C = \frac{A\varepsilon_r(8.85 \times 10^{-12}\text{ F/m})}{d} = \frac{(0.01\text{ m}^2)(5.0)(8.85 \times 10^{-12}\text{ F/m})}{0.02\text{ m}} = 22.1\text{ pF}$$

The calculator sequence is

[•] [0] [1] [×] [5] [×] [8] [•] [8] [5] [EXP] [+/−] [1] [2] [÷] [•] [0] [2] [=]

***Related Exercise*** Determine $C$ where $A = 0.005\text{ m}^2$, $d = 0.008$ m, and ceramic is the dielectric.

**SECTION 13–1 REVIEW**

1. Define *capacitance*.
2. (a) How many microfarads in one farad?
   (b) How many picofarads in one farad?
   (c) How many picofarads in one microfarad?
3. Convert 0.0015 $\mu$F to picofarads. To farads.
4. How much energy in joules is stored by a 0.01 $\mu$F capacitor with 15 V across its plates?
5. (a) When the plate area of a capacitor is increased, does the capacitance increase or decrease?
   (b) When the distance between the plates is increased, does the capacitance increase or decrease?
6. The plates of a ceramic capacitor are separated by 10 mils. What is the typical breakdown voltage?
7. A ceramic capacitor has a plate area of 0.2 $m^2$. The thickness of the dielectric is 0.005 m. What is the capacitance?
8. A capacitor with a value of 2 $\mu$F at 25°C has a positive temperature coefficient of 50 ppm/C°. What is the capacitance value when the temperature increases to 125°C?

## 13–2 ■ TYPES OF CAPACITORS

***Capacitors normally are classified according to the type of dielectric material. The most common types of dielectric materials are mica, ceramic, plastic-film, and electrolytic (aluminum oxide and tantalum oxide). In this section, the characteristics and construction of each of these types of capacitors and variable capacitors are examined.***

***After completing this section, you should be able to***

- **Discuss various types of capacitors**
  - Describe the characteristics of mica, ceramic, plastic-film, and electrolytic capacitors
  - Describe types of variable capacitors
  - Identify capacitor labeling

### Fixed Capacitors

***Mica Capacitors*** Two types of mica capacitors are stacked-foil and silver-mica. The basic construction of the stacked-foil type is shown in Figure 13–8. It consists of alternate layers of metal foil and thin sheets of mica. The metal foil forms the plate, with alternate foil sheets connected together to increase the plate area. More layers are used to increase

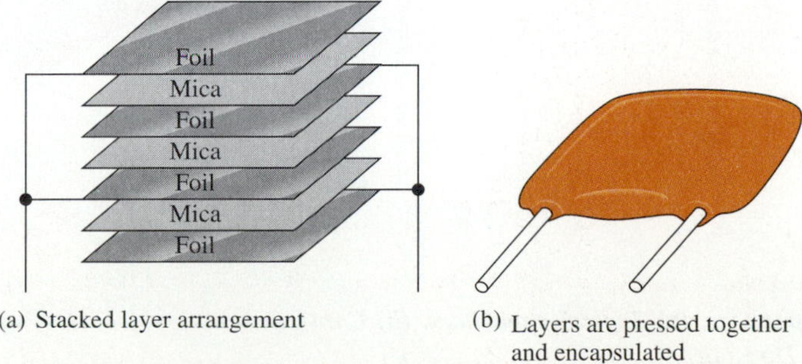


FIGURE 13–8
*Construction of a typical radial-lead mica capacitor.*

the plate area, thus increasing the capacitance. The mica/foil stack is encapsulated in an insulating material such as Bakelite®, as shown in Figure 13–8(b). The silver-mica capacitor is formed in a similar way by stacking mica sheets with silver electrode material screened on them.

Mica capacitors are available with capacitance values ranging from 1 pF to 0.1 $\mu$F and voltage ratings from 100 V dc to 2500 V dc. Common temperature coefficients range from −20 ppm/C° to +100 ppm/C°. Mica has a typical dielectric constant of 5.

***Ceramic Capacitors*** Ceramic dielectrics provide very high dielectric constants (1200 is typical). As a result, comparatively high capacitance values can be achieved in a small physical size. Ceramic capacitors are commonly available in a ceramic disk form, as shown in Figure 13–9, in a multilayer radial-lead configuration, as shown in Figure 13–10, or in a leadless ceramic chip, as shown in Figure 13–11, for surface mounting on PC boards.

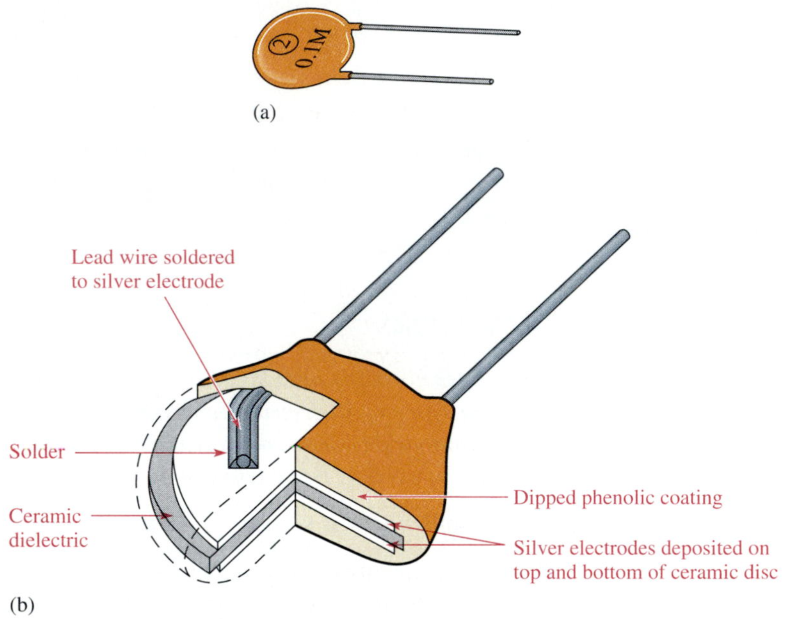

**FIGURE 13–9**
***A ceramic disk capacitor and its basic construction.***

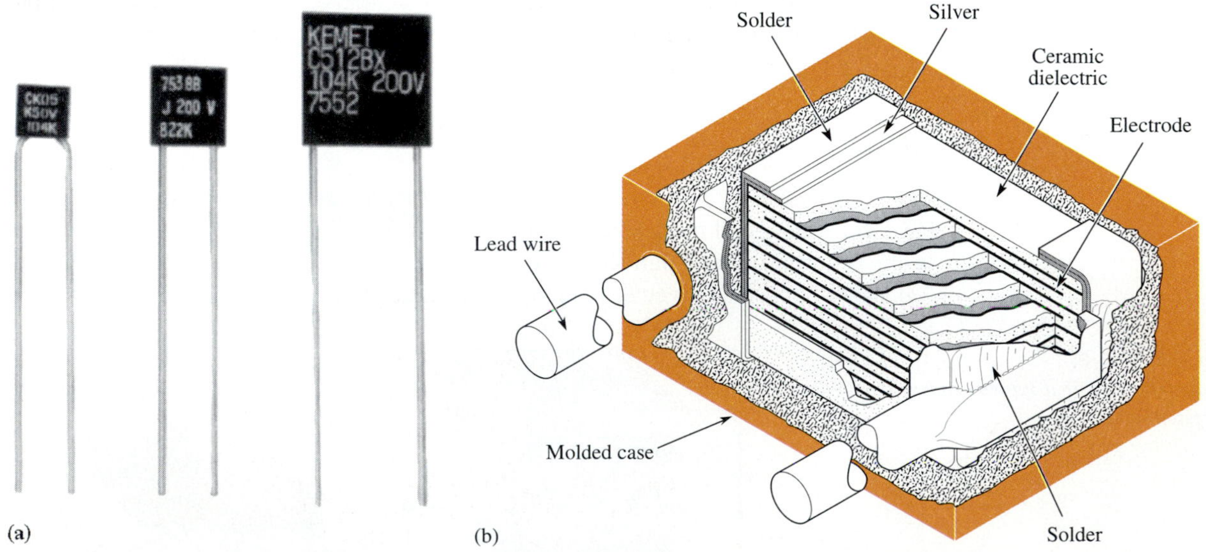

**FIGURE 13–10**
***Ceramic capacitors. (a) Typical capacitors. (b) Construction view (courtesy of Union Carbide Electronics Division).***

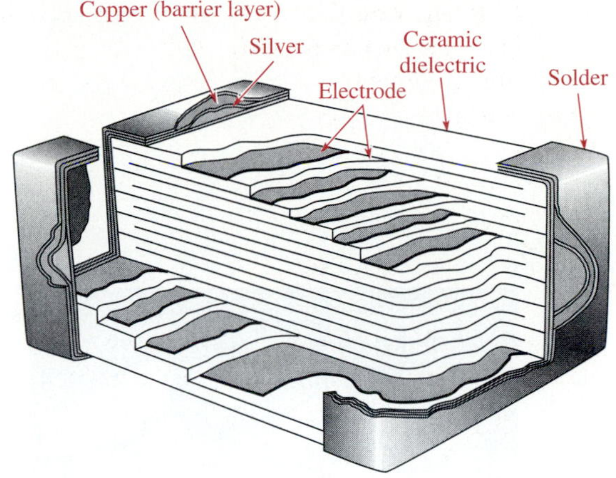


**FIGURE 13–11**
***Construction view of a typical ceramic chip capacitor used for surface mounting on PC boards.***

Ceramic capacitors typically are available in capacitance values ranging from 1 pF to 2.2 $\mu$F with voltage ratings up to 6 kV. A typical temperature coefficient for ceramic capacitors is 200,000 ppm/C°.

***Plastic-Film Capacitors*** There are several types of plastic-film capacitors. Polycarbonate, propylene, parylene, polyester, polystyrene, polypropylene, and mylar are some of the more common dielectric materials used. Some of these types have capacitance values up to 100 $\mu$F.

Figure 13–12 shows a common basic construction used in many plastic-film capacitors. A thin strip of plastic-film dielectric is sandwiched between two thin metal strips that act as plates. One lead is connected to the inner plate and one to the outer plate as indicated. The strips are then rolled in a spiral configuration and encapsulated in a molded case. Thus, a large plate area can be packaged in a relatively small physical size, thereby achieving large capacitance values. Another method uses metal deposited directly on the film dielectric to form the plates.

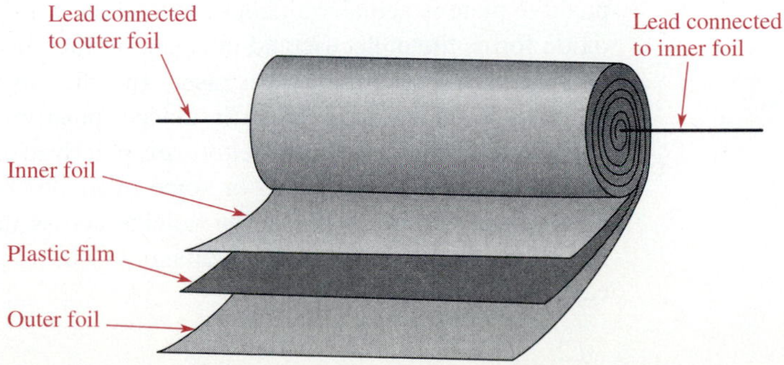


**FIGURE 13–12**
***Basic construction of axial-lead tubular plastic-film dielectric capacitors.***

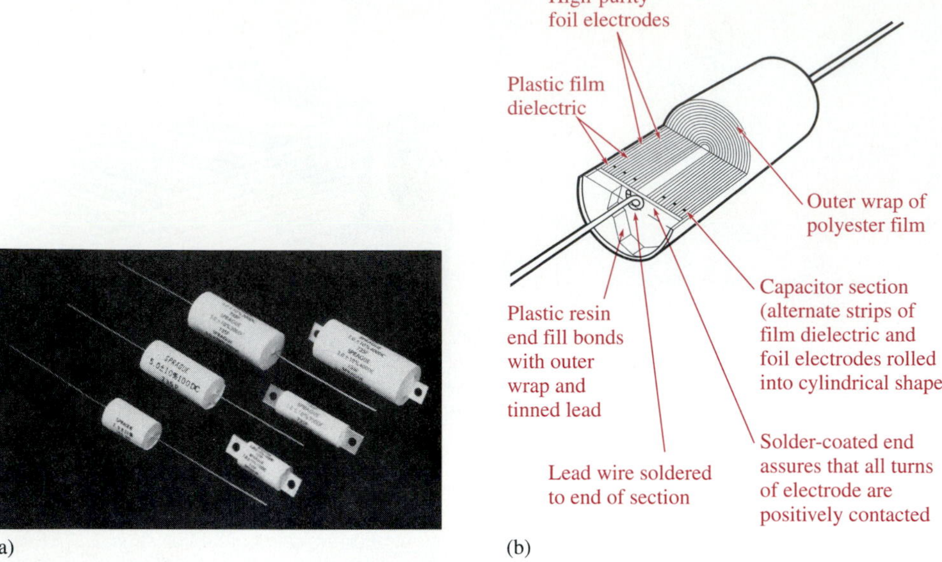


FIGURE 13–13
***Plastic-film capacitors. (a) Typical example (courtesy of Siemens Corp.). (b) Construction view of plastic-film capacitor (courtesy of Union Carbide Electronics Division).***

Figure 13–13(a) shows typical plastic-film capacitors. Figure 13–13(b) shows a construction view for one type of plastic-film capacitor.

***Electrolytic Capacitors*** Electrolytic capacitors are polarized so that one plate is positive and the other negative. These capacitors are used for high capacitance values up to over 200,000 $\mu$F, but they have relatively low breakdown voltages (350 V is a typical maximum) and high amounts of leakage.

Electrolytic capacitors offer much higher capacitance values than mica or ceramic capacitors, but their voltage ratings are typically lower. Aluminum electrolytics are probably the most commonly used type. While other capacitors use two similar plates, the electrolytic consists of one plate of aluminum foil and another plate made of a conducting electrolyte applied to a material such as plastic film. These two "plates" are separated by a layer of aluminum oxide which forms on the surface of the aluminum plate. Figure 13–14(a) illustrates the basic construction of a typical aluminum electrolytic capacitor with axial leads. Other electrolytics with radial leads are shown in Figure 13–14(b); the symbol for an electrolytic capacitor is shown in part (c).

Tantalum electrolytics can be in either a tubular configuration similar to Figure 13–14 or "tear drop" shape as shown in Figure 13–15. In the tear drop configuration, the positive plate is actually a pellet of tantalum powder rather than a sheet of foil. Tantalum oxide forms the dielectric and manganese dioxide forms the negative plate.

Because of the process used for the insulating oxide dielectric, the metallic (aluminum or tantalum) plate is always positive with respect to the electrolyte plate, and, thus all electrolytic capacitors are polarized. The metal plate (positive lead) is usually indicated by a plus sign or some other obvious marking and must always be connected in a dc circuit where the voltage across the capacitor does not change polarity. Reversal of the polarity of the voltage will usually result in complete destruction of the capacitor.

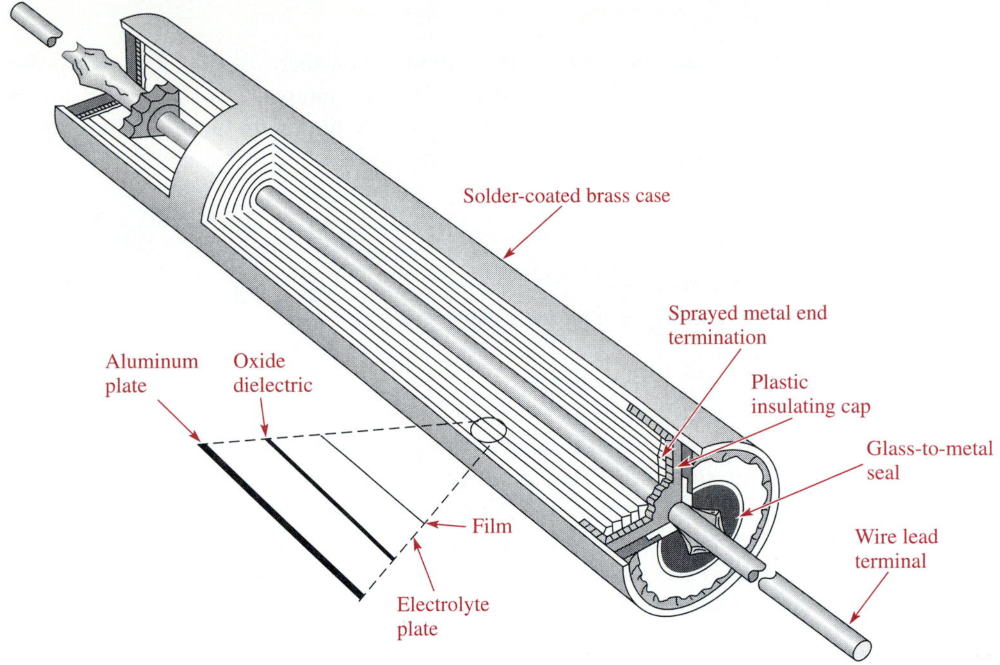


(a) Construction view of an axial-lead electrolytic capacitor

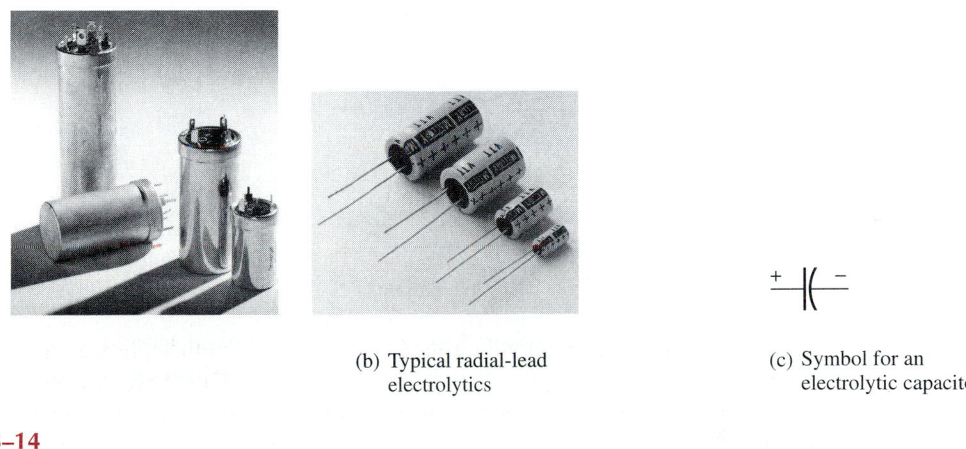

(b) Typical radial-lead electrolytics

(c) Symbol for an electrolytic capacitor

**FIGURE 13–14**
***Electrolytic capacitors.***

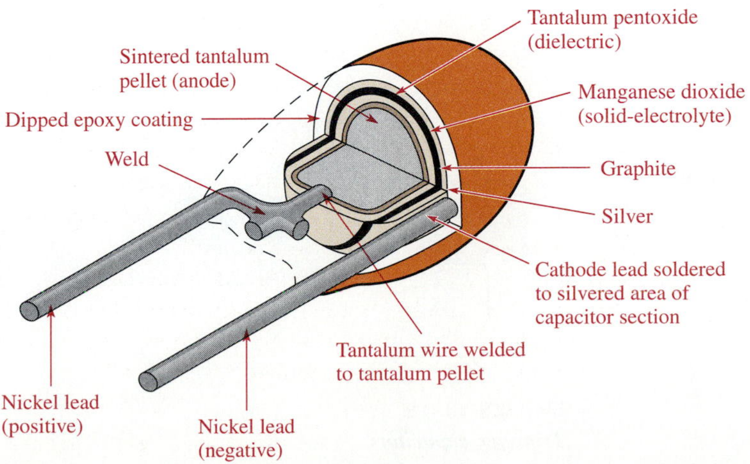


**FIGURE 13–15**
***Construction view of a typical "tear drop" shaped tantalum electrolytic capacitor.***

## Variable Capacitors

Variable capacitors are used in a circuit when there is a need to adjust the capacitance value either manually or automatically, for example, in radio or TV tuners. The major types of variable or adjustable capacitors are now introduced. The schematic symbol for a variable capacitor is shown in Figure 13–16.

FIGURE 13–16
***Schematic symbol for a variable capacitor.***

***Air Capacitor*** Variable capacitors with air dielectrics, such as the one shown in Figure 13–17, are sometimes used as tuning capacitors in applications requiring frequency selection. This type of capacitor is constructed of several plates that mesh together. One set of plates can be moved relative to the other, thus changing the effective plate area and the capacitance. The movable plates are linked together mechanically so that they all move when a shaft is rotated.

FIGURE 13–17
***A typical variable air capacitor.***

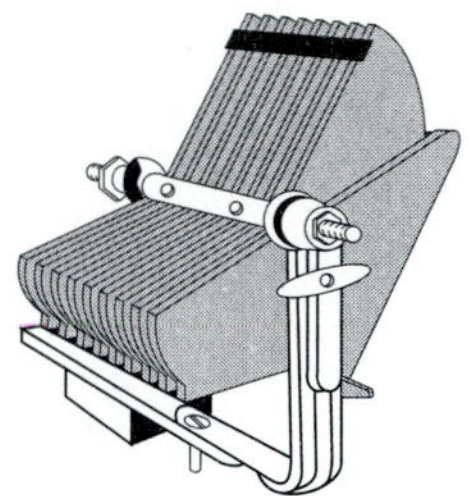

***Trimmers*** Adjustable capacitors that normally have slotted screw-type adjustments and are used for very fine adjustments in a circuit are called **trimmers.** Ceramic or mica is a common dielectric in these types of capacitors, and the capacitance usually is changed by adjusting the plate separation. Figure 13–18 shows some typical devices.

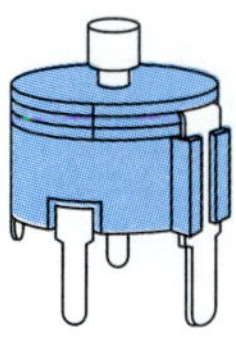
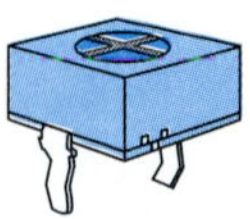
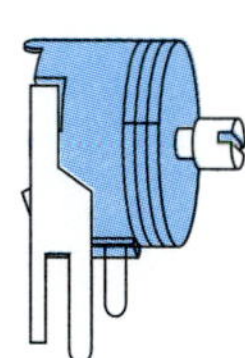

FIGURE 13–18
***Trimmer capacitors.***

***Varactors*** The varactor is a semiconductor device that exhibits a capacitance characteristic that is varied by changing the voltage across its terminals. This device usually is covered in detail in a course on electronic devices.

## Capacitor Labeling

Capacitor values are indicated on the body of the capacitor either by typographical labels or by color codes. Typographical labels consist of letters and numbers that indicate various parameters such as capacitance, voltage rating, and tolerance.

Some capacitors carry no unit designation for capacitance. In these cases, the units are implied by the value indicated and are recognized by experience. For example, a ceramic capacitor marked .001 or .01 has units of microfarads because picofarad values that small are not available. As another example, a ceramic capacitor labeled 50 or 330 has units of picofarads because microfarad units that large normally are not available in this type. In some cases a 3-digit designation is used. The first two digits are the first two digits of the capacitance value. The third digit is the number of zeros after the second digit. For example, 103 means 10,000 pF. In some instances, the units are labeled as pF or $\mu$F; sometimes the microfarad unit is labeled as MF or MFD.

Voltage rating appears on some types of capacitors with WV or WVDC and is omitted on others. When it is omitted, the voltage rating can be determined from information supplied by the manufacturer. The tolerance of the capacitor is usually labeled as a percentage, such as $\pm 10\%$. The temperature coefficient is indicated by a *parts per million* marking. This type of label consists of a P or N followed by a number. For example, N750 means a negative temperature coefficient of 750 ppm/C°, and P30 means a positive temperature coefficient of 30 ppm/C°. An NP0 designation means that the positive and negative coefficients are zero; thus the capacitance does not change with temperature. Certain types of capacitors are color coded. Refer to Appendix D, for color code information.

### SECTION 13–2 REVIEW

1. Name one way capacitors can be classified.
2. What is the difference between a fixed and a variable capacitor?
3. What type of capacitor is polarized?
4. What precautions must be taken when installing a polarized capacitor in a circuit?

## 13–3 ■ SERIES CAPACITORS

***In this section, you will see why the total capacitance of a series connection of capacitors is less than any of the individual capacitances.***

***After completing this section, you should be able to***

- **Analyze series capacitors**
  - ☐ Determine total capacitance
  - ☐ Determine capacitor voltages

### Total Capacitance

When capacitors are connected in series, the total capacitance is less than the smallest capacitance value because the effective plate separation increases. The calculation of total series capacitance is analogous to the calculation of total resistance of parallel resistors.

Consider the generalized circuit in Figure 13–19(a), which has $n$ capacitors in series with a voltage source and a switch. When the switch is closed, the capacitors charge as current is established through the circuit. Since this is a series circuit, the current must be the same at all points, as illustrated. Since current is the rate of flow of charge, the amount of charge stored by each capacitor is equal to the total charge, expressed as follows:

$$Q_T = Q_1 = Q_2 = Q_3 = \cdots = Q_n \tag{13–8}$$

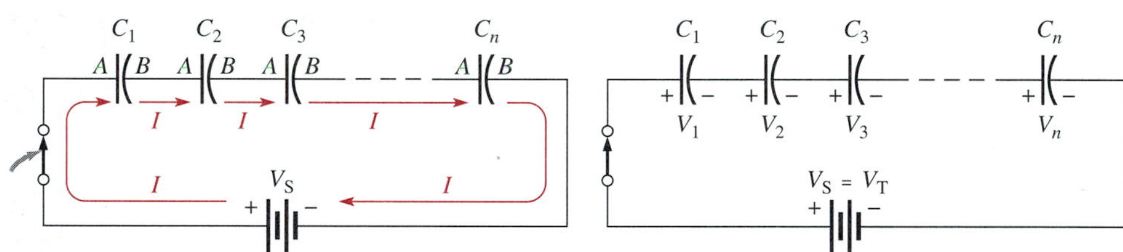


(a) Charging current is same for each capacitor, $I = Q/t$. (b) All capacitors store same amount of charge and $V = Q/C$.

**FIGURE 13–19**
*Series capacitive circuit.*

Next, according to Kirchhoff's voltage law, the sum of the voltages across the charged capacitors must equal the total voltage, $V_T$, as shown in Figure 13–19(b). This is expressed in equation form as

$$V_T = V_1 + V_2 + V_3 + \cdots + V_n$$

From Equation (13–3), $V = Q/C$. When this relationship is substituted into each term of the voltage equation, the following result is obtained:

$$\frac{Q_T}{C_T} = \frac{Q_1}{C_1} + \frac{Q_2}{C_2} + \frac{Q_3}{C_3} + \cdots + \frac{Q_n}{C_n}$$

Since the charges on all the capacitors are equal, the $Q$ terms can be factored and canceled, resulting in

$$\frac{1}{C_T} = \frac{1}{C_1} + \frac{1}{C_2} + \frac{1}{C_3} + \cdots + \frac{1}{C_n} \tag{13–9}$$

Taking the reciprocal of both sides of Equation (13–9) yields the following general formula for total series capacitance:

$$C_T = \frac{1}{\frac{1}{C_1} + \frac{1}{C_2} + \frac{1}{C_3} + \cdots + \frac{1}{C_n}} \tag{13–10}$$

Remember,

**The total series capacitance is always less than the smallest capacitance.**

***Two Capacitors in Series*** When only two capacitors are in series, a special form of Equation (13–9) can be used.

$$\frac{1}{C_T} = \frac{1}{C_1} + \frac{1}{C_2} = \frac{C_1 + C_2}{C_1 C_2}$$

Taking the reciprocal of the left and right terms gives the formula for total capacitance of two capacitors in series.

$$C_T = \frac{C_1 C_2}{C_1 + C_2} \tag{13–11}$$

***Capacitors of Equal Value in Series*** This special case is another in which a formula can be developed from Equation (13–9). When all capacitor values are the same and equal to $C$, the formula is

$$\frac{1}{C_T} = \frac{1}{C} + \frac{1}{C} + \frac{1}{C} + \cdots + \frac{1}{C}$$

Adding all the terms on the right yields

$$\frac{1}{C_T} = \frac{n}{C}$$

where $n$ is the number of equal-value capacitors. Taking the reciprocal of both sides yields

$$C_T = \frac{C}{n} \tag{13–12}$$

The capacitance value of the equal capacitors divided by the number of equal series capacitors gives the total capacitance.

**EXAMPLE 13–5** Determine the total capacitance in Figure 13–20.

FIGURE 13–20

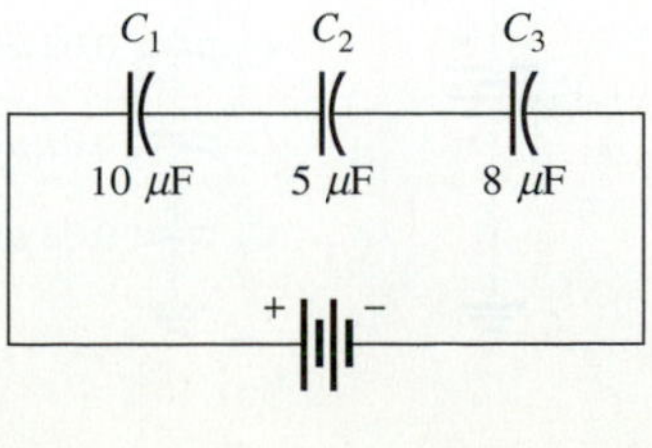

$C_1$ 10 μF, $C_2$ 5 μF, $C_3$ 8 μF

***Solution*** Use Equation (13–10).

$$C_T = \frac{1}{\frac{1}{C_1} + \frac{1}{C_2} + \frac{1}{C_3}} = \frac{1}{\frac{1}{10\ \mu F} + \frac{1}{5\ \mu F} + \frac{1}{8\ \mu F}} = \frac{1}{0.425}\ \mu F = 2.35\ \mu F$$

The calculator sequence is

1 0 EXP +/− 6 2nd 1/x + 5 EXP +/− 6 2nd 1/x +
8 EXP +/− 6 2nd 1/x = 2nd 1/x =

***Related Exercise*** If a 4.7 $\mu$F capacitor is connected in series with the three existing capacitors in Figure 13–20, what is $C_T$?

**EXAMPLE 13–6** Find the total capacitance, $C_T$, in Figure 13–21.

FIGURE 13–21

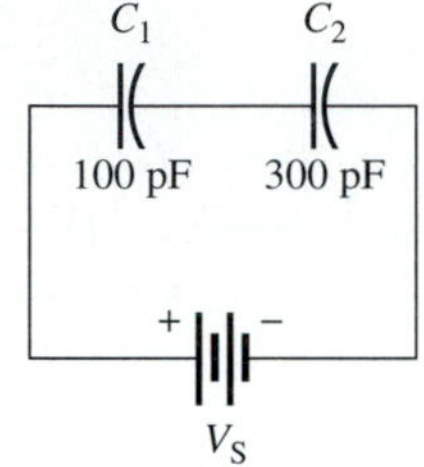


***Solution*** From Equation (13–11),

$$C_T = \frac{C_1 C_2}{C_1 + C_2} = \frac{(100\ pF)(300\ pF)}{400\ pF} = 75\ pF$$

For a calculator solution, use Equation (13–10).

$$C_T = \frac{1}{\frac{1}{100\ pF} + \frac{1}{300\ pF}}$$

1 0 0 EXP +/− 1 2 2nd 1/x + 3 0 0 EXP +/− 1 2
2nd 1/x = 2nd 1/x =

***Related Exercise*** Determine $C_T$ if $C_1 = 470$ pF and $C_2 = 680$ pF in Figure 13–21.

**EXAMPLE 13–7** Determine $C_T$ for the series capacitors in Figure 13–22.

FIGURE 13–22

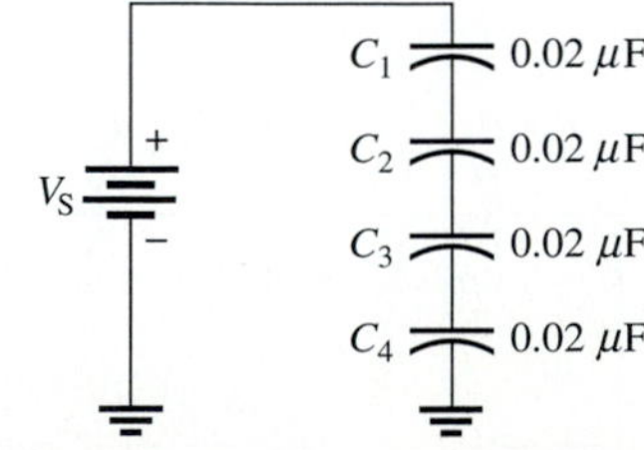

***Solution*** Since $C_1 = C_2 = C_3 = C_4 = C$, use Equation (13–12).

$$C_T = \frac{C}{n} = \frac{0.02\ \mu F}{4} = 0.005\ \mu F$$

***Related Exercise*** Determine $C_T$ if the capacitor values in Figure 13–22 are doubled.

## Capacitor Voltages

A series connection of charged capacitors acts as a voltage divider. The voltage across each capacitor in series is inversely proportional to its capacitance value, as shown by the formula $V = Q/C$.

The voltage across any capacitor in series can be calculated as follows:

$$V_x = \left(\frac{C_T}{C_x}\right)V_T \quad \textbf{(13–13)}$$

where $C_x$ is any capacitor in series, such as $C_1$, $C_2$, or $C_3$, and $V_x$ is the voltage across $C_x$. The derivation is as follows: Since the charge on any capacitor in series is the same as the total charge ($Q_x = Q_T$), and since $Q_x = V_xC_x$ and $Q_T = V_TC_T$, then

$$V_xC_x = V_TC_T$$

Solving for $V_x$ yields

$$V_x = \frac{C_TV_T}{C_x}$$

**The largest-value capacitor in series will have the smallest voltage; and the smallest-value capacitor will have the largest voltage.**

**EXAMPLE 13–8** Find the voltage across each capacitor in Figure 13–23.

FIGURE 13–23

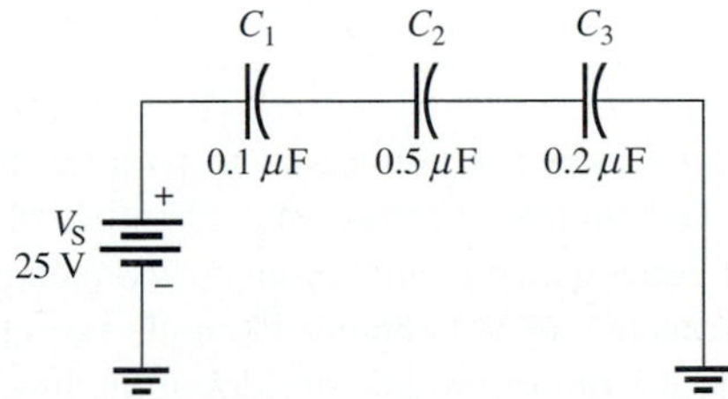


***Solution*** Calculate the total capacitance.

$$\frac{1}{C_T} = \frac{1}{C_1} + \frac{1}{C_2} + \frac{1}{C_3} = \frac{1}{0.1\ \mu F} + \frac{1}{0.5\ \mu F} + \frac{1}{0.2\ \mu F}$$

$$C_T = \frac{1}{17}\ \mu F = 0.0588\ \mu F$$

From Figure 13–23, $V_S = V_T = 25$ V. Therefore, use Equation (13–13) to calculate the voltage across each capacitor.

$$V_1 = \left(\frac{C_T}{C_1}\right)V_T = \left(\frac{0.0588\ \mu\text{F}}{0.1\ \mu\text{F}}\right)25\ \text{V} = 14.7\ \text{V}$$

$$V_2 = \left(\frac{C_T}{C_2}\right)V_T = \left(\frac{0.0588\ \mu\text{F}}{0.5\ \mu\text{F}}\right)25\ \text{V} = 2.94\ \text{V}$$

$$V_3 = \left(\frac{C_T}{C_3}\right)V_T = \left(\frac{0.0588\ \mu\text{F}}{0.2\ \mu\text{F}}\right)25\ \text{V} = 7.35\ \text{V}$$

***Related Exercise*** A 0.47 $\mu$F capacitor is connected in series with the existing capacitors in Figure 13–23. Determine the voltage across the new capacitor.

**SECTION 13–3 REVIEW**

1. Is the total capacitance of a series connection less than or greater than the value of the smallest capacitor?
2. The following capacitors are in series: 100 pF, 250 pF, and 500 pF. What is the total capacitance?
3. A 0.01 $\mu$F and a 0.015 $\mu$F capacitor are in series. Determine the total capacitance.
4. Five 100 pF capacitors are connected in series. What is $C_T$?
5. Determine the voltage across $C_1$ in Figure 13–24.

FIGURE 13–24

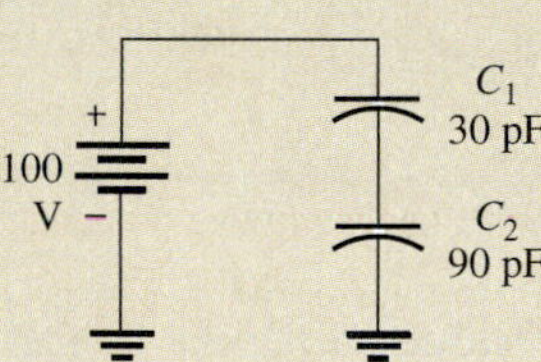


## 13–4 ■ PARALLEL CAPACITORS

***In this section, you will see why capacitances add when they are connected in parallel.***

***After completing this section, you should be able to***

- **Analyze parallel capacitors**
  - Determine total capacitance

When capacitors are connected in parallel, the total capacitance is the sum of the individual capacitances because the effective plate area increases. The calculation of total parallel capacitance is analogous to the calculation of total series resistance.

Consider what happens when the switch in Figure 13–25 is closed. The total charging current from the source divides at the junction of the parallel branches. There is a separate charging current through each branch so that a different charge can be stored by each capacitor. By Kirchhoff's current law, the sum of all of the charging currents is equal to the total current. Therefore, the sum of the charges on the capacitors is equal to the total charge. Also, the voltages across all of the parallel branches are equal. These observations are used to develop a formula for total parallel capacitance as follows for the general case of $n$ capacitors in parallel.

$$Q_T = Q_1 + Q_2 + Q_3 + \cdots + Q_n \quad \textbf{(13–14)}$$

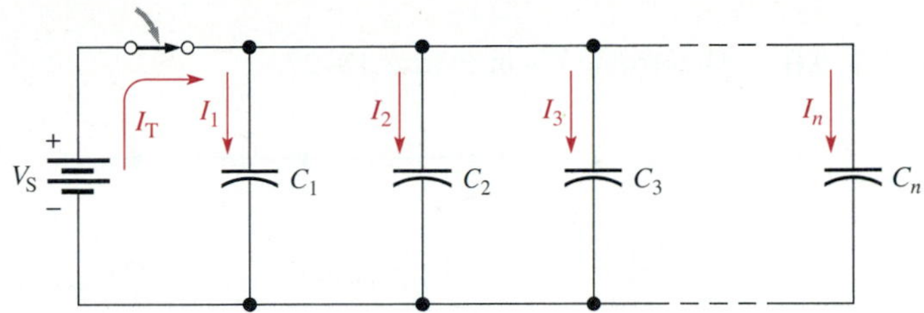


FIGURE 13–25
*Capacitors in parallel.*

From Equation (13–2), $Q = CV$. When this relationship is substituted into each term of Equation (13–14), the following result is obtained:

$$C_TV_T = C_1V_1 + C_2V_2 + C_3V_3 + \cdots + C_nV_n$$

Since $V_T = V_1 = V_2 = V_3 = \cdots = V_n$, the voltages can be factored and canceled, giving

$$C_T = C_1 + C_2 + C_3 + \cdots + C_n \qquad \textbf{(13–15)}$$

Equation (13–15) is the general formula for total parallel capacitance where $n$ is the number of capacitors. Remember,

**The total parallel capacitance is the sum of all the capacitors in parallel.**

For the special case when all of the capacitors have the same value, $C$, multiply the value by the number ($n$) of capacitors in parallel.

$$C_T = nC \qquad \textbf{(13–16)}$$

## EXAMPLE 13–9

What is the total capacitance in Figure 13–26? What is the voltage across each capacitor?

FIGURE 13–26

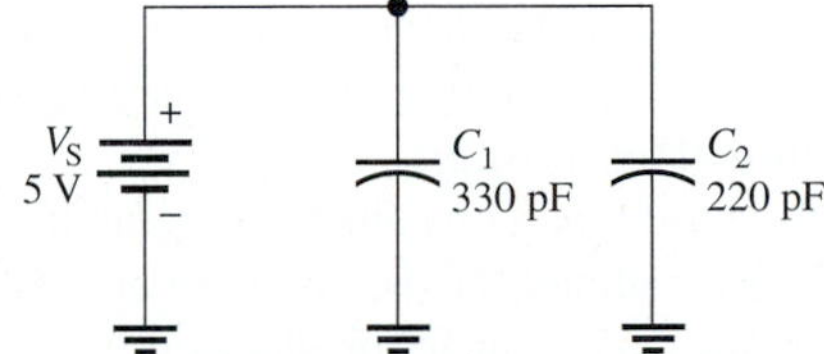


***Solution*** The total capacitance is

$$C_T = C_1 + C_2 = 330\text{ pF} + 220\text{ pF} = 550\text{ pF}$$

The voltage across each capacitor in parallel is equal to the source voltage.

$$V_S = V_1 = V_2 = 5\text{ V}$$

***Related Exercise*** What is $C_T$ if a 100 pF capacitor is connected in parallel with $C_2$ in Figure 13–24?

**EXAMPLE 13–10** Determine $C_T$ in Figure 13–27.

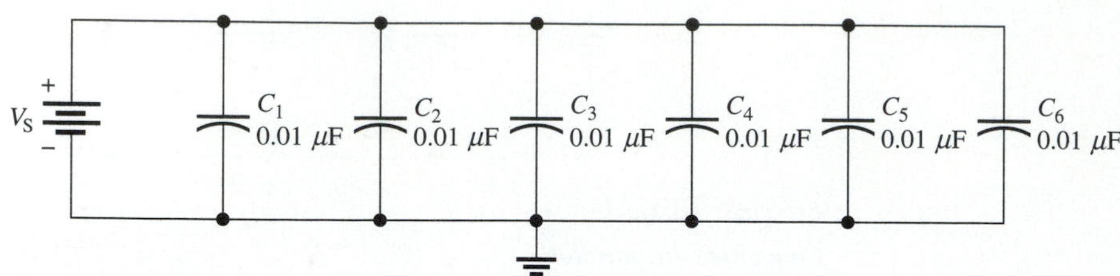


FIGURE 13–27

***Solution*** There are six equal-value capacitors in parallel, so $n = 6$.

$$C_T = nC = (6)(0.01\ \mu F) = 0.06\ \mu F$$

***Related Exercise*** If three more 0.01 μF capacitors are connected in parallel in Figure 13–27, what is the total capacitance?

**SECTION 13–4 REVIEW**

1. How is total parallel capacitance determined?
2. In a certain application, you need 0.05 μF. The only values available are 0.01 μF, which are available in large quantities. How can you get the total capacitance that you need?
3. The following capacitors are in parallel: 10 pF, 5 pF, 33 pF, and 50 pF. What is $C_T$?

## 13–5 ▪ CAPACITORS IN DC CIRCUITS

***A capacitor will charge up when it is connected to a dc voltage source. The buildup of charge across the plates occurs in a predictable manner which is dependent on the capacitance and the resistance in a circuit.***

***After completing this section, you should be able to***

- **Analyze capacitive dc switching circuits**
  - ☐ Describe the charging and discharging of a capacitor
  - ☐ Define *time constant*
  - ☐ Relate the time constant to charging and discharging
  - ☐ Write equations for the charging and discharging curves
  - ☐ Explain why a capacitor blocks dc

### Charging a Capacitor

A capacitor charges when it is connected to a dc voltage source, as shown in Figure 13–28. The capacitor in part (a) of the figure is uncharged; that is, plate *A* and plate *B* have equal numbers of free electrons. When the switch is closed, as shown in part (b), the source moves electrons away from plate *A* through the circuit to plate *B* as the arrows indicate. As plate *A* loses electrons and plate *B* gains electrons, plate *A* becomes positive with respect to plate *B*. As this charging process continues, the voltage across the plates

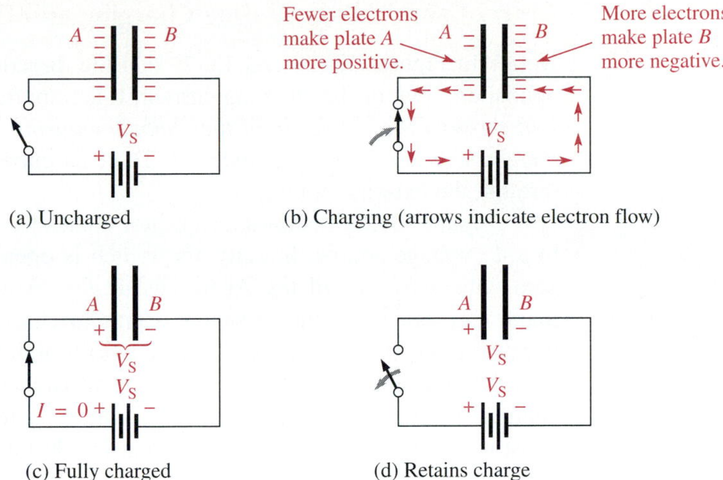


FIGURE 13–28
*Charging a capacitor.*

builds up rapidly until it is equal to the applied voltage, $V_S$, but opposite in polarity, as shown in part (c). When the capacitor is fully charged, there is no current.

**A capacitor blocks constant dc.**

When the charged capacitor is disconnected from the source, as shown in Figure 13–28(d), it remains charged for long periods of time, depending on its leakage resistance, and can cause severe electrical shock. The charge on an electrolytic capacitor generally leaks off more rapidly than in other types of capacitors.

## Discharging a Capacitor

When a wire is connected across a charged capacitor, as shown in Figure 13–29, the capacitor will discharge. In this particular case, a very low resistance path (the wire) is connected across the capacitor with a switch. Before the switch is closed, the capacitor is charged to 50 V, as indicated in part (a). When the switch is closed, as shown in part (b), the excess electrons on plate *B* move through the circuit to plate *A* (indicated by the arrows); as a result of the current through the low resistance of the wire, the energy stored by the capacitor is dissipated in the wire. The charge is neutralized when the numbers of free electrons on both plates are again equal. At this time, the voltage across the capacitor is zero, and the capacitor is completely discharged, as shown in part (c).

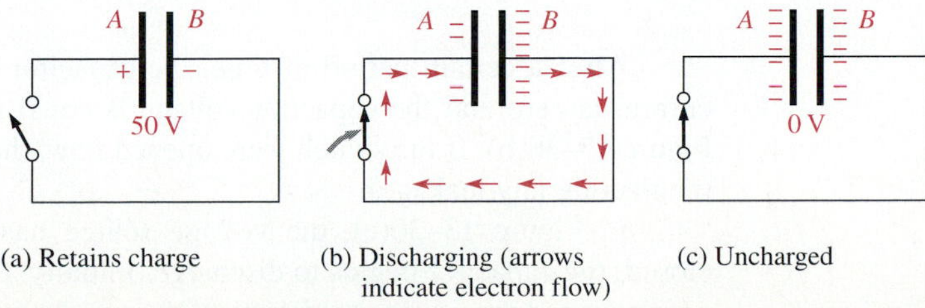


FIGURE 13–29
*Discharging a capacitor.*

## Current and Voltage During Charging and Discharging

Notice in Figures 13–28 and 13–29 that the direction of the current during discharge is opposite to that of the charging current. It is important to understand that *there is no current through the dielectric of the capacitor during charging or discharging because the dielectric is an insulating material.* There is current from one plate to the other only through the external circuit.

Figure 13–30(a) shows a capacitor connected in series with a resistor and a switch to a dc voltage source. Initially, the switch is open and the capacitor is uncharged with zero volts across its plates. At the instant the switch is closed, the current jumps to its maximum value and the capacitor begins to charge. The current is maximum initially because the capacitor has zero volts across it and, therefore, effectively acts as a short; thus, the current is limited only by the resistance. As time passes and the capacitor charges, the current decreases and the voltage across the capacitor ($V_C$) increases. The resistor voltage is proportional to the current during this charging period.

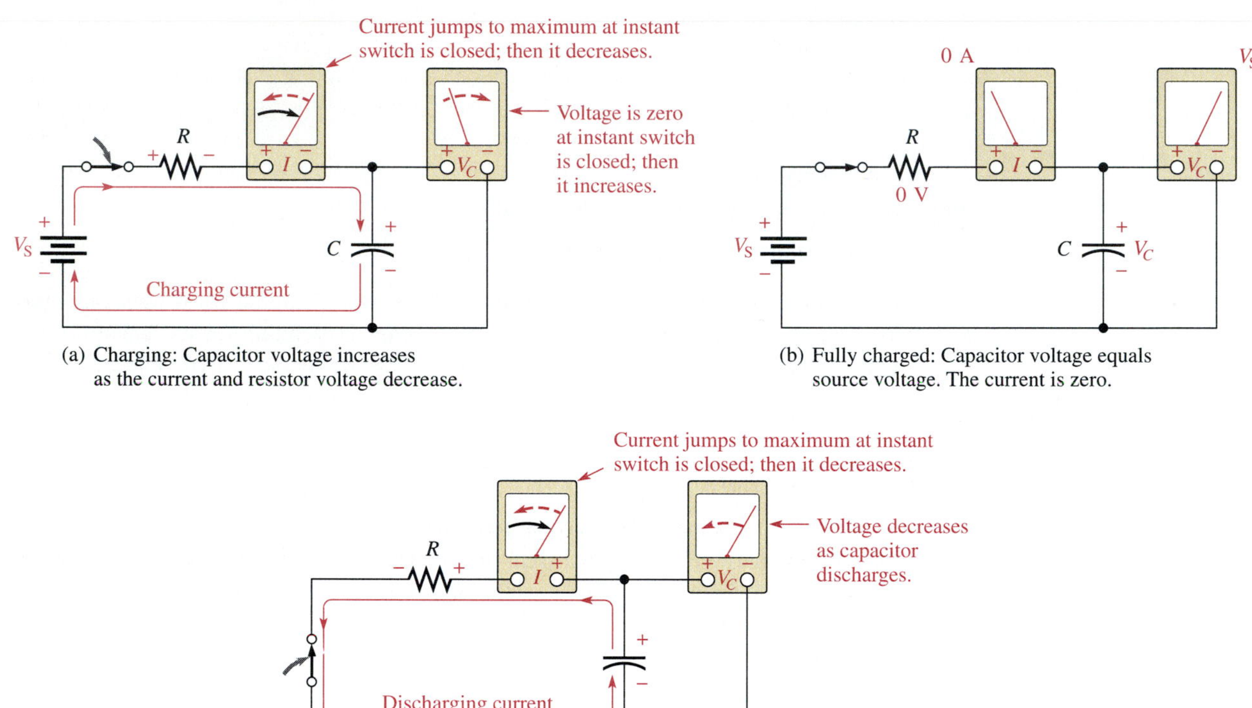


(a) Charging: Capacitor voltage increases as the current and resistor voltage decrease.

(b) Fully charged: Capacitor voltage equals source voltage. The current is zero.

(c) Discharging: Capacitor voltage, resistor voltage, and the current decrease from initial maximums. Note that the discharge current is opposite to the charge current.

**FIGURE 13–30**
***Current and voltage in a charging and discharging capacitor.***

After a certain period of time, the capacitor reaches full charge. At this point, the current is zero and the capacitor voltage is equal to the dc source voltage, as shown in Figure 13–30(b). If the switch were opened now, the capacitor would retain its full charge (neglecting any leakage).

In Figure 13–30(c), the voltage source has been removed. When the switch is closed, the capacitor begins to discharge. Initially, the current jumps to a maximum but in a direction opposite to its direction during charging. As time passes, the current and capacitor voltage decrease. The resistor voltage is always proportional to the current. When the capacitor has fully discharged, the current and the capacitor voltage are zero.

Remember the following rules about capacitors in dc circuits:

1. Voltage across a capacitor *cannot* change instantaneously.
2. Current in a capacitive circuit *can* ideally change instantaneously.
3. A fully charged capacitor appears as an *open* to nonchanging current.
4. An uncharged capacitor appears as a *short* to an instantaneous change in current.

Now we will examine in more detail how the voltage and current change with time in a capacitive circuit.

## The *RC* Time Constant

In a practical situation, there cannot be capacitance without some resistance in a circuit. It may simply be the small resistance of a wire, or it may be a designed-in resistance. Because of this, the charging and discharging characteristics of a capacitor must always be considered in light of the associated resistance. The resistance introduces the element of *time* in the charging and discharging of a capacitor.

When a capacitor charges or discharges through a resistance, a certain time is required for the capacitor to charge fully or discharge fully. The voltage across a capacitor cannot change instantaneously because a finite time is required to move charge from one point to another. The rate at which the capacitor charges or discharges is determined by the **time constant** of the circuit.

**The time constant of a series *RC* circuit is a time interval that equals the product of the resistance and the capacitance.**

The time constant is expressed in seconds when resistance is in ohms and capacitance is in farads. It is symbolized by $\tau$ (Greek letter tau), and the formula is as follows:

$$\tau = RC \qquad \textbf{(13–17)}$$

Recall that $I = Q/t$. The current depends on the amount of charge moved in a given time. When the resistance is increased, the charging current is reduced, thus increasing the charging time of the capacitor. When the capacitance is increased, the amount of charge increases; thus, for the same current, more time is required to charge the capacitor.

**EXAMPLE 13–11** A series *RC* circuit has a resistance of 1 MΩ and a capacitance of 5 μF. What is the time constant?

***Solution*** $\tau = RC = (1 \times 10^6\ \Omega)(5 \times 10^{-6}\ \text{F}) = 5\ \text{s}$

***Related Exercise*** A series *RC* circuit has a 270 kΩ resistor and a 3300 pF capacitor. What is the time constant?

**During one time-constant interval, the charge on a capacitor changes approximately 63%.**

An uncharged capacitor charges to 63% of its fully charged voltage in one time constant. When a capacitor is discharging, its voltage drops to approximately 100% − 63% = 37% of its initial value in one time constant, which is a 63% change.

## The Charging and Discharging Curves

A capacitor charges and discharges following a nonlinear curve, as shown in Figure 13–31. In these graphs, the approximate percentage of full charge is shown at each time-constant interval. This type of curve follows a precise mathematical formula and is called an *exponential curve.* The charging curve is an increasing exponential, and the discharging curve is a decreasing exponential. It takes five time constants to approximately reach the final value. A five time-constant interval is accepted as the time to fully charge or discharge a capacitor and is called the *transient time.*

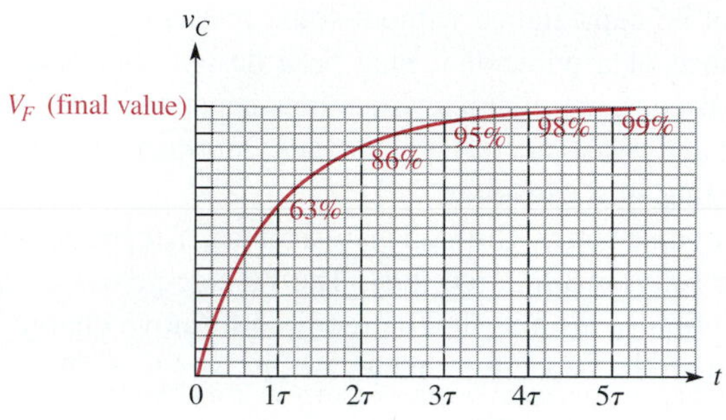


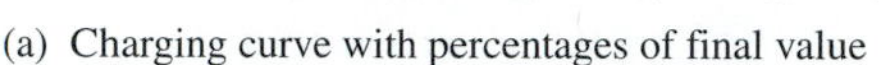

(a) Charging curve with percentages of final value

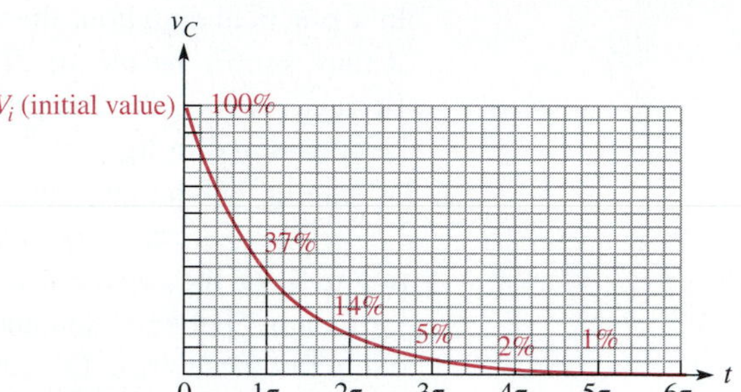


(b) Discharging curve with percentages of initial value

**FIGURE 13–31**

***Charging and discharging exponential curves for an RC circuit.***

***General Formula*** The general expressions for either increasing or decreasing exponential curves are given in the following equations for both instantaneous voltage and instantaneous current.

$$v = V_F + (V_i - V_F)e^{-t/\tau} \quad \textbf{(13–18)}$$

$$i = I_F + (I_i - I_F)e^{-t/\tau} \quad \textbf{(13–19)}$$

where $V_F$ and $I_F$ are the final values, and $V_i$ and $I_i$ are the initial values. $v$ and $i$ are the instantaneous values of the capacitor voltage or current at time $t$, and $e$ is the base of natural logarithms with a value of 2.718. The $e^x$ key or the INV and ln $x$ keys on your calculator make it easy to evaluate this exponential term.

***Charging from Zero*** The formula for the special case in which an increasing exponential voltage curve begins at zero ($V_i = 0$) is given in Equation (13–20). It is developed as follows, starting with the general formula, Equation (13–18).

$$\begin{aligned} v &= V_F + (V_i - V_F)e^{-t/\tau} \\ &= V_F + (0 - V_F)e^{-t/RC} \\ &= V_F - V_F e^{-t/RC} \end{aligned}$$

$$v = V_F(1 - e^{-t/RC}) \quad \textbf{(13–20)}$$

Using Equation (13–20), you can calculate the value of the charging voltage of a capacitor at any instant of time if it is initially uncharged. The same is true for an increasing current.

**EXAMPLE 13–12** In Figure 13–32, determine the capacitor voltage 50 $\mu$s after the switch is closed if the capacitor is initially uncharged. Sketch the charging curve.

FIGURE 13–32

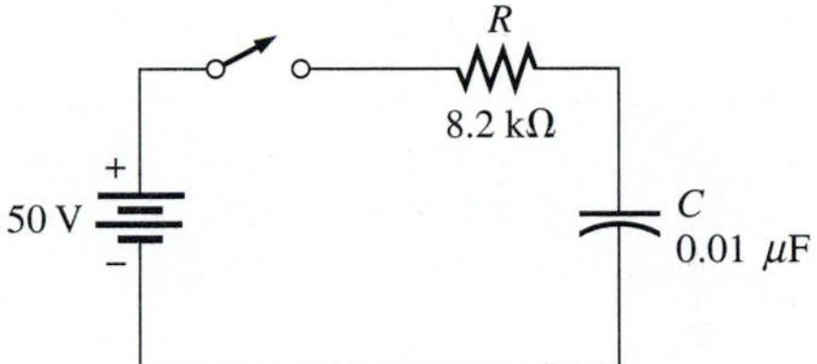


***Solution*** The time constant is $RC = (8.2\text{ k}\Omega)(0.01\ \mu\text{F}) = 82\ \mu\text{s}$. The voltage to which the capacitor will fully charge is 50 V (this is $V_F$). The initial voltage is zero. Notice that 50 $\mu$s is less than one time constant; so the capacitor will charge less than 63% of the full voltage in that time.

$$v_C = V_F(1 - e^{-t/RC}) = (50\text{ V})(1 - e^{-50\mu\text{s}/82\mu\text{s}})$$
$$= (50\text{ V})(1 - e^{-0.61}) = (50\text{ V})(1 - 0.543) = 22.8\text{ V}$$

Determine the value of $e^{-0.61}$ on the calculator by entering –0.61 and then pressing the [2nd] [$e^x$] keys (or [INV] and then [ln x] on some calculators). $e^x$ may or may not be a secondary function on your calculator.

The charging curve for the capacitor is shown in Figure 13–33.

FIGURE 13–33

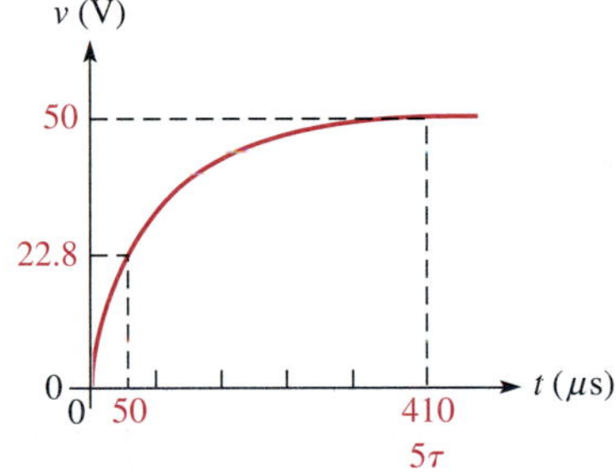


The calculator sequence is

[5] [0] [÷] [8] [2] [=] [+/–] [2nd] [$e^x$] [+/–] [+] [1] [=] [×] [5] [0] [=]

***Related Exercise*** Determine the capacitor voltage 15 $\mu$s after switch closure in Figure 13–32.

***Discharging to Zero*** The formula for the special case in which a decreasing exponential voltage curve ends at zero ($V_F = 0$) is derived from the general formula as follows:

$$v = V_F + (V_i - V_F)e^{-t/\tau}$$
$$= 0 + (V_i - 0)e^{-t/RC}$$

$$v = V_i e^{-t/RC} \quad \textbf{(13–21)}$$

where $V_i$ is the voltage at the beginning of the discharge as shown in Figure 13–31(b). You can use this formula to calculate the discharging voltage at any instant, as Example 13–13 illustrates.

**EXAMPLE 13–13** Determine the capacitor voltage in Figure 13–34 at a point in time 6 ms after the switch is closed. Sketch the discharging curve.

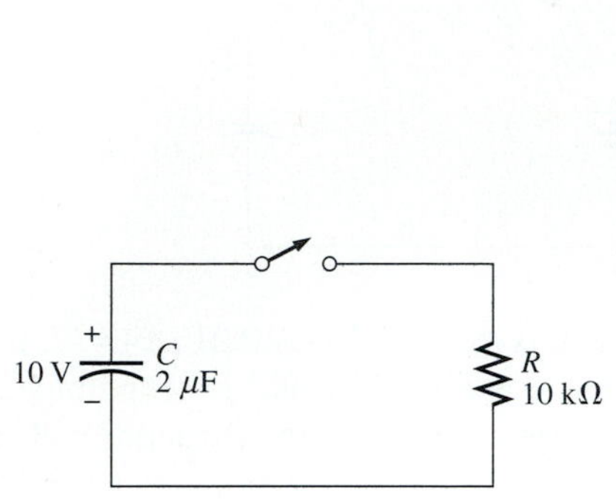


FIGURE 13–34

FIGURE 13–35

***Solution*** The discharge time constant is $RC = (10\text{ k}\Omega)(2\ \mu\text{F}) = 20$ ms. The initial capacitor voltage is 10 V. Notice that 6 ms is less than one time constant, so the capacitor will discharge less than 63%. Therefore, it will have a voltage greater than 37% of the initial voltage at 6 ms.

$$V_C = V_i e^{-t/RC} = 10e^{-6\text{ms}/20\text{ms}} = 10e^{-0.3} = 10(0.741) = 7.41\text{ V}$$

Again, the value of $e^{-0.3}$ can be determined with a calculator.

The discharging curve for the capacitor is shown in Figure 13–35.

***Related Exercise*** In Figure 13–34, change $R$ to 2.2 kΩ and determine the capacitor voltage 1 ms after the switch is closed.

***Graphical Method Using Universal Exponential Curves*** The universal curves in Figure 13–36 provide a graphic solution of the charge and discharge of capacitors. Example 13–14 illustrates this graphical method.

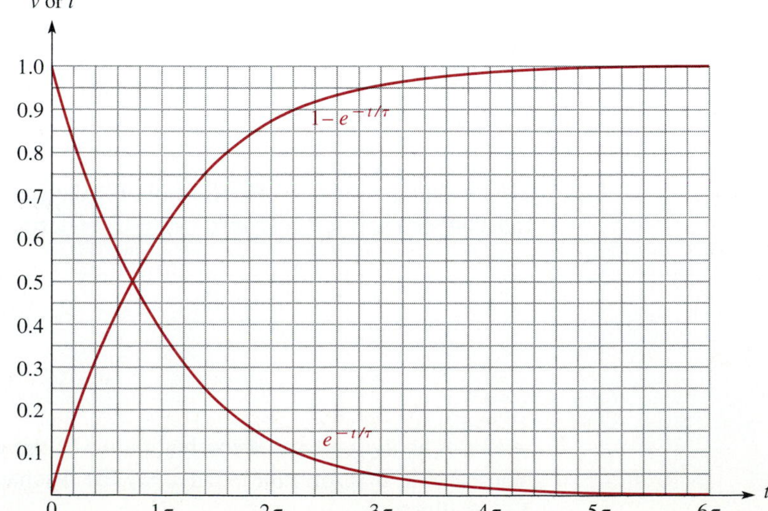


**FIGURE 13–36**
*Normalized universal exponential curves.*

**EXAMPLE 13–14** How long will it take the capacitor in Figure 13–37 to charge to 75 V? What is the capacitor voltage 2 ms after the switch is closed? Use the normalized universal exponential curves in Figure 13–36 to determine the answers.

FIGURE 13–37

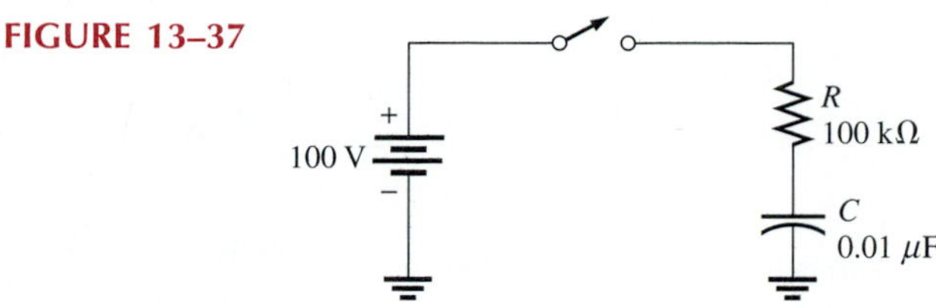


***Solution*** The full charge voltage is 100 V, which is at the 100% level (1.0) on the normalized vertical scale of the graph. The value 75 V is 75% of maximum, or 0.75 on the graph. You can see that this value occurs at 1.4 time constants. One time constant is 1 ms. Therefore, the capacitor voltage reaches 75 V at 1.4 ms after the switch is closed.

The capacitor is at approximately 86 V (0.86 on the vertical axis) in 2 ms. These graphic solutions are shown in Figure 13–38.

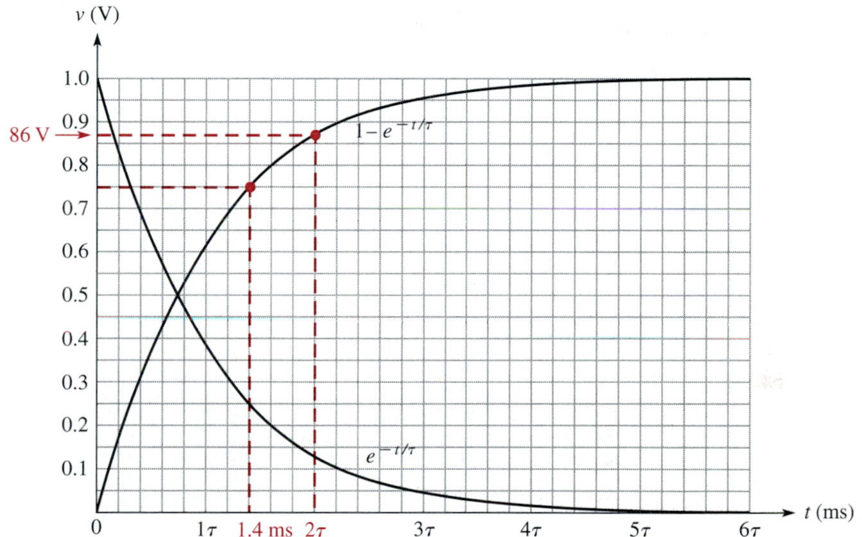


FIGURE 13–38

***Related Exercise*** Using the normalized universal exponential curves, determine how long it will take the capacitor in Figure 13–37 to charge to 50 V? What is the capacitor voltage 3 ms after switch closure?

***Time-Constant Percentage Tables*** The percentages of full charge or discharge at each time-constant interval can be calculated using the exponential formulas, or they can be extracted from the universal exponential curves. The results are summarized in Tables 13–4 and 13–5.

**TABLE 13–4**
*Percentage of final charge after each charging time-constant interval*

| Number of Time Constants | % of Final Charge |
|---|---|
| 1 | 63 |
| 2 | 86 |
| 3 | 95 |
| 4 | 98 |
| 5 | 99 (considered 100%) |

**TABLE 13–5**
*Percentage of initial charge after each discharging time-constant interval*

| Number of Time Constants | % of Initial Charge |
|---|---|
| 1 | 37 |
| 2 | 14 |
| 3 | 5 |
| 4 | 2 |
| 5 | 1 (considered 0) |

## Solving for Time

Occasionally, it is necessary to determine how long it will take a capacitor to charge or discharge to a specified voltage. Equations (13–18) and (13–20) can be solved for $t$ if $v$ is specified. The natural logarithm (abbreviated ln) of $e^{-t/RC}$ is the exponent $-t/RC$. Therefore, taking the natural logarithm of both sides of the equation allows you to solve for time. This procedure is done as follows for the decreasing exponential formula when $V_F = 0$ (Equation 13–21).

$$v = V_i e^{-t/RC}$$

$$\frac{v}{V_i} = e^{-t/RC}$$

$$\ln\left(\frac{v}{V_i}\right) = \ln e^{-t/RC}$$

$$\ln\left(\frac{v}{V_i}\right) = \frac{-t}{RC}$$

$$t = -RC \ln\left(\frac{v}{V_i}\right) \qquad \textbf{(13–22)}$$

The same procedure can be used for the increasing exponential formula in Equation (13–20) as follows:

$$v = V_F(1 - e^{-t/RC})$$

$$\frac{v}{V_F} = 1 - e^{-t/RC}$$

$$1 - \frac{v}{V_F} = e^{-t/RC}$$

$$\ln\left(1 - \frac{v}{V_F}\right) = \ln e^{-t/RC}$$

$$\ln\left(1 - \frac{v}{V_F}\right) = \frac{-t}{RC}$$

$$t = -RC \ln\left(1 - \frac{v}{V_F}\right) \qquad \textbf{(13–23)}$$

**EXAMPLE 13–15** In Figure 13–39, how long will it take the capacitor to discharge to 25 V when the switch is closed?

FIGURE 13–39

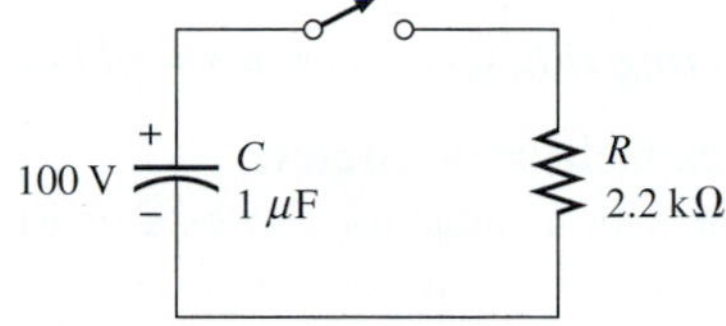


***Solution*** Use Equation (13–22) to find the discharge time.

$$t = -RC \ln\left(\frac{v}{V_i}\right) = -(2.2\ \text{k}\Omega)(1\ \mu\text{F})\ln\left(\frac{25\ \text{V}}{100\ \text{V}}\right)$$
$$= -(2.2\ \text{ms})\ln(0.25) = -(2.2\ \text{ms})(-1.39) = 3.05\ \text{ms}$$

You can determine ln(0.25) with your calculator by first entering 0.25 and then pressing the [ln x] key.

***Related Exercise*** How long will it take the capacitor in Figure 13–39 to discharge to 50 V?

**SECTION 13–5 REVIEW**

1. Determine the time constant when $R = 1.2\ \text{k}\Omega$ and $C = 1000$ pF.
2. If the circuit mentioned in Question 1 is charged with a 5 V source, how long will it take the capacitor to reach full charge? At full charge, what is the capacitor voltage?
3. A certain circuit has a time constant of 1 ms. If it is charged with a 10 V battery, what will the capacitor voltage be at each of the following intervals: 2 ms, 3 ms, 4 ms, and 5 ms?
4. A capacitor is charged to 100 V. If it is discharged through a resistor, what is the capacitor voltage at one time constant?
5. In Figure 13–40, determine the voltage across the capacitor at 2.5 time constants after the switch is closed.
6. In Figure 13–40, how long will it take the capacitor to discharge to 15 V?

FIGURE 13–40

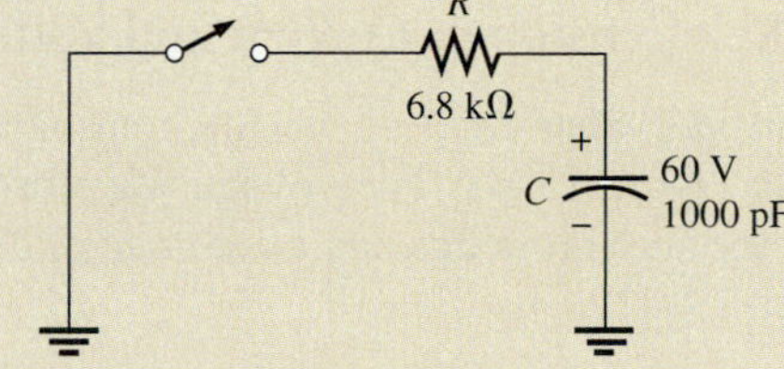

## 13–6 ▪ CAPACITORS IN AC CIRCUITS

***As you saw in the last section, a capacitor blocks dc. You will learn in this section that a capacitor passes ac but with an amount of opposition that depends on the frequency of the ac.***

***After completing this section, you should be able to***

- **Analyze capacitive ac circuits**
  - ☐ Explain why a capacitor causes a phase shift between voltage and current
  - ☐ Define *capacitive reactance*
  - ☐ Determine the value of capacitive reactance in a given circuit
  - ☐ Discuss instantaneous, true, and reactive power in a capacitor

In order to understand fully the action of capacitors in ac circuits, the concept of the derivative must be introduced. *The derivative of a time-varying quantity is the instantaneous rate of change of that quantity.*

Recall that current is the rate of flow of charge (electrons). Therefore, instantaneous current, $i$, can be expressed as the instantaneous rate of change of charge, $q$, with respect to time, $t$.

$$i = \frac{dq}{dt} \quad (13\text{–}24)$$

The term $dq/dt$ is the derivative of $q$ with respect to time and represents the instantaneous rate of change of $q$. Also, in terms of instantaneous quantities, $q = Cv$. Therefore, from a basic rule of differential calculus, the derivative of $q$ is $dq/dt = C(dv/dt)$. Since $i = dq/dt$, we get the following relationship:

$$i = C\left(\frac{dv}{dt}\right) \quad (13\text{–}25)$$

This equation says

> **The instantaneous capacitor current is equal to the capacitance times the instantaneous rate of change of the voltage across the capacitor.**

From this, you can see that the faster the voltage across a capacitor changes, the greater the current. These calculus terms are introduced here for the limited purpose of explaining the phase relationship between current and voltage in a capacitive circuit. Calculus is not used for circuit analysis in this text.

### Phase Relationship of Current and Voltage in a Capacitor

Now consider what happens when a sinusoidal voltage is applied across a capacitor, as shown in Figure 13–41. The voltage waveform has a maximum rate of change ($dv/dt$ = max) at the zero crossings and a zero rate of change ($dv/dt = 0$) at the peaks, as indicated in Figure 13–42.

**FIGURE 13–41**
***Sine wave applied to a capacitor.***

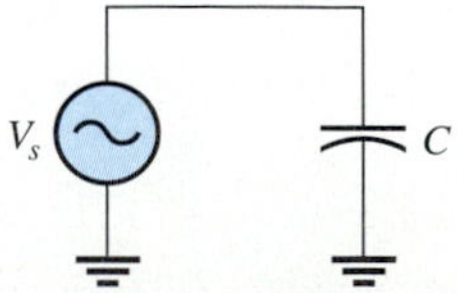

FIGURE 13–42
*The rates of change of a sine wave.*

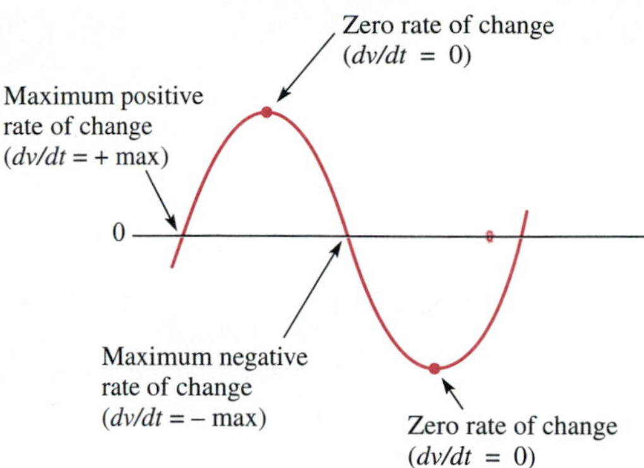


Using Equation (13–25), the **phase** relationship between the current and the voltage for the capacitor can be established. When $dv/dt = 0$, $i$ is also zero because $i = C(dv/dt) = C(0) = 0$. When $dv/dt$ is a positive-going maximum, $i$ is a positive maximum; when $dv/dt$ is a negative-going maximum, $i$ is a negative maximum.

A sinusoidal voltage always produces a sinusoidal current in a capacitive circuit. Therefore, the current can be plotted with respect to the voltage by knowing the points on the voltage curve at which the current is zero and those at which it is maximum. This relationship is shown in Figure 13–43(a). Notice that the current leads the voltage in phase by 90°. This is always true in a purely capacitive circuit. A phasor diagram of this relationship is shown in Figure 13–43(b).

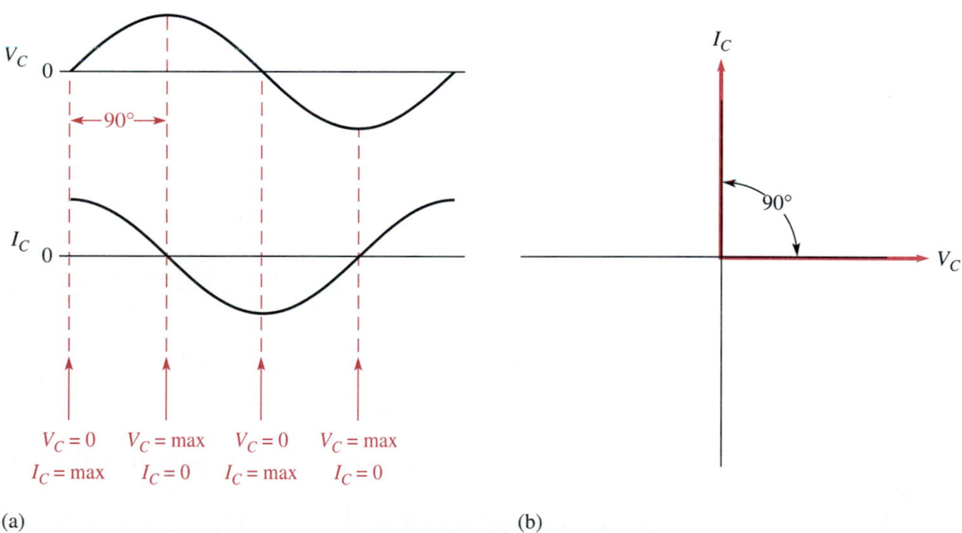


FIGURE 13–43
*Current is always leading the capacitor voltage by 90°.*

## Capacitive Reactance, $X_C$

**Capacitive reactance** is the opposition to sinusoidal current, expressed in ohms. The symbol for capacitive reactance is $X_C$.

To develop a formula for $X_C$, we will use the relationship $i = C(dv/dt)$ and the curves in Figure 13–44. The rate of change of voltage is directly related to frequency. The faster the voltage changes, the higher the frequency. For example, you can see that in

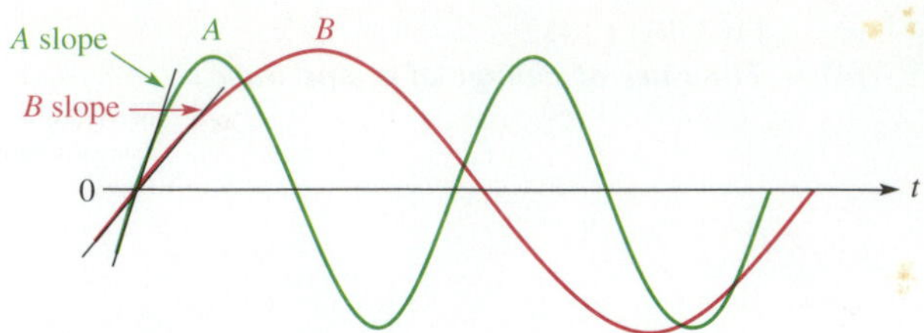


FIGURE 13–44
***A higher frequency waveform has a greater slope at its zero crossings, corresponding to a higher rate of change.***

Figure 13–44 the slope of sine wave *A* at the zero crossings is greater than that of sine wave *B*. The slope of a curve at a point indicates the rate of change at that point. Sine wave *A* has a higher frequency than sine wave *B*, as indicated by a greater maximum rate of change ($dv/dt$ is greater at the zero crossings).

When frequency increases, $dv/dt$ increases, and thus $i$ increases. When frequency decreases, $dv/dt$ decreases, and thus $i$ decreases.

$$\overset{\uparrow}{i} = C(\overset{\uparrow}{dv/dt}) \qquad \text{and} \qquad \underset{\downarrow}{i} = C(\underset{\downarrow}{dv/dt})$$

An increase in $i$ means that there is less opposition to current ($X_C$ is less), and a decrease in $i$ means a greater opposition to current ($X_C$ is greater). Therefore, $X_C$ is inversely proportional to $i$ and thus inversely proportional to frequency.

**$X_C$ is proportional to $\frac{1}{f}$.**

From the same relationship $i = C(dv/dt)$, you can see that if $dv/dt$ is constant and $C$ is varied, an increase in $C$ produces an increase in $i$, and a decrease in $C$ produces a decrease in $i$.

$$\overset{\uparrow}{i} = \overset{\uparrow}{C}(dv/dt) \qquad \text{and} \qquad \underset{\downarrow}{i} = \underset{\downarrow}{C}(dv/dt)$$

Again, an increase in $i$ means less opposition ($X_C$ is less), and a decrease in $i$ means greater opposition ($X_C$ is greater). Therefore, $X_C$ is inversely proportional to $i$ and thus inversely proportional to capacitance.

The capacitive reactance is inversely proportional to both $f$ and $C$.

**$X_C$ is proportional to $\frac{1}{fC}$.**

Thus far, we have determined a proportional relationship between $X_C$ and $1/fC$. We need a formula that tells us what $X_C$ is equal to so that it can be calculated. This important formula is derived in Appendix C and is stated as follows:

$$X_C = \frac{1}{2\pi fC} \qquad \textbf{(13–26)}$$

$X_C$ is in ohms when $f$ is in hertz and $C$ is in farads. Notice that $2\pi$ appears in the denominator as a constant of proportionality. This term is derived from the relationship of a sine wave to rotational motion.

**EXAMPLE 13–16** A sinusoidal voltage is applied to a capacitor, as shown in Figure 13–45. The frequency of the sine wave is 1 kHz. Determine the capacitive reactance.

FIGURE 13–45

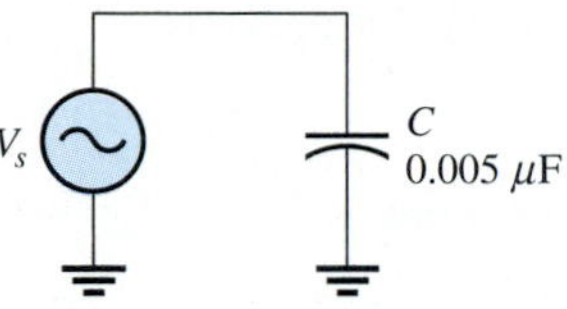


***Solution***

$$X_C = \frac{1}{2\pi fC} = \frac{1}{2\pi(1 \times 10^3\ \text{Hz})(0.005 \times 10^{-6}\ \text{F})} = 31.8\ \text{k}\Omega$$

***Related Exercise*** Determine the frequency required to make the capacitive reactance in Figure 13–45 equal to 10 kΩ.

## Analysis of Capacitive AC Circuits

As you have seen, the current leads the voltage by 90° in purely capacitive ac circuits. If the applied voltage is assigned a reference phase angle of zero, it can be expressed in polar form as $V_s\angle 0°$. The resulting current can be expressed in polar form as $I\angle 90°$ or in rectangular form as $jI$, as shown in Figure 13–46.

FIGURE 13–46

Ohm's law applies to ac circuits containing capacitive reactance with $R$ replaced by $\mathbf{X}_C$ in the Ohm's law formula. Reactance, resistance, voltage, and current are expressed as complex numbers because of the introduction of phase angles as indicated by their boldface, nonitalic format. Applying Ohm's law to the circuit in Figure 13–46 gives the following result:

$$\mathbf{X}_C = \frac{V_s\angle 0°}{I\angle 90°} = \left(\frac{V_s}{I}\right)\angle -90°$$

This shows that $\mathbf{X}_C$ always has a −90° angle attached to its magnitude and is written as $X_C\angle -90°$ or $-jX_C$.

**EXAMPLE 13–17** Determine the rms current in Figure 13–47.

FIGURE 13–47

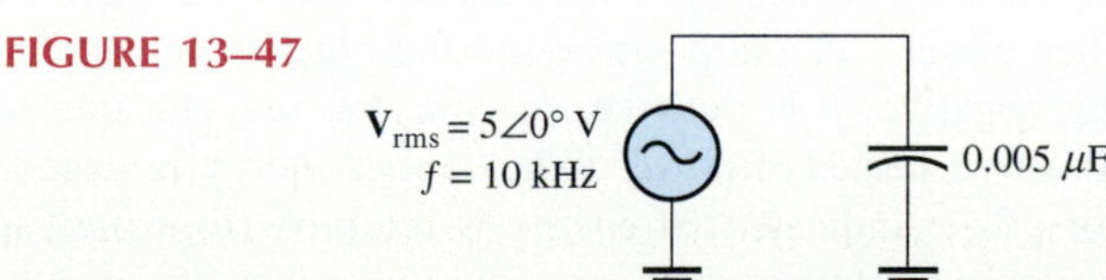

***Solution*** The magnitude of $X_C$ is

$$X_C = \frac{1}{2\pi fC} = \frac{1}{2\pi(10 \times 10^3 \text{ Hz})(0.005 \times 10^{-6} \text{ F})} = 3.18 \text{ k}\Omega$$

Expressed in polar form, $\mathbf{X}_C$ is

$$\mathbf{X}_C = 3.18\angle -90^\circ \text{ k}\Omega$$

Applying Ohm's law,

$$\mathbf{I}_{rms} = \frac{\mathbf{V}_{rms}}{\mathbf{X}_C} = \frac{5\angle 0^\circ \text{ V}}{3.18\angle -90^\circ \text{ k}\Omega} = 1.57\angle 90^\circ \text{ mA}$$

Notice that the current expression has a 90° phase angle indicating that it leads the voltage by 90°.

***Related Exercise*** Change the frequency in Figure 13–47 to 25 kHz and determine the rms current.

## Power in a Capacitor

As discussed earlier in this chapter, a charged capacitor stores energy in the electric field within the dielectric. An ideal capacitor does not dissipate energy; it only stores it. When an ac voltage is applied to a capacitor, energy is stored by the capacitor during a portion of the voltage cycle; then the stored energy is returned to the source during another portion of the cycle. There is no net energy loss. Figure 13–48 shows the power curve that results from one cycle of capacitor voltage and current.

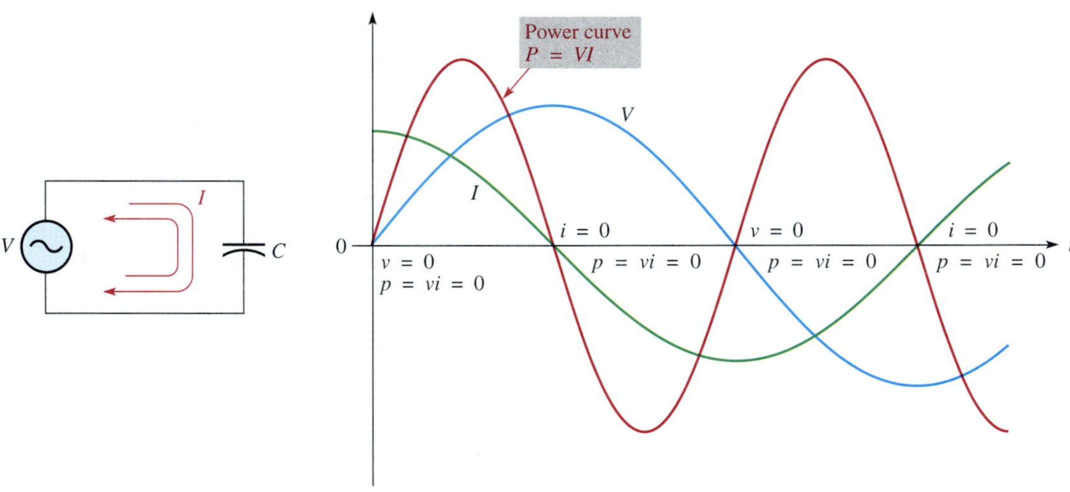


FIGURE 13–48
*Power curve.*

***Instantaneous Power (p)*** The product of $v$ and $i$ gives instantaneous power, $p$. At points where $v$ or $i$ is zero, $p$ is also zero. When both $v$ and $i$ are positive, $p$ is also positive. When either $v$ or $i$ is positive and the other is negative, $p$ is negative. When both $v$ and $i$ are negative, $p$ is positive. As you can see, the power follows a sinusoidal-type curve. Positive values of power indicate that energy is stored by the capacitor. Negative values of power indicate that energy is returned from the capacitor to the source. Note that the power fluctuates at a frequency twice that of the voltage or current as energy is alternately stored and returned to the source.

***True Power ($P_{true}$)*** Ideally, all of the energy stored by a capacitor during the positive portion of the power cycle is returned to the source during the negative portion. No net energy is consumed in the capacitor, so the true power is zero. Actually, because of leakage and foil resistance in a practical capacitor, a small percentage of the total power is dissipated in the form of true power.

***Reactive Power ($P_r$)*** The rate at which a capacitor stores or returns energy is called its **reactive power, $P_r$**. The reactive power is a nonzero quantity, because at any instant in time, the capacitor is actually taking energy from the source or returning energy to it. Reactive power does not represent an energy loss. The following formulas apply:

$$P_r = V_{rms}I_{rms} \tag{13–27}$$

$$P_r = \frac{V_{rms}^2}{X_C} \tag{13–28}$$

$$P_r = I_{rms}^2 X_C \tag{13–29}$$

Notice that these equations are of the same form as those introduced in Chapter 4 for power in a resistor. The voltage and current are expressed in rms. The unit of reactive power is **volt-ampere reactive (VAR).**

**EXAMPLE 13–18** Determine the true power and the reactive power in Figure 13–49.

FIGURE 13–49

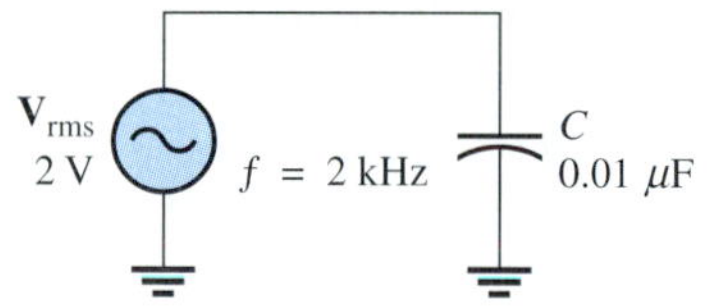

$\mathbf{V}_{rms}$ 2 V, $f$ = 2 kHz, $C$ 0.01 $\mu$F

***Solution*** The true power, $P_{true}$, is *always* zero for an ideal capacitor. The reactive power is determined by first finding the value for the capacitive reactance and then using Equation (13–28).

$$X_C = \frac{1}{2\pi fC} = \frac{1}{2\pi(2 \times 10^3 \text{ Hz})(0.01 \times 10^{-6} \text{ F})} = 7.96 \text{ k}\Omega$$

$$P_r = \frac{V_{rms}^2}{X_C} = \frac{(2 \text{ V})^2}{7.96 \text{ k}\Omega} = 503 \times 10^{-6} \text{ VAR} = 503 \ \mu\text{VAR}$$

***Related Exercise*** If the frequency is doubled in Figure 13–49, what are the true power and the reactive power?

**SECTION 13–6 REVIEW**

1. State the phase relationship between current and voltage in a capacitor.
2. Calculate $X_C$ for $f = 5$ kHz and $C = 50$ pF.
3. At what frequency is the reactance of a 0.1 μF capacitor equal to 2 kΩ?
4. Calculate the rms current in Figure 13–50.

FIGURE 13–50

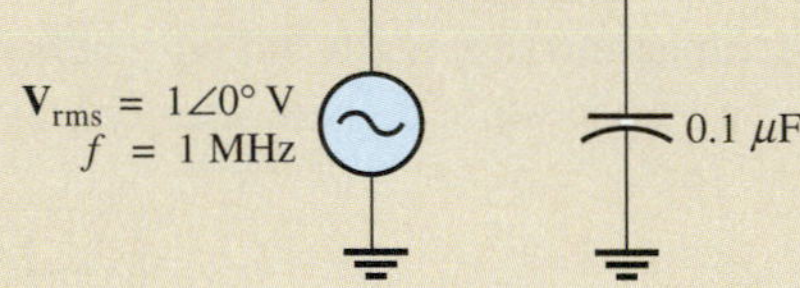


5. A 1 μF capacitor is connected to an ac voltage source of 12 V rms. What is the true power?
6. In Question 5, determine reactive power at a frequency of 500 Hz.

## 13–7 ■ CAPACITOR APPLICATIONS

***Capacitors are widely used in electrical and electronic applications. A few typical applications are discussed in this section to illustrate the usefulness of this component.***

***After completing this section, you should be able to***

- Discuss some capacitor applications
  - ☐ Describe a power supply filter
  - ☐ Explain the purpose of coupling and by-pass capacitors
  - ☐ Discuss the basics of capacitors applied to tuned circuits, timing circuits, and computer memories

If you pick up any circuit board, open any power supply, or look inside any piece of electronic equipment, chances are you will find capacitors of one type or another. These components are used for a variety of reasons in both dc and ac applications.

### Electrical Storage

One of the most basic applications of a capacitor is as a backup voltage source for low-power circuits such as certain types of semiconductor memories in computers. This particular application requires a very high capacitance value and negligible leakage.

The storage capacitor is connected between the dc power supply input to the circuit and ground. When the circuit is operating from its normal power supply, the capacitor remains fully charged to the dc power supply voltage. If the normal power source is disrupted, effectively removing the power supply from the circuit, the storage capacitor temporarily becomes the power source for the circuit.

The capacitor provides voltage and current to the circuit as long as its charge remains sufficient. As current is drawn by the circuit, charge is removed from the capacitor and the voltage decreases. For this reason, the storage capacitor can only be used as a temporary power source. The length of time that the capacitor can provide sufficient power to the circuit depends on the capacitance and the amount of current drawn by the circuit. The smaller the current and the higher the capacitance, the longer the time.

## Power Supply Filtering

A basic dc power supply consists of a circuit known as a **rectifier** followed by a **filter.** The rectifier converts the 110 V, 60 Hz sinusoidal voltage available at a standard outlet to a pulsating dc voltage that can be either a half-wave rectified voltage or a full-wave rectified voltage, depending on the type of rectifier circuit. As shown in Figure 13–51(a), a half-wave rectifier removes each negative half-cycle of the sinusoidal voltage. As shown in Figure 13–51(b), a full-wave rectifier actually reverses the polarity of the negative portion of each cycle. Both half-wave and full-wave rectified voltages are dc because, even though they are changing, they do not alternate polarity.

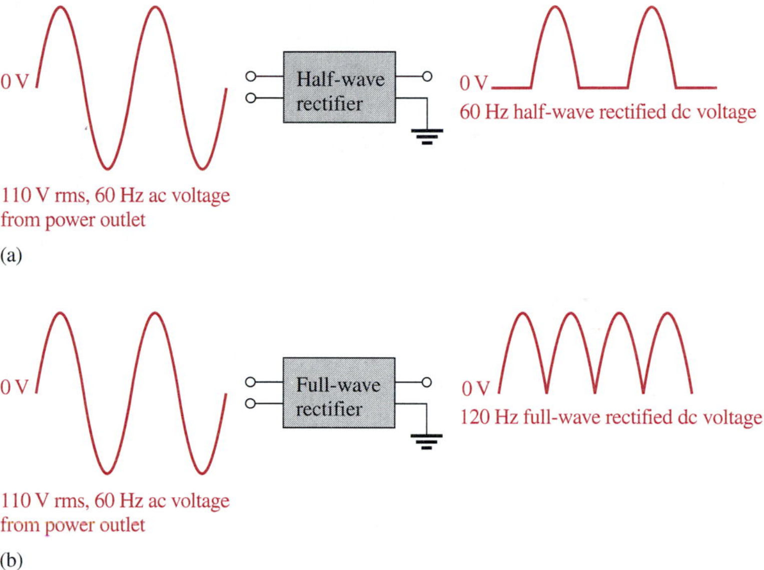


**FIGURE 13–51**
*Half-wave and full-wave rectifier operation.*

To be useful for powering electronic circuits, the rectified voltage must be changed to constant dc voltage because all circuits require constant power. The filter eliminates the fluctuations in the rectified voltage and provides a smooth constant-value dc voltage to the load which is the electronic circuit, as indicated in Figure 13–52.

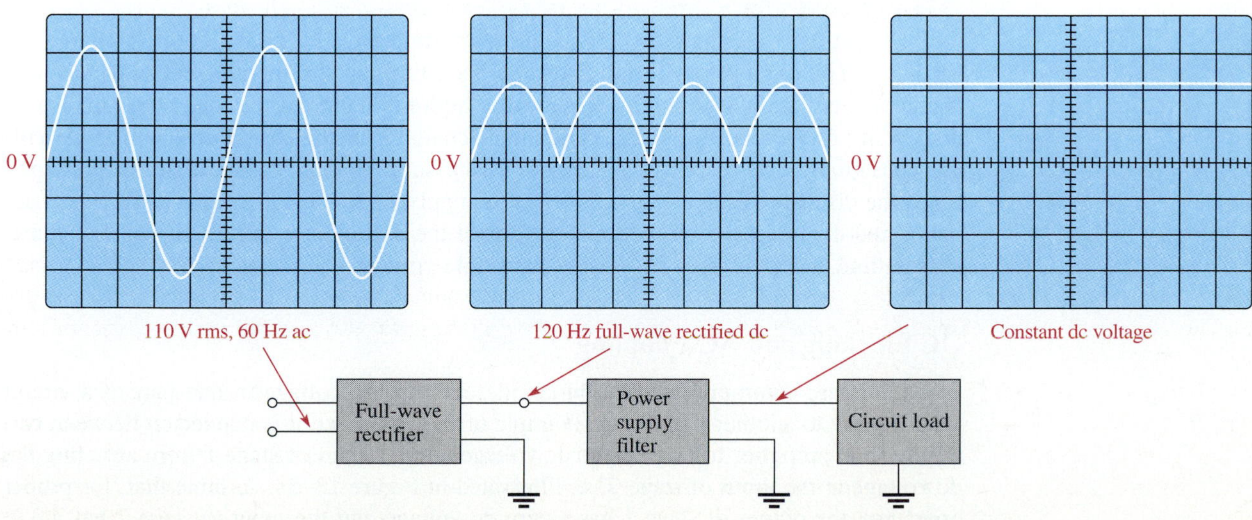


**FIGURE 13–52**
*Basic block diagram and operation of a dc power supply.*

***The Capacitor as a Power Supply Filter*** Capacitors are used as filters in dc power supplies because of their ability to store electrical charge. Figure 13–53(a) shows a dc power supply with a full-wave rectifier and a capacitor filter. The operation can be described from a charging and discharging point-of-view as follows. Assume the capacitor is initially uncharged. When the power supply is first turned on and the first cycle of the rectified voltage occurs, the capacitor will quickly charge through the low resistance of the rectifier. The capacitor voltage will follow the rectified voltage curve up to the peak of the rectified voltage. As the rectified voltage passes the peak and begins to decrease, the capacitor will begin to discharge very slowly through the high resistance of the load circuit, as indicated in Figure 13–53(b). The amount of discharge is typically very small and is exaggerated in the figure for purposes of illustration. The next cycle of the rectified voltage will recharge the capacitor back to the peak value by replenishing the small amount of charge lost since the previous peak. This pattern of a small amount of charging and discharging continues as long as the power is on.

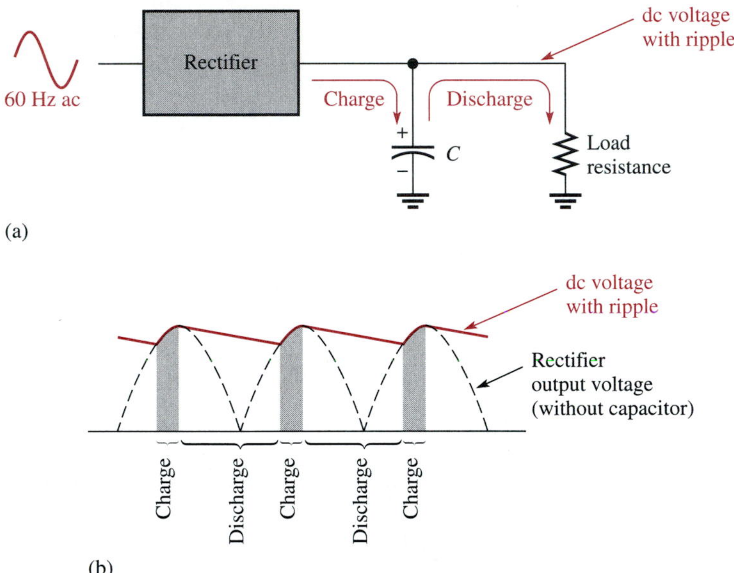


**FIGURE 13–53**
***Basic operation of a power supply filter capacitor.***

A rectifier is designed so that it allows current only in the direction to charge the capacitor. The capacitor will not discharge back through the rectifier but will only discharge a small amount through the relatively high resistance of the load. The small fluctuation in voltage due to the charging and discharging of the capacitor is called the **ripple voltage.** A good dc power supply has a very small amount of ripple on its dc output. Since the discharge time constant of a power supply filter capacitor depends on its capacitance and the resistance of the load, the higher the capacitance value, the longer the discharge time and therefore, the smaller the ripple voltage.

## DC Blocking and AC Coupling

Capacitors are commonly used to block the constant dc voltage in one part of a circuit from getting to another part. As an example of this, a capacitor is connected between two stages of an amplifier to prevent the dc voltage at the output of stage 1 from affecting the dc voltage at the input of stage 2, as illustrated in Figure 13–54. Assume that, for proper operation, the output of stage 1 has a zero dc voltage and the input to stage 2 has a 3 V dc voltage. The capacitor prevents the 3 V dc at stage 2 from getting to the stage 1 output and affecting its zero value, and vice versa.

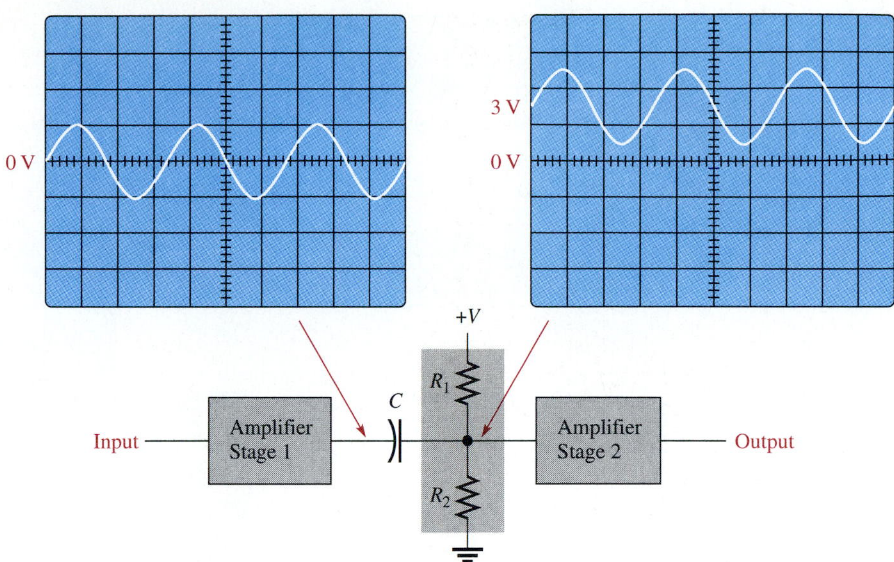


**FIGURE 13–54**
***An application of a capacitor to block dc and couple ac in an amplifier.***

If a sinusoidal signal voltage is applied to the input to stage 1, the signal voltage is increased (amplified) and appears on the ouput of stage 1, as shown in Figure 13–54. The amplified signal voltage is then coupled through the capacitor to the input of stage 2 where it is superimposed on the 3 V dc level and then again amplified by stage 2. In order for the signal voltage to be passed through the capacitor without being reduced, the capacitor must be large enough so that its reactance at the frequency of the signal voltage is negligible. In this type of application, the capacitor is known as a *coupling capacitor* which ideally appears as an open to dc and as a short to ac. As the signal frequency is reduced, the capacitive reactance increases and, at some point, the capacitive reactance becomes large enough to cause a significant reduction in ac voltage between stage 1 and stage 2.

## Power Line Decoupling

Capacitors connected from the dc supply voltage line to ground are used on circuit boards to decouple unwanted voltage transients or spikes that occur on the dc supply voltage because of fast switching digital circuits. A voltage transient contains high frequencies that may affect the operation of the circuits. These transients are shorted to ground through the very low reactance of the decoupling capacitors. Several decoupling capacitors are often used at various points along the supply voltage line on a circuit board.

## Bypassing

Another capacitor application is in bypassing an ac voltage around a resistor in a circuit without affecting the dc voltage across the resistor. In amplifier circuits, for example, dc voltages called *bias voltages* are required at various points. For the amplifier to operate properly, certain bias voltages must remain constant and, therefore, any ac voltages must be removed. A sufficiently large capacitor connected from a bias point to ground provides a low reactance path to ground for ac voltages, leaving the constant dc bias voltage at the given point. At lower frequencies, the bypass capacitor becomes less effective because of its increased reactance. This bypass application is illustrated in Figure 13–55.

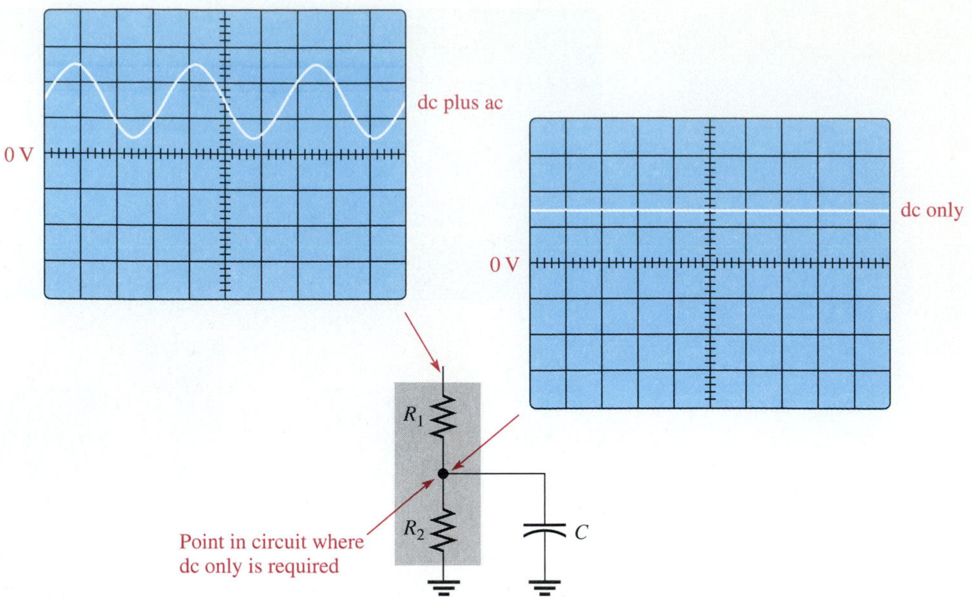


FIGURE 13–55
***Example of the operation of a bypass capacitor.***

## Signal Filters

Capacitors are essential to the operation of a class of circuits called *filters* that are used for selecting one ac signal with a certain specified frequency from a wide range of signals with many different frequencies or for selecting a certain band of frequencies and eliminating all others. A common example of this application is in radio and television receivers where it is necessary to select the signal transmitted from a given station and eliminate or filter out the signals transmitted from all the other stations in the area.

When you tune your radio or TV, you are actually changing the capacitance in the tuner circuit (which is a type of filter) so that only the signal from the station or channel you want passes through to the receiver circuitry. Capacitors are used in conjunction with resistors, inductors (covered in the next chapter), and other components in these types of filters. The topic of filters will be covered in detail in Chapter 19.

The main characteristic of a filter is its frequency selectivity, which is based on the fact that the reactance of a capacitor depends on frequency ($X_C = 1/2\pi fC$).

## Timing Circuits

Another important area in which capacitors are used is in timing circuits that generate specified time delays or produce waveforms with specific characteristics. Recall that the time constant of a circuit with resistance and capacitance can be controlled by selecting appropriate values for $R$ and $C$. The charging time of a capacitor can be used as a basic time delay in various types of circuits. An example is the circuit that controls the turn indicators on your car where the light flashes on and off at regular intervals.

## Computer Memories

Dynamic computer memories use capacitors as the basic storage element for binary information, which consists of two binary digits, 1 and 0. A charged capacitor can represent a stored 1 and a discharged capacitor can represent a stored 0. Patterns of 1s and 0s that make up binary data are stored in a memory that consists of an array of capacitors with associated circuitry. You will study this topic in a computer or digital fundamentals course.

**SECTION 13–7 REVIEW**

1. Explain how half-wave or full-wave rectified dc voltages are smoothed out by a filter capacitor.
2. Explain the purpose of a coupling capacitor.
3. How large must a coupling capacitor be?
4. Explain the purpose of a decoupling capacitor.
5. Discuss how the relationship of frequency and capacitive reactance is important in frequency-selective circuits such as signal filters.
6. What characteristic of a capacitor is most important in time-delay applications.

## 13–8 ■ TESTING CAPACITORS

***Capacitors are very reliable devices but their useful life can be extended significantly by operating them well within the voltage rating and at moderate temperatures. In this section, the basic types of failures are discussed and methods for checking for them are introduced.***

***After completing this section, you should be able to***

- **Test a capacitor**
  - ☐ Perform an ohmmeter check
  - ☐ Explain what an *LC* meter is

Capacitor failures can be categorized into two areas—catastrophic and degradation. The catastrophic failures are usually a short circuit caused by dielectric breakdown or an open circuit caused by connection failure. Degradation usually results in a gradual decrease in leakage resistance, hence an increase in leakage current or an increase in equivalent series resistance or dielectric absorption.

### Ohmmeter Check

When there is a suspected problem, the capacitor can be removed from the circuit and checked with an analog ohmmeter. First, to be sure that the capacitor is discharged, short its leads, as indicated in Figure 13–56(a). Connect the meter—set on a high ohms range such as ×1M—to the capacitor, as shown in part (b), and observe the needle. It should initially indicate near zero ohms. Then it should begin to move toward the high-resistance end of the scale as the capacitor charges from the ohmmeter's battery, as shown in part (c). When the capacitor is fully charged, the meter will indicate an extremely high resistance as shown in part (d).

As mentioned, the capacitor charges from the internal battery of the ohmmeter, and the meter responds to the charging current. The larger the capacitance value, the more slowly the capacitor will charge, as indicated by the needle movement. For very small pF values, the meter response may be insufficient to indicate the fast charging action.

If the capacitor is internally shorted, the meter will go to zero and stay there. If it is leaky, the final meter reading will be much less than normal. Most capacitors have a resistance of several hundred megohms. The exception is the electrolytic, which may normally have less than one megohm of leakage resistance. If the capacitor is open, no charging action will be observed, and the meter will indicate an infinite resistance.

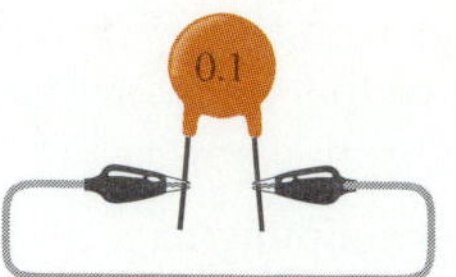


(a) Discharge by shorting leads

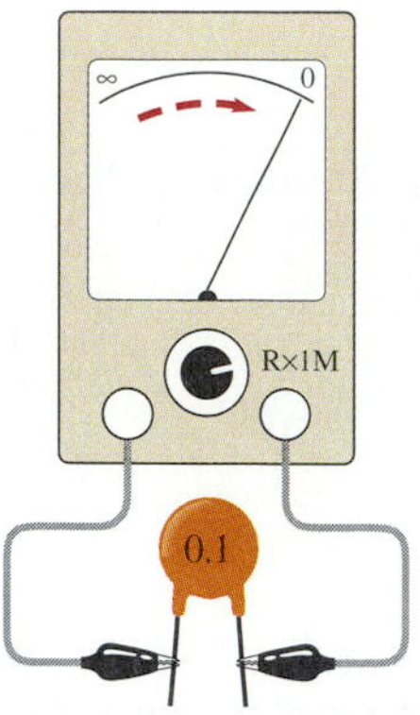


(b) Initially, when the meter is first connected, the pointer jumps immediately to zero.

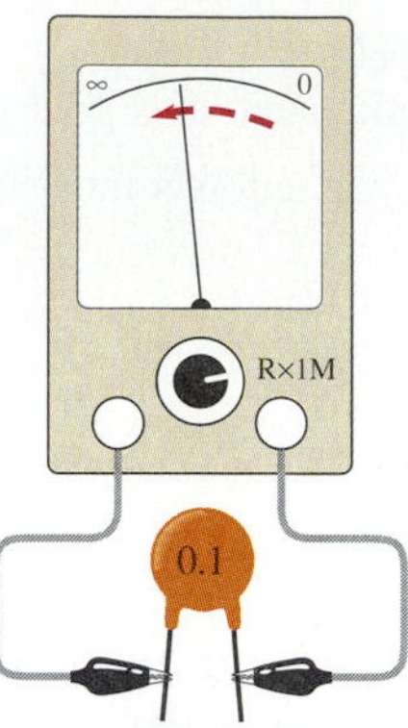


(c) The needle moves slowly toward infinity as the capacitor charges from the ohmmeter battery.

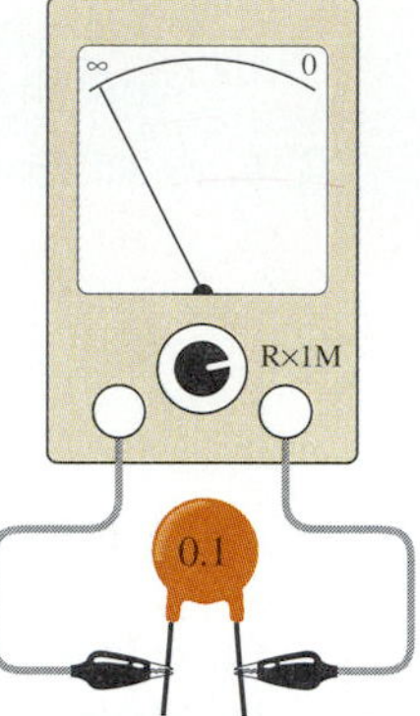


(d) When the capacitor is fully charged the pointer is at infinity.

**FIGURE 13–56**
***Checking a capacitor with an ohmmeter. This check shows a good capacitor.***

## Testing for Capacitance Value and Other Parameters with an *LC* Meter

An *LC* meter such as the one shown in Figure 13–57 can be used to check the value of a capacitor. All capacitors change value over a period of time, some more than others. Ceramic capacitors, for example, often exhibit a 10% to 15% change in value during the first year. Electrolytic capacitors are particularly subject to value change due to drying of the electrolytic solution. In other cases, capacitors may be labeled incorrectly or the wrong value was installed in the circuit. Although a value change represents less than

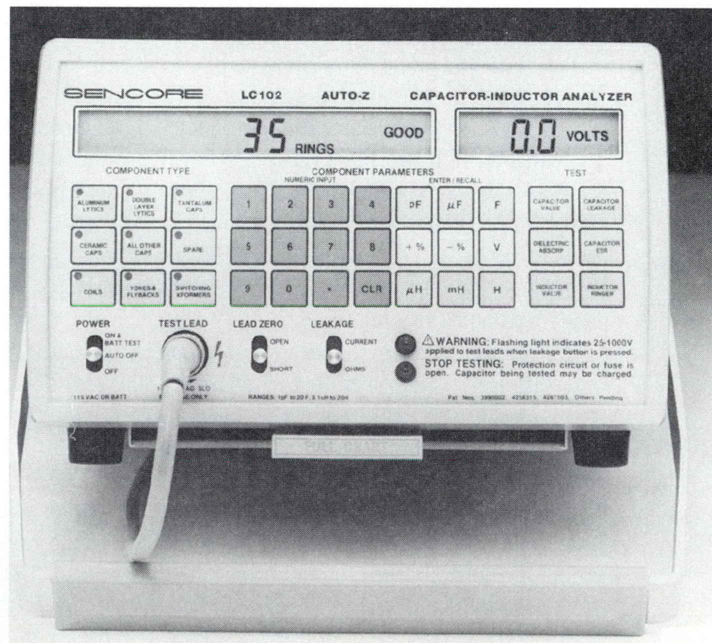


**FIGURE 13–57**
***A typical LC meter (courtesy of SENCORE).***

25% of defective capacitors, a value check should be made to quickly eliminate this as a source of trouble when troubleshooting a circuit.

Typically, values from 1 pF to 200,000 $\mu$F can be measured by simply connecting the capacitor, pushing the appropriate button, and reading the value on the display.

Many *LC* meters can also be used to check for leakage current in capacitors. In order to check for leakage, a sufficient voltage must be applied across the capacitor to simulate operating conditions. This is automatically done by the test instrument. Over 40% of all defective capacitors have excessive leakage current and electrolytics are particularly susceptible to this problem.

The problem of dielectric absorption occurs mostly in electrolytic capacitors when they do not completely discharge during use and retain a residual charge. Approximately 25% of defective capacitors have exhibited this condition.

Another defect sometimes found in capacitors is called equivalent series resistance. This problem may be caused by a defective lead to plate contacts, resistive leads, or resistive plates and shows up only under ac conditions. This is the least common capacitor defect and occurs in less than 10% of all defects.

**SECTION 13–8 REVIEW**

1. How can a capacitor be discharged after removal from the circuit?
2. Describe how the needle of an ohmmeter responds when a good capacitor is checked.
3. List four common capacitor defects.

## 13–9 ■ TECHnology Theory Into Practice

***Capacitors are used in certain types of amplifiers to couple the ac signal while blocking the dc voltage. Capacitors are used in many other applications, but in this TECH TIP, you will focus on the coupling capacitors in an amplifier circuit. This topic was introduced in Section 13–7. A knowledge of amplifier circuits is not necessary for this assignment.***

All amplifier circuits contain transistors that require dc voltages to establish proper operating conditions for amplifying ac signals. These dc voltages are referred to as bias voltages. As indicated in Figure 13–58(a), a common type of dc bias circuit used in amplifiers is the voltage divider formed by $R_1$ and $R_2$, which sets up the proper dc voltage at the input to the amplifier.

When an ac signal voltage is applied to the amplifier, the input coupling capacitor, $C_1$, prevents the internal resistance of the ac source from changing the dc bias voltage. Without the capacitor, the internal source resistance would appear in parallel with $R_2$ and drastically change the value of the dc voltage.

The coupling capacitance is chosen so that its reactance ($X_C$) at the frequency of the ac signal is very small compared to the bias resistor values. The coupling capacitance therefore efficiently couples the ac signal from the source to the input of the amplifier. On the source side of the input coupling capacitor there is only ac but on the amplifier side there is ac plus dc (the signal voltage is riding on the dc bias voltage set by the voltage divider), as indicated in Figure 13–58(a). Capacitor $C_2$ is the output coupling capacitor which couples the amplified ac signal to another amplifier stage that would be connected to the output.

You will check three amplifier boards like the one in Figure 13–58(b) for the proper input voltages using an oscilloscope. If the voltages are incorrect, you will determine the most likely fault. For all measurements, assume the amplifier has no dc loading effect on the voltage-divider bias circuit.

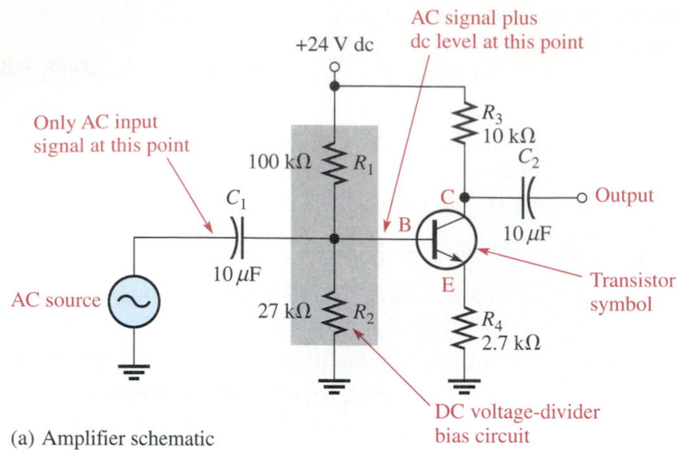

(a) Amplifier schematic

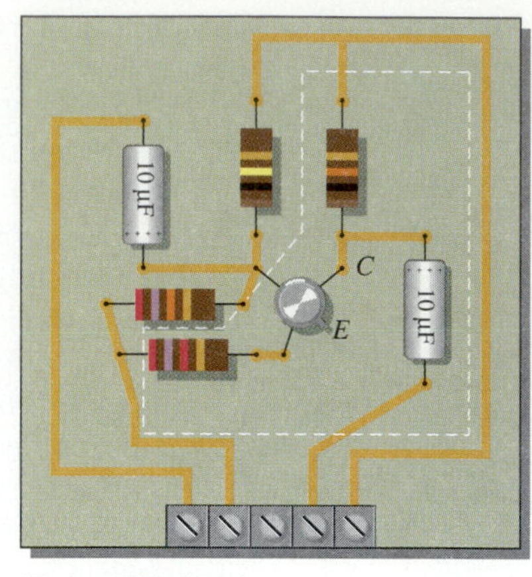

(b) Amplifier board

**FIGURE 13–58**
*Capacitively coupled amplifier.*

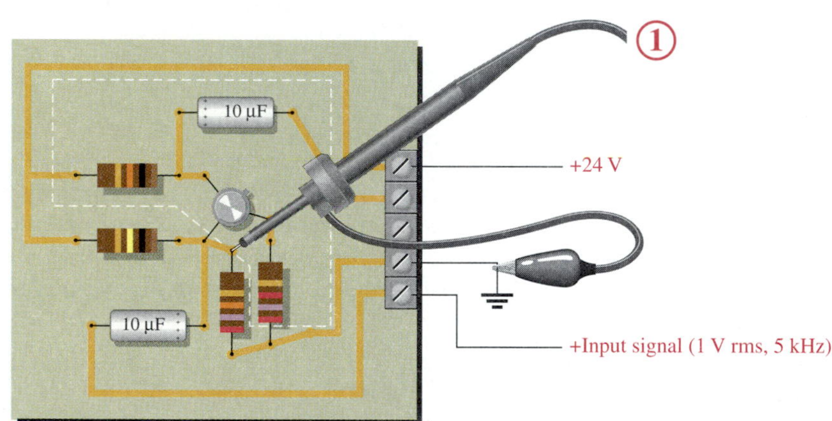


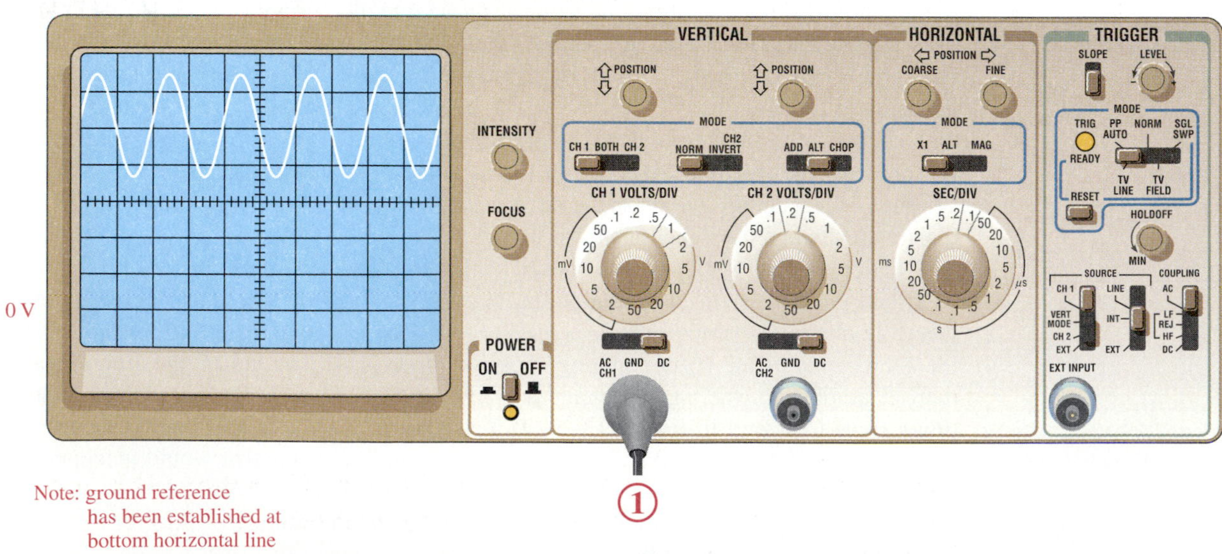


Note: ground reference has been established at bottom horizontal line

**FIGURE 13–59**
*Testing board 1.*

## The Printed Circuit Board and the Schematic

☐ Check the printed circuit board in Figure 13–58(b) to make sure it agrees with the amplifier schematic in part (a).

## Testing Board 1

The oscilloscope probe is connected from channel 1 to the board as shown in Figure 13–59. The input signal from a sinusoidal voltage source is connected to the board and set to a frequency of 5 kHz with an amplitude of 1 V rms.

☐ Determine if the voltage and frequency displayed on the scope are correct. If the scope measurement is incorrect, specify the most likely fault in the circuit.

## Testing Board 2

The oscilloscope probe is connected from channel 1 to board 2 the same as was shown in Figure 13–59 for board 1. The input signal from the sinusoidal voltage source is the same as it was for board 1.

☐ Determine if the scope display in Figure 13–60 is correct. If the scope measurement is incorrect, specify the most likely fault in the circuit.

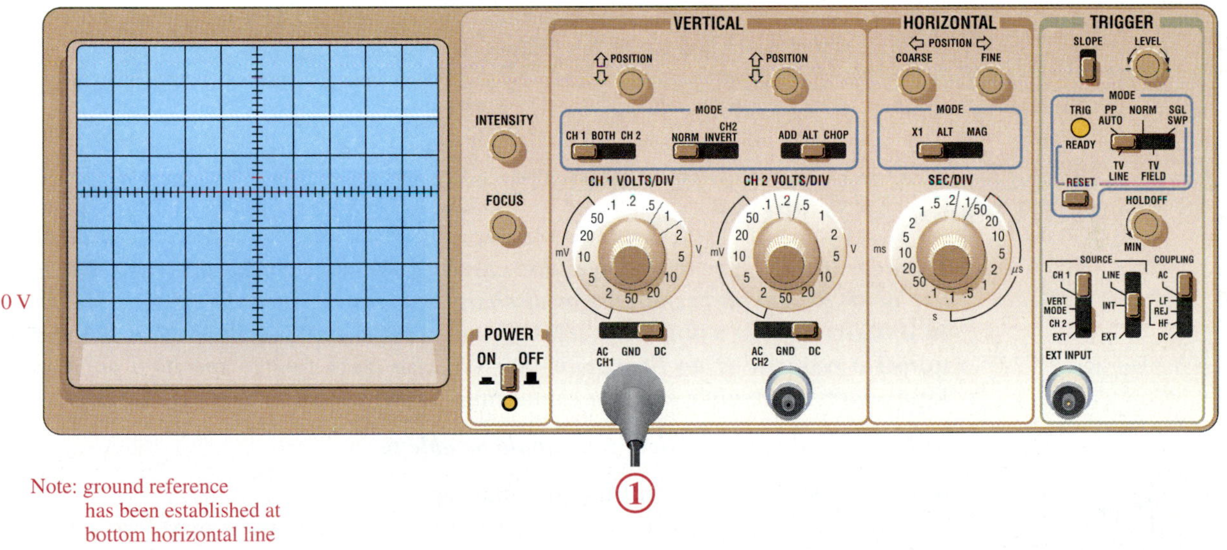


**FIGURE 13–60**
*Testing board 2.*

## Testing Board 3

The oscilloscope probe is connected from channel 1 to board 3 the same as was shown in Figure 13–59 for board 1. The input signal from the sinusoidal voltage source is the same as before.

☐ Determine if the scope display in Figure 13–61 is correct. If the scope measurement is incorrect, specify the most likely fault in the circuit.

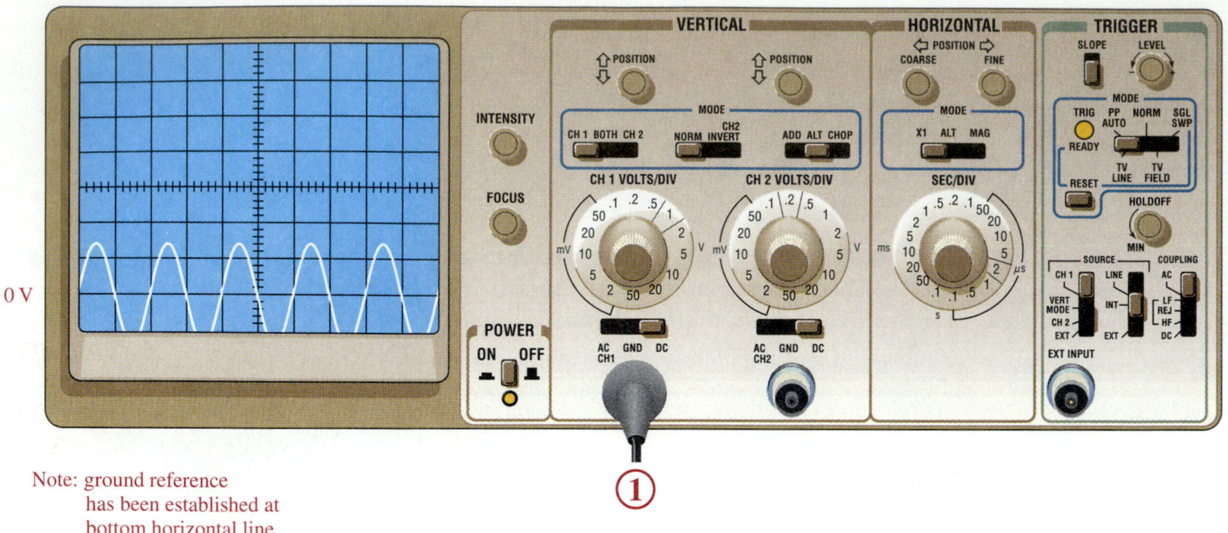


**FIGURE 13–61**
*Testing board 3.*

**SECTION 13–9 REVIEW**

1. Explain why the input coupling capacitor is necessary when connecting an ac source to the amplifier.
2. Capacitor $C_2$ in Figure 13–58 is an output coupling capacitor. Generally, what would you expect to measure at point $C$ and at the output of the circuit when an input signal is applied to the amplifier?

## 13–10 ■ TRANSIENT ANALYSIS OF *RC* CIRCUITS WITH PSpice AND PROBE

***You have learned how the voltages and current change exponentially when the capacitor in an RC circuit is charging or discharging. This is called the transient response. In this section, you will use the .TRAN control statement and Probe to determine the transient response of an RC circuit. You will also learn how to specify a pulse waveform source to simulate a switch connecting the RC circuit to a voltage or ground.***

***After completing this section, you should be able to***

- **Use PSpice and Probe in transient analysis**
  - ☐ Use the PULSE source transient specification to specify pulse waveforms
  - ☐ Simulate a switch closure with a pulse specification
  - ☐ Write circuit files for transient analysis of *RC* circuits
  - ☐ Use Probe to graph charging and discharging curves
  - ☐ Use the Probe cursor to find capacitor voltage and the time constant values
  - ☐ Use the .IC statement to specify initial capacitor voltage

### The PULSE Source Transient Specification

Just as the .AC SIN source transient specification had several parameters that described the sine wave, the PULSE specification has several parameters that describe the pulse waveform. The general format of the PULSE waveform is

PULSE(<Vi> <Vp> <Td> <Tr> <Tf> <Width> <Period>

where Vi is the inital or baseline voltage
Vp is the pulsed voltage
Td is the delay before the pulse starts
Tr is the rise time
Tf is the fall time
Width is the duration of the pulse
Period is the time for the pulses to repeat

For example, the statement

```
VS 1 0 PULSE(0 5 0 0 0 100M 200M)
```

describes a pulse waveform with an inital value of 0 V, a pulsed value of 5 V, a pulse width of 100 ms, and a period of 200 ms. The zero values for Td, Tr, and Tf indicate that the pulse starts at time 0 (no delay) and changes instantaneously from one level to the next (ideal pulse). It is best to use a very small value for Tr and Tf such as 1N (1 nanosecond) rather than 0.

Similarly,

```
VS 1 0 PULSE(0 1 0 10N 10N 80U 1M)
```

describes a waveform with an amplitude of 1 V, rise and fall times of 10 ns, a pulse width of 80 $\mu$s, and a period of 1 ms ($f = 1$ kHz).

## The *RC* Charging Response

When the PULSE specification is applied to the circuit of Figure 13–62, the circuit file is as follows:

```
RC Charging Circuit
* Demonstration of RC Charging
VS 1 0 PULSE(0 5 1N 1N 10M 20M)
R1 1 2 1K
C1 2 0 1U
.TRAN 50U 10M
.PROBE
.END
```

FIGURE 13–62

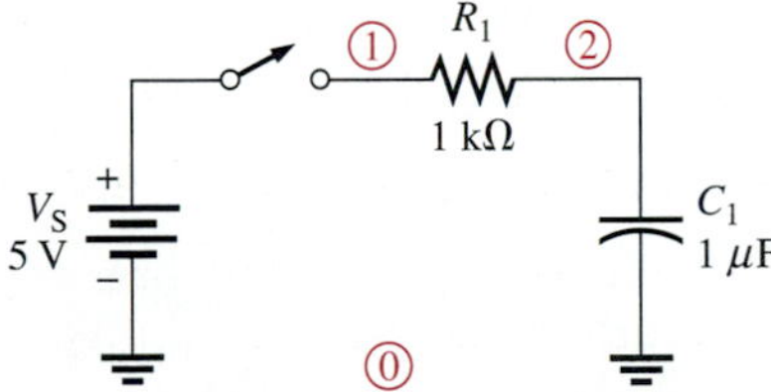


As you can see, PSpice will apply an “ideal” 5 V, 10 ms pulse to simulate a switch connecting a 5 V dc source to the *RC* circuit for 10 ms. Notice that the PULSE specification gives a period of 20 ms; but since Probe will sample the transient response for the first 10 ms (as indicated in the .TRAN statement), any value of period greater than 10 ms will work.

Under the Add_trace or Trace/Add menu option in Probe, enter

```
V(2,0)
```

This specifies the capacitor voltage, since the capacitor is connected between nodes 2 and 0 in the circuit. Probe will draw the capacitor’s charging curve showing that the voltage begins at 0 V and exponentially approaches a final value of 5 V.

***Determining the Time Constant*** Although you can easily calculate the *RC* time constant for the simple circuit in Figure 13–62, you can also do it in Probe, which is useful for more complex circuits. First, delete the curve with the Remove_trace option or, from a Windows platform, select the trace by mouse-clicking on the label V(2,0) and press the DELETE key. Next, select Add_trace or Trace/Add. The prompt "Enter variables or expressions" or "Trace Command:" means that Probe allows you to enter mathematical expressions as well as simple voltage and current variables. Enter the expression,

```
(V(2,0)/5 V)*100
```

Probe will redraw the graph of the capacitor voltage with the vertical axis labeled from 0 to 100 to indicate the percentage of full charge. Position the cursor (refer to Chapter 12 if necessary) at the point on the curve corresponding to 63 on the vertical axis (63%). The corresponding time on the horizontal axis is equal to one time constant. For the circuit in Figure 13–62, this value is approximately 1 ms.

**EXAMPLE 13–19** Write the circuit file for the *RC* circuit in Figure 13–63 using the nodes as numbered. Explain how to determine the final capacitor voltage and the circuit time constant.

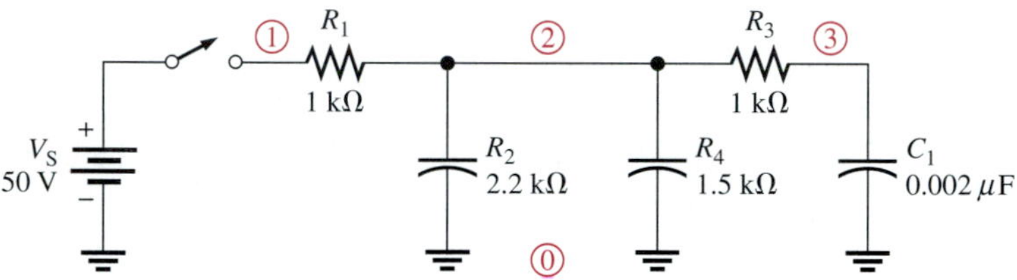


FIGURE 13–63

***Solution*** The circuit file is as follows:

```
RC Circuit
* Figure 13–63
VS  1  0  PULSE(0  50  0  1N  1N  50U  100U)
R1  1  2  1K
R2  2  0  2.2K
R3  2  3  1K
R4  2  0  1.5K
C1  3  0  0.002U
.TRAN  20P  50U
.PROBE
.END
```

After running this file, use Probe to plot the voltage (V(3,0)) across $C_1$. By positioning the cursor at the point where the curve reaches the final voltage, a value of approximately 23.6 V is indicated.

Next, remove this trace and use a trace expression to rescale the graph in terms of percentage charged. Enter the following trace expression:

```
(V(3,0)/23.6 V)*100
```

By positioning the cursor at the 63% point on the graph, a time of about 2.96 μs is read on the horizontal axis. This is the approximate circuit time constant.

***Related Exercise*** Calculate the time constant for the circuit in Figure 13–63.

## The .IC (Initial Condition) Control Statement

Analysis of *RC* discharging is similar to the analysis of *RC* charging response except that there must be an initial voltage across the capacitor. PSpice allows you to set an initial condition using the .IC control statement. The general format for this statement is

```
.IC V(<Label 1>) = <Voltage 1>. . . . . V(<Label N>) = <Voltage N>
```

Labels 1 through N refer to the labels of the nodes to which initial conditions are assigned. Voltages 1 through N are the initial volts for each node. Any node that is not specified in the statement is assumed to start at 0 V.

For example, if you want to begin the analysis of the circuit of Figure 13–62 with an initial voltage on the capacitor of 3 V, you must modify the file by adding the statement,

```
.IC V(2) = 3
```

prior to the .TRAN statement. The resulting graph will show the capacitor voltage as it charges from 3 V to 5 V rather than from 0 V to 5 V.

**EXAMPLE 13–20** Write the circuit file to determine the discharging response of the circuit in Figure 13–64 if the capacitor is initially charged to 10 V.

FIGURE 13–64

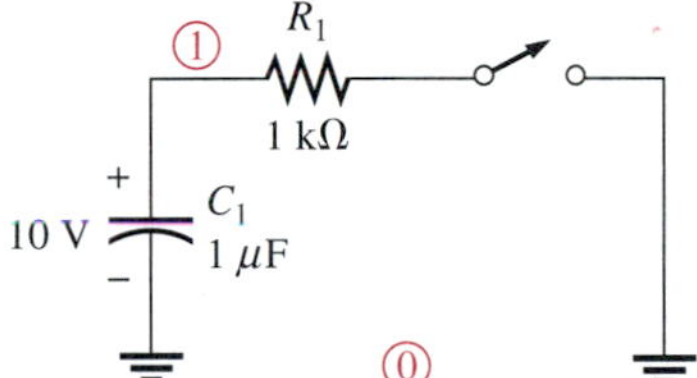


***Solution*** The circuit file is as follows:

```
RC Discharging Circuit
* Demonstration of RC Discharging
R1   1   0   1K
C1   1   0   1U
.IC V(1) = 10
.TRAN   20U   5M
.PROBE
.END
```

***Related Exercise*** Modify the circuit file for an initial voltage of 2.5 V on the capacitor and a capacitor value of 2.2 μF.

**SECTION 13–10 REVIEW**

1. Explain how to determine the time constant of a circuit from a Probe graph of the capacitor voltage.
2. Write an initial condition statement for a circuit in which there are two capacitors. $C_1$ has an inital voltage of 1 V and $C_2$ has an initial voltage of 5 V (The subscript denotes the node to which each capacitor is connected relative to ground.).

## SUMMARY

- A capacitor is composed of two parallel conducting plates separated by a dielectric insulator.
- Energy is stored by a capacitor in the electric field concentrated in the dielectric.
- One farad is the amount of capacitance when one coulomb of charge is stored with one volt across the plates.
- Capacitance is directly proportional to the plate area and inversely proportional to the plate separation.
- The dielectric constant is an indication of the ability of a material to establish an electric field.
- The dielectric strength is one factor that determines the breakdown voltage of a capacitor.
- A capacitor blocks constant dc.
- The time constant for a series *RC* circuit is the resistance times the capacitance.
- In an *RC* circuit, the voltage and current in a charging or discharging capacitor make a 63% change during each time-constant interval.
- Five time constants are required for a capacitor to charge fully or to discharge fully. This is called the *transient time.*
- Charging and discharging follow exponential curves.
- Total series capacitance is less than that of the smallest capacitor in series.
- Capacitance adds in parallel.
- Current leads voltage by 90° in a capacitor.
- $X_C$ is inversely proportional to frequency and capacitance.
- The true power in a capacitor is zero; that is, there is no energy loss in an ideal capacitor.

## GLOSSARY

**Capacitance** The ability of a capacitor to store electrical charge.

**Capacitive reactance** The opposition of a capacitor to sinusoidal current. The unit is the ohm.

**Capacitor** An electrical device consisting of two conductive plates separated by an insulating material and possessing the property of capacitance.

**Coulomb (C)** The unit of electrical charge.

**Coulomb's law** A physical law that states a force exists between two charged bodies that is directly proportional to the product of the two charges and inversely proportional to the square of the distance between them.

**Dielectric** The insulating material between the plates of a capacitor.

**Dielectric constant** A measure of the ability of a dielectric material to establish an electric field.

**Dielectric strength** A measure of the ability of a dielectric material to withstand voltage without breaking down.

**Farad (F)** The unit of capacitance.

**Filter** A type of circuit that passes or blocks certain frequencies to the exclusion of all others.

**Phase** The relative displacement of a time-varying quantity with respect to a given reference.

**Reactive power** The rate at which energy is alternately stored and returned to the source by a capacitor. The unit is the VAR.

**Rectifier** An electronic circuit that converts ac into pulsing dc; one part of a power supply.

**Ripple voltage** The small fluctuation in voltage due to the charging and discharging of a capacitor.

**Temperature coefficient** A constant specifying the amount of change in the value of a quantity for a given change in temperature.

**Time constant** A fixed interval, set by *R* and *C* values, that determines the time response of a circuit.

**Trimmer** A small variable capacitor.

**Volt-ampere reactive (VAR)** The unit of reactive power.

## ■ FORMULAS

| | | |
|---|---|---|
| **(13–1)** | $C = \frac{Q}{V}$ | Capacitance in terms of charge and voltage |
| **(13–2)** | $Q = CV$ | Charge in terms of capacitance and voltage |
| **(13–3)** | $V = \frac{Q}{C}$ | Voltage in terms of charge and capacitance |
| **(13–4)** | $F = \frac{kQ_1Q_2}{d^2}$ | Coulomb's law (force between two charged bodies) |
| **(13–5)** | $W = \frac{1}{2}CV^2$ | Energy stored by a capacitor |
| **(13–6)** | $\varepsilon_r = \frac{\varepsilon}{\varepsilon_0}$ | Dielectric constant (relative permittivity) |
| **(13–7)** | $C = \frac{A\varepsilon_r(8.85 \times 10^{-12}\text{ F/m})}{d}$ | Capacitance in terms of physical parameters |
| **(13–8)** | $Q_T = Q_1 = Q_2 = Q_3 = \cdots = Q_n$ | Total charge of series capacitors (general) |
| **(13–9)** | $\frac{1}{C_T} = \frac{1}{C_1} + \frac{1}{C_2} + \frac{1}{C_3} + \cdots + \frac{1}{C_n}$ | Reciprocal of total series capacitance (general) |
| **(13–10)** | $C_T = \frac{1}{\frac{1}{C_1} + \frac{1}{C_2} + \frac{1}{C_3} + \cdots + \frac{1}{C_n}}$ | Total series capacitance (general) |
| **(13–11)** | $C_T = \frac{C_1C_2}{C_1 + C_2}$ | Total capacitance of two capacitors in series |
| **(13–12)** | $C_T = \frac{C}{n}$ | Total capacitance of equal-value capacitors in series |
| **(13–13)** | $V_x = \left(\frac{C_T}{C_x}\right)V_T$ | Capacitor voltage for capacitors in series |
| **(13–14)** | $Q_T = Q_1 + Q_2 + Q_3 + \cdots + Q_n$ | Total charge of parallel capacitors (general) |
| **(13–15)** | $C_T = C_1 + C_2 + C_3 + \cdots + C_n$ | Total parallel capacitance (general) |
| **(13–16)** | $C_T = nC$ | Total capacitance of equal-value capacitors in parallel |
| **(13–17)** | $\tau = RC$ | Time constant |
| **(13–18)** | $v = V_F + (V_i - V_F)e^{-t/\tau}$ | Exponential voltage (general) |
| **(13–19)** | $i = I_F + (I_i - I_F)e^{-t/\tau}$ | Exponential current (general) |
| **(13–20)** | $v = V_F(1 - e^{-t/RC})$ | Increasing exponential voltage beginning at zero |
| **(13–21)** | $v = V_ie^{-t/RC}$ | Decreasing exponential voltage ending at zero |
| **(13–22)** | $t = -RC \ln\left(\frac{v}{V_i}\right)$ | Time on decreasing exponential ($V_F = 0$) |
| **(13–23)** | $t = -RC \ln\left(1 - \frac{v}{V_F}\right)$ | Time on increasing exponential ($V_i = 0$) |
| **(13–24)** | $i = \frac{dq}{dt}$ | Instantaneous current using charge derivative |
| **(13–25)** | $i = C\left(\frac{dv}{dt}\right)$ | Instantaneous capacitor current using voltage derivative |

| | | |
|---|---|---|
| (13–26) | $X_C = \dfrac{1}{2\pi fC}$ | Capacitive reactance |
| (13–27) | $P_r = V_{rms}I_{rms}$ | Reactive power in a capacitor |
| (13–28) | $P_r = \dfrac{V_{rms}^2}{X_C}$ | Reactive power in a capacitor |
| (13–29) | $P_r = I_{rms}^2 X_C$ | Reactive power in a capacitor |

## ▪ SELF-TEST

1. The following statement(s) accurately describes a capacitor:
   (a) The plates are conductive.
   (b) The dielectric is an insulator between the plates.
   (c) Constant dc flows through a fully charged capacitor.
   (d) A practical capacitor stores charge indefinitely when disconnected from the source.
   (e) none of the above answers
   (d) all of the above answers
   (e) only answers (a) and (b)
2. Which one of the following statements is true?
   (a) There is current through the dielectric of a charging capacitor.
   (b) When a capacitor is connected to a dc voltage source, it will charge to the value of the source.
   (c) An ideal capacitor can be discharged by disconnecting it from the voltage source.
3. A capacitance of 0.01 $\mu$F is larger than
   (a) 0.00001 F (b) 100,000 pF (c) 1000 pF (d) all of the above answers
4. A capacitance of 1000 pF is smaller than
   (a) 0.01 $\mu$F (b) 0.001 $\mu$F (c) 0.00000001 F (d) both (a) and (c)
5. When the voltage across a capacitor is increased, the stored charge
   (a) increases (b) decreases (c) remains constant (d) fluctuates
6. When the voltage across a capacitor is doubled, the stored charge
   (a) stays the same (b) is halved
   (c) increases by four (d) doubles
7. The voltage rating of a capacitor is increased by
   (a) increasing the plate separation (b) decreasing the plate separation
   (c) increasing the plate area (d) answers (b) and (c)
8. The capacitance value is increased by
   (a) decreasing plate area (b) increasing plate separation
   (c) decreasing plate separation (d) increasing plate area
   (e) answers (a) and (b) (f) answers (c) and (d)
9. A 1 $\mu$F, a 2.2 $\mu$F, and a 0.05 $\mu$F capacitor are connected in series. The total capacitance is less than
   (a) 1 $\mu$F (b) 2.2 $\mu$F (c) 0.05 $\mu$F (d) 0.001 $\mu$F
10. Four 0.02 $\mu$F capacitors are in parallel. The total capacitance is
    (a) 0.02 $\mu$F (b) 0.08 $\mu$F (c) 0.05 $\mu$F (d) 0.04 $\mu$F
11. An uncharged capacitor and a resistor are connected in series with a switch and a 12 V battery. At the instant the switch is closed, the voltage across the capacitor is
    (a) 12 V (b) 6 V (c) 24 V (d) 0 V
12. In Question 11, the voltage across the capacitor when it is fully charged is
    (a) 12 V (b) 6 V (c) 24 V (d) –6 V
13. In Question 11, the capacitor will reach full charge in a time equal to approximately
    (a) $RC$ (b) $5RC$ (c) $12RC$ (d) cannot be predicted
14. A sine wave voltage is applied across a capacitor. When the frequency of the voltage is increased, the current
    (a) increases (b) decreases (c) remains constant (d) ceases

15. A capacitor and a resistor are connected in series to a sine wave generator. The frequency is set so that the capacitive reactance is equal to the resistance and, thus, an equal amount of voltage appears across each component. If the frequency is decreased,
(a) $V_R > V_C$ (b) $V_C > V_R$ (c) $V_R = V_C$

16. An ohmmeter is connected across a discharged capacitor and the needle stabilizes at approximately 50 kΩ. The capacitor is
(a) good (b) charged (c) too large (d) leaky

## ■ PROBLEMS

### SECTION 13–1 The Basic Capacitor

1. (a) Find the capacitance when $Q = 50\ \mu C$ and $V = 10$ V.
(b) Find the charge when $C = 0.001\ \mu F$ and $V = 1$ kV.
(c) Find the voltage when $Q = 2$ mC and $C = 200\ \mu F$.

2. Convert the following values from microfarads to picofarads:
(a) 0.1 μF (b) 0.0025 μF (c) 5 μF

3. Convert the following values from picofarads to microfarads:
(a) 1000 pF (b) 3500 pF (c) 250 pF

4. Convert the following values from farads to microfarads:
(a) 0.0000001 F (b) 0.0022 F (c) 0.0000000015 F

5. Calculate the force of repulsion between two electrons 0.001 m apart.

6. What size capacitor is capable of storing 10 mJ of energy with 100 V across its plates?

7. Calculate the absolute permittivity, ε, for each of the following materials:
(a) air (b) oil (c) glass (d) Teflon®

8. A mica capacitor has a plate area of 0.04 $m^2$ and a dielectric thickness of 0.008 m. What is its capacitance?

9. An air capacitor has 0.1 m square plates. The plates are separated by 0.01 m. Calculate the capacitance.

10. At ambient temperature (25°C), a certain capacitor is specified to be 1000 pF. It has a negative temperature coefficient of 200 ppm/C°. What is its capacitance at 75°C?

11. A 0.001 μF capacitor has a positive temperature coefficient of 500 ppm/C°. How much change in capacitance will a 25°C increase in temperature cause?

### SECTION 13–2 Types of Capacitors

12. In the construction of a stacked-foil mica capacitor, how is the plate area increased?

13. What type of capacitor has the highest dielectric constant, mica or ceramic?

14. Show how to connect an electrolytic capacitor across $R_2$ between points $A$ and $B$ in Figure 13–65.

15. Name two types of electrolytic capacitors. How do electrolytics differ from other capacitors?

16. Identify the parts of the ceramic disk capacitor shown in the cutaway view of Figure 13–66.

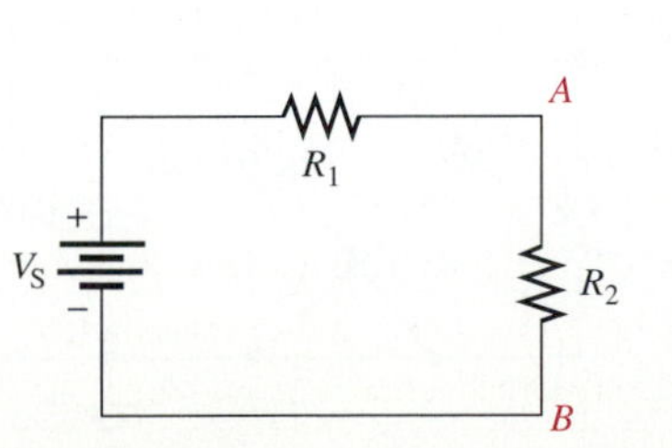


FIGURE 13–65

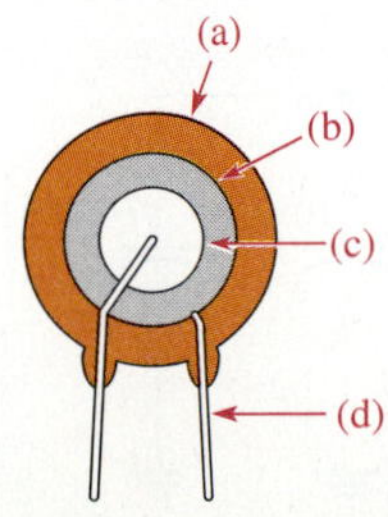


FIGURE 13–66

17. Determine the value of the ceramic disk capacitors in Figure 13–67.

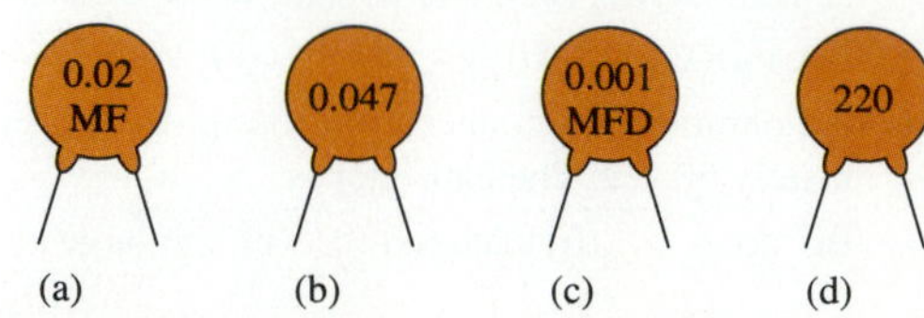


FIGURE 13–67

## SECTION 13–3 Series Capacitors

18. Five 1000 pF capacitors are in series. What is the total capacitance?
19. Find the total capacitance for each circuit in Figure 13–68.

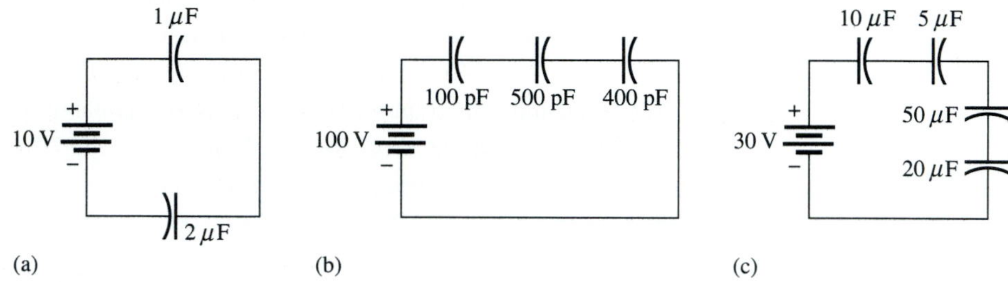


FIGURE 13–68

20. For each circuit in Figure 13–68, determine the voltage across each capacitor.
21. Two series capacitors (one 1 μF, the other of unknown value) are charged from a 12 V source. The 1 μF capacitor is charged to 8 V and the other to 4 V. What is the value of the unknown capacitor?
22. The total charge stored by the series capacitors in Figure 13–69 is 10 μC. Determine the voltage across each of the capacitors.

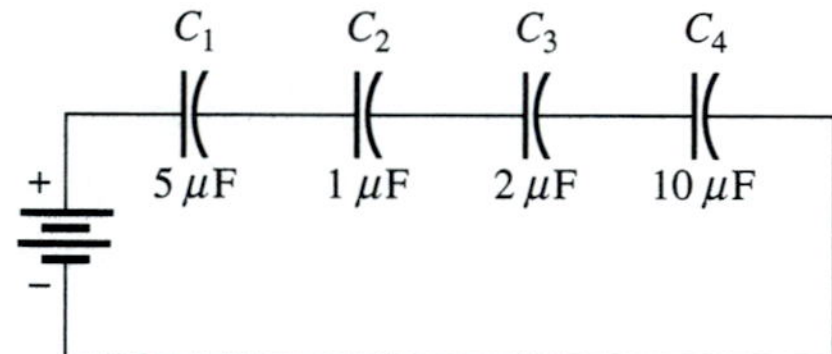


FIGURE 13–69

## SECTION 13–4 Parallel Capacitors

23. Determine $C_T$ for each circuit in Figure 13–70.

FIGURE 13–70

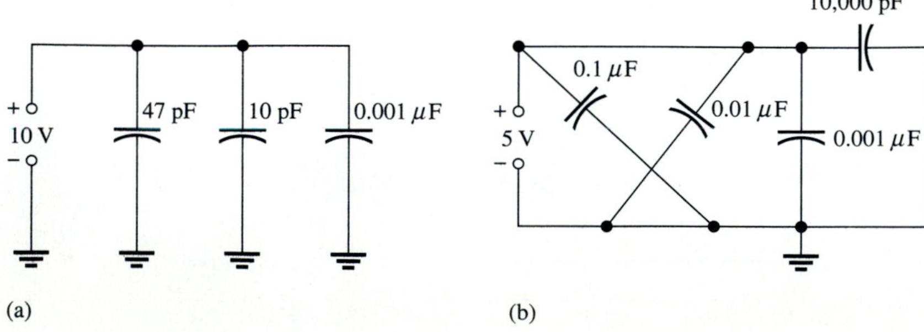

**24.** What is the charge on each capacitor in Figure 13–70?

**25.** Determine $C_T$ for each circuit in Figure 13–71.

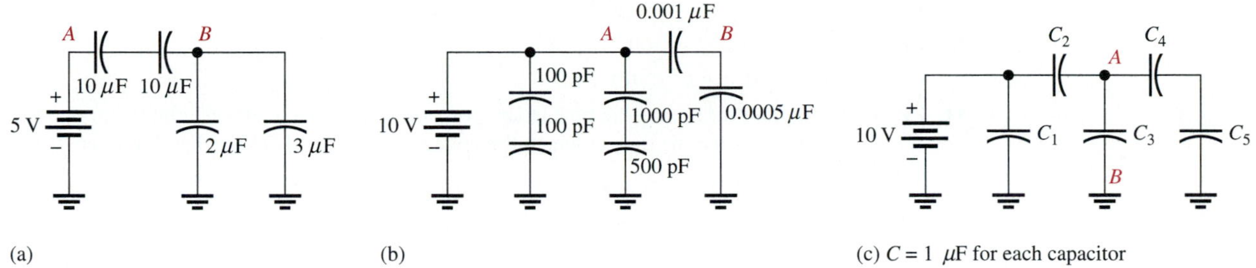


FIGURE 13–71

**26.** What is the voltage between points $A$ and $B$ in each circuit in Figure 13–71?

**27.** How much does the voltage across $C_5$ and the voltage across $C_6$ change when the ganged switch is thrown from position 1 to position 2 in Figure 13–72?

FIGURE 13–72

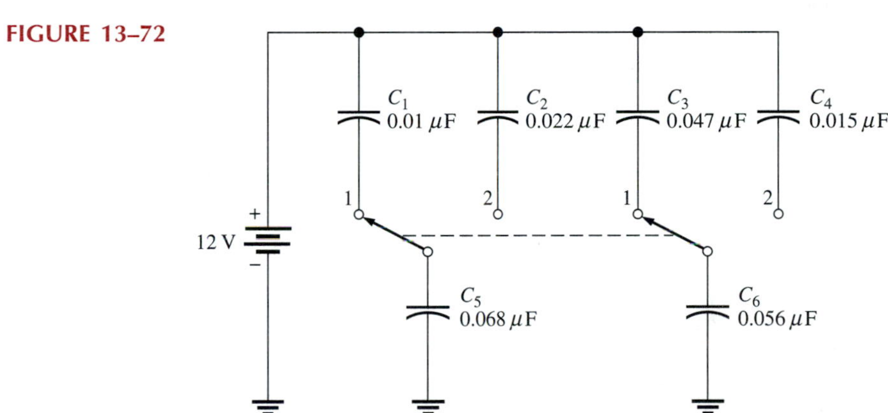


## SECTION 13–5 Capacitors in DC Circuits

**28.** Determine the time constant for each of the following series $RC$ combinations:

**(a)** $R = 100\ \Omega$, $C = 1\ \mu F$ **(b)** $R = 10\ M\Omega$, $C = 50\ pF$

**(c)** $R = 4.7\ k\Omega$, $C = 0.005\ \mu F$ **(d)** $R = 1.5\ M\Omega$, $C = 0.01\ \mu F$

**29.** Determine how long it takes the capacitor to reach full charge for each of the following combinations:

**(a)** $R = 56\ \Omega$, $C = 50\ \mu F$ **(b)** $R = 3300\ \Omega$, $C = 0.015\ \mu F$

**(c)** $R = 22\ k\Omega$, $C = 100\ pF$ **(d)** $R = 5.6\ M\Omega$, $C = 10\ pF$

**30.** In the circuit of Figure 13–73, the capacitor is initially uncharged. Determine the capacitor voltage at the following times after the switch is closed:

**(a)** 10 μs **(b)** 20 μs **(c)** 30 μs **(d)** 40 μs **(e)** 50 μs

FIGURE 13–73

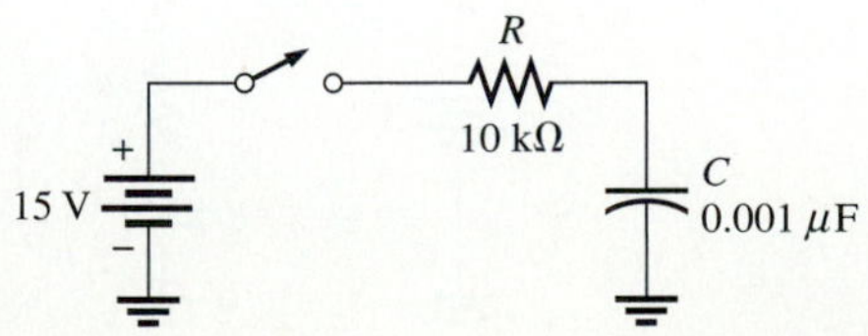

**31.** In Figure 13–74, the capacitor is charged to 25 V. When the switch is closed, what is the capacitor voltage after the following times?

**(a)** 1.5 ms **(b)** 4.5 ms **(c)** 6 ms **(d)** 7.5 ms

**32.** Repeat Problem 30 for the following time intervals:

**(a)** 2 $\mu$s **(b)** 5 $\mu$s **(c)** 15 $\mu$s

**33.** Repeat Problem 31 for the following times:

**(a)** 0.5 ms **(b)** 1 ms **(c)** 2 ms

**34.** Derive the formula for finding the time at any point on an increasing exponential voltage curve. Use this formula to find the time at which the voltage in Figure 13–75 reaches 6 V after switch closure.

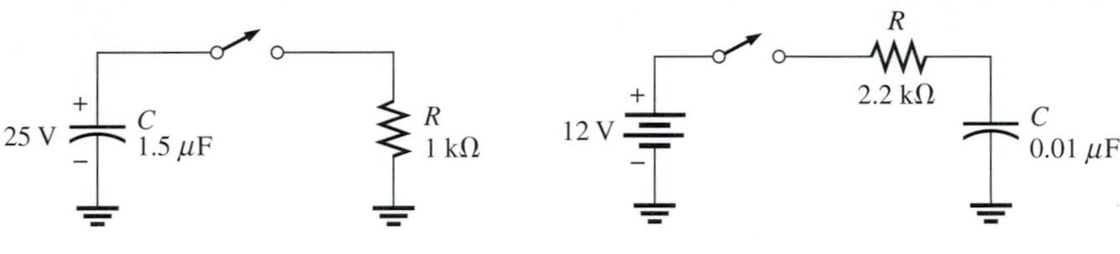


FIGURE 13–74

FIGURE 13–75

**35.** How long does it take $C$ to charge to 8 V in Figure 13–73?

**36.** How long does it take $C$ to discharge to 3 V in Figure 13–74?

**37.** Determine the time constant for the circuit in Figure 13–76.

**38.** In Figure 13–77, the capacitor is initially uncharged. At $t = 10\ \mu$s after the switch is closed, the instantaneous capacitor voltage is 7.2 V. Determine the value of $R$.

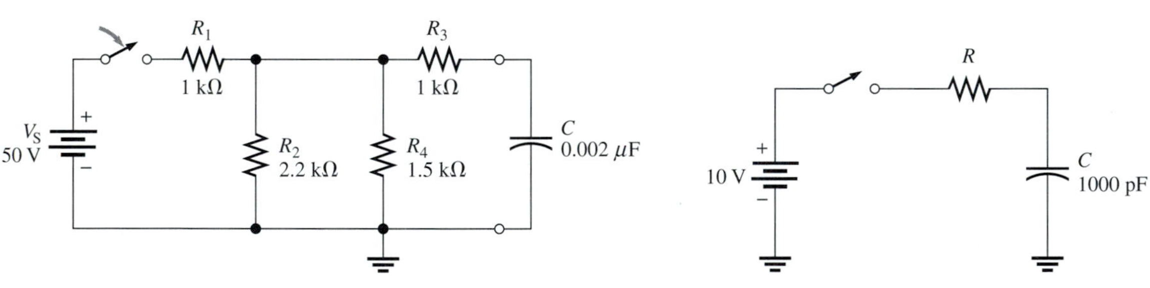


FIGURE 13–76

FIGURE 13–77

**39.** **(a)** The capacitor in Figure 13–78 is uncharged when the switch is thrown into position 1. The switch remains in position 1 for 10 ms and is then thrown into position 2, where it remains indefinitely. Sketch the complete waveform for the capacitor voltage.

**(b)** If the switch is thrown back to position 1 after 5 ms in position 2, and then left in position 1, how would the waveform appear?

FIGURE 13–78

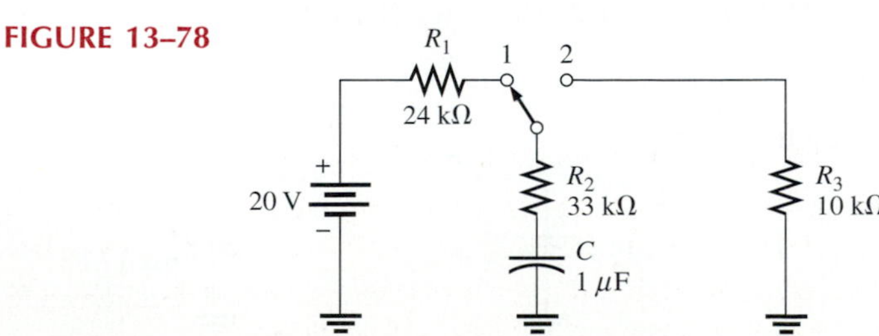

## SECTION 13–6 Capacitors in AC Circuits

**40.** What is the value of the total capacitive reactance in each circuit in Figure 13–79?

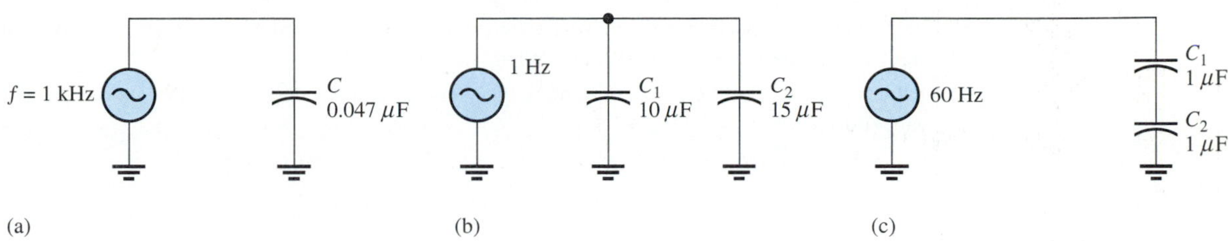


FIGURE 13–79

**41.** In Figure 13–71, each dc voltage source is replaced by a 10 V rms, 2 kHz ac source. Determine the total reactance in each case.

**42.** In each circuit of Figure 13–79, what frequency is required to produce an $X_C$ of 100 Ω? An $X_C$ of 1 kΩ?

**43.** A sinusoidal voltage of 20 V rms produces an rms current of 100 mA when connected to a certain capacitor. What is the reactance?

**44.** A 10 kHz voltage is applied to a 0.0047 μF capacitor, and 1 mA of rms current is measured. What is the value of the voltage?

**45.** Determine the true power and the reactive power in Problem 44.

**46.** Determine the ac voltage across each capacitor and the current in each branch of the circuit in Figure 13–80. What is the phase angle between the current and the voltage in each case?

FIGURE 13–80

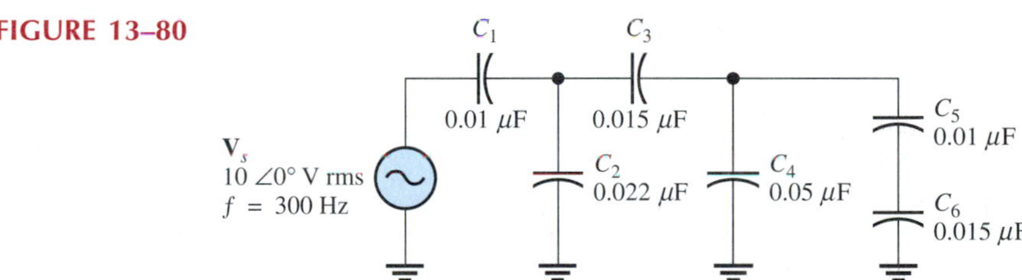


**47.** Find the value of $C_1$ in Figure 13–81.

FIGURE 13–81

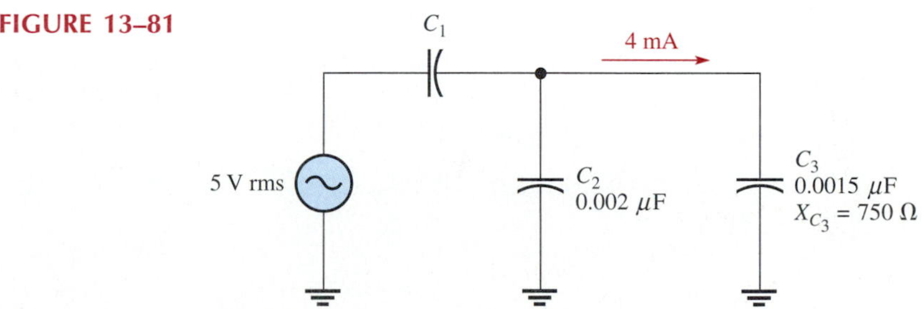


## SECTION 13–7 Capacitor Applications

**48.** If another capacitor is connected in parallel with the existing capacitor in the power supply filter of Figure 13–53 on page 526, how is the ripple voltage affected?

**49.** Ideally, what should the reactance of a bypass capacitor be in order to eliminate a 10 kHz ac voltage at a given point in an amplifier circuit?

## SECTION 13–8 Testing Capacitors

**50.** Assume that you are checking a capacitor with an ohmmeter, and when you connect the leads across the capacitor, the pointer does not move from its left-end scale position. What is the problem?

**51.** In checking a capacitor with the ohmmeter, you find that the pointer goes all the way to the right end of the scale and stays there. What is the problem?

**52.** If $C_4$ in Figure 13–80 opened, determine the voltages that would be measured across the other capacitors.

## ■ ANSWERS TO SECTION REVIEWS

### Section 13–1

**1.** Capacitance is the ability (capacity) to store electrical charge.

**2.** **(a)** 1,000,000 $\mu$F in 1 F **(b)** $1 \times 10^{12}$ pF in 1 F **(c)** 1,000,000 pF in 1 $\mu$F

**3.** 0.0015 $\mu$F = 1500 pF; 0.0015 $\mu$F = 0.0000000015 F

**4.** $W = \frac{1}{2}CV^2 = 1.125\ \mu J$

**5.** **(a)** $C$ increases. **(b)** $C$ decreases.

**6.** (V/mil) mils = 10 kV

**7.** $C = 0.425\ \mu F$

**8.** $C = 2.01\ \mu F$

### Section 13–2

**1.** Capacitors can be classified by the dielectric material.

**2.** A fixed capacitance cannot be changed; a variable capacitor can.

**3.** Electrolytic capacitors are polarized.

**4.** When connecting a polarized capacitor, make sure the voltage rating is sufficient. Connect the positive end to the positive side of the circuit.

### Section 13–3

**1.** Series $C_T$ is less than smallest $C$.

**2.** $C_T = 62.5$ pF

**3.** $C_T = 0.006\ \mu F$

**4.** $C_T = 20$ pF

**5.** $V_1 = 75$ V

### Section 13–4

**1.** The values of the individual capacitors are added in parallel.

**2.** Achieve $C_T$ by using five 0.01 $\mu$F capacitors in parallel.

**3.** $C_T = 98$ pF

### Section 13–5

**1.** $\tau = RC = 1.2\ \mu s$

**2.** $5\tau = 6\ \mu s$; $V_F = 4.97$ V

**3.** $v_{2ms} = 8.65$ V; $v_{3ms} = 9.50$ V; $v_{4ms} = 9.82$ V; $v_{5ms} = 9.93$ V

**4.** $v_C = 36.8$ V

**5.** $v_C = 4.93$ V

**6.** $t = 9.43\ \mu s$

### Section 13–6

**1.** Current leads voltage by 90° in a capacitor.

**2.** $X_C = 1/2\pi fC = 637\ k\Omega$

**3.** $f = 1/2\pi X_C C = 796$ Hz

**4.** $I_{rms} = 628$ mA

**5.** $P_{true} = 0$ W

### Section 13–7

1. Once the capacitor charges to the peak voltage, it discharges very little before the next peak thus smoothing the rectified voltage.
2. A coupling capacitor allows ac to pass from one point to another, but blocks constant dc.
3. A coupling capacitor must be large enough to have a negligible reactance at the frequency that is to be passed without opposition.
4. A decoupling capacitor shorts power line voltage transients to ground.
5. $X_C$ is inversely proportional to frequency and so is the filter's ability to pass ac signals.
6. The time constant is used in delay applications.

### Section 13–8

1. Discharge a capacitor by shorting its leads.
2. Initially, the needle jumps to zero; then it slowly moves to the high-resistance end of the scale when connected across a good capacitor.
3. Shorted, open, leakage, and dielectric absorption are common capacitor defects.

### Section 13–9

1. The coupling capacitor prevents the source from affecting the dc voltage but passes the input signal.
2. An ac voltage riding on a dc voltage is at point *C*.

### Section 13–10

1. Position cursor at 63% point. Read $\tau$ on time axis.
2. .IC V(1) = 1 V(2) = 5

## ■ ANSWERS TO RELATED EXERCISES FOR EXAMPLES

**13–1** 100 kV
**13–2** 0.05 $\mu$F
**13–3** $100 \times 10^6$ pF
**13–4** 6638 pF
**13–5** 1.57 $\mu$F
**13–6** 278 pF
**13–7** 0.01 $\mu$F
**13–8** 2.78 V
**13–9** 650 pF
**13–10** 0.09 $\mu$F
**13–11** 891 $\mu$s
**13–12** 8.36 V
**13–13** 7.97 V
**13–14** ≈0.74 ms; 95 V
**13–15** 1.52 ms
**13–16** 3.18 kHz
**13–17** $3.93\angle 90°$ mA
**13–18** 0 W; 1.01 mVAR
**13–19** 2.95 $\mu$s
**13–20**

```
R1 1 0 1K
C1 1 0 2.2U
.IC V(1)=2.5
.TRAN 20U 5M
.PROBE
.END
```

# 14

# INDUCTORS



## ■ INTRODUCTION

You have already learned about two of the three types of passive electrical components, the resistor and the capacitor. Now you will learn about the inductor, the third type of basic passive component.

In this chapter, you will study the inductor and its characteristics. The basic construction and electrical properties are discussed and the effects of connecting inductors in series and in parallel are analyzed. How an inductor works in both dc and ac circuits is an important part of this coverage and forms the basis for the study of reactive circuits in terms of both frequency response and time response. You will also learn how to check for a faulty inductor. In this chapter and throughout the rest of the book, you will learn the basics of putting technology theory into practice.

The inductor, which is basically a coil of wire, is based on the principle of electromagnetic induction, which was studied in Chapter 10.

Inductance is the property of a coil of wire that opposes a change in current. The basis for inductance is the electromagnetic field that surrounds any conductor when there is current through it. The electrical component designed to have the property of inductance is called an *inductor, coil,* or *choke.* All of these terms refer to the same type of device.

■ **TECHnology Theory Into Practice**

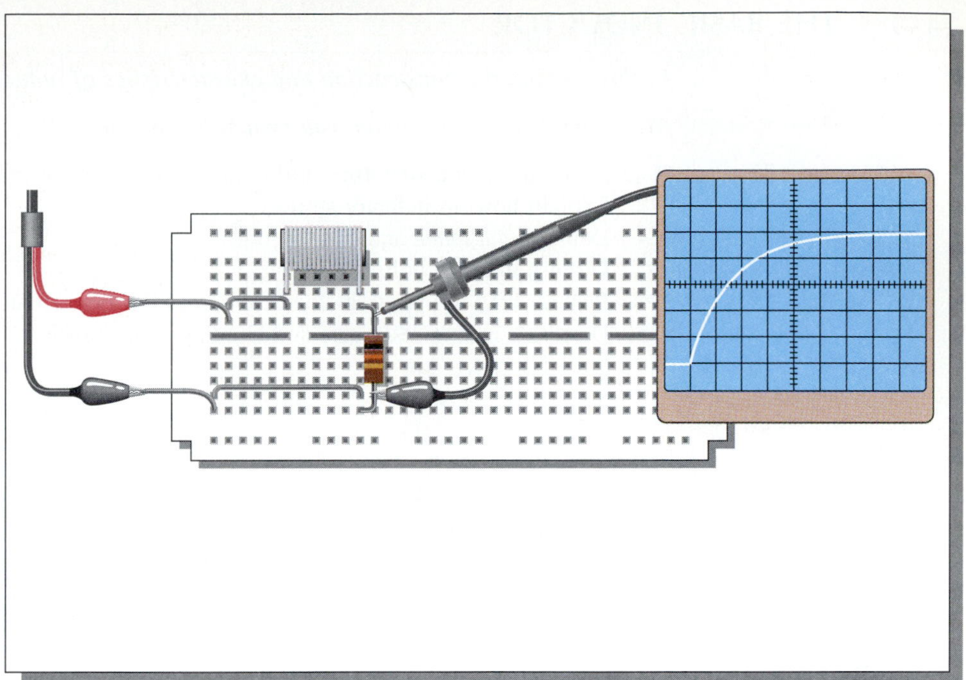

In the TECH TIP assignment in Section 14–9, you will determine the inductance of coils by measuring the time constant of a test circuit using oscilloscope waveforms.

## ■ CHAPTER OBJECTIVES

- ☐ Describe the basic structure and characteristics of an inductor
- ☐ Discuss various types of inductors
- ☐ Analyze series inductors
- ☐ Analyze parallel inductors
- ☐ Analyze inductive dc switching circuits
- ☐ Analyze inductive ac circuits
- ☐ Discuss some inductor applications
- ☐ Test an inductor
- ☐ Change the ranges of the *x* and *y* axes in Probe (optional)

## 14–1 ■ THE BASIC INDUCTOR

***In this section, the construction and characteristics of inductors are examined.***

***After completing this section, you should be able to***

- **Describe the basic structure and characteristics of an inductor**
  - ☐ Explain how an inductor stores energy
  - ☐ Define *inductance* and state its unit
  - ☐ Discuss induced voltage
  - ☐ Specify how the physical characteristics affect inductance
  - ☐ Discuss winding resistance and winding capacitance
  - ☐ State Faraday's law
  - ☐ State Lenz's law

When a length of wire is formed into a coil, as shown in Figure 14–1, it becomes a basic **inductor.** Current through the coil produces an electromagnetic field, as illustrated. The magnetic lines of force around each loop (turn) in the **winding** of the coil effectively add to the lines of force around the adjoining loops, forming a strong electromagnetic field within and around the coil. The net direction of the total electromagnetic field creates a north and a south pole.

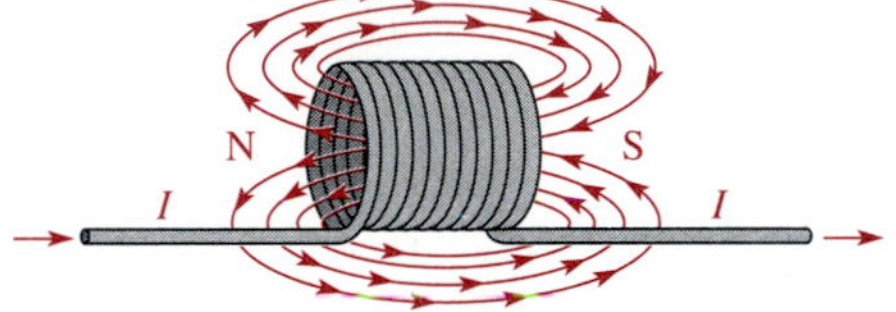


FIGURE 14–1
***A coil of wire forms an inductor. When there is current through it, a three-dimensional electromagnetic field is created, surrounding the coil in all directions.***

To understand the formation of the total electromagnetic field in a coil, consider the interaction of the electromagnetic fields around two adjacent loops. The magnetic lines of force around adjacent loops are each deflected into a single outer path when the loops are brought close together. This effect occurs because the magnetic lines of force are in opposing directions between adjacent loops and therefore cancel out when the loops are close together, as illustrated in Figure 14–2(a). The total electromagnetic field for the two loops is depicted in part (b) of the figure. This effect is additive for many closely adjacent loops in a coil; that is, each additional loop adds to the strength of the electromagnetic field. For simplicity, only single lines of force are shown, although there are many.

FIGURE 14–2
***Interaction of magnetic lines of force in two adjacent loops of a coil.***

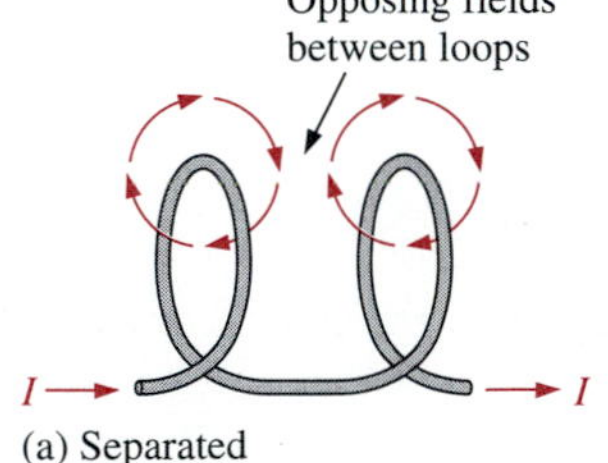


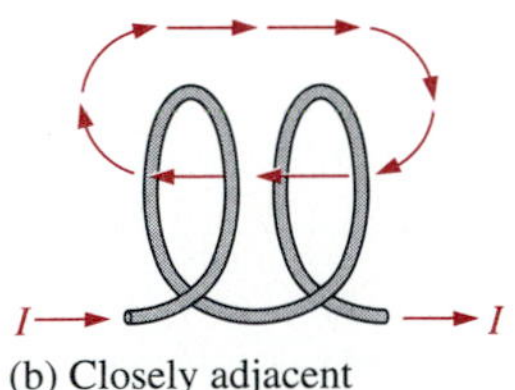


### Self-Inductance

When there is current through an inductor, an electromagnetic field is established. When the current changes, the electromagnetic field also changes. An increase in current expands the electromagnetic field, and a decrease in current reduces it. Therefore, a changing current produces a changing electromagnetic field around the inductor (also known as a **coil** or **choke**). In turn, the changing electromagnetic field causes an **induced voltage** across the coil in a direction to oppose the change in current. This property is called *self-inductance* but is usually referred to as simply **inductance,** symbolized by $L$.

> **Self-inductance is a measure of a coil's ability to establish an induced voltage as a result of a change in its current; and that induced voltage is in a direction to oppose that change in current.**

***The Unit of Inductance*** The **henry, H,** is the basic unit of inductance. By definition, the inductance is one henry when current through the coil, changing at the rate of one ampere per second, induces one volt across the coil. In many practical applications, millihenries (mH) and microhenries ($\mu$H) are the more common units. Figure 14–3 shows a schematic symbol for the inductor.

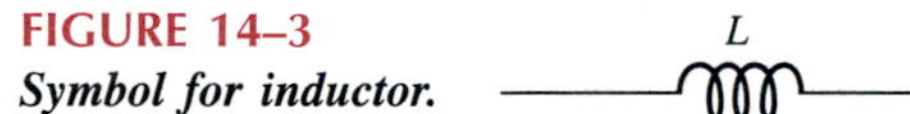


FIGURE 14–3
*Symbol for inductor.*

## The Induced Voltage Depends on $L$ and $di/dt$

The inductance ($L$) of a coil and the time rate of change of the current ($di/dt$) determine the induced voltage ($v_{ind}$). A change in current causes a change in the electromagnetic field, which, in turn, induces a voltage across the coil, as you know. The induced voltage is directly proportional to $L$ and $di/dt$, as stated by the following formula:

$$v_{ind} = L\left(\frac{di}{dt}\right) \qquad \textbf{(14–1)}$$

This formula indicates that the greater the inductance, the greater the induced voltage. Also, it means that the faster the coil current changes (greater $di/dt$), the greater the induced voltage. Notice the similarity of Equation (14–1) to Equation (13–25): $i = C(dv/dt)$.

**EXAMPLE 14–1**

Determine the induced voltage across a 1 henry (1 H) inductor when the current is changing at a rate of 2 A/s.

***Solution***

$$v_{ind} = L\left(\frac{di}{dt}\right) = (1\text{ H})(2\text{ A/s}) = 2\text{ V}$$

***Related Exercise*** Determine the inductance when a current changing at a rate of 10 A/s causes 50 V to be induced.

## Energy Storage

An inductor stores energy in the electromagnetic field created by the current. The energy stored is expressed as follows:

$$W = \frac{1}{2}LI^2 \qquad \textbf{(14–2)}$$

As you can see, the energy stored is proportional to the inductance and the square of the current. When current ($I$) is in amperes and inductance ($L$) is in henries, energy ($W$) is in joules.

## Physical Characteristics of Inductors

The following parameters are important in establishing the inductance of a coil: permeability of the core material, number of turns of wire, core length, and cross-sectional area of the core.

***Core Material*** As discussed earlier, an inductor is basically a coil of wire. The material around which the coil is formed is called the **core.** Coils are wound on either nonmagnetic or magnetic materials. Examples of nonmagnetic materials are air, wood, copper, plastic, and glass. The permeabilities of these materials are the same as for a vacuum. Examples of magnetic materials are iron, nickel, steel, cobalt, or alloys. These materials have permeabilities that are hundreds or thousands of times greater than that of a vacuum and are classified as *ferromagnetic.* A ferromagnetic core provides a better path for the magnetic lines of force and thus permits a stronger magnetic field.

As you learned in Chapter 10, the permeability ($\mu$) of the core material determines how easily a magnetic field can be established. *The inductance is directly proportional to the permeability of the core material.*

***Physical Parameters*** The number of turns of wire, the length, and the cross-sectional area of the core, as indicated in Figure 14–4, are factors in setting the value of inductance. The inductance is inversely proportional to the length of the core and directly proportional to the cross-sectional area. Also, the inductance is directly related to the number of turns squared. This relationship is as follows:

$$L = \frac{N^2 \mu A}{l} \tag{14–3}$$

where $L$ is the inductance in henries, $N$ is the number of turns of wire, $\mu$ is the permeability, $A$ is the cross-sectional area in meters squared, and $l$ is the core length in meters.

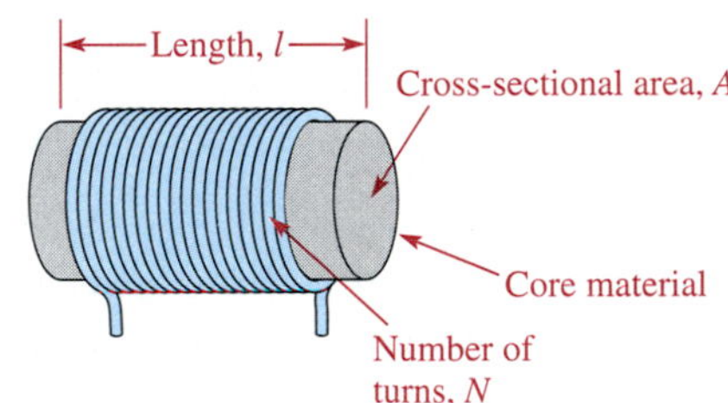


**FIGURE 14–4**
***Physical parameters of an inductor.***

**EXAMPLE 14–2** Determine the inductance of the coil in Figure 14–5. The permeability of the core is $0.25 \times 10^{-3}$.

**FIGURE 14–5**

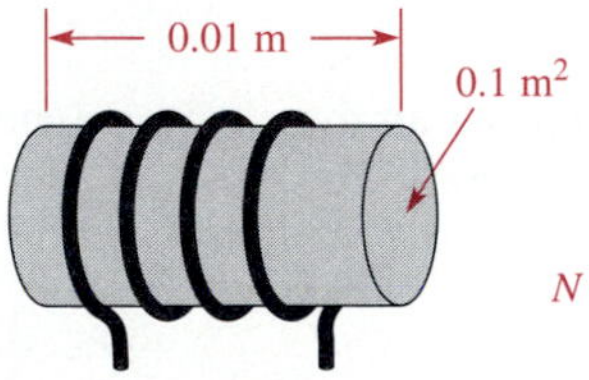

0.01 m

0.1 m$^2$

$N = 4$

***Solution***

$$L = \frac{N^2 \mu A}{l} = \frac{(4)^2(0.25 \times 10^{-3})(0.1)}{0.01} = 40 \text{ mH}$$

The calculator sequence is

[4] [x²] [×] [•] [2] [5] [EXP] [+/−] [3] [×] [•] [1] [÷] [•] [0] [1] [=]

***Related Exercise*** Determine the inductance of a coil with 12 turns around a core that is 0.05 m long and has a cross-sectional area of 0.15 $m^2$. The permeability is $0.25 \times 10^{-3}$.

## Winding Resistance

When a coil is made of a certain material, for example, insulated copper wire, that wire has a certain resistance per unit of length. When many turns of wire are used to construct a coil, the total resistance may be significant. This inherent resistance is called the *dc resistance* or the *winding resistance* ($R_W$). Although this resistance is distributed along the length of the wire, it effectively appears in series with the inductance of the coil, as shown in Figure 14–6. In many applications, the winding resistance may be small enough to be ignored and the coil considered as an ideal inductor. In other cases, the resistance must be considered.

**FIGURE 14–6**
***Winding resistance of a coil.***

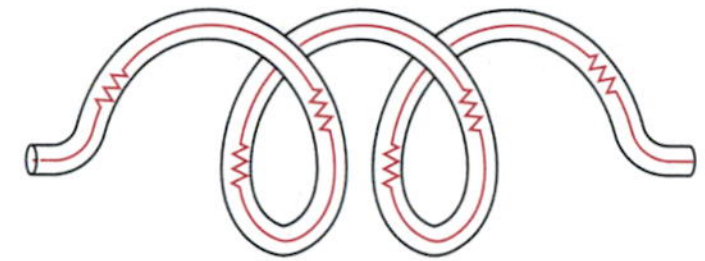

(a) The wire has resistance distributed along its length.

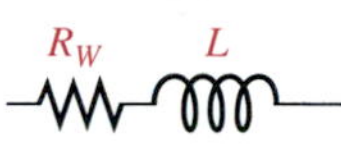


(b) Equivalent circuit

## Winding Capacitance

When two conductors are placed side by side, there is always some capacitance between them. Thus, when many turns of wire are placed close together in a coil, a certain amount of stray capacitance, called *winding capacitance* ($C_W$), is a natural side effect. In many applications, this winding capacitance is very small and has no significant effect. In other cases, particularly at high frequencies, it may become quite important.

The equivalent circuit for an inductor with both its winding resistance ($R_W$) and its winding capacitance ($C_W$) is shown in Figure 14–7. The capacitance effectively acts in parallel.

**FIGURE 14–7**
***Winding capacitance of a coil.***

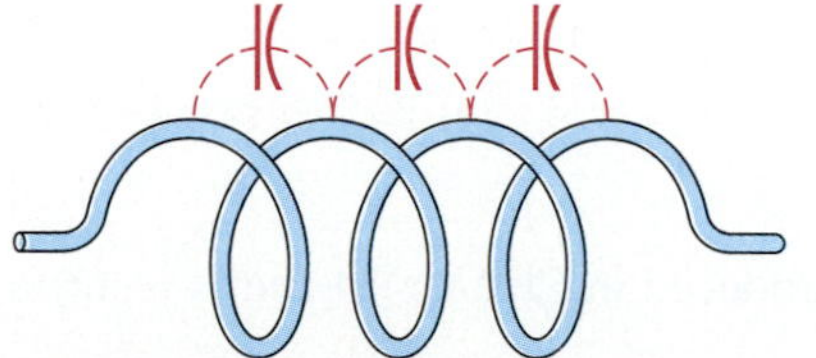

(a) Stray capacitance between each loop appears as a total parallel capacitance.

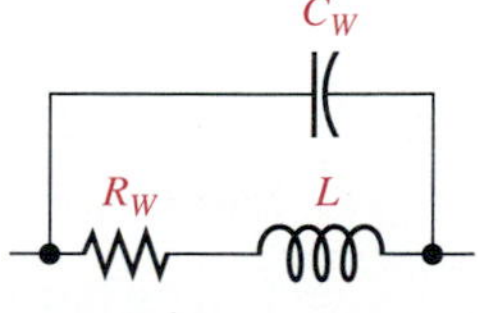


(b) Equivalent circuit

## Review of Faraday's Law

Faraday's law was introduced in Chapter 10 and is reviewed here because of its importance in the study of inductors. Michael Faraday discovered the principle of electromagnetic induction in 1831. He found that by moving a magnet through a coil of wire, a voltage was induced across the coil and that when a complete path was provided, the induced voltage caused an induced current. Faraday observed that

> **The amount of voltage induced in a coil is directly proportional to the rate of change of the magnetic field with respect to the coil.**

This principle is illustrated in Figure 14–8, where a bar magnet is moved through a coil of wire. An induced voltage is indicated by the voltmeter connected across the coil. The faster the magnet is moved, the greater is the induced voltage.

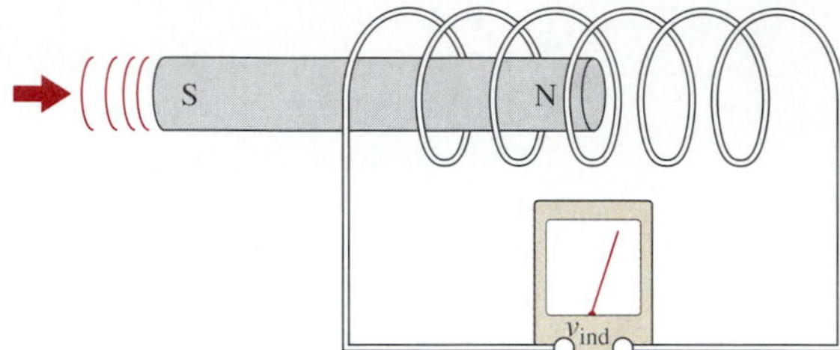


FIGURE 14–8
*Induced voltage created by a changing magnetic field.*

When a wire is formed into a certain number of loops or turns and is exposed to a changing magnetic field, a voltage is induced across the coil. The induced voltage is proportional to the number of turns of wire in the coil, $N$, and to the rate at which the magnetic field changes. The rate of change of the magnetic field is designated $d\phi/dt$, where $\phi$ is the magnetic flux. $d\phi/dt$ is expressed in webers/second (Wb/s). Faraday's law states that the induced voltage across a coil is equal to the number of turns (loops) times the rate of flux change and is expressed in concise form as follows:

$$v_{ind} = N\left(\frac{d\phi}{dt}\right) \tag{14–4}$$

**EXAMPLE 14–3** Apply Faraday's law to find the induced voltage across a coil with 500 turns located in a magnetic field that is changing at a rate of 5 Wb/s.

***Solution***

$$v_{ind} = N\left(\frac{d\phi}{dt}\right) = (500\text{ t})(5\text{ Wb/s}) = 2.5\text{ kV}$$

***Related Exercise*** A 1000 turn coil has an induced voltage of 500 V across it. What is the rate of change of the magnetic field?

## Lenz's Law

**Lenz's law** was introduced in Chapter 10 and is restated here.

> **When the current through a coil changes, an induced voltage is created as a result of the changing electromagnetic field, and the direction of the induced voltage is such that it always opposes the change in current.**

Figure 14–9 illustrates Lenz's law. In part (a), the current is constant and is limited by $R_1$. There is no induced voltage because the electromagnetic field is unchanging. In part (b), the switch suddenly is closed, placing $R_2$ in parallel with $R_1$ and thus reducing the resistance. Naturally, the current tries to increase and the electromagnetic field begins to expand, but the induced voltage opposes this attempted increase in current for an instant.

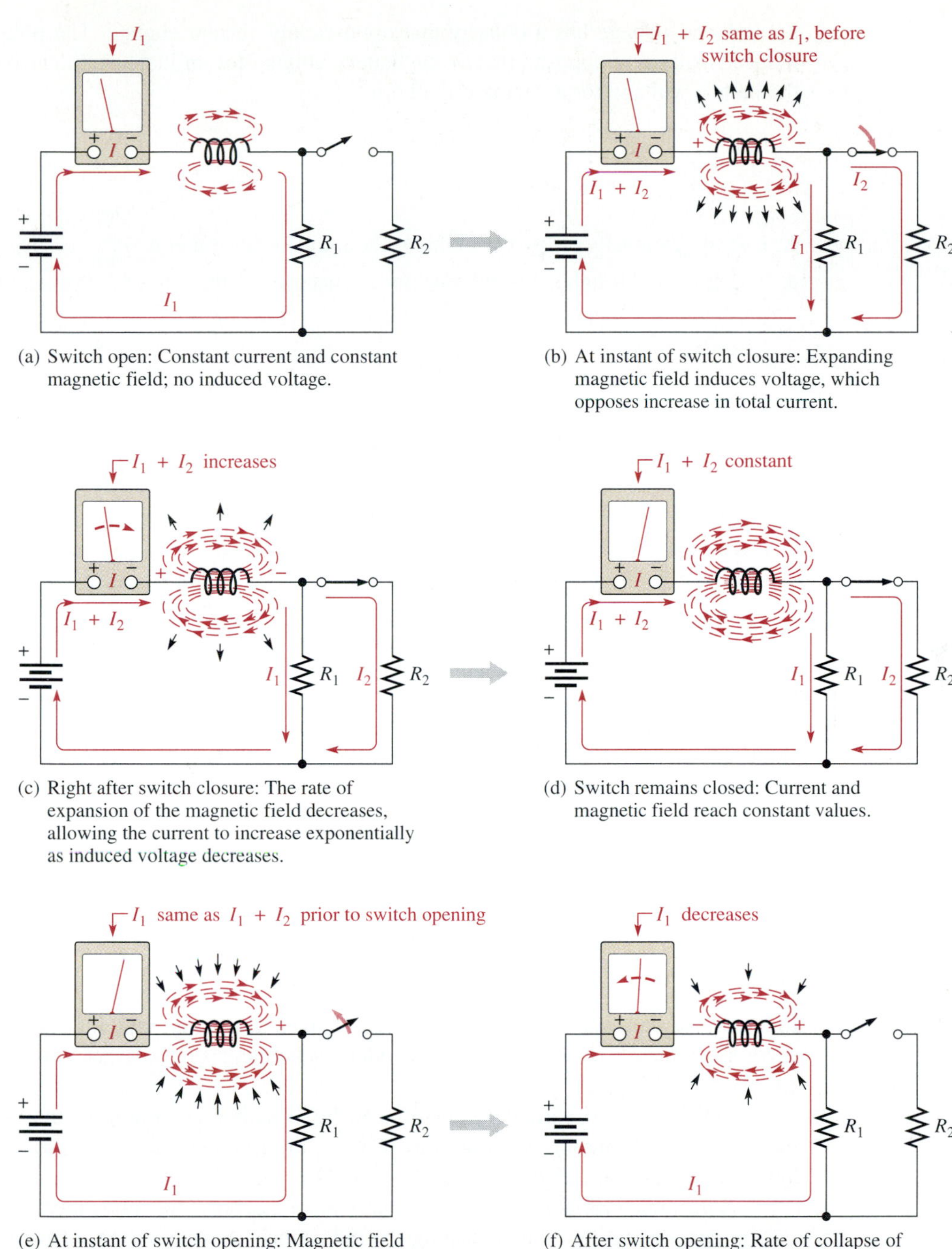

$I_1$
$R_1$
$R_2$
$I_1$
(a) Switch open: Constant current and constant magnetic field; no induced voltage.
$I_1 + I_2$ same as $I_1$, before switch closure
$I_1 + I_2$
$I_2$
$I_1$
$R_1$
$R_2$
(b) At instant of switch closure: Expanding magnetic field induces voltage, which opposes increase in total current.
$I_1 + I_2$ increases
$I_1 + I_2$
$I_1$
$R_1$
$I_2$
$R_2$
(c) Right after switch closure: The rate of expansion of the magnetic field decreases, allowing the current to increase exponentially as induced voltage decreases.
$I_1 + I_2$ constant
$I_1 + I_2$
$I_1$
$R_1$
$I_2$
$R_2$
(d) Switch remains closed: Current and magnetic field reach constant values.
$I_1$ same as $I_1 + I_2$ prior to switch opening
$R_1$
$R_2$
$I_1$
(e) At instant of switch opening: Magnetic field begins to collapse, creating an induced voltage, which opposes decrease in current.
$I_1$ decreases
$R_1$
$R_2$
$I_1$
(f) After switch opening: Rate of collapse of magnetic field decreases, allowing current to decrease exponentially back to original value.

**FIGURE 14–9**

***Demonstration of Lenz's law in an inductive circuit: When the current tries to change suddenly, the electromagnetic field changes and induces a voltage in a direction that opposes that change in current.***

In Figure 14–9(c), the induced voltage gradually decreases, allowing the current to increase. In part (d), the current has reached a constant value as determined by the parallel resistors, and the induced voltage is zero. In part (e), the switch has been suddenly opened, and, for an instant, the induced voltage prevents any decrease in current, and arcing between the switch contacts results. In part (f), the induced voltage gradually decreases, allowing the current to decrease back to a value determined by $R_1$. Notice that the induced voltage has a polarity that opposes any current change. The polarity of the induced voltage is opposite that of the battery voltage for an increase in current and aids the battery voltage for a decrease in current.

**SECTION 14–1 REVIEW**

1. List the parameters that contribute to the inductance of a coil.
2. The current through a 15 mH inductor is changing at the rate of 500 mA/s. What is the induced voltage?
3. Describe what happens to $L$ when
   **(a)** $N$ is increased
   **(b)** The core length is increased
   **(c)** The cross-sectional area of the core is decreased
   **(d)** A ferromagnetic core is replaced by an air core
4. Explain why inductors have some winding resistance.
5. Explain why inductors have some winding capacitance.

## 14–2 ■ TYPES OF INDUCTORS

***Inductors normally are classified according to the type of core material. In this section, the basic types of inductors are examined.***

***After completing this section, you should be able to***

- **Discuss various types of inductors**
  - ☐ Describe the basic types of fixed inductors
  - ☐ Distinguish between fixed and variable inductors

Inductors are made in a variety of shapes and sizes. Basically, they fall into two general categories: fixed and variable. The standard schematic symbols are shown in Figure 14–10.

Both fixed and variable inductors can be classified according to the type of core material. Three common types are the air core, the iron core, and the ferrite core. Each has a unique symbol, as shown in Figure 14–11.

Adjustable (variable) inductors usually have a screw-type adjustment that moves a sliding core in and out, thus changing the inductance. A wide variety of inductors exist, and some are shown in Figure 14–12.

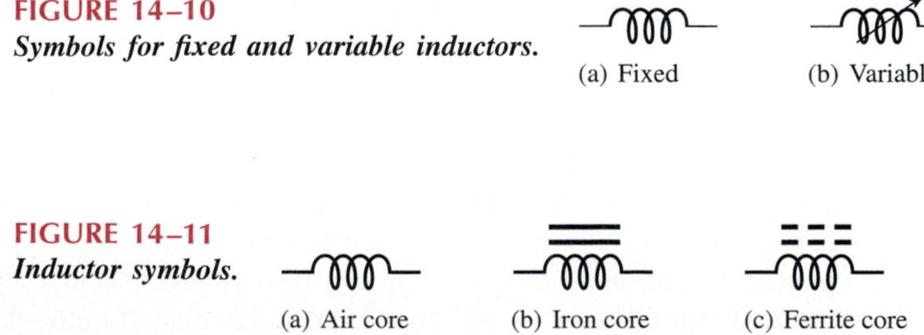


**FIGURE 14–10**
***Symbols for fixed and variable inductors.***

**FIGURE 14–11**
***Inductor symbols.***

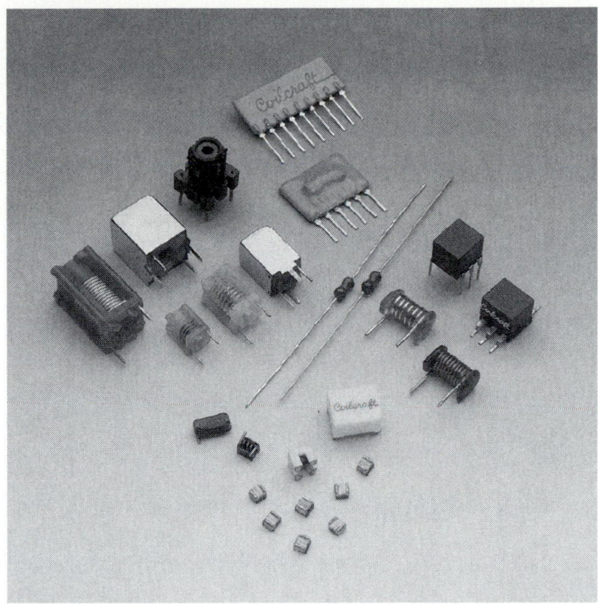

FIGURE 14–12
*Typical inductors. Courtesy of Coilcraft.*

**SECTION 14–2 REVIEW**

1. Name two general categories of inductors.
2. Identify the inductor symbols in Figure 14–13.

FIGURE 14–13

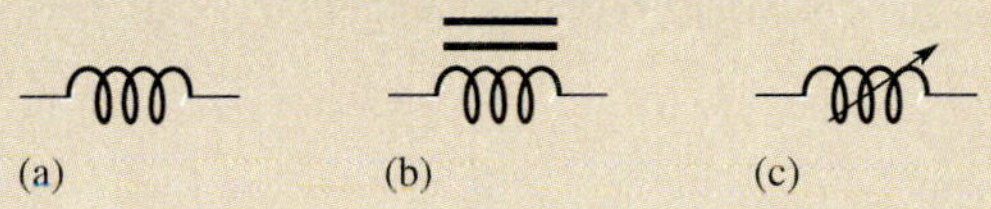

(a) (b) (c)

## 14–3 ■ SERIES INDUCTORS

***In this section, you will see that when inductors are connected in series, the total inductance increases.***

***After completing this section, you should be able to***

- **Analyze series inductors**
  - □ Determine total inductance

When inductors are connected in series, as in Figure 14–14, the total inductance, $L_T$, is the sum of the individual inductances. The formula for $L_T$ is expressed in the following equation for the general case of $n$ inductors in series:

$$L_T = L_1 + L_2 + L_3 + \cdots + L_n \quad (14\text{–}5)$$

Notice that the formula for total inductance in series is similar to the formulas for total resistance in series (Chapter 5) and total capacitance in parallel (Chapter 13).

FIGURE 14–14
*Inductors in series.*

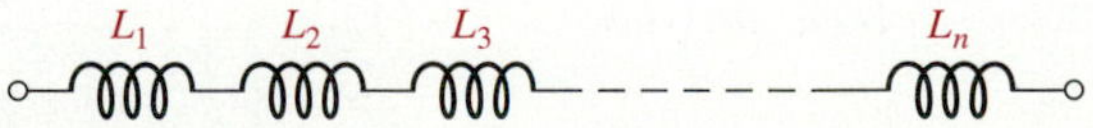

$L_1$ $L_2$ $L_3$ $L_n$

**EXAMPLE 14–4** Determine the total inductance for each of the series connections in Figure 14–15.

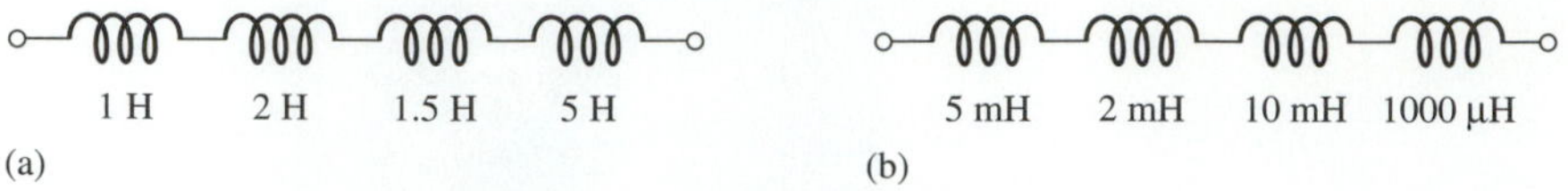

(a) 1 H, 2 H, 1.5 H, 5 H

(b) 5 mH, 2 mH, 10 mH, 1000 μH

FIGURE 14–15

***Solution*** In Figure 14–15(a),

$$L_T = 1\text{ H} + 2\text{ H} + 1.5\text{ H} + 5\text{ H} = 9.5\text{ H}$$

In Figure 14–15(b),

$$L_T = 5\text{ mH} + 2\text{ mH} + 10\text{ mH} + 1\text{ mH} = 18\text{ mH}$$

*Note:* 1000 $\mu$H = 1 mH

***Related Exercise*** What is the total inductance of ten 50 $\mu$H inductors in series?

**SECTION 14–3 REVIEW**

1. State the rule for combining inductors in series.
2. What is $L_T$ for a series connection of 100 $\mu$H, 500 $\mu$H, and 2 mH?
3. Five 100 mH coils are connected in series. What is the total inductance?

## 14–4 ■ PARALLEL INDUCTORS

***In this section, you will see that when inductors are connected in parallel, total inductance is reduced.***

***After completing this section, you should be able to***

- **Analyze parallel inductors**
  - Determine total inductance

### Total Inductance

When inductors are connected in parallel, as in Figure 14–16, the total inductance is less than the smallest inductance. The formula for total inductance in parallel is similar to the formulas for total parallel resistance (Chapter 6) and total series capacitance (Chapter 13).

$$\frac{1}{L_T} = \frac{1}{L_1} + \frac{1}{L_2} + \frac{1}{L_3} + \cdots + \frac{1}{L_n} \qquad \textbf{(14–6)}$$

This general formula states that the reciprocal of the total inductance is equal to the sum of the reciprocals of the individual inductances. $L_T$ can be found by taking the reciprocal of both sides of Equation (14–6).

$$L_T = \frac{1}{\left(\frac{1}{L_1}\right) + \left(\frac{1}{L_2}\right) + \left(\frac{1}{L_3}\right) + \cdots + \left(\frac{1}{L_n}\right)} \qquad \textbf{(14–7)}$$

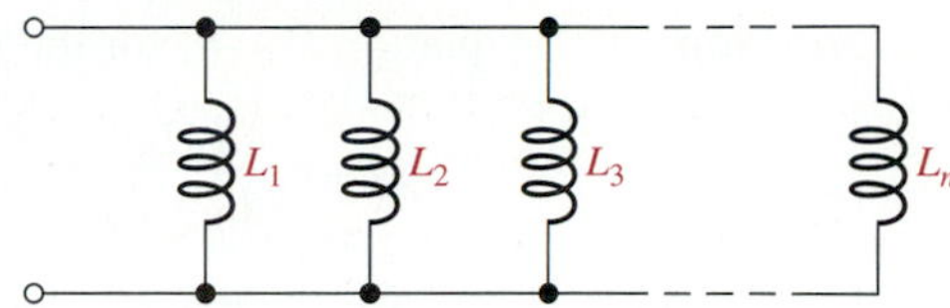


**FIGURE 14–16**
***Inductors in parallel.***

***Two Parallel Inductors*** When only two inductors are in parallel, a special product over sum form of Equation (14–6) can be used.

$$L_T = \frac{L_1 L_2}{L_1 + L_2} \tag{14-8}$$

***Equal-Value Parallel Inductors*** This is another special case in which a short-cut formula can be used. This formula is also derived from the general Equation (14–6) and is stated as follows for $n$ equal-value inductors in parallel:

$$L_T = \frac{L}{n} \tag{14-9}$$

**EXAMPLE 14–5** Determine $L_T$ in Figure 14–17.

**FIGURE 14–17**

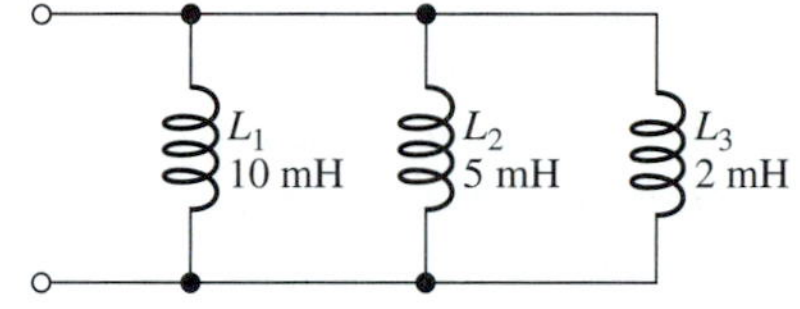


***Solution*** Use Equation (14–7) to determine the total inductance.

$$L_T = \frac{1}{\left(\frac{1}{L_1}\right) + \left(\frac{1}{L_2}\right) + \left(\frac{1}{L_3}\right)} = \frac{1}{\frac{1}{10 \text{ mH}} + \frac{1}{5 \text{ mH}} + \frac{1}{2 \text{ mH}}} = \frac{1}{0.8 \text{ mH}} = 1.25 \text{ mH}$$

The calculator sequence is

1 0 EXP +/− 3 2nd 1/x + 5 EXP +/− 3 2nd 1/x +
2 EXP +/− 3 2nd 1/x = 2nd 1/x

***Related Exercise*** Determine $L_T$ for a parallel connection of 50 $\mu$H, 80 $\mu$H, 100 $\mu$H, and 150 $\mu$H.

**EXAMPLE 14–6** Find $L_T$ for both circuits in Figure 14–18.

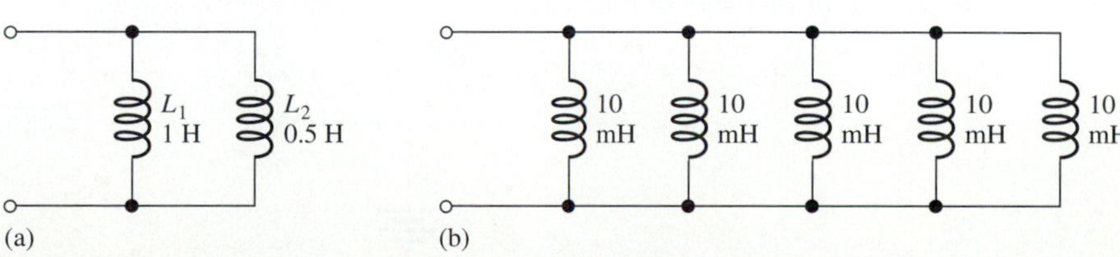


**FIGURE 14–18**

***Solution*** Use Equation (14–8) for the two parallel inductors in Figure 14–18(a).

$$L_T = \frac{L_1 L_2}{L_1 + L_2} = \frac{(1\text{ H})(0.5\text{ H})}{1.5\text{ H}} = 333\text{ mH}$$

Use Equation (14–9) for the equal-value parallel inductors in Figure 14–18(b).

$$L_T = \frac{L}{n} = \frac{10\text{ mH}}{5} = 2\text{ mH}$$

***Related Exercise*** Find $L_T$ for each of the following:
**(a)** $L_1 = 10\ \mu\text{H}$ and $L_2 = 27\ \mu\text{H}$ in parallel **(b)** Five 100 $\mu$H coils in parallel

**SECTION 14–4 REVIEW**

1. Compare the total inductance in parallel with the smallest-value individual inductor.
2. The calculation of total parallel inductance is similar to that for parallel resistance (T or F).
3. Determine $L_T$ for each parallel combination:
   **(a)** 100 mH, 50 mH, and 10 mH **(b)** 40 $\mu$H and 60 $\mu$H
   **(c)** Ten 1 H coils

## 14–5 ■ INDUCTORS IN DC CIRCUITS

***An inductor will energize when it is connected to a dc voltage source. The buildup of current through the inductor occurs in a predictable manner, which is dependent on the inductance and the resistance in a circuit.***

***After completing this section, you should be able to***

- **Analyze inductive dc switching circuits**
  - ☐ Describe the energizing and deenergizing of an inductor
  - ☐ Define *time constant*
  - ☐ Relate the time constant to energizing and deenergizing
  - ☐ Describe induced voltage
  - ☐ Write the exponential equations for current in an inductor

When there is constant direct current in an inductor, there is no induced voltage. There is, however, a voltage drop due to the winding resistance of the coil. The inductance itself appears as a short to dc. Energy is stored in the electromagnetic field according to the formula previously stated in Equation (14–2), $W = \frac{1}{2}LI^2$. The only energy loss occurs in the winding resistance ($P = I^2R_W$). This condition is illustrated in Figure 14–19.

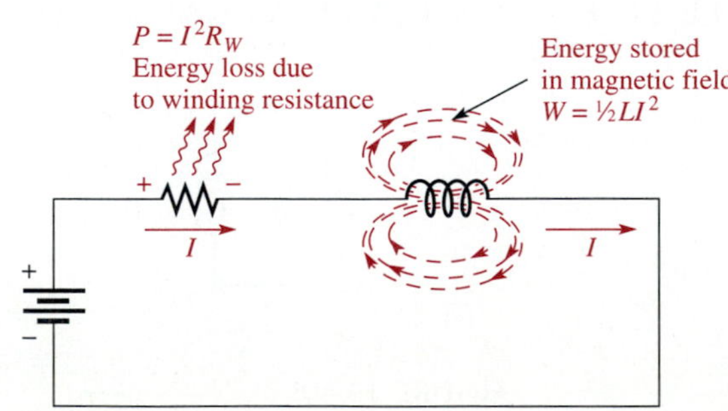


**FIGURE 14–19**
***Energy storage and loss in an inductor in a dc circuit.***

## The *RL* Time Constant

Because the inductor's basic action is to oppose a change in its current, it follows that current cannot change instantaneously in an inductor. A certain time is required for the current to make a change from one value to another. The rate at which the current changes is determined by the **time constant.** The time constant for a series *RL* circuit is

$$\tau = \frac{L}{R} \qquad \textbf{(14–10)}$$

where $\tau$ is in seconds when inductance ($L$) is in henries and resistance ($R$) is in ohms.

**EXAMPLE 14–7** A series *RL* circuit has a resistance of 1 kΩ and an inductance of 1 mH. What is the time constant?

***Solution*** Calculate the time constant as follows:

$$\tau = \frac{L}{R} = \frac{1 \text{ mH}}{1 \text{ k}\Omega} = \frac{1 \times 10^{-3} \text{ H}}{1 \times 10^{3} \ \Omega} = 1 \times 10^{-6} \text{ s} = 1 \ \mu\text{s}$$

***Related Exercise*** Find the time constant for $R = 2.2$ kΩ and $L = 500$ μH.

## Energizing Current in an Inductor

In a series *RL* circuit, the current will increase to approximately 63% of its full value in one time-constant interval after the switch is closed. This buildup of current is analogous to the buildup of capacitor voltage during the charging in an *RC* circuit; they both follow an exponential curve and reach the approximate percentages of final value as indicated in Table 14–1 and as illustrated in Figure 14–20.

**TABLE 14–1**
***Percentage of final current after each time-constant interval during current buildup***

| Number of Time Constants | % Final Value |
|---|---|
| 1 | 63 |
| 2 | 86 |
| 3 | 95 |
| 4 | 98 |
| 5 | 99 (considered 100%) |

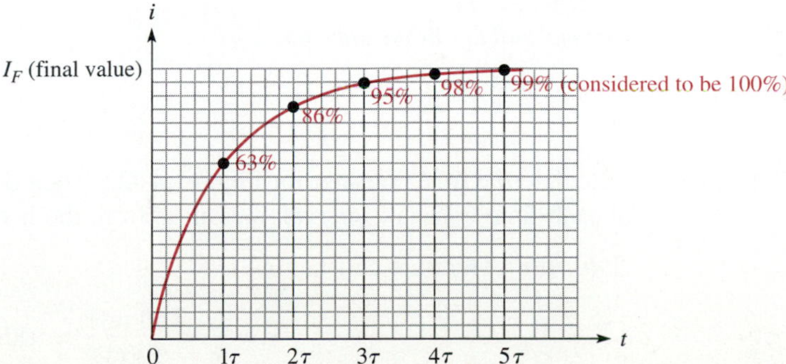


**FIGURE 14–20**
***Energizing current in an inductor.***

The change in current over five time-constant intervals is illustrated in Figure 14–21. When the current reaches its final value at approximately $5\tau$, it ceases to change. At this time, the inductor acts as a short (except for winding resistance) to the constant current. The final value of the current is

$$I_F = \frac{V_S}{R} = \frac{10\text{ V}}{1\text{ k}\Omega} = 10\text{ mA}$$

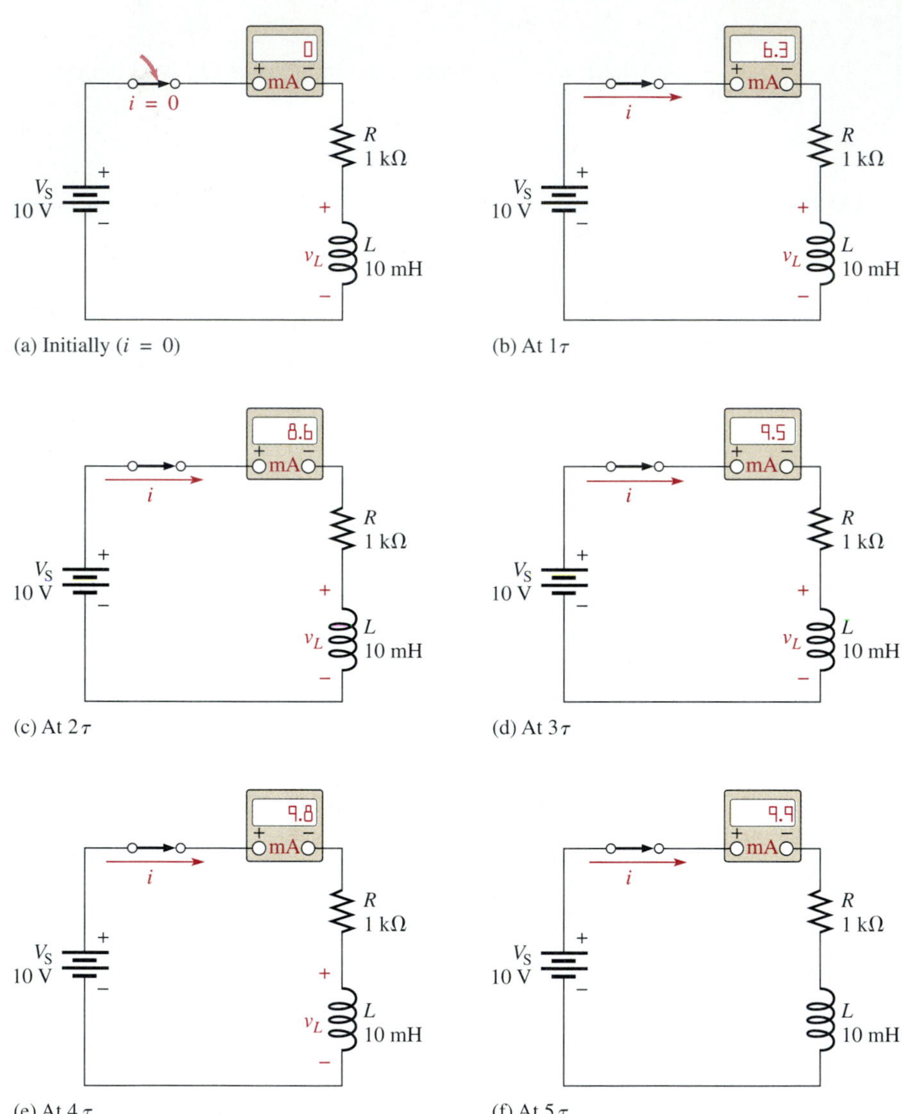


**FIGURE 14–21**
***Current buildup in an inductor.***

**EXAMPLE 14–8** Calculate the time constant for Figure 14–22. Then determine the current and the time at each time-constant interval, measured from the instant the switch is closed.

***Solution*** The time constant is

$$\tau = \frac{L}{R} = \frac{50\text{ mH}}{100\ \Omega} = 500\ \mu\text{s}$$

FIGURE 14–22

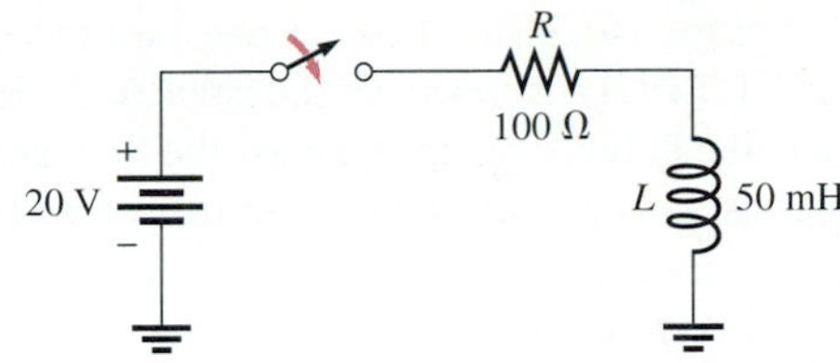


The current at each time-constant interval is a certain percentage of the final current. The final current is

$$I_F = \frac{V_S}{R} = \frac{20\ \text{V}}{100\ \Omega} = 0.2\ \text{A} = 200\ \text{mA}$$

Using the time-constant percentage values from Table 14–1,

$$\text{At } 1\tau = 500\ \mu\text{s:}\quad i = 0.63(200\ \text{mA}) = 126\ \text{mA}$$
$$\text{At } 2\tau = 1\ \text{ms:}\quad i = 0.86(200\ \text{mA}) = 172\ \text{mA}$$
$$\text{At } 3\tau = 1.5\ \text{ms:}\quad i = 0.95(200\ \text{mA}) = 190\ \text{mA}$$
$$\text{At } 4\tau = 2\ \text{ms:}\quad i = 0.98(200\ \text{mA}) = 196\ \text{mA}$$
$$\text{At } 5\tau = 2.5\ \text{ms:}\quad i = 0.99(200\ \text{mA}) = 198\ \text{mA} \cong 200\ \text{mA}$$

***Related Exercise*** Repeat the calculations if $R$ is 680 Ω and $L$ is 100 μH.

## Deenergizing Current in an Inductor

Current in an inductor decreases exponentially according to the approximate percentage values in Table 14–2 and in Figure 14–23.

TABLE 14–2
***Percentage of initial current after each time-constant interval while current is decreasing***

| Number of Time Constants | % of Initial Value |
|---|---|
| 1 | 37 |
| 2 | 14 |
| 3 | 5 |
| 4 | 2 |
| 5 | 1 (considered 0) |

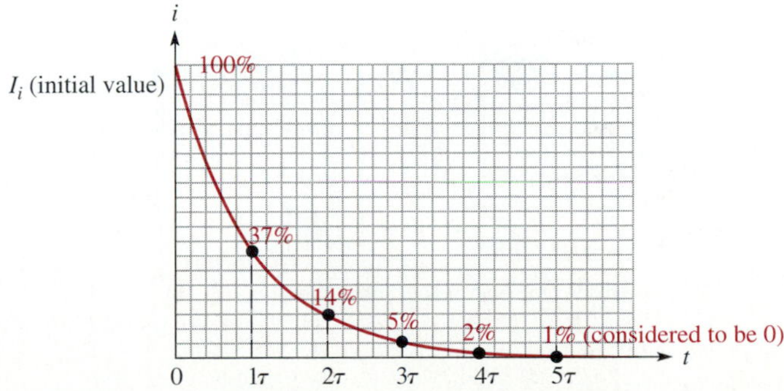


FIGURE 14–23
***Deenergizing current in an inductor.***

Figure 14–24(a) shows a constant current of 1 A (1000 mA) through an inductor. Switch 1 (SW1) is opened and switch 2 (SW2) is closed simultaneously, and for an instant the induced voltage keeps the 1 A current through the inductor. During the first time-constant interval, the current decreases by 63% down to 370 mA (37% of its initial value), as indicated in Figure 14–24(b). During the second time-constant interval, the current decreases by another 63% to 140 mA (14% of its initial value), as shown in Figure 14–24(c). The continued decrease in the current is illustrated by the remaining parts of Figure 14–24. Part (f) shows that only 1% of the initial current is left at the end of five time constants. Traditionally, this value is accepted as the final value and is approximated as zero current. Notice that until after the five time constants have elapsed, there is an induced voltage across the coil which is trying to maintain the current. This voltage follows a decreasing exponential curve, as will be discussed later.

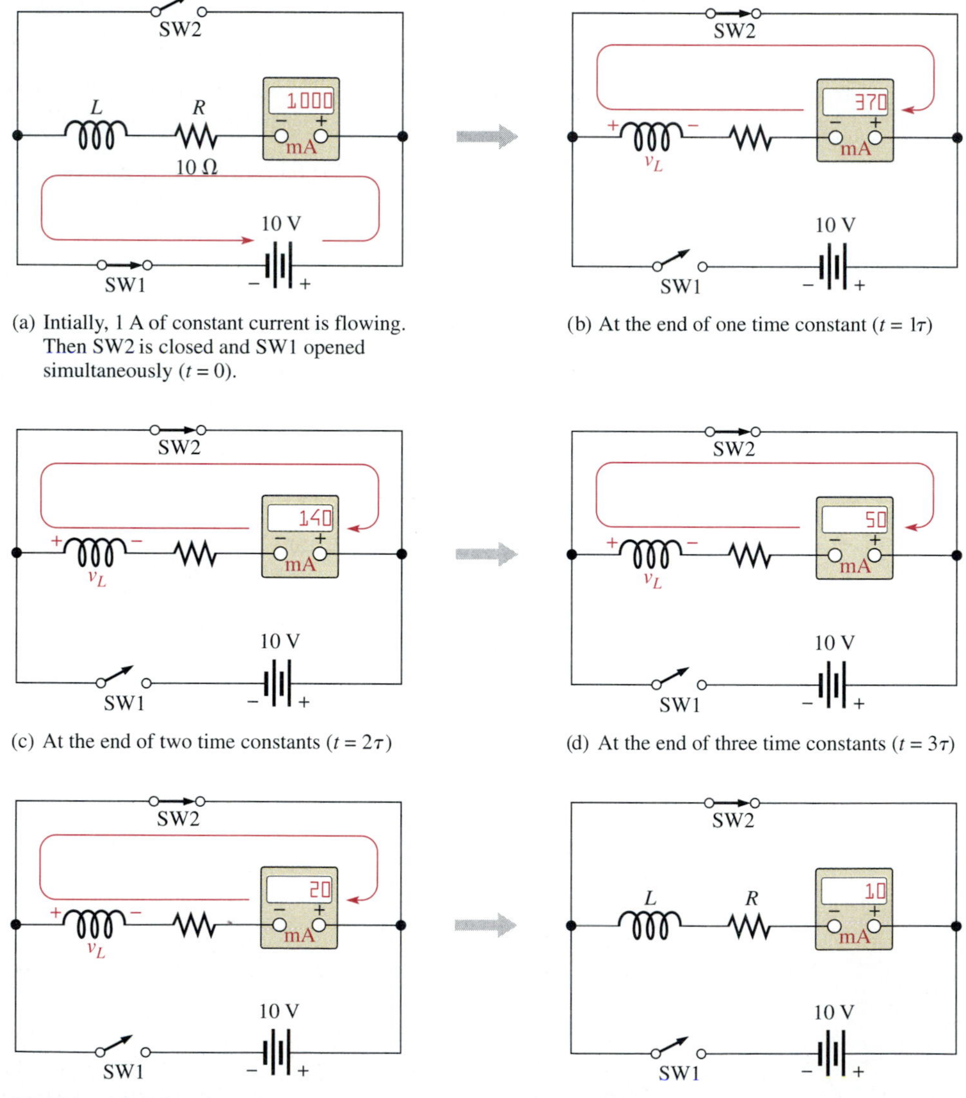


(a) Intially, 1 A of constant current is flowing. Then SW2 is closed and SW1 opened simultaneously ($t = 0$).

(b) At the end of one time constant ($t = 1\tau$)

(c) At the end of two time constants ($t = 2\tau$)

(d) At the end of three time constants ($t = 3\tau$)

(e) At the end of four time constants ($t = 4\tau$)

(f) At the end of five time constants ($t = 5\tau$). Since only 1% of the current is left, this value is taken as the final zero value.

**FIGURE 14–24**

***Illustration of the exponential decrease of current in an inductor. The current decreases another 63% during each time constant interval.***

**EXAMPLE 14–9** In Figure 14–25, SW1 is opened at the instant that SW2 is closed. Assume steady-state current through the coil prior to switch change.

FIGURE 14–25

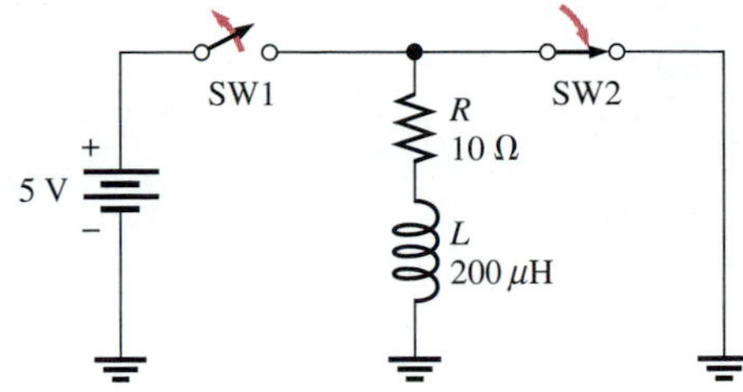


**(a)** What is the time constant?
**(b)** What is the initial coil current at the instant of switching?
**(c)** What is the coil current at $1\tau$?

***Solution***

**(a)** $\tau = \dfrac{L}{R} = \dfrac{200\ \mu\text{H}}{10\ \Omega} = 20\ \mu\text{s}$

**(b)** Current cannot change instantaneously in an inductor. Therefore, the current at the instant of the switch change is the same as the steady-state current.

$$I = \frac{5\text{ V}}{10\ \Omega} = 500\text{ mA}$$

**(c)** At $1\tau$, the current has decreased to 37% of its initial value.

$$i = 0.37(500\text{ mA}) = 185\text{ mA}$$

***Related Exercise*** Change $R$ to 47 Ω and $L$ to 1 mH in Figure 14–25 and repeat each calculation.

## Induced Voltage in the Series *RL* Circuit

As you know, when current changes in an inductor, a voltage is induced. Let's examine what happens to the voltages across the resistor and the coil in a series circuit when a change in current occurs.

Look at the circuit in Figure 14–26(a). When the switch is open, there is no current, and the resistor voltage and the coil voltage are both zero. At the instant the switch is closed, as indicated in part (b), the instantaneous voltage across the resistor ($v_R$) is zero and the instantaneous voltage through the inductor ($v_L$) is 10 V. The reason for this change is that the induced voltage across the coil is equal and opposite to the applied voltage, preventing the current from changing instantaneously. Therefore, *at the instant of switch closure, the inductor effectively acts as an open with all the applied voltage across it.*

During the first five time constants, the current is building up exponentially, and the induced coil voltage is decreasing. The resistor voltage increases with the current, as Figure 14–26(c) illustrates. After five time constants have elapsed, the current has reached its final value, $V_S/R$. At this time, all of the applied voltage is dropped across the resistor and none across the coil. Thus, the inductor effectively acts as a short to nonchanging current, as Figure 14–26(d) illustrates. Keep in mind that the inductor always reacts to a change in current by creating an induced voltage in order to counteract that change in current.

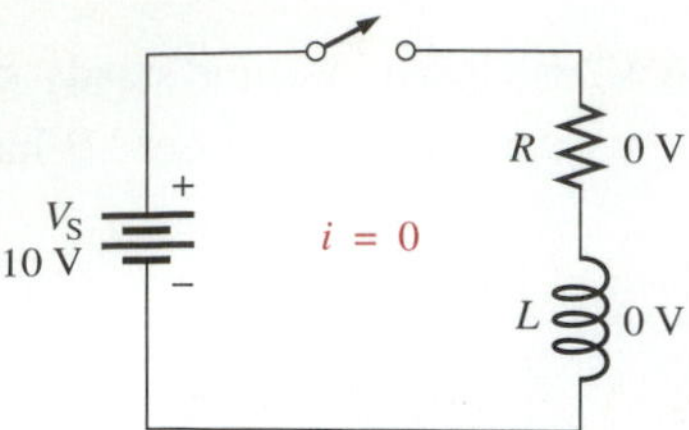

(a) Before switch is closed.

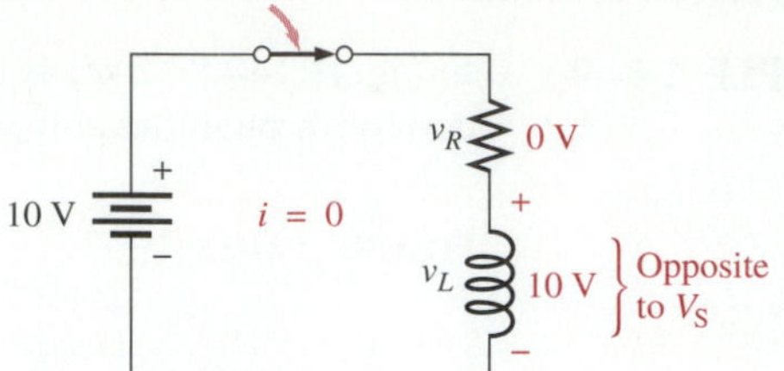

(b) At instant switch is closed, $v_L$ is equal and opposite to $V_S$.

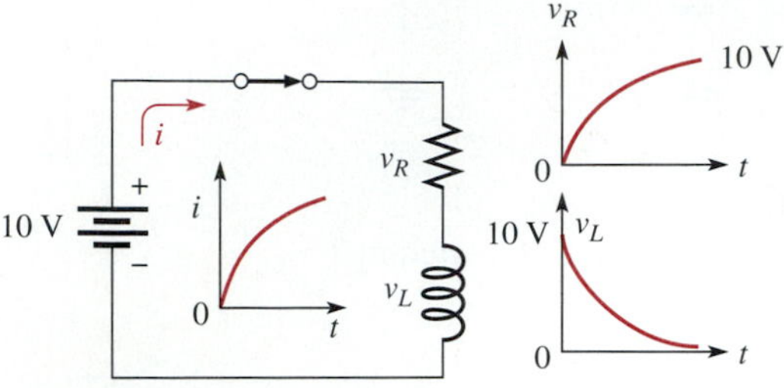

(c) During the first five time constants, $v_R$ increases exponentially with current, and $v_L$ decreases exponentially.

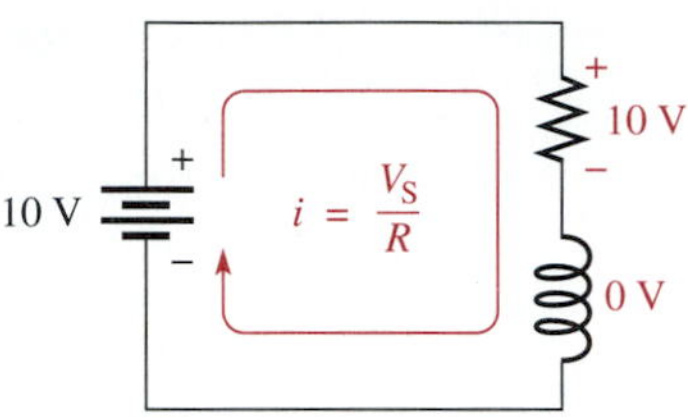

(d) After the first five constants, $v_R = 10$ V and $v_R \approx 0$ V. The current is at a constant maximum value.

**FIGURE 14–26**

***Voltage in an RL circuit as the inductor energizes. The winding resistance is neglected.***

Now let's examine the case illustrated in Figure 14–27, where the steady-state current is switched out, and the inductor discharges through another path. Part (a) shows the steady-state condition, and part (b) illustrates the instant at which the source is removed by opening SW1 and the discharge path is connected with the closure of SW2. There was 1 A through $L$ prior to this. Notice that 10 V are induced in $L$ in the direction to aid the 1 A in an effort to keep it from changing. Then, as shown in part (c), the current decays exponentially, and so do $v_R$ and $v_L$. After $5\tau$, as shown in part (d), all of the energy stored in the magnetic field of $L$ is dissipated, and all values are zero.

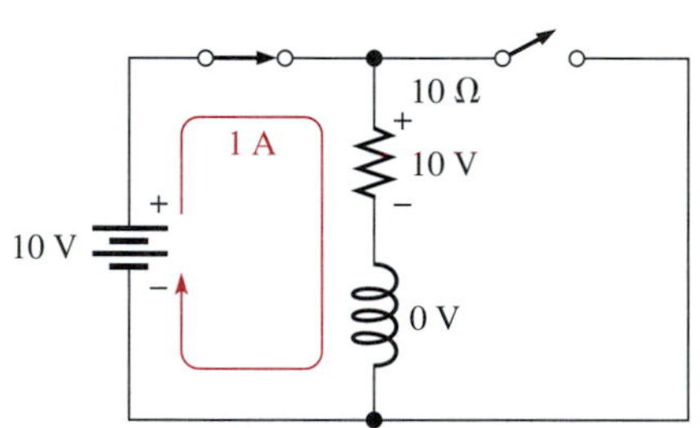

(a) Initially, there is a constant 1 A and the voltage is dropped across $R$.

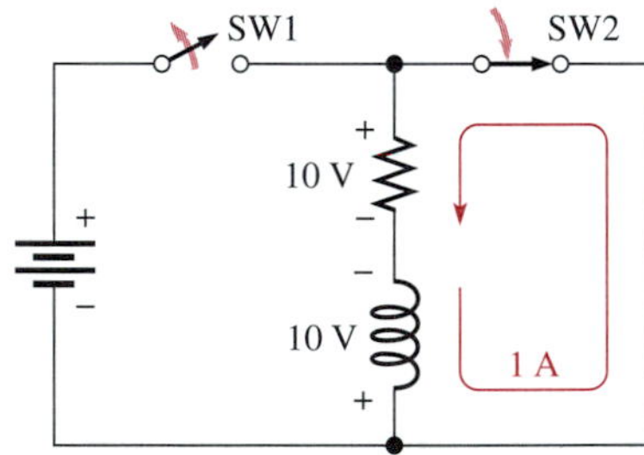

(b) At the instant that SW1 is opened and SW2 closed, 10 V is induced across $L$.

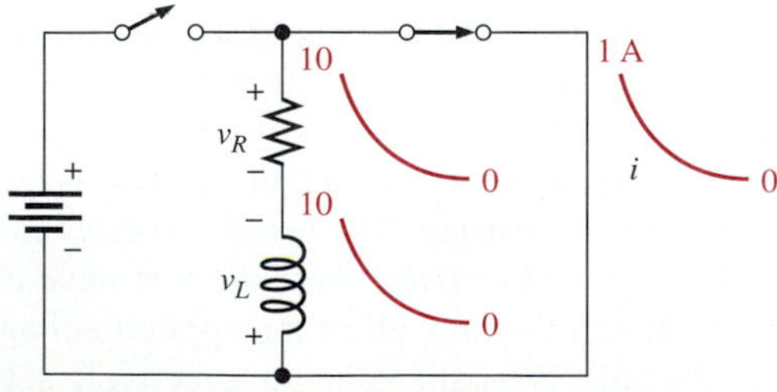

(c) During the five-time-constant interval, $v_R$ and $v_L$ decrease exponentially with the current.

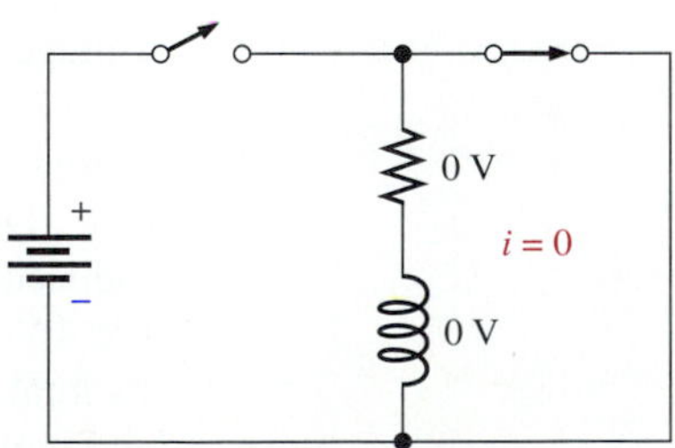

(d) After the five time constants, $v_R$, $v_L$, and $i$ are all zero.

**FIGURE 14–27**

***Voltage in an RL circuit as the inductor deenergizes. The winding resistance is neglected.***

**EXAMPLE 14–10**

**(a)** In Figure 14–28(a), what is $v_L$ at the instant SW1 is closed? What is $v_L$ after $5\tau$?

**(b)** In Figure 14–28(b), what is $v_L$ at the instant SW1 opens and SW2 closes? What is $v_L$ after $5\tau$?

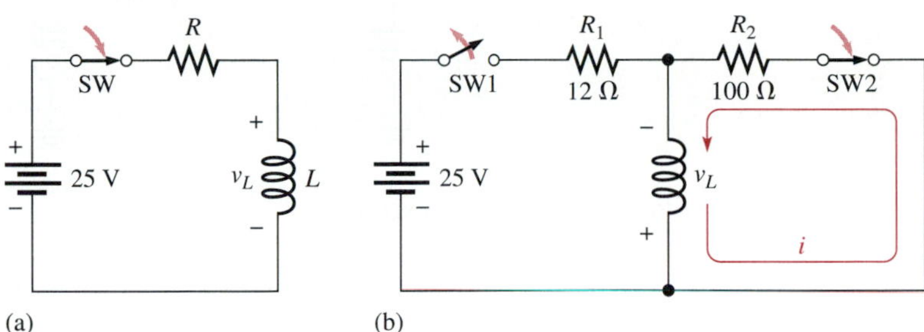


FIGURE 14–28

*Solution*

**(a)** At the instant the switch is closed, all of the source voltage is across $L$. Thus, $v_L = 25$ V, with the polarity as shown. After $5\tau$, $L$ acts as a short, so $v_L = 0$ V.

**(b)** With SW1 closed and SW2 open, the steady-state current is

$$I = \frac{25\text{ V}}{12\ \Omega} = 2.08\text{ A}$$

When the switches are thrown, an induced voltage is created across $L$ sufficient to keep this 2.08 A current for an instant. In this case, it takes

$$v_L = IR_2 = (2.08\text{ A})(100\ \Omega) = 208\text{ V}$$

After $5\tau$, the inductor voltage is zero.

*Related Exercise* Repeat part (a) if the source voltage is 9 V. Repeat part (b) if $R_1$ is 27 Ω with the source voltage at 25 V.

## The Exponential Formulas

The formulas for the exponential current and voltage in an *RL* circuit are similar to those used in the last chapter for the *RC* circuit, and the universal exponential curves in Figure 13–36 apply to inductors as well as capacitors. The general formulas for *RL* circuits are stated as follows:

$$v = V_F + (V_i - V_F)e^{-Rt/L} \quad \textbf{(14–11)}$$

$$i = I_F + (I_i - I_F)e^{-Rt/L} \quad \textbf{(14–12)}$$

where $V_F$ and $I_F$ are the final values, $V_i$ and $I_i$ are the initial values, and $v$ and $i$ are the instantaneous values of the inductor voltage or current at time $t$.

*Increasing Current* The formula for the special case in which an increasing exponential current curve begins at zero ($I_i = 0$) is

$$i = I_F(1 - e^{-Rt/L}) \quad \textbf{(14–13)}$$

Using Equation (14–13), you can calculate the value of the increasing inductor current at any instant of time. The same is true for voltage.

**EXAMPLE 14–11** In Figure 14–29, determine the inductor current 30 μs after the switch is closed.

**FIGURE 14–29**

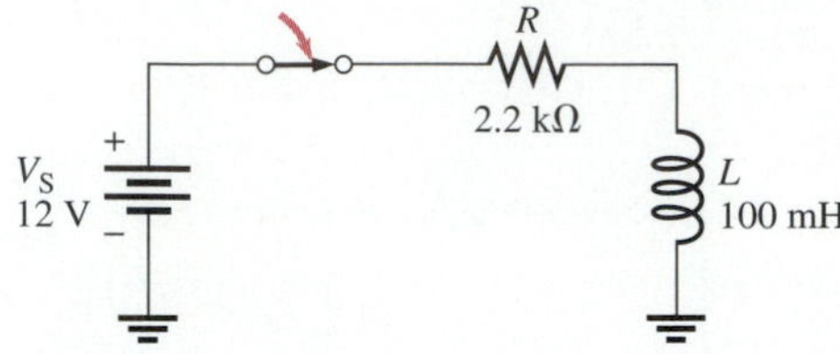


***Solution*** The time constant is

$$\tau = \frac{L}{R} = \frac{100\text{ mH}}{2.2\text{ k}\Omega} = 45.5\ \mu\text{s}$$

The final current is

$$I_F = \frac{V_S}{R} = \frac{12\text{ V}}{2.2\text{ k}\Omega} = 5.45\text{ mA}$$

The initial current is zero. Notice that 30 μs is less than one time constant, so the current will reach less than 63% of its final value in that time.

$$i_L = I_F(1 - e^{-Rt/L}) = 5.45\text{ mA}(1 - e^{-0.66}) = 5.45\text{ mA}(1 - 0.517) = 2.63\text{ mA}$$

The calculator sequence is

3 0 EXP +/− 6 ÷ 4 5 • 5 EXP +/− 6 = +/− 2nd
$e^x$ +/− + 1 = × 5 • 4 5 EXP +/− 3 =

***Related Exercise*** In Figure 14–29, determine the inductor current 55 μs after the switch is closed.

***Decreasing Current*** The formula for the special case in which a decreasing exponential current has a final value of zero ($I_F = 0$) is

$$i = I_i e^{-Rt/L} \qquad \textbf{(14–14)}$$

This formula can be used to calculate the deenergizing current at any instant, as the following example shows.

**EXAMPLE 14–12** Determine the inductor current in Figure 14–30 at a point in time 2 ms after the switches are thrown (SW1 opened and SW2 closed).

**FIGURE 14–30**

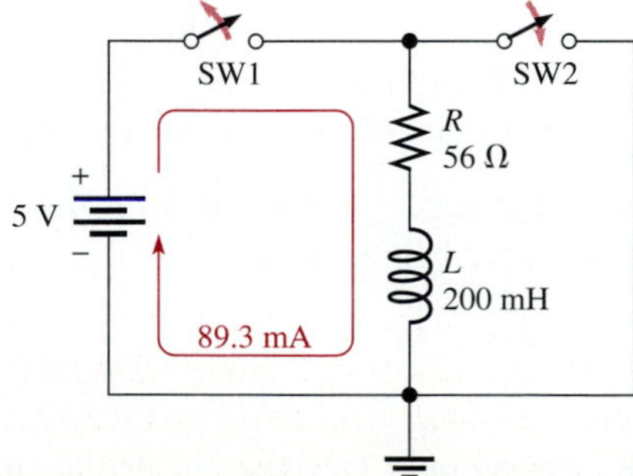

***Solution*** The deenergizing time constant is

$$\tau = \frac{L}{R} = \frac{200 \text{ mH}}{56 \text{ }\Omega} = 3.57 \text{ ms}$$

The initial current in the inductor is 89.3 mA. Notice that 2 ms is less than one time constant, so the current will show a decrease less than 63%. Therefore, the current will be greater than 37% of its initial value at 2 ms after the switches are thrown.

$$i = I_i e^{-Rt/L} = (89.3 \text{ mA})e^{-0.56} = 51.0 \text{ mA}$$

***Related Exercise*** Determine the inductor current in Figure 14–30 at a point in time 6 ms after the switches are thrown if the source voltage is changed to 10 V.

**SECTION 14–5 REVIEW**

1. A 15 mH inductor with a winding resistance of 10 Ω has a constant direct current of 10 mA through it. What is the voltage drop across the inductor?
2. A 20 V dc source is connected to a series *RL* circuit with a switch. At the instant of switch closure, what are the values of $v_R$ and $v_L$?
3. In the same circuit as in Question 2, after a time interval equal to $5\tau$ from switch closure, what are $v_R$ and $v_L$?
4. In a series *RL* circuit where $R = 1$ kΩ and $L = 500$ μH, what is the time constant? Determine the current 0.25 μs after a switch connects 10 V across the circuit.

## 14–6 ■ INDUCTORS IN AC CIRCUITS

***You will learn in this section that an inductor passes ac but with an amount of opposition that depends on the frequency of the ac.***

***After completing this section, you should be able to***

- **Analyze inductive ac circuits**
  - ☐ Explain why an inductor causes a phase shift between voltage and current
  - ☐ Define *inductive reactance*
  - ☐ Determine the value of inductive reactance in a given circuit
  - ☐ Discuss instantaneous, true, and reactive power in an inductor

The concept of the derivative was introduced in Chapter 13. The expression for induced voltage in an inductor was stated earlier in Equation (14–1). This formula is $v_{ind} = L(di/dt)$.

### Phase Relationship of Current and Voltage in an Inductor

From Equation (14–1), the formula for induced voltage, you can see that the faster the current through an inductor changes, the greater the induced voltage will be. For example, if the rate of change of current is zero, the voltage is zero [$v_{ind} = L(di/dt) = L(0) = 0$ V]. When $di/dt$ is a positive-going maximum, $v_{ind}$ is a positive maximum; when $di/dt$ is a negative-going maximum, $v_{ind}$ is a negative maximum.

A sinusoidal current always induces a sinusoidal voltage in inductive circuits. Therefore, the voltage can be plotted with respect to the current by knowing the points on the current curve at which the voltage is zero and those at which it is maximum. This phase relationship is shown in Figure 14–31(a). Notice that the voltage leads the current by 90°. This is always true in a purely inductive circuit. A phasor diagram of this relationship is shown in Figure 14–31(b).

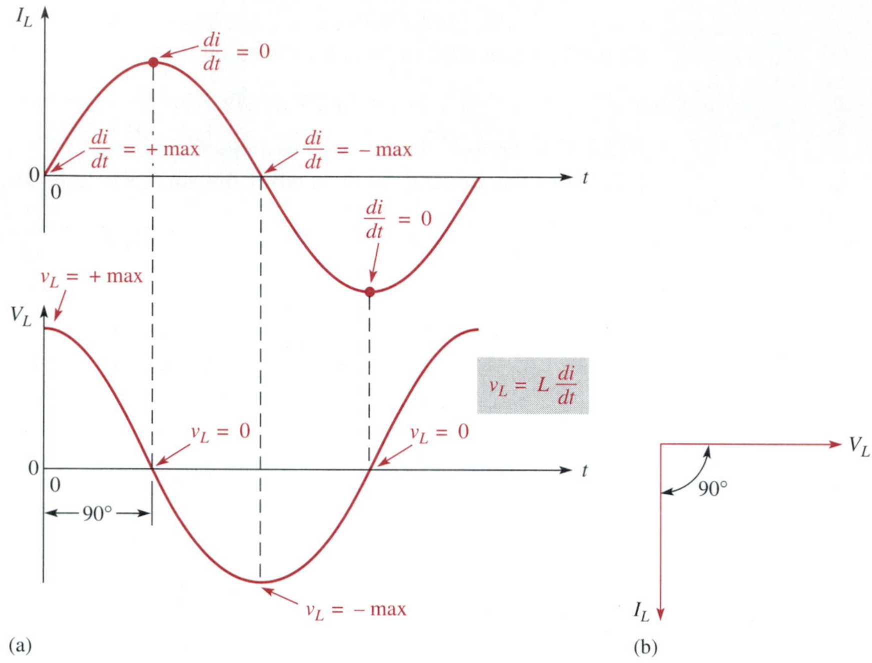


FIGURE 14–31

***Phase relation of $V_{ind}$ and I in an inductor. Current always lags the inductor voltage by 90°.***

## Inductive Reactance, $X_L$

**Inductive reactance** is the opposition to sinusoidal current, expressed in ohms. The symbol for inductive reactance is $X_L$.

To develop a formula for $X_L$, we will use the relationship $v_{ind} = L(di/dt)$ and the curves in Figure 14–32. The rate of change of current is directly related to frequency. The faster the current changes, the higher the frequency. For example, you can see that in Figure 14–32, the slope of sine wave *A* at the zero crossings is greater than that of sine wave *B*. Recall that the slope of a curve at a point indicates the rate of change at that point. Sine wave *A* has a higher frequency than sine wave *B*, as indicated by a greater maximum rate of change (*di/dt* is greater at the zero crossings).

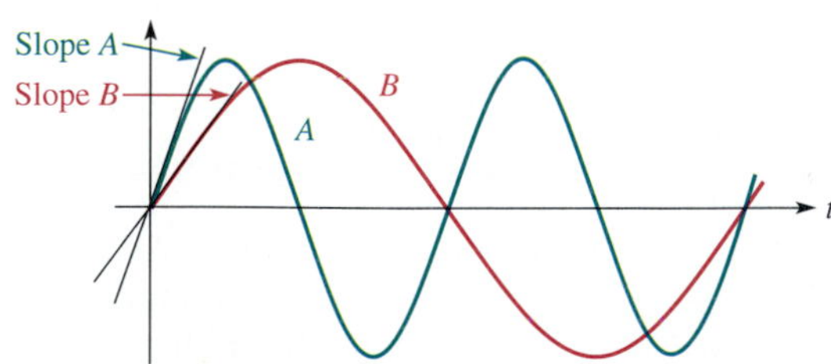


FIGURE 14–32

***Slope indicates rate of change. Sine wave A has a greater rate of change at the zero crossing than B, and thus A has a higher frequency.***

When frequency increases, *di/dt* increases, and thus $v_{ind}$ increases. When frequency decreases, *di/dt* decreases, and thus $v_{ind}$ decreases. The induced voltage is directly dependent on frequency.

$$\overset{\uparrow}{v_{ind}} = \overset{\uparrow}{L(di/dt)} \quad \text{and} \quad \underset{\downarrow}{v_{ind}} = \underset{\downarrow}{L(di/dt)}$$

An increase in induced voltage means more opposition ($X_L$ is greater). Therefore, $X_L$ is directly proportional to induced voltage and thus directly proportional to frequency.

**$X_L$ is proportional to $f$.**

Now, if $di/dt$ is constant and the inductance is varied, an increase in $L$ produces an increase in $v_{ind}$, and a decrease in $L$ produces a decrease in $v_{ind}$, as indicated:

$$\overset{\uparrow}{v_{ind}} = \overset{\uparrow}{L}(di/dt) \quad \text{and} \quad \underset{\downarrow}{v_{ind}} = \underset{\downarrow}{L}(di/dt)$$

Again, an increase in $v_{ind}$ means more opposition (greater $X_L$). Therefore, $X_L$ is directly proportional to induced voltage and thus directly proportional to inductance. The inductive reactance is directly proportional to both $f$ and $L$.

**$X_L$ is proportional to $fL$.**

The complete formula for inductive reactance, $X_L$, is

$$X_L = 2\pi fL \qquad \textbf{(14–15)}$$

Notice that $2\pi$ appears as a constant factor in the equation. This comes from the relationship of a sine wave to rotational motion. $X_L$ is in ohms when $f$ is in hertz and $L$ is in henries.

**EXAMPLE 14–13** A sinusoidal voltage is applied to the circuit in Figure 14–33. The frequency is 1 kHz. Determine the inductive reactance.

FIGURE 14–33

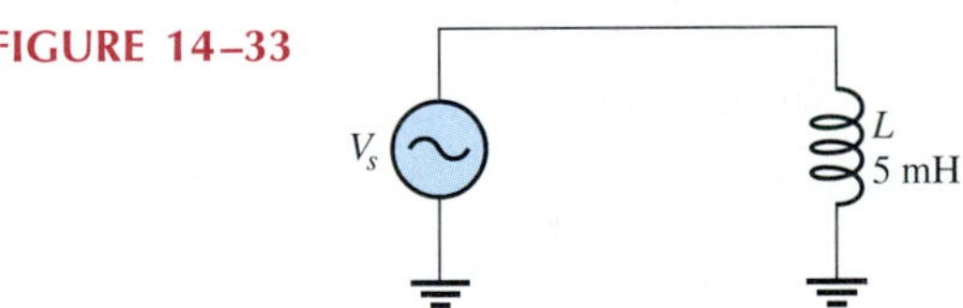


***Solution*** Convert 1 kHz to $1 \times 10^3$ Hz and 5 mH to $5 \times 10^{-3}$ H. Therefore, the inductive reactance is

$$X_L = 2\pi fL = 2\pi(1 \times 10^3\text{ Hz})(5 \times 10^{-3}\text{ H}) = 31.4\ \Omega$$

***Related Exercise*** What is $X_L$ in Figure 14–33 if the frequency is increased to 3.5 kHz?

## Analysis of Inductive AC Circuits

As you have seen, the current lags the voltage by 90° in inductive ac circuits. If the applied voltage is assigned a reference phase angle of 0°, it can be expressed in polar form as $V_s\angle 0°$. The resulting current can be expressed in polar form as $I\angle -90°$ or in rectangular form as $-jI$, as shown in Figure 14–34.

FIGURE 14–34

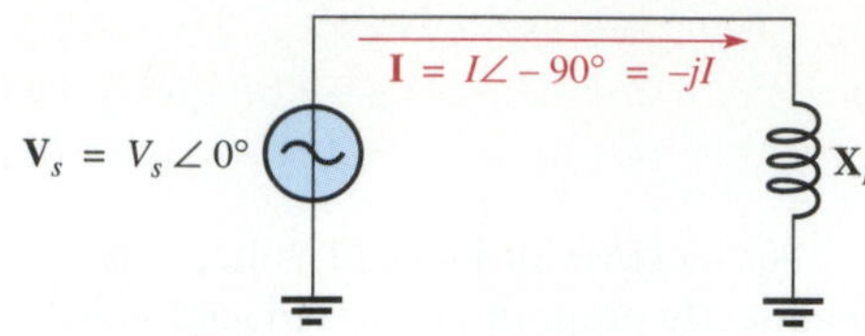


Ohm's law applies to ac circuits with inductive reactance with $R$ replaced by $\mathbf{X}_L$ in the Ohm's law formula. The quantities are expressed as complex numbers because of the introduction of phase angles. Applying Ohm's law to the circuit in Figure 14–34 gives the following result:

$$\mathbf{X}_L = \frac{V_s\angle 0^\circ}{I\angle -90^\circ} = \left(\frac{V_s}{I}\right)\angle 90^\circ$$

This shows that $\mathbf{X}_L$ always has a 90° angle attached to its magnitude and is written as $X_L\angle 90^\circ$ or $jX_L$.

**EXAMPLE 14–14** Determine the rms current in Figure 14–35.

FIGURE 14–35

***Solution*** Convert 10 kHz to $10 \times 10^3$ Hz and 100 mH to $100 \times 10^{-3}$ H. Then calculate the magnitude of $X_L$.

$$X_L = 2\pi fL = 2\pi(10 \times 10^3 \text{ Hz})(100 \times 10^{-3} \text{ H}) = 6283 \ \Omega$$

Expressed in polar form, $X_L$ is

$$\mathbf{X}_L = 6283\angle 90^\circ \ \Omega$$

Apply Ohm's law to determine the current.

$$\mathbf{I} = \frac{\mathbf{V}_{rms}}{\mathbf{X}_L} = \frac{5\angle 0^\circ \text{ V}}{6283\angle 90^\circ \ \Omega} = 796\angle -90^\circ \ \mu\text{A}$$

***Related Exercise*** Determine the rms current in Figure 14–35 for the following values: $\mathbf{V}_{rms} = 12\angle 0^\circ$ V, $f = 4.9$ kHz, and $L = 680 \ \mu$H.

## Power in an Inductor

As discussed earlier, an inductor stores energy in its magnetic field when there is current through it. An ideal inductor (assuming no winding resistance) does not dissipate energy; it only stores it. When an ac voltage is applied to an inductor, energy is stored by the inductor during a portion of the cycle; then the stored energy is returned to the source during another portion of the cycle. There is no net energy loss. Figure 14–36 shows the power curve that results from one cycle of inductor current and voltage.

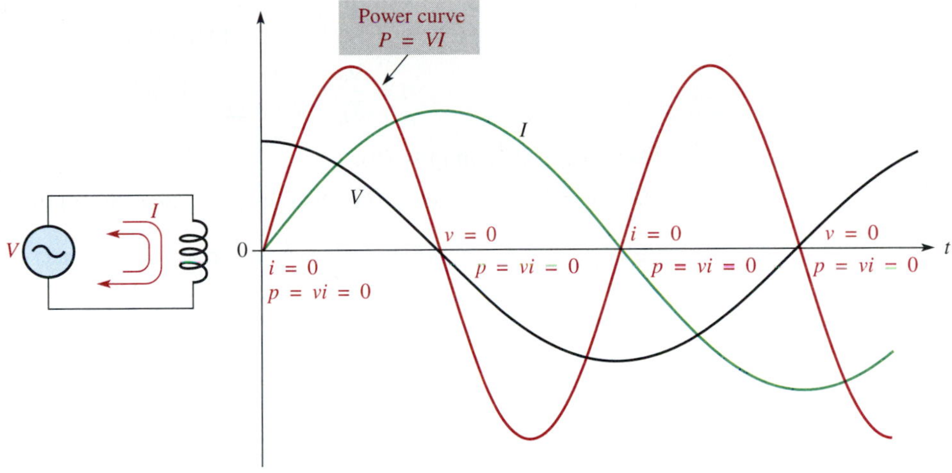


**FIGURE 14–36**
*Power curve.*

***Instantaneous Power (p)*** The product of $v$ and $i$ gives instantaneous power, $p$. At points where $v$ or $i$ is zero, $p$ is also zero. When both $v$ and $i$ are positive, $p$ is also positive. When either $v$ or $i$ is positive and the other negative, $p$ is negative. When both $v$ and $i$ are negative, $p$ is positive. As you can see in Figure 14–36, the power follows a sinusoidal-type curve. Positive values of power indicate that energy is stored by the inductor. Negative values of power indicate that energy is returned from the inductor to the source. Note that the power fluctuates at a frequency twice that of the voltage or current as energy is alternately stored and returned to the source.

***True Power ($P_{true}$)*** Ideally, all of the energy stored by an inductor during the positive portion of the power cycle is returned to the source during the negative portion. No net energy is consumed in the inductance, so the true power is zero. Actually, because of winding resistance in a practical inductor, some power is always dissipated; and there is a very small amount of true power, which can normally be neglected.

$$P_{\text{true}} = (I_{\text{rms}})^2 R_W \qquad \textbf{(14–16)}$$

***Reactive Power ($P_r$)*** The rate at which an inductor stores or returns energy is called its **reactive power, $P_r$.** The reactive power is a nonzero quantity, because at any instant in time, the inductor is actually taking energy from the source or returning energy to it. Reactive power does not represent an energy loss. The following formulas apply:

$$P_r = V_{\text{rms}} I_{\text{rms}} \qquad \textbf{(14–17)}$$

$$P_r = \frac{V_{\text{rms}}^2}{X_L} \qquad \textbf{(14–18)}$$

$$P_r = I_{\text{rms}}^2 X_L \qquad \textbf{(14–19)}$$

**EXAMPLE 14–15** A 10 V rms signal with a frequency of 1 kHz is applied to a 10 mH coil with a negligible winding resistance. Determine the reactive power ($P_r$).

***Solution*** First, calculate the inductive reactance and current values.

$$X_L = 2\pi fL = 2\pi(1\text{ kHz})(10\text{ mH}) = 62.8\ \Omega$$

$$I = \frac{V_S}{X_L} = \frac{10\text{ V}}{62.8\ \Omega} = 159\text{ mA}$$

Then, use Equation (14–19).

$$P_r = I^2X_L = (159\text{ mA})^2(62.8\ \Omega) = 1.59\text{ VAR}$$

***Related Exercise*** What happens to the reactive power if the frequency increases?

## The Quality Factor ($Q$) of a Coil

The quality factor ($Q$) is the ratio of the reactive power in the inductor to the true power in the winding resistance of the coil or the resistance in series with the coil. It is a ratio of the power in $L$ to the power in $R_W$. The quality factor is very important in resonant circuits, which are studied in Chapter 18. A formula for $Q$ is developed as follows:

$$Q = \frac{\text{reactive power}}{\text{true power}} = \frac{I^2X_L}{I^2R_W}$$

In a series circuit, $I$ is the same in $L$ and $R$; thus, the $I^2$ terms cancel, leaving

$$Q = \frac{X_L}{R_W} \qquad \textbf{(14–20)}$$

When the resistance is just the winding resistance of the coil, the circuit $Q$ and the coil $Q$ are the same. Note that $Q$ is a ratio of like units and, therefore, has no unit itself.

**SECTION 14–6 REVIEW**

1. State the phase relationship between current and voltage in an inductor.
2. Calculate $X_L$ for $f = 5$ kHz and $L = 100$ mH.
3. At what frequency is the reactance of a 50 $\mu$H inductor equal to 800 Ω?
4. Calculate the rms current in Figure 14–37.
5. An ideal 50 mH inductor is connected to a 12 V rms source. What is the true power? What is the reactive power at a frequency of 1 kHz?

FIGURE 14–37

## 14–7 ■ INDUCTOR APPLICATIONS

***Inductors are not as versatile as capacitors and tend to be more limited in their applications due, in part, to size and cost factors. However, there are many practical uses for inductors (coils) such as those discussed in Chapter 10. Recall that relay and solenoid coils, recording and pick-up heads, and sensing elements were introduced as electromagnetic applications of coils. In this section, some additional uses of inductors are presented.***

***After completing this section, you should be able to***

- Discuss some inductor applications
  - ☐ Describe a power supply filter
  - ☐ Explain the purpose of an RF choke
  - ☐ Discuss the basics of tuned circuits

### Power Supply Filter

In Chapter 13, you saw that a capacitor is used to filter the pulsating dc in a power supply. The final output voltage was a dc voltage with a small amount of ripple. In many cases, an inductor is used in the filter, as shown in Figure 14–38(a), to smooth out the ripple voltage. The inductor, placed in series with the load as shown, tends to oppose the current fluctuations caused by the ripple voltage, and thus the voltage developed across the load is more constant, as shown in Figure 14–38(b).

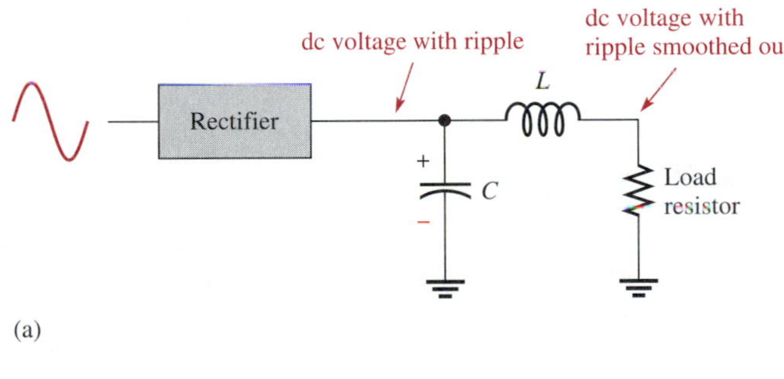


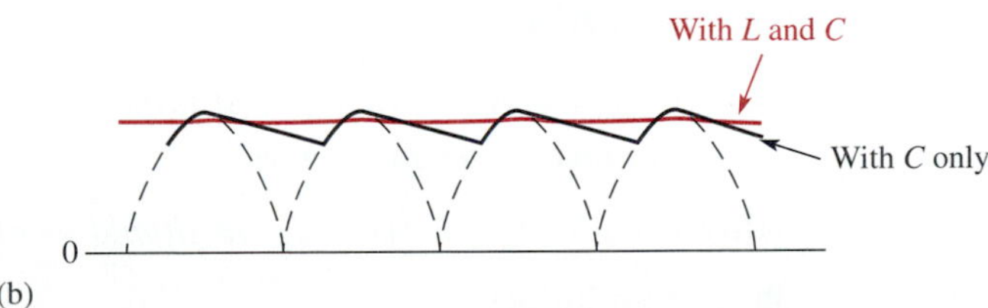


**FIGURE 14–38**
***Basic capacitor power supply filter with a series inductor.***

### RF Choke

Certain types of inductors are used in applications where radio frequencies (RF) must be prevented from getting into parts of a system, such as the power supply or the audio section of a receiver. In these situations, an inductor is used as a series filter and "chokes" off any unwanted RF signals that may be picked up on a line. This filtering action is based on the fact that the reactance of a coil increases with frequency. When the frequency of the current is sufficiently high, the reactance of the coil becomes extremely large and essentially blocks the current. A basic illustration of an inductor used as an RF choke is shown in Figure 14–39.

FIGURE 14–39

*An inductor used as an RF choke to minimize interfering signals on the power supply line.*

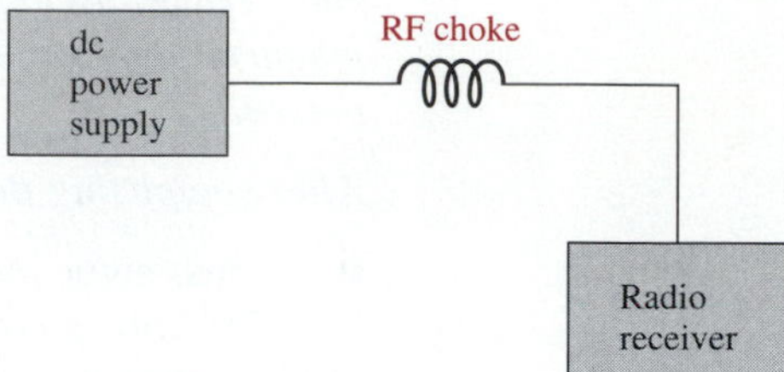


### Tuned Circuits

Inductors are used in conjunction with capacitors to provide frequency selection in communications systems. These tuned circuits allow a narrow band of frequencies to be selected while all other frequencies are rejected. The tuners in your TV and radio receivers are based on this principle and permit you to select one channel or station out of the many that are available.

Frequency selectivity is based on the fact that the reactances of both capacitors and inductors depend on the frequency and on the interaction of these two components when connected in series or parallel. Since the capacitor and the inductor produce opposite phase shifts, their combined opposition to current can be used to obtain a desired response at a selected frequency. Tuned *RLC* circuits are covered in Chapter 18.

**SECTION 14–7 REVIEW**

1. Explain how the ripple voltage from a power supply filter can be reduced through use of an inductor.
2. How does an inductor connected in series act as an RF choke?

## 14–8 ■ TESTING INDUCTORS

***In this section, two basic types of failures in inductors are discussed and methods for testing inductors are introduced.***

***After completing this section, you should be able to***

- Test an inductor
  - ☐ Perform an ohmmeter check for an open winding
  - ☐ Perform an ohmmeter check for shorted windings

The most common failure in an inductor is an open. To check for an open, the coil should be removed from the circuit. If there is an open, an ohmmeter check will indicate infinite resistance, as shown in Figure 14–40(a). If the coil is good, the ohmmeter will show the winding resistance. The value of the winding resistance depends on the wire size and length of the coil. It can be anywhere from one ohm to several hundred ohms. Figure 14–40(b) shows a good reading.

Occasionally, when an inductor is overheated with excessive current, the wire insulation will melt, and two or more turns will short together. This must be tested on an *LC*

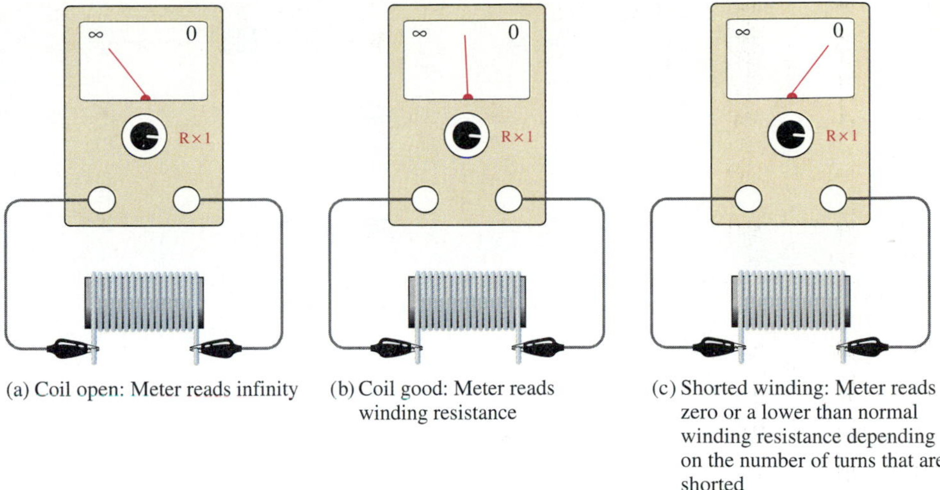


(a) Coil open: Meter reads infinity

(b) Coil good: Meter reads winding resistance

(c) Shorted winding: Meter reads zero or a lower than normal winding resistance depending on the number of turns that are shorted

FIGURE 14–40
***Checking a coil by measuring the resistance.***

meter because, with two shorted turns (or even several), an ohmmeter check may show the coil to be perfectly good from a resistance standpoint. Two shorted turns occur more frequently because the turns are adjacent and can easily short across from poor insulation, voltage breakdown, or simple wear if something is rubbing on them.

**SECTION 14–8 REVIEW**

1. When a coil is checked, a reading of infinity on the ohmmeter indicates a partial short (T or F).
2. An ohmmeter check of a good coil will indicate the value of the inductance (T or F).

## 14–9 ■ TECHnology Theory Into Practice

***In this TECH TIP section, you will test coils for their unknown inductance values using a test setup consisting of a square wave generator and an oscilloscope.***

You are given two coils for which the inductance values are not known. The coils are to be tested using simple laboratory instruments to determine the inductance values. The method is to place the coil in series with a resistor with a known value and measure the time constant. Knowing the time constant and the resistance value, the value of $L$ can be calculated.

The method of determining the time constant is to apply a square wave to the circuit and measure the resulting voltage across the resistor. Each time the square wave input voltage goes high, the inductor is energized and each time the square wave goes back to zero, the inductor is deenergized. The time it takes for the exponential resistor voltage to increase to approximately its final value equals to five time constants. This operation is illustrated in Figure 14–41. To make sure that the winding resistance of the coil can be

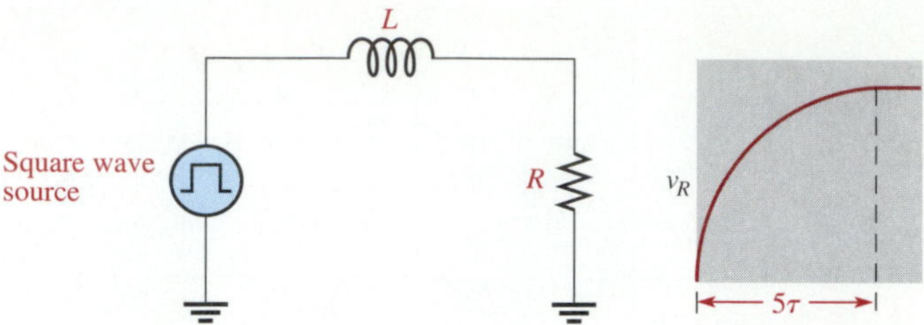


FIGURE 14–41
*Circuit for time-constant measurement.*

neglected, it must be measured and the value of the resistor used in the circuit must be selected to be considerably larger than the winding resistance.

## The Winding Resistance

Assume that the winding resistance of the coil in Figure 14–42 has been measured with an ohmmeter and found to be 85 Ω. To make the winding resistance negligible for time constant measurement, a 10 kΩ series resistor is used in the circuit.

☐ If 10 V dc is connected with the clip leads as shown, how much current is in the circuit after $t = 5\tau$?

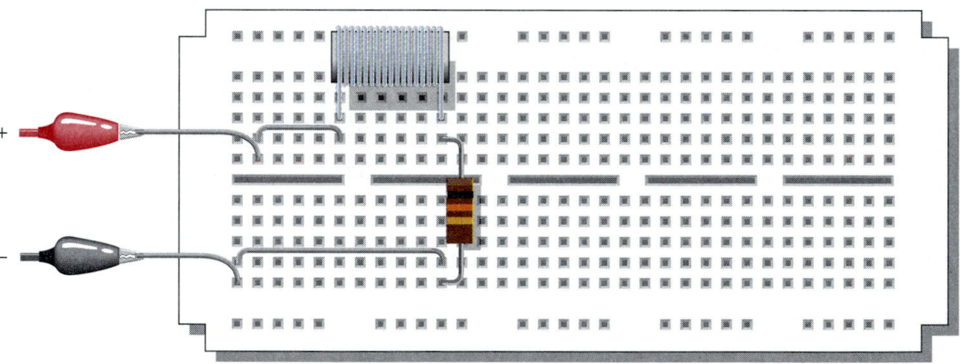


FIGURE 14–42
*Breadboard setup for measuring the time constant.*

## Inductance of Coil 1

Refer to Figure 14–43. To measure the inductance of coil 1, a square wave voltage is applied to the circuit. The amplitude of the square wave is adjusted to 10 V. The frequency is adjusted so that the inductor has time to fully energize during each square wave pulse; the scope is set to view a complete energizing curve as shown.

□ Determine the approximate circuit time constant.

□ Calculate the inductance of coil 1.

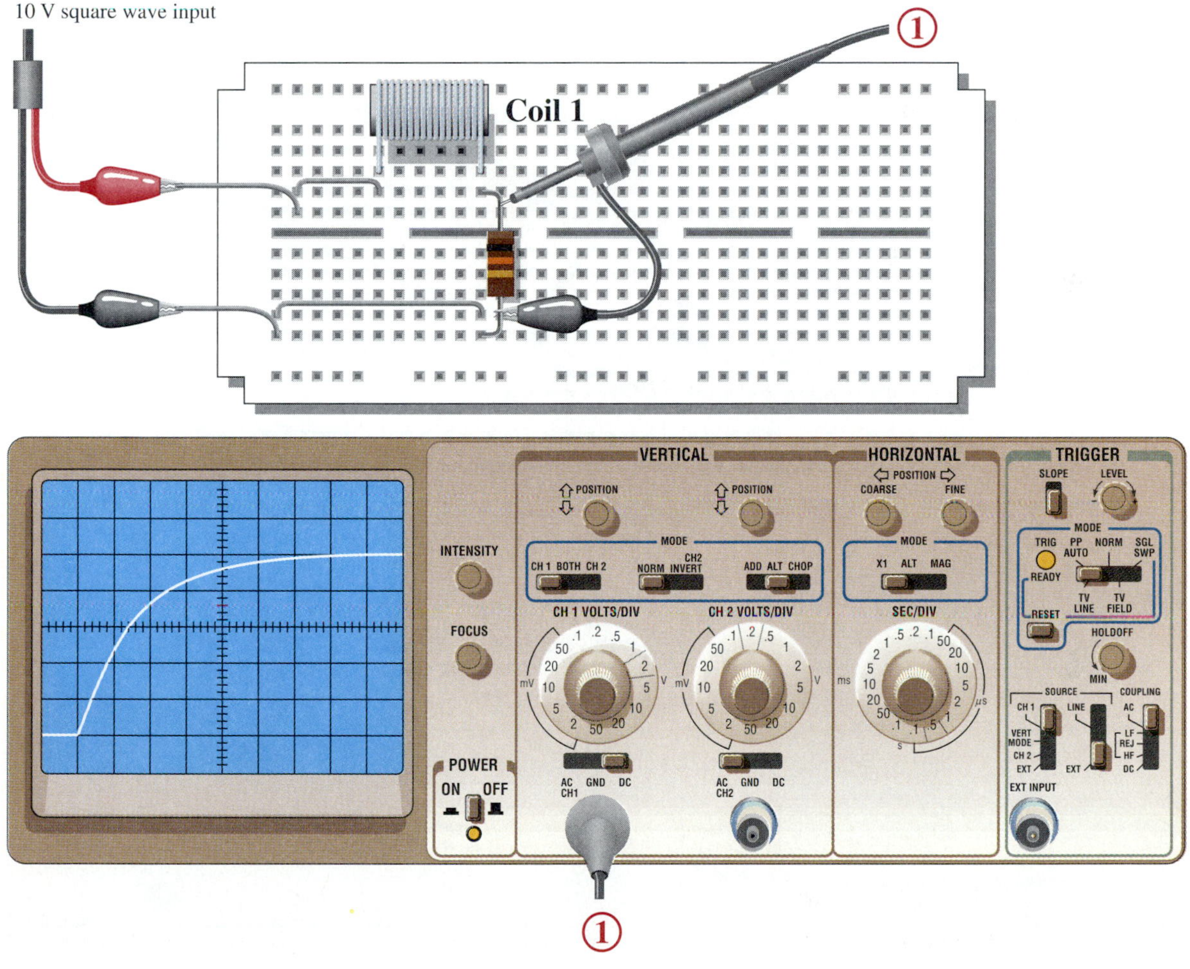


FIGURE 14–43
*Testing coil 1.*

### The Inductance of Coil 2

Refer to Figure 14–44 in which coil 2 replaces coil 1. To determine the inductance, a 10 V square wave is applied to the breadboarded circuit. The frequency of the square wave is adjusted so that the inductor has time to fully energize during each square wave pulse; the scope is set to view a complete energizing curve as shown.

- ☐ Determine the approximate circuit time constant.
- ☐ Calculate the inductance of coil 2.
- ☐ Discuss any difficulty in using this method.
- ☐ Specify how you can use a sinusoidal input voltage instead of a square wave to determine inductance.

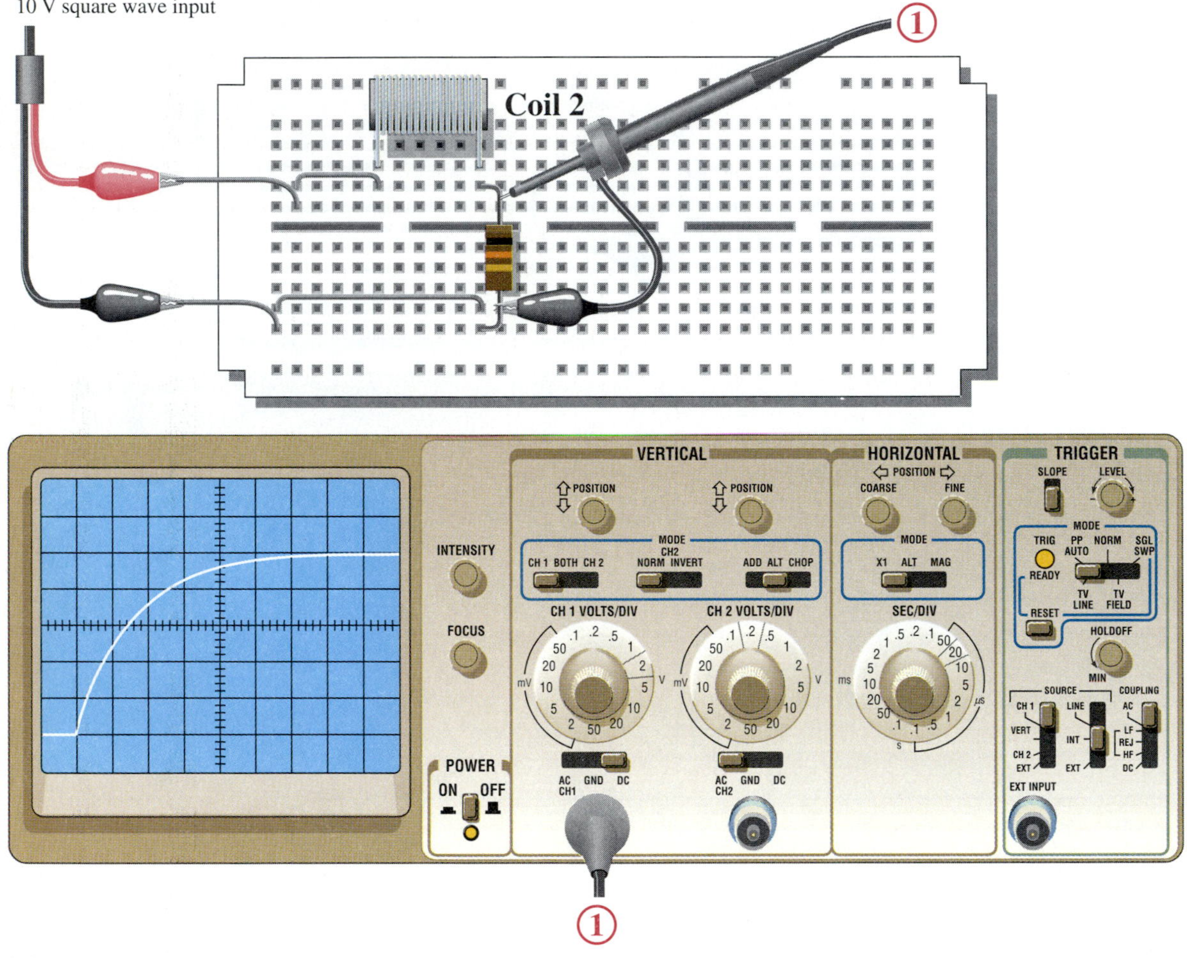


**FIGURE 14–44**
*Testing coil 2.*

**SECTION 14–9 REVIEW**

1. What is the maximum square wave frequency that can be used in Figure 14–43?
2. What is the maximum square wave frequency that can be used in Figure 14–44?
3. What happens if the frequency exceeds the maximum you determined in Questions 1 and 2? Explain how your measurements would be affected.

## 14–10 ■ TRANSIENT ANALYSIS OF *RL* CIRCUITS WITH PSpice AND PROBE

***In the last chapter, you learned how to use PSpice and Probe to analyze the transient response of RC circuits. Analysis of RL circuits is identical except that an inductor rather than a capacitor is the component that stores and releases energy in the circuit. Instead of repeating material from Chapter 13, we will introduce some additional capabilities of Probe.***

***After completing this section, you should be able to***

- **Change the ranges of the *x* and *y* axes in Probe**
  - ☐ More accurately determine the time constant of a circuit

### Setting Ranges in Probe

In Chapter 13, the PULSE and .TRAN specifications were chosen, for purposes of illustration, to display exactly five full time constants and to provide adequate sampling for reasonably accurate results. That approach meant that the time constant was already known, so that Probe was used to find a known value.

In reality, you will not know what the time constant is, so you cannot preset the PULSE and .TRAN parameters to produce nice, neat displays. With Probe, this is not a problem because the scale can be changed to view the part of a trace that is of interest, just as can be done with an oscilloscope. To illustrate, consider the following circuit file for the *RL* circuit in Figure 14–45.

```
RL Transient Circuit
* RL Pulse Demonstration 1
VS  1  0  PULSE(0  5  0  1N  1N  3M  6M)
R1  1  2  1.2K
L1  2  0  560M
.TRAN  50U  6M
.PROBE
.END
```

FIGURE 14–45

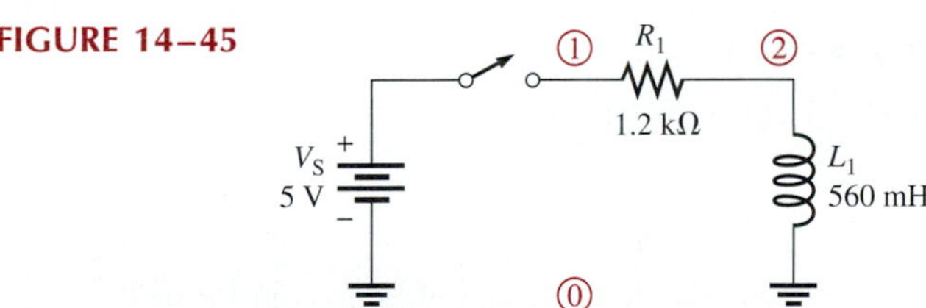


In this circuit the exact time constant is not known, although from the *L* and *R* values you know it is somewhere around 500 μs. Therefore, the transient time (5 time constants) is about 2.5 ms. In order to capture all the data needed, the circuit file specifies a 3 ms pulse and a 6 ms period.

After running PSpice, tell Probe to display the following trace expression:

```
(V(2,0)/5 V)*100
```

This will display the inductor voltage as a percentage of the peak voltage. You will find that the time constant approximation was fairly close because the inductor voltage has time to completely decay from both voltage peaks to zero.

To change the range of the *y*-axis so that you can more accurately observe the positive voltages, the negative portion of the *y*-axis can be eliminated. The sequence of menu and option selections to do this varies slightly depending on whether you are using DOS or Windows, but the effect is the same.

Under DOS, select Y_axis/Set_range and at the "Enter a range:" prompt, type

```
0 100
```

Under Windows, select Plot/Y Axis Settings . . . and mouse-click on User Defined in the Data Range box. Then change −100 to 0 by pressing the 0 key. Finish the command by mouse-clicking on the OK button. Probe will redraw the display so that only the portion of the inductor voltage curve between 0 and 100 is shown.

The trace now shows only the positive portion of the inductor voltage curve which, by inspection, passes through the 37% value at a point between 0 ms and 1 ms on the horizontal axis. So, you can scale the $x$-axis to display only this region of the curve.

Under DOS, select EXIT to get back to the main menu, then select X_axis/Set_range and enter

```
0 1m
```

Under Windows, select Plot/X Axis Settings . . . /User Defined. Then change 6.0 ms to 1.0 ms by pressing TAB on the keyboard or clicking in the 6.0 ms box. Finish the command by clicking the OK button. Probe will redraw the curve with the $x$-axis running from 0 ms to 1 ms. From the main menu select Cursor or Tools/Cursor/Display to determine the time constant as was discussed in Chapter 12. More accurate measurements can usually be achieved by changing the sampling intervals. Previously, we changed the sampling interval by changing the <Print Step> value in the .TRAN statement. For example, the statement

```
.TRAN 50U 6M
```

should take a sample every 50 $\mu$s over an interval of 6 ms while the statement

```
.TRAN 1U 6M
```

should take a sample every 1 $\mu$s over the same interval, thus giving a much more accurate curve.

Unfortunately, PSpice sometimes will ignore the sampling interval specification for saving Probe data if it determines the <Print Step> value is small compared to the <Final Time> value. The solution to this is to completely specify the .TRAN statement by including the values for <No Print Time> and <Max Step> (refer to Chapter 11). For example, to guarantee that Probe will have samples every 1 $\mu$s or the interval from 0 s to 6 ms, you can specify

```
.TRAN 1U 6M 0 1U
```

**SECTION 14–10 REVIEW**

1. How do you change the range of an axis in Probe?
2. Which of the following statements guarantees the sampling interval will be 100 $\mu$s?
   **(a)** .TRAN 100U 1M 100M **(b)** .TRAN 100U 100M 0 100U

## ■ SUMMARY

- Self-inductance is a measure of a coil's ability to establish an induced voltage as a result of a change in its current.
- An inductor opposes a change in its own current.
- Faraday's law states that relative motion between a magnetic field and a coil induces a voltage across the coil.
- The amount of induced voltage is directly proportional to the inductance and to the rate of change in current.
- Lenz's law states that the polarity of induced voltage is such that the resulting induced current is in a direction that opposes the change in the magnetic field that produced it.

- Energy is stored by an inductor in its magnetic field.
- One henry is the amount of inductance when current, changing at the rate of one ampere per second, induces one volt across the inductor.
- Inductance is directly proportional to the square of the number of turns, the permeability, and the cross-sectional area of the core. It is inversely proportional to the length of the core.
- The permeability of a core material is an indication of the ability of the material to establish a magnetic field.
- The time constant for a series *RL* circuit is the inductance divided by the resistance.
- In an *RL* circuit, the voltage and current in an energizing or deenergizing inductor make a 63% change during each time-constant interval.
- Energizing and deenergizing follow exponential curves.
- Inductors add in series.
- Total parallel inductance is less than that of the smallest inductor in parallel.
- Voltage leads current by 90° in an inductor.
- $X_L$ is directly proportional to frequency and inductance.
- The true power in an inductor is zero; that is, there is no energy loss in an ideal inductor, only in its winding resistance.

## GLOSSARY

**Choke** An inductor. The term is more commonly used in connection with inductors used to block or choke off high frequencies.

**Coil** A common term for an inductor.

**Core** The structure around which the winding of an inductor is formed. The core material influences the electromagnetic characteristics of the inductor.

**Henry (H)** The unit of inductance.

**Induced voltage** Voltage produced as a result of a changing magnetic field.

**Inductance** The property of an inductor that produces an opposition to any change in current.

**Inductive reactance** The opposition of an inductor to sinusoidal current. The unit is the ohm.

**Inductor** An electrical device formed by a wire wound around a core having the property of inductance; also known as a coil or a choke.

**Lenz's law** A physical law that states when the current through a coil changes, an induced voltage is created in a direction to oppose the change in current. The current cannot change instantaneously.

**Reactive power ($P_r$)** The rate at which energy is alternately stored and returned to the source by an inductor.

**Time constant** A fixed time interval, set by the *L* and *R* values, that determines the time response of a circuit.

**Winding** The loops or turns of wire in an inductor.

## FORMULAS

| | | |
|---|---|---|
| **(14–1)** | $v_{ind} = L\left(\frac{di}{dt}\right)$ | Induced voltage |
| **(14–2)** | $W = \frac{1}{2}LI^2$ | Energy stored by an inductor |
| **(14–3)** | $L = \frac{N^2 \mu A}{l}$ | Inductance in terms of physical parameters |
| **(14–4)** | $v_{ind} = N\left(\frac{d\phi}{dt}\right)$ | Faraday's law |
| **(14–5)** | $L_T = L_1 + L_2 + L_3 + \cdots + L_n$ | Series inductance |
| **(14–6)** | $\frac{1}{L_T} = \frac{1}{L_1} + \frac{1}{L_2} + \frac{1}{L_3} + \cdots + \frac{1}{L_n}$ | Reciprocal of total parallel inductance |

| | | |
|---|---|---|
| **(14–7)** | $L_T = \dfrac{1}{\left(\dfrac{1}{L_1}\right) + \left(\dfrac{1}{L_2}\right) + \left(\dfrac{1}{L_3}\right) + \cdots + \left(\dfrac{1}{L_n}\right)}$ | Total parallel inductance |
| **(14–8)** | $L_T = \dfrac{L_1 L_2}{L_1 + L_2}$ | Total inductance of two inductors in parallel |
| **(14–9)** | $L_T = \dfrac{L}{n}$ | Total inductance of equal-value inductors in parallel |
| **(14–10)** | $\tau = \dfrac{L}{R}$ | Time constant |
| **(14–11)** | $v = V_F + (V_i - V_F)e^{-Rt/L}$ | Exponential voltage (general) |
| **(14–12)** | $i = I_F + (I_i - I_F)e^{-Rt/L}$ | Exponential current (general) |
| **(14–13)** | $i = I_F(1 - e^{-Rt/L})$ | Increasing exponential current beginning at zero |
| **(14–14)** | $i = I_i e^{-Rt/L}$ | Decreasing exponential current ending at zero |
| **(14–15)** | $X_L = 2\pi f L$ | Inductive reactance |
| **(14–16)** | $P_{true} = (I_{rms})^2 R_W$ | True power |
| **(14–17)** | $P_r = V_{rms} I_{rms}$ | Reactive power |
| **(14–18)** | $P_r = \dfrac{V_{rms}^2}{X_L}$ | Reactive power |
| **(14–19)** | $P_r = I_{rms}^2 X_L$ | Reactive power |
| **(14–20)** | $Q = \dfrac{X_L}{R_W}$ | Quality factor |

## ■ SELF-TEST

**1.** An inductance of 0.05 $\mu$H is larger than
**(a)** 0.0000005 H **(b)** 0.000005 H **(c)** 0.000000008 H **(d)** 0.00005 mH

**2.** An inductance of 0.33 mH is smaller than
**(a)** 33 $\mu$H **(b)** 330 $\mu$H **(c)** 0.05 mH **(d)** 0.0005 H

**3.** When the current through an inductor increases, the amount of energy stored in the electromagnetic field
**(a)** decreases **(b)** remains constant **(c)** increases **(d)** doubles

**4.** When the current through an inductor doubles, the stored energy
**(a)** doubles **(b)** quadruples **(c)** is halved **(d)** does not change

**5.** The winding resistance of a coil can be decreased by
**(a)** reducing the number of turns **(b)** using a larger wire
**(c)** changing the core material **(d)** either answer (a) or (b)

**6.** The inductance of an iron-core coil increases if
**(a)** the number of turns is increased **(b)** the iron core is removed
**(c)** the length of the core is increased **(d)** larger wire is used

**7.** Four 10 mH inductors are in series. The total inductance is
**(a)** 40 mH **(b)** 2.5 mH **(c)** 40,000 $\mu$H **(d)** answers (a) and (c)

**8.** A 1 mH, a 3.3 mH, and a 0.1 mH inductor are connected in parallel. The total inductance is
**(a)** 4.4 mH **(b)** greater than 3.3 mH
**(c)** less than 0.1 mH **(d)** answers (a) and (b)

**9.** An inductor, a resistor, and a switch are connected in series to a 12 V battery. At the instant the switch is closed, the inductor voltage is
**(a)** 0 V **(b)** 12 V **(c)** 6 V **(d)** 4 V

10. A sine wave voltage is applied across an inductor. When the frequency of the voltage is increased, the current
    (a) decreases (b) increases
    (c) does not change (d) momentarily goes to zero
11. An inductor and a resistor are in series with a sine wave voltage source. The frequency is set so that the inductive reactance is equal to the resistance. If the frequency is increased, then
    (a) $V_R > V_L$ (b) $V_L < V_R$ (c) $V_L = V_R$ (d) $V_L > V_R$
12. An ohmmeter is connected across an inductor and the pointer indicates an infinite value. The inductor is
    (a) good (b) open (c) shorted (d) resistive

## ■ PROBLEMS

### SECTION 14–1 The Basic Inductor

1. Convert the following to millihenries:
   (a) 1 H (b) 250 $\mu$H (c) 10 $\mu$H (d) 0.0005 H
2. Convert the following to microhenries:
   (a) 300 mH (b) 0.08 H (c) 5 mH (d) 0.00045 mH
3. What is the voltage across a coil when $di/dt = 10$ mA/$\mu$s and $L = 5$ $\mu$H?
4. Fifty volts are induced across a 25 mH coil. At what rate is the current changing?
5. The current through a 100 mH coil is changing at a rate of 200 mA/s. How much voltage is induced across the coil?
6. How many turns are required to produce 30 mH with a coil wound on a cylindrical core having a cross-sectional area of $10 \times 10^{-5}$ m$^2$ and a length of 0.05 m? The core has a permeability of $1.2 \times 10^{-6}$.
7. A 12 V battery is connected across a coil with a winding resistance of 12 Ω. How much current is there in the coil?
8. How much energy is stored by a 100 mH inductor with a current of 1 A?

### SECTION 14–3 Series Inductors

9. Five inductors are connected in series. The lowest value is 5 $\mu$H. If the value of each inductor is twice that of the preceding one, and if the inductors are connected in order of ascending values, what is the total inductance?
10. Suppose that you require a total inductance of 50 mH. You have available a 10 mH coil and a 22 mH coil. How much additional inductance do you need?
11. Determine the total inductance in Figure 14–46.
12. What is the total inductance between point $A$ and $B$ for each switch position in Figure 14–47?

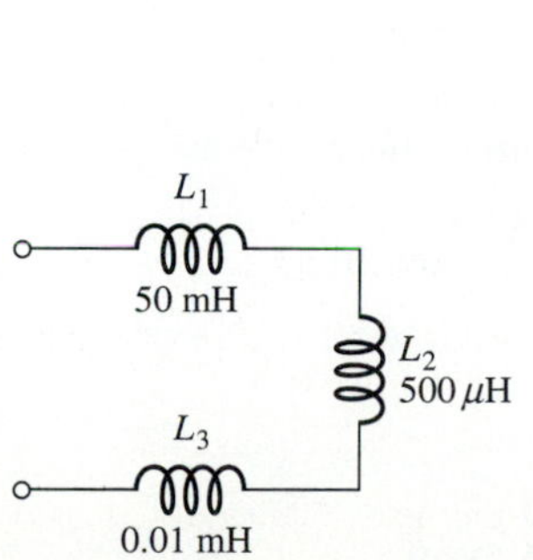


FIGURE 14–46

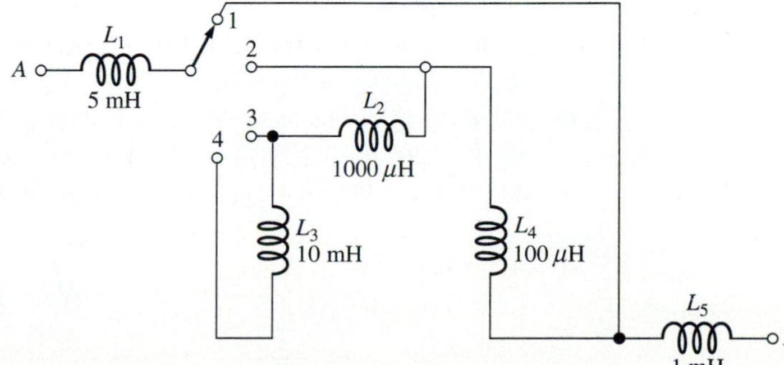


FIGURE 14–47

## SECTION 14–4 Parallel Inductors

**13.** Determine the total parallel inductance for the following coils in parallel: 75 μH, 50 μH, 25 μH, and 15 μH.

**14.** You have a 12 mH inductor, and it is your smallest value. You need an inductance of 8 mH. What value can you use in parallel with the 12 mH to obtain 8 mH?

**15.** Determine the total inductance of each circuit in Figure 14–48.

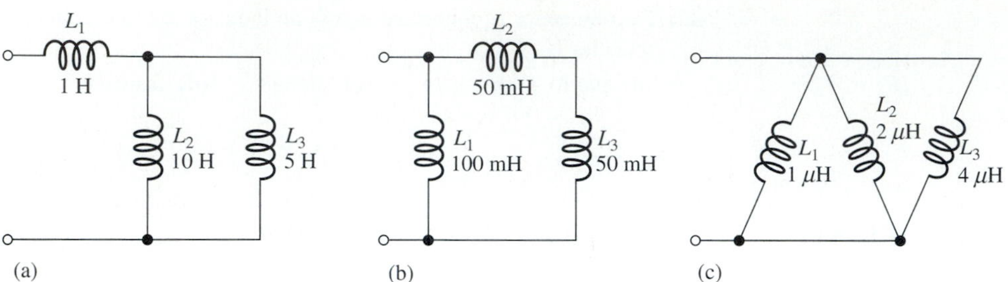


FIGURE 14–48

**16.** Determine the total inductance of each circuit in Figure 14–49.

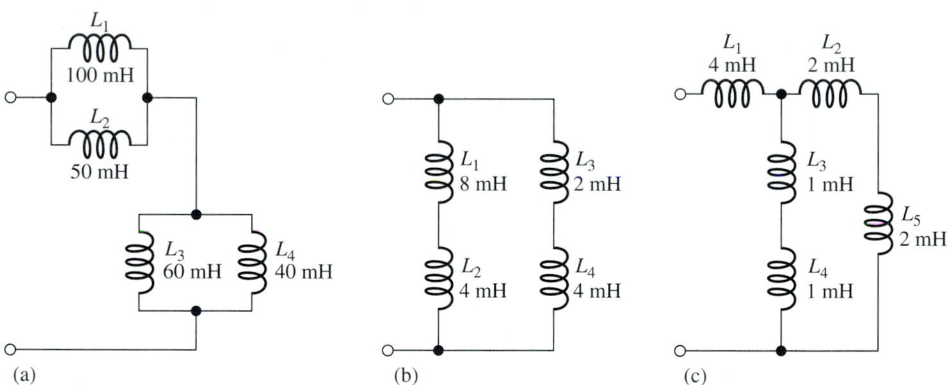


FIGURE 14–49

## SECTION 14–5 Inductors in DC Circuits

**17.** Determine the time constant for each of the following series *RL* combinations:

**(a)** $R = 100\ \Omega$, $L = 100\ \mu\text{H}$ **(b)** $R = 4.7\ \text{k}\Omega$, $L = 10\ \text{mH}$

**(c)** $R = 1.5\ \text{M}\Omega$, $L = 3\ \text{H}$

**18.** In a series *RL* circuit, determine how long it takes the current to build up to its full value for each of the following:

**(a)** $R = 56\ \Omega$, $L = 50\ \mu\text{H}$ **(b)** $R = 3300\ \Omega$, $L = 15\ \text{mH}$

**(c)** $R = 22\ \text{k}\Omega$, $L = 100\ \text{mH}$

**19.** In the circuit of Figure 14–50, there is initially no current. Determine the inductor voltage at the following times after the switch is closed:

**(a)** 10 μs **(b)** 20 μs **(c)** 30 μs **(d)** 40 μs **(e)** 50 μs

FIGURE 14–50

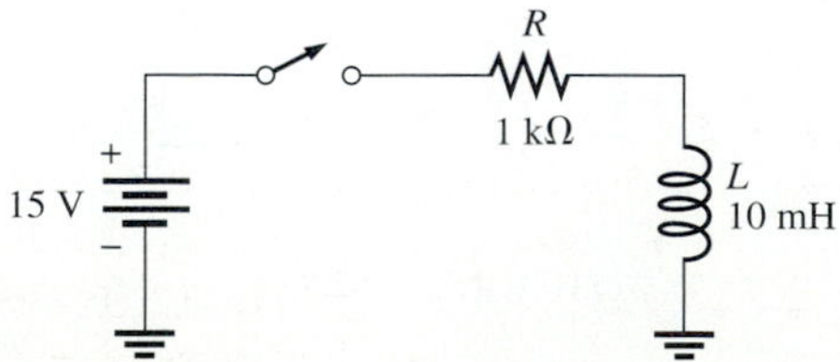

**20.** In Figure 14–51, there are 114 mA through the coil. When SW1 is opened and SW2 simultaneously closed, find the inductor voltage at the following times:

**(a)** Initially **(b)** 1.7 ms **(c)** 5.1 ms **(d)** 6.8 ms

**21.** Repeat Problem 19 for the following times:

**(a)** 2 $\mu$s **(b)** 5 $\mu$s **(c)** 15 $\mu$s

**22.** Repeat Problem 20 for the following times:

**(a)** 0.5 ms **(b)** 1 ms **(c)** 2 ms

**23.** In Figure 14–50, at what time after switch closure does the inductor voltage reach 5 V?

**24.** What is the polarity of the induced voltage in Figure 14–52 when the switch is closed? What is the final value of current if $R_W = 10\ \Omega$?

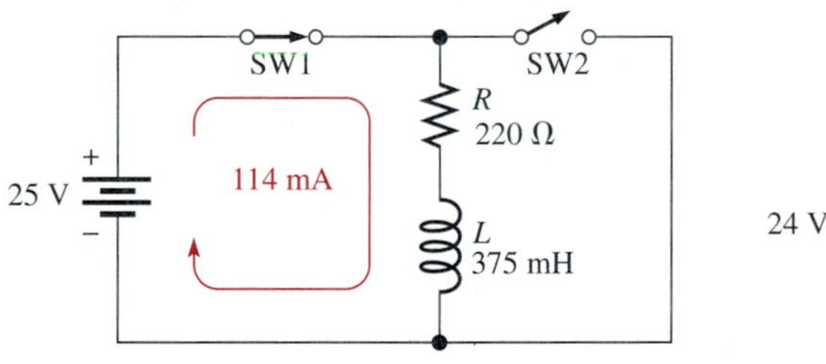


FIGURE 14–51

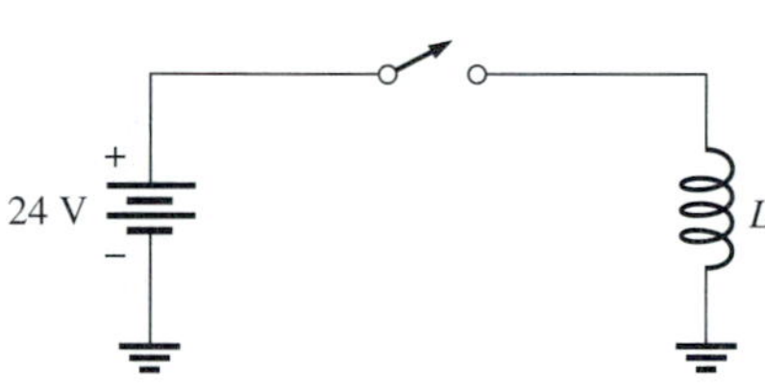


FIGURE 14–52

**25.** Determine the time constant for the circuit in Figure 14–53.

**26.** Find the inductor current at 10 $\mu$s after the switch is thrown from position 1 to position 2 in Figure 14–54. For simplicity, assume that the switch makes contact with position 2 at the same instant it breaks contact with position 1.

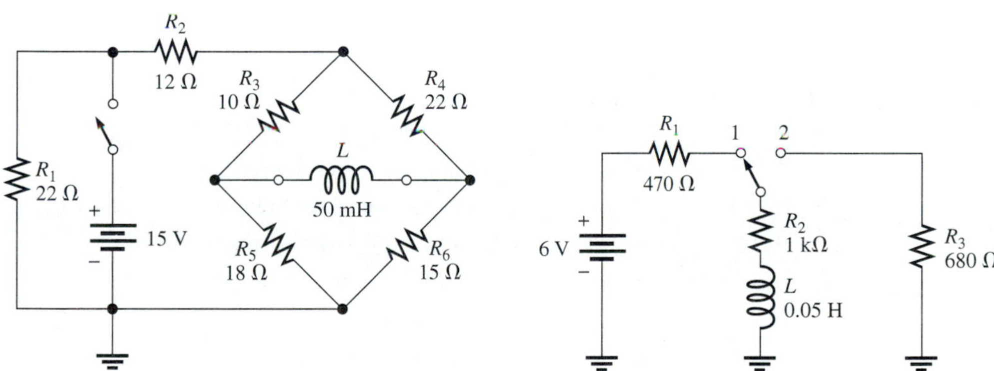


FIGURE 14–53

FIGURE 14–54

**27.** In Figure 14–55, switch 1 is opened and switch 2 is closed at the same instant ($t_0$). What is the instantaneous voltage across $R_2$ at $t_0$?

FIGURE 14–55

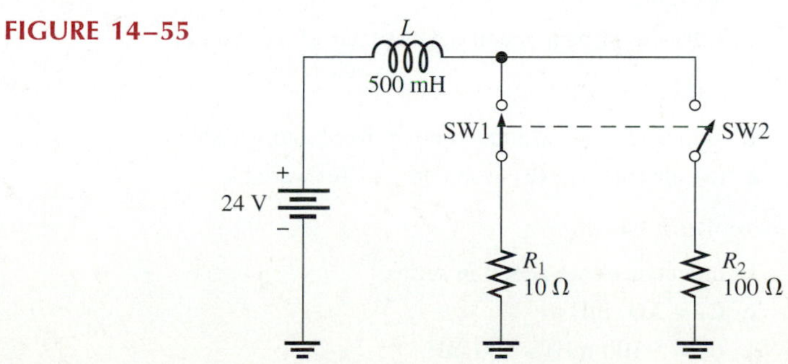

## SECTION 14–6 Inductors in AC Circuits

**28.** Find the total reactance for each circuit in Figure 14–48 when a voltage with a frequency of 5 kHz is applied across the terminals.

**29.** Find the total reactance for each circuit in Figure 14–49 when a 400 Hz voltage is applied.

**30.** Determine the total rms current in Figure 14–56. What are the currents through $L_2$ and $L_3$? Express all currents in polar form.

FIGURE 14–56

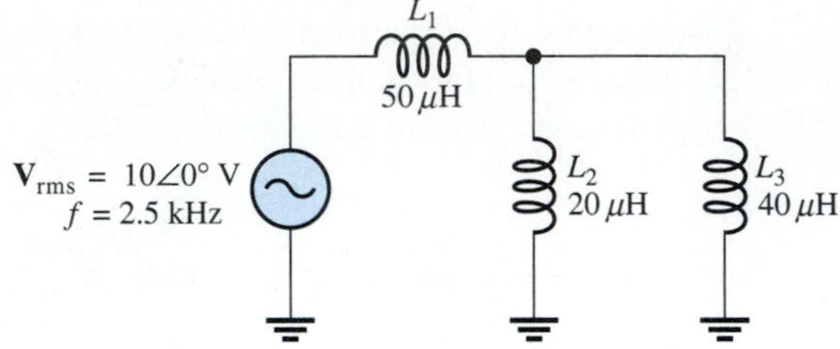


**31.** What frequency will produce 500 mA total rms current in each circuit of Figure 14–49 with an rms input voltage of 10 V?

**32.** Determine the reactive power in Figure 14–56.

**33.** Determine $\mathbf{I}_{L2}$ in Figure 14–57.

FIGURE 14–57

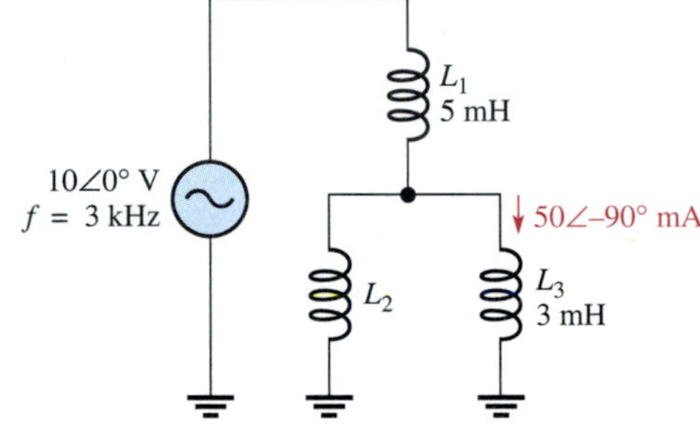


## SECTION 14–8 Testing Inductors

**34.** A certain coil that is supposed to have a 5 Ω winding resistance is measured with an ohmmeter. The meter indicates 2.8 Ω. What is the problem with the coil?

**35.** What is the indication corresponding to each of the following failures in a coil?

**(a)** open **(b)** completely shorted **(c)** some windings shorted

## ▪ ANSWERS TO SECTION REVIEWS

### Section 14–1

**1.** Inductance depends on number of turns of wire, permeability, cross-sectional area; and length.

**2.** $v_{ind} = 7.5$ mV

**3.** **(a)** $L$ increases when $N$ increases. **(b)** $L$ decreases when core length increases.

**(c)** $L$ decreases when core cross-sectional area decreases.

**(d)** $L$ decreases when ferromagnetic core is replaced by air core.

**4.** All wire has some resistance, and since inductors are made from turns of wire, there is always resistance.

**5.** Adjacent turns in a coil act as plates of a capacitor.

### Section 14–2

**1.** Two categories of inductors are fixed and variable.

**2.** **(a)** air core **(b)** iron core **(c)** variable

### Section 14–3

**1.** Inductances are added in series.

**2.** $L_T = 2.60$ mH

**3.** $L_T = 5(100 \text{ mH}) = 500$ mH

## Section 14–4

1. The total parallel inductance is smaller than that of the smallest-value inductor in parallel.
2. True, calculation of parallel inductance is similar to parallel resistance.
3. (a) $L_T = 7.69$ mH (b) $L_T = 24\ \mu$H (c) $L_T = 100$ mH

## Section 14–5

1. $V_L = IR_W = 100$ mV
2. $v_R = 0$ V, $v_L = 20$ V
3. $v_R = 20$ V, $v_L = 0$ V
4. $\tau = 500$ ns, $i_L = 3.93$ mA

## Section 14–6

1. Voltage leads current by 90 degrees in an inductor.
2. $X_L = 2\pi fL = 3.14$ kΩ
3. $f = X_L/2\pi L = 2.55$ MHz
4. $I_{rms} = 15.9\angle{-90°}$ mA
5. $P_{true} = 0$ W; $P_r = 458$ mVAR

## Section 14–7

1. The inductor tends to level out the ripple voltage because of its opposition to changes in current.
2. The inductive reactance is extremely high at radio frequencies, thus blocking signals with these frequencies.

## Section 14–8

1. False, a reading of infinity indicates an open.
2. False, it indicates the winding resistance.

## Section 14–9

1. $f_{max} = 125$ kHz ($5\tau = 4$ ms)
2. $f_{max} = 3.13$ kHz ($5\tau = 160\ \mu$s)
3. If $f > f_{max}$, the inductor will not fully energize because $T/2 < 5\tau$.

## Section 14–10

1. Enter range limits from appropriate menu.
2. Statement (b) guarantees the sampling interval will be 100 μs.

## ■ ANSWERS TO RELATED EXERCISES FOR EXAMPLES

**14–1** 5 H

**14–2** 108 mH

**14–3** 0.5 Wb/s

**14–4** 500 μH

**14–5** 20.3 μH

**14–6** (a) 7.30 μH (b) 20 μH

**14–7** 227 ns

**14–8** $I_F = 29.4$ mA, $\tau = 147$ ns;
at $1\tau$, $i = 18.5$ mA;
at $2\tau$, $i = 25.3$ mA;
at $3\tau$, $i = 27.9$ mA;
at $4\tau$, $i = 28.8$ mA;
at $5\tau$, $i = 29.1$ mA

**14–9** (a) 21.3 μs (b) 106 mA (c) 39.2 mA

**14–10** (a) 9 V, 0 V (b) 92.6 V, 0 V

**14–11** 3.82 mA

**14–12** 33.3 mA

**14–13** 110 Ω

**14–14** 574 mA

**14–15** $P_r$ decreases.

# 15

# TRANSFORMERS



## ■ INTRODUCTION

In Chapter 14, you learned about self-inductance. In this chapter, you will study mutual inductance, which is the basis for the operation of transformers. Transformers are used in all types of applications such as power supplies, electrical power distribution, and signal coupling in communications systems. In this chapter and throughout the rest of the book, you will learn the basics of putting technology theory into practice.

The operation of the transformer is based on the principle of mutual inductance, which occurs when two or more coils are in close proximity. A simple transformer is actually two coils that are electromagnetically coupled by their mutual inductance. Because there is no electrical contact between two magnetically coupled coils, the transfer of energy from one coil to the other can be achieved in a situation of complete electrical isolation. This has many advantages, as you will learn in this chapter.

## TECHnology Theory Into Practice

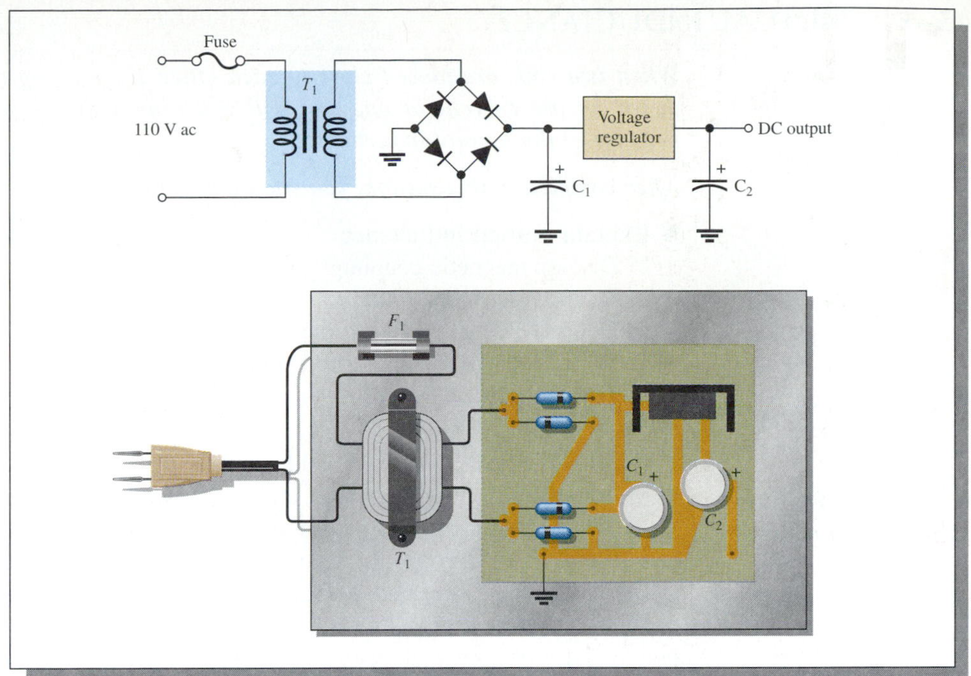


In the TECH TIP assignment in Section 15–12, you will troubleshoot a power supply transformer.

## CHAPTER OBJECTIVES

- ☐ Explain mutual inductance
- ☐ Describe how a transformer is constructed and how it operates
- ☐ Explain how a step-up transformer works
- ☐ Explain how a step-down transformer works
- ☐ Discuss the effect of a resistive load across the secondary winding
- ☐ Discuss the concept of a reflected load in a transformer
- ☐ Discuss impedance matching with transformers
- ☐ Explain how the transformer acts as an isolation device
- ☐ Describe a practical transformer
- ☐ Describe several types of transformers
- ☐ Troubleshoot transformers
- ☐ Use PSpice and Probe for analysis of circuits with transformers (optional)

## 15–1 ▪ MUTUAL INDUCTANCE

***When two coils are placed close to each other, a changing electromagnetic field produced by the current in one coil will cause an induced voltage in the second coil because of the mutual inductance.***

***After completing this section, you should be able to***

- **Explain mutual inductance**
  - ☐ Discuss magnetic coupling
  - ☐ Define *electrical isolation*
  - ☐ Define *coefficient of coupling*
  - ☐ Identify the factors that affect mutual inductance and state the formula

Recall that the electromagnetic field surrounding a coil expands, collapses, and reverses as the current increases, decreases, and reverses.

When a second coil is placed very close to the first coil so that the changing magnetic flux lines cut through the second coil, the coils are magnetically coupled and a voltage is induced, as indicated in Figure 15–1. When two coils are magnetically coupled, they provide **electrical isolation** because there is no electrical connection between them, only a magnetic link. If the current in the first coil is a sine wave, the voltage induced in the second coil is also a sine wave. The amount of voltage induced in the second coil as a result of the current in the first coil is dependent on the **mutual inductance, $L_M$,** between the two coils. The mutual inductance is established by the inductance of each coil ($L_1$ and $L_2$) and by the amount of coupling ($k$) between the two coils.

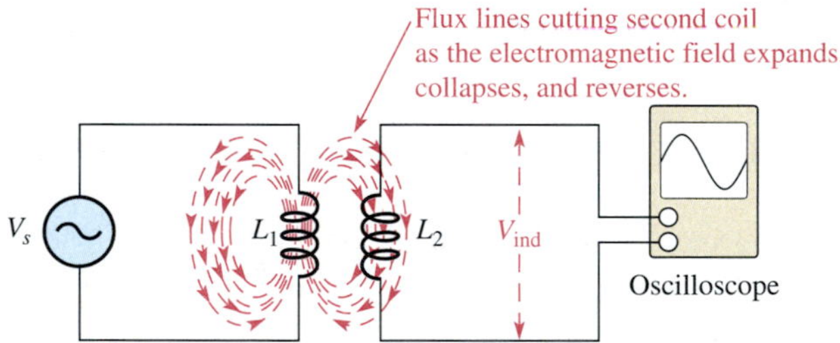


**FIGURE 15–1**
***A voltage is induced in the second coil as a result of the changing current in the first coil producing a changing electromagnetic field that links the second coil.***

### Coefficient of Coupling

The **coefficient of coupling, *k*,** between two coils is the ratio of the lines of force (flux) produced by coil 1 linking coil 2 ($\phi_{1\text{-}2}$) to the total flux produced by coil 1 ($\phi_1$).

$$k = \frac{\phi_{1\text{-}2}}{\phi_1} \qquad \textbf{(15–1)}$$

For example, if half of the total flux produced by coil 1 links coil 2, then $k = 0.5$. A greater value of $k$ means that more voltage is induced in coil 2 for a certain rate of change of current in coil 1. Note that $k$ has no units. Recall that the unit of magnetic lines of force (flux) is the weber, abbreviated Wb.

The coefficient $k$ depends on the physical closeness of the coils and the type of core material on which they are wound. Also, the construction and shape of the cores are factors.

### Formula for Mutual Inductance

The three factors influencing $L_M$ ($k$, $L_1$, and $L_2$) are shown in Figure 15–2. The formula for $L_M$ is

$$L_M = k\sqrt{L_1L_2} \quad (15\text{–}2)$$

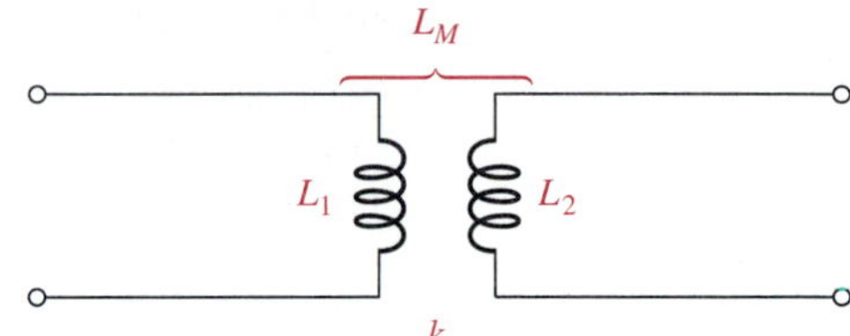


FIGURE 15–2
*The mutual inductance of two coils.*

**EXAMPLE 15–1** One coil produces a total magnetic flux of 50 $\mu$Wb, and 20 $\mu$Wb link coil 2. What is the coefficient of coupling, $k$?

***Solution***

$$k = \frac{\phi_{1\text{-}2}}{\phi_1} = \frac{20\ \mu\text{Wb}}{50\ \mu\text{Wb}} = 0.4$$

***Related Exercise*** Determine $k$ when $\phi_1 = 500\ \mu$Wb and $\phi_{1\text{-}2} = 375\ \mu$Wb.

**EXAMPLE 15–2** Two coils are wound on a single core, and the coefficient of coupling is 0.3. The inductance of coil 1 is 10 $\mu$H, and the inductance of coil 2 is 15 $\mu$H. What is $L_M$?

***Solution*** $L_M = k\sqrt{L_1L_2} = 0.3\sqrt{(10\ \mu\text{H})(15\ \mu\text{H})} = 3.67\ \mu\text{H}$

***Related Exercise*** Determine the mutual inductance when $k = 0.5$, $L_1 = 1$ mH, and $L_2 = 600\ \mu$H.

**SECTION 15–1 REVIEW**

1. Define *mutual inductance.*
2. Two 50 mH coils have $k = 0.9$. What is $L_M$?
3. If $k$ is increased, what happens to the voltage induced in one coil as a result of a current change in the other coil?

## 15–2 ■ THE BASIC TRANSFORMER

***A basic transformer is an electrical device constructed of two coils placed in close proximity to each other so that there is a mutual inductance.***

***After completing this section, you should be able to***

- **Describe how a transformer is constructed and how it operates**
  - ☐ Identify the parts of a basic transformer
  - ☐ Discuss the importance of the core material
  - ☐ Define *primary winding* and *secondary winding*
  - ☐ Define *turns ratio*
  - ☐ Discuss how the direction of windings affects voltage polarities

A schematic of a **transformer** is shown in Figure 15–3(a). As shown, one coil is called the **primary winding,** and the other is called the **secondary winding.** The source voltage is applied to the primary winding, and the load is connected to the secondary winding, as shown in Figure 15–3(b). So, the primary winding is the input winding, and the secondary winding is the output winding.

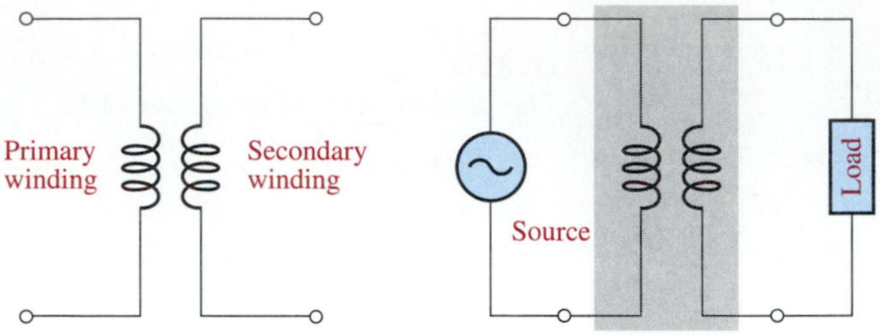

**FIGURE 15–3**
***The basic transformer.***

(a) Schematic symbol (b) Source/load connections

The windings of a transformer are formed around the **core.** The core provides both a physical structure for placement of the windings and a magnetic path so that the magnetic flux lines are concentrated close to the coils. There are three general categories of core material: air, ferrite, and iron. The schematic symbol for each type is shown in Figure 15–4.

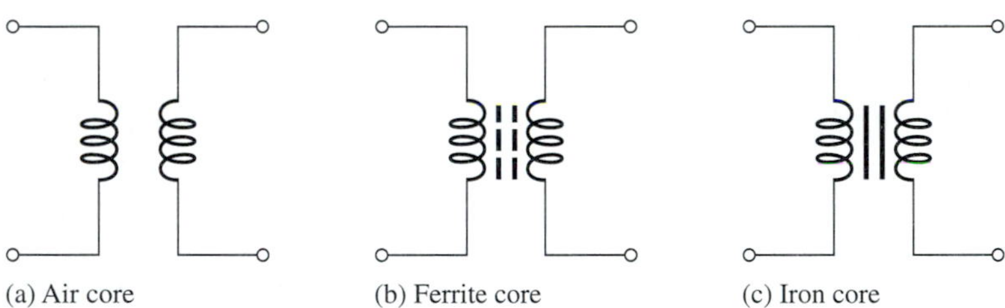

(a) Air core (b) Ferrite core (c) Iron core

**FIGURE 15–4**
***Schematic symbols based on type of core.***

Air-core and ferrite-core transformers generally are used for high-frequency applications and consist of windings on an insulating shell which is hollow (air) or constructed of ferrite, such as depicted in Figure 15–5. The wire is typically covered by a varnish-type coating to prevent the windings from shorting together. The amount of **magnetic**

**FIGURE 15–5**
***Transformers with cylindrical-shaped cores.***

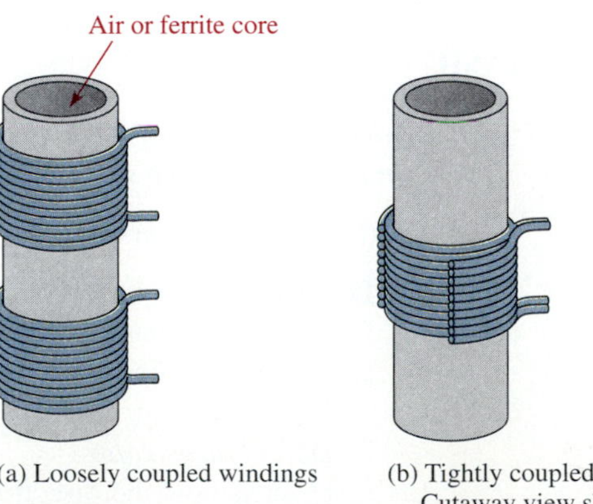


(a) Loosely coupled windings (b) Tightly coupled windings. Cutaway view shows both windings.

**coupling** between the primary winding and the secondary winding is set by the type of core material and by the relative positions of the windings. In Figure 15–5(a), the windings are loosely coupled because they are separated, and in part (b) they are tightly coupled because they are overlapping. The tighter the coupling, the greater the induced voltage in the secondary for a given current in the primary.

Iron-core transformers generally are used for audio frequency (AF) and power applications. These transformers consist of windings on a core constructed from laminated sheets of ferromagnetic material insulated from each other, as shown in Figure 15–6. This construction provides an easy path for the magnetic flux and increases the amount of coupling between the windings. Figure 15–6(a) and 15–6(b) show the basic construction of two major configurations of iron-core transformers. In the core-type construction, shown in part (a), the windings are on separate legs of the laminated core. In the shell-type construction, shown in part (b), both windings are on the same leg. A variety of transformers are shown in Figure 15–7.

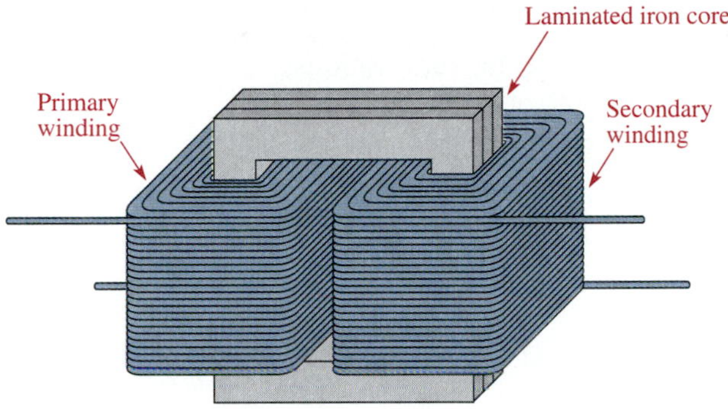


(a) Core-type has each winding on a separate leg.

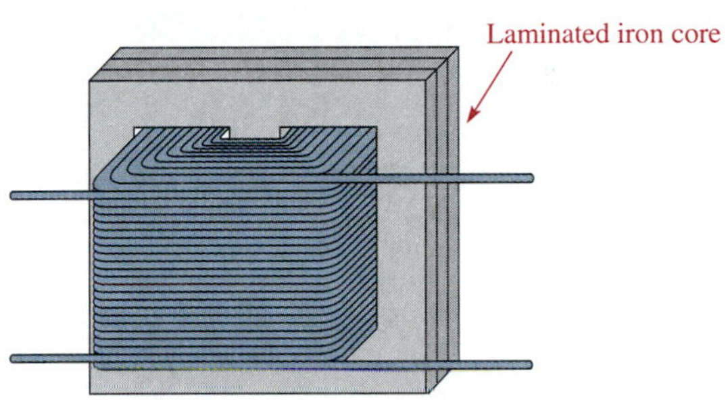


(b) Shell-type has both windings on the same leg.

**FIGURE 15–6**
***Iron-core transformer construction with multilayer windings.***

**FIGURE 15–7**
***Some common types of transformers (courtesy of Litton Triad-Utrad).***

## Turns Ratio

An important parameter of a transformer is its **turns ratio.** The turns ratio ($n$) is the ratio of the number of turns in the secondary winding ($N_{sec}$) to the number of turns in the primary winding ($N_{pri}$).

$$n = \frac{N_{sec}}{N_{pri}} \tag{15–3}$$

In the following sections, you will see how the turns ratio affects the voltages and currents in a transformer. Often, the primary and secondary voltages are specified on a transformer, while the turns ratio is not.

**EXAMPLE 15–3** A transformer primary winding has 100 turns, and the secondary winding has 400 turns. What is the turns ratio?

***Solution*** $N_{sec} = 400$ and $N_{pri} = 100$; therefore, the turns ratio is

$$n = \frac{N_{sec}}{N_{pri}} = \frac{400}{100} = 4$$

***Related Exercise*** A certain transformer has a turns ratio of 10. If $N_{pri} = 50$, what is $N_{sec}$?

## Direction of Windings

Another important transformer parameter is the direction in which the windings are placed around the core. As illustrated in Figure 15–8, the direction of the windings determines the polarity of the voltage across the secondary winding (secondary voltage) with respect to the voltage across the primary winding (primary voltage). Phase dots are used on the schematic symbols to indicate polarities, as shown in Figure 15–9.

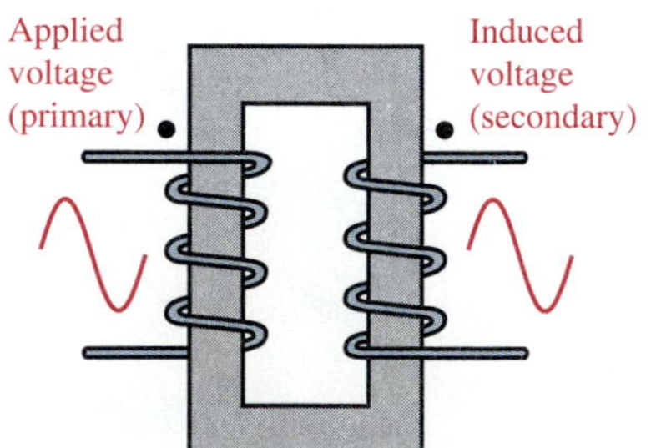


(a) The primary and secondary voltages are in phase when the windings are in the same effective direction around the magnetic path.

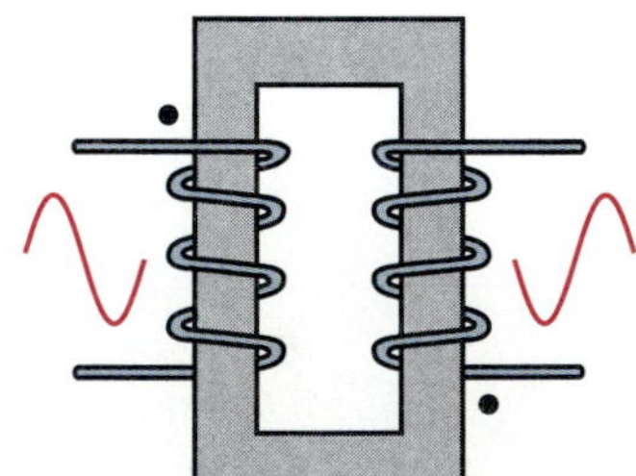

(b) The primary and secondary voltages are 180° out of phase when the windings are in the opposite direction.

FIGURE 15–8

***The direction of the windings determines the relative polarities of the voltages.***

**FIGURE 15–9**
***Phase dots indicate relative polarities of primary and secondary voltages.***

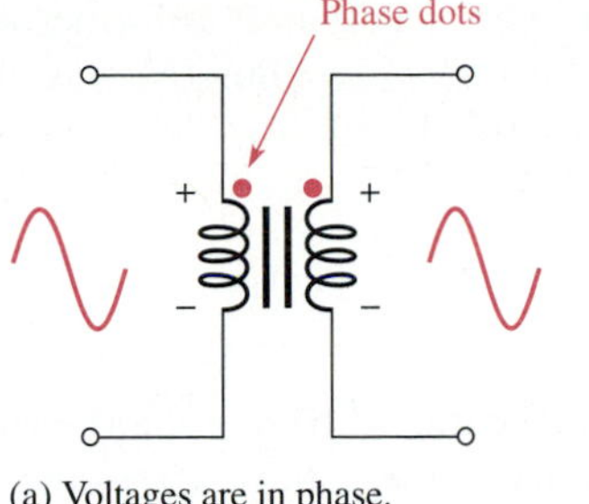


(a) Voltages are in phase.

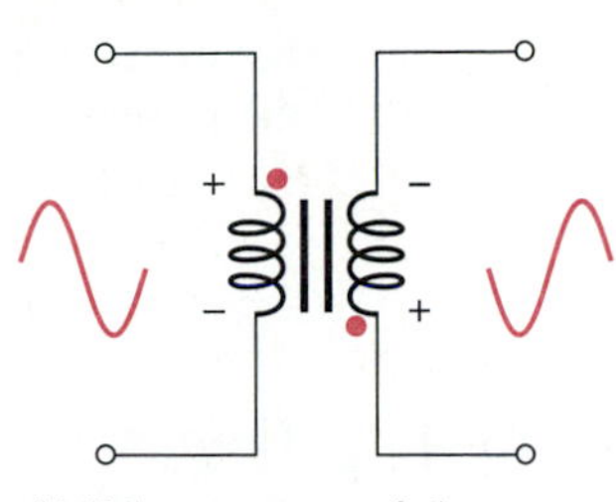


(b) Voltages are out of phase.

**SECTION 15–2 REVIEW**

1. Upon what principle is the operation of a transformer based?
2. Define *turns ratio.*
3. Why are the directions of the windings of a transformer important?
4. A certain transformer has a primary winding with 500 turns and a secondary winding with 250 turns. What is the turns ratio?

## 15–3 ■ STEP-UP TRANSFORMERS

***A step-up transformer has more turns in its secondary winding than in its primary winding and is used to increase ac voltage.***

***After completing this section, you should be able to***

- **Explain how a step-up transformer works**
  - ☐ State the relationship between primary and secondary voltages and the number of turns in the primary and secondary windings
  - ☐ Identify a step-up transformer by its turns ratio

A transformer in which the secondary voltage is greater than the primary voltage is called a **step-up transformer.** The amount that the voltage is stepped up depends on the turns ratio.

> **The ratio of secondary voltage ($V_{sec}$) to primary voltage ($V_{pri}$) is equal to the ratio of the number of turns in the secondary winding ($N_{sec}$) to the number of turns in the primary winding ($N_{pri}$).**

$$\frac{V_{sec}}{V_{pri}} = \frac{N_{sec}}{N_{pri}} \qquad \textbf{(15–4)}$$

From this relationship,

$$V_{sec} = \left(\frac{N_{sec}}{N_{pri}}\right)V_{pri} \qquad \textbf{(15–5)}$$

Equation (15–5) shows that the secondary voltage is equal to the turns ratio times the primary voltage. This condition assumes that the coefficient of coupling is 1, and a good iron core transformer approaches this value.

The turns ratio for a step-up transformer is always greater than 1 because the number of turns in the secondary winding ($N_{sec}$) is always greater than the number of turns in the primary winding ($N_{pri}$).

**EXAMPLE 15–4** The transformer in Figure 15–10 has a 200 turn primary winding and a 600 turn secondary winding. What is the voltage across the secondary winding?

**FIGURE 15–10**

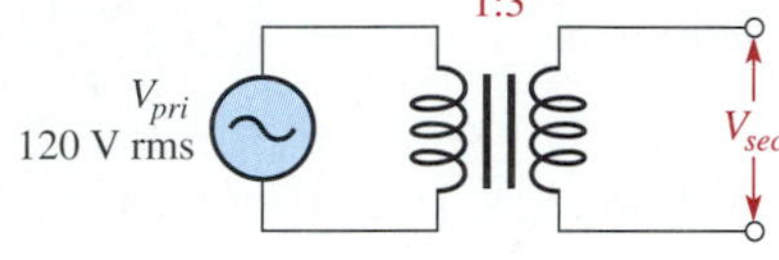


***Solution*** The secondary voltage is

$$V_{sec} = \left(\frac{N_{sec}}{N_{pri}}\right)V_{pri} = \frac{600}{200}(120\text{ V}) = 3(120\text{ V}) = 360\text{ V}$$

Note that the turns ratio of 3 is indicated on the schematic as 1:3, meaning that there are three secondary turns for each primary turn.

***Related Exercise*** The transformer in Figure 15–10 is changed to one with $N_{pri} = 75$ and $N_{sec} = 300$. Determine $V_{sec}$.

**SECTION 15–3 REVIEW**

1. What does a step-up transformer do?
2. If the turns ratio is 5, how much greater is the secondary voltage than the primary voltage?
3. When 240 V ac are applied to the primary winding of a transformer with a turns ratio of 10, what is the secondary voltage?

## 15–4 ■ STEP-DOWN TRANSFORMERS

***A step-down transformer has more turns in its primary winding than in its secondary winding and is used to decrease ac voltage.***

***After completing this section, you should be able to***

- **Explain how a step-down transformer works**
  - ☐ Identify a step-down transformer by its turns ratio

A transformer in which the secondary voltage is less than the primary voltage is called a **step-down transformer.** The amount by which the voltage is stepped down depends on the turns ratio. Equation (15–5) applies also to a step-down transformer.

The turns ratio of a step-down transformer is always less than 1 because the number of turns in the secondary winding is always fewer than the number of turns in the primary winding.

**EXAMPLE 15–5** The transformer in Figure 15–11 has 50 turns in the primary winding and 10 turns in the secondary winding. What is the secondary voltage?

FIGURE 15–11

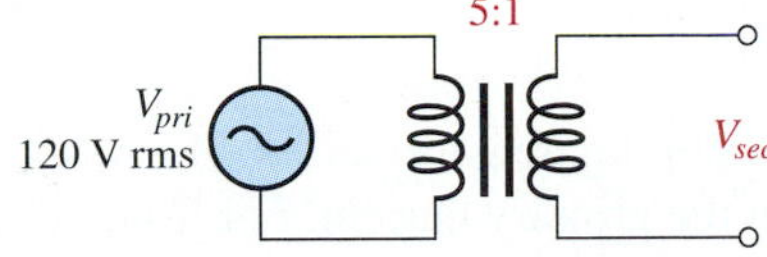


***Solution*** The secondary voltage is

$$V_{sec} = \frac{N_{sec}}{N_{pri}}(V_{pri}) = \frac{10}{50}(120\text{ V}) = 0.2(120\text{ V}) = 24\text{ V}$$

***Related Exercise*** The transformer in Figure 15–11 is changed to one with $N_{pri} = 250$ and $N_{sec} = 120$. Determine the secondary voltage.

**SECTION 15–4 REVIEW**

1. What does a step-down transformer do?
2. A voltage of 120 V ac is applied to the primary winding of a transformer with a turns ratio of 0.5. What is the secondary voltage?
3. A primary voltage of 120 V ac is reduced to 12 V ac. What is the turns ratio?

## 15–5 ■ LOADING THE SECONDARY WINDING

***When a resistive load is connected to the secondary winding of a transformer, the load current depends on both the primary current and the turns ratio.***

***After completing this section, you should be able to***

- **Discuss the effect of a resistive load across the secondary winding**
  - ☐ Determine the secondary current when a step-up transformer is loaded
  - ☐ Determine the secondary current when a step-down transformer is loaded
  - ☐ Discuss power in a transformer

When a load resistor is connected to the secondary winding, as shown in Figure 15–12, there is current through the resulting secondary circuit because of the voltage induced in the secondary coil. It can be shown that the ratio of the primary current, $I_{pri}$, to the secondary current, $I_{sec}$, is equal to the turns ratio, as expressed in the following equation:

$$\frac{I_{pri}}{I_{sec}} = \frac{N_{sec}}{N_{pri}} \quad \textbf{(15–6)}$$

FIGURE 15–12

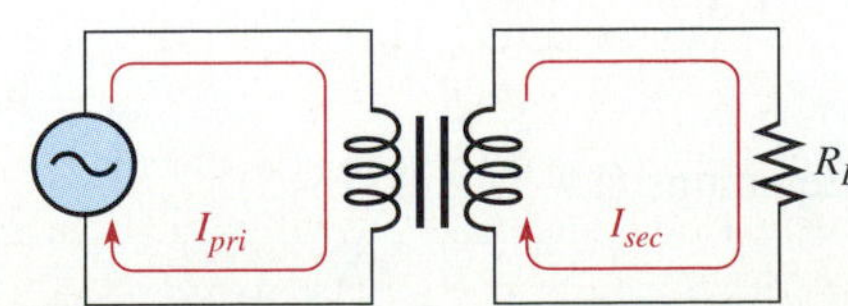

A manipulation of Equation (15–6) gives Equation (15–7), which shows that $I_{sec}$ is equal to $I_{pri}$ times the reciprocal of the turns ratio.

$$I_{sec} = \left(\frac{N_{pri}}{N_{sec}}\right)I_{pri} \tag{15–7}$$

Thus, for a step-up transformer, in which $N_{sec}/N_{pri}$ is greater than 1, the secondary current is less than the primary current. For a step-down transformer, $N_{sec}/N_{pri}$ is less than 1, and $I_{sec}$ is greater than $I_{pri}$.

**EXAMPLE 15–6** The transformers in Figure 15–13(a) and (b) have loaded secondary windings. If the primary current is 100 mA in each case, what is the load current?

FIGURE 15–13

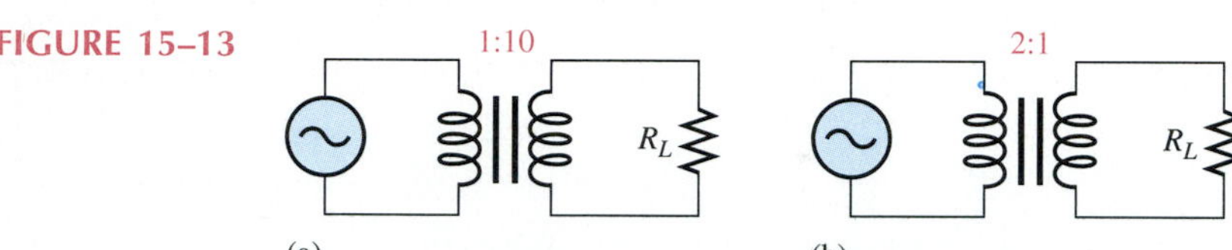


***Solution*** In Figure 15–13(a), the current through the load is

$$I_{sec} = \left(\frac{N_{pri}}{N_{sec}}\right)I_{pri} = 0.1(100 \text{ mA}) = 10 \text{ mA}$$

In Figure 15–13(b), the current through the load is

$$I_{sec} = \left(\frac{N_{pri}}{N_{sec}}\right)I_{pri} = 2(100 \text{ mA}) = 200 \text{ mA}$$

***Related Exercise*** What is the secondary current in Figure 15–13(a) if the turns ratio is doubled? What is the secondary current in Figure 15–13(b) if the turns ratio is halved? Assume $I_{pri}$ remains the same in both circuits.

## Primary Power Equals Secondary Power

When a load is connected to the secondary winding of a transformer, the power transferred to the load can never be greater than the power in the primary winding. For an ideal transformer, the power in the secondary winding ($P_{sec}$) equals the power in the primary winding ($P_{pri}$). When losses are considered, the secondary power is always less.

Power is dependent on voltage and current, and there can be no increase in power in a transformer. Therefore, if the voltage is stepped up, the current is stepped down and vice versa. In an ideal transformer, the secondary power is equal to the primary power regardless of the turns ratio, as the following equations show. The primary power is

$$P_{pri} = V_{pri}I_{pri}$$

and the secondary power is

$$P_{sec} = V_{sec}I_{sec}$$

From Equations (15–7) and (15–5),

$$I_{sec} = \left(\frac{N_{pri}}{N_{sec}}\right)I_{pri} \quad \text{and} \quad V_{sec} = \left(\frac{N_{sec}}{N_{pri}}\right)V_{pri}$$

By substitution,

$$P_{sec} = \left(\frac{N_{pri}}{N_{sec}}\right)\left(\frac{N_{sec}}{N_{pri}}\right)V_{pri}I_{pri}$$

Canceling terms yields

$$P_{sec} = V_{pri}I_{pri} = P_{pri}$$

This result is closely approached in practice because of the very high efficiencies of transformers.

**SECTION 15–5 REVIEW**

1. If the turns ratio of a transformer is 2, is the secondary current greater than or less than the primary current? By how much?
2. A transformer has 100 turns in its primary winding and 25 turns in its secondary winding, and $I_{pri}$ is 0.5 A. What is the value of $I_{sec}$?
3. In Problem 2, how much primary current is necessary to produce a secondary load current of 10 A?

## 15–6 ■ REFLECTED LOAD

***From the viewpoint of the primary circuit, a load connected across the secondary winding of a transformer appears to have a resistance that is not necessarily equal to the actual resistance of the load. The actual load is essentially "reflected" into the primary circuit altered by the turns ratio. This reflected load is what the primary source effectively sees, and it determines the amount of primary current.***

***After completing this section, you should be able to***

- **Discuss the concept of a reflected load in a transformer**
  - ☐ Define *reflected resistance*
  - ☐ Explain how the turns ratio affects the reflected resistance
  - ☐ Calculate reflected resistance

The concept of the **reflected load** is illustrated in Figure 15–14. The load ($R_L$) in the secondary circuit of a transformer is reflected into the primary circuit by transformer action. The load appears to the source in the primary circuit to be a resistance ($R_{pri}$) with a value determined by the turns ratio and the actual value of the load resistance. The resistance $R_{pri}$ is called the **reflected resistance.**

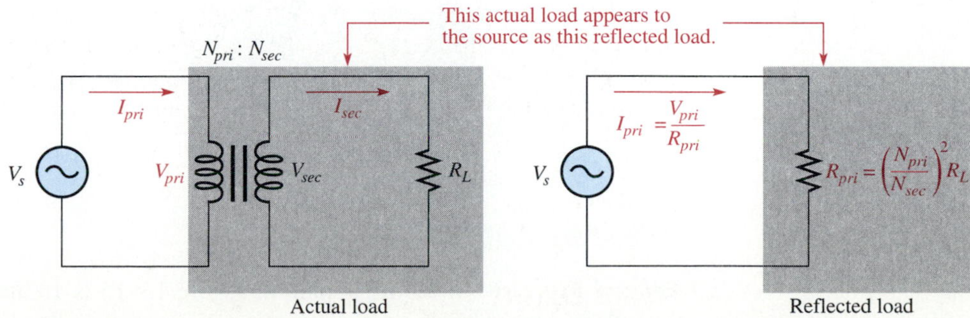


FIGURE 15–14
*Reflected load in a transformer circuit.*

The resistance in the primary circuit of Figure 15–14 is $R_{pri} = V_{pri}/I_{pri}$. The resistance in the secondary circuit is $R_L = V_{sec}/I_{sec}$. From Equations (15–4) and (15–6), you know that $V_{sec}/V_{pri} = N_{sec}/N_{pri}$ and $I_{pri}/I_{sec} = N_{sec}/N_{pri}$. Using these relationships, a formula for $R_{pri}$ in terms of $R_L$ is determined as follows

$$\frac{R_{pri}}{R_L} = \frac{V_{pri}/I_{pri}}{V_{sec}/I_{sec}} = \left(\frac{V_{pri}}{V_{sec}}\right)\left(\frac{I_{sec}}{I_{pri}}\right) = \left(\frac{N_{pri}}{N_{sec}}\right)\left(\frac{N_{pri}}{N_{sec}}\right) = \frac{N_{pri}^2}{N_{sec}^2} = \left(\frac{N_{pri}}{N_{sec}}\right)^2$$

Solving for $R_{pri}$ yields

$$R_{pri} = \left(\frac{N_{pri}}{N_{sec}}\right)^2 R_L \qquad \textbf{(15–8)}$$

Equation (15–8) shows that the resistance reflected into the primary circuit is the square of the reciprocal of the turns ratio times the load resistance.

**EXAMPLE 15–7** Figure 15–15 shows a source that is transformer-coupled to a load resistor of 100 Ω. The transformer has a turns ratio of 4. What is the reflected resistance seen by the source?

**FIGURE 15–15**

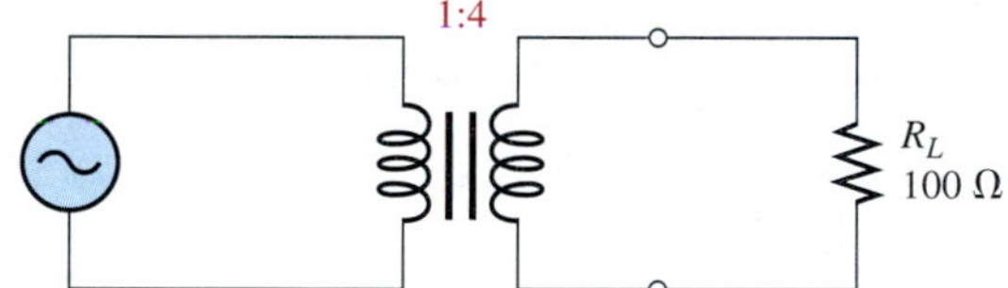


***Solution*** The reflected resistance is determined by Equation (15–8).

$$R_{pri} = \left(\frac{N_{pri}}{N_{sec}}\right)^2 R_L = \left(\frac{1}{4}\right)^2 R_L = \left(\frac{1}{16}\right)(100\ \Omega) = 6.25\ \Omega$$

The source sees a resistance of 6.25 Ω just as if it were connected directly, as shown in the equivalent circuit of Figure 15–16.

**FIGURE 15–16**

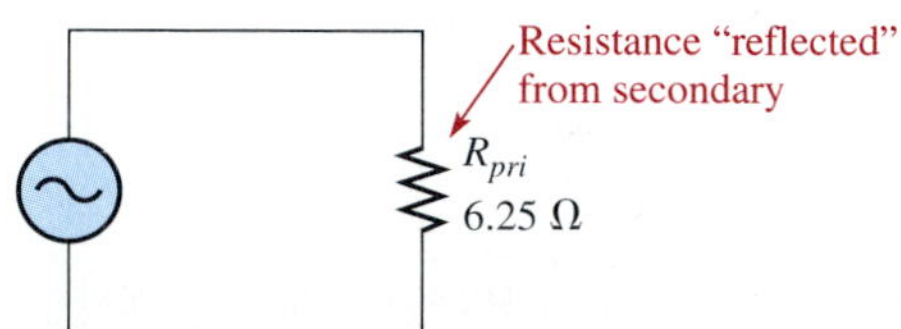


The calculator sequence is

4 2nd 1/x x² × 1 0 0 =

***Related Exercise*** If the turns ratio in Figure 15–15 is 10 and $R_L$ is 600 Ω, what is the reflected resistance?

**EXAMPLE 15–8** If a transformer is used in Figure 15–15 having 40 turns in the primary winding and 10 turns in the secondary winding, what is the reflected resistance?

***Solution*** The reflected resistance is

$$R_{pri} = \left(\frac{N_{pri}}{N_{sec}}\right)^2 R_L = \left(\frac{40}{10}\right)^2 100\ \Omega = (4)^2(100\ \Omega) = 1600\ \Omega$$

This result illustrates the difference that the turns ratio makes.

***Related Exercise*** To achieve a reflected resistance of 800 Ω, what turns ratio is required in Figure 15–15?

**SECTION 15–6 REVIEW**

1. Define *reflected resistance.*
2. What transformer characteristic determines the reflected resistance?
3. A given transformer has a turns ratio of 10, and the load is 50 Ω. How much resistance is reflected into the primary circuit?
4. What is the turns ratio required to reflect a 4 Ω load resistance into the primary circuit as 400 Ω?

## 15–7 ■ MATCHING THE LOAD AND SOURCE RESISTANCES

***One application of transformers is in the matching of a load resistance to a source resistance in order to achieve maximum transfer of power. This technique is called impedance matching. Recall that the maximum power transfer theorem was studied in Chapter 8. In audio systems, transformers are often used to get the maximum amount of power from the amplifier to the speaker by proper selection of the turns ratio.***

***After completing this section, you should be able to***

- **Discuss impedance matching with transformers**
  - ☐ Give a general definition of impedance
  - ☐ Define *impedance matching*
  - ☐ Explain the purpose of impedance matching
  - ☐ Describe a practical application

*Impedance* will be covered thoroughly in Chapter 16. For now, all you need to know is that impedance is a general term for the opposition to current, including the effects of both resistance and reactance combined. However, in this chapter, we will confine our usage to resistance only.

The concept of power transfer in illustrated in the basic circuit of Figure 15–17. Part (a) shows an ac voltage source with a series resistor representing its internal resistance. Some internal resistance is inherent in all sources due to their internal circuitry or physical makeup. When the source is connected directly to a load, as shown in part (b), generally the objective is to transfer as much of the power produced by the source to the load as possible. However, a certain amount of the power produced by the source is dissipated in its internal resistance, and the remaining power goes to the load.

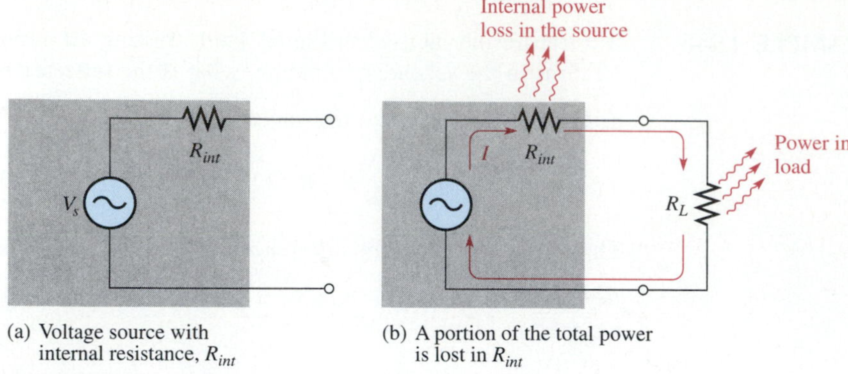


(a) Voltage source with internal resistance, $R_{int}$

(b) A portion of the total power is lost in $R_{int}$

FIGURE 15–17

***Power transfer from a nonideal voltage source to a load.***

In most practical situations, the internal source resistance of various types of sources is fixed. Also, in many cases, the resistance of a device that acts as a load is fixed and cannot be altered. If you need to connect a given source to a given load, remember that only by chance will their resistances match. In this situation a transformer comes in handy. You can use the reflected-resistance characteristic provided by a transformer to make the load resistance appear to have the same value as the source resistance, thereby "fooling" the source into "thinking" that there is a match. This technique is called **impedance matching.**

Let's take a practical, everyday situation to illustrate the concept of impedance matching. The typical resistance of the input to a TV receiver is 300 Ω. An antenna must be connected to this input by a lead-in cable in order to receive TV signals. In this situation, the antenna and the lead-in act as the source, and the input resistance of the TV receiver is the load, as illustrated in Figure 15–18.

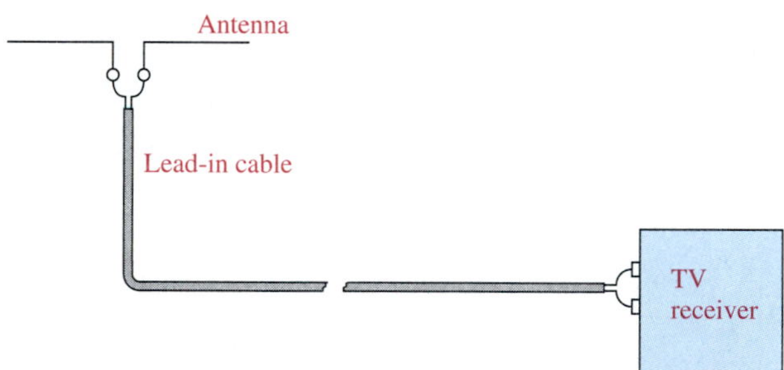


(a) The antenna/lead-in is the source; the TV input is the load.

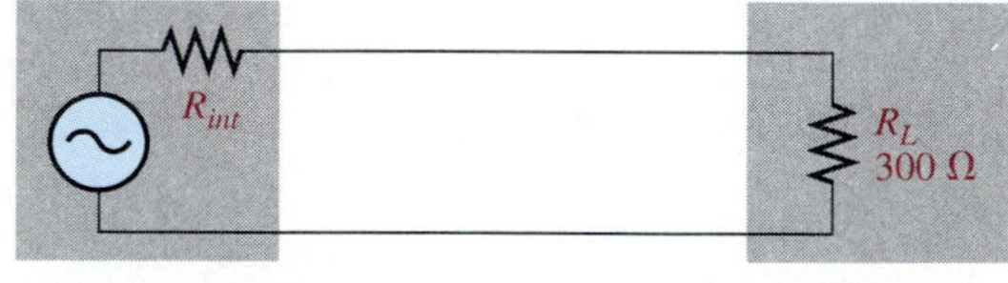


Source–antenna and lead-in

Load–TV receiver

(b) Circuit equivalent of antenna and TV receiver system

FIGURE 15–18

***An antenna directly coupled to a TV receiver.***

It is common for an antenna system to have a characteristic resistance of 75 Ω. Thus, if the 75 Ω source (antenna and lead-in) is connected directly to the 300 Ω TV input, maximum power will not be delivered to the input to the TV, and you will have poor signal reception. The solution is to use a matching transformer, connected as indicated in Figure 15–19, in order to match the 300 Ω load resistance to the 75 Ω source resistance.

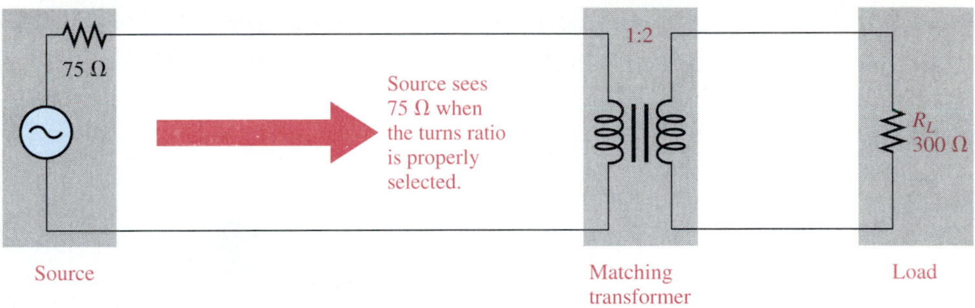


**FIGURE 15–19**
***Example of a load matched to a source by transformer coupling for maximum power transfer.***

To match the resistances, that is, to reflect the load resistance ($R_L$) into the primary circuit so that it appears to have a value equal to the internal source resistance ($R_{int}$), you must select a proper value of turns ratio ($n$). You want the 300 Ω load to look like 75 Ω to the source. You can use Equation (15–8) to determine a formula for the turns ratio, $N_{sec}/N_{pri}$, when you know the values for $R_L$ and $R_{pri}$, the reflected resistance, as follows:

$$R_{pri} = \left(\frac{N_{pri}}{N_{sec}}\right)^2 R_L$$

Transpose terms and divide both sides by $R_L$.

$$\left(\frac{N_{pri}}{N_{sec}}\right)^2 = \frac{R_{pri}}{R_L}$$

Take the square root of both sides.

$$\frac{N_{pri}}{N_{sec}} = \sqrt{\frac{R_{pri}}{R_L}}$$

Invert both sides to get the following formula for the turns ratio:

$$n = \frac{N_{sec}}{N_{pri}} = \sqrt{\frac{R_L}{R_{pri}}} \qquad \textbf{(15–9)}$$

Finally, solve for this particular turns ratio.

$$n = \frac{N_{sec}}{N_{pri}} = \sqrt{\frac{R_L}{R_{pri}}} = \sqrt{\frac{300\ \Omega}{75\ \Omega}} = \sqrt{4} = 2$$

Therefore, a matching transformer with a turns ratio of 2 must be used in this application.

**EXAMPLE 15–9** A certain amplifier has an 800 Ω internal resistance looking from its output. In order to provide maximum power to an 8 Ω speaker, what turns ratio must be used in the coupling transformer?

***Solution*** The reflected resistance must equal 800 Ω. Thus, from Equation (15–9), the turns ratio can be determined.

$$n = \frac{N_{sec}}{N_{pri}} = \sqrt{\frac{R_L}{R_{pri}}} = \sqrt{\frac{8\ \Omega}{800\ \Omega}} = \sqrt{0.01} = 0.1$$

There must be ten primary turns for each secondary turn. The diagram and its equivalent reflected circuit are shown in Figure 15–20.

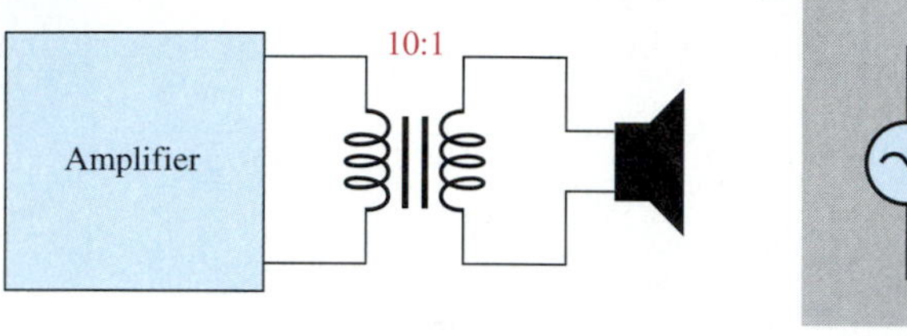


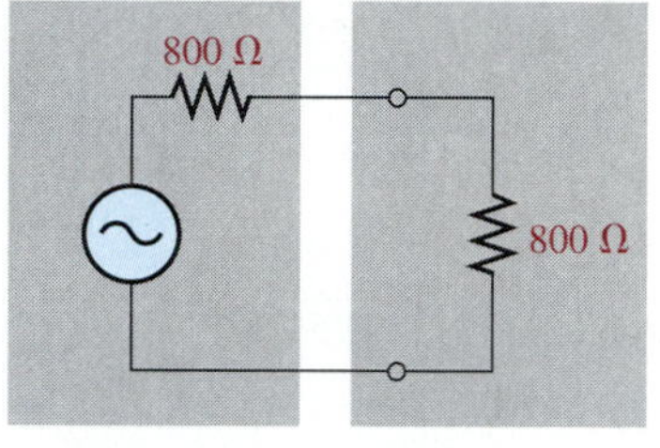


Amplifier equivalent circuit

Speaker/transformer equivalent

FIGURE 15–20

***Related Exercise*** What must be the turns ratio in Figure 15–20 to provide maximum power to two 8 Ω speakers in parallel?

**SECTION 15–7 REVIEW**

1. What does impedance matching mean?
2. What is the advantage of matching the load resistance to the resistance of a source?
3. A transformer has 100 turns in its primary winding and 50 turns in its secondary winding. What is the reflected resistance with 100 Ω across the secondary winding?

## 15–8 ■ THE TRANSFORMER AS AN ISOLATION DEVICE

***Transformers are useful in providing electrical isolation between the primary circuit and the secondary circuit because there is no electrical connection between the two windings. In a transformer, energy is transferred entirely by magnetic coupling.***

***After completing this section, you should be able to***

- **Explain how the transformer acts as an isolation device**
  - ☐ Discuss dc isolation
  - ☐ Discuss power line isolation

### DC Isolation

As illustrated in Figure 15–21(a), if there is a direct current in the primary circuit of a transformer, nothing happens in the secondary circuit. The reason is that a changing current in the primary winding is necessary to induce a voltage in the secondary winding, as

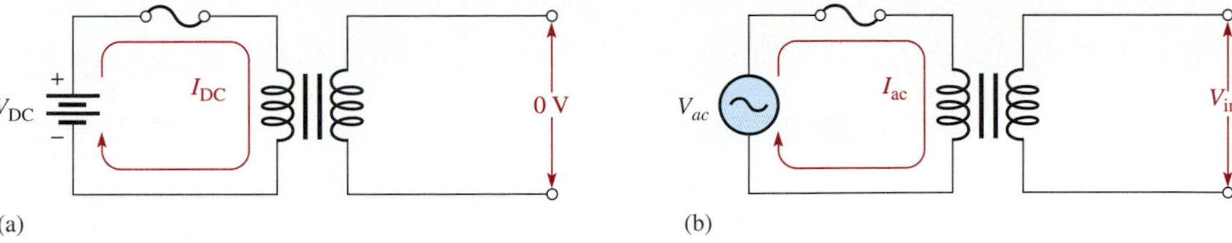


**FIGURE 15–21**
***DC isolation and ac coupling.***

shown in part(b). Therefore, the transformer isolates the secondary circuit from any dc voltage in the primary circuit.

In a typical application, a transformer can be used to keep the dc voltage on the output of an amplifier stage from affecting the dc bias of the next amplifier. Only the ac signal is coupled through the transformer from one stage to the next, as Figure 15–22 illustrates.

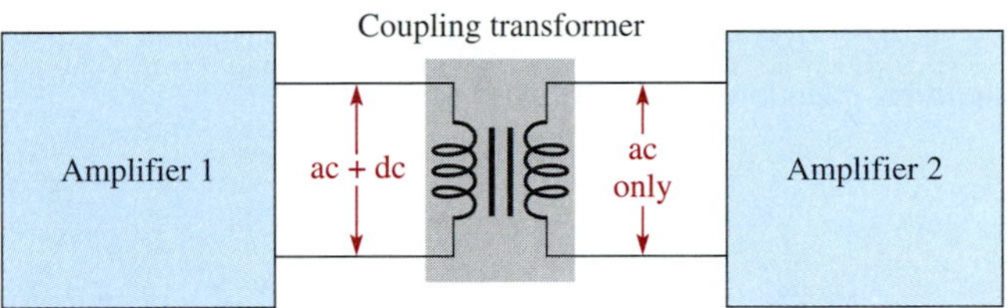


**FIGURE 15–22**
***Amplifier stages with transformer coupling for dc isolation.***

## Power Line Isolation

Transformers are often used to electrically isolate electronic equipment from the 60 Hz, 110 V ac power line. The reason for using an isolation transformer to couple the 60 Hz ac to an instrument is to prevent a possible shock hazard if the 110 V line is connected to the metal chassis of the equipment. This condition is possible if the line cord plug can be inserted into an outlet either way. Incidentally, to prevent this situation, many plugs have keyed prongs so that they can be plugged in only one way.

Figure 15–23 illustrates how a transformer can prevent the metal chassis from being connected to the 110 V line rather than to neutral (ground), no matter how the cord is plugged into the outlet. When an isolation transformer is used, the secondary circuit is said to be "floating" because it is not referenced to the power line ground. Should a person come in contact with the secondary voltage, there is no complete current path back to ground, and therefore there is no shock hazard. There must be current through your body in order for you to receive an electrical shock.

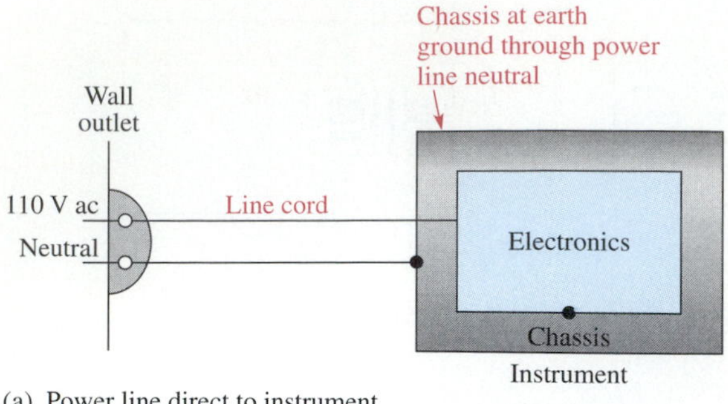


(a) Power line direct to instrument

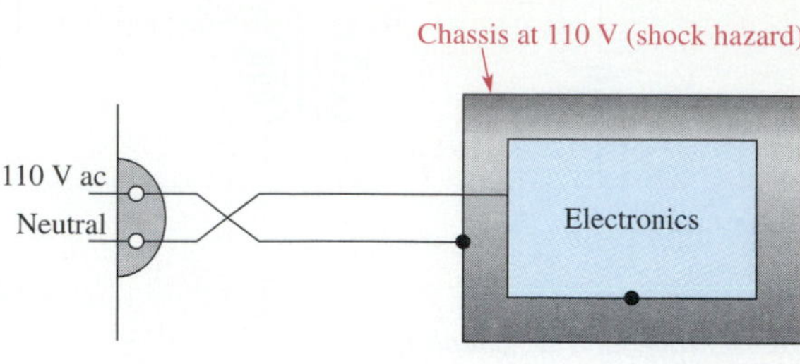


(b) Power line reversed

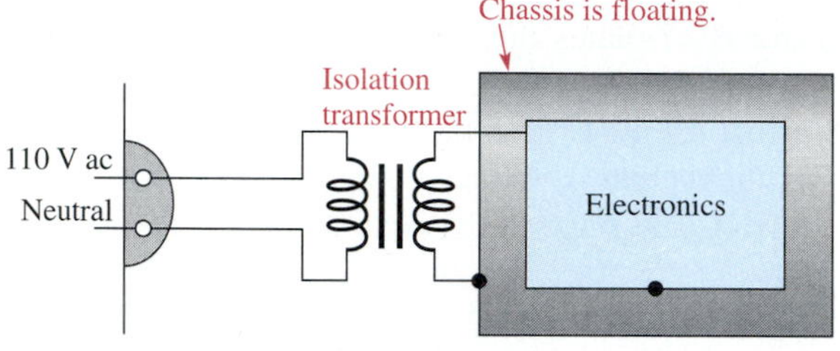


(c) Power line transformer-coupled to instrument

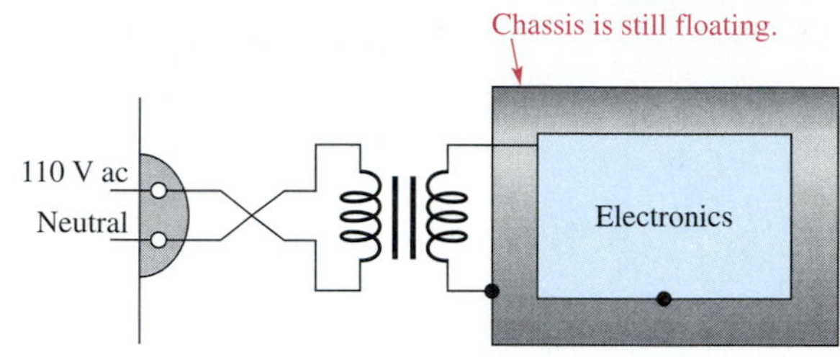


(d) Power line reversed

FIGURE 15–23
*The use of an isolation transformer to prevent a shock hazard.*

**SECTION 15–8 REVIEW**

1. What does the term *electrical isolation* mean?
2. Can a dc voltage be coupled by a transformer?

## 15–9 ■ NONIDEAL TRANSFORMER CHARACTERISTICS

***Up to this point, transformer operation has been discussed from an ideal point of view, and this approach is valid when you are learning new concepts. However, you should be aware of the nonideal characteristics of practical transformers and how they affect performance.***

***After completing this section, you should be able to***

- **Describe a practical transformer**
  - List and describe the nonideal characteristics
  - Explain power rating of a transformer
  - Define *efficiency* of a transformer

So far, we have considered the transformer to be an ideal device. That is, the winding resistance, the winding capacitance, and nonideal core characteristics were all neglected and the transformer was treated as if it had an efficiency of 100%. For studying the basic concepts and in many applications, the ideal model is valid. However, the practical transformer has several nonideal characteristics of which you should be aware.

### Winding Resistance

Both the primary and the secondary windings of a practical transformer have winding resistance. You learned about the winding resistance of inductors in Chapter 14. The winding resistances of a practical transformer are represented as resistors in series with the windings as shown in Figure 15–24.

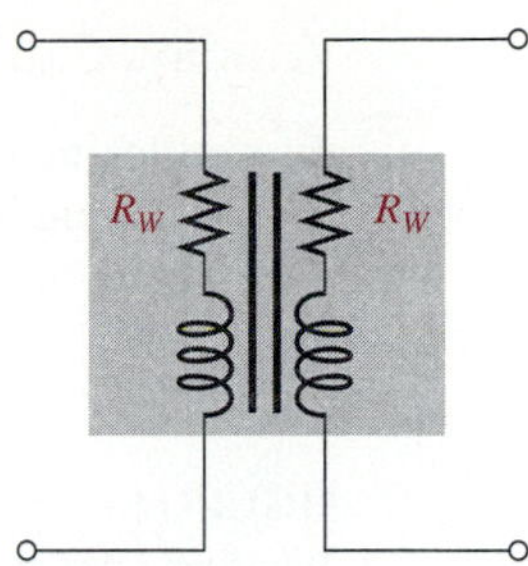


**FIGURE 15–24**
***Winding resistances in a practical transformer.***

Winding resistance in a practical transformer results in less voltage across a secondary load. Voltage drops due to the winding resistance effectively subtract from the primary and secondary voltages and result in load voltage that is less than that predicted by the relationship $V_{sec} = (N_{sec}/N_{pri})V_{pri}$. In many cases, the effect is relatively small and can be neglected.

## Losses in the Core

There is always some energy loss in the core material of a practical transformer. This loss is seen as a heating of ferrite and iron cores, but it does not occur in air cores. Part of this energy is consumed in the continuous reversal of the magnetic field due to the changing direction of the primary current; this energy loss is called *hysteresis loss.* The rest of the energy loss is caused by eddy currents produced when voltage is induced in the core material by the changing magnetic flux, according to Faraday's law. The eddy currents occur in circular patterns in the core resistance, thus causing the energy loss. This loss is greatly reduced by the use of laminated construction of iron cores. The thin layers of ferromagnetic material are insulated from each other to minimize the buildup of eddy currents by confining them to a small area and to keep core losses to a minimum.

## Magnetic Flux Leakage

In an ideal transformer, all of the magnetic flux produced by the primary current is assumed to pass through the core to the secondary winding, and vice versa. In a practical transformer, some of the magnetic flux lines break out of the core and pass through the surrounding air back to the other end of the winding, as illustrated in Figure 15–25 for the magnetic field produced by the primary current. Magnetic flux leakage results in a reduced secondary voltage.

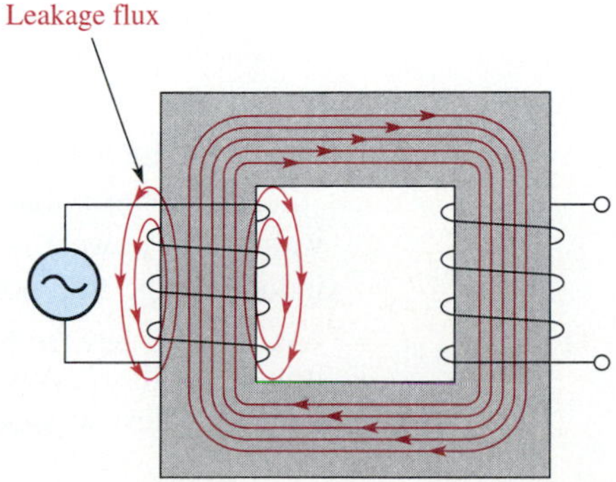


**FIGURE 15–25**
***Flux leakage in a practical transformer.***

The percentage of magnetic flux that actually reaches the secondary winding determines the coefficient of coupling of the transformer. For example, if nine out of ten flux lines remain inside the core, the coefficient of coupling is 0.90 or 90%. Most iron-core transformers have very high coefficients of coupling (greater than 0.99), which ferrite-core and air-core devices have lower values.

## Winding Capacitance

As you learned in Chapter 14, there is always some stray capacitance between adjacent turns of a winding. These stray capacitances result in an effective capacitance in parallel with each winding of a transformer, as indicated in Figure 15–26.

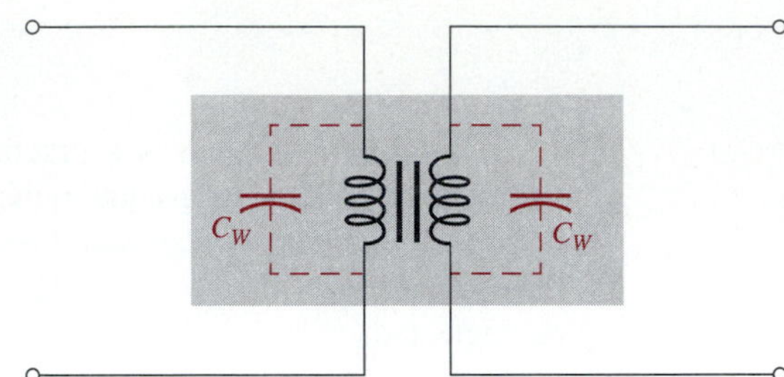


FIGURE 15–26
*Winding capacitance in a practical transformer.*

These stray capacitances have very little effect on the transformer's operation at low frequencies because the reactances ($X_C$) are very high. However, at higher frequencies, the reactances decrease and begin to produce a bypassing effect across the primary winding and across the secondary load. As a result, less of the total primary current is through the primary winding, and less of the total secondary current is through the load. This effect reduces the load voltage as the frequency goes up.

## Transformer Power Rating

A transformer is typically rated in volt-amperes (VA), primary/secondary voltage, and operating frequency. For example, a given transformer rating may be specified as 2 kVA, 500/50, 60 Hz. The 2 kVA value is the **apparent power rating.** The 500 and the 50 can be either secondary or primary voltages. The 60 Hz is the operating frequency.

The transformer rating can be helpful in selecting the proper transformer for a given application. Let's assume, for example, that 50 V is the secondary voltage. In this case the load current is

$$I_L = \frac{P_{sec}}{V_{sec}} = \frac{2\text{ kVA}}{50\text{ V}} = 40\text{ A}$$

On the other hand, if 500 V is the secondary voltage, then

$$I_L = \frac{P_{sec}}{V_{sec}} = \frac{2\text{ kVA}}{500\text{ V}} = 4\text{ A}$$

These are the maximum currents that the secondary can handle in either case.

The reason that the power rating is in volt-amperes (apparent power) rather than in watts (true power) is as follows: If the transformer load is purely capacitive or purely inductive, the true power (watts) delivered to the load is zero. However, the current for $V_{sec} = 500$ V and $X_C = 100\ \Omega$ at 60 Hz, for example, is 5 A. This current exceeds the maximum that the 2 kVA secondary can handle, and the transformer may be damaged. So it is meaningless to specify power in watts.

## Transformer Efficiency

Recall that the secondary power is equal to the primary power in an ideal transformer. Because the nonideal characteristics just discussed result in a power loss in the transformer, the secondary (output) power is always less than the primary (input) power. The

**efficiency** ($\eta$) of a transformer is a measure of the percentage of the input power that is delivered to the output.

$$\eta = \left(\frac{P_{out}}{P_{in}}\right)100\% \qquad (15\text{–}10)$$

Most power transformers have efficiencies in excess of 95%.

**EXAMPLE 15–10** A certain type of transformer has a primary current of 5 A and a primary voltage of 4800 V. The secondary current is 90 A, and the secondary voltage is 240 V. Determine the efficiency of this transformer.

***Solution*** The input power is

$$P_{in} = V_{pri}I_{pri} = (4800 \text{ V})(5 \text{ A}) = 24 \text{ kVA}$$

The output power is

$$P_{out} = V_{sec}I_{sec} = (240 \text{ V})(90 \text{ A}) = 21.6 \text{ kVA}$$

The efficiency is

$$\eta = \left(\frac{P_{out}}{P_{in}}\right)100\% = \left(\frac{21.6 \text{ kVA}}{24 \text{ kVA}}\right)100\% = 90\%$$

***Related Exercise*** A transformer has a primary current of 8 A with a primary voltage of 440 V. The secondary current is 30 A and the secondary voltage is 100 V. What is the efficiency?

**SECTION 15–9 REVIEW**

1. Explain how a practical transformer differs from the ideal model.
2. The coefficient of coupling of a certain transformer is 0.85. What does this mean?
3. A certain transformer has a rating of 10 kVA. If the secondary voltage is 250 V, how much load current can the transformer handle?

## 15–10 ■ OTHER TYPES OF TRANSFORMERS

***There are several important variations of the basic transformer. They include tapped transformers, multiple-winding transformers, and autotransformers.***

***After completing this section, you should be able to***

- **Describe several types of transformers**
  - ☐ Describe center-tapped transformers
  - ☐ Describe multiple-winding transformers
  - ☐ Describe autotransformers

### Tapped Transformers

A schematic of a transformer with a center-tapped secondary winding is shown in Figure 15–27(a). The **center tap (CT)** is equivalent to two secondary windings with half the total voltage across each.

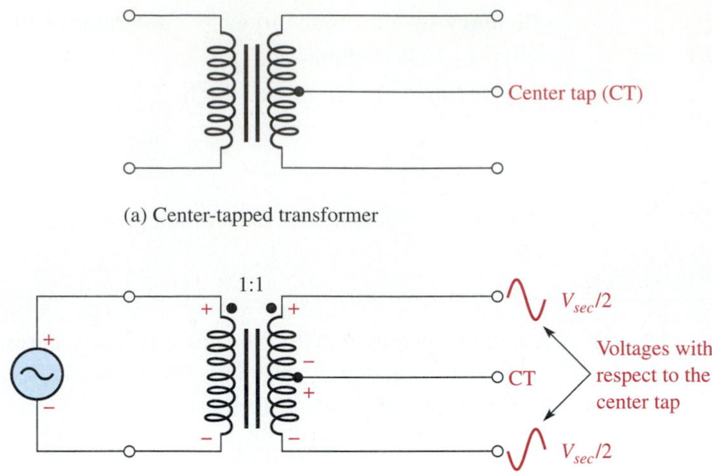


(a) Center-tapped transformer

(b) Output voltages with respect to the center tap are 180° out of phase with each other and are one-half the magnitude of the secondary voltage.

**FIGURE 15–27**
***Operation of a center-tapped transformer.***

The voltages between either end of the secondary winding and the center tap are, at any instant, equal in magnitude but opposite in polarity, as illustrated in Figure 15–27(b). Here, for example, at some instant on the sine wave voltage, the polarity across the entire secondary winding is as shown (top end +, bottom −). At the center tap, the voltage is less positive than the top end but more positive than the bottom end of the secondary. Therefore, measured with respect to the center tap, the top end of the secondary is positive, and the bottom end is negative. This center-tapped feature is used in power supply rectifiers in which the ac voltage is converted to dc, as illustrated in Figure 15–28.

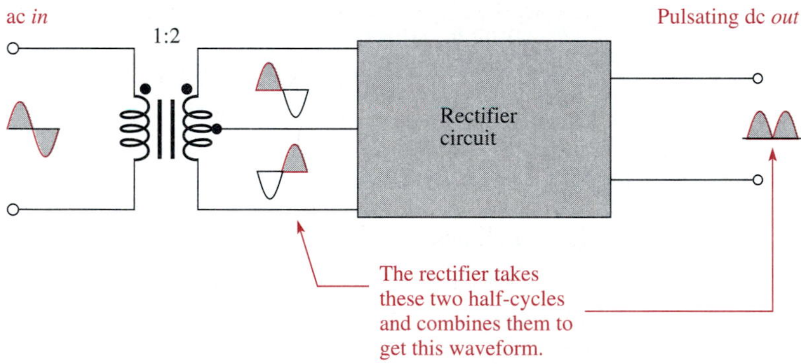


**FIGURE 15–28**
***Application of a center-tapped transformer in ac-to-dc conversion.***

Some tapped transformers have taps on the secondary winding at points other than the electrical center. Also, multiple primary and secondary taps are sometimes used in certain applications. Examples of these types of transformers are shown in Figure 15–29.

One example of a transformer with a multiple-tap primary winding and a center-tapped secondary winding is the utility-pole transformer used by power companies to step down the high voltage from the power line to 110 V/220 V service for residential and commercial customers, as shown in Figure 15–30. The multiple taps on the primary wind-

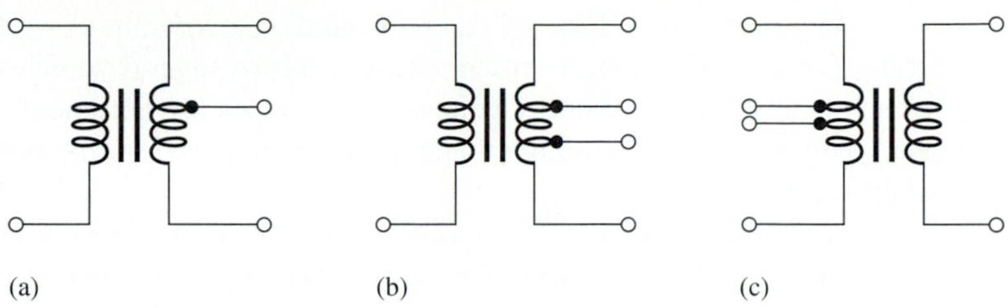


FIGURE 15–29
*Tapped transformers.*

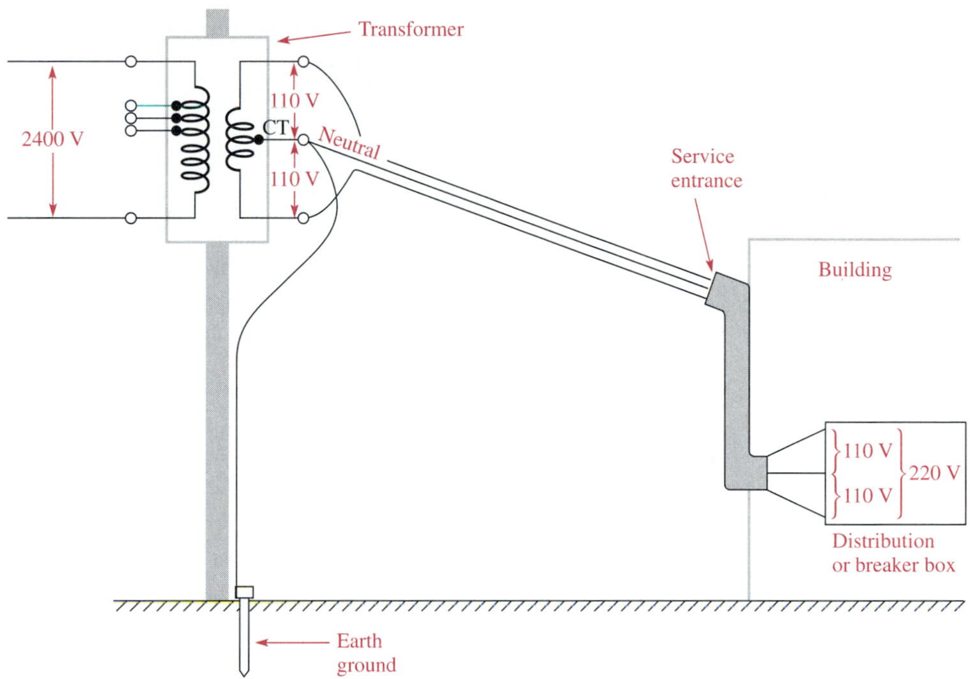


FIGURE 15–30
*Utility-pole transformer in a typical power distribution system.*

ing are used for minor adjustments in the turns ratio in order to overcome line voltages that are slightly too high or too low.

## Multiple-Winding Transformers

Some transformers are designed to operate from either 110 V ac or 220 V ac lines. These transformers usually have two primary windings, each of which is designed for 110 V ac. When the two are connected in series, the transformer can be used for 220 V ac operations, as illustrated in Figure 15–31.

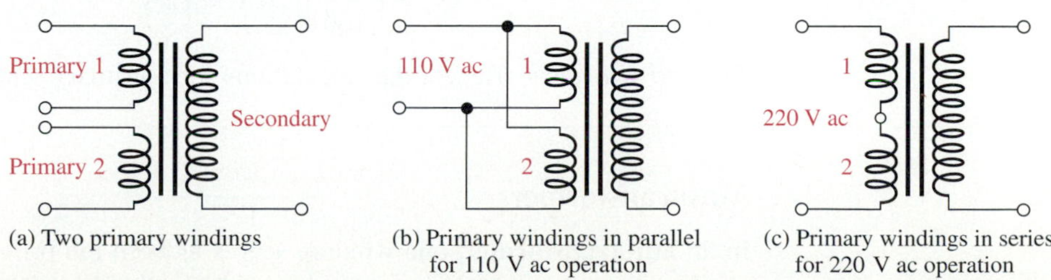


FIGURE 15–31
*Multiple-primary transformer.*

More than one secondary can be wound on a common core. Transformers with several secondary windings are often used to achieve several voltages by either stepping up or stepping down the primary voltage. These types are commonly used in power supply applications in which several voltage levels are required for the operation of an electronic instrument.

A typical schematic of a multiple-secondary transformer is shown in Figure 15–32; this transformer has three secondaries. Sometimes you will find combinations of multiple-primary, multiple-secondary, and tapped transformers all in one unit.

FIGURE 15–32

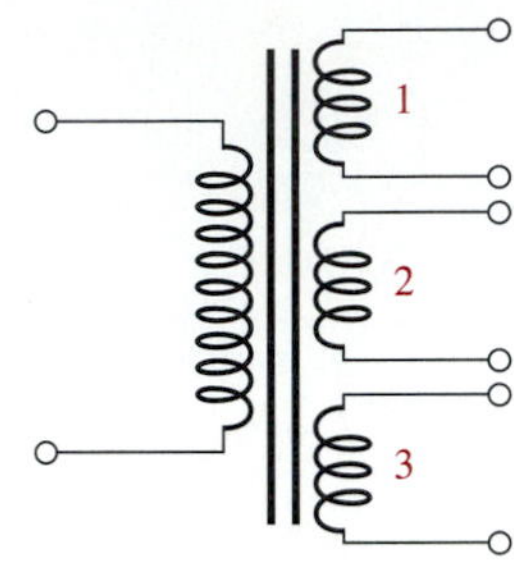


### EXAMPLE 15–11

The transformer shown in Figure 15–33 has the numbers of turns indicated. One of the secondaries is also center tapped. If 120 V ac are connected to the primary, determine each secondary voltage and the voltages with respect to the center tap (CT) on the middle secondary.

FIGURE 15–33

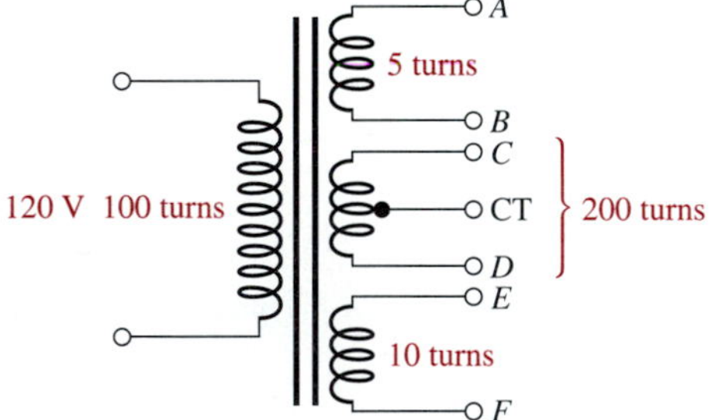


***Solution***

$$V_{AB} = \left(\frac{N_{sec}}{N_{pri}}\right)V_{pri} = \left(\frac{5}{100}\right)120\text{ V} = 6\text{ V}$$

$$V_{CD} = \left(\frac{200}{100}\right)120\text{ V} = 240\text{ V}$$

$$V_{(\text{CT})C} = V_{(\text{CT})D} = \frac{240\text{ V}}{2} = 120\text{ V}$$

$$V_{EF} = \left(\frac{10}{100}\right)120\text{ V} = 12\text{ V}$$

***Related Exercise*** Repeat the calculations for a primary with 50 turns.

## Autotransformers

In an **autotransformer,** one winding serves as both the primary and the secondary. The winding is tapped at the proper points to achieve the desired turns ratio for stepping up or stepping down the voltage.

Autotransformers differ from conventional transformers in that there is no electrical isolation between the primary and the secondary because both are on one winding. Autotransformers normally are smaller and lighter than equivalent conventional transformers because they require a much lower kVA rating for a given load. Many autotransformers provide an adjustable tap using a sliding contact mechanism so that the output voltage can be varied (these are often called *variacs*). Figure 15–34 shows a variable autotransformer and several schematic symbols.

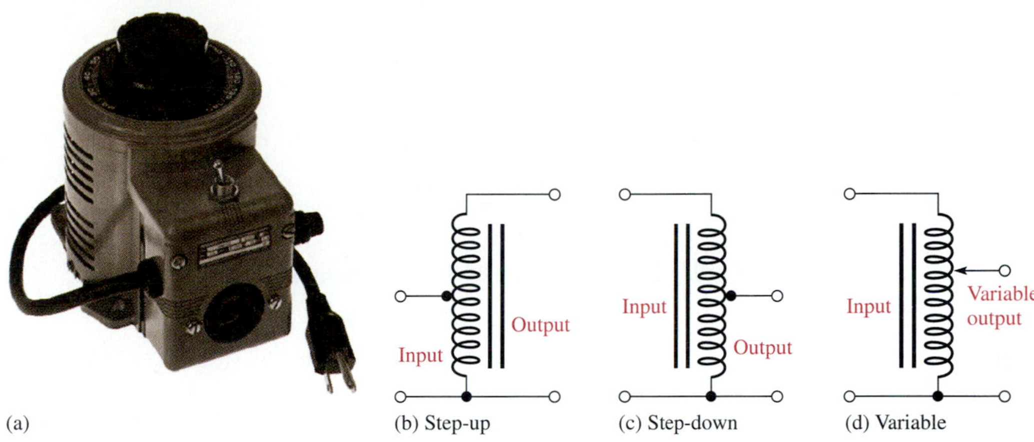


**FIGURE 15–34**
***The variable autotransformer. Part (a) courtesy of Superior Electric Co.***

Example 15–12 illustrates why an autotransformer has a kVA requirement that is less than the input or output kVA.

**EXAMPLE 15–12** A certain autotransformer is used to change a source voltage of 220 V to a load voltage of 160 V across an 8 Ω load resistance. Determine the input and output power in kilovolt-amperes, and show that the actual kVA requirement is less than this value. Assume that this transformer is ideal.

***Solution*** The circuit is shown in Figure 15–35 with voltages and currents indicated. The current directions have been assigned arbitrarily for convenience.

**FIGURE 15–35**

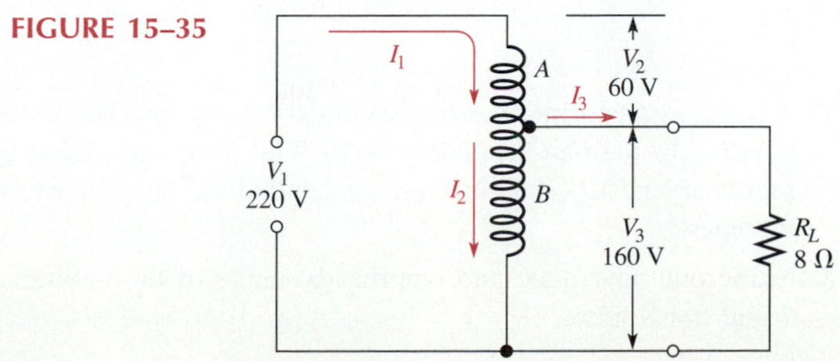

The load current, $I_3$, is determined as

$$I_3 = \frac{V_3}{R_L} = \frac{160\text{ V}}{8\ \Omega} = 20\text{ A}$$

The input power is the total source voltage ($V_1$) times the total current from the source ($I_1$).

$$P_{in} = V_1 I_1$$

The output power is the load voltage, $V_3$, times the load current, $I_3$.

$$P_{out} = V_3 I_3$$

For an ideal transformer, $P_{in} = P_{out}$; thus,

$$V_1 I_1 = V_3 I_3$$

Solving for $I_1$ yields

$$I_1 = \frac{V_3 I_3}{V_1} = \frac{(160\text{ V})(20\text{ A})}{220\text{ V}} = 14.55\text{ A}$$

Applying Kirchhoff's current law at the tap junction,

$$I_1 = I_2 + I_3$$

Solving for $I_2$, the current through winding $B$, yields

$$I_2 = I_1 - I_3 = 14.55\text{ A} - 20\text{ A} = 5.45\text{ A}$$

The minus sign can be dropped because the current directions are arbitrary.

The input and output power are

$$P_{in} = P_{out} = V_3 I_3 = (160\text{ V})(20\text{ A}) = 3.2\text{ kVA}$$

The power in winding $A$ is

$$P_A = V_2 I_1 = (60\text{ V})(14.55\text{ A}) = 873\text{ VA} = 0.873\text{ kVA}$$

The power in winding $B$ is

$$P_B = V_3 I_2 = (160\text{ V})(5.45\text{ A}) = 872\text{ VA} = 0.872\text{ kVA}$$

Thus, the power rating required for each winding is less than the power that is delivered to the load. The slight difference in the calculated powers in windings $A$ and $B$ is due to rounding.

***Related Exercise*** What happens to the kVA requirement if the load is changed to 4 Ω?

## SECTION 15–10 REVIEW

1. A certain transformer has two secondary windings. The turns ratio from the primary winding to the first secondary is 10. The turns ratio from the primary to the other secondary is 0.2. If 240 V ac are applied to the primary, what are the secondary voltages?
2. Name one advantage and one disadvantage of an autotransformer over a conventional transformer.

## 15–11 ■ TROUBLESHOOTING TRANSFORMERS

***The common failures in transformers are opens, shorts, or partial shorts in either the primary or the secondary windings. One cause of such failures is the operation of the device under conditions that exceed its ratings. A few transformer failures and the associated symptoms are covered in this section.***

***After completing this section, you should be able to***

- **Troubleshoot transformers**
  - ☐ Find an open primary or secondary winding
  - ☐ Find a shorted or partially shorted primary or secondary winding

### Open Primary Winding

When there is an open primary winding, there is no primary current and, therefore, no induced voltage or current in the secondary. This condition is illustrated in Figure 15–36(a), and the method of checking with an ohmmeter is shown in part (b).

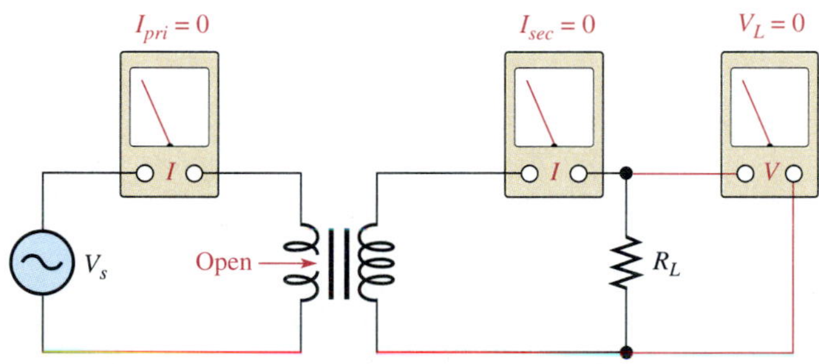


(a) Conditions when the primary winding is open

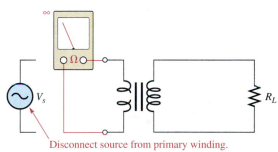

∞
Ω
$V_s$
$R_L$
Disconnect source from primary winding.

(b) Checking the primary winding with an ohmmeter

**FIGURE 15–36**
***Open primary winding.***

### Open Secondary Winding

When there is an open secondary winding, there is no current in the secondary circuit and, as a result, no voltage across the load. Also, an open secondary causes the primary current to be very small (there is only a small magnetizing current). In fact, the primary current may be practically zero. This condition is illustrated in Figure 15–37(a), and the ohmmeter check is shown in part (b).

### Shorted or Partially Shorted Primary Winding

A completely shorted primary winding will draw excessive current from the source; and unless there is a breaker or a fuse in the circuit, either the source or the transformer or both will burn out. A partial short in the primary winding can cause higher than normal or even excessive primary current.

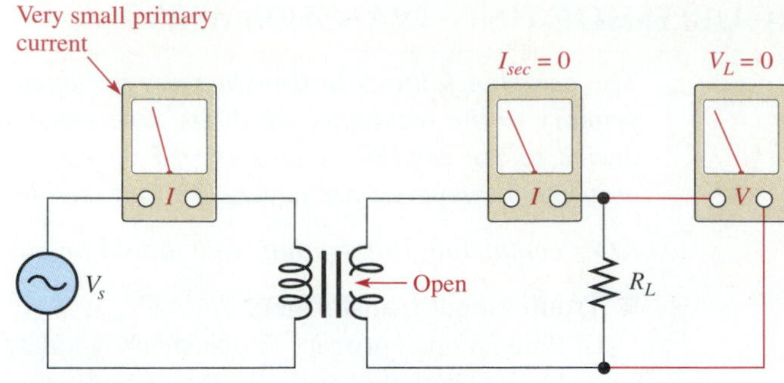


(a) Conditions when the secondary winding is open

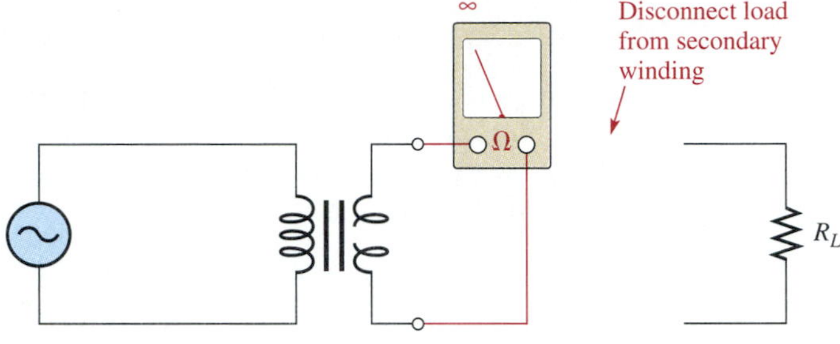


(b) Checking the secondary winding with the ohmmeter

FIGURE 15–37
***Open secondary winding.***

## Shorted or Partially Shorted Secondary Winding

In the case of a shorted or partially shorted secondary winding, there is an excessive primary current because of the low reflected resistance due to the short. Often, this excessive current will burn out the primary winding and result in an open. The short-circuit current in the secondary winding causes the load current to be zero (full short) or smaller than normal (partial short), as demonstrated in Figure 15–38(a) and 15–38(b). The ohmmeter check for this condition is shown in part (c).

Normally, when a transformer fails, it is very difficult to repair, and therefore the simplest procedure is to replace it.

### SECTION 15–11 REVIEW

1. Name two possible failures in a transformer.
2. What is often the cause of transformer failure?

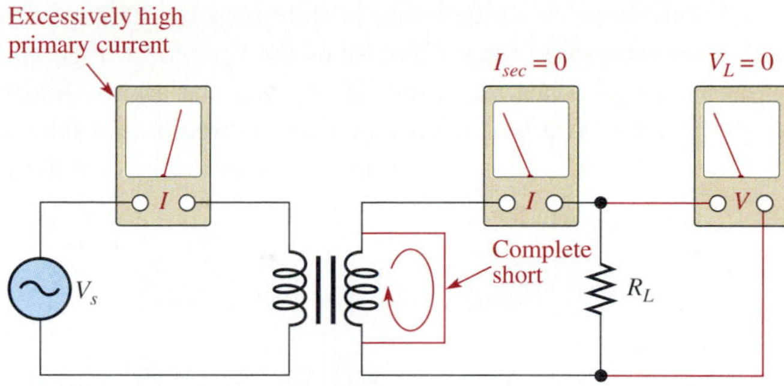


(a) Secondary winding completely shorted

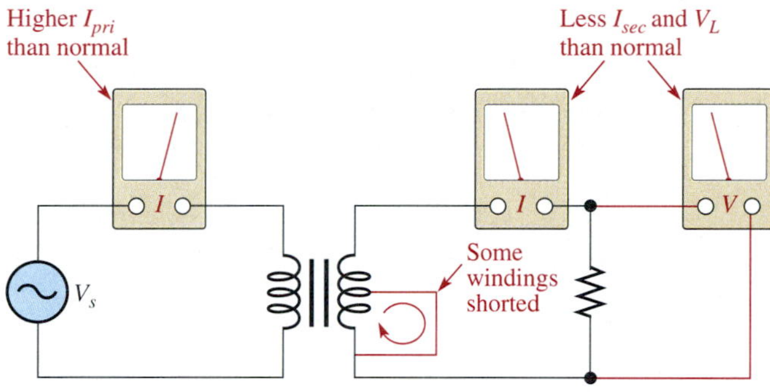


(b) Secondary winding partially shorted

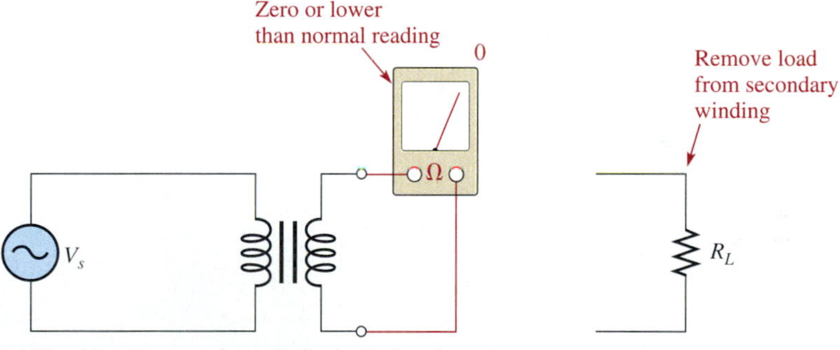


(c) Checking the secondary winding with the ohmmeter

**FIGURE 15–38**
***Shorted secondary winding.***

## 15–12 ■ TECHnology Theory Into Practice

***A common application of the transformer is in dc power supplies. The transformer is used to couple the ac line voltage into the power supply circuitry where it is converted to a dc voltage. Your assignment is to troubleshoot four identical transformer-coupled dc power supplies and, based on a series of measurements, determine the fault, if any, in each.***

The transformer ($T$) in the power supply schematic of Figure 15–39 steps the 110 V rms at the ac outlet down to a level that can be converted by the diode bridge rectifier, filtered, and regulated to obtain a 6 V dc output. The diode rectifier changes the ac to a pulsating full-wave dc voltage that is smoothed by the capacitor filter $C_1$. The voltage regulator is an integrated circuit that takes the filtered voltage and provides a constant 6 V dc over a range of load values and line voltage variations. Additional filtering is provided by capacitor $C_2$. You will learn about these circuits in a later course. The circled numbers in Figure 15–39 correspond to measurement points on the power supply unit.

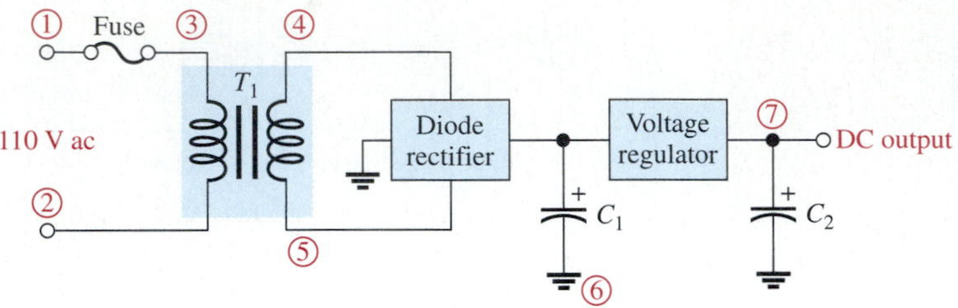


**FIGURE 15–39**
***Basic transformer-coupled dc power supply.***

## The Power Supply

You have four identical power supply units to troubleshoot like the one shown in Figure 15–40. The power line to the primary winding of transformer $T_1$ is fused by $F_1$. The secondary winding is connected to the circuit board containing the rectifier, filter, and regulator. Measurement points are indicated by the circled numbers.

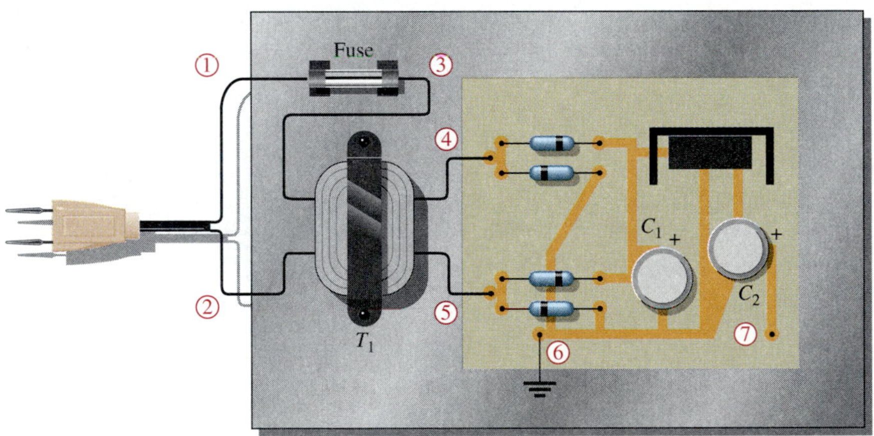


**FIGURE 15–40**
***Power supply unit (top view).***

## Measuring Voltages on Power Supply Unit 1

After plugging the power supply into a standard wall outlet, an autoranging portable multimeter is used to measure the voltages. In an autoranging meter, the appropriate measurement range is automatically selected instead of being manually selected as in a standard multimeter.

- ☐ Determine from the meter readings in Figure 15–41 whether or not the power supply is operating properly. If it is not, isolate the problem to one of the following: the circuit board containing the rectifier, filter, and regulator; the transformer; the fuse; or the power source. The circled numbers on the meter inputs correspond to the numbered points on the power supply in Figure 15–40.

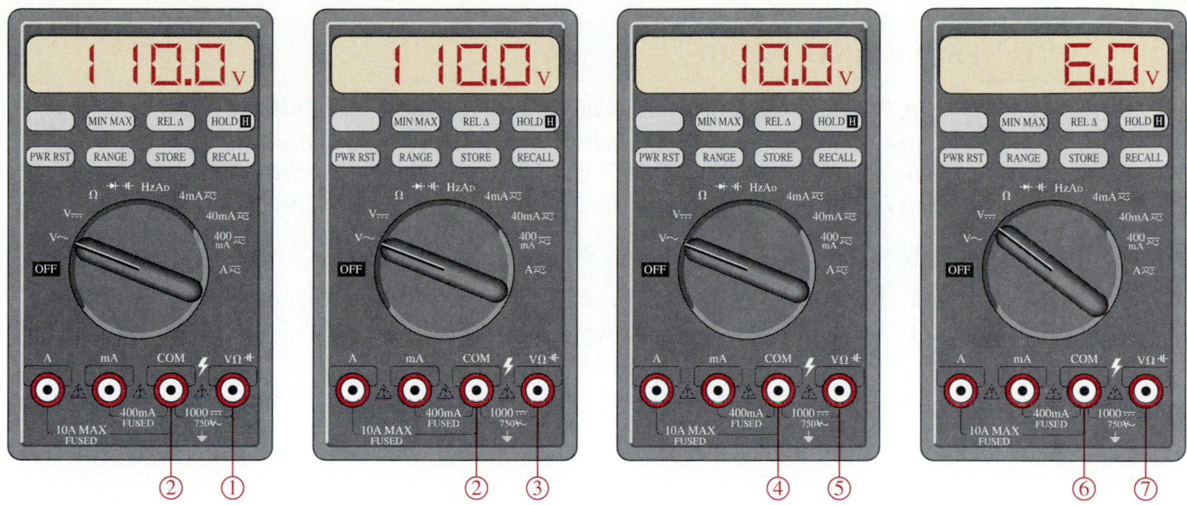


FIGURE 15–41
*Voltage measurements on power supply unit 1.*

## Measuring Voltages on Power Supply Units 2, 3, and 4

☐ Determine from the meter readings for units 2, 3, and 4 in Figure 15–42 whether or not each power supply is operating properly. If it is not, isolate the problem to one of the following: the circuit board containing the rectifier, filter, and regulator; the transformer; the fuse; or the power source. Only the meter displays and corresponding measurement points are shown.

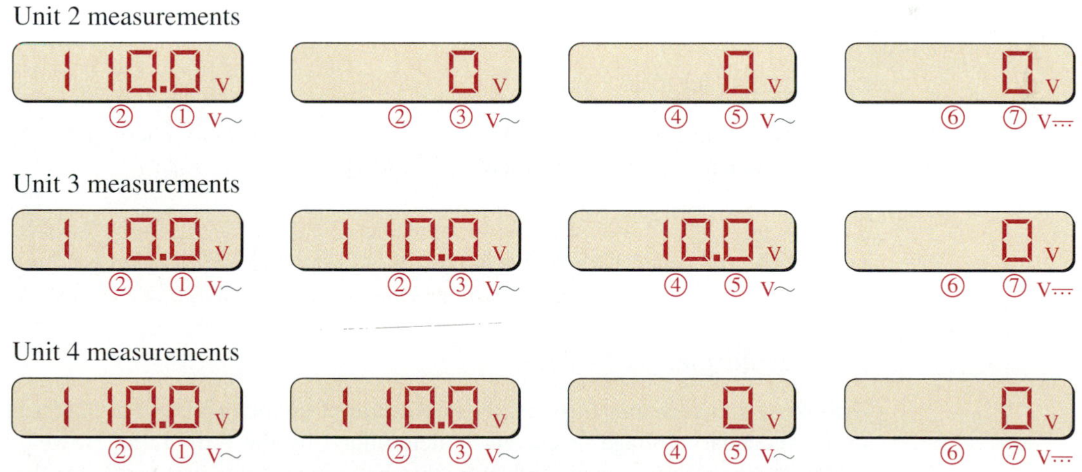


FIGURE 15–42
*Measurements for power supply units 2, 3, and 4.*

**SECTION 15–12 REVIEW**

1. In the case where the transformer was found to be faulty, how can you determine the specific fault (open windings or shorted windings)?
2. What type of fault in the transformer could cause the fuse to blow?

## 15–13 ■ MODELING TRANSFORMERS WITH PSpice AND PROBE

***In PSpice, the transformer is not available as a device. Instead, PSpice uses individual inductors and a coupling coefficient to describe a transformer. Models for describing the coupling of a transformer can be very complex, but the basic model that describes an ideal inductor will work for most applications and is discussed here.***

***After completing this section, you should be able to***

- **Use PSpice and Probe for analysis of circuits with transformers**
  - ☐ Describe a transformer with PSpice statements
  - ☐ Specify the coefficient of coupling
  - ☐ Specify the turns ratio
  - ☐ Specify the polarity

### The PSpice Transformer Core

The circuit file description of a transformer core is similar to other component descriptions except that the core uses inductors instead of nodes to describe the location in the circuit. The general form to describe a transformer core with one primary and one secondary is

```
<Core name><Inductor 1><Inductor 2><Coupling>
```

The core name begins with K, the inductors are specified by their labels in the circuit file, and the coefficient of coupling between the inductors is specified by a number greater than 0 and no greater than 1. As you know, the coefficient of coupling indicates how much of the flux from the primary coil is linked, or coupled, into the secondary. In an ideal transformer, where all of the flux is coupled to the secondary, the coefficient of coupling is 1. In a transformer that couples only half the flux to the secondary, the coefficient of coupling is 0.5.

The statement

```
K12  L1  L2  1
```

describes a transformer core labeled K12 with a maximum coefficient of coupling between inductors (windings) $L_1$ and $L_2$. Similarly,

```
KPS  LP  LS  0.100
```

describes a core called KPS with a coefficient of coupling of 0.100 between windings $L_p$ and $L_s$.

### The Transformer Turns Ratio

With most iron-core transformers, the coefficient of coupling is very high (essentially 1), so the secondary voltage or current is determined by the turns ratio. However, PSpice models the transformer using inductors that have inductance, not turns, for their values. Fortunately, the inductors in PSpice are essentially ideal so that the turns ratio is equal to the square root of the inductance ratio. For example, to describe a transformer with a turns ratio of 2 (the secondary winding has twice as many turns as the primary), you would specify an inductance ratio of 4 (the secondary inductor has an inductance four times that of the primary inductor).

The description lines

```
L1  1  0  25
L2  2  0  1
K12  L1  L2  1
```

represent a step-down transformer with a turns ratio of 1:5 or 0.2. The lines

```
LP  1  0  4
LS  2  0  9
K12  LP  LS  1
```

describe a step-up transformer with a turns ratio of 3:2 or 1.67.

Both transformer windings must have a dc path to ground since PSpice will not allow any "floating" nodes in a circuit and must have some point of reference to relate the primary and the secondary voltages to each other. This may be accomplished by connecting a very large value resistance (for example, 1 GΩ) from the second node of each winding to ground (node 0) if necessary.

## The Transformer Polarity

As you know, transformers can be wound to produce either an in-phase or an out-of-phase (180°) relationship between the primary and secondary voltages. Figure 15–43 shows a transformer circuit with an out-of-phase relationship as indicated by the dot notation on the transformer. The dotted node is listed first in the following component description:

```
L1  2  0  1
L2  0  3  100
```

FIGURE 15–43

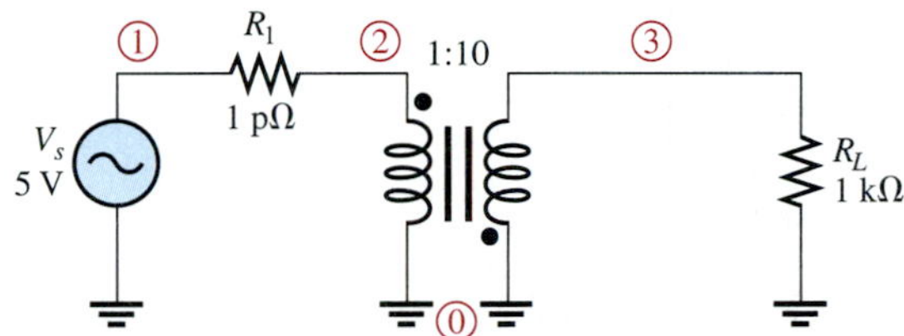


The complete circuit file for Figure 15–43 is as follows:

```
Transformer
*Transformer modeling
VS  1  0  SIN(0 5 1K)
R1  1  2  1P
L1  2  0  1
L2  0  3  100
RL  3  0  1K
K12  L1  L2  1
.TRAN  5U  5M  0  5U
.PROBE
.END
```

You should recognize that the voltage source $V_s$ is a 5 V, 1 kHz ac (sinusoidal) voltage and that the transformer is a step-up with a turns ratio of 10 with the primary and secondary voltages 180° out of phase. The 1 pΩ resistor $R_1$ does not affect the circuit operation, but PSpice requires it to be there. This is because PSpice uses ideal inductors with no winding resistance and without $R_1$, the voltage source would be shorted to ground. PSpice will not allow a short to be placed across a voltage source even though there is no dc voltage involved.

**SECTION 15–13 REVIEW**

1. Modify the circuit file if the transformer in Figure 15–43 is a step-down transformer with a turns ratio of 0.2.
2. Explain why 1 is used in the coefficient of coupling statement in the circuit file.
3. How would you change the circuit file for a transformer having no polarity reversal in Figure 15–43?

## ■ SUMMARY

- A transformer consists of two or more coils that are magnetically coupled on a common core.
- There is mutual inductance between two magnetically coupled coils.
- When current in one coil changes, voltage is induced in the other coil.
- The primary is the winding connected to the source, and the secondary is the winding connected to the load.
- The number of turns in the primary and the number of turns in the secondary determine the turns ratio.
- The relative polarities of the primary and secondary voltages are determined by the direction of the windings around the core.
- A step-up transformer has a turns ratio greater than 1.
- A step-down transformer has a turns ratio less than 1.
- A transformer cannot increase power.
- In an ideal transformer, the power from the source (input power) is equal to the power delivered to the load (output power).
- If the voltage is stepped up, the current is stepped down, and vice versa.
- A load across the secondary winding of a transformer appears to the source as a reflected load having a value dependent on the reciprocal of the turns ratio squared.
- A transformer can match a load resistance to a source resistance to achieve maximum power transfer to the load by selecting the proper turns ratio.
- A transformer does not respond to dc.
- Energy losses in an actual transformer result from winding resistances, hysteresis loss in the core, eddy currents in the core, and flux leakage.

## ■ GLOSSARY

**Apparent power rating** The method of rating transformers in which the power capability is expressed in volt-amperes (VA).

**Autotransformer** A transformer in which the primary and secondary are in a single winding.

**Center tap (CT)** A connection at the midpoint of the secondary winding of a transformer.

**Coefficient of coupling ($k$)** A constant associated with transformers that is the ratio of secondary magnetic flux to primary magnetic flux. The ideal value of 1 indicates that all the flux in the primary winding is coupled into the secondary winding.

**Core** The physical structure around which the windings are placed.

**Efficiency ($\eta$)** The ratio of output power to input power of a circuit expressed as a percentage.

**Electrical isolation** The condition that exists when two coils are magnetically linked but have no electrical connection between them.

**Impedance matching** A technique used to match a load resistance to a source resistance in order to achieve maximum transfer of power.

**Magnetic coupling** The magnetic connection between two coils as a result of the changing magnetic flux lines of one coil cutting through the second coil.

**Mutual inductance ($L_M$)** The inductance between two separate coils, such as a transformer.

**Primary winding** The input winding of a transformer; also called *primary.*

**Reflected load** The load as it appears to the source in the primary of a transformer.

**Reflected resistance** The resistance in the secondary circuit reflected into the primary circuit.

**Secondary winding** The output winding of a transformer; also called *secondary.*

**Step-down transformer** A transformer in which the secondary voltage is less than the primary voltage.

**Step-up transformer** A transformer in which the secondary voltage is greater than the primary voltage.

**Transformer** A device formed by two or more windings that are magnetically coupled to each other and providing a transfer of power electromagnetically from one winding to another.

**Turns ratio** The ratio of turns in the secondary winding to turns in the primary winding.

## ■ FORMULAS

| | | |
|---|---|---|
| **(15–1)** | $k = \dfrac{\phi_{1\text{-}2}}{\phi_1}$ | Coefficient of coupling |
| **(15–2)** | $L_M = k\sqrt{L_1L_2}$ | Mutual inductance |
| **(15–3)** | $n = \dfrac{N_{sec}}{N_{pri}}$ | Turns ratio |
| **(15–4)** | $\dfrac{V_{sec}}{V_{pri}} = \dfrac{N_{sec}}{N_{pri}}$ | Voltage ratio |
| **(15–5)** | $V_{sec} = \left(\dfrac{N_{sec}}{N_{pri}}\right)V_{pri}$ | Secondary voltage |
| **(15–6)** | $\dfrac{I_{pri}}{I_{sec}} = \dfrac{N_{sec}}{N_{pri}}$ | Current ratio |
| **(15–7)** | $I_{sec} = \left(\dfrac{N_{pri}}{N_{sec}}\right)I_{pri}$ | Secondary current |
| **(15–8)** | $R_{pri} = \left(\dfrac{N_{pri}}{N_{sec}}\right)^2 R_L$ | Reflected resistance |
| **(15–9)** | $n = \dfrac{N_{sec}}{N_{pri}} = \sqrt{\dfrac{R_L}{R_{pri}}}$ | Turns ratio for impedance matching |
| **(15–10)** | $\eta = \left(\dfrac{P_{out}}{P_{in}}\right)100\%$ | Transformer efficiency |

## ■ SELF-TEST

**1.** A transformer is used for

**(a)** dc voltages **(b)** ac voltages **(c)** both dc and ac

**2.** Which one of the following is affected by the turns ratio of a transformer?

**(a)** primary voltage **(b)** dc voltage

**(c)** secondary voltage **(d)** none of these

**3.** If the windings of a certain transformer with a turns ratio of 1 are in opposite directions around the core, the secondary voltage is

**(a)** in phase with the primary voltage **(b)** less than the primary voltage

**(c)** greater than the primary voltage **(d)** out of phase with the primary voltage

**4.** When the turns ratio of a transformer is 10 and the primary ac voltage is 6 V, the secondary voltage is

**(a)** 60 V **(b)** 0.6 V **(c)** 6 V **(d)** 36 V

**5.** When the turns ratio of a transformer is 0.5 and the primary ac voltage is 100 V, the secondary voltage is

**(a)** 200 V **(b)** 50 V **(c)** 10 V **(d)** 100 V

**6.** A certain transformer has 500 turns in the primary winding and 2500 turns in the secondary winding. The turns ratio is

**(a)** 0.2 **(b)** 2.5 **(c)** 5 **(d)** 0.5

**7.** If 10 W of power are applied to the primary of an ideal transformer with a turns ratio of 5, the power delivered to the secondary load is

**(a)** 50 W **(b)** 0.5 W **(c)** 0 W **(d)** 10 W

**8.** In a certain loaded transformer, the secondary voltage is one-third the primary voltage. The secondary current is

**(a)** one-third the primary current **(b)** three times the primary current

**(c)** equal to the primary current **(d)** answers (b) and (d)

**9.** When a 1 kΩ load resistor is connected across the secondary winding of a transformer with a turns ratio of 2, the source "sees" a reflected load of

**(a)** 250 Ω **(b)** 2 kΩ **(c)** 4 kΩ **(d)** 1 kΩ

10. In Question 9, if the turns ratio is 0.5, the source "sees" a reflected load of
    (a) 1 kΩ (b) 2 kΩ (c) 4 kΩ (d) 500 Ω
11. The turns required to match a 50 Ω source to a 200 Ω load is
    (a) 0.25 (b) 0.5 (c) 4 (d) 2
12. Maximum power is transferred from a source to a load in a transformer coupled circuit when
    (a) $R_L > R_{int}$ (b) $R_L < R_{int}$ (c) $\left(\frac{1}{n}\right)^2 R_L = R_{int}$ (d) $R_L = nR_{int}$
13. When a 12 V battery is connected across the primary of a transformer with a turns ratio of 4, the secondary voltage is
    (a) 0 V (b) 12 V (c) 48 V (d) 3 V
14. A certain transformer has a turns ratio of 1 and a 0.95 coefficient of coupling. When 1 V ac is applied to the primary, the secondary voltage is
    (a) 1 V (b) 1.95 V (c) 0.95 V

## ▪ PROBLEMS

### SECTION 15–1 Mutual Inductance

1. What is the mutual inductance when $k = 0.75$, $L_1 = 1\ \mu H$, and $L_2 = 4\ \mu H$?
2. Determine the coefficient of coupling when $L_M = 1\ \mu H$, $L_1 = 8\ \mu H$, and $L_2 = 2\ \mu H$.

### SECTION 15–2 The Basic Transformer

3. What is the turns ratio of a transformer having 250 primary turns and 1000 secondary turns? What is the turns ratio when the primary winding has 400 turns and the secondary winding has 100 turns?
4. A certain transformer has 25 turns in its primary winding. In order to double the voltage, how many turns must be in the secondary winding?
5. For each transformer in Figure 15–44, sketch the secondary voltage showing its relationship to the primary voltage. Also indicate the amplitude.

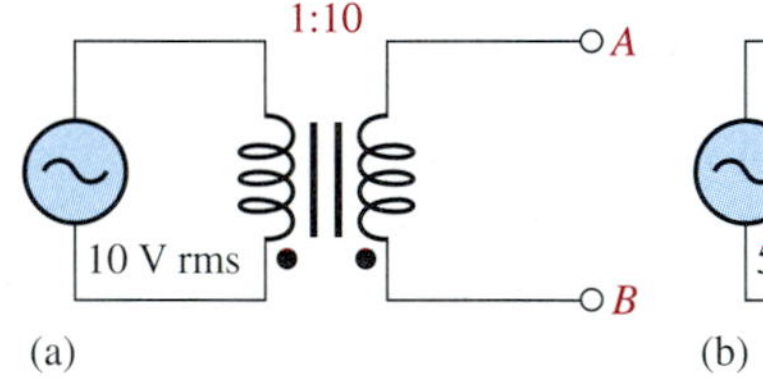

(a)

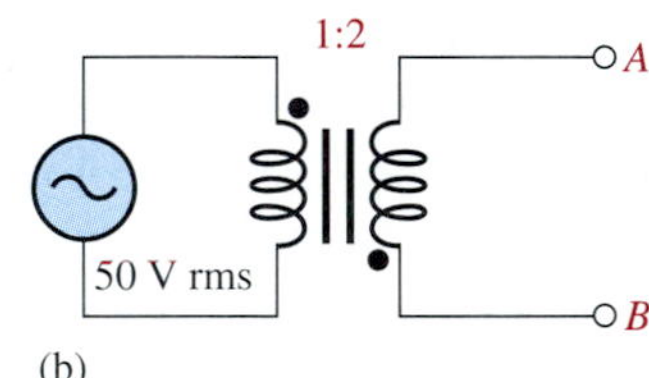

(b)

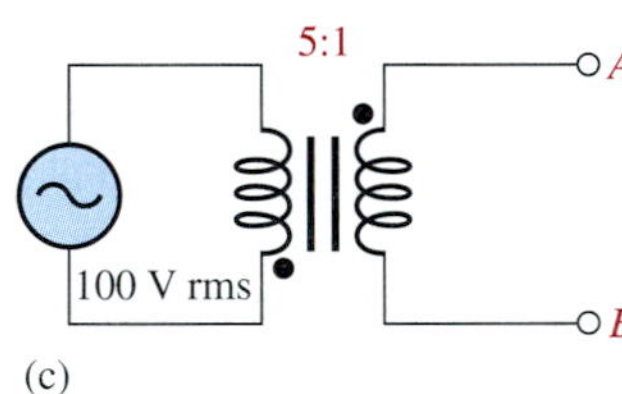

(c)

FIGURE 15–44

### SECTION 15–3 Step-Up Transformers

6. To step 240 V ac up to 720 V, what must the turns ratio be?
7. The primary winding of a transformer has 120 V ac across it. What is the secondary voltage if the turns ratio is 5?
8. How many primary volts must be applied to a transformer with a turns ratio of 10 to obtain a secondary voltage of 60 V ac?

### SECTION 15–4 Step-Down Transformers

9. To step 120 V down to 30 V, what must the turns ratio be?
10. The primary winding of a transformer has 1200 V across it. What is the secondary voltage if the turns ratio is 0.2?
11. How many primary volts must be applied to a transformer with a turns ratio of 0.1 to obtain a secondary voltage of 6 V ac?

## SECTION 15–5 Loading the Secondary Winding

**12.** Determine $I_s$ in Figure 15–45. What is the value of $R_L$?

FIGURE 15–45

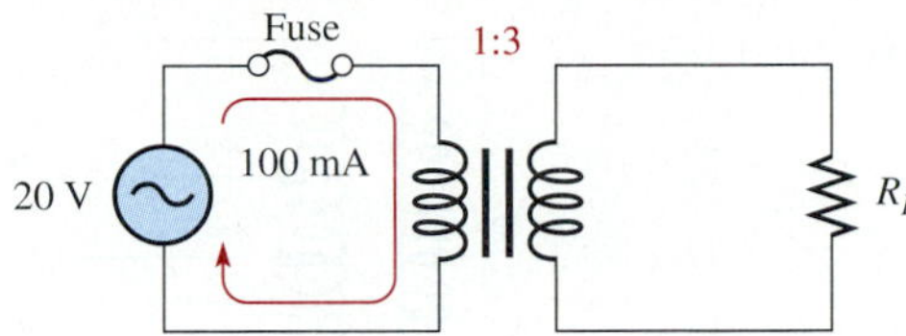


**13.** Determine the following quantities in Figure 15–46.

**(a)** Primary current **(b)** Secondary current

**(c)** Secondary voltage **(d)** Power in the load

FIGURE 15–46

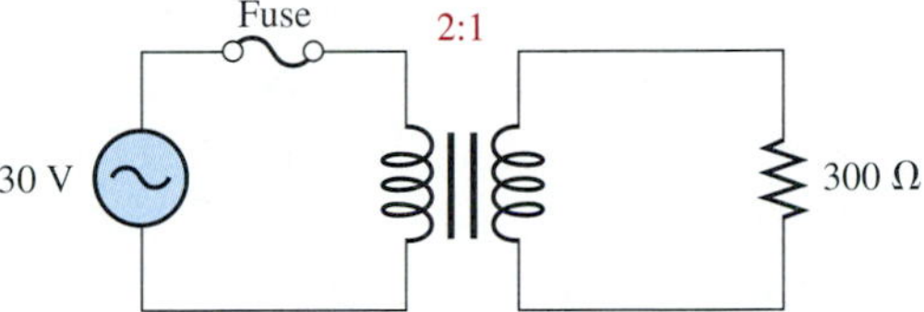


## SECTION 15–6 Reflected Load

**14.** What is the load resistance as seen by the source in Figure 15–47?

FIGURE 15–47

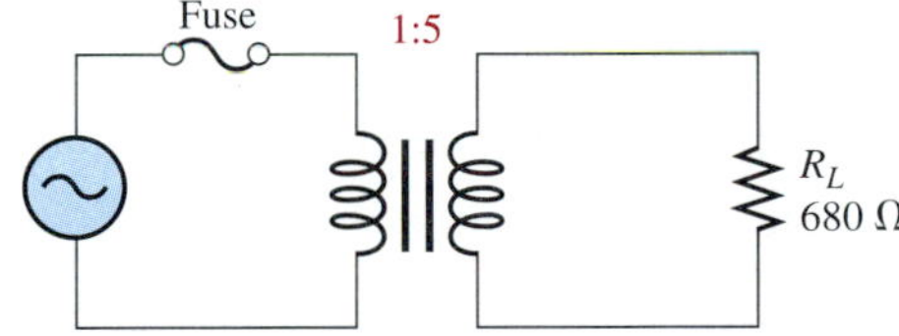


**15.** What must the turns ratio be in Figure 15–48 in order to reflect 300 Ω into the primary circuit?

FIGURE 15–48

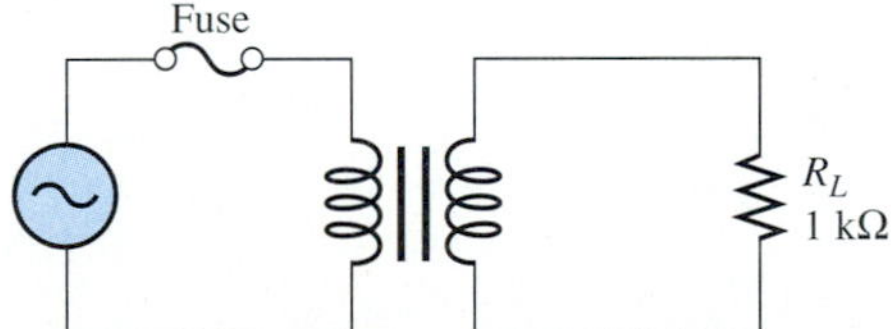


## SECTION 15–7 Matching the Load and Source Resistances

**16.** For the circuit in Figure 15–49, find the turns ratio required to deliver maximum power to the 4 Ω speaker.

**17.** In Figure 15–49, what is the maximum power that can be delivered to the 4 Ω speaker?

FIGURE 15–49

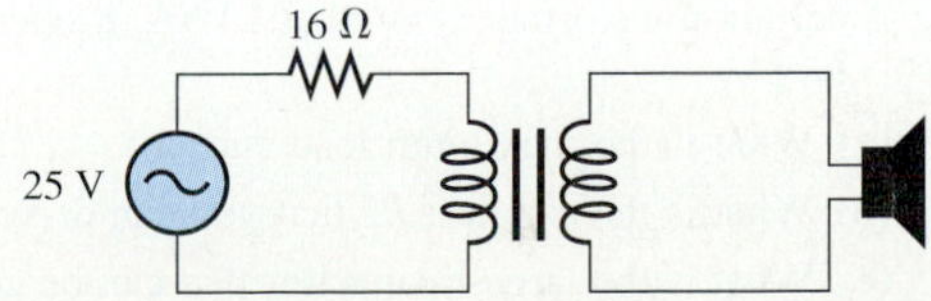

18. Find the appropriate turns ratio for each switch position in Figure 15–50 in order to transfer the maximum power to each load when the source resistance is 10 Ω. Specify the number of turns for the secondary winding if the primary winding has 1000 turns.

FIGURE 15–50

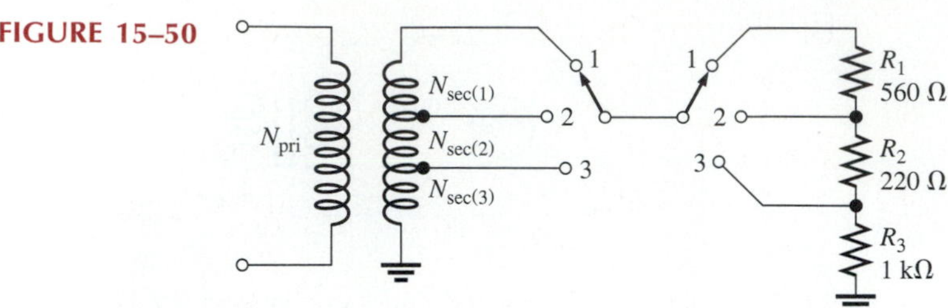


## SECTION 15–8 The Transformer as an Isolation Device

19. What is the voltage across the load in each circuit of Figure 15–51?

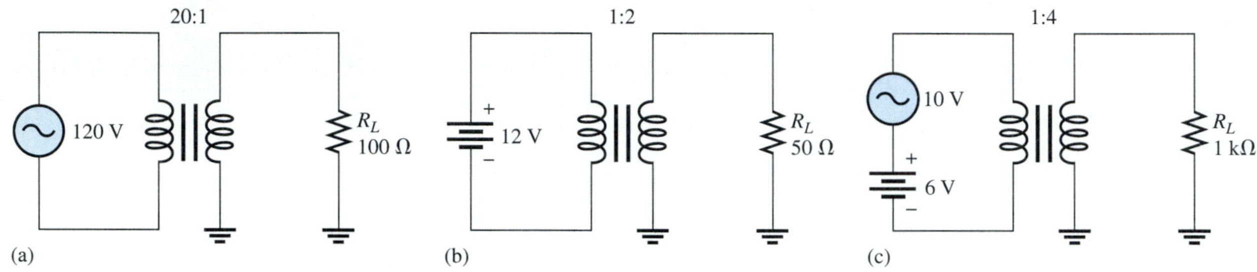


FIGURE 15–51

20. Determine the unspecified meter readings in Figure 15–52.

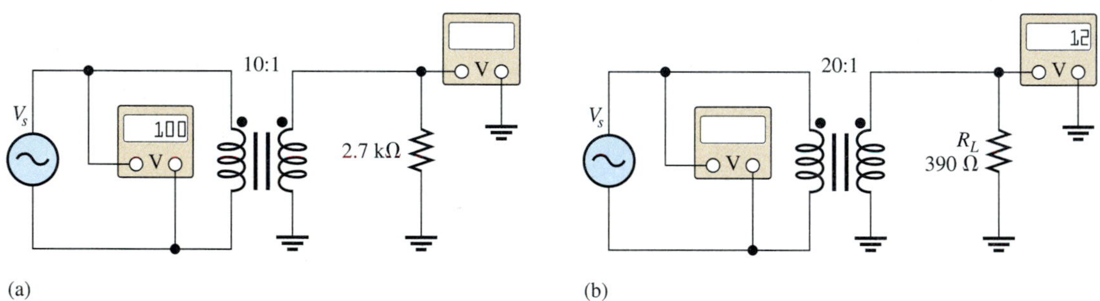


FIGURE 15–52

## SECTION 15–9 Nonideal Transformer Characteristics

21. In a certain transformer, the input power to the primary is 100 W. If 5.5 W are lost in the winding resistances, what is the output power to the load, neglecting any other losses?
22. What is the efficiency of the transformer in Problem 21?
23. Determine the coefficient of coupling for a transformer in which 2% of the total flux generated in the primary does not pass through the secondary.
24. A certain transformer is rated at 1 kVA. It operates on 60 Hz, 120 V ac. The secondary voltage is 600 V.
    **(a)** What is the maximum load current?
    **(b)** What is the smallest $R_L$ that you can drive?
    **(c)** What is the largest capacitor that can be connected as a load?

**25.** What kVA rating is required for a transformer that must handle a maximum load current of 10 A with a secondary voltage of 2.5 kV?

**26.** A certain transformer is rated at 5 kVA, 2400/120 V, at 60 Hz.

**(a)** What is the turns ratio if the 120 V is the secondary voltage?

**(b)** What is the current rating of the secondary if 2400 V is the primary voltage?

**(c)** What is the current rating of the primary winding if 2400 V is the primary voltage?

## SECTION 15–10 Other Types of Transformers

**27.** Determine each unknown voltage indicated in Figure 15–53.

FIGURE 15–53

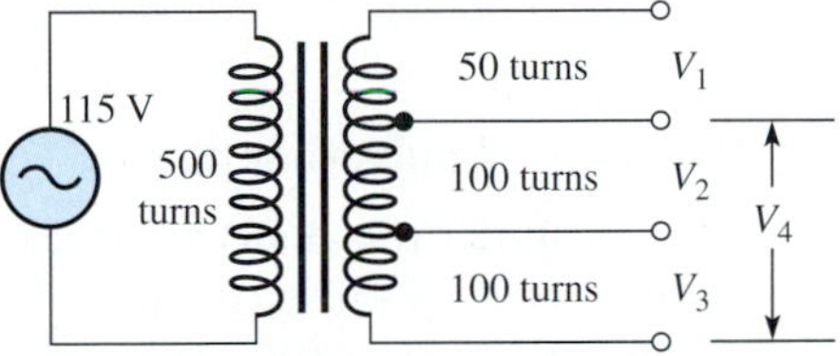


**28.** Using the indicated secondary voltages in Figure 15–54 determine the turns ratio of the primary winding to each tapped section of the secondary winding.

FIGURE 15–54

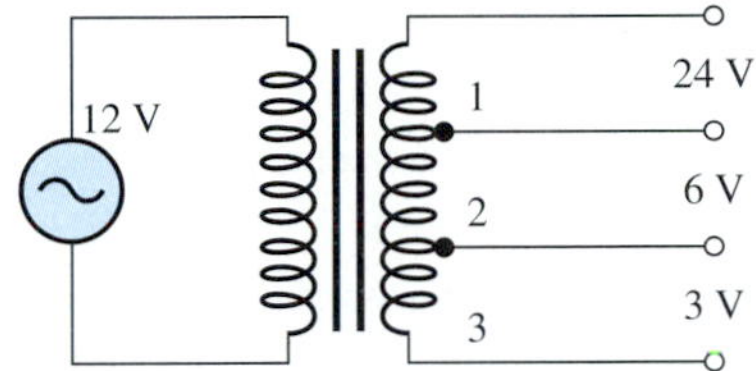


**29.** Find the secondary voltage for each autotransformer in Figure 15–55.

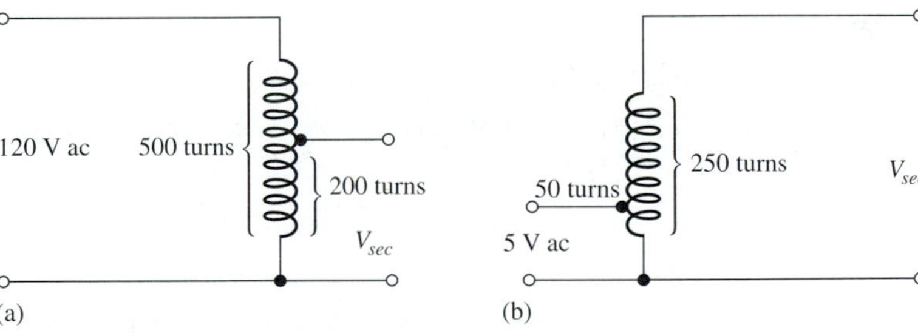


FIGURE 15–55

**30.** In Figure 15–56, each primary can accommodate 120 V ac. How should the primaries be connected for 240 V ac operation? Determine each secondary voltage for 240 V operation.

FIGURE 15–56

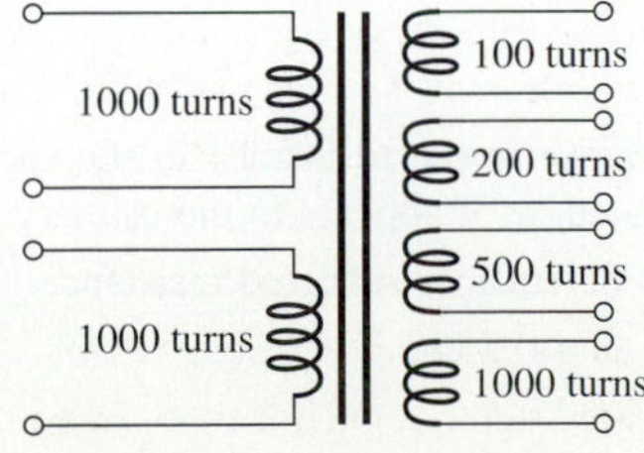

31. For the loaded, tapped-secondary transformer in Figure 15–57, determine the following:
    (a) All load voltages and currents
    (b) The resistance reflected into the primary

FIGURE 15–57

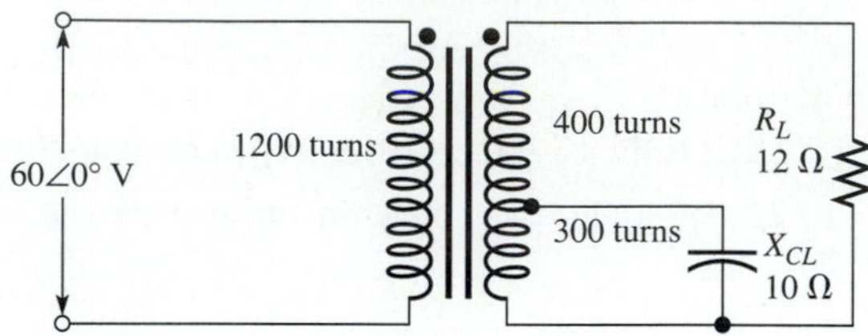


## SECTION 15–11 Troubleshooting

32. When you apply 120 V ac across the primary winding of a transformer and check the voltage across the secondary winding, you get 0 V. Further investigation shows no primary or secondary currents. List the possible faults. What is your next step in investigating the problem?
33. What is likely to happen if the primary winding of a transformer shorts?
34. While checking out a transformer, you find that the secondary voltage is less than it should be although it is not zero. What is the most likely fault?

## ■ ANSWERS TO SECTION REVIEWS

### Section 15–1

1. Mutual inductance is the inductance between two coils.
2. $L_M = k\sqrt{L_1L_2} = 45$ mH
3. The induced voltage increases if $k$ increases.

### Section 15–2

1. Transformer operation is based on mutual inductance.
2. The turns ratio is the ratio of turns in the secondary winding to turns in the primary winding.
3. The directions of the windings determine the relative polarities of the voltages.
4. $n = 250/500 = 0.5$

### Section 15–3

1. A step-up transformer produces a secondary voltage that is greater than the primary voltage.
2. $V_{sec}$ is five times greater than $V_{pri}$.
3. $V_{sec} = (N_{sec}/N_{pri})V_{pri} = 10(240\text{ V}) = 2400$ V

### Section 15–4

1. A step-down transformer produces a secondary voltage that is less than the primary voltage.
2. $V_{sec} = (0.5)120\text{ V} = 60$ V
3. $n = 12\text{ V}/120\text{ V} = 0.1$

### Section 15–5

1. $I_{sec}$ is less than $I_{pri}$ by half.
2. $I_{sec} = (100/25)0.5\text{ A} = 2$ A
3. $I_{pri} = (25/100)10\text{ A} = 2.5$ A

### Section 15–6

1. Reflected resistance is the resistance in the secondary circuit, altered by the reciprocal of the turns ratio squared, as it appears to the primary circuit.
2. The turns ratio determines reflected resistance.
3. $R_{pri} = (0.1)^2 50\ \Omega = 0.5\ \Omega$
4. $n = 0.1$

### Section 15–7

1. Impedance matching makes the load resistance equal the source resistance.
2. Maximum power is delivered to the load when $R_L = R_s$.
3. $R_{pri} = (100/50)^2 100\ \Omega = 400\ \Omega$

### Section 15–8

1. Electrical isolation means no electrical connection between primary and secondary.
2. No, transformers do not couple dc.

### Section 15–9

1. In a practical transformer, energy loss reduces the efficiency. An ideal transformer has an efficiency of 100%.
2. When $k = 0.85$, 85% of the magnetic flux generated in the primary winding passes through the secondary winding.
3. $I_L = 10$ kVA/250 V = 40 A

### Section 15–10

1. $V_{sec} = (10)240$ V = 2400 V, $V_{sec} = (0.2)240$ V = 48 V
2. An autotransformer is smaller and lighter for the same rating than a conventional one. An autotransformer has no electrical isolation.

### Section 15–11

1. Transformer faults: open windings, shorted windings
2. Operating above rated values will cause a failure.

### Section 15–12

1. Use an ohmmeter to check for open windings. Shorted windings are indicated by an incorrect secondary voltage.
2. A short will cause the fuse to blow.

### Section 15–13

1. ```
   L1  2  0  25
   L2  0  3  1
   ```
2. An ideal transformer is assumed.
3. ```
   L1  2  0  1
   L2  3  0  100
   ```

## ■ ANSWERS TO RELATED EXERCISES FOR EXAMPLES

**15–1** 0.75
**15–2** 387 $\mu$H
**15–3** 500 turns
**15–4** 480 V
**15–5** 57.6 V
**15–6** 5 mA; 400 mA
**15–7** 6 $\Omega$
**15–8** 0.354
**15–9** 0.0707 or 14.14:1
**15–10** 85.2%
**15–11** $V_{AB} = 12$ V, $V_{CD} = 480$ V, $V_{(CT)C} = V_{(CT)D} = 240$ V, $V_{EF} = 24$ V
**15–12** Increases to 2.13 kVA

# 16

# *RC* CIRCUITS



## ■ INTRODUCTION

In this chapter, you will study *RC* circuits and in the next chapter you will study *RL* circuits. The analyses of these two types of reactive circuits are similar. The major difference in *RC* and *RL* circuits is that in *RC* circuits, the capacitive reactance decreases with frequency but in *RL* circuits, the inductive reactance increases with frequency. Therefore, *RC* and *RL* circuits have opposite phase responses.

There are two ways to study the material in Chapters 16 and 17. One way is to study all of Chapter 16 first and then study Chapter 17. The other way is to alternate between the two chapters and study corresponding topics. For example, the topic of series *RC* circuits can be followed by a study of series *RL* circuits; then, the topic of parallel *RC* circuits can be followed by a study of parallel *RL* circuits and so on.

In this chapter and throughout the rest of the book, you will learn the basics of putting technology theory into practice.

An *RC* circuit contains both resistance and capacitance. It is one of the basic types of reactive circuits that you will study. In this chapter, basic series and parallel *RC* circuits and their responses to sinusoidal ac voltages are presented. Series-parallel combinations are also analyzed. True, reactive, and apparent power in *RC* circuits are discussed and some basic *RC* applications are introduced. Applications of *RC* circuits include filters, amplifier coupling, oscillators, and wave-shaping circuits. Troubleshooting and PSpice analysis complete the chapter.

## TECHnology Theory Into Practice

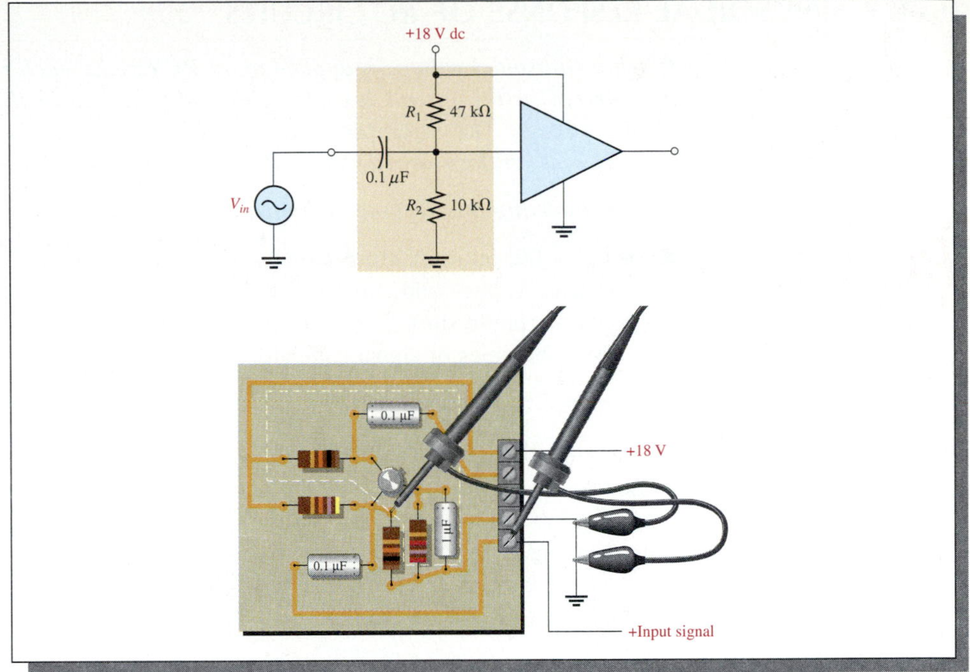


The frequency response of the *RC* input network in an amplifier circuit is similar to the one you worked with in Chapter 13 and is the subject of this chapter's TECH TIP.

## CHAPTER OBJECTIVES

- ☐ Describe the relationship between current and voltage in an *RC* circuit
- ☐ Determine impedance and phase angle in a series *RC* circuit
- ☐ Analyze a series *RC* circuit
- ☐ Determine impedance and phase angle in a parallel *RC* circuit
- ☐ Analyze a parallel *RC* circuit
- ☐ Analyze series-parallel *RC* circuits
- ☐ Determine power in *RC* circuits
- ☐ Discuss some basic *RC* applications
- ☐ Troubleshoot *RC* circuits
- ☐ Use PSpice and Probe to determine the frequency and phase responses of an *RC* circuit (optional)

## 16–1 ■ SINUSOIDAL RESPONSE OF *RC* CIRCUITS

***When a sinusoidal voltage is applied to an RC circuit, each resulting voltage drop and the current in the circuit are also sinusoidal and have the same frequency as the applied voltage. The capacitance causes a phase shift between the voltage and current that depends on the relative values of the resistance and the capacitive reactance.***

***After completing this section, you should be able to***

- **Describe the relationship between current and voltage in an *RC* circuit**
  - ☐ Discuss voltage and current waveforms
  - ☐ Discuss phase shift
  - ☐ Describe types of signal generators

As shown in Figure 16–1, the resistor voltage ($V_R$), the capacitor voltage ($V_C$), and the current ($I$) are all sine waves with the frequency of the source. Phase shifts are introduced because of the capacitance. As you will learn, the resistor voltage and current **lead** the source voltage, and the capacitor voltage **lags** the source voltage. The phase angle between the current and the capacitor voltage is always 90°. These generalized phase relationships are indicated in Figure 16–1.

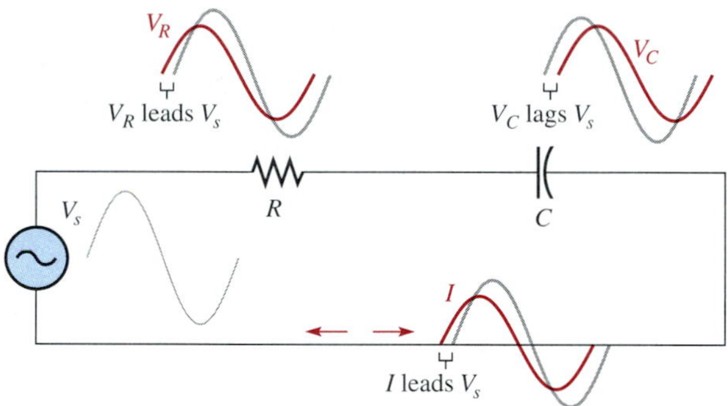


**FIGURE 16–1**

***Illustration of sinusoidal response with general phase relationships of $V_R$, $V_C$, and I relative to the source voltage. $V_R$ leads $V_s$, $V_C$ lags $V_s$, and I leads $V_s$. $V_R$ and I are in phase while $V_R$ and $V_C$ are 90° out of phase.***

The amplitudes and the phase relationships of the voltages and current depend on the values of the resistance and the capacitive reactance. When a circuit is purely resistive, the phase angle between the applied (source) voltage and the total current is zero. When a circuit is purely capacitive, the phase angle between the applied voltage and the total current is 90°, with the current leading the voltage. When there is a combination of both resistance and capacitive reactance in a circuit, the phase angle between the applied voltage and the total current is somewhere between 0° and 90°, depending on the relative values of the resistance and the reactance.

### Signal Generators

When a circuit is hooked up for a laboratory experiment or for troubleshooting, a signal generator similar to those shown in Figure 16–2 is used to provide the source voltage. These instruments, depending on their capability, are classified as sine wave generators, which produce only sine waves; sine/square generators, which produce either sine waves or square waves; or function generators, which produce sine waves, pulse waveforms, or triangular (ramp) waveforms.

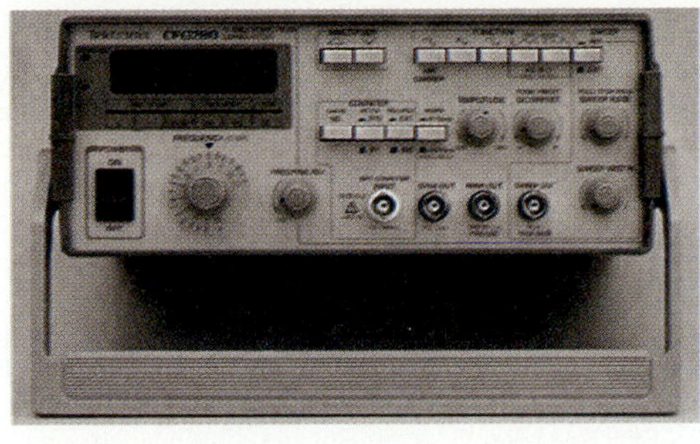
(a)

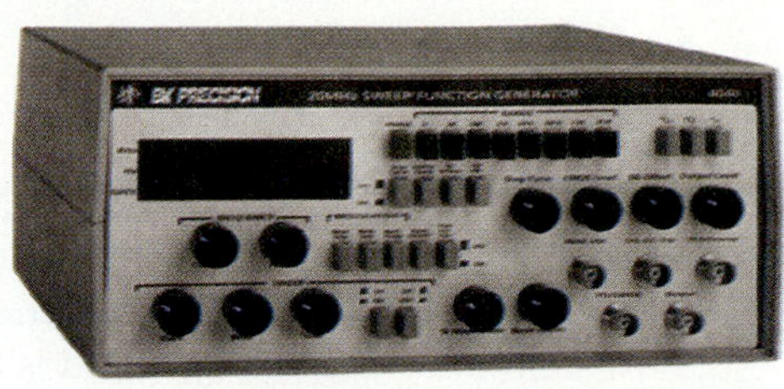
(b)

FIGURE 16–2
*Typical signal (function) generators used in circuit testing and troubleshooting. Part (a) courtesy of Tektronix, Inc. Parts (b) and (c) courtesy of BK Precision, Maxtec International Corp.*

**SECTION 16–1 REVIEW**

1. A 60 Hz sinusoidal voltage is applied to an *RC* circuit. What is the frequency of the capacitor voltage? What is the frequency of the current?
2. What causes the phase shift between $V_s$ and $I$ in a series *RC* circuit?
3. When the resistance in an *RC* circuit is greater than the capacitive reactance, is the phase angle between the applied voltage and the total current closer to 0° or to 90°?

## 16–2 ■ IMPEDANCE AND PHASE ANGLE OF SERIES *RC* CIRCUITS

***The impedance of an RC circuit is the total opposition to sinusoidal current and its unit is the ohm. The phase angle is the phase difference between the total current and the source voltage.***

***After completing this section, you should be able to***

- **Determine impedance and phase angle in a series *RC* circuit**
  - ☐ Define *impedance*
  - ☐ Express capacitive reactance in complex form
  - ☐ Express total impedance in complex form
  - ☐ Draw an impedance triangle
  - ☐ Calculate impedance magnitude and the phase angle

In a purely capacitive circuit, the **impedance** is equal to the total capacitive reactance. The impedance of a series *RC* circuit is determined by both the resistance and the capacitive reactance. These cases are illustrated in Figure 16–3. The magnitude of the impedance is symbolized by *Z*.

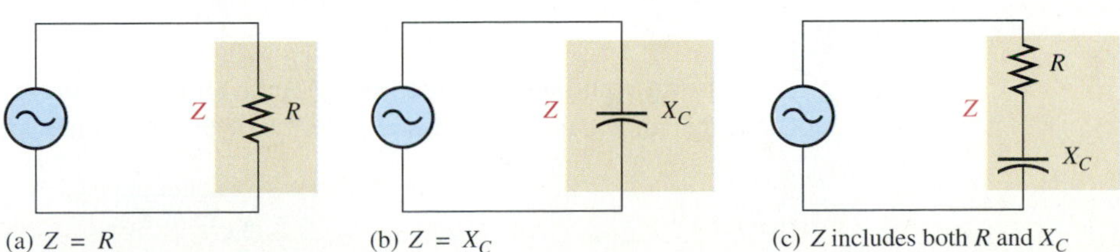


FIGURE 16–3
*Three cases of impedance.*

Recall from Chapter 13 that capacitive reactance is expressed as a complex number in rectangular form as

$$\mathbf{X}_C = -jX_C \qquad \textbf{(16–1)}$$

where boldface $\mathbf{X}_C$ designates a phasor quantity (representing both magnitude and angle) and $X_C$ is just the magnitude.

In the series *RC* circuit of Figure 16–4, the total impedance is the phasor sum of $R$ and $-jX_C$ and is expressed as

$$\mathbf{Z} = R - jX_C \qquad \textbf{(16–2)}$$

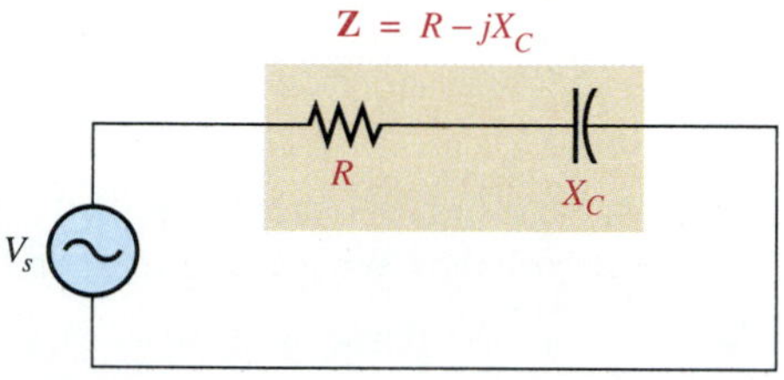

**FIGURE 16–4**
***Series RC circuit.***

$\mathbf{Z} = R - jX_C$

## The Impedance Triangle

In ac analysis, both $R$ and $X_C$ are treated as phasor quantities, as shown in the phasor diagram of Figure 16–5(a), with $X_C$ appearing at a $-90°$ angle with respect to $R$. This relationship comes from the fact that the capacitor voltage in a series *RC* circuit lags the current, and thus the resistor voltage, by 90°. Since $\mathbf{Z}$ is the phasor sum of $R$ and $-jX_C$, its phasor representation is shown in Figure 16–5(b). A repositioning of the phasors, as shown in part (c), forms a right triangle. This is called the *impedance triangle.* The length of each phasor represents the magnitude in ohms, and the angle $\theta$ is the phase angle of the *RC* circuit and represents the phase difference between the applied voltage and the current.

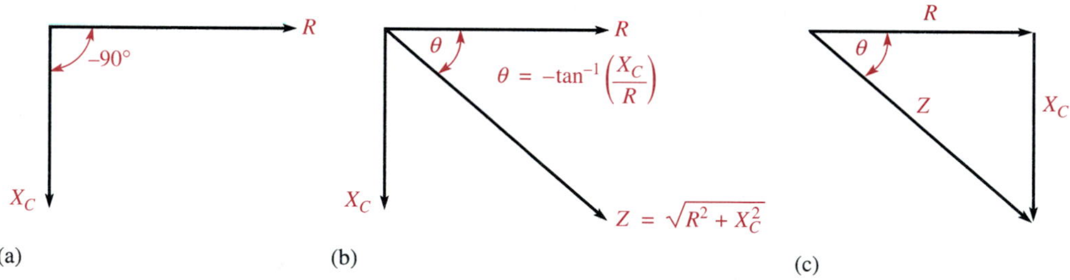


**FIGURE 16–5**
***Development of the impedance triangle for a series RC circuit.***

From right-angle trigonometry (Pythagorean theorem), the magnitude (length) of the impedance can be expressed in terms of the resistance and reactance as

$$Z = \sqrt{R^2 + X_C^2} \qquad \textbf{(16–3)}$$

The italic letter $Z$ represents the magnitude of the phasor quantity $\mathbf{Z}$ and is expressed in ohms.

The phase angle, $\theta$, is expressed as

$$\theta = -\tan^{-1}\left(\frac{X_C}{R}\right) \tag{16–4}$$

The symbol $\tan^{-1}$ stands for inverse tangent and can be found on some calculators by pressing INV, then tan. Combining the magnitude and angle, the phasor expression for impedance in polar form is

$$\mathbf{Z} = \sqrt{R^2 + X_C^2}\angle -\tan^{-1}\left(\frac{X_C}{R}\right) \tag{16–5}$$

**EXAMPLE 16–1** For each circuit in Figure 16–6, write the phasor expression for the impedance in both rectangular form and polar form.

FIGURE 16–6

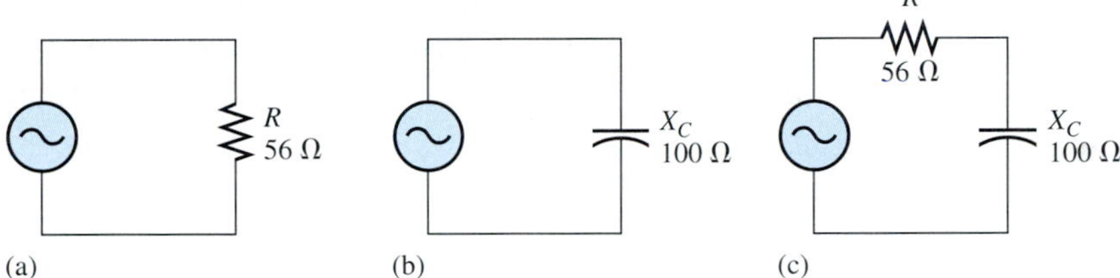


***Solution*** For the circuit in Figure 16–6(a), the impedance is

$\mathbf{Z} = R - j0 = R = 56\ \Omega$ in rectangular form ($X_C = 0$)

$\mathbf{Z} = R\angle 0° = 56\angle 0°\ \Omega$ in polar form

The impedance is simply the resistance, and the phase angle is zero because pure resistance does not cause a phase shift between the voltage and current.

For the circuit in Figure 16–6(b), the impedance is

$\mathbf{Z} = 0 - jX_C = -j100\ \Omega$ in rectangular form ($R = 0$)

$\mathbf{Z} = X_C\angle -90° = 100\angle -90°\ \Omega$ in polar form

The impedance is simply the capacitive reactance, and the phase angle is $-90°$ because the capacitance causes the current to lead the voltage by 90°.

For the circuit in Figure 16–6(c), the impedance in rectangular form is

$$\mathbf{Z} = R - jX_C = 56\ \Omega - j100\ \Omega$$

The impedance in polar form is

$$\mathbf{Z} = \sqrt{R^2 + X_C^2}\angle -\tan^{-1}\left(\frac{X_C}{R}\right)$$

$$= \sqrt{(56\ \Omega)^2 + (100\ \Omega)^2}\angle -\tan^{-1}\left(\frac{100\ \Omega}{56\ \Omega}\right) = 115\angle -60.8°\ \Omega$$

In this case, the impedance is the phasor sum of the resistance and the capacitive reactance. The phase angle is fixed by the relative values of $X_C$ and $R$. Rectangular to polar conversion using a calculator was illustrated in Chapter 12 and can be used in problems like this one.

***Related Exercise*** Use your calculator to convert the impedance in Figure 16–6(c) from rectangular to polar form.

**SECTION 16–2 REVIEW**

1. The impedance of a certain *RC* circuit is 150 Ω – *j*220 Ω. What is the value of the resistance? The capacitive reactance?
2. A series *RC* circuit has a total resistance of 33 kΩ and a capacitive reactance of 50 kΩ. Write the phasor expression for the impedance in rectangular form.
3. For the circuit in Question 2, what is the magnitude of the impedance? What is the phase angle?

## 16–3 ■ ANALYSIS OF SERIES *RC* CIRCUITS

***In the previous section, you learned how to express the impedance of a series RC circuit. Now, Ohm's law and Kirchhoff's voltage law are used in the analysis of RC circuits.***

***After completing this section, you should be able to***

- **Analyze a series *RC* circuit**
  - Apply Ohm's law and Kirchhoff's voltage law to series *RC* circuits
  - Express the voltages and current as phasor quantities
  - Show how impedance and phase angle vary with frequency

### Ohm's Law

The application of Ohm's law to series *RC* circuits involves the use of the phasor quantities of **Z**, **V**, and **I**. Keep in mind that the use of boldface nonitalic letters indicates phasor quantities where both magnitude and angle are included. The three equivalent forms of Ohm's law are as follows:

$$\mathbf{V} = \mathbf{IZ} \tag{16–6}$$

$$\mathbf{I} = \frac{\mathbf{V}}{\mathbf{Z}} \tag{16–7}$$

$$\mathbf{Z} = \frac{\mathbf{V}}{\mathbf{I}} \tag{16–8}$$

From your study of phasor algebra in Chapter 12, you should recall that multiplication and division are most easily accomplished with the polar forms. Since Ohm's law calculations involve multiplications and divisions, the voltage, current, and impedance should be expressed in polar form, as the next two examples show.

**EXAMPLE 16–2** If the current in Figure 16–7 is expressed in polar form as $\mathbf{I} = 0.2\angle 0°$ mA, determine the source voltage expressed in polar form, and draw the phasor diagram.

FIGURE 16–7

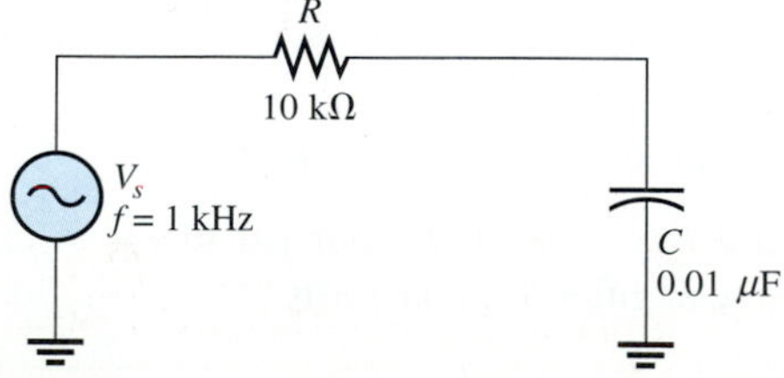

***Solution*** The magnitude of the capacitive reactance is

$$X_C = \frac{1}{2\pi fC} = \frac{1}{2\pi(1000\text{ Hz})(0.01\ \mu\text{F})} = 15.9\text{ k}\Omega$$

The total impedance in rectangular form is

$$\mathbf{Z} = R - jX_C = 10\text{ k}\Omega - j15.9\text{ k}\Omega$$

Converting to polar form yields

$$\mathbf{Z} = \sqrt{R^2 + X_C^2}\angle -\tan^{-1}\left(\frac{X_C}{R}\right)$$

$$= \sqrt{(10\text{ k}\Omega)^2 + (15.9\text{ k}\Omega)^2}\angle -\tan^{-1}\left(\frac{15.9\text{ k}\Omega}{10\text{ k}\Omega}\right) = 18.8\angle -57.8^\circ\text{ k}\Omega$$

Use Ohm's law to determine the source voltage.

$$\mathbf{V}_s = \mathbf{IZ} = (0.2\angle 0^\circ\text{ mA})(18.8\angle -57.8^\circ\text{ k}\Omega) = 3.76\angle -57.8^\circ\text{ V}$$

The magnitude of the source voltage is 3.76 V at an angle of −57.8° with respect to the current; that is, the voltage lags the current by 57.8°, as shown in the phasor diagram of Figure 16–8.

**FIGURE 16–8**

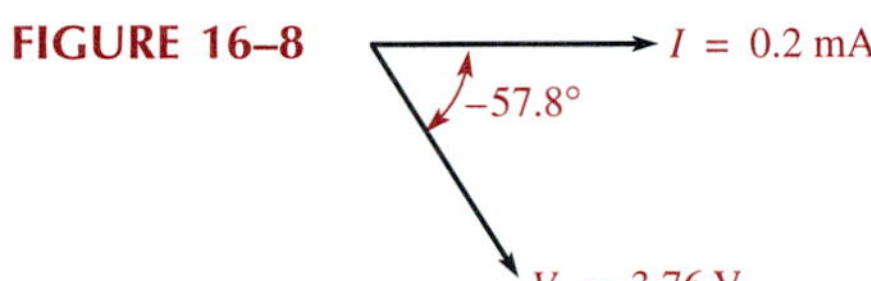


***Related Exercise*** Determine $\mathbf{V}_s$ in Figure 16–7 if $f = 2$ kHz and $\mathbf{I} = 0.2\angle 0^\circ$ A.

**EXAMPLE 16–3** Determine the current in the circuit of Figure 16–9, and draw the phasor diagram.

**FIGURE 16–9**

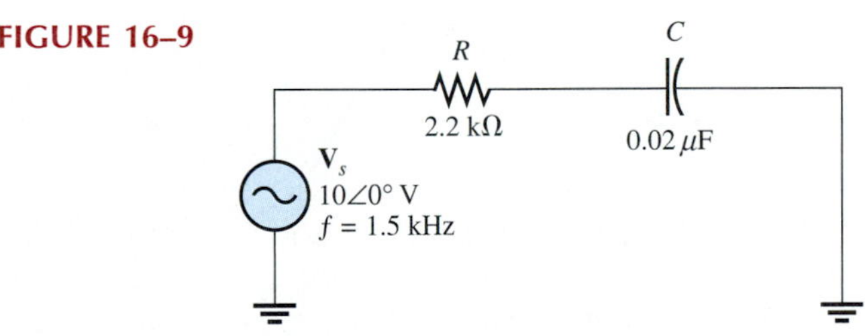


***Solution*** The magnitude of the capacitive reactance is

$$X_C = \frac{1}{2\pi fC} = \frac{1}{2\pi(1.5\text{ kHz})(0.02\ \mu\text{F})} = 5.3\text{ k}\Omega$$

The total impedance in rectangular form is

$$\mathbf{Z} = R - jX_C = 2.2\text{ k}\Omega - j5.3\text{ k}\Omega$$

Converting to polar form yields

$$\mathbf{Z} = \sqrt{R^2 + X_C^2}\angle -\tan^{-1}\left(\frac{X_C}{R}\right)$$

$$= \sqrt{(2.2\text{ k}\Omega)^2 + (5.3\text{ k}\Omega)^2}\angle -\tan^{-1}\left(\frac{5.3\text{ k}\Omega}{2.2\text{ k}\Omega}\right) = 5.74\angle -67.5^\circ\text{ k}\Omega$$

Use Ohm's law to determine the current.

$$\mathbf{I} = \frac{\mathbf{V}}{\mathbf{Z}} = \frac{10\angle 0^\circ \text{ V}}{5.74\angle -67.5^\circ \text{ k}\Omega} = 1.74\angle 67.5^\circ \text{ mA}$$

The magnitude of the current is 1.74 mA. The positive phase angle of 67.5° indicates that the current leads the voltage by that amount, as shown in the phasor diagram of Figure 16–10.

FIGURE 16–10

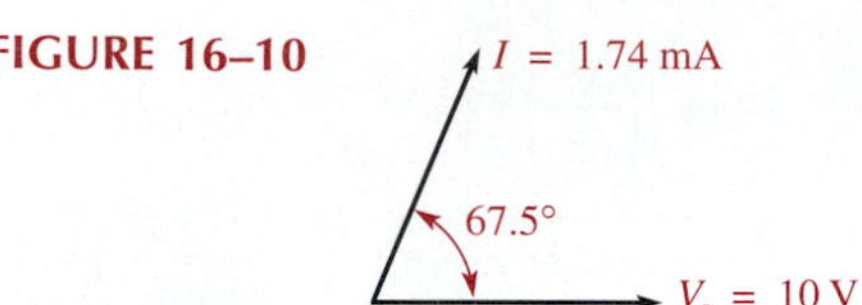


***Related Exercise*** Determine **I** in Figure 16–9 if the frequency is increased to 5 kHz.

## Relationships of the Current and Voltages in a Series *RC* Circuit

In a series circuit, the current is the same through both the resistor and the capacitor. Thus, the resistor voltage is in phase with the current, and the capacitor voltage lags the current by 90°. Therefore, there is a phase difference of 90° between the resistor voltage, $V_R$, and the capacitor voltage, $V_C$, as shown in the waveform diagram of Figure 16–11.

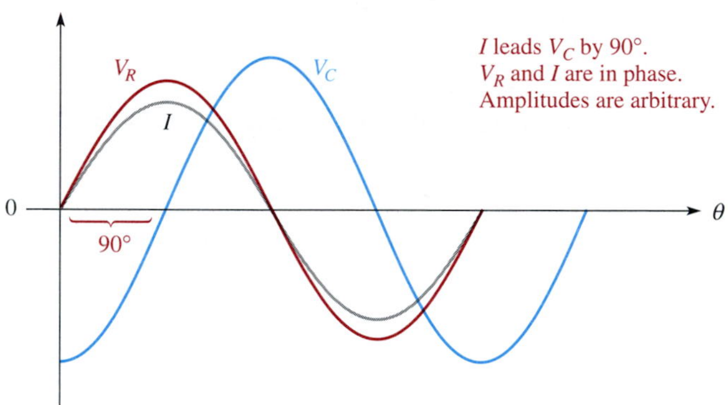


FIGURE 16–11
***Phase relation of voltages and current in a series RC circuit.***

You know from Kirchhoff's voltage law that the sum of the voltage drops must equal the applied voltage. However, since $V_R$ and $V_C$ are not in phase with each other, they must be added as phasor quantities, with $V_C$ lagging $V_R$ by 90°, as shown in Figure 16–12(a). As shown in Figure 16–12(b), $\mathbf{V}_s$ is the phasor sum of $V_R$ and $V_C$, as expressed in rectangular form in the following equation:

$$\mathbf{V}_s = V_R - jV_C \qquad \textbf{(16–9)}$$

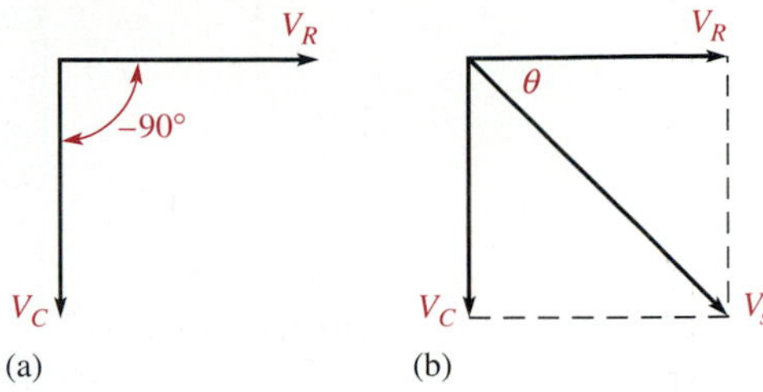


**FIGURE 16–12**
***Voltage phasor diagram for a series RC circuit.***

This equation can be expressed in polar form as

$$\mathbf{V}_s = \sqrt{V_R^2 + V_C^2}\angle -\tan^{-1}\left(\frac{V_C}{V_R}\right) \quad \textbf{(16–10)}$$

where the magnitude of the source voltage is

$$V_s = \sqrt{V_R^2 + V_C^2} \quad \textbf{(16–11)}$$

and the phase angle between the resistor voltage and the source voltage is

$$\theta = -\tan^{-1}\left(\frac{V_C}{V_R}\right) \quad \textbf{(16–12)}$$

Since the resistor voltage and the current are in phase, $\theta$ also represents the phase angle between the source voltage and the current. Figure 16–13 shows a complete voltage and current phasor diagram that represents the waveform diagram of Figure 16–11.

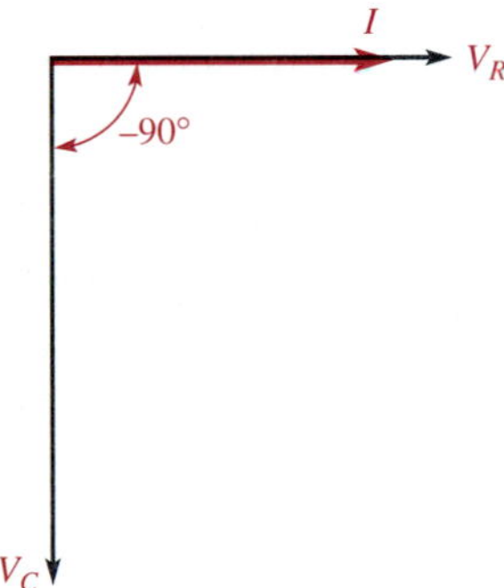


**FIGURE 16–13**
***Voltage and current phasor diagram for the waveforms in Figure 16–11.***

## Variation of Impedance with Frequency

As you know, capacitive reactance varies inversely with frequency. Since $Z = \sqrt{R^2 + X_C^2}$, you can see that when $X_C$ increases, the entire term under the square root sign increases and thus the magnitude of the total impedance also increases; and when $X_C$ decreases, the magnitude of the total impedance also decreases. Therefore, *in an RC circuit, Z is inversely dependent on frequency.*

Figure 16–14 illustrates how the voltages and current in a series *RC* circuit vary as the frequency increases or decreases, with the source voltage held at a constant value. In part (a), as the frequency is increased, $X_C$ decreases; so less voltage is dropped across the capacitor. Also, $Z$ decreases as $X_C$ decreases, causing the current to increase. An increase in the current causes more voltage across $R$.

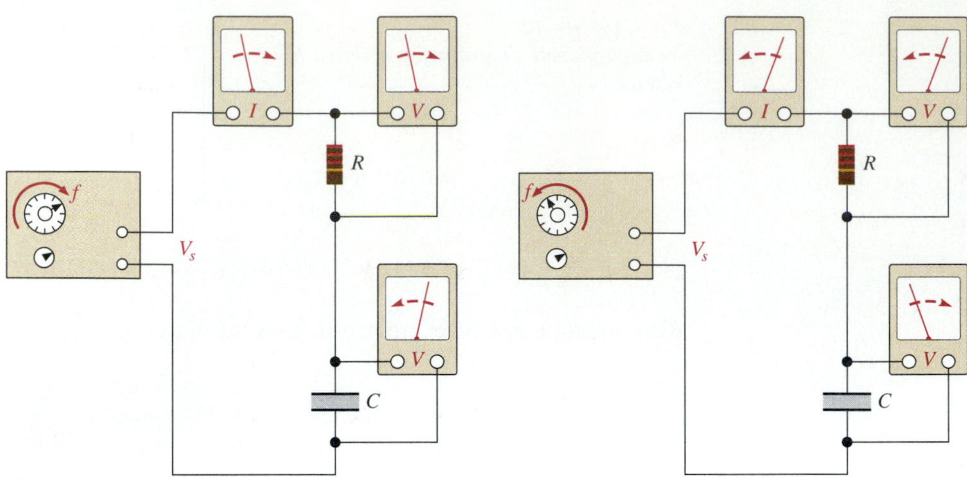


(a) As frequency is increased, $I$ and $V_R$ increase and $V_C$ decreases.

(b) As frequency is decreased, $I$ and $V_R$ decrease and $V_C$ increases.

**FIGURE 16–14**

***An illustration of how the variation of impedance affects the voltages and current as the source frequency is varied. The source voltage is held at a constant amplitude.***

In Figure 16–14(b), as the frequency is decreased, $X_C$ increases; so more voltage is dropped across the capacitor. Also, $Z$ increases as $X_C$ increases, causing the current to decrease. A decrease in the current causes less voltage across $R$.

Changes in $Z$ and $X_C$ can be observed as shown in Figure 16–15. As the frequency increases, the voltage across $Z$ remains constant because $V_s$ is constant. Also, the voltage across $C$ decreases. The increasing current indicates that $Z$ is decreasing. It does so because of the inverse relationship stated in Ohm's law ($Z = V_Z/I$). The increasing current also indicates that $X_C$ is decreasing ($X_C = V_C/I$). The decrease in $V_C$ corresponds to the decrease in $X_C$.

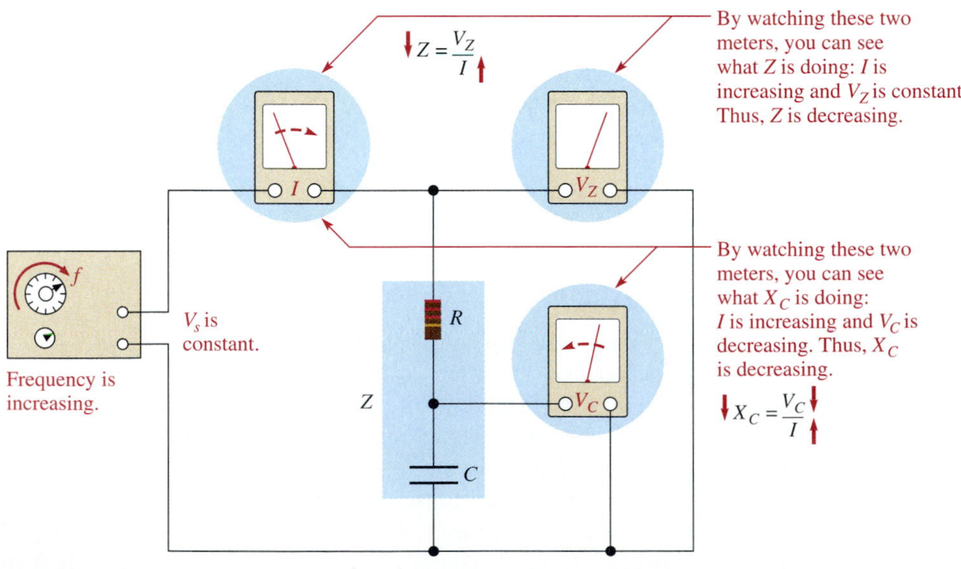


**FIGURE 16–15**

***An illustration of how Z and $X_C$ change with frequency.***

## Variation of the Phase Angle with Frequency

Since $X_C$ is the factor that introduces the phase angle in a series *RC* circuit, a change in $X_C$ produces a change in the phase angle. As the frequency is increased, $X_C$ becomes smaller, and thus the phase angle decreases. As the frequency is decreased, $X_C$ becomes larger, and thus the phase angle increases. This variation is illustrated in Figure 16–16 where a phase meter is connected across the capacitor to measure the angle between $V_s$ and $V_R$. This angle is the phase angle of the circuit because *I* is in phase with $V_R$. By measuring the phase of $V_R$, you are effectively measuring the phase of *I*.

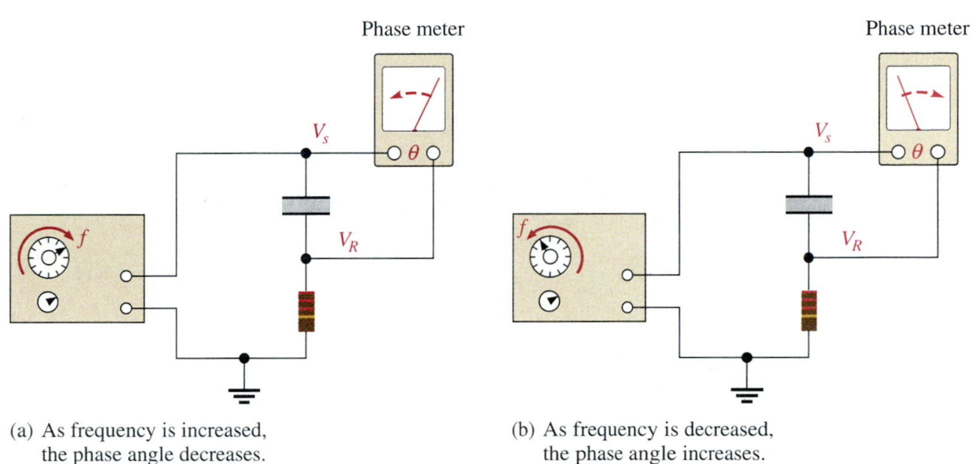


(a) As frequency is increased, the phase angle decreases.

(b) As frequency is decreased, the phase angle increases.

**FIGURE 16–16**

***The phase meter shows the effect of frequency on the phase angle of a circuit. The phase meter indicates the phase angle between two voltages, $V_s$ and $V_R$. Since $V_R$ and I are in phase, this angle is the same as the angle between $V_s$ and I.***

Figure 16–17 uses the impedance triangle to illustrate the variations in $X_C$, *Z*, and $\theta$ as the frequency changes. Of course, *R* remains constant. The key point is that because $X_C$ varies inversely with the frequency, so also do the magnitude of the total impedance and the phase angle. Example 16–4 illustrates this.

**FIGURE 16–17**

***As the frequency increases, $X_C$ decreases, Z decreases, and $\theta$ decreases. Each value of frequency can be visualized as forming a different impedance triangle.***

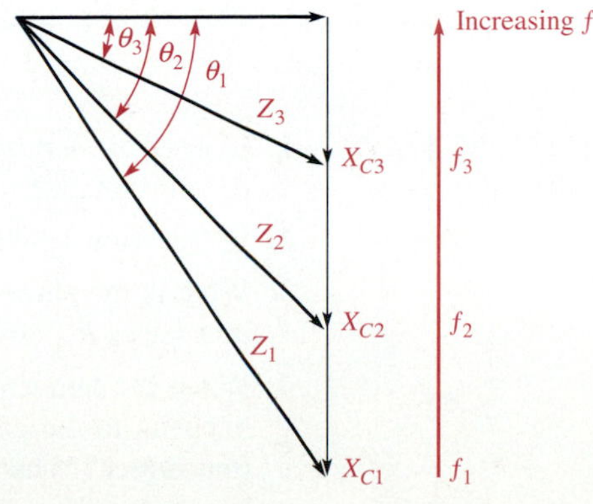

**EXAMPLE 16–4** For the series *RC* circuit in Figure 16–18, determine the magnitude of the total impedance and the phase angle for each of the following values of input frequency:
**(a)** 10 kHz **(b)** 20 kHz **(c)** 30 kHz

FIGURE 16–18

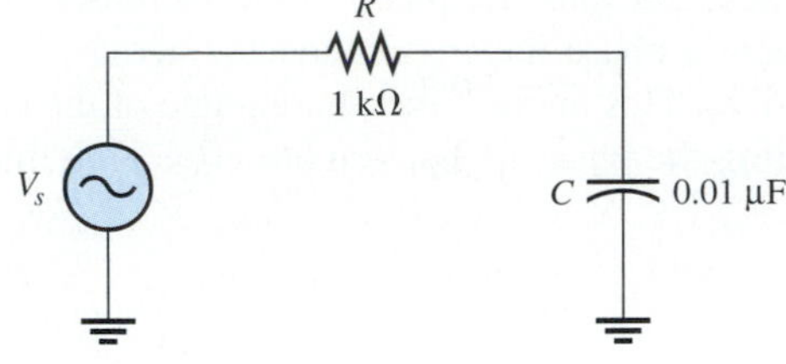


***Solution***

**(a)** For $f = 10$ kHz,

$$X_C = \frac{1}{2\pi fC} = \frac{1}{2\pi(10\text{ kHz})(0.01\ \mu\text{F})} = 1.59\text{ k}\Omega$$

$$\mathbf{Z} = \sqrt{R^2 + X_C^2}\angle -\tan^{-1}\left(\frac{X_C}{R}\right)$$

$$= \sqrt{(1\text{ k}\Omega)^2 + (1.59\text{ k}\Omega)^2}\angle -\tan^{-1}\left(\frac{1.59\text{ k}\Omega}{1\text{ k}\Omega}\right) = 1.88\angle -57.9^\circ\text{ k}\Omega$$

Thus, $Z = 1.88$ kΩ and $\theta = -57.9^\circ$.

**(b)** For $f = 20$ kHz,

$$X_C = \frac{1}{2\pi(20\text{ kHz})(0.01\ \mu\text{F})} = 796\ \Omega$$

$$\mathbf{Z} = \sqrt{(1\text{ k}\Omega)^2 + (796\ \Omega)^2}\angle -\tan^{-1}\left(\frac{796\ \Omega}{1\text{ k}\Omega}\right) = 1.28\angle -38.5^\circ\text{ k}\Omega$$

Thus, $Z = 1.28$ kΩ and $\theta = -38.5^\circ$.

**(c)** For $f = 30$ kHz,

$$X_C = \frac{1}{2\pi(30\text{ kHz})(0.01\ \mu\text{F})} = 531\ \Omega$$

$$\mathbf{Z} = \sqrt{(1\text{ k}\Omega)^2 + (531\ \Omega)^2}\angle -\tan^{-1}\left(\frac{531\ \Omega}{1\text{ k}\Omega}\right) = 1.13\angle -28.0^\circ\text{ k}\Omega$$

Thus, $Z = 1.13$ kΩ and $\theta = -28.0^\circ$.

Notice that as the frequency increases, $X_C$, $Z$, and $\theta$ decrease.

***Related Exercise*** Find the magnitude of the total impedance and the phase angle in Figure 16–18 for $f = 1$ kHz.

**SECTION 16–3 REVIEW**

1. In a certain series *RC* circuit, $V_R = 4$ V, and $V_C = 6$ V. What is the magnitude of the source voltage?
2. In Question 1, what is the phase angle between the source voltage and the current?
3. What is the phase difference between the capacitor voltage and the resistor voltage in a series *RC* circuit?
4. When the frequency of the applied voltage in a series *RC* circuit is increased, what happens to the capacitive reactance? What happens to the magnitude of the total impedance? What happens to the phase angle?

## 16–4 ■ IMPEDANCE AND PHASE ANGLE OF PARALLEL *RC* CIRCUITS

***In this section, you will learn how to determine the impedance and phase angle of a parallel RC circuit. Also, capacitive susceptance and admittance of a parallel RC circuit are introduced.***

***After completing this section, you should be able to***

- **Determine impedance and phase angle in a parallel *RC* circuit**
  - ☐ Express total impedance in complex form
  - ☐ Define and calculate *conductance, capacitive susceptance,* and *admittance*

Figure 16–19 shows a basic parallel *RC* circuit connected to an ac voltage source.

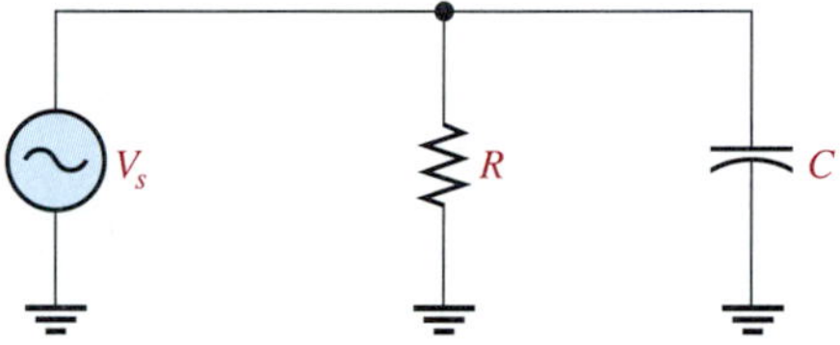


FIGURE 16–19
***Basic parallel RC circuit.***

The expression for the total impedance is developed as follows, using the rules of phasor algebra.

$$\mathbf{Z} = \frac{(R\angle 0°)(X_C\angle -90°)}{R - jX_C}$$

By multiplying the magnitudes, adding the angles in the numerator, and converting the denominator to polar form, we get

$$\mathbf{Z} = \frac{RX_C\angle(0° - 90°)}{\sqrt{R^2 + X_C^2}\angle -\tan^{-1}\left(\frac{X_C}{R}\right)}$$

Now, dividing the magnitude expression in the numerator by that in the denominator, and by subtracting the angle in the denominator from that in the numerator, we get

$$\mathbf{Z} = \left(\frac{RX_C}{\sqrt{R^2 + X_C^2}}\right)\angle\left(-90° + \tan^{-1}\left(\frac{X_C}{R}\right)\right) \quad \textbf{(16–13)}$$

Equation (16–13) is the expression for the total parallel impedance, where the magnitude is

$$Z = \frac{RX_C}{\sqrt{R^2 + X_C^2}} \quad \textbf{(16–14)}$$

and the phase angle between the applied voltage and the total current is

$$\theta = -90° + \tan^{-1}\left(\frac{X_C}{R}\right) \quad \textbf{(16–15)}$$

Equivalently, this expression can be written as

$$\theta = \tan^{-1}\left(\frac{R}{X_C}\right)$$

**EXAMPLE 16–5** For each circuit in Figure 16–20, determine the magnitude of the total impedance and the phase angle.

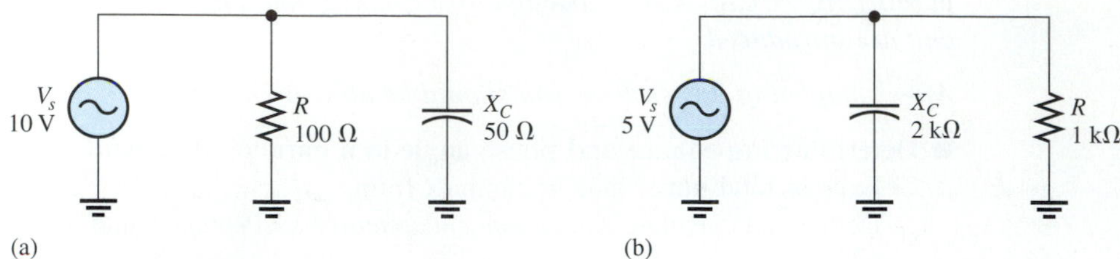


**FIGURE 16–20**

***Solution*** For the circuit in Figure 16–20(a), the total impedance is

$$\mathbf{Z} = \left(\frac{RX_C}{\sqrt{R^2 + X_C^2}}\right)\angle\left(-90^\circ + \tan^{-1}\left(\frac{X_C}{R}\right)\right)$$

$$= \left[\frac{(100\ \Omega)(50\ \Omega)}{\sqrt{(100\ \Omega)^2 + (50\ \Omega)^2}}\right]\angle\left(-90^\circ + \tan^{-1}\left(\frac{50\ \Omega}{100\ \Omega}\right)\right) = 44.7\angle -63.4^\circ\ \Omega$$

Thus, $Z = 44.7\ \Omega$ and $\theta = -63.4^\circ$.

For the circuit in Figure 16–20(b), the total impedance is

$$\mathbf{Z} = \left[\frac{(1\ \text{k}\Omega)(2\ \text{k}\Omega)}{\sqrt{(1\ \text{k}\Omega)^2 + (2\ \text{k}\Omega)^2}}\right]\angle\left(-90^\circ + \tan^{-1}\left(\frac{2\ \text{k}\Omega}{1\ \text{k}\Omega}\right)\right) = 894\angle -26.6^\circ\ \Omega$$

Thus, $Z = 894\ \Omega$ and $\theta = -26.6^\circ$.

***Related Exercise*** Determine **Z** in Figure 16–20(a) if the frequency is doubled.

## Conductance, Susceptance, and Admittance

Recall that **conductance**, *G*, is the reciprocal of resistance. The phasor expression for conductance is expressed as

$$\mathbf{G} = \frac{1}{R\angle 0^\circ} = G\angle 0^\circ \qquad \textbf{(16–16)}$$

Two new terms are now introduced for use in parallel *RC* circuits. **Capacitive susceptance** ($B_C$) is the reciprocal of capacitive reactance and the phasor expression for capacitive susceptance is

$$\mathbf{B}_C = \frac{1}{X_C\angle -90^\circ} = B_C\angle 90^\circ = +jB_C \qquad \textbf{(16–17)}$$

**Admittance** (*Y*) is the reciprocal of impedance and the phasor expression for admittance is

$$\mathbf{Y} = \frac{1}{Z\angle \pm\theta} = Y\angle \mp\theta \qquad \textbf{(16–18)}$$

The unit of each of these terms is the siemens (S), which is the reciprocal of the ohm.

In working with parallel circuits, it is often easier to use $G$, $B_C$, and $Y$ rather than $R$, $X_C$, and $Z$. In a parallel $RC$ circuit, as shown in Figure 16–21, the total admittance is simply the phasor sum of the conductance and the capacitive susceptance.

$$\mathbf{Y} = G + jB_C \quad (16\text{–}19)$$

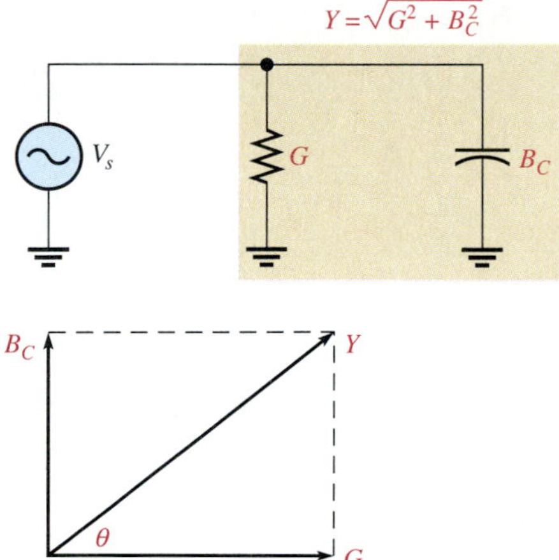


**FIGURE 16–21**
***Admittance in a parallel RC circuit.***

**EXAMPLE 16–6** Determine the admittance (**Y**) and impedance (**Z**) in Figure 16–22. Sketch the admittance phasor diagram.

**FIGURE 16–22**

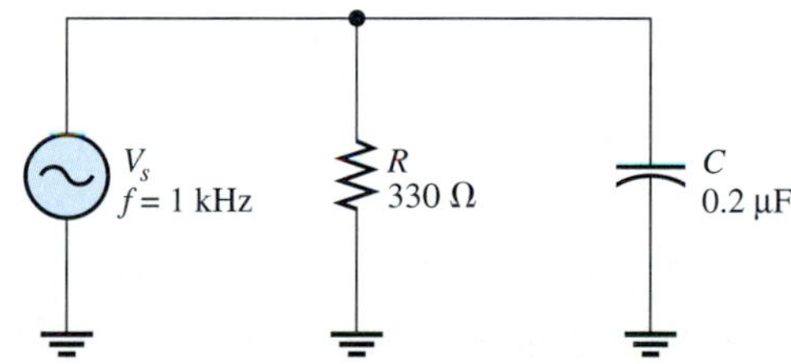


***Solution*** From Figure 16–22, $R = 330\ \Omega$; thus $G = 1/R = 1/330\ \Omega = 3.03$ mS. The capacitive reactance is

$$X_C = \frac{1}{2\pi fC} = \frac{1}{2\pi(1000\text{ Hz})(0.2\ \mu\text{F})} = 796\ \Omega$$

The capacitive susceptance magnitude is

$$B_C = \frac{1}{X_C} = \frac{1}{796\ \Omega} = 1.26\text{ mS}$$

The total admittance is

$$\mathbf{Y} = G + jB_C = 3.03\text{ mS} + j1.26\text{ mS}$$

which can be expressed in polar form as

$$\mathbf{Y} = \sqrt{G^2 + B_C^2}\angle\tan^{-1}\left(\frac{B_C}{G}\right)$$

$$= \sqrt{(3.03\text{ mS})^2 + (1.26\text{ mS})^2}\angle\tan^{-1}\left(\frac{1.26\text{ mS}}{3.03\text{ mS}}\right) = 3.28\angle 22.6^\circ\text{ mS}$$

Admittance is converted to impedance as follows:

$$\mathbf{Z} = \frac{1}{\mathbf{Y}} = \frac{1}{(3.28\angle 22.6^\circ \text{ mS})} = 305\angle -22.6^\circ \ \Omega$$

The admittance phasor diagram is shown in Figure 16–23.

**FIGURE 16–23**

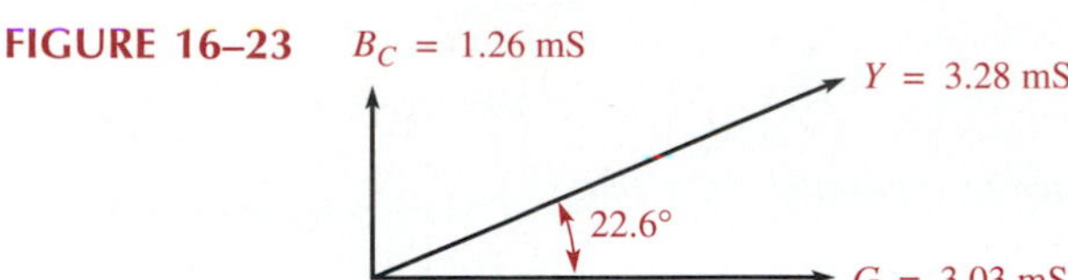


***Related Exercise*** Calculate the admittance in Figure 16–22 if $f$ is increased to 2.5 kHz.

**SECTION 16–4 REVIEW**

1. Define *conductance, capacitive susceptance,* and *admittance.*
2. If $Z = 100\ \Omega$, what is the value of $Y$?
3. In a certain parallel *RC* circuit, $R = 47\ \Omega$ and $X_C = 75\ \Omega$. Determine **Y**.
4. In Question 3, what is the magnitude of **Y**, and what is the phase angle between the total current and the applied voltage?

## 16–5 ■ ANALYSIS OF PARALLEL *RC* CIRCUITS

***In the previous section, you learned how to express the impedance of a parallel RC circuit. Now, Ohm's law and Kirchhoff's current law are used in the analysis of RC circuits. Current and voltage relationships in a parallel RC circuit are examined.***

***After completing this section, you should be able to***

- **Analyze a parallel *RC* circuit**
  - ☐ Apply Ohm's law and Kirchhoff's current law to parallel *RC* circuits
  - ☐ Express the voltages and currents as phasor quantities
  - ☐ Show how impedance and phase angle vary with frequency
  - ☐ Convert from a parallel circuit to an equivalent series circuit

For convenience in the analysis of parallel circuits, the Ohm's law formulas using impedance, previously stated, can be rewritten for admittance using the relation $Y = 1/Z$. Remember, the use of boldface nonitalic letters indicates phasor quantities.

$$\mathbf{V} = \frac{\mathbf{I}}{\mathbf{Y}} \qquad \textbf{(16–20)}$$

$$\mathbf{I} = \mathbf{VY} \qquad \textbf{(16–21)}$$

$$\mathbf{Y} = \frac{\mathbf{I}}{\mathbf{V}} \qquad \textbf{(16–22)}$$

**EXAMPLE 16–7** Determine the total current and phase angle in Figure 16–24. Draw a phasor diagram showing the relationship of $\mathbf{V}_s$ and $\mathbf{I}_{tot}$.

FIGURE 16–24

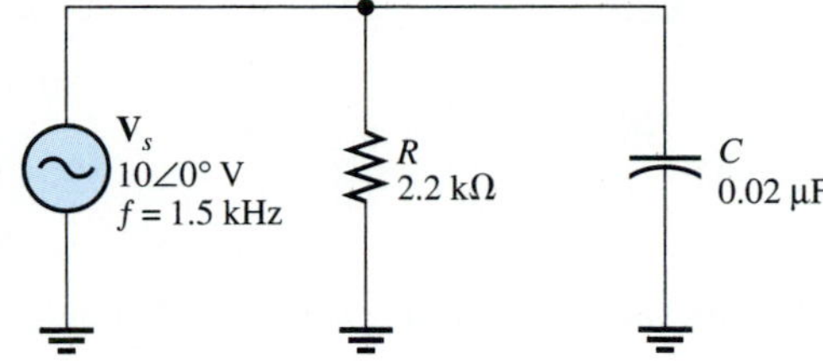


***Solution*** The capacitive reactance is

$$X_C = \frac{1}{2\pi f C} = \frac{1}{2\pi(1.5\text{ kHz})(0.02\ \mu\text{F})} = 5.31\text{ k}\Omega$$

The capacitive susceptance magnitude is

$$B_C = \frac{1}{X_C} = \frac{1}{5.31\text{ k}\Omega} = 188\ \mu\text{S}$$

The conductance magnitude is

$$G = \frac{1}{R} = \frac{1}{2.2\text{ k}\Omega} = 455\ \mu\text{S}$$

The total admittance is

$$\mathbf{Y} = G + jB_C = 455\ \mu\text{S} + j188\ \mu\text{S}$$

Converting to polar form yields

$$\mathbf{Y} = \sqrt{G^2 + B_C^2}\angle\tan^{-1}\left(\frac{B_C}{G}\right)$$

$$= \sqrt{(455\ \mu\text{S})^2 + (188\ \mu\text{S})^2}\angle\tan^{-1}\left(\frac{188\ \mu\text{S}}{455\ \mu\text{S}}\right) = 492\angle 22.5^\circ\ \mu\text{S}$$

The phase angle is 22.5°.

Use Ohm's law to determine the total current.

$$\mathbf{I}_{tot} = \mathbf{V}_s\mathbf{Y} = (10\angle 0^\circ\text{ V})(492\angle 22.5^\circ\ \mu\text{S}) = 4.92\angle 22.5^\circ\text{ mA}$$

The magnitude of the total current is 4.92 mA, and it leads the applied voltage by 22.5°, as the phasor diagram in Figure 16–25 indicates.

FIGURE 16–25

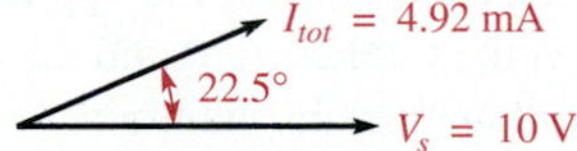


***Related Exercise*** What is the total current (in polar form) if *f* is doubled?

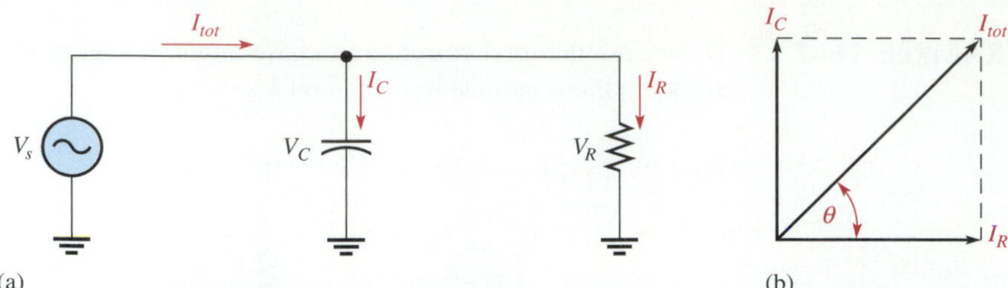


**FIGURE 16–26**
***Currents and voltages in a parallel RC circuit.***

## Relationships of the Currents and Voltages in a Parallel *RC* Circuit

Figure 16–26(a) shows all the currents and voltages in a basic parallel *RC* circuit. As you can see, the applied voltage, $V_s$, appears across both the resistive and the capacitive branches, so $V_s$, $V_R$, and $V_C$ are all in phase and of the same magnitude. The total current, $I_{tot}$, divides at the junction into the two branch currents, $I_R$ and $I_C$.

The current through the resistor is in phase with the voltage. The current through the capacitor leads the voltage, and thus the resistive current, by 90°. By Kirchhoff's current law, the total current is the phasor sum of the two branch currents, as shown by the phasor diagram in Figure 16–26(b). The total current is expressed as

$$\mathbf{I}_{tot} = I_R + jI_C \quad \textbf{(16–23)}$$

This equation can be expressed in polar form as

$$\mathbf{I}_{tot} = \sqrt{I_R^2 + I_C^2}\angle\tan^{-1}\left(\frac{I_C}{I_R}\right) \quad \textbf{(16–24)}$$

where the magnitude of the total current is

$$I_{tot} = \sqrt{I_R^2 + I_C^2} \quad \textbf{(16–25)}$$

and the phase angle between the resistor current and the total current is

$$\theta = \tan^{-1}\left(\frac{I_C}{I_R}\right) \quad \textbf{(16–26)}$$

Since the resistor current and the applied voltage are in phase, $\theta$ also represents the phase angle between the total current and the applied voltage. Figure 16–27 shows a complete current and voltage phasor diagram.

**FIGURE 16–27**
***Current and voltage phasor diagram for a parallel RC circuit (amplitudes are arbitrary).***

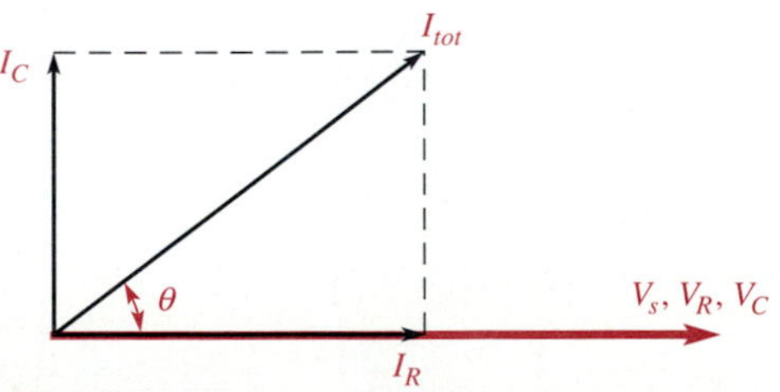

**EXAMPLE 16–8** Determine the value of each current in Figure 16–28, and describe the phase relationship of each with the applied voltage. Draw the phasor diagram.

**FIGURE 16–28**

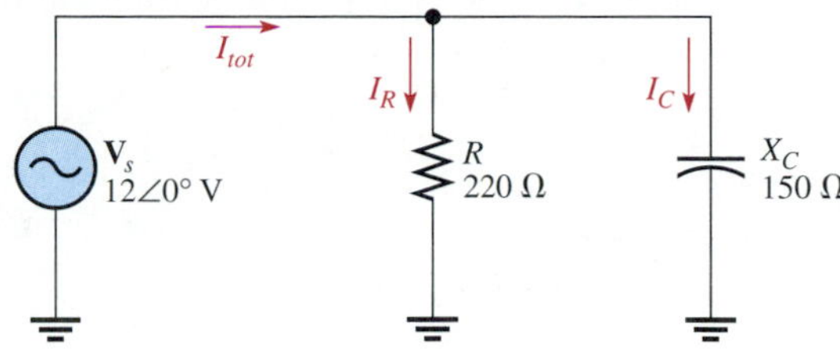


***Solution*** The resistor current, the capacitor current, and the total current are expressed as follows:

$$\mathbf{I}_R = \frac{\mathbf{V}_s}{\mathbf{R}} = \frac{12\angle 0^\circ \text{ V}}{220\angle 0^\circ \ \Omega} = 54.5\angle 0^\circ \text{ mA}$$

$$\mathbf{I}_C = \frac{\mathbf{V}_s}{\mathbf{X}_C} = \frac{12\angle 0^\circ \text{ V}}{150\angle -90^\circ \ \Omega} = 80\angle 90^\circ \text{ mA}$$

$$\mathbf{I}_{tot} = I_R + jI_C = 54.5 \text{ mA} + j80 \text{ mA}$$

Converting $\mathbf{I}_{tot}$ to polar form yields

$$\mathbf{I}_{tot} = \sqrt{I_R^2 + I_C^2}\angle \tan^{-1}\left(\frac{I_C}{I_R}\right)$$

$$= \sqrt{(54.5 \text{ mA})^2 + (80 \text{ mA})^2}\angle \tan^{-1}\left(\frac{80 \text{ mA}}{54.5 \text{ mA}}\right) = 96.8\angle 55.7^\circ \text{ mA}$$

As the results show, the resistor current is 54.5 mA and is in phase with the voltage. The capacitor current is 80 mA and leads the voltage by 90°. The total current is 96.8 mA and leads the voltage by 55.7°. The phasor diagram in Figure 16–29 illustrates these relationships.

**FIGURE 16–29**

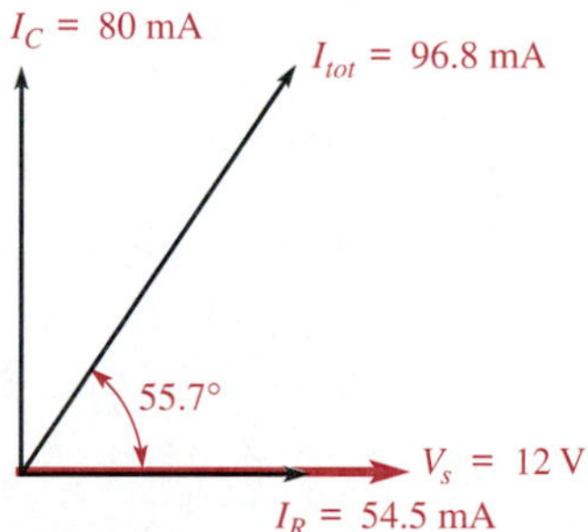


***Related Exercise*** In a parallel circuit, $\mathbf{I}_R = 100\angle 0^\circ$ mA and $\mathbf{I}_C = 60\angle 90^\circ$ mA. Determine the total current.

## Conversion from Parallel to Series Form

For every parallel *RC* circuit, there is an equivalent series *RC* circuit. Two circuits are considered equivalent when they both present an equal impedance at their terminals; that is, the magnitude of impedance and the phase angle are identical.

To obtain the equivalent series circuit for a given parallel $RC$ circuit, first find the impedance and phase angle of the parallel circuit. Then use the values of $Z$ and $\theta$ to construct an impedance triangle shown in Figure 16–30. The vertical and horizontal sides of the triangle represent the equivalent series resistance and capacitive reactance as indicated. These values can be found using the following trigonometric relationships:

$$R_{eq} = Z \cos \theta \quad \textbf{(16–27)}$$

$$X_{C(eq)} = Z \sin \theta \quad \textbf{(16–28)}$$

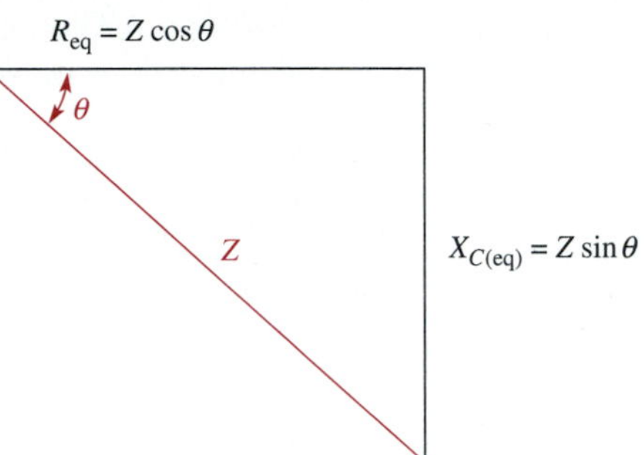


**FIGURE 16–30**
***Impedance triangle for the series equivalent of a parallel RC circuit. Z and θ are the known values for the parallel circuit. $R_{eq}$ and $X_{C(eq)}$ are the series equivalent values.***

## EXAMPLE 16–9

Convert the parallel circuit in Figure 16–31 to a series form.

**FIGURE 16–31**

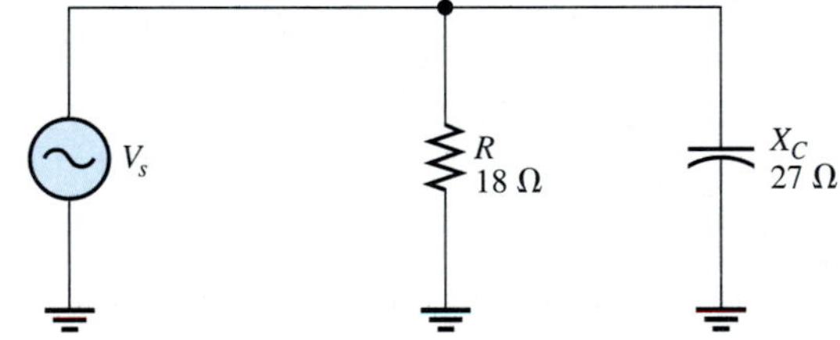


***Solution*** First, find the admittance of the parallel circuit as follows:

$$G = \frac{1}{R} = \frac{1}{18\ \Omega} = 55.6\ \text{mS}$$

$$B_C = \frac{1}{X_C} = \frac{1}{27\ \Omega} = 37.0\ \text{mS}$$

$$\mathbf{Y} = G + jB_C = 55.6\ \text{mS} + j37.0\ \text{mS}$$

Converting to polar form yields

$$\mathbf{Y} = \sqrt{G^2 + B_C^2}\angle\tan^{-1}\left(\frac{B_C}{G}\right)$$

$$= \sqrt{(55.6\ \text{mS})^2 + (37.0\ \text{mS})^2}\angle\tan^{-1}\left(\frac{37.0\ \text{mS}}{55.6\ \text{mS}}\right) = 66.8\angle 33.6^\circ\ \text{mS}$$

Then, the total impedance is

$$\mathbf{Z} = \frac{1}{\mathbf{Y}} = \frac{1}{66.8\angle 33.6^\circ\ \text{mS}} = 15.0\angle -33.6^\circ\ \Omega$$

Converting to rectangular form yields

$$\mathbf{Z} = Z\cos\theta - jZ\sin\theta = R_{eq} - jX_{C(eq)}$$
$$= 15.0\cos(-33.6°) - j15.0\sin(33.6°) = 12.5\ \Omega - j8.31\ \Omega$$

The equivalent series *RC* circuit is a 12.5 Ω resistor in series with a capacitive reactance of 8.31 Ω. This is shown in Figure 16–32.

FIGURE 16–32

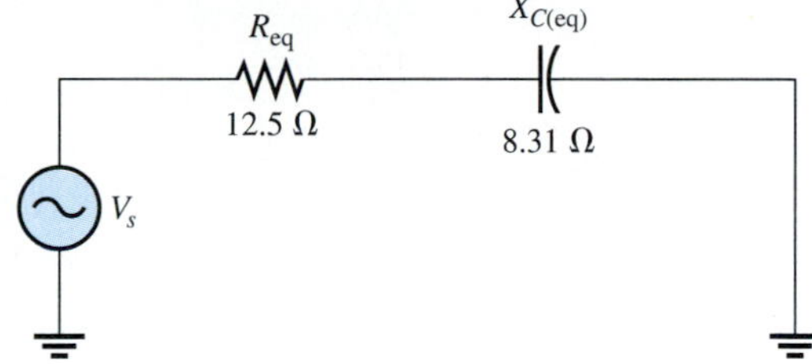


***Related Exercise*** The impedance of a parallel *RC* circuit is $\mathbf{Z} = 10\angle -26°$ kΩ. Convert to an equivalent series circuit.

**SECTION 16–5 REVIEW**

1. The admittance of an *RC* circuit is 3.50 mS, and the applied voltage is 6 V. What is the total current?
2. In a certain parallel *RC* circuit, the resistor current is 10 mA, and the capacitor current is 15 mA. Determine the magnitude and phase angle of the total current. This phase angle is measured with respect to what?
3. What is the phase angle between the capacitor current and the applied voltage in a parallel *RC* circuit?

## 16–6 ■ SERIES-PARALLEL *RC* CIRCUITS

***In this section, the concepts studied in the previous section are used to analyze circuits with combinations of both series and parallel R and C elements.***

***After completing this section, you should be able to***

- **Analyze series-parallel *RC* circuits**
  - ☐ Determine total impedance
  - ☐ Calculate currents and voltages
  - ☐ Measure impedance and phase angle

The following two examples demonstrate how to approach the analysis of complex *RC* networks.

**EXAMPLE 16–10** In the circuit of Figure 16–33, determine the following:
(**a**) total impedance (**b**) total current (**c**) phase angle by which $I_{tot}$ leads $V_s$

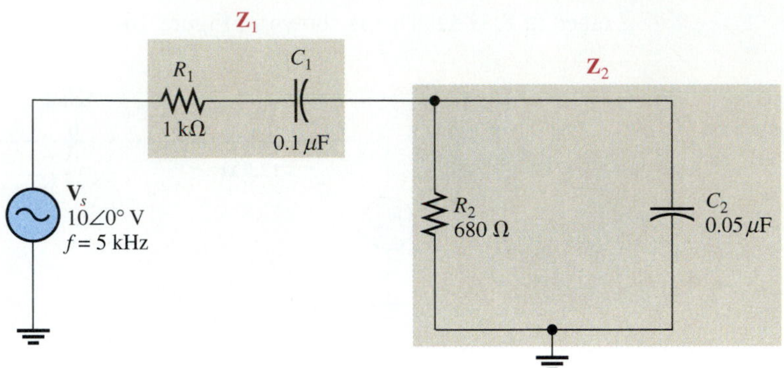


FIGURE 16–33

### *Solution*

(**a**) First, calculate the magnitudes of capacitive reactance.

$$X_{C1} = \frac{1}{2\pi fC} = \frac{1}{2\pi(5\text{ kHz})(0.1\ \mu\text{F})} = 318\ \Omega$$

$$X_{C2} = \frac{1}{2\pi fC} = \frac{1}{2\pi(5\text{ kHz})(0.05\ \mu\text{F})} = 637\ \Omega$$

One approach is to find the impedance of the series portion and the impedance of the parallel portion and combine them to get the total impedance. The impedance of the series combination of $R_1$ and $C_1$ is

$$\mathbf{Z}_1 = R_1 - jX_{C1} = 1\text{ k}\Omega - j318\ \Omega$$

To determine the impedance of the parallel portion, first determine the admittance of the parallel combination of $R_2$ and $C_2$.

$$G_2 = \frac{1}{R_2} = \frac{1}{680\ \Omega} = 1.47\text{ mS}$$

$$B_{C2} = \frac{1}{X_{C2}} = \frac{1}{637\ \Omega} = 1.57\text{ mS}$$

$$\mathbf{Y}_2 = G_2 + jB_{C2} = 1.47\text{ mS} + j1.57\text{ mS}$$

Converting to polar form yields

$$\mathbf{Y}_2 = \sqrt{G_2^2 + B_{C2}^2}\angle\tan^{-1}\left(\frac{B_{C2}}{G_2}\right)$$

$$= \sqrt{(1.47\text{ mS})^2 + (1.57\text{ mS})^2}\angle\tan^{-1}\left(\frac{1.57\text{ mS}}{1.47\text{ mS}}\right) = 2.15\angle 46.9^\circ\text{ mS}$$

Then, the impedance of the parallel portion is

$$\mathbf{Z}_2 = \frac{1}{\mathbf{Y}_2} = \frac{1}{2.15\angle 46.9^\circ\text{ mS}} = 465\angle -46.9^\circ\ \Omega$$

Converting to rectangular form yields

$$\mathbf{Z}_2 = Z_2\cos\theta - jZ_2\sin\theta$$
$$= 465\cos(-46.9^\circ) - j465\sin(-46.9^\circ) = 318\ \Omega - j339\ \Omega$$

The series portion and the parallel portion are in series with each other. Combine $\mathbf{Z}_1$ and $\mathbf{Z}_2$ to get the total impedance.

$$\mathbf{Z}_{tot} = \mathbf{Z}_1 + \mathbf{Z}_2$$
$$= (1\text{ k}\Omega - j318\ \Omega) + (318\ \Omega - j339\ \Omega) = 1318\ \Omega - j657\ \Omega$$

Expressing $\mathbf{Z}_{tot}$ in polar form yields

$$\mathbf{Z}_{tot} = \sqrt{Z_1^2 + Z_2^2}\angle -\tan^{-1}\left(\frac{Z_2}{Z_1}\right)$$
$$= \sqrt{(1318\ \Omega)^2 + (657\ \Omega)^2}\angle -\tan^{-1}\left(\frac{657\ \Omega}{1318\ \Omega}\right) = 1.47\angle -26.5^\circ\text{ k}\Omega$$

**(b)** Use Ohm's law to determine the total current.

$$\mathbf{I}_{tot} = \frac{\mathbf{V}_s}{\mathbf{Z}_{tot}} = \frac{10\angle 0^\circ\text{ V}}{1.47\angle -26.5^\circ\text{ k}\Omega} = 6.80\angle 26.5^\circ\text{ mA}$$

**(c)** The total current leads the applied voltage by 26.5°.

***Related Exercise*** Determine the voltages across $\mathbf{Z}_1$ and $\mathbf{Z}_2$ in Figure 16–33 and express in polar form.

## EXAMPLE 16–11

Determine all currents in Figure 16–34. Sketch a current phasor diagram.

FIGURE 16–34

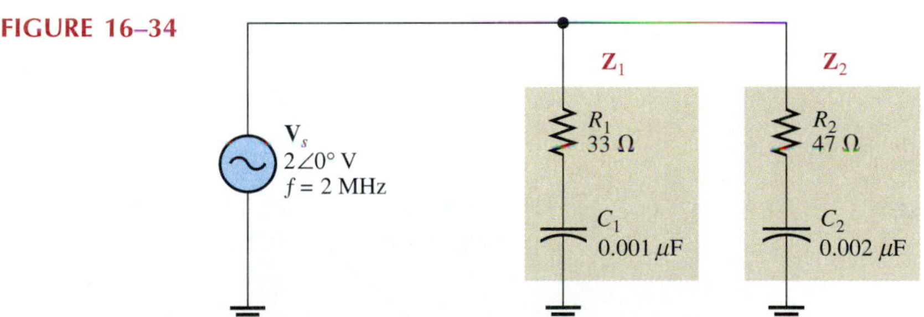


***Solution*** First, calculate $X_{C1}$ and $X_{C2}$.

$$X_{C1} = \frac{1}{2\pi fC} = \frac{1}{2\pi(2\text{ MHz})(0.001\ \mu\text{F})} = 79.6\ \Omega$$
$$X_{C2} = \frac{1}{2\pi fC} = \frac{1}{2\pi(2\text{ MHz})(0.002\ \mu\text{F})} = 39.8\ \Omega$$

Next, determine the impedance of each of the two parallel branches.

$$\mathbf{Z}_1 = R_1 - jX_{C1} = 33\ \Omega - j79.6\ \Omega$$
$$\mathbf{Z}_2 = R_2 - jX_{C2} = 47\ \Omega - j39.8\ \Omega$$

Convert these impedances to polar form.

$$\mathbf{Z}_1 = \sqrt{R_1^2 + X_{C1}^2}\angle -\tan^{-1}\left(\frac{X_{C1}}{R_1}\right)$$
$$= \sqrt{(33\ \Omega)^2 + (79.6\ \Omega)^2}\angle -\tan^{-1}\left(\frac{79.6\ \Omega}{33\ \Omega}\right) = 86.2\angle -67.5^\circ\ \Omega$$

$$\mathbf{Z}_2 = \sqrt{R_2^2 + X_{C2}^2}\angle -\tan^{-1}\left(\frac{X_{C2}}{R_2}\right)$$
$$= \sqrt{(47\ \Omega)^2 + (39.8\ \Omega)^2}\angle -\tan^{-1}\left(\frac{39.8\ \Omega}{47\ \Omega}\right) = 61.6\angle -40.3°\ \Omega$$

Calculate each branch current.

$$\mathbf{I}_1 = \frac{\mathbf{V}_s}{\mathbf{Z}_1} = \frac{2\angle 0°\ \text{V}}{86.2\angle -67.5°\ \Omega} = 23.2\angle 67.5°\ \text{mA}$$
$$\mathbf{I}_2 = \frac{\mathbf{V}_s}{\mathbf{Z}_2} = \frac{2\angle 0°\ \text{V}}{61.6\angle -40.3°\ \Omega} = 32.5\angle 40.3°\ \text{mA}$$

To get the total current, express each branch current in rectangular form so that they can be added.

$$\mathbf{I}_1 = 8.89\ \text{mA} + j21.4\ \text{mA}$$
$$\mathbf{I}_2 = 24.8\ \text{mA} + j21.0\ \text{mA}$$

The total current is

$$\mathbf{I}_{tot} = \mathbf{I}_1 + \mathbf{I}_2$$
$$= (8.89\ \text{mA} + j21.4\ \text{mA}) + (24.8\ \text{mA} + j21.0\ \text{mA}) = 33.7\ \text{mA} + j42.4\ \text{mA}$$

Converting $\mathbf{I}_{tot}$ to polar form yields

$$\mathbf{I}_{tot} = \sqrt{(33.7\ \text{mA})^2 + (42.4\ \text{mA})^2}\angle \tan^{-1}\left(\frac{42.4\ \Omega}{33.7\ \Omega}\right) = 54.2\angle 51.6°\ \text{mA}$$

The current phasor diagram is shown in Figure 16–35.

FIGURE 16–35

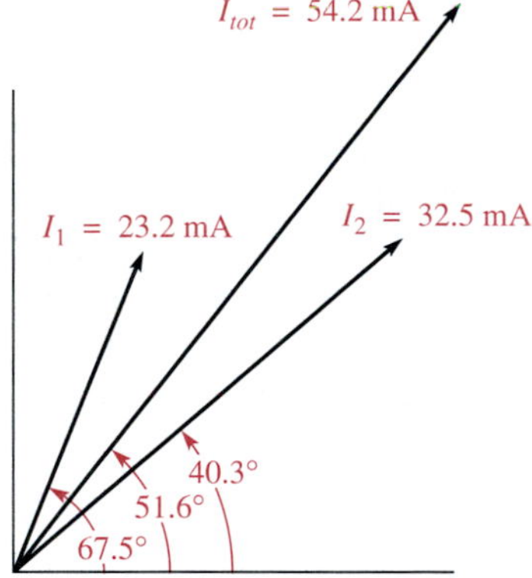


***Related Exercise*** Determine the voltages across each component in Figure 16–34 and sketch a voltage phasor diagram.

## Measurement of $Z_{tot}$ and $\theta$

Now, let's see how the values of $Z_{tot}$ and $\theta$ for the circuit in Example 16–10 can be determined by measurement. First, the total impedance is measured as outlined in the following steps and as illustrated in Figure 16–36 (other ways are also possible):

**Step 1:** Using a sine wave generator, set the source voltage to a known value (10 V) and the frequency to 5 kHz. It is advisable to check the voltage with an ac voltmeter and the frequency with a frequency counter rather than relying on the marked values on the generator controls.

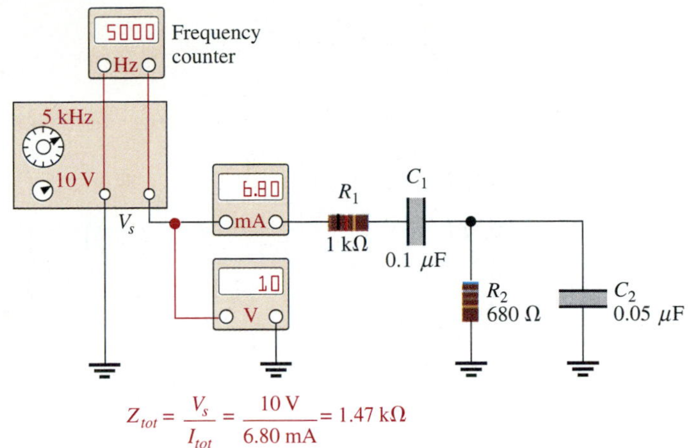


**FIGURE 16–36**
***Determining $Z_{tot}$ by measurement of $V_s$ and $I_{tot}$.***

**Step 2:** Connect an ac ammeter as shown in Figure 16–36, and measure the total current.

**Step 3:** Calculate the total impedance by using Ohm's law.

Although you could use a phase meter to measure the phase angle, we will use an oscilloscope in this illustration because it is more commonly available. The basic method of measuring a phase angle on an oscilloscope was introduced in the TECH TIP in Chapter 12. We will use that method here in an *RC* circuit.

To measure the phase angle, the source voltage and the total current must be displayed on the screen in the proper time relationship. Two basic types of scope probes are available to measure the quantities with an oscilloscope: the voltage probe and the current probe. Although the current probe is a convenient device, it is often not as readily available as a voltage probe. For this reason, we will confine our phase measurement technique to the use of voltage probes in conjunction with the oscilloscope. A typical oscilloscope voltage probe has two points, the probe tip and the ground lead, that are connected to the circuit. Thus, all voltage measurements must be referenced to ground.

Since only voltage probes are to be used, the total current cannot be measured directly. However, for phase measurement, the voltage across $R_1$ is in phase with the total current and can be used to establish the phase angle.

Before proceeding with the actual phase measurement, note that there is a problem with displaying $V_{R1}$. If the scope probe is connected across the resistor, as indicated in Figure 16–37(a), the ground lead of the scope will short point *B* to ground, thus bypassing the rest of the components and effectively removing them from the circuit electrically, as illustrated in Figure 16–37(b) (assuming that the scope is not isolated from power line ground).

To avoid this problem, you can switch the generator output terminals so that one end of $R_1$ is connected to ground, as shown in Figure 16–38(a). This connection does not alter the circuit electrically, because $R_1$ still has the same series relationship with the rest of the circuit. Now the scope can be connected across it to display $V_{R1}$, as indicated in part (b) of the figure. The other probe is connected across the voltage source to display $V_s$ as indicated. Now channel 1 of the scope has $V_{R1}$ as an input, and channel 2 has $V_s$. The trigger source switch on the scope should be on internal so that each trace on the screen will be triggered by one of the inputs and the other will then be shown in the proper time relationship to it. Since amplitudes are not important, the volts/div settings are arbitrary. The sec/div settings should be adjusted so that one half-cycle of the waveforms appears on the screen.

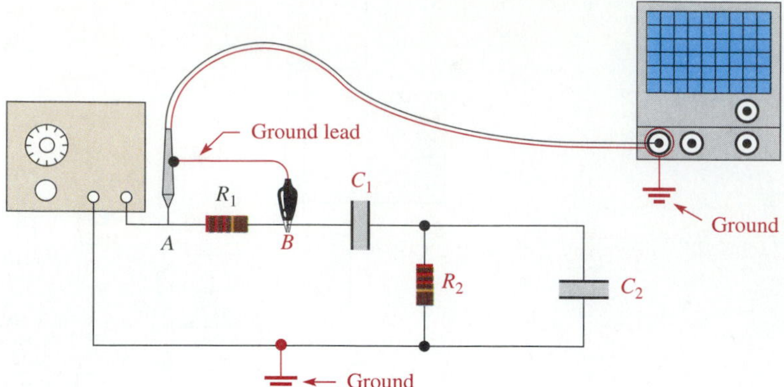


(a) Ground lead on scope probe grounds point $B$.

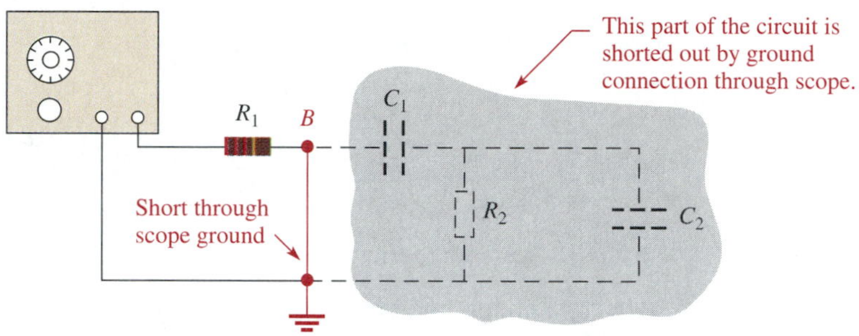


(b) The effect of grounding point $B$ is to short out the rest of the circuit.

**FIGURE 16–37**

***Effects of measuring directly across a component when the instrument and the circuit are grounded.***

Before connecting the probes to the circuit in Figure 16–38, you must align the two horizontal lines (traces) so that they appear as a single line across the center of the screen. To do so, ground the probe tips and adjust the vertical position knobs to move the traces toward the center line of the screen until they are superimposed. This procedure ensures that both waveforms have the same zero crossing so that an accurate phase measurement can be made.

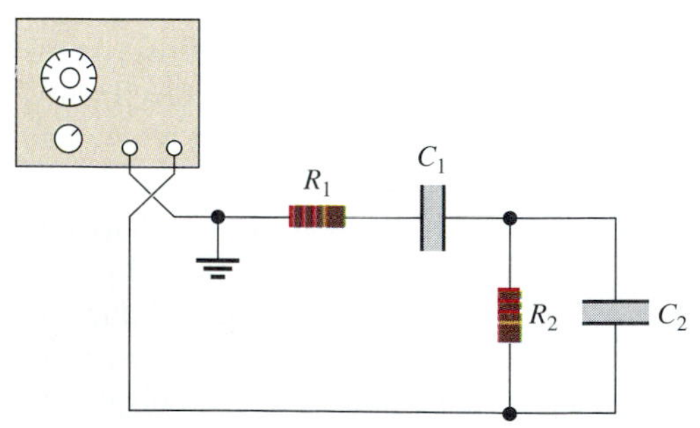


(a) Ground repositioned so that one end of $R_1$ is grounded.

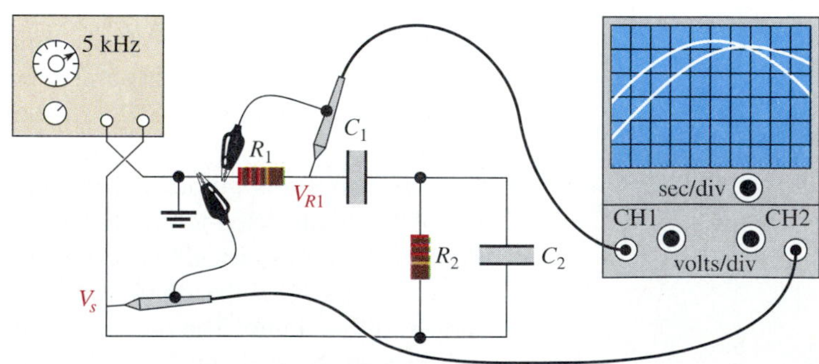


(b) The scope displays a half-cycle of $V_{R1}$ and $V_s$. $V_{R1}$ represents the phase of the total current.

**FIGURE 16–38**

***Repositioning ground so that a direct voltage measurement can be made with respect to ground without shorting out part of the circuit.***

The resulting oscilloscope display is shown in Figure 16–39. Since there are 180° in one half-cycle, each of the ten horizontal divisions across the screen represents 18°. Thus, the horizontal distance between the corresponding points of the two waveforms is the phase angle in degrees as indicated.

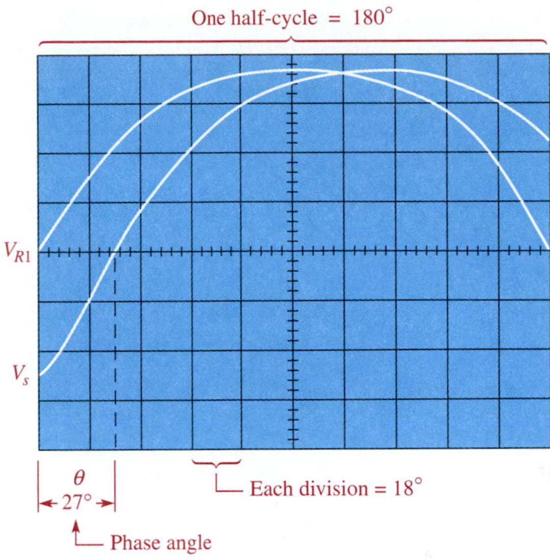


**FIGURE 16–39**
*Measurement of the phase angle on the oscilloscope.*

**SECTION 16–6 REVIEW**

1. What is the equivalent series *RC* circuit for the series-parallel circuit in Figure 16–33?
2. What is the total impedance of the circuit in Figure 16–34?

## 16–7 ■ POWER IN *RC* CIRCUITS

***In a purely resistive ac circuit, all of the energy delivered by the source is dissipated in the form of heat by the resistance. In a purely capacitive ac circuit, all of the energy delivered by the source is stored by the capacitor during a portion of the voltage cycle and then returned to the source during another portion of the cycle so that there is no net energy loss. When there is both resistance and capacitance, some of the energy is alternately stored and returned by the capacitance and some is dissipated by the resistance. The amount of energy loss is determined by the relative values of the resistance and the capacitive reactance.***

***After completing this section, you should be able to***

- Determine power in *RC* circuits
  - ☐ Explain true and reactive power
  - ☐ Draw the power triangle
  - ☐ Define *power factor*
  - ☐ Explain apparent power
  - ☐ Calculate power in an *RC* circuit

It is reasonable to assume that when the resistance is greater than the capacitive reactance, more of the total energy delivered by the source is dissipated by the resistance than is stored by the capacitance. Likewise, when the reactance is greater than the resistance, more of the total energy is stored and returned than is lost.

The formulas for power in a resistor, sometimes called *true power* ($P_{true}$), and the power in a capacitor, called *reactive power* ($P_r$), are restated here. The unit of true power is the watt, and the unit of reactive power is the VAR (volt-ampere reactive).

$$P_{true} = I^2R \quad \textbf{(16–29)}$$

$$P_r = I^2X_C \quad \textbf{(16–30)}$$

## The Power Triangle

The generalized impedance phasor diagram is shown in Figure 16–40(a). A phasor relationship for the powers can also be represented by a similar diagram because the respective magnitudes of the powers, $P_{true}$ and $P_r$, differ from $R$ and $X_C$ by a factor of $I^2$. This is shown in Figure 16–40(b).

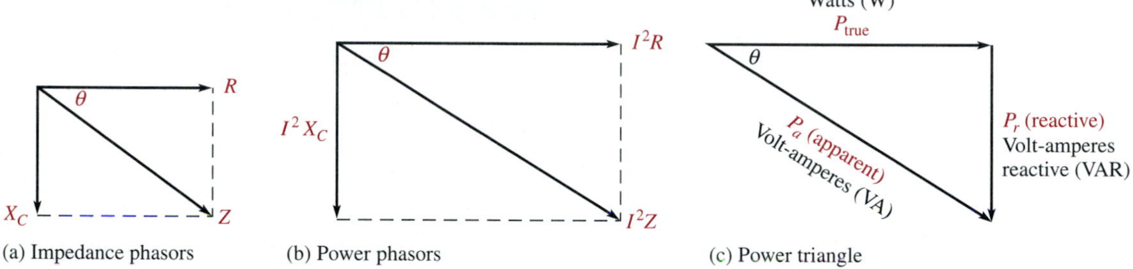


**FIGURE 16–40**
***Development of the power triangle for an RC circuit.***

The resultant power phasor, $I^2Z$, represents the **apparent power**, $P_a$. At any instant in time $P_a$ is the total power that appears to be transferred between the source and the *RC* circuit. The unit of apparent power is the volt-ampere, VA. The expression for apparent power is

$$P_a = I^2Z \quad \textbf{(16–31)}$$

The power phasor diagram in Figure 16–40(b) can be rearranged in the form of a right triangle, as shown in Figure 16–40(c). This is called the *power triangle.* Using the rules of trigonometry, $P_{true}$ can be expressed as

$$P_{true} = P_a\cos\theta \quad \textbf{(16–32)}$$

Since $P_a$ equals $I^2Z$ or $VI$, the equation for the true power dissipation in an *RC* circuit can be written as

$$P_{true} = VI\cos\theta \quad \textbf{(16–33)}$$

where $V$ is the applied voltage and $I$ is the total current.

For the case of a purely resistive current, $\theta = 0°$ and $\cos 0° = 1$, so $P_{true}$ equals $VI$. For the case of a purely capacitive circuit, $\theta = 90°$ and $\cos 90° = 0$, so $P_{true}$ is zero. As you already know, there is no power dissipation in an ideal capacitor.

## The Power Factor

The term *cos* $\theta$ is called the **power factor** and is stated as

$$PF = \cos\theta \tag{16–34}$$

As the phase angle between applied voltage and total current increases, the power factor decreases, indicating an increasingly reactive circuit. The smaller the power factor, the smaller the power dissipation.

The power factor can vary from 0 for a purely reactive circuit to 1 for a purely resistive circuit. In an *RC* circuit, the power factor is referred to as a leading power factor because the current leads the voltage.

**EXAMPLE 16–12** Determine the power factor and the true power in the circuit of Figure 16–41.

FIGURE 16–41

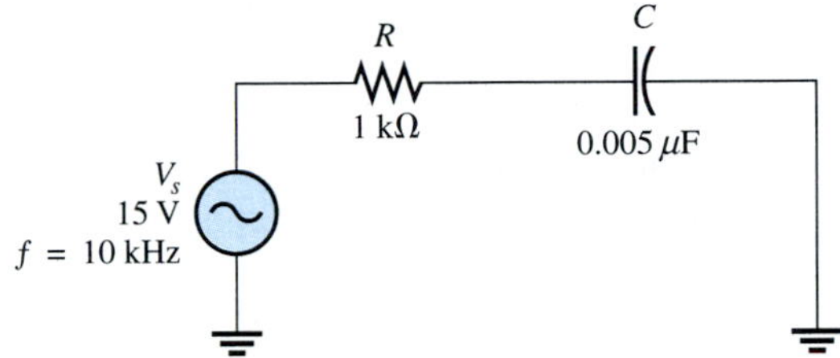


***Solution*** The capacitive reactance is

$$X_C = \frac{1}{2\pi fC} = \frac{1}{2\pi(10\text{ kHz})(0.005\ \mu\text{F})} = 3.18\text{ k}\Omega$$

The total impedance of the circuit in rectangular form is

$$\mathbf{Z} = R - jX_C = 1\text{ k}\Omega - j3.18\text{ k}\Omega$$

Converting to polar form yields

$$\mathbf{Z} = \sqrt{R^2 + X_C^2}\angle -\tan^{-1}\left(\frac{X_C}{R}\right)$$

$$= \sqrt{(1\text{ k}\Omega)^2 + (3.18\text{ k}\Omega^2)}\angle -\tan^{-1}\left(\frac{3.18\text{ k}\Omega}{1\text{ k}\Omega}\right) = 3.33\angle -72.5^\circ\text{ k}\Omega$$

The angle associated with the impedance is $\theta$, the angle between the applied voltage and the total current; therefore, the power factor is

$$PF = \cos\theta = \cos(-72.5^\circ) = 0.301$$

The current magnitude is

$$I = \frac{V_s}{Z} = \frac{15\text{ V}}{3.33\text{ k}\Omega} = 4.50\text{ mA}$$

The true power is

$$P_{\text{true}} = V_s I\cos\theta = (15\text{ V})(4.50\text{ mA})(0.301) = 20.3\text{ mW}$$

***Related Exercise*** What is the power factor if $f$ is reduced by half in Figure 16–41?

## The Significance of Apparent Power

As mentioned, apparent power is the power that appears to be transferred between the source and the load, and it consists of two components—a true power component and a reactive power component.

In all electrical and electronic systems, it is the true power that does the work. The reactive power is simply shuttled back and forth between the source and load. Ideally, in terms of performing useful work, all of the power transferred to the load should be true power and none of it reactive power. However, in most practical situations the load has some reactance associated with it, and therefore you must deal with both power components.

In Chapter 15, we discussed the use of apparent power in relation to transformers. For any reactive load, there are two components of the total current: the resistive component and the reactive component. If you consider only the true power (watts) in a load, you are dealing with only a portion of the total current that the load demands from a source. In order to have a realistic picture of the actual current that a load will draw, you must consider apparent power (in VA).

A source such as an ac generator can provide current to a load up to some maximum value. If the load draws more than this maximum value, the source can be damaged. Figure 16–42(a) shows a 120 V generator that can deliver a maximum current of 5 A to a load. Assume that the generator is rated at 600 W and is connected to a purely resistive load of 24 Ω (power factor of 1). The ammeter shows that the current is 5 A, and the wattmeter indicates that the power is 600 W. The generator has no problem under these conditions, although it is operating at maximum current and power.

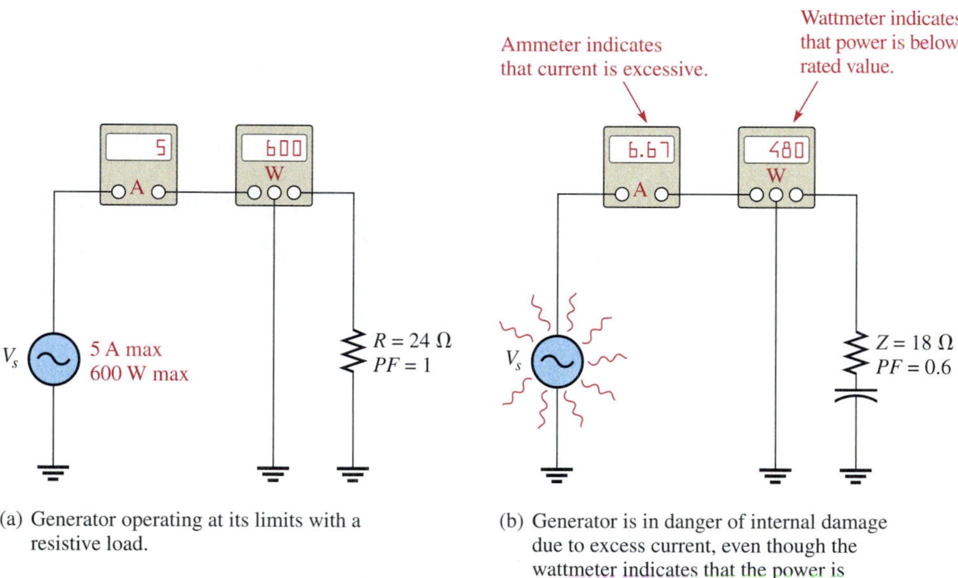


(a) Generator operating at its limits with a resistive load.

(b) Generator is in danger of internal damage due to excess current, even though the wattmeter indicates that the power is below the maximum wattage rating.

**FIGURE 16–42**

***Wattage rating of a source is inappropriate when the load is reactive. The rating should be in VA rather than in watts.***

Now, consider what happens if the load is changed to a reactive one with an impedance of 18 Ω and a power factor of 0.6, as indicated in Figure 16–42(b). The current is 120 V/18 Ω = 6.67 A, which *exceeds* the maximum. Even though the wattmeter reads 480 W, which is less than the power rating of the generator, the excessive current probably will cause damage. This example shows that a true power rating can be deceiving and is inappropriate for ac sources. The ac generator should be rated at 600 VA, a rating that manufacturers generally use, rather than 600 W.

**EXAMPLE 16–13** For the circuit in Figure 16–43, find the true power, the reactive power, and the apparent power. $X_C$ has been determined to be 2 kΩ.

FIGURE 16–43

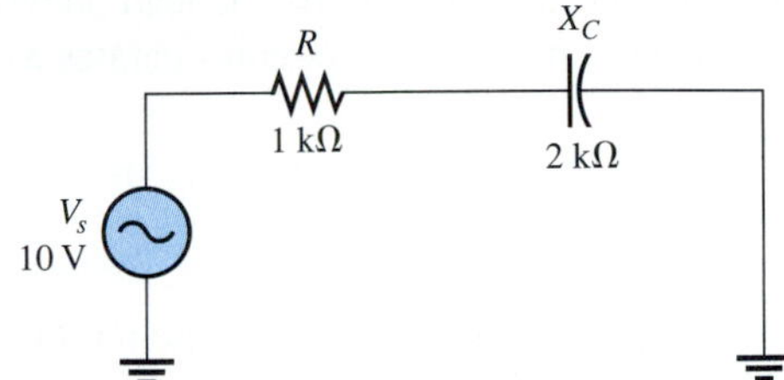


***Solution*** First find the total impedance so that the current can be calculated.

$$\mathbf{Z} = R - jX_C = 1\text{ k}\Omega - j2\text{ k}\Omega$$

$$\mathbf{Z} = \sqrt{R^2 + X_C^2}\angle -\tan^{-1}\left(\frac{X_C}{R}\right)$$

$$= \sqrt{(1\text{ k}\Omega)^2 + (2\text{ k}\Omega)^2}\angle -\tan^{-1}\left(\frac{2\text{ k}\Omega}{1\text{ k}\Omega}\right) = 2.24\angle -63.4^\circ\text{ k}\Omega$$

$$I = \frac{V_s}{Z} = \frac{10\text{ V}}{2.24\text{ k}\Omega} = 4.46\text{ mA}$$

The phase angle, $\theta$, is indicated in the polar expression for impedance.

$$\theta = -63.4^\circ$$

The true power is

$$P_{\text{true}} = V_s I\cos\theta = (10\text{ V})(4.46\text{ mA})\cos(-63.4^\circ) = 20\text{ mW}$$

Note that the same result is realized using the formula $P_{\text{true}} = I^2R$.

The reactive power is

$$P_r = I^2X_C = (4.46\text{ mA})^2(2\text{ k}\Omega) = 39.8\text{ mVAR}$$

The apparent power is

$$P_a = I^2Z = (4.46\text{ mA})^2(2.24\text{ k}\Omega) = 44.6\text{ mVA}$$

The apparent power is also the phasor sum of $P_{\text{true}}$ and $P_r$.

$$P_a = \sqrt{P_{\text{true}}^2 + P_r^2} \cong 44.6\text{ mVA}$$

***Related Exercise*** What is the true power in Figure 16–43 if $X_C = 10$ kΩ?

**SECTION 16–7 REVIEW**

1. To which component in an *RC* circuit is the energy loss due?
2. The phase angle, $\theta$, is 45°. What is the power factor?
3. A certain series *RC* circuit has the following parameter values: $R = 330$ Ω, $X_C = 460$ Ω, and $I = 2$ A. Determine the true power, the reactive power, and the apparent power.

## 16–8 ■ BASIC APPLICATIONS

***RC circuits are found in a variety of applications, often as part of a more complex circuit. Two major applications, phase shift networks and frequency-selective networks (filters), are covered in this section.***

***After completing this section, you should be able to***

- Discuss some basic *RC* applications
  - ☐ Discuss and analyze the *RC* lag network
  - ☐ Discuss and analyze the *RC* lead network
  - ☐ Discuss how the *RC* circuit operates as a filter

### The *RC* Lag Network

The *RC* lag network is a phase shift circuit in which the output voltage lags the input voltage by a specified amount. Figure 16–44(a) shows a series *RC* circuit with the output voltage taken across the capacitor. The source voltage is the input, $V_{in}$. As you know, $\theta$, the phase angle between the current and the input voltage, is also the phase angle between the resistor voltage and the input voltage, because $V_R$ and $I$ are in phase with each other.

**FIGURE 16–44**

***RC lag network ($V_{out} = V_C$).***

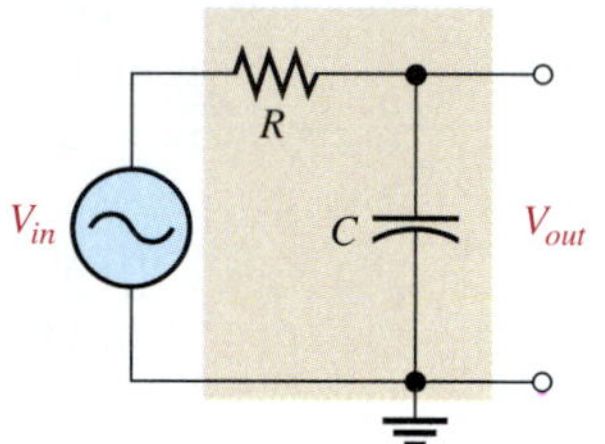


(a) A basic *RC* lag network

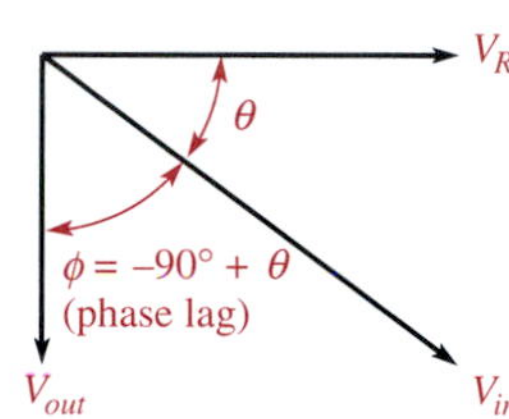


(b) Phasor voltage diagram showing the phase lag between $V_{in}$ and $V_{out}$

Since $V_C$ lags $V_R$ by 90°, the phase angle between the capacitor voltage and the input voltage is the difference between −90° and $\theta$, as shown in Figure 16–44(b). The capacitor voltage is the output, and it lags the input, thus creating a basic lag network.

When the input and output voltage waveforms of the lag network are displayed on an oscilloscope, a relationship similar to that in Figure 16–45 is observed. The amount of phase difference, designated $\phi$, between the input and the output is dependent on the relative sizes of the capacitive reactance and the resistance, as is the magnitude of the output voltage.

**FIGURE 16–45**

***Oscilloscope display of the input and output voltage waveforms of a lag network ($V_{out}$ lags $V_{in}$). The angle shown is arbitrary.***

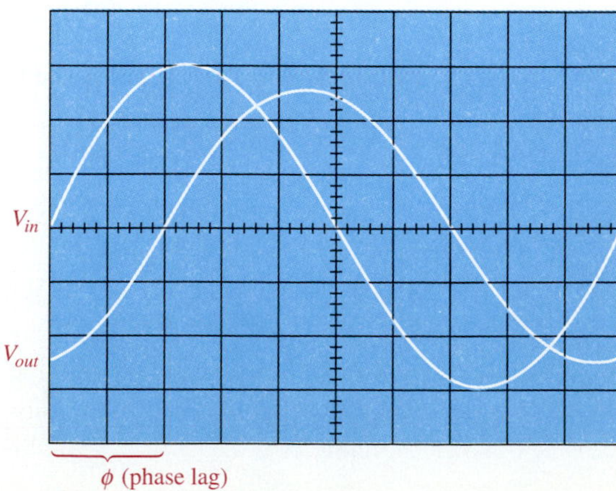

***Phase Difference Between Input and Output*** As already established, $\theta$ is the phase angle between $I$ and $V_{in}$. The angle between $V_{out}$ and $V_{in}$ is designated $\phi$ (phi) and is developed as follows. The polar expressions for the input voltage and the current are $V_{in}\angle 0°$ and $I\angle\theta$, respectively. The output voltage is

$$\mathbf{V}_{out} = (I\angle\theta)(X_C\angle -90°) = IX_C\angle(-90° + \theta)$$

The preceding equation states that the output voltage is at an angle of $-90° + \theta$ with respect to the input voltage. Since $\theta = -\tan^{-1}(X_C/R)$, the angle between the input and output is

$$\phi = -90° + \tan^{-1}\left(\frac{X_C}{R}\right) \quad \textbf{(16–35)}$$

This angle is always negative, indicating that the output voltage lags the input voltage, as shown in Figure 16–46.

FIGURE 16–46

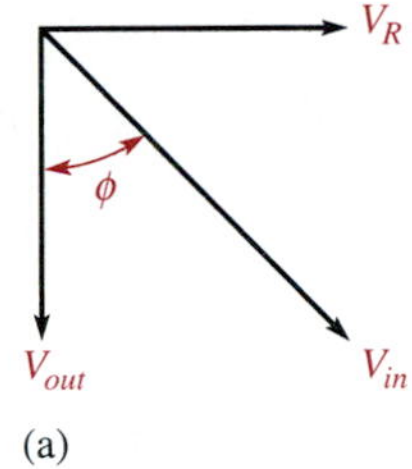


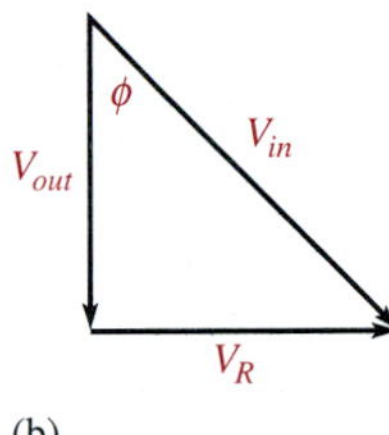


**EXAMPLE 16–14** Determine the amount of phase lag from input to output in each lag network in Figure 16–47.

FIGURE 16–47

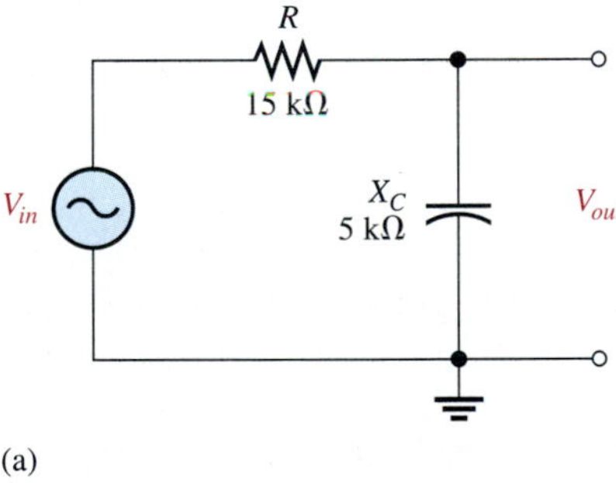


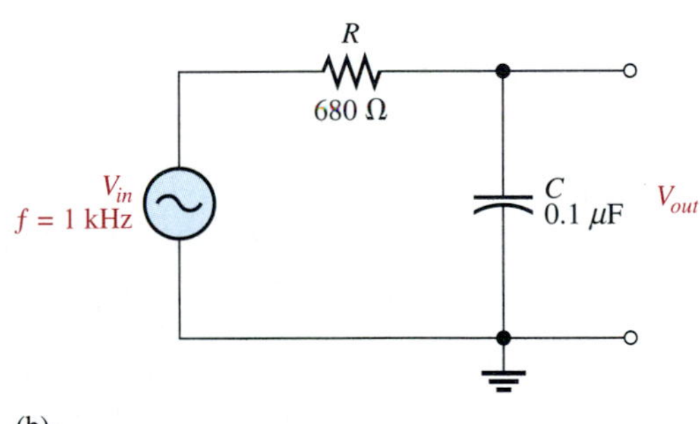


***Solution*** For the lag network in Figure 16–47(a),

$$\phi = -90° + \tan^{-1}\left(\frac{X_C}{R}\right) = -90° + \tan^{-1}\left(\frac{5\text{ k}\Omega}{15\text{ k}\Omega}\right) = -90° + 18.4° = -71.6°$$

The output lags the input by 71.6°.

For the lag network in Figure 16–47(b), first determine the capacitive reactance.

$$X_C = \frac{1}{2\pi fC} = \frac{1}{2\pi(1\text{ kHz})(0.1\ \mu\text{F})} = 1.59\text{ k}\Omega$$

$$\phi = -90° + \tan^{-1}\left(\frac{X_C}{R}\right) = -90° + \tan^{-1}\left(\frac{1.59\text{ k}\Omega}{680\ \Omega}\right) = -23.2°$$

The output lags the input by 23.2°.

***Related Exercise*** In a lag network, what happens to the phase lag if the frequency increases?

***Magnitude of the Output Voltage*** To evaluate the output voltage in terms of its magnitude, visualize the *RC* lag network as a voltage divider. A portion of the total input voltage is dropped across $R$ and a portion across $C$. Since the output voltage is $V_C$, it can be calculated as

$$V_{out} = \left(\frac{X_C}{\sqrt{R^2 + X_C^2}}\right)V_{in} \quad \textbf{(16–36)}$$

Or it can be calculated using Ohm's law as

$$V_{out} = IX_C \quad \textbf{(16–37)}$$

The total phasor expression for the output voltage of a lag network is

$$\mathbf{V}_{out} = V_{out}\angle\phi \quad \textbf{(16–38)}$$

**EXAMPLE 16–15** For the lag network in Figure 16–47(b), determine the output voltage in phasor form when the input voltage has an rms value of 10 V. Sketch the input and output voltage waveforms showing the proper relationships. $X_C$ (1.59 kΩ) and $\phi$ (–23.2°) were found in Example 16–14.

***Solution*** The output voltage in phasor form is

$$\mathbf{V}_{out} = \left(\frac{X_C}{\sqrt{R^2 + X_C^2}}\right)V_{in}\angle\phi$$

$$= \left(\frac{1.59\text{ k}\Omega}{\sqrt{(680\ \Omega)^2 + (1.59\text{ k}\Omega)^2}}\right)10\angle{-23.2°}\text{ V} = 9.20\angle{-23.2°}\text{ V rms}$$

The waveforms are shown in Figure 16–48.

FIGURE 16–48

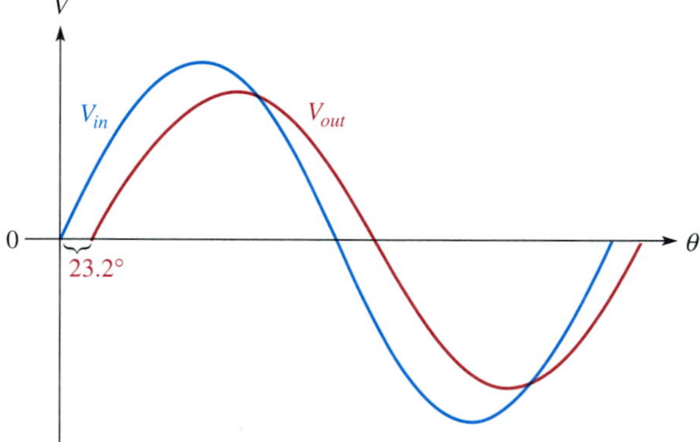


***Related Exercise*** In a lag network, what happens to the output voltage if the frequency increases?

## The *RC* Lead Network

The *RC* lead network is a phase shift circuit in which the output voltage leads the input voltage by a specified amount. When the output of a series *RC* circuit is taken across the resistor rather than across the capacitor, as shown in Figure 16–49(a), it becomes a lead network.

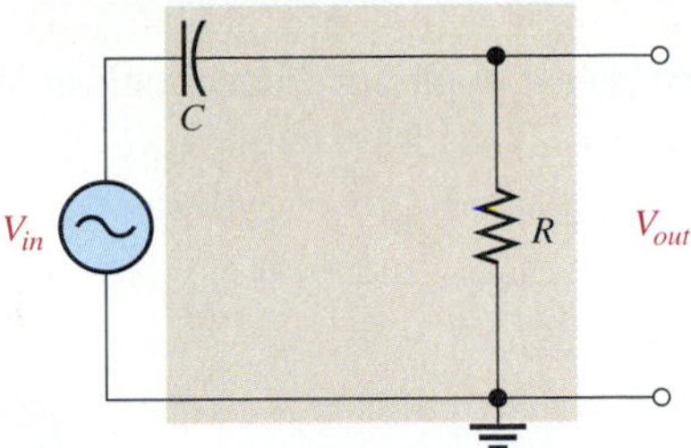


(a) A basic $RC$ lead network

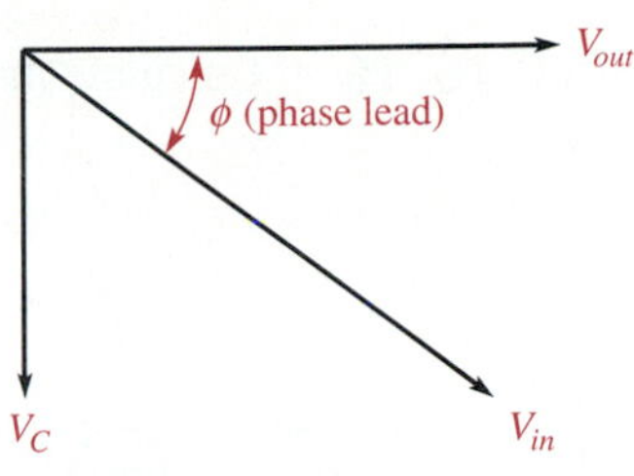


(b) Phasor voltage diagram showing the phase lead between $V_{in}$ and $V_{out}$

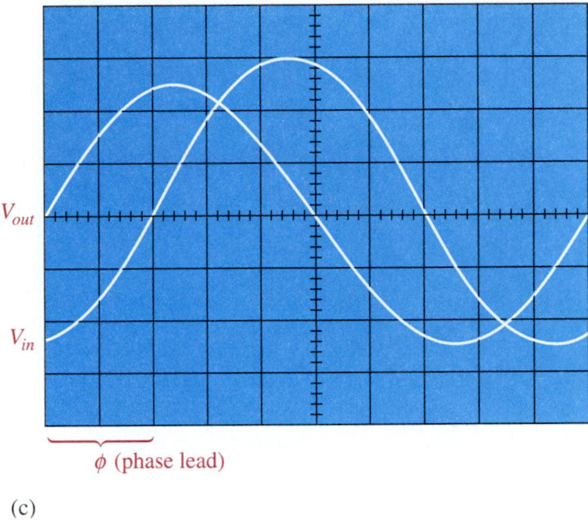


(c)

**FIGURE 16–49**
***RC* lead network ($V_{out} = V_R$).**

***Phase Difference Between Input and Output*** In a series $RC$ circuit, the current leads the input voltage. Also, as you know, the resistor voltage is in phase with the current. Since the output voltage is taken across the resistor, the output leads the input, as indicated by the phasor diagram in Figure 16–49(b). A typical oscilloscope display of the waveforms is shown in Figure 16–49(c).

As in the lag network, the amount of phase difference between the input and output and also the magnitude of the output voltage in the lead network is dependent on the relative values of the resistance and the capacitive reactance. When the input voltage is assigned a reference angle of 0°, the angle of the output voltage is the same as $\theta$ (the angle between total current and applied voltage), because the resistor voltage (output) and the current are in phase with each other. Therefore, since $\phi = \theta$ in this case, the expression is

$$\phi = \tan^{-1}\left(\frac{X_C}{R}\right) \qquad \textbf{(16–39)}$$

This angle is positive, because the output leads the input. The following example illustrates the computation of phase angles for lead networks.

**EXAMPLE 16–16** Calculate the output phase angle for each circuit in Figure 16–50.

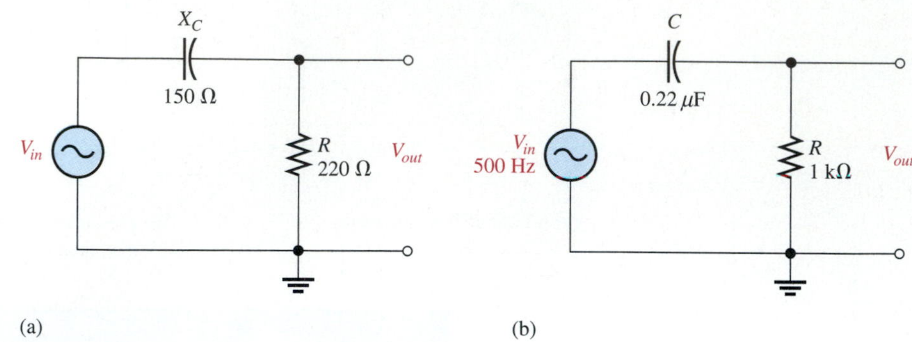


**FIGURE 16–50**

***Solution*** For the lead network in Figure 16–50(a),

$$\phi = \tan^{-1}\left(\frac{X_C}{R}\right) = \tan^{-1}\left(\frac{150\ \Omega}{220\ \Omega}\right) = 34.3°$$

The output leads the input by 34.3°.

For the lead network in Figure 16–50(b), first determine the capacitive reactance.

$$X_C = \frac{1}{2\pi fC} = \frac{1}{2\pi(500\ \text{Hz})(0.22\ \mu\text{F})} = 1.45\ \text{k}\Omega$$

$$\phi = \tan^{-1}\left(\frac{X_C}{R}\right) = \tan^{-1}\left(\frac{1.45\ \text{k}\Omega}{1\ \text{k}\Omega}\right) = 55.4°$$

The output leads the input by 55.4°.

***Related Exercise*** In a lead network, what happens to the phase lead if the frequency increases?

***Magnitude of the Output Voltage*** Since the output voltage of an *RC* lead network is taken across the resistor, the magnitude can be calculated using either the voltage-divider formula or Ohm's law, stated as

$$V_{out} = \left(\frac{R}{\sqrt{R^2 + X_C^2}}\right)V_{in} \quad \textbf{(16–40)}$$

$$V_{out} = IR \quad \textbf{(16–41)}$$

The expression for the output voltage in phasor form is

$$\mathbf{V}_{out} = V_{out}\angle\phi \quad \textbf{(16–42)}$$

**EXAMPLE 16–17** The input voltage in Figure 16–50(b) has an rms value of 10 V. Determine the phasor expression for the output voltage. Sketch the waveform relationships for the input and output voltages showing peak values. The phase angle (55.4°) and $X_C$ (1.45 kΩ) were found in Example 16–16.

***Solution*** The phasor expression for the output voltage is

$$\mathbf{V}_{out} = \left(\frac{R}{\sqrt{R^2 + X_C^2}}\right)V_{in}\angle\phi = \left(\frac{1\text{ k}\Omega}{1.76\text{ k}\Omega}\right)10\angle 55.4^\circ\text{ V} = 5.68\angle 55.4^\circ\text{ V rms}$$

The peak value of the input voltage is

$$V_{in(p)} = 1.414V_{in(rms)} = 1.414(10\text{ V}) = 14.14\text{ V}$$

The peak value of the output voltage is

$$V_{out(p)} = 1.414V_{out(rms)} = 1.414(5.68\text{ V}) = 8.03\text{ V}$$

The waveforms are shown in Figure 16–51.

FIGURE 16–51

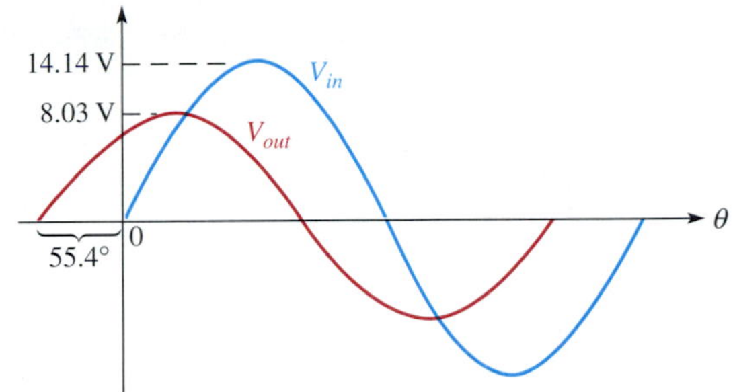


***Related Exercise*** In a lead network, what happens to the output voltage if the frequency is reduced?

## The *RC* Circuit as a Filter

**Filters** are frequency-selective circuits that permit signals of certain frequencies to pass from the input to the output while blocking all others. That is, all frequencies but the selected ones are filtered out. Filters are covered in greater depth in Chapter 19 but are introduced here as an application example.

Series *RC* circuits exhibit a frequency-selective characteristic and therefore act as basic filters. There are two types. The first one that we examine, called a **low-pass filter**, is realized by taking the output across the capacitor, just as in a lag network. The second type, called a **high-pass filter**, is implemented by taking the output across the resistor, as in a lead network.

***Low-Pass Filter*** You have already seen what happens to the output magnitude and phase angle in the lag network. In terms of its filtering action, we are interested primarily in the variation of the output magnitude with frequency.

Figure 16–52 shows the filtering action of a series *RC* circuit using specific values for illustration. In part (a) of the figure, the input is zero frequency (dc). Since the capacitor blocks constant direct current, the output voltage equals the full value of the input voltage, because there is no voltage dropped across *R*. Therefore, the circuit passes all of the input voltage to the output (10 V in, 10 V out).

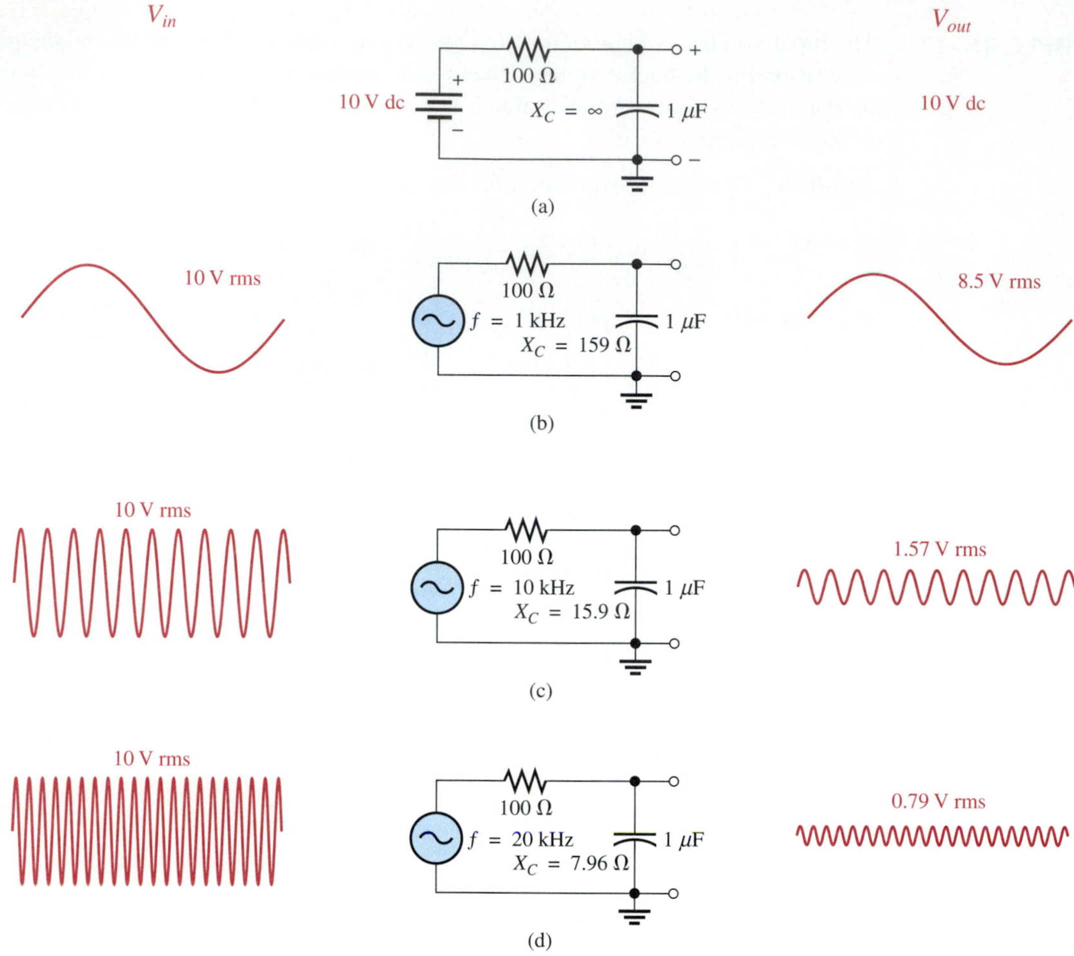


**FIGURE 16–52**
***Low-pass filter action (phase shifts are not indicated).***

In Figure 16–52(b), the frequency of the input voltage has been increased to 1 kHz, causing the capacitive reactance to decrease to 159 Ω. For an input voltage of 10 V rms, the output voltage is approximately 8.5 V rms, which can be calculated using the voltage-divider approach or Ohm's law.

In Figure 16–52(c), the input frequency has been increased to 10 kHz, causing the capacitive reactance to decrease further to 15.9 Ω. For a constant input voltage of 10 V rms, the output voltage is now 1.57 V rms.

As the input frequency is increased further, the output voltage continues to decrease and approaches zero as the frequency becomes very high, as shown in Figure 16–52(d).

A description of the circuit action is as follows: As the frequency of the input increases, the capacitive reactance decreases. Because the resistance is constant and the capacitive reactance decreases, the voltage across the capacitor (output voltage) also decreases according to the voltage-divider principle. The input frequency can be increased until it reaches a value at which the reactance is so small compared to the resistance that the output voltage can be neglected because it is very small compared to the input voltage. At this value of frequency, the circuit is essentially completely blocking the input signal.

As shown in Figure 16–52, the circuit passes dc (zero frequency) completely. As the frequency of the input increases, less of the input voltage is passed through to the output; that is, the output voltage decreases as the frequency increases. It is apparent that the lower frequencies pass through the circuit much better than the higher frequencies. This *RC* circuit is therefore a very basic form of low-pass filter.

The **frequency response** of the low-pass filter circuit in Figure 16–52 is shown in Figure 16–53 with a graph of output voltage magnitude versus frequency. This graph, called a *response curve,* indicates that the output decreases as the frequency increases.

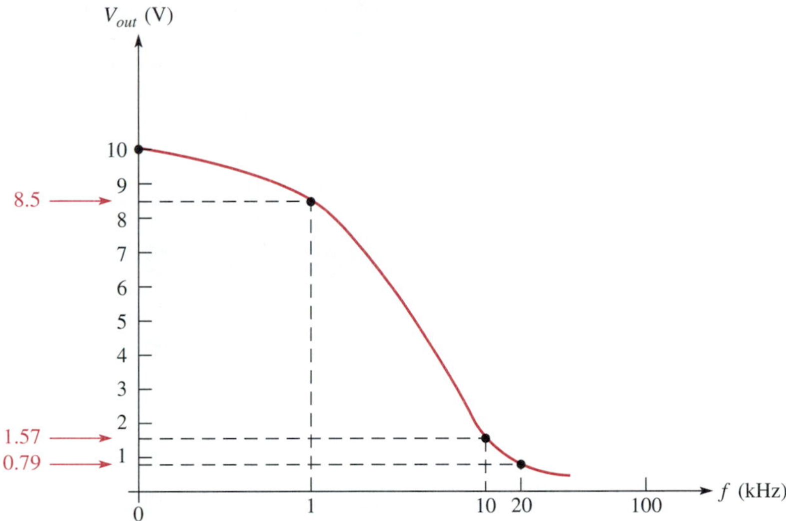


FIGURE 16–53
***Frequency response curve for the low-pass filter in Figure 16–52.***

***High-Pass Filter*** High-pass filter action is illustrated in Figure 16–54, where the output is taken across the resistor, just as in a lead network. When the input voltage is dc (zero frequency) in part (a), the output is zero volts because the capacitor blocks direct current; therefore, no voltage is developed across *R*.

In Figure 16–54(b), the frequency of the input signal has been increased to 100 Hz with an rms value of 10 V. The output voltage is 0.63 V rms. Thus, only a small percentage of the input voltage appears on the output at this frequency.

In Figure 16–54(c), the input frequency is increased further to 1 kHz, causing more voltage to be developed across the resistor because of the further decrease in the capacitive reactance. The output voltage at this frequency is 5.32 V rms. As you can see, the output voltage increases as the frequency increases. A value of frequency is reached at which the reactance is negligible compared to the resistance, and most of the input voltage appears across the resistor, as shown in Figure 16–54(d).

As illustrated, this circuit tends to prevent lower frequencies from appearing on the output but allows higher frequencies to pass through from input to output. Therefore, this *RC* circuit is a very basic form of high-pass filter.

The frequency response of the high-pass filter circuit in Figure 16–54 is shown in Figure 16–55 with a graph of output voltage magnitude versus frequency. This response curve shows that the output increases as the frequency increases and then levels off and approaches the value of the input voltage.

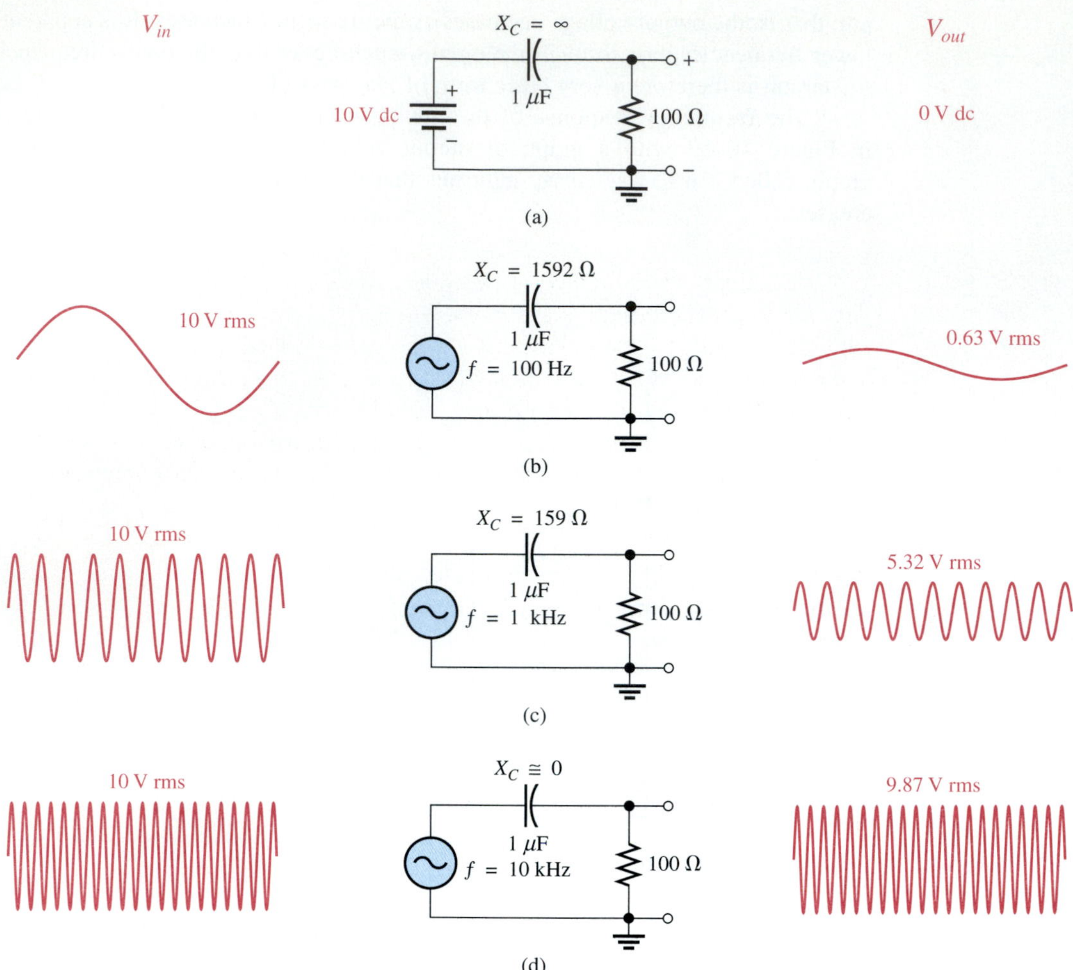


FIGURE 16–54
***High-pass filter action (phase shifts are not indicated).***

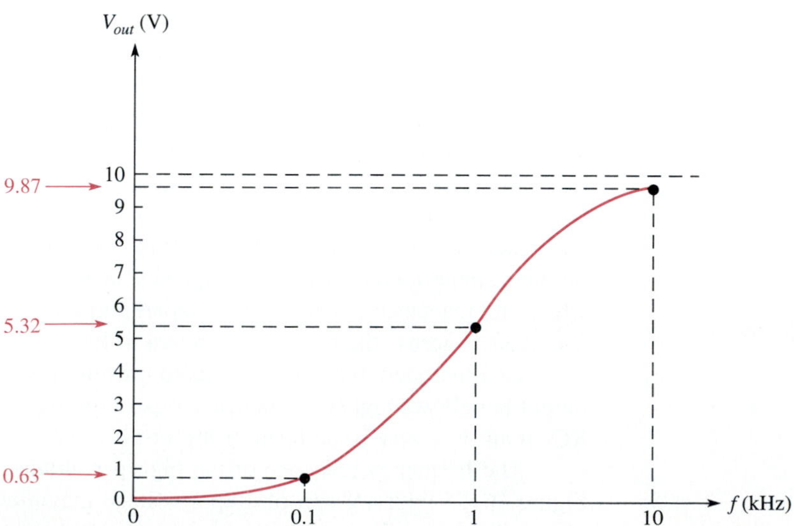


FIGURE 16–55
***Frequency response curve for the high-pass filter in Figure 16–54.***

***The Cutoff Frequency and the Bandwidth of a Filter*** The frequency at which the capacitive reactance equals the resistance in a low-pass or high-pass *RC* filter is called the **cutoff frequency** and is designated $f_c$. This condition is expressed as $1/(2\pi f_c C) = R$. Solving for $f_c$ results in the following formula:

$$f_c = \frac{1}{2\pi RC} \quad \textbf{(16–43)}$$

At $f_c$, the output voltage of the filter is 70.7% of its maximum value. It is standard practice to consider the cutoff frequency as the limit of a filter's performance in terms of passing or rejecting frequencies. For example, in a high-pass filter, all frequencies above $f_c$ are considered to be passed by the filter, and all those below $f_c$ are considered to be rejected. The reverse is true for a low-pass filter.

The range of frequencies that is considered to be passed by a filter is called the **bandwidth**. Figure 16–56 illustrates the bandwidth and the cutoff frequency for a low-pass filter.

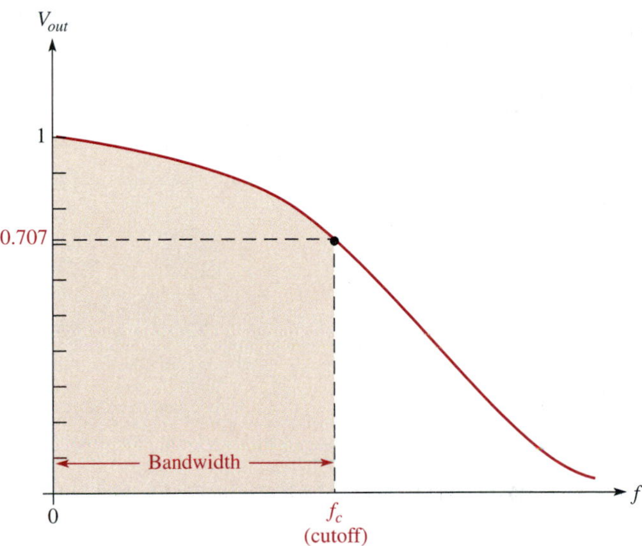


FIGURE 16–56
*Normalized general response curve of a low-pass filter showing the cutoff frequency and the bandwidth.*

## Coupling an AC Signal into a DC Bias Network

Figure 16–57 shows an *RC* network that is used to create a dc voltage level with an ac voltage superimposed on it. This type of circuit is commonly found in amplifiers in which the dc voltage is required to **bias** the amplifier to the proper operating point and the signal voltage to be amplified is coupled through a capacitor and superimposed on the dc level. The capacitor prevents the low internal resistance of the signal source from affecting the dc bias voltage.

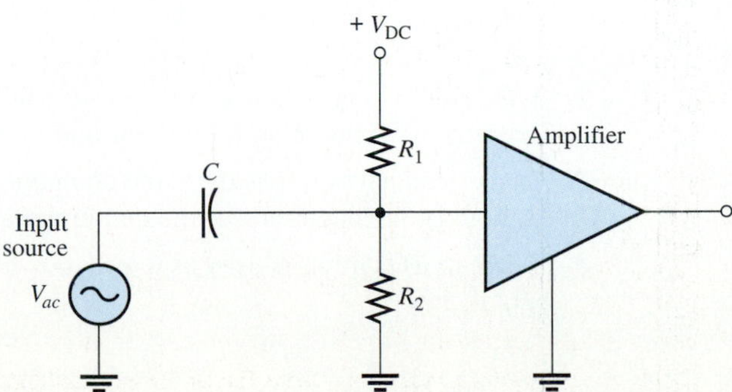


FIGURE 16–57
*Amplifier bias and signal-coupling circuit.*

In this type of application, a relatively high value of capacitance is selected so that for the frequencies to be amplified, the reactance is very small compared to the resistance of the bias network. When the reactance is very small (ideally zero), there is practically no phase shift or signal voltage dropped across the capacitor. Therefore, all of the signal voltage passes from the source to the input to the amplifier.

Figure 16–58 illustrates the application of the superposition principle to the circuit in Figure 16–57. In part (a), the ac source has been effectively removed from the circuit and replaced with a short to represent its ideal internal resistance. Since $C$ is open to dc, the voltage at point $A$ is determined by the voltage-divider action of $R_1$ and $R_2$ and the dc voltage source.

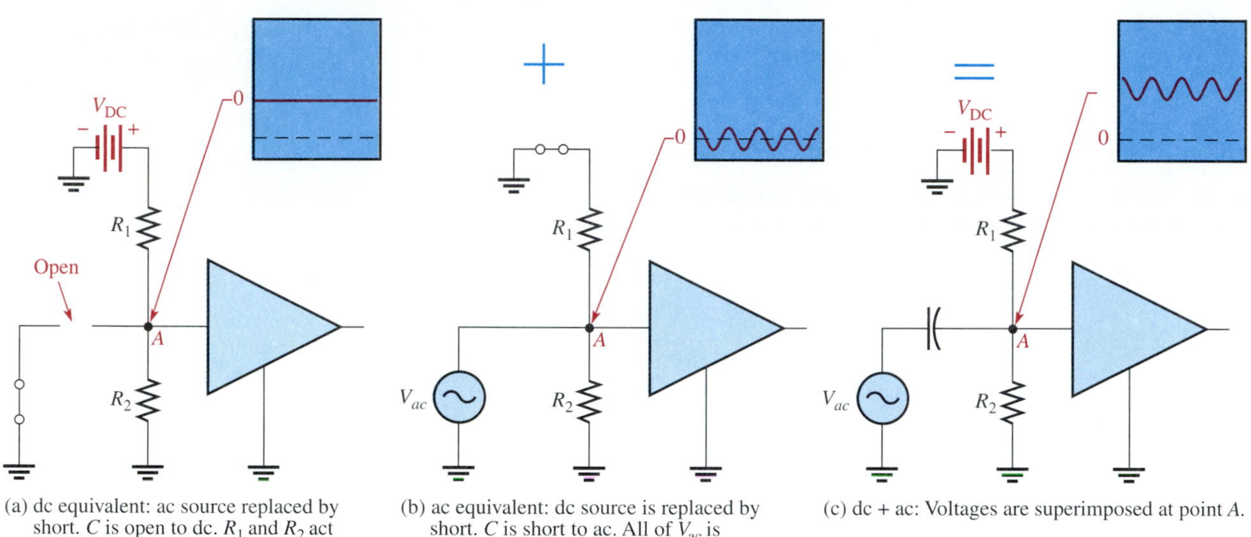


(a) dc equivalent: ac source replaced by short. $C$ is open to dc. $R_1$ and $R_2$ act as dc voltage divider.

(b) ac equivalent: dc source is replaced by short. $C$ is short to ac. All of $V_{ac}$ is coupled to point $A$.

(c) dc + ac: Voltages are superimposed at point $A$.

**FIGURE 16–58**

***The superposition of dc and ac voltages in an RC bias and coupling circuit.***

In Figure 16–58(b), the dc source has been effectively removed from the circuit and replaced with a short to represent its ideal internal resistance. Since $C$ appears as a short at the frequency of the ac, the signal voltage is coupled directly to point $A$ and appears across the parallel combination of $R_1$ and $R_2$. Figure 16–58(c) illustrates that the combined effect of the superposition of the dc and the ac voltages results in the signal voltage "riding" on the dc level.

## SECTION 16–8 REVIEW

1. A certain *RC* lag network consists of a 4.7 kΩ resistor and a 0.022 μF capacitor. Determine the phase shift between input and output at a frequency of 3 kHz.
2. An *RC* lead network has the same component values as the lag network in Question 1. What is the magnitude of the output voltage at 3 kHz when the input is 10 V rms?
3. When an *RC* circuit is used as a low-pass filter, across which component is the output taken?

## 16–9 ▪ TROUBLESHOOTING

***In this section, the effects that typical component failures or degradation have on the response of basic RC circuits are considered.***

***After completing this section, you should be able to***

- **Troubleshoot *RC* circuits**
  - Find an open resistor
  - Find an open capacitor
  - Find a shorted capacitor
  - Find a leaky capacitor

### Effect of an Open Resistor

It is easy to see how an open resistor affects the operation of a basic series *RC* circuit, as shown in Figure 16–59. Obviously, there is no path for current, so the capacitor voltage remains at zero; thus, the total voltage, $V_s$, appears across the open resistor.

### Effect of an Open Capacitor

When the capacitor is open, there is no current; thus, the resistor voltage remains at zero. The total source voltage is across the open capacitor, as shown in Figure 16–60.

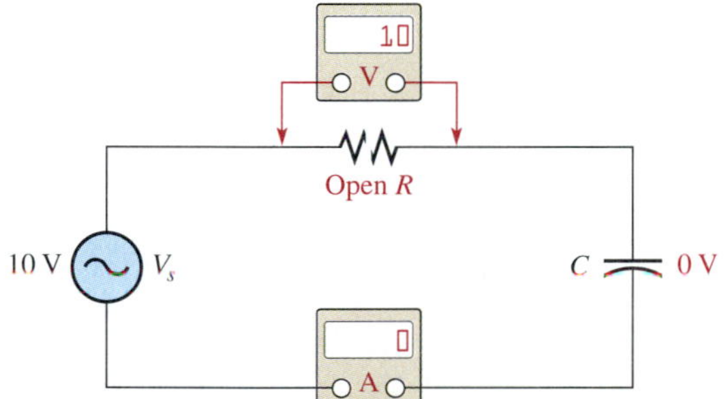


**FIGURE 16–59**
*Effect of an open resistor.*

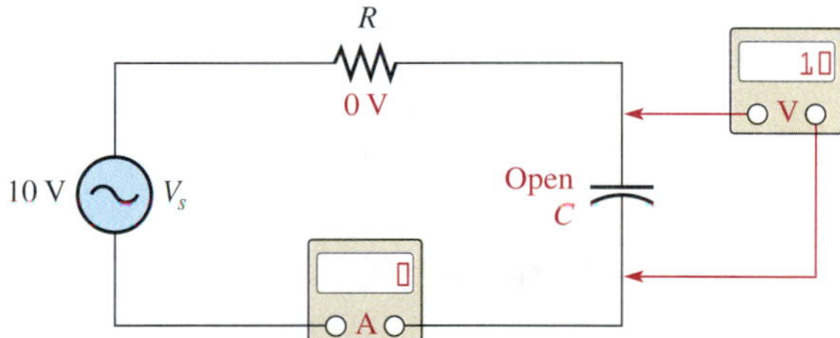


**FIGURE 16–60**
*Effect of an open capacitor.*

### Effect of a Shorted Capacitor

When a capacitor shorts out, the voltage across it is zero, the current equals $V_s/R$, and the total voltage appears across the resistor, as shown in Figure 16–61.

**FIGURE 16–61**
*Effect of a shorted capacitor.*

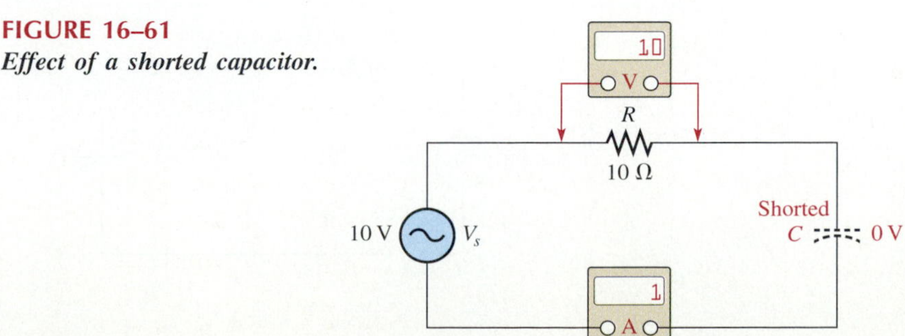

## Effect of a Leaky Capacitor

When a capacitor exhibits a high leakage current, the leakage resistance effectively appears in parallel with the capacitor, as shown in Figure 16–62(a). When the leakage resistance is comparable in value to the circuit resistance, $R$, the circuit response is drastically affected. The circuit, looking from the capacitor toward the source, can be Thevenized, as shown in Figure 16–62(b). The Thevenin equivalent resistance is $R$ in parallel with $R_{leak}$ (the source appears as a short), and the Thevenin equivalent voltage is determined by the voltage-divider action of $R$ and $R_{leak}$.

$$R_{th} = R \parallel R_{leak}$$

$$R_{th} = \frac{RR_{leak}}{R + R_{leak}} \quad \textbf{(16–44)}$$

$$V_{th} = \left(\frac{R_{leak}}{R + R_{leak}}\right)V_s \quad \textbf{(16–45)}$$

As you can see, the voltage to which the capacitor will charge is reduced since $V_{th} < V_s$. Also, the circuit time constant is reduced, and the current is increased. The Thevenin equivalent circuit is shown in Figure 16–62(c).

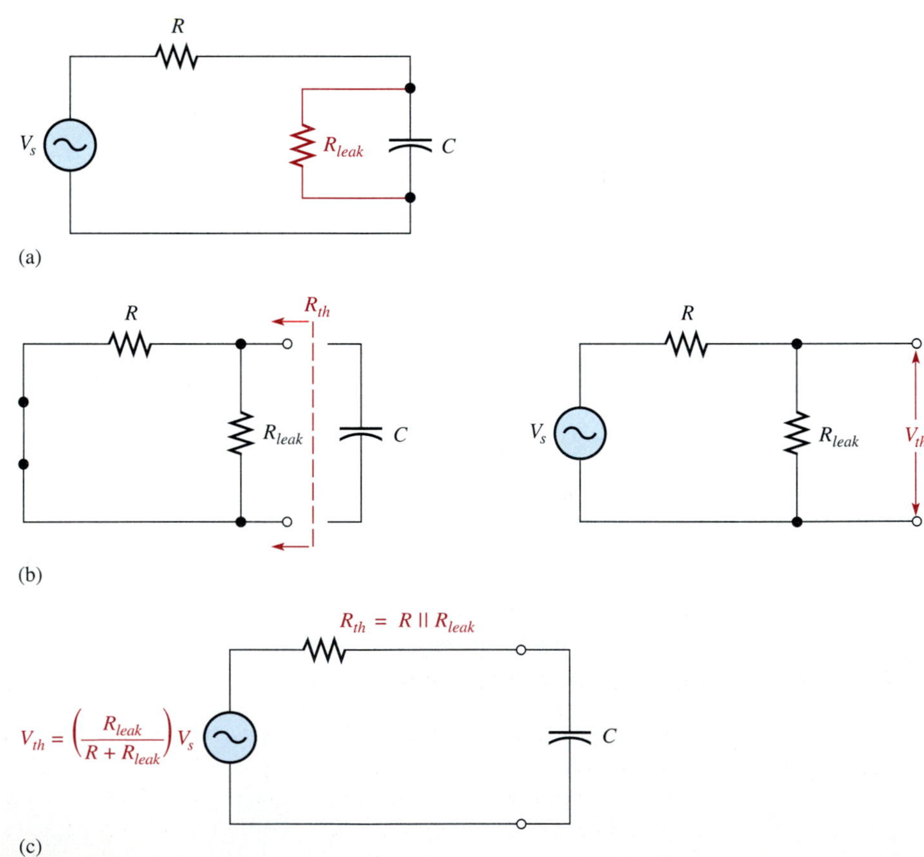


**FIGURE 16–62**
***Effect of a leaky capacitor.***

**EXAMPLE 16–18** Assume that the capacitor in Figure 16–63 is degraded to a point where its leakage resistance is 10 kΩ. Determine the phase shift from input to output and the output voltage under the degraded condition.

FIGURE 16–63

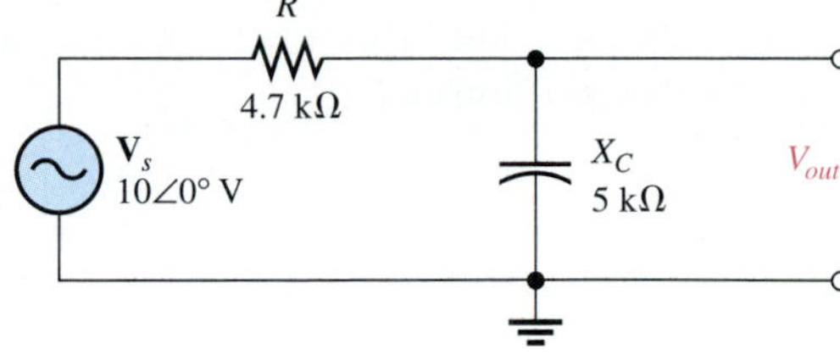


***Solution*** The effective circuit resistance is

$$R_{th} = \frac{RR_{leak}}{R + R_{leak}} = \frac{(4.7\text{ k}\Omega)(10\text{ k}\Omega)}{14.7\text{ k}\Omega} = 3.20\text{ k}\Omega$$

The phase shift is

$$\phi = -90° + \tan^{-1}\left(\frac{X_C}{R_{th}}\right) = -90° + \tan^{-1}\left(\frac{5\text{ k}\Omega}{3.20\text{ k}\Omega}\right) = -32.6°$$

To determine the output voltage, first find the Thevenin equivalent voltage.

$$V_{th} = \left(\frac{R_{leak}}{R + R_{leak}}\right)V_s = \left(\frac{10\text{ k}\Omega}{14.7\text{ k}\Omega}\right)10\text{ V} = 6.80\text{ V}$$

Then,

$$V_{out} = \left(\frac{X_C}{\sqrt{R_{th}^2 + X_C^2}}\right)V_{th} = \left(\frac{5\text{ k}\Omega}{\sqrt{(3.2\text{ k}\Omega)^2 + (5\text{ k}\Omega)^2}}\right)6.80\text{ V} = 5.73\text{ V}$$

***Related Exercise*** What would the output voltage be if the capacitor were not leaky?

**SECTION 16–9 REVIEW**

1. Describe the effect of a leaky capacitor on the response of an *RC* circuit.
2. In a series *RC* circuit, if all of the applied voltage appears across the capacitor, what is the problem?
3. What faults can cause 0 V across a capacitor in a series *RC* circuit?

## 16–10 ■ TECHnology Theory Into Practice

***In Chapter 13, you worked with the capacitively coupled input to an amplifier with voltage-divider bias. In this TECH TIP, you will check the output voltage and phase lag of a similar amplifier's input circuit to determine how they change with frequency. If too much voltage is dropped across the coupling capacitor, the overall performance of the amplifier is adversely affected. You should review the TECH TIP in Chapter 13 before proceeding.***

As you learned in Chapter 13, the coupling capacitor ($C_1$) in Figure 16–64 passes the input signal voltage to the input of the amplifier (point $A$ to point $B$) without affecting the dc level at point $B$ produced by the resistive voltage divider ($R_1$ and $R_2$). If the input frequency is high enough so that the reactance of the coupling capacitor is negligibly small, essentially no ac signal voltage is dropped across the capacitor. As the signal frequency is reduced, the capacitive reactance increases and more of the signal voltage is dropped across the capacitor. This lowers the overall voltage gain of the amplifier and thus degrades its performance.

**FIGURE 16–64**
***A capacitively coupled amplifier.***

The amount of signal voltage that is coupled from the input source (point $A$) to the amplifier input (point $B$) is determined by the values of the capacitor and the dc bias resistors (assuming the amplifier has no loading effect) in Figure 16–64. These components actually form a high-pass $RC$ filter, as shown in Figure 16–65. The voltage-divider bias resistors are effectively in parallel with each other as far as the ac source is concerned. The lower end of $R_2$ goes to ground and the upper end of $R_1$ goes to the dc supply voltage as shown in Figure 16–65(a). Since there is no ac voltage at the +18 V dc termi-

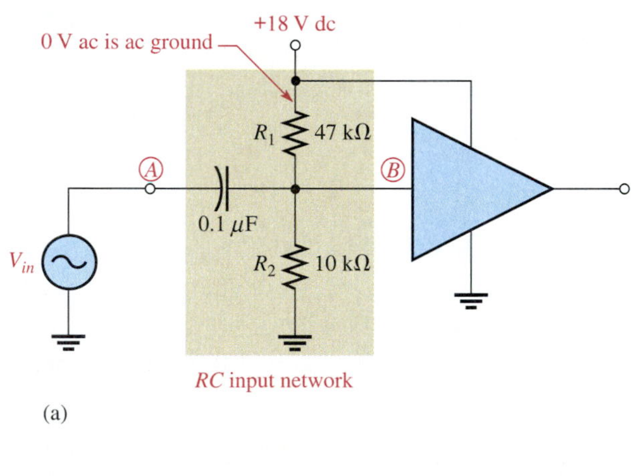


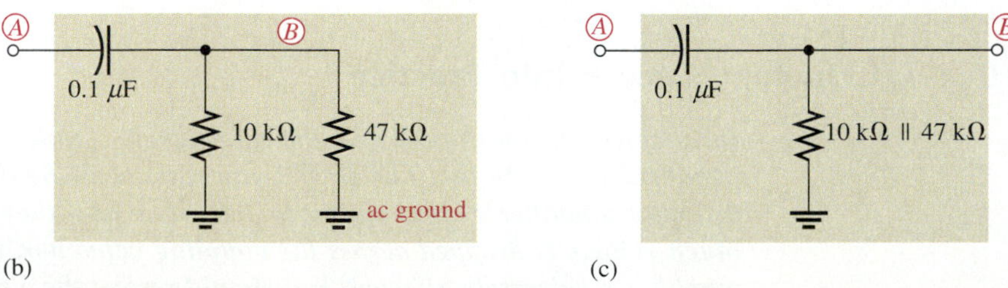


**FIGURE 16–65**
***The RC input circuit acts effectively like a high-pass RC filter.***

nal, the upper end of $R_1$ is at 0 V ac which is referred to as *ac ground.* The development of the circuit into an effective high-pass *RC* filter is shown in parts (b) and (c).

## The Amplifier Input Circuit

□ Determine the value of the equivalent resistance of the input circuit. Assume the amplifier has no loading effect on the input circuit.

## The Response at Frequency $f_1$

Refer to Figure 16–66. The input signal voltage is applied to the amplifier circuit board and displayed on channel 1 of the oscilloscope, and channel 2 is connected to a point on the circuit board.

□ Determine to what point on the circuit the channel 2 probe is connected, the frequency, and the voltage that should be displayed.

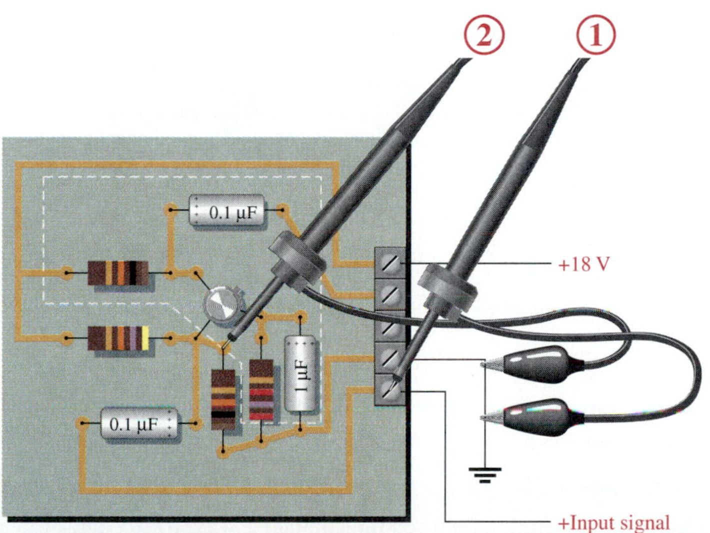


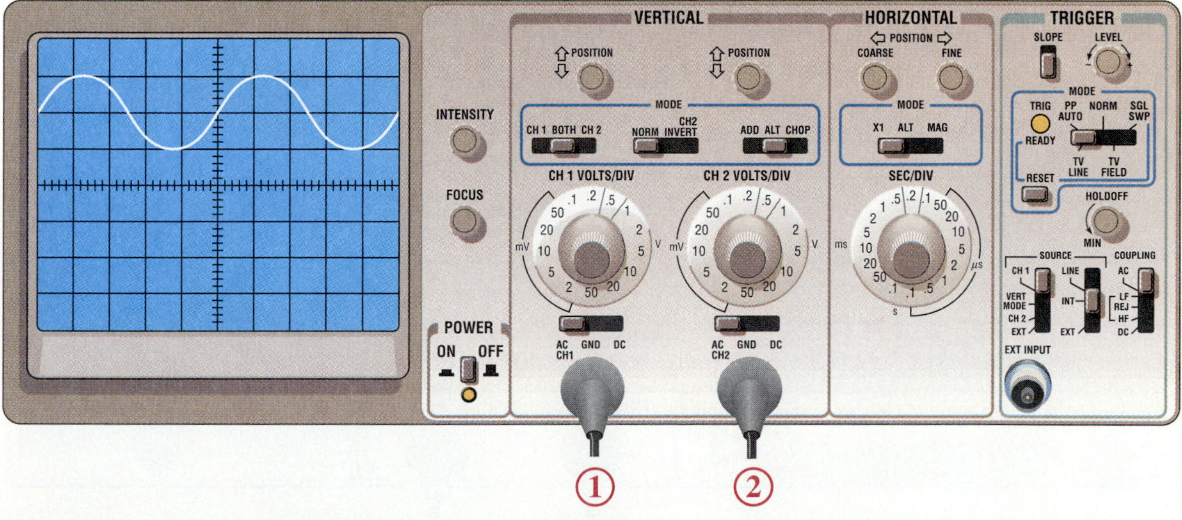


FIGURE 16–66

*Measuring the input circuit response at frequency $f_1$. Circled numbers relate scope inputs to the probes. The channel 1 waveform is shown.*

## The Response at Frequency $f_2$

Refer to Figure 16–67 and the circuit board in Figure 16–66. The input signal voltage displayed on channel 1 of the oscilloscope is applied to the amplifier circuit board.

- ☐ Determine the frequency and the voltage that should be displayed on channel 2.
- ☐ State the difference between the channel 2 waveforms determined for $f_1$ and $f_2$. Explain the reason for the difference.

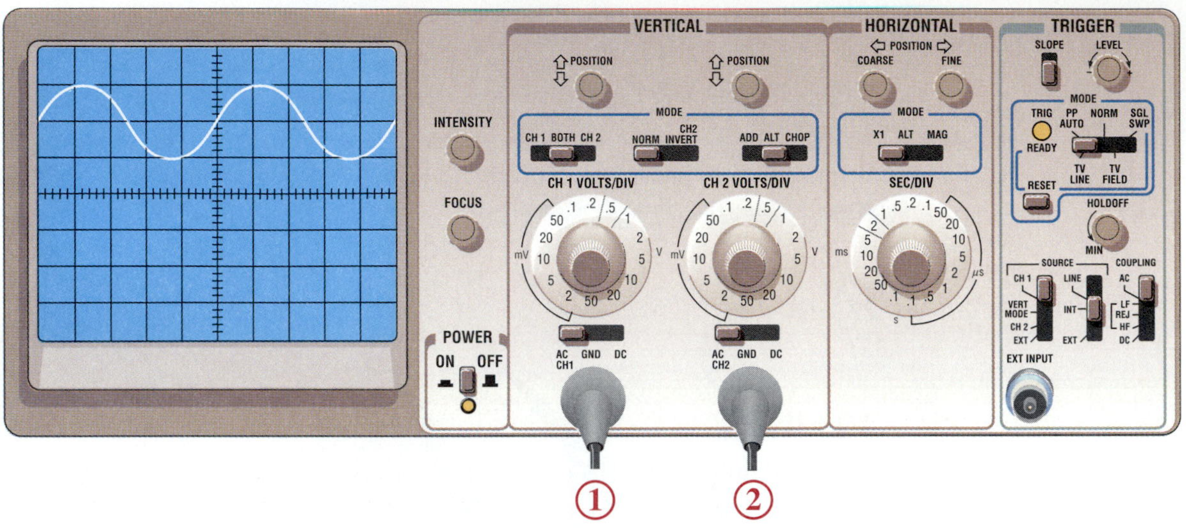


FIGURE 16–67
*Measuring the input circuit response at frequency $f_2$. The channel 1 waveform is shown.*

## The Response at Frequency $f_3$

Refer to Figure 16–68 and the circuit board in Figure 16–66. The input signal voltage displayed on channel 1 of the oscilloscope is applied to the amplifier circuit board.

- ☐ Determine the frequency and the voltage that should be displayed on channel 2.
- ☐ State the difference between the channel 2 waveforms determined for $f_2$ and $f_3$. Explain the reason for the difference.

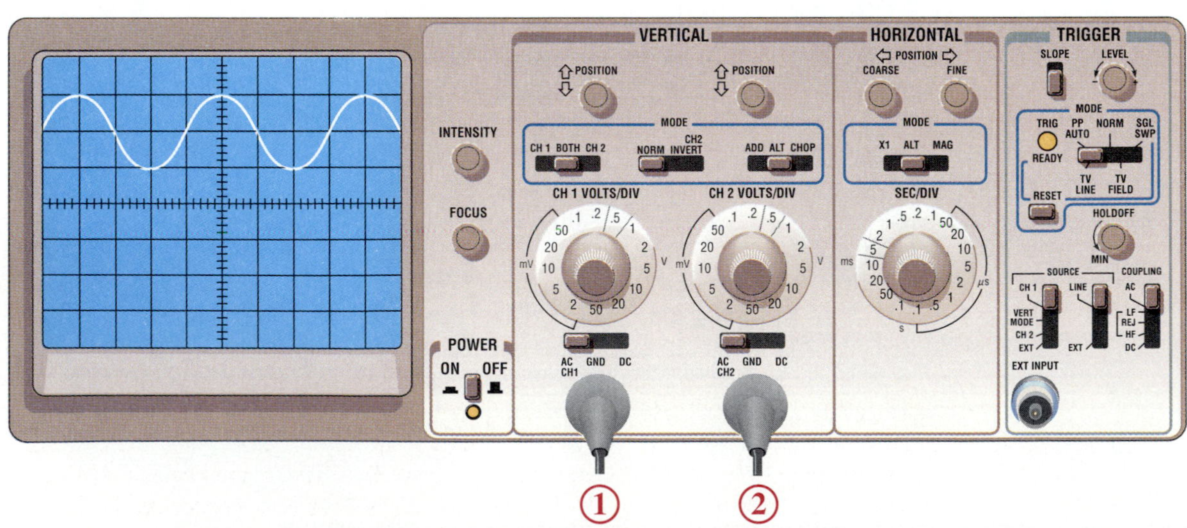


FIGURE 16–68
*Measuring the input circuit response at frequency $f_3$. The channel 1 waveform is shown.*

### Response Curve for the Amplifier Input Circuit

- ☐ Determine the frequency at which the signal voltage at point *B* in Figure 16–64 is 70.7% of its maximum value.
- ☐ Plot the response curve using this voltage value and the values at frequencies $f_1$, $f_2$, and $f_3$.
- ☐ How does this curve show that the input circuit acts as a high-pass filter?
- ☐ What can you do to the circuit to lower the frequency at which the voltage is 70.7% of maximum without affecting the dc bias voltage?

**SECTION 16–10 REVIEW**

1. Explain the effect on the response of the amplifier input circuit of reducing the value of the coupling capacitor.
2. What is the voltage at point *B* in Figure 16–64 if the coupling capacitor opens when the ac input signal is 10 mV rms?
3. What is the voltage at point *B* in Figure 16–64 if resistor $R_1$ is open when the ac input signal is 10 mV rms?

## 16–11 ■ PSpice ANALYSIS OF *RC* CIRCUITS

***So far you have used PSpice to analyze the response of a circuit at one frequency. In this section, you will learn how to determine the frequency response of RC circuits and see how the voltages vary as the frequency changes.***

***After completing this section, you should be able to***

- **Use PSpice and Probe to determine the frequency and phase responses of an *RC* circuit**
  - ☐ Use the .AC control statement
  - ☐ Analyze a series *RC* circuit
  - ☐ Analyze a parallel *RC* circuit
  - ☐ Determine the impedance of a circuit with Probe

### The .AC (AC Analysis) Control Statement

In previous chapters, two control statements, .DC and .TRAN were introduced. The format of the .AC statement is similar to the .DC and .TRAN statements. It specifies a step or increment and range over which PSpice generates data. The basic difference is that while the .DC statement takes samples over a voltage range and the .TRAN statement takes samples over a time range, the .AC statement takes samples over a frequency range. The general format of the .AC control statement is

```
.AC<Sweep type> <Samples> <Start freq> <End freq>
```

The <Sweep type> indicates how to vary or "sweep" the frequency; the <Samples> value specifies the number of samples to take per sweep interval, and the <Start freq> and <End freq> values (in Hertz) determine the range of frequencies.

PSpice uses three types of frequency variations or sweeps: LIN (linear), OCT (by octaves), or DEC (by decades). The <Samples> value specifies the total number of samples taken if the LIN sweep type is specified. It specifies the number of samples taken per octave of frequency if OCT is specified or the number of samples per decade of frequency if DEC is specified. For now, only the linear sweep is discussed. For the AC linear sweep, the format is

```
.AC LIN <Samples> <Start freq> <End freq>
```

For example, the statement

```
.AC LIN 200 100 100K
```

tells PSpice to vary (sweep) the frequency in a circuit from 100 Hz to 100 kHz and take a total of 200 samples within that frequency range.

## Series *RC* Circuit Analysis

Recall from Chapter 11 that the complete general format for an independent voltage or current source was discussed and the AC <value><phase> parameters were introduced. We will now make use of these parameters.

Suppose you want to find the frequency response of the circuit in Figure 16–69 by taking 1000 samples from 1 Hz to 10 kHz using Probe. A circuit file is as follows:

```
RC Circuit Response 1
*Series RC Circuit Frequency Response
VS  1  0  AC  5
R1  1  2  1K
C1  2  0  1U
.AC  LIN  1000  1  10K
.PROBE
.END
```

**FIGURE 16–69**

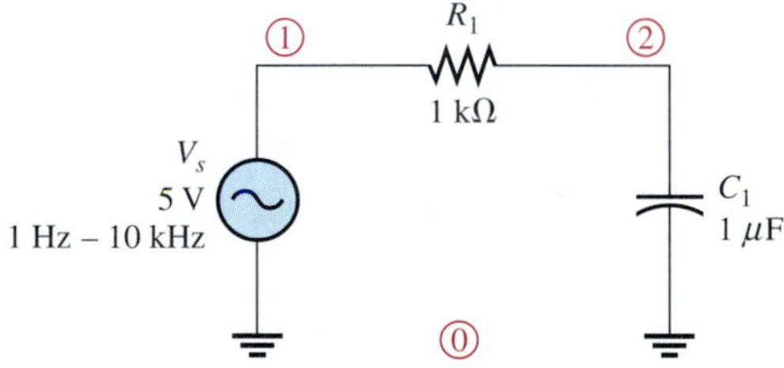


Notice that you do not need to specify a sinusoidal source because PSpice assumes an ac voltage to be sinusoidal. PSpice needs SIN only when the transient response is required. When PSpice is run on this circuit file and Probe started, you will see that the *x*-axis runs from 1 Hz to 10 kHz, as the .AC control statement specified. Think of the "AC" part of the source description as going with the .AC control statement.

Add the following traces to the display:

```
V(1,0)  V(1,2)  V(2,0)
```

in order to view the source voltage, the resistor voltage, and the capacitor voltage as the frequency changes. The source voltage amplitude remains constant at 5 V, while the resistor voltage increases and the capacitor voltage decreases. At a point past 100 Hz, the capacitor and resistor voltages are equal. This point is the *critical frequency,* which is also known as the *half-power, corner,* or *cutoff frequency.* Using the cursors, you will find that the critical frequency is about 159 Hz, and the resistor and capacitor voltages are approximately 3.5 V. By moving the cursors back and forth, the resistor and capacitor voltages at any frequency between 1 Hz and 10 kHz can be found.

## Phase Analysis with Probe

First, remove all traces from the display, and then display the following:

```
VP(1,0)  VP(1,2)  VP(2,0)
```

Probe will again draw three curves for the source voltage, resistor voltage, and capacitor voltage; however, this time the *y*-axis runs from −100d to 100d and the graphs are different than before. Entering VP instead of V for the voltages instructs Probe to display the phase of the voltages. The "d" after the numbers stands for *degrees,* with the phase angle of the source voltage equal to 0°.

From the graph, you can see that the phase of the resistor voltage varies from about 90° to about 0°, while the phase of the capacitor voltage varies from 0° to –90°. These two phases parallel each other as expected, and the angle between the resistor and the capacitor voltages remains 90° with the resistor voltage leading the capacitor voltage. Again, by using the cursor, you can determine the phase angle of any voltage at any frequency between 1 Hz and 10 kHz.

***Phase Angle of the Source*** Since a phase angle was not specified in the circuit file for VS, PSpice assumed a value of 0°. For source phase angles other than 0°, the angle must be specified. For example, if the source voltage is to have a phase angle of 45°, the description in the circuit file would be

```
VS  1  0  AC  5  45
```

## Parallel *RC* Circuit Analysis

The circuit file for the parallel circuit in Figure 16–70 is written as follows:

```
RC Circuit Response 2
*Parallel RC Circuit Frequency Response
VS  1  0  AC  5
R1  1  0  1K
C1  1  0  4.7U
.AC  LIN  1000  1  10K
.PROBE
.END
```

FIGURE 16–70

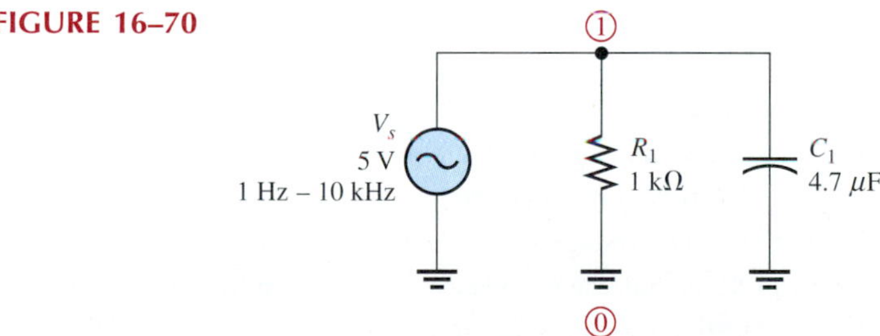


Since the resistor and capacitor are in parallel with the voltage source, the phases and magnitudes of the voltages are identical. The currents, however, are not the same. After running PSpice and starting Probe, select the following traces for display:

```
I(VS)  I(R1)  I(C1)
```

Probe will draw the graphs of the currents through the source, resistor, and capacitor. Because resistance is not dependent on frequency, the current through $R_1$ remains constant, while the total current and the current through $C_1$ increase with frequency. This plot shows what sometimes happens in Probe when traces overlap or when the plot scales "dwarf" a displayed trace. Generally, you can change the trace selection and/or plot scales to let you see a particular trace more clearly. By changing the $x$-axis range to extend from 1 Hz to 100 Hz, you will find that the resistor and capacitor currents are equal at a certain point, which is the critical frequency (a little above 30 Hz).

To see the phase angles, clear the display, reset the $x$-axis range to Auto and enter

```
IP(VS)  IP(R1)  IP(C1)
```

As expected, the phase of the resistor current is constant at 0° and the capacitor current remains at 90°. The total current from the source, however, varies from –180° to –90° as the frequency increases, indicating that the capacitor current affects the circuit phase angle.

### Resistance, Reactance, and Impedance

To determine the magnitudes of the resistance, reactance, or impedance in a circuit, the ratio of voltage to current must be used. To determine the phase angle of the resistance, reactance, or impedance, the difference between the voltage and current phase angles must be used. For example, to find the total impedance magnitude in the circuit of Figure 16–70, display the following trace expression:

V(1,0)/I(VS)

To find the phase angle, display the following trace expression:

VP(1,0) –IP(VS)

The trace will vary with frequency, so you must use the cursor to determine magnitude and phase angle at a specific frequency.

**SECTION 16–11 REVIEW**

1. Rewrite the circuit file for the circuit in Figure 16–69 for the following changes: $R_1$ = 2.2 kΩ, $C_1$ = 5.6 μF, source phase angle is 30°, and the frequency range is 100 Hz to 100 kHz.
2. Rewrite the circuit file for the circuit in Figure 16–70 if a 560 Ω resistor is added in series with the source and connected to the junction of $R_1$ and $C_1$.

## ▪ SUMMARY

- When a sine wave voltage is applied to an *RC* circuit, the current and all the voltage drops are also sine waves.
- Total current in an *RC* circuit always leads the source voltage.
- The resistor voltage is always in phase with the current.
- The capacitor voltage always lags the current by 90°.
- In an *RC* circuit, the impedance is determined by both the resistance and the capacitive reactance combined.
- Impedance is expressed in units of ohms.
- The circuit phase angle is the angle between the total current and the applied (source) voltage.
- The impedance of a series *RC* circuit varies inversely with frequency.
- The phase angle ($\theta$) of a series *RC* circuit varies inversely with frequency.
- For each parallel *RC* circuit, there is an equivalent series circuit for any given frequency.
- The impedance of a circuit can be determined by measuring the applied voltage and the total current and then applying Ohm's law.
- In an *RC* circuit, part of the power is resistive and part reactive.
- The phasor combination of resistive power (true power) and reactive power is called *apparent power.*
- Apparent power is expressed in volt-amperes (VA).
- The power factor (*PF*) indicates how much of the apparent power is true power.
- A power factor of 1 indicates a purely resistive circuit, and a power factor of 0 indicates a purely reactive circuit.
- In a lag network, the output voltage lags the input voltage in phase.
- In a lead network, the output voltage leads the input voltage.
- A filter passes certain frequencies and rejects others.

## ■ GLOSSARY

**Admittance** A measure of the ability of a reactive circuit to permit current; the reciprocal of impedance. The unit is the siemens (S).

**Apparent power** The phasor combination of resistive power (true power) and reactive power. The unit is the volt-ampere (VA).

**Bandwidth** The range of frequencies that is considered to be passed by a filter.

**Bias** The application of a dc voltage to an electronic device to produce a desired mode of operation.

**Capacitive susceptance** The ability of a capacitor to permit current: the reciprocal of capacitive reactance. The unit is the siemens (S).

**Conductance** The reciprocal of resistance. The unit is the siemens (S).

**Cutoff frequency** The frequency at which the output voltage of a filter is 70.7% of the maximum output voltage.

**Filter** A type of circuit that passes certain frequencies and rejects all others.

**Frequency response** In electrical circuits, the variation in the output voltage (or current) over a specified range of frequencies.

**High-pass filter** A certain type of filter whereby higher frequencies are passed and lower frequencies are rejected.

**Impedance** The total opposition to sinusoidal current expressed in ohms.

**Lag** Refers to a condition of the phase or time relationship of waveforms in which one waveform is behind the other in phase or time.

**Lead** Refers to a condition of the phase or time relationship of waveforms in which one waveform is ahead of the other in phase or time.

**Low-pass filter** A certain type of filter whereby lower frequencies are passed and higher frequencies are rejected.

**Power factor** The relationship between volt-amperes and true power or watts. Volt-amperes multiplied by the power factor equals true power.

## ■ FORMULAS

### Series *RC* Circuits

**(16–1)** $\mathbf{X}_C = -jX_C$

**(16–2)** $\mathbf{Z} = R - jX_C$

**(16–3)** $Z = \sqrt{R^2 + X_C^2}$

**(16–4)** $\theta = -\tan^{-1}\left(\frac{X_C}{R}\right)$

**(16–5)** $\mathbf{Z} = \sqrt{R^2 + X_C^2}\angle -\tan^{-1}\left(\frac{X_C}{R}\right)$

**(16–6)** $\mathbf{V} = \mathbf{IZ}$

**(16–7)** $\mathbf{I} = \frac{\mathbf{V}}{\mathbf{Z}}$

**(16–8)** $\mathbf{Z} = \frac{\mathbf{V}}{\mathbf{I}}$

**(16–9)** $\mathbf{V}_s = V_R - jV_C$

**(16–10)** $\mathbf{V}_s = \sqrt{V_R^2 + V_C^2}\angle -\tan^{-1}\left(\frac{V_C}{V_R}\right)$

**(16–11)** $V_s = \sqrt{V_R^2 + V_C^2}$

**(16–12)** $\theta = -\tan^{-1}\left(\frac{V_C}{V_R}\right)$

## Parallel *RC* Circuits

**(16–13)** $\mathbf{Z} = \left(\frac{RX_C}{\sqrt{R^2 + X_C^2}}\right)\angle\left(-90° + \tan^{-1}\left(\frac{X_C}{R}\right)\right)$

**(16–14)** $Z = \frac{RX_C}{\sqrt{R^2 + X_C^2}}$

**(16–15)** $\theta = -90° + \tan^{-1}\left(\frac{X_C}{R}\right)$

**(16–16)** $\mathbf{G} = \frac{1}{R\angle 0°} = G\angle 0°$

**(16–17)** $\mathbf{B}_C = \frac{1}{X_C\angle -90°} = B_C\angle 90° = +jB_C$

**(16–18)** $\mathbf{Y} = \frac{1}{Z\angle \pm\theta} = Y\angle \mp\theta$

**(16–19)** $\mathbf{Y} = G + jB_C$

**(16–20)** $\mathbf{V} = \frac{\mathbf{I}}{\mathbf{Y}}$

**(16–21)** $\mathbf{I} = \mathbf{VY}$

**(16–22)** $\mathbf{Y} = \frac{\mathbf{I}}{\mathbf{V}}$

**(16–23)** $\mathbf{I}_{tot} = I_R + jI_C$

**(16–24)** $\mathbf{I}_{tot} = \sqrt{I_R^2 + I_C^2}\angle\tan^{-1}\left(\frac{I_C}{I_R}\right)$

**(16–25)** $I_{tot} = \sqrt{I_R^2 + I_C^2}$

**(16–26)** $\theta = \tan^{-1}\left(\frac{I_C}{I_R}\right)$

**(16–27)** $R_{eq} = Z\cos\theta$

**(16–28)** $X_{C(eq)} = Z\sin\theta$

## Power in *RC* Circuits

**(16–29)** $P_{true} = I^2R$

**(16–30)** $P_r = I^2X_C$

**(16–31)** $P_a = I^2Z$

**(16–32)** $P_{true} = P_a\cos\theta$

**(16–33)** $P_{true} = VI\cos\theta$

**(16–34)** $PF = \cos\theta$

## Lag Network

**(16–35)** $\phi = -90° + \tan^{-1}\left(\frac{X_C}{R}\right)$

**(16–36)** $V_{out} = \left(\frac{X_C}{\sqrt{R^2 + X_C^2}}\right)V_{in}$

**(16–37)** $V_{out} = IX_C$

**(16–38)** $\mathbf{V}_{out} = V_{out}\angle\phi$

## Lead Network

**(16–39)** $\phi = \tan^{-1}\left(\frac{X_C}{R}\right)$

**(16–40)** $V_{out} = \left(\frac{R}{\sqrt{R^2 + X_C^2}}\right)V_{in}$

**(16–41)** $V_{out} = IR$

**(16–42)** $\mathbf{V}_{out} = V_{out}\angle\phi$

**(16–43)** $f_c = \frac{1}{2\pi RC}$

### Troubleshooting

**(16–44)** $R_{th} = \frac{RR_{leak}}{R + R_{leak}}$

**(16–45)** $V_{th} = \left(\frac{R_{leak}}{R + R_{leak}}\right)V_s$

## ■ SELF-TEST

**1.** In a series *RC* circuit, the voltage across the resistance is
(a) in phase with the source voltage (b) lagging the source voltage by 90°
(c) in phase with the current (d) lagging the current by 90°

**2.** In a series *RC* circuit, the voltage across the capacitor is
(a) in phase with the source voltage (b) lagging the resistor voltage by 90°
(c) in phase with the current (d) lagging the source voltage by 90°

**3.** When the frequency of the voltage applied to a series *RC* circuit is increased, the impedance
(a) increases (b) decreases (c) remains the same (d) doubles

**4.** When the frequency of the voltage applied to a series *RC* circuit is decreased, the phase angle
(a) increases (b) decreases (c) remains the same (d) becomes erratic

**5.** In a series *RC* circuit when the frequency and the resistance are doubled, the impedance
(a) doubles (b) is halved
(c) is quadrupled (d) cannot be determined without values

**6.** In a series *RC* circuit, 10 V rms is measured across the resistor and 10 V rms is also measured across the capacitor. The rms source voltage is
(a) 20 V (b) 14.14 V (c) 28.28 V (d) 10 V

**7.** The voltages in Question 6 are measured at a certain frequency. To make the resistor voltage greater than the capacitor voltage, the frequency
(a) must be increased (b) must be decreased
(c) is held constant (d) has no effect

**8.** When $R = X_C$, the phase angle is
(a) 0° (b) +90° (c) −90° (d) 45°

**9.** To decrease the phase angle below 45°, the following condition must exist:
(a) $R = X_C$ (b) $R < X_C$ (c) $R > X_C$ (d) $R = 10X_C$

**10.** When the frequency of the source voltage is increased, the impedance of a parallel *RC* circuit
(a) increases (b) decreases (c) does not change

**11.** In a parallel *RC* circuit, there is 1 A rms through the resistive branch and 1 A rms through the capacitive branch. The total rms current is
(a) 1 A (b) 2 A (c) 2.28 A (d) 1.414 A

**12.** A power factor of 1 indicates that the circuit phase angle is
(a) 90° (b) 45° (c) 180° (d) 0°

**13.** For a certain load, the true power is 100 W and the reactive power is 100 VAR. The apparent power is
(a) 200 VA (b) 100 VA (c) 141.4 VA (d) 141.4 W

**14.** Energy sources are normally rated in
(a) watts (b) volt-amperes (c) volt-amperes reactive (d) none of these

## ■ PROBLEMS

### SECTION 16–1 Sinusoidal Response of *RC* Circuits

1. An 8 kHz sinusoidal voltage is applied to a series *RC* circuit. What is the frequency of the voltage across the resistor? Across the capacitor?
2. What is the wave shape of the current in the circuit of Problem 1?

### SECTION 16–2 Impedance and Phase Angle of Series *RC* Circuits

3. Express the total impedance of each circuit in Figure 16–71 in both polar and rectangular forms.

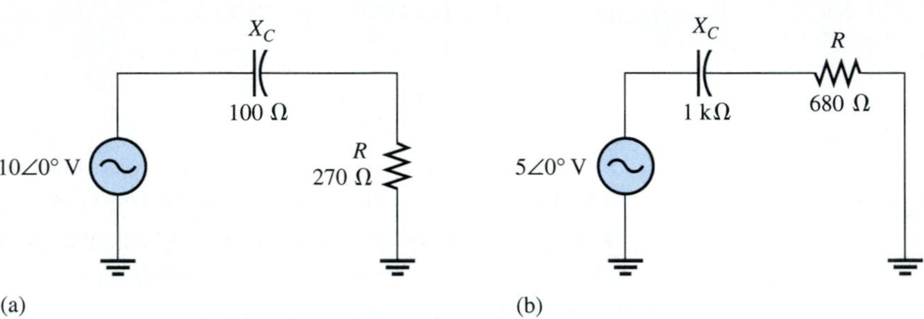


FIGURE 16–71

4. Determine the impedance magnitude and phase angle in each circuit in Figure 16–72.

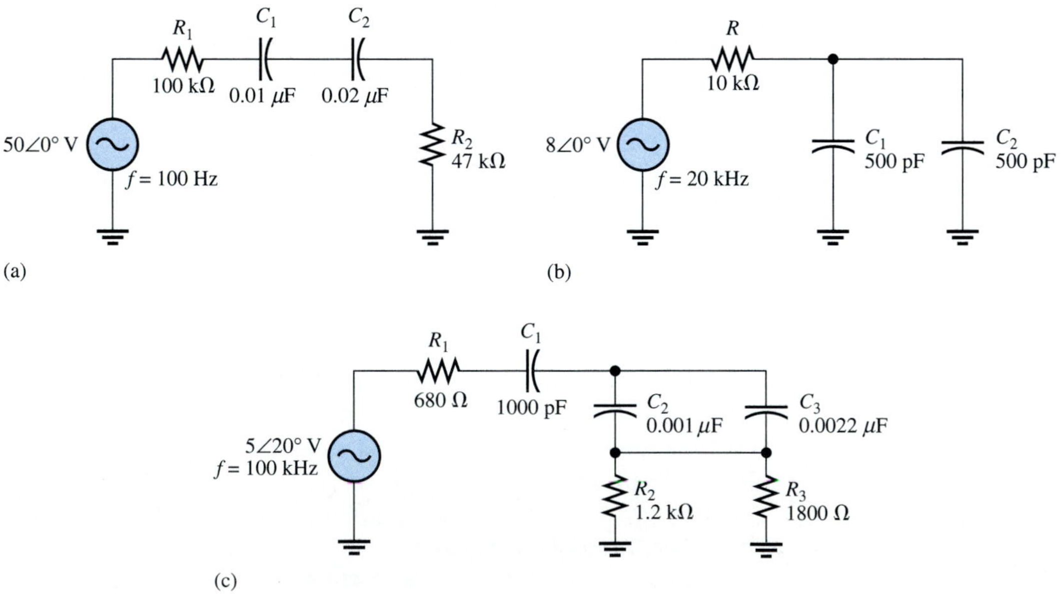


FIGURE 16–72

5. For the circuit of Figure 16–73, determine the impedance expressed in rectangular form for each of the following frequencies:

   **(a)** 100 Hz **(b)** 500 Hz **(c)** 1 kHz **(d)** 2.5 kHz

FIGURE 16–73

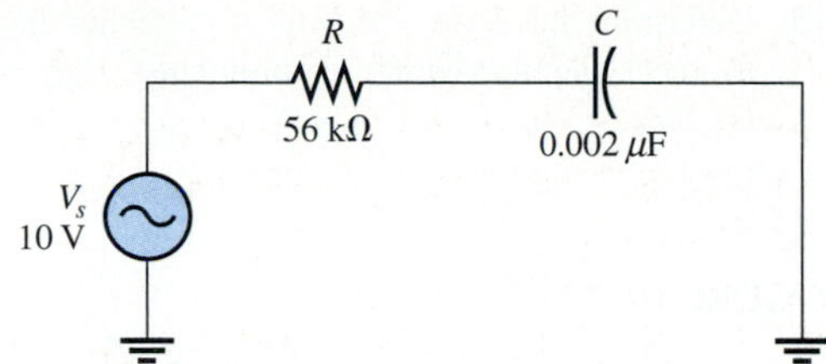


6. Repeat Problem 5 for $C = 0.005\ \mu F$.
7. Determine the values of $R$ and $X_C$ in a series $RC$ circuit for the following values of total impedance:
   (a) $\mathbf{Z} = 33\ \Omega - j50\ \Omega$ (b) $\mathbf{Z} = 300\angle -25°\ \Omega$
   (c) $\mathbf{Z} = 1.8\angle -67.2°\ \text{k}\Omega$ (d) $\mathbf{Z} = 789\angle -45°\ \Omega$

## SECTION 16–3 Analysis of Series *RC* Circuits

8. Express the current in polar form for each circuit of Figure 16–71.
9. Calculate the total current in each circuit of Figure 16–72 and express in polar form.
10. Determine the phase angle between the applied voltage and the current for each circuit in Figure 16–72.
11. Repeat Problem 10 for the circuit in Figure 16–73, using $f = 5$ kHz.
12. For the circuit in Figure 16–74, draw the phasor diagram showing all voltages and the total current. Indicate the phase angles.

FIGURE 16–74

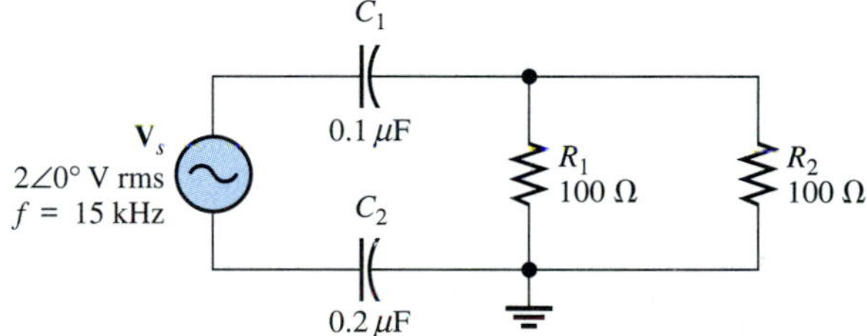


13. For the circuit in Figure 16–75, determine the following in polar form:
    (a) $\mathbf{Z}$ (b) $\mathbf{I}_{tot}$ (c) $\mathbf{V}_R$ (d) $\mathbf{V}_C$

FIGURE 16–75

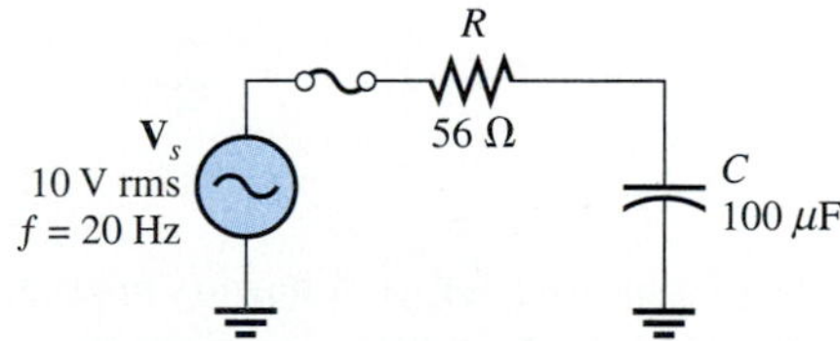


14. To what value must the rheostat be set in Figure 16–76 to make the total current 10 mA? What is the resulting angle?

FIGURE 16–76

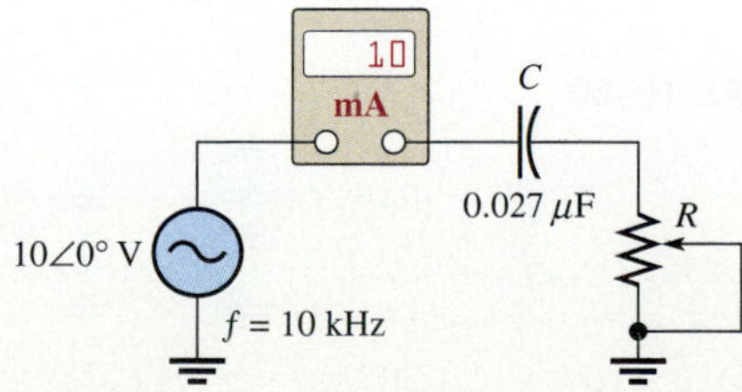

**15.** Determine the series element or elements that must be installed in the block of Figure 16–77 to meet the following requirements: $P_{true} = 400$ W and there is a leading power factor ($I_{tot}$ leads $V_s$).

**FIGURE 16–77**

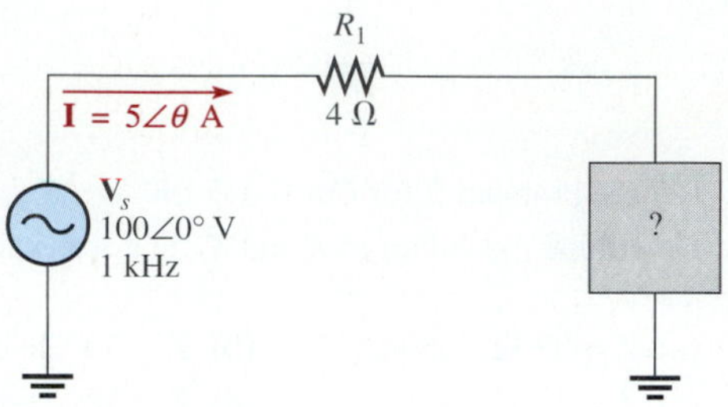

## SECTION 16–4 Impedance and Phase Angle of Parallel *RC* Circuits

**16.** Determine the impedance and express it in polar form for the circuit in Figure 16–78.

**FIGURE 16–78**

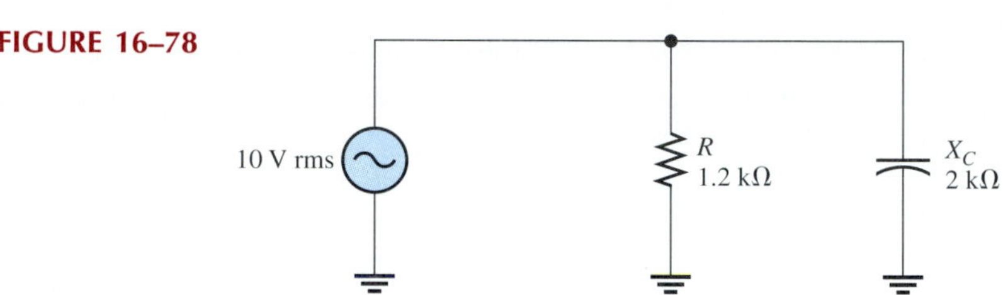


**17.** Determine the impedance magnitude and phase angle in Figure 16–79.

**FIGURE 16–79**

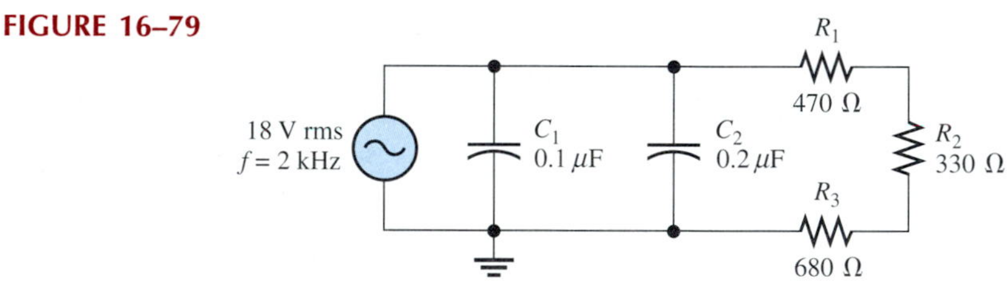


**18.** Repeat Problem 17 for the following frequencies:

**(a)** 1.5 kHz **(b)** 3 kHz **(c)** 5 kHz **(d)** 10 kHz

## SECTION 16–5 Analysis of Parallel *RC* Circuits

**19.** For the circuit in Figure 16–80, find all the currents and voltages in polar form.

**FIGURE 16–80**

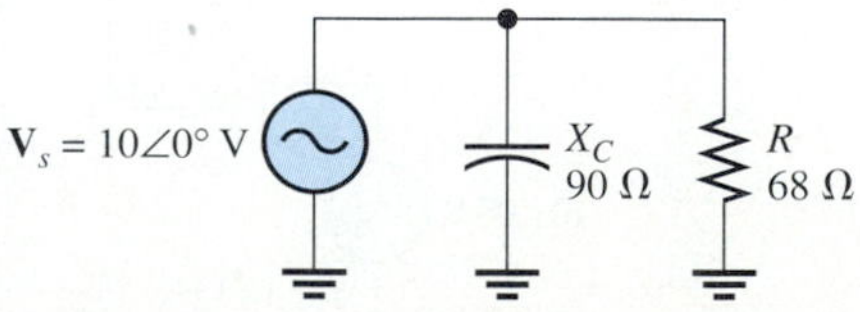

**20.** For the parallel circuit in Figure 16–81, find the magnitude of each branch current and the total current. What is the phase angle between the applied voltage and the total current?

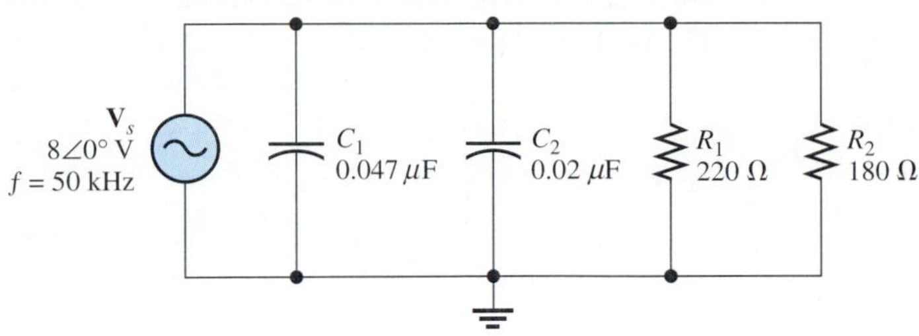


FIGURE 16–81

**21.** For the circuit in Figure 16–82, determine the following:

**(a)** **Z** **(b)** $\mathbf{I}_R$ **(c)** $\mathbf{I}_{C(tot)}$ **(d)** $\mathbf{I}_{tot}$ **(e)** $\theta$

**22.** Repeat Problem 21 for $R = 5.6\ \text{k}\Omega$, $C_1 = 0.05\ \mu\text{F}$, $C_2 = 0.022\ \mu\text{F}$, and $f = 500$ Hz.

FIGURE 16–82

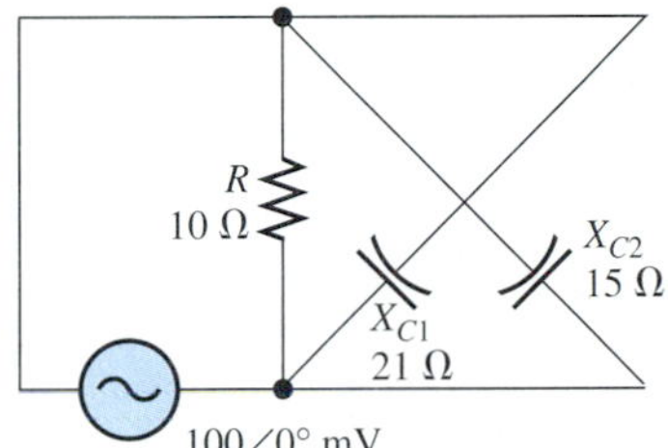


**23.** Convert the circuit in Figure 16–83 to an equivalent series form.

FIGURE 16–83

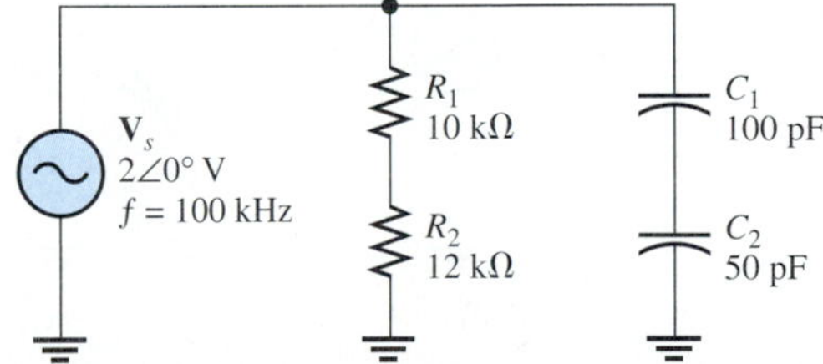


**24.** Determine the value to which $R_1$ must be adjusted to get a phase angle of 30° between the source voltage and the total current in Figure 16–84.

FIGURE 16–84

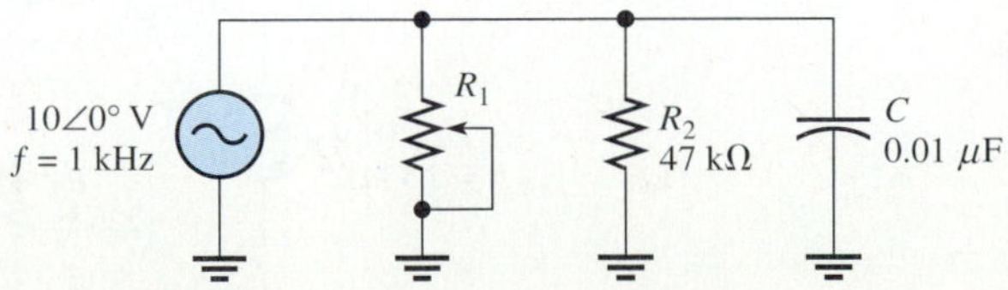

## SECTION 16–6 Series-Parallel *RC* Circuits

**25.** Determine the voltages in polar form across each element in Figure 16–85. Sketch the voltage phasor diagram.

**26.** Is the circuit in Figure 16–85 predominantly resistive or predominantly capacitive?

**27.** Find the current through each branch and the total current in Figure 16–85. Express the currents in polar form. Sketch the current phasor diagram.

FIGURE 16–85

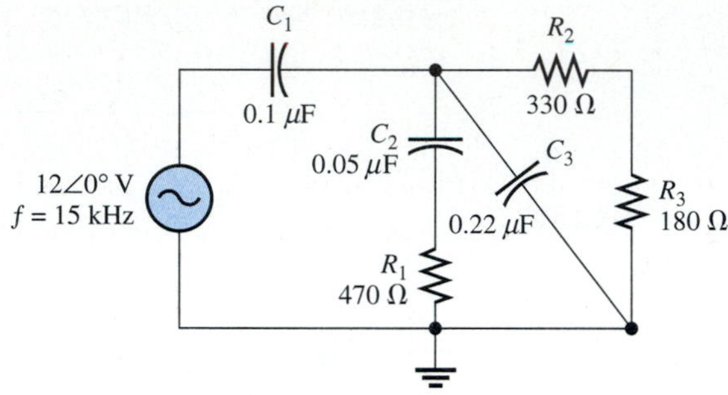


**28.** For the circuit in Figure 16–86, determine the following:

**(a)** $\mathbf{I}_{tot}$ **(b)** $\theta$ **(c)** $\mathbf{V}_{R1}$ **(d)** $\mathbf{V}_{R2}$ **(e)** $\mathbf{V}_{R3}$ **(f)** $\mathbf{V}_C$

FIGURE 16–86

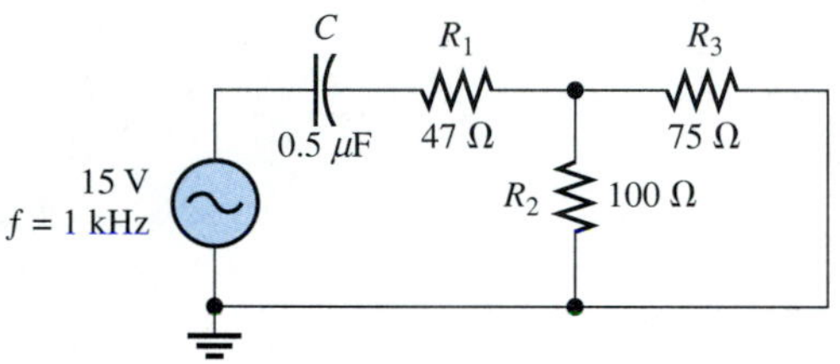


**29.** Determine the value of $C_2$ in Figure 16–87 when $V_A = V_B$.

FIGURE 16–87

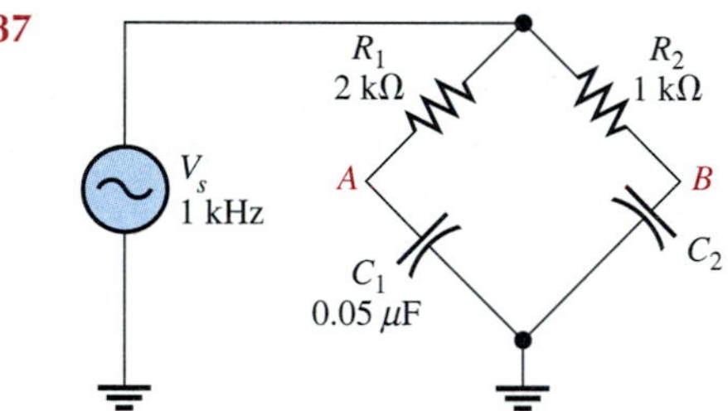


**30.** Determine the voltage and its phase angle at each point labeled in Figure 16–88.

**31.** Find the current through each component in Figure 16–88.

**32.** Sketch the voltage and current phasor diagram for Figure 16–88.

FIGURE 16–88

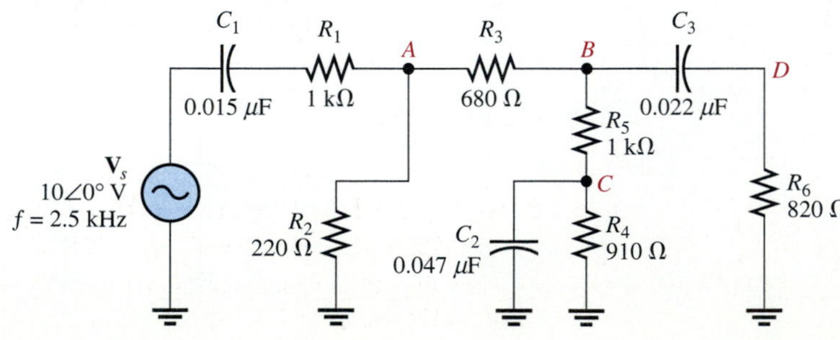

## SECTION 16–7 Power in *RC* Circuits

**33.** In a certain series *RC* circuit, the true power is 2 W, and the reactive power is 3.5 VAR. Determine the apparent power.

**34.** In Figure 16–75, what is the true power and the reactive power?

**35.** What is the power factor for the circuit of Figure 16–83?

**36.** Determine $P_{true}$, $P_r$, $P_a$, and *PF* for the circuit in Figure 16–86. Sketch the power triangle.

**37.** A single 240 V, 60 Hz source drives two loads. Load *A* has an impedance of 50 Ω and a power factor of 0.85. Load *B* has an impedance of 72 Ω and a power factor of 0.95.

**(a)** How much current does each load draw?

**(b)** What is the reactive power in each load?

**(c)** What is the true power in each load?

**(d)** What is the apparent power in each load?

**(e)** Which load has more voltage drop along the lines connecting it to the source?

## SECTION 16–8 Basic Applications

**38.** For the lag network in Figure 16–89, determine the phase shift between the input voltage and the output voltage for each of the following frequencies:

**(a)** 1 Hz **(b)** 100 Hz **(c)** 1 kHz **(d)** 10 kHz

FIGURE 16–89

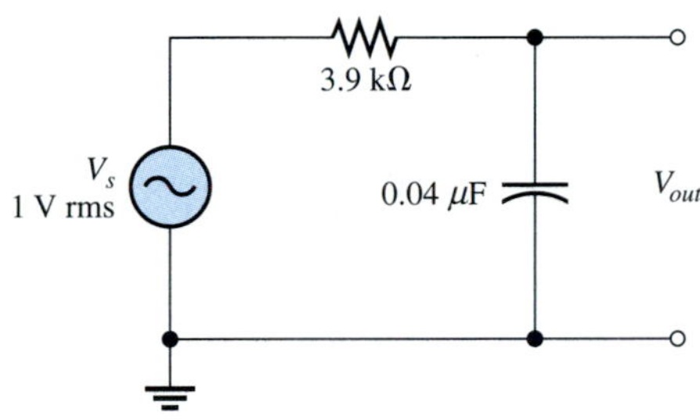


**39.** The lag network in Figure 16–89 also acts as a low-pass filter. Draw a response curve for this circuit by plotting the output voltage versus frequency for 0 Hz to 10 kHz in 1 kHz increments.

**40.** Repeat Problem 38 for the lead network in Figure 16–90.

FIGURE 16–90

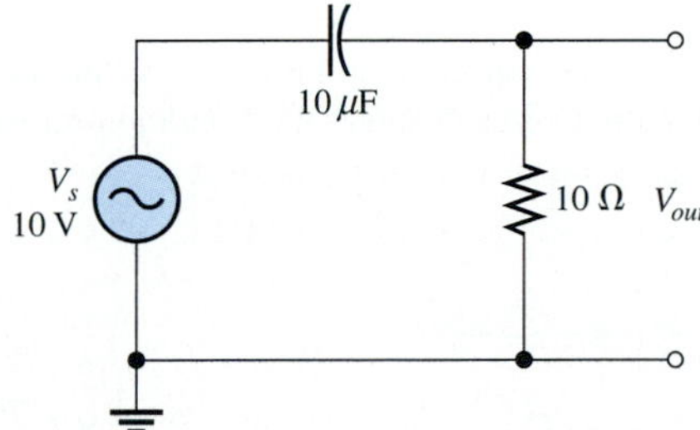


**41.** Plot the frequency response curve of the output amplitude for the lead network in Figure 16–90 for a frequency range of 0 Hz to 10 kHz in 1 kHz increments.

**42.** Draw the voltage phasor diagram for each circuit in Figures 16–89 and 16–90 for a frequency of 5 kHz with $V_s = 1$ V rms.

**43.** What value of coupling capacitor is required in Figure 16–91 so that the signal voltage at the input of amplifier 2 is at least 70.7% of the signal voltage at the output of amplifier 1 when the frequency is 20 Hz?

**FIGURE 16–91**

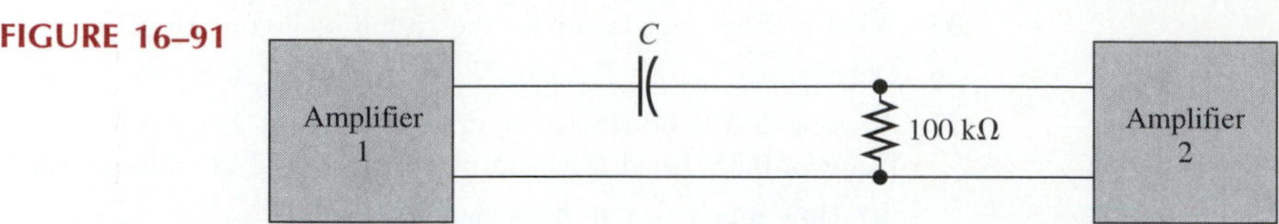


**44.** The rms value of the signal voltage out of amplifier *A* in Figure 16–92 is 50 mV. If the input resistance to amplifier *B* is 10 kΩ, how much of the signal is lost due to the coupling capacitor when the frequency is 3 kHz?

**FIGURE 16–92**

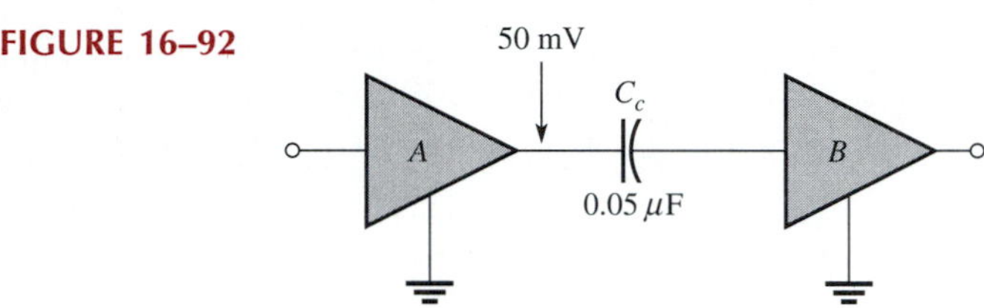


## SECTION 16–9 Troubleshooting

**45.** Assume that the capacitor in Figure 16–93 is excessively leaky. Show how this degradation affects the output voltage and phase angle, assuming that the leakage resistance is 5 kΩ and the frequency is 10 Hz.

**FIGURE 16–93**

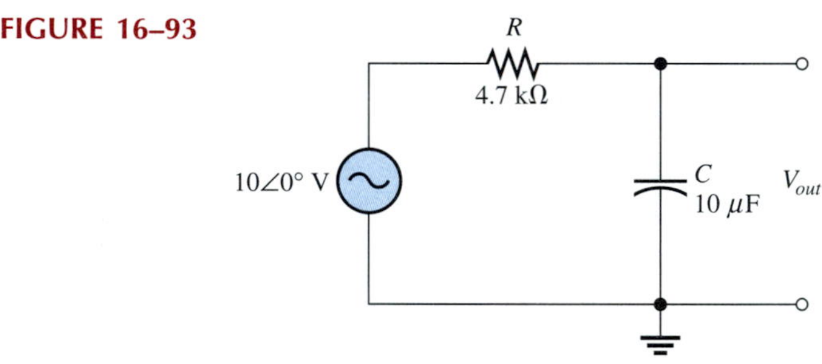


**46.** Each of the capacitors in Figure 16–94 has developed a leakage resistance of 2 kΩ. Determine the output voltages under this condition for each circuit.

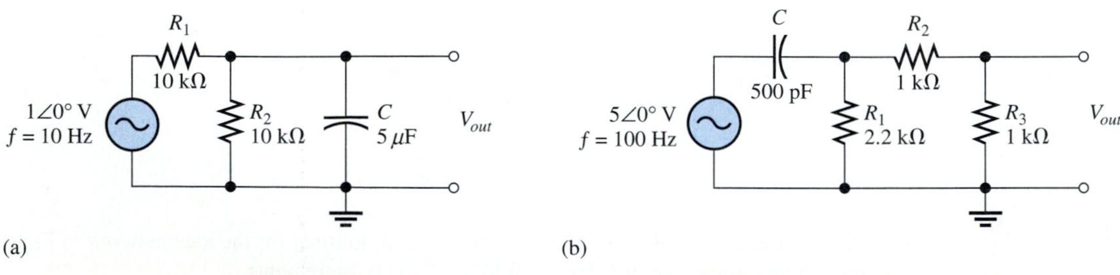


**FIGURE 16–94**

**47.** Determine the output voltage for the circuit in Figure 16–94(a) for each of the following failure modes, and compare it to the correct output:

**(a)** $R_1$ open **(b)** $R_2$ open **(c)** $C$ open **(d)** $C$ shorted

**48.** Determine the output voltage for the circuit in Figure 16–94(b) for each of the following failure modes, and compare it to the correct output:

**(a)** $C$ open **(b)** $C$ shorted **(c)** $R_1$ open **(d)** $R_2$ open **(e)** $R_3$ open

## ▪ ANSWERS TO SECTION REVIEWS

### Section 16–1

**1.** The voltage frequency is 60 Hz. The current frequency is 60 Hz.

**2.** The capacitive reactance causes the phase shift.

**3.** The phase angle is closer to 0°.

### Section 16–2

**1.** $R = 150\ \Omega$; $X_C = 220\ \Omega$

**2.** $\mathbf{Z} = 33\ \text{k}\Omega - j50\ \text{k}\Omega$

**3.** $Z = \sqrt{R^2 + X_C^2} = 59.9\ \Omega$; $\theta = -\tan^{-1}(X_C/R) = -56.6°$

### Section 16–3

**1.** $V_s = \sqrt{V_R^2 + V_C^2} = 7.21\ \text{V}$

**2.** $\theta = -\tan^{-1}(X_C/R) = -56.3°$

**3.** $\theta = 90°$

**4.** When $f$ increases, $X_C$ decreases, $Z$ decreases, and $\theta$ decreases.

### Section 16–4

**1.** Conductance is the reciprocal of resistance, capacitive susceptance is the reciprocal of capacitive reactance, and admittance is the reciprocal of impedance.

**2.** $Y = 1/Z = \dfrac{1}{\sqrt{R^2 + X_C^2}} = 10\ \text{mS}$

**3.** $\mathbf{Y} = 1/\mathbf{Z} = 25.1\angle 32.1°\ \text{mS}$

**4.** $Y = \sqrt{G^2 + B_C^2} = 25.1\ \text{mS}$; $\theta = \tan^{-1}(B_C/G) = 32.1°$

### Section 16–5

**1.** $I_{tot} = V_sY = 21\ \text{mA}$

**2.** $I_{tot} = \sqrt{I_R^2 + I_C^2} = 18\ \text{mA}$; $\theta = \tan^{-1}(I_C/I_R) = 56.3°$; $\theta$ is with respect to applied voltage.

**3.** $\theta = 90°$

### Section 16–6

**1.** See Figure 16–95.

**2.** $\mathbf{Z}_{tot} = \mathbf{V}_s/\mathbf{I}_{tot} = 36.9\angle -51.6°\ \Omega$

FIGURE 16–95

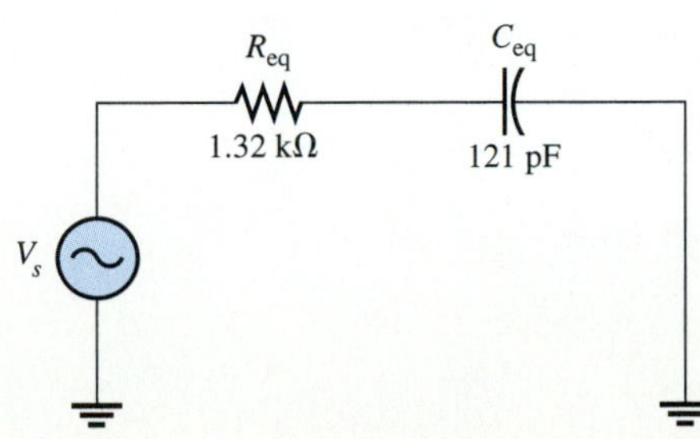

### Section 16–7

1. Energy loss is due to resistance.
2. $PF = \cos\theta = 0.707$
3. $P_{true} = I^2R = 1.32$ kW; $P_r = I^2X_C = 1.84$ kVAR; $P_a = I^2Z = 2.26$ kVA

### Section 16–8

1. $\phi = -90° + \tan^{-1}(X_C/R) = -62.8°$
2. $V_{out} = \left(R/\sqrt{R^2 + X_C^2}\right)V_{in} = 8.90$ V rms
3. The output is across the capacitor.

### Section 16–9

1. The leakage resistance acts in parallel with *C*, which alters the circuit time constant.
2. The capacitor is open.
3. A shorted capacitor or an open resistor can cause 0 V across the capacitor.

### Section 16–10

1. A lower value coupling capacitor will increase the frequency at which a significant drop in voltage occurs.
2. $V_B = 3.16$ V dc
3. $V_B = 10$ mV rms

### Section 16–11

1. Change the following lines;

```
VS  1  0  AC  5  30
R1  1  2  2.2K
C1  2  0  5.6U
.AC  LIN  1000  100  100K
```

2. Change lines for R1 and C1 and add new line for R2.

```
R1  2  0  1K
R2  1  2  560
C1  2  0  4.7U
```

## ■ ANSWERS TO RELATED EXERCISES FOR EXAMPLES

**16–1** [5] [6] [x⇆y] [1] [0] [0] [+/−] [INV] [2nd] [P–R] displays $\theta$. [x⇆y] displays magnitude.

**16–2** $\mathbf{V}_s = 2.56\angle -38.5°$ V

**16–3** $\mathbf{I} = 3.68\angle 35.9°$ mA

**16–4** $Z = 15.9$ kΩ, $\theta = -86.4°$

**16–5** $\mathbf{Z} = 24.3\angle -76.0°$ Ω

**16–6** $\mathbf{Y} = 4.36\angle 46.0°$ mS

**16–7** $\mathbf{I} = 5.91\angle 39.6°$ mA

**16–8** $\mathbf{I}_{tot} = 117\angle 31.0°$ mA

**16–9** $R_{eq} = 8.99$ kΩ, $X_{C(eq)} = 4.38$ kΩ

**16–10** $\mathbf{V}_1 = 7.14\angle 8.9°$ V, $\mathbf{V}_2 = 3.16\angle -20.4°$ V

**16–11** $\mathbf{V}_{R1} = 760\angle 67.5° \text{ mV}$; $\mathbf{V}_{C1} = 1.85\angle -22.5° \text{ V}$; $\mathbf{V}_{R2} = 1.53\angle 40.3° \text{ V}$; $\mathbf{V}_{C2} = 1.29\angle -49.7° \text{ V}$; See Figure 16–96.

FIGURE 16–96

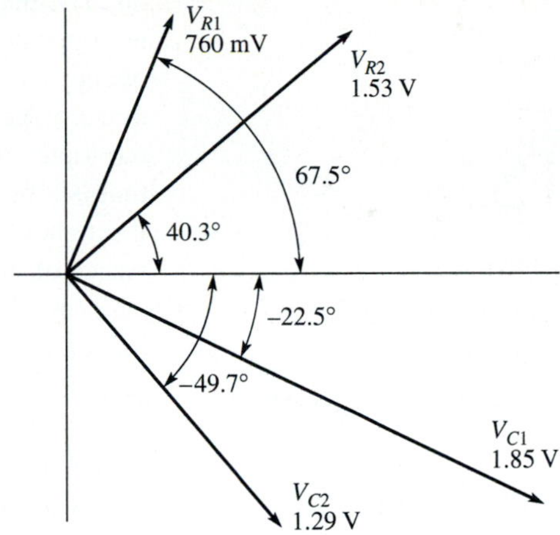


**16–12** $PF = 0.155$

**16–13** $P_{\text{true}} = 990\ \mu\text{W}$

**16–14** The phase lag increases.

**16–15** The output voltage decreases.

**16–16** The phase lead decreases.

**16–17** The output voltage increases.

**16–18** $V_{out} = 7.29\text{ V}$

# 17

# *RL* CIRCUITS



## ■ INTRODUCTION

In this chapter you will study series and parallel *RL* circuits. The analyses of *RL* and *RC* circuits are similar. The major difference is that the phase responses are opposite; inductive reactance increases with frequency, while capacitive reactance decreases with frequency. As discussed in the opening of Chapter 16, the topics in these two chapters may be interwoven if that is a preferred approach. In this chapter and throughout the rest of the book, you will learn the basics of putting technology theory into practice.

An *RL* circuit contains both resistance and inductance. It is one of the basic types of reactive circuits that you will study. In this chapter, basic series and parallel *RL* circuits and their responses to sinusoidal ac voltages are presented. Series-parallel combinations are also analyzed. True, reactive, and apparent power in *RL* circuits are discussed and some basic *RL* applications are introduced. Applications of *RL* circuits include filters, and phase shift networks. Troubleshooting and PSpice analysis complete the chapter.

## TECHnology Theory Into Practice

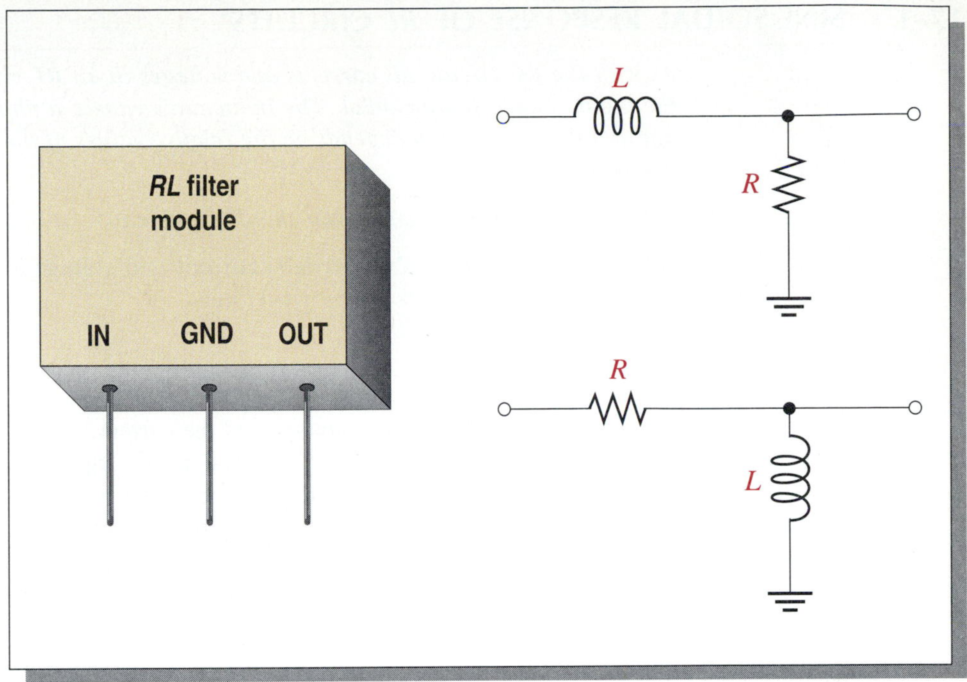


In the TECH TIP assignment in Section 17–10, you will use your knowledge of *RL* circuits to determine, based on parameter measurements, the type of filter circuits and their component values that are encapsulated in sealed modules.

## CHAPTER OBJECTIVES

- Describe the relationship between current and voltage in an *RL* circuit
- Determine impedance and phase angle in a series *RL* circuit
- Analyze a series *RL* circuit
- Determine impedance and phase angle in a parallel *RL* circuit
- Analyze a parallel *RL* circuit
- Analyze series-parallel *RL* circuits
- Determine power in *RL* circuits
- Discuss some basic *RL* applications
- Troubleshoot *RL* circuits
- Use PSpice and Probe to determine the frequency and phase responses of an *RL* circuit (optional)

## 17–1 ▪ SINUSOIDAL RESPONSE OF *RL* CIRCUITS

***As with the RC circuit, all currents and voltages in an RL circuit are sinusoidal when the input voltage is sinusoidal. The inductance causes a phase shift between the voltage and the current that depends on the relative values of the resistance and the inductive reactance.***

***After completing this section, you should be able to***

- **Describe the relationship between current and voltage in an *RL* circuit**
  - ☐ Discuss voltage and current waveforms
  - ☐ Discuss phase shift

As you will learn, the resistor voltage and the current lag the source voltage. The inductor voltage leads the source voltage. The phase angle between the current and the inductor voltage is always 90°. These generalized phase relationships are indicated in Figure 17–1. Notice that they are opposite from those of the *RC* circuit, as discussed in Chapter 16.

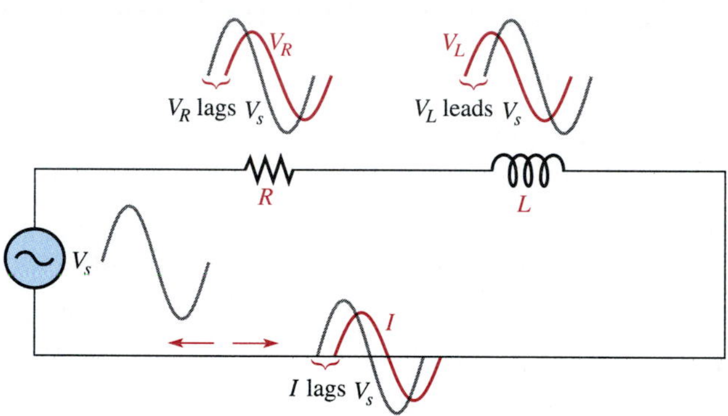


FIGURE 17–1
***Illustration of sinusoidal response with general phase relationships of $V_R$, $V_L$, and I relative to the source voltage. $V_R$ lags $V_s$, $V_L$ leads $V_s$, and I lags $V_s$. $V_R$ and I are in phase, while $V_R$ and $V_L$ are 90° out of phase with each other.***

The amplitudes and the phase relationships of the voltages and current depend on the values of the resistance and the **inductive reactance.** When a circuit is purely inductive, the phase angle between the applied voltage and the total current is 90°, with the current lagging the voltage. When there is a combination of both resistance and inductive reactance in a circuit, the phase angle is somewhere between 0° and 90°, depending on the relative values of the resistance and the reactance.

### SECTION 17–1 REVIEW

1. A 1 kHz sinusoidal voltage is applied to an *RL* circuit. What is the frequency of the resulting current?
2. When the resistance in an *RL* circuit is greater than the inductive reactance, do you think that the phase angle between the applied voltage and the total current is closer to 0° or to 90°?

## 17–2 ■ IMPEDANCE AND PHASE ANGLE OF SERIES *RL* CIRCUITS

***The impedance of an RL circuit is the total opposition to sinusoidal current and its unit is the ohm. The phase angle is the phase difference between the total current and the source voltage.***

***After completing this section, you should be able to***

- **Determine impedance and phase angle in a series *RL* circuit**
  - □ Express inductive reactance in complex form
  - □ Express total impedance in complex form
  - □ Calculate impedance magnitude and the phase angle

The impedance of a series *RL* circuit is determined by the resistance and the inductive reactance. Recall from Chapter 14 that inductive reactance is expressed as a phasor quantity in rectangular form as

$$\mathbf{X}_L = jX_L \quad \text{(17–1)}$$

In the series *RL* circuit of Figure 17–2, the total impedance is the phasor sum of $R$ and $jX_L$ and is expressed as

$$\mathbf{Z} = R + jX_L \quad \text{(17–2)}$$

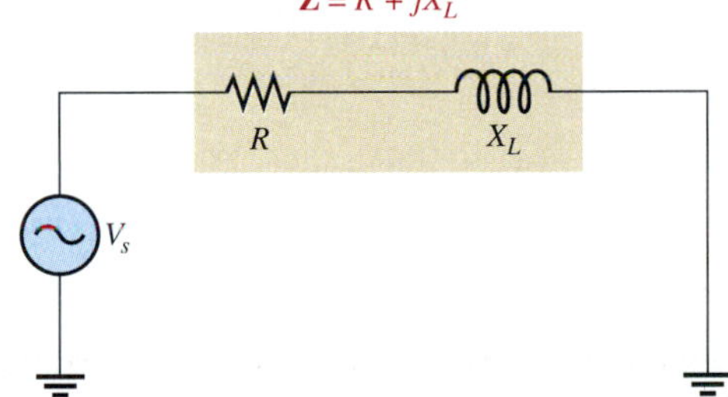


FIGURE 17–2
*Series RL circuit.*

### The Impedance Triangle

In ac analysis, both $R$ and $X_L$ are treated as phasor quantities, as shown in the phasor diagram of Figure 17–3(a), with $X_L$ appearing at a +90° angle with respect to $R$. This relationship comes from the fact that the inductor voltage leads the current, and thus the

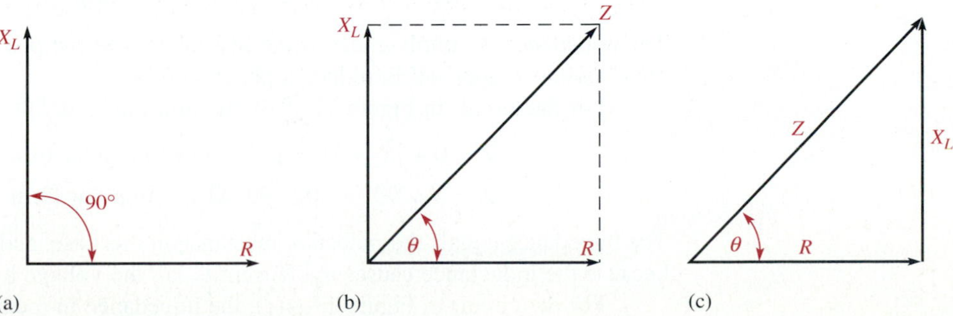


FIGURE 17–3
*Development of the impedance triangle for a series RL circuit.*

resistor voltage, by 90°. Since $\mathbf{Z}$ is the phasor sum of $R$ and $jX_L$, its phasor representation is shown in Figure 17–3(b). A repositioning of the phasors, as shown in part (c), forms a right triangle. This is called the *impedance triangle.* The length of each phasor represents the magnitude of the quantity, and $\theta$ is the phase angle between the applied voltage and the current in the *RL* circuit.

The impedance magnitude of the series *RL* circuit can be expressed in terms of the resistance and reactance as

$$Z = \sqrt{R^2 + X_L^2} \qquad \textbf{(17–3)}$$

The magnitude of the impedance is expressed in ohms.

The phase angle, $\theta$, is expressed as

$$\theta = \tan^{-1}\left(\frac{X_L}{R}\right) \qquad \textbf{(17–4)}$$

Combining the magnitude and the angle, the impedance can be expressed in polar form as

$$\mathbf{Z} = \sqrt{R^2 + X_L^2}\angle\tan^{-1}\left(\frac{X_L}{R}\right) \qquad \textbf{(17–5)}$$

**EXAMPLE 17–1** For each circuit in Figure 17–4, write the phasor expression for the impedance in both rectangular and polar forms.

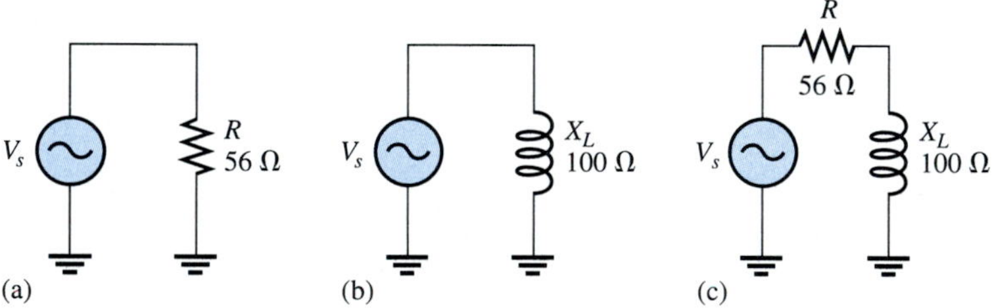


FIGURE 17–4

***Solution*** For the circuit in Figure 17–4(a), the impedance is

$$\mathbf{Z} = R + j0 = R = 56\ \Omega \qquad \text{in rectangular form } (X_L = 0)$$
$$\mathbf{Z} = R\angle 0° = 56\angle 0°\ \Omega \qquad \text{in polar form}$$

The impedance is simply equal to the resistance, and the phase angle is zero because pure resistance does not introduce a phase shift.

For the circuit in Figure 17–4(b), the impedance is

$$\mathbf{Z} = 0 + jX_L = j100\ \Omega \qquad \text{in rectangular form } (R = 0)$$
$$\mathbf{Z} = X_L\angle 90° = 100\angle 90°\ \Omega \qquad \text{in polar form}$$

The impedance equals the inductive reactance in this case, and the phase angle is +90° because the inductance causes the current to lag the voltage by 90°.

For the circuit in Figure 17–4(c), the impedance in rectangular form is

$$\mathbf{Z} = R + jX_L = 56\ \Omega + j100\ \Omega$$

The impedance in polar form is

$$\mathbf{Z} = \sqrt{R^2 + X_L^2}\angle\tan^{-1}\left(\frac{X_L}{R}\right)$$

$$= \sqrt{(56\ \Omega)^2 + (100\ \Omega)^2}\angle\tan^{-1}\left(\frac{100\ \Omega}{56\ \Omega}\right) = 115\angle 60.8°\ \Omega$$

In this case, the impedance is the phasor sum of the resistance and the inductive reactance. The phase angle is fixed by the relative values of $X_L$ and $R$.

Calculator sequences for $Z$ and $\theta$ for the circuit in Figure 17–4(c) are

$Z$: [5] [6] [$x^2$] [+] [1] [0] [0] [$x^2$] [=] [$\sqrt{x}$]
$\theta$: [1] [0] [0] [÷] [5] [6] [=] [INV] [tan]

***Related Exercise*** In a series *RL* circuit, $R = 1.8\ \text{k}\Omega$ and $X_L = 950\ \Omega$. Express the impedance in both rectangular and polar forms.

**SECTION 17–2 REVIEW**

1. The impedance of a certain *RL* circuit is $150\ \Omega + j220\ \Omega$. What is the value of the resistance? The inductive reactance?
2. A series *RL* circuit has a total resistance of 33 kΩ and an inductive reactance of 50 kΩ. Write the phasor expression for the impedance in rectangular form. Convert the impedance to polar form.

## 17–3 ■ ANALYSIS OF SERIES *RL* CIRCUITS

***In the previous section, you learned how to express the impedance of a series RL circuit. Now, Ohm's law and Kirchhoff's voltage law are used in the analysis of RL circuits.***

***After completing this section, you should be able to***

- **Analyze a series *RL* circuit**
  - ☐ Apply Ohm's law and Kirchhoff's voltage law to series *RL* circuits
  - ☐ Express the voltages and current as phasor quantities
  - ☐ Show how impedance and phase angle vary with frequency

### Ohm's Law

The application of Ohm's law to series *RL* circuits involves the use of the phasor quantities of **Z**, **V**, and **I**. The three equivalent forms of Ohm's law were stated in Chapter 16 for *RC* circuits. They apply also to *RL* circuits and are restated here:

$$\mathbf{V} = \mathbf{IZ} \qquad \mathbf{I} = \frac{\mathbf{V}}{\mathbf{Z}} \qquad \mathbf{Z} = \frac{\mathbf{V}}{\mathbf{I}}$$

Recall that since Ohm's law calculations involve multiplication and division operations, the voltage, current, and impedance should be expressed in polar form.

**EXAMPLE 17–2** The current in Figure 17–5 is expressed in polar form as $\mathbf{I} = 0.2\angle 0^\circ$ mA. Determine the source voltage expressed in polar form, and draw the phasor diagram.

FIGURE 17–5

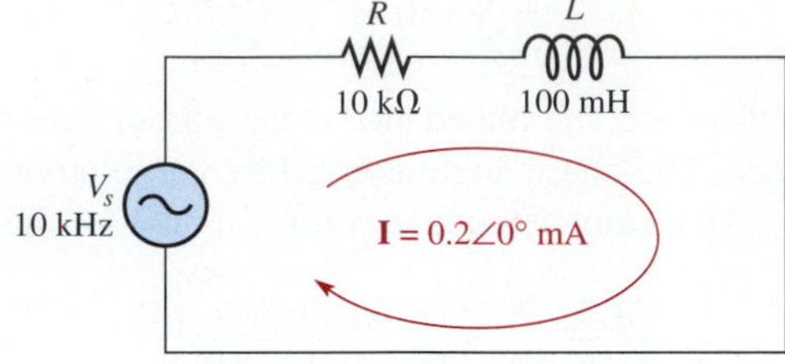


***Solution*** The inductive reactance is

$$X_L = 2\pi fL = 2\pi(10\text{ kHz})(100\text{ mH}) = 6.28\text{ k}\Omega$$

The impedance in rectangular form is

$$\mathbf{Z} = R + jX_L = 10\text{ k}\Omega + j6.28\text{ k}\Omega$$

Converting to polar form yields

$$\mathbf{Z} = \sqrt{R^2 + X_L^2}\angle\tan^{-1}\left(\frac{X_L}{R}\right)$$

$$= \sqrt{(10\text{ k}\Omega)^2 + (6.28\text{ k}\Omega)^2}\angle\tan^{-1}\left(\frac{6.28\text{ k}\Omega}{10\text{ k}\Omega}\right) = 11.8\angle 32.1^\circ\text{ k}\Omega$$

Use Ohm's law to determine the source voltage.

$$\mathbf{V}_s = \mathbf{IZ} = (0.2\angle 0^\circ\text{ mA})(11.8\angle 32.1^\circ\text{ k}\Omega) = 2.36\angle 32.1^\circ\text{ V}$$

The magnitude of the source voltage is 2.36 V at an angle of 32.1° with respect to the current; that is, the voltage leads the current by 32.1°, as shown in the phasor diagram of Figure 17–6.

FIGURE 17–6

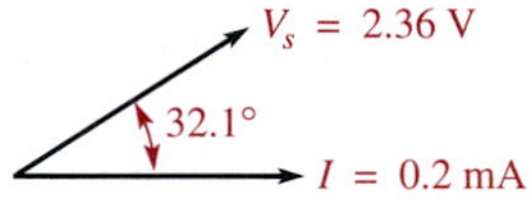

$V_s$ = 2.36 V
32.1°
$I$ = 0.2 mA

***Related Exercise*** If the source voltage in Figure 17–5 were $5\angle 0^\circ$ V, what would be the current expressed in polar form?

## Relationships of the Current and Voltages in a Series *RL* Circuit

In a series *RL* circuit, the current is the same through both the resistor and the **inductor.** Thus, the resistor voltage is in phase with the current, and the inductor voltage leads the current by 90°. Therefore, there is a phase difference of 90° between the resistor voltage, $V_R$, and the inductor voltage, $V_L$, as shown in the waveform diagram of Figure 17–7.

From Kirchhoff's voltage law, the sum of the voltage drops must equal the applied voltage. However, since $V_R$ and $V_L$ are not in phase with each other, they must be added as phasor quantities with $V_L$ leading $V_R$ by 90°, as shown in Figure 17–8(a). As shown in part (b), $\mathbf{V}_s$ is the phasor sum of $V_R$ and $V_L$.

$$\mathbf{V}_s = V_R + jV_L \qquad \textbf{(17–6)}$$

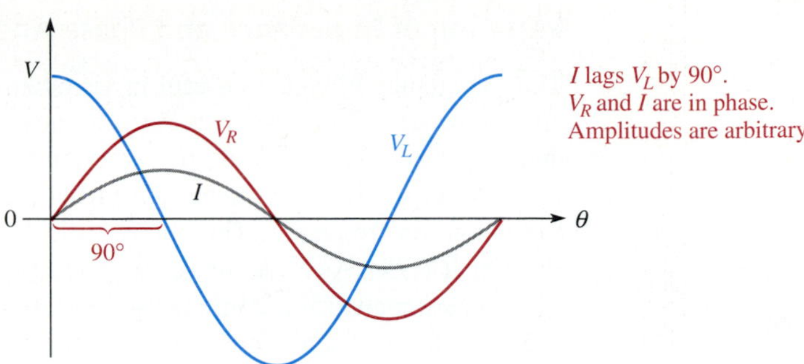


**FIGURE 17–7**
***Phase relation of voltages and current in a series RL circuit.***

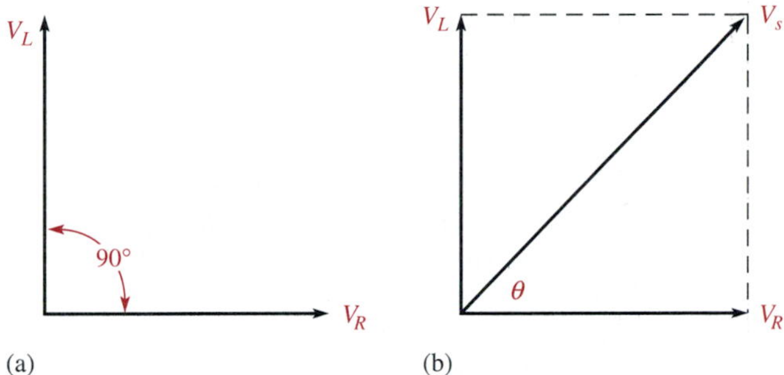


**FIGURE 17–8**
***Voltage phasor diagram for a series RL circuit.***

This equation can be expressed in polar form as

$$\mathbf{V}_s = \sqrt{V_R^2 + V_L^2}\angle\tan^{-1}\left(\frac{V_L}{V_R}\right) \qquad \textbf{(17–7)}$$

where the magnitude of the source voltage is

$$V_s = \sqrt{V_R^2 + V_L^2} \qquad \textbf{(17–8)}$$

and the phase angle between the resistor voltage and the source voltage is

$$\theta = \tan^{-1}\left(\frac{V_L}{V_R}\right) \qquad \textbf{(17–9)}$$

Since the resistor voltage and the current are in phase, $\theta$ is also the phase angle between the source voltage and the current. Figure 17–9 shows a voltage and current phasor diagram that represents the waveform diagram of Figure 17–7.

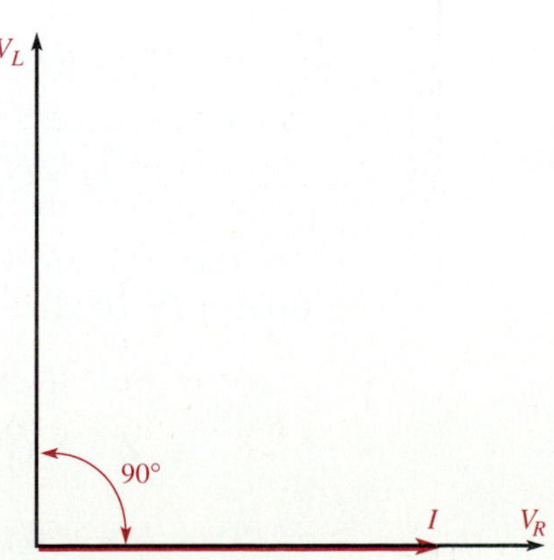


**FIGURE 17–9**
***Voltage and current phasor diagram for the waveforms in Figure 17–7.***

## Variation of Impedance and Phase Angle with Frequency

The impedance triangle is useful in visualizing how the frequency of the applied voltage affects the *RL* circuit response. As you know, inductive reactance varies directly with frequency. When $X_L$ increases, the magnitude of the total impedance also increases; and when $X_L$ decreases, the magnitude of the total impedance decreases. Thus, $Z$ is directly dependent on frequency. The phase angle $\theta$ also varies directly with frequency, because $\theta = \tan^{-1}(X_L/R)$. As $X_L$ increases with frequency, so does $\theta$, and vice versa.

The impedance triangle is used in Figure 17–10 to illustrate the variations in $X_L$, $Z$, and $\theta$ as the frequency changes. Of course, $R$ remains constant. The main point is that because $X_L$ varies directly as the frequency, so also do the magnitude of the total impedance and the phase angle. Example 17–3 illustrates this.

**FIGURE 17–10**
***As the frequency increases, $X_L$ increases, Z increases, and $\theta$ increases. Each value of frequency can be visualized as forming a different impedance triangle.***

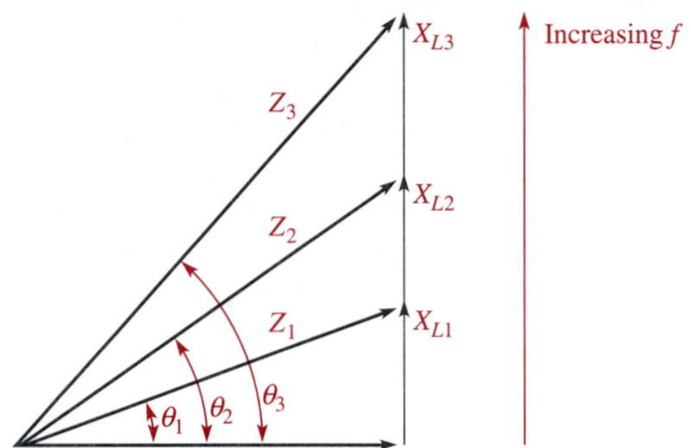


**EXAMPLE 17–3**

For the series *RL* circuit in Figure 17–11, determine the magnitude of the total impedance and the phase angle for each of the following frequencies:
**(a)** 10 kHz **(b)** 20 kHz **(c)** 30 kHz

**FIGURE 17–11**

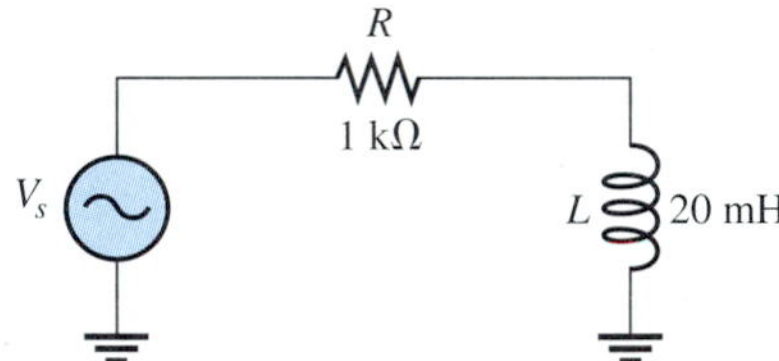


### *Solution*

**(a)** For $f = 10$ kHz,

$$X_L = 2\pi fL = 2\pi(10\text{ kHz})(20\text{ mH}) = 1.26\text{ k}\Omega$$

$$\mathbf{Z} = \sqrt{R^2 + X_L^2}\angle\tan^{-1}\left(\frac{X_L}{R}\right)$$

$$= \sqrt{(1\text{ k}\Omega)^2 + (1.26\text{ k}\Omega)^2}\angle\tan^{-1}\left(\frac{1.26\text{ k}\Omega}{1\text{ k}\Omega}\right) = 1.61\angle 51.6^\circ\text{ k}\Omega$$

Thus, $Z = 1.61$ kΩ and $\theta = 51.6^\circ$.

**(b)** For $f = 20$ kHz,

$$X_L = 2\pi(20\text{ kHz})(20\text{ mH}) = 2.51\text{ k}\Omega$$

$$\mathbf{Z} = \sqrt{(1\text{ k}\Omega)^2 + (2.51\text{ k}\Omega)^2}\angle\tan^{-1}\left(\frac{2.51\text{ k}\Omega}{1\text{ k}\Omega}\right) = 2.70\angle 68.3^\circ\text{ k}\Omega$$

Thus, $Z = 2.70$ kΩ and $\theta = 68.3^\circ$.

**(c)** For $f = 30$ kHz,

$$X_L = 2\pi(30\text{ kHz})(20\text{ mH}) = 3.77\text{ k}\Omega$$

$$\mathbf{Z} = \sqrt{(1\text{ k}\Omega)^2 + (3.77\text{ k}\Omega)^2}\angle\tan^{-1}\left(\frac{3.77\text{ k}\Omega}{1\text{ k}\Omega}\right) = 3.90\angle 75.1°\text{ k}\Omega$$

Thus, $Z = 3.90$ kΩ and $\theta = 75.1°$.

Notice that as the frequency increases, $X_L$, $Z$, and $\theta$ also increase.

***Related Exercise*** Determine $Z$ and $\theta$ in Figure 17–11 if $f$ is 100 kHz.

**SECTION 17–3 REVIEW**

1. In a certain series *RL* circuit, $V_R = 2$ V and $V_L = 3$ V. What is the magnitude of the source voltage?
2. In Question 1, what is the phase angle between the source voltage and the current?
3. When the frequency of the applied voltage in a series *RL* circuit is increased, what happens to the inductive reactance? What happens to the magnitude of the total impedance? What happens to the phase angle?

## 17–4 ■ IMPEDANCE AND PHASE ANGLE OF PARALLEL *RL* CIRCUITS

***In this section, you will learn how to determine the impedance and phase angle of a parallel RL circuit. Also, inductive susceptance and admittance of a parallel RL circuit are introduced.***

***After completing this section, you should be able to***

- **Determine impedance and phase angle in a parallel *RL* circuit**
  - ☐ Express total impedance in complex form
  - ☐ Define and calculate *inductive susceptance* and *admittance*

Figure 17–12 shows a basic parallel *RL* circuit connected to an ac voltage source.

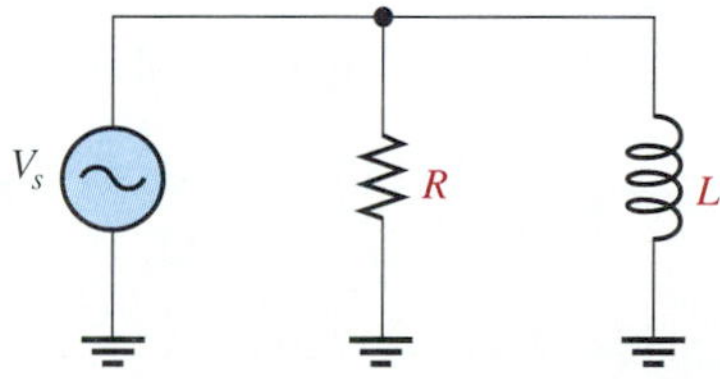


**FIGURE 17–12**
***Parallel RL circuit.***

The expression for the total impedance is developed as follows.

$$\mathbf{Z} = \frac{(R\angle 0°)(X_L\angle 90°)}{R + jX_L} = \frac{RX_L\angle(0° + 90°)}{\sqrt{R^2 + X_L^2}\angle\tan^{-1}\left(\frac{X_L}{R}\right)}$$

$$\mathbf{Z} = \left(\frac{RX_L}{\sqrt{R^2 + X_L^2}}\right)\angle\left(90° - \tan^{-1}\left(\frac{X_L}{R}\right)\right) \quad \textbf{(17–10)}$$

Equation (17–10) is the expression for the total parallel impedance where the magnitude is

$$Z = \frac{RX_L}{\sqrt{R^2 + X_L^2}} \tag{17–11}$$

and the phase angle between the applied voltage and the total current is

$$\theta = 90° - \tan^{-1}\left(\frac{X_L}{R}\right) \tag{17–12}$$

This equation can also be expressed equivalently as $\theta = \tan^{-1}(R/X_L)$.

**EXAMPLE 17–4** For each circuit in Figure 17–13, determine the magnitude of the total impedance and the phase angle.

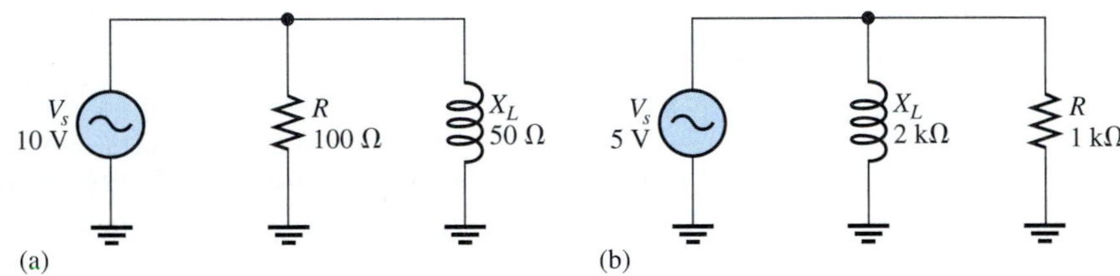


FIGURE 17–13

***Solution*** For the circuit in Figure 17–13(a), the total impedance is

$$\mathbf{Z} = \left(\frac{RX_L}{\sqrt{R^2 + X_L^2}}\right)\angle\left(90° - \tan^{-1}\left(\frac{X_L}{R}\right)\right)$$

$$= \left[\frac{(100\ \Omega)(50\ \Omega)}{\sqrt{(100\ \Omega)^2 + (50\ \Omega)^2}}\right]\angle\left(90° - \tan^{-1}\left(\frac{50\ \Omega}{100\ \Omega}\right)\right) = 44.7\angle 63.4°\ \Omega$$

Thus, $Z = 44.7\ \Omega$ and $\theta = 63.4°$.

For the circuit in Figure 17–13(b), the total impedance is

$$\mathbf{Z} = \left[\frac{(1\ \text{k}\Omega)(2\ \text{k}\Omega)}{\sqrt{(1\ \text{k}\Omega)^2 + (2\ \text{k}\Omega)^2}}\right]\angle\left(90° - \tan^{-1}\left(\frac{2\ \text{k}\Omega}{1\ \text{k}\Omega}\right)\right) = 894\angle 26.6°\ \Omega$$

Thus, $Z = 894\ \Omega$ and $\theta = 26.6°$.

Notice that the positive angle indicates that the voltage leads the current, as opposed to the *RC* case where the voltage lags the current.

The calculator sequences for the circuit in Figure 17–13(a) are

Z: [1] [0] [0] [×] [5] [0] [÷] [(] [1] [0] [0] [x²] [+] [5] [0] [x²] [)] [√x] [=]
θ: [1] [0] [0] [÷] [5] [0] [=] [INV] [tan]

***Related Exercise*** In a parallel circuit, $R = 10\ \text{k}\Omega$ and $X_L = 14\ \text{k}\Omega$. Determine the total impedance in polar form.

## Conductance, Susceptance, and Admittance

As you know from the previous chapter, conductance ($G$) is the reciprocal of resistance, susceptance ($B$) is the reciprocal of reactance, and admittance ($Y$) is the reciprocal of impedance.

For parallel *RL* circuits, the phasor expression for **inductive susceptance** ($B_L$) is

$$\mathbf{B}_L = \frac{1}{X_L\angle 90^\circ} = B_L\angle -90^\circ = -jB_L \tag{17–13}$$

and the phasor expression for **admittance** is

$$\mathbf{Y} = \frac{1}{Z\angle \pm\theta} = Y\angle \mp\theta \tag{17–14}$$

In the basic parallel *RL* circuit shown in Figure 17–14, the total admittance is the phasor sum of the conductance and the inductive susceptance.

$$\mathbf{Y} = G - jB_L \tag{17–15}$$

As with the *RC* circuit, the unit for $G$, $B_L$, and $Y$ is the siemens (S).

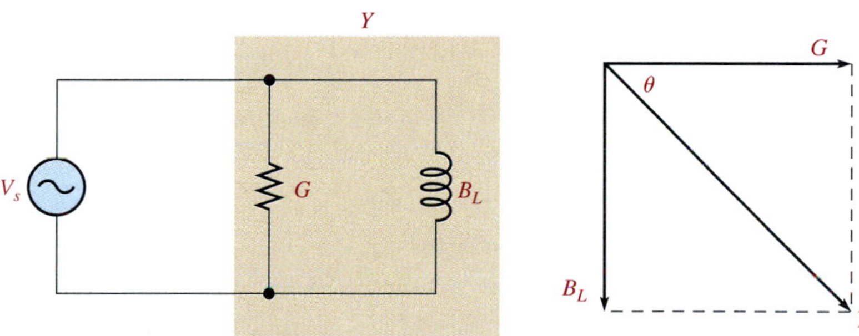


**FIGURE 17–14**
***Admittance in a parallel RL circuit.***

**EXAMPLE 17–5** Determine the total admittance and total impedance in Figure 17–15. Draw the admittance phasor diagram.

**FIGURE 17–15**

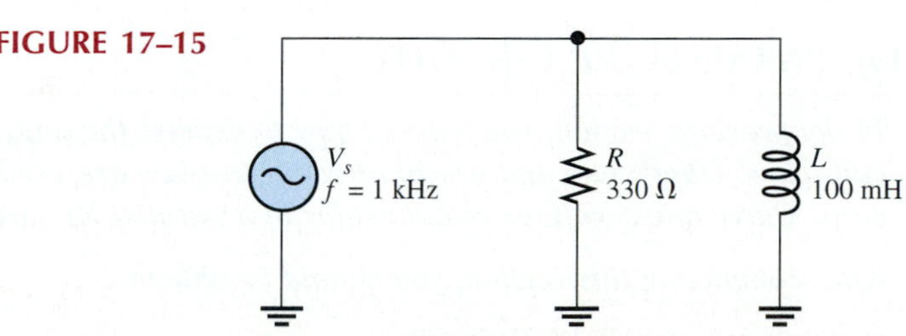


***Solution*** First, determine the conductance magnitude. $R = 330\ \Omega$; thus,

$$G = \frac{1}{R} = \frac{1}{330\ \Omega} = 3.03\text{ mS}$$

Then, determine the inductive reactance.

$$X_L = 2\pi fL = 2\pi(1000 \text{ Hz})(100 \text{ mH}) = 628\ \Omega$$

The inductive susceptance magnitude is

$$B_L = \frac{1}{X_L} = \frac{1}{628\ \Omega} = 1.59 \text{ mS}$$

The total admittance is

$$\mathbf{Y} = G - jB_L = 3.03 \text{ mS} - j1.59 \text{ mS}$$

which can be expressed in polar form as

$$\mathbf{Y} = \sqrt{G^2 + B_L^2}\angle -\tan^{-1}\left(\frac{B_L}{G}\right)$$

$$= \sqrt{(3.03 \text{ mS})^2 + (1.59 \text{ mS})^2}\angle -\tan^{-1}\left(\frac{1.59 \text{ mS}}{3.03 \text{ mS}}\right) = 3.42\angle -27.7^\circ \text{ mS}$$

Admittance is converted to impedance as follows:

$$\mathbf{Z} = \frac{1}{\mathbf{Y}} = \frac{1}{3.42\angle -27.7^\circ \text{ mS}} = 292\angle 27.7^\circ\ \Omega$$

Again, the positive phase angle in the impedance expression indicates that the voltage leads the current. The admittance phasor diagram is shown in Figure 17–16.

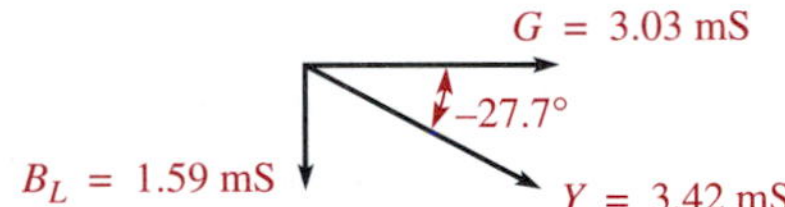


FIGURE 17–16

***Related Exercise*** What is the total admittance of the circuit in Figure 17–15 if $f$ is increased to 2 kHz?

**SECTION 17–4 REVIEW**

1. If $Z = 500\ \Omega$, what is the value of $Y$?
2. In a certain parallel $RL$ circuit, $R = 47\ \Omega$ and $X_L = 75\ \Omega$. Determine the magnitude of the admittance.
3. In the circuit of Question 2, does the total current lead or lag the applied voltage? By what phase angle?

## 17–5 ■ ANALYSIS OF PARALLEL *RL* CIRCUITS

***In the previous section, you learned how to express the impedance of a parallel RL circuit. Now, Ohm's law and Kirchhoff's current law are used in the analysis of RL circuits. Current and voltage relationships in a parallel RL circuit are examined.***

***After completing this section, you should be able to***

- **Analyze a parallel *RL* circuit**
  - ☐ Apply Ohm's law and Kirchhoff's current law to parallel *RL* circuits
  - ☐ Express the voltages and currents as phasor quantities

The following example applies Ohm's law to the analysis of a parallel *RL* circuit.

**EXAMPLE 17–6** Determine the total current and the phase angle in the circuit of Figure 17–17. Draw a phasor diagram showing the relationship of $\mathbf{V}_s$ and $\mathbf{I}_{tot}$.

FIGURE 17–17

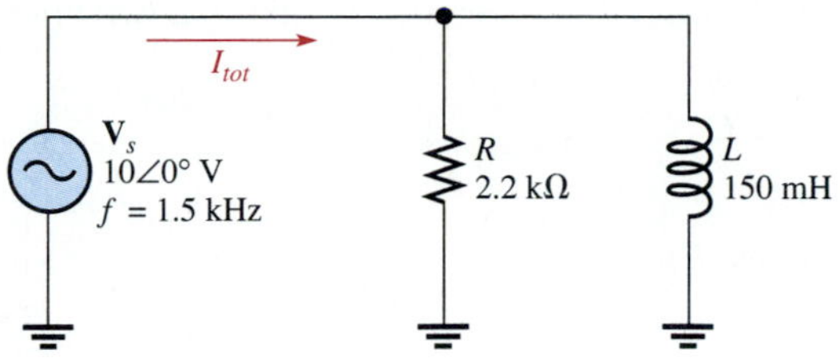

$I_{tot}$

$\mathbf{V}_s$ $10\angle 0°$ V $f$ = 1.5 kHz

$R$ 2.2 kΩ

$L$ 150 mH

***Solution*** The inductive reactance is

$$X_L = 2\pi f L = 2\pi(1.5 \text{ kHz})(150 \text{ mH}) = 1.41 \text{ k}\Omega$$

The inductive susceptance magnitude is

$$B_L = \frac{1}{X_L} = \frac{1}{1.41 \text{ k}\Omega} = 709 \ \mu\text{S}$$

The conductance magnitude is

$$G = \frac{1}{R} = \frac{1}{2.2 \text{ k}\Omega} = 455 \ \mu\text{S}$$

The total admittance is

$$\mathbf{Y} = G - jB_L = 455 \ \mu\text{S} - j709 \ \mu\text{S}$$

Converting to polar form yields

$$\mathbf{Y} = \sqrt{G^2 + B_L^2}\angle -\tan^{-1}\left(\frac{B_L}{G}\right)$$

$$= \sqrt{(455 \ \mu\text{S})^2 + (709 \ \mu\text{S})^2}\angle -\tan^{-1}\left(\frac{709 \ \mu\text{S}}{455 \ \mu\text{S}}\right) = 842\angle -57.3° \ \mu\text{S}$$

The phase angle is −57.3°.

Use Ohm's law to determine the total current.

$$\mathbf{I}_{tot} = \mathbf{V}_s\mathbf{Y} = (10\angle 0° \text{ V})(842\angle -57.3° \ \mu\text{S}) = 8.42\angle -57.3° \text{ mA}$$

The magnitude of the total current is 8.42 mA, and it lags the applied voltage by 57.3°, as indicated by the negative angle associated with it. The phasor diagram in Figure 17–18 shows the voltage-current relationship.

FIGURE 17–18

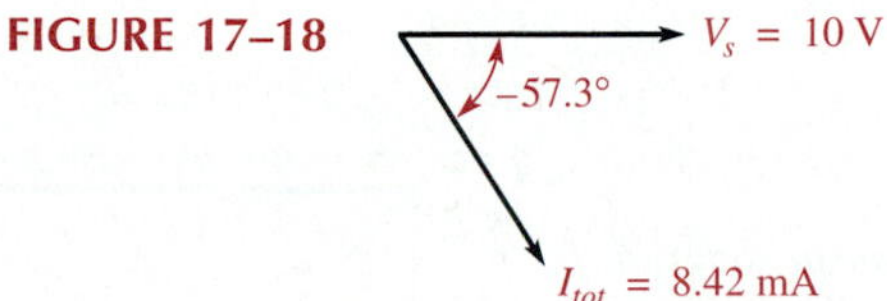


***Related Exercise*** Determine the current in polar form if $f$ is reduced to 800 Hz in Figure 17–17.

## Relationships of the Currents and Voltages in a Parallel *RL* Circuit

Figure 17–19(a) shows all the currents and voltages in a basic parallel *RL* circuit. As you can see, the applied voltage, $V_s$, appears across both the resistive and the inductive branches, so $V_s$, $V_R$, and $V_L$ are all in phase and of the same magnitude. The total current, $I_{tot}$, divides at the junction into the two branch currents, $I_R$ and $I_L$.

**FIGURE 17–19**

***Currents and voltages in a parallel RL circuit.***

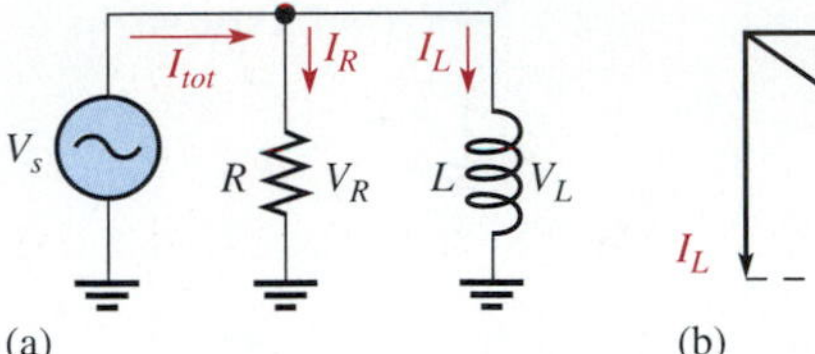


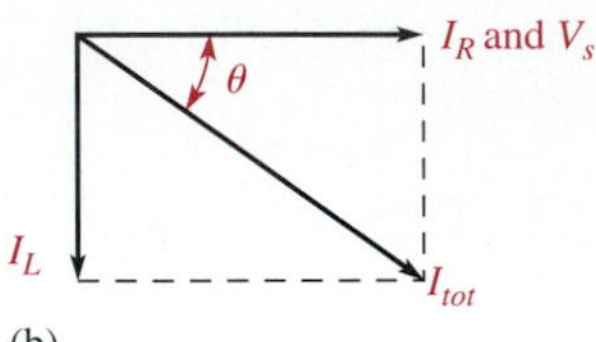


The current through the resistor is in phase with the voltage. The current through the inductor lags the voltage and the resistor current by 90°. By Kirchhoff's current law, the total current is the phasor sum of the two branch currents, as shown by the phasor diagram in Figure 17–19(b). The total current is expressed as

$$\mathbf{I}_{tot} = I_R - jI_L \tag{17–16}$$

This equation can be expressed in polar form as

$$\mathbf{I}_{tot} = \sqrt{I_R^2 + I_L^2}\angle -\tan^{-1}\left(\frac{I_L}{I_R}\right) \tag{17–17}$$

where the magnitude of the total current is

$$\mathbf{I}_{tot} = \sqrt{I_R^2 + I_L^2} \tag{17–18}$$

and the phase angle between the resistor current and the total current is

$$\theta = -\tan^{-1}\left(\frac{I_L}{I_R}\right) \tag{17–19}$$

Since the resistor current and the applied voltage are in phase, $\theta$ also represents the phase angle between the total current and the applied voltage. Figure 17–20 shows a complete current and voltage phasor diagram.

**FIGURE 17–20**

***Current and voltage phasor diagram for a parallel RL circuit (amplitudes are arbitrary).***

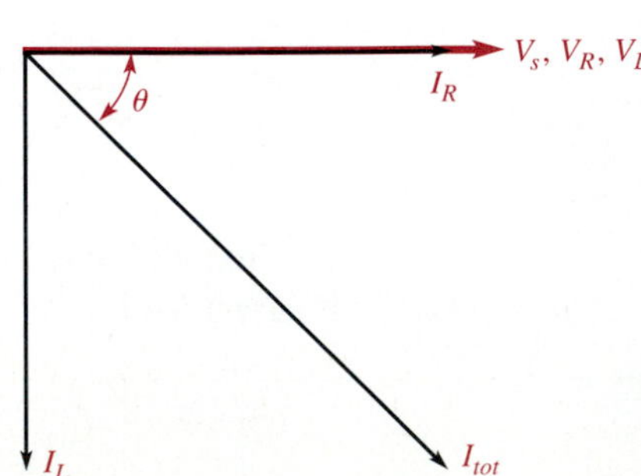

**EXAMPLE 17–7** Determine the value of each current in Figure 17–21, and describe the phase relationship of each with the applied voltage. Draw the phasor diagram.

FIGURE 17–21

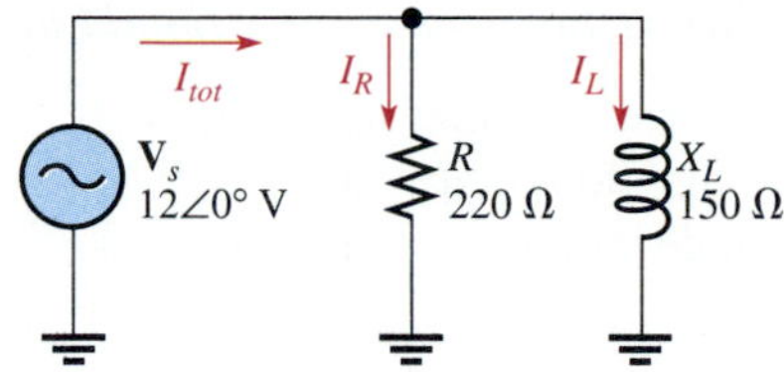


***Solution*** The resistor current, the inductor current, and the total current are expressed as follows:

$$\mathbf{I}_R = \frac{\mathbf{V}_s}{\mathbf{R}} = \frac{12\angle 0^\circ \text{ V}}{220\angle 0^\circ \ \Omega} = 54.5\angle 0^\circ \text{ mA}$$

$$\mathbf{I}_L = \frac{\mathbf{V}_s}{\mathbf{X}_L} = \frac{12\angle 0^\circ \text{ V}}{150\angle 90^\circ \ \Omega} = 80\angle -90^\circ \text{ mA}$$

$$\mathbf{I}_{tot} = I_R - jI_L = 54.5 \text{ mA} - j80 \text{ mA}$$

Converting $\mathbf{I}_{tot}$ to polar form yields

$$\mathbf{I}_{tot} = \sqrt{I_R^2 + I_L^2}\angle -\tan^{-1}\left(\frac{I_L}{I_R}\right)$$

$$= \sqrt{(54.5 \text{ mA})^2 + (80 \text{ mA})^2}\angle -\tan^{-1}\left(\frac{80 \text{ mA}}{54.5 \text{ mA}}\right) = 96.8\angle -55.7^\circ \text{ mA}$$

As the results show, the resistor current is 54.5 mA and is in phase with the applied voltage. The inductor current is 80 mA and lags the applied voltage by 90°. The total current is 96.8 mA and lags the voltage by 55.7°. The phasor diagram in Figure 17–22 shows these relationships.

FIGURE 17–22

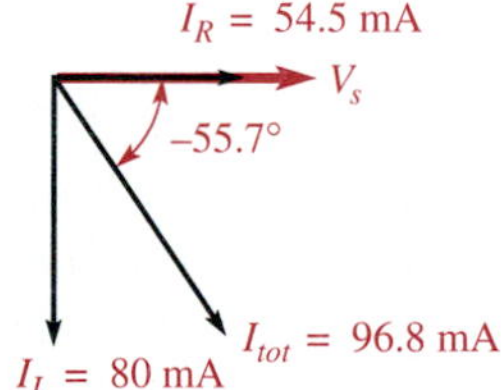


***Related Exercise*** Find the magnitude of $\mathbf{I}_{tot}$ and the circuit phase angle if $X_L$ = 300 Ω in Figure 17–21.

**SECTION 17–5 REVIEW**

1. The admittance of an *RL* circuit is 4 mS, and the applied voltage is 8 V. What is the total current?
2. In a certain parallel *RL* circuit, the resistor current is 12 mA, and the inductor current is 20 mA. Determine the magnitude and phase angle of the total current. This phase angle is measured with respect to what?
3. What is the phase angle between the inductor current and the applied voltage in a parallel *RL* circuit?

## 17–6 ■ SERIES-PARALLEL *RL* CIRCUITS

***In this section, the concepts studied in the previous sections are used to analyze circuits with combinations of both series and parallel R and L elements.***

***After completing this section, you should be able to***

- **Analyze series-parallel *RL* circuits**
  - Determine total impedance
  - Calculate currents and voltages

The following two examples demonstrate how to approach the analysis of complex *RL* networks.

**EXAMPLE 17–8**

In the circuit of Figure 17–23, determine the following values:

**(a)** $\mathbf{Z}_{tot}$ **(b)** $\mathbf{I}_{tot}$ **(c)** $\theta$

FIGURE 17–23

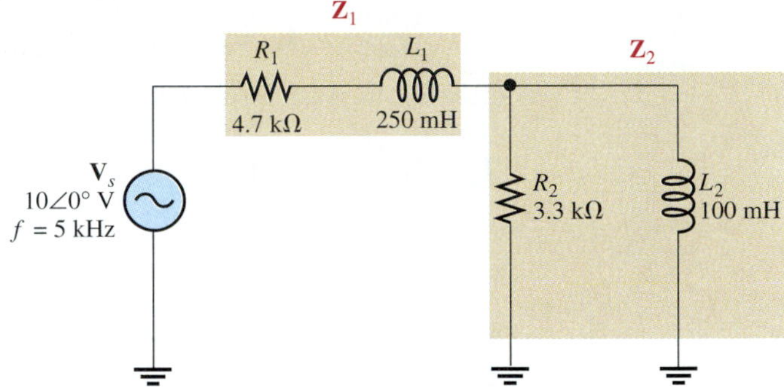


***Solution***

**(a)** First, calculate the magnitudes of inductive reactance.

$$X_{L1} = 2\pi f L_1 = 2\pi(5\text{ kHz})(250\text{ mH}) = 7.85\text{ k}\Omega$$

$$X_{L2} = 2\pi f L_2 = 2\pi(5\text{ kHz})(100\text{ mH}) = 3.14\text{ k}\Omega$$

One approach is to find the impedance of the series portion and the impedance of the parallel portion and combine them to get the total impedance. The impedance of the series combination of $R_1$ and $L_1$ is

$$\mathbf{Z}_1 = R_1 + jX_{L1} = 4.7\text{ k}\Omega + j7.85\text{ k}\Omega$$

To determine the impedance of the parallel portion, first determine the admittance of the parallel combination of $R_2$ and $L_2$.

$$G_2 = \frac{1}{R_2} = \frac{1}{3.3\text{ k}\Omega} = 303\ \mu\text{S}$$

$$B_{L2} = \frac{1}{X_{L2}} = \frac{1}{3.14\text{ k}\Omega} = 318\ \mu\text{S}$$

$$\mathbf{Y}_2 = G_2 - jB_L = 303\ \mu\text{S} - j318\ \mu\text{S}$$

Converting to polar form yields

$$\mathbf{Y}_2 = \sqrt{G_2^2 + B_L^2}\angle -\tan^{-1}\left(\frac{B_L}{G_2}\right)$$

$$= \sqrt{(303\ \mu\text{S})^2 + (318\ \mu\text{S})^2}\angle -\tan^{-1}\left(\frac{318\ \mu\text{S}}{303\ \mu\text{S}}\right) = 439\angle -46.4^\circ\ \mu\text{S}$$

Then, the impedance of the parallel portion is

$$\mathbf{Z}_2 = \frac{1}{\mathbf{Y}_2} = \frac{1}{439\angle -46.4^\circ\ \mu\text{S}} = 2.28\angle 46.4^\circ\ \text{k}\Omega$$

Converting to rectangular form yields

$$\mathbf{Z}_2 = Z_2\cos\theta + jZ_2\sin\theta$$
$$= (2.28\ \text{k}\Omega)\cos(46.4^\circ) + j(2.28\ \text{k}\Omega)\sin(46.4^\circ) = 1.57\ \text{k}\Omega + j1.65\ \text{k}\Omega$$

The series portion and the parallel portion are in series with each other. Combine $\mathbf{Z}_1$ and $\mathbf{Z}_2$ to get the total impedance.

$$\mathbf{Z}_{tot} = \mathbf{Z}_1 + \mathbf{Z}_2$$
$$= (4.7\ \text{k}\Omega + j7.85\ \text{k}\Omega) + (1.57\ \text{k}\Omega + j1.65\ \text{k}\Omega) = 6.27\ \text{k}\Omega + j9.50\ \text{k}\Omega$$

Expressing $\mathbf{Z}_{tot}$ in polar form yields

$$\mathbf{Z}_{tot} = \sqrt{Z_1^2 + Z_2^2}\angle\tan^{-1}\left(\frac{Z_2}{Z_1}\right)$$
$$= \sqrt{(6.27\ \text{k}\Omega)^2 + (9.50\ \text{k}\Omega)^2}\angle\tan^{-1}\left(\frac{9.50\ \text{k}\Omega}{6.27\ \text{k}\Omega}\right) = 11.4\angle 56.6^\circ\ \text{k}\Omega$$

**(b)** Use Ohm's law to find the total current.

$$\mathbf{I}_{tot} = \frac{\mathbf{V}_s}{\mathbf{Z}_{tot}} = \frac{10\angle 0^\circ\ \text{V}}{11.4\angle 56.6^\circ\ \text{k}\Omega} = 877\angle -56.6^\circ\ \mu\text{A}$$

**(c)** The total current lags the applied voltage by 56.6°.

***Related Exercise***

**(a)** Determine the voltage across the series part of the circuit in Figure 17–23.
**(b)** Determine the voltage across the parallel part of the circuit.

**EXAMPLE 17–9** Determine the voltage across each element in Figure 17–24. Sketch a voltage phasor diagram and a current phasor diagram.

FIGURE 17–24

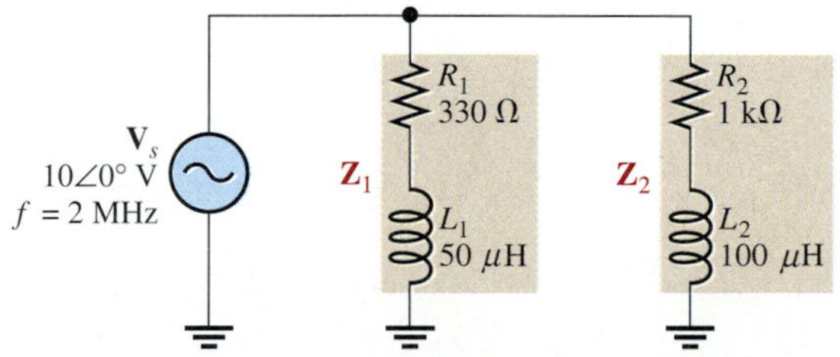

***Solution*** First calculate $X_{L1}$ and $X_{L2}$.

$$X_{L1} = 2\pi fL_1 = 2\pi(2\ \text{MHz})(50\ \mu\text{H}) = 628\ \Omega$$
$$X_{L2} = 2\pi fL_2 = 2\pi(2\ \text{MHz})(100\ \mu\text{H}) = 1.26\ \Omega$$

Next, determine the impedance of each branch.

$$\mathbf{Z}_1 = R_1 + jX_{L1} = 330\ \Omega + j628\ \Omega$$
$$\mathbf{Z}_2 = R_2 + jX_{L2} = 1\ \text{k}\Omega + j1.26\ \text{k}\Omega$$

Convert these impedances to polar form.

$$\mathbf{Z}_1 = \sqrt{R_1^2 + X_{L1}^2}\angle\tan^{-1}\left(\frac{X_{L1}}{R_1}\right)$$

$$= \sqrt{(330\ \Omega)^2 + (628\ \Omega)^2}\angle\tan^{-1}\left(\frac{628\ \Omega}{330\ \Omega}\right) = 709\angle 62.3^\circ\ \Omega$$

$$\mathbf{Z}_2 = \sqrt{R_2^2 + X_{L2}^2}\angle\tan^{-1}\left(\frac{X_{L2}}{R_2}\right)$$

$$= \sqrt{(1\ \text{k}\Omega)^2 + (1.26\ \text{k}\Omega)^2}\angle\tan^{-1}\left(\frac{1.26\ \text{k}\Omega}{1\ \text{k}\Omega}\right) = 1.61\angle 51.6^\circ\ \text{k}\Omega$$

Calculate each branch current.

$$\mathbf{I}_1 = \frac{\mathbf{V}_s}{\mathbf{Z}_1} = \frac{10\angle 0^\circ\ \text{V}}{709\angle 62.3^\circ\ \Omega} = 14.1\angle -62.3^\circ\ \text{mA}$$

$$\mathbf{I}_2 = \frac{\mathbf{V}_s}{\mathbf{Z}_2} = \frac{10\angle 0^\circ\ \text{V}}{1.61\angle 51.6^\circ\ \text{k}\Omega} = 6.21\angle -51.6^\circ\ \text{mA}$$

Now, use Ohm's law to get the voltage across each element.

$$\mathbf{V}_{R1} = \mathbf{I}_1\mathbf{R}_1 = (14.1\angle -62.3^\circ\ \text{mA})(330\angle 0^\circ\ \Omega) = 4.65\angle -62.3^\circ\ \text{V}$$
$$\mathbf{V}_{L1} = \mathbf{I}_1\mathbf{X}_{L1} = (14.1\angle -62.3^\circ\ \text{mA})(628\angle 90^\circ\ \Omega) = 8.85\angle 27.7^\circ\ \text{V}$$
$$\mathbf{V}_{R2} = \mathbf{I}_2\mathbf{R}_2 = (6.21\angle -51.6^\circ\ \text{mA})(1\angle 0^\circ\ \text{k}\Omega) = 6.21\angle -51.6^\circ\ \text{V}$$
$$\mathbf{V}_{L2} = \mathbf{I}_2\mathbf{X}_{L2} = (6.21\angle -51.6^\circ\ \text{mA})(1.26\angle 90^\circ\ \text{k}\Omega) = 7.82\angle 38.4^\circ\ \text{V}$$

The voltage phasor diagram is shown in Figure 17–25, and the current phasor diagram is shown in Figure 17–26.

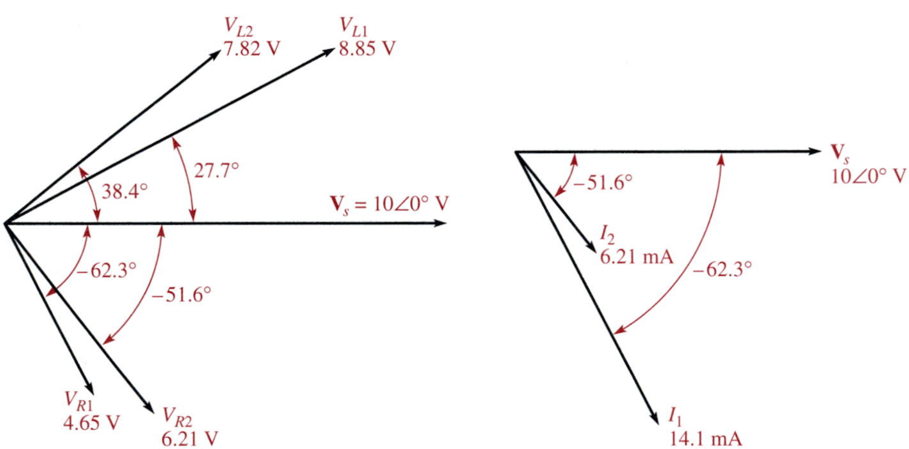


FIGURE 17–25

FIGURE 17–26

***Related Exercise*** What is the total current in polar form in Figure 17–24?

## SECTION 17–6 REVIEW

**1.** What is the total impedance in polar form of the circuit in Figure 17–24?

**2.** Determine the total current in polar form for the circuit in Figure 17–24.

## 17–7 ■ POWER IN *RL* CIRCUITS

***In a purely resistive ac circuit, all of the energy delivered by the source is dissipated in the form of heat by the resistance. In a purely inductive ac circuit, all of the energy delivered by the source is stored by the inductor in its magnetic field during a portion of the voltage cycle and then returned to the source during another portion of the cycle so that there is no net energy loss (conversion to heat). When there is both resistance and inductance, some of the energy is alternately stored and returned by the inductance and some is dissipated by the resistance. The amount of energy loss is determined by the relative values of the resistance and the inductive reactance.***

***After completing this section, you should be able to***

- **Determine power in *RL* circuits**
  - ☐ Explain true and reactive power
  - ☐ Draw the power triangle
  - ☐ Define *power factor*
  - ☐ Explain power factor correction

When the resistance is greater than the inductive reactance, more of the total energy delivered by the source is dissipated by the resistance than is stored by the inductor; and when the reactance is greater than the resistance, more of the total energy is stored and returned than is lost.

As you know, the power loss in a resistance is called the *true power.* The power in an inductor is reactive power and is expressed as

$$P_r = I^2X_L \qquad \textbf{(17–20)}$$

### The Power Triangle

The generalized power triangle for the *RL* circuit is shown in Figure 17–27. The **apparent power**, $P_a$, is the resultant of the average power, $P_{true}$, and the reactive power, $P_r$.

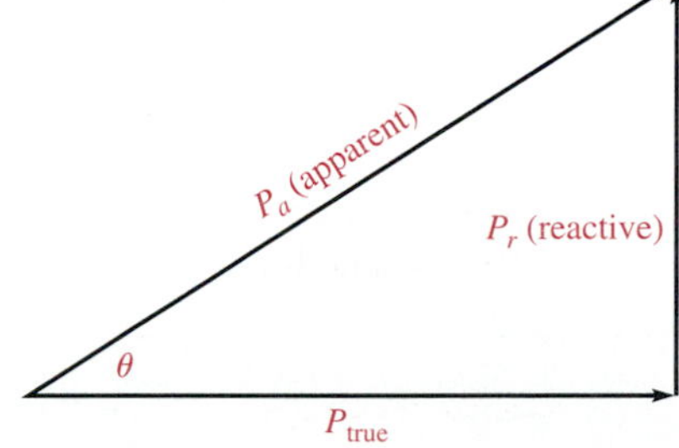


FIGURE 17–27
*Power triangle for an RL circuit.*

Recall that the power factor equals the cosine of $\theta$ ($PF = \cos\theta$). As the phase angle between the applied voltage and the total current increases, the power factor decreases, indicating an increasingly reactive circuit. The smaller the power factor, the smaller the true power is compared to the reactive power.

**EXAMPLE 17–10** Determine the power factor, the true power, the reactive power, and the apparent power in Figure 17–28.

FIGURE 17–28

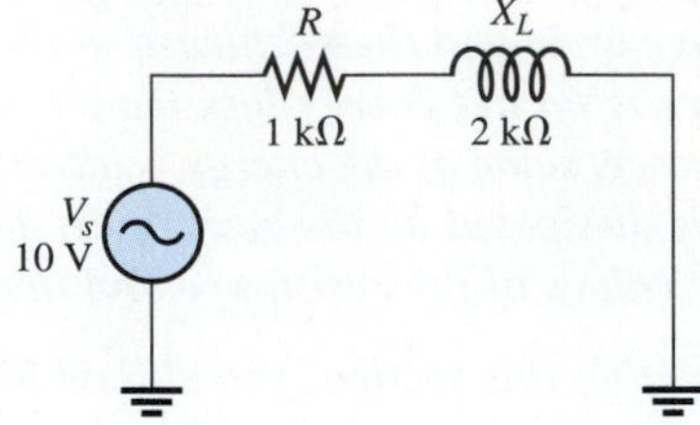


***Solution*** The total impedance of the circuit in rectangular form is

$$\mathbf{Z} = R + jX_L = 1\text{ k}\Omega + j2\text{ k}\Omega$$

Converting to polar form yields

$$\mathbf{Z} = \sqrt{R_2^2 + X_L^2}\angle\tan^{-1}\left(\frac{X_L}{R}\right)$$

$$= \sqrt{(1\text{ k}\Omega)^2 + (2\text{ k}\Omega)^2}\angle\tan^{-1}\left(\frac{2\text{ k}\Omega}{1\text{ k}\Omega}\right) = 2.24\angle 63.4^\circ\text{ k}\Omega$$

The current magnitude is

$$I = \frac{V_s}{Z} = \frac{10\text{ V}}{2.24\text{ k}\Omega} = 4.46\text{ mA}$$

The phase angle is indicated in the expression for **Z**.

$$\theta = 63.4^\circ$$

The power factor is therefore

$$PF = \cos\theta = \cos(63.4^\circ) = 0.448$$

The true power is

$$P_{\text{true}} = V_s I\cos\theta = (10\text{ V})(4.46\text{ mA})(0.448) = 20\text{ mW}$$

The reactive power is

$$P_r = I^2 X_L = (4.46\text{ mA})^2(2\text{ k}\Omega) = 39.8\text{ mVAR}$$

The apparent power is

$$P_a = I^2 Z = (4.46\text{ mA})^2(2.24\text{ k}\Omega) = 44.6\text{ mVA}$$

***Related Exercise*** If the frequency in Figure 17–28 is increased, what happens to $P_{\text{true}}$, $P_r$, and $P_a$?

## Significance of the Power Factor

As you learned in Chapter 16, the **power factor** (*PF*) is important in determining how much useful power (true power) is transferred to a load. The highest power factor is 1, which indicates that all of the current to a load is in phase with the voltage (resistive). When the power factor is 0, all of the current to a load is 90° out of phase with the voltage (reactive).

Generally, a power factor as close to 1 as possible is desirable because then most of the power transferred from the source to the load is useful or true power. True power goes only one way—from source to load—and performs work on the load in terms of energy dissipation. Reactive power simply goes back and forth between the source and the load with no net work being done. Energy must be used in order for work to be done.

Many practical loads have inductance as a result of their particular function, and it is essential for their proper operation. Examples are transformers, electric motors, and speakers, to name a few. Therefore, inductive (and capacitive) loads are important considerations.

To see the effect of the power factor on system requirements, refer to Figure 17–29. This figure shows a representation of a typical inductive load consisting effectively of inductance and resistance in parallel. Part (a) shows a load with a relatively low power factor (0.75), and part (b) shows a load with a relatively high power factor (0.95). Both loads dissipate equal amounts of power as indicated by the wattmeters. Thus, an equal amount of work is done on both loads.

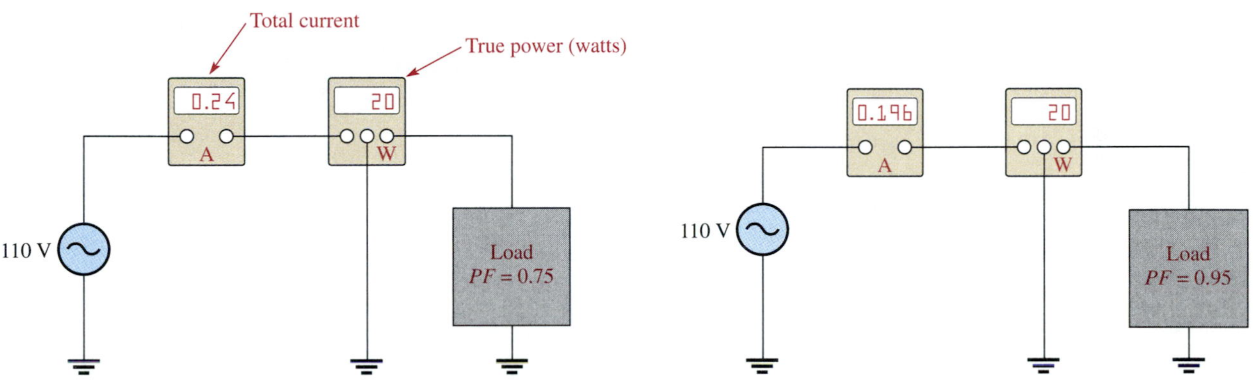


(a) A lower power factor means more total current for a given power dissipation (watts). A larger source is required to deliver the true power (watts).

(b) A higher power factor means less total current for a given power dissipation. A smaller source can deliver the same true power (watts).

**FIGURE 17–29**
***Illustration of the effect of the power factor on system requirements such as source rating (VA) and conductor size.***

Although both loads are equivalent in terms of the amount of work done (true power), the low power factor load in Figure 17–29(a) draws more current from the source than does the high power factor load in Figure 17–29(b), as indicated by the ammeters. Therefore, the source in part (a) must have a higher VA rating than the one in part (b). Also, the lines connecting the source to the load in part (a) must be a larger wire gage than those in part (b), a condition that becomes significant when very long transmission lines are required, such as in power distribution.

Figure 17–29 has demonstrated that a higher power factor is an advantage in delivering power more efficiently to a load.

## Power Factor Correction

The power factor of an inductive load can be increased by the addition of a capacitor in parallel, as shown in Figure 17–30. The capacitor compensates for the phase lag of the total current by creating a capacitive component of current that is 180° out of phase with the inductive component. This has a canceling effect and reduces the phase angle (and power factor) as well as the total current, as illustrated in the figure.

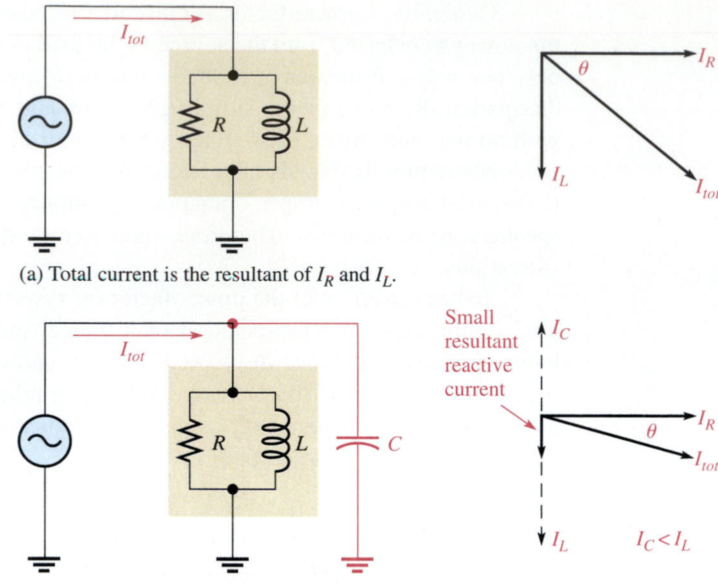
$I_{tot}$
R
L
θ
$I_R$
$I_L$
$I_{tot}$
(a) Total current is the resultant of $I_R$ and $I_L$.
$I_{tot}$
R
L
C
Small
resultant
reactive
current
$I_C$
θ
$I_R$
$I_{tot}$
$I_L$
$I_C < I_L$
(b) $I_C$ subtracts from $I_L$, leaving only a small reactive current, thus decreasing $I_{tot}$ and the phase angle.

FIGURE 17–30
***Example of how the power factor can be increased by the addition of a compensating capacitor.***

**SECTION 17–7 REVIEW**

1. To which component in an *RL* circuit is the energy loss due?
2. Calculate the power factor when $\theta = 50°$.
3. A certain *RL* circuit consists of a 470 Ω resistor and an inductive reactance of 620 Ω at the operating frequency. Determine $P_{true}$, $P_r$, and $P_a$ when $I = 100$ mA.

## 17–8 ■ BASIC APPLICATIONS

***Two basic applications, phase shift networks and frequency-selective networks (filters), are covered in this section.***

***After completing this section, you should be able to***

- **Discuss some basic *RL* applications**
  - ☐ Discuss and analyze the *RL* lead network
  - ☐ Discuss and analyze the *RL* lag network
  - ☐ Discuss how the *RL* circuit operates as a filter

### The *RL* Lead Network

The *RL* lead network is a phase shift circuit in which the output voltage leads the input voltage by a specified amount. Figure 17–31(a) shows a series *RL* circuit with the output voltage taken across the inductor. Note that in the *RC* lead network, the output was taken across the resistor. The source voltage is the input, $V_{in}$. As you know, $\theta$ is the angle between the current and the input voltage; it is also the angle between the resistor voltage and the input voltage, because $V_R$ and $I$ are in phase.

**FIGURE 17–31**
***The RL lead network ($V_{out} = V_L$).***

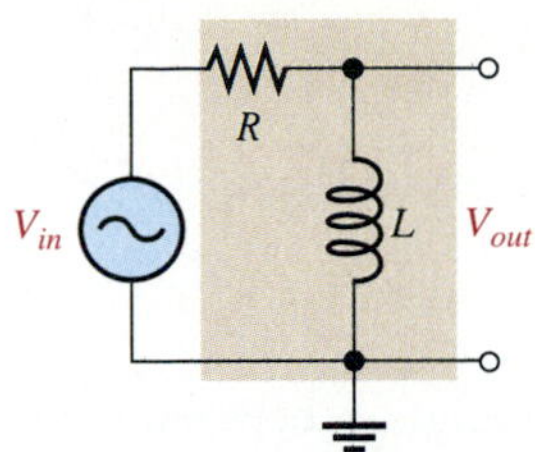

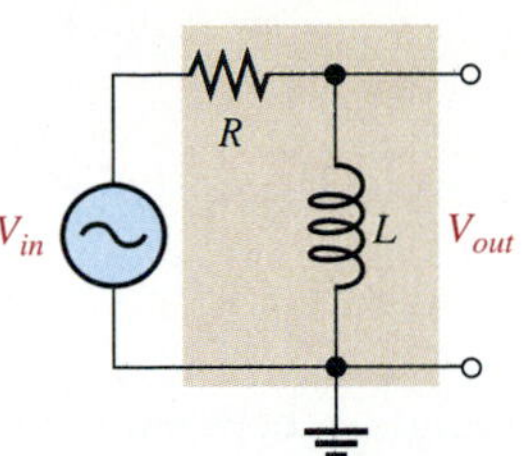

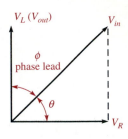

(a) A basic *RL* lead network

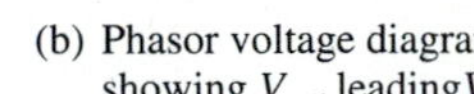


(b) Phasor voltage diagram showing $V_{out}$ leading $V_{in}$

Since $V_L$ leads $V_R$ by 90°, the phase angle between the inductor voltage and the input voltage is the difference between 90° and $\theta$, as shown in Figure 17–31(b). The inductor voltage is the output; it leads the input, thus creating a basic lead network.

When the input and output voltage waveforms of the lead network are displayed on an oscilloscope, a relationship similar to that in Figure 17–32 is observed. The amount of phase difference, designated $\phi$, between the input and the output is dependent on the relative values of the inductive reactance and the resistance, as is the magnitude of the output voltage.

**FIGURE 17–32**
***Input and output voltage waveforms.***

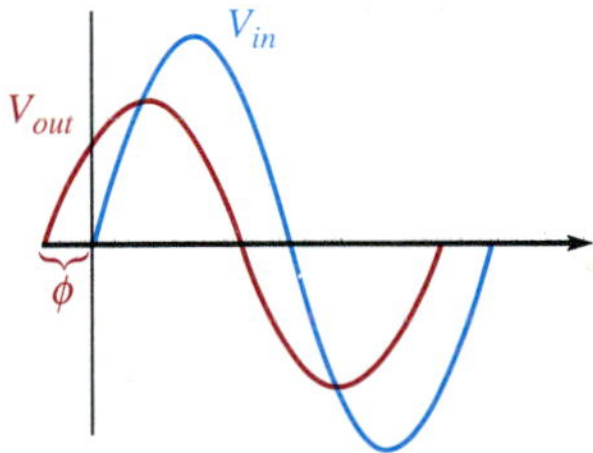


***Phase Difference Between Input and Output*** The angle between $V_{out}$ and $V_{in}$ is designated $\phi$ (phi) and is developed as follows. The polar expressions for the input voltage and the current are $V_{in}\angle 0°$ and $I\angle -\theta$, respectively. The output voltage in polar form is

$$\mathbf{V}_{out} = (I\angle -\theta)(X_L\angle 90°) = IX_L\angle(90° - \theta)$$

This expression shows that the output voltage is at an angle of 90° − $\theta$ with respect to the input voltage. Since $\theta = \tan^{-1}(X_L/R)$, the angle $\phi$ between the input and output is

$$\phi = 90° - \tan^{-1}\left(\frac{X_L}{R}\right) \qquad \textbf{(17–21)}$$

This angle can equivalently be expressed as

$$\phi = \tan^{-1}\left(\frac{R}{X_L}\right) \qquad \textbf{(17–22)}$$

**FIGURE 17–33**

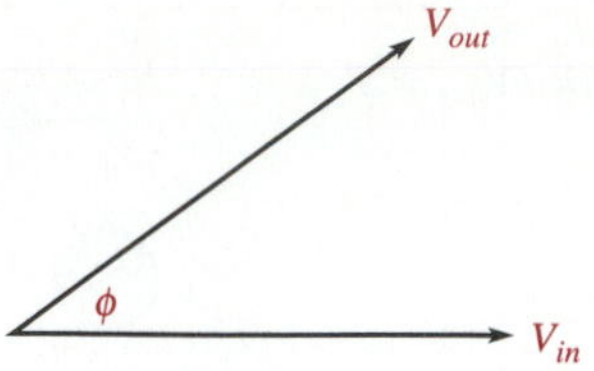

The angle $\phi$ between the output and input is always positive, indicating that the output voltage leads the input voltage, as indicated in Figure 17–33.

**EXAMPLE 17–11** Determine the amount of phase lead from input to output in each lead network in Figure 17–34.

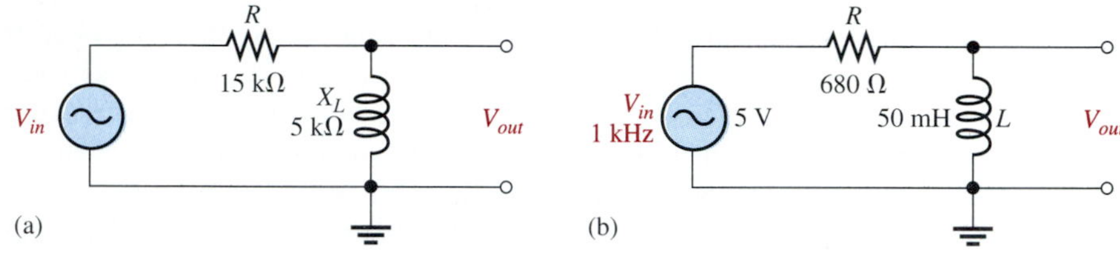


**FIGURE 17–34**

***Solution*** For the lead network in Figure 17–34(a),

$$\phi = 90° - \tan^{-1}\left(\frac{X_L}{R}\right) = 90° - \tan^{-1}\left(\frac{5\text{ k}\Omega}{15\text{ k}\Omega}\right) = 90° - 18.4° = 71.6°$$

The output leads the input by 71.6°.

For the lead network in Figure 17–34(b), first determine the inductive reactance.

$$X_L = 2\pi fL = 2\pi(1\text{ kHz})(50\text{ mH}) = 314\ \Omega$$

$$\phi = 90° - \tan^{-1}\left(\frac{X_L}{R}\right) = 90° - \tan^{-1}\left(\frac{314\ \Omega}{680\ \Omega}\right) = 65.2°$$

The output leads the input by 65.2°.

***Related Exercise*** In a certain lead network, $R = 2.2\text{ k}\Omega$ and $X_L = 1\text{ k}\Omega$. What is the phase lead?

***Magnitude of the Output Voltage*** To evaluate the output voltage in terms of its magnitude, visualize the *RL* lead network as a voltage divider. A portion of the total input voltage is dropped across the resistor and a portion across the inductor. Because the output voltage is the voltage across the inductor, it can be calculated as

$$V_{out} = \left(\frac{X_L}{\sqrt{R^2 + X_L^2}}\right)V_{in} \qquad \textbf{(17–23)}$$

The output voltage can also be found using Ohm's law as

$$V_{out} = IX_L \qquad \textbf{(17–24)}$$

The total phasor expression for the output voltage of an *RL* lead network is

$$\mathbf{V}_{out} = V_{out}\angle\phi \qquad \textbf{(17–25)}$$

**EXAMPLE 17–12** For the lead network in Figure 17–34(b) (Example 17–11), determine the output voltage in phasor form when the input voltage has an rms value of 5 V. Sketch the input and output voltage waveforms showing the proper relationships. $X_L$ (314 Ω) and $\phi$ (65.2°) were found in Example 17–11.

***Solution*** The output voltage in phasor form is

$$\mathbf{V}_{out} = \left(\frac{X_L}{\sqrt{R^2 + X_L^2}}\right)V_{in}\angle\phi$$

$$= \left[\frac{314\ \Omega}{\sqrt{(680\ \Omega)^2 + (314\ \Omega)^2}}\right]5\angle 65.2°\ \text{V} = 2.10\angle 65.2°\ \text{V}$$

The waveforms with their peak values are shown in Figure 17–35. Notice that the output voltage leads the input voltage by 65.2°.

**FIGURE 17–35**

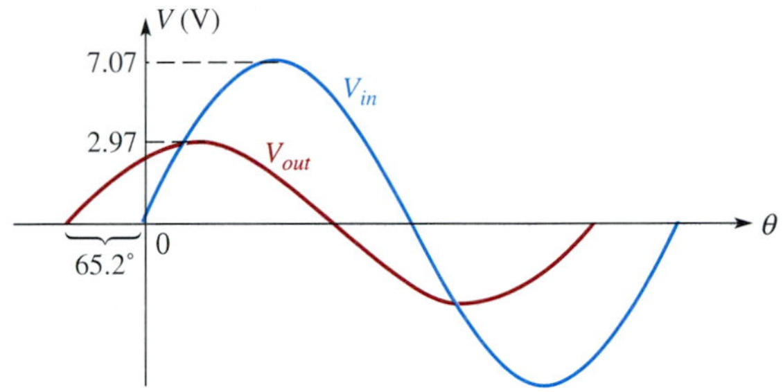


***Related Exercise*** In a lead network, does the output voltage increase or decrease when the frequency increases?

## The *RL* Lag Network

The *RL* lag network is a phase shift circuit in which the output voltage lags the input voltage by a specified amount. When the output of a series *RL* circuit is taken across the resistor rather than the inductor, as shown in Figure 17–36(a), it becomes a lag network.

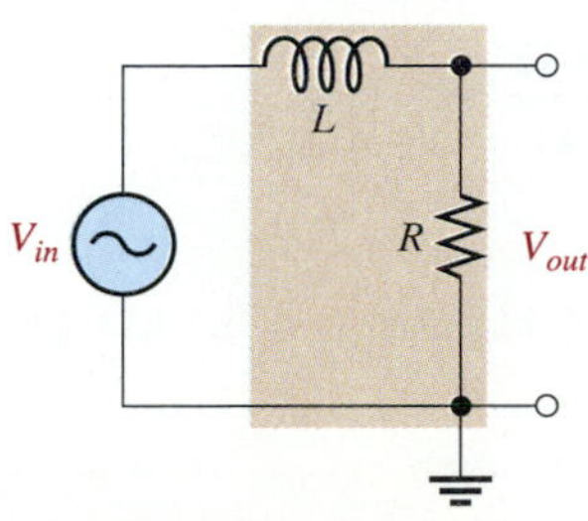


(a) A basic *RL* lag network

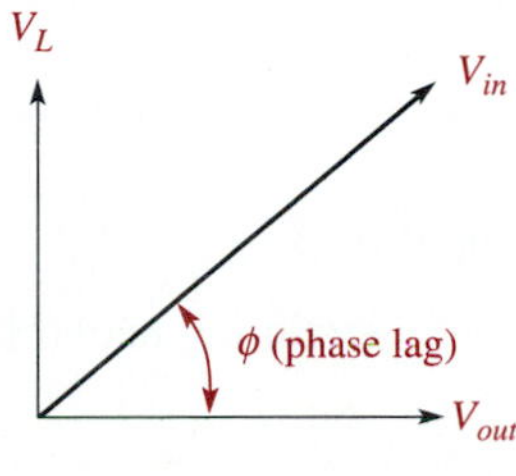


(b) Phasor voltage diagram showing phase lag between $V_{in}$ and $V_{out}$

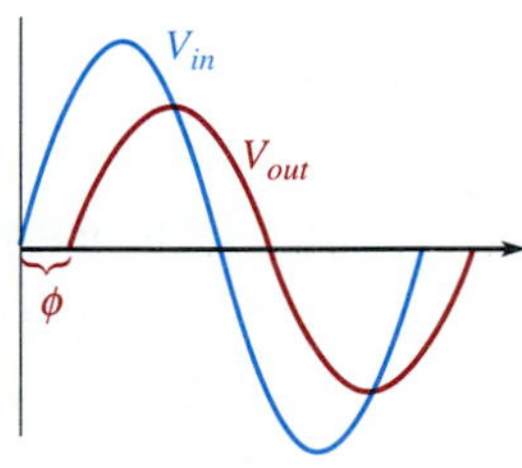


(c) Input and output waveforms

**FIGURE 17–36**
*The RL lag network ($V_{out} = V_R$).*

***Phase Difference Between Input and Output*** In a series *RL* circuit, the current lags the input voltage. Since the output voltage is taken across the resistor, the output lags the input, as indicated by the phasor diagram in Figure 17–36(b). The waveforms are shown in Figure 17–36(c).

As in the lead network, the amount of phase difference between the input and output and the magnitude of the output voltage are dependent on the relative values of the resistance and the inductive reactance. When the input voltage is assigned a reference angle of 0°, the angle of the output voltage ($\phi$) with respect to the input voltage equals $\theta$, because the resistor voltage (output) and the current are in phase with each other. The expression for the angle between the input voltage and the output voltage is

$$\phi = -\tan^{-1}\left(\frac{X_L}{R}\right) \tag{17–26}$$

This angle is negative because the output lags the input.

**EXAMPLE 17–13** Calculate the output phase angle for each circuit in Figure 17–37.

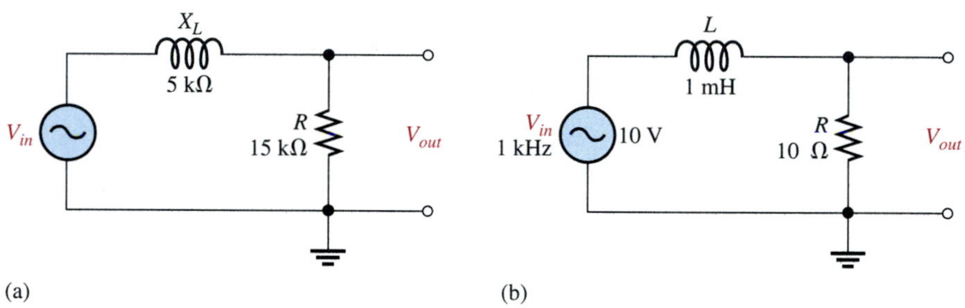


FIGURE 17–37

***Solution*** For the lag network in Figure 17–37(a).

$$\phi = -\tan^{-1}\left(\frac{X_L}{R}\right) = -\tan^{-1}\left(\frac{5\text{ k}\Omega}{15\text{ k}\Omega}\right) = -18.4°$$

The output lags the input by 18.4°.

For the lag network in Figure 17–37(b), first determine the inductive reactance.

$$X_L = 2\pi fL = 2\pi(1\text{ kHz})(1\text{ mH}) = 6.28\ \Omega$$

$$\phi = -\tan^{-1}\left(\frac{X_L}{R}\right) = -\tan^{-1}\left(\frac{6.28\ \Omega}{10\ \Omega}\right) = -32.1°$$

The output lags the input by 32.1°.

***Related Exercise*** In a certain lag network, $R = 5.6$ kΩ and $X_L = 3.5$ kΩ. Determine the phase angle.

***Magnitude of the Output Voltage*** Since the output voltage of an *RL* lag network is taken across the resistor, the magnitude can be calculated using either the voltage-divider approach or Ohm's law.

$$V_{out} = \left(\frac{R}{\sqrt{R^2 + X_L^2}}\right)V_{in} \quad (17\text{–}27)$$

$$V_{out} = IR \quad (17\text{–}28)$$

The expression for the output voltage in phasor form is

$$\mathbf{V}_{out} = V_{out}\angle -\phi \quad (17\text{–}29)$$

**EXAMPLE 17–14** The input voltage in Figure 17–37(b) (Example 17–13) has an rms value of 10 V. Determine the phasor expression for the output voltage. Sketch the waveform relationships for the input and output voltages. The phase angle (–32.1°) and $X_L$ (6.28 Ω) were found in Example 17–13.

***Solution*** The phasor expression for the output voltage is

$$\mathbf{V}_{out} = \left(\frac{R}{\sqrt{R^2 + X_L^2}}\right)V_{in}\angle\phi = \left(\frac{10\ \Omega}{11.8\ \Omega}\right)10\angle -32.1^\circ\ \text{V} = 8.47\angle -32.1^\circ\ \text{V rms}$$

The waveforms are shown in Figure 17–38.

FIGURE 17–38

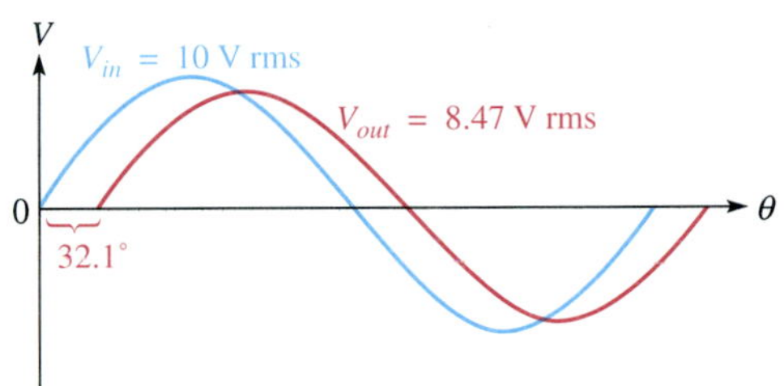


***Related Exercise*** In a lag network, $R = 4.7$ kΩ and $X_L = 6$ kΩ. If the rms input voltage is 20 V, what is the output voltage?

## The *RL* Circuit as a Filter

As with *RC* circuits, series *RL* circuits also exhibit a frequency-selective characteristic and therefore act as basic filters. Filters are introduced here as an application example and will be covered in depth in Chapter 19.

***Low-Pass Filter*** You have seen what happens to the output magnitude and phase angle in the lag network. In terms of the filtering action, the variation of the magnitude of the output voltage as a function of frequency is important.

Figure 17–39 shows the filtering action of a series *RL* circuit using specific values for purposes of illustration. In part (a) of the figure, the input is zero frequency (dc). Since the inductor ideally acts as a short to constant direct current, the output voltage equals the full value of the input voltage (neglecting the winding resistance). Therefore, the circuit passes all of the input voltage to the output (10 V in, 10 V out).

In Figure 17–39(b), the frequency of the input voltage has been increased to 1 kHz, causing the inductive reactance to increase to 62.83 Ω. For an input voltage of 10 V rms, the output voltage is approximately 8.47 V rms, which can be calculated using the voltage divider approach or Ohm's law.

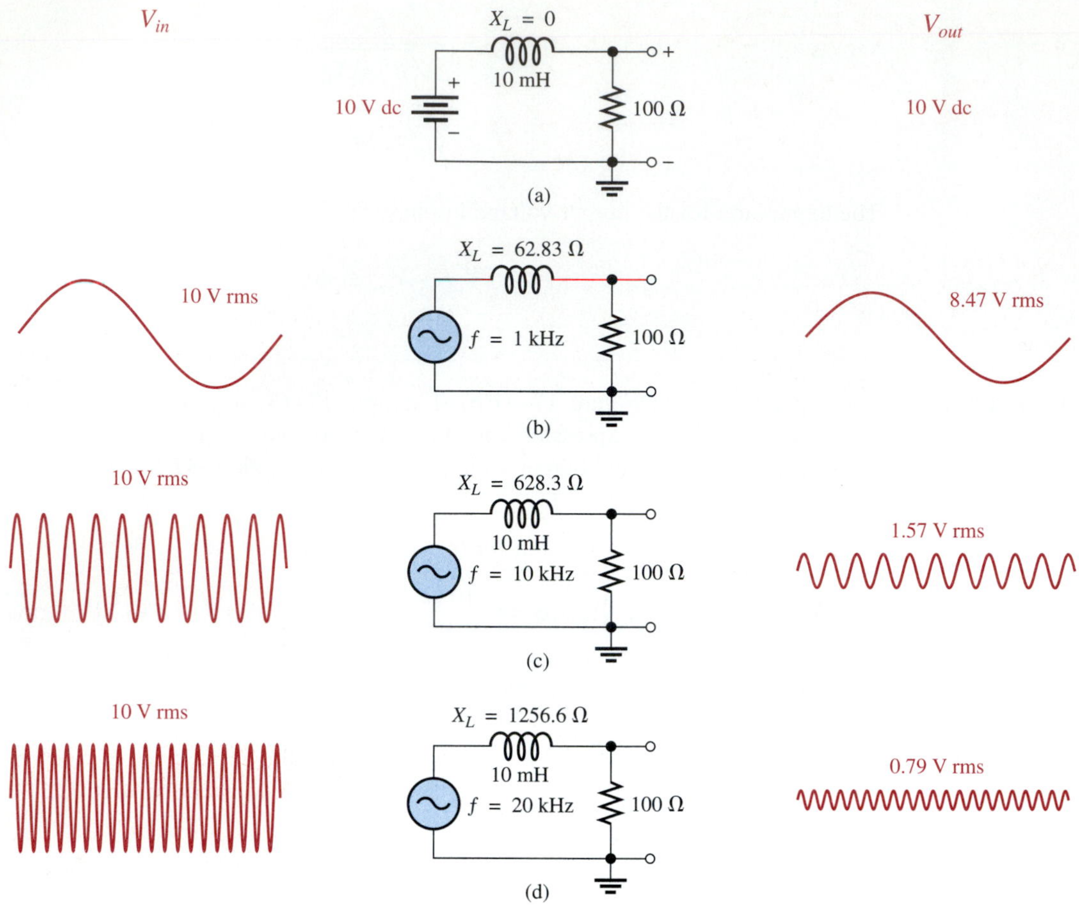


**FIGURE 17–39**

***Low-pass filter action of an RL circuit (phase shift from input to output is not indicated.***

In Figure 17–39(c), the input frequency has been increased to 10 kHz, causing the inductive reactance to increase further to 628.3 Ω. For a constant input voltage of 10 V rms, the output voltage is now 1.57 V rms.

As the input frequency is increased further, the output voltage continues to decrease and approaches zero as the frequency becomes very high, as shown in Figure 17–39(d) for $f = 20$ kHz.

A summary of the circuit action is as follows: As the frequency of the input increases, the inductive reactance increases. Because the resistance is constant and the inductive reactance increases, the voltage across the inductor increases, and that across the resistor (output voltage) decreases. The input frequency can be increased until it reaches a value at which the reactance is so large compared to the resistance that the output voltage can be neglected, because it becomes very small compared to the input voltage.

As shown in Figure 17–39, the circuit passes dc (zero frequency) completely. As the frequency of the input increases, less of the input voltage is passed through to the output. That is, the output voltage decreases as the frequency increases. It is apparent that the lower frequencies pass through the circuit much better than the higher frequencies. This *RL* circuit is therefore a very basic form of low-pass filter.

Figure 17–40 shows a response curve for a low-pass filter.

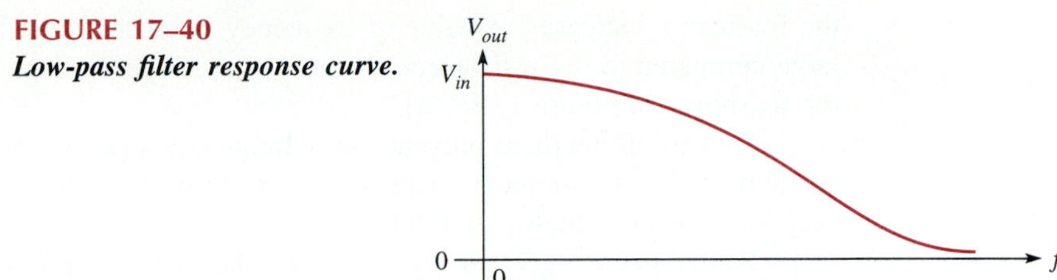


FIGURE 17–40
***Low-pass filter response curve.***

***High-Pass Filter*** Figure 17–41 illustrates high-pass filter action, where the output is taken across the inductor. When the input voltage is dc (zero frequency) in part (a), the output is zero volts because the inductor ideally appears as a short across the output.

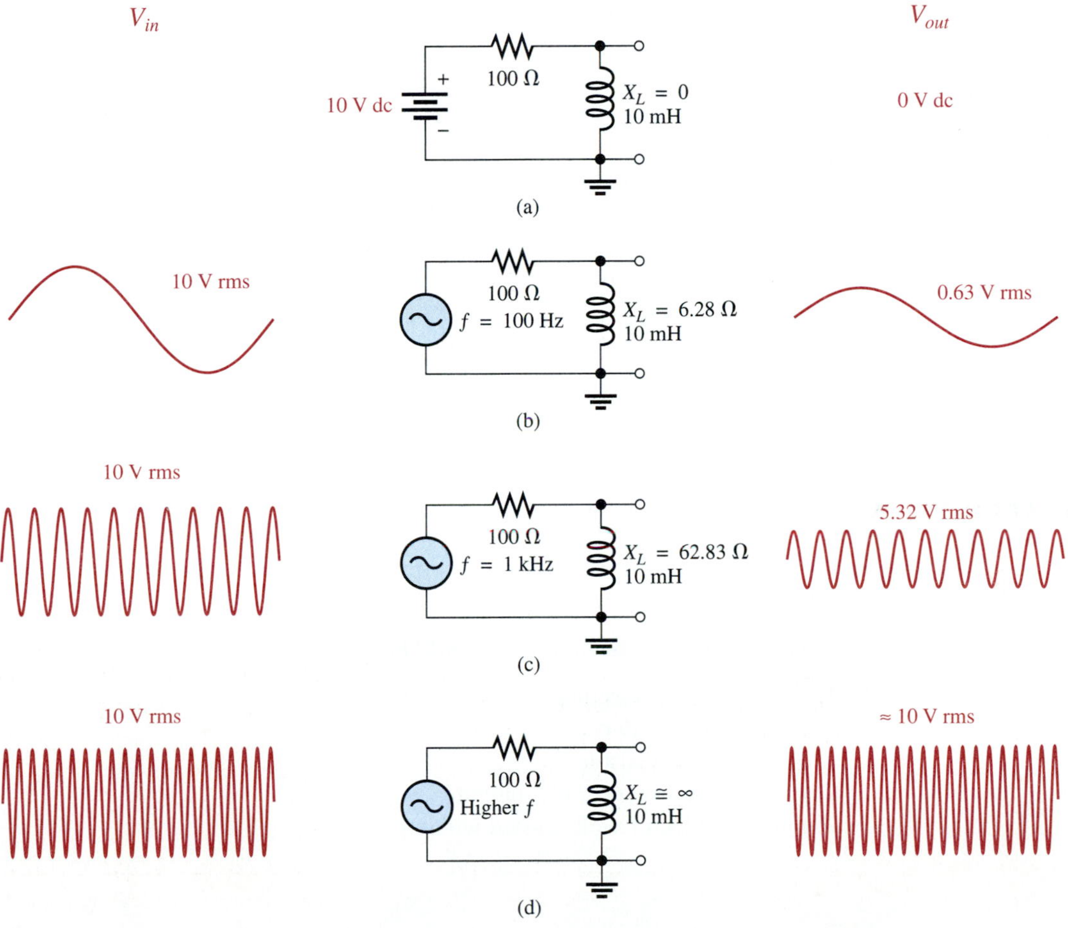


FIGURE 17–41
***High-pass filter action of an RL circuit (phase shift from input to output is not indicated).***

In Figure 17–41(b), the frequency of the input signal has been increased to 100 Hz with an rms value of 10 V. The output voltage is 0.63 V rms. Thus, only a small percentage of the input voltage appears at the output at this frequency.

In Figure 17–41(c), the input frequency is increased further to 1 kHz, causing more voltage to be developed as a result of the increase in the inductive reactance. The output voltage at this frequency is 5.32 V rms. As you can see, the output voltage increases as the frequency increases. A value of frequency is reached at which the reactance is very large compared to the resistance and most of the input voltage appears across the inductor, as shown in Figure 17–41(d).

This circuit tends to prevent lower frequency signals from appearing on the output but permits higher frequency signals to pass through from input to output; thus, it is a very basic form of high-pass filter.

The response curve in Figure 17–42 shows that the output voltage increases and then levels off as it approaches the value of the input voltage as the frequency increases.

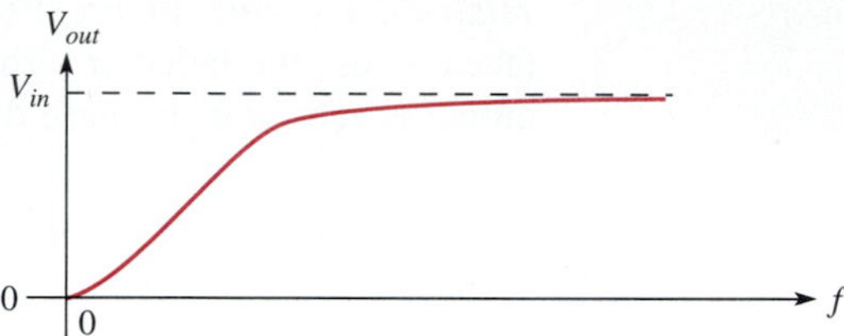


FIGURE 17–42
***High-pass filter response curve.***

**SECTION 17–8 REVIEW**

1. A certain *RL* lead network consists of a 3.3 kΩ resistor and a 15 mH inductor. Determine the phase shift between input and output at a frequency of 5 kHz.
2. An *RL* lag network has the same component values as the lead network in Question 1. What is the magnitude of the output voltage at 5 kHz when the input is 10 V rms?
3. When an *RL* circuit is used as a low-pass filter, across which component is the output taken?

## 17–9 ▪ TROUBLESHOOTING

***In this section, the effects that typical component failures have on the response of basic RL circuits are considered.***

***After completing this section, you should be able to***

- **Troubleshoot *RL* circuits**
  - ☐ Find an open inductor
  - ☐ Find an open resistor
  - ☐ Find an open in a parallel circuit
  - ☐ Find an inductor with shorted windings

### Effect of an Open Inductor

The most common failure mode for inductors occurs when the winding opens as a result of excessive current or a mechanical contact failure. It is easy to see how an open coil affects the operation of a basic series *RL* circuit, as shown in Figure 17–43. Obviously, there is no current path; therefore, the resistor voltage is zero, and the total applied voltage appears across the inductor.

### Effect of an Open Resistor

When the resistor is open, there is no current and the inductor voltage is zero. The total input voltage is across the open resistor, as shown in Figure 17–44.

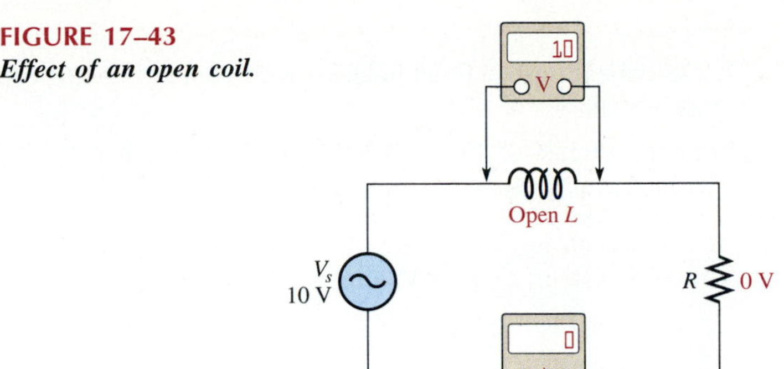


**FIGURE 17–43**
*Effect of an open coil.*

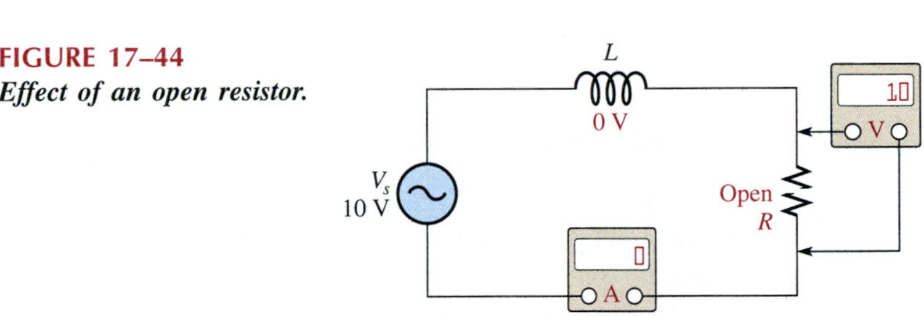


**FIGURE 17–44**
*Effect of an open resistor.*

## Open Components in Parallel Circuits

In a parallel *RL* circuit, an open resistor or inductor will cause the total current to decrease because the total impedance will increase. Obviously, the branch with the open component will have zero current. Figure 17–45 illustrates these conditions.

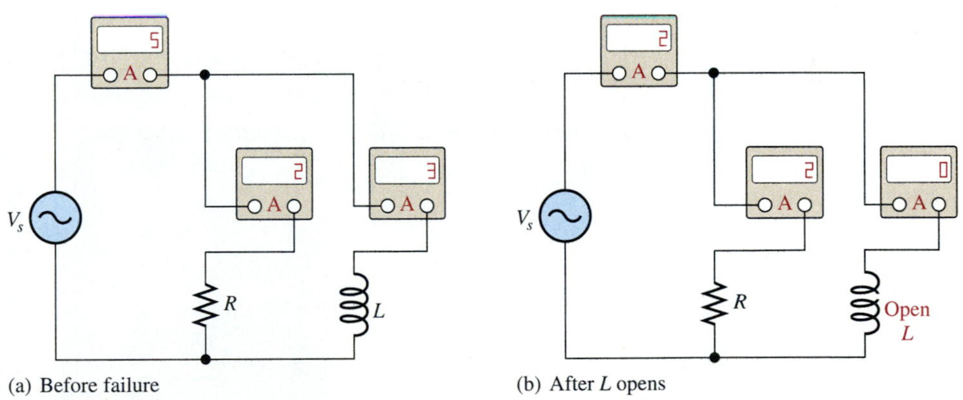


**FIGURE 17–45**
*Effect of an open component in a parallel circuit with $V_s$ constant.*

## Effect of an Inductor with Shorted Windings

It is possible for some of the windings of coils to short together as a result of damaged insulation. This failure mode is much less likely than the open coil. Shorted windings result in a reduction in inductance because the inductance of a coil is proportional to the square of the number of turns. A short between windings effectively reduces the number of turns.

**SECTION 17–9 REVIEW**

1. Describe the effect of an inductor with shorted windings on the response of a series *RL* circuit.
2. In the circuit of Figure 17–46, indicate whether $I_{tot}$, $V_{R1}$, and $V_{R2}$ increase or decrease as a result of $L$ opening.

FIGURE 17–46

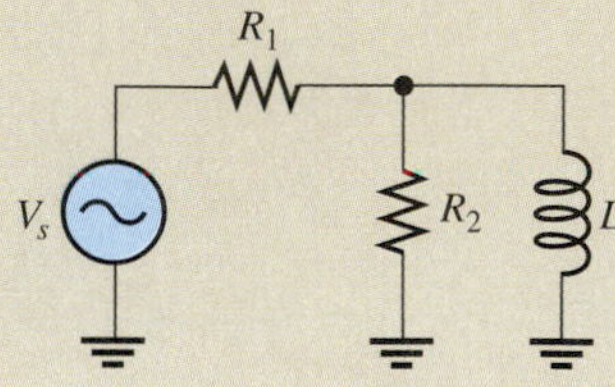


## 17–10 ■ TECHnology Theory Into Practice

***You are given two sealed modules that have been removed from a communications system that is being modified. Each module has three terminals and are labeled as RL filters, but no specifications are given. Your supervisor asks you to test the modules to determine the type of filters and the component values.***

The sealed modules have three terminals labeled IN, GND, and OUT as shown in Figure 17–47. You will apply your knowledge of series *RL* circuits and some basic measurements to determine the internal circuit configuration and the component values.

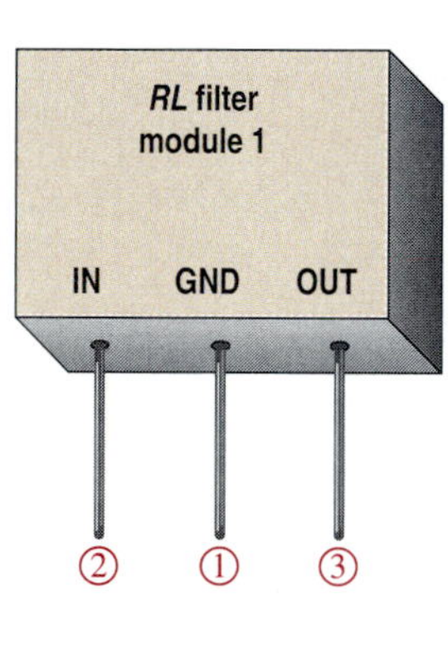


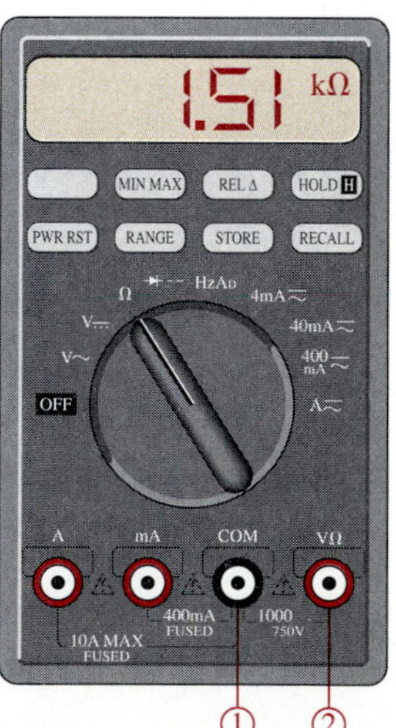


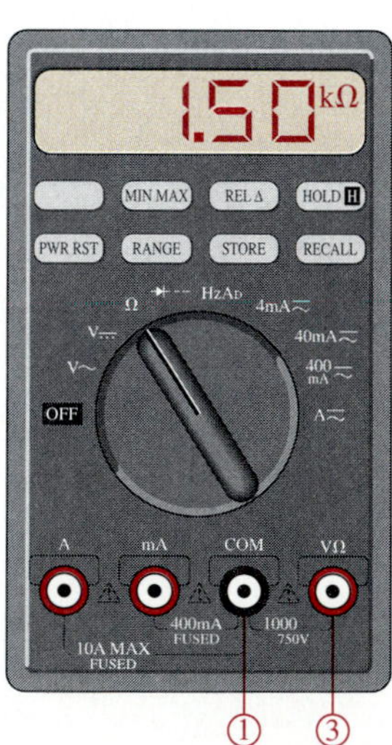


*Ohmmeter measurements of module 1.*

### Ohmmeter Measurement of Module 1

☐ Determine the arrangement of the two components and the values of the resistor and winding resistance for module 1 indicated by the meter readings in Figure 17–47.

## AC Measurement of Module 1

☐ Determine the inductance value for module 1 indicated by the test setup in Figure 17–48.

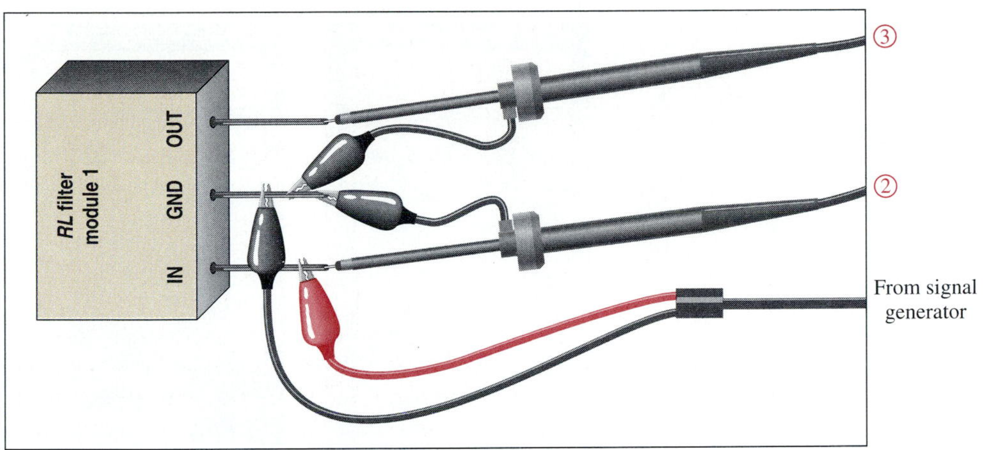


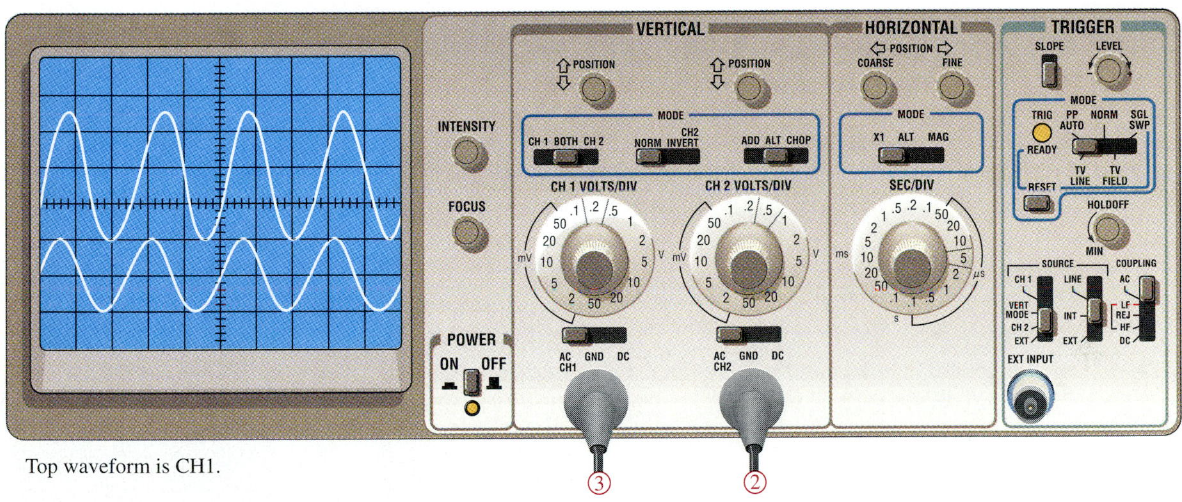


Top waveform is CH1.

**FIGURE 17–48**
***AC measurements for module 1.***

## Ohmmeter Measurement of Module 2

☐ Determine the arrangement of the two components and the values of the resistor and the winding resistance for module 2 indicated by the meter readings in Figure 17–49.

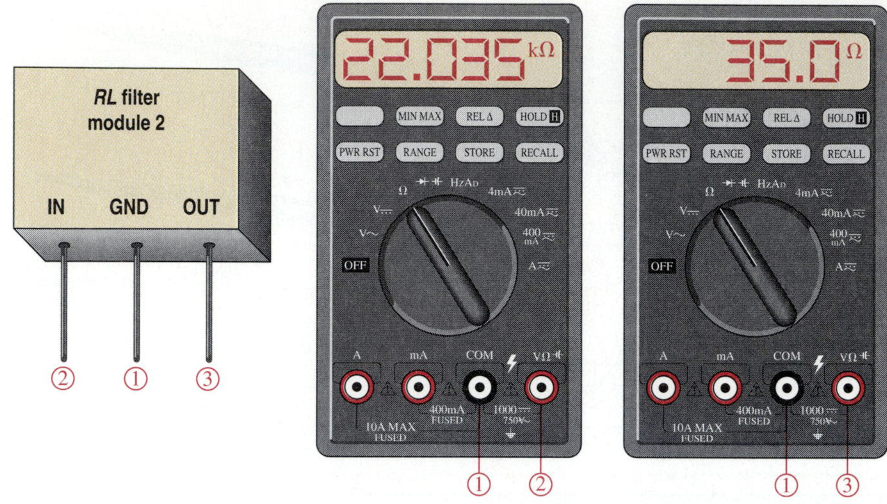


**FIGURE 17–49**
***Ohmmeter measurements of module 2.***

## AC Measurement of Module 2

☐ Determine the inductance value for module 2 indicated by the test setup in Figure 17–50.

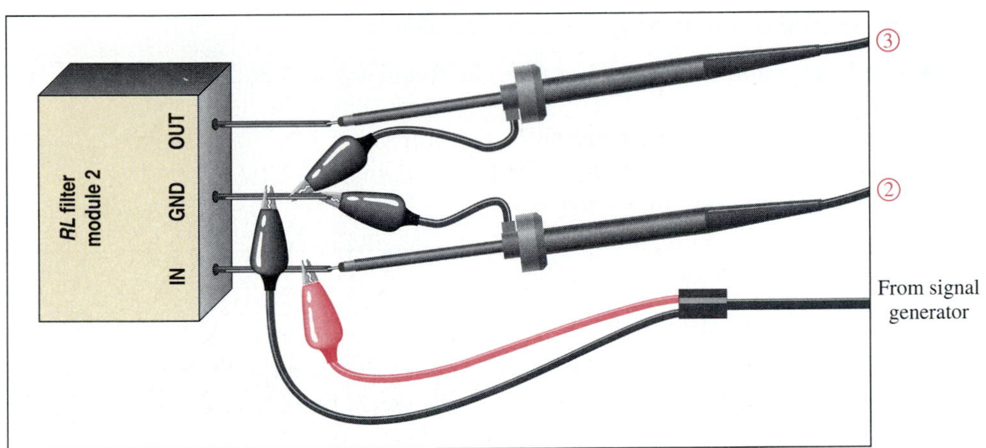


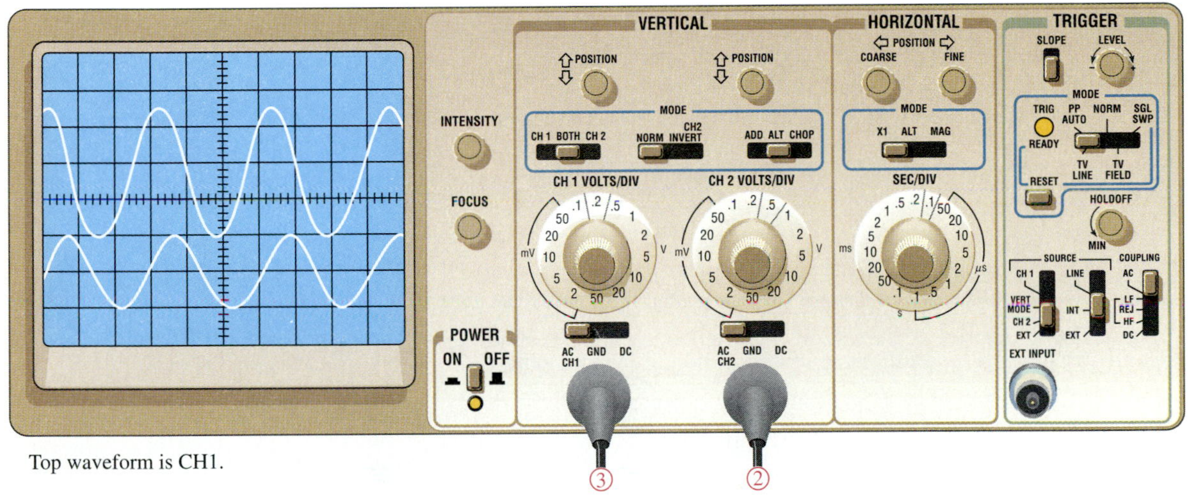


FIGURE 17–50
*AC measurements for module 2.*

**SECTION 17–10 REVIEW**

1. If the inductor in module 1 were open, what would you measure on the output with the test setup of Figure 17–48?
2. If the inductor in module 2 were open, what would you measure on the output with the test setup of Figure 17–50?

## 17–11 ■ PSpice ANALYSIS OF *RL* CIRCUITS

***In the last chapter you saw how to determine the ac response of series and parallel RC circuits. The ac analysis of RL circuits is similar, except that an inductor is the reactive element. Just as in Chapter 16, Probe can be used to determine the magnitude and phase of voltages, currents, and impedances.***

***After completing this section, you should be able to***

- **Use PSpice and Probe to determine the frequency and phase responses of an *RL* circuit**
  - ☐ Use the .AC control statement
  - ☐ Analyze a series *RL* circuit
  - ☐ Analyze a parallel *RL* circuit

### Series *RL* Circuit Analysis

A circuit file for finding the frequency response of the circuit in Figure 17–51 by taking 1000 samples from 1 Hz to 10 kHz using Probe would be as follows:

```
RL Circuit Response 1
*Series RL Circuit Frequency Response
VS  1  0  AC  5
R1  1  2  100
L1  2  0  100M
.AC  LIN  1000  1  10K
.PROBE
.END
```

FIGURE 17–51

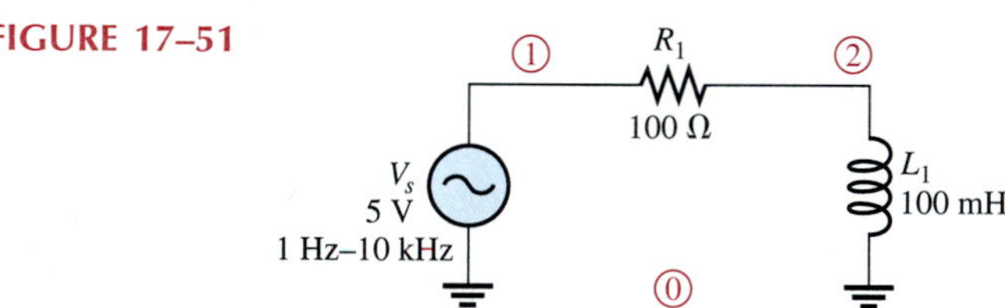


After running PSpice on the file, you can use Probe to compare how the voltage magnitudes and phases differ from the *RC* circuit in Chapter 16. Just as before, there is a critical frequency where the magnitudes of the resistor and inductor voltages are equal, but in this case, the resistor voltage decreases with frequency while the inductor voltage increases. As you would expect, the resistor and inductor voltages remain 90° apart as in the *RC* circuit, but in this case, the phase angle of the resistor voltage decreases from 0° to –90° while the phase angle of the inductor changes from 90° to 0°.

As you know, the responses of *RC* and *RL* circuits are similar in many ways, but unlike in others. At lower frequencies the inductive reactance is low and has less effect on the total impedance; for capacitive reactance the reverse is true. Similarly, at higher frequencies the inductive reactance is high and becomes the dominant effect in the circuit; at higher frequencies capacitive reactance becomes less and less important in the circuit response.

### Parallel *RL* Circuit Analysis

The circuit file for the parallel circuit in Figure 17–52 is written as follows:

```
RL Circuit Response 2
*Parallel RL Circuit Frequency Response
VS  1  0  AC  5
RS  1  2  1P
R1  2  0  100
L1  2  0  100M
.AC  LIN  1000  1  10K
.PROBE
.END
```

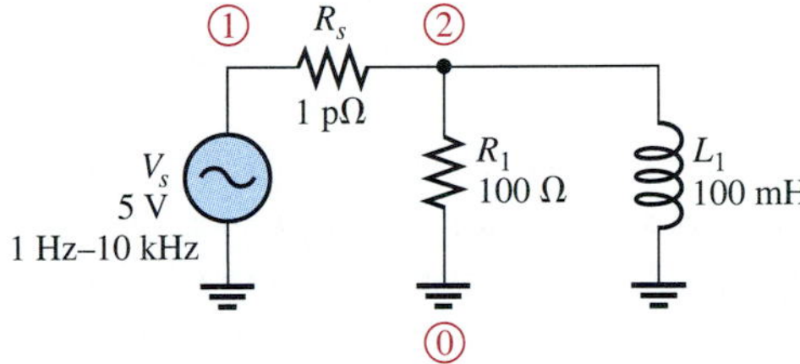


FIGURE 17–52

As was discussed in Chapter 15, the extremely small resistance $R_s$ is added to keep PSpice from seeing $V_s$ shorted to ground through $L_1$. If you use PSpice and Probe to compare the magnitudes and phases of the resistor and inductor currents, you will find the responses similar yet different from those of the parallel *RC* circuit in Chapter 16. The magnitudes of the source and inductor currents decrease with frequency while the resistor current stays constant. The phase angle of the inductor current remains constant at −90° and that of the resistor current remains at 0°. The phase angle of the source (total) current increases from 90° to 180°. Again, this is due to the response of inductors compared to that of capacitors. At low frequencies the inductive current is larger than the resistor current and, therefore, has a greater effect on the circuit response; at higher frequencies the reverse is true.

**SECTION 17–11 REVIEW**

1. Rewrite the circuit file for the circuit in Figure 17–51 for the following changes: $R_1 = 1.8\ \text{k}\Omega$, $L_1 = 10\ \text{mH}$, source phase angle is 30°, and the frequency range is 1 kHz to 100 kHz.
2. Rewrite the circuit file for the circuit in Figure 17–52 if a 470 Ω resistor is added in series with the source and connected to the junction of $R_1$ and $L_1$.

## ■ SUMMARY

- When a sine wave voltage is applied to an *RL* circuit, the current and all the voltage drops are also sine waves.
- Total current in an *RL* circuit always lags the source voltage.
- The resistor voltage is always in phase with the current.
- The inductor voltage always leads the current by 90°.
- In an *RL* circuit, the impedance is determined by both the resistance and the inductive reactance combined.
- Impedance is expressed in units of ohms.
- The impedance of an *RL* circuit varies directly with frequency.
- The phase angle ($\theta$) of a series *RL* circuit varies directly with frequency.

- You can determine the impedance of a circuit by measuring the applied voltage and the total current and then applying Ohm's law.
- In an *RL* circuit, part of the power is resistive and part reactive.
- The power factor indicates how much of the apparent power is true power.
- A power factor of 1 indicates a purely resistive circuit, and a power factor of 0 indicates a purely reactive circuit.
- In a lag network, the output voltage lags the input voltage in phase.
- In a lead network, the output voltage leads the input voltage in phase.
- A filter passes certain frequencies and rejects others.

## GLOSSARY

**Admittance** A measure of the ability of a reactive circuit to permit current; the reciprocal of impedance. The unit is the siemens (S).

**Apparent power** The phasor combination of resistive power (true power) and reactive power. The unit is the volt-ampere (VA).

**Inductive reactance** The opposition of an inductor to sinusoidal current. The unit is the ohm.

**Inductive susceptance** The reciprocal of inductive reactance. The unit is the siemens (S).

**Power factor** The relationship between volt-amperes and true power or watts. Volt-amperes multiplied by the power factor equals true power.

## FORMULAS

### Series *RL* Circuits

**(17–1)** $\mathbf{X}_L = jX_L$

**(17–2)** $\mathbf{Z} = R + jX_L$

**(17–3)** $Z = \sqrt{R^2 + X_L^2}$

**(17–4)** $\theta = \tan^{-1}\left(\frac{X_L}{R}\right)$

**(17–5)** $\mathbf{Z} = \sqrt{R^2 + X_L^2}\angle\tan^{-1}\left(\frac{X_L}{R}\right)$

**(17–6)** $\mathbf{V}_s = V_R + jV_L$

**(17–7)** $\mathbf{V}_s = \sqrt{V_R^2 + V_L^2}\angle\tan^{-1}\left(\frac{V_L}{V_R}\right)$

**(17–8)** $V_s = \sqrt{V_R^2 + V_L^2}$

**(17–9)** $\theta = \tan^{-1}\left(\frac{V_L}{V_R}\right)$

### Parallel *RL* Circuits

**(17–10)** $\mathbf{Z} = \left(\frac{RX_L}{\sqrt{R^2 + X_L^2}}\right)\angle\left(90^\circ - \tan^{-1}\left(\frac{X_L}{R}\right)\right)$

**(17–11)** $Z = \frac{RX_L}{\sqrt{R^2 + X_L^2}}$

**(17–12)** $\theta = 90^\circ - \tan^{-1}\left(\frac{X_L}{R}\right)$

**(17–13)** $\mathbf{B}_L = \frac{1}{X_L\angle 90^\circ} = B_L\angle -90^\circ = -jB_L$

**(17–14)** $\mathbf{Y} = \frac{1}{Z\angle \pm\theta} = Y\angle \pm\mp\theta$

**(17–15)** $\mathbf{Y} = G - jB_L$

(17–16) $\mathbf{I} = I_R - jI_L$

(17–17) $\mathbf{I}_{tot} = \sqrt{I_R^2 + I_L^2}\angle -\tan^{-1}\left(\frac{I_L}{I_R}\right)$

(17–18) $I_{tot} = \sqrt{I_R^2 + I_L^2}$

(17–19) $\theta = -\tan^{-1}\left(\frac{I_L}{I_R}\right)$

### Power in *RL* Circuits

(17–20) $P_r = I^2 X_L$

### Lead Network

(17–21) $\phi = 90° - \tan^{-1}\left(\frac{X_L}{R}\right)$

(17–22) $\phi = \tan^{-1}\left(\frac{R}{X_L}\right)$

(17–23) $V_{out} = \left(\frac{X_L}{\sqrt{R^2 + X_L^2}}\right)V_{in}$

(17–24) $V_{out} = IX_L$

(17–25) $\mathbf{V}_{out} = V_{out}\angle\phi$

### Lag Network

(17–26) $\phi = -\tan^{-1}\left(\frac{X_L}{R}\right)$

(17–27) $V_{out} = \left(\frac{R}{\sqrt{R^2 + X_L^2}}\right)V_{in}$

(17–28) $V_{out} = IR$

(17–29) $\mathbf{V}_{out} = V_{out}\angle -\phi$

## ■ SELF-TEST

**1.** In a series *RL* circuit, the resistor voltage
**(a)** leads the applied voltage **(b)** lags the applied voltage
**(c)** is in phase with the applied voltage **(d)** is in phase with the current
**(e)** answers (a) and (d) **(f)** answers (b) and (d)

**2.** When the frequency of the voltage applied to a series *RL* circuit is increased, the impedance
**(a)** decreases **(b)** increases **(c)** does not change

**3.** When the frequency of the voltage applied to a series *RL* circuit is decreased, the phase angle
**(a)** decreases **(b)** increases **(c)** does not change

**4.** If the frequency is doubled and the resistance is halved, the impedance of a series *RL* circuit
**(a)** doubles **(b)** halves
**(c)** remains constant **(d)** cannot be determined without values

**5.** To reduce the current in a series *RL* circuit, the frequency should be
**(a)** increased **(b)** decreased **(c)** constant

**6.** In a series *RL* circuit, 10 V rms is measured across the resistor, and 10 V rms is measured across the inductor. The peak value of the source voltage is
**(a)** 14.14 V **(b)** 28.28 V **(c)** 10 V **(d)** 20 V

**7.** The voltages in Problem 6 are measured at a certain frequency. To make the resistor voltage greater than the inductor voltage, the frequency is
**(a)** increased **(b)** decreased **(c)** doubled **(d)** not a factor

**8.** When the resistor voltage in a series *RL* circuit becomes greater than the inductor voltage, the phase angle
**(a)** increases **(b)** decreases **(c)** is not affected

**9.** When the frequency is increased, the impedance of a parallel *RL* circuit

**(a)** increases **(b)** decreases **(c)** remains constant

**10.** In a parallel *RL* circuit, there are 2 A rms in the resistive branch and 2 A rms in the inductive branch. The total rms current is

**(a)** 4 A **(b)** 5.656 A **(c)** 2 A **(d)** 2.828 A

**11.** You are observing two voltage waveforms on an oscilloscope. The time base (time/division) of the scope is adjusted so that one-half cycle of the waveforms covers the ten horizontal divisions. The positive-going zero crossing of one waveform is at the leftmost division, and the positive-going zero crossing of the other is three divisions to the right. The phase angle between these two waveforms is

**(a)** 18° **(b)** 36° **(c)** 54° **(d)** 180°

**12.** Which of the following power factors results in less energy loss in an *RL* circuit?

**(a)** 1 **(b)** 0.9 **(c)** 0.5 **(d)** 0.1

**13.** If a load is purely inductive and the reactive power is 10 VAR, the apparent power is

**(a)** 0 VA **(b)** 10 VA **(c)** 14.14 VA **(d)** 3.16 VA

**14.** For a certain load, the true power is 10 W and the reactive power is 10 VAR. The apparent power is

**(a)** 5 VA **(b)** 20 VA **(c)** 14.14 VA **(d)** 100 VA

## ■ PROBLEMS

### SECTION 17–1 Sinusoidal Response of *RL* Circuits

**1.** A 15 kHz sinusoidal voltage is applied to a series *RL* circuit. What is the frequency of $I$, $V_R$, and $V_L$?

**2.** What are the wave shapes of $I$, $V_R$, and $V_L$ in Problem 1?

### SECTION 17–2 Impedance and Phase Angle of Series *RL* Circuits

**3.** Express the total impedance of each circuit in Figure 17–53 in both polar and rectangular forms.

**FIGURE 17–53**

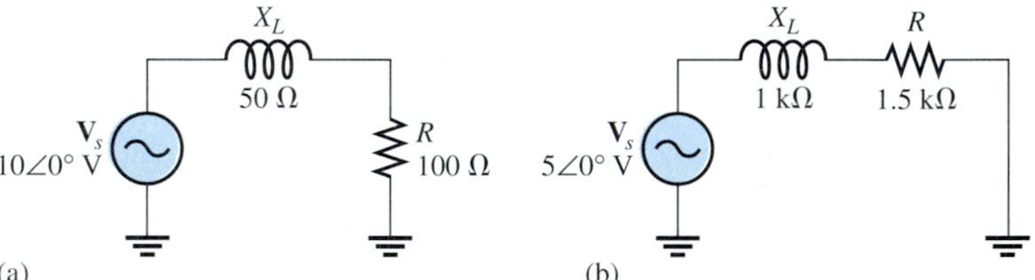


**4.** Determine the impedance magnitude and phase angle in each circuit in Figure 17–54. Sketch the impedance diagrams.

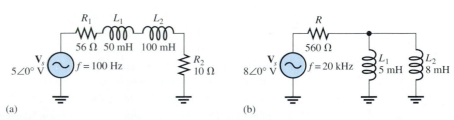

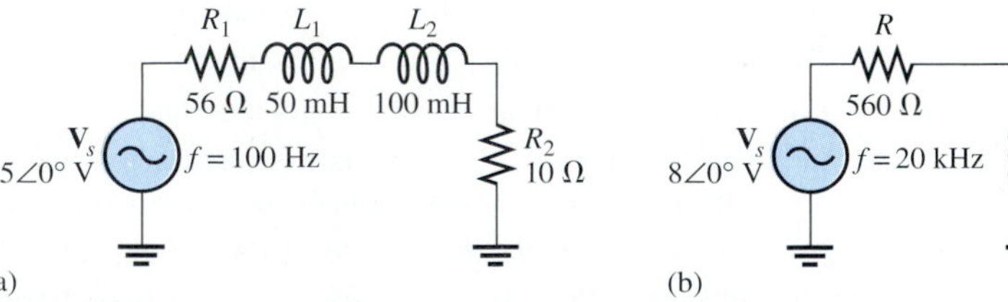


**FIGURE 17–54**

**5.** In Figure 17–55, determine the impedance at each of the following frequencies:

**(a)** 100 Hz **(b)** 500 Hz **(c)** 1 kHz **(d)** 2 kHz

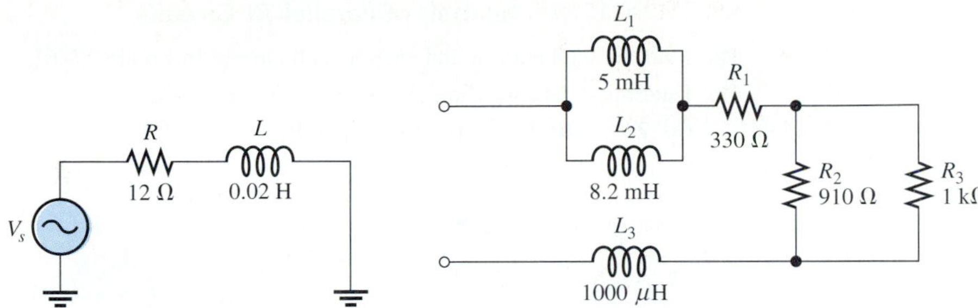


**FIGURE 17–55** **FIGURE 17–56**

6. Determine the values of $R$ and $X_L$ in a series $RL$ circuit for the following values of total impedance:
   (a) $\mathbf{Z} = 20\ \Omega + j45\ \Omega$ (b) $\mathbf{Z} = 500\angle 35°\ \Omega$
   (c) $\mathbf{Z} = 2.5\angle 72.5°\ \text{k}\Omega$ (d) $\mathbf{Z} = 998\angle 45°\ \Omega$
7. Reduce the circuit in Figure 17–56 to a single resistance and inductance in series.

## SECTION 17–3 Analysis of Series *RL* Circuits

8. Express the current in polar form for each circuit of Figure 17–53.
9. Calculate the total current in each circuit of Figure 17–54 and express in polar form.
10. Determine $\theta$ for the circuit in Figure 17–57.
11. If the inductance in Figure 17–57 is doubled, does $\theta$ increase or decrease, and by how many degrees?
12. Sketch the waveforms for $\mathbf{V}_s$, $\mathbf{V}_R$, and $\mathbf{V}_L$ in Figure 17–57. Show the proper phase relationships.
13. For the circuit in Figure 17–58, find $\mathbf{V}_R$ and $\mathbf{V}_L$ for each of the following frequencies:
    (a) 60 Hz (b) 200 Hz (c) 500 Hz (d) 1 kHz
14. Determine the magnitude and phase angle of the source voltage in Figure 17–59.

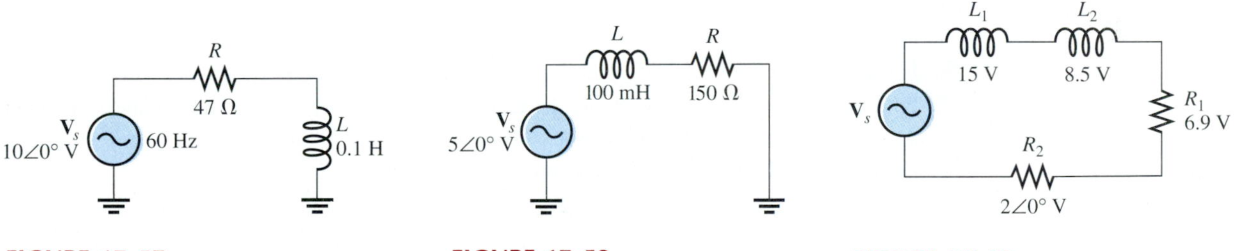


**FIGURE 17–57** **FIGURE 17–58** **FIGURE 17–59**

## SECTION 17–4 Impedance and Phase Angle of Parallel *RL* Circuits

15. What is the impedance expressed in polar form for the circuit in Figure 17–60?

**FIGURE 17–60**

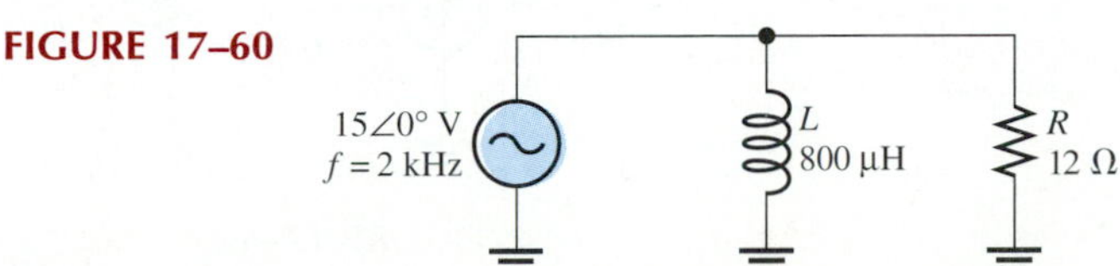


16. Repeat Problem 15 for the following frequencies:
    (a) 1.5 kHz (b) 3 kHz (c) 5 kHz (d) 10 kHz
17. At what frequency does $X_L$ equal $R$ in Figure 17–60?

## SECTION 17–5 Analysis of Parallel *RL* Circuits

**18.** Find the total current and each branch current in Figure 17–61.

**19.** Determine the following quantities in Figure 17–62:

**(a)** $\mathbf{Z}$ **(b)** $\mathbf{I}_R$ **(c)** $\mathbf{I}_L$ **(d)** $\mathbf{I}_{tot}$ **(e)** $\theta$

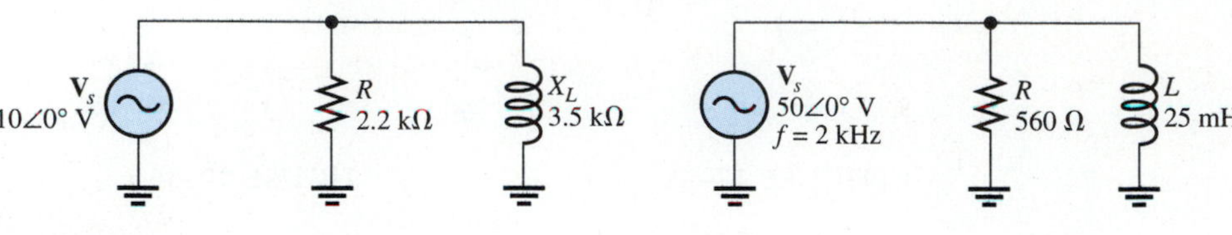


FIGURE 17–61

FIGURE 17–62

**20.** Repeat Problem 19 for $R = 56\ \Omega$ and $L = 330\ \mu\text{H}$.

**21.** Convert the circuit in Figure 17–63 to an equivalent series form.

**22.** Find the magnitude and phase angle of the total current in Figure 17–64.

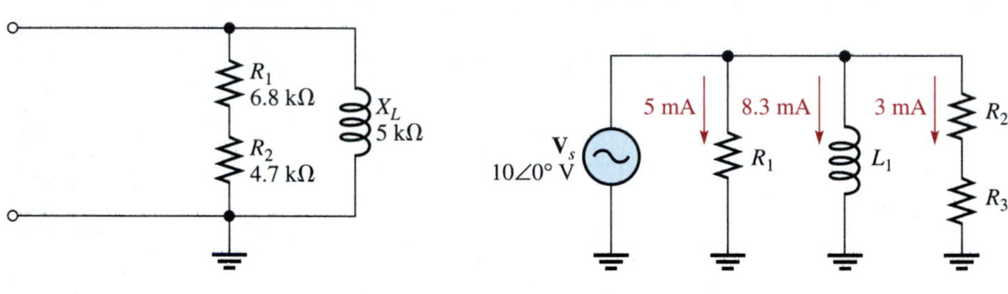


FIGURE 17–63

FIGURE 17–64

## SECTION 17–6 Series-Parallel *RL* Circuits

**23.** Determine the voltages in polar form across each element in Figure 17–65. Sketch the voltage phasor diagram.

**24.** Is the circuit in Figure 17–65 predominantly resistive or predominantly inductive?

**25.** Find the current in each branch and the total current in Figure 17–65. Express the currents in polar form. Sketch the current phasor diagram.

**26.** For the circuit in Figure 17–66, determine the following:

**(a)** $\mathbf{I}_{tot}$ **(b)** $\theta$ **(c)** $\mathbf{V}_{R1}$ **(d)** $\mathbf{V}_{R2}$ **(e)** $\mathbf{V}_{R3}$ **(f)** $\mathbf{V}_{L1}$ **(g)** $\mathbf{V}_{L2}$

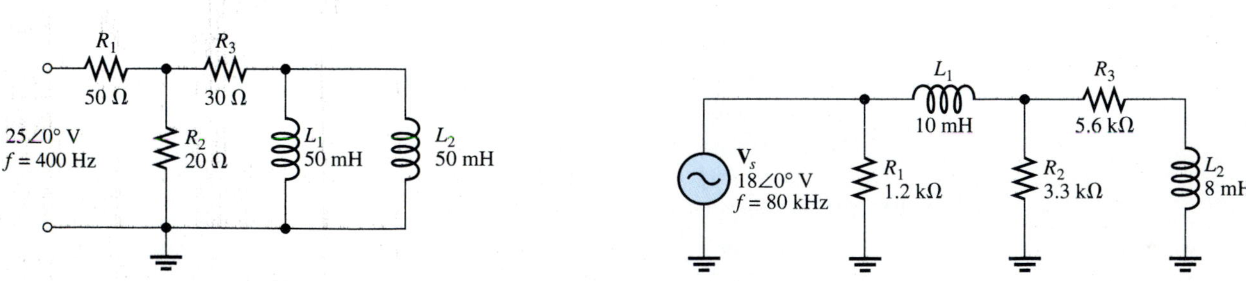


FIGURE 17–65

FIGURE 17–66

**27.** For the circuit in Figure 17–67, determine the following:

**(a)** $\mathbf{I}_{tot}$ **(b)** $\mathbf{V}_{L1}$ **(c)** $\mathbf{V}_{AB}$

28. Draw the phasor diagram of all voltages and currents in Figure 17–67.

29. Determine the phase shift and attenuation (ratio of $V_{out}$ to $V_{in}$) from the input to the output for the network in Figure 17–68.

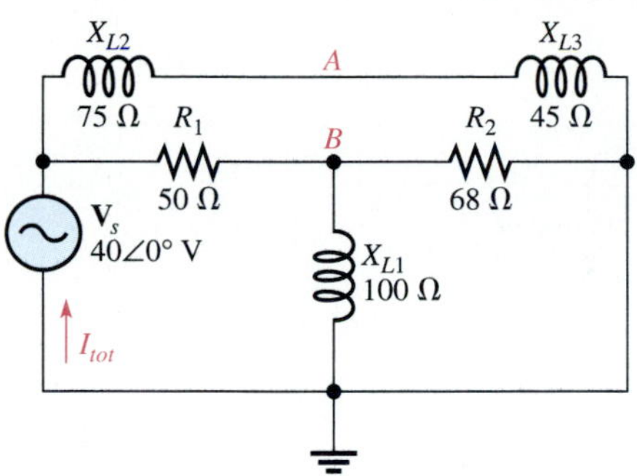


FIGURE 17–67

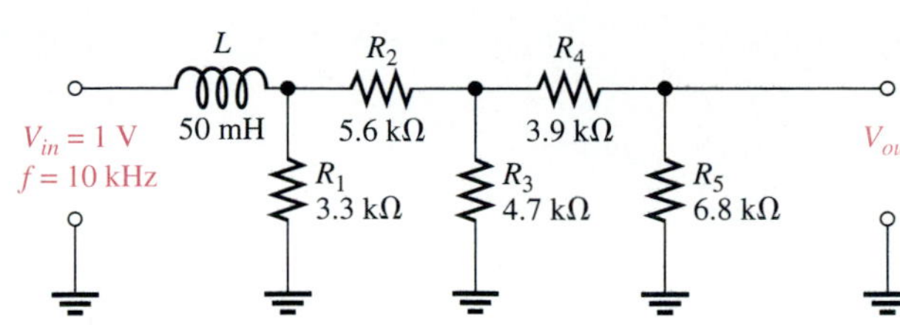


FIGURE 17–68

30. Determine the phase shift and attenuation from the input to the output for the ladder network in Figure 17–69.

FIGURE 17–69

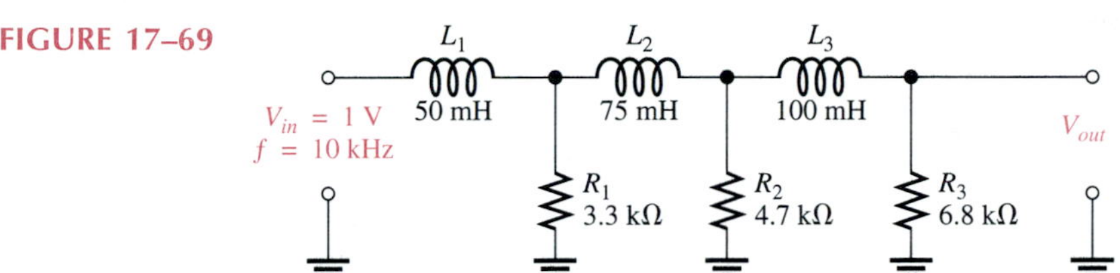


31. Design an ideal inductive switching circuit that will provide a momentary voltage of 2.5 kV from a 12 V dc source when a switch is thrown instantaneously from one position to another. The drain on the source must not exceed 1 A.

## SECTION 17–7 Power in *RL* Circuits

32. In a certain *RL* circuit, the true power is 100 mW, and the reactive power is 340 mVAR. What is the apparent power?

33. Determine the true power and the reactive power in Figure 17–57.

34. What is the power factor in Figure 17–61?

35. Determine $P_{true}$, $P_r$, $P_a$, and *PF* for the circuit in Figure 17–66. Sketch the power triangle.

36. Find the true power for the circuit in Figure 17–67.

## SECTION 17–8 Basic Applications

37. For the lag network in Figure 17–70, determine the phase lag of the output voltage with respect to the input for the following frequencies:

    **(a)** 1 Hz **(b)** 100 Hz **(c)** 1 kHz **(d)** 10 kHz

38. Draw the response curve for the circuit in Figure 17–70. Show the output voltage versus frequency in 1 kHz increments from 0 Hz to 5 kHz.

FIGURE 17–70

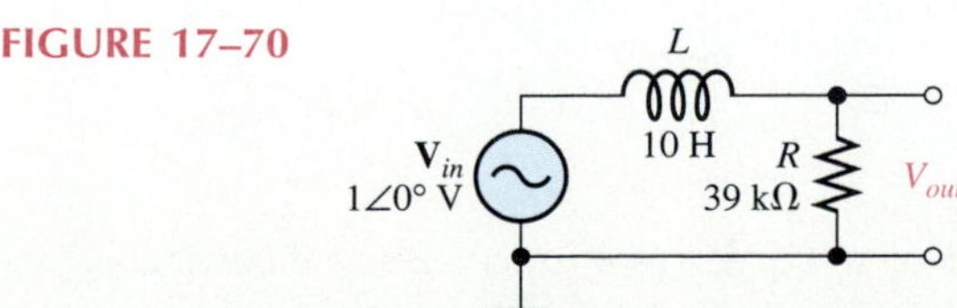

39. Repeat Problem 37 for the lead network to find the phase lead in Figure 17–71.
40. Using the same procedure as in Problem 38, draw the response curve for Figure 17–71.
41. Sketch the voltage phasor diagram for each circuit in Figures 17–70 and 17–71 for a frequency of 8 kHz.

FIGURE 17–71

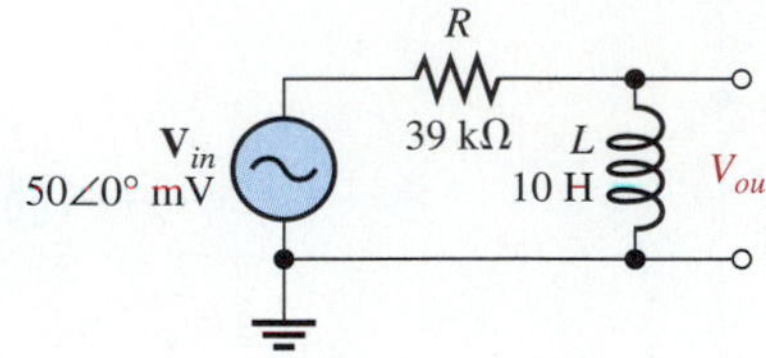


## SECTION 17–9 Troubleshooting

42. Determine the voltage across each element in Figure 17–66 if $L_1$ is open.
43. Determine the output voltage in Figure 17–72 for each of the following failure modes:
**(a)** $L_1$ open **(b)** $L_2$ open **(c)** $R_1$ open **(d)** a short across $R_2$

FIGURE 17–72

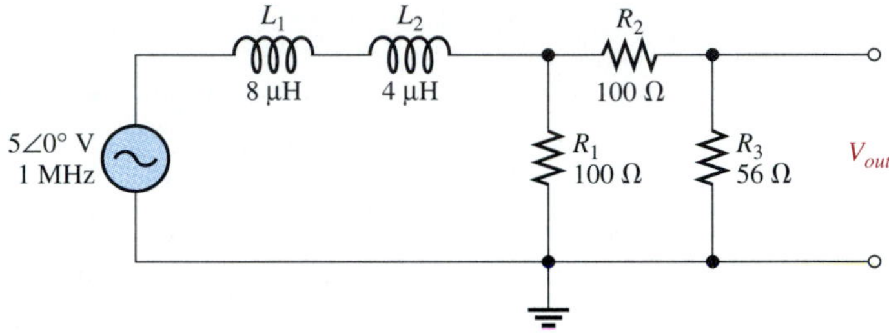


## ■ ANSWERS TO SECTION REVIEWS

### Section 17–1

1. The current frequency is 1 kHz.
2. The phase angle is closer to 0°.

### Section 17–2

1. $R = 150\ \Omega$; $X_L = 220\ \Omega$
2. $\mathbf{Z} = R + jX_L = 33\text{ k}\Omega + j50\text{ k}\Omega$; $\mathbf{Z} = \sqrt{R^2 + X_L^2}\,\angle\tan^{-1}(X_L/R) = 59.9\angle 56.6^\circ\text{ k}\Omega$

### Section 17–3

1. $V_s = \sqrt{V_R^2 + V_L^2} = 3.61\text{ V}$
2. $\theta = \tan^{-1}(V_L/V_R) = 56.3^\circ$
3. When $f$ increases, $X_L$ increases, $Z$ increases, and $\theta$ increases.

### Section 17–4

1. $Y = \dfrac{1}{Z} = \dfrac{1}{\sqrt{R^2 + X_L^2}} = 2\text{ mS}$
2. $Y = \dfrac{1}{Z} = 25.1\text{ mS}$
3. **I** lags $V_s$; $\theta = 32.1^\circ$

### Section 17–5

1. $I_{tot} = 32\text{ mA}$
2. $\mathbf{I}_{tot} = 23.3\angle -59.0^\circ\text{ mA}$; $\theta$ is with respect to the input voltage.
3. $\theta = -90^\circ$

### Section 17–6

1. $\mathbf{Z} = 494\angle 59.0° \ \Omega$
2. $\mathbf{I}_{tot} = 20.2\angle -59.0°$ mA

### Section 17–7

1. Energy loss is due to the resistor.
2. $PF = 0.643$
3. $P_{\text{true}} = 4.7$ W; $P_r = 6.2$ VAR; $P_a = 7.78$ VA

### Section 17–8

1. $\phi = 81.9°$
2. $V_{out} = 9.90$ V
3. The output is across the resistor.

### Section 17–9

1. Shorted windings reduce $L$ and thereby reduce $X_L$ at any given frequency.
2. $I_{tot}$ decreases, $V_{R1}$ decreases, $V_{R2}$ increases.

### Section 17–10

1. $V_{out} = 0$ V
2. $V_{out} = V_{in}$

### Section 17–11

1. Changes lines for VS, R1, L1 and .AC as follows:

```
VS   1   0   AC   5   30
R1   1   2   1.8K
L1   2   0   10M
.AC   LIN   1000   1K   100K
```

2. Change line for RS as follows:

```
RS   1   2   470
```

## ■ ANSWERS TO RELATED EXERCISES FOR EXAMPLES

**17–1** $\mathbf{Z} = 1.8 \text{ k}\Omega + j950 \ \Omega$; $\mathbf{Z} = 2.04\angle 27.8° \text{ k}\Omega$

**17–2** $\mathbf{I} = 423\angle -32.1° \ \mu$A

**17–3** $Z = 1.26 \text{ k}\Omega$; $\theta = 85.5°$

**17–4** $\mathbf{Z} = 8.14\angle 35.5° \text{ k}\Omega$

**17–5** $\mathbf{Y} = 3.03 \text{ mS} - j0.796 \text{ mS}$

**17–6** $\mathbf{I} = 14.0\angle -71.1°$ mA

**17–7** $I_{tot} = 67.6$ mA; $\theta = 36.3°$

**17–8** **(a)** $\mathbf{V}_1 = 8.04\angle 2.52°$ V
**(b)** $\mathbf{V}_2 = 2.00\angle -10.2°$ V

**17–9** $\mathbf{I}_{tot} = 20.2\angle -59.0°$ mA

**17–10** $P_{\text{true}}$, $P_r$, and $P_a$ decrease.

**17–11** $\phi = 65.6°$

**17–12** $V_{out}$ increases.

**17–13** $\phi = -32°$

**17–14** $V_{out} = 12.3$ V rms

# 18

# *RLC* CIRCUITS AND RESONANCE



## ■ INTRODUCTION

In this chapter, the analysis methods learned in Chapters 16 and 17 are extended to the coverage of circuits with combinations of resistive, inductive, and capacitive elements. Series and parallel *RLC* circuits, plus series-parallel combinations, are studied. In this chapter and throughout the rest of the book, you will learn the basics of putting technology theory into practice.

Circuits with both inductance and capacitance can exhibit the property of resonance, which is important in many types of applications. Resonance is the basis for frequency selectivity in communication systems. For example, the ability of a radio or television receiver to select a certain frequency that is transmitted by a particular station and, at the same time, to eliminate frequencies from other stations is based on the principle of resonance. The conditions in *RLC* circuits that produce resonance and the characteristics of resonant circuits are covered in this chapter.

## TECHnology Theory Into Practice

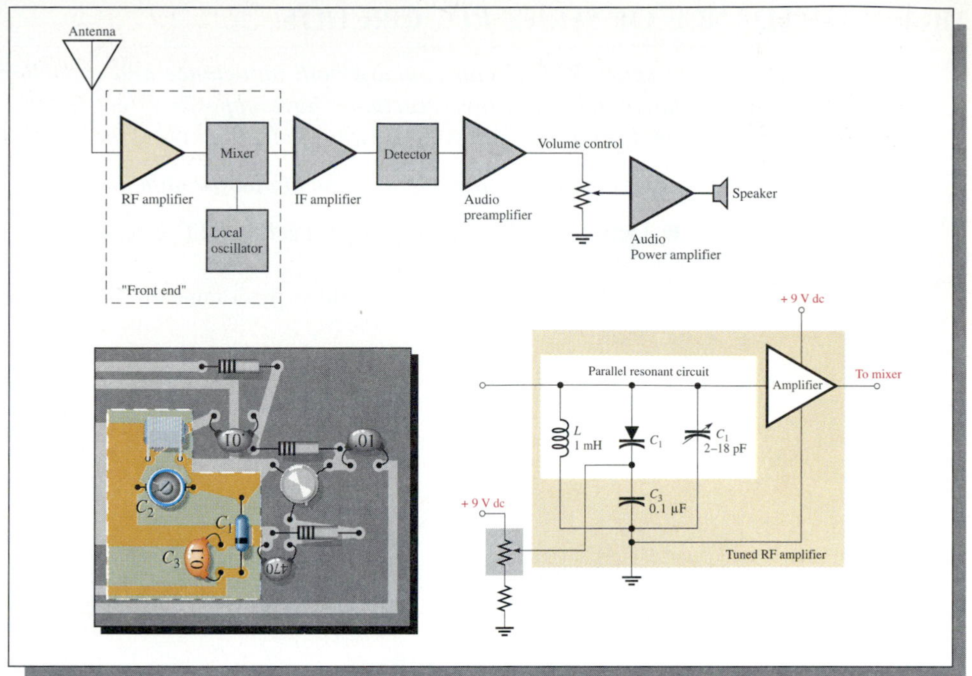


In the TECH TIP assignment in Section 18–9, you will work with the resonant tuning circuit in the RF amplifier of an AM radio receiver. The tuning circuit is used to select any desired frequency within the AM band so that a desired station can be tuned in.

## CHAPTER OBJECTIVES

- ☐ Determine the impedance of a series *RLC* circuit
- ☐ Analyze series *RLC* circuits
- ☐ Analyze a circuit for resonance
- ☐ Determine the impedance of a parallel resonant circuit
- ☐ Analyze parallel and series-parallel *RLC* circuits
- ☐ Analyze a circuit for parallel resonance
- ☐ Determine the bandwidth of resonant circuits
- ☐ Discuss some system applications of resonant circuits
- ☐ Use PSpice and Probe to determine the characteristics of *RLC* circuits (optional)

## 18–1 ■ IMPEDANCE OF SERIES RLC CIRCUITS

***A series RLC circuit contains both inductance and capacitance. Since inductive reactance and capacitive reactance have opposite effects on the circuit phase angle, the total reactance is less than either individual reactance.***

***After completing this section, you should be able to***

- **Determine the impedance of a series *RLC* circuit**
  - ☐ Calculate total reactance
  - ☐ Determine whether a circuit is predominately inductive or capacitive

A series *RLC* circuit is shown in Figure 18–1. It contains resistance, inductance, and capacitance.

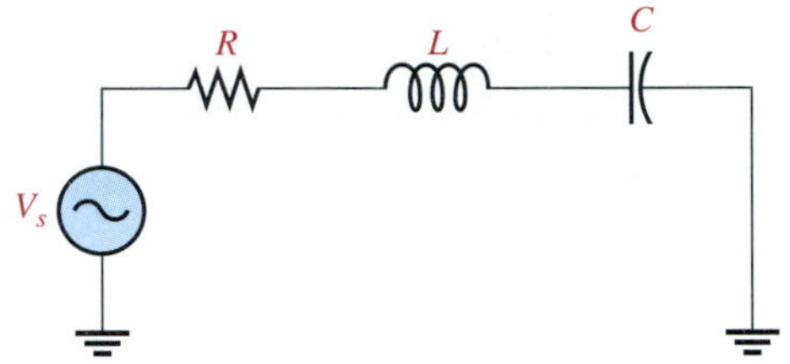

**FIGURE 18–1**
***Series RLC circuit.***



As you know, inductive reactance ($\mathbf{X}_L$) causes the total current to lag the applied voltage. Capacitive reactance ($\mathbf{X}_C$) has the opposite effect: It causes the current to lead the voltage. Thus $\mathbf{X}_L$ and $\mathbf{X}_C$ tend to offset each other. When they are equal, they cancel, and the total reactance is zero. In any case, the magnitude of the total reactance in the series circuit is

$$X_{tot} = |X_L - X_C| \quad \textbf{(18–1)}$$

The term $|X_L - X_C|$ means the absolute value of the difference of the two reactances. That is, the sign of the result is considered positive no matter which reactance is greater. For example, $3 - 7 = -4$, but the absolute value is

$$|3 - 7| = 4$$

When $X_L > X_C$, the circuit is predominantly inductive, and when $X_C > X_L$, the circuit is predominantly capacitive.

The total impedance for the series *RLC* circuit is stated in rectangular form in Equation (18–2) and in polar form in Equation (18–3).

$$\mathbf{Z} = R + jX_L - jX_C \quad \textbf{(18–2)}$$

$$\mathbf{Z} = \sqrt{R^2 + (X_L - X_C)^2}\angle\tan^{-1}\left(\frac{X_{tot}}{R}\right) \quad \textbf{(18–3)}$$

In Equation (18–3), $\sqrt{R^2 + (X_L - X_C)^2}$ is the magnitude and $\tan^{-1}(X_{tot}/R)$ is the phase angle between the total current and the applied voltage.

**EXAMPLE 18–1** Determine the total impedance in Figure 18–2. Express it in both rectangular and polar forms.

FIGURE 18–2

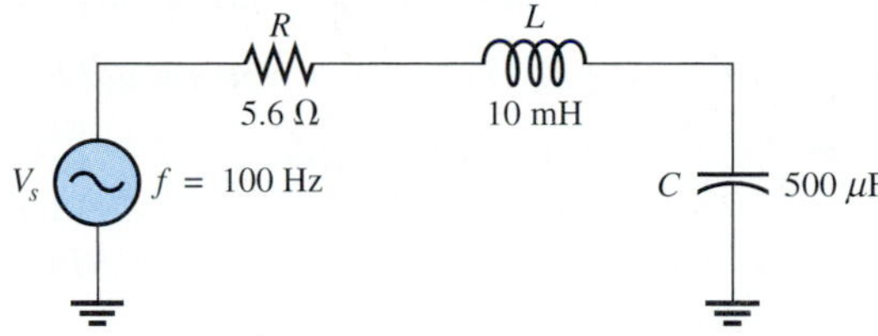


***Solution*** First find $X_C$ and $X_L$.

$$X_C = \frac{1}{2\pi fC} = \frac{1}{2\pi(100\text{ Hz})(500\ \mu\text{F})} = 3.18\ \Omega$$

$$X_L = 2\pi fL = 2\pi(100\text{ Hz})(10\text{ mH}) = 6.28\ \Omega$$

In this case, $X_L$ is greater than $X_C$, and thus the circuit is more inductive than capacitive. The magnitude of the total reactance is

$$X_{tot} = |X_L - X_C| = |6.28\ \Omega - 3.18\ \Omega| = 3.10\ \Omega \qquad \text{inductive}$$

The impedance in rectangular form is

$$\mathbf{Z} = R + (jX_L - jX_C) = 5.6\ \Omega + (j6.28\ \Omega - j3.18\ \Omega) = 5.6\ \Omega + j3.10\ \Omega$$

The impedance in polar form is

$$\mathbf{Z} = \sqrt{R^2 + X_{tot}^2}\angle\tan^{-1}\left(\frac{X_{tot}}{R}\right)$$

$$= \sqrt{(5.6\ \Omega)^2 + (3.10\ \Omega)^2}\angle\tan^{-1}\left(\frac{3.10\ \Omega}{5.6\ \Omega}\right) = 6.40\angle 29.0^\circ\ \Omega$$

The calculator sequence for conversion from the rectangular to the polar form is

Magnitude: [5] [•] [6] [x⇌y] [3] [•] [1] [INV] [2nd] [P–R]
Angle: [x⇌y]

***Related Exercise*** Determine **Z** in polar form if *f* is increased to 200 Hz.

As you have seen, when the inductive reactance is greater than the capacitive reactance, the circuit appears inductive; so the current lags the applied voltage. When the capacitive reactance is greater, the circuit appears capacitive, and the current leads the applied voltage.

**SECTION 18–1 REVIEW**

1. In a given series *RLC* circuit, $X_C$ is 150 Ω and $X_L$ is 80 Ω. What is the total reactance in ohms? Is it inductive or capacitive?
2. Determine the impedance in polar form for the circuit in Question 1 when $R$ = 47 Ω. What is the magnitude of the impedance? What is the phase angle? Is the current leading or lagging the applied voltage?

## 18–2 ■ ANALYSIS OF SERIES *RLC* CIRCUITS

***Recall that capacitive reactance varies inversely with frequency and that inductive reactance varies directly with frequency. In this section, the combined effects of the reactances as a function of frequency are examined.***

***After completing this section, you should be able to***

- **Analyze series *RLC* circuits**
  - ☐ Determine current in a series *RLC* circuit
  - ☐ Determine the voltages in a series *RLC* circuit
  - ☐ Determine the phase angle

Figure 18–3 shows that for a typical series *RLC* circuit the total reactance behaves as follows: Starting at a very low frequency, $X_C$ is high, and $X_L$ is low, and the circuit is predominantly capacitive. As the frequency is increased, $X_C$ decreases and $X_L$ increases until a value is reached where $X_C = X_L$ and the two reactances cancel, making the circuit purely resistive. This condition is **series resonance** and will be studied in Section 18–3. As the frequency is increased further, $X_L$ becomes greater than $X_C$, and the circuit is predominantly inductive. Example 18–2 illustrates how the impedance and phase angle change as the source frequency is varied.

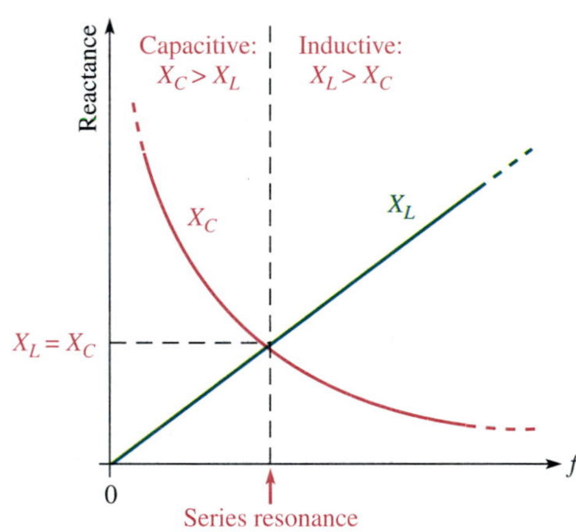


FIGURE 18–3
***How $X_C$ and $X_L$ vary with frequency.***

**EXAMPLE 18–2**

For each of the following input frequencies, find the impedance in polar form for the circuit in Figure 18–4. Note the change in magnitude and phase angle with frequency.
**(a)** $f = 1$ kHz **(b)** $f = 2$ kHz **(c)** $f = 3.5$ kHz **(d)** $f = 5$ kHz

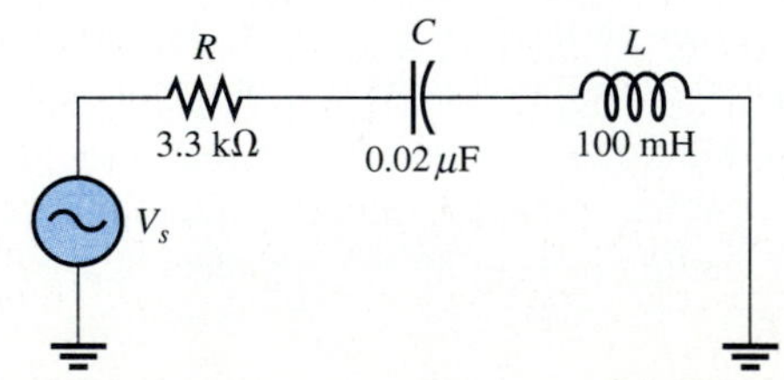


FIGURE 18–4

***Solution***

**(a)** At $f = 1$ kHz,

$$X_C = \frac{1}{2\pi fC} = \frac{1}{2\pi(1\text{ kHz})(0.02\ \mu\text{F})} = 7.96\text{ k}\Omega$$

$$X_L = 2\pi fL = 2\pi(1\text{ kHz})(100\text{ mH}) = 628\ \Omega$$

The circuit is clearly capacitive, and the impedance is

$$\mathbf{Z} = \sqrt{R^2 + (X_L - X_C)^2}\angle -\tan^{-1}\left(\frac{X_{tot}}{R}\right)$$

$$= \sqrt{(3.3\text{ k}\Omega)^2 + (628\ \Omega - 7.96\text{ k}\Omega)^2}\angle -\tan^{-1}\left(\frac{7.33\text{ k}\Omega}{3.3\text{ k}\Omega}\right) = 8.04\angle -65.8^\circ\text{ k}\Omega$$

**(b)** At $f = 2$ kHz,

$$X_C = \frac{1}{2\pi(2\text{ kHz})(0.02\ \mu\text{F})} = 3.98\text{ k}\Omega$$

$$X_L = 2\pi(2\text{ kHz})(100\text{ mH}) = 1.26\text{ k}\Omega$$

The circuit is still capacitive, and the impedance is

$$\mathbf{Z} = \sqrt{(3.3\text{ k}\Omega)^2 + (1.26\text{ k}\Omega - 3.98\text{ k}\Omega)^2}\angle -\tan^{-1}\left(\frac{2.72\text{ k}\Omega}{3.3\text{ k}\Omega}\right)$$

$$= 4.28\angle -39.5^\circ\text{ k}\Omega$$

**(c)** At $f = 3.5$ kHz,

$$X_C = \frac{1}{2\pi(3.5\text{ kHz})(0.02\ \mu\text{F})} = 2.27\text{ k}\Omega$$

$$X_L = 2\pi(3.5\text{ kHz})(100\text{ mH}) = 2.20\text{ k}\Omega$$

The circuit is very close to being purely resistive because $X_C$ and $X_L$ are nearly equal, but is still slightly capacitive. The impedance is

$$\mathbf{Z} = \sqrt{(3.3\text{ k}\Omega)^2 + (2.20\text{ k}\Omega - 2.27\text{ k}\Omega)^2}\angle -\tan^{-1}\left(\frac{0.07\text{ k}\Omega}{3.3\text{ k}\Omega}\right)$$

$$= 3.3\angle -1.22^\circ\text{ k}\Omega$$

**(d)** At $f = 5$ kHz,

$$X_C = \frac{1}{2\pi(5\text{ kHz})(0.02\ \mu\text{F})} = 1.59\text{ k}\Omega$$

$$X_L = 2\pi(5\text{ kHz})(100\text{ mH}) = 3.14\text{ k}\Omega$$

The circuit is now predominantly inductive. The impedance is

$$\mathbf{Z} = \sqrt{(3.3\text{ k}\Omega)^2 + (3.14\text{ k}\Omega - 1.59\text{ k}\Omega)^2}\angle \tan^{-1}\left(\frac{1.55\text{ k}\Omega}{3.3\text{ k}\Omega}\right)$$

$$= 3.65\angle 25.2^\circ\text{ k}\Omega$$

Notice how the circuit changed from capacitive to inductive as the frequency increased. The phase condition changed from the current leading to the current lagging as indicated by the sign of the angle. It is interesting to note that the impedance magnitude decreased to a minimum approximately equal to the resistance and then began increasing again.

***Related Exercise*** Determine **Z** in polar form for $f = 7$ kHz and sketch a graph of impedance vs. frequency using the values in this example.

In a series *RLC* circuit, the capacitor voltage and the inductor voltage are always 180° out of phase with each other. For this reason, $V_C$ and $V_L$ subtract from each other, and thus the voltage across *L* and *C* combined is always less than the larger individual voltage across either element, as illustrated in Figure 18–5 and in the waveform diagram of Figure 18–6.

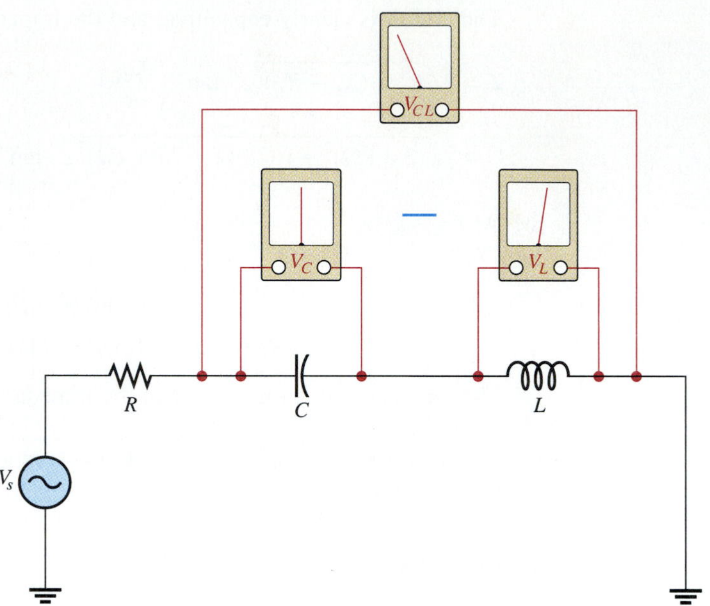


FIGURE 18–5
***The voltage across the series combination of C and L is always less than the larger individual voltage.***

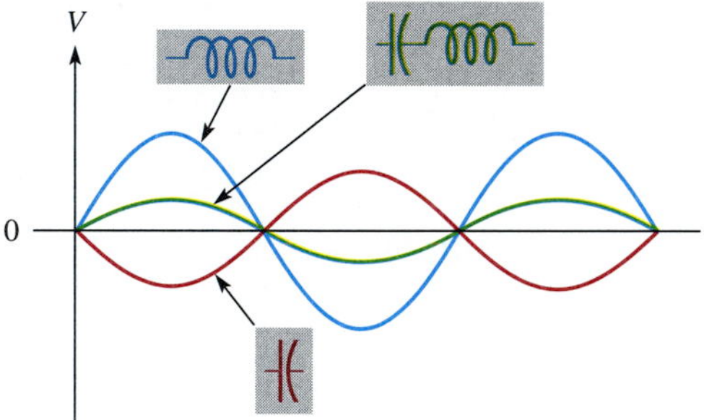


FIGURE 18–6
***$V_L$ and $V_C$ effectively subtract.***

In the next example, Ohm's law is used to find the current and voltages in the series *RLC* circuit.

**EXAMPLE 18–3**

Find the current and the voltages across each element in Figure 18–7. Express each quantity in polar form, and draw a complete voltage phasor diagram.

FIGURE 18–7

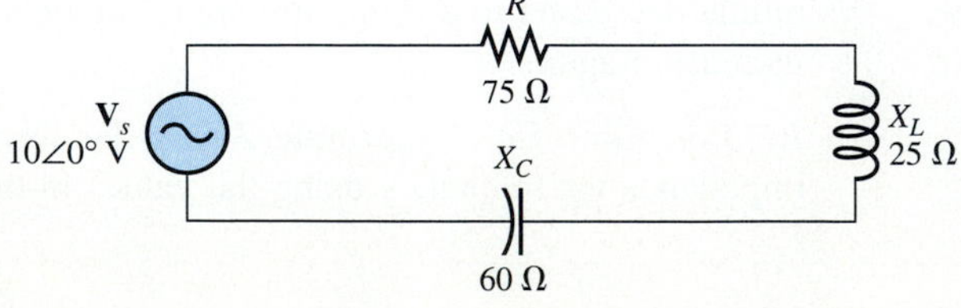

***Solution*** First, find the total impedance.

$$\mathbf{Z} = R + jX_L - jX_C = 75\ \Omega + j25\ \Omega - j60\ \Omega = 75\ \Omega - j35\ \Omega$$

Convert to polar form for convenience in applying Ohm's law.

$$\mathbf{Z} = \sqrt{R^2 + X_{tot}^2}\angle -\tan^{-1}\left(\frac{X_{tot}}{R}\right)$$

$$= \sqrt{(75\ \Omega)^2 + (35\ \Omega)^2}\angle -\tan^{-1}\left(\frac{35\ \Omega}{75\ \Omega}\right) = 82.8\angle -25°\ \Omega$$

Apply Ohm's law to find the current.

$$\mathbf{I} = \frac{\mathbf{V}_s}{\mathbf{Z}} = \frac{10\angle 0°\ \text{V}}{82.8\angle -25°\ \Omega} = 121\angle 25.0°\ \text{mA}$$

Now, apply Ohm's law to find the voltages across $R$, $L$, and $C$.

$$\mathbf{V}_R = \mathbf{IR} = (121\angle 25.0°\ \text{mA})(75\angle 0°\ \Omega) = 9.08\angle 25.0°\ \text{V}$$

$$\mathbf{V}_L = \mathbf{IX}_L = (121\angle 25.0°\ \text{mA})(25\angle 90°\ \Omega) = 3.03\angle 115°\ \text{V}$$

$$\mathbf{V}_C = \mathbf{IX}_C = (121\angle 25.0°\ \text{mA})(60\angle -90°\ \Omega) = 7.26\angle -65.0°\ \text{V}$$

The phasor diagram is shown in Figure 18–8. The magnitudes represent rms values. Notice that $\mathbf{V}_L$ is leading $\mathbf{V}_R$ by 90°, and $\mathbf{V}_C$ is lagging $\mathbf{V}_R$ by 90°. Also, there is a 180° phase difference between $\mathbf{V}_L$ and $\mathbf{V}_C$. If the current phasor were shown, it would be at the same angle as $\mathbf{V}_R$. The current is leading $\mathbf{V}_s$, the source voltage, by 25°, indicating a capacitive circuit ($X_C > X_L$).

**FIGURE 18–8**

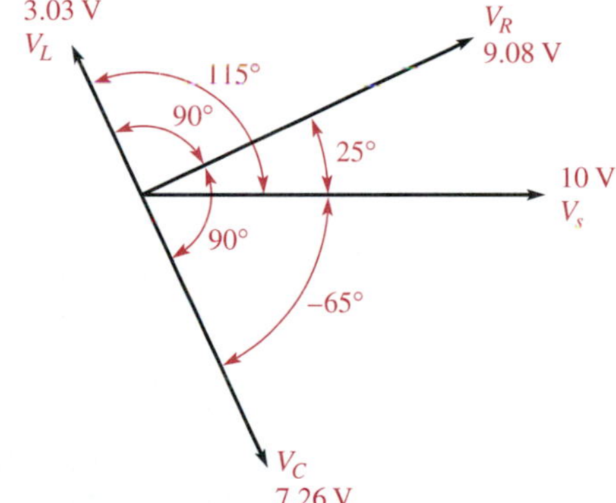


***Related Exercise*** What will happen to the current if the frequency in Figure 18–7 is increased?

## SECTION 18–2 REVIEW

1. The following voltages occur in a certain series *RLC* circuit. Determine the source voltage: $\mathbf{V}_R = 24\angle 30°$ V, $\mathbf{V}_L = 15\angle 120°$ V, and $\mathbf{V}_C = 45\angle -60°$ V.
2. When $R = 10\ \Omega$, $X_C = 18\ \Omega$, and $X_L = 12\ \Omega$, does the current lead or lag the applied voltage?
3. Determine the total reactance in Question 2.

## 18–3 ▪ SERIES RESONANCE

***In a series RLC circuit, series resonance occurs when $X_C = X_L$. The frequency at which resonance occurs is called the resonant frequency and is designated $f_r$.***

***After completing this section, you should be able to***

- **Analyze a circuit for resonance**
  - ☐ Define *resonance*
  - ☐ Determine the impedance at resonance
  - ☐ Explain why the reactances cancel at resonance
  - ☐ Determine the series resonant frequency
  - ☐ Calculate the current, voltages, and phase angle at resonance

Figure 18–9 illustrates the series resonant condition.

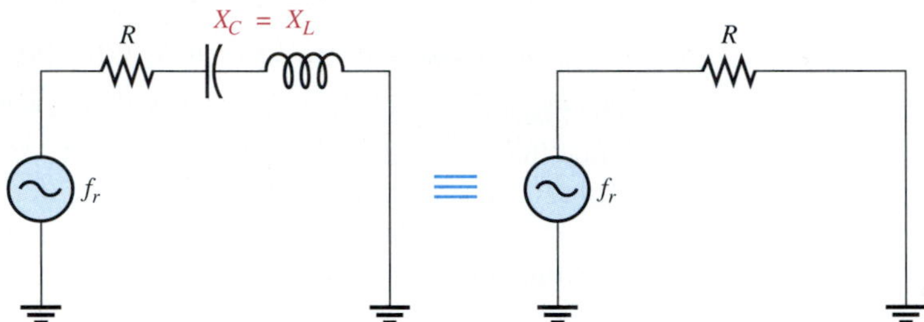


FIGURE 18–9

***Series resonance. $X_C$ and $X_L$ cancel each other resulting in a purely resistive circuit.***

**Resonance** is a condition in a series *RLC* circuit in which the capacitive and inductive reactances are equal in magnitude; thus, they cancel each other and result in a purely resistive impedance. In a series resonant circuit, the total impedance is

$$\mathbf{Z}_r = R + jX_L - jX_C \quad \textbf{(18–4)}$$

Since $X_L = X_C$, the $j$ terms cancel, and the impedance is purely resistive. These resonant conditions are stated in the following equations.

$$X_L = X_C \quad \textbf{(18–5)}$$

$$Z_r = R \quad \textbf{(18–6)}$$

**EXAMPLE 18–4** For the series *RLC* circuit in Figure 18–10, determine $X_C$ and $\mathbf{Z}$ at resonance.

FIGURE 18–10

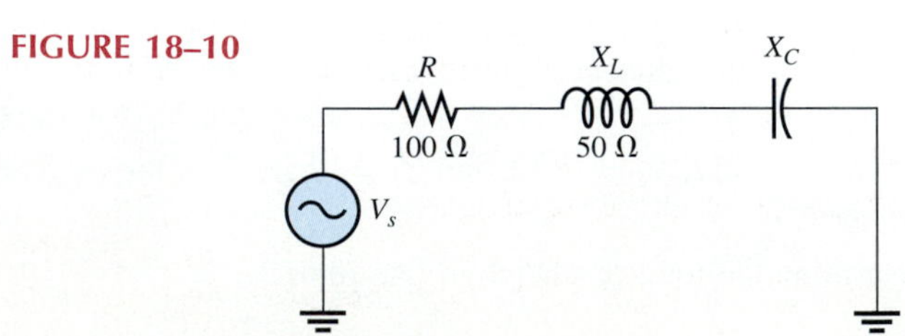

***Solution*** $X_L = X_C$ at the resonant frequency. Thus, $X_C = X_L = 50\ \Omega$. The impedance at resonance is

$$\mathbf{Z}_r = R + jX_L - jX_C = 100\ \Omega + j50\ \Omega - j50\ \Omega = 100\angle 0°\ \Omega$$

The impedance is equal to the resistance because the reactances are equal in magnitude and therefore cancel.

***Related Exercise*** Just below the resonant frequency, is the circuit more inductive or more capacitive?

## $X_L$ and $X_C$ Effectively Cancel at Resonance

At the series **resonant frequency,** the voltages across $C$ and $L$ are equal in magnitude because the reactances are equal and because the same current is through both since they are in series ($IX_C = IX_L$). Also, $V_L$ and $V_C$ are always 180° out of phase with each other.

During any given cycle, the polarities of the voltages across $C$ and $L$ are opposite, as shown in parts (a) and (b) of Figure 18–11. The equal and opposite voltages across $C$ and $L$ cancel, leaving zero volts from point $A$ to point $B$ as shown. Since there is no voltage drop from $A$ to $B$ but there is still current, the total reactance must be zero, as indicated in part (c). Also, the voltage phasor diagram in part (d) shows that $V_C$ and $V_L$ are equal in magnitude and 180° out of phase with each other.

FIGURE 18–11

***At the resonant frequency, $f_r$, the voltages across C and L are equal in magnitude. Since they are 180° out of phase with each other, they cancel, leaving 0 V across the LC combination (point A to point B). The section of the circuit from A to B effectively looks like a short at resonance.***

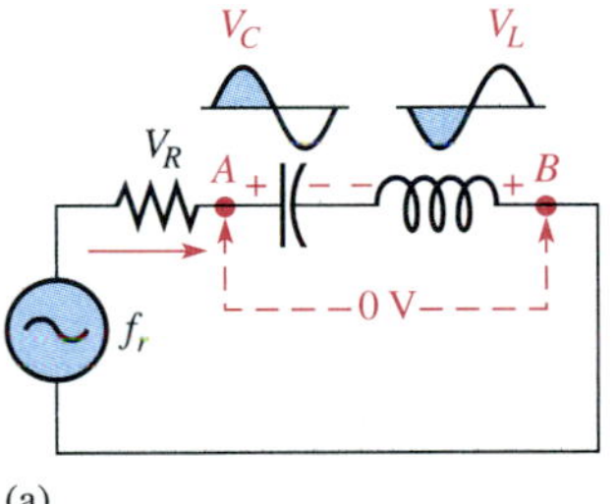

(a)

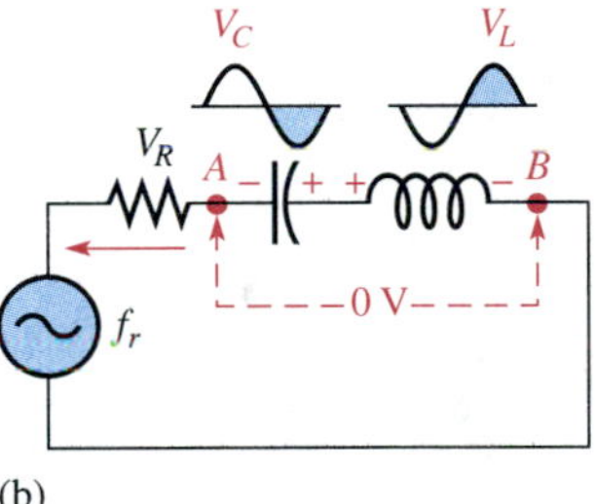

(b)

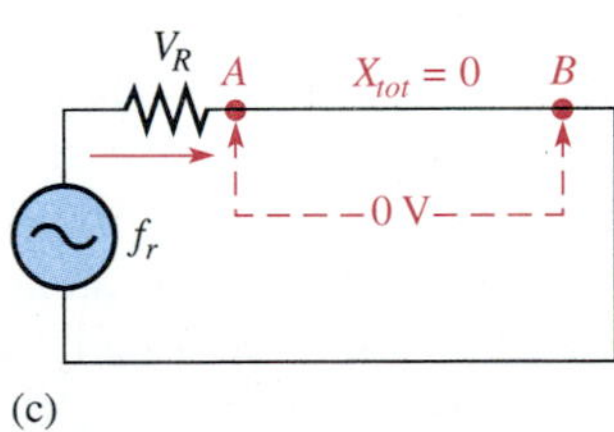

(c)

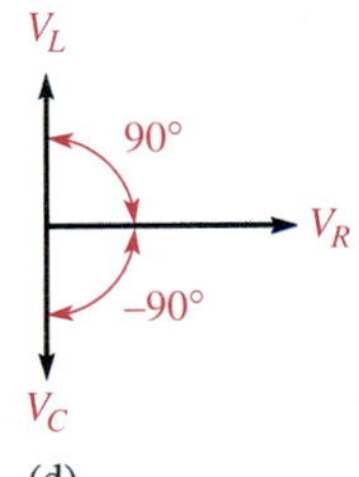

(d)

### Series Resonant Frequency

For a given series $RLC$ circuit, resonance happens at only one specific frequency. A formula for this resonant frequency is developed as follows:

$$X_L = X_C$$

Substitute the reactance formulas:

$$2\pi f_r L = \frac{1}{2\pi f_r C}$$

Solve for $f_r$:

$$(2\pi f_r L)(2\pi f_r C) = 1$$

$$4\pi^2 f_r^2 LC = 1$$

$$f_r^2 = \frac{1}{4\pi^2 LC}$$

Take the square root of both sides. The formula for series resonant frequency is

$$f_r = \frac{1}{2\pi\sqrt{LC}} \quad \textbf{(18–7)}$$

**EXAMPLE 18–5** Find the series resonant frequency for the circuit in Figure 18–12.

**FIGURE 18–12**

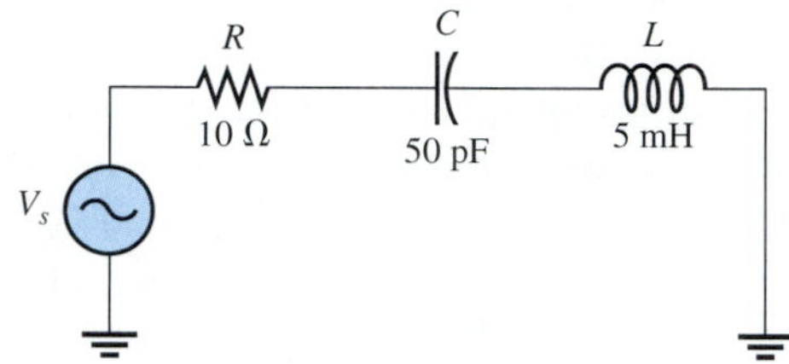


***Solution*** The resonant frequency is

$$f_r = \frac{1}{2\pi\sqrt{LC}} = \frac{1}{2\pi\sqrt{(5\text{ mH})(50\text{ pF})}} = 318\text{ kHz}$$

***Related Exercise*** If $C = 0.01\ \mu$F in Figure 18–12, what is the resonant frequency?

## Series *RLC* Impedance

At frequencies below $f_r$, $X_C > X_L$; thus, the circuit is capacitive. At the resonant frequency, $X_C = X_L$, so the circuit is purely resistive. At frequencies above $f_r$, $X_L > X_C$; thus, the circuit is inductive.

The impedance magnitude is minimum at resonance ($Z = R$) and increases in value above and below the resonant point. The graph in Figure 18–13 illustrates how impedance changes with frequency. At zero frequency, both $X_C$ and $Z$ are infinitely large and $X_L$ is zero because the capacitor looks like an open at 0 Hz and the inductor looks like a short. As the frequency increases, $X_C$ decreases and $X_L$ increases. Since $X_C$ is larger than $X_L$ at frequencies below $f_r$, $Z$ decreases along with $X_C$. At $f_r$, $X_C = X_L$ and $Z = R$. At frequencies above $f_r$, $X_L$ becomes increasingly larger than $X_C$, causing $Z$ to increase.

**FIGURE 18–13**
***Series RLC impedance as a function of frequency.***

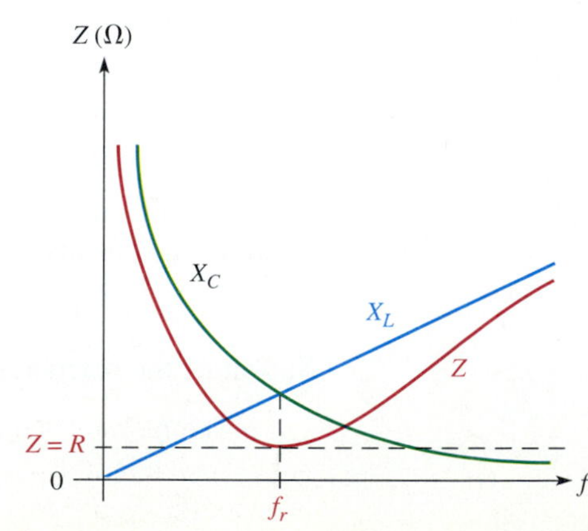

**EXAMPLE 18–6** For the circuit in Figure 18–14, determine the impedance magnitude at the following frequencies:

**(a)** $f_r$ **(b)** 1000 Hz below $f_r$ **(c)** 1000 Hz above $f_r$

FIGURE 18–14

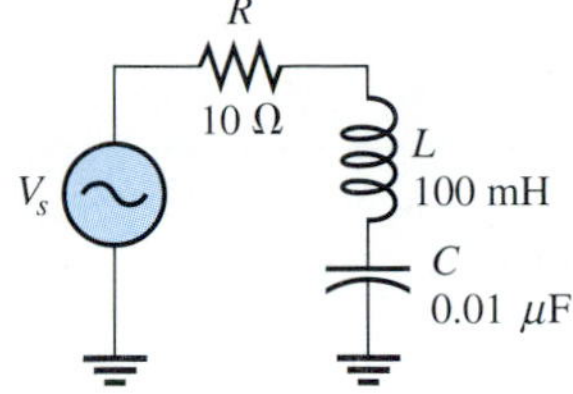


***Solution***

**(a)** At $f_r$, the impedance is equal to $R$:

$$Z = R = 10\ \Omega$$

To determine the impedance above and below $f_r$, first calculate the resonant frequency as follows:

$$f_r = \frac{1}{2\pi\sqrt{LC}} = \frac{1}{2\pi\sqrt{(100\text{ mH})(0.01\ \mu\text{F})}} = 5.03\text{ kHz}$$

**(b)** At 1000 Hz below $f_r$, the frequency and reactances are as follows:

$$f = f_r - 1\text{ kHz} = 5.03\text{ kHz} - 1\text{ kHz} = 4.03\text{ kHz}$$

$$X_C = \frac{1}{2\pi fC} = \frac{1}{2\pi(4.03\text{ kHz})(0.01\ \mu\text{F})} = 3.95\text{ k}\Omega$$

$$X_L = 2\pi fL = 2\pi(4.03\text{ kHz})(100\text{ mH}) = 2.53\text{ k}\Omega$$

Therefore, the impedance at $f_r - 1$ kHz is

$$Z = \sqrt{R^2 + (X_L - X_C)^2} = \sqrt{(10\ \Omega)^2 + (2.53\text{ k}\Omega - 3.95\text{ k}\Omega)^2} = 1.42\text{ k}\Omega$$

**(c)** At 1000 Hz above $f_r$,

$$f = 5.03\text{ kHz} + 1\text{ kHz} = 6.03\text{ kHz}$$

$$X_C = \frac{1}{2\pi(6.03\text{ kHz})(0.01\ \mu\text{F})} = 2.64\text{ k}\Omega$$

$$X_L = 2\pi(6.03\text{ kHz})(100\text{ mH}) = 3.79\text{ k}\Omega$$

Therefore, the impedance at $f_r + 1$ kHz is

$$Z = \sqrt{(10\ \Omega)^2 + (3.79\text{ k}\Omega - 2.64\text{ k}\Omega)^2} = 1.15\text{ k}\Omega$$

In part (b) $Z$ is capacitive, and in part (a) $Z$ is inductive.

***Related Exercise*** What happens to the impedance magnitude if $f$ is decreased below 4.03 kHz? Above 6.03 kHz?

## Current and Voltages in a Series *RLC* Circuit

At the series resonant frequency, the current is maximum ($I_{max} = V_s/R$). Above and below resonance, the current decreases because the impedance increases. A response curve showing the plot of current versus frequency is shown in Figure 18–15(a).

The resistor voltage, $V_R$, follows the current and is maximum (equal to $V_s$) at resonance and zero at $f = 0$ and at $f = \infty$, as shown in Figure 18–15(b). The general shapes of the $V_C$ and $V_L$ curves are indicated in Figure 18–15(c) and (d). Notice that $V_C = V_s$ when

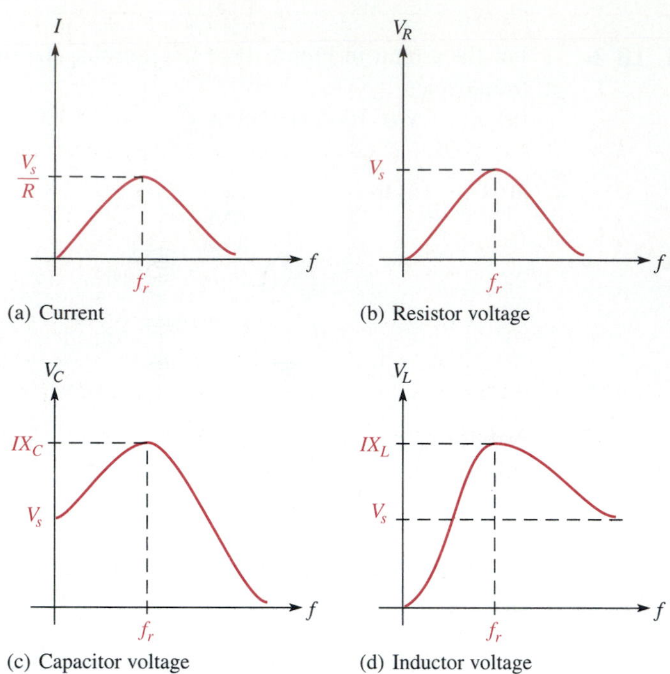


FIGURE 18–15
***Current and voltage magnitudes as a function of frequency in a series RLC circuit. (Note that the peak magnitudes are not shown to scale with respect to each other.)***

$f = 0$, because the capacitor appears open. Also notice that $V_L$ approaches $V_s$ as $f$ approaches infinity, because the inductor appears open.

The voltages are maximum at resonance but drop off above and below $f_r$. The voltages across $L$ and $C$ at resonance are exactly equal in magnitude but 180° out of phase; so they cancel. Thus, the total voltage across both $L$ and $C$ is zero, and $V_R = V_s$ at resonance, as indicated in Figure 18–16. Individually, $V_L$ and $V_C$ can be much greater than the source voltage, as you will see later. Keep in mind that $V_L$ and $V_C$ are always opposite in polarity regardless of the frequency, but only at resonance are their magnitudes equal.

FIGURE 18–16
***Series RLC circuit at resonance.***

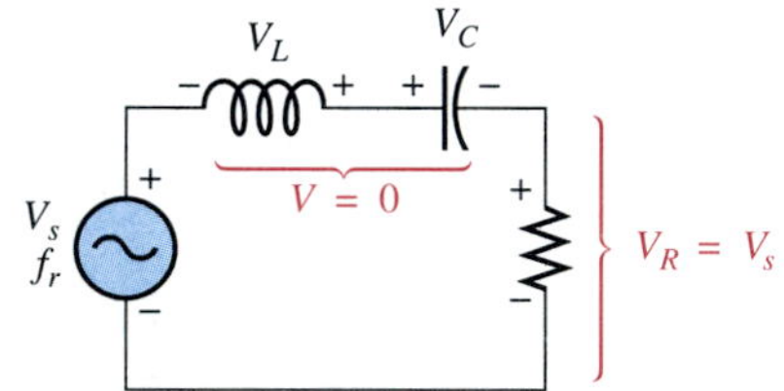


**EXAMPLE 18–7**

Find $I$, $V_R$, $V_L$, and $V_C$ at resonance in Figure 18–17. The resonant values of $X_L$ and $X_C$ are shown.

FIGURE 18–17

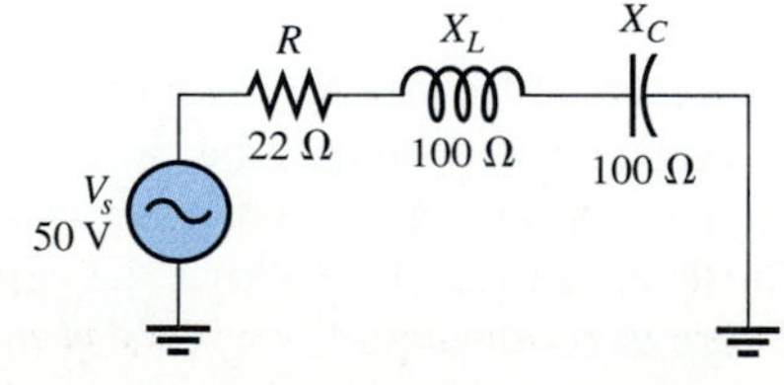

***Solution*** At resonance, $I$ is maximum and equal to $V_s/R$.

$$I = \frac{V_s}{R} = \frac{50\text{ V}}{22\ \Omega} = 2.27\text{ A}$$

Apply Ohm's law to obtain the following voltage magnitudes:

$$V_R = IR = (2.27\text{ A})(22\ \Omega) = 50\text{ V}$$
$$V_L = IX_L = (2.27\text{ A})(100\ \Omega) = 227\text{ V}$$
$$V_C = IX_C = (2.27\text{ A})(100\ \Omega) = 227\text{ V}$$

Notice that all of the source voltage is dropped across the resistor. Also, of course, $V_L$ and $V_C$ are equal in magnitude but opposite in phase. This causes these voltages to cancel, making the total reactive voltage zero.

***Related Exercise*** What is the current at resonance in Figure 18–17 if $X_L = X_C = 1\text{ k}\Omega$?

## The Phase Angle of a Series *RLC* Circuit

At frequencies below resonance, $X_C > X_L$, and the current leads the source voltage, as indicated in Figure 18–18(a). The phase angle decreases as the frequency approaches the resonant value and is 0° at resonance, as indicated in part (b). At frequencies above resonance, $X_L > X_C$, and the current lags the source voltage, as indicated in part (c). As the frequency goes higher, the phase angle approaches 90°. A plot of phase angle versus frequency is shown in part (d) of the figure.

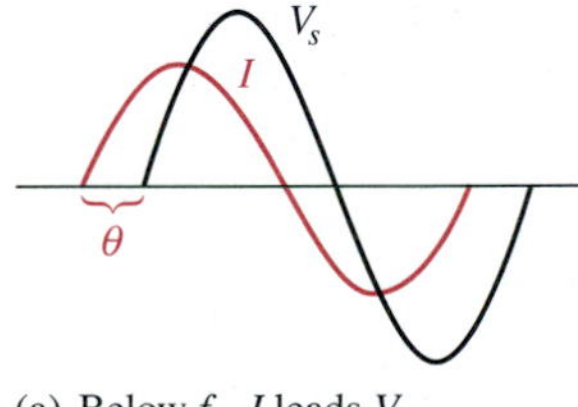


(a) Below $f_r$, $I$ leads $V_s$.

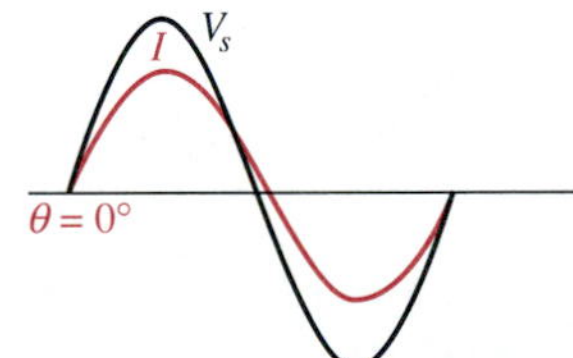


(b) At $f_r$, $I$ is in phase with $V_s$.

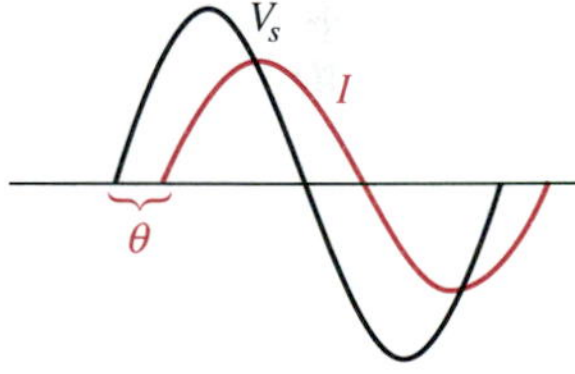


(c) Above $f_r$, $I$ lags $V_s$.

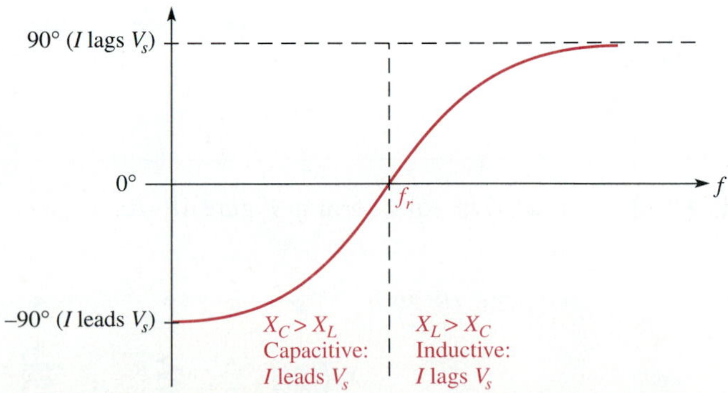


(d) Phase angle versus frequency.

**FIGURE 18–18**

***The phase angle as a function of frequency in a series RLC circuit.***

**SECTION 18–3 REVIEW**

1. What is the condition for series resonance?
2. Why is the current maximum at the resonant frequency?
3. Calculate the resonant frequency for $C = 1000$ pF and $L = 1000$ $\mu$H.
4. In Question 3, is the circuit inductive or capacitive at 50 kHz?

## 18–4 ■ IMPEDANCE OF PARALLEL *RLC* CIRCUITS

***In this section, you will learn how to determine the impedance and phase angle of a parallel RLC circuit. Also, conductance, susceptance, and admittance of a parallel RLC circuit are covered.***

***After completing this section, you should be able to***

- **Determine the impedance of a parallel resonant circuit**
  - ☐ Calculate the conductance, susceptance, and admittance
  - ☐ Determine whether a circuit is predominately inductive or capacitive

Figure 18–19 shows a parallel *RLC* circuit. The total impedance can be calculated using the sum-of-reciprocals method, just as was done for circuits with resistors in parallel.

$$\frac{1}{\mathbf{Z}} = \frac{1}{R\angle 0^\circ} + \frac{1}{X_L\angle 90^\circ} + \frac{1}{X_C\angle -90^\circ}$$

or

$$\mathbf{Z} = \frac{1}{\dfrac{1}{R\angle 0^\circ} + \dfrac{1}{X_L\angle 90^\circ} + \dfrac{1}{X_C\angle -90^\circ}} \qquad \textbf{(18–8)}$$

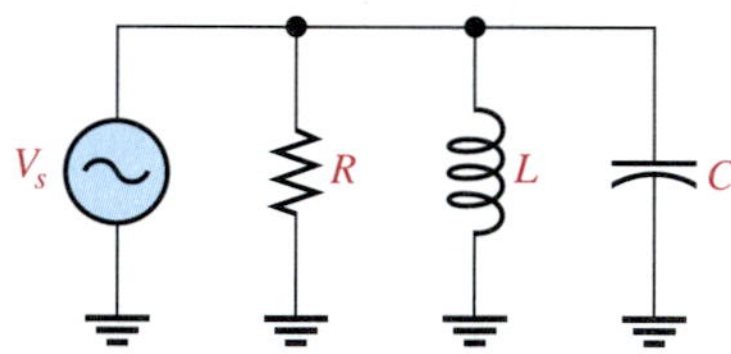


**FIGURE 18–19**
***Parallel RLC circuit.***

**EXAMPLE 18–8** Find **Z** in polar form in Figure 18–20.

**FIGURE 18–20**

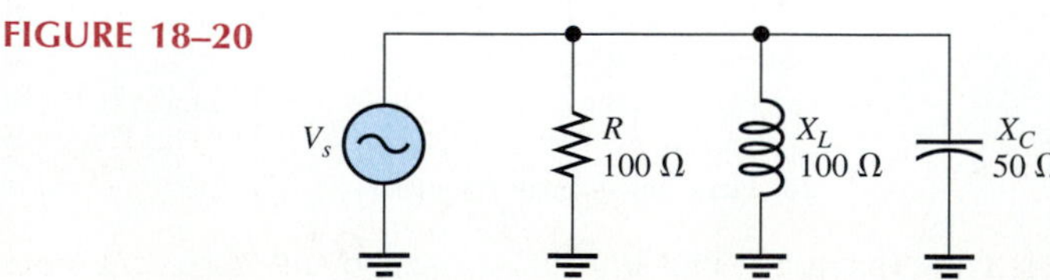

***Solution*** Use the sum-of-reciprocals formula.

$$\frac{1}{\mathbf{Z}} = \frac{1}{R\angle 0^\circ} + \frac{1}{X_L\angle 90^\circ} + \frac{1}{X_C\angle -90^\circ} = \frac{1}{100\angle 0^\circ\ \Omega} + \frac{1}{100\angle 90^\circ\ \Omega} + \frac{1}{50\angle -90^\circ\ \Omega}$$

Apply the rule for division of polar numbers.

$$\frac{1}{\mathbf{Z}} = 10\angle 0^\circ\text{ mS} + 10\angle -90^\circ\text{ mS} + 20\angle 90^\circ\text{ mS}$$

Recall that the sign of the denominator angle changes when dividing.

Now, convert each term to its rectangular equivalent and combine.

$$\frac{1}{\mathbf{Z}} = 10\text{ mS} - j10\text{ mS} + j20\text{ mS} = 10\text{ mS} + j10\text{ mS}$$

Take the reciprocal to obtain **Z** and then convert to polar form.

$$\mathbf{Z} = \frac{1}{10\text{ mS} + j10\text{ mS}} = \frac{1}{\sqrt{(10\text{ mS})^2 + (10\text{ mS})^2}\angle\tan^{-1}\left(\frac{10\text{ mS}}{10\text{ mS}}\right)}$$

$$= \frac{1}{14.14\angle 45^\circ\text{ mS}} = 70.7\angle -45^\circ\ \Omega$$

The negative angle shows that the circuit is capacitive. This may surprise you, since $X_L > X_C$. However, in a parallel circuit, the smaller quantity has the greater effect on the total current. Just as in the case of all resistances in parallel, the smaller reactance draws more current and has the greater effect on the total *R*.

In this circuit, the total current leads the total voltage by a phase angle of 45°.

***Related Exercise*** If the frequency in Figure 18–20 increases, does the impedance increase or decrease?

## Conductance, Susceptance, and Admittance

The concepts of conductance (*G*), capacitive susceptance ($B_C$), inductive susceptance ($B_L$) and admittance (*Y*) were discussed in Chapters 16 and 17. The phasor formulas are restated here.

$$\mathbf{G} = \frac{1}{R\angle 0^\circ} = G\angle 0^\circ \qquad \textbf{(18–9)}$$

$$\mathbf{B}_C = \frac{1}{X_C\angle -90^\circ} = B_C\angle 90^\circ = jB_C \qquad \textbf{(18–10)}$$

$$\mathbf{B}_L = \frac{1}{X_L\angle 90^\circ} = B_L\angle -90^\circ = -jB_L \qquad \textbf{(18–11)}$$

$$\mathbf{Y} = \frac{1}{Z\angle\pm\theta} = Y\angle\pm\theta = G + jB_C - jB_L \qquad \textbf{(18–12)}$$

As you know, the unit of each of these quantities is the siemens (S).

**EXAMPLE 18–9** Determine the conductance, capacitive susceptance, inductive susceptance, and total admittance in Figure 18–21. Also, determine the impedance.

FIGURE 18–21

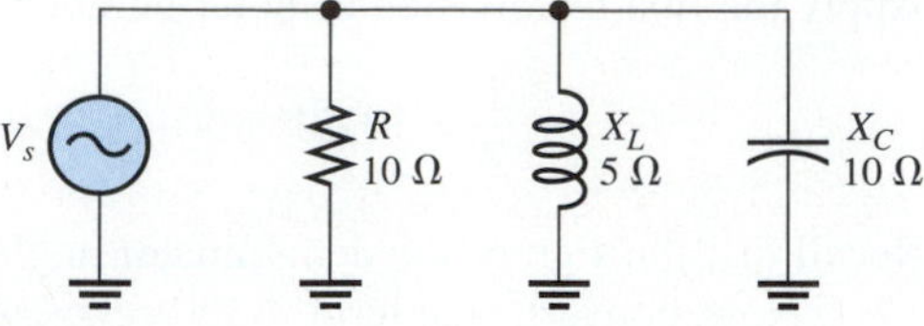


***Solution***

$$\mathbf{G} = \frac{1}{R\angle 0^\circ} = \frac{1}{10\angle 0^\circ\ \Omega} = 100\angle 0^\circ \text{ mS}$$

$$\mathbf{B}_C = \frac{1}{X_C\angle -90^\circ} = \frac{1}{10\angle -90^\circ\ \Omega} = 100\angle 90^\circ \text{ mS}$$

$$\mathbf{B}_L = \frac{1}{X_L\angle 90^\circ} = \frac{1}{5\angle 90^\circ\ \Omega} = 200\angle -90^\circ \text{ mS}$$

$$\mathbf{Y} = G + jB_C - jB_L = 100 \text{ mS} + j100 \text{ mS} - j200 \text{ mS}$$
$$= 100 \text{ mS} - j100 \text{ mS} = 141.4\angle -45^\circ \text{ mS}$$

From **Y**, you can determine **Z.**

$$\mathbf{Z} = \frac{1}{\mathbf{Y}} = \frac{1}{141.4\angle -45^\circ \text{ mS}} = 7.07\angle 45^\circ\ \Omega$$

***Related Exercise*** Is the circuit in Figure 18–21 predominately inductive or predominately capacitive?

**SECTION 18–4 REVIEW**

1. In a certain parallel *RLC* circuit, the capacitive reactance is 60 Ω, and the inductive reactance is 100 Ω. Is the circuit predominantly capacitive or inductive?
2. Determine the admittance of a parallel circuit in which $R = 1$ kΩ, $X_C = 500$ Ω, and $X_L = 1.2$ kΩ.
3. In Question 2, what is the impedance?

## 18–5 ■ ANALYSIS OF PARALLEL AND SERIES-PARALLEL *RLC* CIRCUITS

***As you have seen, the smaller reactance in a parallel circuit dominates because it results in the larger branch current. In this section, you will examine current relationships in parallel and series-parallel circuits and learn how to convert a series-parallel circuit into an equivalent parallel circuit.***

***After completing this section, you should be able to***

- **Analyze parallel and series-parallel *RLC* circuits**
  - ☐ Explain how the currents are related in terms of phase
  - ☐ Calculate impedance, currents, and voltages
  - ☐ Convert from series-parallel to parallel

Recall that capacitive reactance varies inversely with frequency and that inductive reactance varies directly with frequency. In a parallel *RLC* circuit at low frequencies, the inductive reactance is less than the capacitive reactance; therefore, the circuit is inductive. As the frequency is increased, $X_L$ increases and $X_C$ decreases until a value is reached where $X_L = X_C$. This is the point of **parallel resonance.** As the frequency is increased further, $X_C$ becomes smaller than $X_L$, and the circuit becomes capacitive.

## Current Relationships

In a parallel *RLC* circuit, the current in the capacitive branch and the current in the inductive branch are *always* 180° out of phase with each other (neglecting any coil resistance). For this reason, $I_C$ and $I_L$ subtract from each other, and thus the total current into the parallel branches of *L* and *C* is always less than the largest individual branch current, as illustrated in Figure 18–22 and in the waveform diagram of Figure 18–23. Of course, the current in the resistive branch is always 90° out of phase with both reactive currents, as shown in the current phasor diagram of Figure 18–24.

The total current can be expressed as

$$\mathbf{I}_{tot} = \sqrt{I_R^2 + (I_C - I_L)^2}\angle\tan^{-1}\left(\frac{I_{CL}}{I_R}\right) \quad \textbf{(18–13)}$$

where $I_{CL}$ is $I_C - I_L$, the total current into the *L* and *C* branches.

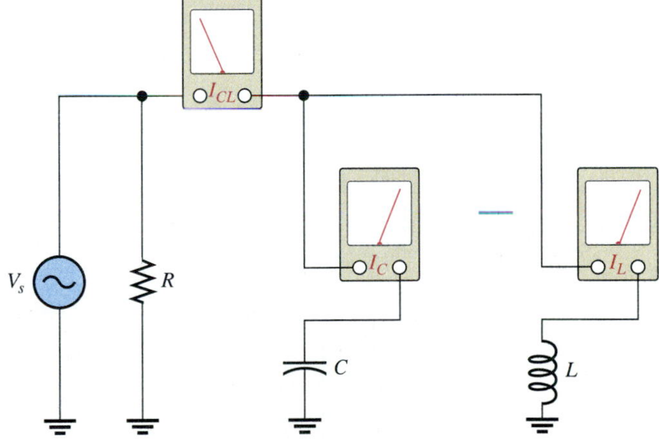


FIGURE 18–22
***The total current into the parallel combination of C and L is the difference of the two branch currents.***

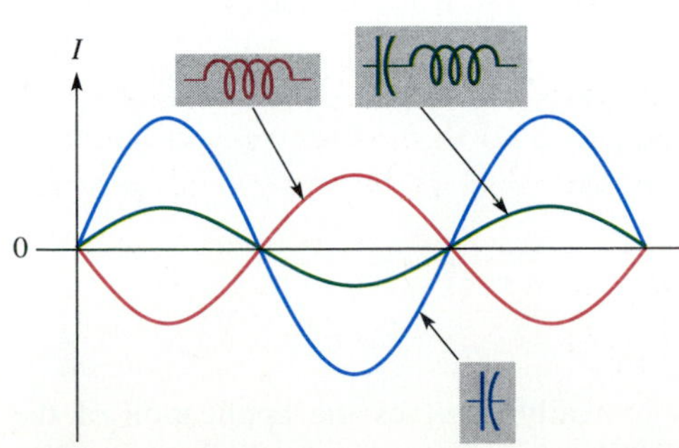


FIGURE 18–23
***$I_C$ and $I_L$ effectively subtract.***

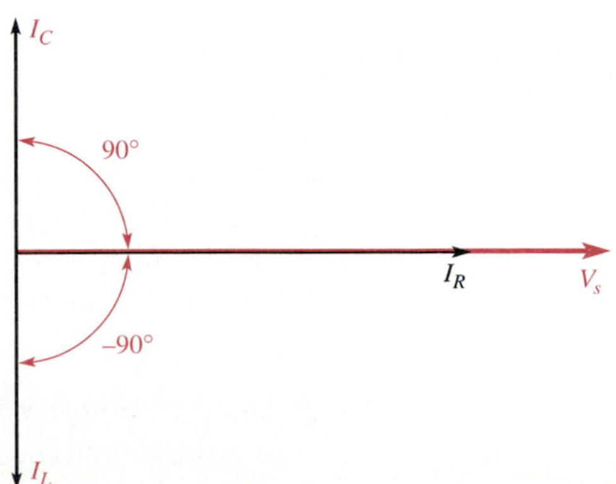


FIGURE 18–24
***Current phasor diagram for a parallel RLC circuit.***

**EXAMPLE 18–10** Find each branch current and the total current in Figure 18–25.

FIGURE 18–25

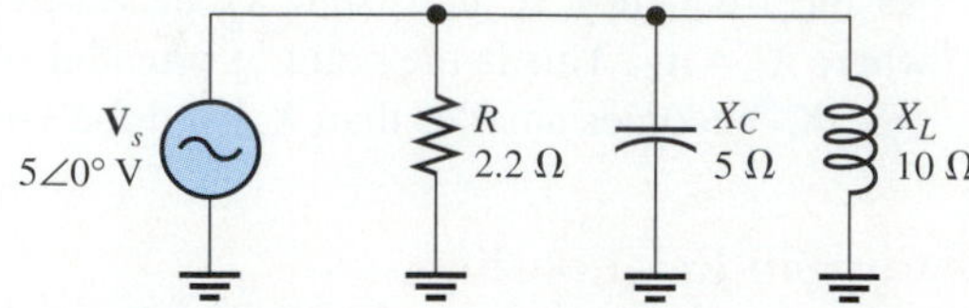


***Solution*** Use Ohm's law to find each branch current in phasor form.

$$\mathbf{I}_R = \frac{\mathbf{V}_s}{\mathbf{R}} = \frac{5\angle 0^\circ \text{ V}}{2.2\angle 0^\circ\ \Omega} = 2.27\angle 0^\circ \text{ A}$$

$$\mathbf{I}_C = \frac{\mathbf{V}_s}{\mathbf{X}_C} = \frac{5\angle 0^\circ \text{ V}}{5\angle -90^\circ\ \Omega} = 1\angle 90^\circ \text{ A}$$

$$\mathbf{I}_L = \frac{\mathbf{V}_s}{\mathbf{X}_L} = \frac{5\angle 0^\circ \text{ V}}{10\angle 90^\circ\ \Omega} = 0.5\angle -90^\circ \text{ A}$$

The total current is the phasor sum of the branch currents. By Kirchhoff's law,

$$\begin{aligned}\mathbf{I}_{tot} &= \mathbf{I}_R + \mathbf{I}_C + \mathbf{I}_L \\ &= 2.27\angle 0^\circ \text{ A} + 1\angle 90^\circ \text{ A} + 0.5\angle -90^\circ \text{ A} \\ &= 2.27 \text{ A} + j1 \text{ A} - j0.5 \text{ A} = 2.27 \text{ A} + j0.5 \text{ A}\end{aligned}$$

Converting to polar form yields

$$\mathbf{I}_{tot} = \sqrt{I_R^2 + (I_C - I_L)^2}\angle \tan^{-1}\left(\frac{I_{CL}}{I_R}\right)$$

$$= \sqrt{(2.27 \text{ A})^2 + (0.5 \text{ A})^2}\angle \tan^{-1}\left(\frac{0.5 \text{ A}}{2.27 \text{ A}}\right) = 2.32\angle 12.4^\circ \text{ A}$$

The total current is 2.32 A leading $V_s$ by 12.4°. Figure 18–26 is the current phasor diagram for the circuit.

FIGURE 18–26

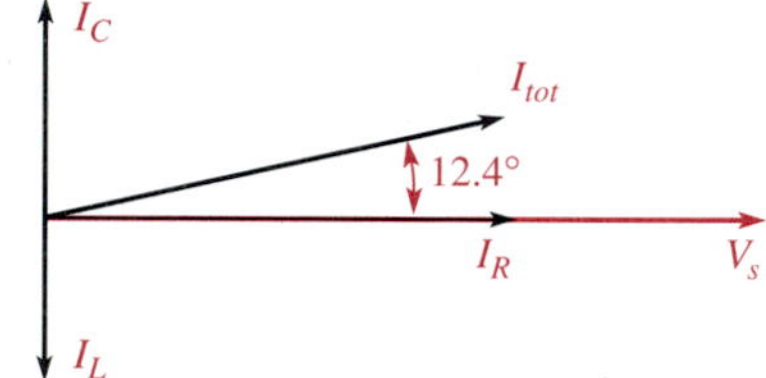


***Related Exercise*** Will total current increase or decrease if the frequency in Figure 18–25 is increased?

## Series-Parallel Analysis

The analysis of series-parallel combinations basically involves the application of the methods covered in Chapters 16 and 17. The following two examples illustrate typical approaches to these more complex circuits.

**EXAMPLE 18–11** In Figure 18–27, find the voltage across the capacitor in polar form. Is this circuit predominantly inductive or capacitive?

FIGURE 18–27

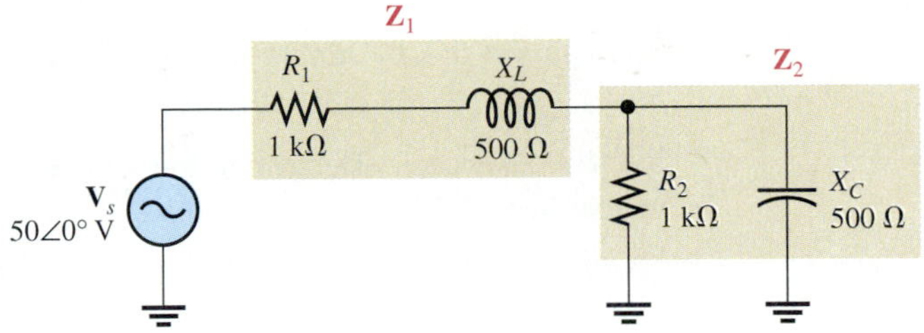

***Solution*** Use the voltage divider formula in this analysis. The impedance of the series combination of $R_1$ and $X_L$ is called $\mathbf{Z}_1$. In rectangular form,

$$\mathbf{Z}_1 = R_1 + jX_L = 1000\ \Omega + j500\ \Omega$$

Converting to polar form yields

$$\mathbf{Z}_1 = \sqrt{R_1^2 + X_L^2}\angle\tan^{-1}\left(\frac{X_L}{R}\right)$$

$$= \sqrt{(1000\ \Omega)^2 + (500\ \Omega)^2}\angle\tan^{-1}\left(\frac{500\ \Omega}{1000\ \Omega}\right) = 1118\angle 26.6^\circ\ \Omega$$

The impedance of the parallel combination of $R_2$ and $X_C$ is called $\mathbf{Z}_2$. In polar form,

$$\mathbf{Z}_2 = \left(\frac{R_2X_C}{\sqrt{R_2^2 + X_C^2}}\right)\angle\left(-90^\circ + \tan^{-1}\left(\frac{X_C}{R_2}\right)\right)$$

$$= \left[\frac{(1000\ \Omega)(500\ \Omega)}{\sqrt{(1000\ \Omega)^2 + (500\ \Omega)^2}}\right]\angle\left(-90^\circ + \tan^{-1}\left(\frac{500\ \Omega}{1000\ \Omega}\right)\right) = 447\angle -63.4^\circ\ \Omega$$

Converting to rectangular form yields

$$\mathbf{Z}_2 = Z_2\cos\theta + jZ_2\sin\theta$$
$$= 447\cos(-63.4^\circ) + j447\sin(-63.4^\circ) = 200\ \Omega - j400\ \Omega$$

The total impedance $\mathbf{Z}_{tot}$ in rectangular form is

$$\mathbf{Z}_{tot} = \mathbf{Z}_1 + \mathbf{Z}_2 = (1000\ \Omega + j500\ \Omega) + (200\ \Omega - j400\ \Omega) = 1200\ \Omega + j100\ \Omega$$

Converting to polar form yields

$$\mathbf{Z}_{tot} = \sqrt{(1200\ \Omega)^2 + (100\ \Omega)^2}\angle\tan^{-1}\left(\frac{100\ \Omega}{1200\ \Omega}\right) = 1204\angle 4.76^\circ\ \Omega$$

Now apply the voltage-divider formula to get $\mathbf{V}_C$.

$$\mathbf{V}_C = \left(\frac{\mathbf{Z}_2}{\mathbf{Z}_{tot}}\right)\mathbf{V}_s = \left(\frac{447\angle -63.4^\circ\ \Omega}{1204\angle 4.76^\circ\ \Omega}\right)50\angle 0^\circ\ \text{V} = 18.6\angle -68.2^\circ\ \text{V}$$

Therefore, $V_C$ is 18.6 V and lags $V_s$ by 68.2°.

The $+j$ term in $\mathbf{Z}_{tot}$, or the positive angle in its polar form, indicates that the circuit is more inductive than capacitive. However, it is just slightly more inductive, because the angle is small. This result may surprise you, because $X_C = X_L = 500\ \Omega$. However, the capacitor is in parallel with a resistor, so the capacitor actually has less effect on the total impedance than does the inductor. Figure 18–28 shows the phasor relationship of $\mathbf{V}_C$ and $\mathbf{V}_s$. Although $X_C = X_L$, this circuit is not at resonance, since the $j$ term of the total impedance is not zero due to the parallel combination of $R_2$ and $X_C$.

FIGURE 18–28

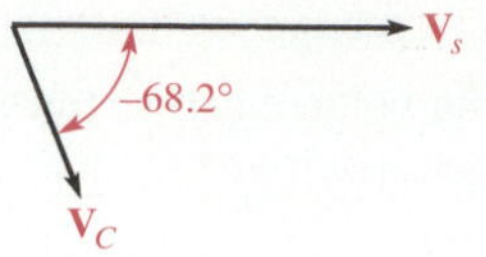

You can see this by noting that the phase angle associated with $\mathbf{Z}_{tot}$ is 4.76° and not zero.

***Related Exercise*** Determine the voltage across the capacitor in polar form if $R_1$ is increased to 2.2 kΩ.

## EXAMPLE 18–12

For the reactive circuit in Figure 18–29, find the voltage at point *B* with respect to ground.

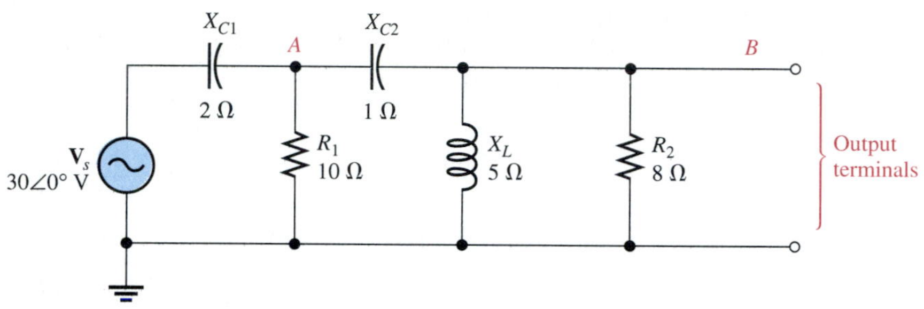


FIGURE 18–29

***Solution*** The voltage ($\mathbf{V}_B$) at point *B* is the voltage across the open output terminals. Use the voltage-divider approach. To do so, you must know the voltage ($\mathbf{V}_A$) at point *A* first; so you need to find the impedance from point *A* to ground as a starting point.

The parallel combination of $X_L$ and $R_2$ is in series with $X_{C2}$. This combination is in parallel with $R_1$. Call this impedance from point *A* to ground, $\mathbf{Z}_A$. To find $\mathbf{Z}_A$, take the following steps. The impedance of the parallel combination of $R_2$ and $X_L$ is called $\mathbf{Z}_1$.

$$\begin{aligned}\mathbf{Z}_1 &= \left(\frac{R_2X_L}{\sqrt{R_2^2 + X_L^2}}\right)\angle\left(90° - \tan^{-1}\left(\frac{X_L}{R}\right)\right)\\ &= \left(\frac{(8\ \Omega)(5\ \Omega)}{\sqrt{(8\ \Omega)^2 + (5\ \Omega)^2}}\right)\angle\left(90° - \tan^{-1}\left(\frac{5\ \Omega}{8\ \Omega}\right)\right)\\ &= \left(\frac{40.0}{9.43}\right)\angle(90° - 32°) = 4.24\angle 58.0°\ \Omega\end{aligned}$$

Next, combine $\mathbf{Z}_1$ in series with $\mathbf{X}_{C2}$ to get an impedance $\mathbf{Z}_2$.

$$\begin{aligned}\mathbf{Z}_2 &= \mathbf{X}_{C2} + \mathbf{Z}_1\\ &= 1\angle -90°\ \Omega + 4.24\angle 58°\ \Omega = -j1\ \Omega + 2.25\ \Omega + j3.6\ \Omega\\ &= 2.25\ \Omega + j2.6\ \Omega\end{aligned}$$

Converting to polar form yields

$$\mathbf{Z}_2 = \sqrt{(2.25\ \Omega)^2 + (2.6\ \Omega)^2}\angle\tan^{-1}\left(\frac{2.6\ \Omega}{2.25\ \Omega}\right) = 3.44\angle 49.1°\ \Omega$$

Finally, combine $\mathbf{Z}_2$ and $\mathbf{R}_1$ in parallel to get $\mathbf{Z}_A$.

$$\mathbf{Z}_A = \frac{\mathbf{R}_1\mathbf{Z}_2}{\mathbf{R}_1 + \mathbf{Z}_2} = \frac{(10\angle 0^\circ)(3.44\angle 49.1^\circ)}{10 + 2.25 + j2.6}$$

$$= \frac{34.4\angle 49.1^\circ}{12.25 + j2.6} = \frac{34.4\angle 49.1^\circ}{12.5\angle 12.0^\circ} = 2.75\angle 37.1^\circ\ \Omega$$

The simplified circuit is shown in Figure 18–30.

FIGURE 18–30

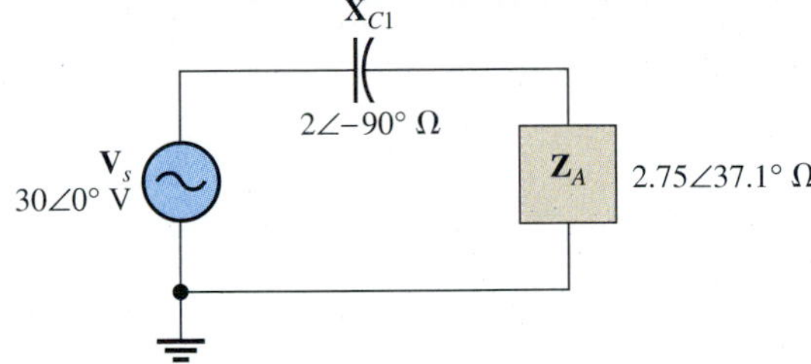


Now, use the voltage-divider principle to find the voltage ($\mathbf{V}_A$) at point $A$ in Figure 18–29. The total impedance is

$$\mathbf{Z}_{tot} = \mathbf{X}_{C1} + \mathbf{Z}_A$$
$$= 2\angle -90^\circ\ \Omega + 2.75\angle 37.1^\circ\ \Omega = -j2\ \Omega + 2.19\ \Omega + j1.66\ \Omega$$
$$= 2.19\ \Omega - j0.340\ \Omega$$

Converting to polar form yields

$$\mathbf{Z}_{tot} = \sqrt{(2.19\ \Omega)^2 + (0.340\ \Omega)^2}\angle -\tan^{-1}\left(\frac{0.340\ \Omega}{2.19\ \Omega}\right) = 2.22\angle -8.82^\circ\ \Omega$$

The voltage at point $A$ is

$$\mathbf{V}_A = \left(\frac{\mathbf{Z}_A}{\mathbf{Z}_{tot}}\right)\mathbf{V}_s = \left(\frac{2.75\angle 37.1^\circ\ \Omega}{2.22\angle -8.82^\circ\ \Omega}\right)30\angle 0^\circ\ \text{V} = 37.2\angle 45.9^\circ\ \text{V}$$

Next, find the voltage ($\mathbf{V}_B$) at point $B$ by dividing $\mathbf{V}_A$ down, as indicated in Figure 18–31. $\mathbf{V}_B$ is the open terminal output voltage.

$$\mathbf{V}_B = \left(\frac{\mathbf{Z}_1}{\mathbf{Z}_2}\right)\mathbf{V}_A = \left(\frac{4.24\angle 58^\circ\ \Omega}{3.44\angle 49.1^\circ\ \Omega}\right)37.2\angle 45.9^\circ\ \text{V} = 45.9\angle 54.8^\circ\ \text{V}$$

Surprisingly, $V_A$ is greater than $V_s$, and $V_B$ is greater than $V_A$! This result is possible because of the out-of-phase relationship of the reactive voltages. Remember that $X_C$ and $X_L$ tend to cancel each other.

FIGURE 18–31

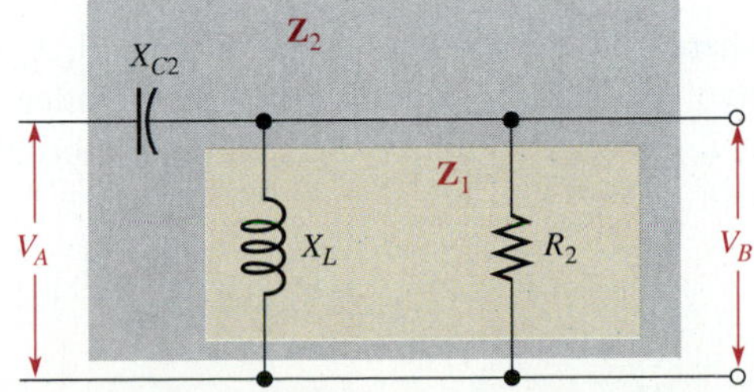


***Related Exercise*** What is the voltage across $C_1$ in Figure 18–29?

## Conversion of Series-Parallel to Parallel

The particular series-parallel configuration shown in Figure 18–32 is important because it represents a circuit having parallel $L$ and $C$ branches, with the winding resistance of the coil taken into account as a series resistance in the $L$ branch.

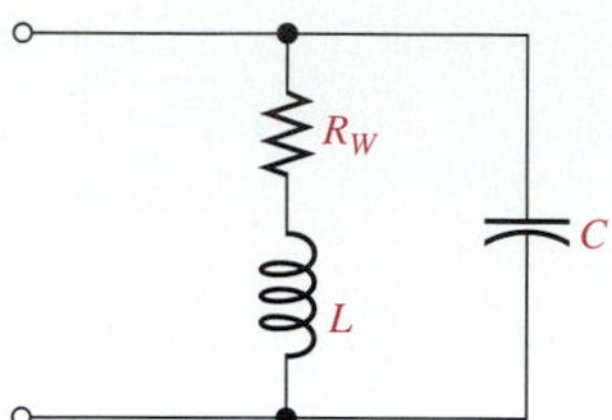


FIGURE 18–32
*A series-parallel RLC circuit ($Q = X_L/R_W$).*

It is helpful to view the series-parallel circuit in Figure 18–32 in an equivalent purely parallel form, as indicated in Figure 18–33. This form will simplify our analysis of parallel resonant characteristics in the next section.

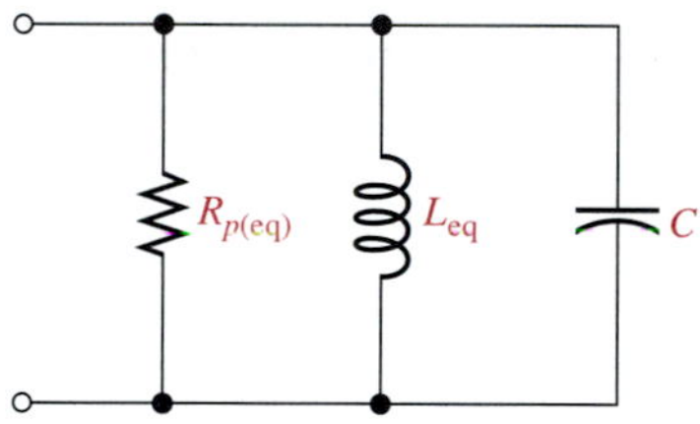


FIGURE 18–33
*Parallel equivalent form of the circuit in Figure 18–32.*

The equivalent inductance, $L_{eq}$, and the equivalent parallel resistance, $R_{p(eq)}$, are given by the following formulas:

$$L_{eq} = L\left(\frac{Q^2 + 1}{Q^2}\right) \tag{18–14}$$

$$R_{p(eq)} = R_W(Q^2 + 1) \tag{18–15}$$

where $Q$ is the **quality factor** of the coil, $X_L/R_W$. Derivations of these formulas are quite involved and thus are not given here. Notice in the equations that for a $Q \geq 10$, the value of $L_{eq}$ is approximately the same as the original value of $L$. For example, if $L = 10$ mH, then

$$L_{eq} = 10\text{ mH}\left(\frac{10^2 + 1}{10^2}\right) = 10\text{ mH}(1.01) = 10.1\text{ mH}$$

The equivalency of the two circuits means that at a given frequency, when the same value of voltage is applied to both circuits, the same total current is in both circuits and the phase angles are the same. Basically, an equivalent circuit simply makes circuit analysis more convenient.

**EXAMPLE 18–13** Convert the series-parallel circuit in Figure 18–34 to an equivalent parallel form at the given frequency.

FIGURE 18–34

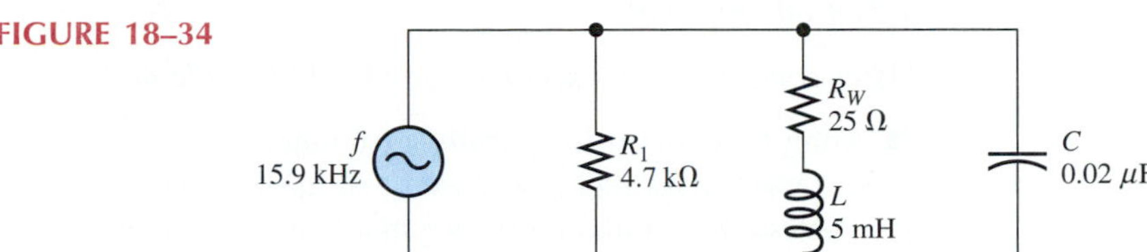


***Solution*** Determine the inductive reactance.

$$X_L = 2\pi fL = 2\pi(15.9 \text{ kHz})(5 \text{ mH}) = 500\ \Omega$$

The $Q$ of the coil is

$$Q = \frac{X_L}{R_W} = \frac{500\ \Omega}{25\ \Omega} = 20$$

Since $Q > 10$, then $L_{eq} \cong L = 5$ mH.

The equivalent parallel resistance is

$$R_{p(\text{eq})} = R_W(Q^2 + 1) = (25\ \Omega)(20^2 + 1) = 10.03 \text{ k}\Omega$$

This equivalent resistance appears in parallel with $R_1$ as shown in Figure 18–35(a). When combined, they give a total parallel resistance ($R_{p(tot)}$) of 3.2 kΩ, as indicated in Figure 18–35(b).

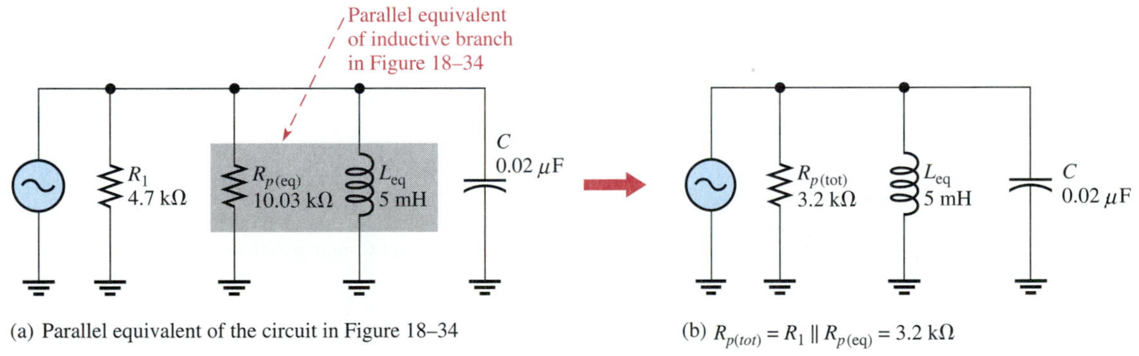


(a) Parallel equivalent of the circuit in Figure 18–34

(b) $R_{p(tot)} = R_1 \parallel R_{p(\text{eq})} = 3.2$ kΩ

FIGURE 18–35

***Related Exercise*** Find the equivalent parallel circuit if $R_W = 10\ \Omega$ in Figure 18–34.

**SECTION 18–5 REVIEW**

1. In a three-branch parallel circuit, $R = 150\ \Omega$, $X_C = 100\ \Omega$, and $X_L = 50\ \Omega$. Determine the current in each branch when $V_s = 12$ V.
2. The impedance of a parallel *RLC* circuit is $2.8\angle -38.9°$ kΩ. Is the circuit capacitive or inductive?
3. Find the equivalent parallel inductance and resistance for a 20 mH coil with a winding resistance of 10 Ω at a frequency of 1 kHz.

## 18–6 ■ PARALLEL RESONANCE

***In this section, we will first look at the resonant condition in an ideal parallel LC circuit. Then, we will examine the more realistic case where the resistance of the coil is taken into account.***

***After completing this section, you should be able to***

- **Analyze a circuit for parallel resonance**
  - Describe parallel resonance in an ideal circuit
  - Describe parallel resonance in a nonideal circuit
  - Explain how impedance varies with frequency
  - Determine current and phase angle at resonance
  - Determine parallel resonant frequency
  - Discuss the effects of loading a parallel resonant circuit

### Condition for Ideal Parallel Resonance

Ideally, parallel resonance occurs when $X_C = X_L$. The frequency at which resonance occurs is called the *resonant frequency,* just as in the series case. When $X_C = X_L$, the two branch currents, $I_C$ and $I_L$, are equal in magnitude, and, of course, they are always 180° out of phase with each other. Thus, the two currents cancel and the total current is zero, as shown in Figure 18–36.

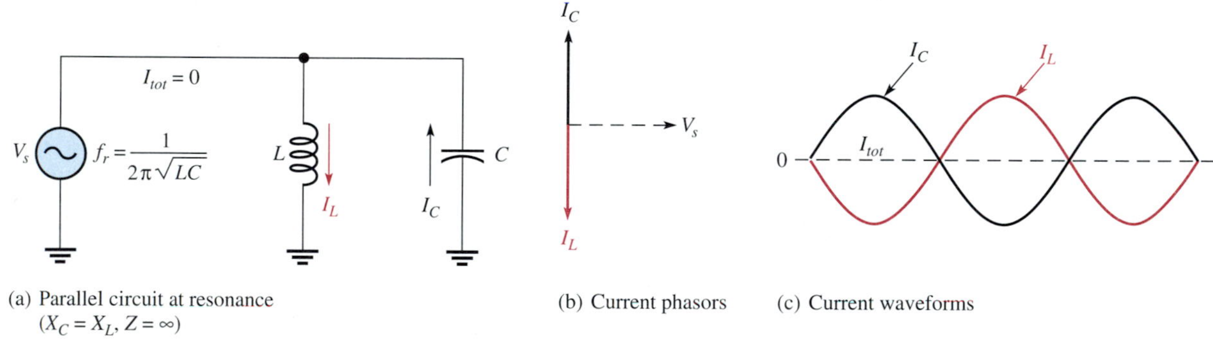


(a) Parallel circuit at resonance ($X_C = X_L$, $Z = \infty$)

(b) Current phasors

(c) Current waveforms

**FIGURE 18–36**
***An ideal parallel LC circuit at resonance.***

Since the total current is zero, the impedance of the parallel *LC* circuit is infinitely large (∞). These ideal resonant conditions are stated as follows:

$$X_L = X_C$$
$$Z_r = \infty$$

### The Ideal Parallel Resonant Frequency

For an ideal (no resistance) parallel resonant circuit, the frequency at which resonance occurs is determined by the same formula as in series resonant circuits; that is,

$$f_r = \frac{1}{2\pi\sqrt{LC}}$$

## The Parallel Resonant *LC* Circuit Is Often Called a Tank Circuit

The term **tank circuit** refers to the fact that the parallel resonant circuit stores energy in the magnetic field of the coil and in the electric field of the capacitor. The stored energy is transferred back and forth between the capacitor and the coil on alternate half-cycles as the current goes first one way and then the other when the inductor deenergizes and the capacitor charges, and vice versa. This concept is illustrated in Figure 18–37.

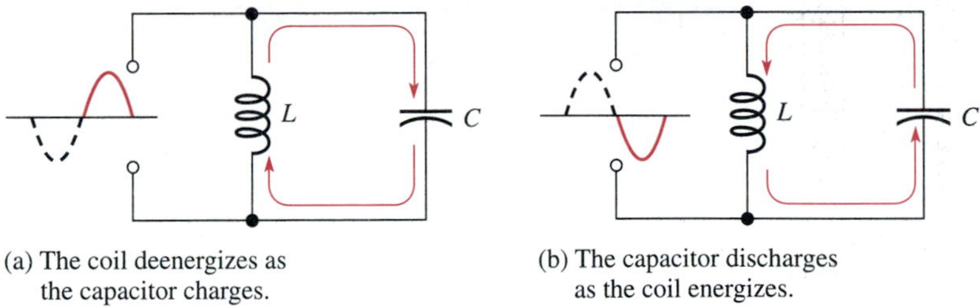


(a) The coil deenergizes as the capacitor charges.

(b) The capacitor discharges as the coil energizes.

FIGURE 18–37
***Energy storage in an ideal parallel resonant tank circuit.***

## Parallel Resonant Conditions in a Nonideal Circuit

So far, the resonance of an ideal parallel *LC* circuit has been examined. Now, let's consider resonance in a tank circuit with the resistance of the coil taken into account. Figure 18–38 shows a nonideal tank circuit and its parallel *RLC* equivalent.

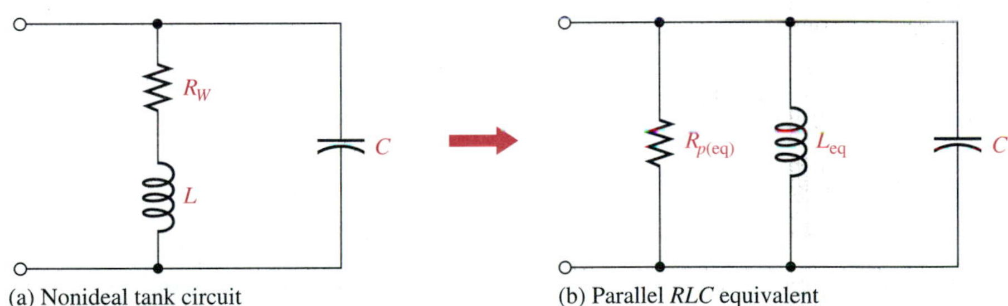


(a) Nonideal tank circuit

(b) Parallel *RLC* equivalent

FIGURE 18–38
***A practical treatment of parallel resonant circuits must include the coil resistance.***

The quality factor, $Q$, of the circuit at resonance is simply the $Q$ of the coil.

$$Q = \frac{X_L}{R_W}$$

The expressions for the equivalent parallel resistance and the equivalent inductance were given in Equations (18–15) and (18–14) as

$$R_{p(eq)} = R_W(Q^2 + 1)$$

$$L_{eq} = L\left(\frac{Q^2 + 1}{Q^2}\right)$$

Recall that for $Q \geq 10$, $L_{eq} \cong L$.

At parallel resonance,

$$X_{L(eq)} = X_C$$

In the parallel equivalent circuit, $R_{p(eq)}$ is in parallel with an ideal coil and a capacitor, so the $L$ and $C$ branches act as an ideal tank circuit which has an infinite impedance at resonance as shown in Figure 18–39. Therefore, the total impedance of the nonideal tank circuit at resonance can be expressed as simply the equivalent parallel resistance.

$$Z_r = R_W(Q^2 + 1) \tag{18–16}$$

A derivation of Equation (18–16) is given in Appendix C.

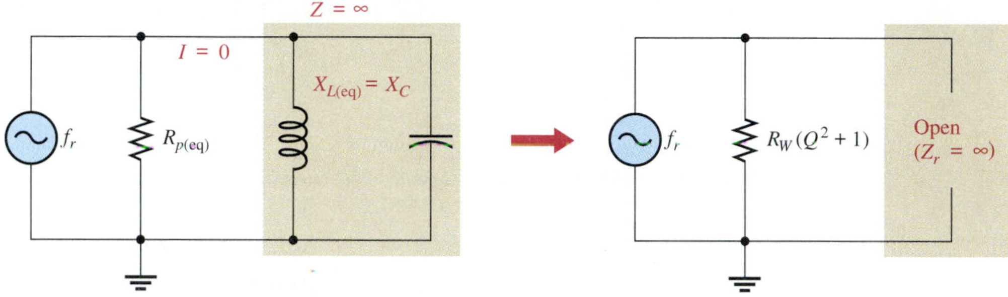


FIGURE 18–39
***At resonance, the parallel LC portion appears open and the source sees only $R_{p(eq)}$.***

**EXAMPLE 18–14** Determine the impedance of the circuit in Figure 18–40 at the resonant frequency ($f_r \cong$ 17,794 Hz).

FIGURE 18–40

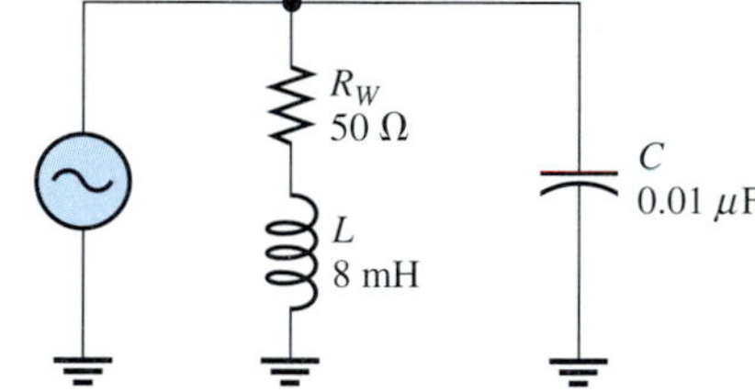


***Solution*** Before you can calculate the impedance using Equation (18–16), you must find the quality factor. To get $Q$, first find the inductive reactance.

$$X_L = 2\pi f_r L = 2\pi(17{,}794 \text{ Hz})(8 \text{ mH}) = 894 \ \Omega$$

$$Q = \frac{X_L}{R_W} = \frac{894 \ \Omega}{50 \ \Omega} = 17.9$$

$$Z_r = R_W(Q^2 + 1) = 50 \ \Omega(17.9^2 + 1) = 16.1 \text{ k}\Omega$$

The calculator sequence is

2 × 2nd π × 1 7 7 9 4 × 8 EXP +/– 3 = ÷ 5 0 =
x² + 1 = × 5 0 =

***Related Exercise*** Determine $Z_r$ for $R_W = 10 \ \Omega$.

## Variation of the Impedance with Frequency

The impedance of a parallel resonant circuit is maximum at the resonant frequency and decreases at lower and higher frequencies, as indicated by the curve in Figure 18–41.

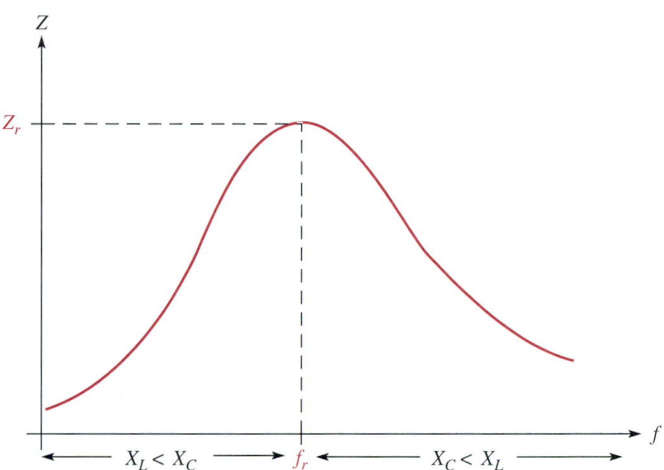


**FIGURE 18–41**
***Generalized impedance curve for a parallel resonant circuit. The circuit is inductive below $f_r$, resistive at $f_r$, and capacitive above $f_r$.***

At very low frequencies, $X_L$ is very small and $X_C$ is very high, so the total impedance is essentially equal to that of the inductive branch. As the frequency goes up, the impedance also increases, and the inductive reactance dominates (because it is less than $X_C$) until the resonant frequency is reached. At this point, of course, $X_L \cong X_C$ (for $Q > 10$) and the impedance is at its maximum. As the frequency goes above resonance, the capacitive reactance dominates (because it is less than $X_L$) and the impedance decreases.

## Current and Phase Angle at Resonance

In the ideal tank circuit, the total current from the source at resonance is zero because the impedance is infinite. In the nonideal case, there is some total current at the resonant frequency, and it is determined by the impedance at resonance.

$$I_{tot} = \frac{V_s}{Z_r} \tag{18–17}$$

The phase angle of the parallel resonant circuit is 0° because the impedance is purely resistive at the resonant frequency.

## Parallel Resonant Frequency in a Nonideal Circuit

As you know, when the coil resistance is considered, the resonant condition is

$$X_{L(\text{eq})} = X_C$$

which can be expressed as

$$2\pi f_r L\left(\frac{Q^2 + 1}{Q^2}\right) = \frac{1}{2\pi f_r C}$$

Solving for $f_r$ in terms of $Q$ yields

$$f_r = \frac{1}{2\pi\sqrt{LC}}\sqrt{\frac{Q^2}{Q^2 + 1}} \tag{18–18}$$

When $Q \geq 10$, the term with the $Q$ factors is approximately 1.

$$\sqrt{\frac{Q^2}{Q^2 + 1}} = \sqrt{\frac{100}{101}} = 0.995 \cong 1$$

Therefore, the parallel resonant frequency is approximately the same as the series resonant frequency as long as $Q$ is equal to or greater than 10.

$$f_r \cong \frac{1}{2\pi\sqrt{LC}} \qquad \text{for } Q \geq 10$$

A precise expression for $f_r$ in terms of the circuit component values is

$$f_r = \frac{\sqrt{1 - (R_W^2 C/L)}}{2\pi\sqrt{LC}} \qquad \textbf{(18–19)}$$

This precise formula is seldom necessary and the simpler equation $f_r = 1/2\pi\sqrt{LC}$ is sufficient for most practical situations. A derivation of Equation (18–19) is given in Appendix C.

**EXAMPLE 18–15** Find the frequency, impedance, and total current at resonance for the circuit in Figure 18–42.

**FIGURE 18–42**

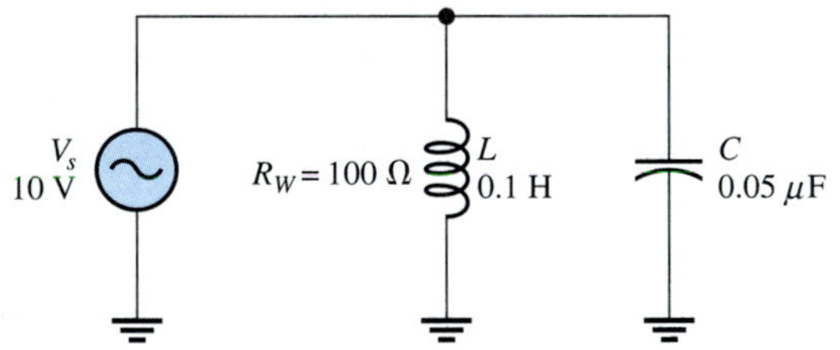

***Solution*** Use Equation (18–19) to find the frequency.

$$f_r = \frac{\sqrt{1 - (R_W^2 C/L)}}{2\pi\sqrt{LC}} = \frac{\sqrt{1 - [(100\ \Omega)^2(0.05\ \mu\text{F})/0.1\ \text{H}]}}{2\pi\sqrt{(0.05\ \mu\text{F})(0.1\ \text{H})}} = 2.25\ \text{kHz}$$

To calculate the impedance, first find $X_L$ and $Q$.

$$X_L = 2\pi f_r L = 2\pi(2.25\ \text{kHz})(0.1\ \text{H}) = 1.41\ \text{k}\Omega$$

$$Q = \frac{X_L}{R_W} = \frac{1.41\ \text{k}\Omega}{100\ \Omega} = 14.1$$

$$Z_r = R_W(Q^2 + 1) = 100\ \Omega(14.1^2 + 1) = 20\ \text{k}\Omega$$

The total current is

$$I_{tot} = \frac{V_s}{Z_r} = \frac{10\ \text{V}}{20\ \text{k}\Omega} = 500\ \mu\text{A}$$

Note that since $Q > 10$, the approximate formula, $f_r \cong 1/2\pi\sqrt{LC}$, could be used.

The calculator sequence for $f_r$ is

1 − ( 1 0 0 x² × • 0 5 EXP +/− 6 ÷ • 1 = ) = √x ÷
2 ÷ 2nd π ÷ ( ( • 0 5 EXP +/− 6 × • 1 ) √x ) =

***Related Exercise*** For a smaller $R_W$, will $I_{tot}$ be less than or greater than 500 $\mu$A?

## An External Parallel Load Resistance Affects a Tank Circuit

There are many practical situations in which an external load resistance appears in parallel with a tank circuit as shown in Figure 18–43(a). Obviously, the external resistor ($R_L$) will dissipate more of the energy delivered by the source and thus will lower the overall $Q$ of the circuit. The external resistor effectively appears in parallel with the equivalent parallel resistance of the coil, $R_{p(\text{eq})}$, and both are combined to determine a total parallel resistance, $R_{p(tot)}$, as indicated in Figure 18–43(b).

$$R_{p(tot)} = R_L \parallel R_{p(\text{eq})}$$

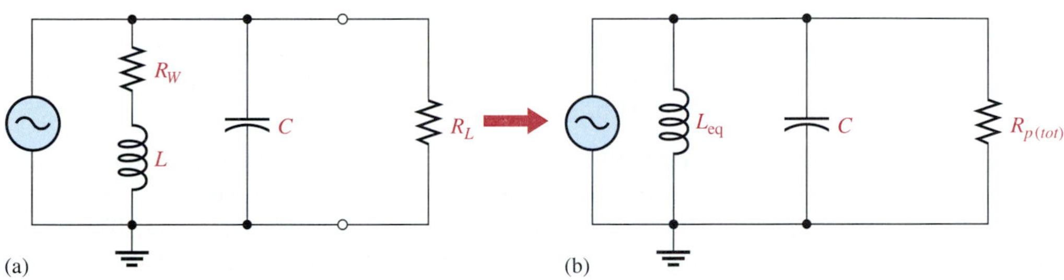


**FIGURE 18–43**
***Tank circuit with a parallel load resistor and its equivalent circuit.***

The overall $Q$, designated $Q_O$, for a parallel $RLC$ circuit is expressed differently from the $Q$ of a series circuit.

$$Q_O = \frac{R_{p(tot)}}{X_{L(\text{eq})}} \qquad \textbf{(18–20)}$$

As you can see, the effect of loading the tank circuit is to reduce its overall $Q$ (which is equal to the coil $Q$ when unloaded).

**SECTION 18–6 REVIEW**

1. Is the impedance minimum or maximum at parallel resonance?
2. Is the current minimum or maximum at parallel resonance?
3. At ideal parallel resonance, $X_L = 1500\ \Omega$. What is $X_C$?
4. A parallel tank circuit has the following values: $R_W = 4\ \Omega$, $L = 50$ mH, and $C =$ 10 pF. Calculate $f_r$ and $Z$ at resonance.
5. If $Q = 25$, $L = 50$ mH, and $C = 1000$ pF, what is $f_r$?
6. In Question 5, if $Q = 2.5$, what is $f_r$?
7. In a certain tank circuit, the coil resistance is 20 Ω. What is the total impedance at resonance if $Q = 20$?

## 18–7 ■ BANDWIDTH OF RESONANT CIRCUITS

***As you have learned, the current in a series RLC is maximum at the resonant frequency because the reactances cancel and the current in a parallel RLC is minimum at the resonant frequency because the inductive and capacitive currents cancel. In this section, you will see how this circuit behavior relates to a characteristic called bandwidth.***

***After completing this section, you should be able to***

- **Determine the bandwidth of resonant circuits**
  - ☐ Discuss the bandwidth of series and parallel resonant circuits
  - ☐ State the formula for bandwidth
  - ☐ Define *half-power frequency*
  - ☐ Define *selectivity*
  - ☐ Explain how the $Q$ affects the bandwidth

### Series Resonant Circuits

The current in a series *RLC* circuit is maximum at the resonant frequency (also known as *center frequency*) and drops off on either side of this frequency. Bandwidth, sometimes abbreviated *BW,* is an important characteristic of a resonant circuit. The bandwidth is the range of frequencies for which the current is equal to or greater than 70.7% of its resonant value.

Figure 18–44 illustrates bandwidth on the response curve of a series *RLC* circuit. Notice that the frequency $f_1$ below $f_r$ is the point at which the current is $0.707I_{max}$ and is commonly called the *lower critical frequency.* The frequency $f_2$ above $f_r$, where the current is again $0.707I_{max}$, is the *upper critical frequency.* Other names for $f_1$ and $f_2$ are *–3 dB frequencies, cutoff frequencies,* and *half-power frequencies.* The significance of the latter term is discussed later in the chapter.

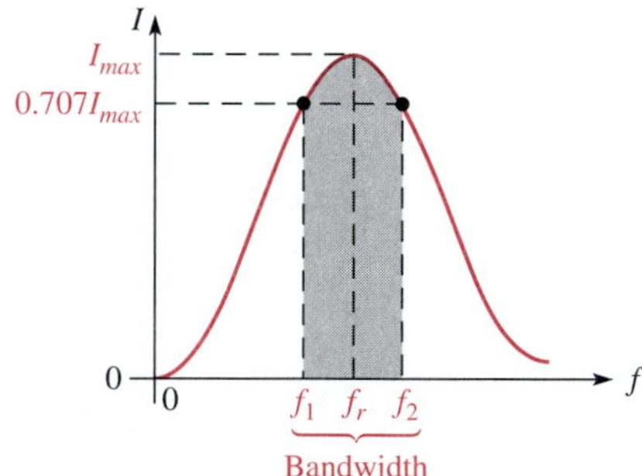


**FIGURE 18–44**
***Bandwidth on series resonant response curve for I.***

**EXAMPLE 18–16** A certain series resonant circuit has a maximum current of 100 mA at the resonant frequency. What is the value of the current at the critical frequencies?

***Solution*** Current at the critical frequencies is 70.7% of maximum.

$$I_{f1} = I_{f2} = 0.707I_{max} = 0.707(100 \text{ mA}) = 70.7 \text{ mA}$$

***Related Exercise*** A certain series resonant circuit has a current of 25 mA at the critical frequencies. What is the current at resonance?

## Parallel Resonant Circuits

For a parallel resonant circuit, the impedance is maximum at the resonant frequency; so the total current is minimum. The bandwidth can be defined in relation to the impedance curve in the same manner that the current curve was used in the series circuit. Of course, $f_r$ is the frequency at which $Z$ is maximum; $f_1$ is the lower critical frequency at which $Z = 0.707Z_{max}$; and $f_2$ is the upper critical frequency at which again $Z = 0.707Z_{max}$. The bandwidth is the range of frequencies between $f_1$ and $f_2$, as shown in Figure 18–45.

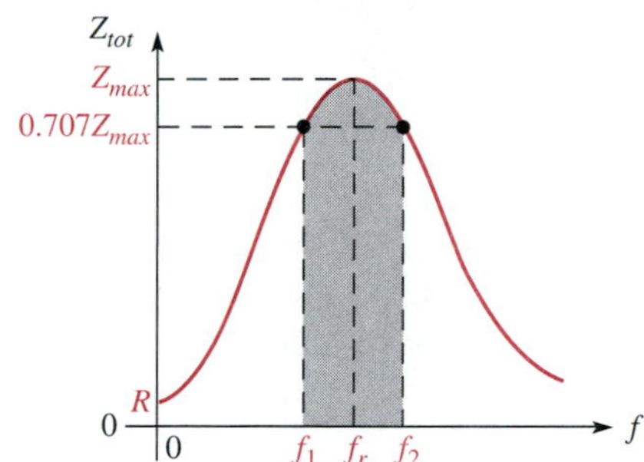


**FIGURE 18–45**
***Bandwidth of the parallel resonant response curve for $Z_{tot}$.***

## Formula for Bandwidth

The bandwidth for either series or parallel resonant circuits is the range of frequencies between the critical frequencies for which the response curve ($I$ or $Z$) is 0.707 of the maximum value. Thus, the bandwidth is actually the difference between $f_2$ and $f_1$.

$$BW = f_2 - f_1 \qquad \textbf{(18–21)}$$

Ideally, $f_r$ is the center frequency and can be calculated as follows:

$$f_r = \frac{f_1 + f_2}{2} \qquad \textbf{(18–22)}$$

**EXAMPLE 18–17** A resonant circuit has a lower critical frequency of 8 kHz and an upper critical frequency of 12 kHz. Determine the bandwidth and center (resonant) frequency.

***Solution***

$$BW = f_2 - f_1 = 12 \text{ kHz} - 8 \text{ kHz} = 4 \text{ kHz}$$

$$f_r = \frac{f_1 + f_2}{2} = \frac{12 \text{ kHz} + 8 \text{ kHz}}{2} = 10 \text{ kHz}$$

***Related Exercise*** If the bandwidth of a resonant circuit is 2.5 kHz and its center frequency is 8 kHz, what are the lower and upper critical frequencies?

## Half-Power Frequencies

As previously mentioned, the upper and lower critical frequencies are sometimes called the **half-power frequencies.** This term is derived from the fact that the power from the source at these frequencies is one-half the power delivered at the resonant frequency. The following steps show that this is true for a series circuit. The same end result also applies to a parallel circuit. At resonance,

$$P_{max} = I_{max}^2 R$$

The power at $f_1$ or $f_2$ is

$$P_{f1} = I_{f1}^2 R = (0.707 I_{max})^2 R = (0.707)^2 I_{max}^2 R = 0.5 I_{max}^2 R = 0.5 P_{max}$$

## Selectivity

The response curves in Figures 18–44 and 18–45 are also called *selectivity curves.* **Selectivity** defines how well a resonant circuit responds to a certain frequency and discriminates against all others. *The narrower the bandwidth, the greater the selectivity.*

We normally assume that a resonant circuit accepts frequencies within its bandwidth and completely eliminates frequencies outside the bandwidth. Such is not actually the case, however, because signals with frequencies outside the bandwidth are not completely eliminated. Their magnitudes, however, are greatly reduced. The further the frequencies are from the critical frequencies, the greater is the reduction, as illustrated in Figure 18–46(a). An ideal selectivity curve is shown in Figure 18–46(b).

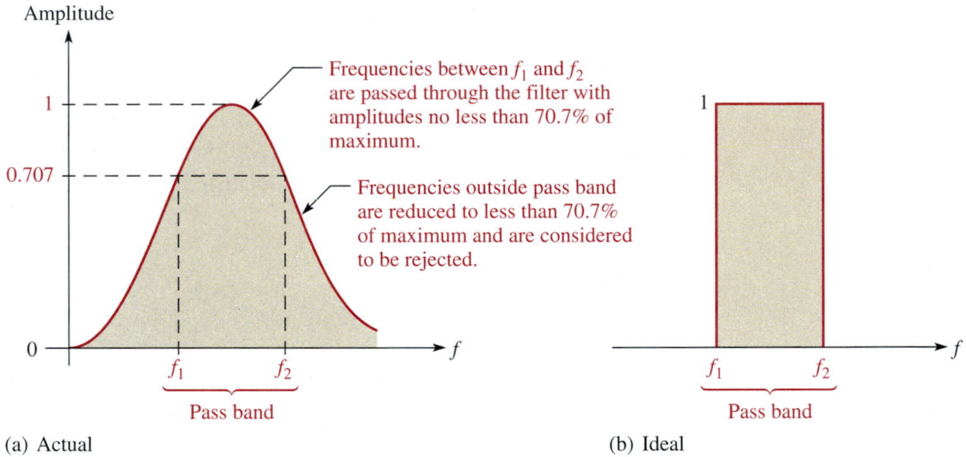


**FIGURE 18–46**
***Generalized selectivity curve of a band-pass filter.***

As you can see in Figure 18–46, another factor that influences selectivity is the sharpness of the slopes of the curve. The faster the curve drops off at the critical frequencies, the more selective the circuit is because it responds only to the frequencies within the bandwidth. Figure 18–47 shows a general comparison of three response curves with varying degrees of selectivity.

## *Q* Affects Bandwidth

A higher value of circuit $Q$ results in a narrower bandwidth. A lower value of $Q$ causes a wider bandwidth. A formula for the bandwidth of a resonant circuit in terms of $Q$ is stated in the following equation:

$$BW = \frac{f_r}{Q} \qquad \textbf{(18–23)}$$

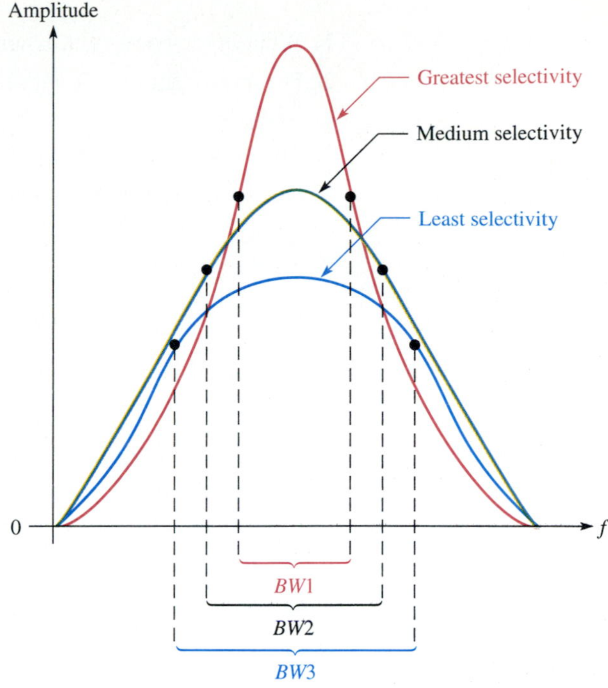


**FIGURE 18–47**
***Comparative selectivity curves.***

## EXAMPLE 18–18

What is the bandwidth of each circuit in Figure 18–48?

**FIGURE 18–48**

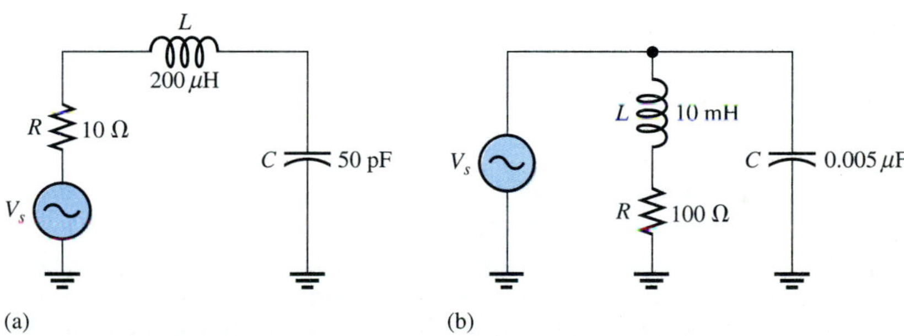


***Solution*** For the circuit in Figure 18–48(a), the bandwidth is found as follows:

$$f_r = \frac{1}{2\pi\sqrt{LC}} = \frac{1}{2\pi\sqrt{(200\ \mu\text{H})(50\ \text{pF})}} = 1.59\ \text{MHz}$$

$$Q = \frac{X_L}{R} = \frac{2\ \text{k}\Omega}{10\ \Omega} = 200$$

$$BW = \frac{f_r}{Q} = \frac{1.59\ \text{MHz}}{200} = 7.95\ \text{kHz}$$

For the circuit in Figure 18–48(b),

$$f_r = \frac{\sqrt{1 - (R_W^2\,C/L)}}{2\pi\sqrt{LC}} \cong \frac{1}{2\pi\sqrt{LC}} = \frac{1}{2\pi\sqrt{(10\ \text{mH})(0.005\ \mu\text{F})}} = 22.5\ \text{kHz}$$

$$Q = \frac{X_L}{R} = \frac{1.41\ \text{k}\Omega}{100\ \Omega} = 14.1$$

$$BW = \frac{f_r}{Q} = \frac{22.5\ \text{kHz}}{14.1} = 1.60\ \text{kHz}$$

***Related Exercise*** Change $C$ in Figure 18–48(a) to 1000 pF and determine the bandwidth.

**SECTION 18–7 REVIEW**

1. What is the bandwidth when $f_2 = 2.2$ MHz and $f_1 = 1.8$ MHz?
2. For a resonant circuit with the critical frequencies in Question 1, what is the center frequency?
3. The power at resonance is 1.8 W. What is the power at the upper critical frequency?
4. Does a larger $Q$ mean a narrower or a wider bandwidth?

## 18–8 ■ SYSTEM APPLICATIONS

***Resonant circuits are used in a wide variety of applications, particularly in communication systems. In this section, we will look briefly at a few common communication systems applications. The purpose in this section is not to explain how the systems work, but to illustrate the importance of resonant circuits in electronic communication.***

***After completing this section, you should be able to***

- **Discuss some system applications of resonant circuits**
  - ☐ Describe a tuned amplifier application
  - ☐ Describe antenna coupling
  - ☐ Describe tuned amplifiers
  - ☐ Describe signal separation in a receiver
  - ☐ Describe a radio receiver

### Tuned Amplifiers

A *tuned amplifier* is a circuit that amplifies signals within a specified band. Typically, a parallel resonant circuit is used in conjunction with an amplifier to achieve the selectivity. In terms of the general operation, input signals with frequencies that range over a wide band are accepted on the amplifier's input and are amplified. The function of the resonant circuit is to allow only a relatively narrow band of those frequencies to be passed on. The variable capacitor allows tuning over the range of input frequencies so that a desired frequency can be selected, as indicated in Figure 18–49.

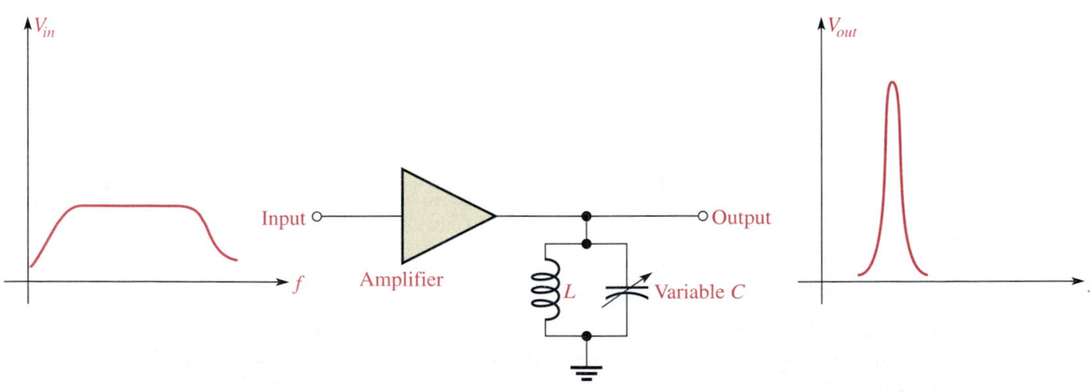


FIGURE 18–49
*A basic tuned band-pass amplifier.*

## Antenna Input to a Receiver

Radio signals are sent out from a transmitter via electromagnetic waves which propagate through the atmosphere. When the electromagnetic waves cut across the receiving antenna, small voltages are induced. Out of all the wide range of electromagnetic frequencies, only one frequency or a limited band of frequencies must be extracted. Figure 18–50 shows a typical arrangement of an antenna coupled to the receiver input by a transformer. A variable capacitor is connected across the transformer secondary to form a parallel resonant circuit.

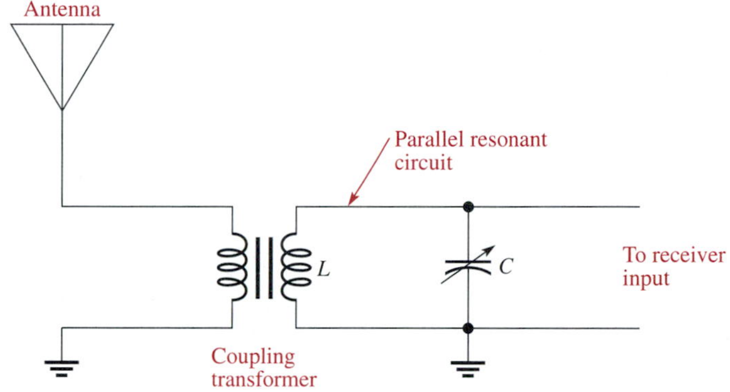


**FIGURE 18–50**
***Resonant coupling from an antenna.***

## Double-Tuned Transformer Coupling in a Receiver

In some types of communication receivers, tuned amplifiers are transformer-coupled together to increase the amplification. Capacitors can be placed in parallel with the primary and secondary windings of the transformer, effectively creating two parallel resonant band-pass filters that are coupled together. This technique, illustrated in Figure 18–51, can result in a wider bandwidth and steeper slopes on the response curve, thus increasing the selectivity for a desired band of frequencies.

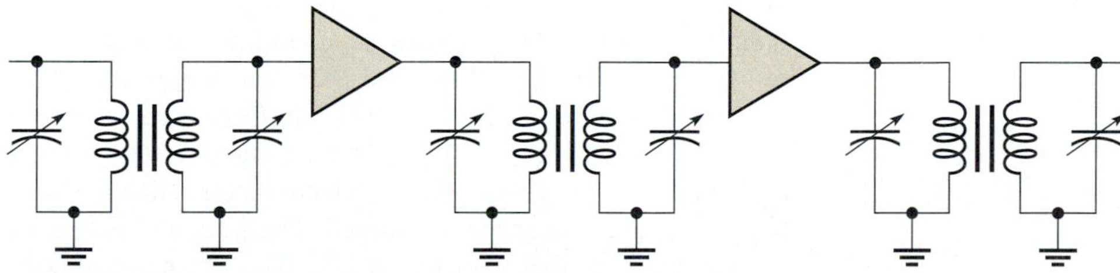

**FIGURE 18–51**
***Double-tuned amplifiers.***

## Signal Reception and Separation in a TV Receiver

A television receiver must handle both video (picture) signals and audio (sound) signals. Each TV transmitting station is allotted a 6 MHz bandwidth. Channel 2 is allotted a band from 54 MHz through 59 MHz, channel 3 is allotted a band from 60 MHz through 65 MHz, on up to channel 13 which has a band from 210 MHz through 215 MHz. You can tune the front end of the TV receiver to select any one of these channels by using tuned amplifiers. The signal output of the front end of the receiver has a bandwidth from 41 MHz through 46 MHz, regardless of the channel that is tuned in. This band, called the *intermediate frequency* (IF) band, contains both video and audio. Amplifiers tuned to the IF band boost the signal and feed it to the video amplifier.

Before the output of the video amplifier is applied to the picture tube, the audio signal is removed by a 4.5 MHz band-stop filter (called a *wave trap*), as shown in Figure 18–52. This trap keeps the sound signal from interfering with the picture. The video amplifier output is also applied to band-pass circuits that are tuned to the sound carrier frequency of 4.5 MHz. The sound signal is then processed and applied to the speaker as indicated in Figure 18–52.

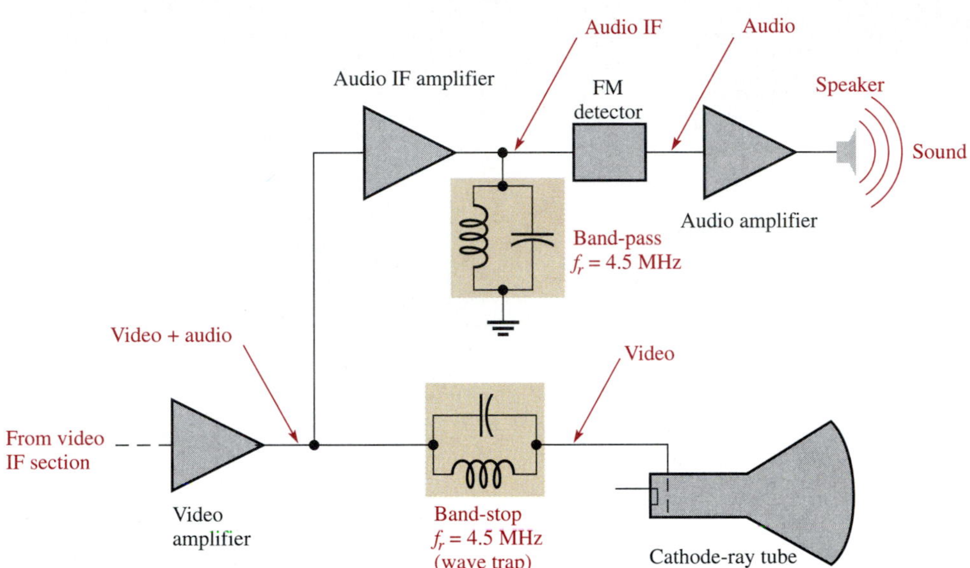


**FIGURE 18–52**
***A simplified portion of a TV receiver showing filter usage.***

## Superheterodyne Receiver

Another good example of filter applications is in the common AM (amplitude modulation) receiver. The AM broadcast band ranges from 535 kHz to 1605 kHz. Each AM station is assigned a certain narrow bandwidth within that range. A simplified block diagram of a superheterodyne AM receiver is shown in Figure 18–53.

In this system, there are basically three parallel resonant band-pass filters in the front end of the receiver. Each of these filters is gang-tuned by capacitors; that is, the capacitors are mechanically or electronically linked together so that they change together as the tuning knob is turned. The front end is tuned to receive a desired station, for example, one that transmits at 600 kHz. The input filter from the antenna and the RF (radio frequency) amplifier filter select only a frequency of 600 kHz out of all the frequencies crossing the antenna. The actual audio (sound) signal is carried by the 600 kHz carrier frequency by modulating the amplitude of the carrier so that it follows the audio signal as indicated. The variation in the amplitude of the carrier corresponding to the audio signal is called the *envelope*. The 600 kHz is then applied to a circuit called the *mixer*. The *local oscillator* (LO) is tuned to a frequency that is 455 kHz above the selected frequency (1055 kHz, in this case). By a process called *heterodyning* or *beating*, the AM signal and the local oscillator signal are mixed together, and the 600 kHz AM signal is converted to a 455 kHz AM signal (1055 kHz − 600 kHz = 455 kHz). The 455 kHz is the intermediate frequency (IF) for standard AM receivers. No matter which

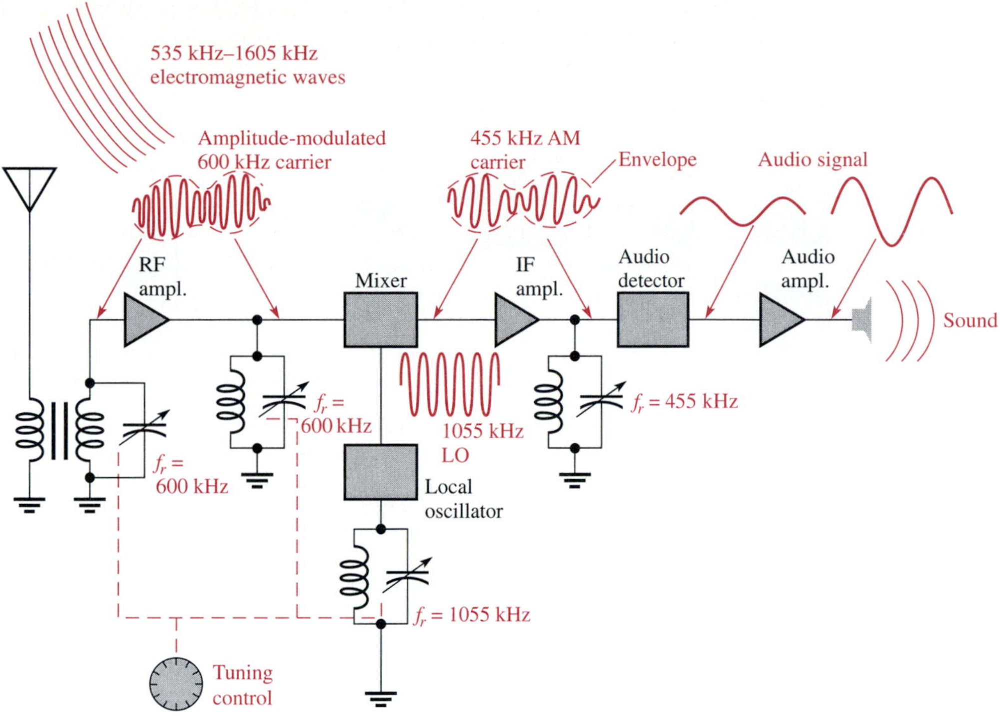


**FIGURE 18–53**

***A simplified diagram of a superheterodyne AM radio broadcast receiver showing an example of the application of tuned resonant circuits.***

station within the broadcast band is selected, its frequency is always converted to the 455 kHz IF. The amplitude-modulated IF is applied to an *audio detector* which removes the IF, leaving only the envelope or audio signal. The audio signal is then amplified and applied to the speaker.

**SECTION 18–8 REVIEW**

1. Generally, why is a tuned filter necessary when a signal is coupled from an antenna to the input of a receiver?
2. What is a wave trap?
3. What is meant by *ganged tuning*?

## 18–9 ■ TECHnology Theory Into Practice

***In the Chapter 11 TECH TIP, you worked with a receiver system to learn basic ac measurements. In this chapter, the receiver is again used to illustrate one application of resonant circuits. We will focus on a part of the "front end" of the receiver system that contains resonant circuits. Generally, the front end includes the RF amplifier, the local oscillator, and the mixer. In this TECH TIP, the RF amplifier is the focus. A knowledge of amplifier circuits is not necessary at this time.***

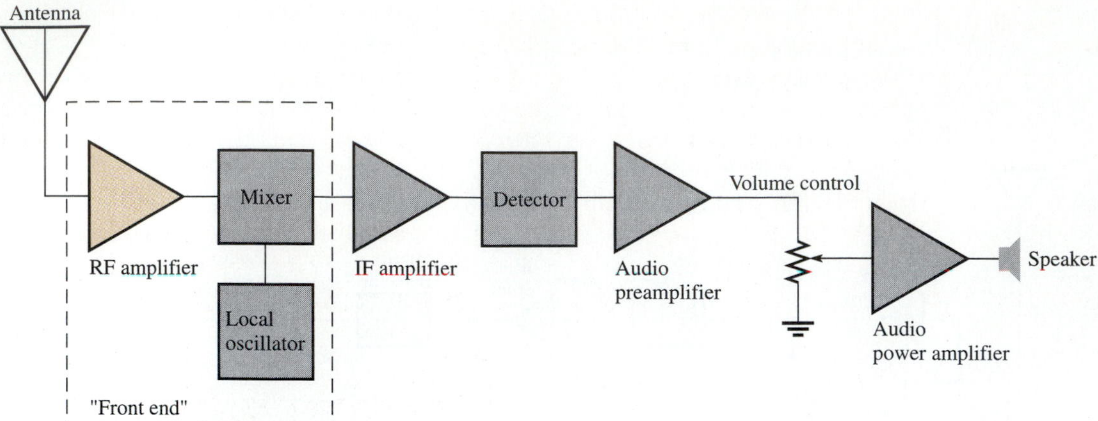


**FIGURE 18–54**
***Simplified block diagram of a basic radio receiver.***

A basic block diagram of an AM radio receiver is shown in Figure 18–54. In this particular system, the "front end" includes the circuitry used for tuning in a desired broadcasting station by frequency selection and then converting that selected frequency to a standard intermediate frequency (IF). AM radio stations transmit in the frequency range from 535 kHz to 1605 kHz. The purpose of the RF amplifier, which is the focus of this TECH TIP, is to take the signals picked up by the antenna, reject all but the signal from the desired station, and amplify it to a higher level.

A schematic of the RF amplifier is shown in Figure 18–55. The parallel resonant tuning circuit consists of $L$, $C_1$ and $C_2$. This particular RF amplifier does not have a resonant circuit on the output. $C_1$ is a varactor, which is a semiconductor device that you will learn more about in a later course. All that you need to know at this point is that the varactor is basically a variable capacitor whose capacitance is varied by changing the dc voltage across it. In this circuit, the dc voltage comes from the wiper of the potentiometer used for tuning the receiver.

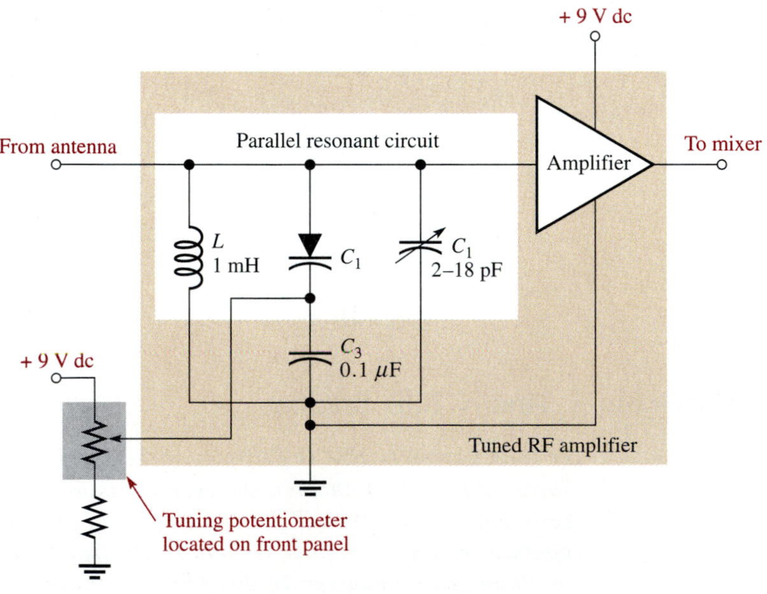


**FIGURE 18–55**
***Partial schematic of the RF amplifier showing the resonant tuning circuit.***

The voltage from the potentiometer can be varied from +1 V to +9 V. The particular varactor used in this circuit can be varied from 200 pF at 1 V to 5 pF at 9 V. The capacitor $C_2$ is a trimmer capacitor that is used for initially adjusting the resonant circuit. Once it is preset, it is left at that value. $C_1$ and $C_2$ are in parallel and their capacitances add to produce the total capacitance for the resonant circuit. $C_3$ has no effect on the resonant circuit. The purpose of $C_3$ is to allow the dc voltage to be applied to the varactor while providing an ac ground.

In this TECH TIP, you will work with the RF amplifier circuit board in Figure 18–56. Although all of the amplifier components are on the board, the part that you are to focus on is the resonant circuit indicated by the highlighted area.

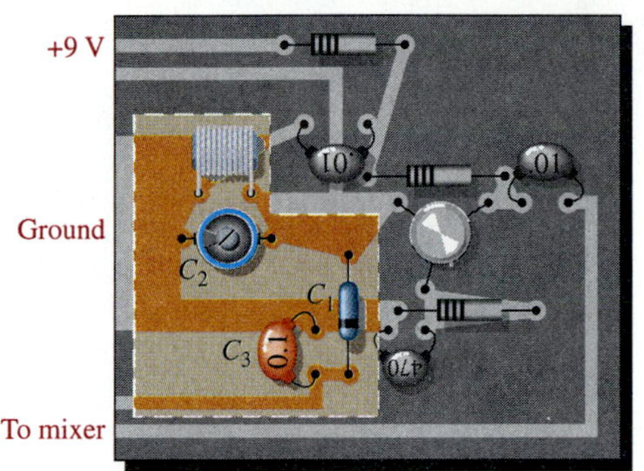


**FIGURE 18–56**
***RF amplifier circuit board.***

## Capacitance in the Resonant Circuit

- ☐ Calculate a capacitance setting for $C_2$ that will ensure a complete coverage of the AM frequency band as the varactor is varied over its capacitance range. The full range of resonant frequencies for the tuning circuit should more than cover the AM band, so that at the maximum varactor capacitance, the resonant frequency will be less than 535 kHz and at the minimum varactor capacitance, the resonant frequency will be greater than 1605 kHz.
- ☐ Using the value of $C_2$ that you have calculated, determine the values of the varactor capacitance that will produce a resonant frequency of 535 kHz and 1605 kHz, respectively.

## Testing the Resonant Circuit

□ Suggest a procedure for testing the resonant circuit using the instruments in the Test Bench setup of Figure 18–57. Develop a test setup by creating a point-to-point hookup of the board and the instruments.

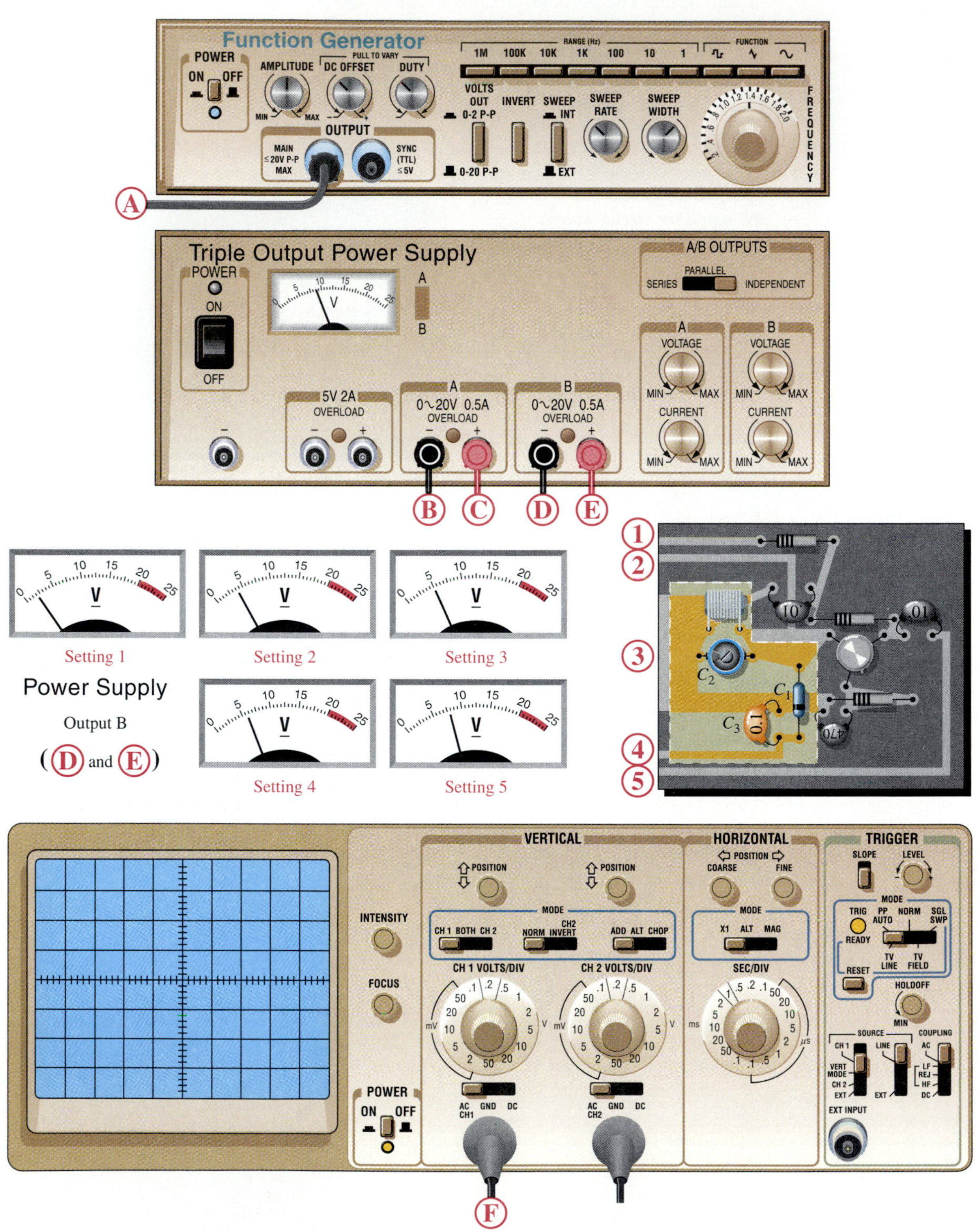


FIGURE 18–57
*Test bench setup. The scope probes are ×1.*

☐ Using the graph in Figure 18–58 that shows the variation in varactor capacitance versus varactor voltage, determine the resonant frequency for each indicated setting from the *B* outputs of the dc power supply (rightmost output terminals). The *B* output of the power supply is used to simulate the potentiometer voltage.

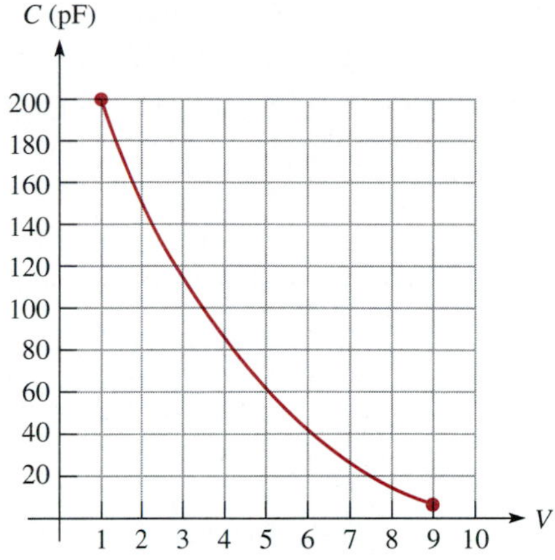


**FIGURE 18–58**
***Varactor capacitance versus voltage.***

> **SECTION 18–9 REVIEW**
>
> **1.** What is the AM frequency range?
>
> **2.** State the purpose of the RF amplifier.
>
> **3.** How is a particular frequency in the AM band selected?

## 18–10 ■ PSpice ANALYSIS OF *RLC* CIRCUITS

***The techniques you have used to analyze RC and RL circuits using PSpice and Probe can also be used to analyze RLC circuits. Both series and parallel RLC circuits will be covered. After completing this section, you should be able to***

- **Use PSpice and Probe to determine the characteristics of *RLC* circuits**
  - ☐ Determine the resonant frequency and bandwidth of a series *RLC* circuit
  - ☐ Determine the resonant frequency and bandwidth of a parallel *RLC* circuit
  - ☐ Analyze the effects of winding resistance on the frequency response of *RLC* circuits

### Series *RLC* Circuit Analysis

A series *RLC* circuit is shown in Figure 18–59. As you know, the reactances of the inductor and capacitor oppose each other and at the resonant frequency they are equal and cancel completely. At the resonant (center) frequency, the circuit has a minimum impedance and, therefore, a maximum current.

**FIGURE 18–59**

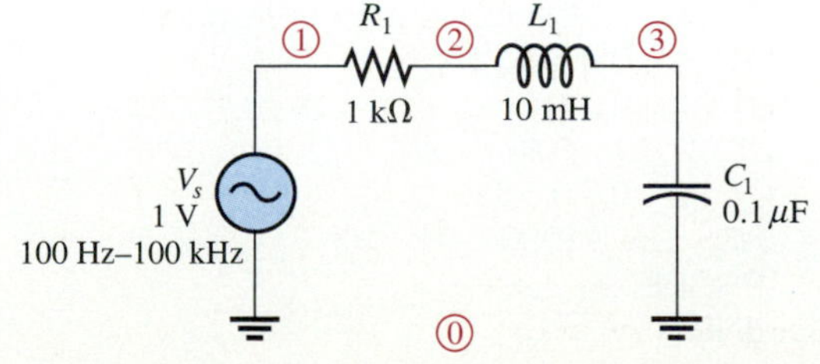

The following circuit file applies to the circuit in Figure 18–59:

```
RLC Frequency Analysis 1
*Series RLC Circuit Frequency Response
VS  1  0  AC  1
R1  1  2  1K
L1  2  3  10 M
C1  3  0  0.1U
.AC  LIN  1000  100  100K
.PROBE
.END
```

After running PSpice on the file, you can use Probe to view the voltages across the resistor, capacitor, and inductor. At approximately 5 kHz, the voltages across the capacitor and inductor are equal and the voltage across the resistor reaches a maximum of 1 V, which is the magnitude of the source voltage. This is the resonant frequency of the circuit at which the reactances cancel each other and the circuit impedance is purely resistive. To see this, remove the traces and then add

```
IP(VS)
```

This should show the phase angle of the total circuit current, but does it? At about 5 kHz, the phase angle of the current should be 0°, indicating a purely resistive circuit because the impedance phase angle is also 0°. Instead, the phase angle appears to be −180°. The reason is that current is out of node 1, which is the first of the two nodes $V_s$ is connected across. PSpice considers positive current into the first node, through the component, and out of the second node. In the ac analysis, this shows up as a 180° phase difference. To correct this, tell Probe to plot the phase of the current through any resistor directly in series with the source. In this case, $R_1$ can be used. Another way is to put a very small value resistor (1 pΩ) in series with the source. This will not affect the operation of a circuit. When you remove the trace and replace it with IP(R1), the correct phase angle of the total circuit current is shown.

***Bandwidth*** As you know, series resonant circuits have a bandwidth, which defines a region on either side of the resonant frequency where the total current is equal to or greater than 70.7% of the maximum current. Remove all traces and then display the voltage source current I(VS). In this case the maximum current is 1 mA, so you can move the cursors (using the RIGHT and LEFT ARROW and SHIFT keys) to find the upper and lower critical frequencies at which the current is equal to 707 μA. If you use cursor 1 to mark the upper critical frequency and cursor 2 to mark the lower critical frequency, Probe will automatically calculate the difference between them, which is the bandwidth. For the circuit in Figure 18–59, the bandwidth is about 16 kHz.

## Parallel *LC* Circuit Analysis

As you know, in a parallel *LC* circuit, the currents through the capacitor and inductor oppose each other because the capacitor current leads the voltage by 90° and the inductor current lags by 90°. These currents cancel at the resonant frequency, and the impedance is maximum (ideally infinite).

A circuit file for the parallel resonant circuit in Figure 18–60 is as follows:

```
RLC Frequency Analysis 2
*Parallel LC Circuit Frequency Response
VS  1  0  AC  1
R1  1  2  1K
L1  2  0  10M
C1  2  0  0.1U
.AC  LIN  1000  1K  10K
.PROBE
.END
```

**FIGURE 18–60**

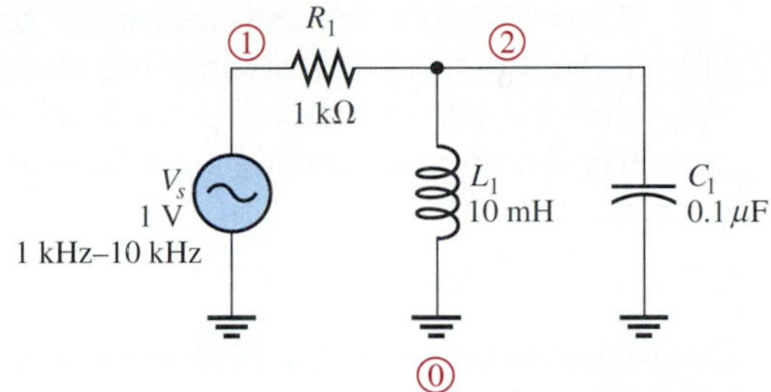


With PSpice and Probe you can view the currents in the resistor (total current), capacitor, and inductor. At the resonant frequency of 5.03 kHz, the inductor and capacitor currents are equal and cancel each other so the total current is minimum (ideally zero). At this point the impedance is maximum, and $R_1$ adds insignificantly to this. When the impedance of the *LC* circuit decreases to a low value at frequencies well above and below the resonant frequency, the current is limited mostly by the resistor.

***Bandwidth*** Again, you can use the cursors to find the bandwidth. For a parallel resonant circuit, the 70.7% rule cannot be applied to the current to determine the critical frequencies as in the series *RLC* circuit. Instead, you must use 70.7% of the voltage across the parallel inductor and capacitor. To do this, remove all traces as before and then display the voltage:

```
V(2,0)
```

Using the cursors, you will see that the voltage peaks at 1 V at approximately 5.03 kHz with a bandwidth of 1.59 kHz.

## Practical Parallel *RLC* Circuits

A major difficulty in the analysis of practical *RLC* circuits is the winding resistance of the inductor. Although, as you know, there are formulas for finding parallel resonant circuit values when the winding resistance is included, Probe allows you to find the resonant frequency and bandwidth of practical circuits using the same methods as for ideal circuits.

Consider the circuit in Figure 18–61, which is the same as the circuit in Figure 18–60 except that the inductor has a 10 Ω winding resistance. The circuit file is as follows:

```
RLC Frequency Analysis 3
*Parallel RLC Circuit Frequency Response
VS  1  0  AC  1
R1  1  2  1K
L1  2  3  10M
RW  3  0  10
C1  2  0  0.1U
.AC  LIN  1000  1K  10K
.PROBE
.END
```

**FIGURE 18–61**

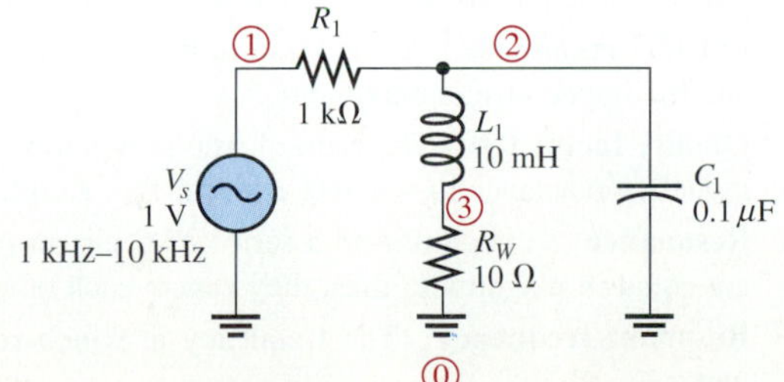

When PSpice is run and the voltage across the tank circuit, V(2,0), is plotted, you will see that the winding resistance has changed the maximum voltage to about 910 mV rather than the 1 V maximum for the ideal circuit. Careful inspection will show that the curve has shifted and flattened out a little as well. To see this, remove the voltage trace and replace it with

```
(V(2,0)/0.910V)*100
```

This scales the response as a percentage of the maximum voltage. Using the cursors, you will find that the resonant frequency has shifted by about 30 Hz to 5.05 kHz and the bandwidth has increased to approximately 1.75 kHz. The larger the winding resistance is in a given *RLC* circuit, the more the resonant frequency will shift and the flatter the response will be. You can observe this effect by running PSpice and Probe for larger values for RW in the circuit file.

## SECTION 18–10 REVIEW

1. Rewrite the circuit file for the circuit in Figure 18–59 for a winding resistance of 20 Ω.
2. Rewrite the circuit file for the circuit in Figure 18–60 for a winding resistance of 15 Ω and the addition of another parallel capacitor with a value of 0.22 μF.

## ■ SUMMARY

- $X_L$ and $X_C$ have opposing effects in an *RLC* circuit.
- In a series *RLC* circuit, the larger reactance determines the net reactance of the circuit.
- At series resonance, the inductive and capacitive reactances are equal.
- The impedance of a series *RLC* circuit is purely resistive at resonance.
- In a series *RLC* circuit, the current is maximum at resonance.
- The reactive voltages $V_L$ and $V_C$ cancel at resonance in a series *RLC* circuit because they are equal in magnitude and 180° out of phase.
- In a parallel *RLC* circuit, the smaller reactance determines the net reactance of the circuit.
- In a parallel resonant circuit, the impedance is maximum at the resonant frequency.
- A parallel resonant circuit is commonly called a *tank circuit.*
- The impedance of a parallel resonant circuit is purely resistive at resonance.
- The bandwidth of a series resonant circuit is the range of frequencies for which the current is $0.707I_{max}$ or greater.
- The bandwidth of a parallel resonant circuit is the range of frequencies for which the impedance is $0.707Z_{max}$ or greater.
- The critical frequencies are the frequencies above and below resonance where the circuit response is 70.7% of the maximum response.
- A higher $Q$ produces a narrower bandwidth.

## ■ GLOSSARY

**Half-power frequency** The frequency at which the output power of a filter is 50% of the maximum (the output voltage is 70.7% of maximum); another name for *critical* or *cutoff frequency.*

**Parallel resonance** A condition in a parallel *RLC* circuit in which the reactances ideally cancel and the impedance is maximum.

**Quality factor ($Q$)** The ratio of true power to reactive power in a resonant circuit or the ratio of inductive reactance to winding resistance in a coil.

**Resonance** A condition in a series *RLC* circuit in which the capacitive and inductive reactances are equal in magnitude; thus, they cancel each other and result in a purely resistive impedance.

**Resonant frequency** The frequency at which resonance occurs; also known as the *center frequency.*

**Selectivity** A measure of how effectively a filter passes certain desired frequencies and rejects all others. Generally, the narrower the bandwidth, the greater the selectivity.

**Series resonance** A condition in a series *RLC* circuit in which the reactances cancel and the impedance is minimum.

**Tank circuit** A parallel resonant circuit.

## ■ FORMULAS

### Series *RLC* Circuits

(18–1) $X_{tot} = |X_L - X_C|$

(18–2) $\mathbf{Z} = R + jX_L - jX_C$

(18–3) $\mathbf{Z} = \sqrt{R^2 + (X_L - X_C)^2}\angle\tan^{-1}\left(\frac{X_{tot}}{R}\right)$

### Series Resonance

(18–4) $\mathbf{Z}_r = R + jX_L - jX_C$

(18–5) $X_L = X_C$

(18–6) $Z_r = R$

(18–7) $f_r = \frac{1}{2\pi\sqrt{LC}}$

### Parallel *RLC* Circuits

(18–8) $\mathbf{Z} = \frac{1}{\frac{1}{R\angle 0^\circ} + \frac{1}{X_L\angle 90^\circ} + \frac{1}{X_C\angle -90^\circ}}$

(18–9) $\mathbf{G} = \frac{1}{R\angle 0^\circ} = G\angle 0^\circ$

(18–10) $\mathbf{B}_C = \frac{1}{X_C\angle -90^\circ} = B_C\angle 90^\circ = jB_C$

(18–11) $\mathbf{B}_L = \frac{1}{X_L\angle 90^\circ} = B_L\angle -90^\circ = -jB_L$

(18–12) $\mathbf{Y} = \frac{1}{Z\angle\pm\theta} = Y\angle\pm\theta = G + jB_C - jB_L$

(18–13) $\mathbf{I}_{tot} = \sqrt{I_R^2 + (I_C - I_L)^2}\angle\tan^{-1}\left(\frac{I_{CL}}{I_R}\right)$

(18–14) $L_{eq} = L\left(\frac{Q^2 + 1}{Q^2}\right)$

(18–15) $R_{p(eq)} = R_W(Q^2 + 1)$

### Parallel Resonance

(18–16) $Z_r = R_W(Q^2 + 1)$

(18–17) $I_{tot} = \frac{V_s}{Z_r}$

(18–18) $f_r = \frac{1}{2\pi\sqrt{LC}}\sqrt{\frac{Q^2}{Q^2 + 1}}$

(18–19) $f_r = \frac{\sqrt{1 - (R_W^2 C/L)}}{2\pi\sqrt{LC}}$

(18–20) $Q_O = \frac{R_{p(tot)}}{X_{L(eq)}}$

(18–21) $$BW = f_2 - f_1$$

(18–22) $$f_r = \frac{f_1 + f_2}{2}$$

(18–23) $$BW = \frac{f_r}{Q}$$

## ■ SELF-TEST

**1.** The total reactance of a series *RLC* circuit at resonance is
**(a)** zero **(b)** equal to the resistance **(c)** infinity **(d)** capacitive

**2.** The phase angle of a series *RLC* circuit at resonance is
**(a)** $-90°$ **(b)** $+90°$ **(c)** $0°$ **(d)** dependent on the reactance

**3.** The impedance at the resonant frequency of a series *RLC* circuit with $L$ = 15 mH, $C$ = 0.015 $\mu$F, and $R_W$ = 80 Ω is
**(a)** 15 kΩ **(b)** 80 Ω **(c)** 30 Ω **(d)** 0 Ω

**4.** In a series *RLC* circuit that is operating below the resonant frequency, the current
**(a)** is in phase with the applied voltage **(b)** lags the applied voltage
**(c)** leads the applied voltage

**5.** If the value of $C$ in a series *RLC* circuit is increased, the resonant frequency
**(a)** is not affected **(b)** increases **(c)** remains the same **(d)** decreases

**6.** In a certain series resonant circuit, $V_C$ = 150 V, $V_L$ = 150 V, and $V_R$ = 50 V. The value of the source voltage is
**(a)** 150 V **(b)** 300 V **(c)** 50 V **(d)** 350 V

**7.** A certain series resonant circuit has a bandwidth of 1 kHz. If the existing coil is replaced with one having a lower value of $Q$, the bandwidth will
**(a)** increase **(b)** decrease **(c)** remain the same **(d)** be more selective

**8.** At frequencies below resonance in a parallel *RLC* circuit, the current
**(a)** leads the source voltages **(b)** lags the source voltage
**(c)** is in phase with the source voltage

**9.** The total current into the $L$ and $C$ branches of a parallel circuit at resonance is ideally
**(a)** maximum **(b)** low **(c)** high **(d)** zero

**10.** To tune a parallel resonant circuit to a lower frequency, the capacitance should be
**(a)** increased **(b)** decreased **(c)** left alone **(d)** replaced with inductance

**11.** The resonant frequency of a parallel circuit is approximately the same as a series circuit when
**(a)** the $Q$ is very low **(b)** the $Q$ is very high
**(c)** there is no resistance **(d)** either answer (b) or (c)

**12.** If the resistance in parallel with a parallel resonant circuit is reduced, the bandwidth
**(a)** disappears **(b)** decreases **(c)** becomes sharper **(d)** increases

## ■ PROBLEMS

### SECTION 18–1 Impedance of Series *RLC* Circuits

**1.** A certain series *RLC* circuit has the following values: $R$ = 10 Ω, $C$ = 0.05 $\mu$F, and $L$ = 5 mH. Determine the impedance in polar form. What is the net reactance? The source frequency is 5 kHz.

**2.** Find the impedance in Figure 18–62, and express it in polar form.

FIGURE 18–62

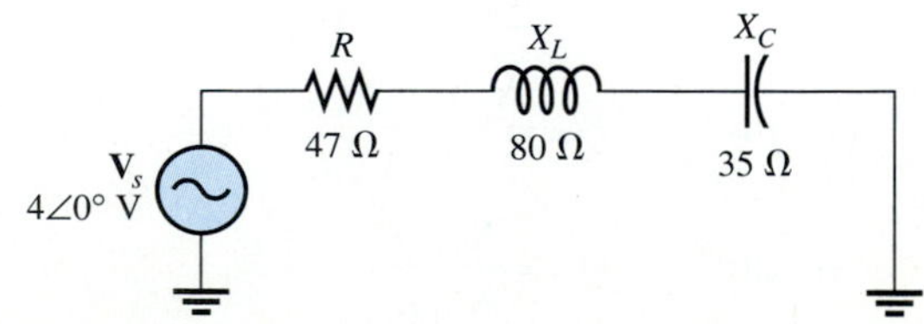

3. If the frequency of the source voltage in Figure 18–62 is doubled from the value that produces the indicated reactances, how does the magnitude of the impedance change?
4. For the circuit of Figure 18–62, determine the net reactance that will make the impedance magnitude equal to 100 Ω.

## SECTION 18–2 Analysis of Series *RLC* Circuits

5. For the circuit in Figure 18–62, find $\mathbf{I}_{tot}$, $\mathbf{V}_R$, $\mathbf{V}_L$, and $\mathbf{V}_C$ in polar form.
6. Sketch the voltage phasor diagram for the circuit in Figure 18–62.
7. Analyze the circuit in Figure 18–63 for the following ($f$ = 25 kHz):
   **(a)** $\mathbf{I}_{tot}$ **(b)** $P_{\text{true}}$ **(c)** $P_r$ **(d)** $P_a$

FIGURE 18–63

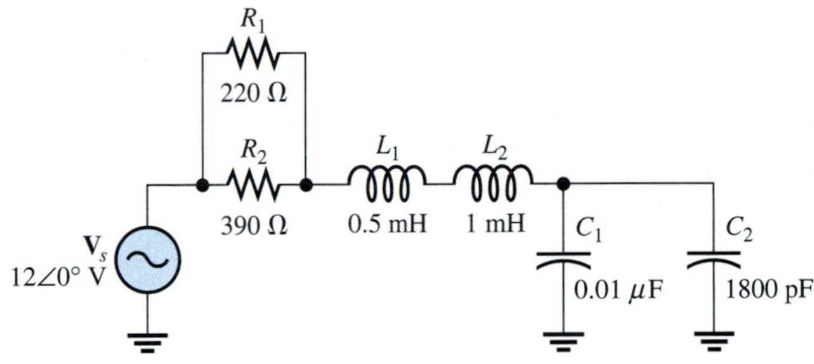


## SECTION 18–3 Series Resonance

8. Find $X_L$, $X_C$, $Z$ and $I$ at the resonant frequency in Figure 18–64.
9. A certain series resonant circuit has a maximum current of 50 mA and a $V_L$ of 100 V. The applied voltage is 10 V. What is $Z$? What are $X_L$ and $X_C$?
10. For the *RLC* circuit in Figure 18–65, determine the resonant frequency.

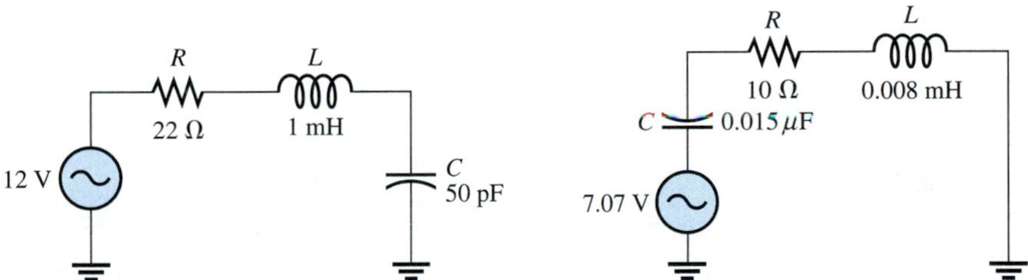


FIGURE 18–64

FIGURE 18–65

11. What is the value of the current at the half-power points in Figure 18–65?
12. Determine the phase angle between the applied voltage and the current at the critical frequencies in Figure 18–65. What is the phase angle at resonance?
13. Design a circuit in which the following series resonant frequencies are switch-selectable:
    **(a)** 500 kHz **(b)** 1000 kHz **(c)** 1500 kHz **(d)** 2000 kHz

## SECTION 18–4 Impedance of Parallel *RLC* Circuits

14. Express the impedance of the circuit in Figure 18–66 in polar form.

FIGURE 18–66

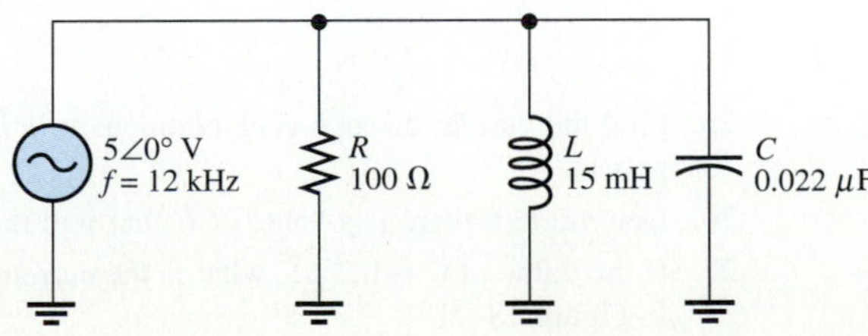

15. Is the circuit in Figure 18–66 capacitive or inductive? Explain.
16. At what frequency does the circuit in Figure 18–66 change its reactive characteristic (from inductive to capacitive or vice versa)?

## SECTION 18–5 Analysis of Parallel and Series-Parallel *RLC* Circuits

17. For the circuit in Figure 18–66, find all the currents and voltages in polar form.
18. Find the total impedance for each circuit in Figure 18–67.

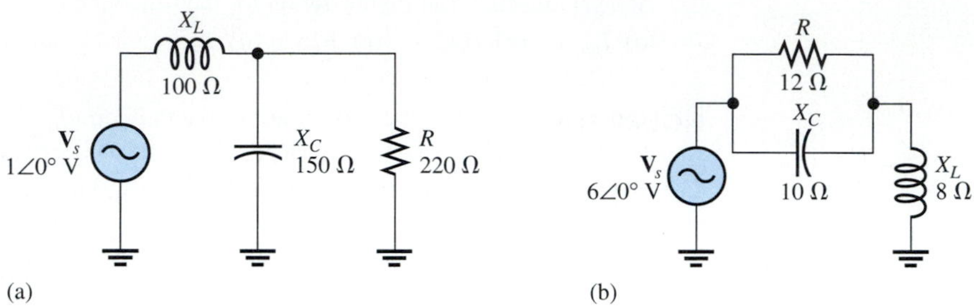


FIGURE 18–67

19. For each circuit in Figure 18–67, determine the phase angle between the source voltage and the total current.
20. Determine the voltage across each element in Figure 18–68, and express each in polar form.
21. Convert the circuit in Figure 18–68 to an equivalent series form.
22. What is the current through $R_2$ in Figure 18–69?

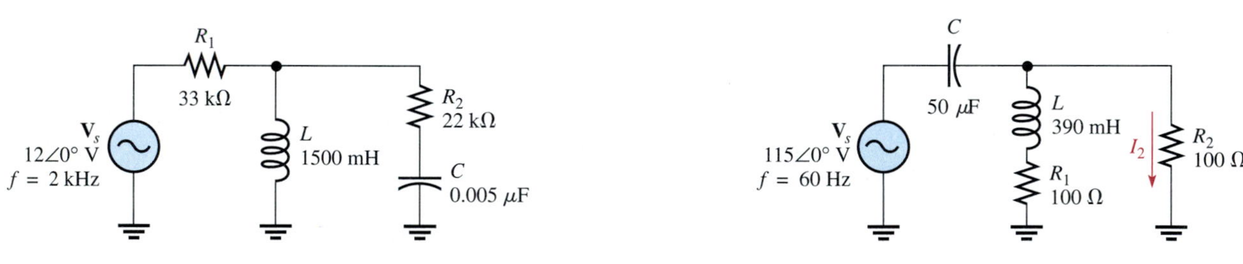


FIGURE 18–68

FIGURE 18–69

23. In Figure 18–69, what is the phase angle between $I_2$ and the source voltage?
24. Determine the total resistance and the total reactance in Figure 18–70.

FIGURE 18–70

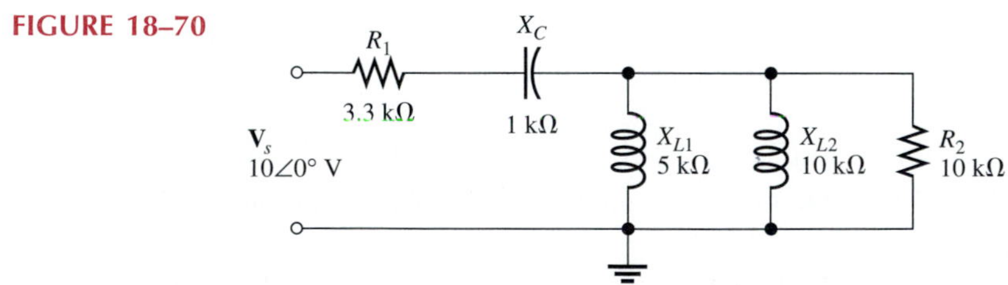


25. Find the current through each component in Figure 18–70. Find the voltage across each component.
26. Determine if there is a value of $C$ that will make $V_{ab} = 0$ V in Figure 18–71. If not, explain.
27. If the value of $C$ is 0.2 μF, what is the current through a 100 Ω resistor connected from $a$ to $b$ in Figure 18–71?

FIGURE 18–71

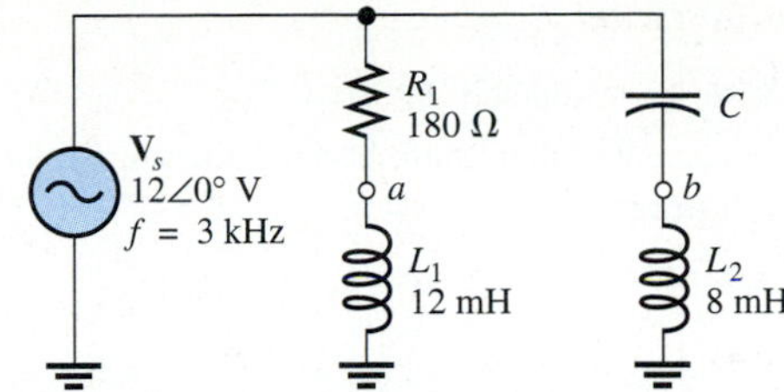


## SECTION 18–6 Parallel Resonance

**28.** What is the impedance of an ideal parallel resonant circuit (no resistance in either branch)?

**29.** Find $Z$ at resonance and $f_r$ for the tank circuit in Figure 18–72.

**30.** How much current is drawn from the source in Figure 18–72 at resonance? What are the inductive current and the capacitive current at the resonant frequency?

**31.** Determine the resonant frequencies and the output voltage at each frequency in Figure 18–73.

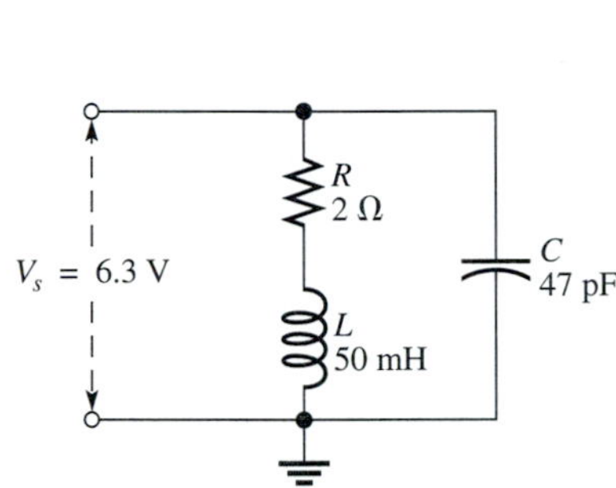


FIGURE 18–72

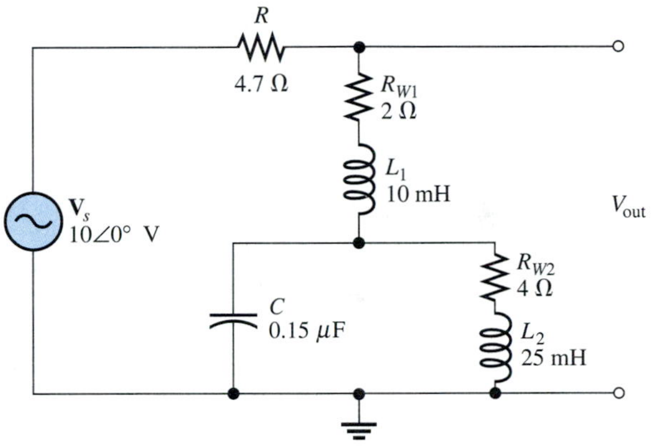

FIGURE 18–73

**32.** Design a parallel-resonant network using a single coil and switch-selectable capacitors to produce the following resonant frequencies: 8 MHz, 9 MHz, 10 MHz, and 11 MHz. Assume a 10 μH coil with a winding resistance of 5 Ω.

## SECTION 18–7 Bandwidth of Resonant Circuits

**33.** At resonance, $X_L = 2\ \text{k}\Omega$ and $R_W = 25\ \Omega$ in a parallel *RLC* circuit. The resonant frequency is 5 kHz. Determine the bandwidth.

**34.** If the lower critical frequency is 2400 Hz and the upper critical frequency is 2800 Hz, what is the bandwidth? What is the resonant frequency?

**35.** In a certain *RLC* circuit, the power at resonance is 2.75 W. What is the power at the lower critical frequency?

**36.** What values of $L$ and $C$ should be used in a tank circuit to obtain a resonant frequency of 8 kHz? The bandwidth must be 800 Hz. The winding resistance of the coil is 10 Ω.

**37.** A parallel resonant circuit has a $Q$ of 50 and a $BW$ of 400 Hz. If $Q$ is doubled, what is the bandwidth for the same $f_r$?

## ▪ ANSWERS TO SECTION REVIEWS

### Section 18–1

**1.** $X_{tot} = 70\ \Omega$; capacitive

**2.** $\mathbf{Z} = 84.3\angle{-56.1°}\ \Omega$; $Z = 84.3\ \Omega$; $\theta = -56.1°$; current is leading $V_s$.

### Section 18–2

**1.** $\mathbf{V}_s = 38.4\angle{-21.3°}$ V

**2.** Current leads the voltage.

**3.** $X_{tot} = 6\ \Omega$

### Section 18–3

1. For series resonance, $X_L = X_C$.
2. The current is maximum because the impedance is minimum.
3. $f_r = 159$ kHz
4. The circuit is capacitive.

### Section 18–4

1. The circuit is capacitive.
2. $\mathbf{Y} = 1.54\angle 49.4°$ mS
3. $\mathbf{Z} = 651\angle -49.4°\ \Omega$

### Section 18–5

1. $I_R = 80$ mA, $I_C = 120$ mA, $I_L = 240$ mA
2. The circuit is capacitive.
3. $L_{eq} = 20.1$ mH, $R_{p(eq)} = 1.59$ kΩ

### Section 18–6

1. Impedance is maximum at parallel resonance.
2. The current is minimum.
3. $X_C = 1500\ \Omega$
4. $f_r = 225$ kHz; $Z = 1250$ MΩ
5. $f_r = 22.5$ kHz
6. $f_r = 20.9$ kHz
7. $Z_{tot} = 8020\ \Omega$

### Section 18–7

1. $BW = f_2 - f_1 = 400$ kHz
2. $f_r = 2$ MHz
3. $P_{f2} = 0.9$ W
4. Larger $Q$ means narrower $BW$.

### Section 18–8

1. A tuned filter is used to select a narrow band of frequencies.
2. A wave trap is a band-stop filter.
3. Ganged tuning is done with several capacitors (or inductors) whose values can be varied simultaneously with a common control.

### Section 18–9

1. The AM frequency range is 535 kHz to 1605 kHz.
2. The RF amplifier rejects all signals but the one from the desired station. It then amplifies the selected signal.
3. A particular AM frequency is selected by varying the varactor capacitance with a dc voltage.

### Section 18–10

1. 
```
RLC Frequency Analysis 1
VS  1  0  AC  1
R1  1  2  1K
RW  2  3  20
L1  3  4  10M
C1  4  0  0.1U
.AC  LIN  1000  100  100K
.PROBE
.END
```

**2.** RLC Frequency Analysis 2

```
VS 1 0 AC 1
R1 1 2 1K
RW 3 0 15
L1 2 3 10M
C1 2 0 0.1U
C2 2 0 0.22U
.AC LIN 1000 1K 10K
.PROBE
.END
```

## ANSWERS TO RELATED EXERCISES FOR EXAMPLES

**18–1** $\mathbf{Z} = 12.3\angle 63.0^\circ\ \Omega$

**18–2** $\mathbf{Z} = 4.64\angle 44.7^\circ\ \text{k}\Omega$. See Figure 18–74.

**18–3** Current will increase with frequency to a certain point and then it will decrease.

**18–4** The circuit is more capacitive.

**18–5** $f_r = 22.5$ kHz

**18–6** $Z$ increases; $Z$ increases.

**18–7** $I = 2.27$ A

**18–8** $Z$ decreases.

**18–9** Inductive

**18–10** $I_{tot}$ increases.

**18–11** $\mathbf{V}_C = 9.30\angle -65.8^\circ$ V

**18–12** $\mathbf{V}_{C1} = 27.1\angle -81.1^\circ$ V

**18–13** $R_{p(\text{eq})} = 25\ \text{k}\Omega$, $L_{\text{eq}} = 5$ mH; $C = 0.02\ \mu$F

**18–14** $Z_r = 79.9\ \text{k}\Omega$

**18–15** Less

**18–16** $I = 35.4$ mA

**18–17** $f_1 = 6.75$ kHz; $f_2 = 9.25$ kHz

**18–18** $BW = 7.96$ kHz

FIGURE 18–74

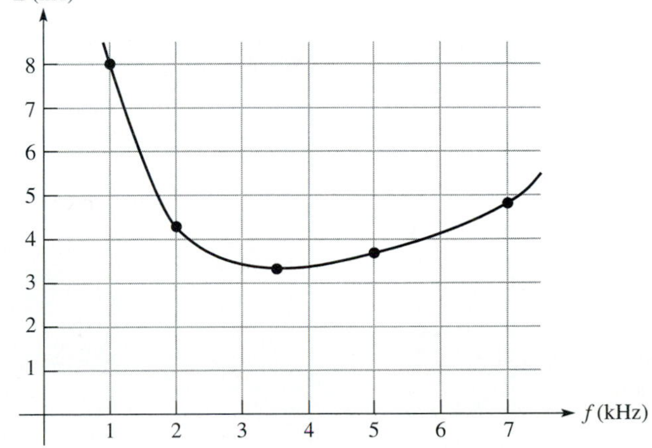

# 19

# BASIC FILTERS



## ■ INTRODUCTION

The concept of filters was introduced in Chapters 16, 17, and 18 to illustrate applications of *RC, RL,* and *RLC* circuits. This chapter is essentially an extension of the earlier material and provides additional coverage of the important topic of filters. In this chapter, you will learn the basics of putting technology theory into practice.

Passive filters are discussed in this chapter. Passive filters use various combinations of resistors, capacitors, and inductors. In a later course, you will study active filters which use passive components combined with amplifiers. You have already seen how basic *RC, RL,* and *RLC* circuits can be used as filters. Now, you will learn that passive filters can be placed in four general categories according to their response characteristics: low-pass, high-pass, band-pass, and band-stop. Within each category, there are several common types that will be examined.

■ **TECHnology Theory Into Practice**

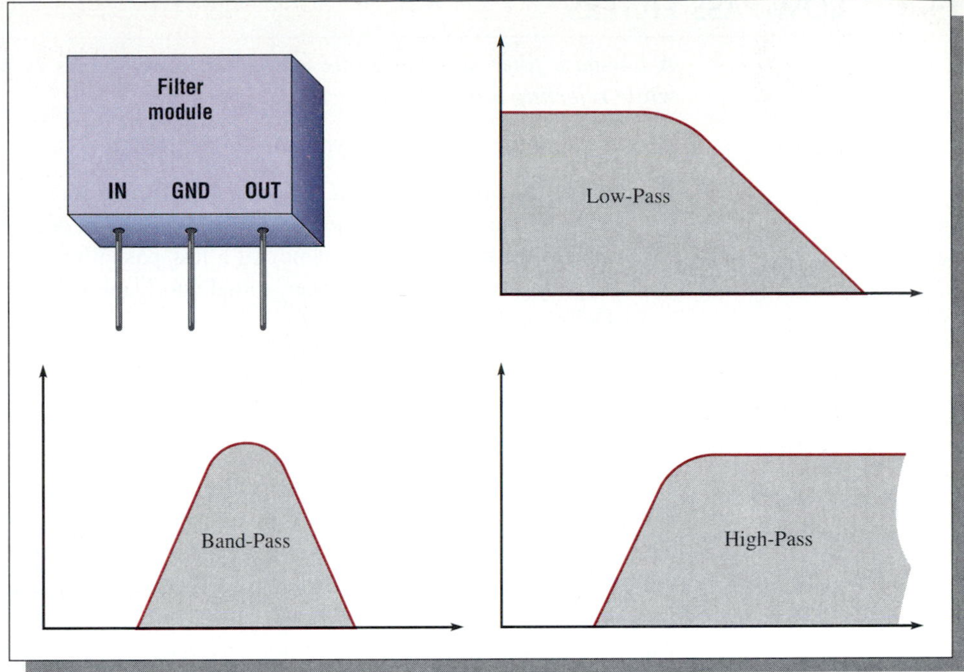


In the TECH TIP assignment in Section 19–5, you will plot the frequency responses of filters based on oscilloscope measurements and identify the types of filters.

## CHAPTER OBJECTIVES

- ☐ Analyze the operation of *RC* and *RL* low-pass filters
- ☐ Analyze the operation of *RC* and *RL* high-pass filters
- ☐ Analyze the operation of band-pass filters
- ☐ Analyze the operation of band-stop filters
- ☐ Use PSpice and Probe to analyze simple filter circuits (optional)

## 19–1 ■ LOW-PASS FILTERS

***A low-pass filter allows signals with lower frequencies to pass from input to output while rejecting higher frequencies.***

***After completing this section, you should be able to***

- **Analyze the operation of *RC* and *RL* low-pass filters**
  - ☐ Express the voltage and power ratios of a filter in decibels
  - ☐ Determine the critical frequency of a low-pass filter
  - ☐ Explain the difference between actual and ideal low-pass response curves
  - ☐ Define *roll-off*
  - ☐ Generate a Bode plot for a low-pass filter
  - ☐ Discuss phase shift in a low-pass filter

Figure 19–1 shows a block diagram and a general response curve for a low-pass filter. The range of low frequencies passed by a low-pass filter within a specified limit is called the **passband** of the filter. The point considered to be the upper end of the passband is at the critical frequency, $f_c$, as illustrated in Figure 19–1(b). The **critical frequency** is the frequency at which the filter's output voltage is 70.7% of the maximum. The filter's critical frequency is also called the *cutoff frequency, break frequency,* or *−3 dB frequency* because the output voltage is down 3 dB from its maximum at this frequency. The term *dB (decibel)* is a commonly used one that you should understand because the decibel unit is used in filter measurements.

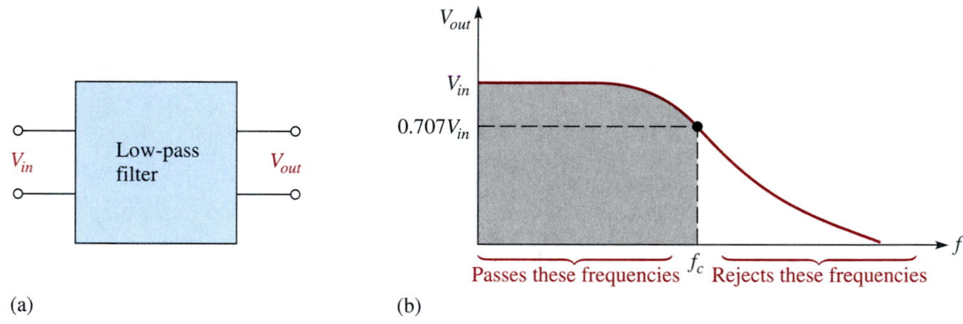


FIGURE 19–1
***Low-pass filter block diagram and general response curve.***

### Decibels

The basis for the decibel unit stems from the logarithmic response of the human ear to the intensity of sound. The **decibel** is a logarithmic measurement of the ratio of one power to another or one voltage to another, which can be used to express the input-to-output relationship of a filter. The following equation expresses a voltage ratio in decibels:

$$\text{dB} = 20 \log\left(\frac{V_{out}}{V_{in}}\right) \qquad \textbf{(19–1)}$$

The following equation is the decibel formula for a power ratio:

$$\text{dB} = 10 \log\left(\frac{P_{out}}{P_{in}}\right) \qquad \textbf{(19–2)}$$

**EXAMPLE 19–1** At a certain frequency, the output voltage of a filter is 5 V and the input is 10 V. Express the voltage ratio in decibels.

***Solution***

$$20 \log\left(\frac{V_{out}}{V_{in}}\right) = 20 \log\left(\frac{5\text{ V}}{10\text{ V}}\right) = 20 \log(0.5) = -6.02\text{ dB}$$

The calculator sequence is

5 ÷ 1 0 = log × 2 0 =

***Related Exercise*** Express the ratio $V_{out}/V_{in} = 0.85$ in decibels.

## *RC* Low-Pass Filter

A basic *RC* low-pass filter is shown in Figure 19–2. Notice that the output voltage is taken across the capacitor.

**FIGURE 19–2**

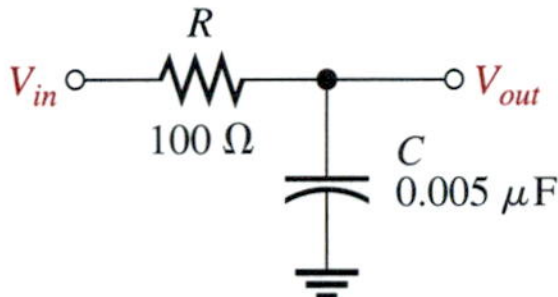


When the input is dc (0 Hz), the output voltage equals the input voltage because $X_C$ is infinitely large. As the input frequency is increased, $X_C$ decreases and, as a result, $V_{out}$ gradually decreases until a frequency is reached where $X_C = R$. This is the critical frequency, $f_c$, of the filter.

$$X_C = R$$

$$\frac{1}{2\pi f_c C} = R$$

$$f_c = \frac{1}{2\pi RC} \qquad \textbf{(19–3)}$$

At the critical frequency, the output voltage magnitude is

$$V_{out} = \left(\frac{X_C}{\sqrt{R^2 + X_C^2}}\right)V_{in}$$

by application of the voltage-divider formula. Since $X_C = R$ at $f_c$, the output voltage can be expressed as

$$V_{out} = \left(\frac{R}{\sqrt{R^2 + R^2}}\right)V_{in} = \left(\frac{R}{\sqrt{2R^2}}\right)V_{in}$$

$$= \left(\frac{R}{R\sqrt{2}}\right)V_{in} = \left(\frac{1}{\sqrt{2}}\right)V_{in} = 0.707V_{in}$$

These calculations show that the output is 70.7% of the input when $X_C = R$. The frequency at which this occurs is, by definition, the critical frequency.

The ratio of output voltage to input voltage at the critical frequency can be expressed in decibels as follows.

$$V_{out} = 0.707V_{in}$$

$$\frac{V_{out}}{V_{in}} = 0.707$$

$$20\log\left(\frac{V_{out}}{V_{in}}\right) = 20\log(0.707) = -3\text{ dB}$$

**EXAMPLE 19–2** Determine the critical frequency for the low-pass $RC$ filter in Figure 19–2.

***Solution***

$$f_c = \frac{1}{2\pi RC} = \frac{1}{2\pi(100\ \Omega)(0.005\ \mu\text{F})} = 318\text{ kHz}$$

The output voltage is 3 dB below $V_{in}$ at this frequency ($V_{out}$ has a maximum value of $V_{in}$).

***Related Exercise*** A certain low-pass $RC$ filter has $R = 1\text{ k}\Omega$ and $C = 0.022\ \mu\text{F}$. Determine its critical frequency.

## "Roll-Off" of the Response Curve

The blue line in Figure 19–3 shows an actual response curve for a low-pass filter. The maximum output is defined to be 0 dB as a reference. Zero decibels corresponds to $V_{out} = V_{in}$, because $20\log(V_{out}/V_{in}) = 20\log 1 = 0$ dB. The output drops from 0 dB to −3 dB at the critical frequency and then continues to decrease at a fixed rate. This pattern of decrease is called the **roll-off** of the frequency response. The red line shows an ideal output response that is considered to be "flat" out to $f_c$. The output then decreases at the fixed rate.

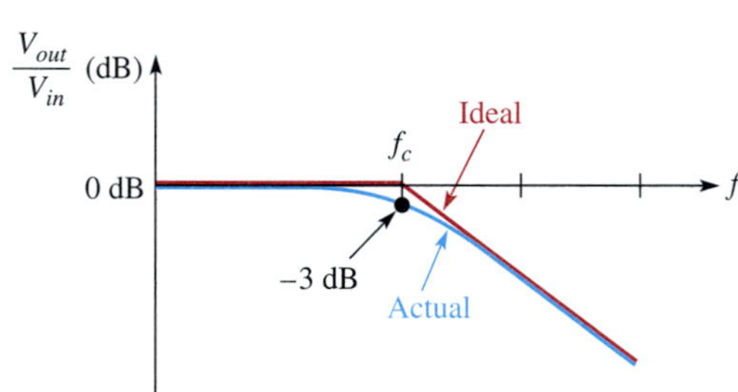


**FIGURE 19–3**
***Actual and ideal response curves for a low-pass filter.***

As you have seen, the output voltage of a low-pass filter decreases by 3 dB when the frequency is increased to the critical value $f_c$. As the frequency continues to increase above $f_c$, the output voltage continues to decrease. In fact, for each tenfold increase in frequency above $f_c$, there is a 20 dB reduction in the output, as shown in the following steps.

Let's take a frequency that is ten times the critical frequency ($f = 10f_c$). Since $R = X_C$ at $f_c$, then $R = 10X_C$ at $10f_c$ because of the inverse relationship of $X_C$ and $f$.

The **attenuation** of the $RC$ circuit is the ratio $V_{out}/V_{in}$ and is developed as follows:

$$\frac{V_{out}}{V_{in}} = \frac{X_C}{\sqrt{R^2 + X_C^2}} = \frac{X_C}{\sqrt{(10X_C)^2 + X_C^2}}$$

$$= \frac{X_C}{\sqrt{100X_C^2 + X_C^2}} = \frac{X_C}{\sqrt{X_C^2(100 + 1)}} = \frac{X_C}{X_C\sqrt{101}} = \frac{1}{\sqrt{101}} \cong \frac{1}{10} = 0.1$$

The dB attenuation is

$$20 \log\left(\frac{V_{out}}{V_{in}}\right) = 20 \log(0.1) = -20 \text{ dB}$$

A tenfold change in frequency is called a **decade.** So, for the *RC* circuit, the output voltage is reduced by 20 dB for each decade increase in frequency. A similar result can be derived for the high-pass circuit. The roll-off is a constant 20 dB/decade for a basic *RC* or *RL* filter. Figure 19–4 shows a frequency response plot on a semilog scale, where each interval on the horizontal axis represents a tenfold increase in frequency. This response curve is called a **Bode plot.**

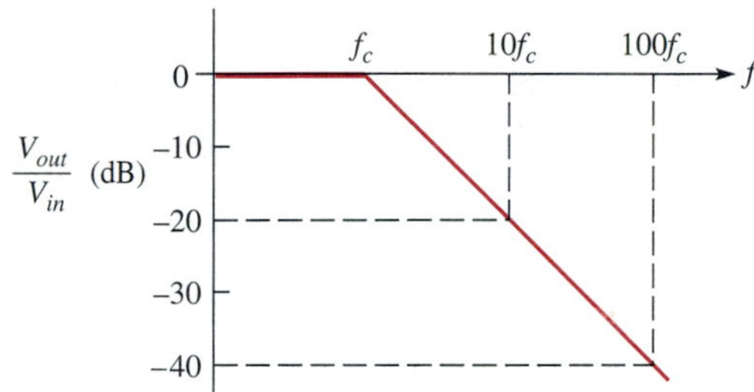


FIGURE 19–4
***Frequency roll-off for a low-pass RC filter (Bode plot).***

**EXAMPLE 19–3**

Make a Bode plot for the filter in Figure 19–5 for three decades of frequency. Use semilog graph paper.

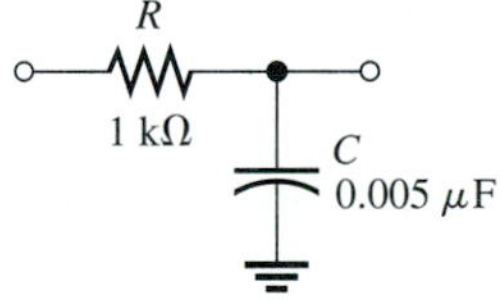


FIGURE 19–5

***Solution*** The critical frequency for this low-pass filter is

$$f_c = \frac{1}{2\pi RC} = \frac{1}{2\pi(1 \text{ k}\Omega)(0.005 \text{ }\mu\text{F})} = 31.8 \text{ kHz}$$

The idealized Bode plot is shown with the red line on the semilog graph in Figure 19–6 on the next page. The approximate actual response curve is shown with the blue line. Notice first that the horizontal scale is logarithmic and the vertical scale is linear. The frequency is on the logarithmic scale, and the filter output in decibels is on the vertical.

The output is flat below $f_c$ (31.8 kHz). As the frequency is increased above $f_c$, the output drops at a 20 dB/decade rate. Thus, for the ideal curve, every time the frequency is increased by ten, the output is reduced by 20 dB. A slight variation from this occurs in actual practice. The output is actually at –3 dB rather than 0 dB at the critical frequency.

***Related Exercise*** What happens to the critical frequency and roll-off rate if *C* is reduced to 0.001 $\mu$F in Figure 19–5?

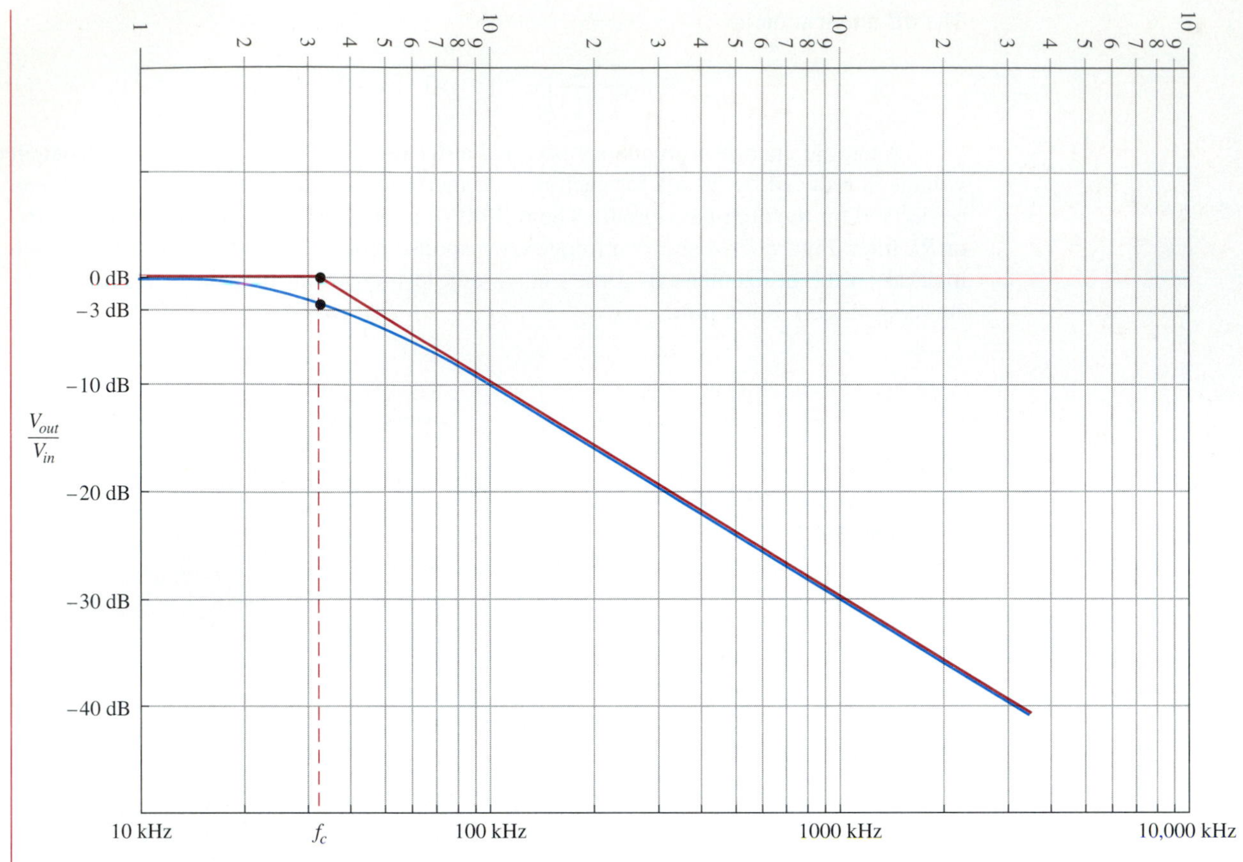


FIGURE 19–6

***Bode plot for Figure 19–5. The red line represents the ideal response curve and the blue line represents the actual response.***

## *RL* Low-Pass Filter

A basic *RL* low-pass filter is shown in Figure 19–7. Notice that the output voltage is taken across the resistor.

FIGURE 19–7

***RL low-pass filter.***



When the input is dc (0 Hz), the output voltage ideally equals the input voltage because $X_L$ is a short (if $R_W$ is neglected). As the input frequency is increased, $X_L$ increases and, as a result, $V_{out}$ gradually decreases until the critical frequency is reached. At this point, $X_L = R$ and the frequency is

$$2\pi f_c L = R$$

$$f_c = \frac{R}{2\pi L}$$

$$f_c = \frac{1}{2\pi(L/R)} \quad \textbf{(19–4)}$$

Just as in the *RC* low-pass filter, $V_{out} = 0.707V_{in}$ and, thus, the output voltage is down 3 dB at the critical frequency.

**EXAMPLE 19–4** Make a Bode plot for the filter in Figure 19–8 for three decades of frequency. Use semilog graph paper.

**FIGURE 19–8**

*L* 4.7 mH, *R* 2.2 kΩ

***Solution*** The critical frequency for this low-pass filter is

$$f_c = \frac{1}{2\pi(L/R)} = \frac{1}{2\pi(4.7\text{ mH}/2.2\text{ k}\Omega)} = 74.5\text{ kHz}$$

The idealized Bode plot is shown with the red line on the semilog graph in Figure 19–9. The approximate actual response curve is shown with the blue line. Notice first that the horizontal scale is logarithmic and the vertical scale is linear. The frequency is on the logarithmic scale, and the filter output in decibels is on the vertical.

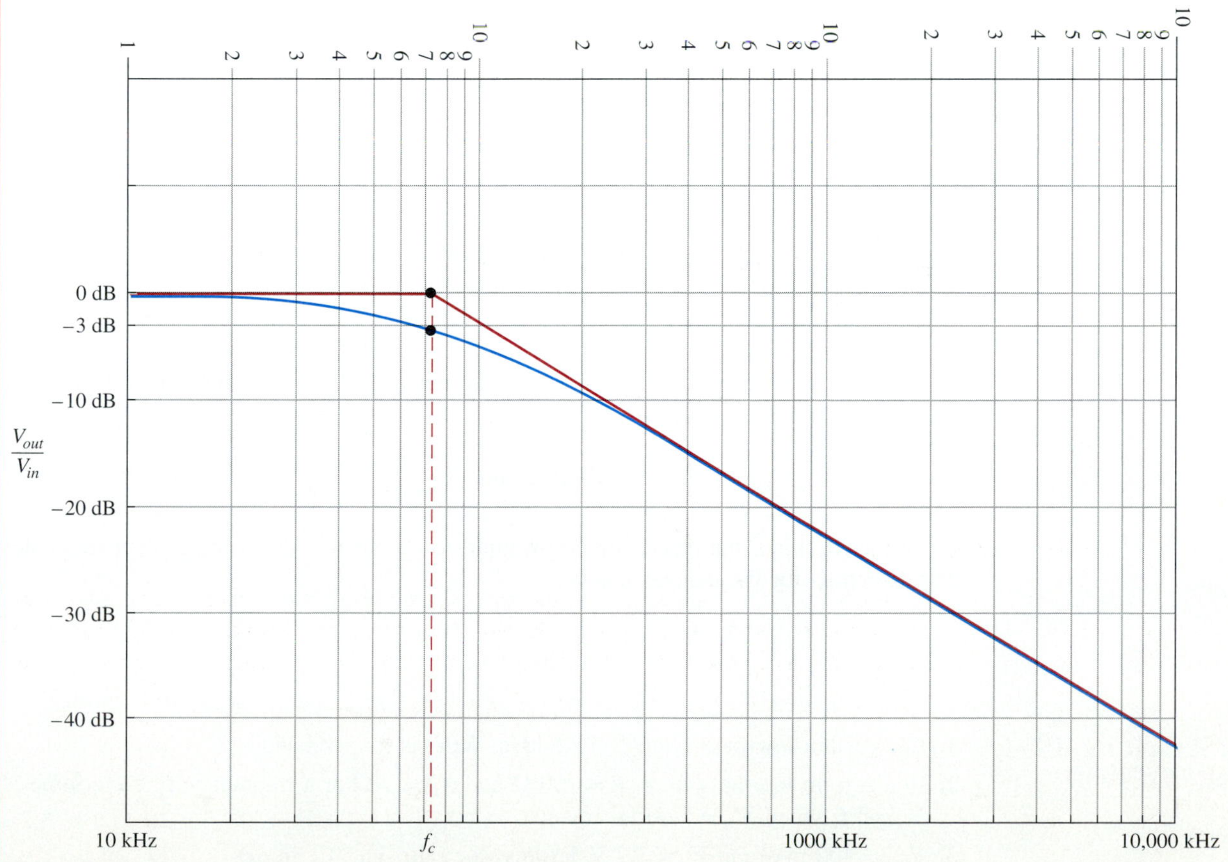


**FIGURE 19–9**
***Bode plot for Figure 19–8. Red line is the ideal and blue line is the actual.***

The output is flat below $f_c$ (74.5 kHz). As the frequency is increased above $f_c$, the output drops at a 20 dB/decade rate. Thus, for the ideal curve, every time the frequency is increased by ten, the output is reduced by 20 dB. A slight variation from this occurs in actual practice. The output is actually at –3 dB rather than 0 dB at the critical frequency.

***Related Exercise*** What happens to the critical frequency and roll-off rate if $L$ is reduced to 1 mH in Figure 19–8?

## Phase Shift in a Low-Pass Filter

The *RC* low-pass filter acts as a lag network. Recall from Chapter 16 that the phase shift from input to output is expressed as

$$\phi = -90° + \tan^{-1}\left(\frac{X_C}{R}\right)$$

At the critical frequency, $X_C = R$ and, therefore, $\phi = -45°$. As the input frequency is reduced, $\phi$ decreases and approaches 0° as the frequency approaches zero, as shown in Figure 19–10.

FIGURE 19–10
***Phase characteristic of a low-pass filter.***

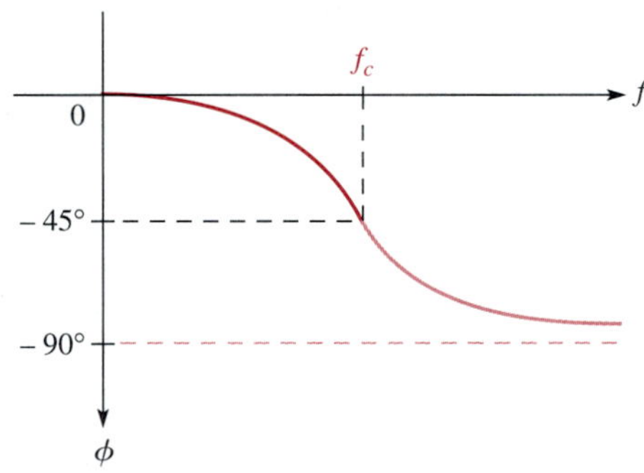


The *RL* low-pass filter also acts as a lag network. Recall from Chapter 17 that the phase shift is expressed as

$$\phi = -\tan^{-1}\left(\frac{X_L}{R}\right)$$

As in the *RC* filter, the phase shift from input to output is –45° at the critical frequency and decreases for frequencies below $f_c$.

**SECTION 19–1 REVIEW**

1. In a certain low-pass filter, $f_c = 2.5$ kHz. What is its passband?
2. In a certain low-pass filter, $R = 100\ \Omega$ and $X_C = 2\ \Omega$ at a frequency, $f_1$. Determine $\mathbf{V}_{out}$ at $f_1$ when $\mathbf{V}_{in} = 5\angle 0°$ V rms.
3. $V_{out} = 400$ mV, and $V_{in} = 1.2$ V. Express the ratio $V_{out}/V_{in}$ in dB.

## 19–2 ■ HIGH-PASS FILTERS

***A high-pass filter allows signals with higher frequencies to pass from input to output while rejecting lower frequencies.***

***After completing this section, you should be able to***

- **Analyze the operation of *RC* and *RL* high-pass filters**
  - □ Determine the critical frequency of a high-pass filter
  - □ Explain the difference between actual and ideal response curves
  - □ Generate a Bode plot for a high-pass filter
  - □ Discuss phase shift in a high-pass filter

Figure 19–11 shows a block diagram and a general response curve for a high-pass filter. The frequency considered to be the lower end of the passband is called the *critical frequency.* Just as in the low-pass filter, it is the frequency at which the output is 70.7% of the maximum, as indicated in the figure.

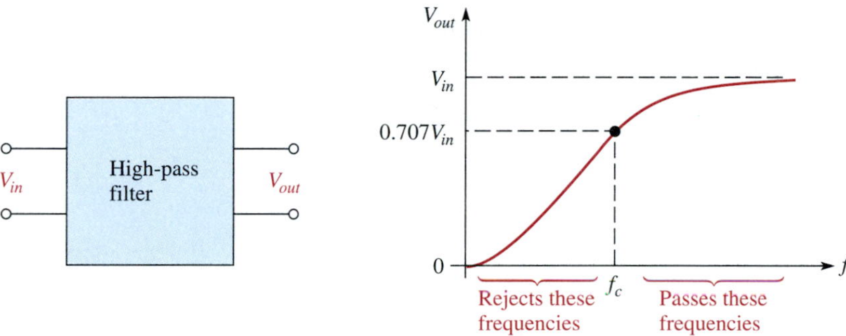


**FIGURE 19–11**
***High-pass filter block diagram and response curve.***

### *RC* High-Pass Filter

A basic *RC* high-pass filter is shown in Figure 19–12. Notice that the output voltage is taken across the resistor.

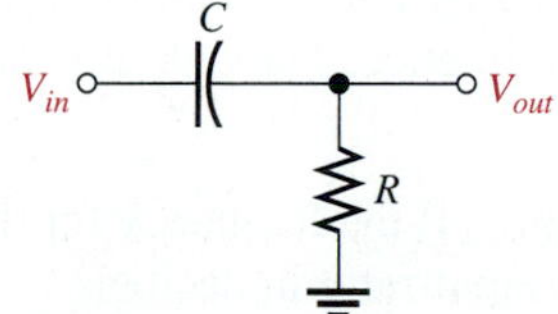


**FIGURE 19–12**
***RC high-pass filter.***

When the input frequency is at its critical value, $X_C = R$ and the output voltage is $0.707V_{in}$, just as in the case of the low-pass filter. As the input frequency increases above $f_c$, $X_C$ decreases and, as a result, the output voltage increases and approaches a value equal to $V_{in}$. The expression for the critical frequency of the high-pass filter is the same as for the low-pass filter.

$$f_c = \frac{1}{2\pi RC}$$

Below $f_c$, the output voltage decreases (rolls off) at a rate of 20 dB/decade. Figure 19–13 shows an actual and an ideal response curve for a high-pass filter.

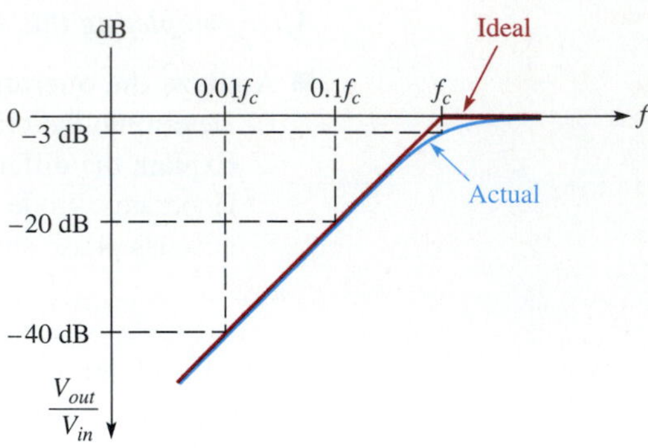


**FIGURE 19–13**
***Actual and ideal response curves for a high-pass filter.***

## EXAMPLE 19–5

Make a Bode plot for the filter in Figure 19–14 for three decades of frequency. Use semilog graph paper.

**FIGURE 19–14**

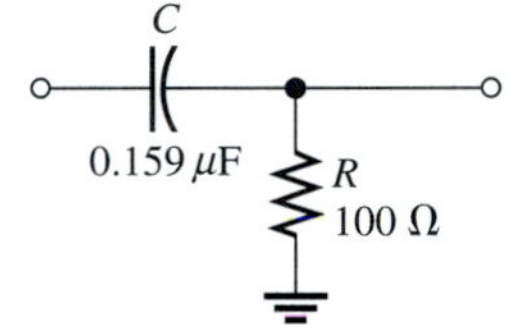


***Solution*** The critical frequency for this high-pass filter is

$$f_c = \frac{1}{2\pi RC} = \frac{1}{2\pi(100\ \Omega)(0.159\ \mu\text{F})} = 10\text{ kHz}$$

The idealized Bode plot is shown with the red line on the semilog graph in Figure 19–15. The approximate actual response curve is shown with the blue line. Notice first that the horizontal scale is logarithmic and the vertical scale is linear. The frequency is on the logarithmic scale, and the filter output in decibels is on the vertical.

The output is flat beyond $f_c$ (10 kHz). As the frequency is reduced below $f_c$, the output drops at a 20 dB/decade rate. Thus, for the ideal curve, every time the frequency is reduced by ten, the output is reduced by 20 dB. A slight variation from this occurs in actual practice. The output is actually at −3 dB rather than 0 dB at the critical frequency.

***Related Exercise*** If the frequency for the high-pass filter is decreased to 10 Hz, what is the output to input ratio in decibels?

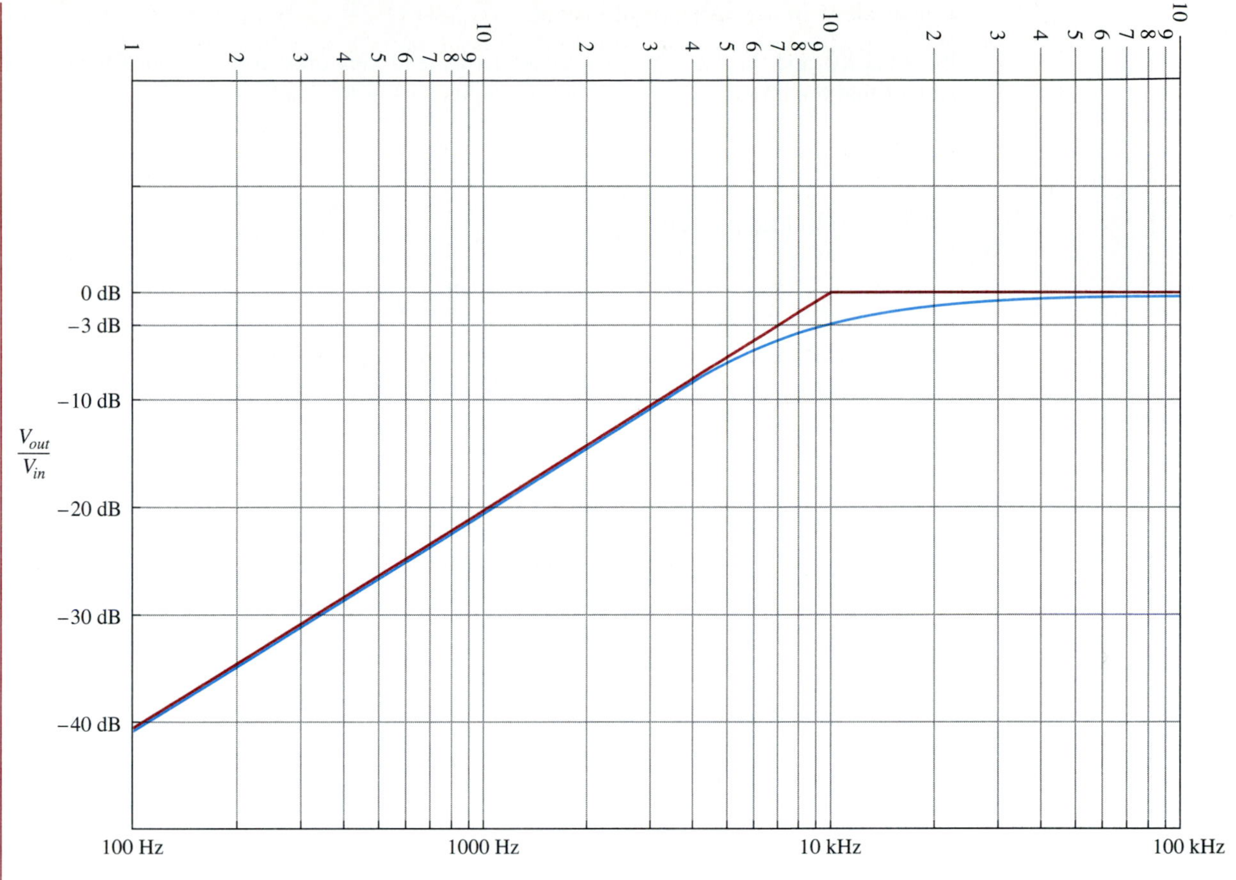


**FIGURE 19–15**
***Bode plot for Figure 19–14. Red line is the ideal and blue line is the actual.***

## *RL* High-Pass Filter

A basic *RL* high-pass filter is shown in Figure 19–16. Notice that the output is taken across the inductor.

**FIGURE 19–16**
***RL high-pass filter.***

When the input frequency is at its critical value, $X_L = R$, and the output voltage is $0.707V_{in}$. As the frequency increases above $f_c$, $X_L$ increases and, as a result, the output voltage increases until it equals $V_{in}$. The expression for the critical frequency of the high-pass filter is the same as for the low-pass filter.

$$f_c = \frac{1}{2\pi(L/R)}$$

## Phase Shift in a High-Pass Filter

Both the *RC* and the *RL* high-pass filters act as lead networks. Recall from Chapters 16 and 17 that the phase shift from input to output for the *RC* lead network is

$$\phi = \tan^{-1}\left(\frac{X_C}{R}\right)$$

and for the *RL* lead network is

$$\phi = 90° - \tan^{-1}\left(\frac{X_L}{R}\right)$$

At the critical frequency, $X_L = R$ and, therefore, $\phi = 45°$. As the frequency is increased, $\phi$ decreases toward 0° as shown in Figure 19–17.

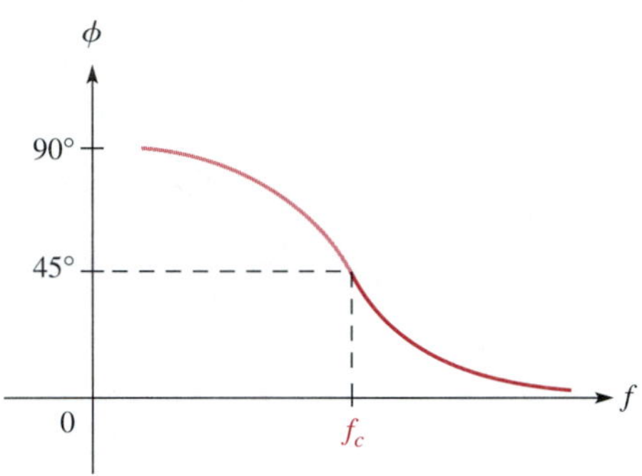


**FIGURE 19–17**
***Phase characteristic of a high-pass filter.***

**EXAMPLE 19–6**

**(a)** In Figure 19–18, find the value of $C$ so that $X_C$ is ten times less than $R$ at an input frequency of 10 kHz.

**(b)** If a 5 V sine wave with a dc level of 10 V is applied, what are the output voltage magnitude and the phase shift?

**FIGURE 19–18**

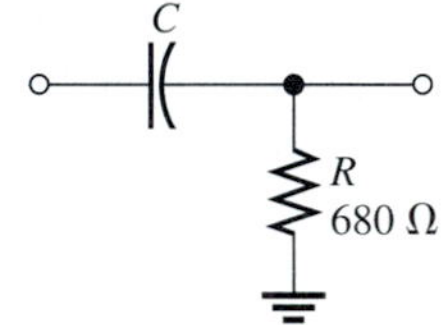


### *Solution*

**(a)** The value of $C$ is determined as follows:

$$X_C = 0.1R = 0.1(680\ \Omega) = 68\ \Omega$$

$$C = \frac{1}{2\pi f X_C} = \frac{1}{2\pi(10\text{ kHz})(68\ \Omega)} = 0.234\ \mu\text{F}$$

**(b)** The magnitude of the sine wave output is determined as follows:

$$V_{out} = \left(\frac{R}{\sqrt{R^2 + X_C^2}}\right)V_{in} = \left(\frac{680\ \Omega}{\sqrt{(680\ \Omega)^2 + (68\ \Omega)^2}}\right)5\text{ V} = 4.98\text{ V}$$

The phase shift is

$$\phi = \tan^{-1}\left(\frac{X_C}{R}\right) = \tan^{-1}\left(\frac{68\ \Omega}{680\ \Omega}\right) = 5.71°$$

At $f = 10$ kHz, which is a decade above the critical frequency, the sinusoidal output is almost equal to the input in magnitude, and the phase shift is very small. The 10 V dc level has been filtered out and does not appear at the output.

***Related Exercise*** Repeat parts (a) and (b) of the example if $R$ is changed to 220 Ω.

**SECTION 19–2 REVIEW**

1. The input voltage of a high-pass filter is 1 V. What is $V_{out}$ at the critical frequency?
2. In a certain high-pass filter, $\mathbf{V}_{in} = 10\angle 0°$ V, $R = 1$ kΩ, and $X_L = 15$ kΩ. Determine $\mathbf{V}_{out}$.

## 19–3 ■ BAND-PASS FILTERS

***A band-pass filter allows a certain band of frequencies to pass and attenuates or rejects all frequencies below and above the passband.***

***After completing this section, you should be able to***

- **Analyze the operation of band-pass filters**
  - Show how a band-pass filter is implemented with low-pass and high-pass filters
  - Define *bandwidth*
  - Explain the series-resonant band-pass filter
  - Explain the parallel-resonant band-pass filter
  - Calculate the bandwidth and output voltage of a band-pass filter

Figure 19–19 shows a typical band-pass response curve.

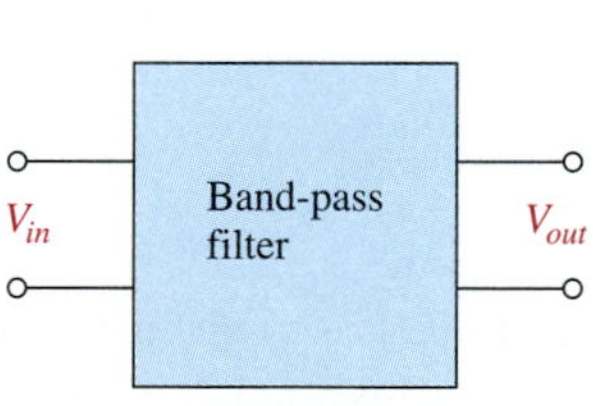


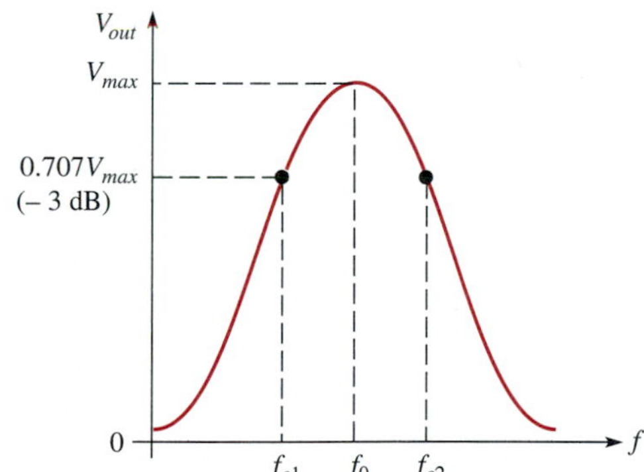


**FIGURE 19–19**
***Typical band-pass response curve.***

### Low-Pass/High-Pass Filter

A combination of a low-pass and a high-pass filter can be used to form a **band-pass filter,** as illustrated in Figure 19–20.

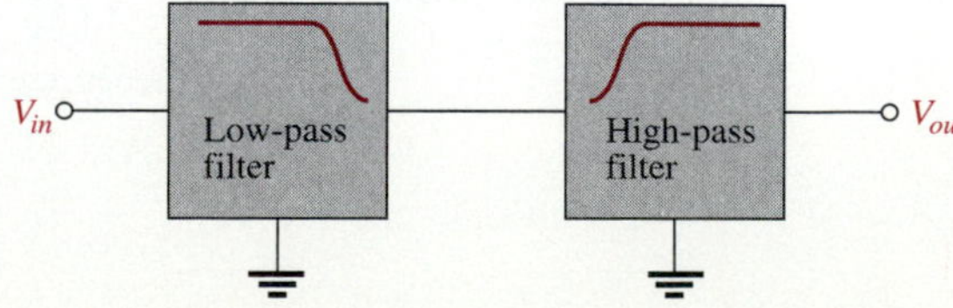


**FIGURE 19–20**
***Low-pass and high-pass filters used to form a band-pass filter.***

If the critical frequency, $f_{c(l)}$, of the low-pass filter is higher than the critical frequency, $f_{c(h)}$, of the high-pass filter, the responses overlap. Thus, all frequencies except those between $f_{c(h)}$ and $f_{c(l)}$ are eliminated, as shown in Figure 19–21.

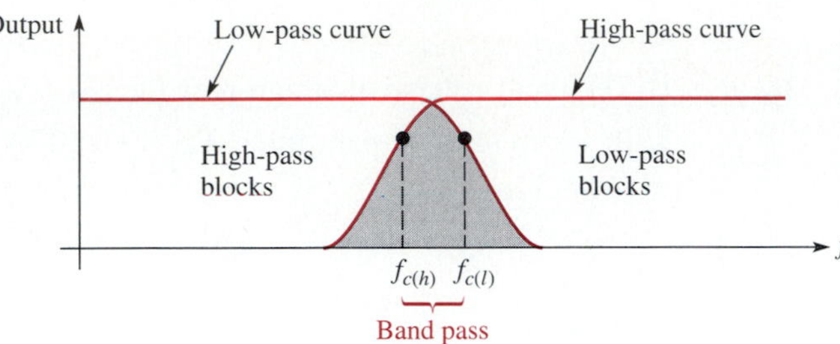


FIGURE 19–21
*Overlapping response curves of a high-pass/low-pass filter.*

**The bandwidth of a band-pass filter is the range of frequencies for which the current, and therefore the output voltage, is equal to or greater than 70.7% of its value at the resonant frequency.**

As you know, bandwidth is often abbreviated *BW*.

**EXAMPLE 19–7** A high-pass filter with $f_c = 2$ kHz and a low-pass filter with $f_c = 2.5$ kHz are used to construct a band-pass filter. What is the bandwidth of the passband?

***Solution*** $BW = f_{c(l)} - f_{c(h)} = 2.5\text{ kHz} - 2\text{ kHz} = 500\text{ Hz}$

***Related Exercise*** If $f_{c(l)} = 9$ kHz and the bandwidth is 1.5 kHz, what is $f_{c(h)}$?

## Series Resonant Band-Pass Filter

A type of series resonant band-pass filter is shown in Figure 19–22. As you learned in Chapter 18, a series resonant circuit has minimum impedance and maximum current at the resonant frequency, $f_r$. Thus, most of the input voltage is dropped across the resistor at the resonant frequency. Therefore, the output across *R* has a band-pass characteristic with a maximum output at the frequency of resonance. The resonant frequency is called the **center frequency, $f_0$.** The bandwidth is determined by the circuit *Q*, as was discussed in Chapter 18.

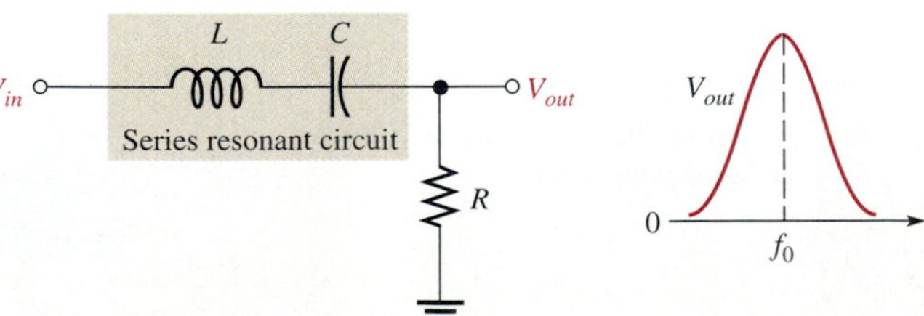


FIGURE 19–22
*Series resonant band-pass filter.*

A higher value of circuit $Q$ results in a smaller bandwidth. A lower value of $Q$ causes a larger bandwidth. A formula for the bandwidth of a resonant circuit in terms of $Q$ is stated in the following equation:

$$BW = \frac{f_0}{Q} \quad (19\text{–}5)$$

**EXAMPLE 19–8** Determine the output voltage magnitude at the center frequency ($f_0$) and the bandwidth for the filter in Figure 19–23.

**FIGURE 19–23**

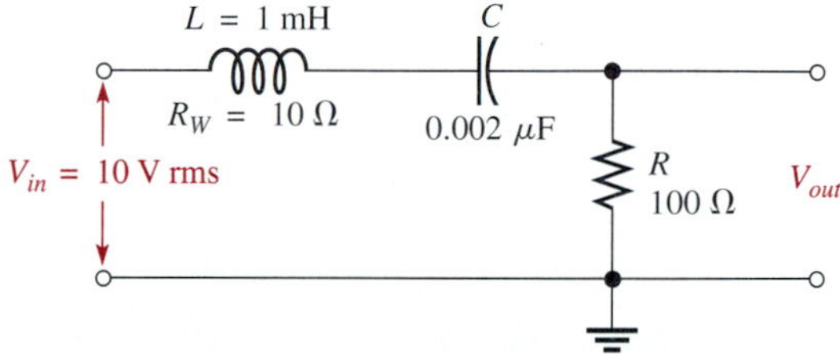


***Solution*** At $f_0$, the impedance of the resonant circuit is equal to the winding resistance, $R_W$. By the voltage-divider formula,

$$V_{out} = \left(\frac{R}{R + R_W}\right)V_{in} = \left(\frac{100\ \Omega}{110\ \Omega}\right)10\ \text{V} = 9.09\ \text{V}$$

The center frequency is

$$f_0 = \frac{1}{2\pi\sqrt{LC}} = \frac{1}{2\pi\sqrt{(1\ \text{mH})(0.002\ \mu\text{F})}} = 113\ \text{kHz}$$

At $f_0$, the inductive reactance is

$$X_L = 2\pi fL = 2\pi(113\ \text{kHz})(1\ \text{mH}) = 710\ \Omega$$

and the total resistance is

$$R_{tot} = R + R_W = 100\ \Omega + 10\ \Omega = 110\ \Omega$$

Therefore, the circuit $Q$ is

$$Q = \frac{X_L}{R_{tot}} = \frac{710\ \Omega}{110\ \Omega} = 6.45$$

The bandwidth is

$$BW = \frac{f_0}{Q} = \frac{113\ \text{kHz}}{6.45} = 17.5\ \text{kHz}$$

***Related Exercise*** If a 1 mH coil with a winding resistance of 18 Ω replaces the existing coil in Figure 19–23, how is the bandwidth affected?

### Parallel Resonant Band-Pass Filter

A type of band-pass filter using a parallel resonant circuit is shown in Figure 19–24. Recall that a parallel resonant circuit has maximum impedance at resonance. The circuit in Figure 19–24 acts as a voltage divider. At resonance, the impedance of the tank is much greater than the resistance. Thus, most of the input voltage is across the tank, producing a maximum output voltage at the resonant (center) frequency.

FIGURE 19–24
***Parallel resonant band-pass filter.***

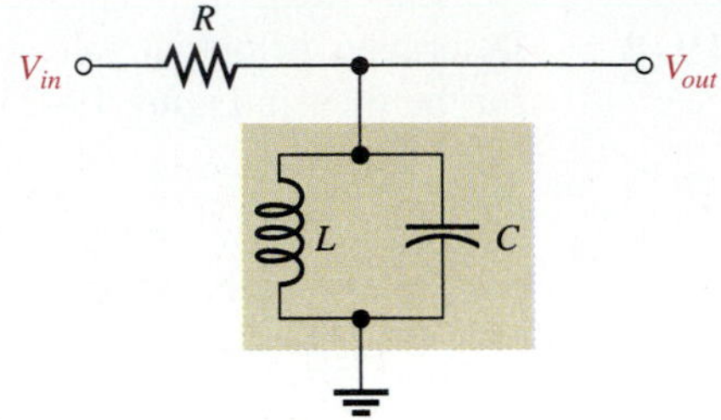


For frequencies above or below resonance, the tank impedance drops off, and more of the input voltage is across $R$. As a result, the output voltage across the tank drops off, creating a band-pass characteristic.

**EXAMPLE 19–9** What is the center frequency of the filter in Figure 19–25? Assume $R_W = 0\ \Omega$.

FIGURE 19–25

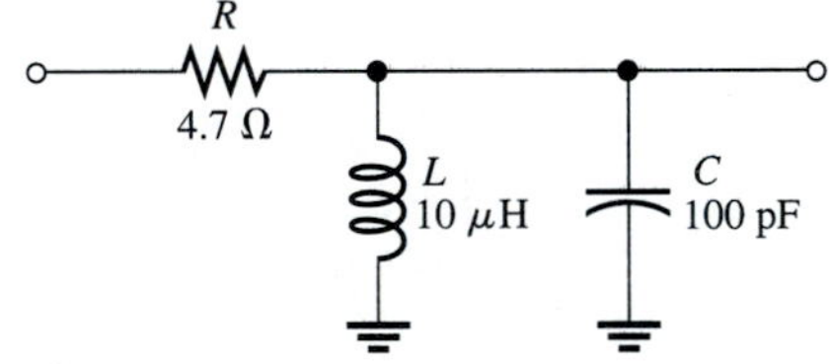


***Solution*** The center frequency of the filter is its resonant frequency.

$$f_0 = \frac{1}{2\pi\sqrt{LC}} = \frac{1}{2\pi\sqrt{(10\ \mu\text{H})(100\ \text{pF})}} = 5.03\ \text{MHz}$$

***Related Exercise*** Determine $f_0$ in Figure 19–25 if $C$ is changed to 1000 pF.

**EXAMPLE 19–10** Determine the center frequency and bandwidth for the band-pass filter in Figure 19–26 if the inductor has a winding resistance of 15 Ω.

FIGURE 19–26

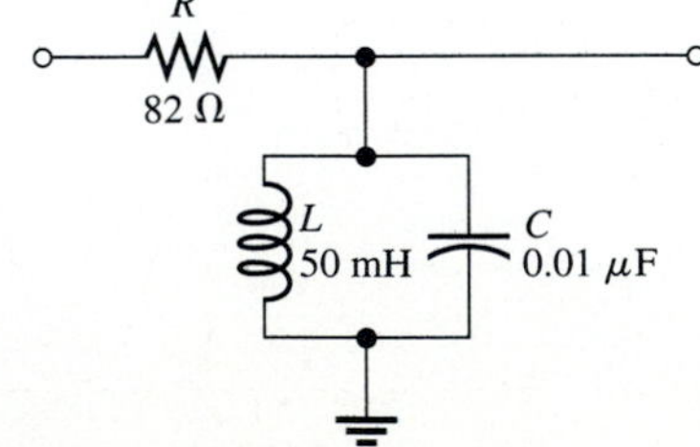

***Solution*** Recall from Chapter 18 that the resonant (center) frequency of a nonideal tank circuit is

$$f_0 = \frac{\sqrt{1 - (R_W^2 C/L)}}{2\pi\sqrt{LC}} = \frac{\sqrt{1 - (15\ \Omega)^2(0.01\ \mu\text{F})/50\ \text{mH}}}{2\pi\sqrt{(50\ \text{mH})(0.01\ \mu\text{F})}} = 7.12\ \text{kHz}$$

The $Q$ of the coil at resonance is

$$Q = \frac{X_L}{R_W} = \frac{2\pi f_0 L}{R_W} = \frac{2\pi(7.12\ \text{kHz})(50\ \text{mH})}{15\ \Omega} = 149$$

The bandwidth of the filter is

$$BW = \frac{f_0}{Q} = \frac{7.12\ \text{kHz}}{149} = 47.8\ \text{Hz}$$

Note that since $Q > 10$, the ideal formula could have been used to calculate $f_0$. However, you did not know this in advance.

***Related Exercise*** Knowing the value of $Q$, recalculate $f_0$ using the ideal formula.

**SECTION 19–3 REVIEW**

1. For a band-pass filter, $f_{c(h)} = 29.8$ kHz and $f_{c(l)} = 30.2$ kHz. What is the bandwidth?
2. A parallel resonant band-pass filter has the following values: $R_W = 15\ \Omega$, $L = 50\ \mu$H, and $C = 470$ pF. Determine the approximate center frequency.

## 19–4 ■ BAND-STOP FILTERS

***A band-stop filter is essentially the opposite of a band-pass filter in terms of the responses. A band-stop filter allows all frequencies to pass except those lying within a certain stop band.***

***After completing this section, you should be able to***

- **Analyze the operation of band-stop filters**
  - ☐ Show how a band-stop filter is implemented with low-pass and high-pass filters
  - ☐ Explain the series-resonant band-stop filter
  - ☐ Explain the parallel-resonant band-stop filter
  - ☐ Calculate the bandwidth and output voltage of a band-stop filter

Figure 19–27 shows a general band-stop response curve.

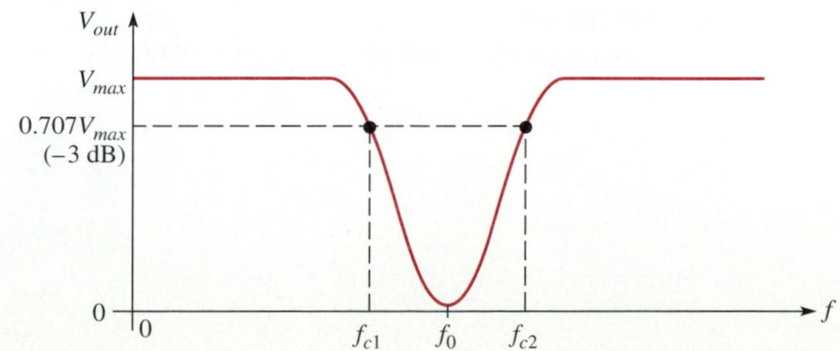


**FIGURE 19–27**
***General band-stop response curve.***

### Low-Pass/High-Pass Filter

A **band-stop filter** can be formed from a low-pass and a high-pass filter, as shown in Figure 19–28.

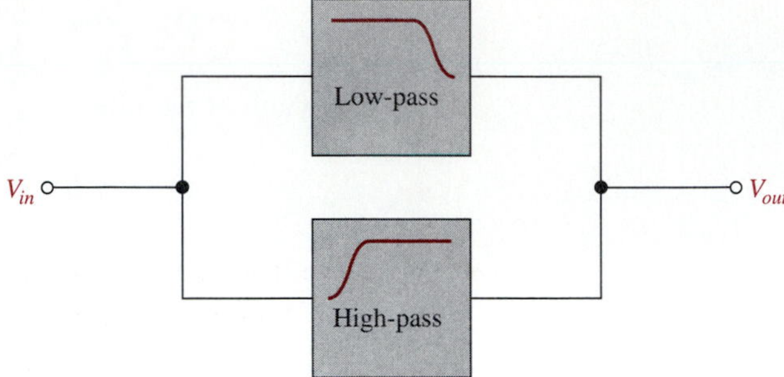


FIGURE 19–28
*Low-pass and high-pass filters used to form a band-stop filter.*

If the low-pass critical frequency, $f_{c(l)}$, is set lower than the high-pass critical frequency, $f_{c(h)}$, a band-stop characteristic is formed as illustrated in Figure 19–29.

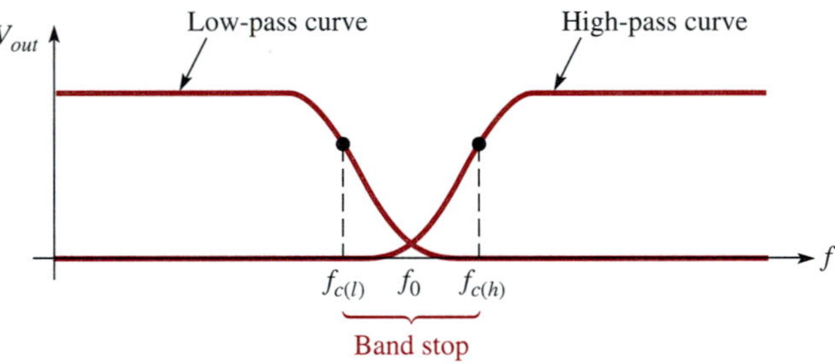


FIGURE 19–29
*Band-stop response curve.*

### Series Resonant Band-Stop Filter

A series resonant circuit used in a band-stop configuration is shown in Figure 19–30. Basically, it works as follows: At the resonant frequency, the impedance is minimum, and therefore the output voltage is minimum. Most of the input voltage is dropped across $R$. At frequencies above and below resonance, the impedance increases, causing more voltage across the output.

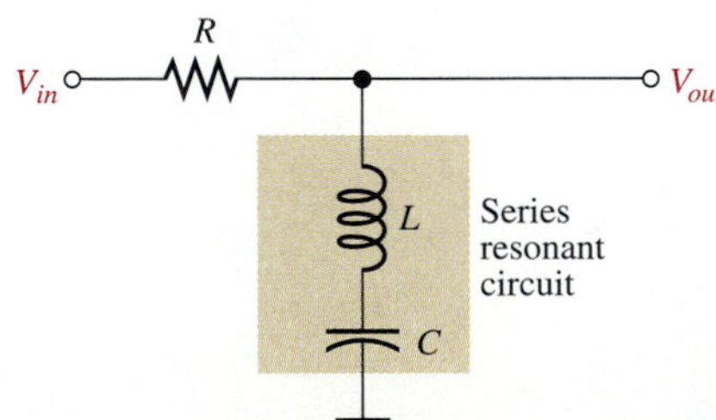


FIGURE 19–30
*Series resonant band-stop filter.*

**EXAMPLE 19–11** Find the output voltage magnitude at $f_0$ and the bandwidth in Figure 19–31.

**FIGURE 19–31**

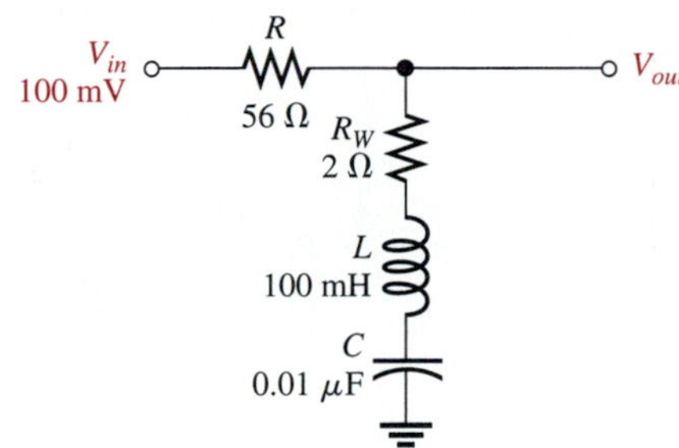


***Solution*** Since $X_L = X_C$ at resonance, the output voltage is

$$V_{out} = \left(\frac{R_W}{R + R_W}\right)V_{in} = \left(\frac{2\ \Omega}{58\ \Omega}\right)100\text{ mV} = 3.45\text{ mV}$$

To determine the bandwidth, calculate the center frequency and $Q$ of the coil.

$$f_0 = \frac{1}{2\pi\sqrt{LC}} = \frac{1}{2\pi\sqrt{(100\text{ mH})(0.01\ \mu\text{F})}} = 5.03\text{ kHz}$$

$$Q = \frac{X_L}{R} = \frac{2\pi f L}{R} = \frac{2\pi(5.03\text{ kHz})(100\text{ mH})}{58\ \Omega} = \frac{3.16\text{ k}\Omega}{58\ \Omega} = 54.5$$

$$BW = \frac{f_0}{Q} = \frac{5.03\text{ kHz}}{54.5} = 92.3\text{ Hz}$$

***Related Exercise*** Assume $R_W = 10\ \Omega$ in Figure 19–31. Determine $V_{out}$ and the bandwidth.

## Parallel Resonant Band-Stop Filter

A parallel resonant circuit used in a band-stop configuration is shown in Figure 19–32. At the resonant frequency, the tank impedance is maximum, and so most of the input voltage appears across it. Very little voltage is across $R$ at resonance. As the tank impedance decreases above and below resonance, the output voltage increases.

**FIGURE 19–32**
***Parallel resonant band-stop filter.***

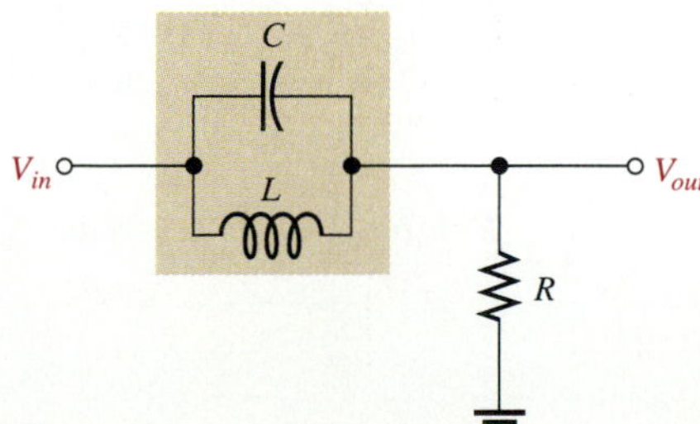

**EXAMPLE 19–12** Find the center frequency of the filter in Figure 19–33. Sketch the output response curve showing the minimum and maximum voltages.

FIGURE 19–33

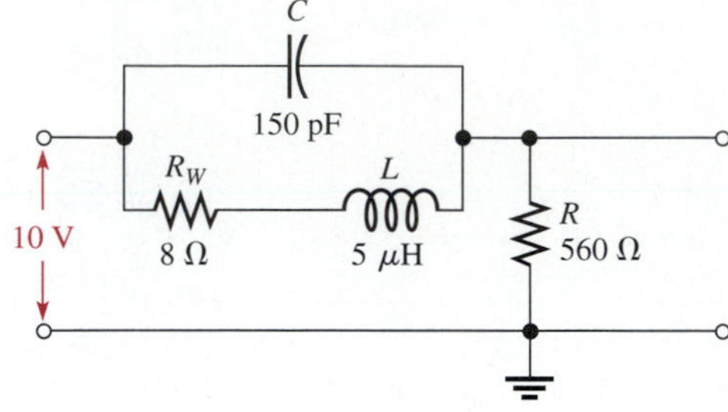


***Solution*** The center frequency is

$$f_0 = \frac{\sqrt{1 - R_W^2 C/L}}{2\pi\sqrt{LC}} = \frac{\sqrt{1 - (8\ \Omega)^2(150\text{ pF})/5\ \mu\text{H}}}{2\pi(5\ \mu\text{H})(150\text{ pF})} = 5.79\text{ MHz}$$

At the center (resonant) frequency,

$$X_L = 2\pi f_0 L = 2\pi(5.79\text{ MHz})(5\ \mu\text{H}) = 182\ \Omega$$

$$Q = \frac{X_L}{R_W} = \frac{182\ \Omega}{8\ \Omega} = 22.8$$

$$Z_r = R_W(Q^2 + 1) = 8\ \Omega(22.8^2 + 1) = 4.17\text{ k}\Omega \quad \text{(purely resistive)}$$

Now, use the voltage-divider formula to find the minimum output voltage magnitude.

$$V_{out(min)} = \left(\frac{R}{R + Z_r}\right)V_{in} = \left(\frac{560\ \Omega}{4.73\text{ k}\Omega}\right)10\text{ V} = 1.18\text{ V}$$

At zero frequency, the impedance of the tank is $R_W$ because $X_C = \infty$ and $X_L = 0\ \Omega$. Therefore, the maximum output voltage below resonance is

$$V_{out(max)} = \left(\frac{R}{R + R_W}\right)V_{in} = \left(\frac{560\ \Omega}{568\ \Omega}\right)10\text{ V} = 9.86\text{ V}$$

As the frequency increases much higher than $f_0$, $X_C$ approaches 0 Ω, and $V_{out}$ approaches $V_{in}$ (10 V). Figure 19–34 shows the response curve.

FIGURE 19–34

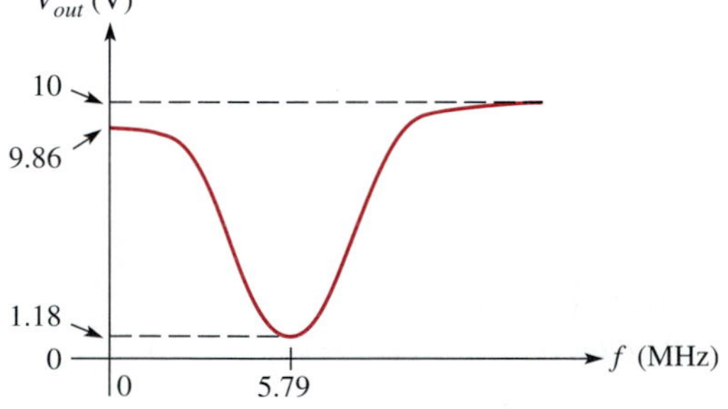


***Related Exercise*** What is the minimum output voltage if $R = 1\text{ k}\Omega$ in Figure 19–33?

**SECTION 19–4 REVIEW**

1. How does a band-stop filter differ from a band-pass filter?
2. Name three basic ways to construct a band-stop filter.

## 19–5 ■ TECHnology Theory Into Practice

***In this TECH TIP, you will plot the frequency responses of two types of filters based on a series of oscilloscope measurements and identify the type of filter in each case.***

The filters are contained in sealed modules as shown in Figure 19–35. You are concerned only with determining the filter response characteristics and not the types of internal components.

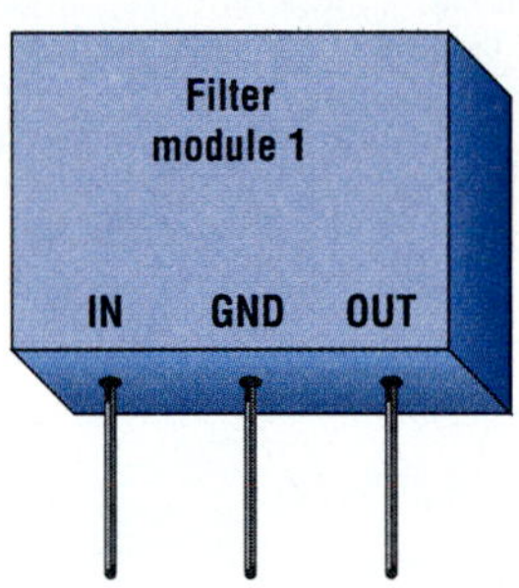


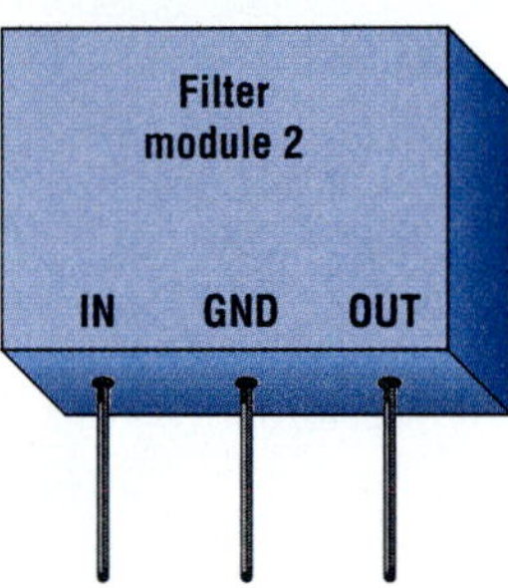


FIGURE 19–35
*Filter modules.*

## Filter Measurement and Analysis

□ Refer to Figure 19–36. Based on the series of four oscilloscope measurements, create a Bode plot for the filter under test, specify applicable frequencies, and identify the type of filter.

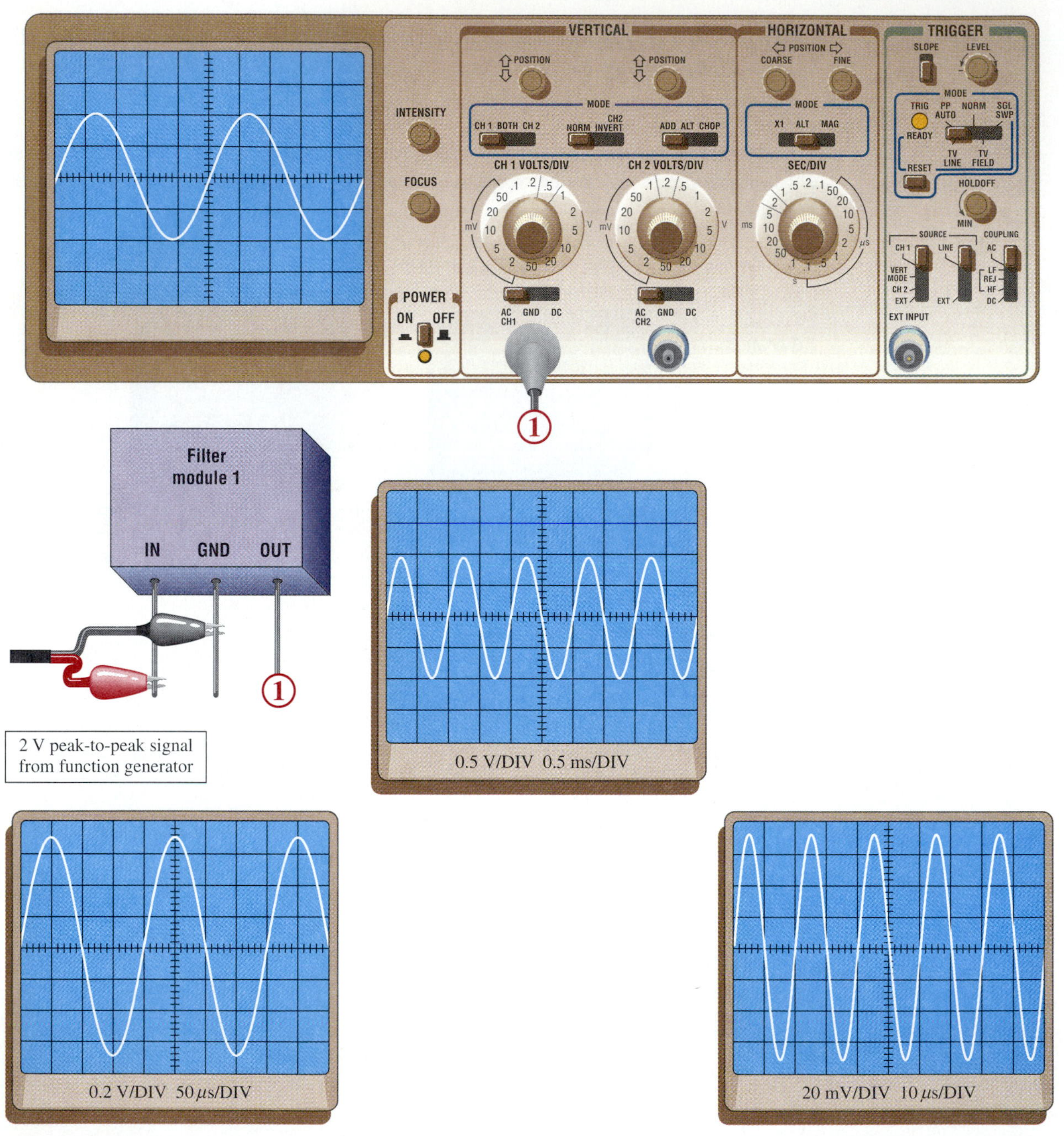


**FIGURE 19–36**
*The scope probes are ×1.*

☐ Refer to Figure 19–37. Based on the series of six oscilloscope measurements, create a Bode plot for the filter under test, specify applicable frequencies, and identify the type of filter.

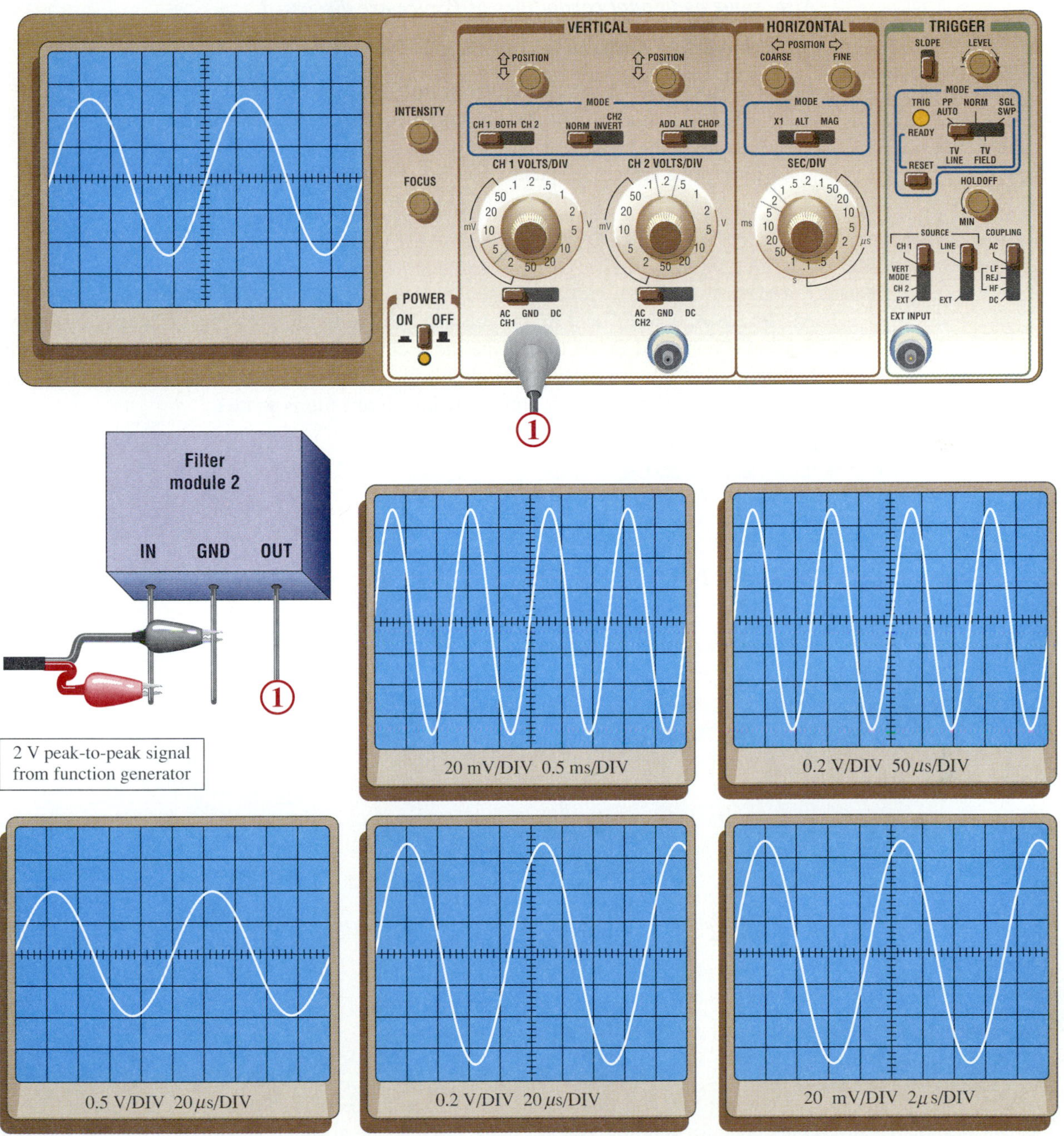


**FIGURE 19–37**
*The scope probes are ×1.*

**SECTION 19–5 REVIEW**

1. Explain how the waveforms in Figure 19–36 indicate the type of filter.
2. Explain how the waveforms in Figure 19–37 indicate the type of filter.

## 19–6 ▪ FILTER ANALYSIS WITH PSpice AND PROBE

***As you have learned, filters are a class of circuits that make use of the frequency-dependent response of RC and RL circuits. In this section, Probe is used to generate a Bode plot from which the critical frequency and roll-off of a filter can be determined. Also, some additional capabilities of PSpice are discussed.***

***After completing this section, you should be able to***

- **Use PSpice and Probe to analyze simple filter circuits**
  - ☐ Use Probe to plot frequency response curves for multiple filter circuits
  - ☐ Use Probe to generate Bode plots
  - ☐ Determine critical frequency and roll-off from Bode plots using cursors

### Low-Pass Filters

As you know, both *RC* and *RL* circuits can be used to form low-pass filters. Although you can use PSpice and Probe to analyze these circuits separately, more than one circuit can be analyzed at one time using one circuit file. For example, if you want to analyze both of the filters in Figure 19–38, the following circuit file is written:

```
Dual Low-Pass Filter Circuit
*Low-Pass RC and RL Circuits
V1  1  0  AC  5
R1  1  2  1K
C1  2  0  0.01U
V2  3  0  AC  5
L1  3  4  10M
R2  4  0  1K
.AC  LIN  1000  1K  1MEG
.PROBE
.END
```

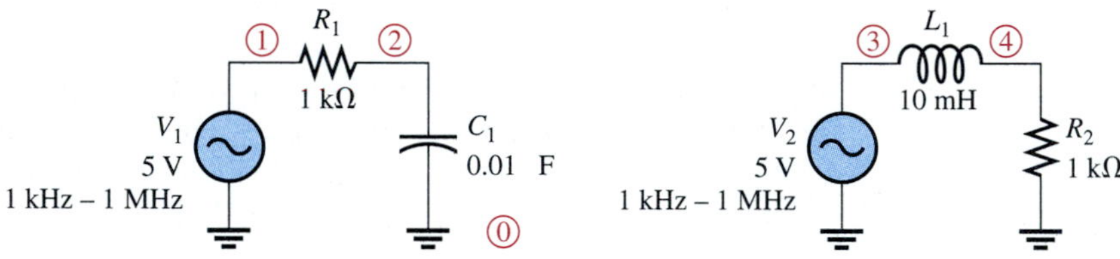


FIGURE 19–38

When you run this circuit file, you will see that PSpice does not object. This is equivalent to building two separate circuits on a single circuit board with the same ground, and it allows the circuits to be analyzed at the same time. This works because the ground node (0) is an absolute reference point, regardless of the circuits that are defined relative to it. To see this, run PSpice on the circuit, start up Probe, and add the following traces to the display:

V(2,0)  V(4,0)

Probe will draw voltage traces for both $C_1$ and $R_2$, which are the output voltages of the two filters. The display may be somewhat difficult to read because both traces, in this case, are identical and appear to be only one trace. Although this makes comparisons easy, it is sometimes better to view the traces on separate displays. Probe allows you to do this with the various Plot_control options which, as stated before, vary slightly with the platform. For DOS Probe users, the following sequence of menu options are selected:

Plot_control
Add_plot
Exit

For Windows Probe users, the following sequence of menu options are selected:

Plot
Add plot

Notice that the plot display area has split into two separate graphs with the original plot in one and an empty plot in the other. Also note that the "new" empty plot area is the "active" plot, as indicated by the SEL>> selected plot indicator.

When a new plot is added, Probe will replace the current display with two smaller displays that are equal in size and have the same capabilities as the original screen for scaling, cursors, etc. However, Probe works with one plot at a time, so it lets you know which plot is active (i.e., selected). To select a different plot to be active, choose Plot_control/Select_plot (if using DOS Probe) or mouse-click anywhere in the plot area of the plot you wish to select. Selecting Plot_control/Remove_plot or Plot/Delete_plot will remove the display currently selected (indicated by SEL next to it) and result in larger displays again.

Using these commands, set up two plots with one displaying the trace for V(2,0) and the other displaying the trace for V(4,0). Probe will draw the traces for the voltage across $R_2$ and $C_1$, allowing you to observe and compare the responses of the two filters. In this case they are identical.

Using two or more displays can be useful in many situations. For example, remove the trace for V(4,0) and add the trace VP(2,0). Probe will draw the phase angle of the voltage, so that you can simultaneously observe how both the magnitude and the phase angle of the voltage across $C_1$ change with frequency.

## Bode Plot

A Bode plot differs from a standard plot in that it plots the amplitude of the voltage in decibels and the frequency scaled in decades. The $x$-axis is already scaled in decades. Probe allows you to plot voltage amplitude in decibels directly. For example, a Bode plot of the voltage across $C_1$ is generated by plotting

VDB(2,0)

A Bode plot can be analyzed using the cursors. To find the critical frequency, a cursor is used to determine where the plotted value is −3 dB from the passband value. To do this, select Cursor and leave cursor 2 at the beginning of the plot. Move cursor 1 along the trace until the difference between them is −3.0 dB. The frequency shown by cursor 1 at this point is the critical frequency. For this example, the passband value is about +14 dB at 1 kHz, and so the critical frequency is about 15.8 kHz at the point where the voltage is down to about +11 dB.

In a similar way, the roll-off of the response can be determined. Cursor 2 is placed at a convenient frequency value, such as 100 kHz in this case, in the linear portion of the roll-off region. Cursor 1 is moved to a point one decade above this (1 MHz in this case) and the dB difference between the cursors is the roll-off rate of the filter in dB/decade. In the case of these simple filters, the roll-off rate is always 20 dB/decade. (Probe may show a value slightly different than this, but it is sufficiently close.)

### High-Pass Filters

The analysis of high-pass *RC* and *RL* filters is identical to that for low-pass filters except that the output voltage is taken across a different component, as you have learned in this chapter.

**SECTION 19–6 REVIEW**

1. Change the filters in Figure 19–38 to high-pass filters and write a combined circuit file to analyze the responses.
2. Explain the difference between V(2,0), VP(2,0), and VDB(2,0).

## ■ SUMMARY

- In an *RC* low-pass filter, the output voltage is taken across the capacitor and the output lags the input.
- In an *RL* low-pass filter, the output voltage is taken across the resistor and the output lags the input.
- In an *RC* high-pass filter, the output is taken across the resistor and the output leads the input.
- In an *RL* high-pass filter, the output is taken across the inductor and the output leads the input.
- The roll-off rate of a basic *RC* or *RL* filter is 20 dB per decade.
- A band-pass filter passes frequencies between the lower and upper critical frequencies and rejects all others.
- A band-stop filter rejects frequencies between its lower and upper critical frequencies and passes all others.
- The bandwidth of a resonant filter is determined by the quality factor ($Q$) of the circuit and the resonant frequency.
- Critical frequencies are also called −3 dB frequencies.
- The output voltage is 70.7% of its maximum at the critical frequencies.

## ■ GLOSSARY

**Attenuation** The ratio with a value of less than 1 of the output voltage to the input voltage of a circuit.

**Band-pass filter** A filter that passes a range of frequencies lying between two critical frequencies and rejects frequencies above and below the range.

**Band-stop filter** A filter that rejects a range of frequencies lying between two critical frequencies and passes frequencies above and below the range.

**Bode plot** The graph of a filter's frequency response showing the change in the output voltage to input voltage ratio expressed in dB as a function of frequency for a constant input voltage.

**Center frequency ($f_0$)** The resonant frequency of a band-pass or band-stop filter.

**Critical frequency** The frequency at which a filter's output voltage is 70.7% of the maximum.

**Decade** A tenfold change in frequency or other parameter.

**Decibel** A logarithmic measurement of the ratio of one power to another or one voltage to another, which can be used to express the input-to-output relationship of a filter.

**Passband** The range of frequencies passed by a filter.

**Roll-off** The rate of decrease of a filter's frequency response.

## ■ FORMULAS

**(19–1)** $$\text{dB} = 20 \log\left(\frac{V_{out}}{V_{in}}\right)$$ Voltage ratio in decibels

**(19–2)** $$\text{dB} = 10 \log\left(\frac{P_{out}}{P_{in}}\right)$$ Power ratio in decibels

**(19–3)** $f_c = \dfrac{1}{2\pi RC}$ Critical frequency

**(19–4)** $f_c = \dfrac{1}{2\pi(L/R)}$ Critical frequency

**(19–5)** $BW = \dfrac{f_0}{Q}$ Bandwidth

## ■ SELF-TEST

**1.** The maximum output voltage of a certain low-pass filter is 10 V. The output voltage at the critical frequency is

**(a)** 10 V **(b)** 0 V **(c)** 7.07 V **(d)** 1.414 V

**2.** A sinusoidal voltage with a peak-to-peak value of 15 V is applied to an *RC* low-pass filter. If the reactance at the input frequency is zero, the output voltage is

**(a)** 15 V peak-to-peak **(b)** zero
**(c)** 10.6 V peak-to-peak **(d)** 7.5 V peak-to-peak

**3.** The same signal in Question 2 is applied to an *RC* high-pass filter. If the reactance is zero at the input frequency, the output voltage is

**(a)** 15 V peak-to-peak **(b)** zero
**(c)** 10.6 V peak-to-peak **(d)** 7.5 V peak-to-peak

**4.** At the critical frequency, the output of a filter is down from its maximum by

**(a)** 0 dB **(b)** −3 dB **(c)** −20 dB **(d)** −6 dB

**5.** If the output of a low-pass *RC* filter is 12 dB below its maximum at $f$ = 1 kHz, then at $f$ = 10 kHz, the output is below its maximum by

**(a)** 3 dB **(b)** 10 dB **(c)** 20 dB **(d)** 32 dB

**6.** In a filter, the ratio $V_{out}/V_{in}$ is called

**(a)** roll-off **(b)** gain **(c)** attenuation **(d)** critical reduction

**7.** For each decade increase in frequency above the critical frequency, the output of a low-pass filter decreases by

**(a)** 20 dB **(b)** 3 dB **(c)** 10 dB **(d)** 0 dB

**8.** At the critical frequency, the phase shift through a high-pass filter is

**(a)** 90° **(b)** 0° **(c)** 45° **(d)** dependent on the reactance

**9.** In a series resonant band-pass filter, a higher value of $Q$ results in

**(a)** a higher resonant frequency **(b)** a smaller bandwidth
**(c)** a higher impedance **(d)** a larger bandwidth

**10.** At series resonance,

**(a)** $X_C = X_L$ **(b)** $X_C > X_L$ **(c)** $X_C < X_L$

**11.** In a certain parallel resonant band-pass filter the resonant frequency is 10 kHz. If the bandwidth is 2 kHz, the lower critical frequency is

**(a)** 5 kHz **(b)** 12 kHz **(c)** 9 kHz **(d)** not determinable

**12.** In a band-pass filter, the output voltage at the resonant frequency is

**(a)** minimum **(b)** maximum
**(c)** 70.7% of maximum **(d)** 70.7% of minimum

**13.** In a band-stop filter, the output voltage at the critical frequencies is

**(a)** minimum **(b)** maximum
**(c)** 70.7% of maximum **(d)** 70.7% of minimum

**14.** At a sufficiently high value of $Q$, the resonant frequency for a parallel resonant filter is ideally

**(a)** much greater than the resonant frequency of a series resonant filter
**(b)** much less than the resonant frequency of a series resonant filter
**(c)** equal to the resonant frequency of a series resonant filter

## ■ PROBLEMS

### SECTION 19–1 Low-Pass Filters

1. In a certain low-pass filter, $X_C = 500\ \Omega$ and $R = 2.2\ \text{k}\Omega$. What is the output voltage ($\mathbf{V}_{out}$) when the input is 10 V rms?
2. A certain low-pass filter has a critical frequency of 3 kHz. Determine which of the following frequencies are passed and which are rejected:
   **(a)** 100 Hz **(b)** 1 kHz **(c)** 2 kHz **(d)** 3 kHz **(e)** 5 kHz
3. Determine the output voltage ($\mathbf{V}_{out}$) of each filter in Figure 19–39 at the specified frequency when $V_{in} = 10$ V.

FIGURE 19–39

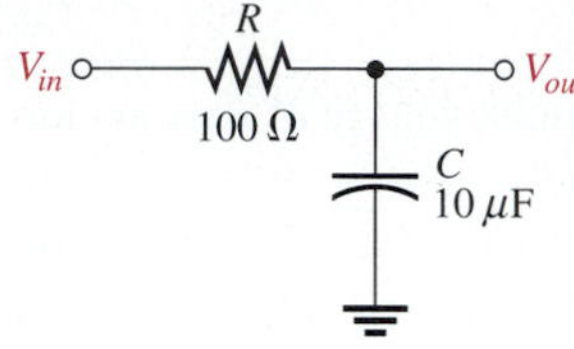

Vin, R 47 Ω, C 8 μF, Vout

(b) $f$ = 400 Hz

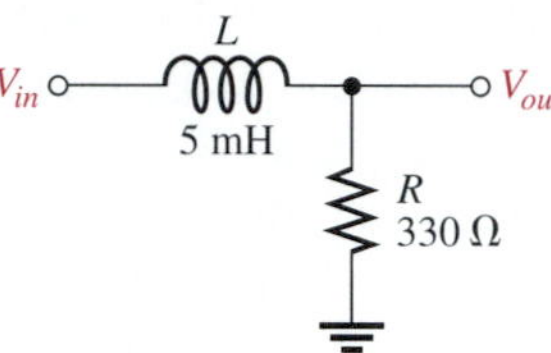


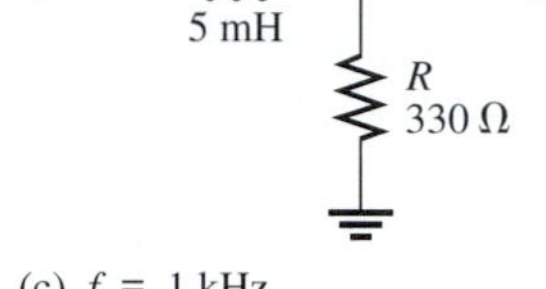

(c) $f$ = 1 kHz

Vin, L 80 μH, R 10 Ω, Vout

(d) $f$ = 2 kHz

4. What is $f_c$ for each filter in Figure 19–39? Determine the output voltage at $f_c$ in each case when $V_{in} = 5$ V.
5. For the filter in Figure 19–40, calculate the value of $C$ required for each of the following critical frequencies:
   **(a)** 60 Hz **(b)** 500 Hz **(c)** 1 kHz **(d)** 5 kHz

FIGURE 19–40

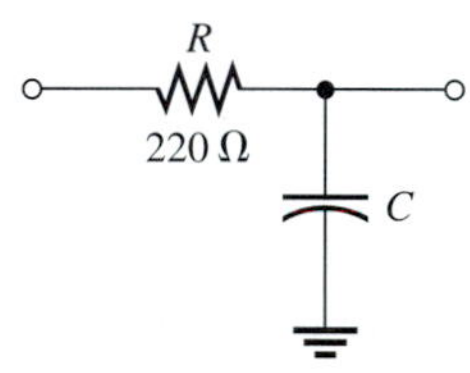


6. Determine the critical frequency for each switch position on the switched filter network of Figure 19–41.

FIGURE 19–41

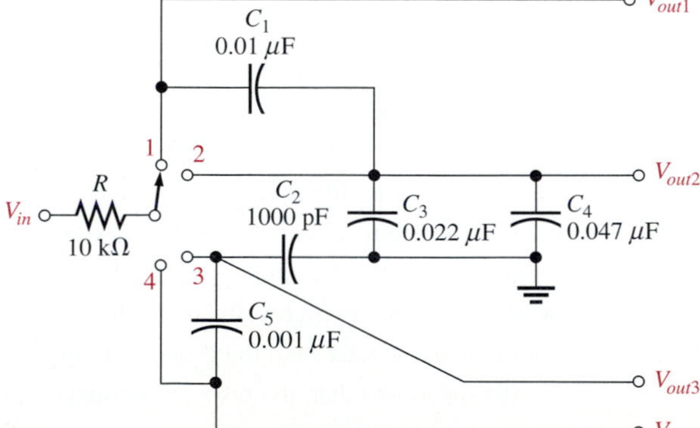

7. Sketch a Bode plot for each part of Problem 5.
8. For each following case, express the voltage ratio in dB:
   (a) $V_{in} = 1$ V, $V_{out} = 1$ V  (b) $V_{in} = 5$ V, $V_{out} = 3$ V
   (c) $V_{in} = 10$ V, $V_{out} = 7.07$ V  (d) $V_{in} = 25$ V, $V_{out} = 5$ V
9. The input voltage to a low-pass *RC* filter is 8 V rms. Find the output voltage at the following dB levels:
   (a) −1 dB  (b) −3 dB  (c) −6 dB  (d) −20 dB
10. For a basic *RC* low-pass filter, find the output voltage in dB relative to a 0 dB input for the following frequencies ($f_c = 1$ kHz):
   (a) 10 kHz  (b) 100 kHz  (c) 1 MHz

## SECTION 19–2 High-Pass Filters

11. In a high-pass filter, $X_C = 500\ \Omega$ and $R = 2.2\ \text{k}\Omega$. What is the output voltage ($\mathbf{V}_{out}$) when $V_{in} = 10$ V rms?
12. A high-pass filter has a critical frequency of 50 Hz. Determine which of the following frequencies are passed and which are rejected:
   (a) 1 Hz  (b) 20 Hz  (c) 50 Hz  (d) 60 Hz  (e) 30 kHz
13. Determine the output voltage of each filter in Figure 19–42 at the specified frequency when $V_{in} = 10$ V.

FIGURE 19–42

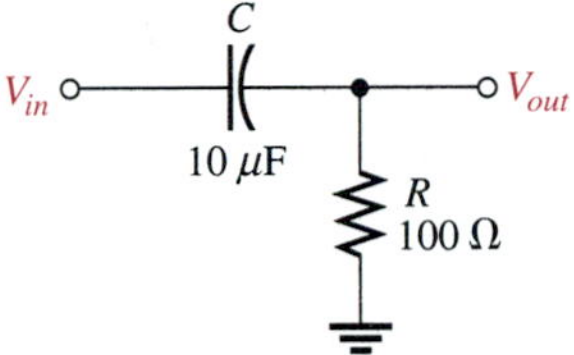

(a) $f = 60$ Hz

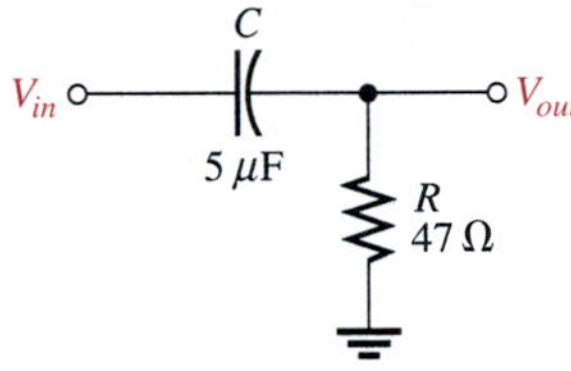

(b) $f = 400$ Hz

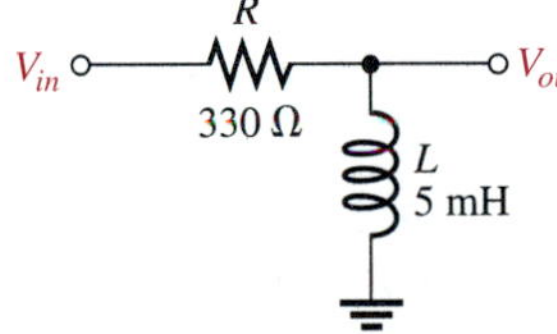

(c) $f = 1$ kHz

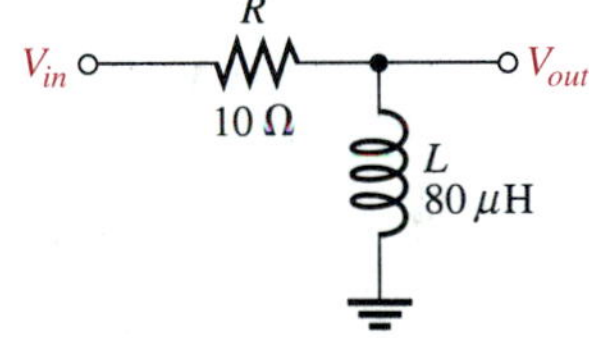

(d) $f = 2$ kHz

14. What is $f_c$ for each filter in Figure 19–42? Determine the output voltage at $f_c$ in each case ($V_{in} = 10$ V).
15. Sketch the Bode plot for each filter in Figure 19–42.
16. Determine $f_c$ for each switch position in Figure 19–43.

FIGURE 19–43

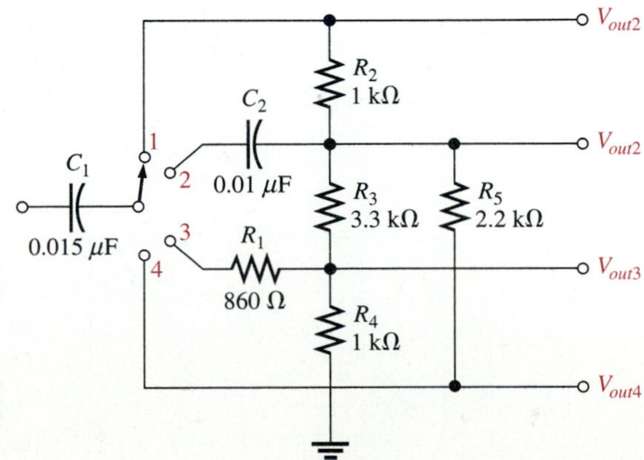

## SECTION 19–3 Band-Pass Filters

17. Determine the center frequency for each filter in Figure 19–44.

FIGURE 19–44

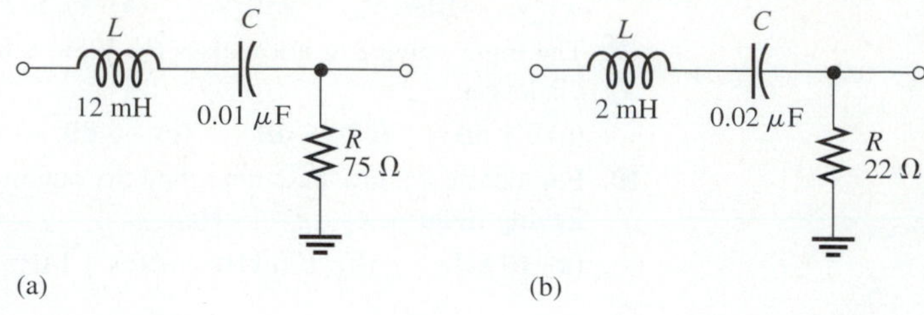


18. Assuming that the coils in Figure 19–44 have a winding resistance of 10 Ω, find the bandwidth for each filter.
19. What are the upper and lower critical frequencies for each filter in Figure 19–44? Assume the response is symmetrical about $f_0$.
20. For each filter in Figure 19–45, find the center frequency of the passband. Neglect $R_W$.

FIGURE 19–45

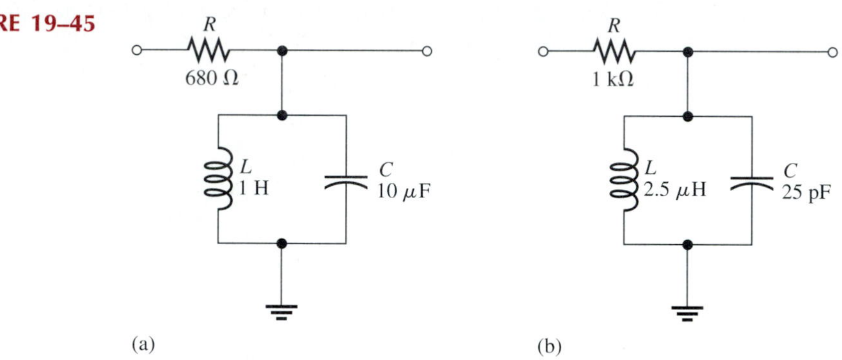


21. If the coils in Figure 19–45 have a winding resistance of 4 Ω, what is the output voltage at resonance when $V_{in} = 120$ V?
22. Determine the separation of center frequencies for all switch positions in Figure 19–46. Do any of the responses overlap? Assume $R_W = 0\ \Omega$ for each coil.

FIGURE 19–46

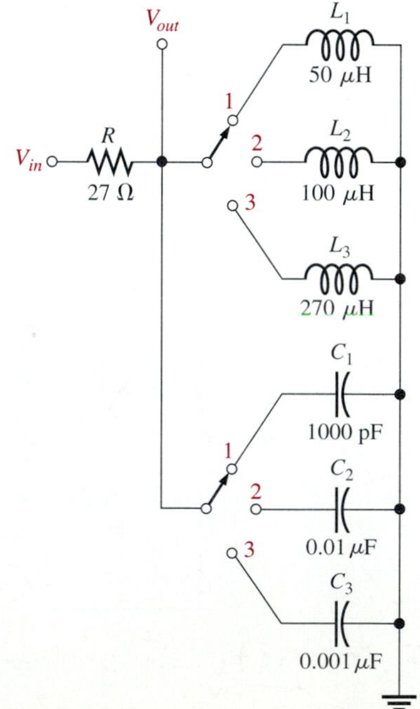

**23.** Design a band-pass filter using a parallel resonant circuit to meet all of the following specifications: $BW = 500$ Hz; $Q = 40$; and $I_{C(max)} = 20$ mA, $V_{C(max)} = 2.5$ V.

## SECTION 19–4 Band-Stop Filters

**24.** Determine the center frequency for each filter in Figure 19–47.

FIGURE 19–47

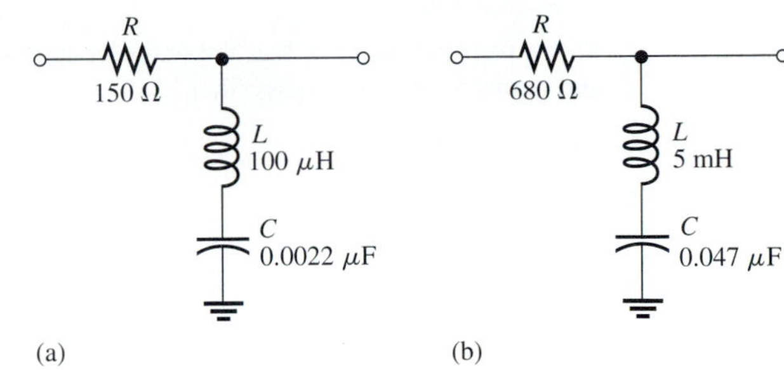


**25.** For each filter in Figure 19–48, find the center frequency of the stop band.

FIGURE 19–48

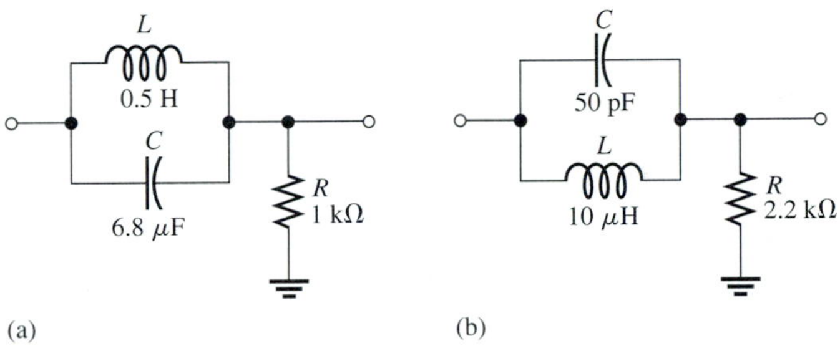


**26.** If the coils in Figure 19–48 have a winding resistance of 8 Ω, what is the output voltage at resonance when $V_{in} = 50$ V?

**27.** Determine the values of $L_1$ and $L_2$ in Figure 19–49 to pass a signal with a frequency of 1200 kHz and stop (reject) a signal with a frequency of 456 kHz.

FIGURE 19–49

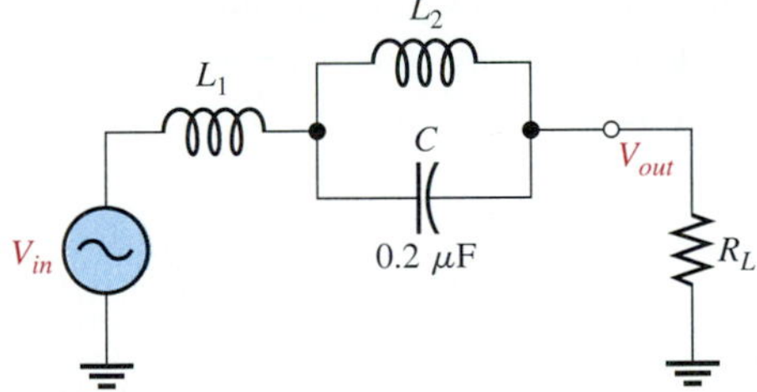


## ■ ANSWERS TO SECTION REVIEWS

### Section 19–1

**1.** The passband is 0 Hz to 2.5 kHz

**2.** $\mathbf{V}_{out} = 100\angle -88.9°$ mV rms

**3.** $20 \log(V_{out}/V_{in}) = -9.54$ dB

### Section 19–2

**1.** $V_{out} = 0.707$ V

**2.** $\mathbf{V}_{out} = 9.98\angle 3.81°$ V

### Section 19–3

**1.** $BW = 30.2$ kHz $- 29.8$ kHz $= 400$ Hz

**2.** $f_0 \cong 1.04$ MHz

### Section 19–4

1. A band-stop filter rejects, rather than passes, a certain band of frequencies.
2. High-pass/low-pass combination, series resonant circuit, and parallel resonant circuit

### Section 19–5

1. The waveforms indicate that the output amplitude decreases with an increase in frequency as in a low-pass filter.
2. The waveforms indicate that the output amplitude is maximum at 10 kHz and drops off above and below as in a band-pass filter.

### Section 19–6

1.
```
V1   1  0  AC  5
C1   1  2  0.01U
R1   2  0  1K
V2   3  0  AC  5
R2   3  4  1K
L1   4  0  10M
.AC  LIN  1000  1K  1MEG
.PROBE
.END
```

2. V(2,0) plots voltage magnitude at node 2 versus frequency, VP(2,0) plots phase angle of the voltage at node 2 versus frequency, and VDB(2,0) plots voltage at node 2 in dB to create a Bode plot.

## ■ ANSWERS TO RELATED EXERCISES FOR EXAMPLES

**19–1** −1.41 dB
**19–2** 7.23 kHz
**19–3** $f_c$ increases to 159 kHz. Roll-off rate remains −20 dB/decade.
**19–4** $f_c$ increases to 350 kHz. Roll-off rate remains −20 dB/decade.
**19–5** −60 dB
**19–6** $C = 0.723\ \mu F$; $V_{out} = 4.98$ V; $\phi = 5.71°$
**19–7** 10.5 kHz
**19–8** $BW$ increases to 18.8 kHz.
**19–9** 1.59 MHz
**19–10** 7.12 kHz (no significant difference)
**19–11** $V_{out} = 15.2$ mV; $BW = 105$ Hz
**19–12** 1.94 V

# 20

# CIRCUIT THEOREMS IN AC ANALYSIS



## INTRODUCTION

Several important theorems were covered in Chapter 8 with emphasis on their applications in the analysis of dc circuits. This chapter is a continuation of that coverage with emphasis on applications in the analysis of ac circuits with reactive elements.

The theorems in this chapter make analysis easier for certain types of circuits. These methods do not replace Ohm's law and Kirchhoff's laws, but they are normally used in conjunction with the laws in certain situations. In this chapter, you will learn the basics of putting technology theory into practice.

Although you are already familiar with the theorems covered in this chapter, a restatement of their purposes may be helpful. The superposition theorem will help you to deal with circuits that have multiple sources. Thevenin's, Norton's, and Millman's theorems provide methods for reducing a circuit to a simple equivalent form for easier analysis. The maximum power transfer theorem is used in applications where it is important for a given circuit to provide maximum power to a load.

## TECHnology Theory Into Practice

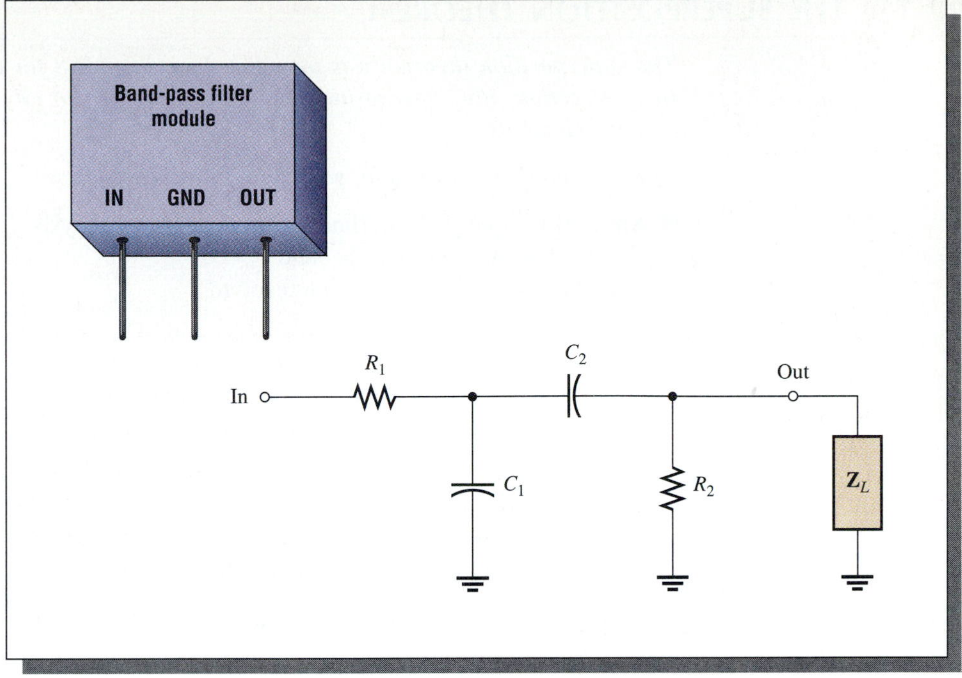


In the TECH TIP assignment in Section 20–6, you will evaluate a band-pass filter module to determine its internal component values, and you will apply Thevenin's theorem to determine an optimum load impedance for maximum power transfer.

## CHAPTER OBJECTIVES

- ☐ Apply the superposition theorem to ac circuit analysis
- ☐ Apply Thevenin's theorem to simplify reactive ac circuits for analysis
- ☐ Apply Norton's theorem to simplify reactive ac circuits
- ☐ Apply Millman's theorem to parallel ac sources
- ☐ Apply the maximum power transfer theorem
- ☐ Use PSpice and Probe to determine Thevenin and Norton equivalent circuits (optional)

## 20–1 ▪ THE SUPERPOSITION THEOREM

***The superposition theorem was introduced in Chapter 8 for use in dc circuit analysis. In this section, the superposition theorem is applied to circuits with ac sources and reactive elements.***

***After completing this section, you should be able to***

- **Apply the superposition theorem to ac circuit analysis**
  - ☐ State the superposition theorem
  - ☐ List the steps in applying the theorem

The **superposition theorem** can be stated as follows:

> **The current in any given branch of a multiple-source circuit can be found by determining the currents in that particular branch produced by each source acting alone, with all other sources replaced by their internal impedances. The total current in the given branch is the phasor sum of the individual source currents in that branch.**

The procedure for the application of the superposition theorem is

**Step 1:** Leave one of the sources in the circuit, and reduce all others to zero. Reduce voltage sources to zero by placing a theoretical short between the terminals; any internal series impedance remains. Reduce current sources to zero by placing an open between the terminals; any internal parallel impedance remains.

**Step 2:** Find the current in the branch of interest produced by the one remaining source.

**Step 3:** Repeat Steps 1 and 2 for each source in turn. When complete, you will have a number of current values equal to the number of sources in the circuit.

**Step 4:** Add the individual current values as phasor quantities.

The following three examples illustrate this procedure.

**EXAMPLE 20–1**

Find the current in $R$ of Figure 20–1 using the superposition theorem. Assume the internal source impedances are zero.

FIGURE 20–1

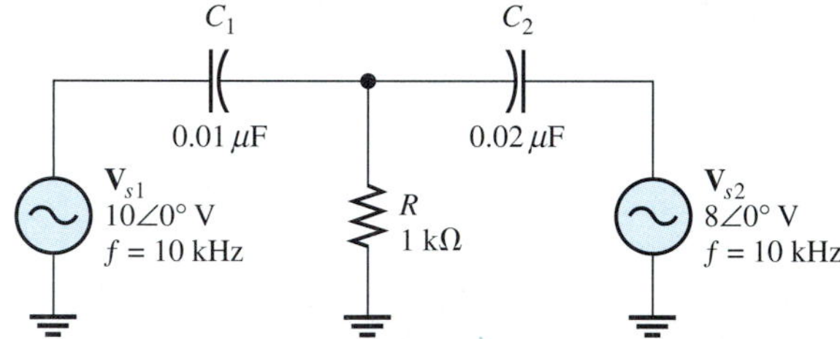


***Solution***

**Step 1:** By zeroing $V_{s2}$, find the current in $R$ due to $V_{s1}$, as indicated in Figure 20–2.

FIGURE 20–2

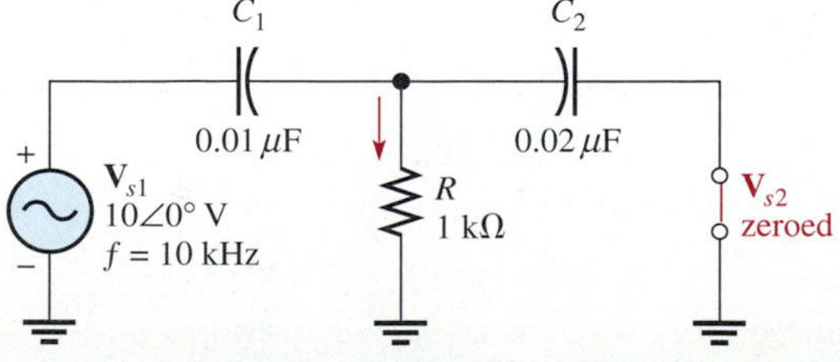

$$X_{C1} = \frac{1}{2\pi f C_1} = \frac{1}{2\pi(10 \text{ kHz})(0.01\ \mu\text{F})} = 1.59 \text{ k}\Omega$$

$$X_{C2} = \frac{1}{2\pi f C_2} = \frac{1}{2\pi(10 \text{ kHz})(0.02\ \mu\text{F})} = 796\ \Omega$$

Looking from $V_{s1}$, the impedance is

$$\mathbf{Z} = \mathbf{X}_{C1} + \frac{\mathbf{R}\mathbf{X}_{C2}}{\mathbf{R} + \mathbf{X}_{C2}} = 1.59\angle{-90^\circ} \text{ k}\Omega + \frac{(1\angle 0^\circ \text{ k}\Omega)(796\angle{-90^\circ}\ \Omega)}{1 \text{ k}\Omega - j796\ \Omega}$$

$$= 1.59\angle{-90^\circ} \text{ k}\Omega + 622\angle{-51.5^\circ}\ \Omega$$

$$= -j1.59 \text{ k}\Omega + 387\ \Omega - j487\ \Omega = 387\ \Omega - j2.08 \text{ k}\Omega$$

Converting to polar form yields

$$\mathbf{Z} = 2.12\angle{-79.5^\circ} \text{ k}\Omega$$

The total current from source 1 is

$$\mathbf{I}_{s1} = \frac{\mathbf{V}_{s1}}{\mathbf{Z}} = \frac{10\angle 0^\circ \text{ V}}{2.12\angle{-79.5^\circ} \text{ k}\Omega} = 4.72\angle 79.5^\circ \text{ mA}$$

Use the current-divider formula. The current through $R$ due to $V_{s1}$ is

$$\mathbf{I}_{R1} = \left(\frac{X_{C2}\angle{-90^\circ}}{R - jX_{C2}}\right)\mathbf{I}_{s1} = \left(\frac{796\angle{-90^\circ}\ \Omega}{1 \text{ k}\Omega - j796\ \Omega}\right)4.72\angle 79.5^\circ \text{ mA}$$

$$= (0.663\angle{-51.5^\circ}\ \Omega)(4.72\angle 79.5^\circ \text{ mA}) = 3.13\angle 28.0^\circ \text{ mA}$$

**Step 2:** Find the current in $R$ due to source $V_{s2}$ by zeroing $V_{s1}$, as shown in Figure 20–3.

FIGURE 20–3

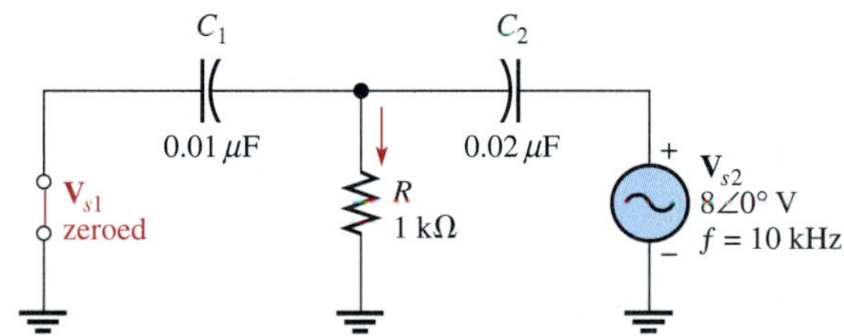


Looking from $V_{s2}$, the impedance is

$$\mathbf{Z} = \mathbf{X}_{C2} + \frac{\mathbf{R}\mathbf{X}_{C1}}{\mathbf{R} + \mathbf{X}_{C1}} = \frac{796\angle{-90^\circ}\ \Omega + (1\angle 0^\circ \text{ k}\Omega)(1.59\angle{-90^\circ} \text{ k}\Omega)}{1 \text{ k}\Omega - j1.59 \text{ k}\Omega}$$

$$= 796\angle{-90^\circ}\ \Omega + 847\angle{-32.2^\circ}\ \Omega$$

$$= -j796\ \Omega + 717\ \Omega - j451\ \Omega = 717\ \Omega - j1247\ \Omega$$

Converting to polar form yields

$$\mathbf{Z} = 1438\angle{-60.1^\circ}\ \Omega$$

The total current from source 2 is

$$\mathbf{I}_{s2} = \frac{\mathbf{V}_{s2}}{\mathbf{Z}} = \frac{8\angle 0^\circ \text{ V}}{1438\angle{-60.1^\circ}\ \Omega} = 5.56\angle 60.1^\circ \text{ mA}$$

Use the current-divider formula. The current through $R$ due to $V_{s2}$ is

$$\mathbf{I}_{R2} = \left(\frac{X_{C1}\angle{-90^\circ}}{R - jX_{C1}}\right)\mathbf{I}_{s2}$$

$$= \left(\frac{1.59\angle{-90^\circ} \text{ k}\Omega}{1 \text{ k}\Omega - j1.59 \text{ k}\Omega}\right)5.56\angle 60.1^\circ \text{ mA} = 4.70\angle 27.9^\circ \text{ mA}$$

**Step 3:** The two individual resistor currents are now converted to rectangular form and added to get the total current through $R$.

$$\mathbf{I}_{R1} = 3.13\angle 28.0^\circ \text{ mA} = 2.76 \text{ mA} + j1.47 \text{ mA}$$
$$\mathbf{I}_{R2} = 4.70\angle 27.9^\circ \text{ mA} = 4.15 \text{ mA} + j2.20 \text{ mA}$$
$$\mathbf{I}_R = \mathbf{I}_{R1} + \mathbf{I}_{R2} = 6.91 \text{ mA} + j3.67 \text{ mA} = 7.82\angle 28.0^\circ \text{ mA}$$

***Related Exercise*** Determine $\mathbf{I}_R$ if $\mathbf{V}_{s2} = 8\angle 180^\circ$ V in Figure 20–1.

## EXAMPLE 20–2

Find the coil current in Figure 20–4. Assume the sources are ideal.

FIGURE 20–4

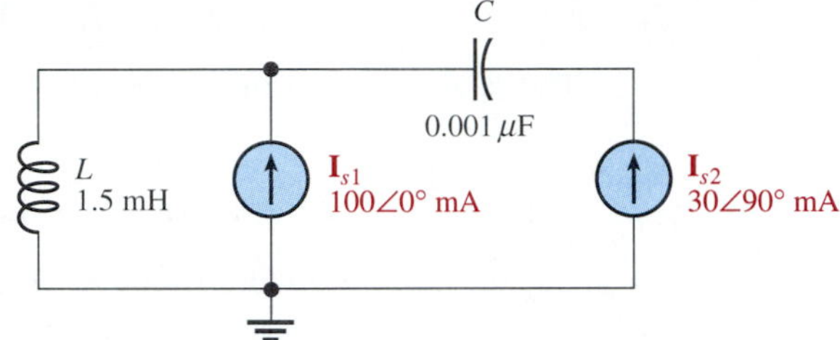


***Solution***

**Step 1:** Find the current through the inductor due to current source $I_{s1}$ by replacing source $I_{s2}$ with an open, as shown in Figure 20–5. As you can see, the entire 100 mA from the current source $I_{s1}$ is through the coil.

FIGURE 20–5

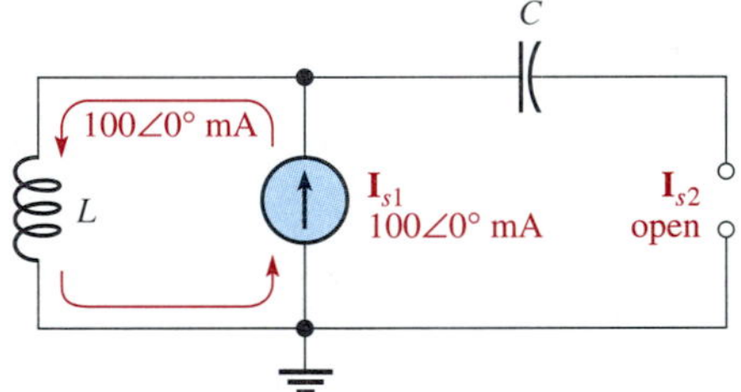


**Step 2:** Find the current through the inductor due to current source $I_{s2}$ by replacing source $I_{s1}$ with an open, as indicated in Figure 20–6. Notice that all of the 30 mA from source $I_{s2}$ is through the coil.

FIGURE 20–6

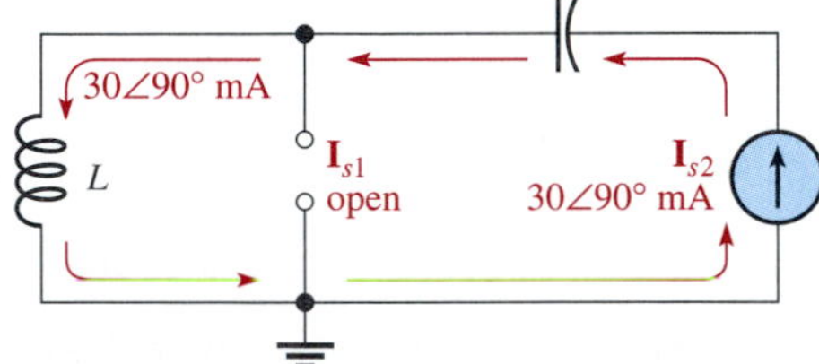


**Step 3:** To get the total inductor current, the two individual currents are superimposed and added as phasor quantities.

$$\mathbf{I}_L = \mathbf{I}_{L1} + \mathbf{I}_{L2}$$
$$= 100\angle 0^\circ \text{ mA} + 30\angle 90^\circ \text{ mA} = 100 \text{ mA} + j30 \text{ mA}$$
$$= 104\angle 16.7^\circ \text{ mA}$$

***Related Exercise*** Find the current through the capacitor in Figure 20–4.

**EXAMPLE 20–3** Find the total current in the resistor $R_L$ in Figure 20–7. Assume the sources are ideal.

FIGURE 20–7

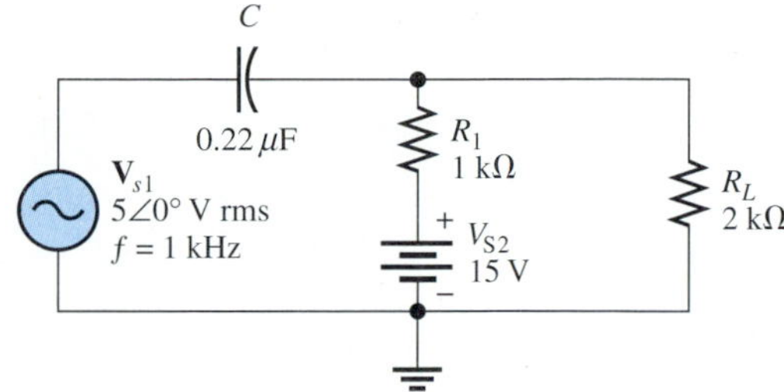


***Solution***

**Step 1:** Find the current through $R_L$ due to source $V_{s1}$ by zeroing the dc source $V_{S2}$, as shown in Figure 20–8. Looking from $V_{s1}$, the impedance is

$$\mathbf{Z} = \mathbf{X}_C + \frac{\mathbf{R}_1\mathbf{R}_L}{\mathbf{R}_1 + \mathbf{R}_L}$$

$$X_C = \frac{1}{2\pi(1\ \text{kHz})(0.22\ \mu\text{F})} = 723\ \Omega$$

$$\mathbf{Z} = 723\angle{-90°}\ \Omega + \frac{(1\angle 0°\ \text{k}\Omega)(2\angle 0°\ \text{k}\Omega)}{3\angle 0°\ \text{k}\Omega}$$

$$= -j723\ \Omega + 667\ \Omega = 984\angle{-47.3°}\ \Omega$$

The total current from source 1 is

$$\mathbf{I}_{s1} = \frac{\mathbf{V}_{s1}}{\mathbf{Z}} = \frac{5\angle 0°\ \text{V}}{984\angle{-47.3°}\ \Omega} = 5.08\angle 47.3°\ \text{mA}$$

Use the current-divider approach. The current in $R_L$ due to $V_{s1}$ is

$$\mathbf{I}_{L(s1)} = \left(\frac{R_1}{R_1 + R_L}\right)\mathbf{I}_{s1} = \left(\frac{1\ \text{k}\Omega}{3\ \text{k}\Omega}\right)5.08\angle 47.3°\ \text{mA} = 1.69\angle 47.3°\ \text{mA}$$

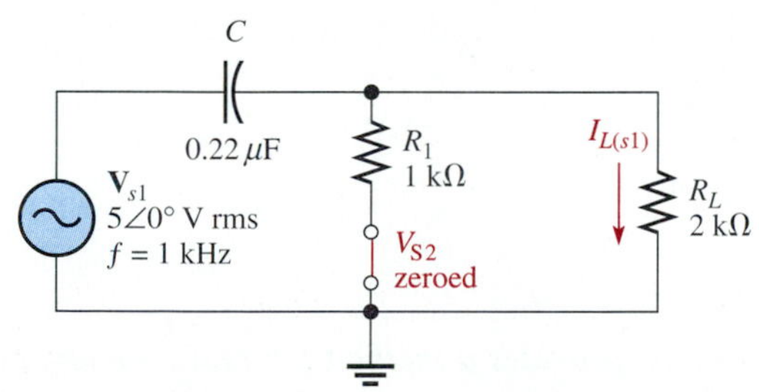


FIGURE 20–8

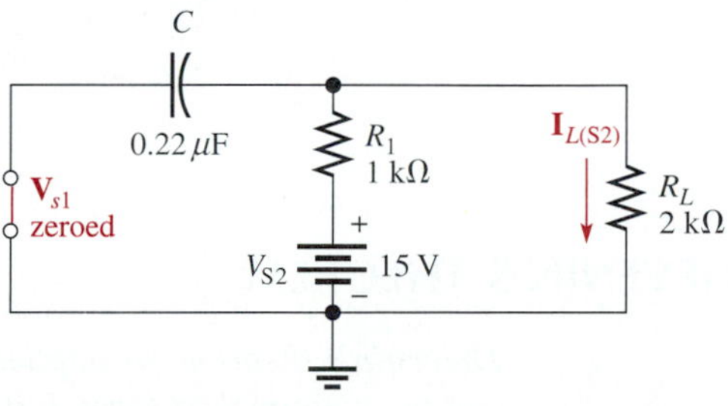


FIGURE 20–9

**Step 2:** Find the current in $R_L$ due to the dc source $V_{S2}$ by zeroing $V_{s1}$, as shown in Figure 20–9.The impedance magnitude as seen by $V_{S2}$ is

$$Z = R_1 + R_L = 3\ \text{k}\Omega$$

The current produced by $V_{S2}$ is

$$I_{L(S2)} = \frac{V_{S2}}{Z} = \frac{15\ \text{V}}{3\ \text{k}\Omega} = 5\ \text{mA dc}$$

**Step 3:** By superposition, the total current in $R_L$ is $1.69\angle 47.3^\circ$ mA riding on a dc level of 5 mA, as indicated in Figure 20–10.

FIGURE 20–10

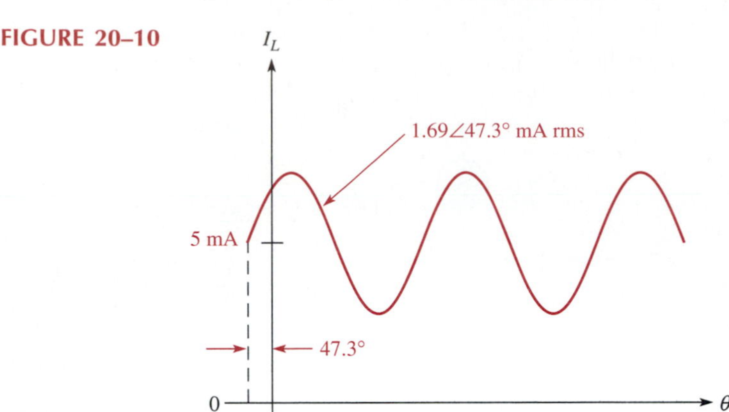


***Related Exercise*** Determine the current through $R_L$ if $V_{S2}$ is changed to 9 V.

**SECTION 20–1 REVIEW**

1. If two equal currents are in opposing directions at any instant of time in a given branch of a circuit, what is the net current at that instant?
2. Why is the superposition theorem useful in the analysis of multiple-source circuits?
3. Using the superposition theorem, find the magnitude of the current through $R$ in Figure 20–11.

FIGURE 20–11

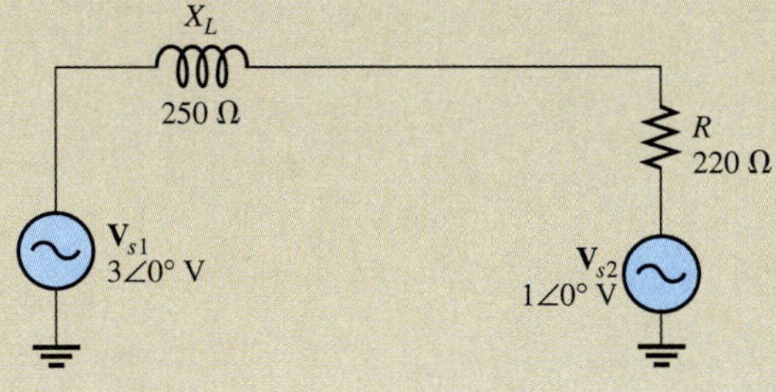


## 20–2 ▪ THEVENIN'S THEOREM

***Thevenin's theorem, as applied to ac circuits, provides a method for reducing any circuit to an equivalent form that consists of an equivalent ac voltage source in series with an equivalent impedance.***

***After completing this section, you should be able to***

- **Apply Thevenin's theorem to simplify reactive ac circuits for analysis**
  - ☐ Describe the form of a Thevenin equivalent circuit
  - ☐ Obtain the Thevenin equivalent ac voltage source
  - ☐ Obtain the Thevenin equivalent impedance
  - ☐ List the steps in applying Thevenin's theorem to an ac circuit

The form of Thevenin's **equivalent circuit** is shown in Figure 20–12. Regardless of how complex the original circuit is, it can always be reduced to this equivalent form. The equivalent voltage source is designated $\mathbf{V}_{th}$; the equivalent impedance is designated $\mathbf{Z}_{th}$ (lowercase italic subscript denotes ac quantity). Notice that the impedance is represented by a block in the circuit diagram. This is because the equivalent impedance can be of several forms: purely resistive, purely capacitive, purely inductive, or a combination of resistance and a reactance.

**FIGURE 20–12**
***Thevenin's equivalent circuit.***

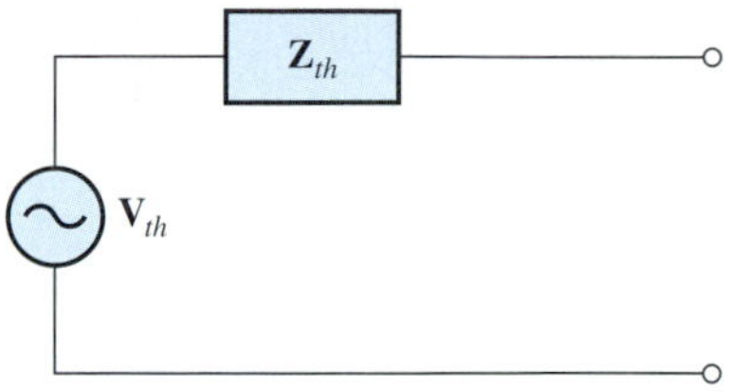

## Equivalency

Figure 20–13(a) shows a block diagram that represents an ac circuit of any given complexity. This circuit has two output terminals, *A* and *B*. A load impedance, $\mathbf{Z}_L$, is connected to the terminals. The circuit produces a certain voltage, $\mathbf{V}_L$, and a certain current, $\mathbf{I}_L$, as illustrated.

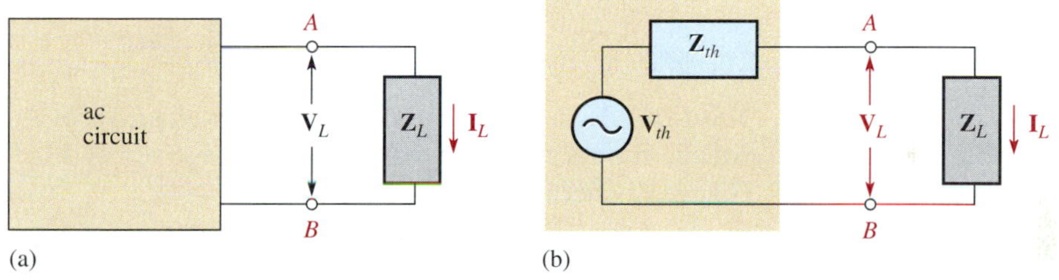


**FIGURE 20–13**
***An ac circuit of any complexity can be reduced to a Thevenin equivalent for analysis purposes.***

By **Thevenin's theorem,** the circuit in the block can be reduced to an equivalent form, as indicated in the beige area of Figure 20–13(b). The term *equivalent* means that when the same value of load is connected to both the original circuit and Thevenin's equivalent circuit, the load voltages and currents are equal for both. Therefore, as far as the load is concerned, there is no difference between the original circuit and Thevenin's equivalent circuit. The load "sees" the same current and voltage regardless of whether it is connected to the original circuit or to the Thevenin equivalent.

## Thevenin's Equivalent Voltage ($\mathbf{V}_{th}$)

As you have seen, the equivalent voltage, $\mathbf{V}_{th}$, is one part of the complete Thevenin equivalent circuit.

> **Thevenin's equivalent voltage is defined as the open circuit voltage between two specified points in a circuit.**

To illustrate, let's assume that an ac circuit of some type has a resistor connected between two defined points, *A* and *B*, as shown in Figure 20–14(a). We wish to find the Thevenin equivalent circuit for the circuit as "seen" by *R*. $\mathbf{V}_{th}$ is the voltage across points *A* and *B*, with *R* removed, as shown in part (b) of the figure. The circuit is viewed from the open terminals *AB*, and *R* is considered external to the circuit for which the Thevenin equivalent is to be found. The following three examples show how to find $\mathbf{V}_{th}$.

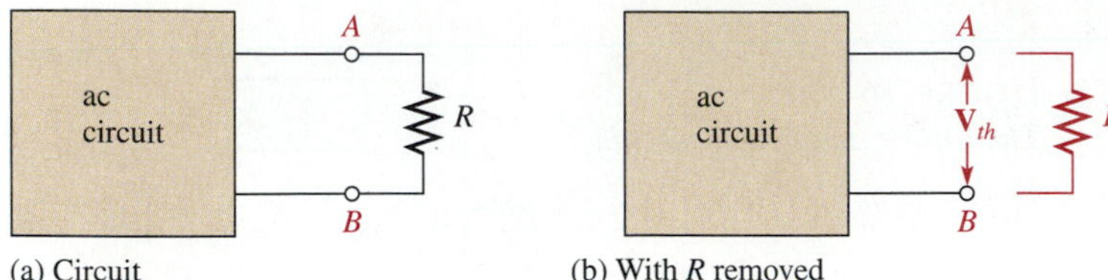


**FIGURE 20–14**
***How $V_{th}$ is determined.***

**EXAMPLE 20–4** Determine $\mathbf{V}_{th}$ for the circuit external to $R_L$ in Figure 20–15. The beige area identifies the portion of the circuit to be thevenized.

**FIGURE 20–15**

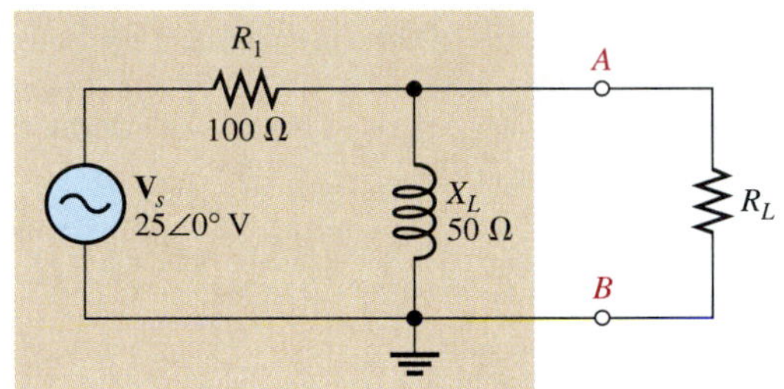


***Solution*** Remove $R_L$ and determine the voltage from *A* to *B* ($\mathbf{V}_{th}$). In this case, the voltage from *A* to *B* is the same as the voltage across $X_L$. This is determined using the voltage-divider method.

$$\mathbf{V}_L = \left(\frac{X_L\angle 90^\circ}{R_1 + jX_L}\right)\mathbf{V}_s = \left(\frac{50\angle 90^\circ\ \Omega}{112\angle 26.6^\circ\ \Omega}\right)25\angle 0^\circ\ \text{V} = 11.2\angle 63.4^\circ\ \text{V}$$

$$\mathbf{V}_{th} = \mathbf{V}_{AB} = \mathbf{V}_L = 11.2\angle 63.4^\circ\ \text{V}$$

***Related Exercise*** Determine $\mathbf{V}_{th}$ if $R_1$ is changed to 47 Ω in Figure 20–15.

**EXAMPLE 20–5** For the circuit in Figure 20–16, determine the Thevenin voltage as seen by $R_L$.

**FIGURE 20–16**

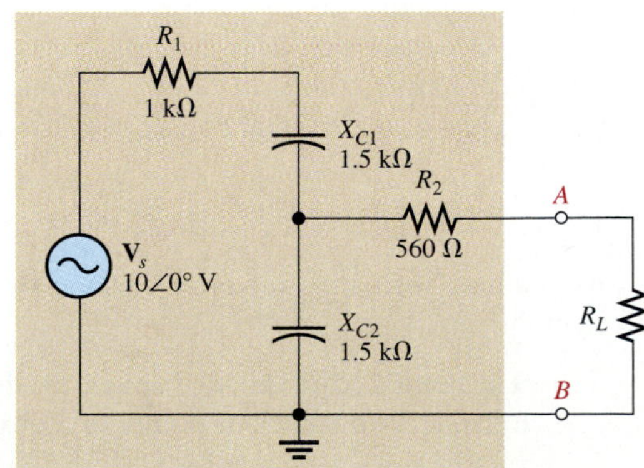

***Solution*** Thevenin's voltage for the circuit between terminals $A$ and $B$ is the voltage that appears across $A$ and $B$ with $R_L$ removed from the circuit.

There is no voltage drop across $R_2$ because the open terminals $AB$ prevent current through it. Thus, $\mathbf{V}_{AB}$ is the same as $\mathbf{V}_{C2}$ and can be found by the voltage-divider formula.

$$\mathbf{V}_{AB} = \mathbf{V}_{C2} = \left(\frac{X_{C2}\angle -90^\circ}{R_1 - jX_{C1} - jX_{C2}}\right)\mathbf{V}_s = \left(\frac{1.5\angle -90^\circ \text{ k}\Omega}{1 \text{ k}\Omega - j3 \text{ k}\Omega}\right)10\angle 0^\circ \text{ V}$$

$$= \left(\frac{1.5\angle -90^\circ \text{ k}\Omega}{3.16\angle -71.6^\circ \text{ k}\Omega}\right)10\angle 0^\circ \text{ V} = 4.75\angle -18.4^\circ \text{ V}$$

$$\mathbf{V}_{th} = \mathbf{V}_{AB} = 4.75\angle -18.4^\circ \text{ V}$$

***Related Exercise*** Determine $\mathbf{V}_{th}$ if $R_1$ is changed to 2.2 kΩ in Figure 20–16.

**EXAMPLE 20–6** For Figure 20–17, find $\mathbf{V}_{th}$ for the circuit external to $R_L$.

FIGURE 20–17

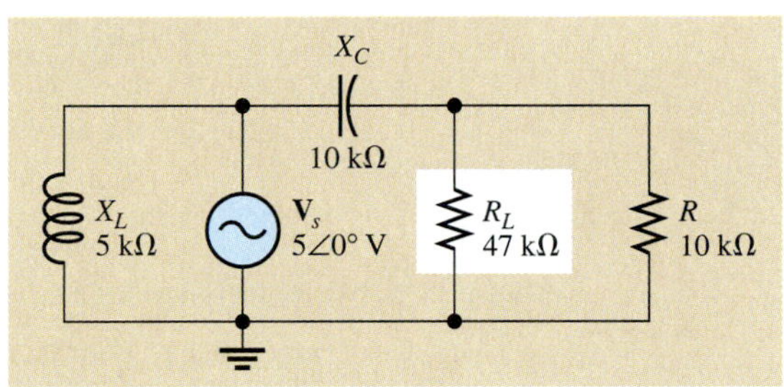


***Solution*** First remove $R_L$ and determine the voltage across the resulting open terminals, which is $\mathbf{V}_{th}$. Find $\mathbf{V}_{th}$ by applying the voltage-divider formula to $X_C$ and $R$.

$$\mathbf{V}_{th} = \mathbf{V}_R = \left(\frac{R\angle 0^\circ}{R - jX_C}\right)\mathbf{V}_s = \left(\frac{10\angle 0^\circ \text{ k}\Omega}{10 \text{ k}\Omega - j10 \text{ k}\Omega}\right)5\angle 0^\circ \text{ V}$$

$$= \left(\frac{10\angle 0^\circ \text{ k}\Omega}{14.14\angle -45^\circ \text{ k}\Omega}\right)5\angle 0^\circ \text{ V} = 3.54\angle 45^\circ \text{ V}$$

Notice the $L$ has no effect on the result, since the 5 V source appears across $C$ and $R$ in combination.

***Related Exercise*** Find $\mathbf{V}_{th}$ if $R$ is 22 kΩ and $R_L$ is 39 kΩ in Figure 20–17.

## Thevenin's Equivalent Impedance ($\mathbf{Z}_{th}$)

The previous examples illustrated how to find only one part of a Thevenin equivalent circuit. Now, let's turn our attention to determining the Thevenin equivalent impedance, $\mathbf{Z}_{th}$. As defined by Thevenin's theorem,

> **Thevenin's equivalent impedance is the total impedance appearing between two specified terminals in a given circuit with all sources zeroed and replaced by their internal impedances (if any).**

Thus, when we wish to find $\mathbf{Z}_{th}$ between any two terminals in a circuit, all the voltage sources are shorted (any internal impedance remains in series). All the current sources are opened (any internal impedance remains in parallel). Then the total impedance between the two terminals is determined. The following three examples illustrate how to find $\mathbf{Z}_{th}$.

**EXAMPLE 20–7** Find $\mathbf{Z}_{th}$ for the part of the circuit in Figure 20–18 that is external to $R_L$. This is the same circuit used in Example 20–4.

FIGURE 20–18

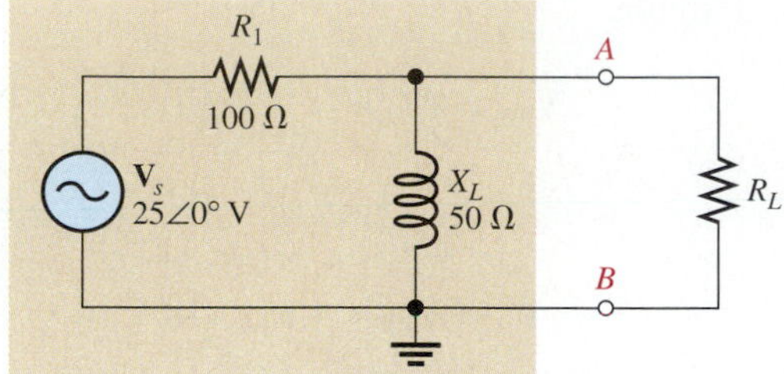


***Solution*** First, reduce $\mathbf{V}_s$ to zero by shorting it, as shown in Figure 20–19.

FIGURE 20–19

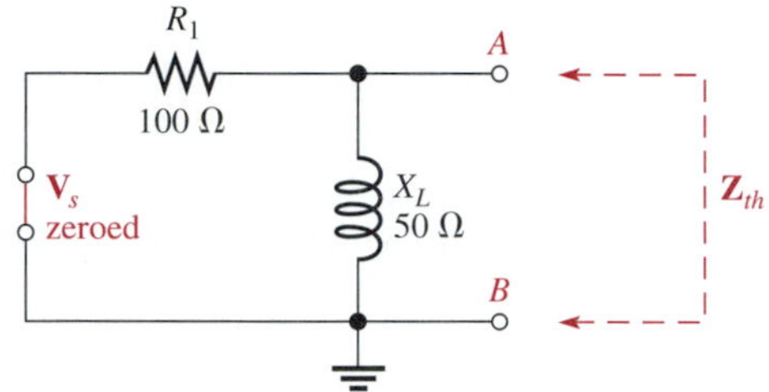


Looking in between terminals $A$ and $B$, $R$ and $X_L$ are in parallel. Thus,

$$\mathbf{Z}_{th} = \frac{(R_1\angle 0^\circ)(X_L\angle 90^\circ)}{R_1 + jX_L} = \frac{(100\angle 0^\circ\ \Omega)(50\angle 90^\circ\ \Omega)}{100\ \Omega + j50\ \Omega}$$

$$= \frac{(100\angle 0^\circ\ \Omega)(50\angle 90^\circ\ \Omega)}{112\angle 26.6^\circ\ \Omega} = 44.6\angle 63.4^\circ\ \Omega$$

***Related Exercise*** Change $R_1$ to 47 Ω and determine $\mathbf{Z}_{th}$.

**EXAMPLE 20–8** For the circuit in Figure 20–20, determine $\mathbf{Z}_{th}$ as seen by $R_L$. This is the same circuit used in Example 20–5.

FIGURE 20–20

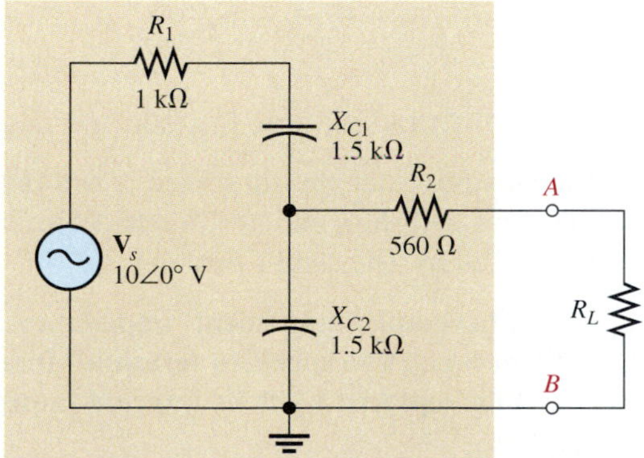


***Solution*** First, zero the voltage source, as shown in Figure 20–21.

FIGURE 20–21

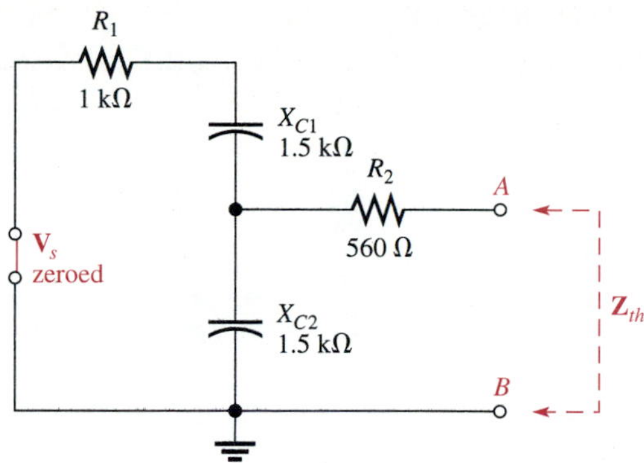


Looking from terminals $A$ and $B$, $C_2$ appears in parallel with the series combination of $R_1$ and $C_1$. This entire combination is in series with $R_2$. The calculation for $\mathbf{Z}_{th}$ is as follows:

$$\mathbf{Z}_{th} = R_2\angle 0^\circ + \frac{(X_{C2}\angle -90^\circ)(R_1 - jX_{C1})}{R_1 - jX_{C1} - jX_{C2}}$$

$$= 560\angle 0^\circ\ \Omega + \frac{(1.5\angle -90^\circ\ \text{k}\Omega)(1\ \text{k}\Omega - j1.5\ \text{k}\Omega)}{1\ \text{k}\Omega - j3\ \text{k}\Omega}$$

$$= 560\angle 0^\circ\ \Omega + \frac{(1.5\angle -90^\circ\ \text{k}\Omega)(1.8\angle -56.3^\circ\ \text{k}\Omega)}{3.16\angle -71.6^\circ\ \text{k}\Omega}$$

$$= 560\angle 0^\circ\ \Omega + 854\angle -74.7^\circ\ \Omega = 560\ \Omega + 225\ \Omega - j824\ \Omega$$

$$= 785\ \Omega - j824\ \Omega = 1138\angle -46.4^\circ\ \Omega$$

***Related Exercise*** Determine $\mathbf{Z}_{th}$ if $R_1$ is changed to 2.2 kΩ in Figure 20–20.

**EXAMPLE 20–9** For the circuit in Figure 20–22, determine $\mathbf{Z}_{th}$ for the portion of the circuit external to $R_L$. This is the same circuit as in Example 20–6.

FIGURE 20–22

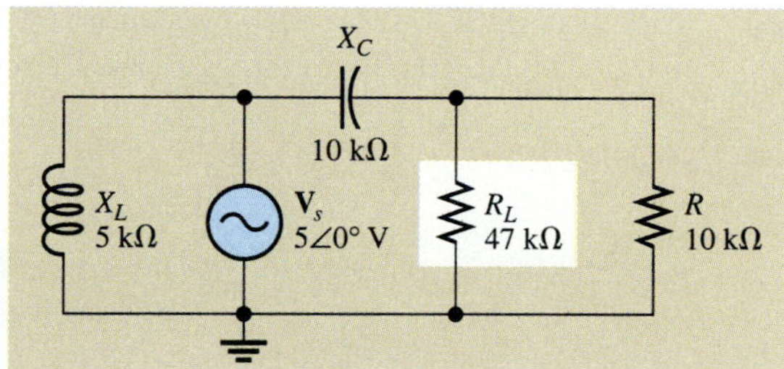


***Solution*** With the voltage source zeroed, $X_L$ is effectively out of the circuit. $R$ and $C$ appear in parallel when viewed from the open terminals, as indicated in Figure 20–23. $\mathbf{Z}_{th}$ is calculated as follows:

$$\mathbf{Z}_{th} = \frac{(R\angle 0^\circ)(X_C\angle -90^\circ)}{R - jX_C} = \frac{(10\angle 0^\circ\ \text{k}\Omega)(10\angle -90^\circ\ \text{k}\Omega)}{14.1\angle -45^\circ\ \text{k}\Omega} = 7.07\angle -45^\circ\ \text{k}\Omega$$

FIGURE 20–23

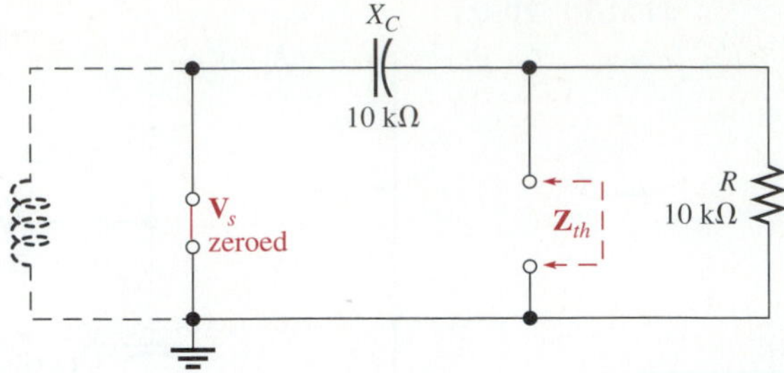


***Related Exercise*** Find $\mathbf{Z}_{th}$ if $R$ is 22 kΩ and $R_L$ is 39 kΩ in Figure 20–22.

## Thevenin's Equivalent Circuit

The previous examples have shown how to find the two equivalent components of a Thevenin circuit, $\mathbf{V}_{th}$ and $\mathbf{Z}_{th}$. Keep in mind that $\mathbf{V}_{th}$ and $\mathbf{Z}_{th}$ can be found for any circuit. Once these equivalent values are determined, they must be connected in series to form the Thevenin equivalent circuit. The following examples use the previous examples to illustrate this final step.

**EXAMPLE 20–10** Draw the Thevenin equivalent circuit for the circuit in Figure 20–24 that is external to $R_L$. This is the circuit used in Examples 20–4 and 20–7.

FIGURE 20–24

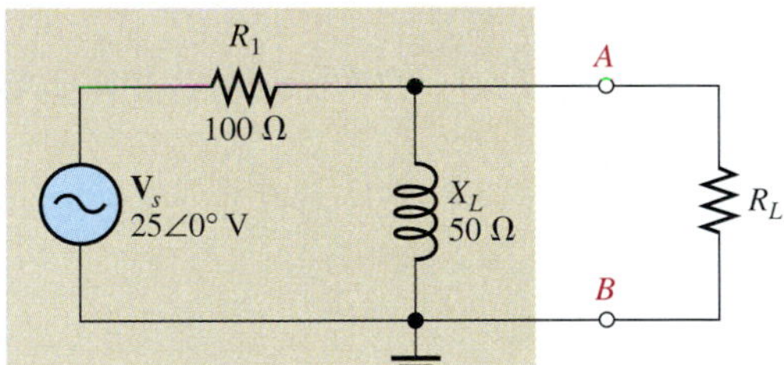


***Solution*** From Examples 20–4 and 20–7 respectively, $\mathbf{V}_{th} = 11.2\angle 63.4°$ V and $\mathbf{Z}_{th} = 44.6\angle 63.4°$ Ω. In rectangular form, the impedance is

$$\mathbf{Z}_{th} = 20\ \Omega + j40\ \Omega$$

This form indicates that the impedance is a 20 Ω resistor in series with a 40 Ω inductive reactance. The Thevenin equivalent circuit is shown in Figure 20–25.

FIGURE 20–25

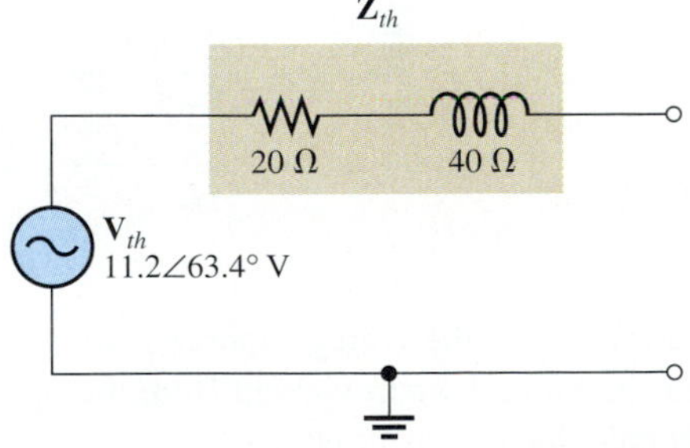


***Related Exercise*** Draw the Thevenin equivalent circuit for Figure 20–24 with $R_1$ = 47 Ω.

**EXAMPLE 20–11** For the circuit in Figure 20–26, sketch the Thevenin equivalent circuit external to $R_L$. This is the circuit used in Examples 20–5 and 20–8.

FIGURE 20–26

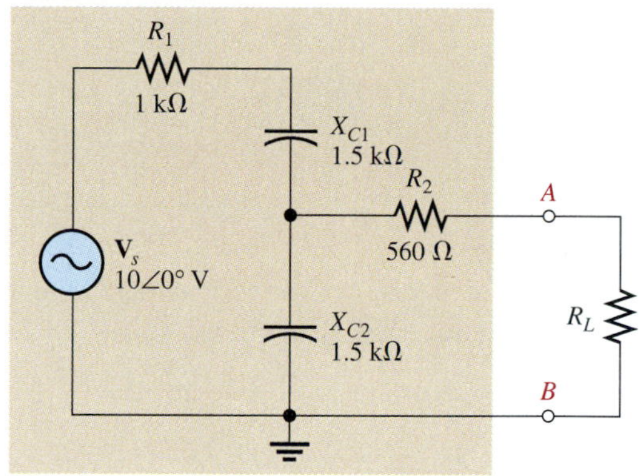


***Solution*** From Examples 20–5 and 20–8 respectively, $\mathbf{V}_{th} = 4.75\angle -18.4°$ V and $\mathbf{Z}_{th} = 1138\angle -46.4°$ Ω. In rectangular form, $\mathbf{Z}_{th} = 785\ \Omega - j824\ \Omega$. The Thevenin equivalent circuit is shown in Figure 20–27.

FIGURE 20–27

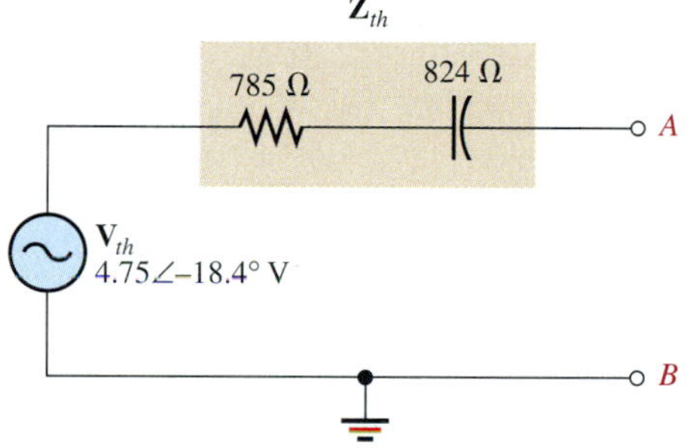


***Related Exercise*** Sketch the Thevenin equivalent for the circuit in Figure 20–26 with $R_1 = 2.2$ kΩ.

**EXAMPLE 20–12** For the circuit in Figure 20–28, determine the Thevenin equivalent circuit as seen by $R_L$. This is the circuit in Examples 20–6 and 20–9.

FIGURE 20–28

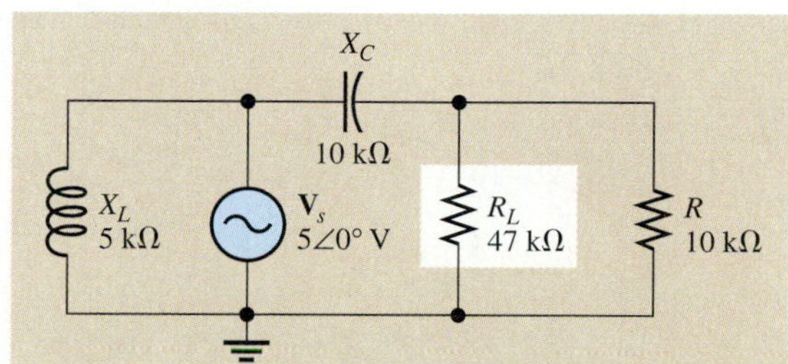


***Solution*** From Examples 20–6 and 20–9 respectively, $\mathbf{V}_{th} = 3.54\angle 45°$ V, and $\mathbf{Z}_{th} = 7.07\angle -45°$ kΩ. The impedance in rectangular form is

$$\mathbf{Z}_{th} = 5\ \text{k}\Omega - j5\ \text{k}\Omega$$

Thus, the Thevenin equivalent circuit is as shown in Figure 20–29.

FIGURE 20–29

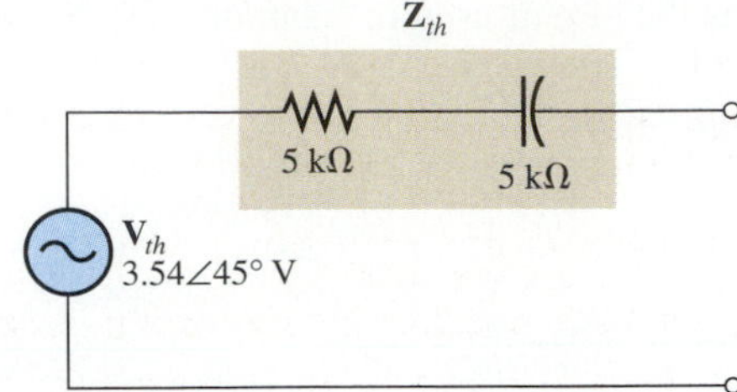


***Related Exercise*** Change $R$ to 22 kΩ and $R_L$ to 39 kΩ in Figure 20–28 and sketch the Thevenin equivalent circuit.

## Summary of Thevenin's Theorem

Remember that the Thevenin equivalent circuit is always of the series form regardless of the original circuit that it replaces. The significance of Thevenin's theorem is that the equivalent circuit can replace the original circuit as far as any external load is concerned. Any load connected between the terminals of a Thevenin equivalent circuit experiences the same current and voltage as if it were connected to the terminals of the original circuit.

A summary of steps for applying Thevenin's theorem follows.

**Step 1:** Open the two terminals between which you want to find the Thevenin circuit. This is done by removing the component from which the circuit is to be viewed.

**Step 2:** Determine the voltage across the two open terminals.

**Step 3:** Determine the impedance viewed from the two open terminals with all sources zeroed (voltage sources replaced with shorts and current sources replaced with opens).

**Step 4:** Connect $\mathbf{V}_{th}$ and $\mathbf{Z}_{th}$ in series to produce the complete Thevenin equivalent circuit.

## SECTION 20–2 REVIEW

1. What are the two basic components of a Thevenin equivalent ac circuit?
2. For a certain circuit, $\mathbf{Z}_{th} = 25\ \Omega - j50\ \Omega$, and $\mathbf{V}_{th} = 5\angle 0°$ V. Sketch the Thevenin equivalent circuit.
3. For the circuit in Figure 20–30, find the Thevenin equivalent looking from terminals *AB*.

FIGURE 20–30

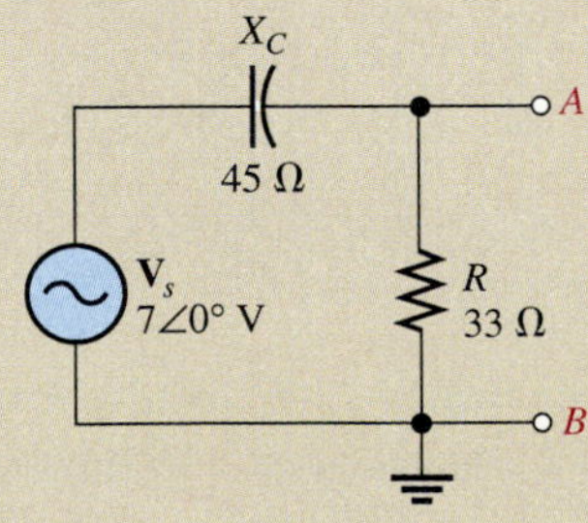

## 20–3 ■ NORTON'S THEOREM

***Like Thevenin's theorem, Norton's theorem provides a method of reducing a more complex circuit to a simpler, more manageable form for analysis. The basic difference is that Norton's theorem gives an equivalent current source (rather than a voltage source) in parallel (rather than in series) with an equivalent impedance.***

***After completing this section, you should be able to***

- **Apply Norton's theorem to simplify reactive ac circuits**
  - ☐ Describe the form of a Norton equivalent circuit
  - ☐ Obtain the Norton equivalent ac current source
  - ☐ Obtain the Norton equivalent impedance

The form of Norton's equivalent circuit is shown in Figure 20–31. Regardless of how complex the original circuit is, it can be reduced to this equivalent form. The equivalent current source is designated $\mathbf{I}_n$, and the equivalent impedance is $\mathbf{Z}_n$ (lowercase italic subscript denotes ac quantity).

FIGURE 20–31
*Norton equivalent circuit.*

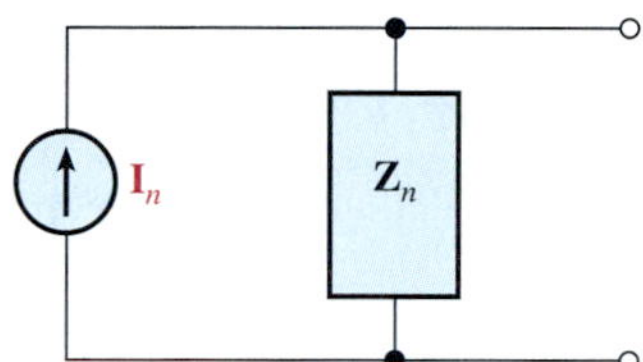


**Norton's theorem** shows you how to find $\mathbf{I}_n$ and $\mathbf{Z}_n$. Once they are known, simply connect them in parallel to get the complete Norton equivalent circuit.

### Norton's Equivalent Current Source ($\mathbf{I}_n$)

$\mathbf{I}_n$ is one part of the Norton equivalent circuit; $\mathbf{Z}_n$ is the other part.

> **Norton's equivalent current is defined as the short circuit current between two specified points in a given circuit.**

Any load connected between these two points effectively "sees" a current source $\mathbf{I}_n$ in parallel with $\mathbf{Z}_n$.

To illustrate, let's suppose that the circuit shown in Figure 20–32 has a load resistor connected to points $A$ and $B$, as indicated in part (a). We wish to find the Norton equivalent for the circuit external to $R_L$. To find $\mathbf{I}_n$, calculate the current between points $A$ and $B$ with those terminals shorted, as shown in part (b). Example 20–13 shows how to find $\mathbf{I}_n$.

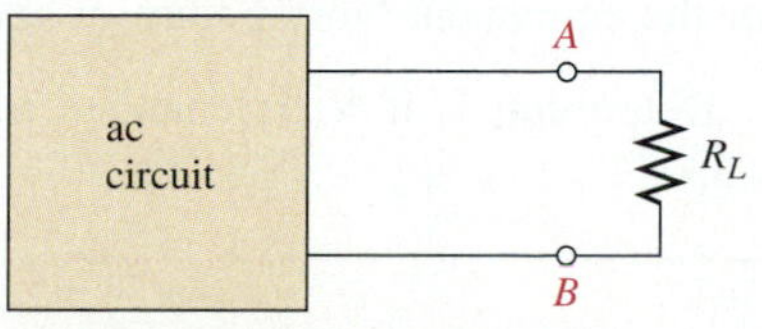


(a) Circuit with load resistor

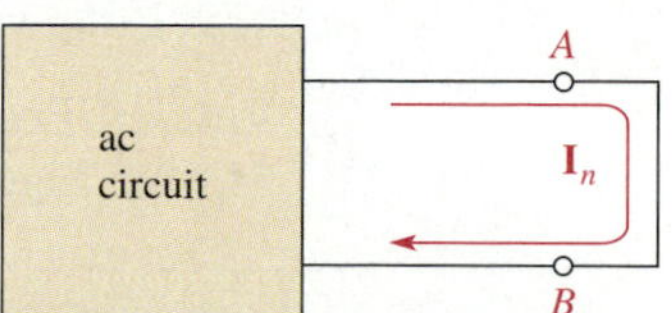


(b) Load is replaced by short and short circuit current is $\mathbf{I}_n$

FIGURE 20–32
***How*** $\mathbf{I}_n$ ***is determined.***

**EXAMPLE 20–13** In Figure 20–33, determine $\mathbf{I}_n$ for the circuit as "seen" by the load resistor. The beige area identifies the portion of the circuit to be nortonized.

FIGURE 20–33

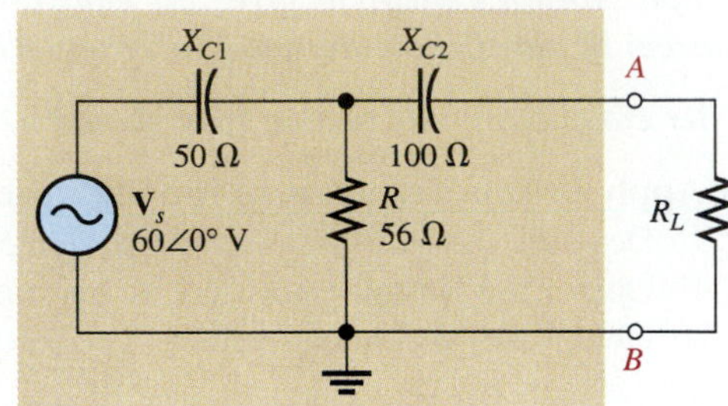


***Solution*** Short the terminals $A$ and $B$, as shown in Figure 20–34.

FIGURE 20–34

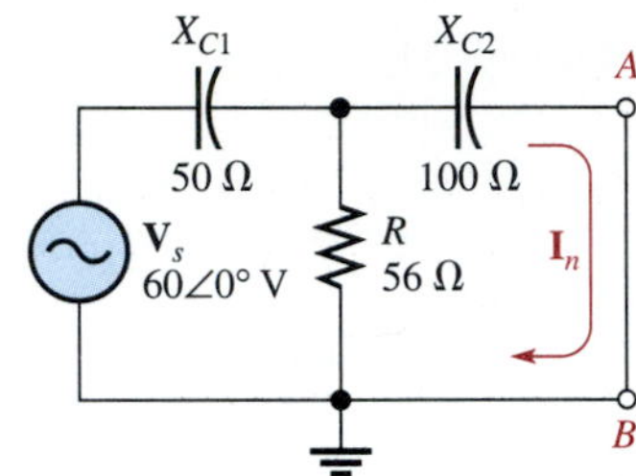


$\mathbf{I}_n$ is the current through the short and is calculated as follows. First, the total impedance viewed from the source is

$$\mathbf{Z} = \mathbf{X}_{C1} + \frac{\mathbf{R}\mathbf{X}_{C2}}{\mathbf{R} + \mathbf{X}_{C2}} = 50\angle{-90^\circ}\ \Omega + \frac{(56\angle 0^\circ\ \Omega)(100\angle{-90^\circ}\ \Omega)}{56\ \Omega - j100\ \Omega}$$

$$= 50\angle{-90^\circ}\ \Omega + 48.9\angle{-29.3^\circ}\ \Omega$$

$$= -j50\ \Omega + 42.6\ \Omega - j23.9\ \Omega = 42.6\ \Omega - j73.9\ \Omega$$

Converting to polar form yields

$$\mathbf{Z} = 85.3\angle{-60.0^\circ}\ \Omega$$

Next, the total current from the source is

$$\mathbf{I}_s = \frac{\mathbf{V}_s}{\mathbf{Z}} = \frac{60\angle 0^\circ\ \text{V}}{85.3\angle{-60.0^\circ}\ \Omega} = 703\angle 60.0^\circ\ \text{mA}$$

Finally, apply the current-divider formula to get $\mathbf{I}_n$ (the current through the short between terminals $A$ and $B$).

$$\mathbf{I}_n = \left(\frac{\mathbf{R}}{\mathbf{R} + \mathbf{X}_{C2}}\right)\mathbf{I}_s = \left(\frac{56\angle 0^\circ\ \Omega}{56\ \Omega - j100\ \Omega}\right)703\angle 60.0^\circ\ \text{mA} = 344\angle 121^\circ\ \text{mA}$$

This is the value for the equivalent Norton current source.

***Related Exercise*** Determine $\mathbf{I}_n$ if $\mathbf{V}_s$ is changed to 25∠0° V and $R$ is changed to 33 Ω in Figure 20–33.

## Norton's Equivalent Impedance ($\mathbf{Z}_n$)

$\mathbf{Z}_n$ is defined the same as $\mathbf{Z}_{th}$: It is the total impedance appearing between two specified terminals of a given circuit viewed from the open terminals with all sources zeroed.

**EXAMPLE 20–14** Find $\mathbf{Z}_n$ for the circuit in Figure 20–33 (Example 20–13) viewed from the open terminals *AB*.

***Solution*** First reduce $\mathbf{V}_s$ to zero, as indicated in Figure 20–35.

FIGURE 20–35

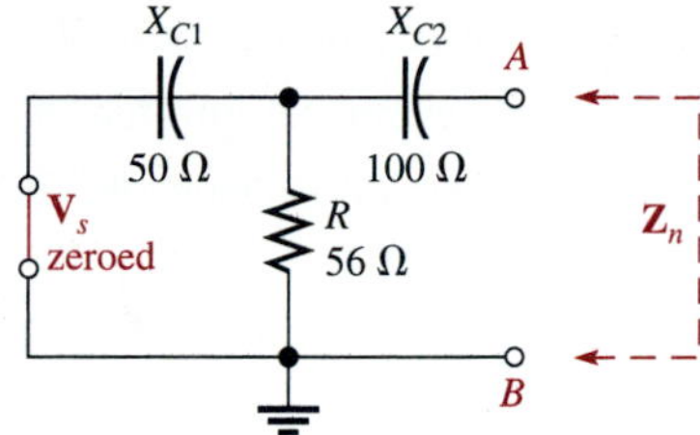


Looking in between terminals *A* and *B*, $C_2$ is in series with the parallel combination of *R* and $C_1$. Thus,

$$\mathbf{Z}_n = \mathbf{X}_{C2} + \frac{\mathbf{R}\mathbf{X}_{C1}}{\mathbf{R} + \mathbf{X}_{C1}} = 100\angle{-90^\circ}\ \Omega + \frac{(56\angle 0^\circ\ \Omega)(50\angle{-90^\circ}\ \Omega)}{56\ \Omega - j50\ \Omega}$$

$$= 100\angle{-90^\circ}\ \Omega + 37.3\angle{-48.2^\circ}\ \Omega$$

$$= -j100\ \Omega + 24.8\ \Omega - j27.8\ \Omega = 24.8\ \Omega - j128\ \Omega$$

The Norton equivalent impedance is a 24.8 Ω resistance in series with a 128 Ω capacitive reactance.

***Related Exercise*** Find $\mathbf{Z}_n$ in Figure 20–33 if $\mathbf{V}_s = 25\angle 0^\circ$ V and $R = 33\ \Omega$.

The previous two examples have shown how to find the two equivalent components of a Norton equivalent circuit. Keep in mind that these values can be found for any given ac circuit. Once these values are known, they are connected in parallel to form the Norton equivalent circuit, as the following example illustrates.

**EXAMPLE 20–15** Sketch the complete Norton equivalent circuit for the circuit in Figure 20–33 (Example 20–13).

***Solution*** From Examples 20–13 and 20–14 respectively, $\mathbf{I}_n = 344\angle 121^\circ$ mA and $\mathbf{Z}_n = 24.8\ \Omega - j128\ \Omega$. The Norton equivalent circuit is shown in Figure 20–36.

FIGURE 20–36

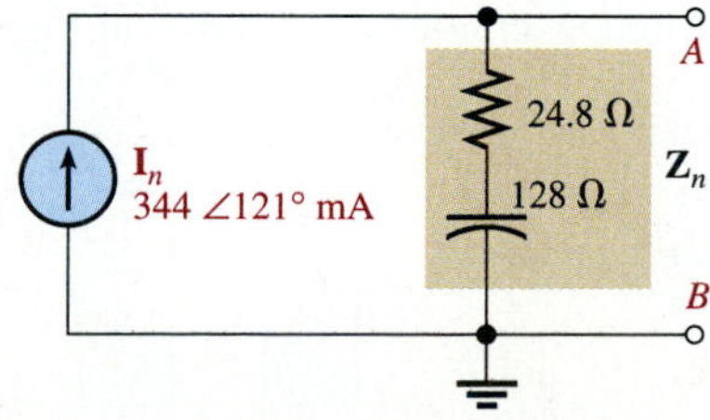


***Related Exercise*** Sketch the Norton equivalent for the circuit in Figure 20–33 if $\mathbf{V}_s = 25\angle 0^\circ$ V and $R = 33\ \Omega$.

### Summary of Norton's Theorem

Any load connected between the terminals of a Norton equivalent circuit will have the same current through it and the same voltage across it as it would when connected to the terminals of the original circuit. A summary of steps for theoretically applying Norton's theorem is as follows:

**Step 1:** Short the two terminals between which the Norton circuit is to be determined.

**Step 2:** Determine the current through the short. This is $\mathbf{I}_n$.

**Step 3:** Determine the impedance between the two open terminals with all sources zeroed. This is $\mathbf{Z}_n$.

**Step 4:** Connect $\mathbf{I}_n$ and $\mathbf{Z}_n$ in parallel.

**SECTION 20–3 REVIEW**

1. For a given circuit, $\mathbf{I}_n = 5\angle 0°$ mA, and $\mathbf{Z}_n = 150\ \Omega + j100\ \Omega$. Draw the Norton equivalent circuit.
2. Find the Norton circuit as seen by $R_L$ in Figure 20–37.

FIGURE 20–37

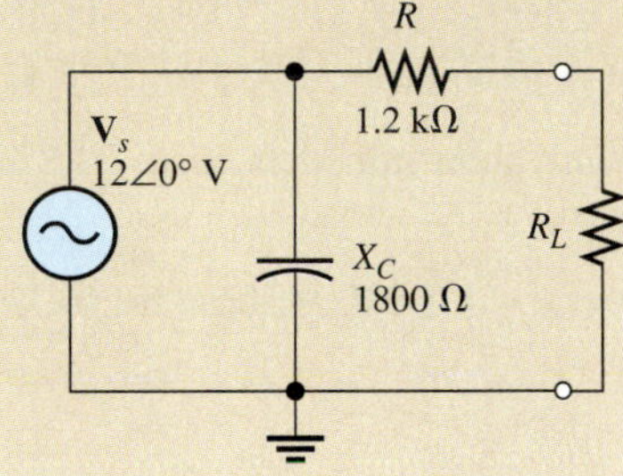


## 20–4 ■ MILLMAN'S THEOREM

***Millman's theorem permits any number of parallel branches consisting of voltage sources and impedances to be reduced to a single equivalent voltage source and equivalent impedance. It can be used as an alternative to Thevenin's theorem for those special cases of all parallel voltage sources with internal impedances.***

***After completing this section, you should be able to***

- **Apply Millman's theorem to parallel ac sources**
  - ☐ Determine the Millman equivalent ac source
  - ☐ Determine the Millman equivalent source impedance

The Millman conversion is illustrated in Figure 20–38.

FIGURE 20–38
***The Millman conversion.***

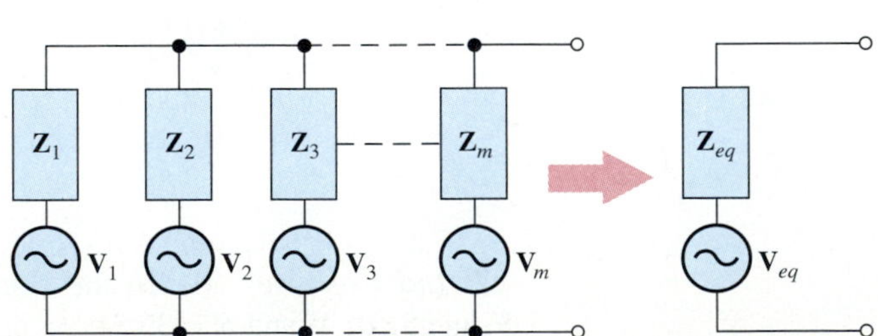

## Millman's Equivalent Voltage ($V_{eq}$) and Equivalent Impedance ($Z_{eq}$)

**Millman's theorem** provides formulas for calculating the equivalent voltage, $\mathbf{V}_{eq}$ and the equivalent impedance, $\mathbf{Z}_{eq}$. To find $\mathbf{V}_{eq}$, convert each of the parallel voltage sources in Figure 20–38 into current sources, as shown in Figure 20–39(a).

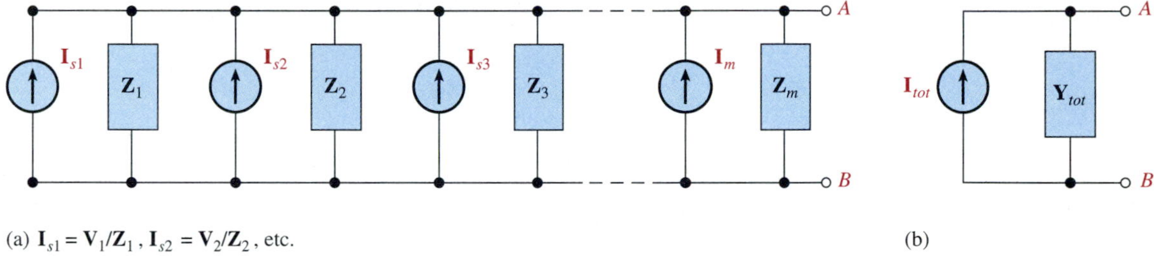


(a) $\mathbf{I}_{s1} = \mathbf{V}_1/\mathbf{Z}_1$, $\mathbf{I}_{s2} = \mathbf{V}_2/\mathbf{Z}_2$, etc.

(b)

FIGURE 20–39

In Figure 20–39(b), the total current from the parallel current sources is

$$\mathbf{I}_{tot} = \mathbf{I}_{s1} + \mathbf{I}_{s2} + \mathbf{I}_{s3} + \cdots + \mathbf{I}_{sm}$$

The total admittance between terminals *A* and *B* is

$$\mathbf{Y}_{tot} = \mathbf{Y}_1 + \mathbf{Y}_2 + \mathbf{Y}_3 + \cdots + \mathbf{Y}_m$$

where $\mathbf{Y}_{tot} = 1/\mathbf{Z}_{tot}$, $\mathbf{Y}_1 = 1/\mathbf{Z}_1$, and so on. (The subscript *m* is the number of parallel branches.) Remember that current sources are effectively open (ideally). Therefore, by Millman's theorem, the equivalent impedance is the total impedance.

$$\mathbf{Z}_{eq} = \frac{1}{\mathbf{Y}_{tot}} = \frac{1}{\dfrac{1}{\mathbf{Z}_1} + \dfrac{1}{\mathbf{Z}_2} + \cdots + \dfrac{1}{\mathbf{Z}_m}} \quad \textbf{(20–1)}$$

Also, by Millman's theorem, the equivalent voltage is $\mathbf{I}_{tot}\mathbf{Z}_{eq}$. An expression for $\mathbf{I}_{tot}$ is as follows:

$$\mathbf{I}_{tot} = \frac{\mathbf{V}_1}{\mathbf{Z}_1} + \frac{\mathbf{V}_2}{\mathbf{Z}_2} + \cdots + \frac{\mathbf{V}_m}{\mathbf{Z}_m}$$

where $\mathbf{V}_1/\mathbf{Z}_1 = \mathbf{I}_{s1}$, $\mathbf{V}_2/\mathbf{Z}_2 = \mathbf{I}_{s2}$, and so on. The following is the formula for the equivalent voltage:

$$\mathbf{V}_{eq} = \frac{\mathbf{V}_1/\mathbf{Z}_1 + \mathbf{V}_2/\mathbf{Z}_2 + \cdots + \mathbf{V}_m/\mathbf{Z}_m}{1/\mathbf{Z}_1 + 1/\mathbf{Z}_2 + \cdots + 1/\mathbf{Z}_m}$$

$$\mathbf{V}_{eq} = \frac{\mathbf{V}_1/\mathbf{Z}_1 + \mathbf{V}_2/\mathbf{Z}_2 + \cdots + \mathbf{V}_m/\mathbf{Z}_m}{\mathbf{Y}_{tot}} \quad \textbf{(20–2)}$$

Equations (20–1) and (20–2) are the two Millman formulas.

**EXAMPLE 20–16** Use Millman's theorem to find the voltage across $R_L$ and the current through $R_L$ in Figure 20–40.

FIGURE 20–40

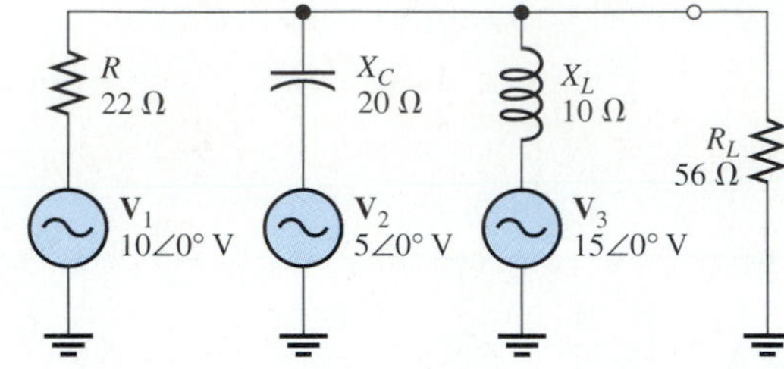


***Solution*** Apply Millman's theorem as follows:

$$\mathbf{Z}_{eq} = \frac{1}{\dfrac{1}{R\angle 0^\circ} + \dfrac{1}{X_C\angle -90^\circ} + \dfrac{1}{X_L\angle 90^\circ}} = \frac{1}{\dfrac{1}{22\angle 0^\circ\ \Omega} + \dfrac{1}{20\angle -90^\circ\ \Omega} + \dfrac{1}{10\angle 90^\circ\ \Omega}}$$

$$= \frac{1}{45\angle 0^\circ\text{ mS} + 50\angle 90^\circ\text{ mS} + 100\angle -90^\circ\text{ mS}} = \frac{1}{45\text{ mS} + j50\text{ mS} - j100\text{ mS}}$$

$$= \frac{1}{45\text{ mS} - j50\text{ mS}} = \frac{1}{67.3\angle -48.0^\circ\text{ mS}} = 14.9\angle 48.0^\circ\ \Omega$$

Converting to rectangular form,

$$\mathbf{Z}_{eq} = 10.0\ \Omega + j11.1\ \Omega$$

The equivalent voltage is then determined as follows:

$$\mathbf{Y}_{eq} = 67.3\angle -48.0^\circ\text{ mS}$$

$$\mathbf{V}_{eq} = \frac{\dfrac{V_1\angle 0^\circ}{R\angle 0^\circ} + \dfrac{V_2\angle 0^\circ}{X_C\angle -90^\circ} + \dfrac{V_3\angle 0^\circ}{X_L\angle 90^\circ}}{\mathbf{Y}_{eq}} = \frac{\dfrac{10\angle 0^\circ\text{ V}}{22\angle 0^\circ\ \Omega} + \dfrac{5\angle 0^\circ\text{ V}}{22\angle -90^\circ\ \Omega} + \dfrac{15\angle 0^\circ\text{ V}}{10\angle 90^\circ\ \Omega}}{67.3\angle -48.0^\circ\text{ mS}}$$

$$= \frac{455\angle 0^\circ\text{ mA} + 227\angle 90^\circ\text{ mA} + 1500\angle -90^\circ\text{ mA}}{67.3\angle -48.0^\circ\text{ mS}} = \frac{1.35\angle -70.3^\circ}{67.3\angle -48.0^\circ\text{ mS}}$$

$$= 20.1\angle -22.3^\circ\text{ V}$$

The single equivalent voltage source is shown in Figure 20–41.

FIGURE 20–41

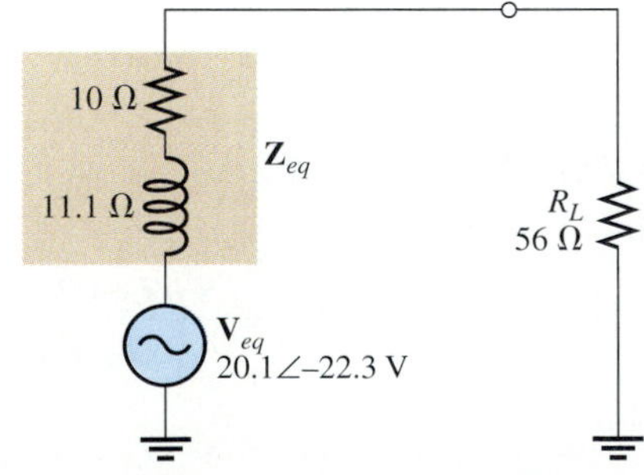

$\mathbf{I}_L$ and $\mathbf{V}_L$ are calculated as follows:

$$\mathbf{I}_L = \frac{\mathbf{V}_{eq}}{\mathbf{Z}_{eq} + \mathbf{R}_L} = \frac{20.1\angle -22.3^\circ\ \text{V}}{10.0\ \Omega + j11.1\ \Omega + 56\ \Omega}$$

$$= \frac{20.1\angle -22.3^\circ\ \text{V}}{66.9\angle 9.55^\circ\ \Omega} = 300\angle -31.9^\circ\ \text{mA}$$

$$\mathbf{V}_L = \mathbf{I}_L\mathbf{R}_L = (300\angle -31.9^\circ\ \text{mA})(56\angle 0^\circ\ \Omega) = 16.8\angle -31.9^\circ\ \text{V}$$

***Related Exercise*** If $X_C = 10\ \Omega$ and $X_L = 20\ \Omega$ in Figure 20–40, is $\mathbf{Z}_{eq}$ predominately inductive or predominately capacitive?

**SECTION 20–4 REVIEW**

1. To what type of circuit does Millman's theorem apply?
2. Find the load current in Figure 20–42 using Millman's theorem.

FIGURE 20–42

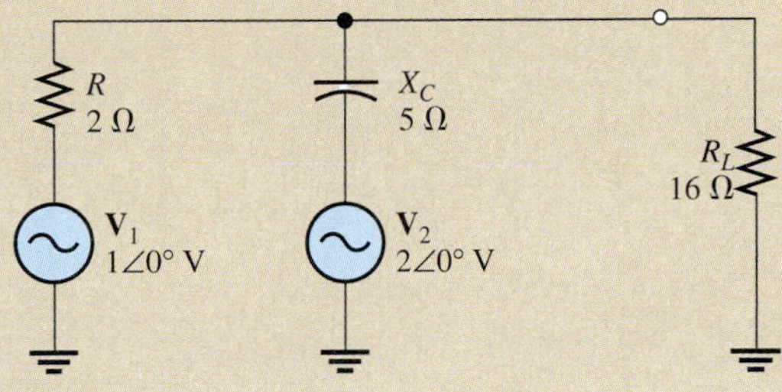


## 20–5 ■ MAXIMUM POWER TRANSFER THEOREM

***When a load is connected to a circuit, maximum power is transferred to the load when the load impedance is the complex conjugate of the circuit's output impedance.***

***After completing this section, you should be able to***

- **Apply the maximum power transfer theorem**
  - ☐ Explain the theorem
  - ☐ Determine the value of load impedance for which maximum power is transferred from a given circuit

The **complex conjugate** of $R - jX_C$ is $R + jX_L$, where the resistances and the reactances are equal in magnitude. The output impedance is effectively Thevenin's equivalent impedance viewed from the output terminals. When $\mathbf{Z}_L$ is the complex conjugate of $\mathbf{Z}_{out}$, maximum power is transferred from the circuit to the load with a power factor of 1. An equivalent circuit with its output impedance and load is shown in Figure 20–43.

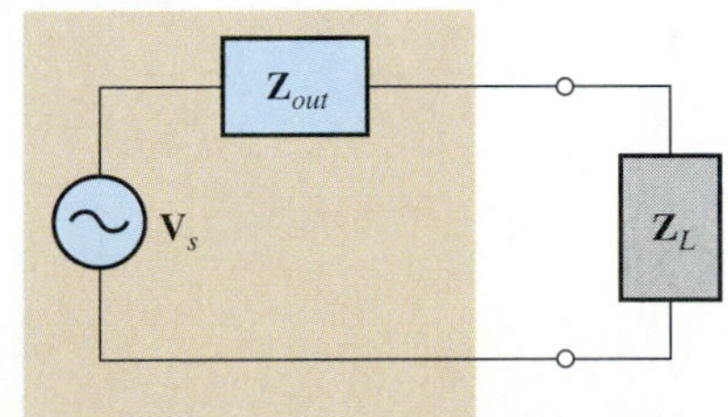


FIGURE 20–43
*Equivalent circuit with load.*

Example 20–17 shows that maximum power occurs when the impedances are conjugately matched.

**EXAMPLE 20–17** The circuit to the left of terminals $A$ and $B$ in Figure 20–44 provides power to the load $\mathbf{Z}_L$. It is the Thevenin equivalent of a more complex circuit. Calculate and plot a graph of the power delivered to the load for each of the following frequencies: 10 kHz, 30 kHz, 50 kHz, 80 kHz, and 100 kHz.

FIGURE 20–44

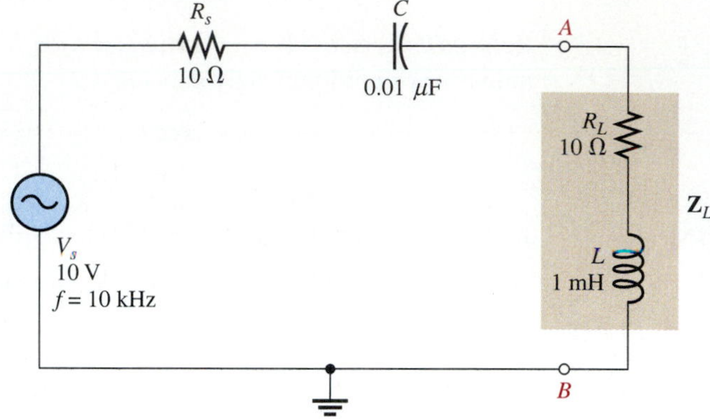


***Solution*** For $f = 10$ kHz,

$$X_C = \frac{1}{2\pi fC} = \frac{1}{2\pi(10 \text{ kHz})(0.01\ \mu\text{F})} = 1.59 \text{ k}\Omega$$

$$X_L = 2\pi fL = 2\pi(10 \text{ kHz})(1 \text{ mH}) = 62.8\ \Omega$$

The magnitude of the total impedance is

$$Z_{tot} = \sqrt{(R_s + R_L)^2 + (X_L - X_C)^2} = \sqrt{(20\ \Omega)^2 + (1.53 \text{ k}\Omega)^2} = 1.53 \text{ k}\Omega$$

The current is

$$I = \frac{V_s}{Z_{tot}} = \frac{10 \text{ V}}{1.53 \text{ k}\Omega} = 6.54 \text{ mA}$$

The load power is

$$P_L = I^2R_L = (6.54 \text{ mA})^2(10\ \Omega) = 428\ \mu\text{W}$$

For $f = 30$ kHz,

$$X_C = \frac{1}{2\pi(30 \text{ kHz})(0.01\ \mu\text{F})} = 531\ \Omega$$

$$X_L = 2\pi(30 \text{ kHz})(1 \text{ mH}) = 189\ \Omega$$

$$Z_{tot} = \sqrt{(20\ \Omega)^2 + (342\ \Omega)^2} = 343\ \Omega$$

$$I = \frac{V_s}{Z_{tot}} = \frac{10 \text{ V}}{343\ \Omega} = 29.2 \text{ mA}$$

$$P_L = I^2R_L = (29.2 \text{ mA})^2(10\ \Omega) = 8.53 \text{ mW}$$

For $f = 50$ kHz,

$$X_C = \frac{1}{2\pi(50 \text{ kHz})(0.01\ \mu\text{F})} = 318\ \Omega$$

$$X_L = 2\pi(50 \text{ kHz})(1 \text{ mH}) = 314\ \Omega$$

Note that $X_C$ and $X_L$ are very close to being equal which makes the impedances approximately complex conjugates. The exact frequency at which $X_L = X_C$ is 50.3 kHz.

$$Z_{tot} = \sqrt{(20\ \Omega)^2 + (4\ \Omega)^2} = 20.4\ \Omega$$

$$I = \frac{V_s}{Z_{tot}} = \frac{10\ \text{V}}{20.4\ \Omega} = 490\ \text{mA}$$

$$P_L = I^2R_L = (490\ \text{mA})^2(10\ \Omega) = 2.40\ \text{W}$$

For $f = 80$ kHz,

$$X_C = \frac{1}{2\pi(80\ \text{kHz})(0.01\ \mu\text{F})} = 199\ \Omega$$

$$X_L = 2\pi(80\ \text{kHz})(1\ \text{mH}) = 503\ \Omega$$

$$Z_{tot} = \sqrt{(20\ \Omega)^2 + (304\ \Omega)^2} = 305\ \Omega$$

$$I = \frac{V_s}{Z_{tot}} = \frac{10\ \text{V}}{305\ \Omega} = 32.8\ \text{mA}$$

$$P_L = I^2R_L = (32.8\ \text{mA})^2(10\ \Omega) = 10.8\ \text{mW}$$

For $f = 100$ kHz,

$$X_C = \frac{1}{2\pi(100\ \text{kHz})(0.01\ \mu\text{F})} = 159\ \Omega$$

$$X_L = 2\pi(100\ \text{kHz})(1\ \text{mH}) = 628\ \Omega$$

$$Z_{tot} = \sqrt{(20\ \Omega)^2 + (469\ \Omega)^2} = 469\ \Omega$$

$$I = \frac{V_s}{Z_{tot}} = \frac{10\ \text{V}}{469\ \Omega} = 21.3\ \text{mA}$$

$$P_L = I^2R_L = (21.3\ \text{mA})^2(10\ \Omega) = 4.54\ \text{mW}$$

As you can see from the results, the power to the load peaks at the frequency at which the load impedance is the complex conjugate of the output impedance (when the reactances are equal in magnitude). A graph of the load power versus frequency is shown in Figure 20–45. Since the maximum power is so much larger than the other values, an accurate plot is difficult to achieve without intermediate values.

FIGURE 20–45

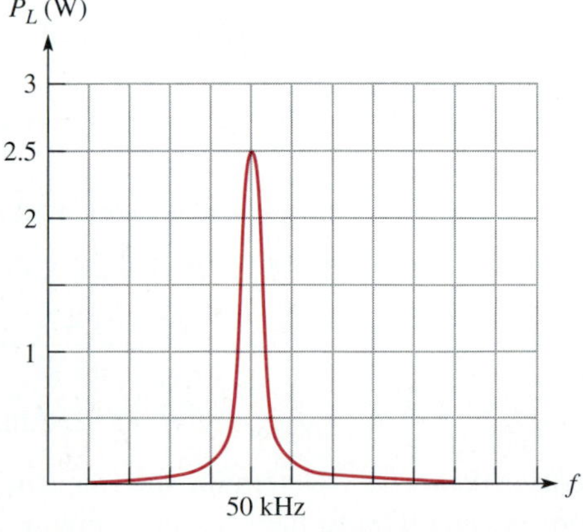


***Related Exercise*** If $R = 47\ \Omega$ and $C = 0.022\ \mu$F in a series $RC$ circuit, what is the complex conjugate of the impedance at 100 kHz?

**EXAMPLE 20–18** (a) Determine the frequency at which maximum power is transferred from the amplifier to the speaker in Figure 20–46(a). The amplifier and coupling capacitor are the source, and the speaker is the load, as shown in the equivalent circuit of Figure 20–46(b).

(b) How many watts of power are delivered to the speaker at this frequency if $V_s$ = 3.8 V rms?

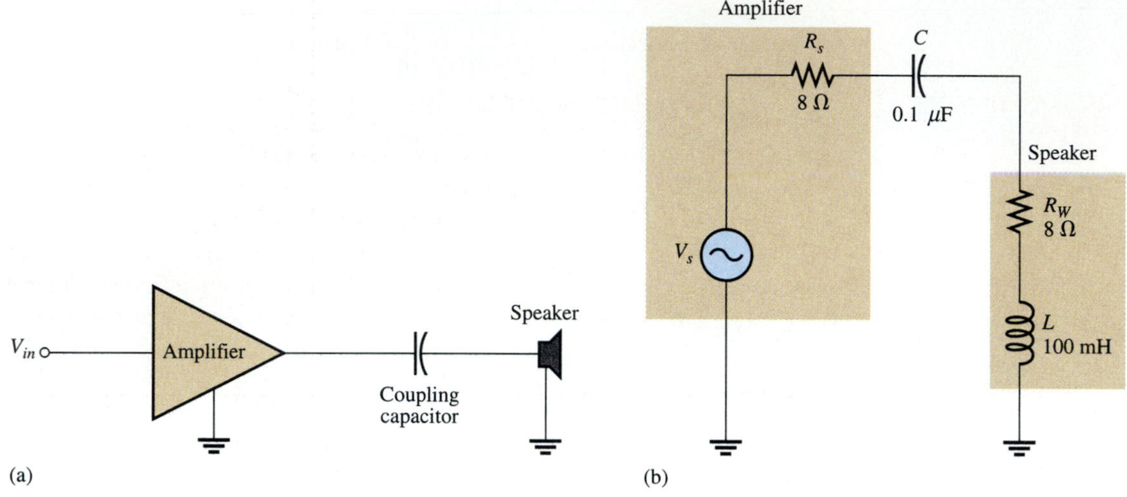


FIGURE 20–46

***Solution***

(a) When the power to the speaker is maximum, the source impedance ($R_s - jX_C$) and the load impedance ($R_W + jX_L$) are complex conjugates, so

$$X_C = X_L$$

$$\frac{1}{2\pi fC} = 2\pi fL$$

Solving for $f$,

$$f^2 = \frac{1}{4\pi^2 LC}$$

$$f = \frac{1}{2\pi\sqrt{LC}} = \frac{1}{2\pi\sqrt{(100\text{ mH})(0.1\ \mu\text{F})}} \cong 1.59\text{ kHz}$$

(b) The power to the speaker is calculated as follows.

$$Z_{tot} = R_s + R_W = 8\ \Omega + 8\ \Omega = 16\ \Omega$$

$$I = \frac{V_s}{Z_{tot}} = \frac{3.8\text{ V}}{16\ \Omega} = 238\text{ mA}$$

$$P_{max} = I^2R_W = (238\text{ mA})^2(8\ \Omega) = 453\text{ mW}$$

***Related Exercise*** Determine the frequency at which maximum power is transferred from the amplifier to the speaker in Figure 20–46 if the coupling capacitor is 1 μF.

**SECTION 20–5 REVIEW**

1. If the output impedance of a certain driving circuit is 50 Ω − $j$10 Ω, what value of load impedance will result in the maximum power to the load?
2. Fir the circuit in Question 1, how much power is delivered to the load when the load impedance is the complex conjugate of the output impedance and when the load current is 2 A?

## 20–6 ■ TECHnology Theory Into Practice

***In this TECH TIP, you are given a sealed band-pass filter module that has been removed from a system and two schematics. Both schematics indicate that the band-pass filter is implemented with a low-pass/high-pass combination. It is uncertain which schematic corresponds to the filter module, but one of them does. By certain measurements, you will determine which schematic represents the filter so that the filter circuit can be reproduced. Also, you will determine the proper load for maximum power transfer.***

The filter circuit contained in a sealed module and two schematics, one of which corresponds to the filter circuit, are shown in Figure 20–47.

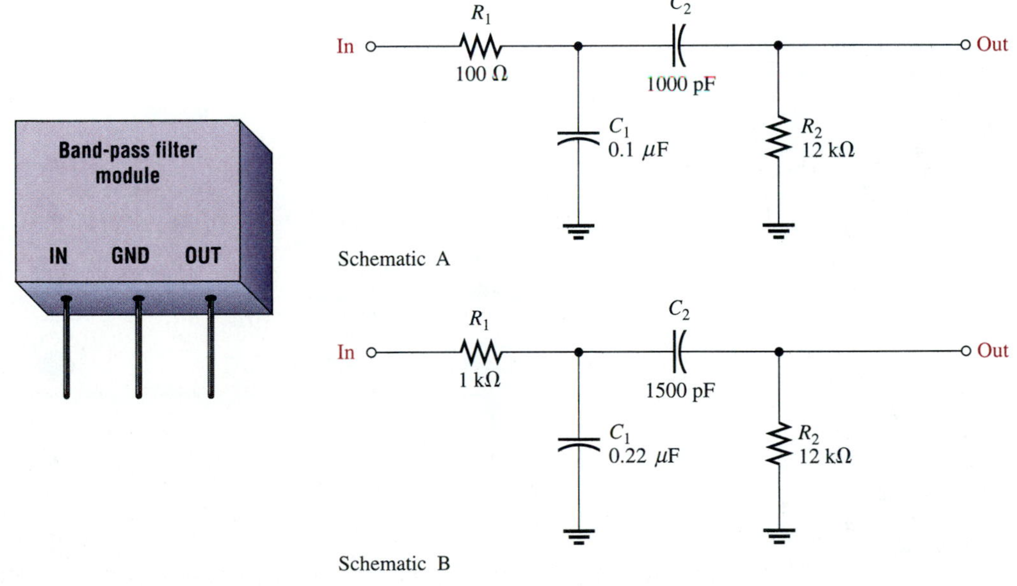


FIGURE 20–47
*Filter module and schematics.*

## Filter Measurement and Analysis

- ☐ Based on the oscilloscope measurement of the filter output shown in Figure 20–48, determine which schematic in Figure 20–47 represents the component values of the filter circuit in the module. A 10 V peak-to-peak voltage is applied to the input.
- ☐ Based on the oscilloscope measurement in Figure 20–48, determine if the filter is operating at its approximate center frequency.
- ☐ Using Thevenin's theorem, determine the load impedance that will provide for maximum power transfer at the center frequency when connected to the output of the filter. Assume the source impedance is zero.

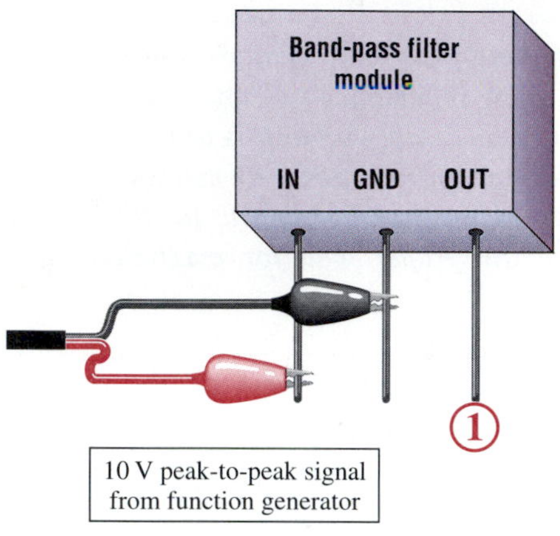


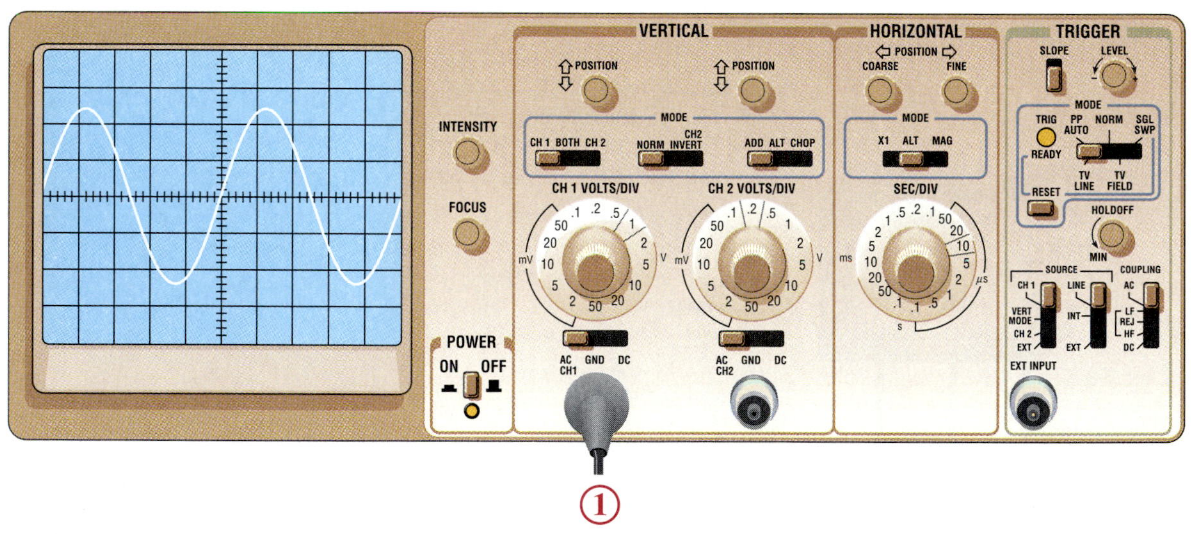


FIGURE 20–48
*The scope probes are ×1.*

**SECTION 20–6 REVIEW**

1. Determine the peak-to-peak output voltage at the frequency shown in Figure 20–48 of the circuit in Figure 20–47 that was determined not to be in the module.
2. Find the center frequency of the circuit in Figure 20–47 that was determined not to be in the module.

## 20–7 ■ AC EQUIVALENT CIRCUITS USING PSpice AND PROBE

***In Chapter 8 you learned how to use PSpice to determine the Thevenin and Norton equivalents for dc circuits. With Probe, you can apply these same techniques to ac circuits. However, with ac circuits, some of the calculations must be done manually. Fortunately, PSpice and Probe allow you to analyze most original circuits directly without the need to reduce them to equivalent forms. However, if you need an equivalent circuit, this is a way to get it.***

***After completing this section, you should be able to***

- **Use PSpice and Probe to determine Thevenin and Norton equivalent circuits**
  - ☐ Determine the Thevenin equivalent voltage
  - ☐ Determine the Norton equivalent current
  - ☐ Use equivalent values to calculate Thevenin and Norton equivalent impedance

An example circuit is shown in Figure 20–49, and the circuit file is as follows:

```
AC Thevenin/Norton Equivalent Example
*Original Circuit
VS  1  0  AC  10
R1  1  2  1K
L1  2  0  100U
R2  2  3  1K
RL  3  0  1K
.AC  LIN  1000  10K  1MEG
.PROBE
.END
```

FIGURE 20–49

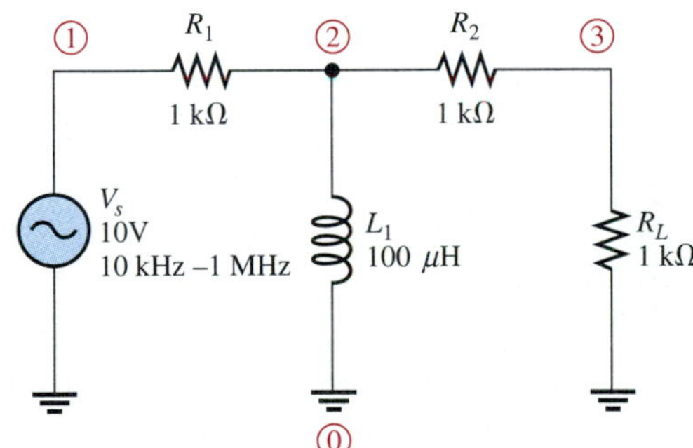


Suppose you want to find the Thevenin or Norton equivalent seen by $R_L$. Recall from Chapter 8 that in the dc circuit procedure, $R_L$ was replaced by a very large resistance to simulate an open in order to find the Thevenin equivalent voltage. Likewise, $R_L$ was replaced by a very small resistance to simulate a short in order to find the Norton equivalent current. The Thevenin voltage was then divided by the Norton current to get the Thevenin (or Norton) equivalent resistance.

The procedure used for dc circuits is the same for ac circuits with two exceptions:

1. The equivalent values must be determined at some specified frequency.
2. Both the magnitude and phase angle of the open circuit voltage and short circuit current must be determined.

For the example circuit in Figure 20–49, assume the frequency of interest is 300 kHz.

## The Thevenin Equivalent Voltage

Change the load resistance from 1 kΩ to 100 GΩ to approximate an open and modify the previous circuit file as follows:

```
AC Thevenin/Norton Equivalent Example
*Open Load Circuit
VS  1  0  AC  10
R1  1  2  1K
L1  2  0  100U
R2  2  3  1K
RL  3  0  100G
.AC  LIN  1000  10K  1MEG
.PROBE
.END
```

Now run PSpice on this new file and start Probe. To find the magnitude of the open load voltage, display the trace for V(3,0).

After Probe draws the trace, select Cursor and move cursor 1 to the right until the frequency is equal to 300 kHz. The magnitude of the open load voltage at this frequency is approximately 1.85 V. To find the phase angle, delete the trace and replace it with VP(3,0) to tell Probe to plot the phase angle versus frequency. Use the cursor to determine the phase angle at 300 kHz which is approximately 79.4°. Combining the magnitude and phase angle, the Thevenin equivalent voltage is

$$\mathbf{V}_{th} = 1.85\angle 79.4^\circ \text{ V}$$

## The Norton Equivalent Current

Exit Probe and change the value of $R_L$ in the circuit file to 1 pΩ to simulate a short.

```
AC Thevenin/Norton Equivalent Example
*Short Circuit Current Circuit
VS  1  0  AC  10
R1  1  2  1K
L1  2  0  100U
R2  2  3  1K
RL  3  0  1P
.AC  LIN  1000  10K  1MEG
.PROBE
.END
```

Run this file and use Probe to view I(RL). Determine the magnitude of the short circuit current at 300 kHz using the cursor. This value should be approximately 1.76 mA.

Now remove the trace and enter IP(RL) to plot the phase angle. Again using the cursor, you will find the phase angle at 300 kHz is approximately 69.4°. Combining these two values gives a Norton equivalent current of

$$\mathbf{I}_n = 1.76\angle 69.4^\circ \text{ mA}$$

## The Equivalent Impedance

The Thevenin and Norton equivalent source impedance is calculated as follows:

$$\mathbf{Z}_{th} = \mathbf{Z}_n = \frac{\mathbf{V}_{th}}{\mathbf{I}_n} = \frac{(1.85\angle 79.4^\circ \text{ V})}{(1.76\angle 69.4^\circ \text{ mA})} = 1.05\angle 10.0^\circ \text{ k}\Omega$$

In this example, the AC sweep range was from 10 kHz to 1 MHz. Increased accuracy can be achieved by changing the AC sweep to a much narrower range about the frequency of interest.

**SECTION 20–7 REVIEW**

1. Explain why $R_L$ was changed to a very high value to determine the Thevenin equivalent voltage.
2. Explain why $R_L$ was changed to a very low value to determine the Norton equivalent current.

## ■ SUMMARY

- The superposition theorem is useful for the analysis of multiple-source circuits.
- Thevenin's theorem provides a method for the reduction of any ac circuit to an equivalent form consisting of an equivalent voltage source in series with an equivalent impedance.
- The term *equivalency,* as used in Thevenin's and Norton's theorems, means that when a given load impedance is connected to the equivalent circuit, it will have the same voltage across it and the same current through it as when it is connected to the original circuit.
- Norton's theorem provides a method for the reduction of any ac circuit to an equivalent form consisting of an equivalent current source in parallel with an equivalent impedance.
- Millman's theorem provides a method for the reduction of parallel voltage sources and impedances to a single equivalent voltage source and an equivalent impedance.
- Maximum power is transferred to a load when the load impedance is the complex conjugate of the output impedance of the driving circuit.

## ■ GLOSSARY

**Complex conjugate** An impedance containing the same resistance and a reactance opposite in phase but equal in magnitude to that of a given impedance.

**Equivalent circuit** A circuit that produces the same voltage and current to a given load as the original circuit that it replaces.

**Millman's theorem** A method for reducing parallel voltage sources to a single equivalent voltage source.

**Norton's theorem** A method for simplifying a given circuit to an equivalent circuit with a current source in parallel with an impedance.

**Superposition theorem** A method for the analysis of circuits with more than one source.

**Thevenin's theorem** A method for simplifying a given circuit to an equivalent circuit with a voltage source in series with an impedance.

## ■ FORMULAS

### MILLMAN'S THEOREM

**(20–1)** $$\mathbf{Z}_{eq} = \frac{1}{\mathbf{Y}_{tot}} = \frac{1}{\dfrac{1}{\mathbf{Z}_1} + \dfrac{1}{\mathbf{Z}_2} + \cdots + \dfrac{1}{\mathbf{Z}_m}}$$

**(20–2)** $$\mathbf{V}_{eq} = \frac{\mathbf{V}_1/\mathbf{Z}_1 + \mathbf{V}_2/\mathbf{Z}_2 + \cdots + \mathbf{V}_m/\mathbf{Z}_m}{\mathbf{Y}_{tot}}$$

## ■ SELF-TEST

**1.** In applying the superposition theorem,

**(a)** all sources are considered simultaneously

**(b)** all voltage sources are considered simultaneously

**(c)** the sources are considered one at a time with all others replaced by a short

**(d)** the sources are considered one at a time with all others replaced by their internal impedances

2. A Thevenin ac equivalent circuit always consists of
   (a) an equivalent ac voltage source and an equivalent capacitance
   (b) an equivalent ac voltage source and an equivalent inductive reactance
   (c) an equivalent ac voltage source and an equivalent impedance
   (d) an equivalent ac voltage source in series with an equivalent capacitive reactance
3. One circuit is equivalent to another, in the context of Thevenin's theorem, when
   (a) the same load has the same voltage and current when connected to either circuit
   (b) different loads have the same voltage and current when connected to either circuit
   (c) the circuits have equal voltage sources and equal series impedances
   (d) the circuits produce the same output voltage
4. The Thevenin equivalent voltage is
   (a) the open circuit voltage
   (b) the short circuit voltage
   (c) the voltage across an equivalent load
   (d) none of the above
5. The Thevenin equivalent impedance is
   (a) the impedance looking from the source with the output shorted
   (b) the impedance looking from the source with the output open
   (c) the impedance looking from any two specified open terminals with all sources replaced by their internal impedances
   (d) the impedance looking from any two specified open terminals with all sources replaced by a short
6. A Norton ac equivalent circuit always consists of
   (a) an equivalent ac current source in series with an equivalent impedance
   (b) an equivalent ac current source in parallel with an equivalent reactance
   (c) an equivalent ac current source in parallel with an equivalent impedance
   (d) an equivalent ac voltage source in parallel with an equivalent impedance
7. The Norton equivalent current is
   (a) the total current from the source
   (b) the short circuit current
   (c) the current to an equivalent load
   (d) none of the above
8. Millman's theorem deals with
   (a) parallel current sources
   (b) parallel voltage sources
   (c) answers (a) and (b)
9. The complex conjugate of $50\ \Omega + j100\ \Omega$ is
   (a) $50\ \Omega - j50\ \Omega$
   (b) $100\ \Omega + j50\ \Omega$
   (c) $100\ \Omega - j50\ \Omega$
   (d) $50\ \Omega - j100\ \Omega$
10. In order to get maximum power transfer from a capacitive source, the load must
   (a) have a capacitance equal to the source capacitance
   (b) have an impedance equal in magnitude to the source impedance
   (c) be inductive
   (d) have an impedance that is the complex conjugate of the source impedance
   (e) answers (a) and (d)

## ▪ PROBLEMS

### SECTION 20–1 The Superposition Theorem

1. Using the superposition method, calculate the current through $R_3$ in Figure 20–50.

FIGURE 20–50

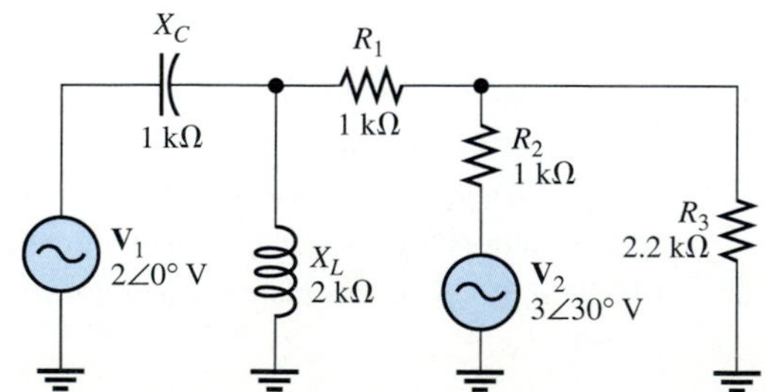

2. Use the superposition theorem to find the current in and the voltage across the $R_2$ branch of Figure 20–50.
3. Using the superposition theorem, solve for the current through $R_1$ in Figure 20–51.

FIGURE 20–51

4. Using the superposition theorem, find the load current in each circuit of Figure 20–52.

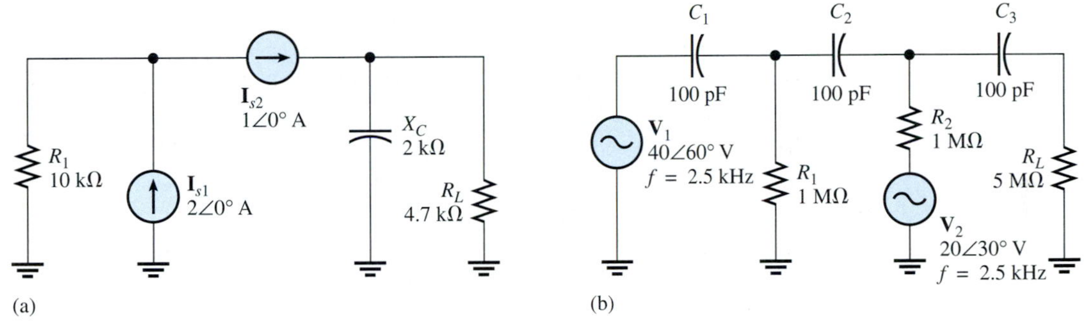


FIGURE 20–52

5. Determine the voltage at each point ($A$, $B$, $C$, $D$) in Figure 20–53. Assume $X_C = 0$ for all capacitors. Sketch the voltage waveforms at each of the points.

FIGURE 20–53

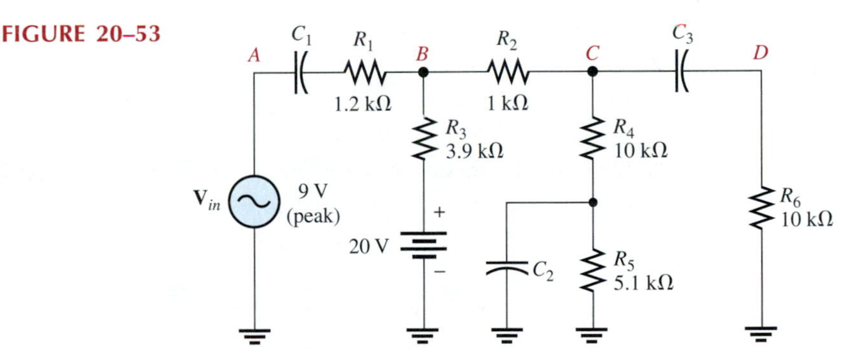


6. Use the superposition theorem to find the capacitor current in Figure 20–54.

FIGURE 20–54

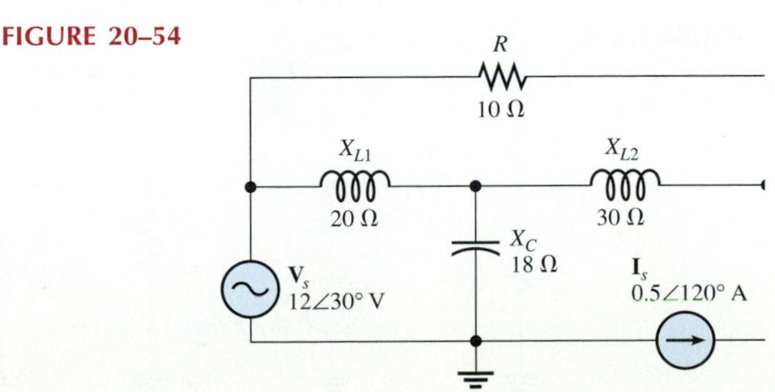

## SECTION 20–2 Thevenin's Theorem

7. For each circuit in Figure 20–55, determine the Thevenin equivalent circuit for the portion of the circuit viewed by $R_L$.

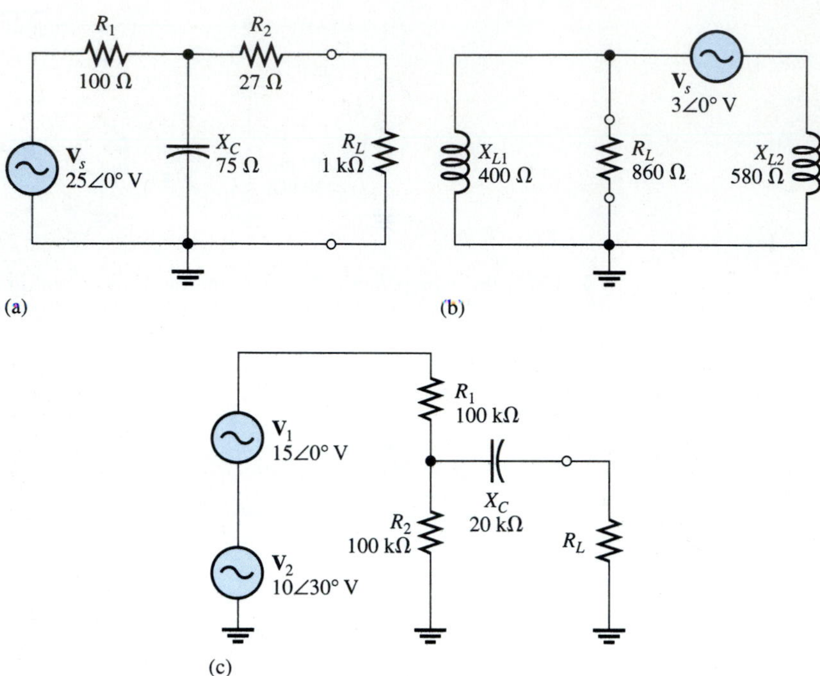


FIGURE 20–55

8. Using Thevenin's theorem, determine the current through the load $R_L$ in Figure 20–56.

FIGURE 20–56

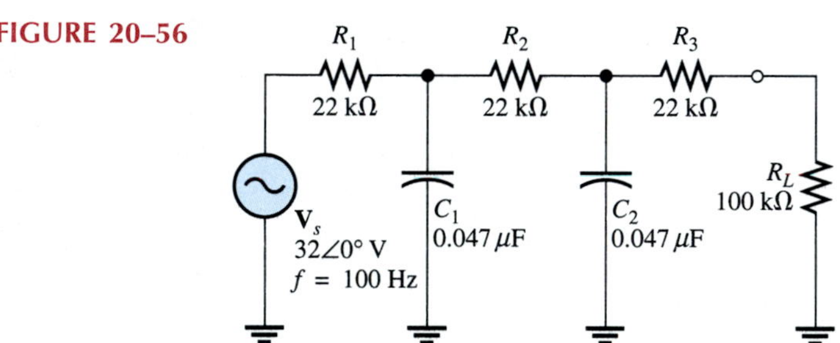


9. Using Thevenin's theorem, find the voltage across $R_4$ in Figure 20–57.

FIGURE 20–57

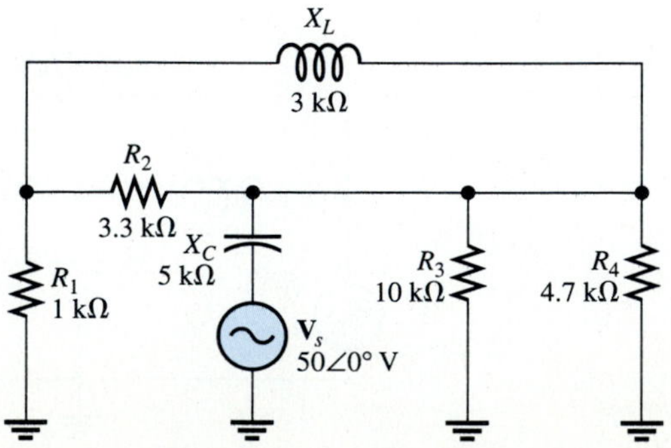

**10.** Simplify the circuit external to $R_3$ in Figure 20–58 to its Thevenin equivalent.

**FIGURE 20–58**

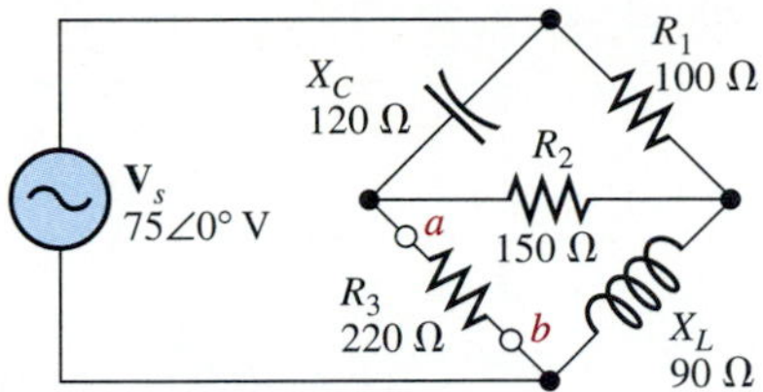


## SECTION 20–3 Norton's Theorem

**11.** For each circuit in Figure 20–55, determine the Norton equivalent as seen by $R_L$.

**12.** Using Norton's theorem, find the current through the load resistor $R_L$ in Figure 20–56.

**13.** Using Norton's theorem, find the voltage across $R_4$ in Figure 20–57.

## SECTION 20–4 Millman's Theorem

**14.** Apply Millman's theorem to the circuit of Figure 20–59.

**FIGURE 20–59**

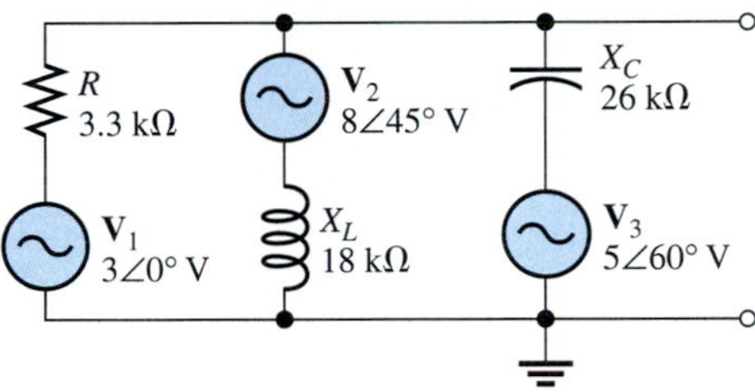


**15.** Use Millman's theorem to reduce the circuit in Figure 20–60 to a single voltage source and impedance.

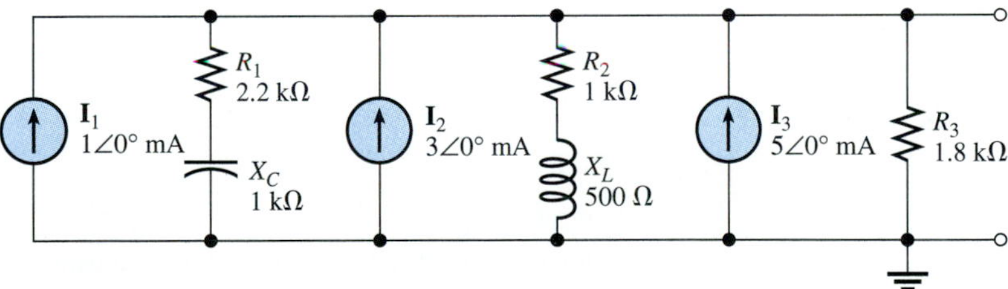


**FIGURE 20–60**

## SECTION 20–5 Maximum Power Transfer Theorem

**16.** For each circuit in Figure 20–61, maximum power is to be transferred to the load $R_L$. Determine the appropriate value for the load impedance in each case.

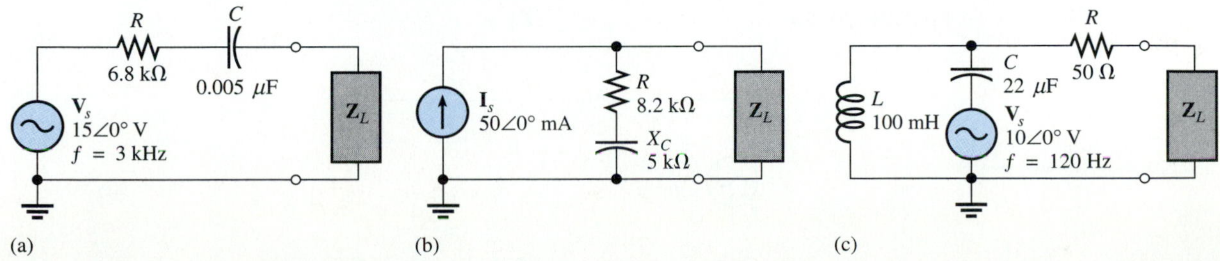


**FIGURE 20–61**

**17.** Determine $\mathbf{Z}_L$ for maximum power in Figure 20–62.

FIGURE 20–62

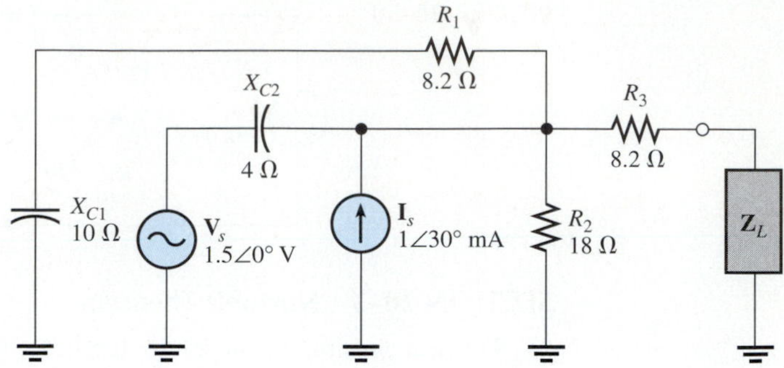


**18.** Find the load impedance required for maximum power transfer to $Z_L$ in Figure 20–63. Determine the maximum true power.

FIGURE 20–63

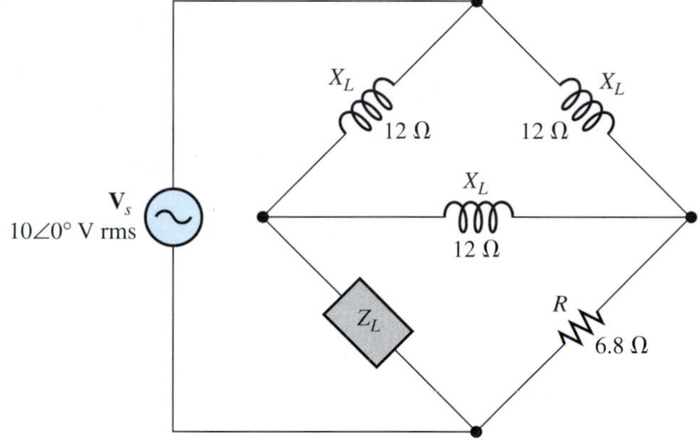


**19.** A load is to be connected in the place of $R_2$ in Figure 20–58 to achieve maximum power transfer. Determine the type of load, and express it in rectangular form.

## ■ ANSWERS TO SECTION REVIEWS

### Section 20–1

1. The net current is zero.
2. The circuit can be analyzed one source at a time using superposition.
3. $I_R = 12$ mA

### Section 20–2

1. The components of a Thevenin equivalent ac circuit are equivalent voltage source and equivalent series impedance.
2. See Figure 20–64.
3. $\mathbf{Z}_{th} = 21.5\ \Omega - j15.7\ \Omega$; $\mathbf{V}_{th} = 4.14\angle 53.8°$ V

FIGURE 20–64

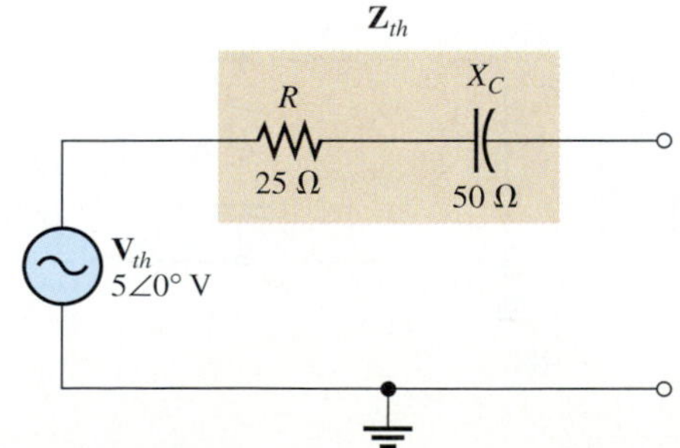

### Section 20–3

**1.** See Figure 20–65.

**2.** $\mathbf{Z}_n = R\angle 0° = 1.2\angle 0°$ kΩ; $\mathbf{I}_n = 10\angle 0°$ mA

FIGURE 20–65

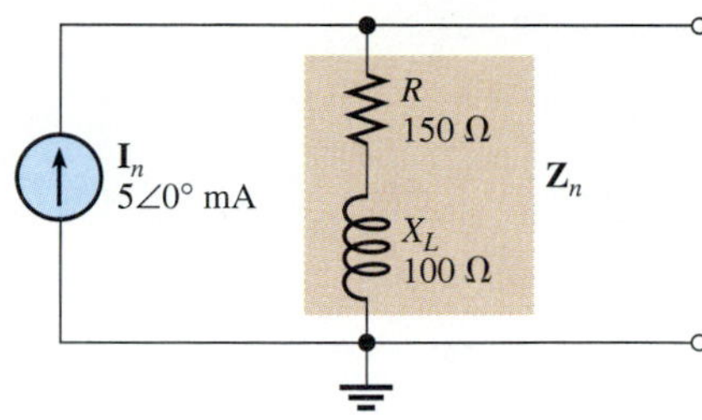


### Section 20–4

**1.** Millman's theorem applies to circuits with parallel voltage sources.

**2.** $\mathbf{I}_L = 67.0\angle 19.1°$ mA

### Section 20–5

**1.** $\mathbf{Z}_L = 50\ \Omega + j10\ \Omega$

**2.** $P_L = 200$ W

### Section 20–6

**1.** $\mathbf{V}_{out} = 166\angle -66.1°$ mV pp

**2.** $f_0 = 4.76$ kHz

### Section 20–7

**1.** $R_L = 100$ GΩ approximates an open.

**2.** $R_L = 1$ pΩ approximates a short.

## ■ ANSWERS TO RELATED EXERCISES FOR EXAMPLES

**20–1** $1.77\angle -152°$ mA

**20–2** $30\angle 90°$ mA

**20–3** $1.69\angle 47.3°$ mA riding on a dc level of 3 mA

**20–4** $18.2\angle 43.2°$ V

**20–5** $4.03\angle -36.3°$ V

**20–6** $4.55\angle 24.4°$ V

**20–7** $34.3\angle 43.2°$ Ω

**20–8** $1.37\angle -47.8°$ kΩ

**20–9** $9.10\angle -65.6°$ kΩ

**20–10** See Figure 20–66.

**20–11** See Figure 20–67.

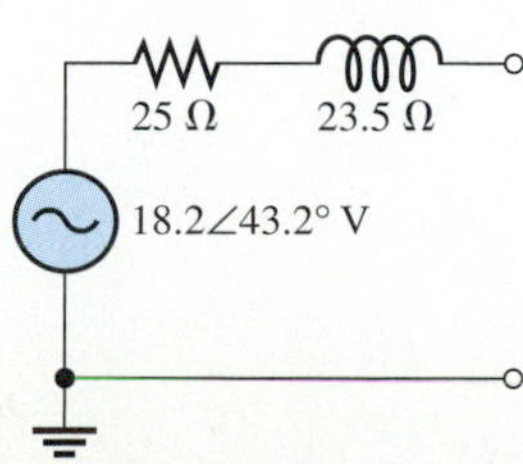


FIGURE 20–66

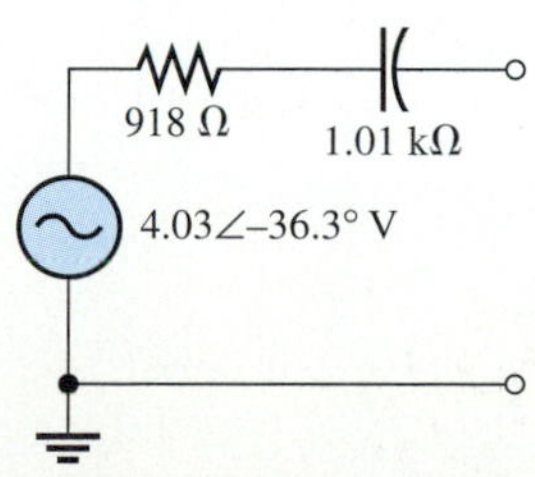


FIGURE 20–67

**20–12** See Figure 20–68.
**20–13** 117∠135° mA
**20–14** 117∠−78.7 Ω
**20–15** See Figure 20–69.
**20–16** Capacitive
**20–17** 47 Ω + $j$72.3 Ω
**20–18** 503 Hz

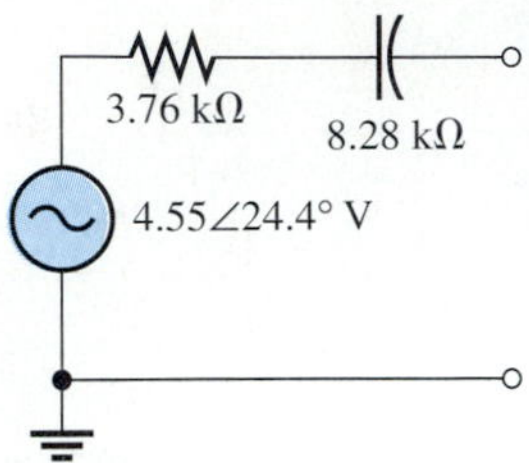

FIGURE 20–68

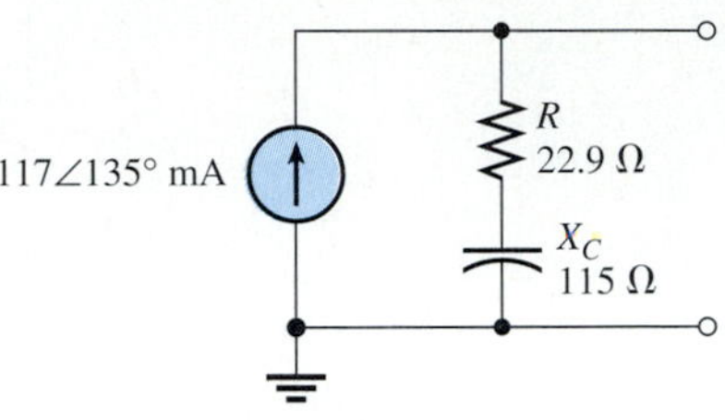

FIGURE 20–69

# 21

# PULSE RESPONSE OF REACTIVE CIRCUITS



## ■ INTRODUCTION

In Chapters 16 and 17, the frequency response of *RC* and *RL* circuits was covered. In this chapter, the response of *RC* and *RL* circuits to pulse inputs is examined. Before starting this chapter, you should review the material in Sections 13–5 and 14–5. Understanding exponential changes in voltages and currents in capacitors and inductors is crucial to the study of pulse response. Throughout this chapter, exponential formulas that were given in Chapters 13 and 14 are used. In this chapter, you will learn the basics of putting technology theory into practice.

With pulse waveform inputs, the time responses of circuits are important. In the areas of pulse and digital circuits, technicians are often concerned with how a circuit responds over an interval of time to rapid changes in voltages or current. The relationship of the circuit time constant to the input pulse characteristics, such as pulse width and period, determines the wave shapes of voltages in the circuit. *Integrator* and *differentiator,* terms used throughout this chapter, refer to mathematical functions that are approximated by these circuits under certain conditions. Mathematical integration is an averaging process, and mathematical differentiation is a process for establishing an instantaneous rate of change of a quantity.

## TECHnology Theory Into Practice

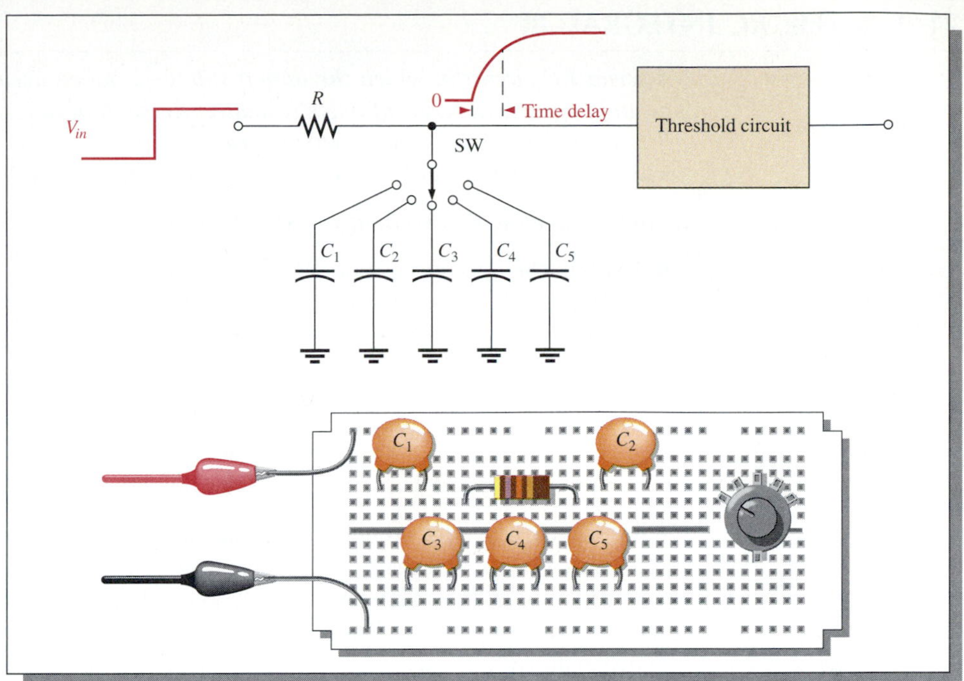


If you think something is missing from the circuit board, you're right. The wiring is missing, so you will have to specify the wiring in the TECH TIP in Section 21–10. You will also determine component values to meet certain specifications and then determine instrument settings to properly test the circuit.

## CHAPTER OBJECTIVES

- ☐ Explain the operation of an *RC* integrator
- ☐ Analyze an *RC* integrator with a single input pulse
- ☐ Analyze an *RC* integrator with repetitive input pulses
- ☐ Analyze an *RC* differentiator with a single input pulse
- ☐ Analyze an *RC* differentiator with repetitive input pulses
- ☐ Analyze the operation of an *RL* integrator
- ☐ Analyze the operation of an *RL* differentiator
- ☐ Explain the relationship of time response to frequency response
- ☐ Troubleshoot integrators and differentiators
- ☐ Use PSpice and Probe to observe pulse responses (optional)

## 21–1 ▪ THE *RC* INTEGRATOR

***A series RC circuit in which the output voltage is taken across the capacitor is known as an integrator in terms of pulse response. Recall that in terms of frequency response, it is a low-pass filter. The term, integrator, is derived from a mathematical function which this type of circuit approximates under certain conditions.***

***After completing this section, you should be able to***

- **Explain the operation of an *RC* integrator**
  - ☐ Describe how the capacitor charges and discharges
  - ☐ Explain how a capacitor reacts to an instantaneous change in voltage or current
  - ☐ Describe the basic output voltage waveform

### How the Capacitor Charges and Discharges with a Pulse Input

When a pulse generator is connected to the input of an *RC* **integrator,** as shown in Figure 21–1, the capacitor will charge and discharge in response to the pulses. When the input goes from its low level to its high level, the capacitor charges toward the high level of the pulse through the resistor. This charging action is analogous to connecting a battery through a switch to the *RC* circuit, as illustrated in Figure 21–2(a). When the pulse goes from its high level back to its low level, the capacitor discharges back through the source. The resistance of the source is assumed to be negligible compared to *R*. This discharging action is analogous to replacing the source with a closed switch, as illustrated in Figure 21–2(b).

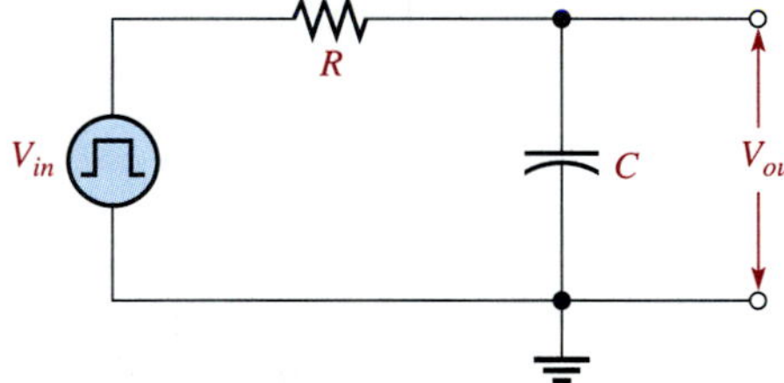


FIGURE 21–1
*An RC integrating circuit.*

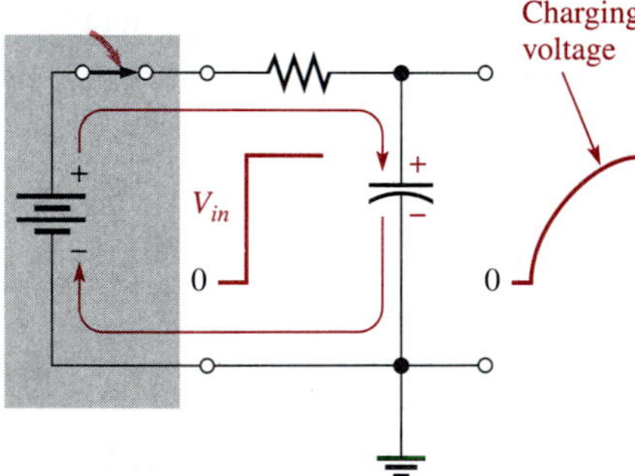


(a) When the input pulse goes high, the source effectively acts as a battery in series with a closed switch, thereby charging the capacitor.

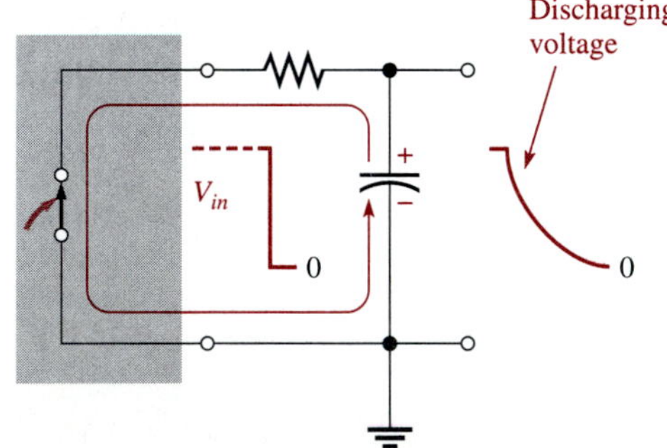


(b) When the input pulse goes back low, the source effectively acts as a closed switch, providing a discharge path for the capacitor.

FIGURE 21–2
*The equivalent action when a pulse source charges and discharges the capacitor.*

As you learned in Chapter 13, the capacitor will charge and discharge following an exponential curve. Its rate of charging and discharging, of course, depends on the *RC* **time constant** ($\tau = RC$).

For an ideal pulse, both edges are considered to occur instantaneously. Two basic rules of capacitor behavior help in understanding the **pulse response** of *RC* circuits.

1. The capacitor appears as a short to an instantaneous change in current and as an open to dc.
2. The voltage across the capacitor cannot change instantaneously—it can change only exponentially.

### The Capacitor Voltage

In an *RC* integrator, the output is the capacitor voltage. The capacitor charges during the time that the pulse is high. If the pulse is at its high level long enough, the capacitor will fully charge to the voltage amplitude of the pulse, as illustrated in Figure 21–3. The capacitor discharges during the time that the pulse is low. If the low time between pulses is long enough, the capacitor will fully discharge to zero, as shown in the figure. Then when the next pulse occurs, it will charge again.

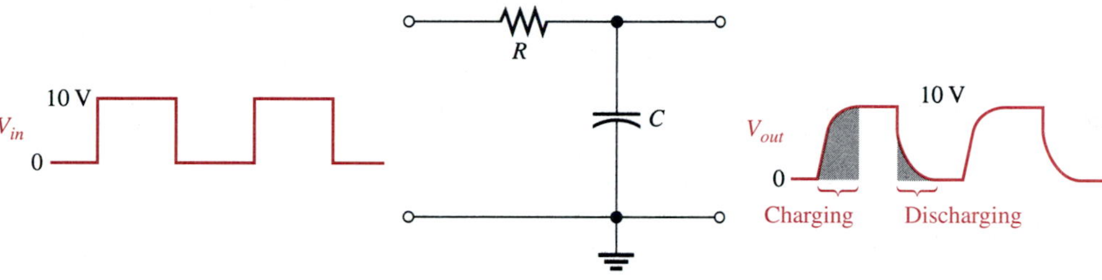


FIGURE 21–3
***Illustration of a capacitor fully charging and discharging in response to a pulse input.***

**SECTION 21–1 REVIEW**

1. Define the term *integrator* in relation to an *RC* circuit.
2. What causes a capacitor in an *RC* circuit to charge and discharge?

## 21–2 ■ SINGLE-PULSE RESPONSE OF *RC* INTEGRATORS

***From the previous section, you have a general idea of how an RC integrator responds to a pulse input. In this section, the response to a single pulse is examined in detail.***

***After completing this section, you should be able to***

- **Analyze an *RC* integrator with a single input pulse**
  - Discuss the importance of the circuit time constant
  - Define *transient time*
  - Determine the response when the pulse width is equal to or greater than five time constants
  - Determine the response when the pulse width is less than five time constants

Two conditions of pulse response must be considered:

1. When the input pulse width ($t_W$) is equal to or greater than five time constants ($t_W \geq 5\tau$)
2. When the input pulse width is less than five time constants ($t_W < 5\tau$)

Recall that five time constants is accepted as the time for a capacitor to fully charge or fully discharge; this time is often called the **transient time.**

### When the Pulse Width Is Equal to or Greater Than Five Time Constants

The capacitor will fully charge if the pulse width is equal to or greater than five time constants ($5\tau$). This condition is expressed as $t_W \geq 5\tau$. At the end of the pulse, the capacitor fully discharges back through the source.

Figure 21–4 illustrates the output waveforms for various *RC* time constants and a fixed input pulse width. Notice that the shape of the output pulse approaches that of the input as the transient time is made small compared to the pulse width. In each case, the output reaches the full amplitude of the input.

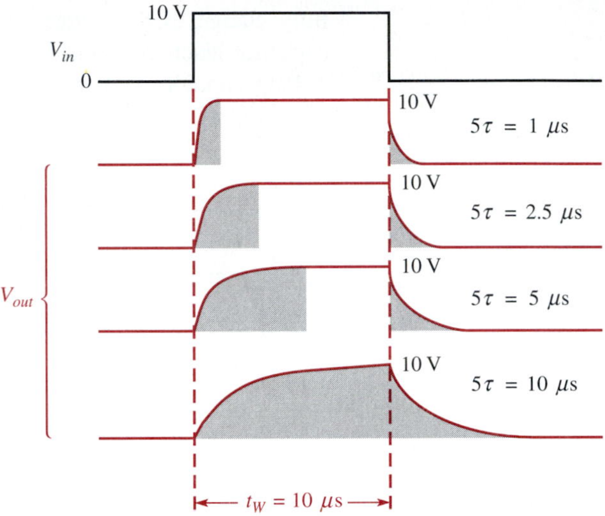


**FIGURE 21–4**
***Variation of an integrator's output pulse shape with time constant. The shaded areas indicate when the capacitor is charging and discharging.***

Figure 21–5 shows how a fixed time constant and a variable input pulse width affect the integrator output. Notice that as the pulse width is increased, the shape of the output pulse approaches that of the input. Again, this means that the transient time is short compared to the pulse width.

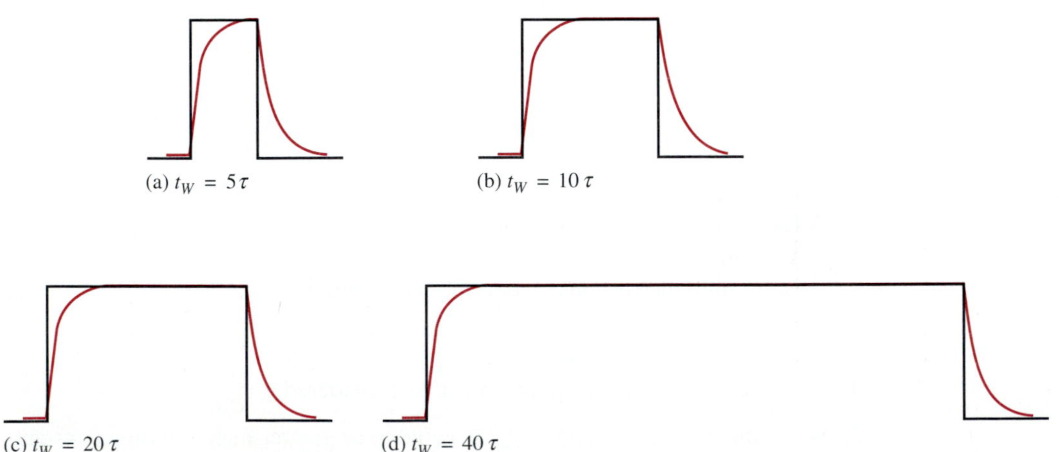


**FIGURE 21–5**
***Variation of an integrator's output pulse shape with input pulse width (the time constant is fixed). Black is input and red is output.***

## When the Pulse Width Is Less Than Five Time Constants

Now let's examine the case in which the width of the input pulse is less than five time constants of the integrator. This condition is expressed as $t_W < 5\tau$.

As before, the capacitor charges for the duration of the pulse. However, because the pulse width is less than the time it takes the capacitor to fully charge ($5\tau$), the output voltage will *not* reach the full input voltage before the end of the pulse. The capacitor only partially charges, as illustrated in Figure 21–6 for several values of *RC* time constants. Notice that for longer time constants, the output reaches a lower voltage because the capacitor cannot charge as much. Of course, in the examples with a single pulse input, the capacitor fully discharges after the pulse ends.

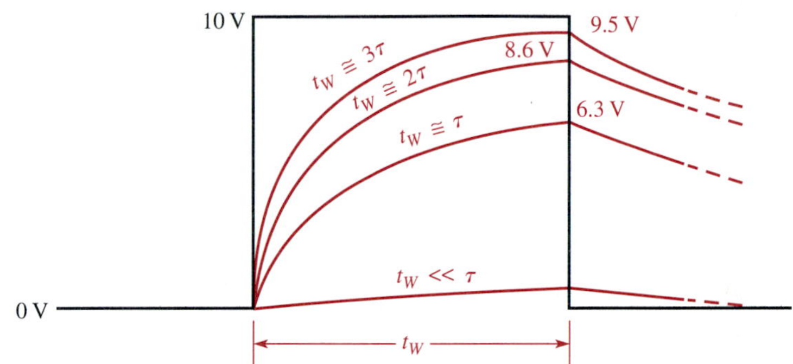


**FIGURE 21–6**
***Capacitor voltage for various time constants that are longer than the input pulse width. Black is input and red is output.***

When the time constant is much greater than the input pulse width, the capacitor charges very little, and, as a result, the output voltage becomes almost negligible, as indicated in Figure 21–6.

Figure 21–7 illustrates the effect of reducing the input pulse width for a fixed time constant value. As the width is reduced, the output voltage becomes smaller because the capacitor has less time to charge. However, it takes the capacitor the same length of time ($5\tau$) to discharge back to zero for each condition after the pulse is removed.

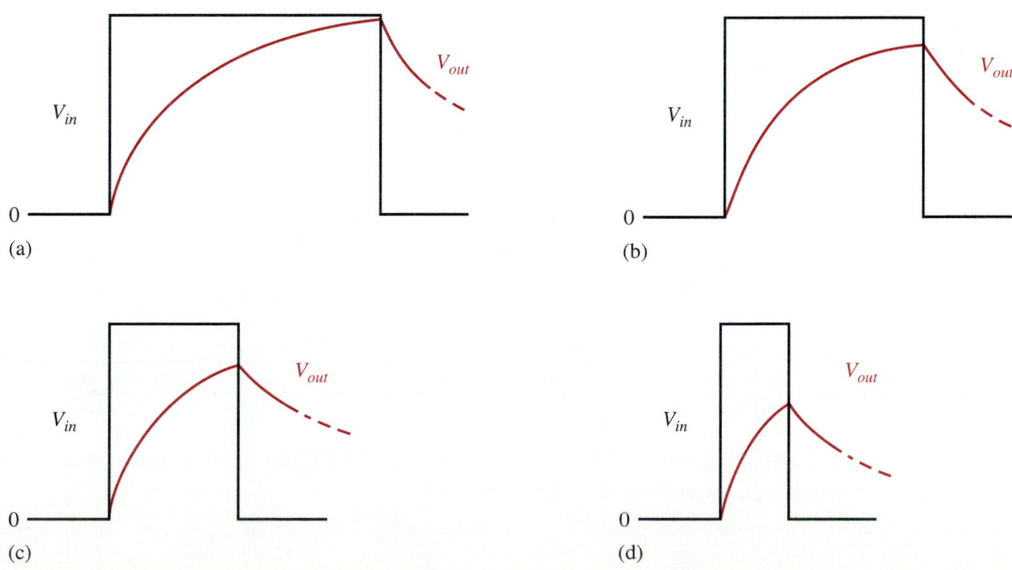


**FIGURE 21–7**
***The capacitor charges less and less as the input pulse width is reduced. The time constant is fixed.***

## EXAMPLE 21–1

A single 10 V pulse with a width of 100 μs is applied to the integrator in Figure 21–8.

**(a)** To what voltage will the capacitor charge?

**(b)** How long will it take the capacitor to discharge if the internal resistance of the pulse source is 50 Ω?

**(c)** Sketch the output voltage.

FIGURE 21–8

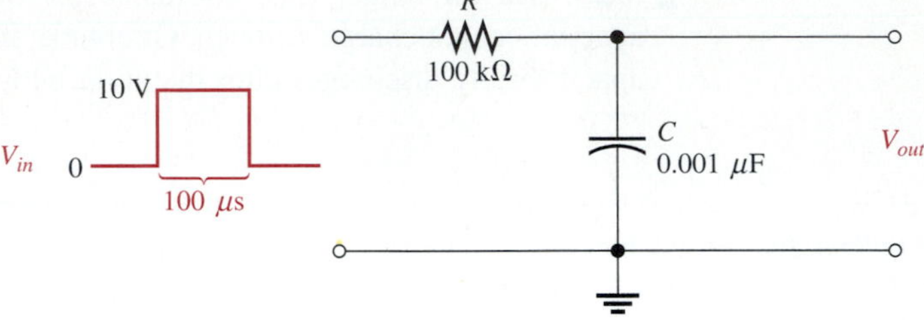


***Solution***

**(a)** The circuit time constant is

$$\tau = RC = (100\ \text{k}\Omega)(0.001\ \mu\text{F}) = 100\ \mu\text{s}$$

Notice that the pulse width is exactly equal to the time constant. Thus, the capacitor will charge approximately 63% of the full input amplitude in one time constant, so the output will reach a maximum voltage of

$$V_{out} = (0.63)10\ \text{V} = 6.3\ \text{V}$$

**(b)** The capacitor discharges back through the source when the pulse ends. You can neglect the 50 Ω source resistance in series with 100 kΩ. The total discharge time, therefore, is

$$5\tau = 5(100\ \mu\text{s}) = 500\ \mu\text{s}$$

**(c)** The output charging and discharging curve is shown in Figure 21–9.

FIGURE 21–9

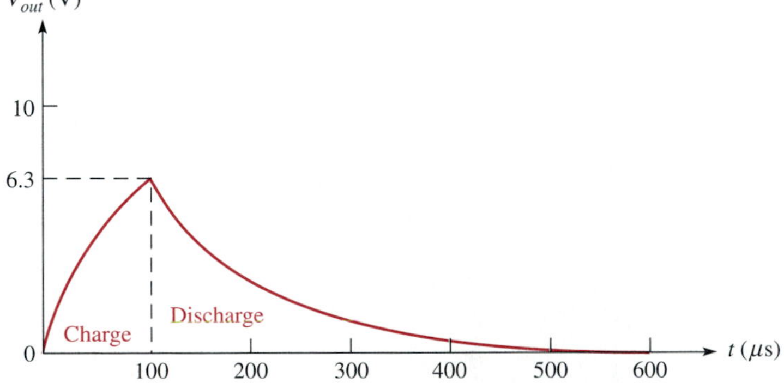


***Related Exercise*** If the input pulse width in Figure 21–8 is increased to 200 μs, to what voltage will the capacitor charge?

**EXAMPLE 21–2** Determine how much the capacitor in Figure 21–10 will charge when the single pulse is applied to the input.

FIGURE 21–10

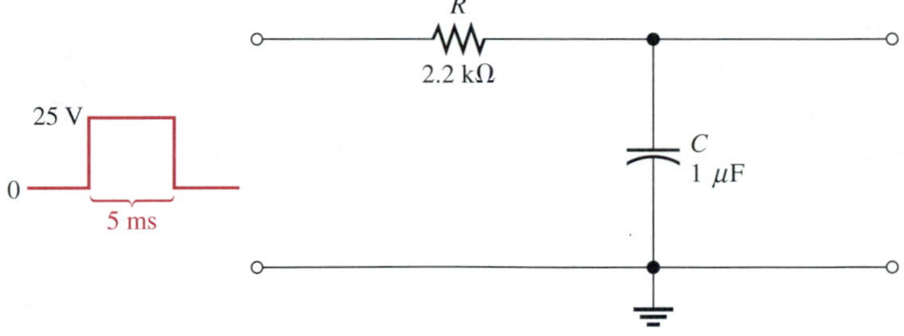


***Solution*** Calculate the time constant.

$$\tau = RC = (2.2\text{ k}\Omega)(1\ \mu\text{F}) = 2.2\text{ ms}$$

Because the pulse width is 5 ms, the capacitor charges for 2.27 time constants (5 ms/2.2 ms = 2.27). Use the exponential formula from Chapter 13 to find the voltage to which the capacitor will charge. With $V_F = 25$ V and $t = 5$ ms, the calculation is done as follows:

$$\begin{aligned} v &= V_F(1 - e^{-t/RC}) \\ &= (25\text{ V})(1 - e^{-5\text{ms}/2.2\text{ms}}) = (25\text{ V})(1 - e^{-2.27}) \\ &= (25\text{ V})(1 - 0.103) = (25\text{ V})(0.897) = 22.4\text{ V} \end{aligned}$$

These calculations show that the capacitor charges to 22.4 V during the 5 ms duration of the input pulse. It will discharge back to zero when the pulse goes back to zero.

***Related Exercise*** Determine how much *C* will charge if the pulse width is increased to 10 ms.

**SECTION 21–2 REVIEW**

1. When an input pulse is applied to an *RC* integrator, what condition must exist in order for the output voltage to reach full amplitude?
2. For the circuit in Figure 21–11, which has a single input pulse, find the maximum output voltage and determine how long the capacitor will discharge.
3. For Figure 21–11, sketch the approximate shape of the output voltage with respect to the input pulse.
4. If the integrator time constant equals the input pulse width, will the capacitor fully charge?
5. Describe the condition under which the output voltage has the approximate shape of a rectangular input pulse.

FIGURE 21–11

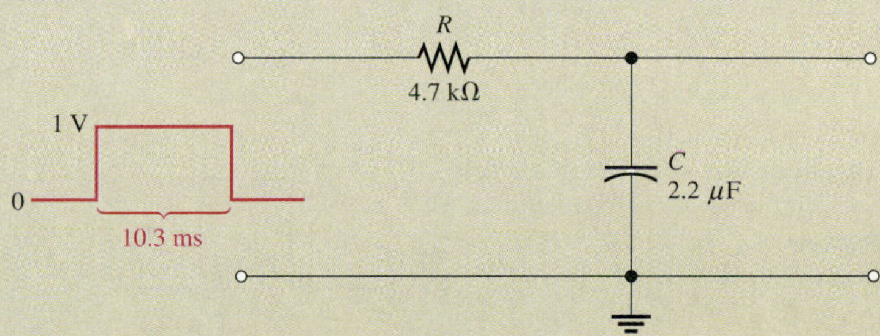

## 21–3 ■ REPETITIVE-PULSE RESPONSE OF *RC* INTEGRATORS

***In the last section, you learned how an RC integrator responds to a single-pulse input. These basic ideas are extended in this section to include the integrator response to repetitive pulses. In electronic systems, you will encounter repetitive-pulse waveforms much more often than single pulses. However, an understanding of the integrator's response to single pulses is necessary in order to understand how these circuits respond to repeated pulses.***

***After completing this section, you should be able to***

- **Analyze an *RC* integrator with repetitive input pulses**
  - ☐ Determine the response when the capacitor does not fully charge or discharge
  - ☐ Define *steady state*
  - ☐ Describe the effect of an increase in time constant on circuit response

If a **periodic** pulse waveform is applied to an *RC* integrator, as shown in Figure 21–12, *the output waveshape depends on the relationship of the circuit time constant and the frequency (period) of the input pulses.* The capacitor, of course, charges and discharges in response to a pulse input. The amount of charge and discharge of the capacitor depends both on the circuit time constant and on the input frequency, as mentioned.

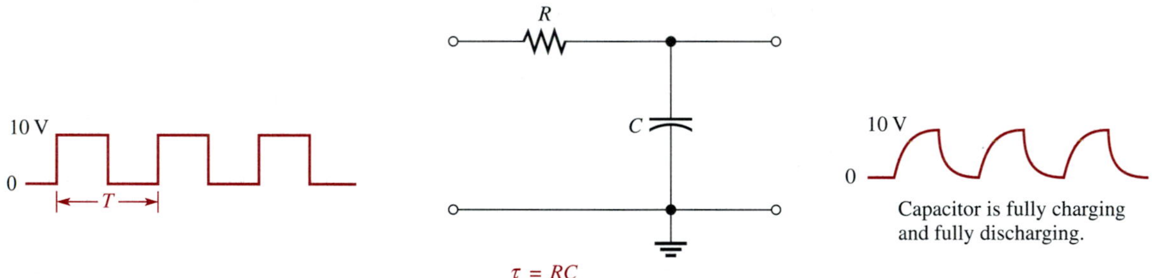


FIGURE 21–12
***RC integrator with a repetitive pulse waveform input.***

If the pulse width and the time between pulses are each equal to or greater than five time constants, the capacitor will fully charge and fully discharge during each period of the input waveform. This case is shown in Figure 21–12.

### When the Capacitor Does Not Fully Charge and Discharge

When the pulse width and the time between pulses are shorter than five time constants, as illustrated in Figure 21–13 for a square wave, the capacitor will *not* completely charge or discharge. We will now examine the effects of this situation on the output voltage of the *RC* integrator.

FIGURE 21–13
***Input waveform that does not allow full charge or discharge of the capacitor in an RC integrator.***

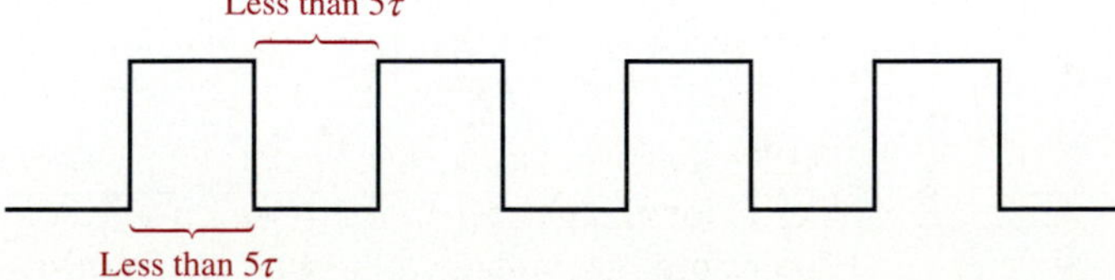

For illustration, let's use an *RC* integrator with a charging and discharging time constant equal to the pulse width of a 10 V square wave input, as in Figure 21–14. This choice will simplify the analysis and will demonstrate the basic action of the integrator under these conditions. At this point, we really do not care what the exact time constant value is because we know from Chapter 13 that an *RC* circuit charges approximately 63% during one time constant interval.

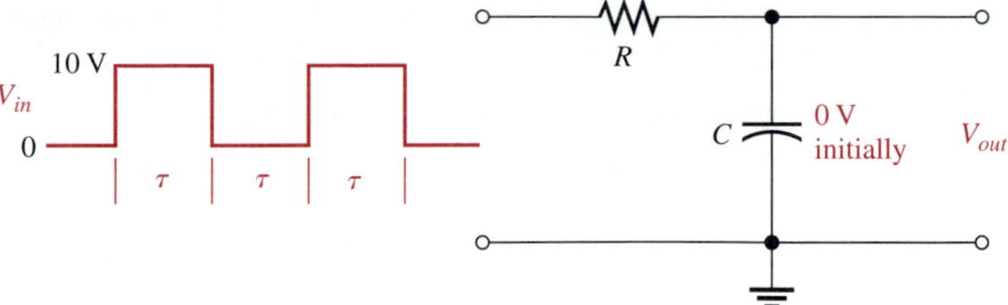


**FIGURE 21–14**
***Integrator with a square wave input having a period equal to two time constants ($T = 2\tau$).***

Let's assume that the capacitor in Figure 21–14 begins initially uncharged and examine the output voltage on a pulse-by-pulse basis. The results of this analysis are shown in Figure 21–15.

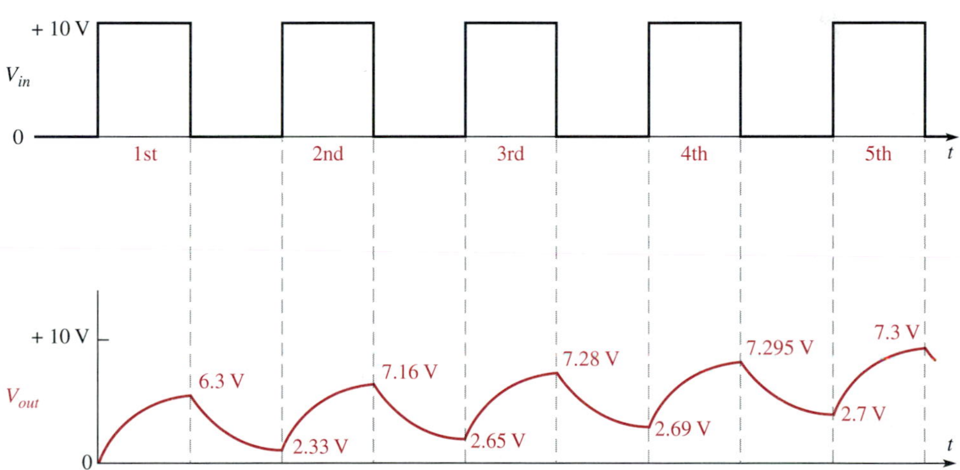


**FIGURE 21–15**
***Input and output for the initially uncharged integrator in Figure 21–14.***

**First pulse** During the first pulse, the capacitor charges. The output voltage reaches 6.3 V (63% of 10 V), as shown in Figure 21–15.

**Between first and second pulses** The capacitor discharges, and the voltage decreases to 37% of the voltage at the beginning of this interval: 0.37(6.3 V) = 2.33 V.

**Second pulse** The capacitor voltage begins at 2.33 V and increases 63% of the way to 10 V. This calculation is as follows: The total charging range is 10 V – 2.33 V = 7.67 V. The capacitor voltage will increase an additional 63% of 7.67 V, which is 4.83 V. Thus, at the end of the second pulse, the output voltage is 2.33 V + 4.83 V = 7.16 V, as shown in Figure 21–15. Notice that the average is building up.

**Between second and third pulses** The capacitor discharges during this time, and therefore the voltage decreases to 37% of the initial voltage by the end of the second pulse: 0.37(7.16 V) = 2.65 V.

**Third pulse** At the start of the third pulse, the capacitor voltage begins at 2.65 V. The capacitor charges 63% of the way from 2.65 V to 10 V: 0.63(10 V – 2.65 V) = 4.63 V. Therefore, the voltage at the end of the third pulse is 2.65 V + 4.63 V = 7.28 V.

**Between third and fourth pulses** The voltage during this interval decreases due to capacitor discharge. It will decrease to 37% of its value by the end of the third pulse. The final voltage in this interval is 0.37(7.28 V) = 2.69 V.

**Fourth pulse** At the start of the fourth pulse, the capacitor voltage is 2.69 V. The voltage increases by 0.63(10 V − 2.69 V) = 4.605 V. Therefore, at the end of the fourth pulse, the capacitor voltage is 2.69 V + 4.605 V = 7.295 V. Notice that the values are leveling off as the pulses continue.

**Between fourth and fifth pulses** Between these pulses, the capacitor voltage drops to 0.37(7.295 V) = 2.7 V.

**Fifth pulse** During the fifth pulse, the capacitor charges 0.63(10 V − 2.7 V) = 4.6 V. Since it started at 2.7 V, the voltage at the end of the pulse is 2.7 V + 4.6 V = 7.3 V.

## Steady-State Response

In the preceding discussion, the output voltage gradually built up and then began leveling off. It takes approximately $5\tau$ for the output voltage to build up to a constant average value. This interval is the transient time of the circuit. Once the output voltage reaches the average value of the input voltage, a **steady-state** condition is reached which continues as long as the periodic input continues. This condition is illustrated in Figure 21–16 based on the values obtained in the preceding discussion.

The transient time for our example circuit is the time from the beginning of the first pulse to the end of the third pulse. The reason for this interval is that the capacitor voltage at the end of the third pulse is 7.28 V, which is about 99% of the final voltage.

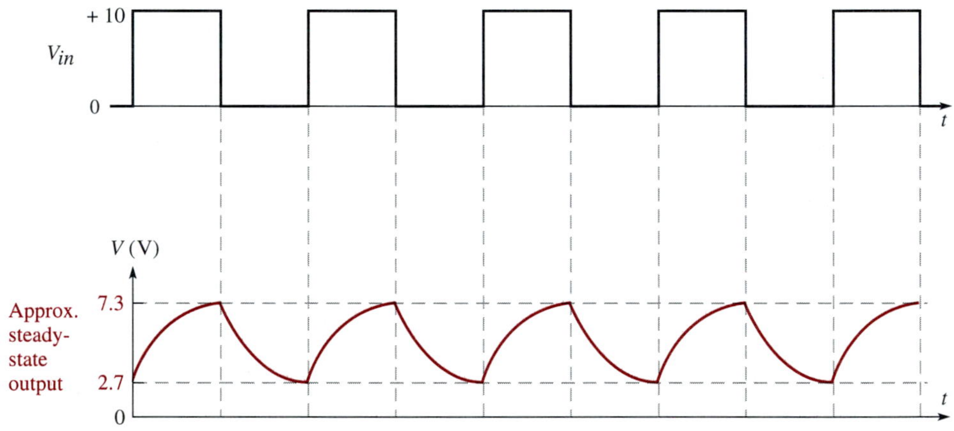


**FIGURE 21–16**
***Output reaches steady state after $5\tau$.***

## The Effect of an Increase in Time Constant

What happens to the output voltage if the *RC* time constant of the integrator is increased with a variable resistor, as indicated in Figure 21–17? As the time constant is increased, the capacitor charges less during a pulse and discharges less between pulses. The result is a smaller fluctuation in the output voltage for increasing values of time constant, as shown in Figure 21–18.

As the time constant becomes extremely long compared to the pulse width, the output voltage approaches a constant dc voltage, as shown in Figure 21–18(c). This value is the average value of the input. For a square wave, it is one-half the amplitude.

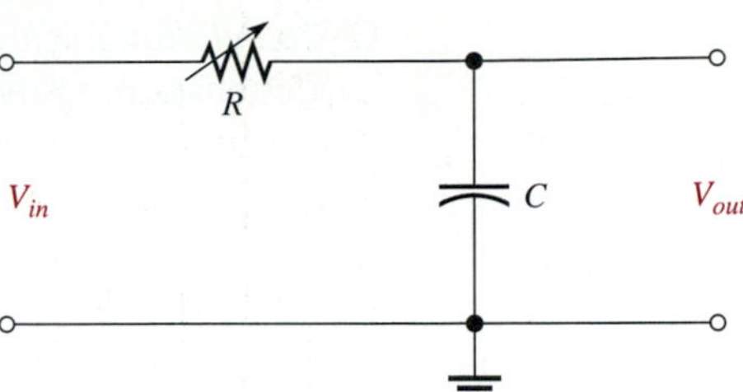


**FIGURE 21–17**
***Integrator with a variable time constant.***

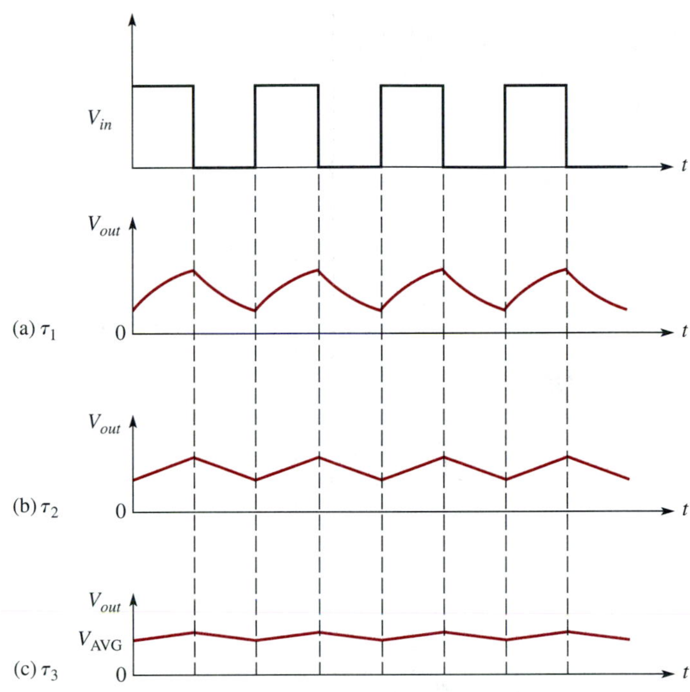


**FIGURE 21–18**
***Effect of longer time constants on the output of an integrator*** $(\tau_3 > \tau_2 > \tau_1)$.

**EXAMPLE 21–3** Determine the output voltage waveform for the first two pulses applied to the integrator circuit in Figure 21–19. Assume that the capacitor is initially uncharged and the rheostat is set to 5 kΩ.

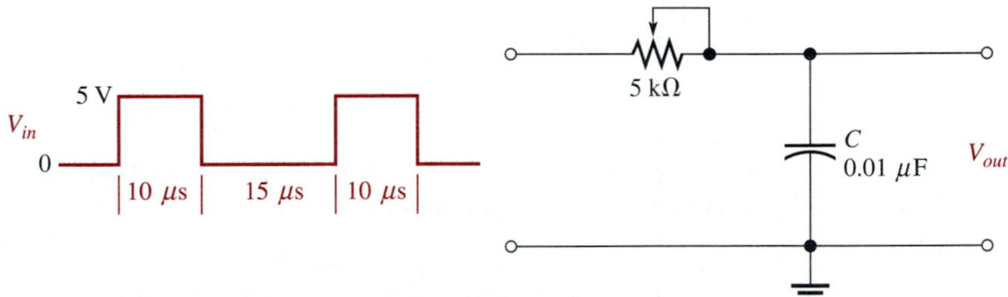


**FIGURE 21–19**

***Solution*** First calculate the circuit time constant.

$$\tau = RC = (5\text{ k}\Omega)(0.01\ \mu\text{F}) = 50\ \mu\text{s}$$

Obviously, the time constant is much longer than the input pulse width or the interval between pulses (notice that the input is not a square wave). Thus, in this case, the exponential formulas must be applied, and the analysis is relatively difficult. Follow the solution carefully.

1. *Calculation for first pulse:* Use the equation for an increasing exponential because $C$ is charging. Note that $V_F$ is 5 V, and $t$ equals the pulse width of 10 μs. Therefore,

$$v_C = V_F(1 - e^{-t/RC}) = (5\text{ V})(1 - e^{-10\mu s/50\mu s})$$
$$= (5\text{ V})(1 - 0.819) = 906\text{ mV}$$

This result is plotted in Figure 21–20(a).

FIGURE 21–20

(a)

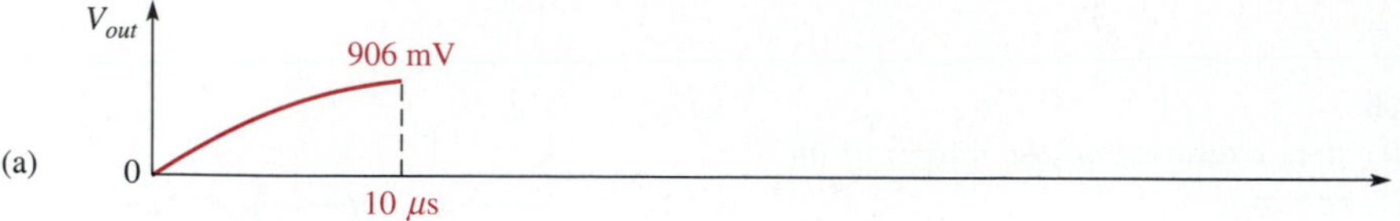


(b)

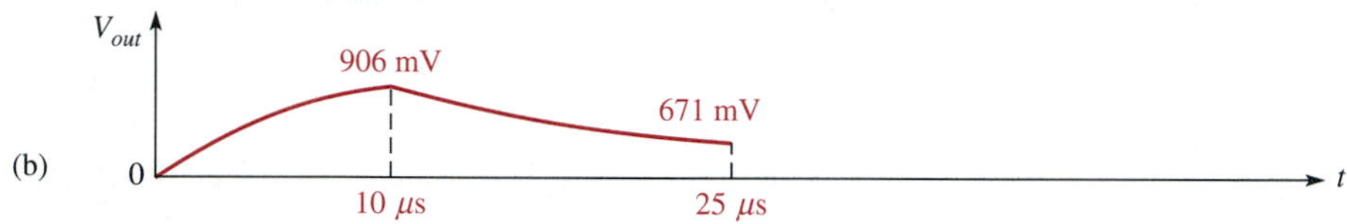


(c)

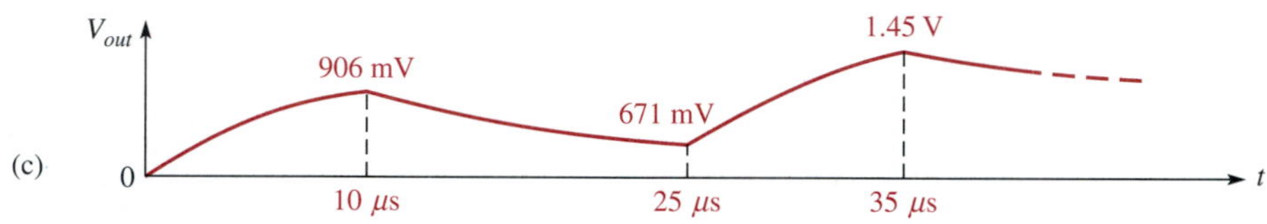


2. *Calculation for interval between first and second pulse:* Use the equation for a decreasing exponential because $C$ is discharging. Note that $V_i$ is 906 mV because $C$ begins to discharge from this value at the end of the first pulse. The discharge time is 15 μs. Therefore,

$$v_C = V_i e^{-t/RC} = (906\text{ mV})e^{-15\mu s/50\mu s}$$
$$= (906\text{ mV})(0.741) = 671\text{ mV}$$

This result is shown in Figure 21–20(b).

3. *Calculation for second pulse:* At the beginning of the second pulse, the output voltage is 671 mV. During the second pulse, the capacitor will again charge. In this case, it does not begin at zero volts. It already has 671 mV from the previous charge and discharge. To handle this situation, you must use the general exponential formula.

$$v = V_F + (V_i - V_F)e^{-t/\tau}$$

Using this equation, you can calculate the voltage across the capacitor at the end of the second pulse as follows:

$$v_C = V_F + (V_i - V_F)e^{-t/RC}$$
$$= 5\text{ V} + (671\text{ mV} - 5\text{ V})e^{-10\mu s/50\mu s}$$
$$= 5\text{ V} + (-4.33\text{ V})(0.819) = 5\text{ V} - 3.55\text{ V} = 1.45\text{ V}$$

This result is shown in Figure 21–20(c).

Notice that the output waveform builds up on successive input pulses. After approximately $5\tau$, it will reach its steady state and will fluctuate between a constant maximum and a constant minimum, with an average equal to the average value of the input. You can see this pattern by carrying the analysis in this example further.

***Related Exercise*** Determine $V_{out}$ at the beginning of the third pulse.

**SECTION 21–3 REVIEW**

1. What conditions allow an *RC* integrator capacitor to fully charge and discharge when a periodic pulse waveform is applied to the input?
2. What will the output waveform look like if the circuit time constant is extremely small compared to the pulse width of a square wave input?
3. When $5\tau$ is greater than the pulse width of an input square wave, the time required for the output voltage to build up to a constant average value is called ______.
4. Define *steady-state response.*
5. What does the average value of the output voltage of an integrator equal during steady state?

## 21–4 ■ SINGLE-PULSE RESPONSE OF *RC* DIFFERENTIATORS

***A series RC circuit in which the output voltage is taken across the resistor is known as a differentiator in terms of pulse response. Recall that in terms of frequency response, it is a high-pass filter. The term, differentiator, is derived from a mathematical function which this type of circuit approximates under certain conditions.***

***After completing this section, you should be able to***

- **Analyze an *RC* differentiator with a single input pulse**
  - ☐ Describe the response at the rising edge of the input pulse
  - ☐ Determine the response during and at the end of a pulse for various pulse width–time constant relationships

Figure 21–21 shows an *RC* **differentiator** with a pulse input. The same action occurs in the differentiator as in the integrator, except the output voltage is taken across the resistor rather than the capacitor. The capacitor charges exponentially at a rate depending on the *RC* time constant. The shape of the differentiator's resistor voltage is determined by the charging and discharging action of the capacitor.

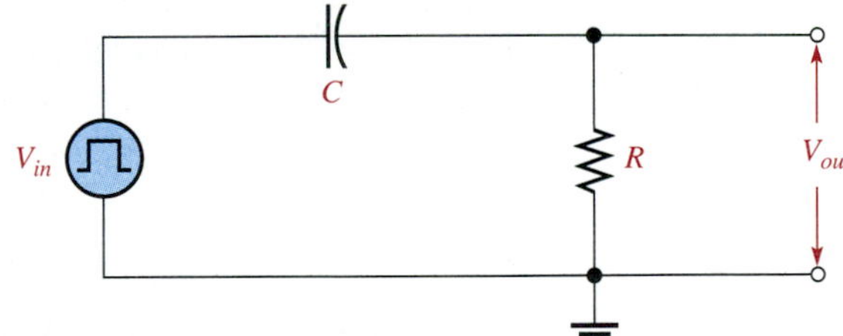


FIGURE 21–21
*An RC differentiating circuit.*

### Pulse Response

To understand how the output voltage is shaped by a differentiator, you must consider the following:

1. The response to the rising pulse edge
2. The response between the rising and falling edges
3. The response to the falling pulse edge

***Response to the Rising Edge of the Input Pulse*** Let's assume that the capacitor is initially uncharged prior to the rising pulse edge. Prior to the pulse, the input is zero volts. Thus, there are zero volts across the capacitor and also zero volts across the resistor, as indicated in Figure 21–22(a).

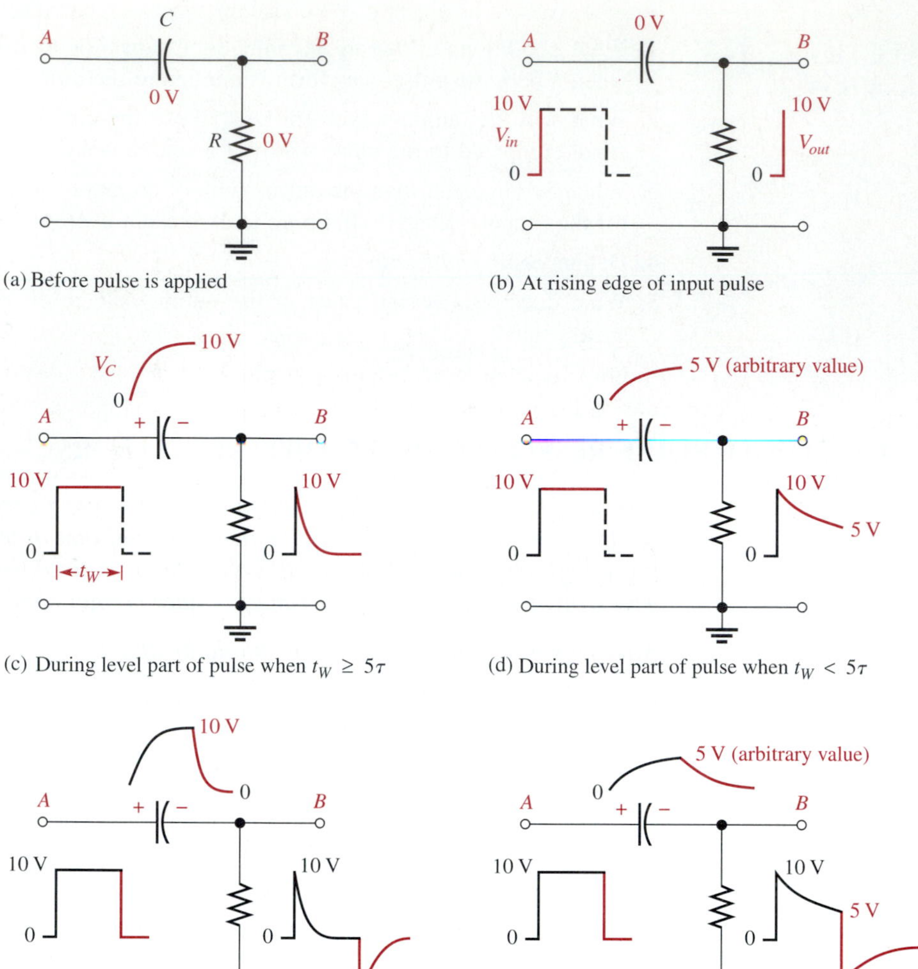


**FIGURE 21–22**

***Examples of the response of a differentiator to a single input pulse under two conditions: $t_W \geq 5\tau$ and $t_W < 5\tau$.***

Now let's assume that a 10 V pulse is applied to the input. When the rising edge occurs, point *A* goes to +10 V. Recall that the voltage across a capacitor cannot change instantaneously, and thus the capacitor appears instantaneously as a short. Therefore, if point *A* goes instantly to +10 V, then point *B must* also go instantly to +10 V, keeping the capacitor voltage zero for the instant of the rising edge. The capacitor voltage is the voltage from point *A* to point *B*.

The voltage at point *B* with respect to ground is the voltage across the resistor (and the output voltage). Thus, the output voltage suddenly goes to +10 V in response to the rising pulse edge, as indicated in Figure 21–22(b).

***Response During Pulse When $t_W \geq 5\tau$*** While the pulse is at its high level between the rising edge and the falling edge, the capacitor is charging. When the pulse width is equal to or greater than five time constants ($t_W \geq 5\tau$), the capacitor has time to fully charge.

As the voltage across the capacitor builds up exponentially, the voltage across the resistor decreases exponentially until it reaches zero volts at the time the capacitor reaches full charge (+10 V in this case). This decrease in the resistor voltage occurs because the sum of the capacitor voltage and the resistor voltage at any instant must be equal to the applied voltage, in compliance with Kirchhoff's voltage law ($v_C + v_R = v_{in}$). This part of the response is illustrated in Figure 21–22(c).

***Response During Pulse When $t_W < 5\tau$*** When the pulse width is less than five time constants ($t_W < 5\tau$), the capacitor does not have time to fully charge. Its partial charge depends on the relation of the time constant and the pulse width.

Because the capacitor does not reach the full +10 V, the resistor voltage will not reach zero volts by the end of the pulse. For example, if the capacitor charges to +5 V during the pulse interval, the resistor voltage will decrease to +5 V, as illustrated in Figure 21–22(d).

***Response to Falling Edge When $t_W \geq 5\tau$*** Let's first examine the case in which the capacitor is fully charged at the end of the pulse ($t_W \geq 5\tau$). Refer to Figure 21–22(e). On the falling edge, the input pulse suddenly goes from +10 V back to zero. An instant before the falling edge, the capacitor is charged to 10 V, so point *A* is +10 V and point *B* is 0 V. Since the voltage across a capacitor cannot change instantaneously, when point *A* makes a transition from +10 V to zero on the falling edge, point *B must* also make a 10 V transition from zero to −10 V. This keeps the voltage across the capacitor at 10 V for the instant of the falling edge.

The capacitor now begins to discharge exponentially. As a result, the resistor voltage goes from −10 V to zero in an exponential curve, as indicated in Figure 21–22(e).

***Response to Falling Edge When $t_W < 5\tau$*** Next, let's examine the case in which the capacitor is only partially charged at the end of the pulse ($t_W < 5\tau$). For example, if the capacitor charges to +5 V, the resistor voltage at the instant before the falling edge is also +5 V, because the capacitor voltage plus the resistor voltage must add up to +10 V, as illustrated in Figure 21–22(d).

When the falling edge occurs, point *A* goes from +10 V to zero. As a result, point *B* goes from +5 V to −5 V, as illustrated in Figure 21–22(f). This decrease occurs, of course, because the capacitor voltage cannot change at the instant of the falling edge. Immediately after the falling edge, the capacitor begins to discharge to zero. As a result, the resistor voltage goes from −5 V to zero, as shown.

## Summary of Differentiator Response to a Single Pulse

Perhaps a good way to summarize this section is to look at the general output waveforms of a differentiator as the time constant is varied from one extreme, when $5\tau$ is much less than the pulse width, to the other extreme, when $5\tau$ is much greater than the pulse width. These situations are illustrated in Figure 21–23. In part (a) of the figure, the output consists of narrow positive and negative "spikes." In part (f), the output approaches the shape of the input. Various conditions between these extremes are illustrated in parts (b) through (e).

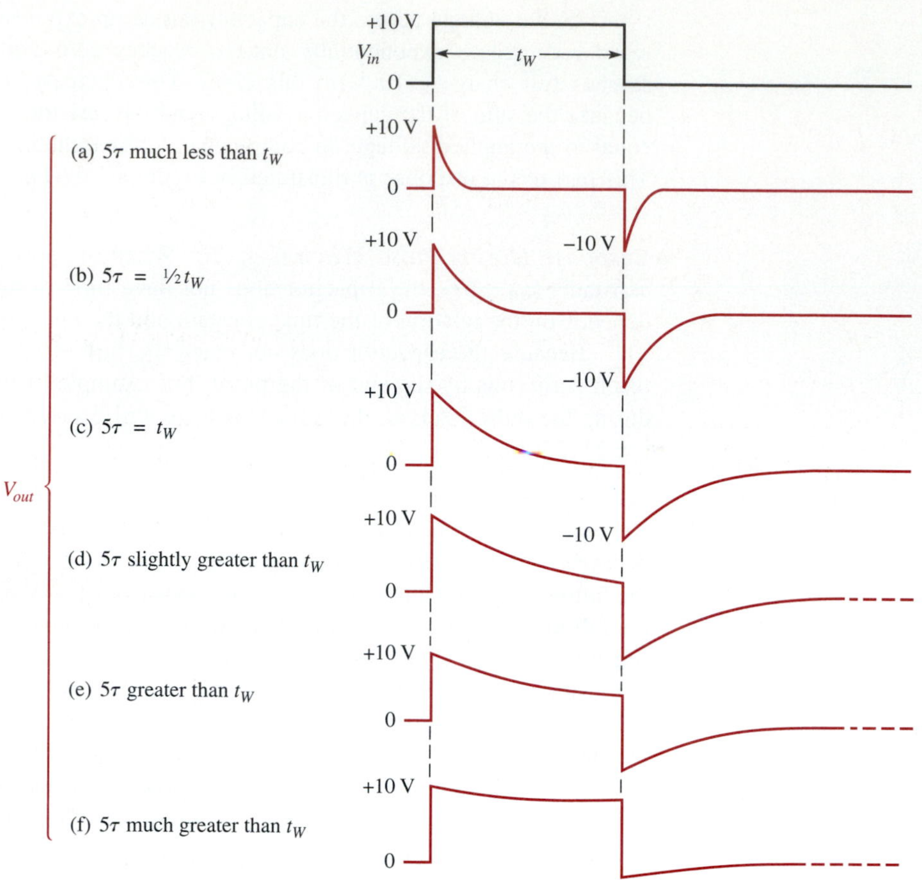


**FIGURE 21–23**
***Effects of a change in time constant on the shape of the output voltage of a differentiator.***

**EXAMPLE 21–4** Sketch the output voltage for the circuit in Figure 21–24.

**FIGURE 21–24**

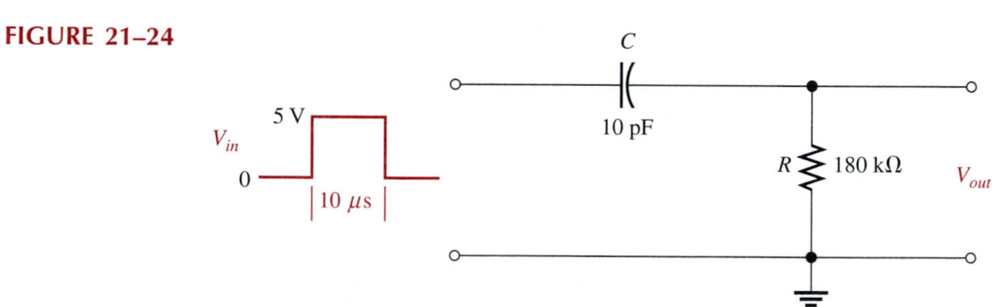


***Solution*** First calculate the time constant.

$$\tau = RC = (180\text{ k}\Omega)(10\text{ pF}) = 1.8\ \mu\text{s}$$

In this case, $t_W > 5\tau$, so the capacitor reaches full charge before the end of the pulse.

On the rising edge, the resistor voltage jumps to +5 V and then decreases exponentially to zero by the end of the pulse. On the falling edge, the resistor voltage jumps to −5 V and then goes back to zero exponentially. The resistor voltage is, of course, the output, and its shape is shown in Figure 21–25.

FIGURE 21–25

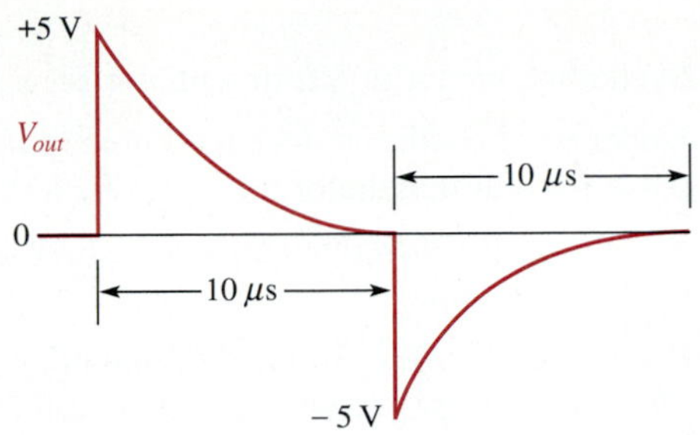


***Related Exercise*** Sketch the output voltage if $C$ is changed to 1 pF in Figure 21–24.

## EXAMPLE 21–5

Determine the output voltage waveform for the differentiator in Figure 21–26 with the rheostat set to 2 kΩ.

FIGURE 21–26

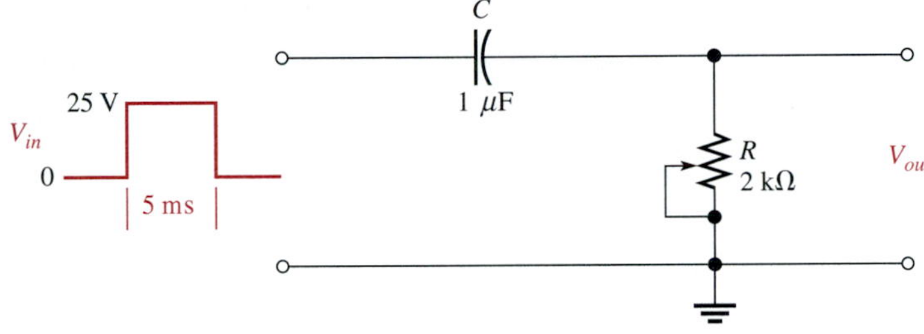


***Solution*** First calculate the time constant.

$$\tau = (2\text{ k}\Omega)(1\ \mu\text{F}) = 2\text{ ms}$$

On the rising edge, the resistor voltage immediately jumps to +25 V. Because the pulse width is 5 ms, the capacitor charges for 2.5 time constants and therefore does not reach full charge. Thus, you must use the formula for a decreasing exponential in order to calculate to what voltage the output decreases by the end of the pulse.

$$v_{out} = V_i e^{-t/RC} = 25e^{-5\text{ms}/2\text{ms}} = 25(0.082) = 2.05\text{ V}$$

where $V_i = 25$ V and $t = 5$ ms. This calculation gives the resistor voltage ($v_{out}$) at the end of the 5 ms pulse width interval.

On the falling edge, the resistor voltage immediately jumps from +2.05 V down to −22.95 V (a 25 V transition). The resulting waveform of the output voltage is shown in Figure 21–27.

FIGURE 21–27

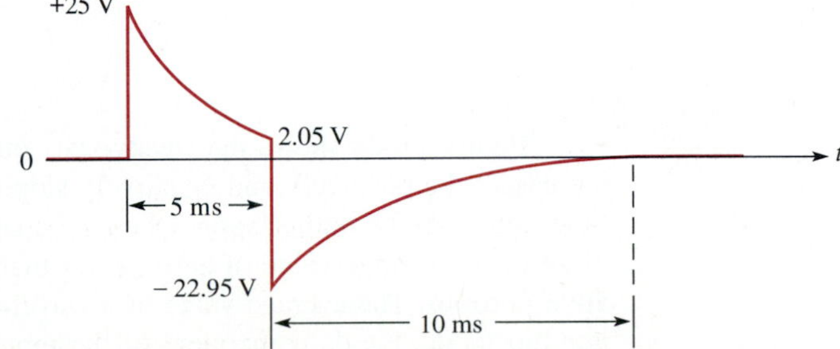


***Related Exercise*** Determine the voltage at the end of the pulse in Figure 21–26 if $R$ is adjusted to 1.5 kΩ.

**SECTION 21–4 REVIEW**

1. Sketch the output of a differentiator for a 10 V input pulse when $5\tau = 0.5t_W$.
2. Under what condition does the output pulse shape most closely resemble the input pulse for a differentiator?
3. What does the differentiator output look like when $5\tau$ is much less than the pulse width of the input?
4. If the resistor voltage in a differentiating circuit is down to +5 V at the end of a 15 V input pulse, to what negative value will the resistor voltage go in response to the falling edge of the input?

## 21–5 ■ REPETITIVE-PULSE RESPONSE OF *RC* DIFFERENTIATORS

***The RC differentiator response to a single pulse, covered in the last section, is extended in this section to repetitive pulses.***

***After completing this section, you should be able to***

- **Analyze an *RC* differentiator with repetitive input pulses**
  - ☐ Determine the response when the pulse width is less than five time constants

If a periodic pulse waveform is applied to an *RC* differentiating circuit, two conditions again are possible: $t_W \geq 5\tau$ or $t_W < 5\tau$. Figure 21–28 shows the output when $t_W = 5\tau$. As the time constant is reduced, both the positive and the negative portions of the output become narrower. Notice that the average value of the output is zero.

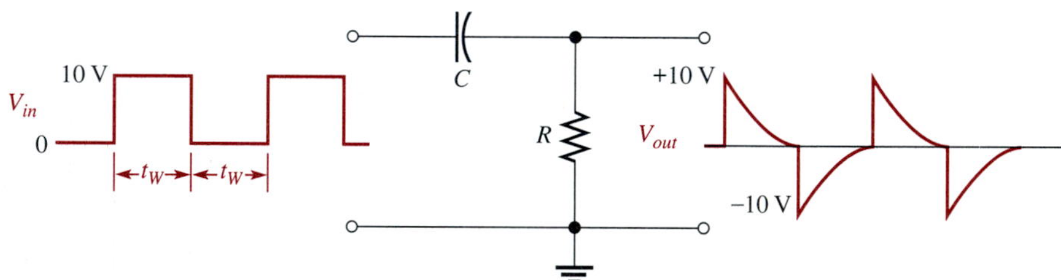


**FIGURE 21–28**
***Example of differentiator response when $t_W = 5\tau$.***

Figure 21–29 shows the steady-state output when $t_W < 5\tau$. As the time constant is increased, the positively and negatively sloping portions become flatter. For a very long time constant, the output approaches the shape of the input, but with an average value of zero. An average value of zero means that the waveform has equal positive and negative portions. The average value of a waveform is its **dc component.** Because a capacitor blocks dc, the dc component of the input is prevented from passing through to the output.

Like the integrator, the differentiator output takes time ($5\tau$) to reach steady state. To illustrate the response, let's take an example in which the time constant equals the input pulse width.

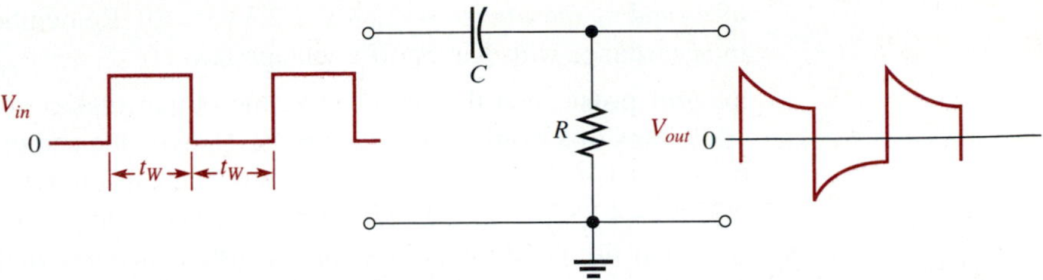


**FIGURE 21–29**
*Example of differentiator response when $t_W < 5\tau$.*

## Analysis of a Repetitive Waveform

At this point, we do not care what the circuit time constant is, because we know that the resistor voltage will decrease to approximately 37% of its maximum value during one pulse ($1\tau$). Let's assume that the capacitor in Figure 21–30 begins initially uncharged and then examine the output voltage on a pulse-by-pulse basis. The results of the analysis to follow are shown in Figure 21–31.

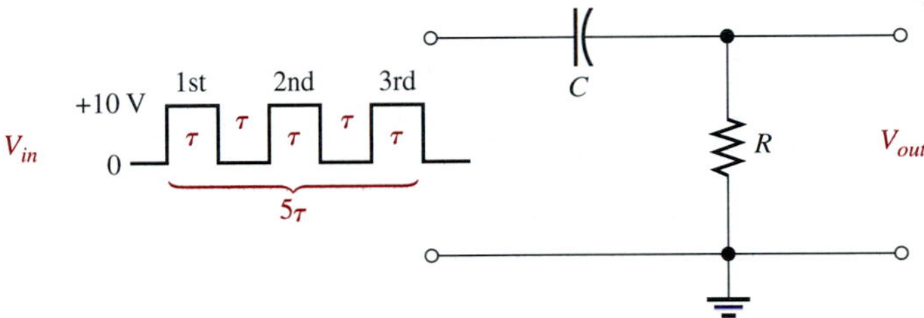


**FIGURE 21–30**
*RC differentiator with $\tau = t_W$.*

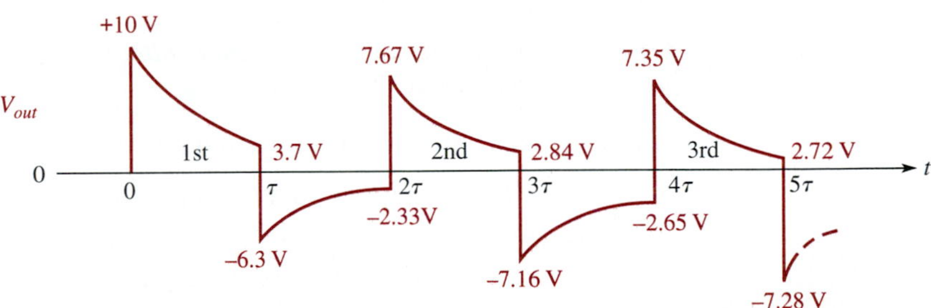


**FIGURE 21–31**
*Differentiator output waveform during transient time for the circuit in Figure 21–30.*

**First pulse** On the rising edge, the output instantaneously jumps to +10 V. Then the capacitor partially charges to 63% of 10 V, which is 6.3 V. Thus, the output voltage must decrease to 3.7 V, as shown in Figure 21–31.

On the falling edge, the output instantaneously makes a negative-going 10 V transition to –6.3 V (–10 V + 3.7 V = –6.3 V).

**Between first and second pulses** The capacitor discharges to 37% of 6.3 V, which is 2.33 V. Thus, the resistor voltage, which starts at –6.3 V, must increase to –2.33 V. Why? Because at the instant prior to the next pulse, the input voltage is zero. Therefore, the sum of $v_C$ and $v_R$ must be zero (2.33 V – 2.33 V = 0). Remember that $v_C + v_R = v_{in}$ at all times, in accordance with Kirchhoff's voltage law.

**Second pulse** On the rising edge, the output makes an instantaneous, positive-going, 10 V transition from –2.33 V to 7.67 V. Then by the end of the pulse the capacitor charges 0.63 × (10 V – 2.33 V) = 4.83 V. Thus, the capacitor voltage increases from 2.33 V to 2.33 V + 4.83 V = 7.16 V. The output voltage drops to 0.37 × 7.67 V = 2.84 V.

On the falling edge, the output instantaneously makes a negative-going transition from 2.84 V to –7.16 V, as shown in Figure 21–31.

**Between second and third pulses** The capacitor discharges to 37% of 7.16 V, which is 2.65 V. Thus, the output voltage starts at –7.16 V and increases to –2.65 V, because the capacitor voltage and the resistor voltage must add up to zero at the instant prior to the third pulse (the input is zero).

**Third pulse** On the rising edge, the output makes an instantaneous 10 V transition from –2.65 V to +7.35 V. Then the capacitor charges 0.63 × (10 V – 2.65 V) = 4.63 V to 2.65 V + 4.63 V = 7.28 V. As a result, the output voltage drops to 0.37 × 7.35 V = 2.72 V. On the falling edge, the output instantly goes from +2.72 V down to –7.28 V.

After the third pulse, five time constants have elapsed, and the output voltage is close to its steady state. Thus, it will continue to vary from a positive maximum of about +7.3 V to a negative maximum of about –7.3 V, with an average value of zero.

**SECTION 21–5 REVIEW**

1. What conditions allow an *RC* differentiator to fully charge and discharge when a periodic pulse waveform is applied to the input?
2. What will the output waveform look like if the circuit time constant is extremely small compared to the pulse width of a square wave input?
3. What does the average value of the differentiator output voltage equal during steady state?

## 21–6 ■ PULSE RESPONSE OF *RL* INTEGRATORS

***A series RL circuit in which the output voltage is taken across the resistor is known as an integrator in terms of pulse response.***

***After completing this section, you should be able to***

- **Analyze the operation of an *RL* integrator**
  - ☐ Determine the response to a single input pulse

Figure 21–32 shows an *RL* integrator. The output waveform is taken across the resistor and, under equivalent conditions, is the same shape as that for the *RC* integrator. Recall that in the *RC* case, the output was across the capacitor.

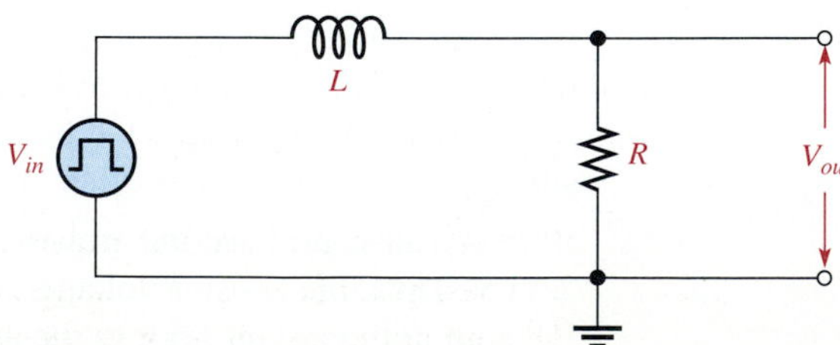


FIGURE 21–32
*An RL integrating circuit.*

As you know, each edge of an ideal pulse is considered to occur instantaneously. Two basic rules for inductor behavior will aid in analyzing *RL* circuit responses to pulse inputs:

1. The inductor appears as an open to an instantaneous change in current and as a short (ideally) to dc.
2. The current in an inductor cannot change instantaneously—it can change only exponentially.

## Response of the Integrator to a Single Pulse

When a pulse generator is connected to the input of the integrator and the voltage pulse goes from its low level to its high level, the inductor prevents a sudden change in current. As a result, the inductor acts as an open, and all of the input voltage is across it at the instant of the rising pulse edge. This situation is indicated in Figure 21–33(a).

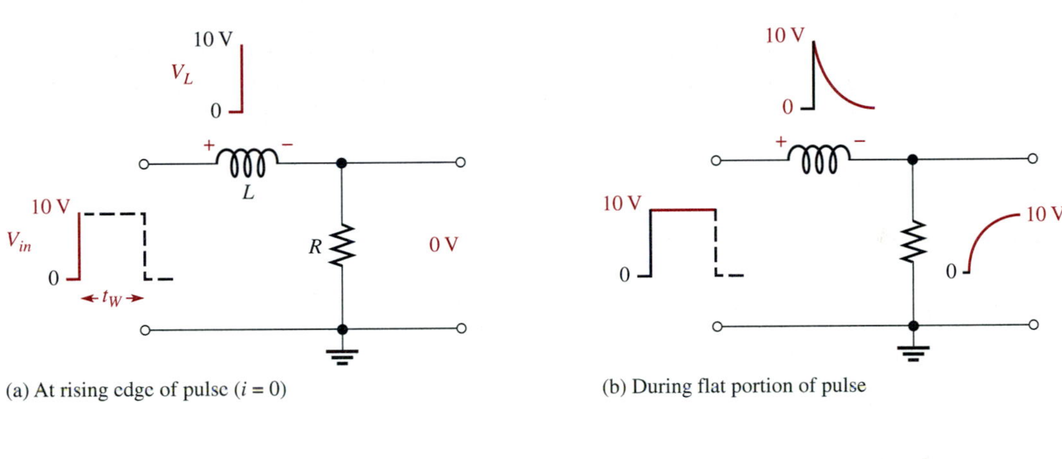


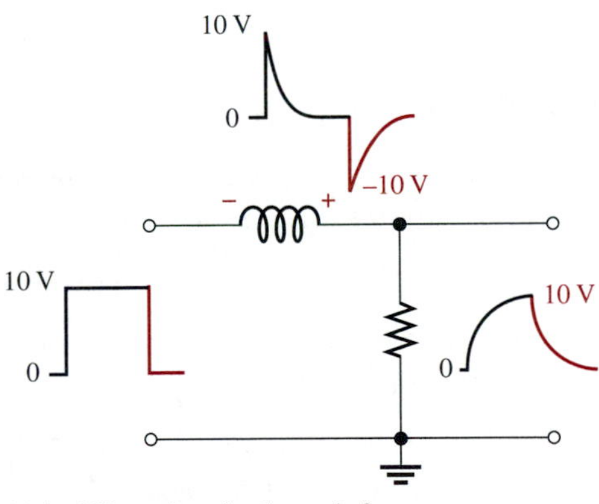


**FIGURE 21–33**
***Illustration of the pulse response of an RL integrator ($t_W > 5\tau$).***

After the rising edge, the current builds up, and the output voltage follows the current as it increases exponentially, as shown in Figure 21–33(b). The current can reach a maximum of $V_p/R$ if the transient time is shorter than the pulse width ($V_p$ = 10 V in this example).

When the pulse goes from its high level to its low level, an induced voltage with reversed polarity is created across the coil in an effort to keep the current equal to $V_p/R$. The output voltage begins to decrease exponentially, as shown in Figure 21–33(c).

The exact shape of the output depends on the $L/R$ time constant as summarized in Figure 21–34 for various relationships between the time constant and the pulse width. You should note that the response of this $RL$ circuit in terms of the shape of the output is identical to that of the $RC$ integrator. The relationship of the $L/R$ time constant to the input pulse width has the same effect as the $RC$ time constant that we discussed earlier in this chapter. For example, when $t_W < 5\tau$, the output voltage will not reach its maximum possible value.

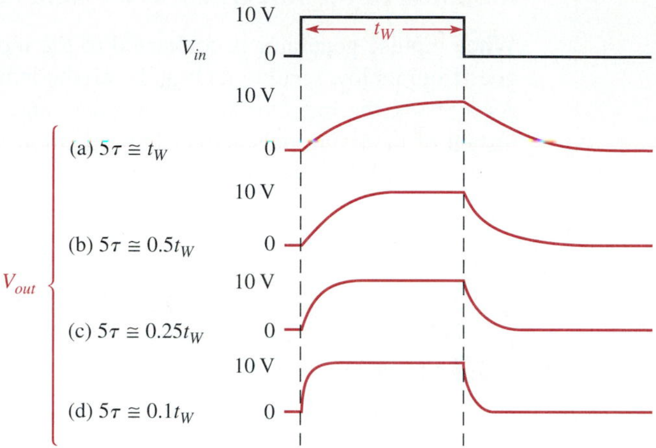


**FIGURE 21–34**
***Illustration of the variation in integrator output pulse shape with time constant.***

## EXAMPLE 21–6

Determine the maximum output voltage for the integrator in Figure 21–35 when a single pulse is applied as shown. The rheostat is set to 50 Ω.

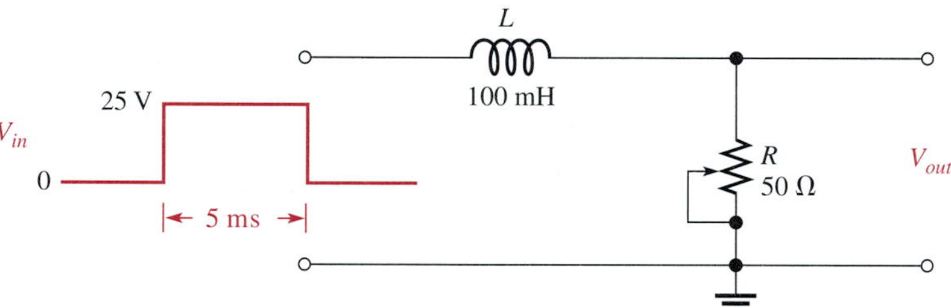


**FIGURE 21–35**

***Solution*** Calculate the time constant.

$$\tau = \frac{L}{R} = \frac{100 \text{ mH}}{50 \ \Omega} = 2 \text{ ms}$$

Because the pulse width is 5 ms, the inductor charges for $2.5\tau$. Use the exponential formula to calculate the voltage:

$$v_{out} = V_F(1 - e^{-t/\tau}) = 25(1 - e^{-5\text{ms}/2\text{ms}})$$
$$= 25(1 - e^{-2.5}) = 25(1 - 0.082) = 25(0.918) = 22.9 \text{ V}$$

***Related Exercise*** To what resistance must $R$ be set for the output voltage to reach 25 V by the end of the pulse in Figure 21–35?

**EXAMPLE 21–7** A pulse is applied to the *RL* integrator circuit in Figure 21–36. Determine the complete waveshapes and the values for $I$, $V_R$, and $V_L$.

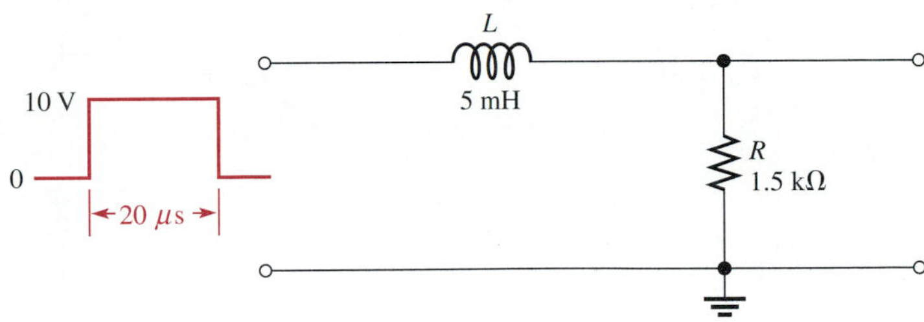


FIGURE 21–36

***Solution*** The circuit time constant is

$$\tau = \frac{L}{R} = \frac{5\text{ mH}}{1.5\text{ k}\Omega} = 3.33\ \mu\text{s}$$

Since $5\tau = 16.7\ \mu$s is less than $t_W$, the current will reach its maximum value and remain there until the end of the pulse.

At the rising edge of the pulse,

$$i = 0\text{ A}$$
$$v_R = 0\text{ V}$$
$$v_L = 10\text{ V}$$

Since the inductor is initially an open, all of the input voltage appears across $L$.

During the pulse,

$$i \text{ increases exponentially to } \frac{V_p}{R} = \frac{10\text{ V}}{1.5\text{ k}\Omega} = 6.67\text{ mA in } 16.7\ \mu\text{s}$$

$v_R$ increases exponentially to 10 V in 16.7 $\mu$s

$v_L$ decreases exponentially to zero in 16.7 $\mu$s

At the falling edge of the pulse,

$$i = 6.67\text{ mA}$$
$$v_R = 10\text{ V}$$
$$v_L = -10\text{ V}$$

After the pulse,

$i$ decreases exponentially to zero in 16.7 $\mu$s

$v_R$ decreases exponentially to zero in 16.7 $\mu$s

$v_L$ decreases exponentially to zero in 16.7 $\mu$s

The waveforms are shown in Figure 21–37.

FIGURE 21–37

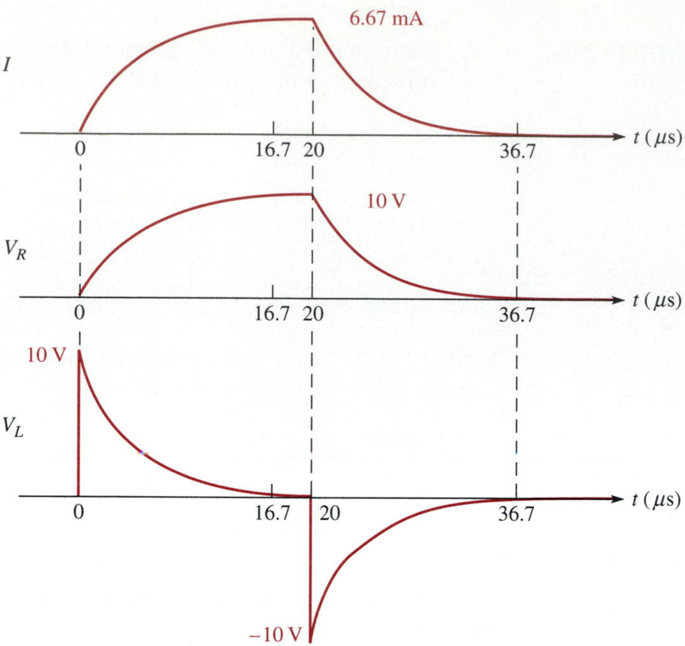


***Related Exercise*** What will be the maximum output voltage if the amplitude of the input pulse is increased to 20 V in Figure 21–36?

**EXAMPLE 21–8** A 10 V pulse with a width of 1 ms is applied to the integrator in Figure 21–38. Determine the voltage level that the output will reach during the pulse. If the source has an internal resistance of 30 Ω, how long will it take the output to decay to zero? Sketch the output voltage waveform.

FIGURE 21–38

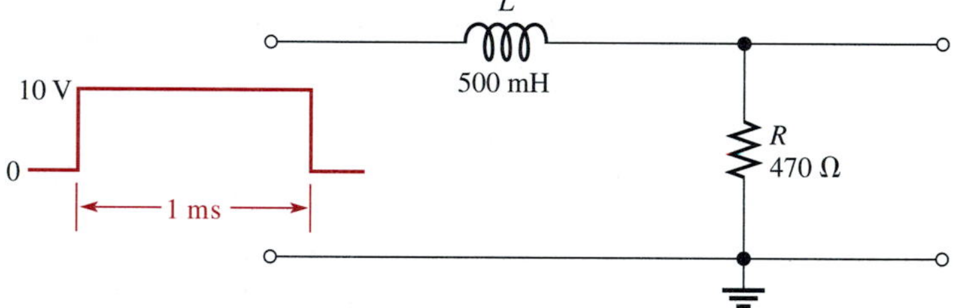


***Solution*** The coil charges through the 30 Ω source resistance plus the 470 Ω external resistor. The time constant is

$$\tau = \frac{L}{R_{tot}} = \frac{500\text{ mH}}{470\ \Omega + 30\ \Omega} = \frac{500\text{ mH}}{500\ \Omega} = 1\text{ ms}$$

Notice that in this case the pulse width is exactly equal to $\tau$. Thus, the output $V_R$ will reach 63% of the full input amplitude in $1\tau$. Therefore, the output voltage gets to 6.3 V at the end of the pulse.

After the pulse is gone, the inductor discharges back through the 30 Ω source and the 470 Ω resistor. The source takes $5\tau$ to completely discharge.

$$5\tau = 5(1\text{ ms}) = 5\text{ ms}$$

The output voltage is shown in Figure 21–39.

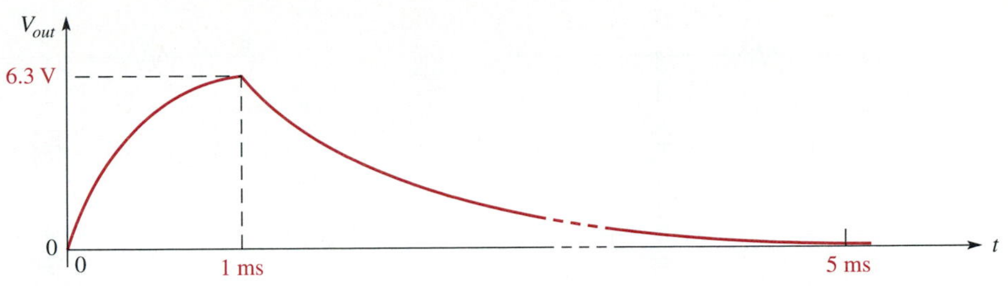


FIGURE 21–39

***Related Exercise*** To what value must $R$ be changed to allow the output voltage to reach the input level during the pulse?

**SECTION 21–6 REVIEW**

1. In an *RL* integrator, across which component is the output voltage taken?
2. When a pulse is applied to an *RL* integrator, what condition must exist in order for the output voltage to reach the amplitude of the input?
3. Under what condition will the output voltage have the approximate shape of the input pulse?

## 21–7 ■ PULSE RESPONSE OF *RL* DIFFERENTIATORS

***A series RL circuit in which the output voltage is taken across the inductor is known as a differentiator.***

***After completing this section, you should be able to***

- **Analyze the operation of an *RL* differentiator**
  - ☐ Determine the response to a single input pulse

### Response of the Differentiator to a Single Pulse

Figure 21–40 shows an *RL* differentiator with a pulse generator connected to the input. Initially, before the pulse, there is no current in the circuit. When the input pulse goes from its low level to its high level, the inductor prevents a sudden change in current. It does so, as you know, with an induced voltage equal and opposite to the input. As a result, $L$ looks like an open, and all of the input voltage appears across it at the instant of the rising edge, as shown in Figure 21–41(a) with a 10 V pulse.

FIGURE 21–40
***An RL differentiating circuit.***

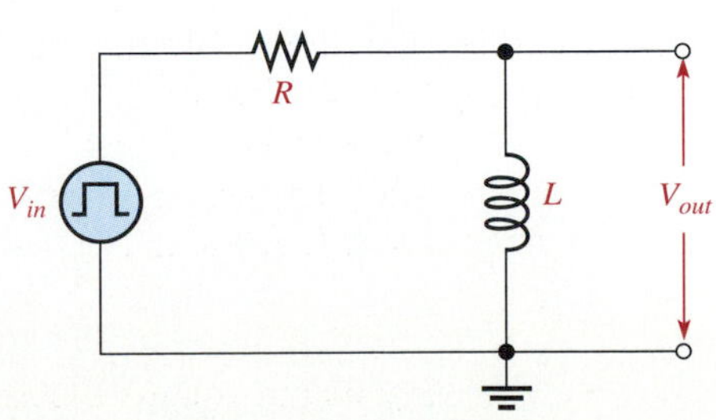

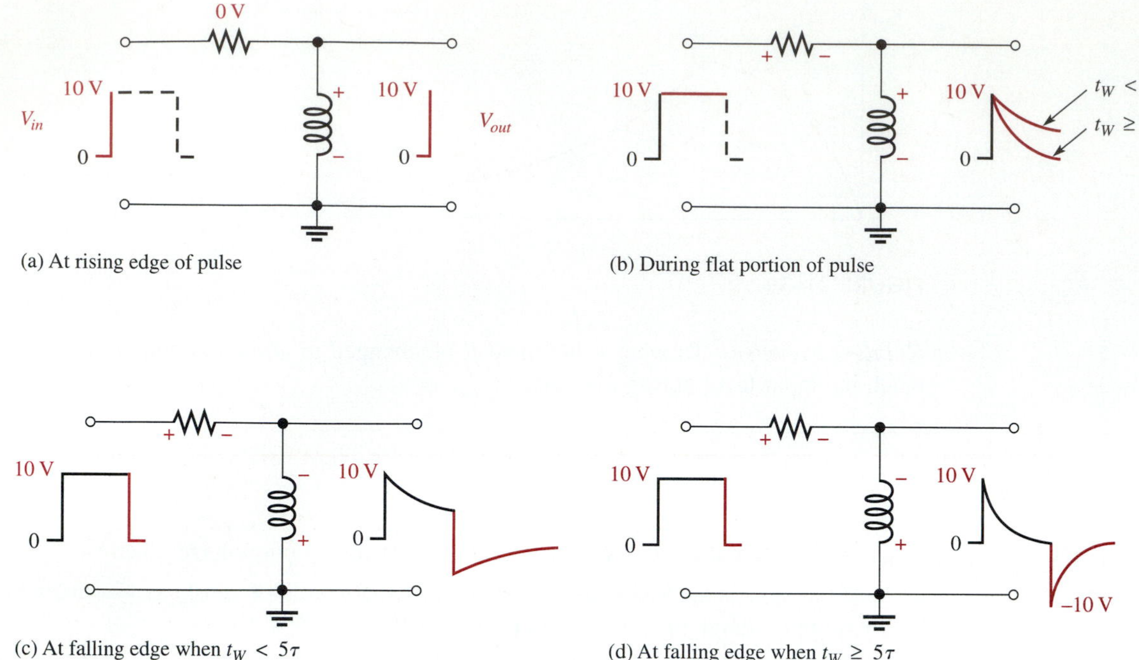


**FIGURE 21–41**
***Illustration of the response of an RL differentiator for both time constant conditions.***

During the pulse, the current exponentially builds up. As a result, the inductor voltage decreases, as shown in Figure 21–41(b). The rate of decrease, as you know, depends on the *L*/*R* time constant. When the falling edge of the input appears, the inductor reacts to keep the current as is, by creating an induced voltage in a direction as indicated in Figure 21–41(c). This reaction is seen as a sudden negative-going transition of the inductor voltage, as indicated in Figure 21–41(c) and (d).

Two conditions are possible, as indicated in Figure 21–41(c) and (d). In part (c), $5\tau$ is greater than the input pulse width, and the output voltage does not have time to decay to zero. In part (d), $5\tau$ is less than the pulse width, and so the output decays to zero before the end of the pulse. In this case a full, negative, 10 V transition occurs at the trailing edge.

Keep in mind that as far as the input and output waveforms are concerned, the *RL* integrator and differentiator perform the same as their *RC* counterparts.

A summary of the *RL* differentiator response for relationships of various time constants and pulse widths is shown in Figure 21–42.

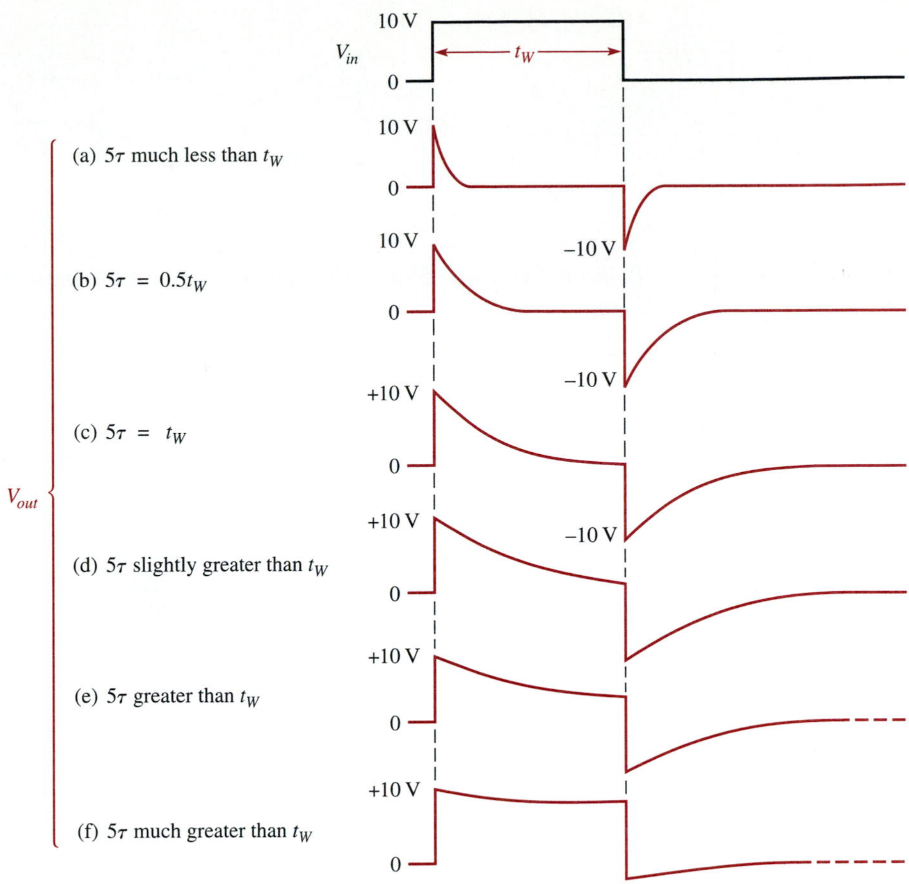


**FIGURE 21–42**
***Illustration of the variation in output pulse shape with the time constant.***

**EXAMPLE 21–9** Sketch the output voltage for the circuit in Figure 21–43.

**FIGURE 21–43**

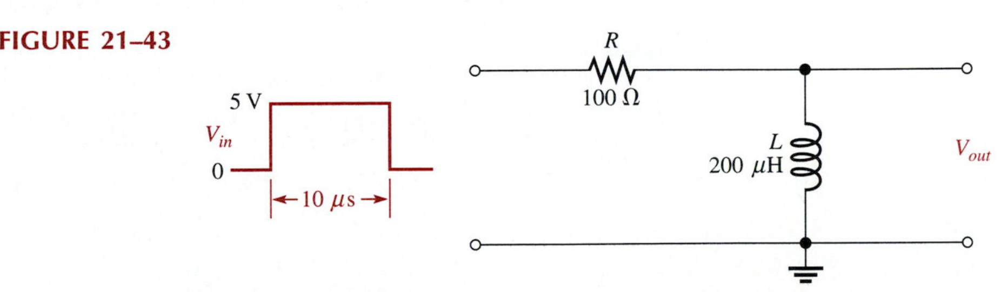


***Solution*** First calculate the time constant.

$$\tau = \frac{L}{R} = \frac{200\ \mu\text{H}}{100\ \Omega} = 2\ \mu\text{s}$$

In this case, $t_W = 5\tau$, so the output will decay to zero at the end of the pulse.

On the rising edge, the inductor voltage jumps to +5 V and then decays exponentially to zero. It reaches approximately zero at the instant of the falling edge. On the falling edge of the input, the inductor voltage jumps to −5 V and then goes back to zero. The output waveform is shown in Figure 21–44.

FIGURE 21–44

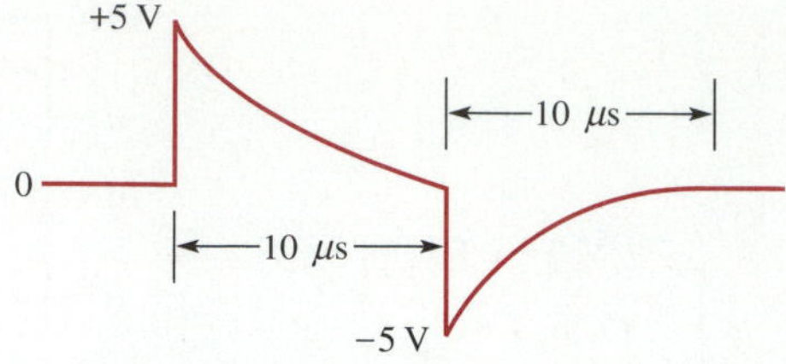


***Related Exercise*** Sketch the output voltage if the pulse width is reduced to 5 μs in Figure 21–43.

**EXAMPLE 21–10** Determine the output voltage waveform for the differentiator in Figure 21–45.

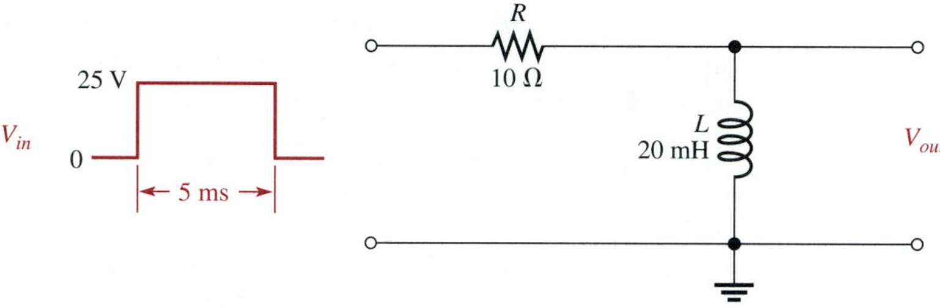


FIGURE 21–45

***Solution*** First calculate the time constant.

$$\tau = \frac{L}{R} = \frac{20\text{ mH}}{10\ \Omega} = 2\text{ ms}$$

On the rising edge, the inductor voltage immediately jumps to +25 V. Because the pulse width is 5 ms, the inductor charges for only 2.5$\tau$, so you must use the formula for a decreasing exponential.

$$v_L = V_i e^{-t/\tau} = 25e^{-5\text{ms}/2\text{ms}} = 25e^{-2.5} = 25(0.082) = 2.05\text{ V}$$

This result is the inductor voltage at the end of the 5 ms input pulse.

On the falling edge, the output immediately jumps from +2.05 V down to −22.95 V (a 25 V negative-going transition). The complete output waveform is sketched in Figure 21–46.

FIGURE 21–46

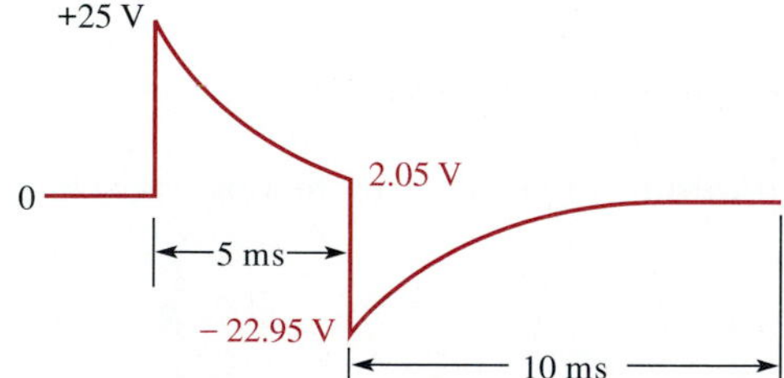


***Related Exercise*** What must be the value of $R$ for the output voltage to reach zero by the end of the pulse in Figure 21–45?

**SECTION 21–7 REVIEW**

1. In an *RL* differentiator, across which component is the output taken?
2. Under what condition does the output pulse shape most closely resemble the input pulse?
3. If the inductor voltage in an *RL* differentiator is down to +2 V at the end of a +10 V input pulse, to what negative voltage will the output go in response to the falling edge of the input?

## 21–8 ■ RELATIONSHIP OF TIME (PULSE) RESPONSE TO FREQUENCY RESPONSE

***A definite relationship exists between time (pulse) response and frequency response. The fast rising and falling edges of a pulse waveform contain the higher frequency components. The flatter portions of the pulse waveform, which are the tops of the pulses, represent slow changes or lower frequency components. The average value of the pulse waveform is its dc component.***

***After completing this section, you should be able to***

- **Explain the relationship of time response to frequency response**
  - ☐ Describe a pulse waveform in terms of its frequency components
  - ☐ Explain how *RC* and *RL* integrators act as filters
  - ☐ Explain how *RC* and *RL* differentiators act as filters
  - ☐ State the formulas that relate rise and fall times to frequency

The relationships of pulse characteristics and frequency content of pulse waveforms are indicated in Figure 21–47.

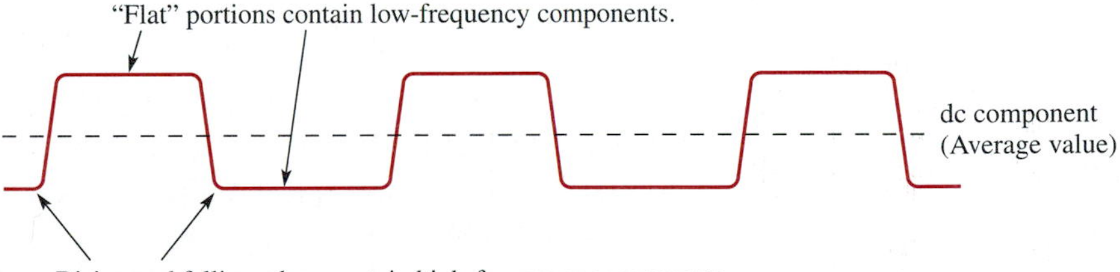
"Flat" portions contain low-frequency components.
dc component
(Average value)
Rising and falling edges contain high-frequency components.

FIGURE 21–47
*General relationship of a pulse waveform to frequency content.*

### The *RC* Integrator Acts as a Low-Pass Filter

As you learned, the integrator tends to exponentially "round off" the edges of the applied pulses. This rounding off occurs to varying degrees, depending on the relationship of the time constant to the pulse width and period. The rounding off of the edges indicates that the integrator tends to reduce the higher frequency components of the pulse waveform, as illustrated in Figure 21–48.

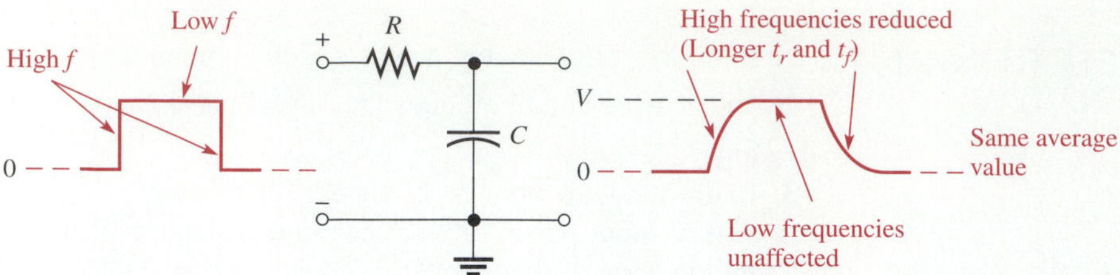


**FIGURE 21–48**
***Time and frequency response relationship in an integrator (one pulse in a repetitive waveform shown).***

## The *RL* Integrator Acts as a Low-Pass Filter

Like the *RC* integrator, the *RL* integrator also acts as a basic low-pass filter, because *L* is in series between the input and output. $X_L$ is small for low frequencies and offers little opposition. It increases with frequency, so at higher frequencies most of the total voltage is dropped across *L* and very little across *R*, the output. If the input is dc, *L* is like a short ($X_L = 0$). At high frequencies, *L* becomes like an open, as illustrated in Figure 21–49.

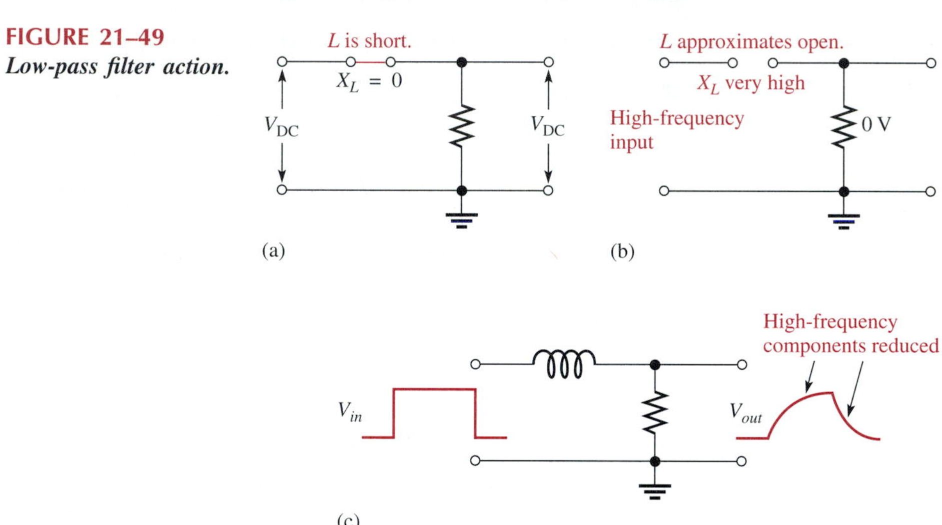


**FIGURE 21–49**
***Low-pass filter action.***

## The *RC* Differentiator Acts as a High-Pass Filter

As you know, the differentiator tends to introduce tilt to the flat portion of a pulse. That is, it tends to reduce the lower frequency components of a pulse waveform. Also, it completely eliminates the dc component of the input and produces a zero average-value output. This action is illustrated in Figure 21–50.

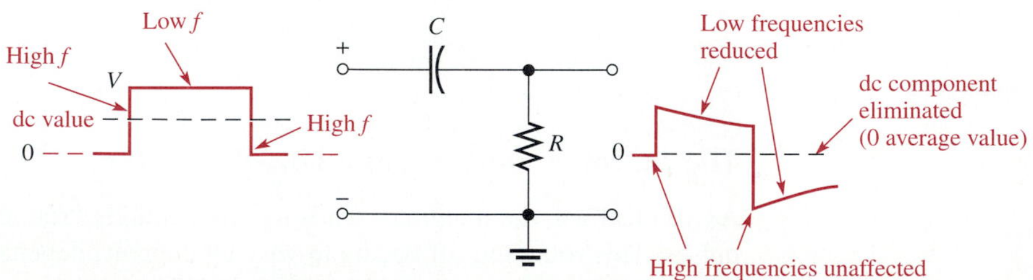


**FIGURE 21–50**
***Time and frequency response relationship in a differentiator (one pulse in a repetitive waveform shown).***

## The *RL* Differentiator Acts as a High-Pass Filter

Again like the *RC* differentiator, the *RL* differentiator also acts as a basic high-pass filter. Because *L* is connected across the output, less voltage is developed across it at lower frequencies than at higher ones. There are zero volts across the output for dc. For high frequencies, most of the input voltage is dropped across the output coil ($X_L = 0$ for dc; $X_L \cong$ open for high frequencies). Figure 21–51 shows high-pass filter action.

**FIGURE 21–51**
***High-pass filter action.***

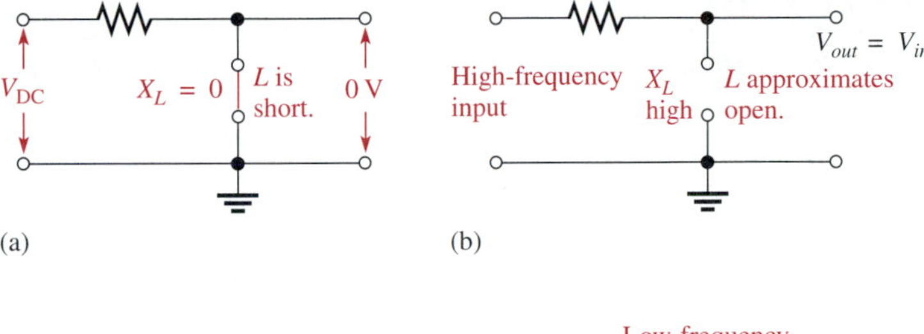


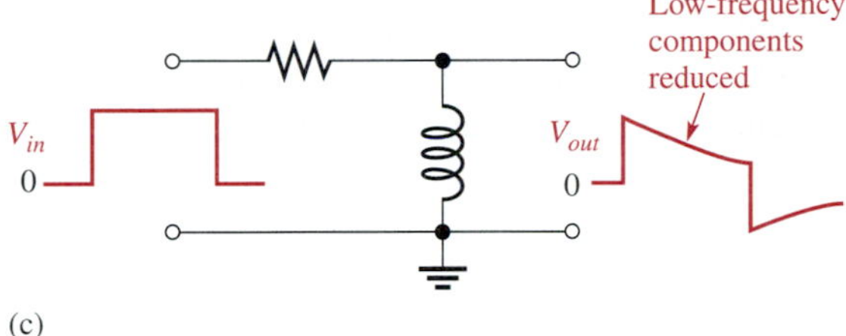


## Formula Relating Rise Time and Fall Time to Frequency

It can be shown that the fast transitions of a pulse (rise time, $t_r$, and fall time, $t_f$) are related to the highest frequency component, $f_h$, in that pulse by the following formula:

$$t_r = \frac{0.35}{f_h} \qquad \textbf{(21–1)}$$

This formula also applies to fall time, and the fastest transition determines the highest frequency in the pulse waveform.

Equation (21–1) can be rearranged to give the highest frequency as follows:

$$f_h = \frac{0.35}{t_r} \qquad \textbf{(21–2)}$$

or

$$f_h = \frac{0.35}{t_f} \qquad \textbf{(21–3)}$$

**EXAMPLE 21–11** What is the highest frequency contained in a pulse that has rise and fall times equal to 10 nanoseconds (10 ns)?

***Solution***

$$f_h = \frac{0.35}{t_r} = \frac{0.35}{10 \times 10^{-9}\text{ s}} = 0.035 \times 10^9\text{ Hz}$$
$$= 35 \times 10^6\text{ Hz} = 35\text{ MHz}$$

***Related Exercise*** What is the highest frequency in a pulse with $t_r$ = 20 ns and $t_f$ = 15 ns?

**SECTION 21–8 REVIEW**

1. What type of filter is an integrator?
2. What type of filter is a differentiator?
3. What is the highest frequency component in a pulse waveform having $t_r$ and $t_f$ equal to 1 $\mu$s?

## 21–9 ■ TROUBLESHOOTING

***In this section, RC circuits with pulse inputs are used to demonstrate the effects of common component failures in selected cases. The concepts can then be easily related to RL circuits.***

***After completing this section, you should be able to***

- **Troubleshoot integrators and differentiators**
  - ☐ Recognize the effect of an open capacitor
  - ☐ Recognize the effect of a leaky capacitor
  - ☐ Recognize the effect of a shorted capacitor
  - ☐ Recognize the effect of an open resistor

### Open Capacitors

If the capacitor in an integrator opens, the output has the same waveshape as the input, as shown in Figure 21–52(a). If the capacitor in a differentiator opens, the output is zero because it is held at ground through the resistor, as illustrated in part (b).

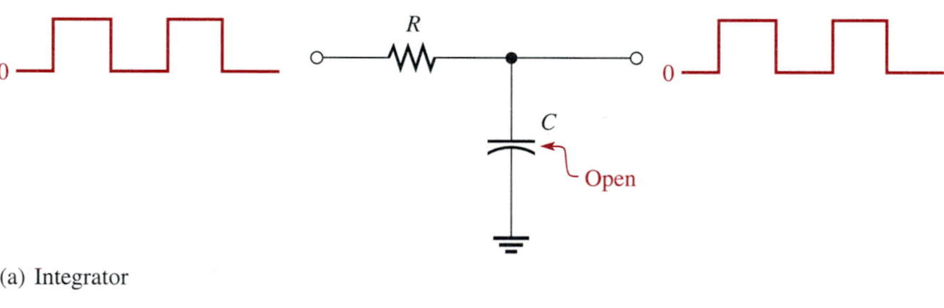


(a) Integrator

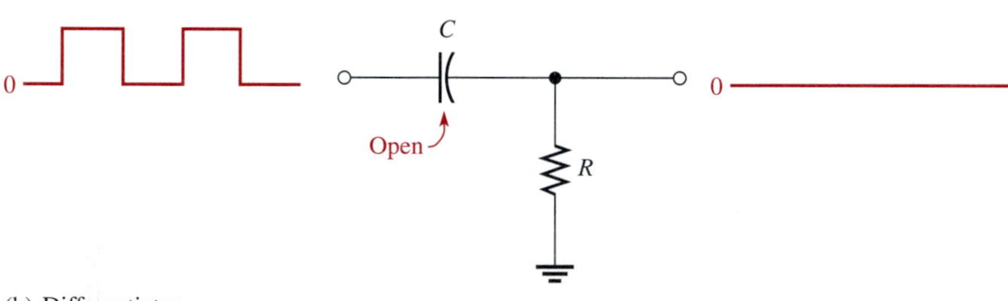


(b) Differentiator

FIGURE 21–52
*Examples of the effect of an open capacitor.*

### Leaky Capacitor

If the capacitor in an integrator becomes leaky, three things happen: (a) the time constant will be effectively reduced by the leakage resistance (when thevenized, looking from $C$ it appears in parallel with $R$); (b) the waveshape of the output voltage (across $C$) is altered from normal by a shorter charging time; and (c) the amplitude of the output is reduced because $R$ and $R_{leak}$ effectively act as a voltage divider. These effects are illustrated in Figure 21–53(a).

If the capacitor in a differentiator becomes leaky, the time constant is reduced, just as in the integrator (they are both simply series $RC$ circuits). When the capacitor reaches full charge, the output voltage (across $R$) is set by the effective voltage-divider action of $R$ and $R_{leak}$, as shown in Figure 21–53(b).

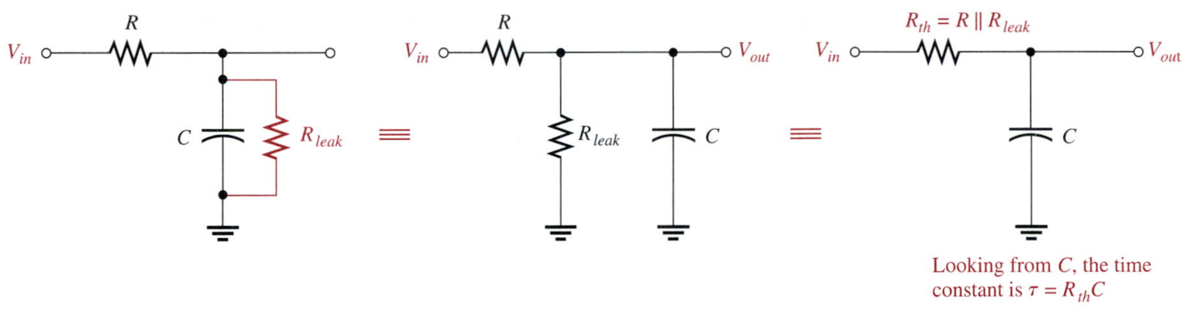


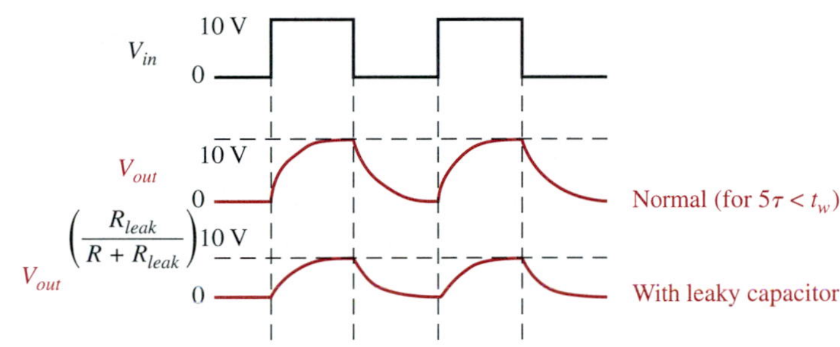


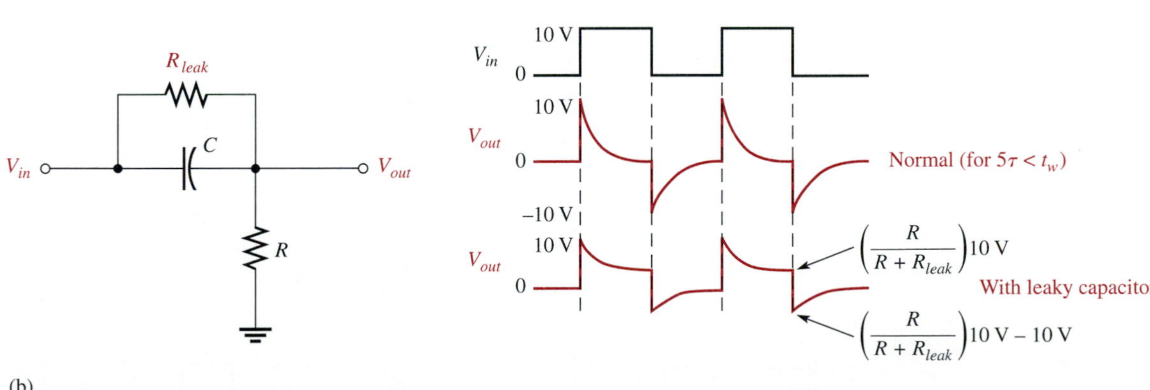


**FIGURE 21–53**
***Examples of the effect of a leaky capacitor.***

## Shorted Capacitor

If the capacitor in an integrator shorts, the output is at ground as shown in Figure 21–54(a). If the capacitor in a differentiator shorts, the output is the same as the input, as shown in part (b).

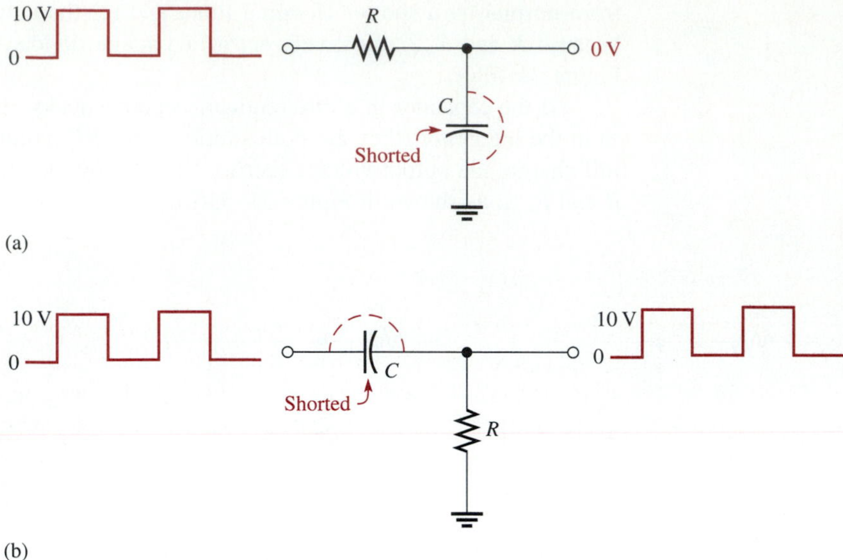


**FIGURE 21–54**
*Examples of the effect of a shorted capacitor.*

## Open Resistor

If the resistor in an integrator opens, the capacitor has no discharge path, and, ideally, it will hold its charge. In an actual situation, the charge will gradually leak off or the capacitor will discharge slowly through a measuring instrument connected to the output. This is illustrated in Figure 21–55(a).

If the resistor in a differentiator opens, the output looks like the input except for the dc level because the capacitor now must charge and discharge through the extremely high resistance of the oscilloscope, as shown in Figure 21–55(b).

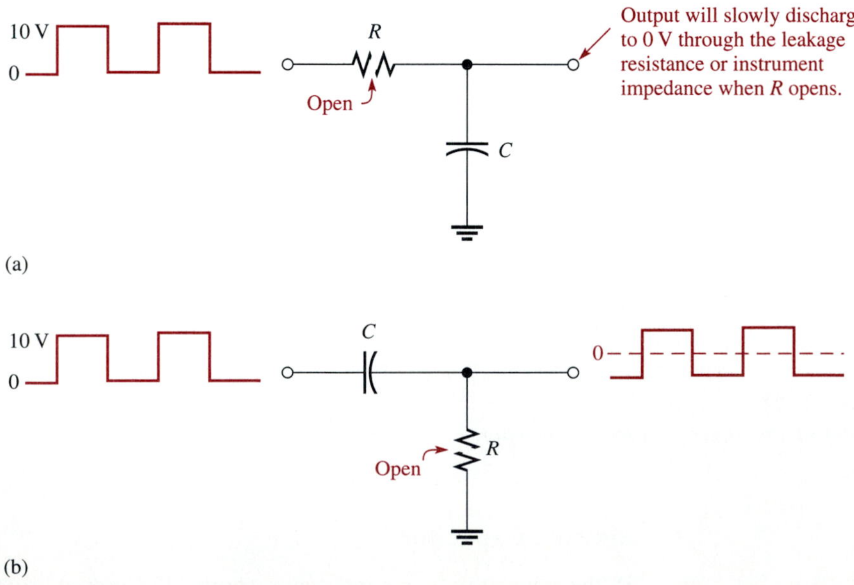


**FIGURE 21–55**
*Examples of the effects of an open resistor.*

**SECTION 21–9 REVIEW**

1. An *RC* integrator has a zero output with a square wave input. What are the possible causes of this problem?
2. If the capacitor in a differentiator is shorted, what is the output for a square wave input?

## 21–10 ■ TECHnology Theory Into Practice

***In this TECH TIP section, you are asked to build and test a time-delay circuit that will provide five switch-selectable delay times. An RC integrator is selected for this application. The input is a 5 V pulse of long duration, and the output goes to a threshold trigger circuit that is used to turn the power on to a portion of a system at any of the five selected time intervals after the occurrence of the original pulse.***

A schematic of the selectable time-delay integrator circuit is shown in Figure 21–56. The *RC* integrator is driven by a pulse input; and the output is an exponentially increasing voltage that is used to trigger a threshold circuit at the 3.5 V level, which then turns power on to part of a system. The basic concept is shown in Figure 21–57. In this application, the delay time of the integrator is specified to be the time from the rising edge of the input pulse to the point where the output voltage reaches 3.5 V. The specified delay times are as listed in Table 21–1 on the next page.

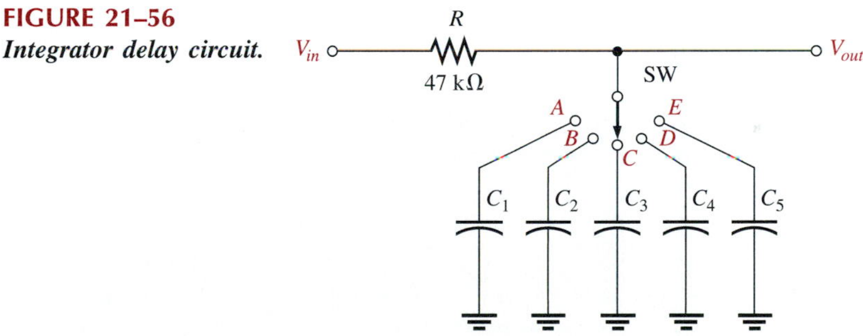


**FIGURE 21–56**
***Integrator delay circuit.***

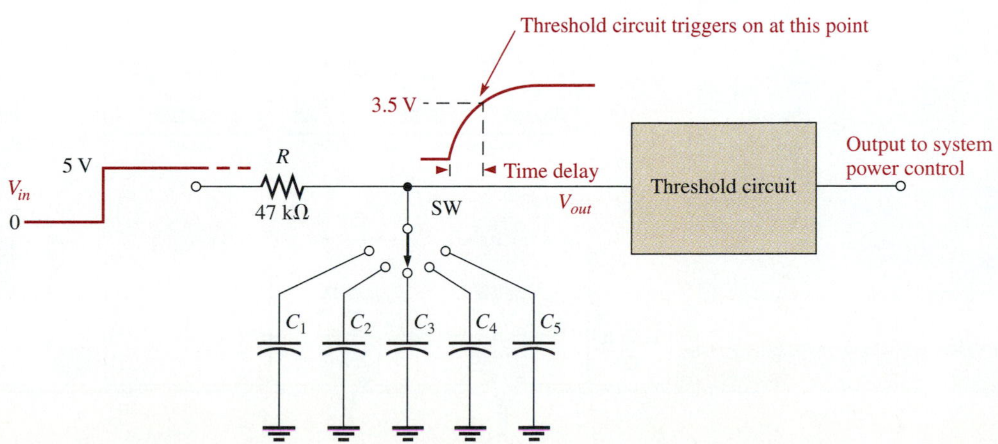


**FIGURE 21–57**
***Illustration of the time-delay application.***

**TABLE 21–1**

| Switch Position | Delay Time |
|---|---|
| A | 10 ms |
| B | 25 ms |
| C | 40 ms |
| D | 65 ms |
| E | 85 ms |

## Capacitor Values

☐ Determine a value for each capacitor that will provide the specified delay times within 10%. Select from the following list of standard values: 0.1 μF, 0.12 μF, 0.15 μF, 0.18 μF, 0.22 μF, 0.27 μF, 0.33 μF, 0.39 μF, 0.47 μF, 0.56 μF, 0.68 μF, 0.82 μF, 1.0 μF, 1.2 μF, 1.5 μF, 1.8 μF, 2.2 μF, 2.7 μF, 3.3 μF, 3.9 μF, 4.7 μF, 5.6 μF, 6.8 μF, 8.2 μF.

## Circuit Connections

Refer to Figure 21–58. The components for the *RC* integrator circuit in Figure 21–56 are assembled, but not interconnected, on the circuit board.

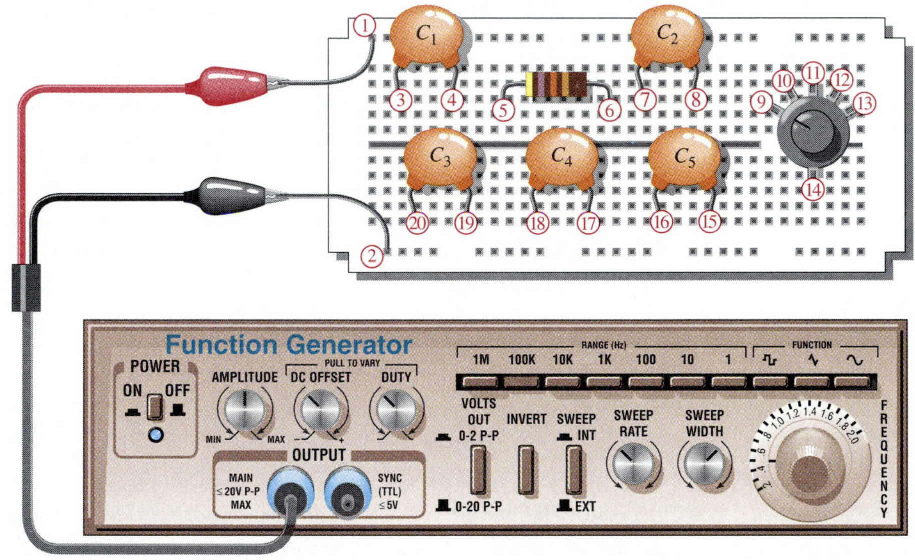


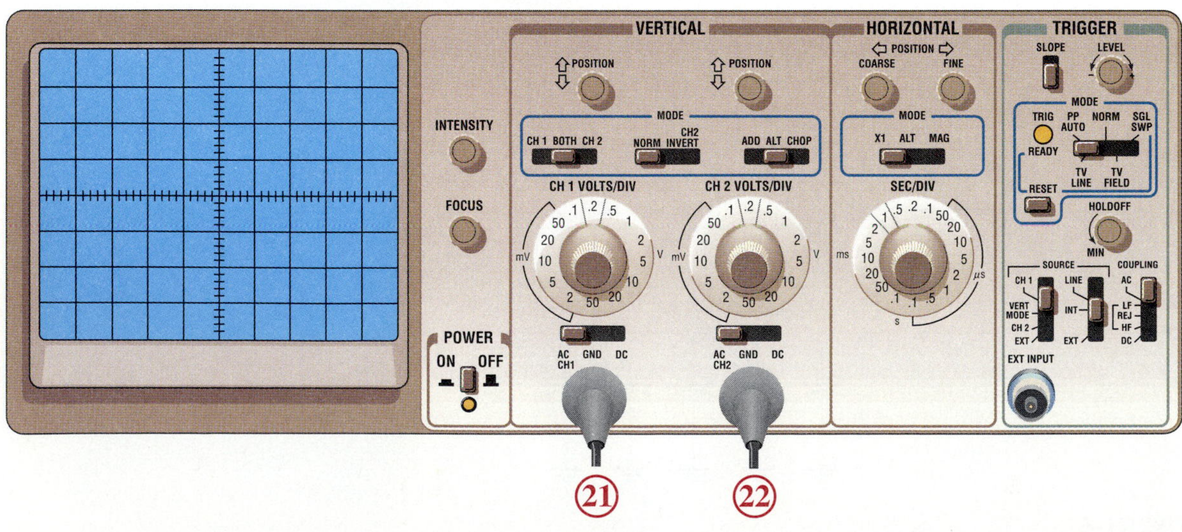


**FIGURE 21–58**
*The scope probes are ×1.*

□ Using the circled numbers, develop a point-to-point wiring list to properly connect the circuit on the board.

□ Indicate, using the appropriate circled numbers, how you would connect the instruments to test the circuit.

### Test Procedure

□ Specify the function, amplitude, and minimum frequency settings for the function generator in order to test all output delay times in Figure 21–58.

□ Specify the minimum oscilloscope settings for measuring each of the specified delay times in Figure 21–58.

**SECTION 21–10 REVIEW**

1. To add an additional time delay to the circuit of Figure 21–57, what changes must be made?
2. An additional time delay of 100 ms is required for the time-delay circuit. Determine the capacitor value that should be added.

## 21–11 ■ APPLYING PSpice AND PROBE TO PULSED CIRCUITS

***In Chapters 13 and 14, the transient responses of RC and RL circuits were observed using Probe and the .TRAN control statement. In those chapters, only single pulses with pulse widths of at least 5τ were used. In this section, you will learn how to use PSpice and Probe to observe the responses of differentiators and integrators to repetitive pulse waveforms.***

***After completing this section, you should be able to***

- **Use PSpice and Probe to observe pulse responses**
  - □ Write circuit files for *RC* and *RL* circuits with pulse inputs

Refer to Chapter 13 for a review of the PULSE source transient specification. An example *RC* differentiating circuit is shown in Figure 21–59, and the circuit file is as follows:

```
RC Differentiator Circuit
*Pulse Waveform Demonstration
VS  1  0  PULSE(0  10  0  1N  1N  500U  1M)
C1  1  2  1U
R1  2  0  1K
.PROBE
.TRAN  50P  5M
.END
```

FIGURE 21–59

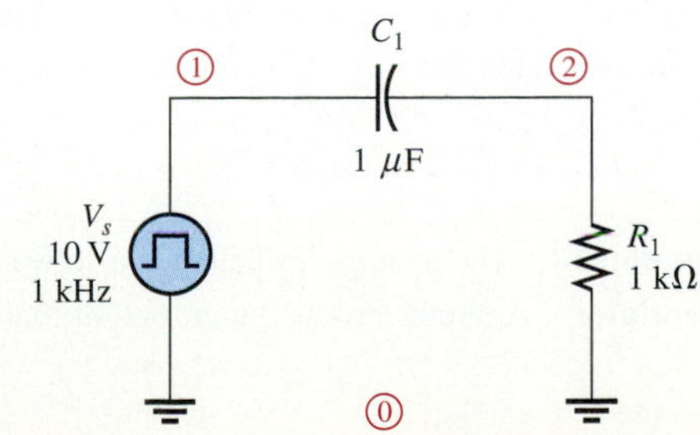

In this circuit, the voltage source produces 1 ms wide pulses that repeat every 2 ms ($T$ = 500 $\mu$s). Since the time constant of the circuit is 1 ms, the capacitor is unable to completely charge or discharge. The .TRAN statement samples data every 50 ps over a 5 ms interval so that you can see the effects of three consecutive pulses.

The source voltage and the voltages across the capacitor and the resistor can be observed by using Probe to display V(1,0), V(1,2), and V(2,0). The initial 10 V pulse produces 10 V across the resistor, which then begins to decrease exponentially as the capacitor charges exponentially. Before the capacitor can fully charge, the source voltage drops from 10 V to 0 V at the end of the pulse, forcing the resistor voltage to drop 10 V below its current value to a negative value. During the time between the first and second pulses as the capacitor discharges, its voltage is decreasing toward 0 V and the resistor voltage is increasing toward 0 V. The cursors can be used to determine values at any point on the waveforms.

At the beginning of the second pulse, the source voltage again jumps from 0 V to 10 V. This causes the resistor voltage to jump 10 V above its current value; as the capacitor again begins to charge, the resistor voltage decreases, and the capacitor voltage increases from its current value. This process of charging and discharging continues for successive pulses. With each pulse, the average value of the resistor voltage gets closer to 0 V, and the average value of the capacitor voltage approaches the average value of the pulse waveform of the source.

If Probe is allowed to sample the voltage long enough, you will observe the waveforms reach their final average values. This can be done by changing the .TRAN parameter from 5 ms to 50 ms or more.

## SECTION 21–11 REVIEW

1. Rewrite the circuit file for an *RC* integrator with the same values as in Figure 21–59.
2. Rewrite the .TRAN statement in order to guarantee taking a sample every 10 ps over a 25 ms interval.

## ■ SUMMARY

- In an *RC* integrating circuit, the output voltage is taken across the capacitor.
- In an *RC* differentiating circuit, the output voltage is taken across the resistor.
- In an *RL* integrating circuit, the output voltage is taken across the resistor.
- In an *RL* differentiating circuit, the output voltage is taken across the inductor.
- In an integrator, when the pulse width ($t_W$) of the input is much less than the transient time, the output voltage approaches a constant level equal to the average value of the input.
- In an integrator, when the pulse width of the input is much greater than the transient time, the output voltage approaches the shape of the input.
- In a differentiator, when the pulse width of the input is much less than the transient time, the output voltage approaches the shape of the input but with an average value of zero.
- In an differentiator, when the pulse width of the input is much greater than the transient time, the output voltage consists of narrow, postive-going and negative-going spikes occurring on the leading and trailing edges of the input pulses.
- The rising and falling edges of a pulse waveform contain the higher frequency components.
- The flat portion of the pulse contains the lower frequency components.

## ■ GLOSSARY

**DC component** The average value of a pulse waveform.

**Differentiator** A circuit producing an output that approaches the mathematical derivative of the input.

**Integrator** A circuit producing an output that approaches the mathematical integral of the input.

**Periodic** Characterized by a repetition at fixed-time intervals.

**Pulse response** In electric circuits, the reaction of a circuit to a given pulse input.

**Steady state** The equilibrium condition of a circuit that occurs after an initial transient time.

**Time constant** A fixed-time interval, set by *R* and *C*, or *R* and *L* values, that determines the time response of a circuit.

**Transient time** An interval equal to approximately five time constants.

## ■ FORMULAS

**(21–1)** $t_r = \dfrac{0.35}{f_h}$ Rise time

**(21–2)** $f_h = \dfrac{0.35}{t_r}$ Highest frequency in relation to rise time

**(21–3)** $f_h = \dfrac{0.35}{t_f}$ Highest frequency in relation to fall time

## ■ SELF-TEST

**1.** The output of an *RC* integrator is taken across the
**(a)** resistor **(b)** capacitor **(c)** source **(d)** coil

**2.** When a 10 V input pulse with a width equal to one time constant is applied to an *RC* integrator, the capacitor charges to
**(a)** 10 V **(b)** 5 V **(c)** 6.3 V **(d)** 3.7 V

**3.** When a 10 V pulse with a width equal to one time constant is applied to an *RC* differentiator, the capacitor charges to
**(a)** 6.3 V **(b)** 10 V **(c)** 0 V **(d)** 3.7 V

**4.** In an *RC* integrator, the output pulse closely resembles the input pulse when
**(a)** $\tau$ is much larger than the pulse width **(b)** $\tau$ is equal to the pulse width
**(c)** $\tau$ is less than the pulse width **(d)** $\tau$ is much less than the pulse width

**5.** In an *RC* differentiator, the output pulse closely resembles the input pulse when
**(a)** $\tau$ is much larger than the pulse width **(b)** $\tau$ is equal to the pulse width
**(c)** $\tau$ is less than the pulse width **(d)** $\tau$ is much less than the pulse width

**6.** The positive and negative portions of a differentiator's output voltage are equal when
**(a)** $5\tau < t_W$ **(b)** $5\tau > t_W$ **(c)** $5\tau = t_W$ **(d)** $5\tau > 0$

**7.** The output of an *RL* integrator is taken across the
**(a)** resistor **(b)** coil **(c)** source **(d)** capacitor

**8.** The maximum possible current in an *RL* integrator is
**(a)** $I = \dfrac{V_p}{X_L}$ **(b)** $I = \dfrac{V_p}{Z}$ **(c)** $I = \dfrac{V_p}{R}$

**9.** The current in an *RL* differentiator reaches its maximum possible value when
**(a)** $5\tau = t_W$ **(b)** $5\tau < t_W$ **(c)** $5\tau > t_W$ **(d)** $\tau = 0.5t_W$

**10.** If you have an *RC* and an *RL* differentiator with equal time constants sitting side-by-side and you apply the same input pulse to both,
**(a)** the *RC* has the widest output pulse
**(b)** the *RL* has the most narrow spikes on the output
**(c)** the output of one is an increasing exponential and the output of the other is a decreasing exponential
**(d)** you can't tell the difference by observing the output waveforms

## ■ PROBLEMS

### SECTION 21–1 The *RC* Integrator

1. An integrating circuit has $R = 2.2\text{ k}\Omega$ in series with $C = 0.05\ \mu\text{F}$. What is the time constant?
2. Determine how long it takes the capacitor in an integrating circuit to reach full charge for each of the following series *RC* combinations:
   **(a)** $R = 56\ \Omega$, $C = 50\ \mu\text{F}$ **(b)** $R = 3300\ \Omega$, $C = 0.015\ \mu\text{F}$
   **(c)** $R = 22\text{ k}\Omega$, $C = 100\text{ pF}$ **(d)** $R = 5.6\text{ M}\Omega$, $C = 10\text{ pF}$

### SECTION 21–2 Single-Pulse Response of *RC* Integrators

3. A 20 V pulse is applied to an *RC* integrator. The pulse width equals one time constant. To what voltage does the capacitor charge during the pulse? Assume that it is initially uncharged.
4. Repeat Problem 3 for the following values of $t_W$:
   **(a)** $2\tau$ **(b)** $3\tau$ **(c)** $4\tau$ **(d)** $5\tau$
5. Sketch the approximate shape of an integrator output voltage where $5\tau$ is much less than the pulse width of a 10 V square-wave input. Repeat for the case in which $5\tau$ is much larger than the pulse width.
6. Determine the output voltage for an integrator with a single input pulse, as shown in Figure 21–60. For repetitive pulses, how long will it take this circuit to reach steady state?

FIGURE 21–60

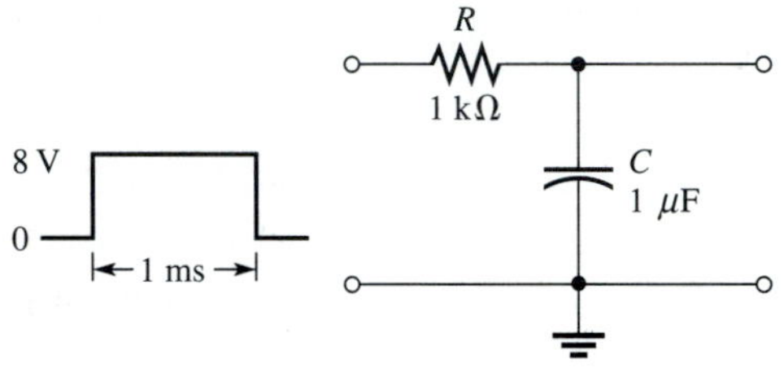


7. **(a)** What is $\tau$ in Figure 21–61? **(b)** Sketch the output voltage.

FIGURE 21–61

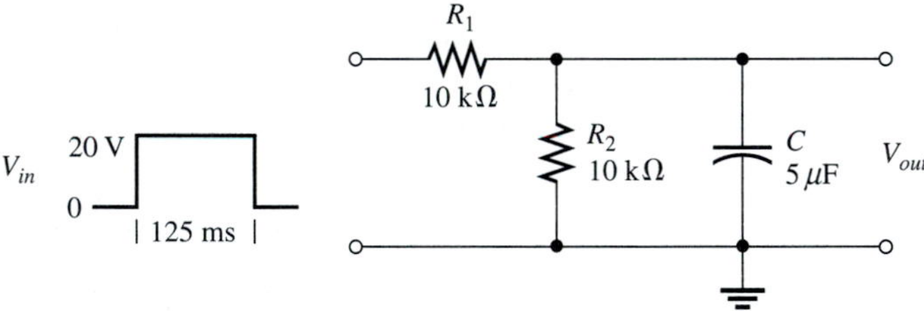


### SECTION 21–3 Repetitive-Pulse Response of *RC* Integrators

8. Sketch the integrator in Figure 21–62, showing maximum voltages.
9. A 1 V, 10 kHz pulse waveform with a duty cycle of 25% is applied to an integrator with $\tau = 25\ \mu\text{s}$. Graph the output voltage for three initial pulses. *C* is initially uncharged.

FIGURE 21–62

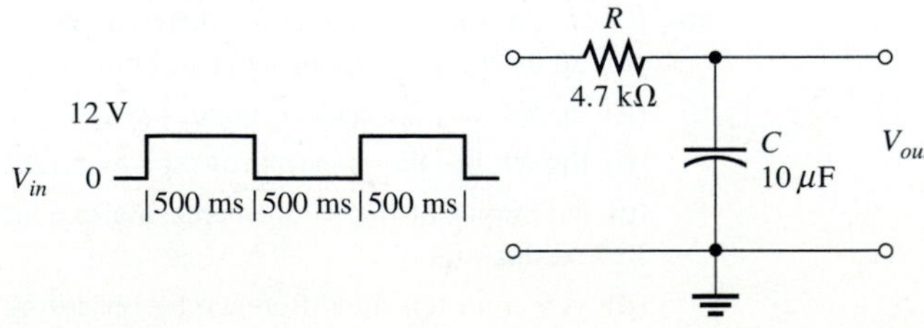

**10.** What is the steady-state output voltage of the integrator with a square-wave input shown in Figure 21–63?

FIGURE 21–63

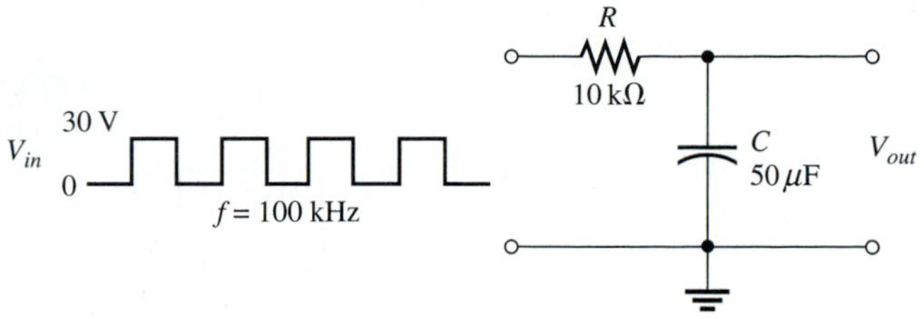

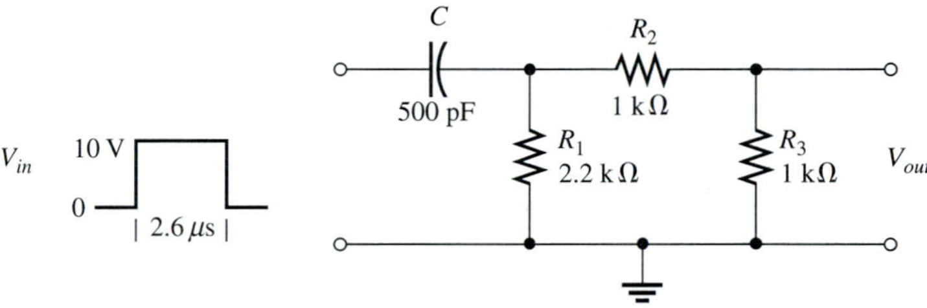


## SECTION 21–4 Single-Pulse Response of *RC* Differentiators

**11.** Repeat Problem 5 for an *RC* differentiator.

**12.** Redraw the circuit in Figure 21–60 to make it a differentiator, and repeat Problem 6.

**13.** **(a)** What is $\tau$ in Figure 21–64? **(b)** Sketch the output voltage.

FIGURE 21–64

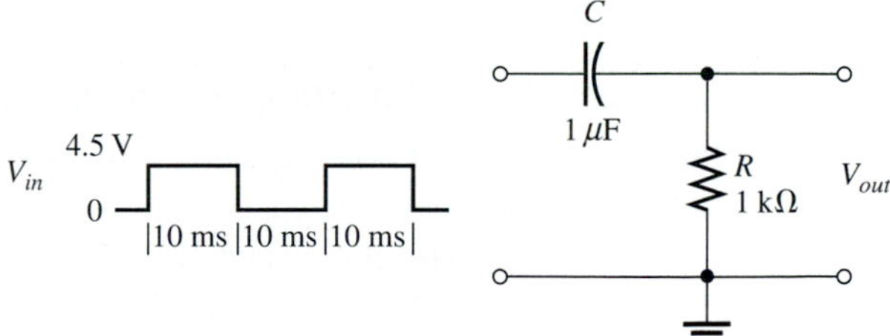


## SECTION 21–5 Repetitive-Pulse Response of *RC* Differentiators

**14.** Sketch the differentiator output in Figure 21–65, showing maximum voltages.

FIGURE 21–65

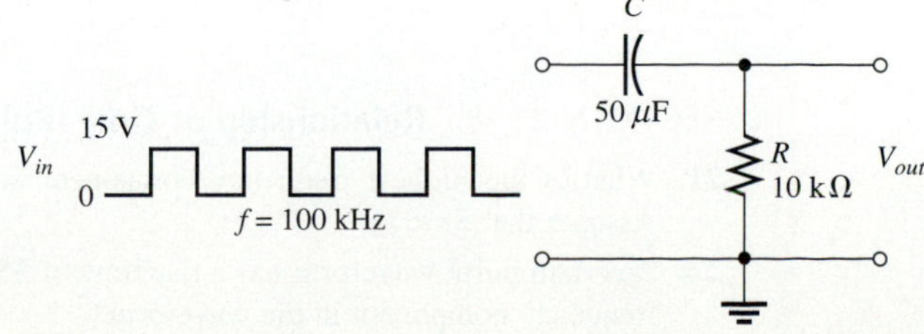


**15.** What is the steady-state output voltage of the differentiator with the square-wave input shown in Figure 21–66?

FIGURE 21–66

## SECTION 21–6 Pulse Response of *RL* Integrators

**16.** Determine the output voltage for the circuit in Figure 21–67. A single input pulse is applied as shown.

FIGURE 21–67

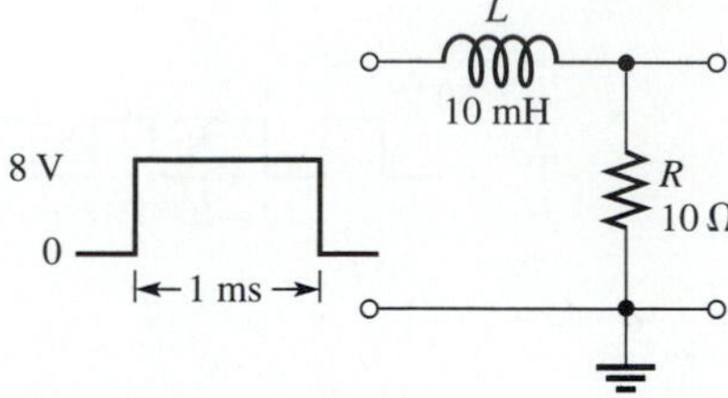


**17.** Sketch the integrator in Figure 21–68, showing maximum voltages.

FIGURE 21–68

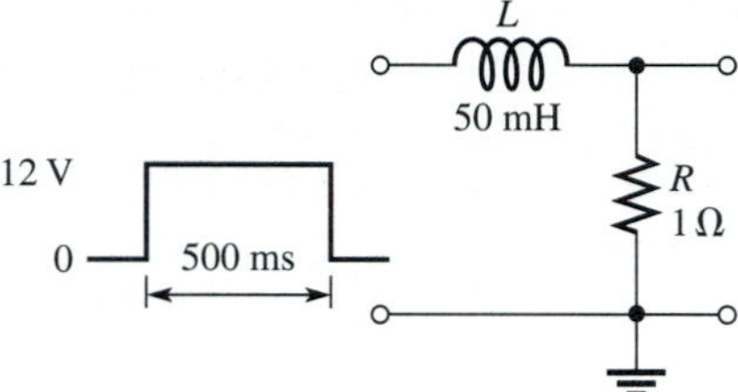


**18.** Determine the time constant in Figure 21–69. Is this circuit an integrator or a differentiator?

FIGURE 21–69

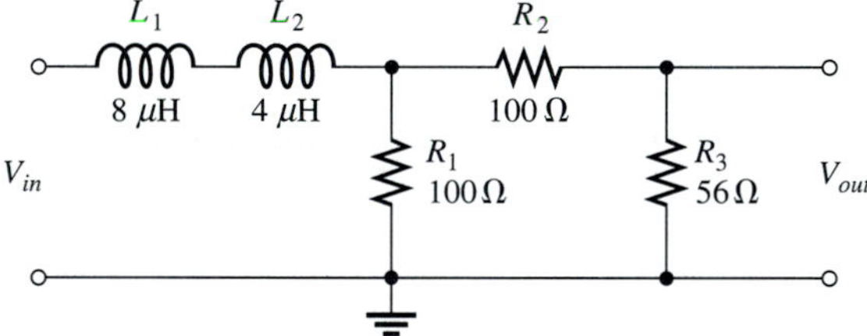


## SECTION 21–7 Pulse Response of *RL* Differentiators

**19.** **(a)** What is $\tau$ in Figure 21–70? **(b)** Sketch the output voltage.

**20.** Draw the output waveform if a periodic pulse waveform with $t_W = 25\ \mu s$ and $T = 60\ \mu s$ is applied to the circuit in Figure 21–70.

FIGURE 21–70

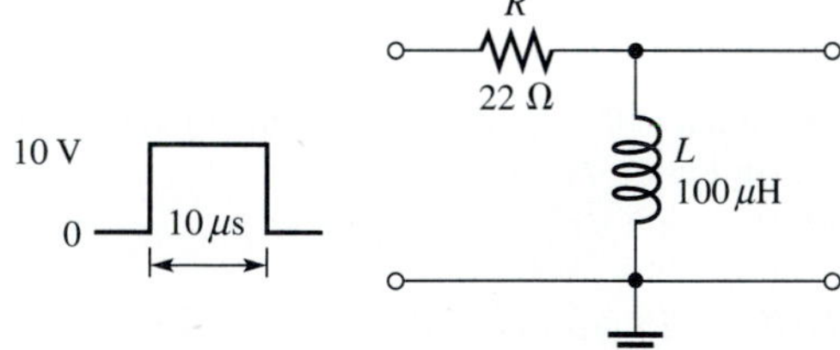


## SECTION 21–8 Relationship of Time (Pulse) Response to Frequency Response

**21.** What is the highest frequency component in the output of an integrator with $\tau = 10\ \mu s$? Assume that $5\tau < t_W$.

**22.** A certain pulse waveform has a rise time of 55 ns and a fall time of 42 ns. What is the highest frequency component in the waveform?

## SECTION 21–9 Troubleshooting

**23.** Determine the most likely fault(s), in the circuit of Figure 21–71(a) for each set of waveforms in parts (b) through (d). $V_{in}$ is a square wave with a period of 8 ms.

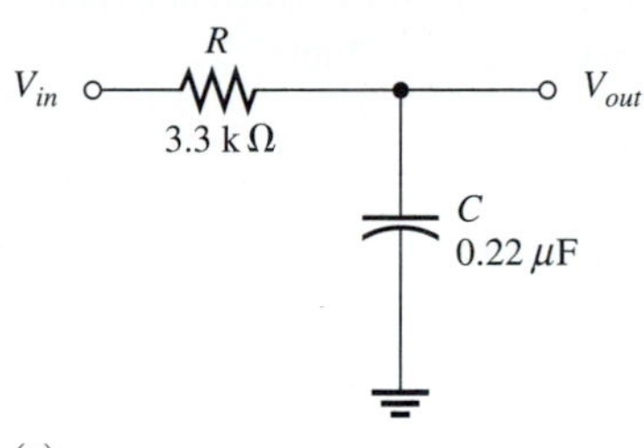


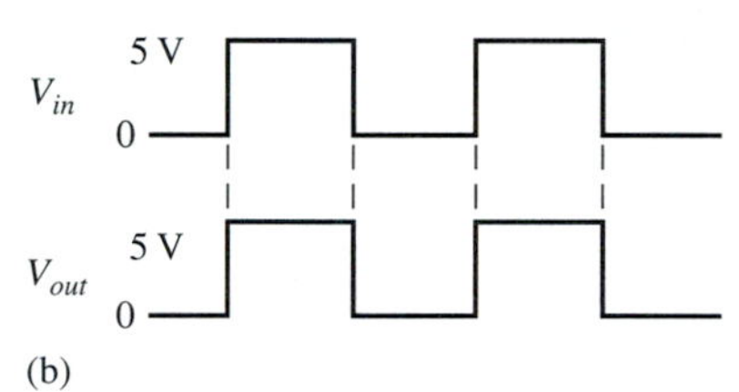


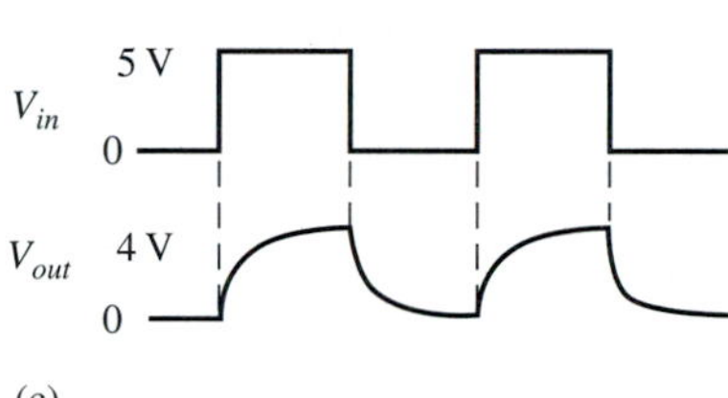


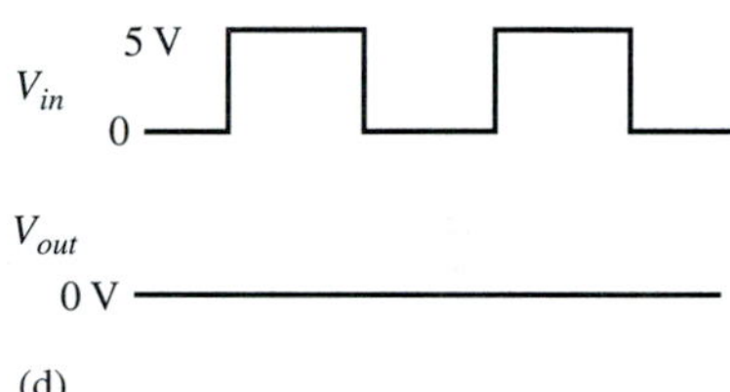


FIGURE 21–71

**24.** Determine the most likely fault(s), if any, in the circuit of Figure 21–72(a) for each set of waveforms in parts (b) through (d). $V_{in}$ is a square wave with a period of 8 ms.

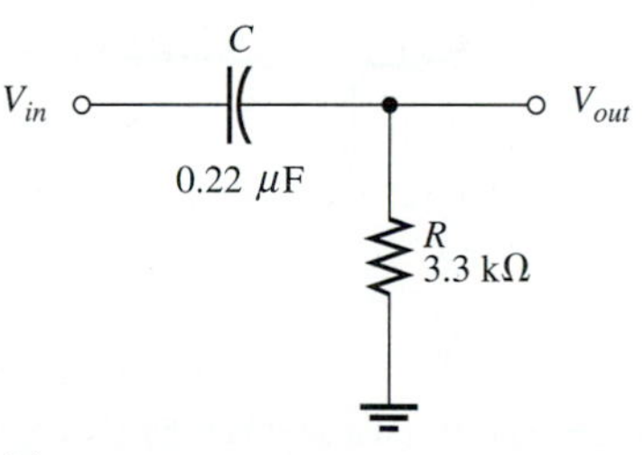


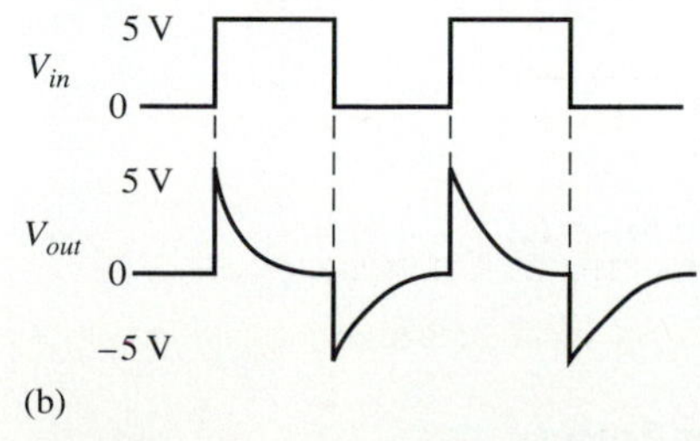


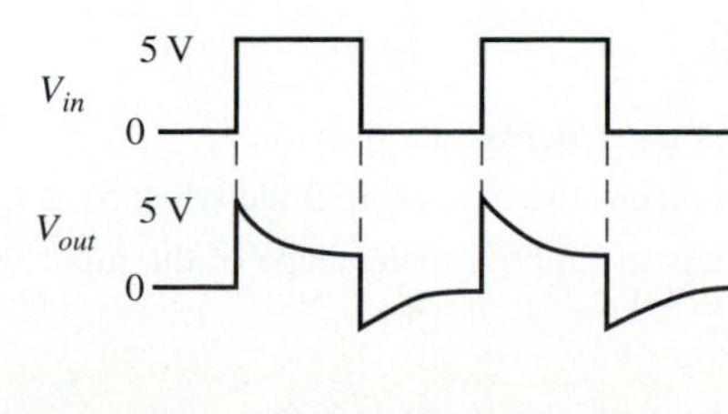


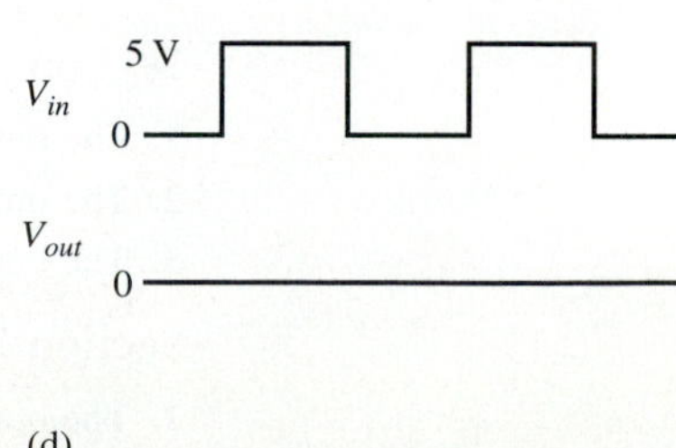


FIGURE 21–72

## ■ ANSWERS TO SECTION REVIEWS

### Section 21–1

1. An integrator is a series *RC* circuit in which the output is across the capacitor.
2. A voltage applied to the input causes the capacitor to charge. A short across the input causes the capacitor to discharge.

### Section 21–2

1. For the output of an integrator to reach amplitude, $5\tau \leq t_W$.
2. $V_{out(max)} = 630$ mV; $t_{\text{disch}} = 51.7$ ms
3. See Figure 21–73.
4. No, *C* will not fully charge.
5. The output has approximately the shape of the input when $5\tau << t_W$ ($5\tau$ much less than $t_W$).

FIGURE 21–73

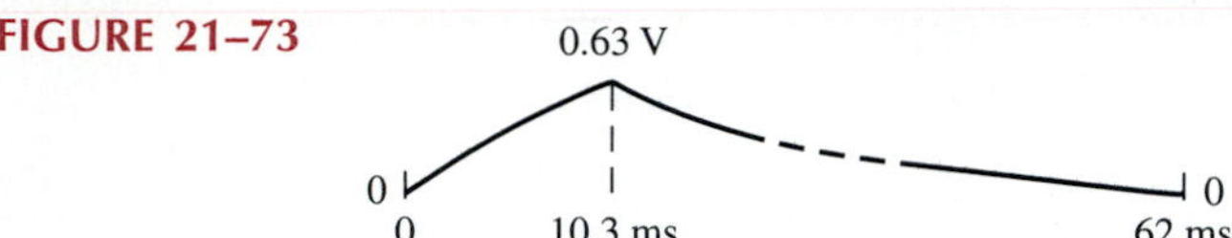


### Section 21–3

1. *C* will fully charge and discharge when $5\tau \leq t_W$, and $5\tau \leq$ time between pulses.
2. When $\tau << t_W$, the output is approximately like the input.
3. Transient time
4. Steady-state response is the response after the transient time has passed.
5. The average value of the output equals the average value of the input voltage.

### Section 21–4

1. See Figure 21–74.
2. The output resembles the input when $5\tau >> t_W$.
3. The output appears to be positive and negative spikes.
4. $V_R$ will go to −10 V.

FIGURE 21–74

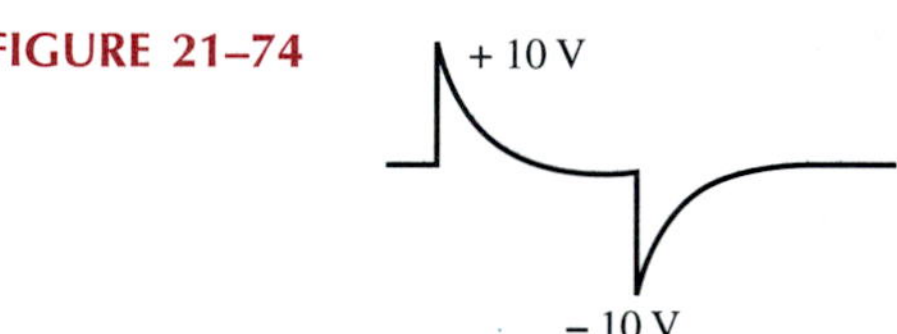


### Section 21–5

1. *C* will fully charge and discharge when $5\tau \leq t_W$ *and* $5\tau \leq$ time between pulses.
2. The output appears to be positive and negative spikes.
3. The average value is 0 V.

### Section 21–6

1. The output is taken across the resistor.
2. The output reaches the input amplitude when $5\tau \leq t_W$.
3. The output has the approximate shape of the input when $5\tau << t_W$.

### Section 21–7

1. The output is taken across the inductor.
2. The output has the approximate shape of the input when $5\tau >> t_W$.
3. $V_L$ will go to −8 V.

### Section 21–8

**1.** An integrator is a low-pass filter.

**2.** A differentiator is a high-pass filter.

**3.** $f_{max} = 350$ kHz

### Section 21–9

**1.** A 0 V output may be caused by an open resistor or shorted capacitor.

**2.** If $C$ is shorted, the output is the same as the input.

### Section 21–10

**1.** A capacitor must be added and the switch changed to one with six positions.

**2.** $C_6 = 100\text{ ms}/[(0.357)(47\text{ k}\Omega)] = 5.96\ \mu\text{F}$ (use 5.6 $\mu$F)

### Section 21–11

**1.**

```
RC Integrator Circuit
*Pulse Waveform Demonstration
VS  1  0  PULSE(0  10  0  1N  1N  500U  1M)
R1  1  2  1K
C1  2  0  1U
.PROBE
.TRAN  50P  5M
.END
```

**2.**

```
.TRAN  10P  25M  0  10P
```

## ■ ANSWERS TO RELATED EXERCISES FOR EXAMPLES

**21–1** 8.65 V

**21–2** 24.7 V

**21–3** 1.08 V

**21–4** See Figure 21–75.

**21–5** 892 mV

**21–6** Impossible with a 50 Ω rheostat

**21–7** 20 V

**21–8** 2.5 kΩ

**21–9** See Figure 21–76.

**21–10** 20 Ω

**21–11** 23.3 MHz

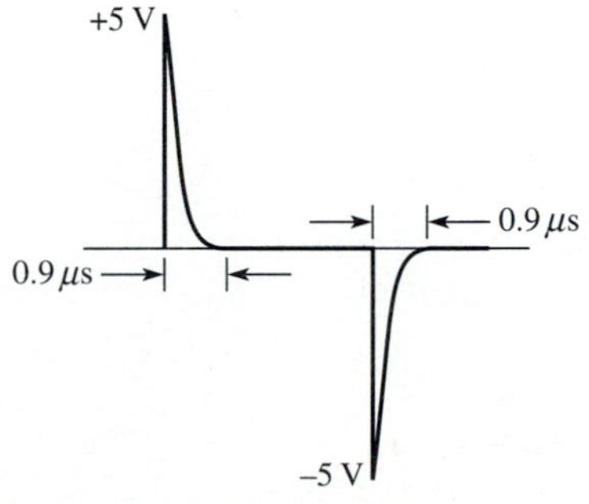


FIGURE 21–75

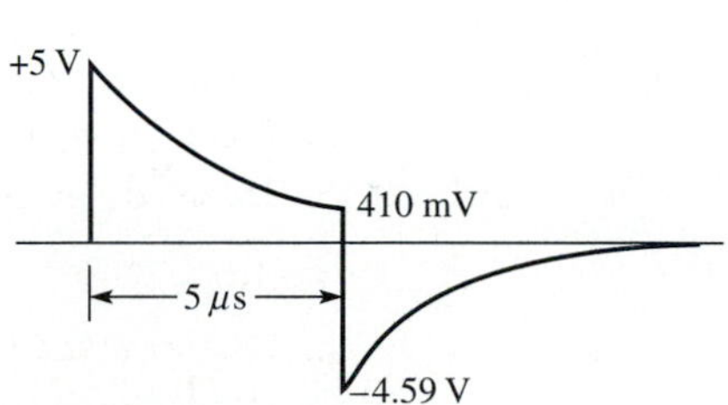


FIGURE 21–76

# 22

# POLYPHASE SYSTEMS IN POWER APPLICATIONS



## ■ INTRODUCTION

In the coverage of ac analysis in previous chapters, only single-phase sinusoidal sources have been considered. In Chapter 11, you learned how a sinusoidal voltage can be generated by the rotation of a conductor at a constant velocity in a magnetic field, and the basic concepts of ac generators were introduced.

In this chapter, the basic generation of polyphase sinusoidal waveforms is examined. The advantages of polyphase systems in power applications are covered, and various types of three-phase connections and power measurement are introduced.

## TECHnology Theory Into Practice

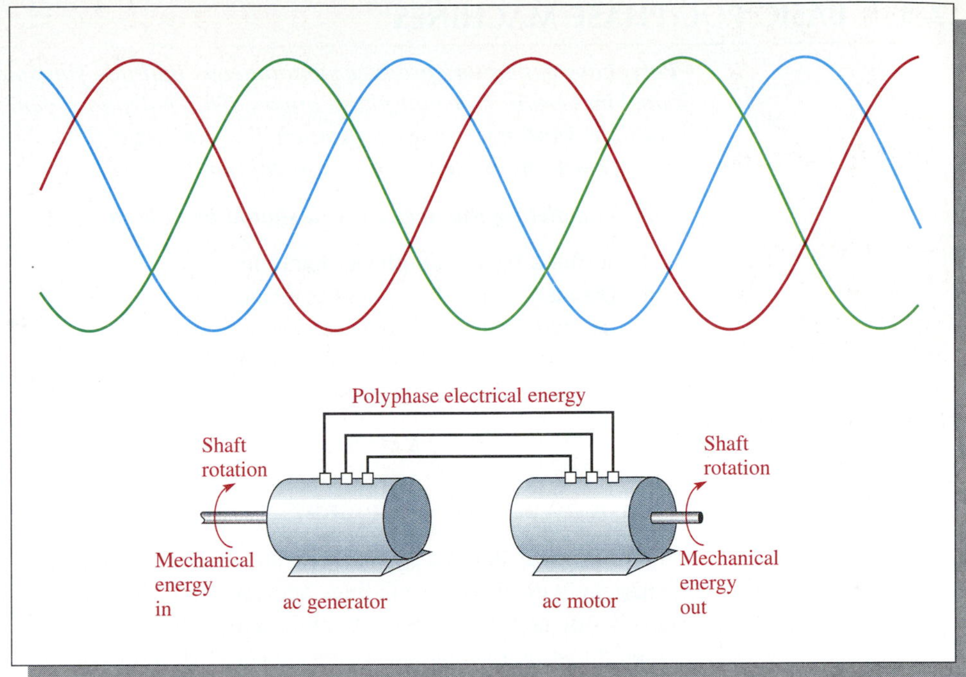


## CHAPTER OBJECTIVES

- ☐ Describe a basic polyphase machine
- ☐ Discuss the advantages of polyphase in power applications
- ☐ Analyze three-phase generator configurations
- ☐ Analyze three-phase generators with three-phase loads
- ☐ Discuss power measurements in three-phase systems
- ☐ Use PSpice and Probe to analyze three-phase circuits (optional)

## 22–1 ■ BASIC POLYPHASE MACHINES

***Polyphase generators produce simultaneous multiple sinusoidal voltages that are separated by certain constant phase angles. This polyphase generation is accomplished by multiple windings rotating through the magnetic field. Similarly, polyphase motors operate with multiple-phase sinusoidal inputs.***

***After completing this section, you should be able to***

- **Describe a basic polyphase machine**
  - ☐ Discuss a basic two-phase generator
  - ☐ Discuss a basic three-phase generator
  - ☐ Describe the construction of a three-phase generator
  - ☐ Describe a basic three-phase induction motor

### A Basic Two-Phase Generator

Figure 22–1(a) shows a second separate conductor winding added to a basic single-phase generator with the two windings separated by 90°. Both windings are mounted on the same **rotor** and therefore rotate at the same speed. Winding $A$ is 90° ahead of winding $B$ in the direction of rotation. As they rotate, two induced sinusoidal voltages are produced that are 90° apart in phase, as shown in part (b) of the figure.

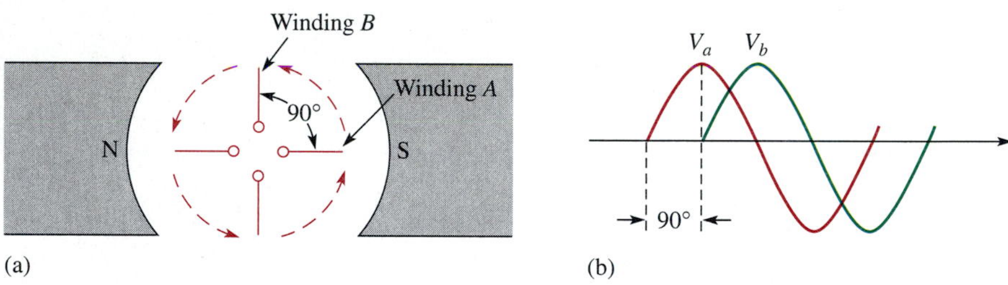


FIGURE 22–1
*Basic two-phase generator.*

### A Basic Three-Phase Generator

Figure 22–2(a) shows a **polyphase** generator with three separate conductor windings placed at 120° intervals around the rotor. This configuration generates three sinusoidal voltages that are separated from each other by phase angles of 120°, as shown in part (b).

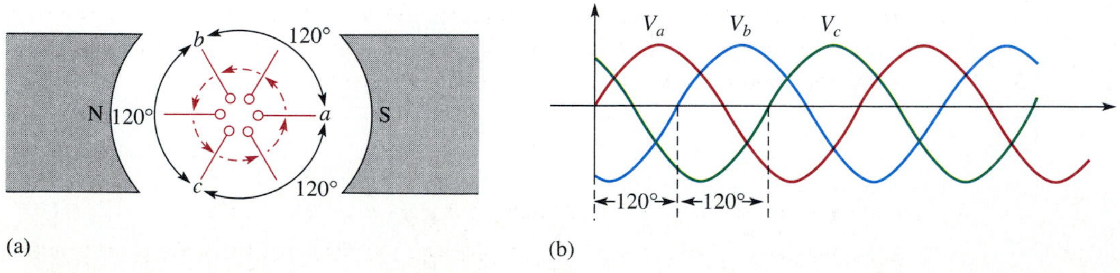


FIGURE 22–2
*Basic three-phase generator.*

## A Practical Three-Phase Generator

The purpose of using the simple ac generators (**alternators**) presented up to this point was to illustrate the basic concept of the generation of sinusoidal voltages from the rotation of a conductor in a magnetic field. Many ac generators produce three-phase ac using a configuration that is somewhat different from that previously discussed. The basic principle, however, is the same.

A basic two-pole, three-phase generator is shown in Figure 22–3. Most practical generators are of this basic form. Rather than using a permanent magnet in a fixed position, a rotating electromagnet is used. The electromagnet is created by passing direct current ($I_F$) through a winding around the rotor, as shown. This winding is called the **field winding.** The direct current is applied through a brush and slip ring assembly. The stationary outer portion of the generator is called the **stator.** Three separate windings are placed 120° apart around the stator; three-phase voltages are induced in these windings as the magnetic field rotates, as indicated in Figure 22–2(b).

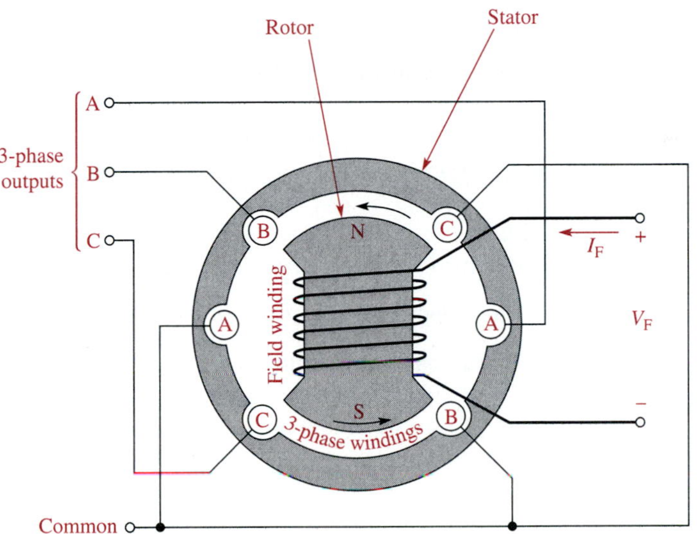


FIGURE 22–3
***Basic two-pole, three-phase generator.***

## A Basic Three-Phase Motor

The most common type of ac motor is the three-phase induction motor. Basically, it consists of a stator with stator windings and a rotor assembly constructed as a cylindrical frame of metal bars arranged in a **squirrel-cage** type configuration. A basic end-view diagram is shown in Figure 22–4 on the next page.

When the three-phase voltages are applied to the stator windings, a rotating magnetic field is established. As the magnetic field rotates, currents are induced in the conductors of the squirrel-cage rotor. The interaction of the induced currents and the magnetic field produces forces that cause the rotor to also rotate.

**SECTION 22–1 REVIEW**

1. Describe the basic principle used in ac generators.
2. A two-pole, single-phase generator rotates at 400 rps. What frequency is produced?
3. How many separate armature windings are required in a three-phase alternator?

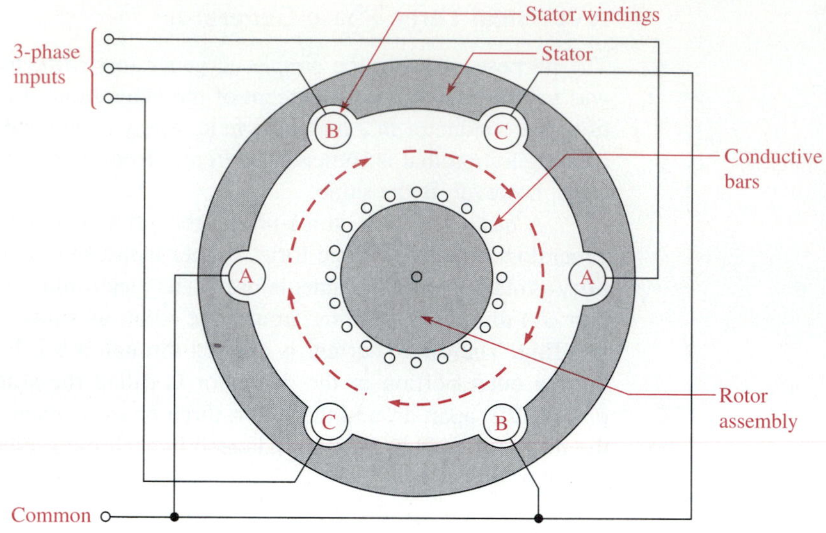


FIGURE 22–4
*Basic three-phase induction motor.*

## 22–2 ■ POLYPHASE GENERATORS IN POWER APPLICATIONS

***There are several advantages of using polyphase generators to deliver power to a load over using a single-phase machine. These advantages are discussed in this section.***

***After completing this section, you should be able to***

- **Discuss the advantages of polyphase in power applications**
  - Explain the copper advantage
  - Compare single-phase, two-phase, and three-phase systems in terms of the copper advantage
  - Explain the advantage of constant power
  - Explain the advantage of a constant rotating magnetic field

### The Copper Advantage

The size of the copper wire required to carry current from a generator to a load can be reduced when a polyphase rather than a single-phase generator is used.

***A Single-Phase System*** Figure 22–5 is a simplified representation of a single-phase generator connected to a resistive load. The coil symbol represents the generator winding.

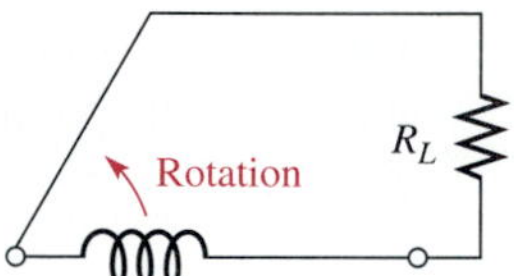


FIGURE 22–5
*Simplified representation of a single-phase generator connected to a resistive load.*

For example, a single-phase sinusoidal voltage is induced in the winding and applied to a 60 Ω load, as indicated in Figure 22–6. The resulting load current is

$$\mathbf{I}_{RL} = \frac{120\angle 0^\circ \text{ V}}{60\angle 0^\circ \ \Omega} = 2\angle 0^\circ \text{ A}$$

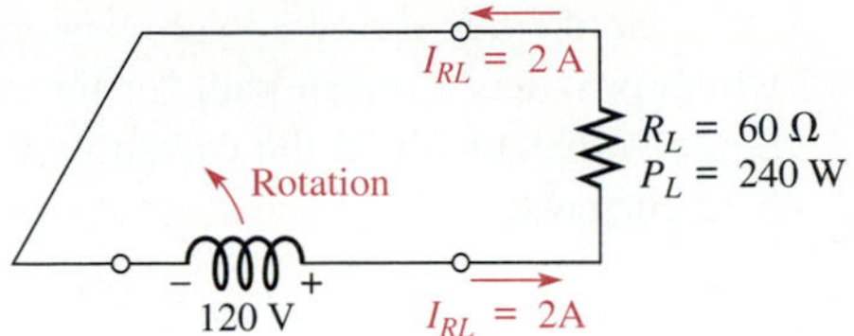


FIGURE 22–6
*Single-phase example.*

The total current that must be delivered by the generator to the load is 2∠0° A. This means that the two conductors carrying current to and from the load must each be capable of handling 2 A; thus, the total copper cross section must handle 4 A. (The copper cross section is a measure of the total amount of wire required based on its physical size as related to its diameter.) The total load power is

$$P_{L(tot)} = I_{RL}^2 R_L = 240 \text{ W}$$

***A Two-Phase System*** Figure 22–7 shows a simplified representation of a two-phase generator connected to two 120 Ω load resistors. The coils drawn at a 90° angle represent the armature windings spaced 90° apart. An equivalent single-phase system would be required to feed two 120 Ω resistors in parallel, thus creating a 60 Ω load.

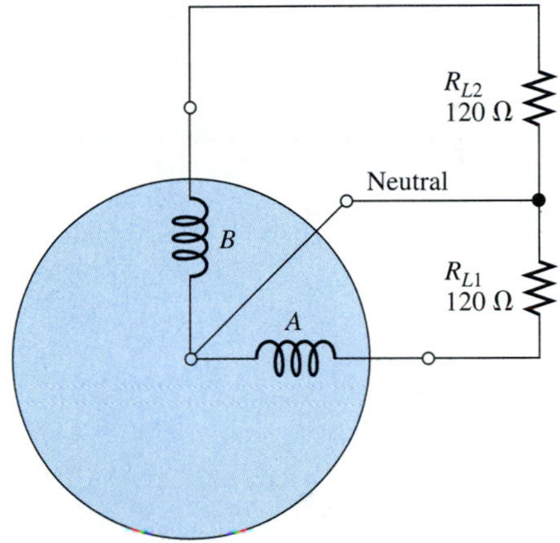


FIGURE 22–7
*Simplified representation of a two-phase generator with each phase connected to a 120 Ω load.*

The voltage across load $R_{L1}$ is 120∠0° V, and the voltage across load $R_{L2}$ is 120∠90° V, as indicated in Figure 22–8(a). The current in $R_{L1}$ is

$$\mathbf{I}_{RL1} = \frac{120\angle 0^\circ \text{ V}}{120\angle 0^\circ \ \Omega} = 1\angle 0^\circ \text{ A}$$

and the current in $R_{L2}$ is

$$\mathbf{I}_{RL2} = \frac{120\angle 90^\circ \text{ V}}{120\angle 0^\circ \ \Omega} = 1\angle 90^\circ \text{ A}$$

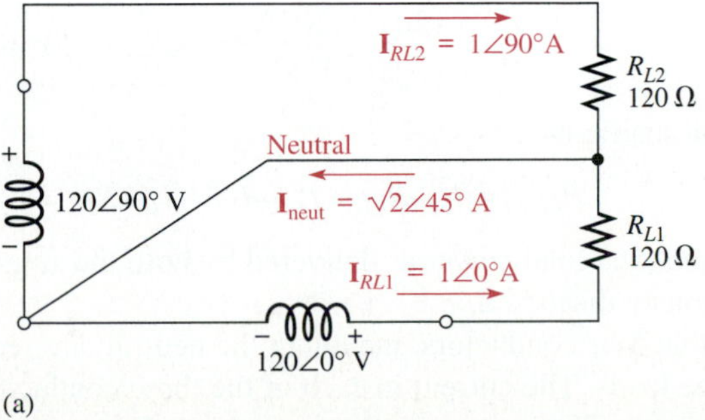


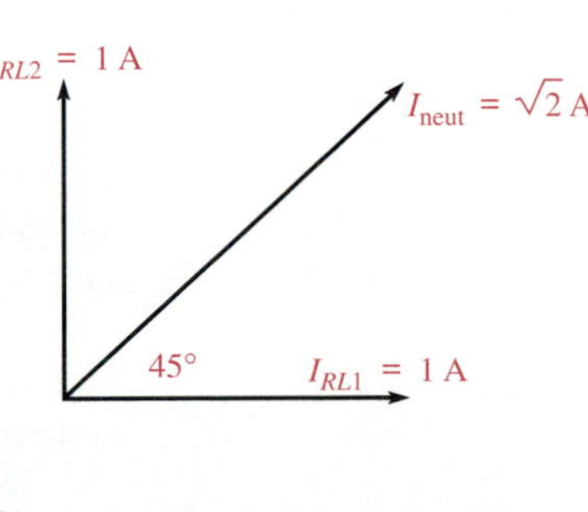


FIGURE 22–8
*Two-phase example.*

Notice that the two load resistors are connected to a common or neutral conductor, which provides a return path for the currents. Since the two load currents are 90° out of phase with each other, the current in the neutral conductor ($I_{\text{neut}}$) is the phasor sum of the load currents.

$$I_{\text{neut}} = \sqrt{I_{RL1}^2 + I_{RL2}^2} = \sqrt{2}\text{ A} = 1.414\text{ A}$$

Figure 22–8(b) shows the phasor diagram for the currents in the two-phase system.

Three conductors are required in this system to carry current to and from the loads. Two of the conductors must be capable of handling 1 A each, and the neutral conductor must handle 1.414 A. The total copper cross section must handle 1 A + 1 A + 1.414 A = 3.414 A. This is a smaller copper cross section than was required in the single-phase system to deliver the same amount of total load power.

$$P_{L(tot)} = P_{RL1} + P_{RL2} = I_{RL1}^2 R_{L1} = I_{RL2}^2 R_{L2} = 120\text{ W} + 120\text{ W} = 240\text{ W}$$

***A Three-Phase System*** Figure 22–9 shows a simplified representation of a three-phase generator connected to three 180 Ω resistive loads. An equivalent single-phase system would be required to feed three 180 Ω resistors in parallel, thus creating an effective load resistance of 60 Ω. The coils represent the generator windings separated by 120°.

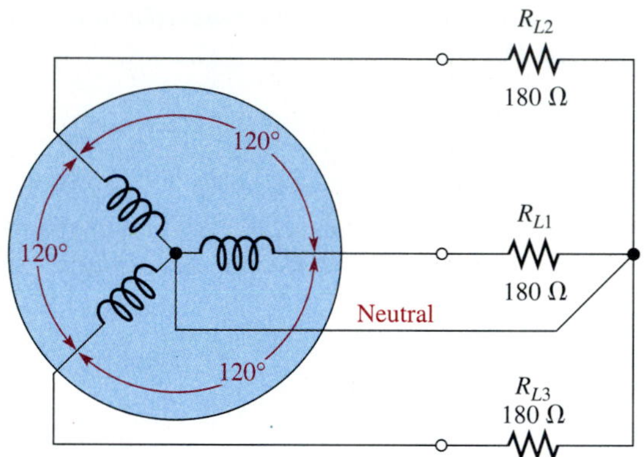


FIGURE 22–9
*A simplified representation of a three-phase generator with each phase connected to a 180 Ω load.*

The voltage across $R_{L1}$ is 120∠0° V, the voltage across $R_{L2}$ is 120∠120° V, and the voltage across $R_{L3}$ is 120∠−120° V, as indicated in Figure 22–10(a). The current from each winding to its respective load is as follows:

$$\mathbf{I}_{RL1} = \frac{120\angle 0^\circ\text{ V}}{180\angle 0^\circ\ \Omega} = 667\angle 0^\circ\text{ mA}$$

$$\mathbf{I}_{RL2} = \frac{120\angle 120^\circ\text{ V}}{180\angle 0^\circ\ \Omega} = 667\angle 120^\circ\text{ mA}$$

$$\mathbf{I}_{RL3} = \frac{120\angle -120^\circ\text{ V}}{180\angle 0^\circ\ \Omega} = 667\angle -120^\circ\text{ mA}$$

The total load power is

$$P_{L(tot)} = I_{RL1}^2 R_{L1} + I_{RL2}^2 R_{L2} + I_{RL3}^2 R_{L3} = 240\text{ W}$$

This is the same total load power as delivered by both the single-phase and the two-phase systems previously discussed.

Notice that four conductors, including the neutral, are required to carry the currents to and from the loads. The current in each of the three conductors is 667 mA, as indicated in Figure 22–10(a). The current in the neutral conductor is the phasor sum of the three load currents and is equal to zero, as shown in the following equation, with reference to the phasor diagram in Figure 22–10(b).

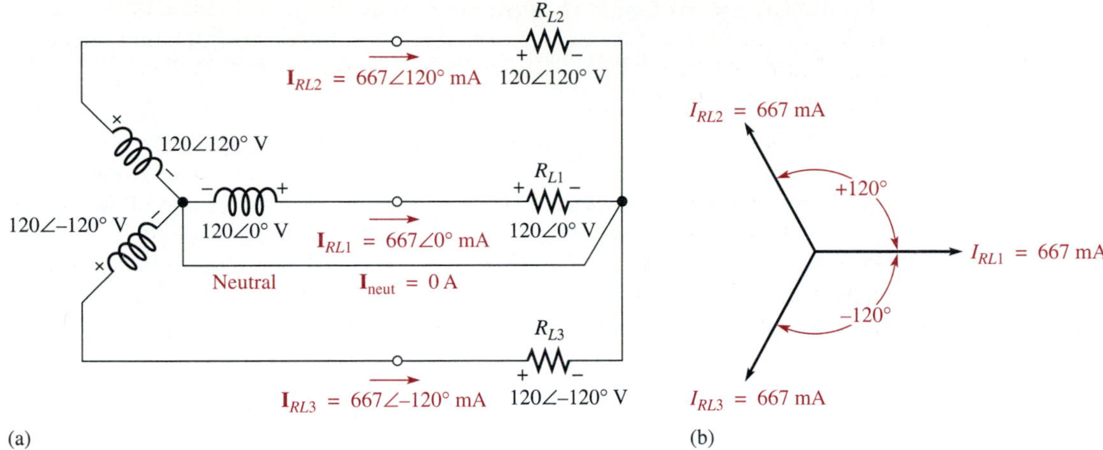


**FIGURE 22–10**
***Three-phase example.***

$$\begin{aligned}\mathbf{I}_{RL1} + \mathbf{I}_{RL2} + \mathbf{I}_{RL3} &= 667\angle 0^\circ \text{ mA} + 667\angle 120^\circ \text{ mA} + 667\angle -120^\circ \text{ mA} \\ &= 667 \text{ mA} - 333.5 \text{ mA} + j578 \text{ mA} - 333.5 \text{ mA} - j578 \text{ mA} \\ &= 667 \text{ mA} - 667 \text{ mA} = 0 \text{ A}\end{aligned}$$

This condition, where all load currents are equal and the neutral current is zero, is called a **balanced load** condition.

The total copper cross section must handle 667 mA + 667 mA + 667 mA + 0 mA = 2 A. This result shows that considerably less copper is required to deliver the same load power with a three-phase system than is required for either the single-phase or the two-phase systems. The amount of copper is an important consideration in power distribution systems.

## EXAMPLE 22–1

Compare the total copper cross sections in terms of current-carrying capacity for single-phase and three-phase 120 V systems with effective load resistances of 12 Ω.

***Solution***

*Single-phase system:* The load current is

$$I_{RL} = \frac{120 \text{ V}}{12\ \Omega} = 10 \text{ A}$$

The conductor to the load must carry 10 A, and the conductor from the load must also carry 10 A.

The total copper cross section, therefore, must be sufficient to handle 20 A.

*Three-phase system:* For an effective load resistance of 12 Ω, the three-phase generator feeds three load resistors of 36 Ω each. The current in each load resistor is

$$I_{RL} = \frac{120 \text{ V}}{36\ \Omega} = 3.33 \text{ A}$$

Each of the three conductors feeding the balanced load must carry 3.33 A, and the neutral current is zero.

Therefore, the total copper cross section must be sufficient to handle 10 A. This is significantly less than for the single-phase system with an equivalent load.

***Related Exercise*** Compare the total copper cross sections in terms of current-carrying capacity for single-phase and three-phase 240 V systems with effective load resistances of 100 Ω.

## The Advantage of Constant Power

A second advantage of polyphase systems over single-phase is that polyphase systems produce a constant amount of power in the load.

***A Single-Phase System*** As shown in Figure 22–11, the load power fluctuates as the square of the sinusoidal voltage divided by the resistance. It changes from a maximum of $V^2_{RL(\text{max})}/R_L$ to a minimum of zero at a frequency equal to twice that of the voltage.

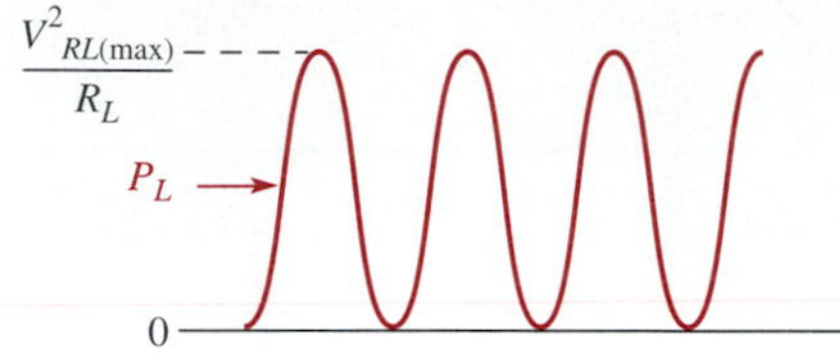


**FIGURE 22–11**
***Single-phase load power (sin² curve).***

***A Polyphase System*** The power waveform across one of the load resistors in a two-phase system is 90° out of phase with the power waveform across the other, as shown in Figure 22–12. When the instantaneous power to one load is maximum, the other is minimum. In between the maximum and minimum points, one power is increasing while the other is decreasing. Careful examination of the power waveforms shows that when two instantaneous values are added, the sum is always constant and equal to $V^2_{RL(\text{max})}/R_L$.

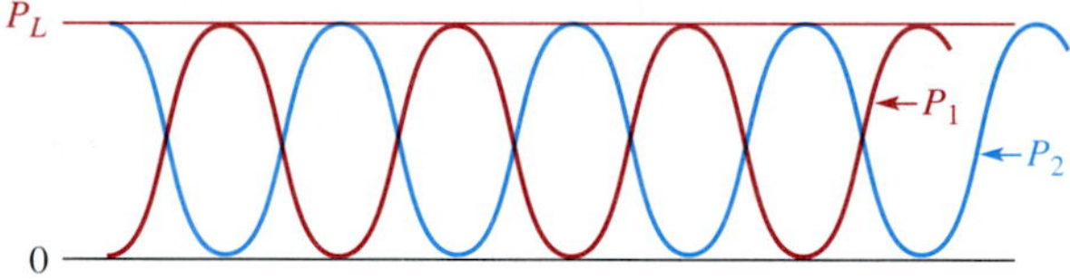


**FIGURE 22–12**
***Two-phase power ($P_L = V^2_{RL(max)}/R_L$).***

In a three-phase system, the total power delivered to the load resistors is also constant for the same basic reason as for the two-phase system. The sum of the instantaneous voltages is always the same; therefore, the power has a constant value. A constant load power means a uniform conversion of mechanical to electrical energy, which is an important consideration in many power applications.

## The Advantage of a Constant, Rotating Magnetic Field

In many applications, ac generators are used to drive ac motors for conversion of electrical energy to mechanical energy in the form of shaft rotation in the motor. The original energy for operation of the generator can come from any of several sources such as hydroelectric or steam. Figure 22–13 illustrates the basic concept.

When a polyphase generator is connected to the motor windings as depicted in Figure 22–14, where a two-phase system is used for purposes of illustration, a magnetic field

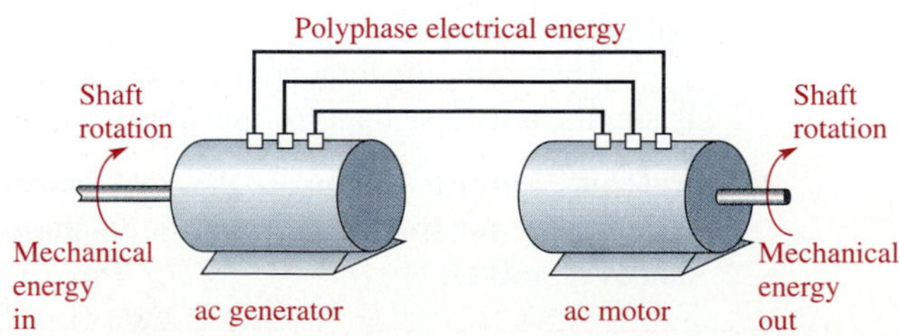


**FIGURE 22–13**
***Simple example of mechanical-to-electrical-to-mechanical energy conversion.***

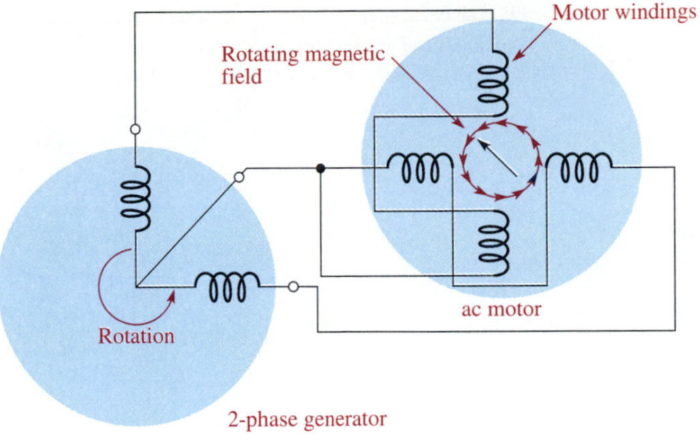


**FIGURE 22–14**
*A two-phase generator producing a constant rotating magnetic field in an ac motor.*

is created within the motor that has a constant flux density and that rotates at the frequency of the two-phase sine wave. The motor's rotor is pulled around at a constant rotational velocity by the rotating magnetic field, producing a constant shaft rotation.

A three-phase system, of course, has the same advantage as a two-phase system. A single-phase system is unsuitable for many applications, because it produces a magnetic field that fluctuates in flux density and reverses direction during each cycle without providing the advantage of constant rotation.

**SECTION 22–2 REVIEW**

1. List three advantages of polyphase systems over single-phase systems.
2. Which advantage(s) is/are most important in mechanical-to-electrical energy conversions?
3. Which advantage(s) is/are most important in electrical-to-mechanical energy conversions?

## 22–3 ■ THREE-PHASE GENERATORS

***In the previous sections, the Y-connection was used for illustration. In this section, the Y-connection is examined further and a second type, the Δ-configuration, is introduced.***

***After completing this section, you should be able to***

- **Analyze three-phase generator configurations**
  - ☐ Analyze the Y-connected generator
  - ☐ Analyze the Δ-connected generator

### The Y-Connected Generator

A Y-connected system can be either a three-wire or, when the neutral is used, a four-wire system, as shown in Figure 22–15 on the next page, connected to a generalized load represented by the block. Recall that when the loads are perfectly balanced, the neutral current is zero; therefore, the neutral conductor is unnecessary. However, in cases where the loads are not equal (unbalanced), a neutral wire is essential to provide a return current path, because the neutral current has a nonzero value.

The voltages across the generator windings are called **phase voltages** ($V_\theta$), and the currents through the windings are called **phase currents** ($I_\theta$). Also, the currents in the lines connecting the generator windings to the load are called **line currents** ($I_L$), and the

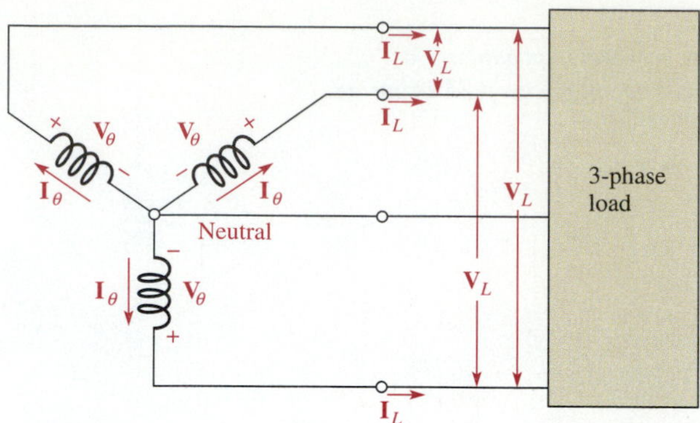


**FIGURE 22–15**
*Y-connected generator.*

voltages across the lines are called the **line voltages** ($V_L$). Note that the magnitude of each line current is equal to the corresponding phase current in the Y-connected circuit.

$$I_L = I_\theta \tag{22–1}$$

In Figure 22–16, the line terminations of the windings are designated *a, b,* and *c,* and the neutral point is designated *n.* These letters are added as subscripts to the phase and line currents to indicate the phase with which each is associated. The phase voltages are also designated in the same manner. Notice that the phase voltages are always positive at the terminal end of the winding and are negative at the neutral point. The line voltages are from one winding terminal to another, as indicated by the double-letter subscripts. For example, $\mathbf{V}_{L(ba)}$ is the line voltage from $b$ to $a$.

**FIGURE 22–16**
*Phase voltages and line voltages in a Y-connected system.*

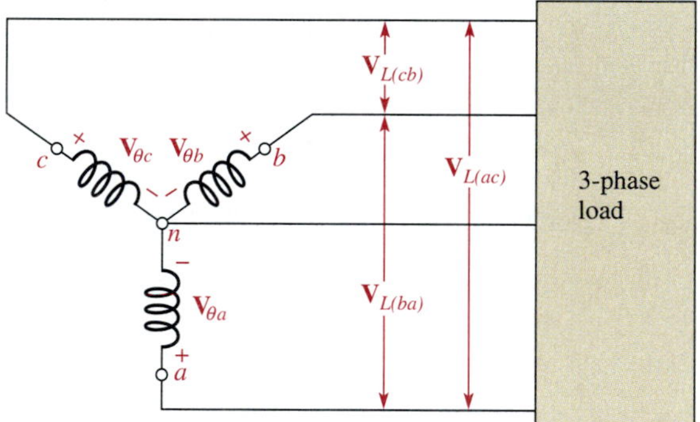


Figure 22–17(a) shows a phasor diagram for the phase voltages. By rotation of the phasors, as shown in part (b), $V_{\theta a}$ is given a reference angle of zero, and the polar expressions for the phasor voltages are as follows:

$$\mathbf{V}_{\theta a} = V_{\theta a}\angle 0^\circ$$
$$\mathbf{V}_{\theta b} = V_{\theta b}\angle 120^\circ$$
$$\mathbf{V}_{\theta c} = V_{\theta c}\angle -120^\circ$$

There are three line voltages: one between *a* and *b,* one between *a* and *c,* and another between *b* and *c.* It can be shown that the magnitude of each line voltage is equal to $\sqrt{3}$ times the magnitude of the phase voltage and that there is a phase angle of 30° between each line voltage and the nearest phase voltage.

$$V_L = \sqrt{3}V_\theta \tag{22–2}$$

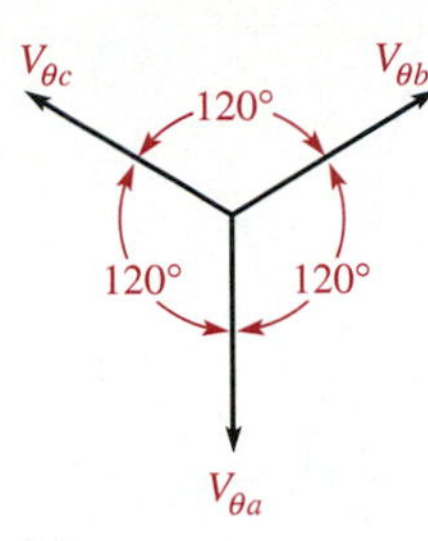


(a)

$V_{\theta b}$
120°
$V_{\theta a}$
120°
$V_{\theta c}$

(b)

**FIGURE 22–17**
***Phase voltage diagram.***

Since all phase voltages are equal in magnitude,

$$\mathbf{V}_{L(ba)} = \sqrt{3}V_{\theta}\angle 150°$$
$$\mathbf{V}_{L(ac)} = \sqrt{3}V_{\theta}\angle 30°$$
$$\mathbf{V}_{L(cb)} = \sqrt{3}V_{\theta}\angle -90°$$

The line voltage phasor diagram is shown in Figure 22–18 superimposed on the phasor diagram for the phase voltages. Notice that there is a phase angle of 30° between each line voltage and the nearest phase voltage and that the line voltages are 120° apart.

**FIGURE 22–18**
***Phasor diagram for the phase voltages and line voltages in a Y-connected, three-phase system.***

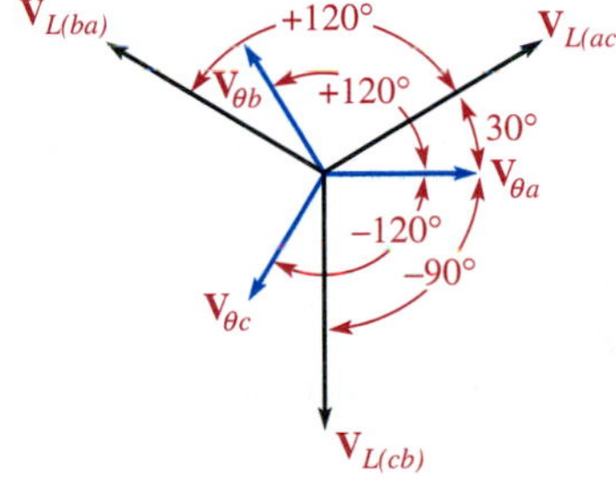


**EXAMPLE 22–2** The instantaneous position of a certain Y-connected ac generator is shown in Figure 22–19. If each phase voltage has a magnitude of 120 V rms, determine the magnitude of each line voltage, and sketch the phasor diagram.

**FIGURE 22–19**

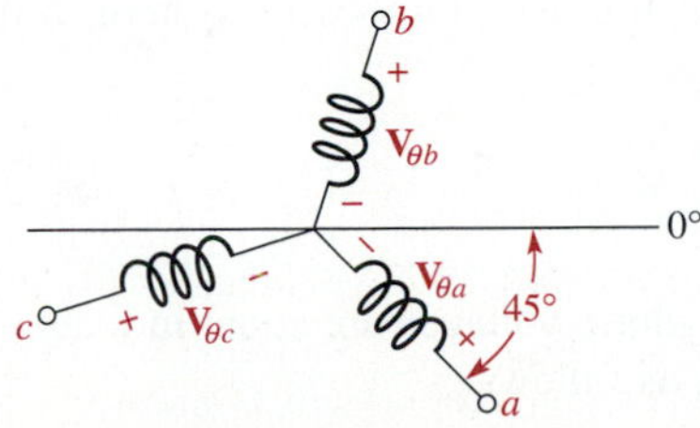


***Solution*** The magnitude of each line voltage is

$$V_L = \sqrt{3}V_{\theta} = \sqrt{3}(120\text{ V}) = 208\text{ V}$$

FIGURE 22–20

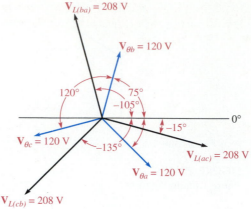

The phasor diagram for the given instantaneous generator position is shown in Figure 22–20.

***Related Exercise*** Determine the line voltage magnitude if the generator position indicated in Figure 22–19 is rotated another 45° clockwise.

## The Δ-Connected Generator

In the Y-connected generator, two voltage magnitudes are available at the terminals in the four-wire system: the phase voltage and the line voltage. Also, in the Y-connected generator, the line current is equal to the phase current. Keep these characteristics in mind as you examine the Δ-connected generator.

The windings of a three-phase generator can be rearranged to form a Δ-connected generator, as shown in Figure 22–21. By examination of this diagram, it is apparent that the magnitudes of the line voltages and phase voltages are equal, but the line currents do not equal the phase currents.

FIGURE 22–21
***Δ-connected generator.***

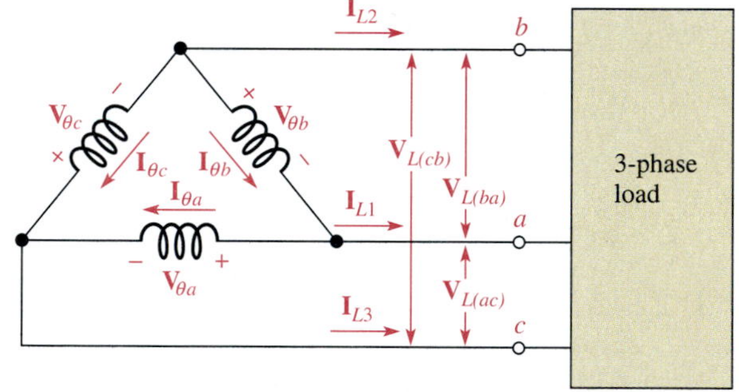


Since this is a three-wire system, only a single voltage magnitude is available, expressed as

$$V_L = V_\theta \qquad \textbf{(22–3)}$$

All of the phase voltages are equal in magnitude; thus, the line voltages are expressed in polar form as follows:

$$\mathbf{V}_{L(ac)} = V_\theta\angle 0^\circ$$
$$\mathbf{V}_{L(ba)} = V_\theta\angle 120^\circ$$
$$\mathbf{V}_{L(cb)} = V_\theta\angle -120^\circ$$

The phasor diagram for the phase currents is shown in Figure 22–22, and the polar expressions for each current are as follows:

$$\mathbf{I}_{\theta a} = I_{\theta a}\angle 0^\circ$$
$$\mathbf{I}_{\theta b} = I_{\theta b}\angle 120^\circ$$
$$\mathbf{I}_{\theta c} = I_{\theta c}\angle -120^\circ$$

It can be shown that the magnitude of each line current is equal to $\sqrt{3}$ times the magnitude of the phase current and that there is a phase angle of 30° between each line current and the nearest phase current.

$$I_L = \sqrt{3}I_\theta \qquad \textbf{(22–4)}$$

Since all phase currents are equal in magnitude,

$$\mathbf{I}_{L1} = \sqrt{3}I_\theta\angle -30^\circ$$
$$\mathbf{I}_{L2} = \sqrt{3}I_\theta\angle 90^\circ$$
$$\mathbf{I}_{L3} = \sqrt{3}I_\theta\angle -150^\circ$$

The current phasor diagram is shown in Figure 22–23.

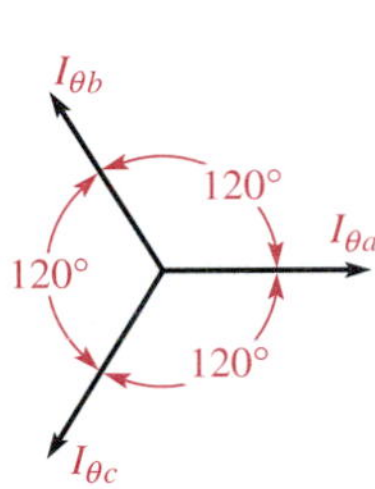


**FIGURE 22–22**
***Phase current diagram for the Δ-connected system.***

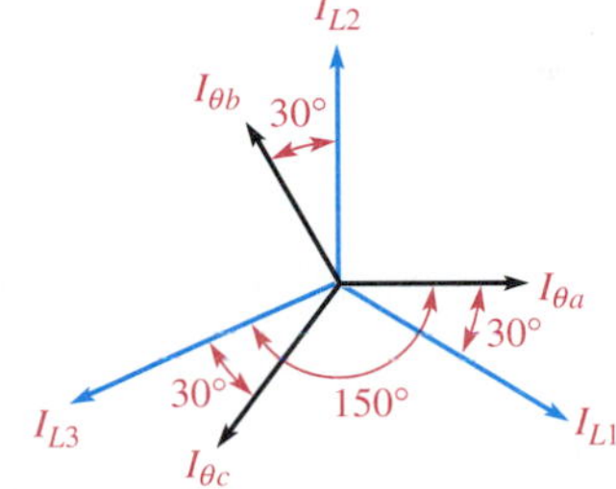


**FIGURE 22–23**
***Phasor diagram of phase currents and line currents.***

**EXAMPLE 22–3**

The three-phase Δ-connected generator represented in Figure 22–24 is driving a balanced load such that each phase current is 10 A in magnitude. When $\mathbf{I}_{\theta a} = 10\angle 30^\circ$ A, determine the following:

**(a)** The polar expressions for the other phase currents
**(b)** The polar expressions for each of the line currents
**(c)** The complete current phasor diagram

**FIGURE 22–24**

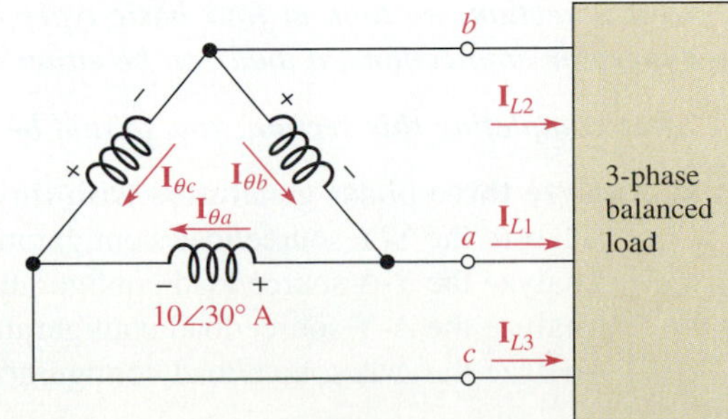

***Solution***

**(a)** The phase currents are separated by 120°; therefore,

$$\mathbf{I}_{\theta b} = 10\angle(30° + 120°) = 10\angle 150° \text{ A}$$
$$\mathbf{I}_{\theta c} = 10\angle(30° - 120°) = 10\angle -90° \text{ A}$$

**(b)** The line currents are separated from the nearest phase current by 30°; therefore,

$$\mathbf{I}_{L1} = \sqrt{3}I_{\theta a}\angle(30° - 30°) = 17.3\angle 0° \text{ A}$$
$$\mathbf{I}_{L2} = \sqrt{3}I_{\theta b}\angle(150° - 30°) = 17.3\angle 120° \text{ A}$$
$$\mathbf{I}_{L3} = \sqrt{3}I_{\theta c}\angle(-90° - 30°) = 17.3\angle -120° \text{ A}$$

**(c)** The phasor diagram is shown in Figure 22–25.

**FIGURE 22–25**

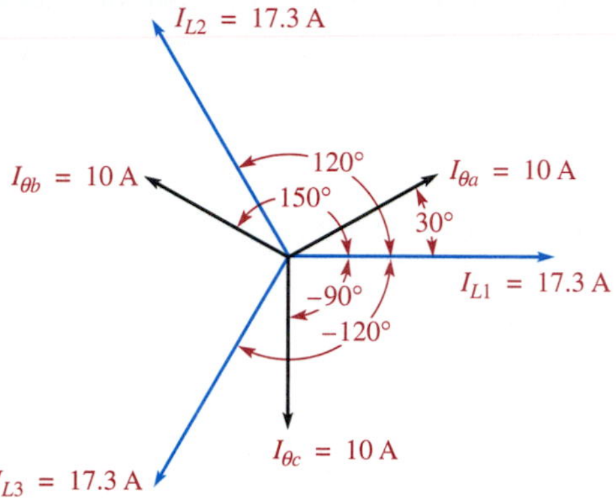


***Related Exercise*** Repeat parts (a) and (b) of the example if $\mathbf{I}_{\theta a} = 8\angle 60°$ A.

## SECTION 22–3 REVIEW

1. In a certain three-wire, Y-connected generator, the phase voltages are 1 kV. Determine the magnitude of the line voltages.
2. In the Y-connected generator mentioned in Question 1, all the phase currents are 5 A. What are the line current magnitudes?
3. In a Δ-connected generator, the phase voltages are 240 V. What are the line voltages?
4. In a Δ-connected generator, a phase current is 2 A. Determine the magnitude of the line current.

## 22–4 ■ THREE-PHASE SOURCE/LOAD ANALYSIS

***In this section, we look at four basic types of source/load configurations. As with the generator connections, a load can be either a Y or a Δ configuration.***

***After completing this section, you should be able to***

- **Analyze three-phase generators with three-phase loads**
  - ☐ Analyze the Y-Y source/load configuration
  - ☐ Analyze the Y-Δ source/load configuration
  - ☐ Analyze the Δ-Y source/load configuration
  - ☐ Analyze the Δ-Δ source/load configuration

A Y-connected load is shown in Figure 22–26(a), and a Δ-connected load is shown in part (b). The blocks $Z_a$, $Z_b$, and $Z_c$ represent the load impedances, which can be resistive, reactive, or both.

In this section, four source/load configurations are examined: a Y-connected source driving a Y-connected load (Y-Y system), a Y-connected source driving a Δ-connected load (Y-Δ system), a Δ-connected source driving a Y-connected load (Δ-Y system), and a Δ-connected source driving a Δ-connected load (Δ-Δ system).

**FIGURE 22–26**
***Three-phase loads.***

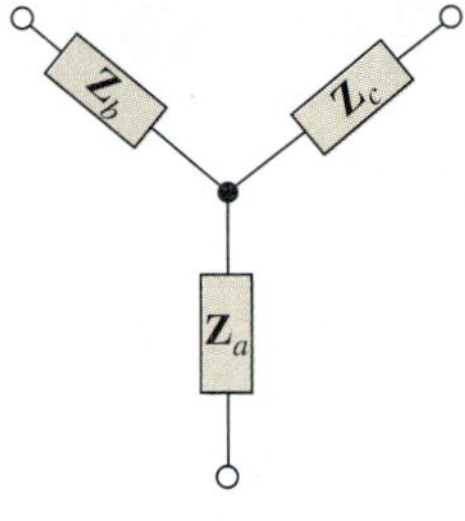


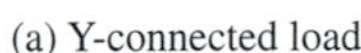
(a) Y-connected load

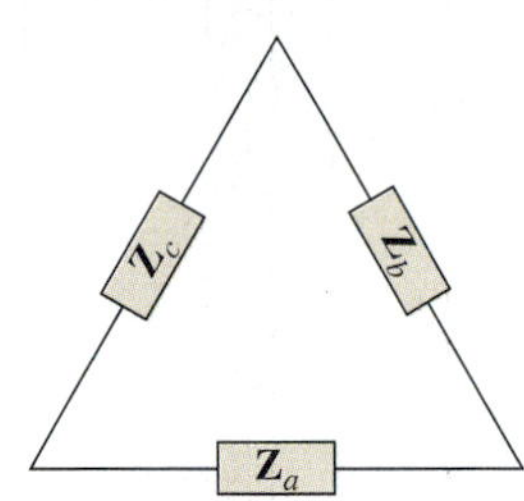

(b) Δ-connected load

## The Y-Y System

Figure 22–27 shows a Y-connected source driving a Y-connected load. The load can be a balanced load, such as a three-phase motor where $Z_a = Z_b = Z_c$, or it can be three independent single-phase loads where, for example, $Z_a$ is a lighting circuit, $Z_b$ is a heater, and $Z_c$ is an air-conditioning compressor.

**FIGURE 22–27**
***A Y-connected source feeding a Y-connected load.***

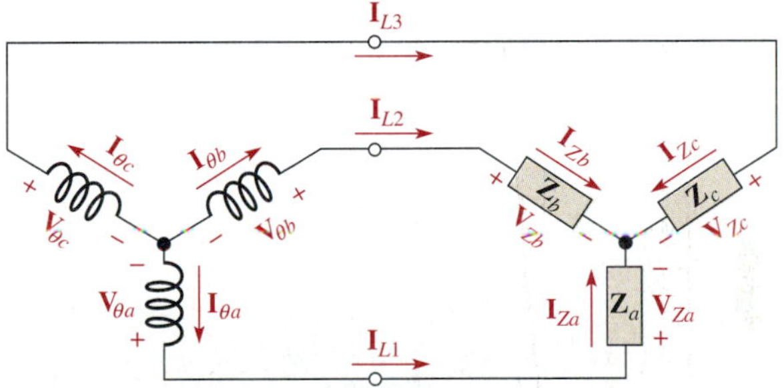


An important feature of a Y-connected source is that two different values of three-phase voltage are available: the phase voltage and the line voltage. For example, in the standard power distribution system, a three-phase transformer can be considered a source of three-phase voltage supplying 120 V and 208 V. In order to utilize a phase voltage of 120 V, the loads are connected in the Y configuration. Later, you will see that a Δ-connected load is used for the 208 V line voltages.

Notice in the Y-Y system in Figure 22–27 that the phase current, the line current, and the load current are all equal in each phase. Also, each load voltage equals the corresponding phase voltage. These relationships are expressed as follows and are true for either a balanced or an unbalanced load.

$$I_\theta = I_L = I_Z \qquad \textbf{(22–5)}$$

$$V_\theta = V_Z \qquad \textbf{(22–6)}$$

where $V_Z$ and $I_Z$ are the load voltage and current.

For a balanced load, all the phase currents are equal, and the neutral current is zero. For an unbalanced load, each phase current is different, and the neutral current is, therefore, nonzero.

**EXAMPLE 22–4** In the Y-Y system of Figure 22–28, determine the following:

**(a)** Each load current **(b)** Each line current **(c)** Each phase current

**(d)** Neutral current **(e)** Each load voltage

FIGURE 22–28

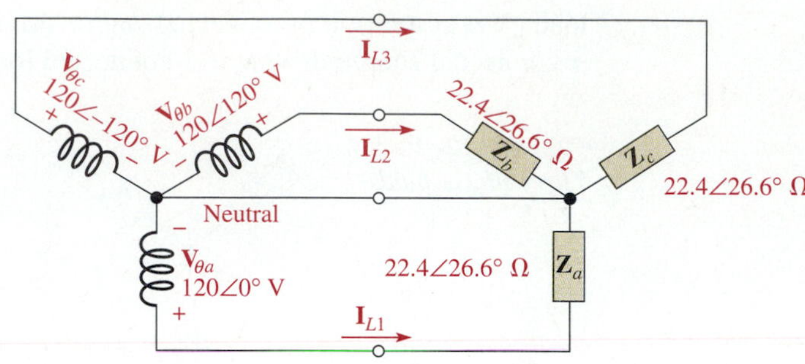


***Solution*** This system has a balanced load.

**(a)** $\mathbf{Z}_a = \mathbf{Z}_b = \mathbf{Z}_c = 22.4\angle 26.6°\ \Omega$

$$\mathbf{I}_{Za} = \frac{\mathbf{V}_{\theta a}}{\mathbf{Z}_a} = \frac{120\angle 0°\text{ V}}{22.4\angle 26.6°\ \Omega} = 5.36\angle -26.6°\text{ A}$$

$$\mathbf{I}_{Zb} = \frac{\mathbf{V}_{\theta b}}{\mathbf{Z}_b} = \frac{120\angle 120°\text{ V}}{22.4\angle 26.6°\ \Omega} = 5.36\angle 93.4°\text{ A}$$

$$\mathbf{I}_{Zc} = \frac{\mathbf{V}_{\theta c}}{\mathbf{Z}_c} = \frac{120\angle -120°\text{ V}}{22.4\angle 26.6°\ \Omega} = 5.36\angle -147°\text{ A}$$

**(b)** $\mathbf{I}_{L1} = 5.36\angle -26.6°\text{ A}$

$\mathbf{I}_{L2} = 5.36\angle 93.4°\text{ A}$

$\mathbf{I}_{L3} = 5.36\angle -147°\text{ A}$

**(c)** $\mathbf{I}_{\theta a} = 5.36\angle -26.6°\text{ A}$

$\mathbf{I}_{\theta b} = 5.36\angle 93.4°\text{ A}$

$\mathbf{I}_{\theta c} = 5.36\angle -147°\text{ A}$

**(d)**
$$\begin{aligned}\mathbf{I}_{\text{neut}} &= \mathbf{I}_{Za} + \mathbf{I}_{Zb} + \mathbf{I}_{Zc}\\ &= 5.36\angle -26.6°\text{ A} + 5.36\angle 93.4°\text{ A} + 5.36\angle -147°\text{ A}\\ &= (4.80\text{ A} - j2.40\text{ A}) + (-0.33\text{ A} + j5.35\text{ A}) + (-4.47\text{ A} - j2.95\text{ A}) = 0\text{ A}\end{aligned}$$

If the load impedances were not equal (balanced load), the neutral current would have a nonzero value.

**(e)** The load voltages are equal to the corresponding source phase voltages.

$$\mathbf{V}_{Za} = 120\angle 0°\text{ V}$$
$$\mathbf{V}_{Zb} = 120\angle 120°\text{ V}$$
$$\mathbf{V}_{Zc} = 120\angle -120°\text{ V}$$

***Related Exercise*** Determine the neutral current if $\mathbf{Z}_a$ and $\mathbf{Z}_b$ are the same as in Figure 22–28, but $\mathbf{Z}_c = 50\angle 26.6°\ \Omega$.

## The Y-Δ System

Figure 22–29 shows a Y-connected source feeding a Δ-connected load. An important feature of this configuration is that each phase of the load has the full line voltage across it.

$$V_Z = V_L \tag{22–7}$$

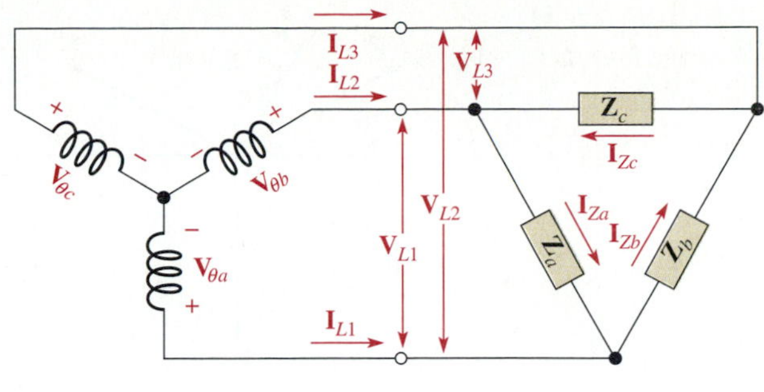


**FIGURE 22–29**
*A Y-connected source feeding a Δ-connected load.*

The line currents equal the corresponding phase currents, and each line current divides into two load currents, as indicated. For a balanced load ($Z_a = Z_b = Z_c$), the expression for the current in each load is

$$I_L = \sqrt{3}I_Z \tag{22–8}$$

**EXAMPLE 22–5** Determine the load voltages and load currents in Figure 22–30, and show their relationship in a phasor diagram.

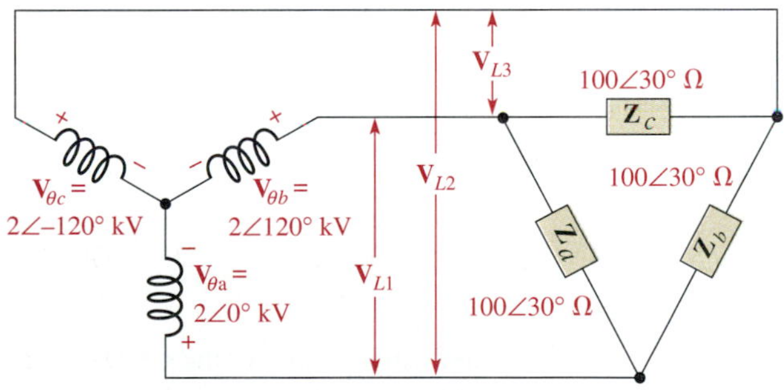


**FIGURE 22–30**

***Solution***

$$\mathbf{V}_{Za} = \mathbf{V}_{L1} = 2\sqrt{3}\angle 150° \text{ kV} = 3.46\angle 150° \text{ kV}$$
$$\mathbf{V}_{Zb} = \mathbf{V}_{L2} = 2\sqrt{3}\angle 30° \text{ kV} = 3.46\angle 30° \text{ kV}$$
$$\mathbf{V}_{Zc} = \mathbf{V}_{L3} = 2\sqrt{3}\angle -90° \text{ kV} = 3.46\angle -90° \text{ kV}$$

The load currents are

$$\mathbf{I}_{Za} = \frac{\mathbf{V}_{Za}}{\mathbf{Z}_a} = \frac{3.46\angle 150° \text{ kV}}{100\angle 30° \ \Omega} = 34.6\angle 120° \text{ A}$$
$$\mathbf{I}_{Zb} = \frac{\mathbf{V}_{Zb}}{\mathbf{Z}_b} = \frac{3.46\angle 30° \text{ kV}}{100\angle 30° \ \Omega} = 34.6\angle 0° \text{ A}$$
$$\mathbf{I}_{Zc} = \frac{\mathbf{V}_{Zc}}{\mathbf{Z}_c} = \frac{3.46\angle -90° \text{ kV}}{100\angle 30° \ \Omega} = 34.6\angle -120° \text{ A}$$

The phasor diagram is shown in Figure 22–31.

FIGURE 22–31

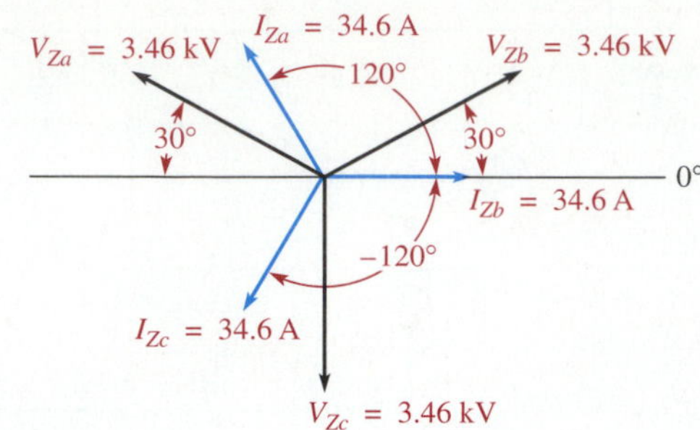


***Related Exercise*** Determine the load currents in Figure 22–30 if the phase voltages have a magnitude of 240 V.

## The Δ-Y System

Figure 22–32 shows a Δ-connected source feeding a Y-connected balanced load. By examination of the figure, you can see that the line voltages are equal to the corresponding phase voltages of the source. Also, each phase voltage equals the difference of the corresponding load voltages, as you can see by the polarities.

FIGURE 22–32

*A Δ-connected source feeding a Y-connected load.*

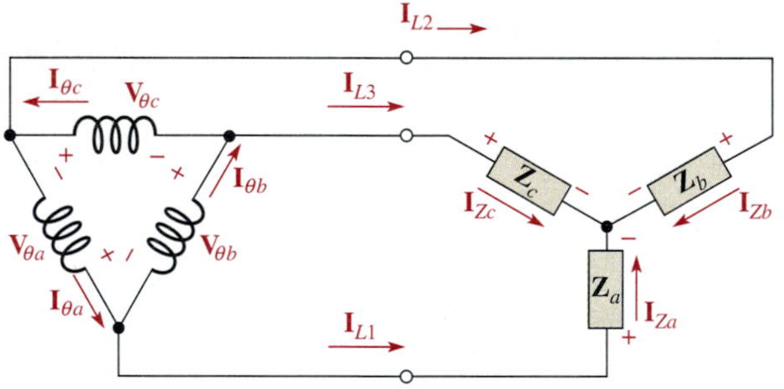


Each load current equals the corresponding line current. The sum of the load currents is zero, because the load is balanced; thus, there is no need for a neutral return.

The relationship between the load voltages and the corresponding phase voltages (and line voltages) is

$$V_\theta = \sqrt{3}V_Z \tag{22–9}$$

The line currents and corresponding load currents are equal, and for a balanced load, the sum of the load currents is zero.

$$\mathbf{I}_L = \mathbf{I}_Z \tag{22–10}$$

As you can see in Figure 22–32, each line current is the difference of two phase currents.

$$\mathbf{I}_{L1} = \mathbf{I}_{\theta a} - \mathbf{I}_{\theta b}$$
$$\mathbf{I}_{L2} = \mathbf{I}_{\theta c} - \mathbf{I}_{\theta a}$$
$$\mathbf{I}_{L3} = \mathbf{I}_{\theta b} - \mathbf{I}_{\theta c}$$

**EXAMPLE 22–6** Determine the currents and voltages in the balanced load and the magnitude of the line voltages in Figure 22–33.

FIGURE 22–33

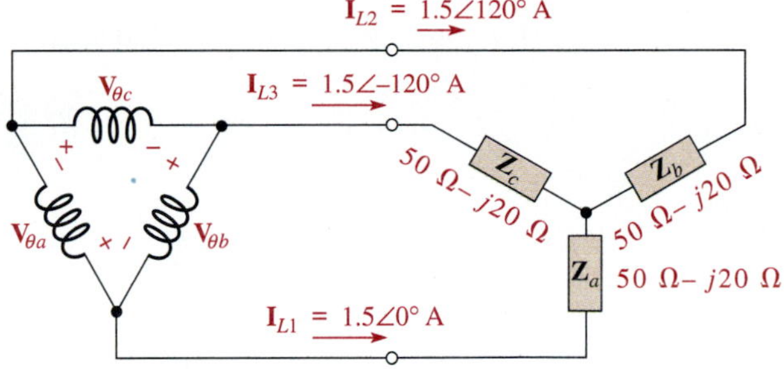


***Solution*** The load currents equal the specified line currents.

$$\mathbf{I}_{Za} = \mathbf{I}_{L1} = 1.5\angle 0^\circ \text{ A}$$
$$\mathbf{I}_{Zb} = \mathbf{I}_{L2} = 1.5\angle 120^\circ \text{ A}$$
$$\mathbf{I}_{Zc} = \mathbf{I}_{L3} = 1.5\angle -120^\circ \text{ A}$$

The load voltages are

$$\mathbf{V}_{Za} = \mathbf{I}_{Za}\mathbf{Z}_a$$
$$= (1.5\angle 0^\circ \text{ A})(50\ \Omega - j20\ \Omega)$$
$$= (1.5\angle 0^\circ \text{ A})(53.9\angle -21.8^\circ\ \Omega) = 80.9\angle -21.8^\circ \text{ V}$$
$$\mathbf{V}_{Zb} = \mathbf{I}_{Zb}\mathbf{Z}_b$$
$$= (1.5\angle 120^\circ \text{ A})(53.9\angle -21.8^\circ\ \Omega) = 80.9\angle 98.2^\circ \text{ V}$$
$$\mathbf{V}_{Zc} = \mathbf{I}_{Zc}\mathbf{Z}_c$$
$$= (1.5\angle -120^\circ \text{ A})(53.9\angle -21.8^\circ\ \Omega) = 80.9\angle -142^\circ \text{ V}$$

The magnitude of the line voltages is

$$V_L = V_\theta = \sqrt{3}V_Z = \sqrt{3}(80.9 \text{ V}) = 140 \text{ V}$$

***Related Exercise*** If the magnitudes of the line currents are 1 A, what are the load currents?

## The Δ-Δ System

Figure 22–34 shows a Δ-connected source driving a Δ-connected load. Notice that the load voltage, line voltage, and source phase voltage are all equal for a given phase.

$$V_{\theta a} = V_{L1} = V_{Za}$$
$$V_{\theta b} = V_{L2} = V_{Zb}$$
$$V_{\theta c} = V_{L3} = V_{Zc}$$

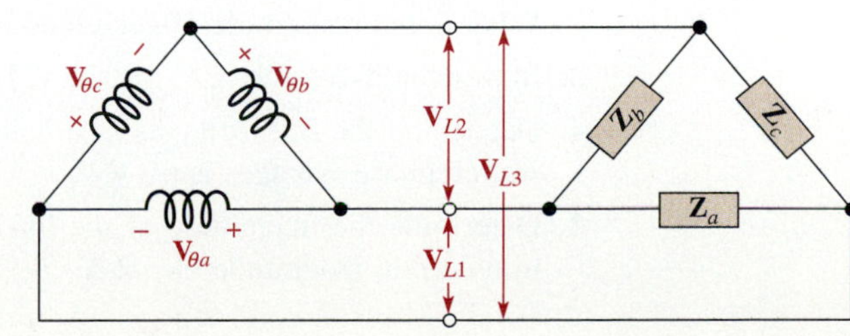


FIGURE 22–34
*A Δ-connected source feeding a Δ-connected load.*

Of course, when the load is balanced, all the voltages are equal, and a general expression can be written

$$V_\theta = V_L = V_Z \quad (22\text{–}11)$$

For a balanced load and equal source phase voltages, it can be shown that

$$I_L = \sqrt{3}I_Z \quad (22\text{–}12)$$

**EXAMPLE 22–7** Determine the magnitude of the load currents and the line currents in Figure 22–35.

FIGURE 22–35

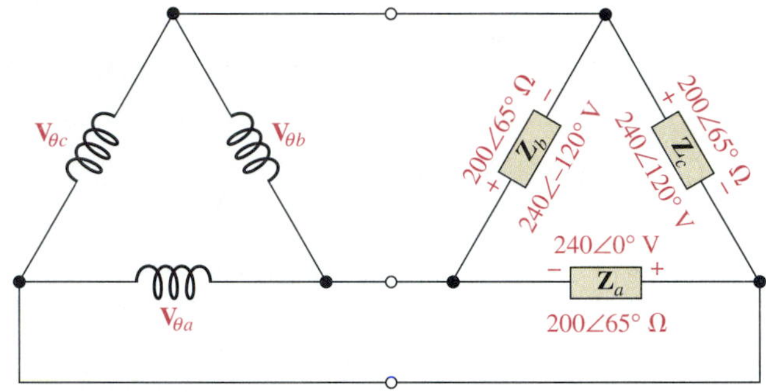


***Solution***

$$V_{Za} = V_{Zb} = V_{Zc} = 240\text{ V}$$

The magnitude of the load currents is

$$I_{Za} = I_{Zb} = I_{Zc} = \frac{V_{Za}}{Z_a} = \frac{240\text{ V}}{200\ \Omega} = 1.20\text{ A}$$

The magnitude of the line currents is

$$I_L = \sqrt{3}I_Z = \sqrt{3}(1.20\text{ A}) = 2.08\text{ A}$$

***Related Exercise*** Determine the magnitude of the load and line currents in Figure 22–35 if the magnitude of the load voltages is 120 V and the impedances are 600 Ω.

**SECTION 22–4 REVIEW**

1. List the four types of three-phase source/load configurations.
2. In a certain Y-Y system, the source phase currents each have a magnitude of 3.5 A. What is the magnitude of each load current for a balanced load condition?
3. In a given Y-Δ system, $V_L = 220$ V. Determine $V_Z$.
4. Determine the line voltages in a balanced Δ-Y system when the magnitude of the source phase voltages is 60 V.
5. Determine the magnitude of the load currents in a balanced Δ-Δ system having a line current magnitude of 3.2 A.

## 22–5 ■ THREE-PHASE POWER

***In this section, power in three-phase systems is studied and methods of power measurement are introduced.***

***After completing this section, you should be able to***

- **Discuss power measurements in three-phase systems**
  - ☐ Describe the three-wattmeter method
  - ☐ Describe the two-wattmeter method

Each phase of a balanced three-phase load has an equal amount of power. Therefore, the total true load power is three times the power in each phase of the load.

$$P_{L(tot)} = 3V_Z I_Z \cos\theta \qquad \textbf{(22–13)}$$

where $V_Z$ and $I_Z$ are the voltage and current associated with each phase of the load, and $\cos\theta$ is the power factor.

Recall that in a balanced Y-connected system, the line voltage and line current were

$$V_L = \sqrt{3}V_Z \qquad \text{and} \qquad I_L = I_Z$$

and in a balanced Δ-connected system, the line voltage and line current were

$$V_L = V_Z \qquad \text{and} \qquad I_L = \sqrt{3}I_Z$$

When either of these relationships is substituted into Equation (22–13), the total true power for both Y- and Δ-connected systems is

$$P_{L(tot)} = \sqrt{3}V_L I_L \cos\theta \qquad \textbf{(22–14)}$$

**EXAMPLE 22–8**

In a certain Δ-connected balanced load, the line voltages are 250 V and the impedances are $50\angle 30^\circ\ \Omega$. Determine the total load power.

***Solution*** In a Δ-connected system, $V_Z = V_L$ and $I_L = \sqrt{3}I_Z$. The load current magnitudes are

$$I_Z = \frac{V_Z}{Z} = \frac{250\text{ V}}{50\ \Omega} = 5\text{ A}$$

and

$$I_L = \sqrt{3}I_Z = \sqrt{3}(5\text{ A}) = 8.66\text{ A}$$

The power factor is

$$\cos\theta = \cos 30^\circ = 0.866$$

The total load power is

$$P_{L(tot)} = \sqrt{3}V_L I_L \cos\theta = \sqrt{3}(250\text{ V})(8.66\text{ A})(0.866) = 3.25\text{ kW}$$

***Related Exercise*** Determine the total load power if $V_L = 120$ V and $\mathbf{Z} = 100\angle 30^\circ\ \Omega$.

### Power Measurement

Power is measured in three-phase systems using wattmeters. The wattmeter uses a basic electrodynamometer-type movement consisting of two coils. One coil is used to measure the current, and the other is used to measure the voltage. The needle of the meter is deflected proportionally to the current through a load and the voltage across the load, thus indicating power. Figure 22–36 shows a basic wattmeter symbol and the connections for measuring power in a load. The resistor in series with the voltage coil limits the current through the coil to a small amount proportional to the voltage across the coil.

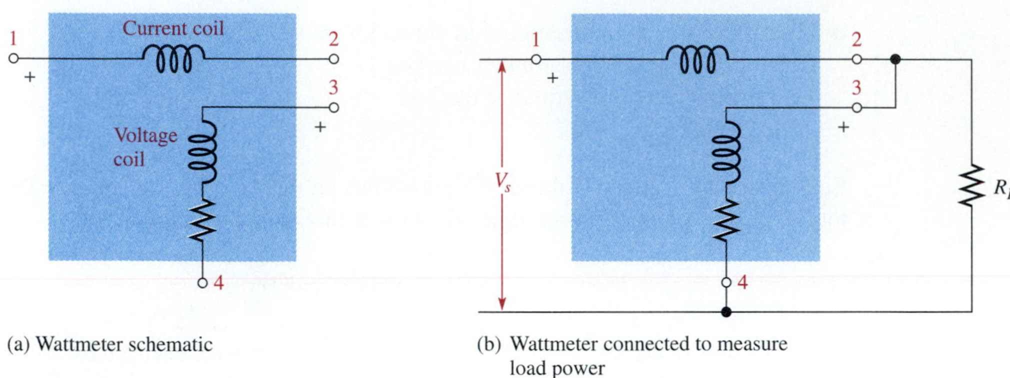


FIGURE 22–36

***Three-Wattmeter Method*** Power can be measured easily in a balanced or unbalanced three-phase load of either the Y or the Δ type by using three wattmeters connected as shown in Figure 22–37. This is sometimes known as the *three-wattmeter method.*

The total power is determined by summing the three wattmeter readings.

$$P_{tot} = P_1 + P_2 + P_3 \tag{22–15}$$

If the load is balanced, the total power is simply three times the reading on any one wattmeter.

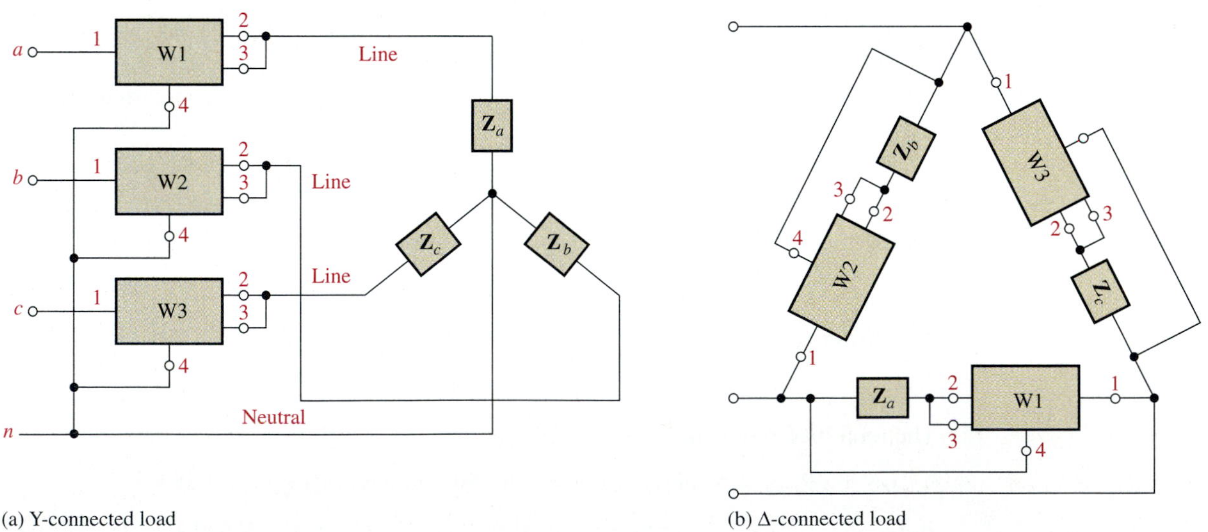


**FIGURE 22–37**
***Three-wattmeter method of power measurement.***

In many three-phase loads, particularly the Δ configuration, it is difficult to connect a wattmeter such that the voltage coil is across the load or such that the current coil is in series with the load because of inaccessibility of points within the load.

***Two-Wattmeter Method*** Another method of three-phase power measurement uses only two wattmeters. The connections for this two-wattmeter method are shown in Figure 22–38. Notice that the voltage coil of each wattmeter is connected across a line voltage and that the current coil has a line current through it. It can be shown that the algebraic sum of the two wattmeter readings equals the total power in the Y- and Δ-connected load.

$$P_{tot} = P_1 \pm P_2 \qquad (22\text{–}16)$$

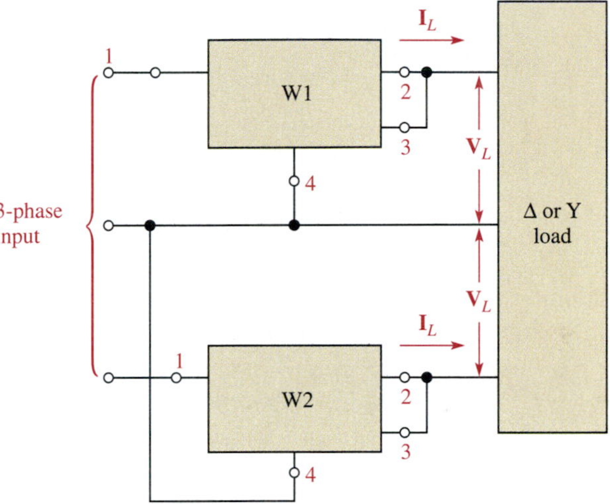


**FIGURE 22–38**
***Two-wattmeter method.***

**SECTION 22–5 REVIEW**

1. $V_L = 30$ V, $I_L = 1.2$ A, and the power factor is 0.257. What is the total power in a balanced Y-connected load? In a balanced Δ-connected load?
2. Three wattmeters connected to measure the power in a certain balanced load indicate a total of 2678 W. How much power does each meter measure?

## 22–6 ■ POLYPHASE CIRCUIT MODELING WITH PSpice

***All of the PSpice circuit analyses to this point have been for single-phase circuits. Although PSpice was not primarily intended to study three-phase circuits, it can be used to model and analyze them. In this section, you will learn how to model wye and delta sources and loads.***

***After completing this section, you should be able to***

- **Use PSpice and Probe to analyze three-phase circuits**
  - ☐ Determine currents and powers in a Y-Y circuit
  - ☐ Determine currents and powers in a Δ-Δ circuit

The circuit in Figure 22–39 is a wye-wye (Y-Y) connection because both the source and the load are in a Y configuration. Notice that all of the three sources are identical in magnitude and frequency but are 120° apart in phase and the three loads are the same. This is called a balanced circuit because of the equal sources and the equal loads.

FIGURE 22–39

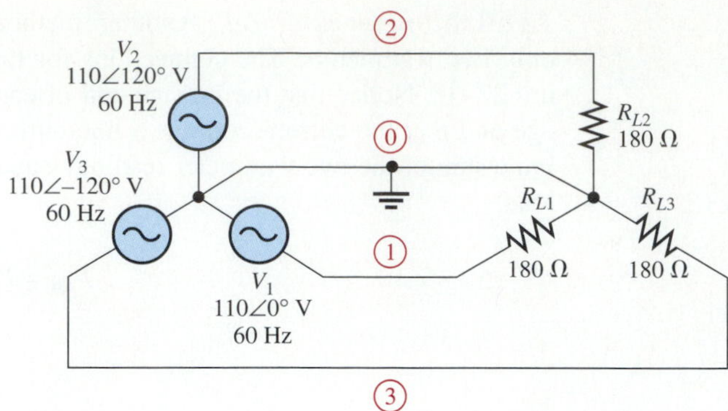


The circuit file for the balanced Y-Y circuit is as follows:

```
Three-Phase Circuit
*Y-Y Connection
V1  1  0  SIN(0  110  60  0  0  0)
V2  2  0  SIN(0  110  60  0  0  -120)
V3  3  0  SIN(0  110  60  0  0  -240)
RL1  1  0  180
RL2  2  0  180
RL3  3  0  180
.PROBE
.TRAN  100U  50M
.END
```

Recall from Chapter 12 that the transient phase specification uses the convention where the angle at the positive-going zero crossing is written with a negative sign. Thus V2 should have a positive-going zero crossing at 120° and V3 should have a positive-going zero crossing at 240°, which is the same as saying –120°.

Run PSpice on this circuit and display the following with Probe:

```
I(RL1)  I(RL2)  I(RL3)
```

You will see that the currents have equal magnitudes of 611 mA but are shifted in phase. You should be able to tell by inspection that they are 120° apart, but if you wish to confirm this, the $x$-axis can be changed to degrees. To do this, change the $x$-axis variable as was done in Chapter 12. At the prompt for variables or expressions, or X-axis Variable, enter

```
(TIME*60)*(360/1s)
```

This will convert the scale from seconds to degrees. Probe will redraw the $x$-axis so that values run from 0° to 360°. The trace for $I_{RL1}$ begins (positive-going zero crossing) at 0°, the trace for $I_{RL2}$ begins at 120°, and the trace for $I_{RL3}$ begins at 240°. The waveforms are 120° out of phase with each other as expected.

A delta-delta (Δ-Δ) three-phase circuit is shown in Figure 22–40. Both the sources and the loads are connected in a delta arrangement. The circuit file for this circuit is

```
Three-Phase Circuit
*Δ-Δ Connection
VA  2  1  SIN(0  110  60  0  0  0)
VB  4  3  SIN(0  110  60  0  0  -120)
VC  0  5  SIN(0  110  60  0  0  -240)
RSA  0  1  1P
RSB  2  3  1P
RSC  4  5  1P
RLA  2  0  180
RLB  4  2  180
```

FIGURE 22–40

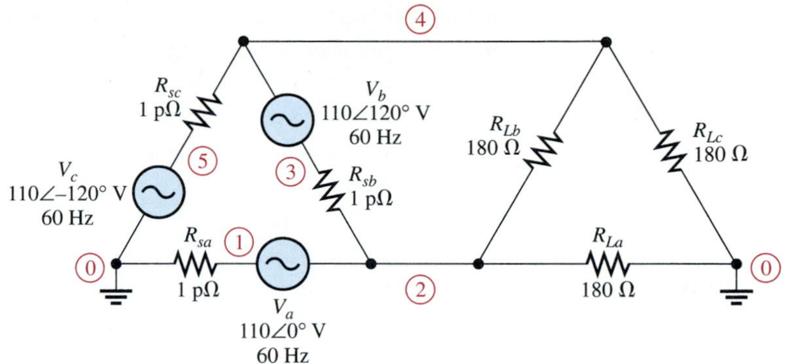


```
RLC  0  4  180
.PROBE
.TRAN  100U  50M
.END
```

There are special requirements to use PSpice in this circuit. First, each voltage source must have a series resistor to keep it balanced and prevent PSpice from finding a short across any source. A very small value resistance (1 pΩ) is added in series with each source to satisfy this requirement. Second, node 0 is a reference ground only so that PSpice can perform its analysis on the circuit and not an actual circuit ground. In actual circuits, unless a ground is placed at some point, the entire circuit will float.

Use Probe as before to view the currents through $R_{La}$, $R_{Lb}$, and $R_{Lc}$. Again, the currents are all 611 mA and 120° apart in phase just as in the Y-Y circuit.

## Power in Three-Phase Circuits

To find the power in three-phase circuits, you simply tell Probe to find the product of the voltage and the current in the load. For example, if you want the power in $R_{La}$ for the Δ-Δ circuit in Figure 22–40, enter the following trace expression:

```
V(2,0)*I(RLA)
```

Probe will draw the power curve for the specified load resistor. By selecting Cursor, you can find the peak or peak-to-peak magnitude of the power in the load resistor. In this case, the power is approximately 67.2 W.

**SECTION 21–11 REVIEW**

1. Rewrite the circuit file if the three-phase load in Figure 22–39 is changed from a wye to a delta configuration.
2. What effect do the series resistors in the three-phase delta source connection have on the load currents? Explain.

## ■ SUMMARY

- A simple two-phase generator consists of two conductive loops, separated by 90°, rotating in a magnetic field.
- A simple three-phase generator consists of three conductive loops separated by 120°.
- Three advantages of polyphase systems over single-phase systems are a smaller copper cross section for the same power delivered to the load, constant power delivered to the load, and a constant, rotating magnetic field.
- In a Y-connected generator, $I_L = I_\theta$ and $V_L = \sqrt{3}V_\theta$.
- In a Y-connected generator, there is a 30° difference between each line voltage and the nearest phase voltage.

- In a Δ-connected generator, $V_L = V_\theta$ and $I_L = \sqrt{3}I_\theta$.
- In a Δ-connected generator, there is a 30° difference between each line current and the nearest phase current.
- A balanced load is one in which all the impedances are equal.
- Power is measured in a three-phase load using either the three-wattmeter method or the two-wattmeter method.

## ■ GLOSSARY

**Alternator** An electromechanical ac generator.

**Balanced load** A condition where all the load currents are equal and the neutral current is zero.

**Field winding** The winding on the rotor of an ac generator.

**Line current ($I_L$)** The current through a line feeding a load.

**Line voltage ($V_L$)** The voltage between lines feeding a load.

**Phase current ($I_\theta$)** The current through a generator winding.

**Phase voltage ($V_\theta$)** The voltage across a generator winding.

**Polyphase** Characterized by two or more sinusoidal voltages, each having a different phase angle.

**Rotor** The rotating assembly in a generator or motor.

**Squirrel-cage** A type of ac induction motor.

**Stator** The stationary outer part of a generator or motor.

## ■ FORMULAS

### Y Generator

**(22–1)** $I_L = I_\theta$

**(22–2)** $V_L = \sqrt{3}V_\theta$

### Δ Generator

**(22–3)** $V_L = V_\theta$

**(22–4)** $I_L = \sqrt{3}I_\theta$

### Y-Y System

**(22–5)** $I_\theta = I_L = I_Z$

**(22–6)** $V_\theta = V_Z$

### Y-Δ System

**(22–7)** $V_Z = V_L$

**(22–8)** $I_L = \sqrt{3}I_Z$

### Δ-Y System

**(22–9)** $V_\theta = \sqrt{3}V_Z$

**(22–10)** $\mathbf{I}_L = \mathbf{I}_Z$

### Δ-Δ System

**(22–11)** $V_\theta = V_L = V_Z$

**(22–12)** $I_L = \sqrt{3}I_Z$

### Three-Phase Power

**(22–13)** $P_{L(tot)} = 3V_Z I_Z \cos\theta$

**(22–14)** $P_{L(tot)} = \sqrt{3}V_L I_L \cos\theta$

### Three-Wattmeter Method

**(22–15)** $P_{tot} = P_1 + P_2 + P_3$

### Two-Wattmeter Method

**(22–16)** $P_{tot} = P_1 \pm P_2$

## ■ SELF-TEST

1. In a three-phase system, the voltages are separated by
   (a) 90° (b) 30° (c) 180° (d) 120°
2. The term *squirrel-cage* applies to a type of
   (a) three-phase ac generator (b) single-phase ac generator
   (c) a three-phase ac motor (d) a dc motor
3. An alternator is a
   (a) three-phase ac generator (b) single-phase ac generator
   (c) three-phase ac motor (d) dc generator
4. Two major parts of an ac generator are
   (a) rotor and stator (b) rotor and stabilizer
   (c) regulator and slip-ring (d) magnets and brushes
5. Advantages of a three-phase system over a single-phase system are
   (a) smaller cross-sectional area for the copper conductors
   (b) slower rotor speed
   (c) constant power
   (d) smaller chance of overheating
   (e) both (a) and (c)
   (f) both (b) and (c)
6. The phase current produced by a certain 240 V, Y-connected generator is 12 A. The corresponding line current is
   (a) 36 A (b) 4 A (c) 12 A (d) 6 A
7. A certain Δ-connected generator produces phase voltages of 30 V. The magnitude of the line voltages are
   (a) 10 V (b) 30 V (c) 90 V (d) none of the above
8. A certain Δ-Δ system produces phase currents of 5 A. The line currents are
   (a) 5 A (b) 15 A (c) 8.66 A (d) 2.87 A
9. A certain Y-Y system produces phase currents of 15 A. Each line and load current is
   (a) 26 A (b) 8.66 A (c) 5 A (d) 15 A
10. If the source phase voltages of a Δ-Y system are 220 V, the magnitude of the load voltages is
    (a) 220 V (b) 381 V (c) 127 V (d) 73.3 V

## ■ PROBLEMS

### SECTION 22–1 Basic Polyphase Machines

1. The output of an ac generator has a maximum value of 250 V. At what angle is the instantaneous value equal to 75 V?
2. A certain two-pole single-phase generator has a speed of rotation of 3600 rpm. What is the frequency of the voltage produced by this generator?

### SECTION 22–2 Polyphase Generators in Power Applications

3. A single-phase generator feeds a load consisting of a 200 Ω resistor and a capacitor with a reactance of 175 Ω. The generator produces a voltage of 100 V. Determine the magnitude of the load current and its phase relation to the generator voltage.
4. In a two-phase system, the two currents through the lines connecting the generator to the load are 3.8 A. Determine the current in the neutral line.
5. A certain three-phase unbalanced load in a four-wire system has currents of $2\angle 20^\circ$ A, $3\angle 140^\circ$ A, and $1.5\angle -100^\circ$ A. Determine the current in the neutral line.

## SECTION 22–3 Three-Phase Generators

**6.** Determine the line voltages in Figure 22–41.

FIGURE 22–41

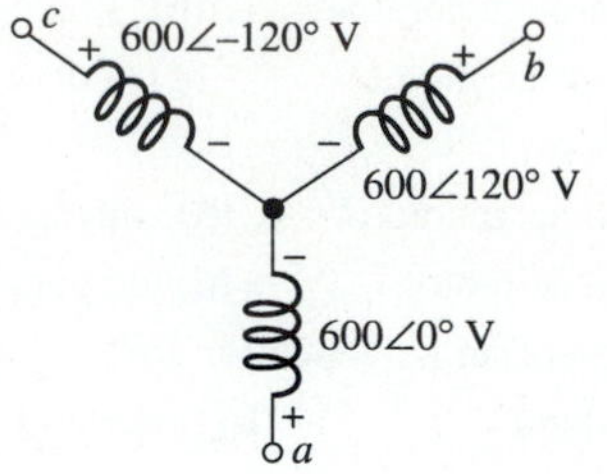

**7.** Determine the line currents in Figure 22–42.

**8.** Develop a complete current phasor diagram for Figure 22–42.

FIGURE 22–42

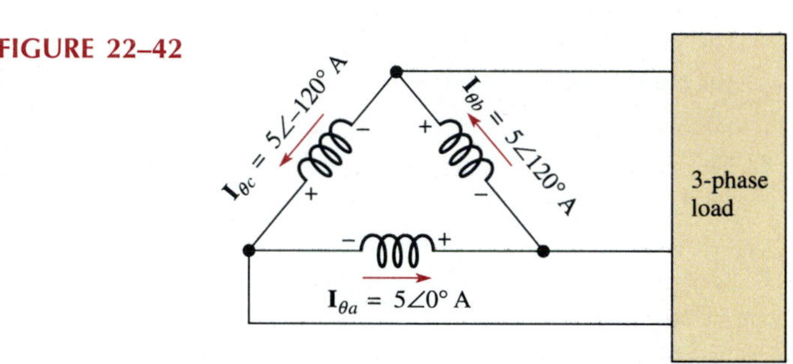


## SECTION 22–4 Three-Phase Source/Load Analysis

**9.** Determine the following quantities for the Y-Y system in Figure 22–43:

**(a)** Line voltages **(b)** Phase currents **(c)** Line currents

**(d)** Load currents **(e)** Load voltages

FIGURE 22–43

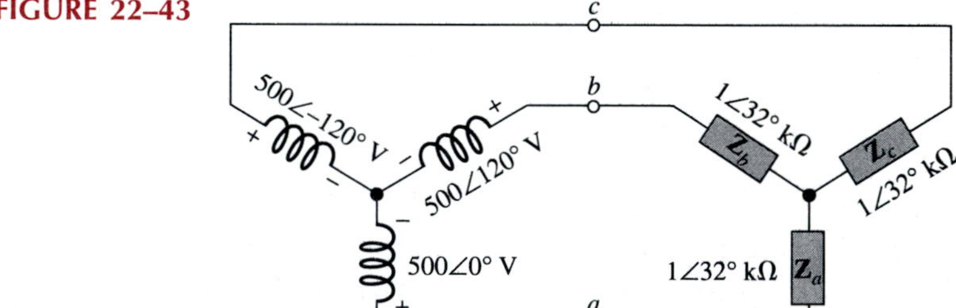


**10.** Repeat Problem 9 for the system in Figure 22–44, and also find the neutral current.

FIGURE 22–44

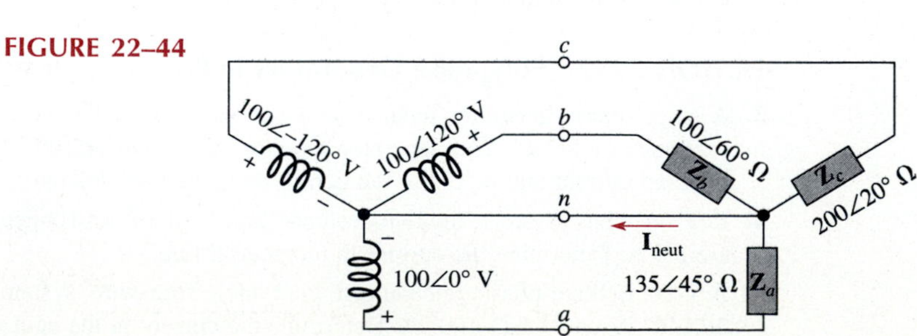

**11.** Repeat Problem 9 for the system in Figure 22–45.

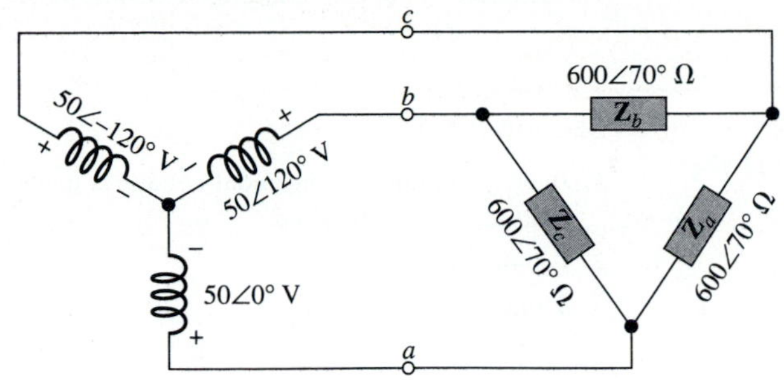


FIGURE 22–45

**12.** Repeat Problem 9 for the system in Figure 22–46.

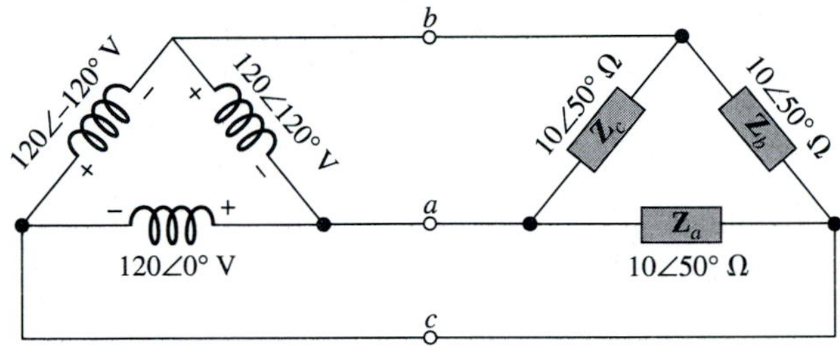


FIGURE 22–46

**13.** Determine the line voltages and load currents for the system in Figure 22–47.

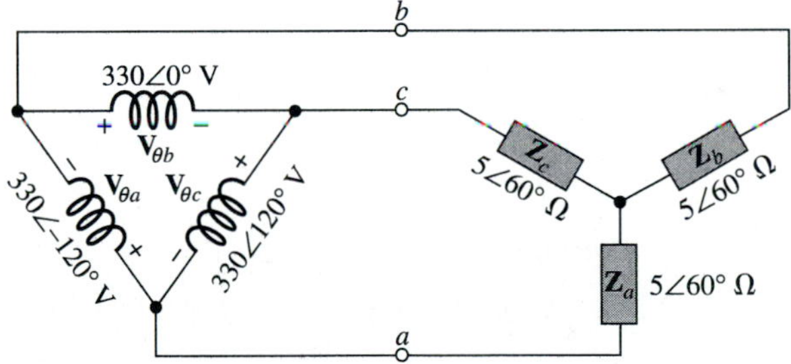


FIGURE 22–47

## SECTION 22–5 Three-Phase Power

**14.** The power in each phase of a balanced three-phase system is 1200 W. What is the total power?

**15.** Determine the load power in Figures 22–43 through 22–47.

**16.** Find the total load power in Figure 22–48.

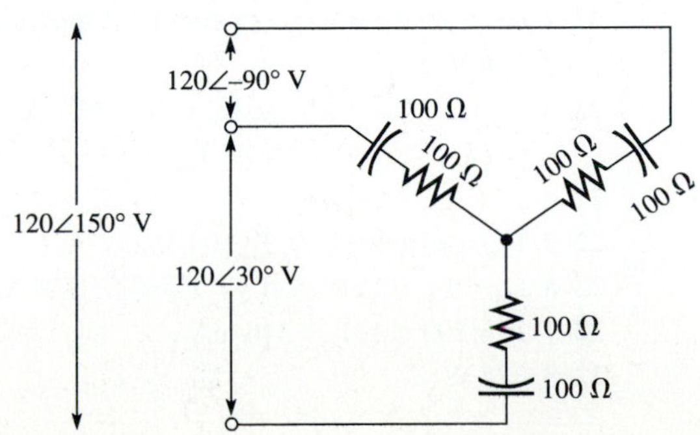


FIGURE 22–48

**17.** Using the three-wattmeter method for the system in Figure 22–48, how much power does each wattmeter indicate?

**18.** Repeat Problem 17 using the two-wattmeter method.

## ■ ANSWERS TO SECTION REVIEWS

### Section 22–1

**1.** In ac generators, a sinusoidal voltage is induced when a conductive loop is rotated in a magnetic field at a constant speed.

**2.** $f = 400$ Hz

**3.** Three armature windings

### Section 22–2

**1.** The advantages of polyphase systems are less copper cross section to conduct current; constant power to load; and constant, rotating magnetic field.

**2.** Constant power

**3.** Constant magnetic field

### Section 22–3

**1.** $V_L = 1.73$ kV

**2.** $I_L = 5$ A

**3.** $V_L = 240$ V

**4.** $I_L = 3.46$ A

### Section 22–4

**1.** The source/load configurations are Y-Y, Y-Δ, Δ-Y, and Δ-Δ.

**2.** $I_L = 3.5$ A

**3.** $V_Z = 220$ V

**4.** $V_L = 60$ V

**5.** $I_Z = 1.85$ A

### Section 22–5

**1.** $P_Y = 16.0$ W; $P_\Delta = 16.0$ W

**2.** $P = 893$ W

### Section 22–6

**1.** Change only the resistor lines as follows:

```
RL1  1  2  180
RL2  2  3  180
RL3  1  3  180
```

**2.** The series resistors included with each source have no effect on load currents because of their extremely small value.

## ■ ANSWERS TO RELATED EXERCISES FOR EXAMPLES

**22–1** 4.8 A total for single phase; 2.4 A total for three-phase

**22–2** 208 V

**22–3** (a) $\mathbf{I}_{\theta b} = 8\angle 180°$ A, $\mathbf{I}_{\theta c} = 8\angle -60°$ A
(b) $\mathbf{I}_{L1} = 13.9\angle 30°$ A, $\mathbf{I}_{L2} = 13.9\angle 150°$ A, $\mathbf{I}_{L3} = 13.9\angle -90°$ A

**22–4** $2.96\angle 33.4°$ A

**22–5** $\mathbf{I}_{Za} = 4.16\angle 120°$ A, $\mathbf{I}_{Zb} = 4.16\angle 0°$ A, $\mathbf{I}_{Zc} = 4.16\angle -120°$ A

**22–6** $\mathbf{I}_{L1} = \mathbf{I}_{Za} = 1\angle 0°$ A, $\mathbf{I}_{L2} = \mathbf{I}_{Zb} = 1\angle 120°$ A, $\mathbf{I}_{L3} = \mathbf{I}_{Zc} = 1\angle -120°$ A

**22–7** $I_Z = 200$ mA, $I_L = 346$ mA

**22–8** 374 W

# TABLE OF STANDARD RESISTOR VALUES

Resistance Tolerance (±%)

| 0.1%<br>0.25%<br>0.5% | 1% | 2%<br>5% | 10% | 0.1%<br>0.25%<br>0.5% | 1% | 2%<br>5% | 10% | 0.1%<br>0.25%<br>0.5% | 1% | 2%<br>5% | 10% | 0.1%<br>0.25%<br>0.5% | 1% | 2%<br>5% | 10% | 0.1%<br>0.25%<br>0.5% | 1% | 2%<br>5% | 10% | 0.1%<br>0.25%<br>0.5% | 1% | 2%<br>5% | 10% |
|---|---|---|---|---|---|---|---|---|---|---|---|---|---|---|---|---|---|---|---|---|---|---|---|
| 10.0 | 10.0 | 10 | 10 | 14.7 | 14.7 | — | — | 21.5 | 21.5 | — | — | 31.6 | 31.6 | — | — | 46.4 | 46.4 | — | — | 68.1 | 68.1 | 68 | 68 |
| 10.1 | — | — | — | 14.9 | — | — | — | 21.8 | — | — | — | 32.0 | — | — | — | 47.0 | — | 47 | 47 | 69.0 | — | — | — |
| 10.2 | 10.2 | — | — | 15.0 | 15.0 | 15 | 15 | 22.1 | 22.1 | 22 | 22 | 32.4 | 32.4 | — | — | 47.5 | 47.5 | — | — | 69.8 | 69.8 | — | — |
| 10.4 | — | — | — | 15.2 | — | — | — | 22.3 | — | — | — | 32.8 | — | — | — | 48.1 | — | — | — | 70.6 | — | — | — |
| 10.5 | 10.5 | — | — | 15.4 | 15.4 | — | — | 22.6 | 22.6 | — | — | 33.2 | 33.2 | 33 | 33 | 48.7 | 48.7 | — | — | 71.5 | 71.5 | — | — |
| 10.6 | — | — | — | 15.6 | — | — | — | 22.9 | — | — | — | 33.6 | — | — | — | 49.3 | — | — | — | 72.3 | — | — | — |
| 10.7 | 10.7 | — | — | 15.8 | 15.8 | — | — | 23.2 | 23.2 | — | — | 34.0 | 34.0 | — | — | 49.9 | 49.9 | — | — | 73.2 | 73.2 | — | — |
| 10.9 | — | — | — | 16.0 | — | 16 | — | 23.4 | — | — | — | 34.4 | — | — | — | 50.5 | — | — | — | 74.1 | — | — | — |
| 11.0 | 11.0 | 11 | — | 16.2 | 16.2 | — | — | 23.7 | 23.7 | — | — | 34.8 | 34.8 | — | — | 51.1 | 51.1 | 51 | — | 75.0 | 75.0 | 75 | — |
| 11.1 | — | — | — | 16.4 | — | — | — | 24.0 | — | 24 | — | 35.2 | — | — | — | 51.7 | — | — | — | 75.9 | — | — | — |
| 11.3 | 11.3 | — | — | 16.5 | 16.5 | — | — | 24.3 | 24.3 | — | — | 35.7 | 35.7 | — | — | 52.3 | 52.3 | — | — | 76.8 | 76.8 | — | — |
| 11.4 | — | — | — | 16.7 | — | — | — | 24.6 | — | — | — | 36.1 | — | 36 | — | 53.0 | — | — | — | 77.7 | — | — | — |
| 11.5 | 11.5 | — | — | 16.9 | 16.9 | — | — | 24.9 | 24.9 | — | — | 36.5 | 36.5 | — | — | 53.6 | 53.6 | — | — | 78.7 | 78.7 | — | — |
| 11.7 | — | — | — | 17.2 | — | — | — | 25.2 | — | — | — | 37.0 | — | — | — | 54.2 | — | — | — | 79.6 | — | — | — |
| 11.8 | 11.8 | — | — | 17.4 | 17.4 | — | — | 25.5 | 25.5 | — | — | 37.4 | 37.4 | — | — | 54.9 | 54.9 | — | — | 80.6 | 80.6 | — | — |
| 12.0 | — | 12 | 12 | 17.6 | — | — | — | 25.8 | — | — | — | 37.9 | — | — | — | 56.2 | — | — | — | 81.6 | — | — | — |
| 12.1 | 12.1 | — | — | 17.8 | 17.8 | — | — | 26.1 | 26.1 | — | — | 38.3 | 38.3 | — | — | 56.6 | 56.6 | 56 | 56 | 82.5 | 82.5 | 82 | 82 |
| 12.3 | — | — | — | 18.0 | — | 18 | 18 | 26.4 | — | — | — | 38.8 | — | — | — | 56.9 | — | — | — | 83.5 | — | — | — |
| 12.4 | 12.4 | — | — | 18.2 | 18.2 | — | — | 26.7 | 26.7 | — | — | 39.2 | 39.2 | 39 | 39 | 57.6 | 57.6 | — | — | 84.5 | 84.5 | — | — |
| 12.6 | — | — | — | 18.4 | — | — | — | 27.1 | — | 27 | 27 | 39.7 | — | — | — | 58.3 | — | — | — | 85.6 | — | — | — |
| 12.7 | 12.7 | — | — | 18.7 | 18.7 | — | — | 27.4 | 27.4 | — | — | 40.2 | 40.2 | — | — | 59.0 | 59.0 | — | — | 86.6 | 86.6 | — | — |
| 12.9 | — | — | — | 18.9 | — | — | — | 27.7 | — | — | — | 40.7 | — | — | — | 59.7 | — | — | — | 87.6 | — | — | — |
| 13.0 | 13.0 | 13 | — | 19.1 | 19.1 | — | — | 28.0 | 28.0 | — | — | 41.2 | 41.2 | — | — | 60.4 | 60.4 | — | — | 88.7 | 88.7 | — | — |
| 13.2 | — | — | — | 19.3 | — | — | — | 28.4 | — | — | — | 41.7 | — | — | — | 61.2 | — | — | — | 89.8 | — | — | — |
| 13.3 | 13.3 | — | — | 19.6 | 19.6 | — | — | 28.7 | 28.7 | — | — | 42.2 | 42.2 | — | — | 61.9 | 61.9 | 62 | — | 90.9 | 90.9 | 91 | — |
| 13.5 | — | — | — | 19.8 | — | — | — | 29.1 | — | — | — | 42.7 | — | — | — | 62.6 | — | — | — | 92.0 | — | — | — |
| 13.7 | 13.7 | — | — | 20.0 | 20.0 | 20 | — | 29.4 | 29.4 | — | — | 43.2 | 43.2 | 43 | — | 63.4 | 63.4 | — | — | 93.1 | 93.1 | — | — |
| 13.8 | — | — | — | 20.3 | — | — | — | 29.8 | — | — | — | 43.7 | — | — | — | 64.2 | — | — | — | 94.2 | — | — | — |
| 14.0 | 14.0 | — | — | 20.5 | 20.5 | — | — | 30.1 | 30.1 | 30 | — | 44.2 | 44.2 | — | — | 64.9 | 64.9 | — | — | 95.3 | 95.3 | — | — |
| 14.2 | — | — | — | 20.8 | — | — | — | 30.5 | — | — | — | 44.8 | — | — | — | 65.7 | — | — | — | 96.5 | — | — | — |
| 14.3 | 14.3 | — | — | 21.0 | 21.0 | — | — | 30.9 | 30.9 | — | — | 45.3 | 45.3 | — | — | 66.5 | 66.5 | — | — | 97.6 | 97.6 | — | — |
| 14.5 | — | — | — | 21.3 | — | — | — | 31.2 | — | — | — | 45.9 | — | — | — | 67.3 | — | — | — | 98.8 | — | — | — |

Note: These values are generally available in multiples of 0.1, 1, 10, 100, 1k, and 1M.

# B

# BATTERIES

Batteries are an important source of dc voltage. They are available in two basic categories: the wet cell and the dry cell. A battery generally is made up of several individual cells.

A cell consists basically of two electrodes immersed in an electrolyte. A voltage is developed between the electrodes as a result of the chemical action between the electrodes and the electrolyte. The electrodes typically are two dissimilar metals, and the electrolyte is a chemical solution.

## Simple Wet Cell

Figure B–1 shows a simple copper-zinc (Cu-Zn) chemical cell. One electrode is made of copper, the other of zinc. These electrodes are immersed in a solution of water and hydrochloric acid (HCl), which is the electrolyte.

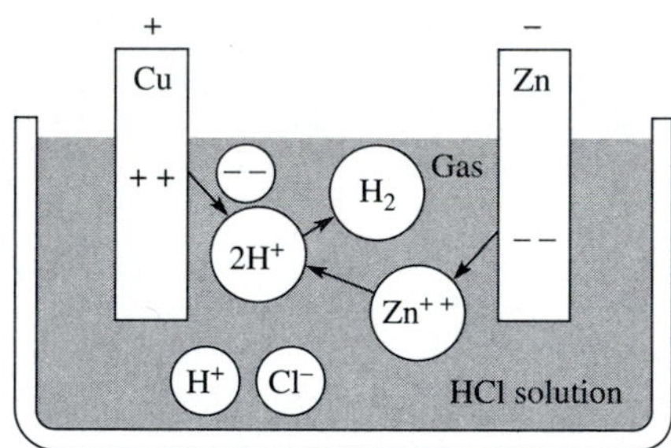


**FIGURE B–1**
***Simple chemical cell.***

Positive hydrogen ions ($H^+$) and negative chlorine ions ($Cl^-$) are formed when the HCl ionizes in the water. Since zinc is more active than hydrogen, zinc atoms leave the zinc electrode and form zinc ions ($Zn^{++}$) in the solution. When a zinc ion is formed, two excess electrons are left on the zinc electrode, and two hydrogen ions are displaced from the solution. These two hydrogen ions will migrate to the copper electrode, take two electrons from the copper, and form a molecule of hydrogen gas ($H_2$). As a result of this reaction, a negative charge develops on the zinc electrode, and a positive charge develops on the copper electrode, creating a potential difference or voltage between the two electrodes.

In this copper-zinc cell, the hydrogen gas given off at the copper electrode tends to form a layer of bubbles around the electrodes, insulating the copper from the electrolyte. This effect, called *polarization,* results in a reduction in the voltage produced by the cell. Polarization can be remedied by the addition of an agent to the electrolyte to remove hydrogen gas or by the use of an electrolyte that does not form hydrogen gas.

***Lead-Acid Cell*** The positive electrode of a lead-acid cell is lead peroxide ($PbO_2$), and the negative electrode is spongy lead (Pb). The electrolyte is sulfuric acid ($H_2SO_4$) in water. Thus, the lead-acid cell is classified as a wet cell.

Two positive hydrogen ions ($2H^+$) and one negative sulfate ion ($SO_4^{--}$) are formed when the sulfuric acid ionizes in the water. Lead ions ($Pb^{++}$) from both electrodes displace the hydrogen ions in the electrolyte solution. When the lead ion from the spongy lead electrode enters the solution, it combines with a sulfate ion ($SO_4^{--}$) to form lead sulfate ($PbSO_4$), and it leaves two excess electrons on the electrode.

When a lead ion from the lead peroxide electrode enters the solution, it also leaves two excess electrons on the electrode and forms lead sulfate in the solution. However, because this electrode is lead peroxide, two free oxygen atoms are created when a lead atom leaves and enters the solution as a lead ion. These two oxygen atoms take four electrons from the lead peroxide electrode and become oxygen ions ($O^{--}$). This process creates a deficiency of two electrons on this electrode (there were initially two excess electrons).

The two oxygen ions ($2O^{--}$) combine in the solution with four hydrogen ions ($4H^+$) to produce two molecules of water ($2H_2O$). This process dilutes the electrolyte over a period of time. Also, there is a buildup of lead sulfate on the electrodes. These two factors result in a reduction in the voltage produced by the cell and necessitate periodic recharging.

As you have seen, for each departing lead ion, there is an excess of two electrons on the spongy lead electrode, and there is a deficiency of two electrons on the lead peroxide electrode. Therefore, the lead peroxide electrode is positive, and the spongy lead electrode is negative. This chemical reaction is pictured in Figure B–2.

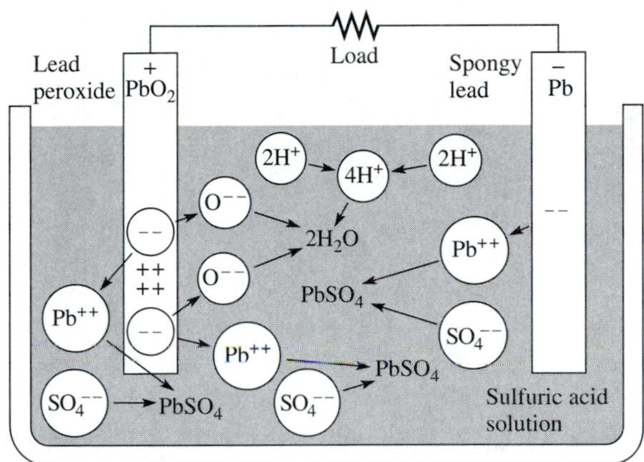


**FIGURE B–2**
*Chemical reaction in a discharging lead-acid cell.*

As mentioned, the dilution of the electrolyte by the formation of water and lead sulfate requires that the lead-acid cell be recharged to reverse the chemical process. A chemical cell that can be recharged is called a *secondary cell.* One that cannot be recharged is called a *primary cell.*

The cell is recharged by the connection of an external voltage source to the electrodes, as shown in Figure B–3. The formula for the chemical reaction in a lead-acid cell is

$$Pb + PbO_2 + 2H_2SO_4 \rightarrow 2PbSO_4 + 2H_2O$$

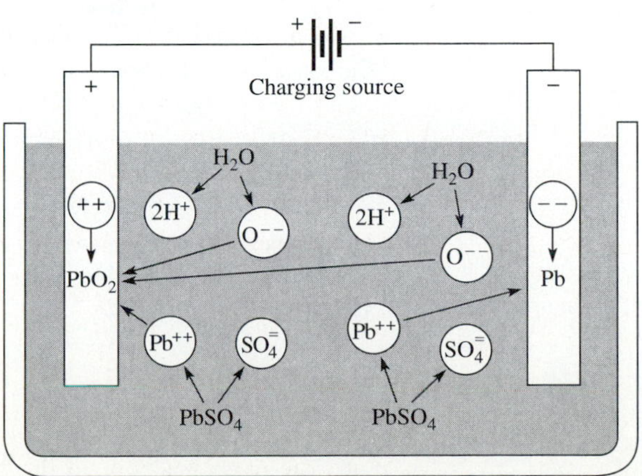


**FIGURE B–3**
*Recharging a lead-acid cell.*

## Dry Cell

In a dry cell, some of the disadvantages of a liquid electrolyte are overcome. Actually, the electrolyte in a typical dry cell is not dry but rather is in the form of a moist paste. This electrolyte is a combination of granulated carbon, powdered manganese dioxide, and ammonium chloride solution.

A typical carbon-zinc dry cell is illustrated in Figure B–4. The zinc container or can is dissolved by the electrolyte. As a result of this reaction, an excess of electrons accumulates on the container, making it the negative electrode.

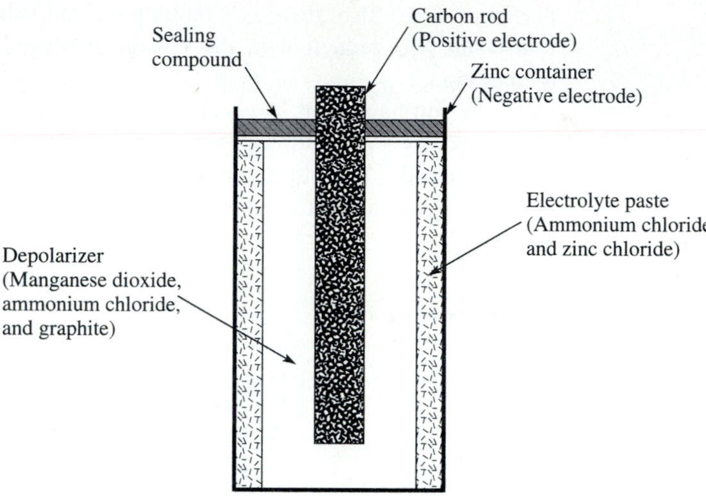


**FIGURE B–4**
*Simplified construction of a dry cell.*

The hydrogen ions in the electrolyte take electrons from the carbon rod, making it the positive electrode. Hydrogen gas is formed near the carbon electrode, but this gas is eliminated by reaction with manganese dioxide (called a *depolarizing agent*). This depolarization prevents bursting of the container due to gas formation. Because the chemical reaction is not reversible, the carbon-zinc cell is a primary cell.

## Types of Chemical Cells

Although only two common types of battery cells have been discussed, there are several types, listed in Table B–1.

**TABLE B–1**
*Types of battery cells*

| Type | + electrode | – electrode | Electrolyte | Volts | Comments |
|---|---|---|---|---|---|
| Carbon-zinc | Carbon | Zinc | Ammonium and zinc chloride | 1.5 | Dry, primary |
| Lead-acid | Lead peroxide | Spongy lead | Sulfuric acid | 2.0 | Wet, secondary |
| Manganese-alkaline | Manganese dioxide | Zinc | Potassium hydroxide | 1.5 | Dry, primary or secondary |
| Mercury | Zinc | Mercuric oxide | Potassium hydroxide | 1.3 | Dry, primary |
| Nickel-cadmium | Nickel | Cadmium hydroxide | Potassium hydroxide | 1.25 | Dry, secondary |
| Nickel-iron (Edison cell) | Nickel oxide | Iron | Potassium hydroxide | 1.36 | Wet, secondary |

# C DERIVATIONS

## Equation (11–6) RMS (Effective) Value of a Sine Wave

The abbreviation "rms" stands for the root mean square process by which this value is derived. In the process, we first square the equation of a sine wave.

$$v^2 = V_p^2 \sin^2\theta$$

Next, we obtain the mean or average value of $v^2$ by dividing the area under a half-cycle of the curve by $\pi$ (see Figure C–1). The area is found by integration and trigonometric identities.

$$V_{\text{avg}}^2 = \frac{\text{area}}{\pi} = \frac{1}{\pi}\int_0^{\pi} V_p^2 \sin^2\theta \, d\theta$$

$$= \frac{V_p^2}{2\pi}\int_0^{\pi}(1 - \cos 2\theta)d\theta = \frac{V_p^2}{2\pi}\int_0^{\pi} 1 \, d\theta - \frac{V_p^2}{2\pi}\int_0^{\pi}(-\cos 2\theta)\, d\theta$$

$$= \frac{V_p^2}{2\pi}(\theta - \tfrac{1}{2}\sin 2\theta)_0^{\pi} = \frac{V_p^2}{2\pi}(\pi - 0) = \frac{V_p^2}{2}$$

Finally, the square root of $V_{\text{avg}}^2$ is $V_{\text{rms}}$.

$$V_{\text{rms}} = \sqrt{V_{\text{avg}}^2} = \sqrt{V_p^2/2} = \frac{V_p}{\sqrt{2}} = 0.707V_p$$

FIGURE C–1

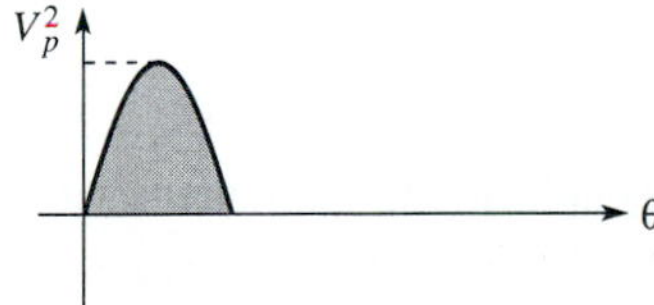


## Equation (11–12) Average Value of a Half-Cycle Sine Wave

The average value of a sine wave is determined for a half-cycle because the average over a full cycle is zero.

The equation for a sine wave is

$$v = V_p \sin \theta$$

The average value of the half-cycle is the area under the curve divided by the distance of the curve along the horizontal axis (see Figure C–2).

$$V_{\text{avg}} = \frac{\text{area}}{\pi}$$

FIGURE C–2

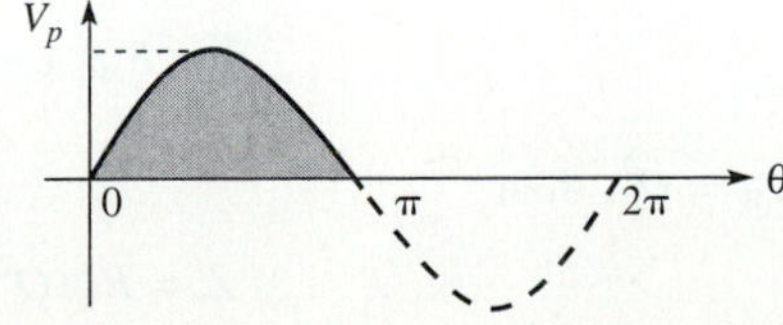

To find the area, we use integral calculus.

$$V_{avg} = \frac{1}{\pi}\int_0^{\pi} V_p \sin\theta \, d\theta = \frac{V_p}{\pi}(-\cos\theta)\Big|_0^{\pi}$$
$$= \frac{V_p}{\pi}[-\cos\pi - (-\cos 0)] = \frac{V_p}{\pi}[-(-1) - (-1)]$$
$$= \frac{V_p}{\pi}(2) = \frac{2}{\pi}V_p = 0.637V_p$$

## Equations (13–26) and (14–15) Reactance Derivations

### *Derivation of Capacitive Reactance*

$$\theta = 2\pi ft = \omega t$$
$$i = C\frac{dv}{dt} = C\frac{d(V_p \sin\theta)}{dt} = C\frac{d(V_p \sin\omega t)}{dt} = \omega C(V_p \cos\omega t)$$
$$I_{rms} = \omega C V_{rms}$$
$$X_C = \frac{V_{rms}}{I_{rms}} = \frac{V_{rms}}{\omega C V_{rms}} = \frac{1}{\omega C} = \frac{1}{2\pi fC}$$

### *Derivation of Inductive Reactance*

$$v = L\frac{di}{dt} = L\frac{d(I_p \sin\omega t)}{dt} = \omega L(I_p \cos\omega t)$$
$$V_{rms} = \omega L I_{rms}$$
$$X_L = \frac{V_{rms}}{I_{rms}} = \frac{\omega L I_{rms}}{I_{rms}} = \omega L = 2\pi fL$$

### *Equation (18–16) Impedance of Nonideal Tank Circuit at Resonance*

$$\frac{1}{\mathbf{Z}} = \frac{1}{-jX_C} + \frac{1}{R_W + jX_L}$$
$$= j\left(\frac{1}{X_C}\right) + \frac{R_W - jX_L}{(R_W + jX_L)(R_W - jX_L)} = j\left(\frac{1}{X_C}\right) + \frac{R_W - jX_L}{R_W^2 + X_L^2}$$

The first term plus splitting the numerator of the second term yields

$$\frac{1}{\mathbf{Z}} = j\left(\frac{1}{X_C}\right) - j\left(\frac{X_L}{R_W^2 + X_L^2}\right) + \frac{R_W}{R_W^2 + X_L^2}$$

At resonance, $\mathbf{Z}$ is purely resistive; so it has no $j$ part (the $j$ terms in the last expression cancel). Thus, only the real part is left, as stated in the following equation for $Z$ at resonance:

$$Z_r = \frac{R_W^2 + X_L^2}{R_W}$$

Splitting the denominator, we get

$$Z_r = \frac{R_W^2}{R_W} + \frac{X_L^2}{R_W} = R_W + \frac{X_L^2}{R_W}$$

Factoring out $R_W$ gives

$$Z_r = R_W\left(1 + \frac{X_L^2}{R_W^2}\right)$$

Since $X_L^2/R_W^2 = Q^2$, then

$$Z_r = R_W(Q^2 + 1)$$

## Equation (18–19) Resonant Frequency for a Nonideal Parallel Resonant Circuit

The $j$ terms that were part of the derivation of Equation (18–16) are equal.

$$\frac{1}{X_C} = \frac{X_L}{R_W^2 + X_L^2}$$

Thus,

$$R_W^2 = X_L^2 = X_L X_C$$

$$R_W^2 + (2\pi f_r L)^2 = \frac{2\pi f_r L}{2\pi f_r C}$$

$$R_W^2 + 4\pi^2 f_r^2 L^2 = \frac{L}{C}$$

$$4\pi^2 f_r^2 L^2 = \frac{L}{C} - R_W^2$$

Solving for $f_r^2$,

$$f_r^2 = \frac{\left(\frac{L}{C}\right) - R_W^2}{4\pi^2 L^2}$$

Multiply both numerator and denominator by $C$,

$$f_r^2 = \frac{L - R_W^2 C}{4\pi^2 L^2 C} = \frac{L - R_W^2 C}{L(4\pi^2 LC)}$$

Factoring an $L$ out of the numerator and canceling gives

$$f_r^2 = \frac{1 - (R_W^2 C/L)}{4\pi^2 LC}$$

Taking the square root of both sides yields $f_r$,

$$f_r = \frac{\sqrt{1 - (R_W^2 C/L)}}{2\pi\sqrt{LC}}$$

# D CAPACITOR COLOR CODING

Some capacitors have color-coded designations. The color code used for capacitors is basically the same as that used for resistors. Some variations occur in tolerance designation. The basic color codes are shown in Table D–1, and some typical color-coded capacitors are illustrated in Figure D–1.

**TABLE D–1**
*Typical composite color codes for capacitors (picofarads)*

| Color | Digit | Multiplier | Tolerance |
|---|---|---|---|
| Black | 0 | 1 | 20% |
| Brown | 1 | 10 | 1% |
| Red | 2 | 100 | 2% |
| Orange | 3 | 1000 | 3% |
| Yellow | 4 | 10000 | |
| Green | 5 | 100000 | 5% (EIA) |
| Blue | 6 | 1000000 | |
| Violet | 7 | | |
| Gray | 8 | | |
| White | 9 | | |
| Gold | | 0.1 | 5% (JAN) |
| Silver | | 0.01 | 10% |
| No color | | | 20% |

NOTE: EIA stands for Electronic Industries Association, and JAN stands for Joint Army-Navy, a military standard.

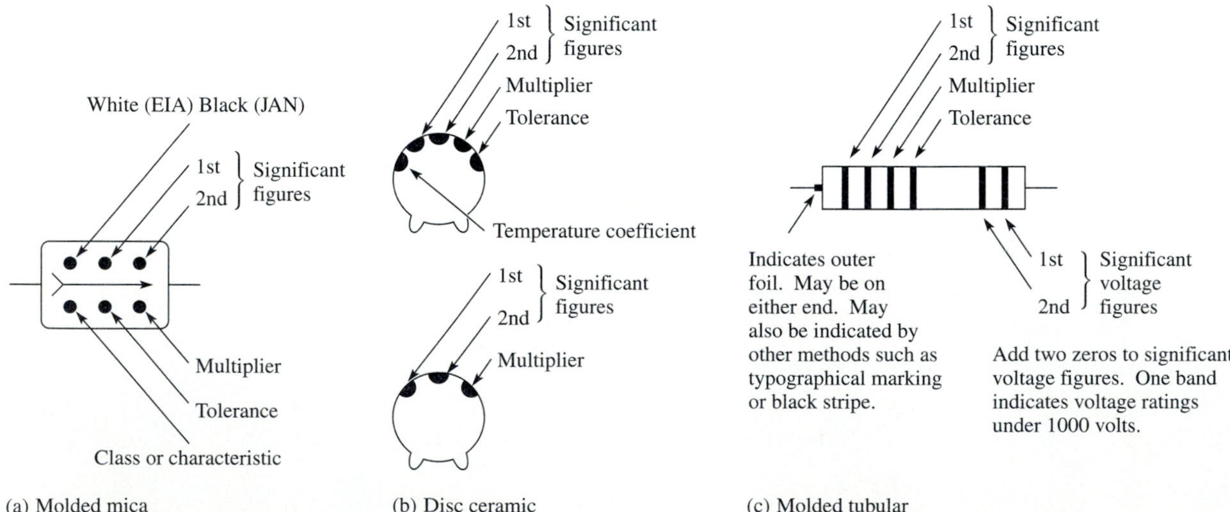


**FIGURE D–1**
*Typical color-coded capacitors.*

# ANSWERS TO SELF-TESTS

### Chapter 1

**1.** (c) **2.** (c) **3.** (c) **4.** (d) **5.** (b) **6.** (b) **7.** (a) **8.** (d)
**9.** (d) **10.** (d) **11.** (c) **12.** (c) **13.** (c)

### Chapter 2

**1.** (b) **2.** (a) **3.** (c) **4.** (c) **5.** (b) **6.** (b) **7.** (c) **8.** (d)
**9.** (b) **10.** (b) **11.** (e) **12.** (b) **13.** (b) **14.** (c)

### Chapter 3

**1.** (b) **2.** (c) **3.** (b) **4.** (d) **5.** (a) **6.** (d) **7.** (b) **8.** (d)
**9.** (c) **10.** (b)

### Chapter 4

**1.** (c) **2.** (c) **3.** (a) **4.** (d) **5.** (b) **6.** (d) **7.** (a) **8.** (a)
**9.** (d) **10.** (d) **11.** (b) **12.** (c) **13.** (c) **14.** (a) **15.** (c) **16.** (a)
**17.** (b) **18.** (c)

### Chapter 5

**1.** (a) **2.** (d) **3.** (b) **4.** (d) **5.** (d) **6.** (a) **7.** (b) **8.** (c)
**9.** (b) **10.** (c) **11.** (a) **12.** (d) **13.** (d) **14.** (d)

### Chapter 6

**1.** (b) **2.** (c) **3.** (b) **4.** (d) **5.** (a) **6.** (c) **7.** (c) **8.** (c)
**9.** (d) **10.** (b) **11.** (a) **12.** (d) **13.** (c) **14.** (b)

### Chapter 7

**1.** (e) **2.** (c) **3.** (c) **4.** (c) **5.** (b) **6.** (a) **7.** (b) **8.** (b)
**9.** (d) **10.** (b) **11.** (b) **12.** (b) **13.** (b) **14.** (a) **15.** (d)

### Chapter 8

**1.** (b) **2.** (c) **3.** (a) **4.** (b) **5.** (d) **6.** (c) **7.** (b) **8.** (d)
**9.** (c) **10.** (d) **11.** (b)

### Chapter 9

**1.** (e) **2.** (d) **3.** (b) **4.** (c) **5.** (c) **6.** (b) **7.** (d) **8.** (a)
**9.** (e) **10.** (b)

### Chapter 10

**1.** (b) **2.** (c) **3.** (a) **4.** (d) **5.** (b) **6.** (c) **7.** (a) **8.** (d)
**9.** (c) **10.** (c) **11.** (b) **12.** (c)

### Chapter 11

**1.** (b) **2.** (b) **3.** (c) **4.** (b) **5.** (d) **6.** (a) **7.** (a) **8.** (a)
**9.** (c) **10.** (b) **11.** (a) **12.** (d) **13.** (c) **14.** (b) **15.** (d)

### Chapter 12

**1.** (b) **2.** (b) **3.** (a) **4.** (d) **5.** (c) **6.** (c) **7.** (a) **8.** (b)
**9.** (c) **10.** (d)

## Chapter 13

**1.** (g) **2.** (b) **3.** (c) **4.** (d) **5.** (a) **6.** (d) **7.** (a) **8.** (f)
**9.** (c) **10.** (b) **11.** (d) **12.** (a) **13.** (b) **14.** (a) **15.** (b) **16.** (d)

## Chapter 14

**1.** (c) **2.** (d) **3.** (c) **4.** (b) **5.** (d) **6.** (a) **7.** (d) **8.** (c)
**9.** (b) **10.** (a) **11.** (d) **12.** (b)

## Chapter 15

**1.** (b) **2.** (c) **3.** (d) **4.** (a) **5.** (b) **6.** (c) **7.** (d) **8.** (b)
**9.** (a) **10.** (c) **11.** (d) **12.** (c) **13.** (a) **14.** (c)

## Chapter 16

**1.** (c) **2.** (b) **3.** (b) **4.** (a) **5.** (d) **6.** (b) **7.** (a) **8.** (d)
**9.** (c) **10.** (b) **11.** (d) **12.** (d) **13.** (c) **14.** (b)

## Chapter 17

**1.** (f) **2.** (b) **3.** (a) **4.** (d) **5.** (a) **6.** (d) **7.** (b) **8.** (b)
**9.** (a) **10.** (d) **11.** (c) **12.** (d) **13.** (b) **14.** (c)

## Chapter 18

**1.** (a) **2.** (c) **3.** (b) **4.** (c) **5.** (d) **6.** (c) **7.** (a) **8.** (b)
**9.** (d) **10.** (a) **11.** (d) **12.** (d)

## Chapter 19

**1.** (c) **2.** (b) **3.** (a) **4.** (b) **5.** (d) **6.** (c) **7.** (a) **8.** (c)
**9.** (b) **10.** (a) **11.** (c) **12.** (b) **13.** (c) **14.** (c)

## Chapter 20

**1.** (d) **2.** (c) **3.** (a) **4.** (a) **5.** (c) **6.** (c) **7.** (b) **8.** (c)
**9.** (d) **10.** (d)

## Chapter 21

**1.** (b) **2.** (c) **3.** (a) **4.** (d) **5.** (a) **6.** (a) **7.** (a) **8.** (c)
**9.** (b) **10.** (d)

## Chapter 22

**1.** (d) **2.** (c) **3.** (a) **4.** (a) **5.** (e) **6.** (c) **7.** (b) **8.** (c)
**9.** (d) **10.** (c)

# ANSWERS TO SELECTED ODD-NUMBERED PROBLEMS

## Chapter One

1. (a) $3 \times 10^3$ (b) $7.5 \times 10^4$ (c) $2 \times 10^6$
3. (a) $8.4 \times 10^3 = 0.84 \times 10^4 = 0.084 \times 10^5$
   (b) $99 \times 10^3 = 9.9 \times 10^4 = 0.99 \times 10^5$
   (c) $200 \times 10^3 = 20 \times 10^4 = 2 \times 10^5$
5. (a) 0.0000025 (b) 5000 (c) 0.39
7. (a) $126 \times 10^6$ (b) $855 \times 10^{-3}$
   (c) $606 \times 10^{-8}$
9. (a) $20 \times 10^8$ (b) $36 \times 10^{14}$
   (c) $15.4 \times 10^{-15}$
11. (a) $2370 \times 10^{-6}$ (b) $18.56 \times 10^{-6}$
    (c) $0.00574389 \times 10^{-6}$
    (d) $100{,}000{,}000{,}000 \times 10^{-6}$
13. (a) $12.25 \times 10^{14}$ (b) $5 \times 10^3$
    (c) $10.575 \times 10^{-6}$ (d) $2 \times 10^{10}$
15. (a) 3 $\mu$F (b) 3.3 M$\Omega$
    (c) 350 nA or 0.35 $\mu$A
17. (a) 24 $\mu$A (b) 9.7 M$\Omega$ (c) 3 pW
19. (a) 82 $\mu$W (b) 450 W
21. (a) 1000 $\mu$A (b) 50,000 mV (c) $2 \times 10^{-5}$ M$\Omega$
    (d) $1.55 \times 10^{-4}$ kW

## Chapter Two

1. $80 \times 10^{12}$ C
3. (a) 10 V (b) 2.5 V (c) 4 V
5. 20 V
7. 33.3 V
9. 0.2 A
11. 0.15 C
13. (a) 27 k$\Omega$ ± 5% (b) 1.8 k$\Omega$ ± 10%
15. 330 $\Omega$: orange, orange, brown
    2.2 k$\Omega$: red, red, red
    56 k$\Omega$: green, blue, orange
    100 k$\Omega$: brown, black, yellow
    39 k$\Omega$: orange, white, orange
17. (a) 200 mS (b) 40 mS (c) 10 mS
19. AWG #27
21. Through lamp 2
23. Circuit (b)
25. See Figure P–1.

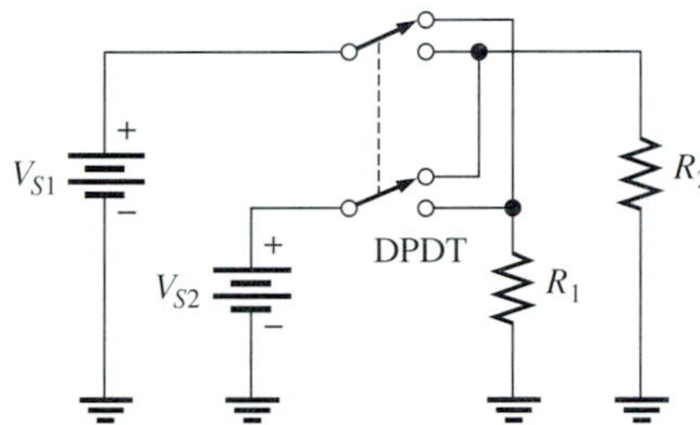


FIGURE P–1

27. See Figure P–2.

FIGURE P–2

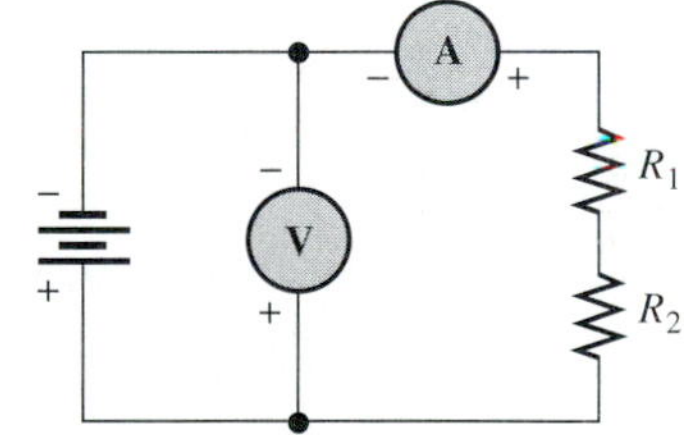


29. Position 1: V1 = 0 V, V2 = $V_S$
    Position 2: V1 = $V_S$, V2 = 0 V
31. See Figure P–3.

FIGURE P–3

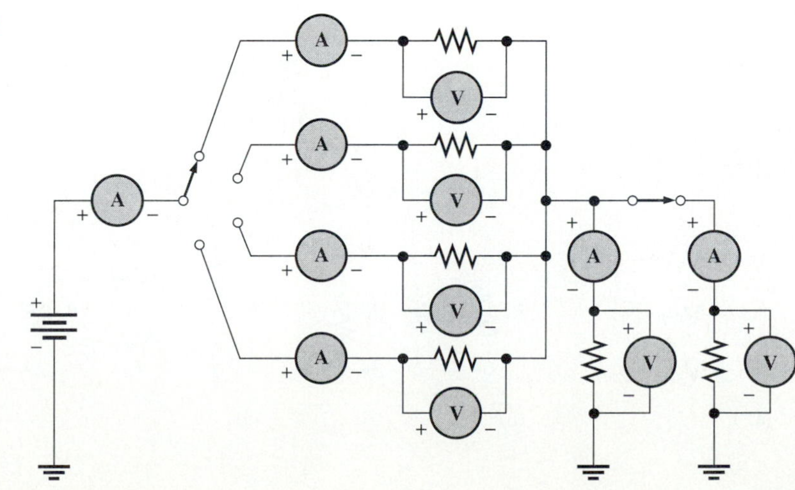

**33.** **(a)** 0.25 V **(b)** 250 V
**35.** **(a)** 200 Ω **(b)** 150 MΩ **(c)** 4500 Ω
**37.** See Figure P–4.

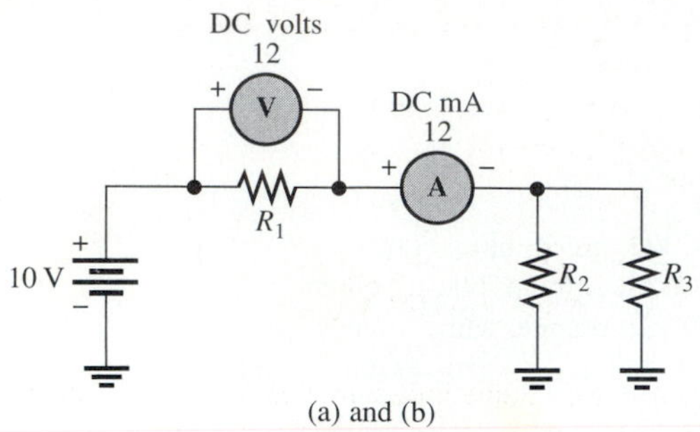


(a) and (b)

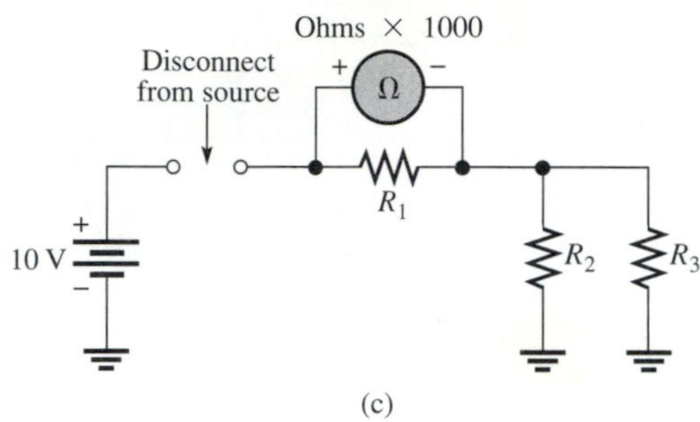


(c)

FIGURE P–4

## Chapter Three

**1.** **(a)** Current triples. **(b)** Current is reduced 75%. **(c)** Current is halved. **(d)** Current increases 35%. **(e)** Current quadruples. **(f)** Current is unchanged.
**3.** $V = IR$
**5.** **(a)** 5 A **(b)** 1.5 A **(c)** 500 mA **(d)** 2 mA **(e)** 44.6 μA
**7.** 1.2 A
**9.** 532 μA
**11.** **(a)** 36 V **(b)** 280 V **(c)** 1700 V **(d)** 28.2 V **(e)** 56 V
**13.** 81 V
**15.** **(a)** 59.9 mA **(b)** 5.99 V **(c)** 4.59 mV
**17.** **(a)** 2 kΩ **(b)** 3.5 kΩ **(c)** 2 kΩ **(d)** 100 kΩ **(e)** 1 MΩ
**19.** 150 Ω
**21.** 133 Ω, 100 Ω
**23.** The graph is a straight line, indicating a linear relationship between $V$ and $I$.
**25.** $R_1 = 0.5\ \Omega$, $R_2 = 1\ \Omega$, $R_3 = 2\ \Omega$

## Chapter Four

**1.** 350 W
**3.** 20 kW
**5.** **(a)** 1 MW **(b)** 3 MW **(c)** 150 MW **(d)** 8.7 MW
**7.** **(a)** 2,000,000 μW **(b)** 500 μW **(c)** 250 μW **(d)** 6.67 μW
**9.** 8640 J
**11.** 2.02 kW/day
**13.** 0.00186 kWh
**15.** 37.5 Ω
**17.** 360 W
**19.** 100 μW
**21.** 40.2 mW
**23.** **(a)** 0.480 Wh **(b)** Equal
**25.** At least 12 W, to allow a safety margin
**27.** 7.07 V
**29.** 8 A
**31.** 100 mW, 80%
**33.** 0.032 kWh

## Chapter Five

**1.** See Figure P–5.
**3.** $R_1, R_7, R_8, R_{10}$
$R_2, R_4, R_6, R_{11}$
$R_3, R_5, R_9, R_{12}$
**5.** 5 mA
**7.** See Figure P–6.
**9.** **(a)** 1560 Ω **(b)** 103 Ω **(c)** 13.7 kΩ **(d)** 3.671 MΩ
**11.** 67.2 kΩ
**13.** 3.9 kΩ
**15.** 17.8 MΩ
**17.** **(a)** 625 μA **(b)** 4.26 μA
**19.** 355 mA
**21.** $R_1 = 330\ \Omega$, $R_2 = 220\ \Omega$, $R_3 = 100\ \Omega$, $R_4 = 470\ \Omega$
**23.** Position $A$: 5.45 mA
Position $B$: 6.06 mA
Position $C$: 7.95 mA
Position $D$: 12 mA
**25.** 14 V
**27.** **(a)** 23 V **(b)** 35 V **(c)** 0 V
**29.** 4 V
**31.** 22 Ω

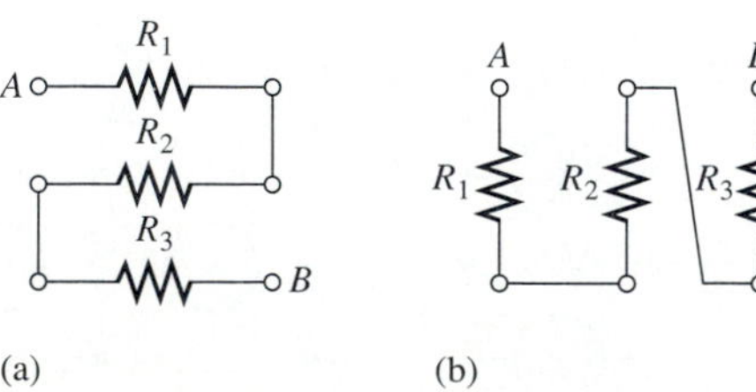


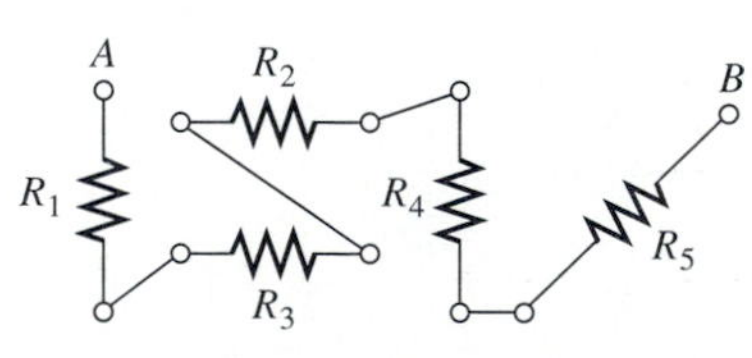


FIGURE P–5

FIGURE P–6

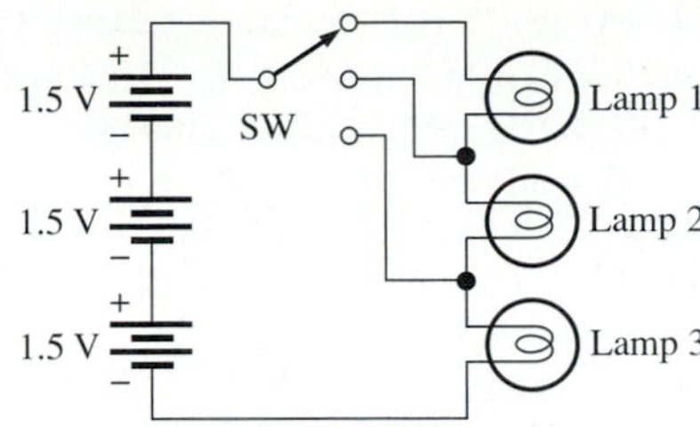


**33.** Position $A$: 4.0 V
Position $B$: 4.5 V
Position $C$: 5.4 V
Position $D$: 7.2 V

**35.** 4.82%

**37.** $V_R = 6$ V, $V_{2R} = 12$ V, $V_{3R} = 18$ V, $V_{4R} = 24$ V, $V_{5R} = 30$ V

**39.** $V_2 = 1.79$ V, $V_3 = 1$ V, $V_4 = 17.9$ V

**41.** See Figure P–7.

FIGURE P–7

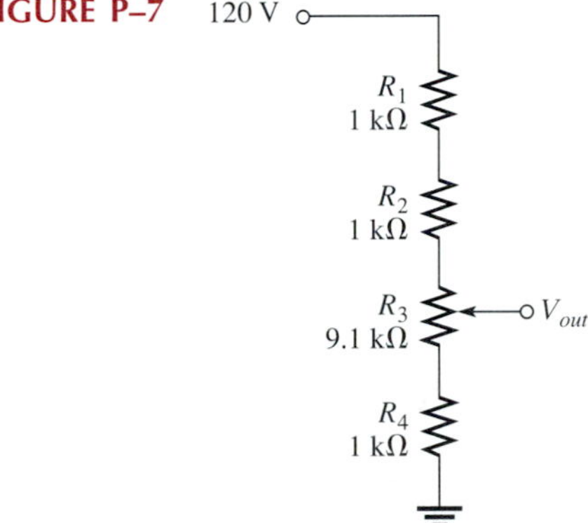


**43.** 54.9 mW

**45.** 12.5 MΩ

**47.** $V_{AG} = 100$ V, $V_{BG} = 57.7$ V, $V_{CG} = 15.2$ V, $V_{DG} = 7.58$ V

**49.** $V_{AG} = 14.82$ V, $V_{BG} = 12.97$ V, $V_{CG} = 12.64$ V, $V_{DG} = 9.34$ V

**51.** **(a)** $R_4$ is open.
**(b)** Short from $A$ to $B$

**53.** Table 5–1 is correct.

**55.** Yes. There is a short between pin 4 and the upper side of $R_{11}$.

## Chapter Six

**1.** See Figure P–8.

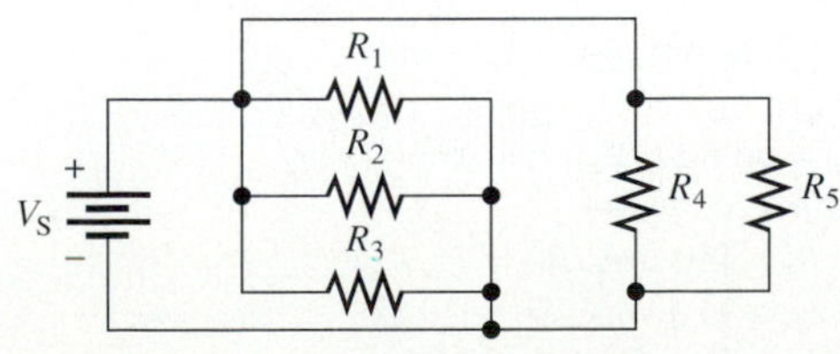


FIGURE P–8

**3.** $R_1, R_2, R_5, R_9, R_{10}, R_{12}$
$R_4, R_6, R_7, R_8$
$R_3, R_{11}$

**5.** 100 V

**7.** 1.35 A

**9.** $R_2 = 22\ \Omega$, $R_3 = 100\ \Omega$, $R_4 = 33\ \Omega$

**11.** 11.4 mA

**13.** **(a)** 360 Ω **(b)** 25.6 Ω
**(c)** 819 Ω **(d)** 997 Ω

**15.** 567 Ω

**17.** 2.46 Ω

**19.** **(a)** 510 kΩ **(b)** 245 kΩ
**(c)** 510 kΩ **(d)** 193 kΩ

**21.** 10 A

**23.** 50 mA; When one bulb burns out the others remain on.

**25.** 53.7 Ω

**27.** $I_2 = 167$ mA, $I_3 = 83.3$ mA, $I_T = 300$ mA, $R_1 = 2$ kΩ, $R_2 = 600\ \Omega$

**29.** Position $A$: 2.25 A
Position $B$: 4.75 A
Position $C$: 7 A

**31.** **(a)** $I_1 = 6.88\ \mu$A, $I_2 = 3.12\ \mu$A,
**(b)** $I_1 = 5.25$ mA, $I_2 = 2.39$ mA, $I_3 = 1.59$ mA, $I_4 = 772\ \mu$A

**33.** $R_1 = 3.3$ kΩ, $R_2 = 1.8$ kΩ, $R_3 = 5.6$ kΩ, $R_4 = 3.9$ kΩ

**35.** 0.05 Ω

**37.** **(a)** 68.8 μW **(b)** 52.5 mW

**39.** $P_1 = 1.25$ W, $I_2 = 75$ mA, $I_1 = 125$ A, $V_S = 10$ V, $R_1 = 80\ \Omega$, $R_2 = 133\ \Omega$

**41.** 682 mA, 3.41 A

**43.** The 8.2 kΩ resistor is open.

**45.** Connect ohmmeter between the following pins:

Pins 1-2
Correct reading: $R = 1\ \text{k}\Omega \parallel 3.3\ \text{k}\Omega = 767\ \Omega$
$R_1$ open: $R = 3.3$ kΩ
$R_2$ open: $R = 1$ kΩ

Pins 3-4
Correct reading: $R = 270\ \Omega \parallel 390\ \Omega = 159.5\ \Omega$
$R_3$ open: $R = 390\ \Omega$
$R_4$ open: $R = 270\ \Omega$

Pins 5-6
Correct reading: $R = 1\ \text{M}\Omega \parallel 1.8\ \text{M}\Omega \parallel 680\ \text{k}\Omega \parallel 510\ \text{k}\Omega = 201\ \text{k}\Omega$
$R_5$ open: $R = 1.8\ \text{M}\Omega \parallel 680\ \text{k}\Omega \parallel 510\ \text{k}\Omega = 251\ \text{k}\Omega$
$R_6$ open: $R = 1\ \text{M}\Omega \parallel 680\ \text{k}\Omega \parallel 510\ \text{k}\Omega = 227\ \text{k}\Omega$
$R_7$ open: $R = 1\ \text{M}\Omega \parallel 1.8\ \text{M}\Omega \parallel 510\ \text{k}\Omega = 284\ \text{k}\Omega$
$R_8$ open: $R = 1\ \text{M}\Omega \parallel 1.8\ \text{M}\Omega \parallel 680\ \text{k}\Omega = 330\ \text{k}\Omega$

**47.** Short between pins 3 and 4:
**(a)** $R_{1\text{-}2} = (R_1 \parallel R_2 \parallel R_3 \parallel R_4 \parallel R_{11} \parallel R_{12}) + (R_5 \parallel R_6 \parallel R_7 \parallel R_8 \parallel R_9 \parallel R_{10}) = 940\ \Omega$
**(b)** $R_{2\text{-}3} = R_5 \parallel R_6 \parallel R_7 \parallel R_8 \parallel R_9 \parallel R_{10} = 518\ \Omega$
**(c)** $R_{2\text{-}4} = R_5 \parallel R_6 \parallel R_7 \parallel R_8 \parallel R_9 \parallel R_{10} = 518\ \Omega$
**(d)** $R_{1\text{-}4} = R_1 \parallel R_2 \parallel R_3 \parallel R_4 \parallel R_{11} \parallel R_{12} = 422\ \Omega$

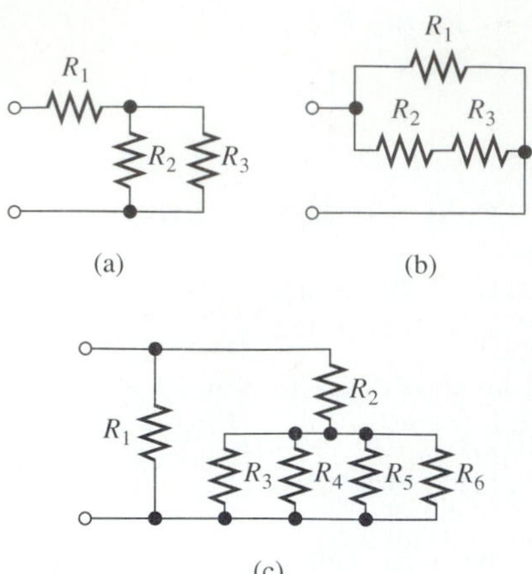


FIGURE P–9

## Chapter Seven

1. See Figure P–9.
3. (a) $R_1$ and $R_4$ are in series with the parallel combination of $R_2$ and $R_3$.
   (b) $R_1$ is in series with the parallel combination of $R_2$, $R_3$, and $R_4$.
   (c) The parallel combination of $R_2$ and $R_3$ is in series with the parallel combination of $R_4$ and $R_5$. This is all in parallel with $R_1$.
5. See Figure P–10.
7. See Figure P–11.

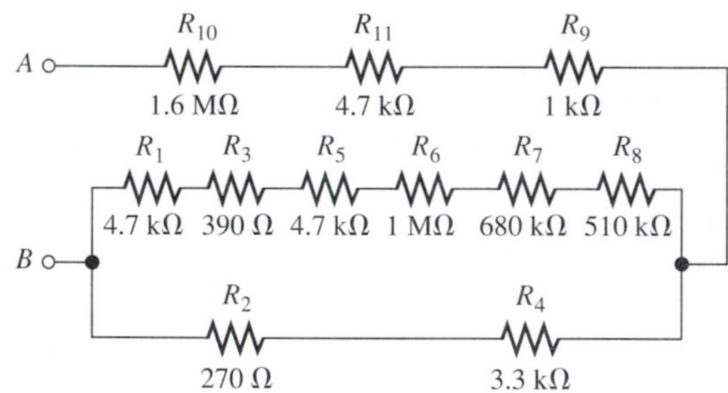


FIGURE P–10

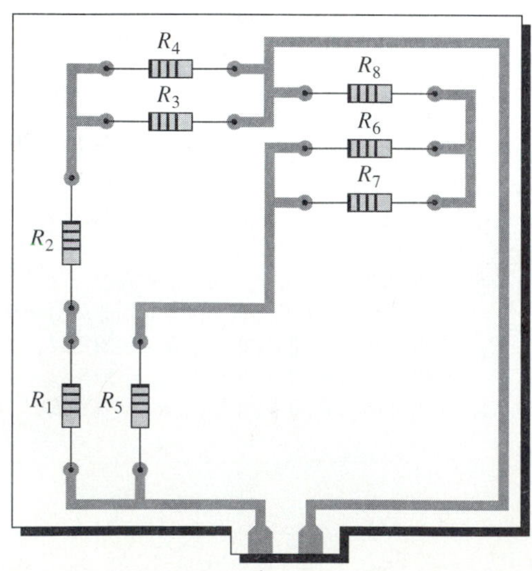


FIGURE P–11

9. (a) 133 Ω (b) 779 Ω (c) 852 Ω
11. (a) $I_1 = I_4 = 11.3$ mA, $I_2 = I_3 = 5.64$ mA, $V_1 = 633$ mV, $V_2 = V_3 = 564$ mV, $V_4 = 305$ mV
    (b) $I_1 = 3.85$ mA, $I_2 = 563$ μA, $I_3 = 1.16$ mA, $I_4 = 2.13$ mA, $V_1 = 2.62$ V, $V_2 = V_3 = V_4 = 383$ mV
    (c) $I_1 = 5$ mA, $I_2 = 303$ μA, $I_3 = 568$ μA, $I_4 = 313$ μA, $I_5 = 558$ μA, $V_1 = 5$ V, $V_2 = V_3 = 1.88$ V, $V_4 = V_5 = 3.13$ V
13. SW1 closed, SW2 open: 220 Ω
    SW1 closed, SW2 closed: 200 Ω
    SW1 open, SW2 open: 320 Ω
    SW1 open, SW2 closed: 300 Ω
15. $V_{AG} = 100$ V, $V_{BG} = 61.5$ V, $V_{CG} = 15.7$ V, $V_{DG} = 7.87$ V
17. Measure the voltage at $A$ with respect to ground and the voltage at $B$ with respect to ground. The difference is $V_{R2}$.
19. 110 Ω
21. $R_{AB} = 1.65$ kΩ
    $R_{BC} = 1.65$ kΩ
    $R_{CD} = 0$ Ω
23. 7.5 V unloaded, 7.29 V loaded
25. 47 kΩ
27. $R_1 = 1000$ Ω; $R_2 = R_3 = 500$ Ω;
    lower tap loaded: $V_{lower} = 1.82$ V, $V_{upper} = 4.55$ V
    upper tap loaded: $V_{lower} = 1.67$ V, $V_{upper} = 3.33$ V
29. (a) $V_G = 1.75$ V, $V_S = 0.25$ V
    (b) $I_1 = I_2 = 6.48$ μA, $I_D = I_S = 167$ μA
    (c) $V_{DS} = 7.92$ V, $V_{DG} = 6.42$ V
31. 1000 V
33. (a) 0.5 V range (b) Approximately 1 mV
35. (a) 271 Ω (b) 221 mA
    (c) 58.7 mA (d) 12 V
37. 621 Ω, $I_1 = I_9 = 16.1$ mA, $I_2 = 8.27$ mA, $I_3 = I_8 = 7.84$ mA, $I_4 = 4.06$ mA, $I_5 = I_6 = I_7 = 3.78$ mA
39. 971 mA
41. (a) 9 V (b) 3.75 V (c) 11.25 V
43. 2184 Ω
45. No, it should be 4.39 V.
47. The 2.2 kΩ resistor ($R_3$) is open.
49. The 3.3 kΩ resistor ($R_4$) is open.

## Chapter Eight

1. $I_S = 6$ A, $R_S = 50$ Ω
3. $V_S = 720$ V, $R_S = 1.2$ kΩ
5. 845 μA
7. 1.6 mA
9. 90.7 V
11. $I_{S1} = 2.28$ mA, $I_{S2} = 1.35$ mA
13. 116 μA
15. $R_{TH} = 88.6$ Ω, $V_{TH} = 1.09$ V
17. 100 μA
19. (a) $I_N = 110$ mA, $R_N = 76.7$ Ω
    (b) $I_N = 11.1$ mA, $R_N = 73$ Ω
    (c) $I_N = 50$ μA, $R_N = 35.9$ kΩ
    (d) $I_N = 68.8$ mA, $R_N = 1.3$ kΩ

**21.** 17.9 V
**23.** $I_N = 953\ \mu A$, $R_N = 1175\ \Omega$
**25.** $R_{EQ} = 56.9\ \Omega$, $V_{EQ} = -2.74$ V
**27.** SW1, SW2 closed: $I_L = 448\ \mu A$
SW1, SW3 closed: $I_L = 506\ \mu A$
SW2, SW3 closed: $I_L = 315\ \mu A$
SW1, SW2, SW3 closed: $I_L = 459\ \mu A$
**29.** 11.1 Ω
**31.** $R_{TH} = 48\ \Omega$, $R_4 = 160\ \Omega$
**33.** **(a)** $R_A = 39.8\ \Omega$, $R_B = 73\ \Omega$, $R_C = 48.7\ \Omega$
**(b)** $R_A = 21.2$ kΩ, $R_B = 10.3$ kΩ, $R_C = 14.9$ kΩ

## Chapter Nine

**1.** Six possible loops
**3.** $I_1 - I_2 - I_3 = 0$
**5.** $V_1 = 5.66$ V, $V_2 = 6.33$ V, $V_3 = 325$ mV
**7.** $I_1 = 738$ mA, $I_2 = -527$ mA, $I_3 = -469$ mA
**9.** −1.84 V
**11.** $I_1 = 0$ A, $I_2 = 2$ A
**13.** **(a)** −16,470 **(b)** −1.59
**15.** $I_1 = 1.24$ A, $I_2 = 2.05$ A, $I_3 = 1.89$ A
**17.** $I_1 = -5.11$ mA, $I_2 = -3.52$ mA
**19.** $V_{1k} = 5.11$ V, $V_{560} = 890$ mV, $V_{820} = 2.89$ V
**21.** $I_1 = 15.6$ mA, $I_2 = -61.3$ mA, $I_3 = 61.5$ mA
**23.** −11.2 mV
**25.** 2.7 mA
**27.** $I_{R1} = 20.6$ mA, $I_{R3} = 193$ mA, $I_{R2} = -172$ mA
**29.** $V_A = 1.5$ V, $V_B = -5.65$ V
**31.** $I_{R1} = 193\ \mu A$, $I_{R2} = 370\ \mu A$, $I_{R3} = 179\ \mu A$, $I_{R4} = 328\ \mu A$, $I_{R5} = 1.46$ mA, $I_{R6} = 522\ \mu A$, $I_{R7} = 2.16$ mA, $I_{R8} = 1.64$ mA, $V_A = -3.70$ V, $V_B = -5.85$ V, $V_C = -15.7$ V

## Chapter Ten

**1.** Decreases
**3.** 37.5 μWb
**5.** 597
**7.** 150 At
**9.** **(a)** Electromagnetic field **(b)** Spring
**11.** Forces produced by the interaction of the electromagnetic field and the permanent magnetic field
**13.** Change the current.
**15.** Material A
**17.** 1 mA
**19.** Lenz's law defines the polarity of the induced voltage.
**21.** The commutator and brush arrangement electrically connects the loop to the external circuit.
**23.** Figure P–12.

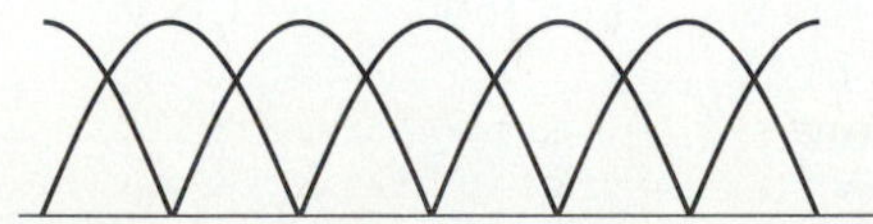

FIGURE P–12

## Chapter Eleven

**1.** **(a)** 1 Hz **(b)** 5 Hz
**(c)** 20 Hz **(d)** 1 kHz
**(e)** 2 kHz **(f)** 100 kHz
**3.** 2 μs
**5.** 250 Hz
**7.** 200 rps
**9.** **(a)** 7.07 mA **(b)** 0 A (full cycle), 4.5 mA (half-cycle) **(c)** 14.14 mA
**11.** **(a)** 0.524 or π/6 rad **(b)** 0.785 or π/4 rad
**(c)** 1.361 or 39π/90 rad **(d)** 2.356 or 3π/4 rad
**(e)** 3.491 or 10π/9 rad **(f)** 5.236 or 5π/3 rad
**13.** 15°, *A* leading
**15.** See Figure P–13.

FIGURE P–13

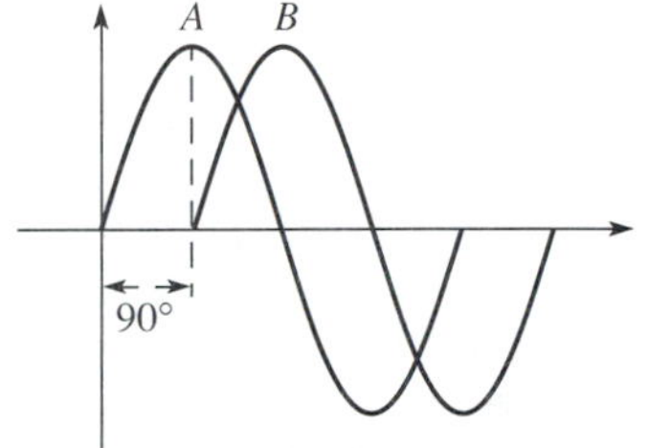


**17.** **(a)** 57.4 mA **(b)** 99.6 mA
**(c)** −17.4 mA **(d)** −57.4 mA
**(e)** −99.6 mA **(f)** 0 mA
**19.** 30°: 13.0 V
45°: 14.5 V
90°: 13.0 V
180°: −7.5 V
200°: −11.5 V
300°: −7.5 V
**21.** 22.1 V
**23.** $V_{R1(avg)} = 40.5$ V, $V_{R2(avg)} = 31.5$ V
**25.** $V_{max} = 39$ V, $V_{min} = 9$ V
**27.** −1 V
**29.** $t_r \cong 3.5$ ms, $t_f \cong 3.5$ ms, $t_W \cong 12.5$ ms, Ampl. ≅ 5 V
**31.** 5.84 V
**33.** **(a)** −0.375 V **(b)** 3.01 V
**35.** **(a)** 50 kHz **(b)** 10 Hz
**37.** 75 kHz, 125 kHz, 175 kHz, 225 kHz, 275 kHz, 325 kHz
**39.** $V_p = 600$ mV, $T = 500$ ms
**41.** Ampl = 1.4 V, $t_W = 120$ ms, $T = 400$ ms, %dc = 30%
**43.** $V_{p(in)} = 1.89$ V, $f_{in} = 2$ Hz

## Chapter Twelve

**1.** See Figure P–14.

FIGURE P–14

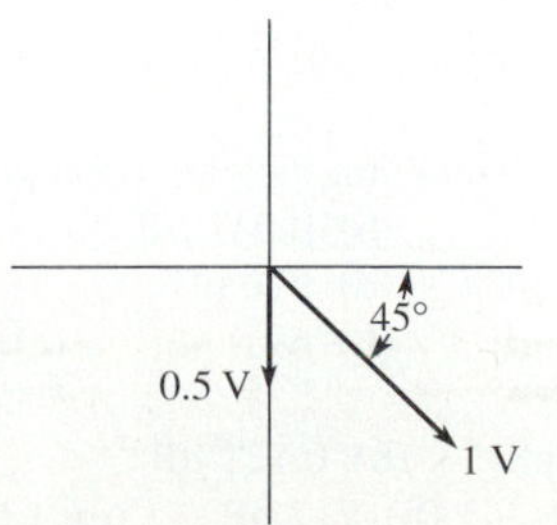

3. (a) 9.55 Hz (b) 57.3 Hz
(c) 0.318 Hz (d) 200 Hz
5. 54.5 $\mu$s
7. See Figure P–15.

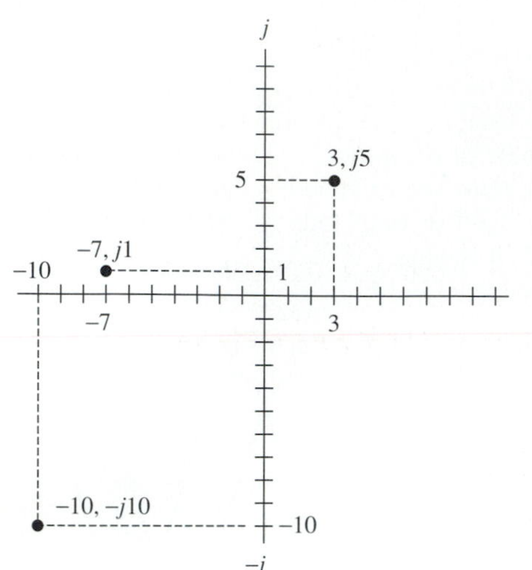


**FIGURE P–15**

9. (a) −5, +j3 and 5, −j3
(b) −1, −j7 and 1, +j7
(c) −10, +j10 and 10, −j10
11. 18.0
13. (a) 643 − j766 (b) −14.1 + j5.13
(c) −17.7 − j17.7 (d) −3 + j0
15. (a) Fourth (b) Fourth (c) Fourth (d) First
17. (a) $12\angle 115°$ (b) $20\angle 230°$
(c) $100\angle 190°$ (d) $50\angle 160°$
19. (a) 1.1 + j0.7 (b) −81 − j35
(c) 5.28 − j5.27 (d) −50.4 + j62.5
21. (a) $3.2\angle 11°$ (b) $7\angle -101°$
(c) $1.52\angle 70.6°$ (d) $2.79\angle -63.5°$
23. $9.87\angle -8.80°$ V, $1.97\angle -8.80°$ mA

## Chapter Thirteen

1. (a) 5 $\mu$F (b) 1 $\mu$C (c) 10 V
3. (a) 0.001 $\mu$F (b) 0.0035 $\mu$F
(c) 0.00025 $\mu$F
5. $2.3 \times 10^{-22}$ newton
7. (a) $8.85 \times 10^{-12}$ F/m
(b) $35.4 \times 10^{-12}$ F/m
(c) $66.4 \times 10^{-12}$ F/m
(d) $17.7 \times 10^{-12}$ F/m
9. 8.85 pF
11. 12.5 pF increase
13. Ceramic
15. Aluminum, tantalum; they are polarized.
17. (a) 0.02 $\mu$F (b) 0.047 $\mu$F
(c) 0.001 $\mu$F (d) 220 pF
19. (a) 0.667 $\mu$F (b) 69.0 pF (c) 2.70 $\mu$F
21. 2 $\mu$F
23. (a) 1057 pF (b) 0.121 $\mu$F
25. (a) 2.5 $\mu$F (b) 717 pF (c) 1.6 $\mu$F
27. $\Delta V_5 = +1.39$ V, $\Delta V_6 = -2.96$ V
29. (a) 14 ms (b) 247.5 $\mu$s
(c) 11 $\mu$s (d) 280 $\mu$s
31. (a) 9.20 V (b) 1.24 V
(c) 0.458 V (d) 0.168 V
33. (a) 17.9 V (b) 12.8 V (c) 6.59 V
35. 7.62 $\mu$s
37. 2.94 $\mu$s
39. See Figure P–16.

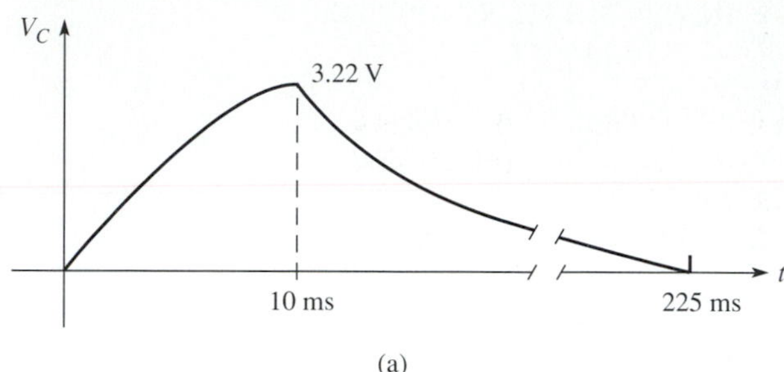


(a)

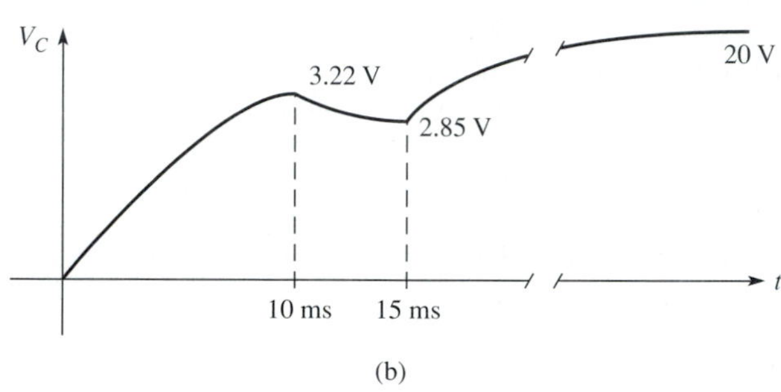


(b)

**FIGURE P–16**

41. (a) 31.8 Ω (b) 111 kΩ (c) 49.7 Ω
43. 200 Ω
45. 0 W, 3.39 mVAR
47. 0.00525 $\mu$F
49. $X_C$ of the bypass capacitor should ideally be 0 Ω.
51. The capacitor is shorted.

## Chapter Fourteen

1. (a) 1000 mH (b) 0.25 mH
(c) 0.01 mH (d) 0.5 mH
3. 50 mV
5. 20 mV
7. 1 A
9. 155 $\mu$H
11. 50.51 mH
13. 7.14 $\mu$H
15. (a) 4.33 H (b) 50 mH (c) 0.571 $\mu$H
17. (a) 1 $\mu$s (b) 2.13 $\mu$s (c) 2 $\mu$s
19. (a) 5.52 V (b) 2.03 V (c) 747 mV
(d) 275 mV (e) 101 mV
21. (a) 12.3 V (b) 9.10 V (c) 3.35 V
23. 11.0 $\mu$s
25. 3.18 ms
27. 240 V
29. (a) 144 Ω (b) 10.1 Ω (c) 13.4 Ω
31. (a) 55.5 Hz (b) 796 Hz (c) 597 Hz

**33.** $26.1\angle{-90°}$ mA

**35.** **(a)** Infinite resistance **(b)** Zero resistance
**(c)** Lower $R_W$

## Chapter Fifteen

**1.** 1.5 $\mu$H

**3.** 4; 0.25

**5.** **(a)** 100 V rms; in phase
**(b)** 100 V rms; out of phase
**(c)** 20 V rms; out of phase

**7.** 600 V

**9.** 0.25 (4:1)

**11.** 60 V

**13.** **(a)** 22.7 mA **(b)** 45.4 mA
**(c)** 15 V **(d)** 681 mW

**15.** 1.83

**17.** 9.76 W

**19.** **(a)** 6 V **(b)** 0 V **(c)** 40 V

**21.** 94.5 W

**23.** 0.98

**25.** 25 kVA

**27.** $V_1 = 11.5$ V, $V_2 = 23.0$ V, $V_3 = 23.0$ V, $V_4 = 46.0$ V

**29.** **(a)** 48 V **(b)** 25 V

**31.** **(a)** $\mathbf{V}_{RL} = 35\angle 0°$ V, $\mathbf{I}_{RL} = 2.92\angle 0°$ A,
$\mathbf{V}_C = 15\angle 0°$ V, $\mathbf{I}_C = 1.5\angle 90°$ A
**(b)** $34.4\angle{-12.5°}$ Ω

**33.** Excessive primary current is drawn, potentially burning out the source and/or the transformer unless the primary is protected by a fuse.

## Chapter Sixteen

**1.** 8 kHz, 8 kHz

**3.** **(a)** 270 Ω – $j$100 Ω, $288\angle{-20.3°}$ Ω
**(b)** 680 Ω – $j$1000 Ω, $1.21\angle{-55.8°}$ kΩ

**5.** **(a)** 56 kΩ – $j$796 kΩ
**(b)** 56 kΩ – $j$159 kΩ
**(c)** 56 kΩ – $j$79.6 kΩ
**(d)** 56 kΩ – $j$31.8 kΩ

**7.** **(a)** $R = 33$ Ω, $X_C = 50$ Ω
**(b)** $R = 272$ Ω, $X_C = 127$ Ω
**(c)** $R = 698$ Ω, $X_C = 1.66$ kΩ
**(d)** $R = 558$ Ω, $X_C = 558$ Ω

**9.** **(a)** $179\angle 58.2°$ $\mu$A **(b)** $625\angle 38.5°$ $\mu$A
**(c)** $1.98\angle 76.2°$ mA

**11.** 15.9°

**13.** **(a)** $97.3\angle{-54.9°}$ Ω **(b)** $103\angle 54.9°$ mA
**(c)** $5.76\angle 54.9°$ V **(d)** $8.18\angle{-35.1°}$ V

**15.** $R_X = 12$ Ω, $C_X = 13.3$ $\mu$F in series.

**17.** 261 Ω, –79.8°

**19.** $\mathbf{V}_C = \mathbf{V}_R = 10\angle 0°$ V
$\mathbf{I}_{tot} = 184\angle 37.1°$ mA
$\mathbf{I}_R = 147\angle 0°$ mA
$\mathbf{I}_C = 111\angle 90°$ mA

**21.** **(a)** $6.58\angle{-48.8°}$ Ω **(b)** $10\angle 0°$ mA
**(c)** $11.4\angle 90°$ mA **(d)** $15.2\angle 48.8°$ mA
**(e)** –48.8° ($I_{tot}$ leading $V_s$)

**23.** 18.2 kΩ resistor in series with 190 pF capacitor.

**25.** $\mathbf{V}_{C1} = 8.42\angle{-3.0°}$ V, $\mathbf{V}_{C2} = 1.50\angle{-58.9°}$ V
$\mathbf{V}_{C3} = 3.65\angle 6.8°$ V, $\mathbf{V}_{R1} = 3.32\angle 31.1°$ V
$\mathbf{V}_{R2} = 2.36\angle 6.8°$ V, $\mathbf{V}_{R3} = 1.29\angle 6.8°$ V

**27.** $\mathbf{I}_{tot} = 79.5\angle 87°$ mA, $\mathbf{I}_{C2R1} = 7.07\angle 31.1°$ mA
$\mathbf{I}_{C3} = 75.7\angle 96.8°$ mA, $\mathbf{I}_{R2R3} = 7.16\angle 6.8°$ mA

**29.** 0.1 $\mu$F

**31.** $\mathbf{I}_{C1} = \mathbf{I}_{R1} = 22.7\angle 74.5°$ mA
$\mathbf{I}_{R2} = 20.4\angle 72.0°$ mA
$\mathbf{I}_{R3} = 2.46\angle 84.3°$ mA
$\mathbf{I}_{R4} = 1.49\angle 41.2°$ mA
$\mathbf{I}_{R5} = 1.80\angle 75.1°$ mA
$\mathbf{I}_{R6} = \mathbf{I}_{C3} = 1.01\angle 135°$ mA
$\mathbf{I}_{C2} = 1.01\angle 131°$ mA

**33.** 4.03 VA

**35.** 0.909

**37.** **(a)** $I_{LA} = 4.8$ A, $I_{LB} = 3.33$ A
**(b)** $P_{rA} = 606$ VAR, $P_{rB} = 250$ VAR
**(c)** $P_{\text{true}A} = 979$ W, $P_{\text{true}B} = 759$ W
**(d)** $P_{aA} = 1151$ VA, $P_{aB} = 799$ VA
**(e)** Load $A$

**39.**

| | |
|---|---|
| 0 Hz | 1 V |
| 1 kHz | 706 mV |
| 2 kHz | 445 mV |
| 3 kHz | 315 mV |
| 4 kHz | 242 mV |
| 5 kHz | 195 mV |
| 6 kHz | 164 mV |
| 7 kHz | 141 mV |
| 8 kHz | 123 mV |
| 9 kHz | 110 mV |
| 10 kHz | 99.0 mV |

**41.**

| | |
|---|---|
| 0 Hz | 0 V |
| 1 kHz | 5.32 V |
| 2 kHz | 7.82 V |
| 3 kHz | 8.83 V |
| 4 kHz | 9.29 V |
| 5 kHz | 9.53 V |
| 6 kHz | 9.66 V |
| 7 kHz | 9.76 V |
| 8 kHz | 9.80 V |
| 9 kHz | 9.84 V |
| 10 kHz | 9.87 V |

**43.** 0.0796 $\mu$F

**45.** Reduces $V_{out}$ to 2.83 V and $\theta$ to –56.7°

**47.** **(a)** No output voltage
**(b)** $303\angle{-72.3°}$ mV
**(c)** $500\angle 0°$ mV
**(d)** 0 V

## Chapter Seventeen

**1.** 15 kHz

**3.** **(a)** 100 Ω + $j$50 Ω;
$112\angle 26.6°$ Ω
**(b)** 1.5 kΩ + $j$1 kΩ;
$1.80\angle 33.7°$ kΩ

**5.** **(a)** $17.4\angle 46.4°$ Ω
**(b)** $64.0\angle 79.2°$ Ω
**(c)** $127\angle 84.6°$ Ω
**(d)** $251\angle 87.3°$ Ω

**7.** 806 Ω, 4.11 mH

**9.** **(a)** $435\angle{-55°}$ mA
**(b)** $11.8\angle{-34.6°}$ mA

**11.** $\theta$ increases from 38.7° to 58.1°.

**13.** **(a)** $\mathbf{V}_R = 4.85\angle -14.1°$ V
$\mathbf{V}_L = 1.22\angle 75.9°$ V
**(b)** $\mathbf{V}_R = 3.83\angle -40.0°$ V
$\mathbf{V}_L = 3.21\angle 50.0°$ V
**(c)** $\mathbf{V}_R = 2.16\angle -64.5°$ V
$\mathbf{V}_L = 4.51\angle 25.5°$ V
**(d)** $\mathbf{V}_R = 1.16\angle -76.6°$ V
$\mathbf{V}_L = 4.86\angle 13.4°$ V
**15.** $7.75\angle 49.9°\ \Omega$
**17.** 2.39 kHz
**19.** **(a)** $274\angle 60.7°\ \Omega$
**(b)** $89.3\angle 0°$ mA
**(c)** $159\angle -90°$ mA
**(d)** $182\angle -60.7°$ mA
**(e)** 60.7° ($I_{tot}$ lagging $V_s$)
**21.** 1.83 kΩ resistor in series with 4.21 kΩ inductive reactance
**23.** $\mathbf{V}_{R1} = 18.6\angle -3.39°$ V
$\mathbf{V}_{R2} = 6.52\angle 9.71°$ V
$\mathbf{V}_{R3} = 2.81\angle -54.8°$ V
$\mathbf{V}_{L1} = \mathbf{V}_{L2} = 5.88\angle 35.2°$ V
**25.** $\mathbf{I}_{R1} = 372\angle -3.39°$ mA
$\mathbf{I}_{R2} = 326\angle 9.71°$ mA
$\mathbf{I}_{R3} = 93.7\angle -54.8°$ mA
$\mathbf{I}_{L1} = \mathbf{I}_{L2} = 46.8\angle -54.8°$ mA
**27.** **(a)** $588\angle -50.5°$ mA **(b)** $22.0\angle 16.1°$ V
**(c)** $8.63\angle -135°$ V
**29.** $\theta = 52.5°$ ($V_{out}$ lags $V_{in}$), 0.143
**31.** See Figure P–17.

**FIGURE P–17**

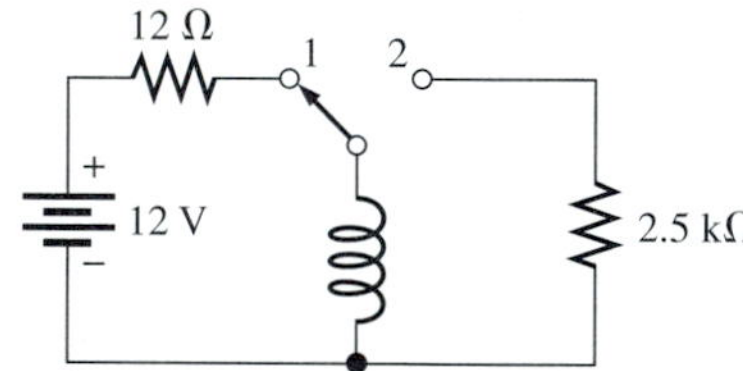


**33.** 1.29 W, 1.04 VAR
**35.** $P_{true} = 290$ mW; $P_r = 50.8$ mVAR;
$P_a = 296$ mV; $PF = 0.985$
**37.** **(a)** −0.0923° **(b)** −9.15°
**(c)** −58.2° **(d)** −86.4°
**39.** **(a)** 89.9° **(b)** 80.9°
**(c)** 31.8° **(d)** 3.60°
**41.** See Figure P–18.

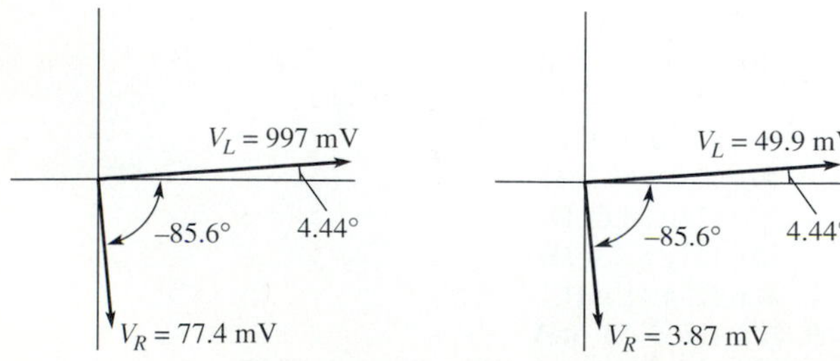


**FIGURE P–18**

**43.** **(a)** 0 V **(b)** 0 V **(c)** $1.62\angle -25.8°$ V
**(d)** $2.15\angle -64.5°$ V

## Chapter Eighteen

**1.** $480\angle -88.8°\ \Omega$; 480 Ω capacitive
**3.** Impedance increases to 150 Ω
**5.** $\mathbf{I}_{tot} = 61.4\angle -43.8°$ mA
$\mathbf{V}_R = 2.89\angle -43.8°$ V
$\mathbf{V}_L = 4.91\angle 46.2°$ V
$\mathbf{V}_C = 2.15\angle -134°$ V
**7.** **(a)** $35.8\angle 65.1°$ mA **(b)** 181 mW
**(c)** 390 mVAR **(d)** 430 mVA
**9.** $Z = 200\ \Omega$, $X_C = X_L = 2$ kΩ
**11.** 500 mA
**13.** See Figure P–19.

**FIGURE P–19**

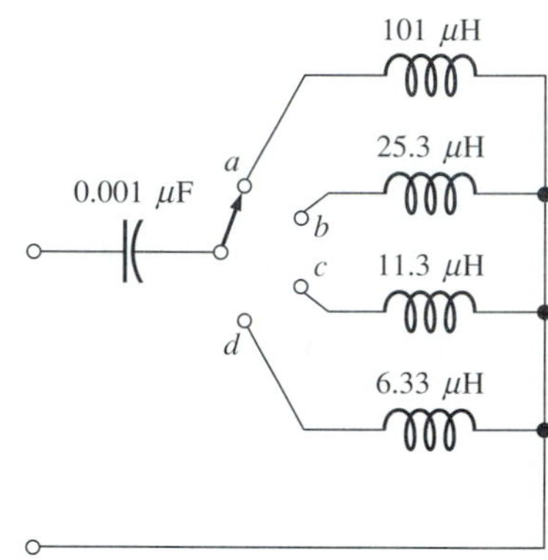


**15.** The phase angle of −4.43° indicates a slightly capacitive circuit.
**17.** $\mathbf{I}_R = 50\angle 0°$ mA
$\mathbf{I}_L = 4.42\angle -90°$ mA
$\mathbf{I}_C = 8.29\angle 90°$ mA
$\mathbf{I}_{tot} = 50.2\angle 4.43°$ mA
$\mathbf{V}_R = \mathbf{V}_L = \mathbf{V}_C = 5\angle 0°$ V
**19.** **(a)** −1.97° ($V_s$ lags $I_{tot}$)
**(b)** 23.0° ($V_s$ leads $I_{tot}$)
**21.** 49.0 kΩ resistor in series with 1.33 H inductor
**23.** 42.8° ($I_2$ leads $V_s$)
**25.** $\mathbf{I}_{R1} = \mathbf{I}_{C1} = 1.09\angle -25.7°$ mA
$\mathbf{I}_{R2} = 767\angle 19.3°\ \mu$A
$\mathbf{I}_{C2} = 767\angle 109.3°\ \mu$A
$\mathbf{I}_L = 1.53\angle -70.7°$ mA
$\mathbf{V}_{R2} = \mathbf{V}_{C2} = \mathbf{V}_L = 7.67\angle 19.3°$ V
$\mathbf{V}_{R1} = 3.60\angle -25.7°$ V
$\mathbf{V}_{C1} = 1.09\angle -116°$ V
**27.** $48.9\angle 131°$ mA
**29.** 486 MΩ, 104 kHz
**31.** $f_{r(series)} = 4.1$ kHz
$\mathbf{V}_{out} = 9.97\angle -1.65°$ V
$f_{r(parallel)} = 2.6$ kHz
$\mathbf{V}_{out} \cong 10\angle 0°$ V
**33.** 62.5 Hz
**35.** 1.38 W
**37.** 200 Hz

## Chapter Nineteen

**1.** $2.22\angle -77.2°$ V rms

**3.** **(a)** $9.36\angle -20.7°$ V **(b)** $7.29\angle -43.2°$ V
**(c)** $9.96\angle -5.44°$ V **(d)** $9.95\angle -5.74°$ V

**5.** **(a)** 12.1 μF **(b)** 1.45 μF
**(c)** 0.723 μF **(d)** 0.144 μF

**9.** **(a)** 7.13 V **(b)** 5.67 V
**(c)** 4.01 V **(d)** 0.800 V

**11.** $9.75\angle 12.8°$ V

**13.** **(a)** 3.53 V **(b)** 5.08 V
**(c)** 947 mV **(d)** 995 mV

**17.** **(a)** 14.5 kHz **(b)** 25.2 kHz

**19.** **(a)** 15.06 kHz, 13.94 kHz
**(b)** 26.48 kHz, 23.93 kHz

**21.** **(a)** 117 V **(b)** 115 V

**23.** $C = 0.064\ \mu$F, $L = 989\ \mu$H, $f_r = 20$ kHz

**25.** **(a)** 86.3 Hz **(b)** 7.12 MHz

**27.** $L_1 = 0.088\ \mu$H, $L_2 = 0.609\ \mu$H

## Chapter Twenty

**1.** $1.22\angle 28.6°$ mA

**3.** $80.5\angle -11.3°$ mA

**5.** $V_{A(\text{dc})} = 0$ V, $V_{B(\text{dc})} = 16.1$ V, $V_{C(\text{dc})} = 15.1$ V, $V_{D(\text{dc})} = 0$ V, $V_{A(\text{peak})} = 9$ V, $V_{B(\text{peak})} = 5.96$ V, $V_{C(\text{peak})} = V_{D(\text{peak})} = 4.96$ V

**7.** **(a)** $\mathbf{V}_{th} = 15\angle -53.1°$ V
$\mathbf{Z}_{th} = 63\ \Omega - j48\ \Omega$
**(b)** $\mathbf{V}_{th} = 1.22\angle 0°$ V
$\mathbf{Z}_{th} = j237\ \Omega$
**(c)** $\mathbf{V}_{th} = 12.1\angle 11.9°$ V
$\mathbf{Z}_{th} = 50\ \text{k}\Omega - j20\ \text{k}\Omega$

**9.** $16.9\angle 88.2°$ V

**11.** **(a)** $\mathbf{I}_n = 189\angle -15.8°$ mA
$\mathbf{Z}_n = 63\ \Omega - j48\ \Omega$
**(b)** $\mathbf{I}_n = 5.15\angle -90°$ mA
$\mathbf{Z}_n = j237\ \Omega$
**(c)** $\mathbf{I}_n = 224\angle 33.7°\ \mu$A
$\mathbf{Z}_n = 50\ \text{k}\Omega - j20\ \text{k}\Omega$

**13.** $16.8\angle 88.5°$ V

**15.** $\mathbf{V}_{eq} = 5.12\angle 7.54°$ V
$\mathbf{Z}_{eq} = 571\angle 7.54°\ \Omega$

**17.** $9.18\ \Omega + j2.90\ \Omega$

**19.** $95.2\ \Omega + j42.7\ \Omega$

## Chapter Twenty-One

**1.** 110 μs

**3.** 12.6 V

**5.** See Figure P–20.

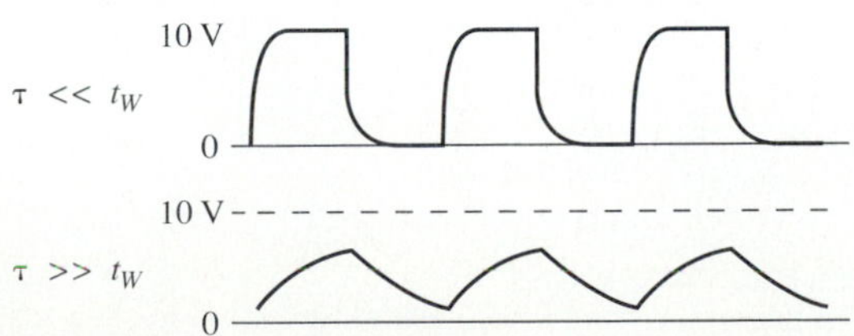


FIGURE P–20

**7.** **(a)** 25 ms
**(b)** See Figure P–21.

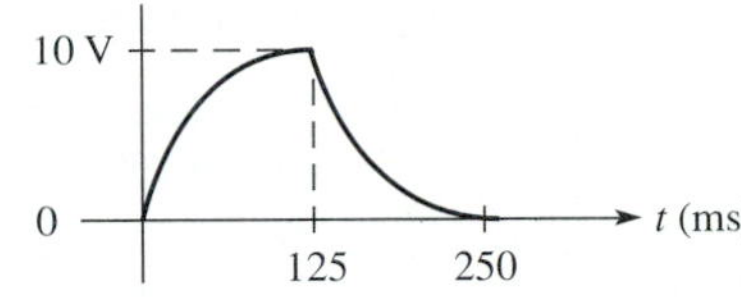


FIGURE P–21

**9.** See Figure P–22.

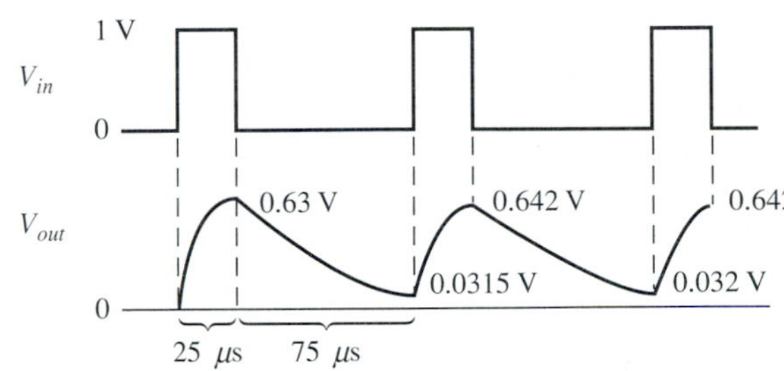


FIGURE P–22

**11.** See Figure P–23.

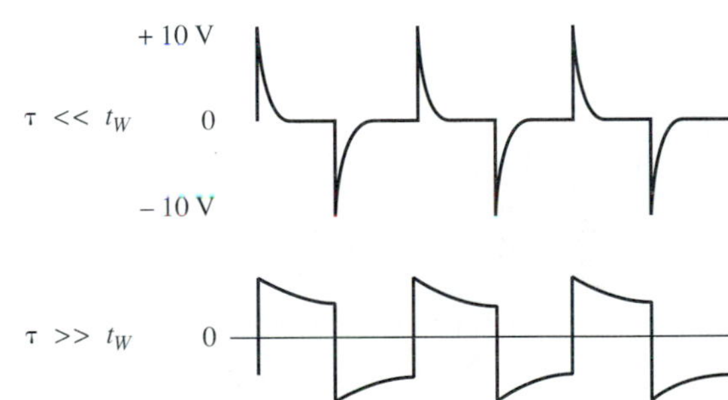


FIGURE P–23

**13.** **(a)** 525 ns
**(b)** See Figure P–24.

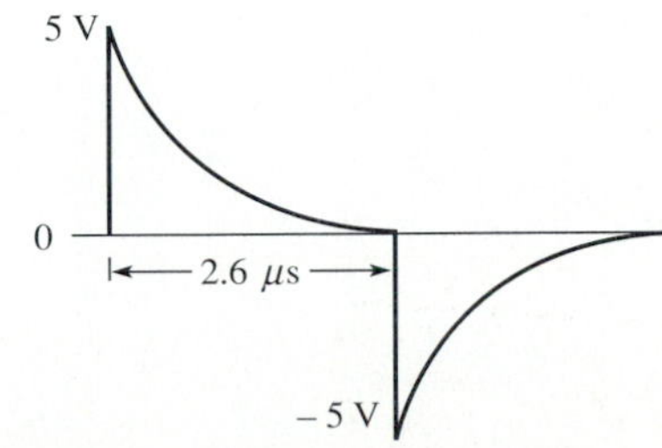


FIGURE P–24

**15.** An approximate square wave with an average value of zero.

FIGURE P–25

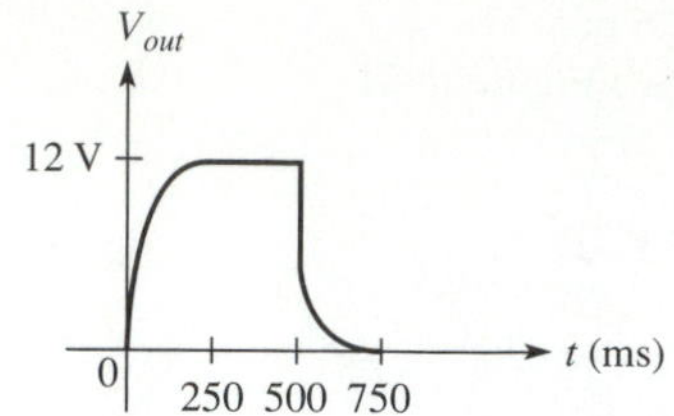


**17.** See Figure P–25.
**19.** **(a)** 4.55 μs
**(b)** See Figure P–26.

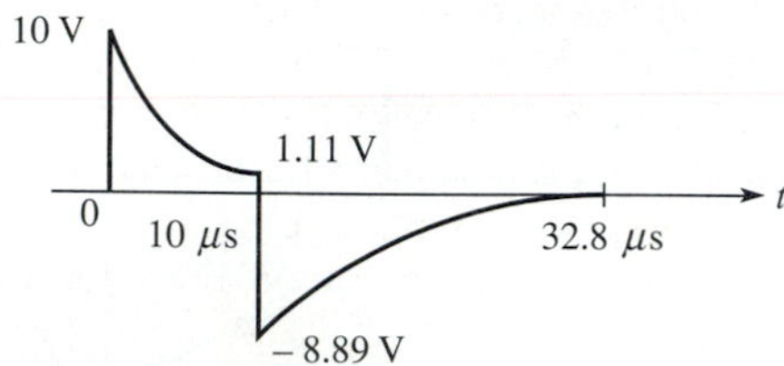


FIGURE P–26

**21.** 15.9 kHz
**23.** **(b)** Capacitor open
**(c)** *C* leaky or $R > 3.3$ kΩ or $C > 0.22$ μF
**(d)** Resistor open

## Chapter Twenty-Two

**1.** 17.5°
**3.** $376\angle 41.2°$ mA
**5.** $1.32\angle 121°$ A
**7.** $\mathbf{I}_{La} = 8.66\angle -30°$ A
$\mathbf{I}_{Lb} = 8.66\angle 90°$ A
$\mathbf{I}_{Le} = 8.66\angle -150°$ A
**9.** **(a)** $\mathbf{V}_{L(ab)} = 866\angle -30°$ V
$\mathbf{V}_{L(ca)} = 866\angle -150°$ V
$\mathbf{V}_{L(bc)} = 866\angle 90°$ V
**(b)** $\mathbf{I}_{\theta a} = 500\angle -32°$ mA
$\mathbf{I}_{\theta b} = 500\angle 88°$ mA
$\mathbf{I}_{\theta c} = 500\angle -152°$ mA
**(c)** $\mathbf{I}_{La} = 500\angle -32°$ mA
$\mathbf{I}_{Lb} = 500\angle 88°$ mA
$\mathbf{I}_{Lc} = 500\angle -152°$ mA
**(d)** $\mathbf{I}_{Za} = 500\angle -32°$ mA
$\mathbf{I}_{Zb} = 500\angle 88°$ mA
$\mathbf{I}_{Zc} = 500\angle -152°$ mA
**(e)** $\mathbf{V}_{Za} = 500\angle 0°$ V
$\mathbf{V}_{Zb} = 500\angle 120°$ V
$\mathbf{V}_{Zc} = 500\angle -120°$ V
**11.** **(a)** $\mathbf{V}_{L(ab)} = 86.6\angle -30°$ V
$\mathbf{V}_{L(ca)} = 86.6\angle -150°$ V
$\mathbf{V}_{L(bc)} = 86.6\angle 90°$ V
**(b)** $\mathbf{I}_{\theta a} = 250\angle 110°$ mA
$\mathbf{I}_{\theta b} = 250\angle -130°$ mA
$\mathbf{I}_{\theta c} = 250\angle -10°$ mA
**(c)** $\mathbf{I}_{La} = 250\angle 110°$ mA
$\mathbf{I}_{Lb} = 250\angle -130°$ mA
$\mathbf{I}_{Lc} = 250\angle -10°$ mA
**(d)** $\mathbf{I}_{Za} = 144\angle 140°$ mA
$\mathbf{I}_{Zb} = 144\angle 20°$ mA
$\mathbf{I}_{Zc} = 144\angle -100°$ mA
**(e)** $\mathbf{V}_{Za} = 86.6\angle -150°$ V
$\mathbf{V}_{Zb} = 86.6\angle 90°$ V
$\mathbf{V}_{Zc} = 86.6\angle -30°$ V
**13.** $\mathbf{V}_{L(ab)} = 330\angle -120°$ V
$\mathbf{V}_{L(ca)} = 330\angle 120°$ V
$\mathbf{V}_{L(bc)} = 330\angle 0°$ V
$\mathbf{I}_{Za} = 38.2\angle -150°$ A
$\mathbf{I}_{Zb} = 38.2\angle -30°$ A
$\mathbf{I}_{Zc} = 38.2\angle 90°$ A
**15.** Figure 22–43: 636 W
Figure 22–44: 149 W
Figure 22–45: 12.8 W
Figure 22–46: 2.78 kW
Figure 22–47: 10.9 kW
**17.** 24 W

# GLOSSARY

**Admittance** A measure of the ability of a reactive circuit to permit current; the reciprocal of impedance. The unit is the siemens (S).

**Alternating current (ac)** Current that reverses direction in response to a change in source voltage polarity.

**Alternator** An electromechanical ac generator.

**American wire gage (AWG)** A standardization based on wire diameter.

**Ammeter** An electrical instrument used to measure current.

**Ampere (A)** The unit of electrical current.

**Ampere-hour rating** A number given in ampere-hours (Ah) determined by multiplying the current (A) times the length of time (h) a battery can deliver that current to a load.

**Ampere-turn** The unit of magnetomotive force (mmf).

**Amplitude** The maximum value of a voltage or current.

**Angular velocity** The rotational velocity of a phasor which is related to the frequency of the sine wave that the phasor represents.

**Apparent power** The phasor combination of resistive power (true power) and reactive power. The unit is the volt-ampere (VA).

**Apparent power rating** The method of rating transformers in which the power capability is expressed in volt-amperes (VA).

**Atom** The smallest particle of an element possessing the unique characteristics of that element.

**Atomic number** The number of protons in a nucleus.

**Atomic weight** The number of protons and neutrons in the nucleus of an atom.

**Attenuation** The ratio with a value of less than 1 of the output voltage to the input voltage of a circuit.

**Autotransformer** A transformer in which the primary and secondary are in a single winding.

**Average value** The average of a sine wave over one-half cycle. It is 0.637 times the peak value.

**Balanced load** A condition where all the load currents are equal and the neutral current is zero.

**Band-pass filter** A filter that passes a range of frequencies lying between two critical frequencies and rejects frequencies above and below that range.

**Band-stop filter** A filter that rejects a range of frequencies lying between two critical frequencies and passes frequencies above and below that range.

**Bandwidth** The range of frequencies for which the current (or output voltage) is equal to or greater than 70.7% of its value at the resonant frequency.

**Baseline** The normal level of a pulse waveform; the voltage level in the absence of a pulse.

**Battery** An energy source that uses a chemical reaction to convert chemical energy into electrical energy.

**Bias** The application of a dc voltage to an electronic device to produce a desired mode of operation.

**Bleeder current** The current left after the total load current is subtracted from the total current into the circuit.

**Bode plot** The graph of a filter's frequency response showing the change in the output voltage to input voltage ratio expressed in dB as a function of frequency for a constant input voltage.

**Branch** One current path in a parallel circuit; a current path that connects two nodes.

**Branch current** The actual current in a branch.

**Capacitance** The ability of a capacitor to store electrical charge.

**Capacitive reactance** The opposition of a capacitor to sinusoidal current. The unit is the ohm (Ω).

**Capacitive susceptance** The ability of a capacitor to permit current; the reciprocal of capacitive reactance. The unit is the siemens (S).

**Capacitor** An electrical device consisting of two conductive plates separated by an insulating material and possessing the property of capacitance.

**Cathode-ray tube (CRT)** A vacuum tube device containing an electron gun that emits a narrow focused beam of electrons onto a phosphor-coated screen.

**Center frequency** The resonant frequency of a band-pass or band-stop filter.

**Center tap (CT)** A connection at the midpoint of the secondary winding of a transformer.

**Charge** An electrical property of matter that exists because of an excess or a deficiency of electrons. Charge can be either positive or negative.

**Choke** An inductor. The term is more commonly used in connection with inductors used to block or choke off high frequencies.

**Circuit** An interconnection of electrical components designed to produce a desired result. A basic circuit consists of a source, a load, and an interconnecting current path.

**Circuit breaker** A resettable protective device used for interrupting excessive current in an electric circuit.

**Circuit ground** A method of grounding whereby a large conducive area on a printed circuit board or the metal chassis that houses the assembly is used as the common or reference point; also called *chassis ground.*

**Circular mil (CM)** The unit of the cross-sectional area of a wire.

**Closed circuit** A circuit with a complete current path.

**Coefficient** The constant number that appears in front of a variable.

**Coefficient of coupling** A constant associated with transformers that is the ratio of secondary magnetic flux to primary magnetic flux. The ideal value of 1 indicates that all the flux in the primary winding is coupled into the secondary winding.

**Coil** A common term for an inductor or for the primary or secondary winding of a transformer.

**Complex conjugate** An impedance containing the same resistance and a reactance opposite in phase but equal in magnitude to that of a given impedance.

**Complex plane** An area consisting of four quadrants on which a quantity containing both magnitude and direction can be represented.

**Component** In a circuit, a device such as a resistor, a capacitor, or an inductor that affects some electrical characteristics of that circuit.

**Conductance** The ability of a circuit to allow current; the reciprocal of resistance. The unit is the siemens (S).

**Conductor** A material in which electric current is easily established. An example is copper.

**Core** The physical structure around which the winding of an inductor is formed. The core material influences the electromagnetic characteristics of the inductor.

**Coulomb (C)** The unit of electrical charge.

**Coulomb's law** A physical law that states a force exists between two charged bodies that is directly proportional to the product of the two charges and inversely proportional to the square of the distance between them.

**Critical frequency** The frequency at which a filter's output voltage is 70.7% of the maximum.

**Current** The rate of flow of charge (electrons).

**Current source** A device that ideally provides a constant value of current regardless of the load.

**Cutoff frequency ($f_c$)** The frequency at which the output voltage of a filter is 70.7% of the maximum output voltage; another term for critical frequency.

**Cycle** One repetition of a periodic waveform.

**DC component** The average value of a pulse waveform.

**Decade** A tenfold change in frequency or other parameter.

**Decibel** A logarithmic measurement of the ratio of one power to another or one voltage to another, which can be used to express the input-to-output relationship of a filter.

**Degree** The unit of angular measure corresponding to 1/360 of a complete revolution.

**Determinant** An array of coefficients and constants in a given set of simultaneous equations.

**Dielectric** The insulating material between the plates of a capacitor.

**Dielectric constant** A measure of the ability of a dielectric material to establish an electric field.

**Dielectric strength** A measure of the ability of a dielectric material to withstand voltage without breaking down.

**Differentiator** A circuit producing an output that approaches the mathematical derivative of the input.

**Duty cycle** A characteristic of a pulse waveform that indicates the percentage of time that a pulse is present during a cycle; the ratio of pulse width to period.

**Earth ground** A method of grounding whereby one side of a power line is neutralized by connecting it to a water pipe or a metal rod driven into the ground.

**Effective value** A measure of the heating effect of a sine wave; also known as the rms (root mean square) value.

**Efficiency** The ratio of the output power to the input power of a circuit expressed as a percentage.

**Electrical** Related to the use of electrical voltage and current to achieve desired results.

**Electrical isolation** The condition that exists when two coils are magnetically linked but have no electrical connection between them.

**Electromagnetic field** A formation of a group of magnetic lines of force surrounding a conductor created by electrical current in the conductor.

**Electromagnetic induction** The phenomenon or process by which a voltage is produced in a conductor when there is relative motion between the conductor and a magnetic or electromagnetic field.

**Electron** The basic particle of electrical charge in matter. The electron possesses negative charge.

**Electronic** Related to the movement and control of free electrons in semiconductors or vacuum devices.

**Element** One of the unique substances that make up the known universe. Each element is characterized by a unique atomic structure.

**Energy** The fundamental ability to do work.

**Equivalent circuit** A circuit that produces the same voltage and current to a given load as the original circuit that it replaces.

**Falling edge** The negative-going transition of a pulse.

**Fall time** The time interval required for a pulse to change from 90% to 10% of its full amplitude.

**Farad (F)** The unit of capacitance.

**Faraday's law** A law stating that the voltage induced across a coil of wire equals the number of turns in the coil times the rate of change of the magnetic flux.

**Field winding** The winding on the rotor of an ac generator.

**Filter** A type of circuit that passes certain frequencies and rejects all others.

**Free electron** A valence electron that has broken away from its parent atom and is free to move from atom to atom within the atomic structure of a material.

**Frequency** A measure of the rate of change of a periodic function; the number of cycles completed in 1 s. The unit of frequency is the hertz.

**Frequency response** In electric circuits, the variation in the output voltage (or current) over a specified range of frequencies.

**Fundamental frequency** The repetition rate of a waveform.

**Fuse** A protective device that burns open when there is excessive current in a circuit.

**Generator** An energy source that produces electrical signals.

**Ground** In electric circuits, the common or reference point.

**Half-power frequency** The frequency at which the output power of a filter is 50% of the maximum (the output voltage is 70.7% of maximum); another name for critical or cutoff frequency.
**Harmonics** The frequencies contained in a composite waveform, which are integer multiples of the repetition frequency (fundamental).
**Henry (H)** The unit of inductance.
**Hertz (Hz)** The unit of frequency. One hertz equals one cycle per second.
**High-pass filter** A certain type of filter whereby higher frequencies are passed and lower frequencies are rejected.
**Hysteresis** A characteristic of a magnetic material whereby a change in magnetization lags the application of a magnetizing force.

**Imaginary number** A number that exists on the vertical axis of the complex plane.
**Impedance** The total opposition to sinusoidal current expressed in ohms.
**Impedance matching** A technique used to match a load resistance to a source resistance in order to achieve maximum transfer of power.
**Induced voltage** Voltage produced as a result of a changing magnetic field.
**Inductance** The property of an inductor whereby a change in current causes the inductor to produce a voltage that opposes the change in current.
**Inductive reactance** The opposition of an inductor to sinusoidal current. The unit is the ohm (Ω).
**Inductive susceptance** The reciprocal of inductive reactance. The unit is the siemens (S).
**Inductor** An electrical device formed by a wire wound around a core having the property of inductance; also known as a coil or a choke.
**Instantaneous value** The voltage or current value of a waveform at a given instant in time.
**Insulator** A material that does not allow current under normal conditions.
**Integrator** A circuit producing an output that approaches the mathematical integral of the input.

**Joule (J)** The unit of energy.
**Junction** A point at which two or more components are connected.

**Kilowatt-hour (kWh)** A common unit of energy used mainly by utility companies.
**Kirchhoff's current law** A law stating that the total current into a junction equals the total current out of the junction.

**Lag** Refers to a condition of the phase or time relationship of waveforms in which one waveform is behind the other in phase or time.
**Lead** Refers to a condition of the phase or time relationship of waveforms in which one waveform is ahead of the other in phase or time; also, a wire or cable connection to a device or instrument.
**Leading edge** The first step or transition of a pulse.
**Lenz's law** A physical law that states when the current through a coil changes, an induced voltage is created in a direction to oppose the change in current. The current cannot change instantaneously.
**Linear** Characterized by a straight-line relationship.
**Line current** The current through a line feeding a load.
**Line voltage** The voltage between lines feeding a load.
**Load** An element (resistor or other component) connected across the output terminals of a circuit that draws current from the circuit; the device in a circuit upon which work is done.
**Loop** A closed current path in a circuit.
**Low-pass filter** A certain type of filter whereby lower frequencies are passed and higher frequencies are rejected.

**Magnetic coupling** The magnetic connection between two coils as a result of the changing magnetic flux lines of one coil cutting through the second coil.
**Magnetic flux** The lines of force between the north and south poles of a permanent magnet or an electromagnet.
**Magnetic flux density** The number of lines of force per unit area in a magnetic field.
**Magnetizing force** The amount of mmf per unit length of magnetic material.
**Magnetomotive force (mmf)** The force that produces the magnetic field.
**Magnitude** The value of a quantity, such as the number of volts of voltage or the number of amperes of current.
**Maximum power transfer theorem** A theorem that states the maximum power is transferred from a source to a load when the load resistance equals the internal source resistance.
**Millman's theorem** A method for reducing parallel voltage sources to a single equivalent voltage source.
**Multimeter** An instrument that measures voltage, current, and resistance.
**Mutual inductance** The inductance between two separate coils, such as a transformer.

**Neutron** An atomic particle having no electrical charge.
**Node** A unique point in a circuit where two or more components are connected.
**Norton's theorem** A method for simplifying a given circuit to an equivalent circuit with a current source in parallel with a resistance or impedance.

**Ohm (Ω)** The unit of resistance.
**Ohmmeter** An instrument for measuring resistance.
**Ohm's law** A law stating that current is directly proportional to voltage and inversely proportional to resistance.
**Open circuit** A circuit in which there is not a complete current path.
**Oscillator** An electronic circuit that produces a time-varying signal without an external input signal using positive feedback.
**Oscilloscope** A measurement instrument that displays signal waveforms on a screen.

**Parallel** The relationship in electric circuits in which two or more current paths are connected between the same two points.

**Parallel resonance** A condition in a parallel *RLC* circuit in which the reactances ideally cancel and the impedance is maximum.
**Passband** The range of frequencies passed by a filter.
**Peak-to-peak value** The voltage or current value of a waveform measured from its minimum to its maximum points.
**Peak value** The voltage or current value of a waveform at its maximum positive or negative points.
**Period (*T*)** The time interval of one complete cycle of a periodic waveform.
**Periodic** Characterized by a repetition at fixed-time intervals.
**Permeability** The measure of ease with which a magnetic field can be established in a material.
**Phase** The relative displacement of a time-varying quantity with respect to a given reference.
**Phase current** The current through a generator winding.
**Phase voltage** The voltage across a generator winding.
**Phasor** A representation of a sine wave in terms of its magnitude (amplitude) and direction (phase angle).
**Photoconductive cell** A type of variable resistor that is light-sensitive.
**Polar form** One form of a complex number made up of a magnitude and an angle.
**Polyphase** Characterized by two or more sinusoidal voltages, each having a different phase angle.
**Potentiometer** A three-terminal variable resistor.
**Power** The rate of energy usage.
**Power factor** The relationship between volt-amperes and true power or watts. Volt-amperes multiplied by the power factor equals true power.
**Power rating** The maximum amount of power that a resistor can dissipate without being damaged by excessive heat buildup.
**Power supply** An electronic instrument that produces voltage, current and power from the ac power line or batteries in the form suitable for use in powering electronic equipment.
**Primary winding** The input winding of a transformer; also called *primary.*
**Proton** A positively charged atomic particle.
**Pulse** A type of waveform that consists of two equal and opposite steps in voltage or current separated by a time interval.
**Pulse repetition frequency** The fundamental frequency of a repetitive pulse waveform; the rate at which the pulses repeat expressed in either hertz or pulses per second.
**Pulse response** In electric circuits, the reaction of a circuit to a given pulse input.
**Pulse width** The time interval between the opposite steps of an ideal pulse. For a nonideal pulse, the time between the 50% points of the leading and trailing edges.

**Quality factor (*Q*)** The ratio of true power to reactive power in a resonant circuit or the ratio of inductive reactance to winding resistance in a coil.

**Radian** A unit of angular measurement. There are $2\pi$ radians in one complete 360° revolution. One radian equals 57.3°.
**Ramp** A type of waveform characterized by a linear increase or decrease in voltage or current.
**Reactive power** The rate at which energy is alternately stored and returned to the source by a capacitor. The unit is the VAR.
**Rectangular form** One form of a complex number made up of a real part and an imaginary part.
**Rectifier** An electronic circuit that converts ac into pulsating dc; one part of a power supply.
**Reflected load** The load as it appears to the source in the primary of a transformer.
**Reflected resistance** The resistance in the secondary circuit reflected into the primary circuit.
**Relay** An electromagnetically controlled mechanical device in which electrical contacts are opened or closed by a magnetizing current.
**Reluctance** The opposition to the establishment of a magnetic field in a material.
**Resistance** Opposition to current. The unit is the ohm (Ω).
**Resistor** An electrical component designed specifically to provide resistance.
**Resonance** A condition in a series *RLC* circuit in which the capacitive and inductive reactances are equal in magnitude; thus, they cancel each other and result in a purely resistive impedance.
**Resonant frequency** The frequency at which resonance occurs; also known as *center frequency.*
**Retentivity** The ability of a material, once magnetized, to maintain a magnetized state without the presence of a magnetizing force.
**Rheostat** A two-terminal variable resistor.
**Ripple voltage** The variation in the dc voltage on the output of a filtered rectifier caused by the slight charging and discharging action of the filter capacitor.
**Rise time** The time interval required for a pulse to change from 10% to 90% of its amplitude.
**Rising edge** The positive-going transition of a pulse.
**Roll-off** The rate of decrease of a filter's frequency response.
**Root mean square (rms)** The value of a sine wave that indicates its heating effect, also known as the effective value. It is equal to 0.707 times the peak value.
**Rotor** The rotating assembly in a generator or motor.

**Sawtooth waveform** A type of electrical waveform composed of ramps; a special case of a triangular waveform in which one ramp is much shorter than the other.
**Schematic** A symbolized diagram of an electrical or electronic circuit.
**Secondary winding** The output winding of a transformer; also called *secondary.*
**Selectivity** A measure of how effectively a filter passes certain desired frequencies and rejects all others. Generally, the narrower the bandwidth, the greater the selectivity.
**Semiconductor** A material that has a conductance value between that of a conductor and an insulator. Silicon and germanium are examples.
**Sensitivity factor** The ohms-per-volt rating of a voltmeter.
**Series** In an electrical circuit, a relationship of components in which the components are connected such that they provide a single current path between two points.

**Series resonance** A condition in a series *RLC* circuit in which the reactances cancel and the impedance is minimum.
**Shell** The orbit in which an electron revolves.
**Short circuit** A circuit in which there is a zero or abnormally low resistance path between two points; usually an inadvertent condition.
**Solenoid** An electromagnetically controlled device in which the mechanical movement of a shaft or plunger is activated by a magnetizing current.
**Source** A device that produces electrical energy.
**Squirrel-cage** A type of ac induction motor.
**Stator** The stationary outer part of a generator or motor.
**Steady state** The equilibrium condition of a circuit that occurs after an initial transient time.
**Step-down transformer** A transformer in which the secondary voltage is less than the primary voltage.
**Step-up transformer** A transformer in which the secondary voltage is greater than the primary voltage.
**Superposition theorem** A method for the analysis of circuits with more than one source.
**Switch** An electrical device for opening and closing a current path.

**Tank circuit** A parallel resonant circuit.
**Tapered** Nonlinear, such as a tapered potentiometer.
**Temperature coefficient** A constant specifying the amount of change in the value of a quantity for a given change in temperature.
**Terminal equivalency** The concept that when any given load resistance is connected to two sources, the same load voltage and load current are produced by both sources.
**Tesla** The unit of flux density.
**Thermistor** A temperature-sensitive resistor.
**Thevenin's theorem** A method for simplifying a given circuit to an equivalent circuit with a voltage source in series with a resistance or impedance.
**Time constant** A fixed-time interval, set by $R$ and $C$, or $R$ and $L$ values, that determines the time response of a circuit.
**Tolerance** The limits of variation in the value of a component.
**Trailing edge** The second step of transition of a pulse.
**Transformer** A device formed by two or more windings that are magnetically coupled to each other and provide a transfer of power electromagnetically from one winding to another.
**Transient time** An interval equal to approximately five time constants.
**Triangular waveform** A type of electrical waveform that consists of two ramps.
**Trigger** The activating unit of some electronic devices or instruments.
**Trimmer** A small variable capacitor.
**Turns ratio** The ratio of turns in the secondary winding to turns in the primary winding.

**Valance** Related to the outer shell or orbit of an atom.
**Valence electron** An electron that is present in the outermost shell of an atom.
**Volt-ampere reactive (VAR)** The unit of reactive power.
**Volt** The unit of voltage or electromotive force.
**Voltage** The amount of energy available to move a certain number of electrons from one point to another in an electric circuit.
**Voltage drop** The drop in energy level through a resistor.
**Voltage source** A device that ideally provides a constant value of voltage regardless of the load.
**Voltmeter** An instrument used to measure voltage.

**Watt (W)** The unit of power.
**Watt's law** A law that states the relationships of power to current, voltage, and resistance.
**Waveform** The pattern of variations of a voltage or current showing how the quantity changes with time.
**Weber** The unit of magnetic flux.
**Winding** The loops or turns of wire in an inductor.
**Wiper** The sliding contact in a potentiometer.

# INDEX